CRC

Handbook of Chromatography

Volume I

EDITORS

GUNTER ZWEIG, Ph.D.

Syracuse University Research Corporation
Syracuse, New York

JOSEPH SHERMA, Ph.D.

Lafayette College
Easton, Pennsylvania

published by:

CRC PRESS

A DIVISION OF
THE CHEMICAL RUBBER CO.
18901 Cranwood Parkway • Cleveland, Ohio 44128

This book presents data obtained from authentic and highly regarded sources. Reprinted material is quoted with permission, and sources are indicated. A wide variety of references are listed. Every reasonable effort has been made to give reliable data and information but the editor and the publisher cannot assume responsibility for the validity of all materials or for the consequences of their use.

Library of Congress Card No. 76-163067
Printed in the United States of America
International Standard Book Number 0-87819-560-2 Complete set
International Standard Book Number 0-87819-561-0 Volume One

HANDBOOK SERIES

Handbook of Chemistry and Physics, 53rd edition
Standard Mathematical Tables, 20th edition
Handbook of tables for Mathematics, 4th edition
Handbook of tables for Organic Compound Identification, 3rd edition
Handbook of Biochemistry, selected data for Molecular Biology, 2nd edition
Handbook of Clinical Laboratory Data, 2nd edition
Manual for Clinical Laboratory Procedures, 2nd edition
Manual of Nuclear Medicine Procedures, 1st edition
Handbook of Laboratory Safety, 2nd edition
Handbook of tables for Probability and Statistics, 2nd edition
Fenaroli's Handbook of Flavor Ingredients, 1st edition
Handbook of Analytical Toxicology, 1st edition
Manual of Laboratory Procedures in Toxicology, 1st edition
Handbook of tables for Applied Engineering Science, 2nd edition
Handbook of Chromatography, 1st edition
Handbook of Environmental Control, 1st edition
Handbook of Food Additives, 2nd edition
Handbook of Radioactive Nuclides, 1st edition
Handbook of Lasers, 1st edition
Handbook of Microbiology, 1st edition
Handbook of Engineering in Medicine and Biology, 1st edition
*Handbook of Marine Sciences, 1st edition
*Handbook of Material Science, 1st edition
*Handbook of Spectroscopy, 1st edition
*Atlas of Spectral Data and Physical Constants for Organic Compounds, 1st edition

*Currently in preparation.

PREFACE

Volumes One and Two of the *Handbook of Chromatography* represent over three years of intensive labor by the editors and members of the Editorial Advisory Board. When the Publishers first suggested the production of a Handbook for Chromatographic Data, it seemed to the editors to be just a simple compilation of R_F-tables and retention times or volumes. However, a quick perusal of the literature of chromatography, accumulated during approximately twenty-five years of modern chromatography, revealed that the task would be a formidable one and could not be accomplished in one year, as was first estimated.

The Handbook, as it finally appears, should serve chemists in all fields as a working manual and reference book which should aid in their search for identification of unknowns as well as suggest quantitative methods of analysis.

Volume One of the Handbook contains over 549 tables of chromatographic data expressed in uniform terms of R_F values, retention times and retention volumes as well as other terms used in chromatography. Each table has literature citations which will refer to the primary and sometimes secondary sources. Over 12,000 compounds are cross-indexed to direct the reader to appropriate tables in this volume. Thus, for example, the amino acid alanine will have entries in tables under gas, liquid-column, paper, and thin-layer chromatography.

Volume Two of the Handbook has been designed to give the researcher, even the novice in chromatography, a working knowledge of the theory and practices of the various fields of chromatography–gas, liquid-column, paper, and thin-layer chromatography.

Volume Two also contains two useful sections. One is on detection reagents for paper and thin-layer chromatography, with an alphabetical index for chemical classes and "name" reagents. This section is very helpful for the interpretation of the tables in Volume One. The other section describes selected methods for sample preparations, which will be expanded in future editions of the Handbook. Subsequent sub-sections in **PRACTICAL APPLICATIONS** supply the researcher with useful information on commercial sources of all types of chromatographic supplies.

A Book Directory at the close of Volume Two should serve as a good source for more detailed reading in chromatography.

The editors would be ungrateful if they did not acknowledge the invaluable help and advice from the members of the Advisory Editorial Board.

Special thanks are due to the industrious group of compilers, especially Mrs. Ellen Burton, Miss Susan Rodems, Miss Irene Zweig, Mrs. Frances K. Zweig, and Dr. and Mrs. Coleman Hamel. Special thanks are also due to Dr. Irving Sunshine whose initiative and perseverance are in no small manner responsible for the writing and creation of this Handbook, and to Mrs. Florence Thomas and Mrs. Ruth Pokorney of The Chemical Rubber Co. for their unstinting effort in the production and final editing of the books.

The editors would greatly appreciate suggestions for improvements and additions in future editions from any interested persons.

Gunter Zweig

Joseph Sherma

November 1972

Advisory Board

TABLE OF CONTENTS

VOLUME I

CHROMATOGRAPHIC DATA

Abbreviations and Symbols, inside covers

SECTION I LIST OF TABLES FOR

I.I **GAS** Chromatography
I.II **LIQUID** Chromatography
I.III **PAPER** Chromatography
I.IV **THIN-LAYER** Chromatography

SECTION II. TABLES FOR

II.I **GAS** Chromatography, 1
II.II **LIQUID** Chromatography, 179
II.III **PAPER** Chromatography, 283
II.IV **THIN-LAYER** Chromatography, 437

SECTION III.I COMPOUND Index, 659

VOLUME II

PRINCIPLES AND TECHNIQUES

Abbreviations and Symbols, inside covers

SECTION I PRINCIPLES AND TECHNIQUES

I.I Introduction, 1
I.II **GAS** Chromatography, 7
I.III **LIQUID COLUMN** Chromatography, 25
I.IV **PAPER** Chromatography, 73
I.V **THIN-LAYER** Chromatography, 89

PRACTICAL APPLICATIONS

SECTION II PRACTICAL APPLICATIONS

II.I Detection Reagents for Paper and Thin-Layer Chromatography, 103
II.II Selected Methods of Sample Preparation, 191
II.III Products and Sources of Chromatographic Materials, 255
II.IV International Chromatography Book Directory, 319
II.V Subject Index, 337

SECTION I.I

GAS CROMATOGRAPHY TABLES

	Table No. GC
Acids, Aliphatic; Dibasic Esters	1
Acids, Aliphatic; Methyl Esters	2,3
Acids, Aliphatic; Unsaturated Esters	91
Acids, Aromatic, Trimethylsilyl Derivatives	4,5
Acids, Krebs Cycle1	89
Acids, Phenolic	90
Alcohols and Derivatives	92,94
Aldehydes, Dimethyl Acetals	6
Aldehydes and Ketones	7
Aliphatics, Miscellaneous	93
Alkaloids	8,96
Alkaloids, Tobacco	95
Amines	11,98
Amines, Acetylated Derivatives	9
Amines, Aliphatic	97
Amines, Aliphatic, Trifluoroacetyl Derivatives	10
Amines, Pentafluoropropionamides	99
Amino Acid Derivatives	12
Amino Acid Esters	101
Anilines, Substituted	100,102
Antihistamines	54
Aromatic Nitro Compounds and Anilines	46
Carbohydrates, Acetals and Ketals	13
Carbohydrates, Alditol Acetates	14
Carbohydrates, Trimethylsilyl Derivatives	15
Dipeptide Derivatives	16
Drugs	17-23,103-111
Esters	24,25
Gases	26,27,112-114
Gases: Isotopes of Hydrogen	115
Glycol Ethers	116
Glycopyranosides, Methyl Ethers	28
Hydrocarbons	118
Hydrocarbons, Aliphatic, Unsaturated	29
Hydrocarbons, Alkyl Paraffins	30
Hydrocarbons, Aromatic	119
Hydrocarbons, Aromatic and Aliphatic	120
Hydrocarbons, Chlorinated Benzenes	31
Hydrocarbons, Chlorinated and Brominated Aliphatics	34
Hydrocarbons, Chlorinated Olefins	32
Hydrocarbons, Cyclic	41
Hydrocarbons, Cyclohexanes	35
Hydrocarbons, Cyclopentanes	36
Hydrocarbons, Iodinated Alkanes	37
Hydrocarbons, Naphthalenes	38
Hydrocarbons, Octanes	33
Hydrocarbons, Octenes	39
Hydrocarbons, Olefins	122
Hydrocarbons and Oxygenated Compounds	121
Hydrocarbons, Polynuclear	40
Hydrocarbons, Saturated	42
Inorganic Chlorides	50
Inorganics, Cation Acetylacetonates	43
Isocyanates	44
Ketones	117
Ketones, Acetophenones	48
Miscellaneous Organics, Aliphatic and Aromatic	126
Nitriles	123
Nitrogen Compounds	45
Oligosaccharides, Trimethylsilyl Ether Derivatives	124
Organics	47,49,51-53,55,56,127
Organics, Aliphatics	58
Organics, Various Compounds	125
Organics, Volatiles	57
Perhalomethyl Sulfenyl and Thiocarbonyl Compounds	59
Pesticides	60-65,68,128,129
Pesticides, Carbamates and Enols	130
Pesticides, Organophosphates	67
Pesticides, Triazines	66
Phenols	69-75
Plasticizers	132
Polyols and Methylglycosides-Trifluoroacetyl Derivatives	76
Purines, Pyrimidines and Related Compounds	134
Purines, Pyrimidines, (TMS Derivatives)	77
Pyridines	131
Pyrroles	133
Quinolines and Pyridines	78
S Compounds	85
Steroids	79-81
Steroids, Androstanes and Derivatives	137
Steroids, Bile Acids and TMS Derivatives	138
Steroids and Derivatives	136
Steroids, Estrogens, TMS Derivatives	82
Steroids, Heptafluorobutyrates	83
Steroids, Plant Sterols and Derivatives	84
Sulfur Compounds, Dialkylsulfides	135
Sulfur Compounds, Sulfones	139
Terpenes	86,87
Terpenes, Sesquiterpenes	140,141
Vitamins	88,142

SECTION I.II

LIQUID CHROMATOGRAPHY TABLES

	Table No. LC
Anion Exchange Chromatography of Benzoic Acid Derivatives and Related Compounds	30
Anion Exchange Chromatography of Hydroxy Acids	23
Anion Exchange Chromatography of Purines and Pyrimidines	31
Anion Exchange Chromatography of Saccharides	25
Automatic Ion Exchange Chromatography of Amino Acids	33,41
Automatic Ion Exchange Chromatography of Carbohydrates and Related Compounds	29
Automatic Ion Exchange Chromatography of Organic Acids	40
Cation Exchange Chromatography of (Uncommon) Amino Acids	59
Cation Exchange Chromatography of Nucleotides, Nucleosides and Nucleic Bases	22
Chromatography of Food Preservatives on Polyamide Columns	74
Chromatography of Hydroxy- and Ketoprogesterones on Lipophilic Sephadex	10
Chromatography of Primary Aromatic Amines on Alginic Acid and Carboxymethylcellulose Columns	35

	Table No. LC
Chromatography of Purines and Related Compounds on Columns of Sephadex G-10	46
Chromatography of Steroids and Derivatives on Lipophilic Sephadex Gels	6
Column Chromatography of Alkaloid Drugs	55
Column Chromatography of Amino Acids on Polystyrene	12
Column Chromatography of Iodotyrosines, on Sephadex and Derivatives	76
Column Chromatography of Nucleic Acid Components on Sephadex Gels	3
Column Chromatography of Nucleotides on Poly-*N*-Vinyl Pyrrolidone	17
Column Chromatography of Phenols, Flavonoids and Derivatives on Sephadex	1
Column Chromatography of Proteins used to Calibrate Sephadex Gels for Molecular Weight Determinations	19
Column Chromatography of Steroids on Sephadex LH-20	62
Column Gel Chromatography of Amides on Polyacrylamides	73
Column Gel Chromatography of Carboxylic Acids on Polyacrylamides	58
Column Gel Chromatography of Phenols on Polyacrylamides	51
Column Partition Chromatography of Rare Earths	39
Column Polyacrylamide Gel Chromatography of Sugars and Alcohols	71
Gel Chromatography of Flavonoid Compounds	28
Gel Chromatography of Hydrocarbons	27
Gel Chromatography of NAD-Glycohydrolase and other Proteins	64
Gel Chromatography of Nucleic Acid Components	15
Gel Chromatography of Organophosphorus Pesticides	43
Gel Chromatography of Peptides	14
Gel Chromatography of Phosphate Esters	18
Gel Chromatography of Protected Peptides on Methylated Sephadex	9
Gel Filtration of Amino Acids	56
Gel Filtration of Amino Acids on Sephadex G-10	4
Gel Filtration of Aromatic Acids and Others	53
Gel Filtration of Aromatic and Heterocyclic Acids and Glycols	72
Gel Filtration of Buffer Components	34
Gel Filtration of Heterocyclic Amines	77
Gel Filtration of Phosphate Esters and Related Compounds	47
Gel Filtration of Purines and Pyrimidines	13
Gel Filtration of Vitamins and Alkaloids	20
Gel Permeation Chromatography of Alcohols	54
Gel Permeation Chromatography of Carbonyls	70
Gel Permeation Chromatography of Hydrocarbons and Ethers	21
Gel Permeation Chromatography of Low Molecular Weight Compounds: Alcohols and Glycols	37
Gel Permeation Chromatography of Low Molecular Weight Compounds: Aliphatic Hydrocarbons	38
Gel Permeation Chromatography of Low Molecular Weight Compounds: Carboxylic Acids	57
Gel Permeation Chromatography of Low Molecular Weight Compounds: Miscellaneous Organics	36
Gel Permeation Chromatography of Miscellaneous Compounds	60
Gel Permeation Chromatography of Monomers and Oligomers in Nylon 6, Nylon 66, and Nylon 12	49
Gel Permeation Chromatography of Organic Halides	75
Gel Permeation Chromatography of Phenols and Organic Acids	61
Gel Permeation Chromatography of Resin and Fatty Acid Methyl Esters	67
Gel Permeation Chromatography of Triglycerides	78
Ion Exchange Chromatography of α- and β-Aspartyl Oligo Peptides	2
Ion Exchange Chromatography of Carboxylic Acids	65
Ion Exchange Chromatography of Penicillin and Related Compounds	32
Ion Exchange Chromatography of Polyamines and Related Compounds	66
Ion Exchange Chromatography of Polyols	69
Ion Exchange Chromatography of Compounds Related to Urea	63
Liquid Chromatography of Insecticides	44
Metal Ions	79-89
Metal Ions Figures, following Table 91	
Absorption of the Elements	
From HBr Solutions By A Cation Exchange Resin LC-1	
From $HC1O_4$ Solutions By A Cation Exchange Resin LC-2	
From HCl Solutions By A Cation Exchange Resin LC-3	
From HCl Solutions By An Anion Exchange Resin LC-4	
Metal Ions, Rare Earth	90,91
Partition Chromatography of Hydrocarbons	42
Partition Chromatography of Sugars on Ion Exchange Resins	48
Purines and Pyrimidines	50
Salting-Out Chromatography of Alcohols	5
Salting-Out Chromatography of Aldehydes and Ketones	8
Salting-Out Chromatography of Amines	26
Sephadex Gel Chromatography of Monosubstituted Phenols	52
Solubilization Chromatography of Alcohols and Phenols	11
Solubilization Chromatography of Ethers, Carboxylic Acids and Hydrocarbons	7
Solubilization Chromatography of Ketones	16
Steroids, Gibberellins, Pigments and Esters	45
Sterol Acetates	24
Supercritical Fluid Chromatography of Fatty Acids, Lipids, and Misc. Compounds	68

SECTION I.III

PAPER CHROMATOGRAPHY TABLES

	Table No. PC
Alcohols	53
Alcohols, Alcohol Derivatives	52
Alcohols, Polyhydric	1
Alkaloids	2-4,54

	Table No. TLC
Alcohols, Glycol Esters	112
Alcohols, Polyalcohols	113
Alkaloids	2,3
Alkaloids, Ergot Alkaloids and Hydrogenated Derivatives	114
Alkaloids, Opium Alkaloids	4
Alkaloids, Opium Alkaloids and Synthetic Morphine Derivatives	116
Alkaloids, Rauwolfia	115
Alkaloids, Tropine	117
Amines	5
Amines, Adrenaline Derivatives	9
Amines, Aromatic	119
Amines, Arylamine Metabolites	6
Amines, 4-(Phenylazo)Benzene Sulfonamide Derivatives	10
Amines, Catecholamines	121
Amines, Epinephrine (Adrenaline) and Related Compounds	8
Amines and Hydrocarbons	7
Amines and Imines	120
Amines, Ureas	122
Amino Acids	11,15,16
Amino Acids, derivatives of, *N*-2,4-Dinitro-5-Aminophenyl (DNAP)	14
Amino Acids, 1-Dimethylaminonaphthalene-5-Sulfonylamino	123
Amino Acids, Dinitrophenyl Derivatives	17,18
Amino Acids, Iodoamino Acids and Derivatives	19
Amino Acids and Peptides, Carbobenzoxy Derivatives	12
Amino Acids, 3-Phenyl-2-Thiohydantoin (PTH)	124
Amino Acids and Related Compounds	13
Amino Acids, Two-Dimensional	118
Antibiotics, Drugs	20,21
Antibiotics, Macrolide Group	153
Antioxidants	125
Antioxidants and Accelerators in Vulcanized Rubbers	22
Carbohydrates	23
Carbohydrates, Acetyl Derivatives	25
Carbohydrates-Derivatives	24
Carbohydrates, Methyl Derivatives	26
Carbohydrates, Mono- and Disaccharides	126
Carbohydrates, Sugar Alcohols	127
Carbohydrates, Sugar Phosphates	128
Carboxylic Acid Esters, Plasticizers and "Softeners"	27
Carboxylic Acids	130,134
Carboxylic Acids, Aliphatic Acids	29
Carboxylic Acids, Amino Benzoates and Salicylates	129
Carboxylic Acids, Cyclic Dicarboxylic Anhydride Derivatives	30
Carboxylic Acids, Dicarboxylic Acids	31
Carboxylic Acids, 2,4-Dinitrophenylhydrazones of Keto Acids	32
Carboxylic Acids, Fatty Acids, Esters and Alcohols	33
Carboxylic Acids, Hydroxamic Acid, Derivatives of Fatty Acids	132
Carboxylic Acids, Lichen Substances	28,34,35
Carboxylic Acids, Lichen Substances (Aliphatic Lichen Acids)	131
Carboxylic Acids, Organic Acids	36
Carboxylic Acids, Phenolic Acids	133
Carboxylic Acids, Phenolic Dicarboxylic Acids	135
Carboxylic Acids, Substituted Benzoic Acids and Phenolic Compounds	37

	Table No. TLC
Drugs	39,45
Drugs (Analgesics and Antipyretics)	40
Drugs, (Local) Anesthetics	43
Drugs, Animal	136
Drugs, Barbitals	137
Drugs, Diuretics and Antihypertensives	41
Drugs, Medicinal Carbamates	38
Drugs, Narcotics, Analgesics	138
Drugs, Phenothiazines	139
Drugs, Psychotropic	42
Drugs, Sulfonamides	44,140
Dyes, Azo Dyestuffs	47,142
Dyes, Fat-Soluble	46
Dyes, Oil-Soluble	48
Dyes, Synthetic Food Dyes	49
Food Dyes, Synthetic	141
Gibberellins	144
Gibberellins, Methyl Esters of	145
Hydrocarbons, Anthracene Derivatives	50
Hydrocarbons, Aromatic	51,53
Hydrocarbons, Polycyclic Aromatic	146
Hydrocarbons, Terpene	54,143
Inorganic Ions	150
Inorganics, Anions	147
Inorganics, Cations	52
Inorganics, Condensed Phosphates	151
Inorganics, Metal Ions	148,149
Inorganics, Noble Metals	55
Lipids	56,57
Lipids, Brain	152
Lipids, Cholesterol and Methyl Esters of Fatty Acids	154
Nitrogen Compounds, Explosive Materials, Components of	157
Nitrogen Heterocyclics	58
Nitrogen Heterocyclics, Imidazoles	155
Nitrogen Heterocyclics, Indoles	156
Nitrogen Heterocyclics, Indoles, 2,4-Dinitrophenylthio Derivatives	59
Nitrogen Heterocyclics, Pyridine Carboxylic Acid Derivatives	158
Nitrogen Heterocyclics, Pyridine Derivatives	61
Nitrogen Heterocyclics, Pyridines	60
Nucleic Acid, Free Bases and Nucleosides	159
Nucleosides, Nucleotides and Oligonucleotides	62
Nucleotides	63,160
Organic Iodine Compounds	64
Organic Peroxides	65
Organic Sulfur Compounds, Accelerators	66
Organic Sulfur Compounds, Mercaptans and Thiophenols	162
Organic Sulfur Compounds, Sulfides	163
Organic Sulfur Compounds, Sulfoxides	164
Organomercury Compounds, Dithizonates of	161
Organo-Metallic Tin Compounds	67
Oxo Compounds, Aliphatic and Terpene 2,4-Dinitrophenylhydrazones	76
Oxo Compounds, Aminoanthraquinones	69
Oxo Compounds, Aromatic Aldehydes	68
Oxo Compounds, Benzaldehyde and Related Compounds	71
Oxo Compounds, Derivatives	165
Oxo Compounds, 2,4-Dinitrophenylhydrazones	72,73,167
Oxo Compounds, DNPH-Derivatives	166
Oxo Compounds, 2-Hydrazinebenzothiazole Derivatives	168

	Table No. PC
Amines, Adrenaline and Related Compounds	5
Amines, Aliphatic	56
Amines, Aliphatic, 3,5-Dinitrobenzoyl	55
Amines, Aliphatic Polyamines and Amino Alcohols	7
Amines, Amino Alcohols	57
Amines, Aminobenzene Derivatives	59
Amines, Aromatic	6
Amines, Catecholamines	58
Amines, Choline Esters	8
Amines, Guanidines	9,10
Amines, Hydrazines	61
Amines, Nitrogen Compounds, Biologically Active	60
Amino Acids	11,63
Amino Acids, Derivatives	62
Amino Acids, Dinitrophenyl Derivatives	65
Amino Acids and Related Compounds (Indoles, Imidazoles, Guanidines)	64
Antibiotics, Macrolide	66
Antibiotics, Streptomycins, Neomycins, Kanamycins	67
Antibiotics, Tetracycline	68
Antioxidants	69
Carbohydrates, Lactones, Glycosides, and Sugar Acids	72
Carbohydrates, Methylated Sugars	73
Carbohydrates, Oligosaccharides and Derivatives	74
Carbohydrates, Sugar Acids and Lactones	76
Carbohydrates, Sugar Esters	71
Carbohydrates, Sugar Phosphates	12
Carbohydrates, Sugars	70
Carbohydrates, Sugars and Glucosides	75
Carboxylic Acids	13
Carboxylic Acids, Benzoic Acid Derivatives	78
Carboxylic Acids and Derivatives	77
Carboxylic Acids, Hippuric Acid Derivatives	80
Carboxylic Acids, α-Keto Acids and Derivatives	81
Carboxylic Acids, Lower Fatty Acids	79
Carboxylic Acids, Miscellaneous	83
Carboxylic Acids, Phenols, and Related Compounds	82
Drugs, Antihistamines and Tranquilizers	85
Drugs, Barbiturates	86
Drugs and Miscellaneous Compounds	84
Drugs, Sulfa	87
Dyes, Indicator	88
Hydrocarbons, Aromatic	89,90
Inorganics, Anions	14,91,93
Inorganics, Cationic Complexes	94
Inorganics, Cations	92,97
Inorganics, Cations and Anions	15
Inorganics, Metal Ions	95,96
Inorganic, Phosphates	98
Lipids and Related Compounds	99
Nitro Compounds	100
Nitrogen Heterocyclics, Aminohydroxypyrimidines and Imidazoles	16
Nitrogen Heterocyclics, Imidazoles	17
Nitrogen Heterocyclics, Indoles	101,103
Nitrogen Heterocyclics, Indoles and Related Compounds	102
Nitrogen Heterocyclics, Naphthalene and Quinoline Derivatives	105
Nitrogen Heterocyclics, Nicotinic Acid and Related Compounds	18
Nitrogen Heterocyclics, Quinoline Derivatives	104,106
Nucleic Acid Derivatives	19
Nucleotides, Nucleosides and Bases	108
Oxo Compounds, Aldehydes, Ketones and Derivatives	20
Oxo Compounds, Anthraquinone Derivatives	110
Oxo Compounds, Anthraquinones	109
Oxo Compounds, Azines of Aromatic Carbonyls	23
Oxo Compounds, Ketones	21
Oxo Compounds, Naphthoquinones, Substituted	111
Oxo Compounds, 4-Nitrophenylhydrazones of Aromatic Carbonyl Compounds	22
Oxygen Heterocyclics, Anthocyanins	24,25
Oxygen Heterocyclics, Coumarins and Related Compounds	113
Oxygen Heterocyclics, Flavonoids	26
Peptides, Tyrosine, Diiodotyrosine and Thyroxine	114
Peroxides, Isomeric Monosubstituted Peroxybenzoic Acids	107
Peroxides, Organic	27
Pesticides, Chlorinated	116
Pesticides, Dithiocarbamate Fungicides	117
Pesticides, Esters of 2,4-Dichloro-Phenoxyacetic acid	115
Pesticides, Herbicides	118
Pesticides, Insecticides, Organophosphorus	28
Pesticides, Pyrethrins	119
Phenols	31
Phenols, Chlorinated, and Related Compounds	29
Phenols and Derivatives	120,121
Phenols, Polyphenols	30
Phosphorus Compounds, Alkyl Phosphates	34
Phosphorus Compounds, Aromatic and Pyridyl Phosphate Esters, and Related Compounds	112
Phosphorus Compounds, Ethyl Esters of Condensed Phosphoric Acids	35
Phosphorus Compounds, Phosphate Esters	32
Pigments, Algal	122
Pigments, Porphyrin Methyl Esters	123
Pigments, Porphyrins	33
Purines, Pyrimidines and Derivatives	36
Purines, Pyrimidines, Nucleotides and Related Compounds	37
Silicon Compounds, Methyl Phenyl Polysiloxanes	38
Steroid Glycosides, Cardiac Glycosides, Aglycones, and their Acetates	127
Steroids	124
Steroids, Bile Acids	39
Steroids, C_{20-21}	43
Steroids, C_{21}	44,45,125
Steroids, 17-Ketosteroids	41
Steroids, C_{19-20} (17-Oxysteroids)	40
Steroids and Steroid Esters	42
Steroids, Sterols and Vitamins	126
Sulfur Compounds, Sulfonic Acids	46
Sulfur Compounds, Thioureas	47
Vitamins and Related Compounds	48,50,51
Vitamins A_1, A_2 and their derivatives	49

SECTION I.IV

THIN-LAYER CHROMATOGRAPHY TABLES

	Table No. TLC
Alcohols	1
Alcohols, DNB and ONPU Derivatives	111

	Table No. TLC
Oxo Compounds, Ketones	70
Oxo Compounds, 4-Nitrophenylhydrazones	74
Oxo Compounds, Polynuclear Ring-Carbonyl Compounds	75
Oxo Compounds, Quinones, Natural	169
Oxo Compounds, Terpene Essential Oils, Carbonyls	77
Oxygen Heterocyclics, Anthocyanins and Anthocyanidins	78
Oxygen Heterocyclics, Coumarins	79
Pesticides	80,81
Pesticides, Chlorinated	82
Pesticides, Insecticides	170
Pesticides, Insecticides, Carbamate	83
Pesticides, Insecticides, Phosphorus	171
Pesticides, Organophosphorus	84,172
Pesticides, Pyrethrins	85
Pesticides, Triazine Herbicides	86
Pesticides, Urea Herbicides	173
Phenolic Compounds	174
Phenols	87,88
Phenols, 4-Acetyl-2-Nitrophenol Derivatives (ANP)	91
Phenols, Acyl	89
Phenols, Antioxidants	176
Phenols and Phenol Ethers	90
Phenols, Polynuclear	92
Phenols, Rubber Compounding Ingredients	175
Pigments, Carotenoids	93,94
Steroid 2,4-Dinitrophenylhydrazones	97
Steroid Glycosides, Digitalis Glycosides and Anglycones	177
Steroid Sapogenins	103
Steroids	95,96,98,100,101,178,179
Steroids, Cardiac Glycosides	102
Steroids, Cholanic Acids	180
Steroids, Estrogens	181
Steroids, β-Sitosteryl Esters	182
Steroids, Sterols	99
Sulfur Compounds, Thiophene Derivatives	104
Terpene Alcohols	183
Terpene Esters	105
Terpenes	106
Terpenes, Triterpene Acids	184
Toxins, Mycotoxins	185
Vitamin A and Related Compounds	108
Vitamin B_6 Compounds	109
Vitamins	189
Vitamins, Fat-Soluble	107,186,187
Vitamins, Tocopherols	110
Vitamins, Water-Soluble	188

Volume One

SECTION II.I

CHROMATOGRAPHIC DATA
Gas Chromatography Tables

Prepared by Dr. Coleman Hamel
Kutztown State College
Kutztown, Pennsylvania

Wherever possible, tables were arranged according to classes of chemical compounds. However, this was not always possible when diverse compounds were chromatographed under the same experimental conditions. The reader is referred to the master compound index for specific compounds which may appear in diverse tables.

Techniques

n.a. = not available from original source.
Solvent composition always v/v unless otherwise stated; if solid is dissolved in solvent, g/v.

TABLE GC 1
ACIDS, ALIPHATIC; DIBASIC ESTERS

Column Packing	P1	P1	P1	P1	P1	P2	P3	P4
Temperature (°C)	150	170	180	200	220	150	150	175
Gas; Flow Rate (ml/min)	13,He	13,He	13,He	13,He	13,He	150,He	150,He	10.3,He
Column								
length, cm	190.5	190.5	190.5	190.5	190.5	228.6	228.6	150 ft
diameter, cm, I.D.^{a}								
cm, O.D.^{b}	0.8^a	0.8^a	0.8^a	0.8^a	0.8^a	0.635^b	0.635^b	0.01 in.
form	U-tube	U-tube	U-tube	U-tube	U-tube	coiled	coiled	capillary
material	SS	SS	SS	SS	SS	copper	copper	na
Detector	TC	TC	TC	TC	TC	TC	TC	na
Literature	1	1	1	1	1	2	2	3
Compound				t_r				r*
Diethyl malonate	8.8	5.6	5.0	3.5	—	1.00	1.00	—
Diethyl succinate	13.4	8.6	7.5	5.4	—	1.48	1.30	—
Diethyl glutarate	20.7	12.1	10.0	—	—	—	—	—
Diethyl adipate	31.8	17.8	13.8	12.7	7.3	—	—	—
Diethyl pimelate	52.1	26.7	19.6	18.4	11.3	—	—	—
Diethyl suberate	—	—	—	24.5	14.8	—	—	—
Diethyl azelate	—	—	—	33.8	19.8	—	—	—
Diethyl fumarate	—	—	—	—	—	1.48	1.30	—
Diethyl maleate	—	—	—	—	—	2.06	2.04	—
Diethyl malate	—	—	—	—	—	—	6.81	—
Dimethyl oxalate	—	—	—	—	—	—	—	0.23
Dimethyl malonate	—	—	—	—	—	—	—	0.35
Dimethyl fumarate	—	—	—	—	—	—	—	0.38
Dimethyl succinate	—	—	—	—	—	—	—	0.47
Dimethyl itaconate	—	—	—	—	—	—	—	0.59
Dimethyl maleate	—	—	—	—	—	—	—	0.64
Dimethyl glutarate	—	—	—	—	—	—	—	0.70
Dimethyl adipate	—	—	—	—	—	—	—	1.00
Dimethyl pimelate	—	—	—	—	—	—	—	1.34
Dimethyl suberate	—	—	—	—	—	—	—	1.93
Dimethyl isosebacate	—	—	—	—	—	—	—	1.94
Dimethyl azelate	—	—	—	—	—	—	—	2.67
Dimethyl sebacate	—	—	—	—	—	—	—	3.44
Dimethyl orthophthalate	—	—	—	—	—	—	—	5.94
Dimethyl dodecanedioate	—	—	—	—	—	—	—	6.69

* Relative to dimethyl adipate.

Column Packing
P1 = 10% DC-550 on acid-washed, size-graded Celite 545 (40/120 mesh).
P2 = 25% Poly(1,4-butanediol succinate) on C-22 firebrick (60/80 mesh).
P3 = 25% Poly(diethylene glycol succinate) on C-22 firebrick (60/80 mesh).
P4 = Diethylene glycol succinate (coating).

Literature
1. Nowakowska, J., Melvin, E. H., and R. Wiebe, *J. Amer. Oil Chem. Soc.*, **34**, 411 (1957).
2. Mirocha, C. J. and J. E. DeVay, *Phytopathology*, **51**, 274 (1961).
3. Afremow, L. C., *J. Gas Chromatogr.*, **1**, (6), 29 (1963).

TABLE GC 2
ACIDS, ALIPHATIC; METHYL ESTERS

Column Packing	P1	P2	P3	P4
Temperature (°C)	150	150	150	150
Gas; Flow Rate (ml/min)	65,He	65,He	65,He	65,He
Column				
length, ft	12	12	12	12
diameter, in., O.D.	¼	¼	¼	¼
form	na	na	na	na
material	aluminum	aluminum	aluminum	aluminum
Detector	FI and TC	FI and TC	FI and TC	FI and TC

Compound	r*			
Methyl formate	0.075	0.123	0.170	0.350
Ethyl formate	0.129	0.220	0.260	0.560
Propyl formate	0.216	0.380	0.450	0.810
Isopropyl formate	0.162	0.125	0.250	0.560
Butyl formate	0.375	0.630	0.774	1.310
2-Methylpropyl formate	0.301	0.370	0.566	1.040
Pentyl formate	0.645	1.03	1.28	2.060
3-Methylbutyl formate	0.512	0.650	1.00	1.660
Hexyl formate	—	1.79	2.04	3.060
Octyl formate	—	4.93	5.50	7.310
Methyl acetate	0.140	0.260	0.330	0.500
Ethyl acetate	0.200	0.360	0.440	0.69
Propyl acetate	0.338	0.580	0.780	1.07
Isopropyl acetate	0.259	0.390	0.490	0.78
Butyl acetate	0.585	0.970	1.21	1.63
2-Methylpropyl acetate	0.468	0.800	0.490	1.32
Pentyl acetate	0.980	1.60	2.00	2.50
3-Methylbutyl acetate	0.798	1.30	1.56	2.10
Hexyl acetate	—	2.70	3.60	3.88
Octyl acetate	—	7.96	—	9.31
Methyl propionate	0.235	0.390	0.512	0.76
Ethyl propionate	0.335	0.550	0.700	1.00
Propyl propionate	0.570	0.950	1.18	1.55
Isopropyl propionate	0.405	0.700	0.770	1.19
Butyl propionate	0.955	1.53	1.95	2.44
2-Methylpropyl propionate	0.790	1.26	1.50	2.00
Pentyl propionate	1.602	2.56	3.14	3.75
3-Methylbutyl propionate	1.304	2.07	2.50	3.15
Hexyl propionate	—	4.16	5.05	5.82
Octyl propionate	—	—	12.95	14.50
Methyl butyrate	0.397	0.630	0.78	1.13
Ethyl butyrate	0.565	0.930	1.07	1.44
Propyl butyrate	0.920	1.57	1.80	2.25
Isopropyl butyrate	0.680	1.07	1.19	1.66
Butyl butyrate	1.48	2.47	2.93	3.35
2-Methylpropyl butyrate	1.18	1.97	2.35	2.72
Pentyl butyrate	2.35	4.11	4.78	5.12
3-Methylbutyl butyrate	1.97	3.32	3.80	4.28
Hexyl butyrate	—	6.704	7.58	7.80
Octyl butyrate	—	18.10	19.20	18.37
Methyl isobutyrate	0.319	—	—	—
Ethyl isobutyrate	0.445	0.710	0.83	1.19
Propyl isobutyrate	0.722	1.130	1.40	1.69
Isopropyl isobutyrate	0.565	0.770	0.86	1.23
Butyl isobutyrate	1.17	1.83	2.21	2.80
2-Methylpropyl isobutyrate	0.997	1.490	1.73	2.25
Pentyl isobutyrate	1.87	3.10	3.54	4.10
3-Methybutyl isobutyrate	1.59	2.46	2.86	3.44
Hexyl isobutyrate	—	4.94	5.70	6.57
Octyl isobutyrate	—	13.28	14.55	15.50
Methyl pentanoate	0.650	1.07	1.40	2.00
Ethyl pentanoate	0.900	1.50	1.90	2.31
Propyl pentanoate	1.480	2.46	3.00	3.63
Isopropyl pentanoate	1.110	1.68	2.00	2.50

TABLE GC 2—(Continued)
ACIDS, ALIPHATIC; METHYL ESTERS

Column Packing	P1	P2	P3	P4
Temperature (°C)	150	150	150	150
Gas; Flow Rate (ml/min)	65,He	65,He	65,He	65,He
Column				
length, ft	12	12	12	12
diameter, in., O.D.	¼	¼	¼	¼
form	na	na	na	na
material	aluminum	aluminum	aluminum	aluminum
Detector	FI and TC	FI and TC	FI and TC	FI and TC
Compound		r*		
Butyl pentanoate	2.34	4.01	4.86	5.67
2-Methylpropyl pentanoate	1.97	3.30	3.74	4.44
Pentyl pentanoate	3.72	6.40	7.50	8.40
3-Methylbutyl pentanoate	3.12	5.28	6.21	6.65
Hexyl pentanoate		10.20	12.0	12.46
Octyl pentanoate		—	29.8	28.15
Methyl 3-methylbutyrate	0.508	0.900	1.10	1.41
Ethyl 3-methylbutyrate	0.740	1.130	1.44	1.72
Propyl 3-methylbutyrate	1.19	1.900	2.25	2.62
Isopropyl 3-methylbutyrate	0.92	1.230	1.49	1.91
Butyl 3-methylbutyrate	1.89	3.06	3.58	4.05
2-Methylpropyl 3-methylbutyrate	1.58	2.43	2.86	3.25
Pentyl 3-methylbutyrate	2.99	5.30	5.45	6.12
3-Methylbutyl 3-methylbutyrate	2.52	3.95	4.70	4.96
Hexyl 3-methylbutyrate	—	8.35	9.14	9.16
Octyl 3-methylbutyrate	—	—	—	21.42
Methyl hexanoate	1.06	1.83	2.25	2.70
Ethyl hexanoate	1.47	2.68	3.08	3.26
Propyl hexanoate	2.36	4.35	4.90	5.19
Isopropyl hexanoate	1.78	3.07	3.30	3.63
Butyl hexanoate	3.75	7.30	7.64	7.88
2-Methylpropyl hexanoate	3.08	5.65	6.02	6.51
Pentyl hexanoate	5.96	11.72	12.12	12.20
3-Methylbutyl hexanoate	4.93	9.18	9.52	9.79
Hexyl hexanoate	—	—	19.68	18.09
Octyl hexanoate	—	—	—	41.40
Methyl 4-methylpentanoate	0.88	1.65	1.84	2.25
Ethyl 4-methylpentanoate	1.25	2.16	2.56	2.84
Propyl 4-methylpentanoate	1.96	3.50	3.96	4.38
Isopropyl 4-methylpentanoate	1.51	2.49	2.66	3.20
Butyl 4-methylpentanoate	3.20	5.52	6.37	6.63
2-Methylpropyl 4-methylpentanoate	2.65	4.70	4.80	5.35
Pentyl 4-methylpentanoate	5.02	9.45	10.00	10.10
3-Methylbutyl 4-methylpentanoate	4.25	7.70	7.79	8.03
Hexyl 4-methylpentanoate	—	13.25	15.80	15.30
Octyl 4-methylpentanoate	—	—	40.60	35.20

* Relative to nonane.

Column Packing P1 = 10% SE-30 on acid-washed silanized Celite 560 (60/80 mesh).
P2 = 10% OV-17 on acid-washed silanized Celite 560 (60/80 mesh).
P3 = 10% OV-25 on acid-washed silanized Celite 560 (60/80 mesh).
P4 = 10% XE-60 on acid-washed silanized Celite 560 (60/80 mesh).

Literature Allen, I. D. and J. K. Haken, *J. Chromatogr.*, **49**, 409 (1970).

TABLE GC 3
ACIDS, ALIPHATIC; METHYL ESTERS

Column Packing	P1	P2
Temperature (°C)	197	180
Gas; Flow Rate (ml/min)	Ar	Ar
Column		
length, ft	4	4
diameter, mm, I.D.	4 to 5	4 to 5
form	na	na
material	na	na
Detector	EC	EC

Compound	r*	
Methyl *n*-pentanoate	0.019	—
Methyl *n*-hexanoate	0.031	—
Methyl 4-methylhexanoate	0.041	—
Methyl *n*-heptanoate	0.046	—
Methyl 6-methylheptanoate	0.057	—
Methyl *n*-octanoate	0.071	0.141
Methyl 6-methyloctanoate	0.097	—
Methyl *n*-nonanoate	0.11	0.193
Methyl 8-methylnonanoate	0.144	—
Methyl *n*-decanoate	0.173	0.264
Methyl 8-methyldecanoate	0.235	—
Methyl *n*-undecanoate	0.274	0.360
Methyl 10-methylundecanoate	0.354	—
Methyl *n*-dodecanoate	0.426	0.510
Methyl 10-methyldodecanoate	0.567	—
Methyl *n*-tridecanoate	0.580	0.710
Methyl *cis*- Δ^{9}-tetradecenoate	0.92	1.2
Methyl *n*-tetradecanoate	1.0	1.0
Methyl 12-methyltetradecanoate	1.35	—
Methyl *n*-pentadecanoate	1.54	1.38
Methyl 14-methylpentadecanoate	2.04	—
Methyl *cis*- Δ^{9}-hexadecenoate	2.09	2.26
Methyl *trans*- Δ^{9}-hexadecenoate	2.16	2.26
Methyl *n*-hexadecanoate	2.34	1.96
Methyl 14-methylhexadecanoate	3.24	—
Methyl *n*-heptadecanoate	3.59	2.75
Methyl *cis*- $\Delta^{9,12}$-octadecadienoate	4.55	5.27
Methyl $\Delta^{9,12,15}$-octadecatrienoate	4.60	6.88
Methyl *cis*- Δ^{9}-octadecenoate	4.75	4.33
Methyl *trans*- Δ^{9}-octadecenoate	4.95	4.33
Methyl *cis*- Δ^{6}-octadecenoate	4.87	4.35
Methyl *cis*- Δ^{4}-octadecenoate	4.95	4.36
Methyl *trans*- Δ^{4}-octadecenoate	5.15	—
Methyl *n*-octadecanoate	5.5	3.86
Methyl $\Delta^{5,8,11,14}$-eicostatetraenoate	8.25	12.85
Methyl eicosapentaenoate	8.25	16.7
Methyl *n*-eicosanoate	12.1	7.54
Methyl docosahexaenoate	16.8	35.7
Methyl Δ^{13}-docosaenoate	25.6	16.0
Methyl *n*-docosanoate	26.3	—

* Relative to myristate.

Column Packing P1 = 20% Apiezon L grease on Celite (100/210 mesh).
P2 = 20% Polyethylene glycol adipate on Celite (100/210 mesh).

Literature James, A. T., *J. Chromatogr.*, **2**, 552 (1959).

TABLE GC 4
ACIDS, AROMATIC, TRIMETHYLSILYL DERIVATIVES

Column Packing	P1
Temperature (°C)	T1
Gas; Flow Rate (ml/min)	170,N_2
Column	
length, ft	8
diameter, mm, I.D.	5.5
form	U-tube
material	—
Detector	FI

Compound	Methylene unit*
Benzoic acid	12.33
Phenylacetic acid	12.81
3-Phenylpropionic acid	13.98
***m*-Methoxyphenylacetic acid**	15.01
***p*-Methoxyphenylacetic acid**	15.17
Cinnamic acid	15.27
***o*-Hydroxyphenylacetic acid**	15.60
***m*-Hydroxybenzoic acid**	15.60
β-Phenyllactic acid	15.85
3-(*o*-Methoxyphenyl)propionic acid	15.90
***m*-Hydroxyphenylacetic acid**	15.98
***p*-Hydroxybenzoic acid**	16.21
***p*-Hydroxyphenylacetic acid**	16.28
3-(*p*-Methoxyphenyl)propionic acid	16.37
2,5-Dimethoxyphenylacetic acid	16.77
3,4-Dimethoxyphenylacetic acid	16.92
Phenylpyruvic acid	17.05
3-(*p*-Hydroxyphenyl)propionic acid	17.51
***o*-Methoxycinnamic acid**	17.54
***m*-Methoxycinnamic acid**	17.66
β-(*p*-Methoxyphenyl)lactic acid	17.97
***p*-Methoxycinnamic acid**	18.03
3,4,5-Trimethoxyphenylacetic acid	18.12
***o*-Methoxyphenylpyruvic acid**	18.60
β-(*p*-Hydroxyphenyl)lactic acid	19.06
3-(3,4-Dihydroxyphenyl)propionic acid	19.46
***p*-Methoxyphenylpyruvic acid**	19.50
3,5-Dimethoxycinnamic acid	19.94
3,4-Dimethoxycinnamic acid	19.98
***m*-Hydroxyphenylpyruvic acid**	20.00
2,4-Dimethoxycinnamic acid	20.12
***p*-Hydroxyphenylpyruvic acid**	20.51
4-Hydroxy-3-methoxycinnamic acid	20.80
3,4,5-Trimethoxycinnamic acid	21.19
3,4-Dimethoxyphenylpyruvic acid	21.30
3,4-Dihydroxycinnamic acid	21.36
3,4-Dihydroxyphenylpyruvic acid	22.09
3,5-Dimethoxy-4-hydroxycinnamic acid	22.12

* 1 methylene unit = retention time difference between an *n*-alkane with *m* and one with *m* + 1 carbons.

Column Packing P1 = 13% SE-30 on acid-washed Chromosorb W (silanized glass wool used at exit end of column).

Temperature T1 = programmed; held at 100 °C for 5 minutes after sample injection, then to 300 °C at 5°/minute.

Literature Hoffman, E., Milling, Jr., A., and D. Parmelee, *Anal. Biochem.*, **32**, 386 (1969).

TABLE GC 5
ACIDS, AROMATIC, TRIMETHYLSILYL DERIVATIVES

Column Packing	P1	P2	P1	P2
Temperature (°C)	T1	T1	T2	T2
Gas; Flow Rate (ml/min)	50,Ar	50,Ar	50,Ar	50,Ar
Column				
length, m	1.5	1.5	4	4
diameter, mm, I.D.	4	4	4	4
form	coiled	coiled	coiled	coiled
material	—	—	—	—
Detector	FI	FI	FI	FI

Compound	Methylene units			
Benzoic	13.65	12.26	13.96	12.31
Hippuric	21.05	18.02	21.13	18.04
Phenylacetic	14.40	12.64	14.40	12.72
Phenaceturic	21.67	18.47	21.77	18.43
	—	18.62	—	18.57
Phenylacetylglutamic	24.68	20.67	24.91	20.68
	26.06	22.39	26.00	22.35
	—	22.96	—	22.95
Phenylacetylglutamine	24.67	20.68	24.93	20.68
	24.97	20.80	25.21	20.82
	27.96	21.95	*ca.* 28.4	22.00
	—	—	*ca.* 28.5	—
β-Phenylpropionic	15.52	14.05	15.65	14.06
Cinnamic	17.04	14.97	17.17	15.21
Mandelic	15.91	14.69	15.90	14.57
β-Phenyllactic	16.88	15.87	16.83	15.82
Phenylpyruvic	18.28	17.02	18.32	16.99
***N*-Acetylphenylalanine**	19.39	17.88	19.31	17.90
	20.54	—	20.65	—
***o*-Hydroxybenzoic**	16.26	15.01	16.30	15.00
***o*-Hydroxyhippuric**	22.92	20.41	23.02	20.51
***o*-Hydroxyphenylacetic**	16.92	15.56	16.85	15.53
***o*-Hydroxyphenylpyruvic, lactone**	18.95	16.50	19.32	16.74
***o*-Hydroxycinnamic**	19.57	18.00	19.62	18.03
***m*-Hydroxybenzoic**	16.66	15.57	16.76	15.51
***m*-Hydroxyhippuric**	23.80	21.20	23.78	21.20
***m*-Hydroxyphenylacetic**	17.36	15.97	17.25	15.97
β-(*m*-Hydroxyphenyl)propionic	18.52	17.17	18.48	17.13
***m*-Hydroxycinnamic**	20.05	18.63	20.00	18.57
β-(*m*-Hydroxyphenyl)hydracrylic	19.41	18.63	19.22	18.50
***p*-Hydroxybenzoic**	17.32	16.20	17.25	16.20
***p*-Hydroxyphenylacetic**	17.65	16.29	17.59	16.25
β-(*p*-Hydroxyphenyl)propionic	18.88	17.53	18.83	17.48
***p*-Hydroxycinnamic**	20.84	19.28	20.84	19.30
***p*-Hydroxymandelic**	18.88	17.94	18.65	17.74
β-(*p*-Hydroxyphenyl)lactic	20.00	19.13	19.86	19.05
***p*-Hydroxyphenylpyruvic**	21.59	20.61	21.49	20.54
***N*-Acetyltyrosine**	24.86	21.29	24.90	21.24
***p*-Methoxybenzoic**	16.96	14.95	17.01	15.04
***p*-Methoxyhippuric**	24.47	20.79	24.58	20.83
***p*-Methoxyphenylacetic**	17.28	15.13	17.33	15.10
***p*-Methoxyphenaceturic**	24.75	20.82	24.82	20.85
***p*-Methoxyphenylacetylglutamic**	27.80	23.19	*ca.* 28.1	23.16
	ca. 28.8	24.44	*ca.* 28.8	24.42
β-(*p*-Methoxyphenyl)propionic	18.46	16.30	18.54	16.36
***p*-Methoxycinnamic**	20.58	17.96	20.69	18.08
***p*-Methoxymandelic**	18.75	16.91	18.64	16.81
β-(*p*-Methoxyphenyl)lactic	19.62	17.95	19.60	18.01
***p*-Methoxyphenylpyruvic**	21.42	19.54	21.46	19.54
***N*-Acetyl-*p*-methoxyphenylalanine**	23.51	20.09	23.56	20.13
2,3-Dihydroxybenzoic	18.48	17.49	18.45	17.51
2,4-Dihydroxybenzoic	19.20	18.23	19.07	18.18
2,5-Dihydroxybenzoic	18.78	17.85	18.60	17.74

TABLE GC 5—(Continued)
ACIDS, AROMATIC, TRIMETHYLSILYL DERIVATIVES

Column Packing	P1	P2	P1	P2
Temperature (°C)	T1	T1	T2	T2
Gas; Flow Rate (ml/min)	50,Ar	50,Ar	50,Ar	50,Ar
Column				
length, m	1.5	1.5	4	4
diameter, mm, I.D.	4	4	4	4
form	coiled	coiled	coiled	coiled
material	—	—	—	—
Detector	FI	FI	FI	FI

Compound	Methylene units			
2,5-Dihydroxyphenylacetic	19.40	18.45	19.27	18.32
2,5-Dihydroxyphenylacetic, lactone	18.70	17.30	18.71	17.30
2,5-Dihydroxyphenylpyruvic, lactone	22.03	20.21	22.27	20.35
2,6-Dihydroxybenzoic	18.86	17.68	18.81	17.64
3,4-Dihydroxybenzoic	19.17	18.25	19.05	18.21
3,4-Dihydroxyphenylacetic	19.39	18.35	19.31	18.30
3,4-Dihydroxyphenylcinnamic	22.48	21.49	22.43	21.40
3,4-Dihydroxymandelic	20.19	19.48	20.02	19.33
3,5-Dihydroxybenzoic	19.28	18.28	18.99	18.11
3-Hydroxy-4-methoxybenzoic	19.33	17.60	19.23	17.52
3-Hydroxy-4-methoxyhippuric	26.34	23.26	26.31	23.18
3-Hydroxy-4-methoxyphenylacetic	19.39	17.58	19.30	17.48
β-(3-Hydroxy-4-methoxyphenyl)propionic	20.45	18.71	20.42	18.68
3-Hydroxy-4-methoxycinnamic	22.57	20.54	22.55	20.59
β-(3-Hydroxy-4-methoxyphenyl)hydracrylic	21.18	20.04	21.06	19.88
4-Hydroxy-3-methoxybenzoic	19.11	17.53	19.05	17.49
4-Hydroxy-3-methoxyhippuric	24.88	23.27	24.71	23.32
4-Hydroxy-3-methoxyphenylacetic	19.38	17.61	19.33	17.54
β-(4-Hydroxy-3-methoxyphenyl)propionic	20.53	18.84	20.57	18.83
4-Hydroxy-3-methoxycinnamic	22.76	20.78	22.75	20.78
4-Hydroxy-3-methoxymandelic	20.20	18.92	20.06	18.74
β-(4-Hydroxy-3-methoxyphenyl)lactic	21.45	20.27	21.36	20.23
4-Hydroxy-3-methoxyphenylpyruvic	23.34	21.67	23.29	21.76
4-Hydroxy-3-ethoxybenzoic	19.55	18.04	19.46	18.02
3,4-Dimethoxybenzoic	19.47	16.84	19.44	16.76
3,4-Dimethoxyphenylacetic	19.61	16.87	19.56	16.78
4-Hydroxy-3,5-dimethoxybenzoic	20.90	18.89	20.83	18.80
3-Indolylcarboxylic	22.60	19.95	22.78	20.01
3-Indolylacetic (i)	21.82	18.65	22.44	18.74
(ii)	21.71	19.34	22.00	19.41
β-(3-Indolyl)propionic (i)	23.15	19.71	23.55	19.86
(ii)	22.76	20.53	23.07	20.81
β-(3-Indolyl)lactic (i)	23.92	21.07	24.17	21.18
(ii)	23.34	21.68	23.55	21.79
3-Indolylpyruvic	26.33	24.30	26.41	24.34
***N*-Acetyltryptophan**	25.65	24.00	25.82	24.11
	27.60	—	27.62	—
3-(5-Hydroxyindolyl)acetic	24.31	22.06	24.29	22.03

Column Packing P1 = 10% OV-17 on Diatoport S (80/100 mesh).
P2 = 10% OV-1 on Diatoport S (80/100 mesh).

Temperature T1 = initial temperature 100 °C, 2°/minute.
T2 = initial temperature 170 °C, 1°/minute.

Literature Coward, R. F. and P. Smith, *J. Chromatogr.*, **45**, 230 (1969).

TABLE GC 6
ALDEHYDES, DIMETHYL ACETALS

Column Packing	P1	P2	P3
Temperature (°C)	190	150	150
Gas; Flow Rate (ml/min)	100,Ar	50,N_2	50,N_2
Column			
length, cm	120	150	150
diameter, mm	4	4	4
form	na	coiled	coiled
material	glass	glass	glass
Detector	na	na	na

Compound	r (hexadecanal acetal derivative)		
Dodecanal	0.170	0.214	0.251
11-Methyldodecanal	0.221	0.265	0.298
10-Methyldodecanal	0.229	0.279	0.324
Tridecanal	0.267	0.314	0.353
12-Methyltridecanal	0.345	0.392	0.417
11-Methyltridecanal	0.363	0.412	0.450
Tetradecanal (myristaldehyde)	0.412	0.460	0.500
13-Methyltetradecanal	0.541	0.570	0.593
12-Methyltetradecanal	0.563	0.596	0.628
Pentadecanal	0.645	0.68	0.705
14-Methylpentadecanal	0.840	0.846	0.84
13-Methylpentadecanal	0.895	0.890	0.90
Hexadecanal (palmitaldehyde)	1.00	1.00	1.00
Hexadecenal (palmitoleic aldehyde)	0.840	1.13	1.15
15-Methylhexadecanal	1.30	1.25	1.18
14-Methylhexadecanal	1.36	1.32	1.28
Heptadecanal	1.55	1.47	1.43
Octadecanal (stearaldehyde)	2.39	2.16	2.01
***cis*-Octadeca-9-Dienol (oleylaldehyde)**	2.02	2.43	2.33
Octadecenal	2.14	—	—
***cis,cis*-Octadeca-9,12-Dienol (linoleylaldehyde)**	1.89	2.92	—
Octadeca-9,12,15-enal (linolenylaldehyde)	1.89	3.67	—
Methyl hexadecanoate (palmitate)	0.775	1.16	1.25

Column Packing P1 = 12.5% Apiezon L on "Alkaline" Gas Chrom P (80/100 mesh).
P2 = 10% EGA on Gas Chrom CLH (80/100 mesh).
P3 = 15% EGSSX on Gas Chrom CLH (80/100 mesh).

Literature Gray, G. M., *J. Chromatogr.*, **6**, 236 (1961).

TABLE GC 7
ALDEHYDES AND KETONES

Column Packing	P1	P2	P3	P4	P5	P6	P7	P7
Temperature (°C)	100	100	100	70	100	70	90	150
Gas; Flow Rate (ml/min.)	—	—	—	—	—	—	32,He	32,He
Column								
length, ft	—	—	—	—	—	—	10	10
diameter, in., O.D.	—	—	—	—	—	—	$\frac{1}{4}$	$\frac{1}{4}$
form	—	—	—	—	—	—	na	na
material	—	—	—	—	—	—	SS	SS
Detector	—	—	—	—	—	—	TC	TC
Literature	1	1	1	1	1	2	3	3
Compound	r (pentane or benzene)						t_r	
Acetaldehyde	—	—	—	0.16	—	0.43	4.5	—
Propionaldehyde	0.80	1.36	1.70	0.34	10.5	0.72	6.8	—
Acrolein	0.74	1.49	1.96	0.46	13.9	0.96	9.2	—
Isobutyraldehyde	1.48	2.26	2.82	0.46	12.2	0.78	7.7	—
n-Butyraldehyde	1.97	3.12	4.06	0.69	19.3	1.2	10.4	—
Crotonaldehyde	3.22	6.00	8.43	1.9	58.4	4.0	—	6.9
Paraldehyde	12.8	26.8	13.2	—	61.7	—	—	—
Isovaleraldehyde	—	—	—	—	—	—	13.1	—
n-Valeraldehyde	—	—	—	—	—	—	17.0	—
2-Methyl-1-butanal	—	—	—	—	—	—	12.0	—
Crotonaldehyde	—	—	—	—	—	—	26.9	—
2,4-Pentadienal	—	—	—	—	—	—	—	12.2
2-Hexenal	—	—	—	—	—	—	—	12.7
Benzaldehyde	—	—	—	—	—	—	—	>32
Acetone	0.96	1.34	1.92	0.41	13.7	1.1	7.6	—
Methyl ethyl ketone	2.07	3.21	4.44	0.79	22.7	1.7	11.7	—
Diethyl ketone	4.29	6.68	9.01	1.5	37.9	2.4	—	—
Acetylacetone	8.10	17.5	22.9	—	159	—	—	—
Ethanol	0.69	1.49	1.68	0.73	31.5	1.1	—	—
2-Pentanone	—	—	—	—	—	—	17.8	4.6
3-Pentanone	—	—	—	—	—	—	19.1	5.3
3-Methyl-2-butanone	—	—	—	—	—	—	14.0	—
2-Hexanone	—	—	—	—	—	—	30.0	6.6
4-Methyl-3-penten-2-one	—	—	—	—	—	—	—	12.3
Cyclohexanone	—	—	—	—	—	—	—	24.6
Diacetyl (bis)	—	—	—	—	—	—	>25	—
n-Pentane	1.00	1.00	1.00	—	1.00	—	—	—
Benzene	—	—	—	1.00	—	1.00	—	—

Column Packing
P1 = Silicone.
P2 = Dioctyl sebacate.
P3 = Didecyl phthalate.
P4 = Hexamethylene glycol dimethyl ether.
P5 = Polyethylene glycol.
P6 = β,β'-Oxydipropionitrile.
P7 = 30% Carbowax 1540 on firebrick.

Literature
1. Raupp, G., *Z. Anal. Chemie.*, **164**, 135 (1958).
2. Kelker, H., *Angew. Chemie.*, **71**, 218 (1959).
3. Ralls, J. W., *Anal. Chem.*, **32**, 332 (1960).

TABLE GC 8
ALKALOIDS

Column Packing	P1		P1		P1		P1	
Temperature (°C)	210		230		250		270	
Gas; Flow Rate (cc/min)	30.7,N_2		30.7,N_2		30.7,N_2		30.7,N_2	
Column								
length, ft.	5		5		5		5	
diameter, in.	0.093		0.093		0.093		0.093	
form	na		na		na		na	
material	SS		SS		SS		SS	
Detector	FI		FI		FI		FI	
Compound	t_r	r*	t_r	r	t_r	r	t_r	r
Aconitine	—	—	10.6	1.08	5.1	1.06	3.0	1.00
Apomorphine	21.3	1.38	12.3	1.28	5.1	1.06	3.2	1.07
Atropine (*d*-hyoscyamine)	9.2	0.60	5.7	0.59	3.0	0.63	2.1	0.70
Berberine	—	—	3.9	0.41	2.1	0.44	—	—
Betaine	24.0	1.56	11.3	1.18	5.2	1.08	3.1	1.03
Caffeine	3.4	0.22	2.1	0.22	1.2	0.25	—	—
Cinchonine (*d*-cinchonine)	—	—	13.9	1.45	7.9	1.65	5.3	1.77
Cinchonidine (*l*-cinchonine)	—	—	13.9	1.45	7.9	1.65	5.2	1.73
Cocaine	9.2	0.60	5.5	0.57	2.8	0.58	1.9	0.63
Codeine	15.4	1.00	9.6	1.00	4.8	1.00	3.0	1.00
Colchicine	—	—	11.0	1.15	5.2	1.08	3.0	1.00
Cotarnine	3.1	0.20	2.2	0.23	1.2	0.25	—	—
Diacetylmorphine	32.0	2.08	13.3	1.39	6.3	1.31	3.9	1.30
Dibucaine	—	—	20.3	2.12	9.4	1.96	5.5	1.83
Dihydrocodeinone	—	—	11.3	1.18	5.4	1.12	—	—
Dihydrohydroxycodeinone	—	—	1.8	0.19	1.1	0.23	—	—
Dihydromorphinone	25.0	1.62	14.1	1.47	6.1	1.27	1.2	0.40
Emetine	—	—	11.6	1.21	5.1	1.06	—	—
Ethylmorphine	19.5	1.27	11.6	1.21	5.0	1.04	3.2	1.07
Gelsemine	—	—	27.0	2.81	10.0	2.08	6.1	2.04
Homatropine	7.3	0.47	4.2	0.44	2.4	0.50	1.6	0.53
Hydrastine	—	—	11.3	1.18	5.2	1.08	3.2	1.07
l-Hyoscyamine	9.4	0.61	5.7	0.59	2.9	0.60	2.1	0.70
Meperidine	—	—	1.7	0.18	1.1	0.23	—	—
Mescaline	2.3	0.15	2.0	0.21	1.0	0.21	—	—
d,l-Methadone	—	—	4.7	0.49	—	—	—	—
Morphine	17.0	1.10	11.6	1.21	5.8	1.19	4.2	1.40
N-Allylnormorphine	—	—	17.8	1.85	8.0	1.67	5.0	1.67
Nicotine	2.6	0.17	0.8	0.08	—	—	—	—
Papaverine	—	—	35.6	3.71	13.2	2.75	7.8	2.60
Pilocarpine	—	—	7.0	0.73	2.3	0.48	1.7	0.57
Piperine	—	—	28.5	2.97	13.6	2.84	7.8	2.60
Procaine	6.1	0.40	3.5	0.36	2.0	0.42	1.4	0.47
Quinine	—	—	27.0	2.81	13.0	2.70	8.2	2.74
Scopolamine	13.6	0.88	8.1	0.84	3.8	0.79	2.6	0.87
Sparteine	2.8	0.18	1.9	0.20	1.6	0.33	—	—
Strychnine	—	—	—	—	26.0	5.40	—	—
Thebaine	25.6	1.66	13.8	1.44	6.0	1.25	3.8	1.27
Theobromine	24.3	1.58	10.4	1.08	5.0	1.04	3.2	1.06
d-Tubocurarine	—	—	11.2	1.17	5.0	1.04	—	—
Yohimbine	—	—	3.6	0.38	1.9	0.40	—	—

* Retention time relative to codeine.

Column Packing P1 = 5% (w/w) SE-30 on acid washed Chromosorb W (60/80 mesh).

Literature Parker, K. D., Fontan, C. R., and P. L. Kirk, *Anal. Chem.*, **35**, 356 (1963).

TABLE GC 9
AMINES, ACETYLATED DERIVATIVES

Column Packing	P1	P1	P1	P1	P2
Temperature (°C)	198	216	165	183	215
Gas; Flow Rate	Ar,na	Ar,na	Ar,na	Ar,na	Ar,na
Column,					
length, ft	6	6	6	6	6
diameter, mm, I.D.	4	4	4	4	4
form	U-tube	U-tube	U-tube	U-tube	U-tube
material	glass	glass	glass	glass	glass
Detector	EC	EC	EC	EC	EC
Acetylated derivatives	r (anthracene)				
β-Phenylethylamine	0.39	—	0.11	0.12	—
β-Hydroxyphenylethylamine	1.08	—	0.18	0.20	0.22
Tyramine	2.20	1.77	0.49	0.46	1.14
Octopamine	5.91	4.39	1.00	0.88	3.3
Homovanillylamine	4.44	3.43	0.73	0.68	0.86
Normetanephrine	11.0	7.43	1.28	1.15	2.24
Dopamine	7.53	5.44	—	—	—
Norepinephrine	19.5	12.05	—	—	—
Homoveratrylamine	2.35	2.00	—	—	—
Metanephrine	9.81	6.80	—	—	—
Epinephrine	17.3	11.2	—	—	—
Amphetamine	0.37	—	0.09	0.11	0.06
Norephedrine	0.96	0.84	—	—	—
Vanillylamine	3.45	2.75	—	—	—
Mescaline	4.34	—	—	—	—
3,4-Dihydroxynorephedrine	14.1	9.18	—	—	—
Deoxyephedrine	0.49	—	—	—	—
Ephedrine	1.12	1.05	—	—	—
Synephrine	5.35	4.17	—	—	—
Phenylephrine	4.71	3.36	—	—	—
Hordenine	0.36	—	—	—	—
4-Methoxy-β-phenylethylamine	1.20	—	—	—	—
Tryptamine	—	—	—	1.8	3.4
***N*-Methyltryptamine**	2.34	—	—	—	—
***N,N*-Dimethyltryptamine**	1.00	0.98	—	—	—
5-Hydroxy-*N,N*-dimethyltryptamine (bufotenin)	—	3.71	—	—	—
5-Methoxytryptamine	2.84	—	—	4.5	—
5-Methoxy-*N,N*-dimethyltryptamine	2.60	2.22	—	—	—
***N*-Acetyltryptamine**	8.73	6.35	—	—	—
***N*-Acetyl-*N*-methyltryptamine**	7.43	5.80	—	—	—
***N*-Acetyl-5-acetoxytryptamine (Diacetylserotonin)**	—	36.0	—	—	—
5-Acetoxy-*N,N*-dimethyltryptamine	5.70	4.16	—	—	—
***N*-Acetyl-5-methoxytryptamine**	22.5	15.0	—	—	—
***N*-Acetyl-*N*-methylhistamine**	1.23	1.01	—	—	—
***N,N*-Diacetylhistamine**	2.31	1.94	—	—	—
Phenylalanine	0.78	—	—	—	—
Tyrosine	3.89	—	—	—	—
Histidine	4.67	—	—	—	—
Tryptophan	16.7	—	—	—	—

Column Packing P1 = 7% Silicone oil F-60 and 1% ethylene glycol succinate-phenylmethylsiloxane Copolymer EGSS-Z on acid washed, dichlorodimethylsilane treated Gas Chrom P (80/100 or 100/120 mesh).

P2 — 10% NGS (neopentyl glycol succinate) polyester on acid washed, dichlorodimethylsilane treated Gas Chrom P (80/100 or 100/120 mesh). Preconditioned 210°-225°C in argon 12–24 hours.

Literature Brooks, C. J. W. and E. C. Horning, *Anal. Chem.*, **36**, 1540 (1964).

TABLE GC 10
AMINES, ALIPHATIC, TRIFLUOROACETYL DERIVATIVES

Column Packing	P1	P2	P3	P4	P5	P5
Temperature (°C)	T1	T2	T3	T4	T5	220 °C
Gas; Flow Rate (ml/min)	He	He	He	He	80,He	80,He
Column						
length, m	50	50	30	15	2	2
diameter, mm, I.D.	0.25	0.25	0.25	0.5	3	3
form	capillary	capillary	capillary	—	na	na
material	SS	SS	SS	SS	SS	SS
Detector	FI	FI	FI	FI	TC	TC
Literature	1	1	1	1	2	2
Trifluoroacetyl derivatives of:			$r_{(di-n-butvlamine)}$			
Ethylamine	0.12	0.05	0.08	0.05	—	—
***n*-Propylamine**	0.20	0.13	0.14	0.73	—	—
Isopropylamine	0.09	0.05	0.07	0.32	—	—
***n*-Butylamine**	0.40	0.26	0.28	0.96	—	—
Isobutylamine	0.24	0.18	0.20	0.77	—	—
***sec*-Butylamine**	0.19	0.14	0.17	0.61	—	—
***tert*-Butylamine**	0.06	0.05	0.09	0.23	—	—
***n*-Amylamine**	0.66	0.46	0.48	1.19	—	—
Isoamylamine	0.55	0.38	0.32	1.07	—	—
***n*-Hexylamine**	0.94	0.72	0.73	1.39	—	—
***n*-Heptylamine**	1.22	1.00	1.00	1.68	—	—
***n*-Octylamine**	1.52	1.28	1.39	1.86	—	—
Allylamine	0.20	0.10	0.12	—	—	—
Dimethylamine	0.09	0.05	0.07	0.12	—	—
Diethylamine	0.19	0.18	0.20	0.27	—	—
Diisopropylamine	0.27	0.33	0.35	—	—	—
Di-*n*-proplyamine	0.48	0.52	0.53	0.50	—	—
Diisobutylamine	0.66	0.75	0.76	—	—	—
Di-*n*-butylamine	1.00	1.00	1.00	1.00	—	—
Di-*sec*-butylamine	0.69	0.81	—	0.70	—	—
Diisoamylamine	1.22	1.28	1.29	—	—	—
Di-*n*-amylamine	1.37	1.39	1.50	—	—	—
***N*-Methyl-ethylamine**	0.13	0.13	0.12	0.20	—	—
***N*-Methyl-isopropylamine**	0.19	0.18	0.20	0.19	—	—
***N*-Methyl-*n*-propylamine**	0.24	0.22	0.23	—	—	—
***N*-Methyl-isobutylamine**	0.31	0.32	0.32	0.35	—	—
***N*-Methyl-*n*-butylamine**	0.43	0.41	0.41	0.46	—	—
***N*-Methyl-isoamylamine**	0.55	0.53	0.54	0.60	—	—
***N*-Methyl-*n*-amylamine**	0.71	0.65	0.64	—	—	—
Diallylamine	0.40	0.68	0.39	—	—	—
Aniline	1.34	0.94	0.85	2.08	—	—
***o*-Toluidine**	1.44	1.09	1.00	1.90	—	—
***m*-Toluidine**	1.60	1.21	1.13	—	—	—
***p*-Toluidine**	1.64	1.24	1.17	—	—	—
2,3-Dimethylaniline	1.87	1.47	1.35	2.30	—	—
2,4-Dimethylaniline	1.77	1.39	0.88	—	—	—
2,5-Dimethylaniline	1.68	1.34	1.24	2.17	—	—
2,6-Dimethylaniline	1.74	1.34	0.97	2.27	—	—
3,4-Dimethylaniline	2.02	1.58	1.47	2.45	—	—
***o*-Ethylaniline**	1.57	1.24	1.17	2.02	—	—
***p*-Ethylaniline**	1.89	1.49	1.50	—	—	—
Pyrrole	0.31	0.52	0.53	—	—	—
Pyrroline	0.66	0.43	0.39	0.79	—	—
Pyrrolidine	0.80	0.52	0.48	0.80	—	—
Piperidine	0.80	0.65	0.58	0.76	—	—
(±)-1-Phenethylamine	1.60	1.15	1.13	2.18	—	—
2-Phenethylamine	1.93	1.39	1.35	2.39	—	—
1,2,5,6-Tetrahydropryridine	0.81	0.63	—	0.89	—	—
Trimethylenediamine	—	—	—	—	18.0	4.2
Tetramethylenediamine	—	—	—	—	19.9	5.3
Pentamethylenediamine	—	—	—	—	21.5	6.6
Hexamethylenediamine	—	—	—	—	23.0	8.1
Heptamethylenediamine	—	—	—	—	24.9	9.9

TABLE GC 10—(Continued)
AMINES, ALIPHATIC, TRIFLUOROACETYL DERIVATIVES

Column Packing	P1	P2	P3	P4	P5	P5
Temperature (°C)	T1	T2	T3	T4	T5	220 °C
Gas; Flow Rate (ml/min)	He	He	He	He	80,He	80,He
Column						
length, m	50	50	30	15	2	2
diameter, mm, I.D.	0.25	0.25	0.25	0.5	3	3
form	capillary	capillary	capillary	—	na	na
material	SS	SS	SS	SS	SS	SS
Detector	FI	FI	FI	FI	TC	TC
Literature	1	1	1	1	2	2
Trifluoroacetyl derivatives of:			$r_{\text{di-}n\text{-butylamine}}$			
Octamethylenediamine	—	—	—	—	27.9	12.9
Nonamethylenediamine	—	—	—	—	30.8	15.5
Decamethylenediamine	—	—	—	—	35.8	20.6
Dodecamethylenediamine	—	—	—	—	49.2	33.7

Column Packing
P1 = PPE OS 124 (polyphenyl ether).
P2 = Apiezon L.
P3 = SE-30 (silicone rubber).
P4 = Polyethylene glycol succinate.
P5 = 5% NPGS on Celite 545 (80/100 mesh).

Temperature
Programmed:
T1 = 85 ° to 185 °C at 4 °/minute.
T2 = 95 ° to 195 °C at 4 °/minute.
T3 = 90 ° to 180 °C at 4 °/minute.
T4 = 50 ° to 180 °C at 3 °/minute.
T5 = 140 ° to 220° C at 4 °/minute.

Literature
1. Irvine, W. J. and M. J. Saxby, *J. Chromatogr.*, **43**, 129 (1969).
2. Mori, S., Furusawa, M, and T. Takeuchi, *Anal. Chem.*, **42**, 138 (1970).

TABLE GC 11
AMINES

Column Packing	P1	P2	P3	P4
Temperature (°C)	160	160	100	100
Gas; Flow Rate (ml/min.)	25,N_2	25,N_2	25,N_2	25,N_2
Column				
length, m	1.6	1.6	1.6	1.6
diameter, mm, I.D.	4	4	4	4
form	na	na	na	na
material	glass	glass	glass	glass
Detector	FID	FID	FID	FID

Compound	t_r			
n-Propylamine	1.9	2.2	—	0.9
n-Butylamine	3.4	4.9	—	1.6
N,N-Dimethyldiaminoethane	6.6	7.4	3.3	3.0
n-Pentylamine	5.9	6.7	—	2.4
N-Methyldiaminoethane	7.7	8.5	3.0	6.1
Diaminoethane	6.5	8.8	2.1	6.2
1,2-Diaminopropane	7.5	9.8	2.7	6.2
N,N'-Dimethyldiaminoethane	8.5	10.1	4.5	4.3
N,N,N',N'-Tetramethyldiaminoethane	6.0	7.6	6.0	2.6
n-Hexylamine	10.0	—	—	2.9
N,N-Dimethyl-1,3-diaminopropane	11.9	13.3	6.7	6.1
1,3-Diaminopropane	12.5	19.3	4.4	11.5
N,N,N'-Trimethyldiaminoethane	6.6	8.3	5.1	2.6
N-Metnyl-1,3-diaminopropane	14.7	18.5	7.0	10.2
N,N,N',N'-Tetramethyl-1,3-diaminopropane	12.0	12.0	11.5	3.4
N,N'-Dimethyl-1,3-diaminopropane	15.2	17.7	9.7	8.7
N,N'-Diethyldiaminoethane	15.1	17.0	13.0	7.3
1,4-Diaminobutane	21.6	29.3	9.6	22.6
N,N,N',N'-Tetramethyl-1,3-diaminobutane	17.6	18.5	21.9	5.8
N,N,N',N'-Tetramethyl-1,4-diaminobutane	18.8	19.6	24.1	11.8
Triethylenediamine	45.5	57.4	30.5	28.7
1,5-Diaminopentane	36.3	44.5	21.3	43.5
1,6-Diaminohexane	59.3	72.3	42.8	75.5
1,7-Diaminoheptane	97.6	117.3	90.0	132.5
1,8-Diaminooctane	159.1	189.5	—	—

Column Packing P1 = PEI (polyethylene imine) 10% w/w + PAR-1 (80/120 mesh).
P2 = PEI (polyethylene imine) 20% w/w + PAR-1 (80/120 mesh).
P3 = Apiezon L 20% w/w + Celite (100/120 mesh) (5% w/w KOH).
P4 = Carbowax 20M 20% w/w + Celite (100/120 mesh) (5% w/w KOH).

Literature Smith, J. R. L. and D. J. Waddington, *J. Chromatogr.*, **42**, 195 (1969).

TABLE GC 12
AMINO ACID DERIVATIVES

Column Packing		P1	P1	P1	P1
Temperature (°C)		110	110	110	100
Gas; Flow Rate		He	He	He	He
Column					
length, ft		(1) 500 (2) 100	— 500	(1) 400 (2) 200	(1) 400 (2) 200
diameter, in.		0.02	0.02	0.02	0.02
form		—	—	—	—
material		—	—	—	—
Amino Acid Derivative		1	2	3	4
Derivative	Isomer	Relative Retention Volume			
Alanine	**D**	0.280 (1)	0.245	0.298 (1)	0.300 (1)
	L	0.307	0.262	0.324 (1)	0.339
α-Amino-*n*-butyric acid	**D**	0.405 (1)	0.348	—	0.425 (1)
	L	0.441	0.368	—	0.475
Valine	**D**	0.455 (1)	0.392	0.451 (1)	0.462 (1)
	L	0.491	0.404	0.476 (1)	0.504
Glycine	—	0.514 (1)	0.433	0.556 (1)	0.543 (1)
Threonine	**D**	0.556 (1)	0.535	0.533	0.543 (1)
	L	0.614	0.564	0.584	0.597 (1)
Norvaline	**D**	0.662 (1)	0.600	—	0.664
	L	0.725	0.631	—	0.745
Alloisoleucine	**D**	0.633 (1)	0.600	0.654 (1)	0.642 (1)
	L	0.727	0.633	0.708 (1)	0.716
Isoleucine	**D**	0.727 (1)	0.633	0.708	0.716 (1)
	L	0.783	0.670	0.749	0.787
β-Alanine	—	0.839 (1)	—	—	—
Leucine	**D**	0.898 (1)	0.858	0.912 (1)	0.874 (1)
	L	1.000	0.906	1.000	1.000
Serine	**D**	1.000 (1)	1.000	1.000 (1) (2)	0.968 (1) (2)
	L	1.080	1.043	1.079	1.088
Allothreonine	**D**	1.000	—	—	—
	L	1.080	—	—	—
Norleucine	**D**	1.075 (1)	1.000	—	1.062 (1)
	L	1.188	1.052	—	1.201
Proline	**D**	1.314 (1)	1.179	1.298 (1)	1.331 (1)
	L	1.370 (1) (2)	1.232	1.356	1.383
Aspartic acid	**D**	3.45	2.101	8.500 (2)	8.547 (2)
	L	3.66	2.282	9.790	9.254
Methionine	**D**	5.45 (1)	5.270	5.975 (2)	—
	L	6.08	5.680	6.630	—
Phenylalanine	**D**	7.88	7.260	8.240 (2)	9.254 (2)
	L	8.97	7.861	9.240	10.173
Glutamic acid	**D**	10.23 (2)	6.625	—	—
	L	11.56	7.168	—	—

( ) refers to different flow rates.

Column Packing P1 = coated with N-TFA-L-valyl-L-valine cyclohexyl ester.

Derivatives *N*-trifluoroacetyl amino acid;
(1) isopropyl ester;
(2) methyl;
(3) *n*-butyl;
(4) 3-pentyl.
Conditioning: 100 °C for 24 hours and 110 °C for 24 hours with 20 psi Helium.

Literature Nakaparksin, S., Birrell, P., Gil-av, E. and J. Oro, *J. Chromatogr. Sci.*, **8,** 177 (1970).

TABLE GC 13
CARBOHYDRATES, ACETALS AND KETALS

Column Packing	P1
Temperature (°C)	206
Gas; Flow Rate	—
Column	
length, cm	117
diameter, cm, I.D.	0.5
form	—
material	glass
Detector	—

Acetals or Ketals of	r*
1,2:4,5-Di-*O*-isopropylideneribitol	0.32
5-Deoxy-1,2-*O*-isopropylidene-L-arabinose	0.42
3-*O*-Methyl-1,2-*O*-isopropylidene-D-fructose	0.47
Methyl 2,3-di-*O*-methyl-4,6-*O*-isopropylidene-α-D-glucopyranoside	0.51
1,3:2,4-Di-*O*-methyleneribitol	0.58
1,2-*O*-Isopropylidene-α-D-glucofuranose-5,6-carbonate	0.58
Methyl 2,3-di-*O*-methyl-4,6-*O*-ethylidene-β-D-mannopyranoside	0.65
Methyl 2,3:4,6-di-*O*-isopropylidene-α-D-mannopyranoside	0.66
3-*O*-Acetyl-1,2:4,5-di-*O*-isopropylidene-D-fructose	0.72
2,3-*O*-Isopropylidene-L-rhamnose	0.79
1,2:4,5-Di-*O*-isopropylidene-D-fructose	0.82
Methyl 4,6-*O*-ethylidene-2,3-di-*O*-methyl-β-D-galactopyranoside	0.84
1,2-*O*-Isopropylidene-D-xylulose	0.97
1,2:5,6-Di-*O*-isopropylidene-D-glucose	1.00
2,3:4,5-Di-*O*-isopropylidene-D-fructose	1.04
3,4-*O*-Isopropylidene-β-D-galactosan	1.12
2,3:4,5-Di-*O*-isopropylidenegalactitol	1.13
1,2:4,5:Di-*O*-isopropylidenegalactitol	1.16
1,2:3,4:5,6-Tri-*O*-isopropylidene-D-mannitol	1.20
2,3:5,6-Di-*O*-isopropylidene-D-mannose	1.20
1,2:5,6-Di-*O*-isopropylidene-D-mannitol	1.40
1,2:4,5-Di-*O*-isopropylidene-D-mannitol	1.40
5-Deoxy-5-S-ethyl-1,2-*O*-isopropylidene-D-xylose	1.63
4,6-*O*-Ethylidene-1,2-*O*-isopropylidene-D-galactose	1.67
2,3:5,6-Di-*O*-isopropylidene-D-mannono-1→ 4-lactone	1.81
4,6-*O*-Ethylidene-2,3-*O*-isopropylidene-D-mannose	1.86
3,6-Di-*O*-acetyl-5-*O*-methyl-1,2-*O*-isopropylidene-D-glucose	1.91
5-Deoxy-5-S-ethyl-1,2-*O*-isopropylidene-L-arabinose	2.00
Methyl 3,4-*O*-isopropylidene-2-*O*-methanesulfonyl-β-D-arabinoside	2.24
2,3:4,5-Di-*O*-isopropylidene-1,6-di-*O*-methanesulfonylgalactitol	2.27
1,2-*O*-Isopropylidene-D-fructose	2.70
3,4:5,6-Di-*O*-isopropylidene-D-glucose dimethylacetal	3.13
2,3-*O*-Isopropylidene-D-ribono-1→ 4-lactone	4.05
3-*O*-Benzyl-1,2:4,5-di-*O*-isopropylidene-D-fructose	5.40
Methyl 4,6-*O*-benzylidene-2,3-di-*O*-methyl-α-D-glucoside	5.82
1,2:4,5-Di-*O*-isopropylidene-3-*O*-methanesulfonyl-D-fructose	6.54
1,2-*O*-Isopropylidene-*O*-glucofuranose	6.60
1,2:3,4-Di-*O*-isopropylidene-6-*O*-methanesulfonyl-D-galactose	6.64
2-*O*-Benzyl-3,4:5,6-di-*O*-isopropylidene-D-mannose dimethylacetal	7.46
1,3:2,4-Di-*O*-methylene-D-glucitol	8.88
2,3:5,6-Di-*O*-isopropylidene-4-*O*-toluene-*p*-sulfonyl-D-glucose dimethylacetal	8.92
3,5-Di-*O*-acetyl-2,4-*O*-benzylidene-D-xylose dimethylacetal	12.8
2,3:4,5-Di-*O*-isopropylidene-1-*O*-toluene-*p*-sulfonyl-D-fructose	16.7
1,3,5-Tri-*O*-acetyl-2,4-*O*-benzylidenexylitol	17.8
Methyl 3,4-*O*-isopropylidene-2-*O*-toluene-*p*-sulfonyl-β-D-arabinoside	18.0
1,2:5,6-Di-*O*-isopropylidene-3-*O*-toluene-*p*-sulfonyl-D-glucose	25.6
Methyl 4,6-*O*-isopropylidene-2,3-di-*O*-methanesulfonyl-α-D-glucoside	40.1
Methyl 4,6-*O*-isopropylidene-2,3-di-*O*-benzoyl-α-D-glucoside	42.7
4,6-*O*-Ethylidene-1,2-*O*-isopropylidene-3-*O*-toluene-*p*-sulfonyl-D-galactose	54.0
3,4-*O*-Ethylidene-1,2-*O*-isopropylidene-6-*O*-toluene-*p*-sulfonyl-D-galactose	56.8

* t_r relative to 1,2:5,6-di-O-isopropylidene-D-glucose.

Column Packing P1 = 20% (w/w) Apiezon M on Chromosorb W (60/80 mesh) and 20% (w/w) butanediol succinate polyester on Chromosorb W (60/80 mesh), placed on top of 1% SE-30 on glass beads (60 plus mesh).

Literature Jones, H. G., Jones, J. K. N., and M. B. Perry, *Can. J. Chem.*, **40**, 1559 (1962).

TABLE GC 14
CARBOHYDRATES, ALDITOL ACETATES

Column Packing	P1	P2	P3	P4	P5	P6
Temperature (°C)	213	213	213	190	199	140
Gas; Flow Rate (ml/min)	na	na	na	na	N_2, 40	N_2, 70
Column						
length, ft	na	na	na	10	9	1.8 m
diameter, in.	na	na	na	¼	3.16 mm i.d.	4 mm i.d.
form	na	na	na	na	na	na
material	na	na	na	na	glass	na
Detector	na	na	na	na	FI	FI
Literature	1	1	1	2	3	4

Acetates of		r (arabitol) pentaacetate		r (xylitol) penta-acetate	r (glucitol hexa-acetate)	r (xylitol TFA[a])
Glycerol	0.120	0.118	0.122	0.04	—	—
Erythritol	0.378	0.358	0.393	0.19	0.131	—
D-Threitol	0.403	0.429	0.433	0.26	—	—
Ribitol	0.918	0.895	0.944	0.64	0.272	0.70
L-Arabitol	1.000	1.000	1.000	—	—	—[b]
Xylitol	1.18	1.18	1.20	1.00 (30.2 min)	0.370	1.00
Allitol	1.78	1.89	1.90	—	0.728	—
D-Talitol	2.07	2.16	2.03	—	—	—
D-Mannitol	2.13	2.25	2.16	1.89	0.840	1.44
D-Glucitol	2.39	2.59	2.73	2.56	1.000	1.92
Galactitol	2.42	2.59	2.75	2.18	0.930	2.12
D-Iditol	2.74	2.97	2.85	—	1.142	—
6-Deoxy-L-talitol	0.707	0.712	0.698	—	—	—
L-Rhamnitol	0.715	0.733	0.730	0.41	0.221	0.42
D-Fucitol	0.750	0.778	0.750	0.46	0.235	0.57
Pentaerythritol	0.807	0.798	0.790	—	—	—
meso-Glycero-alloheptitol	—	—	3.56	—	—	—
D-Glycero-D-altroheptitol	—	—	4.06	—	—	—
D-Glycero-D-mannoheptitol	—	—	4.73	—	—	—
meso-Glycero-guloheptitol	—	—	4.90	—	—	—
D-Glycero-D-galactoheptitol	5.60	—	5.05	—	—	—
D-Glycero-D-glucoheptitol	—	—	5.10	—	—	—
L-Glycero-D-glucoheptitol	—	—	6.11	—	—	—
meso-Glycero-idoheptitol	—	—	6.60	—	—	—
D-Erythro-L-galactooctitol	—	—	11.50	—	—	—
D-Threo-L-galactooctitol	—	—	12.61	—	—	—
L-Arabinitol	—	—	—	0.72	0.297	0.85
2-Deoxyribitol	—	—	—	—	0.146	0.55
2-Deoxyglucitol	—	—	—	—	0.463	1.23
2-Deoxygalactinol	—	—	—	—	0.475	1.57
Altritol	—	—	—	—	0.842	—

[a] Trifluoroacetate derivatives.
[b] t_r = 6.91 min.

Column Packing

P1 = 1 : 1 w/w mixture of 20% w/w butanediol succinate on 60–80 mesh Chromosorb W and 20% w/w Apiezon M grease on 60–80 mesh silver-coated Chromosorb W.
P2 = 1 : 1 w/w mixture of 20% w/w butanediol succinate polyester on 60–80 mesh Chromosorb W and 20% w/w Dow Corning grease on 60–80 mesh Chromosorb W.
P3 = 1 : 1 w/w mixture of P1 and 0.3% w/w Apiezon M grease on 60 plus mesh silver-coated glass beads.
P4 = 3% ECNSS-M coated on 100–120 mesh Gas-Chrom Q.
P5 = 2% EGSS-X on non-acid washed Chromosorb W (60–80).
P6 = 2% XF-1105 on Gas-Chrom P (80-100).

Literature

1. Gunner, S. W., Jones, J. K. N., and M. B. Perry, *Can. J. Chem.* **39**, 1892 (1961).
2. Sawardeker, J. S., Sloneker, J. N., and A. Jeanes, *Anal. Chem.*, **37**, 1602 (1965).
3. Shaw, D. H. and G. W. Moss, *J. Chromatogr.*, **41**, 350 (1969).
4. Imanari, T., Arakawa, Y., and Z. Tamura, *Chem. Pharm. Bull.*, **17**, 1967 (1969).

TABLE GC 15
CARBOHYDRATES, TRIMETHYLSILYL DERIVATIVES

Column Packing	P1	P1	P1	P1	P1	P2	P3	P4	P5
Temperature (°C)	145	155	165	175	185	145	T1	T2	175
Gas; Flow Rate (ml/min)	40	40	40	40	40	N_2,na	N_2,na	N_2,na	na
Column									
length, ft	5	5	5	5	5	6	6	6	na
diameter, in., I.D.	$\frac{1}{8}$	$\frac{1}{8}$	$\frac{1}{8}$	$\frac{1}{8}$	$\frac{1}{8}$	5mm	5mm	5mm	na
form	tubing	tubing	tubing	tubing	tubing	U-shaped	U-shaped	U-shaped	na
material	SS	SS	SS	SS	SS	glass	glass	glass	na
Detector	FI	FI	FI	FI	FI	FI	FI	FI	na
Literature	1	1	1	1	1	2	2	2	3
Trimethylsilyl derivatives of:	$r_{D\text{-ribose}}$					t_r	t_r	t_r	$r_{D\text{-glucose}}$
Glycolaldehyde	0.127	0.139	0.155	0.170	0.182	—	—	—	—
D-Erythrose	0.234	0.234	0.234	0.234	0.234	3.5	2	3.3	—
L-Arabinose	0.825	0.797	0.820	0.817	0.798	9.9	3.5	7.1	0.21
D-Ribulose	0.836	0.831	0.838	0.835	0.838	—	—	—	—
D-Ribose	1.000	1.000	1.000	1.000	1.000	11.0	4.2	7.8	—
D-Xylose	1.300	1.240	1.290	1.296	1.262	12.4	5.7	8.7	0.29
DL-Glyceraldehyde	1.640	1.610	1.550	1.530	1.470	—	—	—	—
Dihydroxyacetone	1.976	1.880	1.840	1.770	1.720	—	—	—	—
D-Fructose	2.350	2.260	2.300	2.200	2.170	16.2	7.1	11.7	—
D-Sorbose	2.970	2.745	2.780	2.740	2.620	—	—	—	—
D-Galactose	3.130	3.070	2.760	2.690	2.560	17.7	11.0	15.7	0.59[a]
α-D-Glucose	3.620	3.370	3.170	3.060	2.860	18.5	11.7	14.9	1.00
D-Sedoheptulosan	3.550	3.360	3.360	3.200	3.130	—	—	—	—
D-Mannose	3.830	3.740	3.500	3.310	3.300	16.1	7.7	12.8	0.48
Deoxyribose	—	—	—	—	—	6.3	2.7	4.5	0.11
Erythritol	—	—	—	—	—	7.1	2	4.0	—
Rhamnose	—	—	—	—	—	10.0	3.3	6.6	—
Xylitol	—	—	—	—	—	12.7	5.0	9.7	0.33
Sorbitol	—	—	—	—	—	20.4	14.9	18.0	—
β-D-Glucose	—	—	—	—	—	21.8	21.9	19.1	—
2-Deoxyglucose	—	—	—	—	—	—	—	—	0.43
L-Fucose	—	—	—	—	—	—	—	—	0.26
L-Arabitinol	—	—	—	—	—	—	—	—	0.33
Ribitol	—	—	—	—	—	—	—	—	0.34
D-Glucitol	—	—	—	—	—	—	—	—	0.85
Inositol	—	—	—	—	—	—	—	—	1.53

[a] t_r = 11.4 min.

Column Packing
P1 = 5% SE-30 on Chromosorb W (60/80) mesh.
P2 = 15% Carbowax 20M on Chromosorb W (80/100) mesh.
P3 = 3% TCEPE (tetracyanoethylated pentaerythritol) on Aeropak 30 (70/80) mesh.
P4 = 3% DV-1 on Gas Chrom Q (60/80) mesh.
P5 = 3% JXR on Gas Chrom Q (100/120) mesh, purchased from Applied Science Laboratory, State College, Pennsylvania.

Temperature
T1 = Programmed from 100–140 °C (2 °C/min.).
T2 = Programmed from 120–190 °C (3 °C/min.).

Literature
1. Arreguin, B. and J. Taboada, *J. Chromatogr. Sci.*, **8**, 187 (1970).
2. Farshtchi, D. and C. W. Moss, *J. Chromatogr.*, **42**, 108 (1969).
3. Supina, W. R., Kruppa, R. F., and R. S. Henly, *J. Amer. Oil Chem. Soc.*, **44**, 74 (1967).

TABLE GC 16
DIPEPTIDE DERIVATIVES

Column Packing	P1	P1	P1	P2	P2
Temperature (°C)	206	196	196	115	165
Gas; Flow Rate (ml/min)	Ar,40	Ar,40	Ar,40	Ar,40	Ar,40
Column					
length, ft	6	6	6	3	3
diameter, m	3.4	3.4	3.4	3.4	3.4
form	coiled	coiled	coiled	coiled	coiled
material	glass	glass	glass	glass	glass
Detector	FI	FI	FI	FI	FI
Derivatives	D1	D1	D2	D3	D3
Derivative			t_r		
Glycyl-glycine	25.3	41.5	5.8	4.1	—
Glycyl-alanine	15.5	25.6	9.5	—	—
Glycyl-valine	17.1	28.7	5.3	—	—
Glycyl-proline	22.8	37.5	13.8	—	—
Alanyl-proline (*cis* + *trans*)	10.4	17.4	17.2	—	—
Alanyl-leucine (*cis* + *trans*)	13.4	22.1	10.5	10.6	—
Valyl-valine (*cis*)	10.6	17.2	7.0	11.3	—
Valyl-valine (*trans*)	9.6	15.4	5.4	9.7	—
Valyl-proline (*cis*)	12.3	19.5	3.5	—	—
Valyl-proline (*trans*)	16.7	26.8	—	—	—
Isoleucyl-proline (*cis*)	16.3	25.7	—	—	—
Isoleucyl-proline (*trans*)	21.8	35.2	—	—	—
Alanyl-valine (*cis*)	—	—	—	5.4	—
Alanyl-valine (*trans*)	—	—	—	4.9	—
Glycyl-phenylalanine	—	—	—	—	4.5
Alanyl-phenylalanine (*cis*)	—	—	—	—	4.9
Alanyl-phenylalanine (*trans*)	—	—	—	—	4.5
Leucyl-phenylalanine (*cis* + *trans*)	—	—	—	—	>5.0

Column Packing P1 = 3% EGSP-Z on Gas Chrom Q (100/120 mesh).
P2 = 3% SE-30 on Gas Chrom Q (100/120 mesh).

Derivatives D1 = Diketopiperazines.
D2 = *N*-Methyl diketopiperazines.
D3 = *N*-Trimethylsilyl diketopiperazines.

Literature Mauger, A. B., *J. Chromatogr.*, **37**, 315 (1968).

TABLE GC 17
DRUGS

Column Packing	P1	P1	P1	P1
Column temperature (°C)	150	165	180	210
Gas; Flow Rate (ml/min)	Ar,40	Ar,40	Ar,40	Ar,40
Column				
length, m	1.22	1.22	1.22	1.22
diameter, mm, I.D.	5	5	5	5
form	—	—	—	—
material	glass	glass	glass	glass
Detector	D1	D1	D1	D1

Compound	t_r			
Free form				
Phenothiazine Derivatives				
Chlorpromazine	—	5.7	2.7	0.9
Ethopropazine	—	3.7	—	—
Fluphenazine	—	[a]	—	6.6
Mepazine	—	7.3	3.3	—
Methoxypromazine	—	8.5	3.8	—
Perphenazine	—	[a]	—	[a]
Prochlorperazine	—	32.0	12.5	2.9
Promazine	—	3.3	—	—
Promethazine	—	2.9	—	—
Pyrathiazine	—	7.3	3.3	—
Thiopropazate	—	[a]	[a]	[a]
Thioridazine	—	61.0	22.7	4.7
Trifluoperazine	—	12.1	5.3	1.4
Triflupromazine	5.2	2.2	—	—
Trimeprazine	—	3.1	—	—
Rauwolfia Alkaloids				
Deserpidine	—	[a]	[a]	[a]
Rescinnamine	—	[a]	[a]	[a]
Reserpine	—	[a]	[a]	[a]
Diphenylmethane Derivatives				
Adiphenine	5.2	2.7	—	—
Azacyclonol	—	4.2	—	—
Buclizine	—	[a]	41.0	7.3
Captodiamine	—	18.8	7.5	1.8
Chlorcyclizine	5.0	2.7	—	—
Diphenylpyraline	3.2	1.5	—	—
Hydroxyzine	—	[a]	—	5.1
Meclizine	—	41.0	16.8	3.4
Phenyltoloxamine	1.8	0.9	—	—
Pipradol	3.8	2.1	—	—
Piperilate	[a]	2.3	—	—
Butanediols, Propanediols, and Related Compounds				
Carisoprodol	3.3	1.8	—	—
Mephenesin	1.3	0.8	—	—
Meprobamate	5.4	2.7	—	—
Methocarbamol	—	6.1	2.0	—
Phenaglycodol	0.6	[a]	[a]	[a]
Ureides, Amides, Hydrazines				
α-Methylphenethylhydrazine	2.8	1.3	[a]	—
Dextroamphetamine	[a]	[a]	—	—
Ectylurea	1.0	[a]	—	—
Glutethimide	3.6	2.0	—	—
Iproniazid	—	1.4	—	—
Isocarboxazide	[a]	[a]	[a]	[a]
Nialamide	[a]	[a]	[a]	[a]
Phenelzine	3.3	1.8	—	—
Acetylcarbromal	[a]	[a]	[a]	[a]

TABLE 17—(Continued)
DRUGS

Column Packing	P1	P1	P1	P1
Column temperature (°C)	150	165	180	210
Gas; Flow Rate (ml/min)	Ar,40	Ar,40	Ar,40	Ar,40
Column				
length, m	1.22	1.22	1.22	1.22
diameter, mm, I.D.	5	5	5	5
form	—	—	—	—
material	glass	glass	glass	glass
Detector	D1	D1	D1	D1

Compound	t_r			
Miscellaneous				
Betazole	[a]	[a]	[a]	[a]
Chlormethazanone	—	3.6	[a]	[a]
Chlorzoxazone	[a]	[a]	—	—
Imipramine	5.0	2 5	—	—
Methylphenidate	0.8	[a]	—	—
Pyrrobutamine	—	5.0	2.3	—
Thonzylamine	5.2	2.7	—	—
Salt form				
Phenothiazine Derivatives				
Chlorpromazine hydrochloride	—	6.2	3.0	—
Ethopropazine hydrochloride	—	4.1	2.0	—
Fluphenazine dihydrochloride	—	[a]	[a]	7.3
Mepazine hydrochloride	—	7.7	3.5	—
Methoxypromazine maleate	—	9.5	4.3	—
Prochlorperazine dimaleate	—	37.0	13.8	3.0
Prochlorperazine ethane disulfonate	—	[a]	—	2.8
Prochlorperazine dihydrochloride	—	[a]	—	3.1
Promazine hydrochloride	—	3.7	—	—
Promethazine hydrochloride	6.5	3.0	—	—
Pyrathiazine hydrochloride	—	7.6	3.5	—
Thiopropazate dihydrochloride	[a]	[a]	—	[a]
Thioridazine hydrochloride	—	70.0	23.3	4.7
Trifluoperazine dihydrochloride	—	12.4	5.7	1.4
Triflupromazine hydrochloride	5.3	2.6	—	—
Trimeprazine tartrate	—	3.2	—	—
Diphenylmethane Derivatives				
Adiphenine hydrochloride	5.3	3.0	—	—
Azacyclonol hydrochloride	—	6.5	2.5	—
Buclizine dihydrochloride	—	[a]	[a]	6.7
Captodiamine hydrochloride	—	17.9	7.4	1.7
Chlorocyclizine dihydrochloride	5.4	2.6	—	—
Diphenylpyraline hydrochloride	3.5	1.9	—	—
Hydroxyzine dihydrochloride	—	[a]	[a]	5.6
Meclizine dihydrochloride	—	49.0	17.3	3.3
Phenyltoloxamine dihydrogen citrate	1.9	1.0	—	—
Pipradol hydrochloride	4.5	2.8	—	—
Piperilate hydrochloride	[a]	2.4	—	—
Ureides, Amides, Hydrazines				
α-Methylphenethylhydrazine hydrochloride	[a]	[a]	—	—
Dextroamphetamine sulfate	—	[a]	—	—
Iproniazid phosphate	[a]	1.4	—	—
Phenelzine dihydrogen sulfate	—	[a]	—	—
Miscellaneous				
Betazole dihydrochloride	—	[a]	[a]	—
Imipramine hydrochloride	5.3	2.6	—	—
Methylphenidate hydrochloride	1.2	[a]	—	—
Pyrrobutamine diphosphate	—	5.8	2.5	—
Thonzylamine hydrochloride	5.8	2.9	—	—

TABLE GC 17—(Continued)
DRUGS

[a] Compound was injected but no response was observed.

Column Packing P1 = 60/80 mesh glass microbeads, 0.05% SE-30.
Detection D1 = Electron capture.
Literature Parker, K. D., Fontan, C. R., and P. L. Kirk, *Anal. Chem.*, **34**, 757 (1962).

TABLE GC 18
DRUGS

Column Packing	P1	P1
Temperature (°C)	T1	T1
Gas; Flow Rate (ml/min)	Ar,47	Ar,47
Column		
length, m	1.83	1.83
diameter, mm, I.D.	3	3
form	coiled tube	coiled tube
material	glass	glass
Detector	D1	D1

Compound	t_r	$r_{codeine}$
l-Acetylmethadol	3.65	0.61
dl-Alphaprodine	1.22	0.20
Codeine	6.00	1.00[a]
Dihydrocodeinone (Hydrocodone)	6.94	1.16
Dihydromorphine	6.94	1.16
Dihydromorphinone (Hydromorphone)	7.31	1.22
Dihydrohydroxycodeinone (Oxycodone)	8.62	1.44
Dihydrohydroxymorphinone (Oxymorphone)	8.81	1.47
Ethylmorphine	6.71	1.12
Heroin	11.34	1.89
2′-Hydroxy-5,9-dimethyl-2-cyclopropylmethyl-6,7-benzomorphan	3.78	0.63
2′-Hydroxy-5,9-dimethyl-2-(3,3-dimethylallyl)-6,7-benzomorphan	4.27	0.71
dl-2′-Hydroxy-5,9-dimethyl-2-phenethyl-6,7-benzomorphan	12.66	2.11
1,3-Hydroxy-*N*-methylmorphinian (Levorphanol)	4.12	0.69
1,3-Hydroxymorphinan	4.12	0.69
1,3-Hydroxy-*N*-allylmorphinian (Levallorphan)	5.72	0.95
l-2′-Hydroxy-2,5,9-trimethyl-6,7-benzomorphan	1.87	0.31
Ketobemidone	2.53	0.42
dl-Methadone	3.32	0.55
l-3-Methoxy-*N*-methylmorphinan	3.19	0.53
l-3-Methoxymorphinan	3.19	0.53
Methyldihydromorphinone (Metopon)	7.50	1.25
6-Monacetylmorphine	8.62	1.44
Morphine	6.94	1.16
Nalorphine	9.84	1.64
Norcodeine	6.00	1.00
Normorphine	6.94	1.16
Norpethidine	1.22	0.20
Pethidine (Meperidine)	1.22	0.20
Pimonodine	21.94	3.66
d-Propoxyphene	3.56	5.9

[a] Reference retention time of codeine, 6.1 min.

Column Packing P1 = 80/100 mesh Gas Chrom S, 2% SE-30.
Temperature T1 = injection 260 °C, column 215 °C, detector 215 °C.
Detector D1 = Argon β-ionization (Ra 226).
Literature Mulé, S. J., *Anal. Chem.*, **36**, 1907 (1964).

TABLE GC 19
DRUGS

Column Packing	P1	P2	P3	P4	P5
Temperature (°C)	150	137	170	165	210
Gas; Flow Rate (ml/min)	na, 37.5	na, 68	na, 40	na, 25	Ar,50
Column					
length, m	1.83	1.83	1.83	1.83	1.525
diameter, mm	3 I.D.	3 I.D.	3 I.D.	3 I.D.	6 O.D.
form	U-shaped	U-shaped	U-shaped	U-shaped	na
Material	glass	glass	glass	glass	glass
Detector	D1	D1	D1	D1	HFI
Literature	1	1	1	1	2

Compound	$r_{pentobarbital}$				$r_{barbitone}$
Allobarbitone (Allobarbital, diallylbarbituric acid)	0.46	0.52	0.97	0.57	1.47
Allylisopropylacetylurea	0.40	0.45	—	—	—
Amobarbital (Amylobarbitone)	0.82	0.92	0.92	0.96	2.14
Aprobarbital	0.55	0.56	0.85	0.63	—
Barbital (Barbitone)	0.29	0.32	0.53	0.47	1.00
5(2-Bromallyl)-5-isopropyl barbituric acid	1.77	1.91	4.30	1.80	—
Butabarbitone (Butabarbital)	—	—	—	—	1.94
Butethal	0.65	0.71	0.84	0.80	1.84
Cabromal	—	—	—	—	0.31
Caffeine	—	—	—	—	4.38
Cyclobarbital (Cyclobarbitone)	3.42	3.03	4.32	2.50	7.20
Cyclopal	2.06	1.82	3.14	1.47	—
Ethinamate	—	—	—	—	0.49
Glutethimide	1.98	1.36	0.53	1.67	4.10
Heptobarbital (Medomin)	—	—	—	—	10.50
Hexobarbital (Hexobarbitone)	1.88	1.61	0.62	1.54	4.04
Hexethal	2.15	1.94	1.66	1.51	—
3-Hydroxyamylobarbitone	—	—	—	—	4.40
Itobarbital (Butalbital)	0.68	0.71	0.91	0.74	—
Megimide (Bemigride)	—	—	—	—	0.62
Mephobarbital (Methylphenobarbitone)	2.30	1.91	1.04	1.67	4.79
Methaqualone	—	—	—	—	10.80
Metharbital	0.24	0.23	0.11	0.34	—
Methyl-*p*-hydroxybenzoic acid	—	—	—	—	0.93
Methyprylone	—	—	—	—	1.25
Nealbarbitone	—	—	—	—	2.35
p-Nitrophenol	—	—	—	—	1.90
Pentobarbital (Pentabarbitone)	1.00[a]	1.00[b]	1.00[c]	1.00[d]	2.55
Phenacetin	—	—	—	—	2.25
Phenazone	—	—	—	—	5.20
Phenobarbital (Phenobarbitone)	3.10	2.88	8.78	2.50	6.93
Probarbital	0.44	0.42	0.65	0.57	—
Rutonal	—	—	—	—	5.60
Salicylamide	—	—	—	—	1.10
Secobarbital (Quinolbarbitone)	1.26	1.29	1.38	1.09	3.02
Thiamylal	2.82	2.05	1.41	1.09	—
Thiopental	2.21	1.59	1.08	1.01	—
Tybamate	—	—	—	—	5.60
Vanillic acid diethylamide (Ethamivan)	—	—	—	—	3.67
Vinbarbital	0.99	1.17	1.45	1.12	—

[a] Retention time of pentobarbital, 8.4 min.
[b] Retention time of pentobarbital, 6.6 min.
[c] Retention time of pentobarbital, 7.4 min.
[d] Retention time of pentobarbital, 7.6 min.

Column Packing P1 = 60/80 mesh Chromosorb W, 1.4% Apiezon L.
P2 = 60/80 mesh Chromosorb W, 1.2% SE-30.
P3 = 60/80 mesh Chromosorb W, 0.6% PEG 4000.
P4 = 60/80 mesh Chromosorb W, 2.8% QF-1-0065.
P5 = 80/100 mesh Chromosorb W, 10% Apiezon L.

Detector D1 = argon β-ionization.

Literature
1. Brochmann-Hanssen, E. and A. Baerheim Svendsen, *J. Pharm. Sci.*, **51**, 318 (1962).
2. Leach, H. and P. A. Toseland, *Clin. Chim. Acta*, **20**, 195 (1968).

TABLE GC 20
DRUGS

Column Packing	P1	P2	P3	P3	P4	P5
Temperature (°C)	T1	T2	T3	T4	T5	T6
Gas; Flow Rate (ml/min)	N_2,79	N_2,200	—	—	Ar,60	Ar,60
Column						
length, m	1.5	1.2	1.83	1.83	1.83	1.83
diameter, mm	3.2 O.D.	4 I.D.	5 I.D.	5 I.D.	na	3 I.D.
form	Tube	na	na	na	na	U-shaped
material	S	na	G	G	G	G
Detector	HFI	HFI	D1	D1	D2	D2
Literature	1	2	3	3	4	5

Compound	$r_{methapyrilene\ HCl}$	$r_{diphenylpyraline\ HCl}$	$r_{chlorpromazine}$		— t_r —	
Antazoline	0.603	—	—	—	3.10	65.80
Bromophenhydramine HCl	1.82	—	—	—	—	—
Carbinoxamine	—	—	—	—	14.20	8.50
Carbinoxamine maleate	1.53	—	—	—	—	—
Chlorcyclizine HCl	—	1.66	—	—	—	—
Chlorothen citrate	1.84	1.45	—	—	—	—
Chlorpheniramine	—	—	—	—	9.00	5.20
					9.70[a]	
Chlorpheniramine maleate	0.984	0.76	—	—	—	—
Chlorpromazine	—	—	1.00[b]	1.00[c]	—	—
Clemizole HCl	no peak	—	—	—	—	—
Cyclizine HCl	—	0.64	—	—	—	—
Diphenylhydramine HCl	0.460	—	—	—	—	—
Diphenylpyraline HCl	—	1.00	—	—	—	—
Doxylamine succinate	0.579	—	—	—	—	—
Ethopropazine	—	—	0.70	—	—	—
Ethyl isobutrazine	—	—	0.91	—	—	—
Mepazine	—	—	1.13	—	—	—
Methapyrilene HCl	1.00[d]	0.77	—	—	—	—
Meclizine	—	—	—	—	no response	no response
Methdilazine HCl	8.00	—	—	—	—	—
Methdilazine	—	—	0.92	—	—	—
Methopromazine (methoxypromazine)	—	—	1.20	—	—	—
Parabromdylamine maleate (brompheniramine maleate)	1.55	—	—	—	—	—
Phenylpropanolamine HCl	—	0.13	—	—	—	—
Phenirimine maleate	—	0.31	—	—	—	—
Pyrilamine maleate	3.33	2.51	—	—	—	—
Pyribenzamine citrate (tripelennamine citrate)	—	0.67	—	—	—	—
Thonzylamine HCl	—	2.08	—	—	—	—
Tripelennamine base	0.857	—	—	—	—	—
Tripelennamine	—	—	—	—	7.00	5.00
Pyrilamine	—	—	—	—	25.90	20.30
Triprolidine HCl	3.02	—	—	—	—	—
	5.18[e]	—	—	—	—	—
Phenyltoloxamine hydrogen citrate	—	0.53	—	—	—	—
Dextramethorphan HBr	—	1.05	—	—	—	—
Dexbrompheniramine maleate (brompheniramine maleate)	—	1.21	—	—	—	—
Phenindamine tartrate	—	1.49	—	—	—	—
Thenyldiamine	—	—	—	—	8.20	6.90
Trifluomeprazine	—	—	0.43	—	—	—
Trifluoperazine	—	—	1.69	—	—	—
Triflupromazine	—	—	0.46	—	—	—
Promethazine	—	—	0.53	—	4.00	21.70
	—	—	—	—	7.20	—
	—	—	—	—	7.20[e]	—
Trimeprazine	—	—	0.58	—	—	—
Promazine	—	—	0.60	—	—	—

TABLE GC 20—(Continued)
DRUGS

Column Packing	P1	P2	P3	P3	P4	P5
Temperature (°C)	T1	T2	T3	T4	T5	T6
Gas; Flow Rate (ml/min)	N_2,79	N_2,200	—	—	Ar,60	Ar,60
Column						
length, m	1.5	1.2	1.83	1.83	1.83	1.83
diameter, mm	3.2 O.D.	4 I.D.	5 I.D.	5 I.D	na	3 I.D.
form	Tube	na	na	na	na	U-shaped
material	S	na	G	G	G	G
Detector	HFI	HFI	D1	D1	D2	D2
Literature	1	2	3	3	4	5

Compound	$r_{\text{methapyrilene HCl}}$	$r_{\text{diphenylpyraline HCl}}$	$r_{\text{chlorpromazine}}$		— t_r	—
Pyrathiazine	—	—	1.11	—	—	—
Propiopromazine	—	—	2.38	—	—	—
Prochlorperazine	—	—	—	2.72	—	—
Thioridazine	—	—	—	3.47	—	—
Thiopropazate	—	—	—	6.97	—	—
Pipamazine	—	—	—	8.83	—	—
Perphenazine	—	—	—	>23.90	—	—
Fluphenazine	—	—	—	>23.90	—	—
Cyclazine	—	—	—	—	1.90	4.90
	—	—	—	—	8.90[e]	—
Diphenhydramine	—	—	—	—	6.30	2.50
Doxylamine	—	—	—	—	8.20	3.90
Pheniramine	—	—	—	—	8.20	2.10
Chlorothen (chloromethapyrilene)	—	—	—	—	14.70	10.50
Chlorcyclizine	—	—	—	—	20.00	12.90
	—	—	—	—	22.90[e]	—
Thonzylamine	—	—	—	—	20.90	18.00
Methapyrilene	—	—	—	—	—	5.80

[a] Major peaks .
[b] Retention time of chlorpromazine is 24.9 min.
[c] Retention time of chlorpromazine is 3.80 min.
[d] Retention time of methapyrilene HCl is 6.3 min.
[e] Larger of the 2 peaks, and asymmetrical.

Column Packing
P1 = 60/80 mesh Chromosorb W, 2% Carbowax 20M.
P2 = 80/90 mesh Anakrom ABS, 2% SE-30 and 2% Carbowax 20M.
P3 = 80/100 mesh Gas Chrom S, 2% Se-30.
P4 = 100/120 mesh Gas Chrom P, 1% SE-30.
P5 = 120/170 mesh glass beads, 0.08% PDEAS.

Temperature
T1 = injection 230 °C, column 190 °C, detector na.
T2 = injection 350 °C, column 185 °C, detector 250 °C.
T3 = injection 225 °C, column 205 °C, detector na.
T4 = injection 305 °C, column 270 °C, detector na.
T5 = injection 256 °C, column 173 °C, detector 228 °C.
T6 = injection 260 °C, column 175 °C, detector 230 °C.

Detector
D1 = argon β-ionization (Sr).
D2 = argon β-ionization.

Literature
1. Fontan, C. R., Smith, W. C., and P. L. Kirk, *Anal. Chem.*, **35**, 591 (1963).
2. Celeste, A. C. and M. V. Polito, *J. Ass. Offic. Agr. Chem.*, **49**, 543 (1966).
3. Anders, M. W. and G. J. Mannering, *J. Chromatogr.*, **7**, 259 (1962).
4. MacDonald, Jr., A. and R. T. Pflaum, *J. Pharm. Sci.*, **52**, 816 (1963).
5. MacDonald, Jr., A. and R. T. Pflaum, *J. Pharm. Sci.*, **53**, 887 (1964).

TABLE GC 21
DRUGS

Column Packing	P1	P1	P1	P1	P1	P1	P1	P1	P1
Temperature (°C)	T1	T2	T3	T4	T5	T6	T7	T8	T9
Gas; Flow Rate (ml/min)	Ar,50	Ar,50	Ar,60	Ar,60	Ar,65	Ar,65	Ar,70	Ar,80	Ar,80
Column									
length, m	1.83	1.83	1.83	1.83	1.83	1.83	1.83	1.83	1.83
diameter, mm, I.D.	4	4	4	4	4	4	4	4	4
form	U-shaped	U-shaped	U-shaped	U-shaped	U-shaped	U-shaped	U-shaped	U-shaped	U-shaped
material	G	G	G	G	G	G	G	G	G
Detector	D1	D1	D1	D1	D1	D1	D1	D1	D1

Compound	t_r								
Ethchlorvynol (Placidyl®)	1.4	1.2	—	—	—	—	—	—	—
Phenethylamine	2.1	1.7	—	—	—	—	—	—	—
Amphetamine	2.4	1.9	—	—	—	—	—	—	—
Desoxyephedrine (Methamphetamine)	3.0	2.1	1.3	—	—	—	—	—	—
Nicotine	6.7	4.1	2.2	—	—	—	—	—	—
Ephedrine	7.5	4.2	2.3	1.2	—	—	—	—	—
Ethynylcyclohexyl carbamate (Valmid®)	9.5	4.9	2.5	—	—	—	—	—	—
Warfarin	—	—	—	1.7	1.2	0.8		—	—
Methyprylon (Noludar®)	—	10.8	4.3	2.9	1.8	—	—	—	—
Barbital	—	—	4.3	2.8	1.6	—	—	—	—
Probarbital (Ipral®)	—	—	5.6	3.5	1.9	—	—	—	—
Acetophenetidin (Phenacetin)	—	—	8.6	5.7	2.6	—	—	—	—
Meperidine (Demerol®)	—	—	9.7	5.9	3.0	1.6	—	—	—
Amobarbital (Amytal®)	—	—	10.9	6.0	3.0	—	—	—	—
Hydroxyphenamate (Listica®)	—	—	11.3	6.5	3.2	1.6	—	—	—
Pentobarbital	—	—	13.2	6.5	3.2	—	—	—	—
Pheniramine (Trimeton®)	—	—	13.0	7.5	3.6	—	—	—	—
Caffeine	—	—	14.7	8.1	3.8	—	—	—	—
Secobarbital (Seconal®)	—	—	16.8	8.2	3.8	—	—	—	—
Diphenhydramine (Benadryl®)	—	—	16.9	9.3	4.2	1.9	—	—	—
Glutethimide (Doridene®)	—	—	—	9.1	4.2	2.0	1.0	—	—
Meprobamate (Miltown®)	—	—	—	10.4	4.4	1.8	—	—	—
Lidocaine (Xylocaine®)	—	—	18.0	9.9	4.5	2.1	—	—	—
Prominal (Mebaral®)	—	—	23.1	11.2	5.2	—	—	—	—
Antipyrine	—	—	20.8	11.8	5.3	2.3	1.9	—	—
Aminopyrine	—	—	—	12.0	5.7	2.6	—	—	—
Tripelennamine (Pyribenzamine®)	—	—	—	14.7	6.7	3.0	—	—	—
Methapyrilene (Histadyl®)	—	—	—	14.8	6.8	3.0	—	—	—
Chlorpheniramine (Chlor-Trimeton®)	—	—	—	16.3	6.8	3.0	—	—	—
Phenobarbital	—	—	—	19.2	7.8	3.3	—	—	—
Procaine	—	—	—	18.6	7.9	3.4	—	—	—
Methadone (Dolophin®)	—	—	—	—	12.1	4.9	3.3	—	—
Bromodiphenhydramine (Ambodryl®)	—	—	—	—	12.3	4.7	3.3	2.0	—
Propoxyphene (Darvon®)	—	—	—	—	13.8	5.5	4.0	—	—
Atropine	—	—	—	—	14.0	5.6	4.0	2.3	—
Thonzylamine (Anahist®)	—	—	—	—	14.7	5.7	4.0	2.3	—
Chlorcyclizine (Perazil)	—	—	—	—	15.5	6.3	4.4	2.4	1.3
α-Cyclohexyl-α-phenyl-1-piperidine-propanol (Artane®)	—	—	—	—	16.4	6.5	4.3	—	—
Tetracaine (Pontocaine®)	—	—	—	—	17.2	6.4	4.2	—	—
Promazine (Sparine®)	—	—	—	—	—	—	5.5	3.0	1.6
Librium	—	—	—	—	—	—	5.6	3.0	1.6
Scopolamine	—	—	—	—	—	8.1	5.7	3.1	1.6
DDT (Dichlorodiphenyltrichloroethane)	—	—	—	—	26.0	8.5	5.5	—	—
Antazoline (Antistine®)	—	—	—	—	20.3	8.6	6.0	3.2	1.6
Codeine	—	—	—	—	—	9.2	6.8	3.5	1.8
Ethylmorphine (Dionin®)	—	—	—	—	—	10.7	7.4	3.8	2.0
Diphenylhydantoin (Dilantin®)	—	—	—	—	—	—	9.0	3.9	1.9
Chlorpromazine (Thorazine®)	—	—	—	—	—	—	9.7	4.6	2.2
Morphine	—	—	—	—	—	—	9.9	4.7	2.3

TABLE GC 21—(Continued)
DRUGS

Column Packing	P1	P1	P1	P1	P1	P1	P1	P1	P1
Temperature (°C)	T1	T2	T3	T4	T5	T6	T7	T8	T9
Gas; Flow Rate (ml/min)	Ar,50	Ar,50	Ar,60	Ar,60	Ar,65	Ar,65	Ar,70	Ar,80	Ar,80
Column									
length, m	1.83	1.83	1.83	1.83	1.83	1.83	1.83	1.83	1.83
diameter, mm, I.D.	4	4	4	4	4	4	4	4	4
form	U-shaped	U-shaped	U-shaped	U-shaped	U-shaped	U-shaped	U-shaped	U-shaped	U-shaped
material	G	G	G	G	G	G	G	G	G
Detector	D1	D1	D1	D1	D1	D1	D1	D1	D1
Compound					t_r				
Cinchonine	—	—	—	—	—	—	13.6	6.2	2.8
Cinchonidine	—	—	—	—	—	—	14.6	6.4	2.8
Diacetylmorphine (Heroin)	—	—	—	—	—	—	14.5	6.9	3.0
Chloroquin	—	—	—	—	—	—	14.7	6.9	2.6
Dibucaine (Nupercaine®)	—	—	—	—	—	—	20.9	9.1	3.1
Quinidine	—	—	—	—	—	—	—	11.0	4.4
Quinine	—	—	—	—	—	—	—	11.2	4.4
Anileridine (Leritine)	—	—	—	—	—	—	32.2	13.9	4.4
Meclizine (Bonamine)	—	—	—	—	—	—	—	20.3	6.2
Strychnine	—	—	—	—	—	—	—	23.7	8.7

Column Packing P1 = Chromosorb W, 1% SE-30.
Temperature T1 = column 115 °C.
T2 = column 130 °C.
T3 = column 150 °C.
T4 = injection 215 °C, column 165 °C, detector 190 °C.
T5 = injection 230 °C, column 180 °C, detector 205 °C.
T6 = injection 250 °C, column 200 °C, detector 225 °C.
T7 = injection 260 °C, column 210 °C, detector 235 °C.
T8 = column 225 °C.
T9 = column 250 °C.
Detector D1 = argon β-ionization.
Literature Kazyak, L. and E. C. Knoblock, *Anal. Chem.*, **35**, 1448 (1963).

TABLE GC 22
DRUGS

Column Packing	P1	P1	P1	P1	P1	P1	P1	P1	P1	P1
Temperature (°C)	130	140	160	180	205	215	235	270	290	310
Gas; Flow Rate (ml/min)	N_2, 50–60	N_2, 50–60	N_2, 50–60	N_2, 50–60	N_2, 50–60	N_2, 50–60	N_2, 50–60	N_2, 50–60	N_2, 50–60	N_2, 50–60
Column										
length, ft	6	6	6	6	6	6	6	6	6	6
diameter, in., O.D.	$\frac{1}{8}$	$\frac{1}{8}$	$\frac{1}{8}$	$\frac{1}{8}$	$\frac{1}{8}$	$\frac{1}{8}$	$\frac{1}{8}$	$\frac{1}{8}$	$\frac{1}{8}$	$\frac{1}{8}$
in., I.D.	0.085	0.085	0.085	0.085	0.085	0.085	0.085	0.085	0.085	0.085
form	helix	helix	helix	helix	helix	helix	helix	helix	helix	helix
material	SS	SS	SS	SS	SS	SS	SS	S	SS	SS
Detector	FI	FI	FI	FI	FI	FI	FI	FI	FI	FI
Compound	t_r									
Amphetamine	1.9	1.5	—	—	—	—	—	—	—	—
Phentermine	—	1.7	—	—	—	—	—	—	—	—
Methylamphetamine	2.45	1.85	—	—	—	—	—	—	—	—
Mephentermine	—	2.65	1.5	—	—	—	—	—	—	—
Chlorphentermine	—	4.1	2.25	1.25	—	—	—	—	—	—
Ephedrine	6.4	4.4	2.3	1.3	—	—	—	—	—	—
Phenmetrazine	—	6.0	2.65	1.7	—	—	—	—	—	—
Hordenine	—	—	3.2	—	—	—	—	—	—	—
Meclofenoxate	—	—	4.4	2.3	1.2	—	—	—	—	—
Metaraminol	—	—	—	—	1.6	—	—	—	—	—
Pheniramine	—	—	—	—	2.85	—	—	—	—	—
Metyrapone	—	—	—	—	3.5	2.65	—	—	—	—
Orphenadrine	—	—	—	9.1	—	2.85	—	—	—	—
Aribine	—	—	—	10.0	4.25	3.2	—	—	—	—
Chlorpheniramine	—	—	—	—	—	3.9	2.1	—	—	—
Adiphenine	—	—	—	—	—	6.85	3.5	—	—	—
Nortriptyline	—	—	—	—	—	—	3.7	—	—	—
Chlorcyclizine	—	—	—	—	—	7.6	3.9	—	—	—
Diazepam	—	—	—	—	—	—	6.6	2.4	—	—
Chlorpromazine	—	—	—	—	—	—	7.9	2.75	—	—
Pentaquine	—	—	—	—	—	—	9.3	2.95	—	—
Acetylpromazine	—	—	—	—	—	—	13.5	4.3	—	—
Quinine	—	—	—	—	—	—	16.4	5.05	—	—
Papaverine	—	—	—	—	—	—	18.4	5.15	—	—
Prochlorperazine	—	—	—	—	—	—	—	7.6	4.0	—
Octaverine	—	—	—	—	—	—	33.2	8.3	4.05	—
Prolixin	—	—	—	—	—	—	—	8.8	4.5	—
Strychnine	—	—	—	—	—	—	—	—	—	3.1
Brucine	—	—	—	—	—	—	—	—	—	5.7

Column Packing P1 = SE-30 on Chromosorb W (100/120 mesh).
Literature Street, H. V., *J. Chromatogr.*, **29**, 68 (1967).

TABLE GC 23
DRUGS

Column Packing	P1	P1	P1	P1
Temperature (°C)	82	104	135	179
Gas; Flow Rate (ml/min)	na 33	na 22.5	na 30	na 25
Column				
length, m	2.4	2.4	2.4	2.4
diameter, mm, I.D.	3	3	3	3
form	U-shaped	U-shaped	U-shaped	U-shaped
material	G	G	G	G
Detector	D1	D1	D1	D1

Compound	t_r			
Amphetamine	7.2	3.3	—	—
Benzphetamine	—	—	15.0	—
Cyclopentamine	5.6	2.6	—	—
Ephedrine	—	11.4	—	—
Epinephrine	—	—	—	13.4[a]
Hydroxyamphetamine	—	—	2.9	—
Levarterenol (*l*-norepinephrine)	—	—	—	12.4[a]
Mephentermine	16.7	6.1	—	—
Metaraminol	—	—	[b]	—
Methamphetamine	10.6	4.6	—	—
Phenylephrine	—	—	9.0[c]	—
Phenylpropanolamine	—	9.1	—	—
Propylhexedrine	10.3	4.0	—	—
Pseudoephedrine	—	11.6	—	—
Synephrine	—	—	9.1[c]	—
Tuaminoheptane	1.9	with solvent peak	—	—

[a] Triacetyl trimethylsilyl ether.
[b] No elution peak observed.
[c] Adsorptive effects caused a badly tailing peak. Retention time dependent on sample size.

Column Packing P1 = 100/140 mesh Gas Chrom P, 1.15% SE-30.
Detector D1 = Argon β-ionization (Ra 226).
Literature Brochmann-Hanssen, E. and A. B. Svendsen, *J. Pharm. Sci.*, **51**, 939 (1962).

TABLE GC 24
ESTERS

Column Packing	P1	P2
Temperature (°C)	T1	T1
Gas; Flow Rate (ml/min)	80	80
Column		
length	na	na
diameter	na	na
form	na	na
material	na	na
Detector	TC	TC

Compound	I	
Methyl ester of		
Propionic acid	—	885–885
Butyric acid	—	970–972–971
Isobutyric acid		903
Valeric acid	808	1081–1081–1085
Isovaleric acid	—	1013
Hexanoic acid	907	1183–1182–1183
Isohexanoic acid	—	1094
Heptanoic acid	1008	1282–1281
Octanoic acid	1109	1378–1380
Nonanoic acid	1211	1484–1487
Decanoic acid	1310	1584–1588
Undecanoic acid	1410	1694–1696
Dodecanoic acid	1513	1800–1801
Myristic acid	1714	2002–1998
Hexadecanoic acid	1911	2190
Octadecanoic acid	2098	—
Benzoic acid	1080	1631
Phenylacetic acid	1156–1154	1759–1761–1762
Salicylic acid	1181–1181	1794
***p*-Hydroxybenzoic acid**	1435	—
***o*-Hydroxyphenylacetic acid**	1260–1269	—
***p*-Hydroxyphenylacetic acid**	1460	—
Cinnamic acid	1363	2065
***o*-Hydroxycinnamic acid**	1430	—
***m*-Hydroxycinnamic acid**	1690	—
***p*-Hydroxycinnamic acid**	1498	—
Anthranilic acid	1325	2259
β-Hydroxybutyric acid	1320	1464–1457
2-Hydroxyisobutyric acid	1118	—
2-Methylbutyric acid	758	980
2-Keto-octanoic acid	1200	—
Crotonic acid	—	1102
Furoic acid	950	—
Dimethyl ester of		
Oxalic acid	837	1381–1383
Malonic acid	895	1489–1489–1490
Succinic acid	1000–1004	1576
Glutaric acid	1105	1686–1687–1689
Adipic acid	1213	1804
Pimelic acid	1313	1908–1909
Suberic acid	1416	2010
Azelaic acid	1519	2102–2100
Sebacic acid	—	2213–2210
Ethyl ester of		
Formic acid	—	822
Acetic acid	—	866
Propionic acid	700	940
Butyric acid	787	1032
Isobutyric acid	—	950
Valeric acid	884	1130–1128
Isovaleric acid	—	1064
Hexanoic acid	979–983	1228–1227
Heptanoic acid	1080–1081	1324–1327

TABLE GC 24—(Continued)
ESTERS

Column Packing	P1	P2
Temperature (°C)	T1	T1
Gas; Flow Rate (ml/min)	80	80
Column		
length	na	na
diameter	na	na
form	na	na
material	na	na
Detector	TC	TC

Compound	I	
Ethyl ester of		
Octanoic acid	1181	1422–1427
Decanoic acid	1379–1379	1631
Lauric acid	1579	1840
Myristic acid	1780	2038
Palmitic acid	1979	2238
Stearic acid	2175	—
Salicylic acid	1261	1828
Cinnamic acid	1447	2108
Lactic acid	801	—
Diethyl ester of		
Oxalic acid	948	—
Malonic acid	1035	—
Succinic acid	1139	—
Propyl ester of		
Acetic acid	704	—
Isopropyl ester of		
Acetic acid	—	866–866
Butyl ester of		
Acetic acid	802	1065
Benzoic acid	1360	1871
Isobutyl ester of		
Acetic acid	—	1002–1002
Isobutyric acid	901	1090
Cinnamic acid	1598	—
Benzoic acid	1318	1799
Amyl ester of		
Acetic acid	896–900	1169
Benzoic acid	1462	—
Salicylic acid	1535	—
Isoamyl ester of		
Acetic acid	853	1116–1116
Benzoic acid	1425	1921
Hexyl ester of		
Formic acid	913	1216
Acetic acid	993–993	1264
Butyric acid	1177	1406
Isobutyric acid	—	1337
Benzoic acid	1565	2068–2070
Salicylic acid	1684	2208
Hexanoic acid	1371	1606
Isohexyl ester of		
Acetic acid	—	1208
Heptyl ester of		
Acetic acid	1096	—
Octyl ester of		
Salicylic acid	1895	—
Nonyl ester of		
Acetic acid	1296	1569
Decyl ester of		
Acetic acid	1395	1674
Dodecyl ester of		
Acetic acid	1595	—

TABLE GC 24—(Continued)
ESTERS

Column Packing	P1	P2
Temperature (°C)	T1	T1
Gas; Flow Rate (ml/min)	80	80
Column		
length	na	na
diameter	na	na
form	na	na
material	na	na
Detector	TC	TC

Compound	I	
Benzyl ester of		
Formic acid	1057	1687
Acetic acid	1141	1728
Propionic acid	1237	1791
Butyric acid	1325	1870
Cinnamic acid	1682	—
Phenylethyl ester of		
Cinnamic acid	2143	—
Anthranilic acid	2088	—
Cinnamyl ester of		
Formic acid	1332	—
Acetic acid	1422	2125
Propionic acid	1519	2194
Isobutyric acid	1562	2179
Isovaleric acid	1663	2289
Cinnamic acid	2052	—
Allyl ester of		
Hexanoic acid	1062–1060	1360
Heptanoic acid	1163	1463
Octanoic acid	1262	1566

Column Packing P1 = 25% Silicone rubber SE-30 on Celite.
P2 = 25% Carbowax 20M on Celite.

Temperature T1 = programmed from 75° to 228 °C at 4.6°/min.

Literature Van den Dool, H. and P. D. Kratz, *J. Chromatogr.*, **11**, 463 (1963).

TABLE GC 25
ESTERS

Column Packing	P1	P1	P1	P2	P2	P2
Temperature (°C)	200	220	240	200	220	240
Gas; Flow Rate (ml/min)	H_2,50	H_2,50	H_2,50	H_2,50	H_2,50	H_2,50
Column						
length, ft	6	6	6	6	6	6
diameter, mm, I.D.	4	4	4	4	4	4
form	U-shaped	U-shaped	U-shaped	U-shaped	U-shaped	U-shaped
material	—	—	—	—	—	—
Detector	—	—	—	—	—	—
Compound	log V_R					
Succinates						
Dimethyl	1.003	0.723	0.554	1.560	1.329	1.186
Diethyl	1.262	1.091	0.952	1.775	1.509	1.395
Di-*n*-propyl	1.648	1.470	1.330	2.071	1.827	1.682
Di-*n*-butyl	2.075	1.848	1.642	2.514	2.218	2.055
n-Propyl-*n*-hexyl	2.185	1.969	1.775	2.632	2.334	2.184
Di-*n*-amyl	2.631	2.128	1.932	2.803	2.484	2.315
Di-*n*-hexyl	2.723	2.485	2.247	—	2.826	2.663
n-Amyl-*n*-octyl	2.886	2.642	2.406	3.381	2.956	2.806
Di-*n*-heptyl	3.046	2.781	2.546	—	3.152	2.947
Di-*n*-octyl	3.391	3.110	2.845	—	3.472	3.255
n-Heptyl-*n*-nonyl	3.416	3.122	2.862	—	3.487	3.270
Di-*n*-nonyl	—	3.403	3.122	—	—	—
Adipates						
Dimethyl	1.401	1.247	1.035	1.982	1.777	1.637
Diethyl	1.629	1.569	1.225	2.168	1.960	1.789
Di-*n*-propyl	1.996	1.896	1.565	2.499	2.272	2.083
n-Propyl-*n*-hexyl	2.500	2.387	2.000	3.034	2.766	2.544
Di-*n*-amyl	2.685	2.554	2.151	3.223	2.945	2.703
Di-*n*-hexyl	3.026	2.849	2.418	—	3.271	3.035
Di-*n*-heptyl	3.343	3.178	2.714	—	—	3.421
Di-*n*-octyl	—	3.536	3.012	—	—	—
Suberates						
Diethyl	2.138	2.231	1.707	2.600	2.293	2.131
Di-*n*-propyl	—	2.215	—	2.828	—	2.374
n-Propyl-*n*-hexyl	—	—	2.398	—	—	2.808
Di-*n*-amyl	3.071	2.788	2.552	—	3.212	2.993
Di-*n*-hexyl	3.401	3.115	2.852	—	3.536	3.292
n-Hexyl-*n*-heptyl	3.397	3.118	2.854	—	3.557	3.293
Di-*n*-heptyl	—	—	3.151	—	—	3.612
Sebacates						
Dimethyl	2.222	1.974	1.778	2.777	2.455	2.290
Diethyl	2.502	2.231	2.052	2.969	2.636	2.465
Di-*n*-propyl	2.753	2.510	2.291	3.260	2.936	2.713
Di-*n*-butyl	3.150	2.843	2.625	3.655	3.272	3.054
n-Propyl-*n*-hexyl	3.282	2.998	2.745	—	3.432	3.189
Di-*n*-amyl	—	3.129	2.847	—	—	3.433
Di-*n*-hexyl	—	3.445	3.151	—	—	3.592
n-Amyl-*n*-heptyl	—	3.460	3.162	—	—	—
Phthalates						
Dimethyl	1.831	1.660	1.495	2.602	2.390	2.226
Diethyl	2.097	1.883	1.702	2.805	2.567	2.392
Di-*n*-propyl	2.362	2.214	1.982	3.069	2.806	2.619
Di-*n*-butyl	2.687	2.451	2.229	3.400	3.123	2.917
n-Propyl-*n*-hexyl	2.873	2.628	2.437	3.560	3.278	3.056
Di-*n*-hexyl	3.330	3.047	2.768	—	—	3.494
Di-*n*-heptyl	—	3.346	3.055	—	—	—
Phosphates						
Triethyl	1.138	1.024	0.855	1.716	1.502	1.351
Tri-*n*-butyl	2.256	1.986	1.803	2.667	2.363	2.204
Tri-*n*-hexyl	3.216	2.957	2.704	—	3.407	3.223

Column Packing P1 = 0.5% silicone SE-30 on hydrofluoric acid-etched beads.
P2 = 0.5% neopentylglycol polysebacate (Analabs) on acid-etched beads.

Literature Zulaica, J. and G. Guiochon, *Bull. Soc. Chim. France*, 1242 (1963).

TABLE GC 26
GASES

Column Packing	P1	P2	P3	P4	P5	P6	P7	P8
Temperature (°C)	26	48	26	18	28.5	0	T1	130
Gas; Flow Rate (ml/min)	He,33	Ar,16.5	He,33	N_2,40	N_2,40	He,95	He,na	He,na
Column								
length, ft (a), **cm** (b)	10(a)	4½(a)	10(a)	320(b)	300(b)	20(a)	10(a)	200–400 (b)
diameter, in. (a), **mm** (b)	0.25(a)	0.25(a)	0.25(a)	0.7(b)	3.5(b)	0.25(a)	0.25(a)	⅛(a)
form	na	na	na	na	na	na	na	tubing
material	monel metal	monel metal	SS	na	na	copper	copper	SS
Detector	TC	gas density	TC	interfer-ometer cell	TC	TC	TC	TC
Literature	1	2	1	3	4	5	6	7
Compound				t_r				I
Silicon tetrafluoride	2.8	—	2.8	—	—	—	—	—
Carbon tetrafluoride	4.5	—	2.9	—	—	2.0	5.8	—
Nitrogen trifluoride	4.6	—	2.9	—	—	—	—	—
Hydrogen fluoride	7.0	1.70	—	—	—	—	—	—
Sulfur hexafluoride	12.5	—	4.3	—	—	—	—	—
Trifluorochloromethane	14.0	—	3.3	2	—	—	—	—
Fluorine	15.5	—	—	—	—	—	—	—
Trifluorobromomethane	31.5	—	—	—	—	—	—	—
Chlorine trifluoride	46.5	2.80	—	—	—	—	—	—
Chlorine	50.0	2.00	5.5	—	1.00	—	—	—
Difluorodichloromethane	56.0	—	5.8	3	—	—	—	318.2
Trifluorochloroethylene	67.5	—	6.3	—	—	—	—	—
Difluorodibromomethane	430.0	—	19.7	—	—	—	—	—
Fluorodichloromethane	448.0	—	12.2	30	—	—	—	411.1
Nitrogen	—	1.20	—	—	—	—	—	—
Chlorine monofluoride	—	1.40	—	—	—	—	—	—
Bromine	—	4.55	—	—	—	—	—	—
Bromine pentafluoride	—	6.10	—	—	—	—	—	—
Uranium hexafluoride	—	10.50	—	—	—	—	—	—
Oxygen	—	—	—	—	—	—	—	—
Freon 23	—	—	—	5	—	—	—	—
Freon 22 ($CHClF_2$)	—	—	—	12	—	—	—	277.6
Fluorotrichloromethane	—	—	—	—	—	—	—	458.0
Carbon tetrachloride	—	—	—	—	—	—	—	610.1
Hydrogen chloride	—	—	—	—	0.51	—	—	—
Phosgene	—	—	—	—	1.70	—	—	—
Boron trichloride	—	—	—	—	1.98	—	—	—
Silicon tetrachloride	—	—	—	—	7.70	—	—	—
Air	—	—	—	—	—	2.0	—	—
Hexafluoroethane	—	—	—	—	—	2.4	13.0	161.4
Tetrafluoroethene	—	—	—	—	—	2.8	16.0	—
Octafluoropropane	—	—	—	—	—	3.2	23.0	—
Difluoroacetylene	—	—	—	—	—	4.7	19.0	—
Hexafluoropropene	—	—	—	—	—	5.6	31.0	—
Octafluorocyclobutane	—	—	—	—	—	8.0	37.0	—
***trans*-Octafluorobutene-2**	—	—	—	—	—	9.6	43.0	—
***cis*-Octafluorobutene-2**	—	—	—	—	—	10.5	43.0	—
Octafluorobutene-1	—	—	—	—	—	14.2	57.0	—
Octafluoroisobutene	—	—	—	—	—	16.0	50.0	—
Hexafluorocyclopropane	—	—	—	—	—	—	28.4	—
Fluoromethane	—	—	—	—	—	—	—	174.9
Chloromethane	—	—	—	—	—	—	—	307.2
Bromomethane	—	—	—	—	—	—	—	372.8
Dichloromethane	—	—	—	—	—	—	—	446.4
Chlorofluoromethane	—	—	—	—	—	—	—	321.5
Difluoromethane	—	—	—	—	—	—	—	173.1
Chloroform	—	—	—	—	—	—	—	550.1
Trifluoromethane	—	—	—	—	—	—	—	160.2

TABLE GC 26—(Continued)
GASES

Column Packing	P1	P2	P3	P4	P5	P6	P7	P8
Temperature (°C)	26	48	26	18	28.5	0	T1	130
Gas; Flow Rate (ml/min)	He,33	Ar,16.5	He,33	N_2,40	N_2,40	He,95	He,na	He,na
Column								
length, ft (a), cm (b)	10(a)	4½(a)	10(a)	320(b)	300(b)	20(a)	10(a)	200–400 (b)
diameter, in. (a), mm (b)	0.25(a)	0.25(a)	0.25(a)	0.7(b)	3.5(b)	0.25(a)	0.25(a)	⅛(a)
form	na	na	na	na	na	na	na	tubing
material	monel metal	monel metal	SS	na	na	copper	copper	SS
Detector	TC	gas density	TC	interfer-ometer cell	TC	TC	TC	TC
Literature	1	2	1	3	4	5	6	7
Compound	t_r							I
Trifluorochloromethane	—	—	—	—	—	—	—	188.9
1,1,2-Trichloro-1,2,2-trifluoro-ethane	—	—	—	—	—	—	—	477.2
1,2-Dichloro-1,1,2,2-tetra-fluoroethane	—	—	—	—	—	—	—	393.9
1-Chloro-1,1,2,2,2-penta-fluoroethane	—	—	—	—	—	—	—	268.0
Ethylene (ethene)	—	—	—	—	—	—	—	181.2
Fluoroethylene (vinyl fluoride)	—	—	—	—	—	—	—	212.6
Vinyl chloride	—	—	—	—	—	—	—	352.2
Vinyl bromide	—	—	—	—	—	—	—	417.1
1,1-Difluoroethene	—	—	—	—	—	—	—	191.9

Column Packing
P1 = 50% halocarbon oil plasticized w/Kel-F 300 (30/50 mesh).
P2 = 17% w/w Kel-F 40 on Kel-F polymer (36/72 mesh).
P3 = 30% halocarbon oil on Chromosorb W (80/100 mesh).
P4 = 20% dibutylphthalate on diatomite.
P5 = Hexadecane on Celite 545.
P6 = 30% w/w $CH_2{=}CH{\cdot}CO_2{\cdot}CH_2{-}(CF_2CF_2)_3H$ on Chromosorb W.
P7 = Silica gel (50/80 mesh).
P8 = Porapak Q (80/100 mesh).

Temperature
T1 = from room temperature to 180 °C in 60 minutes.

Literature

1. Lysyj, I. and P. R. Newton, *Anal. Chem.*, **35**, 90 (1963).
2. Ellis, J. F. and G. Iveson, *Proc. 2nd Symp. Gas Chromatography*, 1958, Amsterdam, D. H. Desty, Ed., Butterworths, London, 1958, p. 300.
3. Gol'dinov, A. L., Lukhovitskii, V. I., and N. A. Svinina, *Zavod. Lab.*, **28**, 150 (1962).
4. Green, S. W., "*Vapor Phase Chromatography*", London, 1956, D. H. Desty, Ed., Butterworths, London, 1957, p. 388.
5. Turkel'taub, N. N., Ainshtein, S. A., and S. V. Syavtsillo, *Zavod. Lab.*, **28**, 141 (1962).
6. Greene, S. A. and F. M. Wachi, *Anal. Chem.*, **35**, 928 (1963).
7. Foris, A. and J. G. Lehman, *Separ. Sci.*, **4**, 230 (1969).

TABLE GC 27
GASES

Column Packing	P1	P2	P3	P4	P5	P6	P7	P8
Temperature (°C)	0	−78	40	25	27	60	0	30
Gas; Flow Rate (ml/min)	na	na	na	na	He,111	He,24	He,2.73	He
Column								
length, ft[a], cm[b]	33[a]	10[a]	18[a]	4[a]	10[a]	10[a]	125[b]	—
diameter, in.[a], cm[b]	na	na	na	na	0.25[a]	0.25[a]	0.4[b]	na
form	na	na	na	na	na	na	na	na
material	na	na	na	na	copper	copper	na	na
Detector	na	na	na	na	TC	TC	na	na
Literature	1	1	1	1	2	2	3	4
Compound	V_R	V_R	V_R	V_R	t_r	t_r	t_r	V_R
Hydrogen	0.46	—	—	0.95	—	—	—	—
Diborane	0.62	—	0.97	—	2	3.5	8.0	—
Hydrogen chloride	1.00	1.00	1.00	1.00	—	—	—	—
Chlorodiborane	4.05	—	2.65	—	—	—	—	—
Dichloroborine	—	1.81	4.70	—	—	—	—	—
Boron trichloride	—	4.54	6.30	2.72	—	—	—	—
B_4H_{10}	—	—	—	—	7	—	—	—
B_5H_9	—	—	—	—	19	—	—	—
B_5H_{11}	—	—	—	—	26	—	—	—
Ethane	—	—	—	—	—	2.8	—	—
Diethyl ether	—	—	—	—	—	15.0	—	—
Methyldiborane	—	—	—	—	—	—	12.5	—
Trimethylboron	—	—	—	—	—	—	18.0	—
1,1-Dimethylborane	—	—	—	—	—	—	27.0	—
Silane	—	—	—	—	—	—	—	0.27
Methylsilane	—	—	—	—	—	—	—	0.36
Dimethylsilane	—	—	—	—	—	—	—	0.59
Trimethylsilane	—	—	—	—	—	—	—	1.0
Tetramethylsilane	—	—	—	—	—	—	—	1.63

Column Packing
P1 = Silicone oil dispersed on Chromosorb (60/80 mesh).
P2 = Teflon.
P3 = Hexadecane dispersed on Chromosorb (60/80 mesh).
P4 = Fluorolube on Teflon.
P5 = 30% Paraffin oil on Celite 545 (60/100 or 60/80 mesh).
P6 = 30% Octoil on Celite 545 (60/100 or 60/80 mesh).
P7 = 38% Mineral oil on firebrick (52/65 mesh).
P8 = Tetra-iso-butene on C-22 firebrick.

Literature
1. Myers, H. W. and R. F. Putnam, *Anal. Chem.*, **34**, 664 (1962).
2. Kaufman, J. J., Todd, J. E., and W. S. Koski, *Anal. Chem.*, **29**, 1032 (1957).
3. Seely, G. R., Oliver, J. P., and D. M. Ritter, *Anal. Chem.*, **31**, 1993 (1959).
4. Russell, G. A., *J. Amer. Chem. Soc.*, **81**, 4825 (1959).

ERRATA

HANDBOOK OF CHROMATOGRAPHY

VOLUMES I AND II

Addition to **ADVISORY BOARD**

Ivor Smith, Ph.D., F.R.I.C.
Courtauld Institute of Biochemistry
The Middlesex Hospital
Medical School
London, England

VOLUME I

Page 96 – “^{r}quinoline” found in Table GC-78.
Page 659 – “Abietate” found in Table LC-67.
Page 659 – Delete “Ac.amines GC-98”

SECTION I LIST OF TABLES FOR (5 pages – unnumbered)

Section I.III Paper Chromatography Tables (PC) Pages 2 and 4
Section I.IV Thin-Layer Chromatography Tables (TLC) Pages 4, 3, and 5

VOLUME II

Page 73 – “Section I.IIIB3e” should be “Section I.IIIB3c.”
Pages 109 and 157 – Detection Reagent No. 350 should be “Nitrobenzenediazonium fluoborate”
Page 319 – should be “Zweig and Sherma, Analytical Methods for Pesticides and Plant Growth Regulators.”

ERRATA

HANDBOOK OF CHROMATOGRAPHY

VOLUMES I AND II

Addition to **ADVISORY BOARD**

Ivor Smith, Ph.D., F.R.I.C.
Courtauld Institute of Biochemistry
The Middlesex Hospital
Medical School
London, England

VOLUME I

Page 96 – “^{r}quinoline” found in Table GC-78.
Page 659 – “Abietate” found in Table LC-67.
Page 659 – Delete “Ac.amines GC-98”

SECTION I LIST OF TABLES FOR (5 pages – unnumbered)

Section I.III Paper Chromatography Tables (PC) Pages 2 and 4
Section I.IV Thin-Layer Chromatography Tables (TLC) Pages 4, 3, and 5

VOLUME II

Page 73 – “Section I.IIIB3e” should be “Section I.IIIB3c.”
Pages 109 and 157 – Detection Reagent No. 350 should be “Nitrobenzenediazonium fluoborate”
Page 319 – should be “Zweig and Sherma, Analytical Methods for Pesticides and Plant Growth Regulators.”

TABLE GC 28
GLYCOPYRANOSIDES, METHYL ETHERS

Column Packing	P1	P2	P3	P4
Temperature (°C)	150	200	212	170
Gas; Flow Rate (ml/min)	Ar,60	Ar,60	N_2,na	He,130
Column				
length, ft	4	4	3	8
diameter	5 I.D.	5 I.D.	1/8″	7 mm
form	na	na	coiled	na
material	na	na	SS	glass
Detector	EC	EC	FI	TC
Literature	1	1	2	3

Compound	r (methyl-2,3,4,6-tetra-O-methyl-α-D glucopyranoside		r (β-isomer)	r (quinoline)
Glucopyranoside methyl ether of:				
β-2,3,4,6	0.67	0.82	1.00	—
α-2,3,4,6	1.000 (reference)	1.000 (reference)	1.37	—
β-2,3,4	1.78	0.98	—	—
α-2,3,4	2.19	1.12	—	—
β-2,4,6	2.34	1.09	—	—
β-2,3,6	2.48	1.12	2.95	—
β-3,4,6	2.62	1.13	—	—
α-3,4,6	2.65	1.13	—	—
α-2,3,6	3.56	1.26	3.65	—
α-2,4,6	3.56	1.38	—	—
α,β-3,6	—	—	6.95	—
α-4,6	—	—	6.96	—
β-2,3	—	—	8.05	—
α-2,3	—	—	9.70	—
α,β-2,6	—	—	10.50	—
Methyl 2,3,4-tri-O-methyl-β-D-xylopyranoside	—	—	—	0.528
Methyl 2,3,4-tri-O-methyl-β-L-arabopyranoside	—	—	—	0.785
Methyl 2,3,4,6-tetra-O-methyl-α-D-mannopyranoside	—	—	—	1.468
Methyl 2,3,4,6-tetra-O-methyl-α-D-glucopyranoside	—	—	—	1.594
Methyl 2,3,4,6-tetra-O-methyl-α-D-galactopyranoside	—	—	—	1.708
Quinoline (reference)	—	—	—	1.000

Column Packing P1 = >100 mesh, alkali-washed Celite 545 with butanediol succinate polyester.
P2 = >100 mesh, alkali-washed Celite 545 with butanediol succinate polyester on Apiezon M.
P3 = 25% LAC-2-R-446 and 2% H_3PO_4 on Chromosorb W (80/100 mesh).
P4 = 20% Apiezon M on Celite 545.

Literature
1. Bishop, C. T. and F. P. Cooper, *Can. J. Chem.*, **38**, 388 (1960).
2. Neely, W. B., Nott, J., and C. B. Roberts, *Anal. Chem.*, **34**, 1423 (1962).
3. McInnes, A. G., Ball, D. H., Cooper, F. P., and C. T. Bishop, *J. Chromatogr.*, **1**, 556 (1958).

TABLE GC 29
HYDROCARBONS, ALIPHATIC, UNSATURATED

Column Packing	P1	P1	P1	P2	P2	P2
Temperature (°C)	35.0	22.25	15.0	30.0	22.0	15.0
Gas; Flow Rate (ml/min)	He	He	He	He	He	He
Column						
length, ft	18	18	18	11.4	11.4	11.4
diameter, in.	$\frac{1}{4}$	$\frac{1}{4}$	$\frac{1}{4}$	$\frac{1}{4}$	$\frac{1}{4}$	$\frac{1}{4}$
form	na	na	na	na	na	na
material	SS	SS	SS	SS	SS	SS
Detector	na	na	na	na	na	na

Compound	$r_{isoprene}$					
Acetylene	0.137	0.116	0.106	—	—	—
Isopentane	0.217	0.200	0.184	—	—	—
***n*-Pentane**	0.284	0.263	0.247	0.111	0.103	0.093
3-Methylbutene-1	0.294	0.260	0.244	0.452	0.457	0.457
Butadiene-1,3	0.369	0.329	0.311	0.366	0.353	0.336
Neohexane	0.382	0.359	0.343	—	—	—
Propyne	0.384	0.356	0.350	—	—	—
Pentene-1	0.414	0.393	0.378	—	—	—
2-Methylbutene-1	0.482	0.461	0.447	—	—	—
***trans*-Pentene-2**	0.503	0.486	0.476	0.371	0.355	0.352
***cis*-Pentene-2**	0.551	0.530	0.518	0.810	0.832	0.851
Pentadiene-1,4	0.615	0.592	0.574	1.571	1.648	1.735
2-Methylbutene-2	0.634	0.618	0.608	0.393	0.381	0.366
***n*-Hexane**	0.706	0.698	0.692	0.268	0.254	0.246
Cyclopentane	0.756	0.717	0.712	0.367	0.342	0.325
Butyne-1	0.800	0.789	0.774	—	—	—
Isoprene	1.00	1.00	1.00	1.00	1.00	1.00
Isopropylacetylene	1.055	1.075	1.096	—	—	—
Cyclopentene	1.075	1.04	1.04	2.461	2.611	2.778
2-Methylpentene-1	1.094	1.108	1.115	1.122	1.155	1.194
***trans*-1,3-Pentadiene**	1.335	1.358	1.375	1.377	1.390	1.407
Butyne-2	1.41	1.435	1.456	1.619	1.608	1.606
***cis*-1,3-Pentadiene**	1.48	1.514	1.541	1.825	1.874	1.924
***n*-Heptane**	1.724	1.823	1.887	0.649	0.633	0.638
Cyclopentadiene	1.90	1.954	1.995	2.464	2.460	2.50
Pentyne-1	1.96	2.09	2.18	0.597	0.609	0.626
Isopropenylacetylene	2.21	2.38	2.50	—	—	—
Pentyne-2	2.66	2.86	3.01	3.236	3.353	3.490
***n*-Butane**	—	—	—	0.041	0.036	0.034
Ethylene	—	—	—	0.042	0.041	0.038
Propylene	—	—	—	0.100	0.095	0.091
Propadiene	—	—	—	0.115	0.103	0.094
***trans*-Butene-2**	—	—	—	0.154	0.1465	0.136
Isobutylene	—	—	—	0.195	0.1875	0.180
Butene-1	—	—	—	0.287	0.287	0.284
***cis*-Butene-2**	—	—	—	0.345	0.330	0.327
Cyclohexane	—	—	—	0.811	0.766	0.750

Column Packing P1 = 20% DMS (dimethylsulfolane) on Chromosorb (30/50 mesh).
P2 = 30% BC-$AgNO_3$ (benzyl cyanide) on Chromosorb (30/50 mesh).

Literature Armitage, F., *J. Chromatogr.*, **2**, 655 (1959).

TABLE GC 30
HYDROCARBONS, ALKYL PARAFFINS

Column Packing	P1	P2	P2	P3	P4	P5	P6	P7	P8
Temperature (°C)	80	80	25	25	78.5	78.5	95	78.5	25
Gas; Flow Rate (ml/min)	N_2	N_2	N_2	N_2	N_2	N_2	N_2	N_2	N_2
Column									
length, ft	20–30	20–30	20–30	20–30	20–30	20–30	20–30	20–30	20–30
diameter, in., O.D.	$\frac{1}{8}$	$\frac{1}{8}$	$\frac{1}{8}$	$\frac{1}{8}$	$\frac{1}{8}$	$\frac{1}{8}$	$\frac{1}{8}$	$\frac{1}{8}$	$\frac{1}{8}$
form	U-tube	U-tube	U-tube	U-tube	U-tube	U-tube	U-tube	U-tube	U-tube
material	copper	copper	copper	copper	copper	copper	copper	copper	copper
Detector	FI	FI	FI	FI	FI	FI	FI	FI	FI

Compound	I								
2,2-Dimethylbutane	530	539	535	534	531	527	—	532	536
2,3-Dimethylbutane	562	569	566	563	569	566	—	571	565
2-Methylpentane	568	570	570	568	569	568	571	569	573
3-Methylpentane	583	585	584	581	585	585	586	586	583
2,2-Dimethylpentane	622	626	625	624	625	613	—	628	626
2,4-Dimethylpentane	628	630	630	629	628	616	—	633	631
2,2,3-Trimethylbutane	643	643	637	632	642	629	—	—	635
3,3-Dimethylpentane	660	662	656	653	662	662	—	664	654
2,3-Dimethylpentane	672	672	670	667	674	668	—	675	668
2-Methylhexane	667	667	667	666	666	658	670	667	667
3-Methylhexane	677	677	676	674	677	671	679	678	674
3-Ethylhexane	688	688	685	682	689	684	—	688	682
2,2,4-Trimethylpentane	692	694	687	687	691	674	—	691	690
2,2,3,3-Tetramethylbutane	—	733	720	—	—	—	—	—	—
2,2-Dimethylhexane	721	722	718	716	715	705	725	722	719
2,5-Dimethylhexane	728	729	728	727	725	714	733	729	731
2,4-Dimethylhexane	731	734	731	729	730	720	—	734	733
2,2,3-Trimethylpentane	739	742	734	730	740	733	—	741	733
3,3-Dimethylhexane	746	747	740	739	745	737	—	747	743
2,3,4-Trimethylpentane	754	758	748	745	755	751	—	757	750
2,3,3-Trimethylpentane	762	765	755	—	765	760	—	763	755
2,3-Dimethylhexane	761	761	758	—	760	757	—	761	761
2-Methyl-3-ethylhexane	763	764	759	—	763	762	—	773	760
2-Methylheptane	764	765	764	—	762	758	767	764	770
4-Methylheptane	769	768	767	—	766	762	—	769	771
3,4-Dimethylhexane	773	774	768	—	773	773	—	774	770
3-Methyl-3-ethylpentane	781	780	769	—	770	788	—	779	770
3-Ethylhexane	776	775	771	—	772	773	—	782	774
3-Methylheptane	775	773	770	—	771	758	776	—	775
2,2,4,4-Tetramethylpentane	—	—	768	—	—	—	—	—	—
2,2,5-Trimethylhexane	783	778	—	—	—	—	—	—	—
2,2,4-Trimethylhexane	794	—	—	—	—	—	—	—	—
2,3,5-Trimethylhexane	818	—	810	—	—	—	—	—	—
2,2-Dimethylheptane	820	—	808	—	—	—	822	—	—
2,4-Dimethylheptane	823	822	—	—	—	—	—	—	—
2,6-Dimethylheptane	828	827	—	—	—	—	831	—	—
2,5-Dimethylheptane	834	833	—	—	—	—	—	—	—
1,1,3-Trimethylhexane	842	—	834	—	—	—	—	—	—
2,3-Dimethylheptane	858	856	—	—	—	—	—	—	—
2-Methyloctane	865	864	—	—	—	—	865	—	—
3-Methyloctane	871	870	—	—	—	—	872	—	—
4-Methyloctane	860	859	—	—	—	—	—	—	—
4-Methylnonane	961	—	—	—	—	—	—	—	—

Column Packing P1 = 5% SE-30 Silicone gum rubber on acid-washed silanized Chromosorb W (60/80 mesh).
P2 = 5% Squalane on acid-washed silanized Chromosorb W (60/80 mesh).
P3 = 5% Hexadecane on acid-washed silanized Chromosorb W (60/80 mesh).
P4 = 5% Hexatricontane on acid-washed silanized Chromosorb W (60/80 mesh).
P5 = 5% Benzyl diphenyl on acid-washed silanized Chromosorb W (60/80 mesh).
P6 = 5% High vacuum silicone grease on acid-washed silanized Chromosorb W (60/80 mesh).
P7 = 5% Paraffin wax on acid-washed silanized Chromosorb W (60/80 mesh).
P8 = 5% Hexadecane-fluolubrol mixed phase on acid-washed silanized Chromosorb W (60/80 mesh).

Literature Mitra, G. D. and N. C. Saha, *J. Chromatogr. Sci.*, **8**, 95 (1970).

TABLE GC 31
HYDROCARBONS, CHLORINATED BENZENES

Column Packing	P1	P1	P1
Temperature (°C)	70	100	130
Gas; Flow Rate (ml/min)	N_2	N_2	N_2
Column			
length, m	10	10	10
diameter, mm, I.D.	0.25	0.25	0.25
form	capillary	capillary	capillary
material	SS	SS	SS
Detector	FI	FI	FI

Compound	$r_{1,2,4\text{-trichlorobenzene}}$		
Benzene	0.02	—	—
Chlorobenzene	0.10	—	—
1,3-Dichlorobenzene	0.30	0.36	—
1,4-Dichlorobenzene	0.34	0.42	—
1,2-Dichlorobenzene	0.45	0.52	0.54
1,3,5-Trichlorobenzene	0.54	0.58	0.62
1,2,4-Trichlorobenzene	1.00	1.00	1.00
1,2,3-Trichlorobenzene	1.53	1.45	1.38
1,2,3,5-Tetrachlorobenzene	2.23	1.98	1.75
1,2,4,5-Tetrachlorobenzene	2.42	2.12	1.84
1,2,3,4-Tetrachlorobenzene	—	3.36	2.87
Pentachlorobenzene	—	5.68	4.42
Hexachlorobenzene	—	—	9.99

Column Packing P1 = Diethylene glycol succinate.
Literature Hrivňák, J. and M. Michalek, *Chromatographia*, **3**, 123 (1970).

TABLE GC 32
HYDROCARBONS, CHLORINATED OLEFINS

Column Packing	P1	P2	P3
Temperature (°C)	30	25	25
Gas; Flow Rate (cc/min)	1–2,Ar	1–2,Ar	1–2,Ar
Column			
length, ft	100	100	200
diameter, in., I.D.	0.01	0.01	0.01
form	capillary	capillary	capillary
material	304SS	304SS	321SS
Detector	Triode argon	Triode argon	Triode argon

Compound	$r_{\text{chloroform}}$		
Vinyl chloride	0.330	0.417	0.64
Vinylidene chloride	0.558	0.520	0.79
Methylene chloride	0.496	0.660	0.76
1,1-Dichloroethane	0.720	0.709	0.86
***trans*-Dichloroethylene**	0.742	0.643	0.88
***cis*-Dichloroethylene**	0.840	0.895	0.92
Chloroform	1.00	1.00	1.00
1,1,1-Trichloroethane	1.39	0.843	1.19
Ethylene dichloride	1.19	1.44	1.08
Carbon tetrachloride	1.78	0.843	1.38
Trichloroethylene	2.47	1.31	1.68
Propylene dichloride	2.24	1.66	1.49
DL-3-Chloro-1-butene	0.84	0.675	0.93
2-Chloro-1-butene	0.877	—	0.94
2-Chloro-1,3-butadiene	0.925	—	0.97

TABLE GC 32—(Continued)
HYDROCARBONS, CHLORINATED OLEFINS

cis-1-Chloro-1-butene	1.02	—	1.01
trans-1-Chloro-1-butene	1.20	—	1.11
cis-2-Chloro-2-butene	1.25	0.747	1.09
trans-2-Chloro-2-butene	1.06	0.636	1.04
trans-1-Chloro-2-butene	1.52	1.14	1.26
3-Methylpentane	1.07	0.497	1.06
3-Methylhexane	2.58	0.637	—

Column Packing P1 = Squalane.
P2 = 25% GE 96 silicone and 75% oxybis(2-ethyl benzoate).
P3 = Hexadecane.
Literature Hollis, O. L. and W. V. Hayes, *Anal. Chem.*, **34**, 1223 (1962).

TABLE GC 33
HYDROCARBONS, OCTANES

Column Packing	P1	P1
Temperature (°C)	30	50
Gas; Flow Rate (ml/min)	0.3,N_2	0.3,N_2
Column		
length, m	55	55
diameter, mm	0.25	0.25
form	capillary	capillary
material	SS	SS
Detector	FID	FID

Compound	I	
Paraffins		
2-Methylbutane	474	474
n-Pentane	500	500
2,3-Dimethylbutane	565	567.5
2-Methylpentane	569	569.5
3-Methylpentane	582.5	584.5
n-Hexane	600	600
2,2-Dimethylpentane	624.5	626
2,4-Dimethylpentane	629	630
2,2,3-Trimethylbutane	636.5	639.5
2-Methylhexane	666	666.5
2,3-Dimethylpentane	669.5	671
3-Methylhexane	675	676
3-Ethylpentane	684.5	686
2,2,4-Trimethylpentane	688	690
n-Heptane	700	700
2,2-Dimethylhexane	718.5	719.5
2,5-Dimethylhexane	728	729
2,4-Dimethylhexane	731.5	732.5
2,2,3-Trimethylpentane	734	737
3,3-Dimethylhexane	741	743.5
2,3,4-Trimethylpentane	750.5	753
2,3,3-Trimethylpentane	756	760
2,3-Dimethylhexane	759.5	760.5
2-Methylheptane	764.5	765
4-Methylheptane	767.5	767.5
3,4-Dimethylhexane	769	771
3-Methylheptane	773	773.5
2,2,5-Trimethylhexane	775	776.5
2,2,4-Trimethylhexane	786.5	789
n-Octane	800	800
2,3,5-Trimethylhexane	810.5	812

Column Packing P1 = Squalane.
Literature Tourres, D. A., *J. Gas Chromatogr.*, **5**, 35 (967).

TABLE GC 34

HYDROCARBONS, CHLORINATED AND BROMINATED ALIPHATICS

Column Packing	P1	P2	P3	P3	P3
Temperature (°C)	40	40	50	80	120
Gas; Flow Rate (ml/min)	13–15,N_2	25–30	na	na	na
Column					
length, m	4	4	45	45	45
diameter, mm	1.4 O.D.	2.8 O.D.	0.2 I.D.	0.2 I.D.	0.2 I.D.
form	na	na	capillary	capillary	capillary
material	SS	SS	SS	SS	SS
Detector	FI	FI	FI	FI	FI
Literature	1	1	2	2	2

Compound	$r_{\text{1-chloro-2,2 dimethylpropane}}$			r_{pentane}	
1-Chloro-2,2-dimethylpropane	1.00	1.00	—	—	—
2-Chloro-2-methylbutane	1.00	1.13	—	—	—
2-Chloro-3-methylbutane	1.37	1.47	—	—	—
2-Chloropentane	1.52	1.80	—	—	—
3-Chloropentane	1.63	1.86	—	—	—
1-Chloro-3-methylbutane	1.68	1.70	—	—	—
1-Chloro-2-methylbutane	1.83	1.83	3.34	1.92	1.48
1-Chloropentane	2.41	2.66	—	6.80	3.33
Chloroethane	—	—	0.78	0.9	0.98
1-Chloropropane	—	—	1.25	1.15	1.10
2-Chloropropane	—	—	1.04	1.05	1.11
1-Chlorobutane	—	—	2.32	1.47	1.21
2-Chloro-2-methylpropane	—	—	1.73	1.33	1.19
1-Chlorohexane	—	—	—	4.24	2.17
1-Chlorooctane	—	—	—	16.70	5.44
1-Chlorononane	—	—	—	18.75	9.45
1-Chlorodecane	—	—	—	35.5	17.0
***n*-Pentane**	—	—	2.01 sec.	1.83 sec.	1.56 sec.
1-Bromo-2,2-dimethylpropane	1.00	1.00	—	—	—
2-Bromo-2-methylbutane	1.00	(1.15)	—	—	—
2-Bromo-3-methylbutane	1.33	1.52	—	—	—
2-Bromopentane	1.42	1.86	—	—	—
3-Bromopentane	1.55	1.93	—	—	—
1-Bromo-3-methylbutane	1.55	1.67	—	—	—
1-Bromo-2-methylbutane	1.68	1.82	—	—	—
1-Bromopentane	2.15	2.69	—	—	—

Column Packing P1 = 10% Squalane on Chromosorb W (80/100 mesh).
P2 = 11.5% Bentone 34 and 11.5% MS-555 (silicone) on Chromosorb W (80/100 mesh).
P3 = Squalene.

Literature
1. Chaudri, B. A., Hudson, H. R., and W. S. Murphy, *J. Chromatogr.*, **29**, 218 (1967).
2. Drabkin, A. E., Yaminov, M. S., and M. G. Rudin, *Zh. Prikl. Khim.*, **42**, 425 (1969).

TABLE GC 35
HYDROCARBONS, CYCLOHEXANES

Column Packing	P1	P1	P2	P3	P4	P2
Temperature (°C)	80	130	25	78.5	25	70
Gas; Flow Rate (ml/min)	N_2	N_2	N_2	N_2	N_2	N_2
Column						
length, ft	20–30	20–30	20–30	20–30	20–30	20–30
diameter, in., O.D.	$\frac{1}{8}$	$\frac{1}{8}$	$\frac{1}{8}$	$\frac{1}{8}$	$\frac{1}{8}$	$\frac{1}{8}$
form	U-tube	U-tube	U-tube	U-tube	U-tube	U-tube
material	copper	copper	copper	copper	copper	copper
Detector	FI	FI	FI	FI	FI	FI

Compound			I			
Cyclohexane	667	676	662	672	645	668
Methylcyclohexane	727	739	720	735	711	730
***trans*-1,4-Dimethylcyclohexane**	781	796	780	—	779	789
1,1-Dimethylcyclohexane	784	—	780	796	776	792
***cis*-1,3-Dimethylcyclohexane**	783	798	779	795	779	789
***trans*-1,2-Dimethylcyclohexane**	802	817	793	—	789	807
***cis*-1,4-Dimethylcyclohexane**	804	820	798	—	790	810
***trans*-1,3-Dimethylcyclohexane**	806	823	799	—	790	810
***cis*-1,2-Dimethylcyclohexane**	831	850	—	—	811	836
Ethylcyclohexane	825	842	828	845	818	839
1,3,5-Trimethylcyclohexane (*trans*-)	—	840	—	—	—	—
1,3,5-Trimethylcyclohexane (*cis*-)	—	870	—	—	—	—
1,2,3-Trimethylcyclohexane (*trans*-)	—	954	—	—	—	—
1,2,3-Trimethylcyclohexane (*cis*-)	—	982	—	—	—	—
1,2,4-Trimethylcyclohexane (*trans*-)	—	868	—	—	—	—
1,2,4-Trimethylcyclohexane (*cis*-)	—	895	—	—	—	—
Isopentylcyclohexane	—	934	—	—	—	923
***n*-Pentylcyclohexane**	—	944	—	—	—	931
***trans*-1-Methyl-4-isopentylcyclohexane**	—	997	—	—	—	—
1,2,4,5-Tetramethylcyclohexane	—	1036	—	—	—	—
***cis*-1-Methyl-4-isopentylcyclohexane**	—	1000	—	—	—	—
1,2,4,5-Tetramethylcyclohexane	—	1054	—	—	—	—
Isobutylcyclohexane	—	1005	—	—	—	983
1,4-Diethylcyclohexane	—	1072	—	—	—	—
		1105				
1,3-Diethylcyclohexane	—	1017	—	—	—	—
***n*-Butylcyclohexane**	—	1061	—	—	—	1029

Column Packing P1 = 5% SE-30 Silicone gum rubber on acid-washed silanized Chromosorb W (60/80 mesh).
P2 = 5% Squalane on acid-washed silanized Chromosorb W (60/80 mesh).
P3 = 5% Hexatriacontane on acid-washed silanized Chromosorb W (60/80 mesh).
P4 = 5% Hexadecane-fluolubrol mixed phase on acid-washed silanized Chromosorb W (60/80 mesh).

Literature Mitra, G. D. and N. C. Saha, *J. Chromatogr. Sci.*, **8**, 95 (1970).

TABLE GC 36
HYDROCARBONS, CYCLOPENTANES

Column Packing	P1	P2	P3	P4	P5	P2
Temperature (°C)	130	25	25	78.5	25	70
Gas; Flow Rate (ml/min)	N_2	N_2	N_2	N_2	N_2	N_2
Column						
length, ft	20–30	20–30	20–30	20–30	20–30	20–30
diameter, in., O.D.	$\frac{1}{8}$	$\frac{1}{8}$	$\frac{1}{8}$	$\frac{1}{8}$	$\frac{1}{8}$	$\frac{1}{8}$
form	U-tube	U-tube	U-tube	U-tube	U-tube	U-tube
material	copper	copper	copper	copper	copper	copper
Detector	FI	FI	FI	FI	FI	FI
Compound				I		
Cyclopentane	572	563	555	573	551	570
Methylcyclopentane	634	625	617	634	617	631
1,1-Dimethylcyclopentane	683	670	663	679	661	677
1,3-*cis*-Dimethylcyclopentane	694	679	674	689	672	686
1,3-*trans*-Dimethylcyclopentane	698	683	678	—	676	690
1,2-*trans*-Dimethylcyclopentane	702	686	680	696	678	—
1,2-*cis*-Dimethylcyclopentane	734	716	708	727	705	725
Ethylcyclopentane	749	729	722	741	721	738
1,1,3-Trimethylcyclopentane	740	719	715	769	716	—
1,2-*trans*-(1)4-*cis*-Trimethylcyclopentane	759	738	733	—	736	—
1,2-*trans*-(1)3-*cis*-Trimethylcyclopentane	766	744	738	—	742	—
1,1,2-Trimethylcyclopentane	779	757	749	—	753	—
1,2-*cis*-(1)4-*trans*-Trimethylcyclopentane	791	768	—	—	765	—
1,2-*cis*-(1)3-*trans*-Trimethylcyclopentane	798	773	—	—	767	—
***trans*-1-Methyl-3-ethylcyclopentane**	810	785	—	—	783	—
***trans*-1-Methyl-2-ethylcyclopentane**	811	786	—	—	783	—
1-Methyl-1-ethylcyclopentane	814	788	—	801	783	—
1,2-*cis*-(1)3-*cis*-Trimethylcyclopentane	821	794	—	—	789	—
***cis*-1-Methyl-3-ethylcyclopentane**	808	783	—	—	781	—
Isopentylcyclopentane	835	806	—	823	798	—
***n*-Pentylcyclopentane**	851	—	—	838	819	834
1,2-*cis*-(1)4-*cis*-Trimethylcyclopentane	797	—	—	—	770	—
***cis*-1-Methyl-2-ethylcyclopentane**	846	—	—	—	807	—

Column Packing
P1 = 5% SE-30 silicone gum rubber on acid washed silanized Chromosorb W (60/80 mesh).
P2 = 5% Squalane on acid washed silanized Chromosorb W (60/80 mesh).
P3 = 5% Hexadecane on acid washed silanized Chromosorb W (60/80 mesh).
P4 = 5% Hexatriacontane on acid washed silanized Chromosorb W (60/80 mesh).
P5 = 5% Hexadecane-fluolubrol mixed phase on acid washed silanized Chromosorb W (60/80 mesh).

Literature
Mitra, G. D. and N. C. Saha, *J. Chromatogr. Sci.*, **8**, 95 (1970).

TABLE GC 37
HYDROCARBONS, IODINATED ALKANES

Column Packing	P1	P1	P1	P1	P1	P1	P2	P3
Temperature (°C)	60	100	120	60	100	120	40	40
Gas; Flow Rate (ml/min)	N_2,25	N_2,25	N_2,25	N_2,25	N_2,25	N_2,25	N_2,14	N_2,28
Column								
length, m	1.2	1.2	1.2	3.0	3.0	3.0	4.0	4.0
diameter, mm, I.D.	1.5	1.5	1.5	1.5	1.5	1.5	—	—
in., O.D.	$\frac{1}{8}$	$\frac{1}{8}$	$\frac{1}{8}$	$\frac{1}{8}$	$\frac{1}{8}$	$\frac{1}{8}$	$\frac{1}{16}$	$\frac{1}{8}$
form	helix	helix	helix	helix	helix	helix	na	na
material	pyrex glass	pyrex glass	pyrex glass	pyrex glass	pyrex glass	pyrex glass	SS	SS
Detector	EC	EC	EC	EC	EC	EC	na	na
Literature	1	1	1	1	1	1	2	2

Compound	t_r[a]						$r_{1\text{-chloro-2,2-dimethylpropane}}$	
Iodomethane	0.77	0.38	0.23	1.15	0.64	0.56	—	—
Iodoethane	1.74	0.71	0.39	2.04	1.32	1.02	—	—
1-Iodopropane	3.67	1.33	0.76	5.82	2.64	1.81	—	—
2-Iodopropane	2.36	1.00	0.59	—	1.84	—	—	—
1-Iodobutane	7.95	2.52	1.42	14.64	5.16	3.31	—	—
1-Iodopentane	18.35	4.84	2.65	34.63	10.32	5.98	2.14	2.51
1-Iodohexane	41.25	9.54	4.90	78.23	20.32	10.79	—	—
1-Iodoheptane	89.88	18.43	8.96	170.12	39.76	19.29	—	—
1-Iodooctane	204.00	35.57	16.23	—	79.81	34.40	—	—
1-Iodononane	—	68.32	29.27	—	155.00	60.87	—	—
1-Iododecane	—	134.90	53.01	—	—	106.92	—	—
1-Iodo-2-methylpropane	5.89	2.12	1.10	—	3.96	—	—	—
2-Iodo-2-methylpropane	2.87	1.09	0.64	—	2.16	—	—	—
2-Iodo-2-methylbutane	8.27	2.73	1.88	—	7.04	—	0.97	(1.08)[b]
1-Iodo-3,3-dimethylbutane	18.07	5.20	2.74	—	10.02	—	—	—
2-Iodo-2,3-dimethylbutane	18.40	5.88	2.88	—	10.36	—	—	—
2-Iodo-3,3-dimethylbutane	18.15	5.64	2.80	—	10.15	—	—	—
2-Iodo-2,4-dimethylpentane	41.60	9.60	5.31	—	20.96	—	—	—
1-Iodo-2,2,4-trimethylpentane	65.30	15.75	7.49	—	31.36	—	—	—
Diiodomethane	47.00	12.23	6.04	—	24.52	—	—	—
1,2-Diiodoethane	91.25	21.95	10.14	—	43.84	—	—	—
1,3-Diiodopropane	126.40	44.83	26.40	—	80.10	—	—	—
3-Iodopropene	4.00	1.47	0.89	—	2.96	—	—	—
5-Iodohexene-1	23.60	6.56	3.66	—	14.68	—	—	—
1-Iodo-2,2-dimethylpropane	—	—	—	—	—	—	1.00	1.00
2-Iodo-3-methylbutane	—	—	—	—	—	—	1.33	1.54
2-Iodopentane	—	—	—	—	—	—	1.39	1.77
3-Iodopentane	—	—	—	—	—	—	1.53	1.90
1-Iodo-3-methylbutane	—	—	—	—	—	—	1.53	1.58
1-Iodo-2-methylbutane	—	—	—	—	—	—	1.66	1.78

[a] Adjusted t_r for holdup of carrier gas.
[b] Determined in 1 m glass column at 20°.

Column Packing
P1 = 15% Tricresyl phosphate on DMCS-treated Chromosorb W (80/100 mesh).
P2 = 10% Squalane on Chromosorb W (80/100 mesh).
P3 = 11.5% Bentone 34 and 11.5% MS-555 silicone on Chromosorb W (80/100 mesh).

Literature

1. Castello, G., D'Amato, G., and E. Biagini, *J. Chromatogr.*, **41**, 313 (1969).
2. Chaudri, B. A., Hudson, H. R., and W. S. Murphy, *J. Chromatogr.*, **29**, 218 (1967).

TABLE GC 38
HYDROCARBONS, NAPHTHALENES

Column Packing	P1	P2	P3	P4	P4
Temperature (°C)	180	200	190	215	215
Gas; Flow Rate(ml/min)	50,He	60,He	46,He	68,He	28,He
Column					
length, ft	5	10	5	10	6
diameter, in., O.D.	$\frac{1}{4}$	$\frac{1}{4}$	$\frac{1}{4}$	$\frac{1}{4}$	$\frac{1}{8}$
form	na	na	na	na	na
material	copper or SS	copper or SS	copper or SS	copper or SS	copper or SS
Detector	na	na	na	na	na

Compound	$r_{naphthalene}$				
***n*-Tetradecane**	1.53	1.30	0.30	0.24	0.20
***n*-Hexadecane**	4.18	2.45	0.59	0.38	0.39
Naphthalene	1.00	1.00	1.00	1.00	1.00
2-Methylnaphthalene	1.53	1.30	1.89	1.34	1.46
1-Methylnaphthalene	1.53	1.42	2.04	1.48	1.68
2-Ethylnaphthalene	2.18	1.64	3.72	1.69	2.00
1-Ethylnaphthalene	2.18	1.64	3.72	1.78	2.19
2,6-Dimethylnaphthalene	2.18	1.64	3.72	1.78	2.19
2,7-Dimethylnaphthalene	2.18	1.64	3.72	1.78	2.19
1,3-Dimethylnaphthalene	2.18	1.64	3.72	1.94	2.40
1,7-Dimethylnaphthalene	2.18	1.64	3.72	1.94	2.40
1,6-Dimethylnaphthalene	2.18	1.64	3.72	1.99	2.48
1,4-Dimethylnaphthalene	2.18	1.82	5.00	2.25	2.79
2,3-Dimethylnaphthalene	2.18	1.82	5.00	2.25	2.79
1,5-Dimethylnaphthalene	2.18	1.82	5.00	2.25	2.79
1,2-Dimethylnaphthalene	2.18	1.82	5.00	2.32	3.04
1,8-Dimethylnaphthalene	2.18	1.97	5.00	2.83	3.57
Biphenyl	—	—	—	—	2.30
Acenaphthene	—	—	—	—	3.88

Column Packing
P1 = 20% (w/w) G.E. Silicone gum rubber on Chromosorb W (30/60 mesh).
P2 = 30% (w/w) G.E. Silicone oil S-96 on Chromosorb W (30/60 mesh).
P3 = 3% (w/w) 2,4,7-trinitro-9-fluorenone on Chromosorb R (30/60 mesh).
P4 = 20% (w/w) Craig polyester succinate on Chromosorb W (30/60 mesh).

Literature
King, R. W., Fabrizio, F. A., and A. R. Donnell in "*Gas Chromatography*" by Brenner, N., Collen, J. E., and M. D. Weiss, Eds., Academic Press, Inc., N.Y., 1962.

TABLE GC 39
HYDROCARBONS, OCTENES

Column Packing	P1	P1
Temperature (°C)	30	50
Gas; Flow Rate (ml/min)	0.3,N_2	0.3,N_2
Column		
length, m	55	55
diameter, mm	0.25	0.25
form	capillary	capillary
material	SS	SS
Detector	FID	FID

Compound	I	
2-Methyl-2-butene	515	515
3-Methyl-*cis*-2-pentene	612.5	613
4,4-Dimethyl-*trans*-2-pentene	614.5	615
2,4-Dimethyl-1-pentene	636	637.5
2,4-Dimethyl-2-pentene	641	641
2-Methyl-*trans*-3-hexene	647	647
2,3-Dimethyl-1-pentene	648	650
4-Methyl-1-hexene	655.5	656.5
4-Methyl-*trans*-2-hexene	658	659
3,4-Dimethyl-*trans*-2-pentene	677	677.5
3-Methyl-*trans*-3-hexene	684	684
2-Methyl-2-hexene	690.5	691
2,2-Dimethyl-*trans*-3-hexene	693	693
3-Methyl-*cis*-3-hexene	695	695
3,4,4-Trimethyl-1-pentene	698.5	700
2,3-Dimethyl-2-pentene	702	704
2,4,4-Trimethyl-1-pentene	702	704
5,5-Dimethyl-*trans*-2-hexene	706	707
3-Methyl-2-isopropyl-1-butene	708.5	710.5
2,4,4-Trimethyl-2-pentene	714.5	715
5,5-Dimethyl-*cis*-2-hexene	721.5	723.5
2,3,4-Trimethyl-1-pentene	724.5	726.5
2,4-Dimethyl-*cis-trans*-3-hexane	730	730
2,3,3-Trimethyl-1-pentene	730.5	734
4,5-Dimethyl-1-hexene	734	735
2,4-Dimethyl-1-hexene	737.5	738.5
2,3-Dimethyl-2-hexene	740	741
2,5-Dimethyl-1-hexene	740	741
3,4,4-Trimethyl-*cis-trans*-2-pentene	745.5	746.5
2-Isopropyl-1-pentene	748.5	750
2,5-Dimethyl-2-hexene	749.5	750
3,5-Dimethyl-*cis-trans*-2-hexene	751.5	752
2,3-Dimethyl-*cis-trans*-3-hexene	754.5	756
3,4-Dimethyl-1-hexene	755.5	756
3,4-Dimethyl-*cis-trans*-2-hexene	759	760.5
2,3,4-Trimethyl-2-pentene	765.5	766
2-Methyl-*x*-heptene	772	772
3-Methyl-*x*-heptene	to	to
4-Methyl-*x*-heptene		
3,4-Dimethyl-*x*-hexene	782.5	782.5
2,3-Dimethyl-2-hexene	787.5	788.5
2-Methyl-2-heptene	789.5	789.5
2,5,5-Trimethyl-*x*-hexene	794.5	795.5
3-Methyl-*cis-trans*-2-heptene	797	797.5

Column Packing P1 = Squalane.
Literature Tourres, D. A., *J. Gas Chromatogr.*, **5**, 35 (1967).

TABLE GC 40

HYDROCARBONS, POLYNUCLEAR

Column Packing	P1		P1	P2	P3	P3	P4
Temperature (°C)	T1		200	230	260	260	T2
Gas; Flow Rate (ml/min)	N_2,1.0		N_2,1.0	H_2,97	H_2,72.5	H_2,125	N_2,30–75
Column							
length, m	65		65	1.0	1.5	0.5	3.65
diameter, mm^1, cm I.D.^{2}, in. O.D.^{3}	0.30^1		0.30^1	0.4^2	0.4^2	0.4^2	$\frac{1}{8}^3$
form	capillary		capillary	na	na	na	na
material	G		G	Cu	Cu	Cu	SS
Detector	FI or EC		FI or EC	Te	Te	Te	FI
Literature	1		1	2	2	2	3
Compound	t_r	I	r^a	$r_{\times 100}{}^b$	r^c	r^d	r^e
Naphthalene	105	1150	2.0	20	—	—	—
Azulene	111.5	1292	2.8	27	—	—	—
Diphenyl	116	1373	3.3	—	—	—	—
Acetylnaphthalene	121	1437	4.8	—	—	—	—
Acenaphthene	123	1458	5.3	40	—	—	—
Fluorene	132.5	1555	7.6	52	—	—	—
2-Methylfluorene	145	1673	7.6	—	—	—	—
1-Methylfluorene	145.5	1677	7.6	—	—	—	—
trans-Stilbene	146.5	1686	10.9	—	—	—	—
Phenanthrene	152.5	1741	14.5	100	—	—	—
Anthracene	153.5	1750	15.1	100	—	—	—
2-Methylanthracene	166	1870	22.0	—	—	—	—
4,5-Methylenephenanthrene	166.5	1875	22.0	—	—	—	—
1-Methylphenanthrene	168	1890	22.9	—	—	—	—
9-Methylanthracene	171	1920	26.3	—	—	—	—
Fluoranthene	181	2020	36.3	216	—	—	112
Pyrene	186	2070	43.0	265	49	—	100
1,2-Benzofluorene	196.5	2179	58.4	—	57	21	—
2,3-Benzofluorene	198	2195	61.5	—	—	—	—
3,4-Benzofluorene	198	2195	61.5	—	—	—	—
1-Methylpyrene	200	2215	67.6	—	—	—	—
3-Methylpyrene	200.5	2220	69.3	436	70	—	—
Benzo(g,h,i)fluoranthene	211	2326	100	—	—	—	—
3,4-Benzophenanthrene	211.5	2332	100	—	—	—	—
1,2-Benzoanthracene	217	2389	120	—	94	39	—
Chrysene	217.5	2395	124	—	100	48	192
Triphenylene	217.5	2395	124	—	—	—	195
Naphthacene	220	2425	134	—	—	—	—
7-Methyl-1,2-benzanthracene	231.5	2575	209	—	—	—	—
2,3-Benzofluoranthene	242.5	2700	308	—	—	—	—
3,4-Benzofluoranthene	242.5	2700	308	—	—	—	—
11,12-Benzofluoranthene	243	2706	313	—	—	—	(261)
7,12-Dimethyl-1,2-benzanthracene	243.5	2713	318	—	—	—	—
1,2-Benzpyrene	247	2760	364	—	255	100	253
3,4-Benzpyrene	248	2773	378	—	255	100	248
Perylene	250	2800	404	—	288	100	252
3-Methylcholanthrene	258	2906	553	—	—	—	—
20-Methylcholanthrene	258	2906	553	—	—	—	—
1,2,7,8-Dibenzanthracene	270.5	3078	—	—	—	—	—
1,2,3,4-Dibenzanthracene	273	3114	—	—	—	—	—
1,2,5,6-Dibenzanthracene	273	3114	—	—	—	251	(343)
Benzotetraphene	274.5	3136	—	—	—	—	—
Benzo(g,h,i)perylene	275.5	3150	—	—	—	—	308
Picene	275.5	3150	—	—	—	—	—
Anthanthrene	278	3186	—	—	—	271	—
1,2,3,4-Dibenzopyrene	297	3477	—	—	—	—	—
Coronene	300.5	3544	—	—	—	—	—
1,2,4,5-Dibenzopyrene	301.5	3567	—	—	—	—	—
3,4,9,10-Dibenzopyrene	303	3600	—	—	—	664	—
3,4,8,9-Dibenzopyrene	303.5	3620	—	—	—	—	188
Benz(a)anthracene	—	—	—	—	—	—	—

TABLE GC 40—(Continued)
HYDROCARBONS, POLYNUCLEAR

[a] $r_{(benzo(g,h,i,)fluoranthene)}$.
[b] anthracene.
[c] chrysene.
[d] perylene.
[e] pyrene.
( ) isomers not verified.

Column Packing P1 = SE-52.
P2 = 20% Apiezon L on Celite (30/60 mesh).
P3 = 9.1% Apiezon L on Celite (30/60 mesh).
P4 = 20% LiCl on Anakrom P (40/60 mesh) fired at 700 °C.

Temperature T1 = 100 °–300 °C at 1.8°/min (programmed).
T2 = 200 °–325 °C at 2°/min (programmed).

Literature
1. Carugno, N. and S. Rossi, *J. Gas Chromatogr.*, **5**, 103 (1967).
2. Carugno, N. and G. Giovanozzi-Sermanni, *Tobacco*, **63**, 285 (1959).
3. Gump, B. H., *J. Chromatogr. Sci.*, **7**, 755 (1969).

TABLE GC 41
HYDROCARBONS, CYCLIC

Column Packing	P1	P1	P1	P1	P1
Temperature (°C)	30	30	30	30	30
Gas; Flow Rate (ml/min)	60,He	52,He	46,He	36,He	42,He
Column					
length, m	2	2	2	2	2
diameter, mm, I.D.	4.2	4.2	4.2	4.2	4.2
form	U-shaped	U-shaped	U-shaped	U-shaped	U-shaped
material	glass	glass	glass	glass	glass
Detector	TC	TC	TC	TC	TC
Literature	1	1	1	2	2

Compound	$r_{toluene}$				
Vinylcyclohexane	3.72	—	—	—	—
Ethylidenecyclohexane	1.70	—	—	—	—
1-Ethylcyclohexene	0.85	—	—	—	—
3-Ethylcyclohexene	2.43	—	—	—	—
4-Ethylcyclohexene	2.37	—	—	—	—
Isopropylidenecyclohexane	1.51	—	—	—	—
1-Isopropylcyclohexene	—	1.00	—	—	—
3-Isopropylcyclohexene	—	2.70	—	—	—
4-Isopropylcyclohexene	—	3.85	—	—	—
1(7)-*p*-Menthene	—	8.17	—	—	—
8-*p*-Menthene	—	—	4.34	—	—
4(8)-*p*-Menthene	—	—	1.92	—	—
1-*p*-Menthene	—	—	2.03	—	—
***trans*-2-*p*-Menthene**	—	—	3.17	—	—
3-*p*-Menthene	—	—	1.21	—	—
1-Methylcyclopentene	—	—	—	0.75	—
3-Methylcyclopentene	—	—	—	2.35	—
4-Methylcyclopentene	—	—	—	1.23	—
Methylenecyclopentane	—	—	—	1.66	—
1-Ethylcyclopentene	—	—	—	—	1.30
3-Ethylcyclopentene	—	—	—	—	3.37
4-Ethylcyclopentene	—	—	—	—	2.12
Ethylidenecyclopentane	—	—	—	—	0.68

Column Packing P1 = 25% silver nitrate/glycol on acid-washed C-22 firebrick (50/80 mesh).

Literature
1. Herling, J., Shabtai, J., and E. Gil-Av, *J. Chromatogr.*, **8**, 349 (1962).
2. Shabtai, J., Herling, J., and E. Gil-Av, *J. Chromatogr.*, **2**, 406 (1959).

TABLE GC 42
HYDROCARBONS, SATURATED

Column Packing	P1	P1
Temperature (°C)	25	51
Gas; Flow Rate (cc/min)	0.6	0.6
Column		
length, ft	300	300
diameter, in.	0.012	0.012
form	coiled capillary	coiled capillary
material	glass	glass
Detector	FI	FI

Compound	$r_{n\text{-hexane}}$	
Methane	0.005	0.01
Ethane	0.01	0.02
Propane	0.03	0.07
2-Methylpropane	0.07	0.17
***n*-Butane**	0.11	0.21
2,2-Dimethylpropane	0.13	0.23
2-Methylbutane	0.26	0.36
***n*-Pentane**	0.35	0.44
2,2-Dimethylbutane	0.51	0.59
Cyclopentane	0.68	0.75
2,3-Dimethylbutane	0.69	0.75
2-Methylpentane	0.72	0.77
3-Methylpentane	0.84	0.87
***n*-Hexane**	1.00	1.00
2,2-Dimethylpentane	1.28	1.23
Methylcyclopentane	1.30	1.27
2,4-Dimethylpentane	1.37	1.29
2,2,3-Trimethylbutane	1.43	1.37
3,3-Dimethylpentane	1.75	1.61
Cyclohexane	1.78	1.68
2-Methylhexane	2.02	1.77
2,3-Dimethylpentane	2.02	1.81
1,1-Dimethylcyclopentane	2.06	1.87
3-Methylhexane	2.19	1.91
1,*cis*-3-Dimethylcyclopentane	2.29	2.04
1,*trans*-3-Dimethylcyclopentane	2.35	2.09
1,*trans*-2-Dimethylcyclopentane	2.43	2.16
3-Ethylpentane	2.43	2.09
2,2,4-Trimethylpentane	2.48	2.16
***n*-Heptane**	2.82	2.35
1,*cis*-2-Dimethylcyclopentane	3.29	2.79
Methylcyclohexane	3.29	2.81
2,2,3,3-Tetramethylbutane	3.29	2.81
1,1,3-Trimethylcyclopentane	3.42	2.88
2,2-Dimethylhexane	3.49	~2.84
Ethylcyclopentane	3.71	3.09
2,2,3-Trimethylpentane	3.81	3.14
2,5-Dimethylhexane	3.87	3.09
2,4-Dimethylhexane	3.93	3.14
1,*trans*-2,*cis*-4-Trimethylcyclopentane	4.06	3.31
3,3-Dimethylhexane	4.14	3.34
1,*trans*-2,*cis*-3-Trimethylcyclopentane	4.36	3.52
2,3,4-Trimethylpentane	4.49	3.60
2,3,3-Trimethylpentane	4.61	3.73
1,1,2-Trimethylcyclopentane	4.93	3.98
2,3-Dimethylhexane	5.09	3.93
3-Ethyl-2-methylpentane	5.09	3.98
1,1,3,3-Tetramethylcyclopentane	5.09	4.07
3-Ethyl-3-methylpentane	5.40	4.26
2-Methylheptane	5.46	4.14
4-Methylheptane	5.53	4.19
3,4-Dimethylhexane	5.53	4.26
1,*cis*-2,*trans*-4-Trimethylcyclopentane	5.53	4.34

TABLE GC 42—(Continued)
HYDROCARBONS, SATURATED

Column Packing	P1	P1
Temperature (°C)	25	51
Gas; Flow Rate (cc/min)	0.6	0.6
Column		
length, ft	300	300
diameter, in.	0.012	0.012
form	coiled capillary	coiled capillary
material	glass	glass
Detector	FI	FI

Compound	$r_{n\text{-hexane}}$	
2,2,4,4-Tetramethylpentane	5.53	4.36
1,*cis*-3-Dimethylcyclohexane	5.67	4.50
1,1,*cis*-3,*trans*-4-Tetramethylcyclopentane	5.74	4.45
1,*cis*-2,*trans*-3-Trimethylcyclopentane	5.74	4.50
1,*trans*-4-Dimethylcyclohexane	5.78	4.55
3-Methylheptane	5.85	4.41
3-Ethylhexane	5.88	4.45
1,1-Dimethylcyclohexane	6.02	4.76
1-Ethyl-*trans*-3-methylcyclopentane	6.37	4.90
Cycloheptane	6.37	5.04
2,2,5-Trimethylhexane	6.49	4.84
1-Ethyl-*cis*-3-methylcyclopentane	6.49	4.99
1-Ethyl-*trans*-2-methylcyclopentane	6.56	5.04
1-Ethyl-1-methylcyclopentane	6.62	5.13
2,2,4-Trimethylhexane	6.76	5.13
1,*trans*-2-Dimethylcyclohexane	6.76	5.25
1,*cis*-2,*cis*-3-Trimethylcyclopentane	7.12	5.54
1,*cis*-4-Dimethylcyclohexane	7.26	5.54
1,*trans*-3-Dimethylcyclohexane	7.26	5.59
***n*-Octane**	7.66	5.54
Isopropylcyclopentane	7.75	5.87
1,1,*cis*-2-*trans*-4-Tetramethylcyclopentane	7.86	5.91
1,1,*cis*-2,*cis*-4-Tetramethylcyclopentane	7.86	5.91
2,4,4-Trimethylhexane	7.86	5.87
1,1,*cis*-3,*cis*-4-Tetramethylcyclopentane	8.11	6.03
1-Ethyl-*cis*-2-methylcyclopentane	8.49	6.41
2,2,3,4-Tetramethylpentane	8.49	6.41
2,3,5-Trimethylhexane	8.75	6.33
3-Ethyl-2,2-dimethylpentane	8.91	6.60
1,*cis*-2-Dimethylcyclohexane	8.91	6.72
2,2-Dimethylheptane	9.16	6.51
2,2,3-Trimethylhexane	9.28	6.72
Ethylcyclohexane	9.44	7.02
1,*cis*-3,*cis*-5-Trimethylcyclohexane	9.52	7.02
***n*-Propylcyclopentane**	9.57	7.02

Column Packing P1 = 6E Silicone fluid (SF-96) (from 10% in *n*-hexane).
Literature Polgar, A. G., Holst, J. J., and S. Groennings, *Anal. Chem.*, **34**, 1226 (1962).

TABLE GC 43

INORGANICS, CATION ACETYLACETONATES

Column Packing	P1	P2
Temperature (°C)	115	115
Gas; Flow Rate (ml/min)	80,CCl_2F_2	163,CCl_2F_2
Pressure, psi.	950	1100
Column		
length, in.	46	46
diameter, in., I.D.	$\frac{1}{8}$	$\frac{1}{8}$
form	na	na
material	titanium	titanium
Detector	UV monochromator	UV monochromator

Compound	$r_{Be\ II}$	
Li(I)	1.28	1.31
Na(I)	1.17	1.17
K(I)	1.19	1.17
Be(II)	1.00	1.00[a]
Mg(II)	1.24	1.19
Ca(II)	[b]	1.22
Sr(II)	[b]	1.24
Ba(II)	[b]	1.24
Al(III)	1.14	1.12
Ga(III)	1.17	1.17
In(III)	1.39	1.25
Tl(I)	1.20	1.18
Sc(III)	1.10	1.08
Y(III)	1.13	1.11
La(III)	1.18	1.15
Ce(III)	1.18	1.14
Pr(III)	1.18	1.15
Nd(III)	1.17	1.14
Tb(III)	1.17	—
TiO	1.13	—
$TiCl_2$	1.27	1.22
Zr(IV)	1.19	1.18
VO	1.10	1.09
Cr(III)	1.28	1.22
MoO_2	1.17	1.19
Mn(II)	1.32	1.23
Mn(III)	1.18	1.17
Fe(III)	1.11	1.13
Ru(III)	1.43	1.25
Co(II)	1.14	1.15
Co(III)	1.48	1.32
Rh(III)	1.50	1.34
Ir(III)	1.42	1.25
Ni(II)	1.12	1.13
Pd(II)	1.83	1.59
Pt(II)	2.32	1.88
Cu(II)	1.12	1.13
Zn(II)	1.23	1.20
Cd(II)	1.39	1.25
Pb(II)	1.35	1.24
Th(IV)	1.17	1.17
U(IV)	1.39	1.28
UO_2	1.27	1.22

[a] t_r Be(II) chelate: 19.95 min. (flow rate 80 ml/min.); 8.55 min.
[b] broad peak.

Column Packing P1 = 10% Epon 1001 Resin on Chromosorb W (60/80 mesh).
P2 = 5% Epon 1009 Resin on AW-HMDS-treated Chromosorb P (80/100 mesh).

Literature Karayannis, N. M. and A. H. Corwin, *J. Chromatogr. Sci.*, **8**, 251 (1970).

TABLE GC 44
ISOCYANATES

Column Packing	P1	P2	P3	P1	P2	P3
Temperature (°C)	100	100	100	170	170	150
Gas; Flow Rate (ml/min)	na,N_2	32,N_2	8,N_2	na,N_2	32,N_2	8,N_2
Column						
length, m	2.00	2.00	50	2.00	2.00	50
diameter, cm, O.D.	0.3	0.3	—	0.3	0.3	—
form	na	na	na	na	na	na
material	SS	SS	SS	SS	SS	SS
Detector	FID	FID	FID	FID	FID	FID

Compound	t_r					
Isocyanates						
Methyl	0.16	0.42	0.29	—	—	—
Ethyl	0.24	0.40	0.42	—	—	—
Allyl	0.45	0.85	0.77	—	—	—
Isopropyl	0.26	0.29	0.38	—	—	—
***n*-Propyl**	0.47	0.60	0.62	—	—	—
***tert*-Butyl**	0.27	0.23	0.29	—	—	—
***sec*-Butyl**	0.58	0.48	0.56	—	—	—
Isobutyl	0.70	0.68	0.72	—	—	—
***n*-Butyl**	1.00	1.00	1.00	0.26	0.25	2.27
Cyclohexyl	6.06	4.50	4.14	1.03	0.83	0.82
***n*-Hexyl**	4.40	3.04	2.85	0.73	0.56	0.61
Phenyl	6.05	6.59	5.16	1.00	1.00	1.00
***o*-Tolyl**	—	—	—	1.56	1.33	1.35
***m*-Tolyl**	—	—	—	1.72	1.43	1.48
***p*-Tolyl**	—	—	—	1.79	1.54	1.58
Benzyl	—	—	—	2.20	3.82	3.97
***n*-Decyl**	—	—	—	5.52	2.59	3.04

Column Packing P1 = 15% Apiezon M on Celite 545 (60/80 mesh).
P2 = 20% Carbowax 20M on Chromosorb W.
P3 = Tween 80.

Literature Binder, H., *J. Chromatogr.*, **41**, 448 (1969).

TABLE GC 45
NITROGEN COMPOUNDS

Column Packing	P1	P1	P2	P2	P3
Temperature (°C)	180	220	180	220	220
Gas; Flow Rate (ml/min)	73,He	73,He	73,He	73,He	73,He
Column					
length, ft	10	10	10	10	10
diameter, in.	0.1	0.1	0.1	0.1	0.1
form	na	na	na	na	na
material	SS	SS	SS	SS	SS
Detector	FI	FI	FI	FI	FI
Compound			$r_{quinoline}$		
Hydrocarbons					
Indene	0.221	0.255	0.212	0.256	—
Naphthalene	0.544	0.585	0.528	0.581	0.855
Decalin (*trans*)	0.084	0.107	0.072	0.108	—
(*cis*)	0.111	0.142	0.095	0.133	—
Eicosane	—	1.49	—	1.10	—
Phenanthrene	—	7.74	—	7.64	9.31
Anthracene	10.8	8.09	—	7.83	9.67
Amines					
Di-*n*-amylamine	0.119	—	—	—	—
Benzylamine	0.290	—	—	—	—
Aniline	0.437	—	0.481	—	—
p-Toluidine	0.578	—	—	—	—
4-Ethylaniline	0.818	—	—	—	—
1,2,3,4-Tetrahydroisoquinoline	0.912	—	—	—	—
2,3-Dimethylaniline	0.950	—	—	—	—
1,2,3,4-Tetrahydroquinoline	1.50	—	—	—	—
1-Naphthylamine	—	6.34	—	6.99	—
2-Naphthylamine	—	7.32	—	8.06	—
Pyridines					
2-(*n*-Pentyl)pyridine	0.297	0.318	—	—	—
4-(*n*-Pentyl)pyridine	0.491	0.491	—	—	—
2-(5-Nonyl)pyridine	0.665	0.641	—	0.541	1.53
4-(5-Nonyl)pyridine	1.23	1.10	0.970	0.890	2.51
2-Benzylpyridine	2.81	2.03	—	—	—
4-Benzylpyridine	3.64	2.92	—	2.75	2.88
Quinolines					
Quinoline	1.00	1.00	1.00	1.00	1.00
8-Methylquinoline	1.10	1.08	1.08	1.06	1.34
6-Methylquinoline	1.51	1.44	—	1.39	1.56
2,6-Dimethylquinoline	1.51	1.50	—	1.45	—
4-Methylquinoline	1.80	1.68	—	1.63	1.78
2,4-Dimethylquinoline	1.91	1.82	—	1.72	—
2,4,6,8-Tetramethylquinoline	2.78	2.32	—	—	—
Acridine	14.2	10.2	—	10.0	9.32
Phenanthridine	—	11.8	—	11.6	10.5
3,4-Benzacridine	—	84.0	—	79.0	77.6
1,2-Benzacridine	—	110.0	—	104.0	88.2
Isoquinolines					
Isoquinoline	1.16	1.15	—	1.15	1.06
3-Methylisoquinoline	1.25	1.22	—	1.18	1.44
Indoles					
1-Methylindole	—	1.06	—	—	—
1,2-Dimethylindole	2.01	1.79	2.00	1.78	1.80
Indole	4.31	3.18	4.95	3.66	0.954
3-Methylindole	5.10	3.77	5.63	4.16	1.42
2-Methylindole	5.16	3.80	—	—	—
2,7-Dimethylindole	5.56	3.86	—	—	—
5-Methylindole	5.75	4.16	—	—	
2,3-Dimethylindole	—	4.77	—	—	—
2,5-Dimethylindole	6.95	4.86	—	—	—
Carbazole	—	30.6	—	34.5	9.24

TABLE GC 45—(Continued)
NITROGEN COMPOUNDS

Column Packing P1 = 5% (w/w) Triton X-305 (octylphenoxypolyethoxy ethanol) on acid-washed Chromosorb G (70/80 mesh).
P2 = 5% Carbowax 20M (polyethylene glycol) on Chromosorb G (70/80 mesh).
P3 = 5% Apiezon L doped with Carbowax 20M-TPA on Chromosorb G (70/80 mesh).
Literature Poulson, R. E., *J. Chromatogr. Sci.*, 7, 152 (1969).

TABLE GC 46
AROMATIC NITRO COMPOUNDS AND ANILINES

Column Packing	P1		P1
Temperature (°C)	200–210		200–210
Gas; Flow Rate (ml/min)	20,He		20,He
Column			
length, ft	8		8
diameter, in., O.D.	¼		¼
form	tube		tube
material	copper		copper
Detector	TC		TC

Nitro compound	t_r	Amine compound	t_r
2-Nitrotoluene	11.5	**2-Methylaniline**	8
4-Nitrotoluene	14.5	**3-Methylaniline**	8
1-Nitro-2-ethylbenzene	14.5	**4-Methylaniline**	8
1-Nitro-2,6-dimethylbenzene	11.5	**2-Ethylaniline**	10.5
1-Nitro-2,5-dimethylbenzene	18.5	**4-Ethylaniline**	11
1-Nitro-2,3-dimethylbenzene	19.0	**2,6-Dimethylaniline**	11.5
1-Nitro-2,4,6-trimethylbenzene	18.0	**2,4-Dimethylaniline**	11.5
1-Nitro-2,3,6-trimethylbenzene	21.0	**3,5-Dimethylaniline**	11.5
1-Nitro-2,4,5-trimethylbenzene	33.0	**2,5-Dimethylaniline**	11.5
1,Nitro-2,3,5-trimethylbenzene	28.5	**2,3-Dimethylaniline**	13.5
1-Nitro-2,3,4-trimethylbenzene	36.5	**3,4-Dimethylaniline**	13.5
1-Nitro-3,4,5-trimethylbenzene	45.0	**2,4,6-Trimethylaniline**	15.5
1-Nitro-2-methyl-5-ethylbenzene	24.0	**2,3,6-Trimethylaniline**	19.5
1-Nitro-5-methyl-2-ethylbenzene	22.5	**2,4,5-Trimethylaniline**	19.5
1-Nitro-2-methyl-6-ethylbenzene	16.5	**2,3,5-Trimethylaniline**	16.5
1-Nitro-2-methyl-4-ethylbenzene	26.0	**2,3,4-Trimethylaniline**	23.5
1-Nitro-4-methyl-2-ethylbenzene	22.5	**3,4,5-Trimethylaniline**	23
1-Nitro-2-methyl-3-ethylbenzene	26.0	**2-Methyl-5-ethylaniline**	15
1-Nitro-3-methyl-2-ethylbenzene	24.0	**5-Methyl-2-ethylaniline**	14
1-Nitro-3-methyl-4-ethylbenzene	34.5	**2-Methyl-6-ethylaniline**	13.5
1-Nitro-4-methyl-3-ethylbenzene	35.0	**2-Methyl-4-ethylaniline**	16
		4-Methyl-2-ethylaniline	15
		2-Methyl-3-ethylaniline	17
		3-Methyl-2-ethylaniline	14
		3-Methyl-4-ethylaniline	16
		4-Methyl-3-ethylaniline	16
		2,4,5,6-Tetramethylaniline	32
		2,3,5,6-Tetramethylaniline	32

Column Packing P1 = 30% DC-710 silicone fluid on firebrick C-22 (30/60 mesh).
Literature Jones, J. H., Ritchie, C. D., and K. S. Heine, Jr., *J. Ass. Offic. Agr. Chem.*, **41**, 749 (1958).

TABLE GC 47
ORGANICS

Column Packing	P1	P2	P3	P4	P5	P6	P7	P8
Temperature (°C)	80	80	80	80	80	80	80	80
Gas; Flow Rate	—	—	—	—	—	—	—	—
Column								
length, m.	2	2	2	2	2	2	2	2
diameter, in.	$\frac{3}{8}$	$\frac{3}{8}$	$\frac{3}{8}$	$\frac{3}{8}$	$\frac{3}{8}$	$\frac{3}{8}$	$\frac{3}{8}$	$\frac{3}{8}$
form	helix	helix	helix	helix	helix	helix	helix	helix
material	copper	copper	copper	copper	copper	copper	copper	copper
Detector	TC	TC	TC	TC	TC	TC	TC	TC

Compound	V_g							
Isopentane	21.4	16.9	16.7	13.9	10.4	—	—	—
n-Pentane	26.0	21.0	21.7	15.9	11.5	—	2.4	—
2,2-Dimethylbutane	36.2	28.9	—	21.5	17.4	—	2.9	—
2,3-Dimethylbutane	47.6	37.6	—	27.6	22.6	—	4.3	—
2-Methylpentane	49.8	39.2	38.6	28.9	23.1	12.3	4.1	—
3-Methylpentane	56.6	44.3	43.5	33.0	26.7	—	4.2	—
n-Hexane	63.9	50.3	48.3	36.9	29.3	—	5.1	—
2,4-Dimethylpentane	84.4	68.5	65.0	51.2	44.2	—	5.4	—
3-Methylhexane	123	95	—	69.9	54.6	—	8.5	—
n-Heptane	150	115	107	84.0	65.1	36.0	10.1	—
3-Methylheptane	278	213	—	153	119	—	16.6	—
n-Octane	347	269	240	190	148	77.0	20.1	—
Cyclohexane	115	93.6	86.7	70.5	58.8	—	16.0	—
2-Methyl-1-butane	24.1	21.2	28.0	17.8	15.8	—	4.1	—
2-Methyl-2-butane	29.4	25.9	36.1	22.1	19.1	—	5.3	—
1-Pentene	22.9	—	—	—	—	—	3.1	—
1-Hexene	55.6	—	—	—	—	—	6.9	—
1-Heptene	131	—	—	—	—	—	13.7	—
Cyclopentene	41.8	37.4	—	31.6	28.8	—	10.3	—
Cyclohexene	123	108	—	88.8	82.1	—	25.5	—
Benzene	94.2	104	124	102	121	84.5	63.0	6.1
Toluene	237	256	291	245	277	185	117	8.8
Dichloromethane	24.4	32.1	51.2	45.7	51.2	—	52.0	—
Chloroform	58.6	79.6	122	95.6	115	—	95.6	—
Carbon tetrachloride	102	94.8	106	86.1	88.2	—	38.8	—
cis-Dichloroethylene	40.5	44.2	59.8	46.2	49.2	—	33.4	4.6
trans-Dichloroethylene	48.0	65.1	98.2	86.5	103	—	79.9	9.4
Chlorobenzene	432	509	—	—	—	—	—	28.7
Ethyl bromide	28.6	31.1	25.6	80.2	34.9	—	20.2	—
n-Propyl bromide	67.5	73.6	84.4	90.5	83.2	—	36.4	—
sec-Butyl bromide	118	128	141	130	137	—	46.9	—
n-Butyl bromide	162	174	116	182	187	—	65.3	—
n-Propyl chloride	32.5	34.7	—	37.6	40.3	—	17.3	—
Allyl chloride	27.2	31.9	40.8	—	44.7	—	25.0	—
Ethyl iodide	66.2	73.6		74.5	81.9	—	42.4	5.9
n-Propyl iodide	159	176	—	170	183	—	75.0	—
Diethyl ether	20.7	31.9	22.9	21.3	22.1	—	8.1	—
Dioxan	103	225	185	184	220	—	179	—
Acetone	15.4	38.6	29.9	34.5	46.6	—	35.0	—
Methyl ethyl ketone	40.2	88.8	68.8	75.4	97.0	—	59.9	—
Methyl *n*-propyl ketone	86.0	190	144	156	196	—	98.0	—
Methyl formate	7.6	14.1	15.7	15.4	20.2	—	20.6	—
Ethyl formate	18.0	31.9	33.0	32.6	41.6	—	31.6	8.7
n-Propyl formate	46.4	77.8	79.4	77.5	95.5	—	58.6	12
n-Butyl formate	113	183	182	180	216	—	109	16.0
iso-Pentyl formate	200	314	316	303	360	—	150	—
Ethyl acetate	41.5	80.0	—	68.3	69.4	54.8	48.7	—
Ethyl propionate	99.0	173	—	139	164	—	97.9	—
Dimethyl carbonate	40.8	—	—	87.8	116	—	95.6	22.1
Ethyl chloroformate	65.2	91.7	124	127	155	—	—	—
Nitromethane	19.5	35	—	104	158	—	264	—
Methanol	5.3	52.5	—	—	25.5	26.9	75.1	—

TABLE GC 47—(Continued)
ORGANICS

Column Packing	P1	P2	P3	P4	P5	P6	P7	P8
Temperature (°C)	80	80	80	80	80	80	80	80
Gas; Flow Rate	—	—	—	—	—	—	—	—
Column								
length, m	2	2	2	2	2	2	2	2
diameter, in.	$\frac{3}{8}$	$\frac{3}{8}$	$\frac{3}{8}$	$\frac{3}{8}$	$\frac{3}{8}$	$\frac{3}{8}$	$\frac{3}{8}$	$\frac{3}{8}$
form	helix	helix	helix	helix	helix	helix	helix	helix
material	copper	copper	copper	copper	copper	copper	copper	copper
Detector	TC	TC	TC	TC	TC	TC	TC	TC
Compound	V_g							
Ethanol	10.3	90.4	43.2	39.0	41.9	41.2	95.7	72.1
n-Propanol	27.9	208	96.5	84.2	99.0	95	179	93.5
n-Butanol	71.4	503	233	193	235	—	387	131
n-Pentanol	174	—	—	—	—	—	694	193
Allyl alcohol	29.8	196	98.3	—	129	—	277	—

Column Packing P1 = 20% Squalane on acid washed Celite (80/100 mesh).
P2 = 20% Dodecanal on acid washed Celite (80/100 mesh).
P3 = 20% Dioctyl sebacate on acid washed Celite (80/100 mesh).
P4 = 20% Dinonyl phthalate on acid washed Celite (80/100 mesh).
P5 = 20% Dibutyl phthalate on acid washed Celite (80/100 mesh).
P6 = 20% Tricresyl phosphate on acid washed Celite (80/100 mesh).
P7 = 20% Polyethylene glycol 400 on acid washed Celite (80/100 mesh).
P8 = 20% Diglycerol on acid washed Celite (80/100 mesh).

Literature Littlewood A. B., *J. Gas Chromatogr.*, **1**, 6 (1963).

TABLE GC 48
KETONES, ACETOPHENONES

Column Packing	P1	P2	P3	P4
Temperature (°C)	170	170	170	170
Gas; Flow Rate, (ml/min)	540,H_2	736,H_2	824,H_2	1250,H_2
Column				
length, m	1	1	1	1
diameter, mm, I.D.	5	6	10	5
form	na	na	na	na
material	na	na	na	na
Detector	TC	TC	TC	TC
Compound	r 6′-hydroxy-2′,3′,4′,-trimethoxy acetophenone			
2′-Hydroxy-4′-methoxyacetophenone	0.274	0.327	0.334	0.282
6′-Hydroxy-2′,3′-dimethoxyacetophenone	0.356	0.404	0.432	0.328
6′-Hydroxy-2′,3′,4′-trimethoxyacetophenone	1.00	1.00	1.00	1.00
2′-Hydroxy-3′,4′,6′-trimethoxyacetophenone	1.64	1.56	1.58	2.50
2′-Hydroxy-2,4′,6′-trimethoxyacetophenone	2.19	2.27	2.06	—
6′-Hydroxy-2,2′,3′,4′-tetramethoxyacetophenone	2.54	2.34	2.32	—
2′-Hydroxy-2,3′,4′,6′-tetramethoxyacetophenone	4.56	4.28	3.88	—
2′-Hydroxy-2,3′,4′,5′,6′-pentamethoxyacetophenone	3.00	2.58	2.90	—
2′,4′-Dimethoxyacetophenone	0.810	0.924	0.844	0.932
2′,3′,4′,6′-Tetramethoxyacetophenone	0.987	0.654	0.943	1.28
2,2′,3′,4′,6′-Pentamethoxyacetophenone	2.30	1.41	1.96	4.03
6′-Hydroxy-2′,3′,4′-trimethoxyacetophenone (min)	5.82	4.15	4.07	10.45

Column Packing P1 = 20% D-550 Silicone oil on Chromosorb W (60/80 mesh).
P2 = 8% Apiezon L on Chromosorb G (60/80 mesh).
P3 = 8% SE-30 on Chromosorb G (60/80 mesh).
P4 = 8% Epikote 728 on Chromosorb G (60/80 mesh).

Literature Békássy, S. and M. Nógrádi, *Acta Chim. Acad. Sci. Hung.*, **59**, 3–4 **425–29** (1969), from *J. Chromatogr.* **51**, D39 (1970).

TABLE GC 49
ORGANICS

Column Packing	P1	P1	P2	P2	P3	P3
Temperature (°C)	150	100	150	100	150	100
Gas; Flow Rate (ml/min)	40,N_2	40,N_2	25,N_2	25,N_2	25,N_2	25,N_2
Column						
length, m	1	1	1	1	1	1
diameter, mm, I.D.	2.3	2.3	2.3	2.3	2.3	2.3
form	na	na	na	na	na	na
material	SS	SS	SS	SS	SS	SS
Detector	FI	FI	FI	FI	FI	FI

Compound	r_{octane}					
Pentane	0.343	0.194	0.367	0.198	0.190	0.375
trans-1,2-Dichloroethylene	—	0.270	—	0.259	0.253	—
Methyl acetate	0.440	0.292	0.912[a]	0.842	>50	1.75[a]
Acetone	0.449	0.299	1.210	1.230	>50	2.230[a]
Formaldehyde	0.388	0.300	0.536	—	—	1.030[a]
Hexane	0.460	0.315	0.483	0.320	0.309	0.490
Methanol	0.412	0.318	0.481	0.655 (severe tailing)	>50	—
Dichloromethane (CH_2Cl_2)	—	0.332	—	0.295	0.261	—
Isobutyraldehyde	0.480	0.364	1.150[a]	—	—	1.270
Tetrachloromethane (CCl_4)	—	0.388	—	0.369	0.372	—
1,1-Dichloroethane	—	0.388	—	0.428	0.361	—
Ethanol	0.492	0.425	0.996	—	—	—
Butyraldehyde	0.544	0.432	1.520[a]	—	—	1.580
Benzene	0.560	0.440	0.700	0.560	0.630	0.720
Ethyl acetate	0.640	0.476	1.510	1.700	—	3.350[a]
2-Butanone	0.612	0.482	1.790	2.130	—	3.370[a]
cis-1,2-Dichloroethylene	—	0.489	—	0.428	0.373	—
Trichloroethylene	—	0.541	—	0.489	0.499	—
Heptane	0.660	0.546	0.657	0.550	0.548	0.680
1,2-Dichloroethane	—	0.621	—	0.648	0.656	—
Trichloromethane	—	0.621	—	0.489	0.372	—
Isopropyl acetate	0.768	0.688	2.170	2.790	—	5.000[a]
Propanol	0.748	0.800	—	—	—	—
Toluene	0.868	0.828	1.100	1.130	1.410	1.170
Propyl acetate	0.904	0.870	2.460	3.320	—	—
Octane	1.000 (0.984 min)	1.000 (2.60 min)	1.000 (1.63 min)	1.000 (4.80 min)	1.000 (2.49 min)	1.000 (0.788 min)
2-Ethyl butyraldehyde	1.060	1.120	—	—	—	3.490
Isobutanol	1.060	1.370	—	—	—	—
Isobutyl acetate	1.250	1.380	3.630	5.480	—	—
Ethylbenzene	1.370	1.500	1.700	2.050	2.650	1.800
m-Xylene	1.370	1.580	1.800	2.260	3.140	2.000
p-Xylene	1.370	1.580	1.800	2.260	3.140	2.000
Butanol	1.210	1.640	—	—	—	—
Butyl acetate	1.420	1.650	4.110	6.530	—	—
o-Xylene	1.510	1.690	1.970	2.530	3.620	2.220
Nonane	1.540	1.860	1.550	1.870	1.860	1.490
Isopentanol	1.940	3.000	—	—	—	—
Pentanol	2.010	3.320	—	—	—	—
1,2,4-Trimethylbenzene	2.460	3.380	3.280	5.050	7.910	3.800
Decane	2.470	3.490	2.470	3.520	3.500	2.290
1,2,3-Trimethylbenzene	2.720	3.830	3.650	5.740	9.420	4.400
Acetic acid	2.300	—	—	—	—	—
Hexanol	3.360	—	—	—	—	—
Undecane	3.960	—	3.920	—	—	3.570
Heptanol	5.700	—	—	—	—	—
Dodecane	6.480	—	6.640	—	—	5.620
Octanol	9.560	—	—	—	—	—
Tridecane	10.600	—	11.100	—	—	8.950
Nonanol	26.600	—	—	—	—	—
Tetradecane	17.300	—	18.500	—	—	14.200

[a] Tailing.
All packings available from Waters Assoc. as Durapak®.

TABLE GC 49—(Continued)
ORGANICS

Column Packing	P1 = Carbowax 400 on Porasil C (100/120 mesh). P2 = OPN (oxypropionitrile) on Porasil C (80/100 mesh). P3 = Octane on Porasil C (120/150 mesh).
Literature	Little, J. N., Dark, W. A., Farlinger, P. W., and K. J. Bombaugh, *J. Chromatogr. Sci.*, **8**, 647 (1970).

TABLE GC 50
INORGANIC CHLORIDES

Column Packing	P1	P1	P1	P1
Temperature (°C)	75	100	125	150
Gas; Flow Rate (ml/min)	83,He	83,He	83,He	83,He
Column				
length, m	2.00	2.00	2.00	2.00
diameter, mm, I.D.	8	8	8	8
form	W-shaped	W-shaped	W-shaped	W-shaped
material	G	G	G	G
Detector	gas density balance	gas density balance	gas density balance	gas density balance

Compound		$p_{\text{carbon tetrachloride}}$		
Carbon tetrachloride	100	100	100	100
Silicon tetrachloride	54	55	56	57
Phosphorus trichloride	78	79	80	—
Phosphorus oxychloride	227	189	172	—
Sulfur dichloride	42	44	49	—
Sulfur chloride (mono)	264	233	216	—
Thionyl chloride	79	76	74	—
Sulfuryl chloride	85	85	85	—
Titanium tetrachloride	379	313	287	—
Vanadium tetrachloride	—	445	386	319
Vanadium oxychloride	—	260	222	209
Chromium oxychloride	251	220	204	—
Digallium hexachloride	—	—	—	755
Germanium tetrachloride	104	100	99	—
Arsenic trichloride	227	202	189	174
Molybdenum dioxydichloride	—	285	248	218
Stannic chloride	183	161	150	139
Antimony trichloride	—	666	546	460
Antimony pentachloride	—	1180	886	683

Column Packing	P1 = 15 wt % Kel-F 40 on Haloport-F (30/80 mesh).
Literature	Sie, S. T., Bleumer, J. P. A., and G. W. A. Rijnders, *Separ. Sci.*, **1**(1), 41 (1966).

TABLE GC 51
ORGANICS

Column Packing	P1	P1	P1	P2	P2	P2
Temperature (°C)	70	130	190	70	130	190
Gas; Flow Rate (ml/min)	60,He	60,He	60,He	60,He	60,He	60,He
Column						
length, m	2.25	2.25	2.25	2.25	2.25	2.25
diameter, mm, O.D.	6	6	6	6	6	6
form	na	na	na	na	na	na
material	glass	glass	glass	glass	glass	glass
Detector	na	na	na	na	na	na

Compound			I			
Pentane	500	500	—	500	500	—
Heptane	700	700	—	700	700	—
Nonane	—	900	900	—	900	900
Undecane	—	1100	1100	—	1100	1100
2-Methylpentane	570	571	—	568	570	—
2,3-Dimethylbutane	570	572	—	567	570	—
2,2,4-Trimethylpentane	690	694	—	685	693	—
Pentene-1	485	479	—	520	526	—
Hexene-1	587	583	—	622	627	—
Heptene-1	687	684	—	721	726	—
Octene-1	—	783	779	818	827	—
2-Methylbutene-1	500	499	—	541	544	—
3-Methylbutene-1	450	448	—	484	489	—
2-Methylbutene-2	527	526	—	566	570	—
***trans*-4-Methylpentene-2**	557	559	—	590	592	—
***cis*-4-Methylpentene-2**	558	560	—	590	594	—
Isoprene	511	514	—	595	599	—
Diethyl ether	476	484	—	574	578	—
Dipropyl ether	661	664	—	737	742	—
Dibutyl ether	—	858	862	—	934	938
Diamyl ether	—	1052	1057	—	1131	1132
Diisopropyl ether	570	568	—	638	630	—
2-Ethylbutanol-1	—	802	806	—	1160	1149
2-Ethylhexanol-1	—	986	990	—	1341	1330
Acetone	447	450	—	702	708	—
Butanone-2	547	551	—	784	792	—
Pentanone-2	638	644	—	862	878	—
Hexanone-2	—	747	752	—	979	1001
Heptanone-2	—	846	853	—	1079	1103
Octanone-2	—	947	953	—	1178	1202
Nonanone-2	—	1047	1055	—	1277	1301
Decanone-2	—	1148	1154	—	—	—
Undecanone-2	—	1247	1256	—	—	—
Pentanone-3	647	651	—	873	884	—
Hexanone-3	—	746	751	—	966	973
Heptanone-3	—	847	852	—	1061	1078
Octanone-3	—	948	952	—	1159	1180
Heptanone-4	—	833	838	—	1043	1055
Nonanone-5	—	1035	1039	—	1238	1256
Undecanone-6	—	1231	1237	—	1436	1452
3-Methylbutanone-2	605	627	—	834	849	—
4-Methylpentanone-2	688	706	—	908	923	—
5-Methylhexanone-2	807	812	—	—	1050	—
3,3-Dimethylbutanone-2	665	672	—	865	883	—
2,4-Dimethylpentanone-2	749	756	—	—	939	948
2,6-Dimethylheptanone-4	—	935	943	—	1108	1121
Acetonitrile	439	447	—	849	853	—
Propionitrile	510	524	—	880	892	—
Butyronitrile	—	619	—	—	970	990
Valeronitrile	—	721	—	—	1066	1086
Capronitrile	—	822	829	—	1164	1187
Nitromethane	483	487	—	964	967	—

TABLE GC 51—(Continued)
ORGANICS

Column Packing	P1	P1	P1	P2	P2	P2
Temperature (°C)	70	130	190	70	130	190
Gas: Flow Rate (ml/min)	60,He	60,He	60,He	60,He	60,He	60,He
Column						
length, m	2.25	2.25	2.25	2.25	2.25	2.25
diameter, mm, O.D.	6	6	6	6	6	6
form	na	na	na	na	na	na
material	glass	glass	glass	glass	glass	glass
Detector	na	na	na	na	na	na
Compound			I			
Nitroethane	583	590	—	—	992	997
1-Nitropropane	677	686	—	—	1064	1080
1-Nitrobutane	—	791	800	—	1149	1167
1-Nitropentane	—	896	905	—	1237	1257
Formic acid ester of:						
Methanol	370	362	—	643	637	—
Ethanol	468	455	—	717	714	—
Propanol-1	570	574	—	804	814	—
Propanol-2	532	525	—	748	754	—
Butanol-1	679	680	—	910	913	—
Butanol-2	632	631	—	850	855	—
2-Methylpropanol-1	639	642	—	862	868	—
2-Methylpropanol-2	585	589	—	790	795	—
Pentanol-1	781	787	—	—	1014	1017
3-Methylbutanol-1	739	748	—	—	975	986
Acetic acid ester of:						
Methanol	475	468	—	719	717	—
Ethanol	557	547	—	784	779	—
Propanol-1	654	648	—	867	874	—
Propanol-2	596	588	—	793	798	—
Butanol-1	757	755	—	—	971	981
Butanol-2	697	692	—	—	895	—
2-Methylpropanol-1	711	713	—	—	921	—
2-Methylpropanol-2	640	636	—	817	820	—
Pentanol-1	—	855	856	—	1072	1084
3-Methylbutanol-1	—	819	—	—	1028	1039
Propionic acid ester of:						
Methanol	578	572	—	800	802	—
Ethanol	654	645	—	859	859	—
Propanol-1	752	751	—	—	951	—
Propanol-2	685	683	—	872	876	—
Butanol-1	—	851	853	—	1047	1058
Butanol-2	—	784	—	—	970	973
2-Methylpropanol-1	—	808	—	—	999	1002
2-Methylpropanol-2	727	725	—	886	889	—
Pentanol-1	—	949	950	—	1148	1156
3-Methylbutanol-1	—	914	918	—	1104	1115
Butyric acid ester of:						
Methanol	669	665	—	882	885	—
Ethanol	741	738	—	—	944	—
Propanol-1	—	840	—	—	1034	1039
Propanol-2	776	774	—	953	959	—
Butanol-1	—	940	937	—	1127	1142
Butanol-2	—	877	874	—	1052	1059
2-Methylpropanol-1	—	900	900	—	1080	1090
2-Methylpropanol-2	—	815	—	—	971	979
Pentanol-1	—	1039	1039	—	1226	1239
3-Methylbutanol-1	—	1002	1007	—	1185	1193
Isobutyric acid ester of:						
Methanol	616	610	—	826	825	—
Ethanol	699	693	—	—	881	—
Propanol-1	—	790	—	—	974	
Propanol-2	729	723	—	—	897	—
Butanol-1	—	894	898	—	1065	1079

TABLE GC 51—(Continued)
ORGANICS

Column Packing	P1	P1	P1	P2	P2	P2
Temperature (°C)	70	130	190	70	130	190
Gas: Flow Rate (ml/min)	60,He	60,He	60,He	60,He	60,He	60,He
Column						
length, m	2.25	2.25	2.25	2.25	2.25	2.25
diameter, mm, O.D.	6	6	6	6	6	6
form	na	na	na	na	na	na
material	glass	glass	glass	glass	glass	glass
Detector	na	na	na	na	na	na
Compound			I			
Butanol-2	—	829	827	—	987	993
2-Methylpropanol-1	—	855	856	—	1021	1025
2-Methylpropanol-2	—	763	—	—	902	910
Pentanol-1	—	989	996	—	1164	1175
3-Methylbutanol-1	—	954	960	—	1126	1135
Cyclopentane	571	587	—	604	627	—
Cyclohexane	676	700	—	704	734	—
Cycloheptane	—	846	874	—	883	919
Cyclooctane	—	979	1011	—	1020	1060
Cyclononane	—	1093	1132	—	1137	1186
Cyclodecane	—	1198	1238	—	1245	1293
Cycloundecane	—	1292	1334	—	1338	1387
Cyclododecane	—	1384	1430	—	1431	1482
trans-Decalin	—	1125	1166	—	1165	1207
cis-Decalin	—	1170	1213	—	1219	1267
Methylcyclopentane	635	651	—	661	684	—
Methylcyclohexane	738	760	—	762	788	—
Methylcycloheptane	—	908	938	—	939	975
Methylcyclooctane	—	1019	1052	—	1054	1091
Cyclopentene	565	581	—	642	659	—
Cyclohexene	695	714	—	764	787	—
Cycloheptene	—	840	864	—	918	942
Cyclooctene	—	945	985	—	1026	1059
1-Methylcyclopentene-1	—	675	683	—	740	753
1-Methylcyclohexene-1	—	807	816	—	866	887
1-Methylcycloheptene-1	—	909	930	—	976	1003
1-Methylcyclooctene-1	—	1023	1048	—	1092	1121
Chlorocyclopentane	—	804	827	—	967	995
Chlorocyclohexane	—	922	952	—	1085	1119
Chlorocycloheptane	—	1069	1106	—	1237	1278
Chlorocyclooctane	—	1197	1237	—	1371	1411
Bromocyclopentane	—	905	933	—	1075	1105
Bromocyclohexane	—	1023	1056	—	1189	1231
Bromocycloheptane	—	1157	1203	—	1333	1386
Bromocyclooctane	—	1258	1304	—	1432	1486
Cyclopentanone	—	766	785	—	1072	1101
Cyclohexanone	—	886	909	—	1171	1206
Cycloheptanone	—	1009	1038	—	1299	1343
Cyclooctanone	—	1115	1158	—	1405	1455
2-Methylcyclohexanone	—	937	971	—	1208	1246
3-Methylcyclohexanone	—	943	965	—	1205	1244
4-Methylcyclohexanone	—	950	973	—	1231	1278
Cyclopentanol	—	768	773	—	1135	1135
Cyclohexanol	—	880	898	—	1242	1247
Cycloheptanol	—	1022	1052	—	1385	1403
Cyclooctanol	—	1155	1189	—	1520	1548
1-Methylcyclopentanol-1	—	770	782	—	1091	1096
1-Methylcyclohexanol-1	—	882	906	—	1192	1203
1-Methylcycloheptanol-1	—	1009	1036	—	1335	1359
1-Methylcyclooctanol-1	—	1121	1156	—	1445	1474
Nitrocyclohexane	—	1083	1119	—	1385	1466
Tetrahydrofuran	618	631	—	780	800	—
Tetrahydropyran	697	714	—	—	870	897
Benzene	—	691	—	—	862	884

TABLE GC 51—(Continued)
ORGANICS

Column Packing	P1	P1	P1	P2	P2	P2
Temperature (°C)	70	130	190	70	130	190
Gas: Flow Rate (ml/min)	60,He	60,He	60,He	60,He	60,He	60,He
Column						
length, m	2.25	2.25	2.25	2.25	2.25	2.25
diameter, mm, O.D.	6	6	6	6	6	6
form	na	na	na	na	na	na
material	glass	glass	glass	glass	glass	glass
Detector	na	na	na	na	na	na

Compound	I					
Naphthalene	—	1263	1292	—	1531	1585
Azulene	—	1399	1419	—	1707	1763
Toluene	—	798	816	—	963	984
Ethylbenzene	—	893	913	—	1053	1076
Propylbenzene	—	978	998	—	1136	1161
o-Xylene	—	930	953	—	1090	1128
m-Xylene	—	904	924	—	1056	1086
p-Xylene	—	904	923	—	1051	1084
Isopropylbenzene	—	947	966	—	1104	1125
p-Cymene	—	1051	1072	—	1199	1223
Styrene	—	929	949	—	1128	1169
Fluorobenzene	—	681	—	—	879	—
Chlorobenzene	—	855	914	—	1099	1134
Bromobenzene	—	982	1019	—	1210	1257
Iodobenzene	—	1104	1152	—	1352	1405
o-Dichlorobenzene	—	1076	1117	—	1126	1375
m-Dichlorobenzene	—	1058	1097	—	1275	1320
p-Dichlorobenzene	—	1060	1096	—	1288	1335
o-Chlorotoluene	—	996	1024	—	1194	1231
m-Chlorotoluene	—	995	1025	—	1205	1241
p-Chlorotoluene	—	996	1026	—	1206	1241
o-Bromotoluene	—	1091	1129	—	1306	1352
m-Bromotoluene	—	1095	1130	—	1319	1362
p-Bromotoluene	—	1094	1132	—	1321	1364
Benzylbromide	—	1125	1161	—	—	—
2-Brom-1-phenylethane	—	1219	1257	—	1529	1585
Phenylcyanide	—	965	993	—	1406	1444
Phenylacetonitrile	—	1097	1127	—	1646	1686
Anisole	—	930	949	—	1200	1231
Phenetole	—	996	1013	—	1246	1276
Nitrobenzene	—	1088	1131	—	1499	1536
o-Nitrotoluene	—	1155	1190	—	1547	1588
m-Nitrotoluene	—	1201	1237	—	1592	1639
p-Nitrotoluene	—	1212	1258	—	1615	1668
Benzaldehyde	—	965	996	—	1336	1371
Phenylacetaldehyde	—	1029	1057	—	1430	1465
Hydrocinnamaldehyde	—	1146	1174	—	1546	1584
m-Tolualdehyde	—	1081	1108	—	1429	1470
p-Tolualdehyde	—	1089	1122	—	1449	1490
Acetophenone	—	1067	1095	—	1445	1479
Propiophenone	—	1164	1189	—	1518	1556
Butyrophenone	—	1240	1271	—	1583	1621
1-Phenylpropanone-2	—	1098	1127	—	1499	1538
1-Phenylbutanone-2	—	1192	1223	—	1583	1623
1-Phenylbutanone-3	—	1214	1243	—	1622	1654
Benzyl alcohol	—	1010	1044	—	1573	1598
2-Phenylethanol	—	1089	1128	—	1610	1644
3-Phenylpropanol-1	—	1207	1234	—	1716	1757
1-Phenylethanol-1	—	1042	1071	—	1558	1577
2-Phenylpropanol-2	—	1076	1099	—	1540	1562

Column Packing P1 = 40% Apiezon L on Celite (200/250 mμ).
P2 = 40% Emulphor-O on Celite (200/250 mμ).

Literature Wehrli, A. and E. Kovats, *Helv. Chim. Acta* **42**, 2709 (1959).

TABLE GC 52
ORGANICS

Column Packing	P1	P2
Temperature (°C)	70	70
Gas; Flow Rate (ml/min)	165	300
Column		
length, m	4	4
diameter, mm, I.D.	6	10
form	na	na
material	Cu	Cu
Detector	TC	TC

Organic compounds	$r_{benzene}$	
***n*-Heptane**	0.07_5	0.29
***n*-Octane**	0.13	0.62
***n*-Decane**	0.47	—
Cyclohexane	0.11	0.30
Methylcyclohexane	0.14	0.46
Benzene	1.00	1.00
Methyl chloride	0.08_5	0.06_5
Methylene chloride	0.64	0.75
Chloroform	0.91	1.6_3
Carbon tetrachloride	0.36	0.70
Ethyl chloride	0.14	0.14
Vinyl chloride	0.07	0.08
1,1-Dichloroethane	0.58	0.68
1,2-Dichloroethane	2.1_5	2.0_6
1,1-Dichloroethylene	0.16	0.24
***trans*-1,2-Dichloroethylene**	0.35	0.54
***cis*-1,2-Dichloroethylene**	0.87	1.2_4
1,1,1-Trichloroethane	0.49	0.74
1,1,2-Trichloroethane	5.8	>4
Trichloroethylene	0.77	1.4_3
Perchloroethylene	0.98	1.9_3
Ethylene oxide	0.37	0.16
Propylene oxide	0.48	0.24
Isobutylene oxide	0.52	0.31
***cis*-2,3-Butylene oxide**	0.78	0.46
***trans*-2,3-Butylene oxide**	0.57	0.35
1,2-Butylene oxide	0.78	0.52
Acetaldehyde	0.43	0.16
Propionaldehyde	0.72	0.34
Isobutyraldehyde	0.78	0.46
***n*-Butyraldehyde**	1.1_9	0.69
Acrolein	0.96	0.46
Crotonaldehyde	3.9_6	1.8_8
Acetone	1.1_0	0.41
Methyl ethyl ketone	1.6_7	0.79
Diethyl ketone	2.4_0	1.4_6
Methyl isopropyl ketone	1.8_9	1.0_8
Methyl *n*-propyl ketone	2.4_7	1.4_6
Formaldehyde dimethyl acetal	0.40	0.25
Formaldehyde diethyl acetal	0.69	0.69
Acetaldehyde dimethyl acetal	0.59	0.41
Acetaldehyde diethyl acetal	0.84	0.97
Propionaldehyde dimethyl acetal	0.87	0.78
Butyraldehyde dimethyl acetal	1.3_8	1.5_8
Methyl formate	0.50	0.24
Ethyl formate	0.75	0.43
Methyl acetate	0.82	0.43
Ethyl acetate	1.0_9	0.73
Isopropyl acetate	1.0_2	0.88
***n*-Propyl acetate**	1.7_8	1.4_7
***n*-Butyl acetate**	3.0_0	3.1_0
Vinyl acetate	0.97	0.73

TABLE GC 52—(Continued)

ORGANICS

Column Packing	P1	P2
Temperature (°C)	70	70
Gas; Flow Rate (ml/min)	165	300
Column		
length, m	4	4
diameter, mm, I.D.	6	10
form	na	na
material	Cu	Cu
Detector	TC	TC

Organic compounds	$r_{benzene}$	
Methyl propionate	1.2_4	0.82
Methyl acrylate	1.4_1	0.96
Methyl isobutyrate	—	1.0_1
Ethyl butyrate	2.3_3	2.4_7
Vinyl butyrate	2.0_8	2.3_2
Methyl butyrate	—	1.5
Methanol	0.91	0.53
Ethanol	1.1_3	0.73
Isopropanol	1.0_3	0.78
***n*-Propanol**	1.9_6	1.6_3
***n*-Butanol**	3.4_7	3.6_0
Isobutanol	2.4_4	2.5_6
***sec*-Butanol**	1.7_9	1.6_6
***tert*-Butanol**	0.8_9	—
Dimethyl ether	0.07	0.03_5
Diethyl ether	0.12	0.12
Diisopropyl ether	0.12	0.20
Di-*n*-propyl ether	0.24	0.43
Di-*n*-butyl ether	0.71	1.9_5
***sym.*-Hexafluorodiethyl ether**	—	0.38
Water	2.2	1.5
Carbon disulfide	0.13	0.23
Dioxan	4.4_5	2.0_4
Tetrahydrofuran	0.91	0.58
Diacetyl	—	1.3

Column Packing P1 = β,β'-Hydroxydipropionitrate, 30% on Sterchamol®[a] (0.25–0.35 mm).

[a] Purchased from Sterchamol, Dortmund, Germany.

P2 = Hexaethylene glycol dimethyl ether, 30% on Sterchamol® (0.25–0.35 mm).

Literature Kelker, H. *Angew. Chem.*, **71**, 218 (1959).

TABLE GC 53
ORGANICS

Column Packing	P1
Temperature (°C)	T1
Gas; Flow Rate (ml/min)	100,He
Column	
length, m	4
diameter, in., O.D.	$\frac{1}{4}$
form	—
material	aluminum
Detector	TC

Compound	$r_{n\text{-nonane}}$
2-Methylbutane (isopentane)	0.116
***n*-Pentane**	0.158
Pentene-1	0.171
2,2-Dimethylbutane (neohexane)	0.180
Pentene-2 (*trans*)	0.218
2,3-Dimethylbutane	0.256
Diethyl ether	0.256
2-Methylpentane	0.275
2-Methyl-1,3-butadiene	0.285
3-Methylpentane	0.296
Cyclopentane	0.312
1,3-Pentadiene	0.341
***n*-Hexane**	0.350
Methyl formate	0.354
***tert*-Butyl chloride**	0.366
Hexene-1	0.373
2,4-Dimethylpentane	0.388
Dimethoxymethane (methylal)	0.393
Methylcyclopentane	0.443
Propionaldehyde	0.460
2,3-Dimethylpentane	0.498
Acetone (2-propanone)	0.501
2,2,4-Trimethylpentane (isooctane)	0.511
Ethyl formate	0.520
Cyclohexane	0.529
2-Chlorobutane (*sec*-butyl chloride)	0.563
***n*-Heptane**	0.582
Heptene-1	0.606
Methanol	0.614
Diisobutylene	0.637
Cyclohexene	0.650
Vinyl acetate	0.664
***n*-Butyraldehyde**	0.664
1-Chlorobutane (*n*-butyl chloride)	0.673
Methylcylclohexane	0.679
Methylene chloride	0.684
Ethyl acetate	0.686
2,3,4-Trimethylpentane	0.691
Methyl ethyl ketone (2-butanone)	0.705
Ethanol	0.708
***tert*-Butyl alcohol (2-methyl-2-propanol)**	0.719
Carbon tetrachloride	0.721
Isopropyl alcohol (2-propanol)	0.732
Isopropyl acetate	0.739
3-Methyl-3-butenyl methyl ether	0.788
***n*-Octane**	0.799
Water	0.810
Benzene	0.813
Octene-1	0.822
Fluorobenzene	0.864
2-Pentanone	0.868
***n*-Propyl acetate**	0.875
3-Pentanone (diethyl ketone)	0.882
Chloroform	0.887

TABLE GC 53—(Continued)
ORGANICS

Column Packing	P1
Temperature (°C)	T1
Gas; Flow Rate (ml/min)	100,He
Column	
length, m	4
diameter, in., O.D.	¼
form	—
material	aluminum
Detector	TC

Compound	$r_{n\text{-nonane}}$
3-Methyl-2-butenyl methyl ether	0.893
Valeraldehyde (pentanol)	0.894
Methyl butyrate	0.895
Propanol	0.903
sec-Butyl alcohol	0.917
tert-Amyl chloride	0.930
Crotonaldehyde	0.932
4-Methyl-2-pentanone	0.960
Allyl alcohol (2-propen-1-ol)	0.983
n-Nonane	1.000
Isobutyl alcohol (2-methyl-1-propanol)	1.011
Ethyl butyrate	1.019
Nonene-1	1.027
Toluene (methylbenzene)	1.039
Neopentyl alcohol (2,2-dimethyl-1-propanol)	1.071
n-Butyl acetate	1.087
Butanol	1.120
Mesityl oxide (4-methyl-3-penten-2-one)	1.205
Isoamyl acetate	1.240
n-Decane	1.242
3-Methyl-3-chlorobutyl methyl ether	1.248
Ethylbenzene	1.263
4-Heptanone	1.281
Decene-1	1.287
Isoamyl alcohol (3-methyl-1-butanol)	1.298
m-Xylene	1.308
p-Xylene	1.308
3-Heptanone	1.346
4-Methyl-3-chlorobutyl methyl ether	1.408
2-Heptanone	1.415
Amyl alcohol	1.418
Cumene (isopropylbenzene)	1.431
Chlorobenzene	1.433
o-Xylene	1.435
Diisobutyl ketone	1.541
Propylbenzene	1.578
n-Undecane (hendecane)	1.596
N,*N*-Dimethylformamide	1.652
Undecene-1	1.663
Mesitylene (1,3,5-trimethylbenzene)	1.714
tert-Butylbenzene	1.740
sec-Butylbenzene	1.803
Isobutylbenzene	1.822
Hexanol	1.858
Pseudocumene (1,2,4-trimethylbenzene)	1.898
Diacetone alcohol (4-methyl-4-hydroxy-2-pentanone)	1.966
Bromobenzene	2.005
n-Butylbenzene	2.142
n-Dodecane	2.155
Dodecene-1	2.267
2-Cyclohexene-1-one	2.427

TABLE GC 53—(Continued)
ORGANICS

Column Packing	P1
Temperature (°C)	T1
Gas; Flow Rate (ml/min)	100,He
Column	
length, m	4
diameter, in., O.D,	$\frac{1}{4}$
form	—
material	aluminum
Detector	TC

Compound	$r_{n\text{-nonane}}$
Heptanol	2.604
2,5-Hexanedione (acetonylacetone)	2.791
Benzaldehyde	2.911
Iodobenzene	3.406

Column Packing P1 = 10% (w/w) Ethofat (25/60 mesh) on Teflon® 6 (40/60 mesh).
Temperature T1 = 30 ° to 140 °C at 7.5 °/min linear.
Literature Mowry, D. P. *J. Gas Chromatogr.*, **4**, 433 (1966).

TABLE GC 54
ANTIHISTAMINES

Column Packing	P1
Temperature (°C)	185
Gas; Flow Rate (ml/min)	200,N_2
Column	
length, m	4
diameter, mm, I.D.	4
form	—
material	—
Detector	F1

Compound	$r_{\text{diphenylpyraline HCl}}$
Phenylpropanolamine HCl	0.13
Pheniramine maleate	0.31
Phenyltoloxamine hydrogen citrate	0.53
Cyclizine HCl	0.64
Pyribenzamine citrate	0.67
Chlorpheniramine maleate	0.76
Methapyrilene HCl	0.77
Diphenylpyraline HCl	1.000
Dextromethorphan HBr	1.05
Dexbrompheniramine maleate	1.21
Chlorothen citrate	1.45
Phenindamine tartrate	1.49
Chlorcyclizine HCl	1.66
Thonzylamine HCl	2.08
Pyrilamine maleate	2.51

Column Packing P1 = 2% SE-30 and 2% Carbowax 20M on Anakrom ABS, conditioned overnight at 200 °C and 200 ml/min. of N_2.
Literature Celeste, A. C. and M. V. Polito, *J. Ass. Offic. Agr. Chem.*, **49,** 541 (1966).

TABLE GC 55
ORGANICS

Column Packing	P1
Temperature (°C)	80
Gas; Flow Rate (ml/min)	81,H_2
Column	
length, m	5
diameter, in.	¼
form	na
material	Cu
Detector	TC

Compound	$r_{n\text{-decane}}$
Isobutyl acetate	1.088
tert-Amyl alcohol	1.103
n-Butyl formate	1.116
Thiophene	1.132
Allyl acetate	1.186
sec-Butyl alcohol	1.195
Propionitrile	1.253
Toluene	1.283
1-Propanol	1.309
Crotonaldehyde	1.347
Cyclobutanone	1.364
1,4-Dioxane	1.562
n-Butyl acetate	1.603
Isobutyl alcohol	1.833
3-Methyl 2-butanol	1.837
Crotononitrile	1.872
Butyronitrile	1.956
n-Undecane	1.976
3-Pentanol	2.039
Allyl alcohol	2.115
Isoamyl acetate	2.218
Ethylbenzene	2.220
2-Pentanol	2.238
m-Xylene	2.436
n-Butyl alcohol	2.646
Isopropylbenzene	2.972
o-Xylene	3.161
Cyclopentanone	3.489
n-Propylbenzene	3.743
n-Dodecane	3.845
n-Hexane	0.068
Cyclopentane	0.084
Isoprene	0.090
1,3-Pentadiene	0.107
Methylcyclopentane	0.116
n-Heptane	0.135
Acetaldehyde	0.141
Cyclohexane	0.165
Methyl formate	0.196
Methylcyclohexane	0.228
Furan	0.244
Propanal	0.254
n-Octane	0.262
Cyclohexene	0.282
Pivalaldehyde	0.284
Isobutanal	0.295
Acetone	0.305
Ethyl formate	0.306
Methyl acetate	0.315
Acrolein	0.366
Isopropyl formate	0.370
2-Methylfuran	0.404
Tetrahydrofuran	0.430

TABLE GC 55—(Continued)
ORGANICS

Column Packing	P1
Temperature (°C)	80
Gas; Flow Rate (ml/min)	81,H_2
Column	
length, m	5
diameter, in.	¼
form	na
material	Cu
Detector	TC

Compound	$r_{n\text{-decane}}$
Butanal	0.456
Methacrolein	0.462
Ethyl acetate	0.467
Isopropyl acetate	0.505
***n*-Nonane**	0.513
***tert*-Butyl alcohol**	0.529
Methyl propionate	0.532
Methyl ethyl ketone	0.533
Methanol	0.536
***n*-Propyl formate**	0.586
Isovaleraldehyde	0.603
Tetrahydropyran	0.624
Isopropyl alcohol	0.633
Isopropyl methyl ketone	0.659
Methyl acrylate	0.661
Ethanol	0.669
Benzene	0.684
2,5-Dimethylfuran	0.704
Vinyl methyl ketone	0.723
Ethyl propionate	0.749
Biacetyl	0.829
Allyl formate	0.837
***n*-Propyl acetate**	0.843
Diethyl ketone	0.879
***n*-Propyl methyl ketone**	0.887
Valeraldehyde	0.901
Methyl butyrate	0.905
Methacrylonitrile	0.912
Acrylonitrile	0.954
***n*-Decane**	1.000
Methyl methacrylate	1.014
Isobutyronitrile	1.070
Acetonitrile	1.074

Column Packing P1 = 15% Carbowax 1540 on acid-washed Sil-O-Cell C-22 (0.21-0.25 mesh).

Literature Caroff, J. N., Bahurel, Y., and J. Veron, *J. Gas Chromatogr.*, **4**, 234 (1966).

TABLE GC 56
ORGANICS

Column Packing	P1	P1	P1
Temperature (°C)	20	65	100
Gas; Flow Rate	na	na	na
Column			
length, ft	5	5	5
diameter, mm, I.D.	4	4	4
form	na	na	na
material	glass	glass	glass
Detector	FI	FI	FI

Compound	$r_{n\text{-nonane}}$		
1-Chloropropane	0.033	0.182	0.250
1-Chlorobutane	0.106	0.451	0.525
1-Chloropentane	0.333	1.091	1.104
1-Chlorohexane	—	2.643	2.310
1-Chloroheptane	—	—	4.77
1-Chlorooctane	—	—	20.15
Bromoethane	0.027	0.156	—
1-Bromopropane	0.084	0.381	0.470
1-Bromobutane	0.259	0.933	0.993
1-Bromopentane	—	2.233	2.054
1-Bromohexane	—	5.40	4.275
1-Bromoheptane	—	—	8.80
Iodomethane	0.027	0.148	0.222
Iodoethane	0.076	0.355	0.463
1-Iodopropane	0.228	0.852	0.957
1-Iodobutane	0.680	2.033	1.960
1-Iodopentane	—	4.81	4.050
1-Iodohexane	—	—	8.40
1-Iodoheptane	—	—	17.08
Methyl cyanide	0.048	0.194	—
Ethyl cyanide	0.114	0.426	0.506
***n*-Propyl cyanide**	0.316	1.056	1.077
***n*-Butyl cyanide**	0.966	2.572	2.267
***n*-Pentyl cyanide**	—	—	4.665
Ethanethiol	0.021	0.129	—
***n*-Propanethiol**	0.065	0.320	0.401
***n*-Butanethiol**	0.206	0.777	0.845
***n*-Pentanethiol**	0.626	1.863	1.759
***n*-Hexanethiol**	—	4.479	3.642
***n*-Heptanethiol**	—	—	7.47
Pent-1-ene	0.011	0.072	—
Hex-1-ene	0.034	0.177	0.237
Hept-1-ene	0.106	0.420	0.492
Oct-1-ene	0.321	2.420	2.092
Dec-1-ene	—	—	8.75

Column Packing P1 = 20% w/w mixture of dinonyl phthalate (B.D.H.) and acid-washed (1% phosphoric acid), 60–72 mesh Celite (J. J.'s, Ewell, Surrey) carrier gas: 3 : 1 (by volume) mixture of hydrogen and nitrogen.

Literature Evans, M. B. and J. F. Smith, *J. Chromatogr.*, **8**, 303 (1962).

TABLE GC 57
ORGANICS, VOLATILES

Column Packing	P1	P2	P3
Temperature (°C)	70	60	55
Gas; Flow Rate (ml/min)	51.3,Ar	51.3,Ar	51.3,Ar
Column			
length, m	2	2	2
diameter, mm, I.D.	4.8	4.8	4.8
form	na	na	na
material	copper	copper	copper
Detector	FID	FID	FID

Compound	r (a)*n* undecane (b)*n*-decane (a)	(b)	(a)
Acetaldehyde	0.008	0.016	0.016
Methyl formate	0.013	0.021	0.023
Propionaldehyde	0.017	0.037	0.030
Acetone	0.022	0.042	0.041
Ethyl formate	0.024	0.043	0.039
Methyl acetate	0.025	0.048	0.041
Isobutyraldehyde	0.026	0.057	0.040
Isopropyl formate	0.034	0.062	0.048
Butyraldehyde	0.037	0.082	0.063
Isopropanol	0.044	0.055	0.098
Methyl ethyl ketone	0.045	0.093	0.076
Ethyl acetate	0.045	0.089	0.067
Ethanol	0.046	0.057	0.099
2,3-Butanedione	0.050	0.098	0.110
Methyl propionate	0.052	0.104	0.073
Propyl formate	0.053	0.098	0.075
Methyl isobutyrate	0.071	0.148	0.090
***tert*-Butyl acetate**	0.071	0.151	0.090
Methyl propyl ketone	0.088	0.186	0.140
Ethyl propionate	0.089	0.189	0.116
Valeraldehyde	0.089	0.194	0.144
***n*-Propanol**	0.094	0.117	0.206
2-Butanol	0.096	0.129	0.200
***n*-Propyl acetate**	0.098	0.199	0.136
2-Methyl-2-butanol	0.101	0.135	0.210
Methyl *n*-butyrate	0.103	0.212	0.137
Ethyl isobutyrate	0.117	0.256	0.144
***n*-Butyl formate**	0.119	0.277	0.163
***iso*-Butyl acetate**	0.143	0.308	0.192
***iso*-Butanol**	0.152	0.178	0.317
Methyl *iso*-valerate	0.153	0.315	0.206
Ethyl *n*-butyrate	0.183	0.392	0.232
***n*-Propyl *n*-propionate**	0.190	0.416	0.234
***iso*-Amyl formate**	0.195	0.373	0.268
2-Pentanol	0.198	0.272	0.430
Methyl butyl ketone	0.198	0.421	0.305
Caproaldehyde	0.204	0.447	0.328
3-Pentanol	0.212	0.257	0.378
***n*-Butyl acetate**	0.218	0.456	0.293
***n*-Butanol**	0.218	0.267	0.462
Amyl formate	0.269	0.514	,0.358
***iso*-Amyl acetate**	0.345	0.734	0.476
***n*-Propyl *n*-butyrate**	0.370	0.855	0.461
***iso*-Butyl *iso*-butyrate**	0.370	0.868	0.481
2-Methyl-1-butanol	0.375	0.442	0.757
3-Methyl-1-butanol	0.378	0.448	0.796
Ethyl *n*-valerate	0.402	0.896	0.510
2-Hexenal	0.405	0.882	0.716
3-Hexanol	0.418	0.558	0.790
Heptanal	0.457	1.00	0.727
2-Hexenol	0.481	0.619	1.00

TABLE GC 57—(Continued)
ORGANICS, VOLATILES

Column Packing	P1	P2	P3
Temperature (°C)	70	60	55
Gas: Flow Rate (ml/min)	51.3,Ar	51.3,Ar	51.3,Ar
Column			
length, m	2	2	2
diameter, mm, I.D.	4.8	4.8	4.8
form	na	na	na
material	copper	copper	copper
Detector	FID	FID	FID

Compound	r (a)*n*-undecane (b)*n*-decane (a)	(b)	(a)
Amyl acetate	0.484	1.00	0.644
1-Hexen-3-ol	0.493	0.593	1.024
***n*-Pentanol**	0.505	0.608	1.00
Methyl *n*-caproate	0.518	1.131	0.641
***n*-Butyl *iso*-butyrate**	0.527	1.249	0.653
***iso*-Butyl *n*-butyrate**	0.553	1.303	0.715
***n*-Hexyl formate**	0.597	1.159	0.786
***iso*-Amyl propionate**	0.654	1.523	0.812
***iso*-Butyl *iso*-valerate**	0.782	1.881	1.112
***n*-Butyl *n*-butyrate**	0.802	1.919	0.992
***iso*-Amyl *iso*-butyrate**	0.827	2.005	1.076
Ethyl *n*-caproate	0.881	2.019	1.107
Amyl *n*-propionate	0.916	—	1.112
***cis*-3-Hexen-1-ol**	0.992	1.318	2.226
Octanal	1.00	—	1.610
***n*-Hexyl acetate**	1.058	—	1.402
Amyl *iso*-butyrate	1.137	—	1.447
***n*-Hexanol**	1.141	1.357	2.290
***iso*-Amyl *n*-butyrate**	1.244	—	1.643
***trans*-3-Hexen-1-ol**	1.350	1.813	—
Amyl *n*-butyrate	1.756	—	2.192
Ethyl *n*-heptanoate	1.932	—	—

Column Packing P1 = 2% UCON 50-LB-1200X on DMCS treated Chromosorb G.
P2 = 5% DC-710 Carbowax 20M on acid washed Chromosorb P.
P3 = 5% Carbowax 20M on non-acid washed Chromosorb W.

Literature Gasco L., Barrera, R., and F. de la Cruz, *J. Chromatogr. Sci.*, **7**, 228 (1969).

TABLE GC 58
ORGANICS, ALIPHATICS

Column Packing	P1	P2	P3
Temperature (°C)	—	—	—
Gas; Flow Rate (ml/min)	40,N_2	25,N_2	25,N_2
Column			
length, m	1	1	1
diameter, mm, I.D.	2.3	2.3	2.3
form	na	na	na
material	SS	SS	SS
Detector	FI	FI	FI

Compound	r_{hexane}		
Methane	—	0.069	0.059
Ethane	0.084	0.086	0.073
Ethylene	0.084	0.086	0.073
Acetylene	0.108	0.122	0.079
Propane	0.125	0.132	0.097
Propylene	0.125	0.132	0.125
Isobutane	0.200	0.194	0.171
Butane	0.226	0.212	0.195
Butene-1	0.226	0.238	0.248
Isobutene	—	0.238	0.284
***trans*-Butene-2**	0.248	0.256	0.302
***cis*-Butene-2**	0.248	0.256	0.326
Isopentane	—	—	0.369
Pentane	0.453	0.434	0.427
Pentene-1	0.470	0.491	0.524
***trans*-Pentene-2**	0.504	0.532	0.712
***cis*-Pentene-2**	0.670	0.532	0.750
***trans*-1,2-Dichloroethylene**	0.834	0.703	0.826
2-Methylpentane	0.853	0.832	0.847
Acetone	0.953	7.450	—
Methyl acetate	0.984	4.530	—
Hexane	1.000 (2.54 min)	1.000 (4.71 min)	1.000 (5.34 min)
Methanol	1.280	>6 (severe tailing)	—
Isobutyraldehyde	1.290	—	—
Dichloromethane (methylene chloride)	1.300	0.866	0.897
Tetrachloromethane (CCl_4)	1.530	1.160	1.270
1,1-Dichloroethane	1.530	1.510	1.530
Benzene	1.630	2.110	6.140
Butyraldehyde	1.730	—	—
2-Butanone	1.970	>12	—
Ethyl acetate	2.030	>9	—
Ethanol	2.250	—	—
Heptane	2.290	2.320	2.420
***cis*-1,2-Dichloroethylene**	2.300	1.510	1.530
Trichloroethylene	2.400	1.870	2.270
1,2-Dichloroethane	3.080	2.680	3.870
Trichloromethane (chloroform)	3.150	1.960	1.510
Isopropyl acetate	3.380	>13	—
Toluene	4.000	>4	—
Propyl acetate	4.630	>15	—
Octane	5.350	5.430	5.960
Propanol	5.620	—	—
2-Ethyl butyraldehyde	6.240	—	—
Isobutanol	12.000	—	—
Butanol	15.000	—	—

All packings available from Waters Assoc. as Durapak® phases.

Column Packing P1 = Carbowax 400 on Porasil C (100/120 mesh).
P2 = OPN (oxypropionitrile) on Porasil C (80/100 mesh).
P3 = Octane on Porasil C (120/150 mesh).

Literature Little, J. N., Dark, W. A., Farlinger, P. W., and K. J. Bombaugh, *J. Chromatogr. Sci.*, **8**, 647 (1970).

TABLE GC 59

PERHALOMETHYL SULFENYL AND THIOCARBONYL COMPOUNDS

Column Packing	P1	P2	P3	P4	P4
Temperature (°C)	70	70	70	70	100
Gas; Flow Rate (ml/min)	37,He	30,He	25,He	25,He	25,He
Column					
length, cm	473	360	180	180	180
diameter, mm	6	6.3	6.3	6.3	6.3
form	na	na	na	na	na
material	copper	SS	SS	SS	SS
Detector	TC	TC	TC	TC	TC

Compound			r_{CCl_4}		
CO_2	—[a]	0.01	—[a]	0.01	0.01
COS	0.02	0.05	0.04	0.04	0.05
CF_3SCl	0.11	0.14	—	0.05	0.06
CSFCl	0.05	0.16	—[b]	0.14	0.15
CF_2ClSCF_3	—	0.23	0.04	0.04	0.07
CF_3SO_2Cl	0.27	0.30	0.12	0.09	0.13
CF_3SSCF_3	0.25	0.31	0.04	0.09	0.10
CF_3SCN	0.18	0.22	0.39	0.13	0.14
CF_3SCSF	0.38	0.52	—[b]	0.26	0.23
CF_3SNH_2	0.32	0.26	1.48	0.36	0.37
COFSCl	0.33	0.36	—[b]	0.31	0.41
F_2CSSCF_2	0.45	0.51	0.64	0.31	0.32
$CF_3SOC(O)CF_3$	—[b]	—[b]	0.22[c]	—[b]	—[b]
CF_2ClSCl	0.44	0.64	—	0.51	0.43
CCl_2FSCF_3	0.65	0.72	0.16	0.29	0.31
$CF_3SO_2SCF_3$	0.72	0.87	0.13	0.17	0.19
CF_3SSCF_2Cl	—	1.10	0.19	0.41	—
$(CF_3S)_2NH$	0.99	1.17	—	0.85	0.75
$CClF_2SOC(O)CF_3$	—	1.37	0.4[c]	—	0.48
$(CF_3)_2CClSCl$	1.71	1.75	0.56	0.63	0.64
$(CF_3S)_3N$	1.87	2.45	—	—	—
$(CF_3)_2CSSC(CF_3)_2$	2.54	—	—	—	—
$CCl_2FSOC(O)CF_3$	—	—	—	—	0.74
$CF_3SCFClSCl$	—	—	—	—	2.10
$CF(O)SSCF_2Cl$	—	—	—[b]	—	1.43

[a] No separation from air peak.
[b] Decomposition.
[c] Decomposition was not complete in one test.

Column Packing P1 = 15% Kel F oil No. 10 on Haloport F (30/60 mesh).
P2 = 15% Fluorolube on Teflon T6 (42/60 mesh).
P3 = 30% Polyethylene glycol 1000 on C-22 firebricks (42/60 mesh).
P4 = 50% Silicone oil DC-550 on C-22 firebricks (42/60 mesh).

Literature Benseler, E. and A. Haas, *J. Chromatogr. Sci.*, **8**, 727 (1970).

TABLE GC 60
PESTICIDES

Column Packing	P1
Temperature (°C)	200
Gas; Flow Rate (ml/min)	120,N_2
Column	
length, ft	6
diameter, mm, I.D. (coiled to 4″ dia)	4
form	coiled
material	all-glass
Detector	EC

Compound	r_{aldrin}
Phosdrin®	0.44
α-BHC	0.51
Isopropyl 2,4-D	0.62
Diazinon®	0.64
Naled	0.65
Lindane	0.71
Di-Syston®	0.77
Heptachlor	0.83
Isopropyl 2,4,5-T	0.93
β-BHC	0.96
Aldrin	1.00[a]
Demeton	*1.00*,[b] 0.82
Butyl 2,4-D	1.02
Ronnel	1.12
Dimethoate	1.25
Dicofol	*1.59*, 2.12, 2.78
Heptachlor epoxide	1.63
Methyl parathion	1.78
Malathion	1.80
Chlorbenzide	1.84
Endosulfan	2.08, 3.74
***p,p′*-DDE**	2.12
Parathion	2.25
Perthane®	*2.28*, 2.92
Dieldrin	2.49
***o,p′*-DDD**	2.50
Captan	2.85
***o,p′*-DDT**	2.90
Endrin	*3.02*, 5.18, 8.47
Chlorobenzilate®	3.18
***p,p′*-DDD**	3.37
Methyl trithion®	3.58
***p,p′*-DDT**	3.92
Thiram	3.98
Ethion	4.12
Trithion®	4.40
Mirex	5.33
Methoxychlor	7.00
Guthion®	15.0
Chlordane	0.82, 1.68, 1.84, others
Strobane®	3.73, 4.39, others
Toxaphene	*4.34*, others

[a] 2.5 min.
[b] Major peaks in italics.

Column Packing P1 = 8% QF-1 and 2% OV-17 on Gas Chrom Q (80/100 mesh).
Literature Windham, E. S., *J. Ass. Offic. Agr. Chem.*, **52**, 1237 (1969).

TABLE GC 61
PESTICIDES

Column Packing	P1
Temperature (°C)	See table
Gas; Flow Rate	na
Column	
length, ft	2
diameter, in., I.D.	$\frac{3}{16}$
form	na
material	borosilicate glass
Detector	na

Compound	Column Temp., °C	r_{thimet} Retention Time, min. Actual	Relative
Ciodrin®	160	34.8	9.7
Co-Ral oxygen analog	210	35.1	9.8
Dibrom	160	3.6	1.0
Dimethoate	160	6.6	1.8
Dimethoate oxygen analog	160	3.6	1.0
Guthion®	210	26.0	7.2
Guthion oxygen analog	210	20.8	5.8
Imidan®	210	18.6	5.2
Imidoxon	210	14.8	4.2
Malathion	158	22.2	6.2
Malathion oxygen analog	158	18.7	5.2
Parathion	158	21.7	6.0
Parathion oxygen analog	158	18.0	5.0
Prefar	212	32.0	8.9
Phosdrin®	130	4.6	1.3
Phosphamidon	160	11.5	3.2
Thimet®	160	3.6	1.0
Thimet oxygen analog	160	2.8	0.8
Thimet sulfoxide	160	18.1	5.0
Thimet sulfone	160	19.8	5.5
Thimet oxygen analog sulfoxide	160	14.5	4.0
Thimet oxygen analog sulfone	160	14.8	4.1
Ciba 10573 (O-(1,3-dithiolan-2-yl)-phenyl-*N*-methylcarbamate)	172	60.5	16.8
Mesurol (Bayer 37344) (*N*-methyl-3,5-dimethyl-4-thiomethylphenylcarbamate)	153	23.0	6.4
Mesurol phenol	153	2.4	0.7
Mobam (4-benzothienyl *N*-methylcarbamate)	172	9.8	2.7
Monuron	142	25.2	7.0
Bromacil	185	6.3	1.7
Terbacil	185	3.2	0.9
DNOC	160	1.8	0.5
DNOSBP	160	3.6	1.0
Aldrin	172	6.1	1.7
***p,p'*-DDT**	172	40.3	11.2
***p,p'*-DDD**	172	32.3	8.9
***p,p'*-DDE**	172	19.1	5.3
Dieldrin	160	15.6	4.3
Lindane	142	11.8	3.3
Dacthal mono acid	158	4.1	1.1
Temik	115	19.7	5.5
Thiram	185	48.8	10.8

Column Packing P1 = 10% OV-17 on silyated Gas Chrom Q (80/100 mesh).
Literature Bache, C. A. and D. J. Lisk, *J. Ass. Offic. Agr. Chem.*, **51**, 1270 (1968).

TABLE GC 62
PESTICIDES

Column Packing	P1
Temperature (°C)	150
Gas; Flow Rate (ml/min)	53,N_2
Column	
length, ft	3
diameter, in.	$\frac{1}{8}$
form	coiled
material	glass
Detector	EC

Compound	$r_{o,p'\text{-DDD}}$
PCNB®	0.18
Vancide PB	0.19
***p,p'*-DDE**	0.97
Botran	0.18
Ovex	0.85
Tedion®	3.6
***p,p'*-DDD**	1.3
***o,p'*-DDD**	1.0
***o,p'*-DDT**	1.3
Dicryl	0.41
Dichlone	0.23
Captan	0.60
Unads	0.58
Perthane®	1.2
Antu®	1.8
Atrazine	0.19
Redax	2.3

Column Packing P1 = 4% Dow-11 on HMDS-treated Chromosorb W (40/60 mesh); conditioned 18 hours at 200 °C with N_2.

Literature Zielinski, Jr., W. L., Fishbein, L., and L. Martin, Jr., *J. Gas Chromatogr.*, **5**, 552 (1967).

TABLE GC 63
PESTICIDES

Column Packing	P1	P2	P3
Temperature (°C)	T1	T1	T1
Gas; Flow Rate (ml/min)	N_2,1.0 Kp/cm sq or 1.5 Kp/cm sq	N_2,1.0 Kp/cm sq or 1.5 Kp/cm sq	N_2,1.0 Kp/cm sq or 1.5 Kp/cm sq
Column			
length, m	0.80	0.80	0.80
diameter, mm, I.D.	3	3	3
form	U-shaped	U-shaped	U-shaped
material	glass	glass	glass
Detector	D1	D1	D1

Compound	t_r		
Bromoxynol	1.81	2.70	2.35
DNOSBP (Dinitrobutylphenol)	1.10	1.45	1.12
DNOC (Dinitrocresol)	0.82	1.04	0.93
DNOCHP (Dinitrocyclohexylphenol)	4.25	6.18	4.49
DNOPP(2,4-Dinitro-6-phenylphenol	9.22	14.54	11.93
Ioxynil	5.40	9.22	7.40

TABLE GC 63—(Continued)
PESTICIDES

Column Packing	P1 = 60/80 mesh Chromosorb W, 3 % neopentylglycol succinate and 1 % H_3PO_4.
	P2 = 60/80 mesh Chromosorb W, 3 % cyclohexanedimethanol succinate and 1 % H_3PO_4.
	P3 = 60/80 mesh Chromosorb W, 3 % butanediol succinate and 1 % H_3PO_4.
Temperature	T1 = injection, 220 °C; column 200 °C; detector, na.
Detector	D1 = FI and EC, with tritium source.
Literature	Hrivnak, J. and Stota, Zdenek, *J. Gas Chromatogr.*, **6**, 9 (1968).

TABLE GC 64
PESTICIDES

Column Packing	P1	P2
Temperature (°C)	210	210
Gas; Flow Rate (ml/min)	N_2,120	N_2,120
Column		
length, ft	6	6
diameter, mm, I.D.	4.5	4.5
form	na	na
material	Al	Al
Detector	D1	D1

Compound	r	μg[a]	r	μg[a]
Carbophenothion, sulfoxide	0.13, 1.52, 1.61, 2.10, *3.90*[b]	16	—	—
Carbophenothion, thiol	0.13, *1.61*	4	—	—
Ethion	0.16, *1.83*	1	1.18	3.5
Thiono demeton	0.27	1	—	—
Demeton	0.27, *0.37*	0.75	0.16, 0.27	1
Dioxathion	*0.41*, 0.67	4	0.13, *0.26*, 0.44	5
Diazinon®	0.44	1	0.23	3
Dimethoate	0.51	0.75	0.38	1
Methyl parathion	0.60	2	0.55	2
Ronnel	0.65	2.5	0.39	5
Malathion	0.71	0.75	0.55	0.75
Carbophenothion, thiolsulfone	0.72, 1.61, 2.77	60	—	—
Parathion	0.78	1.5	0.71	1.5
Chlorthion®	0.81	1	—	—
Captan	0.95	6	0.80	6
Sulphenone	1.00[c]	2.5[c]	1.00	2.5
Ovex	1.23	3.5	1.03	3.5
Endosulfan	*1.23*, 1.64	10	*0.71*, 1.09	10
Aramite®	1.48, 1.58	10	0.96, 1.02	20.0
Carbophenothion, thiolsulfoxide	1.53, 1.62, 2.03	60	—	—
Carbophenothion	2.10	3	1.27	3
EPN	3.15	17	2.65	17
Tetradifon	3.67	20	3.07	20
Guthion®	3.75	20	3.21	75
Co-Ral	6.6	6	8.7	100
Thimet®	—	—	0.19	1
Imidan®	—	—	2.71	20

[a] Response in μg for ½ F.S.D., 6452.
[b] When more than one peak is present, the major peak(s) appear in italic figures.
[c] Sulphenone reference. Retention time approxmately 5.9 minutes.

Column Packing	P1 = 10 % DC-200 on 80–100 mesh Gas-Chrom Q.
	P2 = Equal portions of 15 % QF-1 on 80–100 mesh Gas-Chrom Q and 10 % DC-200 on 80–100 mesh Gas-Chrom Q.
Detector	D1 = Dohrmann Coulometer Model C-100, sulfur cell, Model T-200-P.
Literature	Pesticide Analytical Manual, Tables 312.52 A, B, Vol. 1, F.D.A.

TABLE GC 65
PESTICIDES

Column Packing	P1	P2	P3	P4
Temperature (°C)	T1	T1	T1	T1
Gas; Flow Rate (ml/min)	160,N_2	160,N_2	160,N_2	160,N_2
Column				
length, cm	240	240	240	240
diameter, mm, O.D.	6	6	6	6
mm, I.D.	4	4	4	4
form	na	na	na	na
material	G	G	G	G
Detector	D1	D1	D1	D1

		$r_{parathion}$							
Compound	PS in molecule	Isoth	Progr	Isoth	Progr	Isoth	Progr	Isoth	Progr
Abate®	P_2S_3	20.9	2.67	29.6	3.06	12.5	2.43	[d]	[d]
Accothion®	PS	0.84	0.93	0.98	1.00	0.86	0.93	0.98	1.00
Accothion O-analog	P	0.69	0.81	0.81	0.91	1.18	1.08	1.01	1.00
Aldicarb	S	0.03	0.07	0.03	0.06	0.03	0.07	0.02	0.06
Aldicarb sulfoxide	S	0.05	0.11	0.06	0.12	0.06	0.14	0.06	0.18
Aldicarb sulfone	S	0.14	0.27	0.08	0.17	0.11	0.24	0.10	0.27
Amidithion	PS_2	0.90	0.95	1.15	1.04	1.03	1.01	1.48	1.15
Apholate	P_3	4.60	1.83	6.94	1.83	2.14	1.41	[d]	[d]
Azinphosethyl	PS_2	4.77	1.85	7.11	1.85	3.74	1.75	4.93	1.72
Azinphosmethyl	PS_2	4.00	1.75	6.39	1.79	3.52	1.70	[d]	[d]
Azodrin®	P	0.37	0.55	0.52	0.73	0.67	0.82	0.84	0.95
Bay 30911	PS	0.47	0.65	0.48	0.67	0.23	0.42	0.30	0.61
Bay 37289	PS	1.16	1.08	0.93	0.96	0.48	0.68	0.56	0.80
Bay 37342	PS_2	1.04	1.01	1.17	1.05	0.53	0.73	0.78	0.92
Bay 77488	PS	1.43	1.19	1.74	1.22	1.90	1.33	[d]	[d]
Bidrin®	P	0.37	0.55	0.48	0.67	0.64	0.81	0.51	0.78
Captan	S	1.16	1.08	1.56	1.18	1.07	1.02	1.59	1.17
Carbophenothion	PS_3	2.28	1.48	2.64	1.41	1.18	1.08	1.84	1.22
Carbophenothion O-analog	PS_2	1.85	1.35	2.23	1.33	1.44	1.18	1.82	1.21
Chipman RP-11783	PS_2	2.10	1.42	3.25	1.49	2.37	1.45	3.18	1.45
Chlorbenside	S	1.34	1.15	1.38	1.13	0.51	0.71	0.89	0.96
2-Chlorovinyl diethyl phosphate	P	0.08	0.18	0.08	0.17	0.08	0.18	0.06	0.18
Chlorthion®	PS	1.02	1.00	1.16	1.05	1.08	1.03	1.25	1.08
Ciba C-2307	P	0.34	0.53	0.42	0.64	0.50	0.69	0.40	0.70
Ciba C-7019	S	0.58	0.74	0.58	0.77	0.36	0.53	0.45	0.73
Ciba C-8874	PS	1.97	1.39	2.05	1.30	0.87	0.93	1.26	1.08
Ciba C-9491	PS	1.57	1.25	1.80	1.23	0.74	0.86	1.20	1.06
Ciba C-9491 O-analog	P	1.32	1.15	1.58	1.18	1.04	1.01	1.30	1.09
Ciba C-14421	S_2	2.14	1.45	1.17	1.05	1.33	1.14	1.31	1.09
Ciodrin®	P	1.27	1.14	1.44	1.16	1.33	1.14	1.23	1.07
Compound 4072	P	1.25	1.13	1.29	1.10	0.94	0.98	0.97	1.00
Coumaphos	PS	6.02	1.97	7.88	1.88	7.66	2.10	6.62	1.81
Coumaphos O-analog	P	5.07	1.90	6.91	1.83	9.85	2.29	7.02	1.86
Dasanit®	PS_2	1.92	1.36	2.77	1.43	2.84	1.56	2.91	1.42
Dasanit O-analog	PS	1.59	1.27	2.39	1.36	3.63	1.72	3.03	1.43
Dazomet	S_2	0.40	0.58	0.69	0.83	0.63	0.80	[d]	[d]
DEF®	PS_3	1.64	1.32	1.54	1.16	0.78	0.89	0.87	0.95
Demeton	PS_2	0.31	0.48	0.29	0.50	0.16	0.31	—	—
		0.43	0.62	0.43	0.67	0.38	0.55	0.31	0.63
Diazinon®	PS	0.57	0.73	0.50	0.71	0.24	0.41	0.26	0.58
Diazoxon	P	0.54	0.69	0.49	0.70	0.42	0.60	0.31	0.63
Dicapthon	PS	1.04	1.01	1.10	1.03	0.95	0.98	1.11	1.03
Dichlorvos	P	0.08	0.17	0.08	0.18	0.08	0.17	0.06	0.21
Difolatan®	S	2.58	1.53	3.92	1.55	2.18	1.42	[d]	[d]
Dimethoate	PS_2	0.43	0.61	0.61	0.78	0.52	0.72	0.88	0.96
Dioxathion	P_2S_4	0.14	0.23	0.12	0.23	0.04	0.16	0.06	0.20
		0.49	0.67	0.57	0.76	0.21	0.51	0.42	0.71
		6.10	2.01	—	—	3.32	1.67	—	—
Disulfoton	PS_3	0.59	0.75	0.55	0.74	0.30	0.47	0.34	0.66
Disulfoton sulfoxide	PS_3	1.38	1.18	1.86	1.25	2.15	1.42	2.43	1.36

TABLE GC 65—(Continued)
PESTICIDES

Column Packing	P1	P2	P3	P4
Temperature (°C)	T1	T1	T1	T1
Gas: Flow Rate (ml/min)	160,N_2	160,N_2	160,N_2	160,N_2
Column				
length, cm	240	240	240	240
diameter, mm, O.D.	6	6	6	6
mm, I.D.	4	4	4	4
form	na	na	na	na
material	G	G	G	G
Detector	D1	D1	D1	D1

		$r_{parathion}$							
Compound	Ps in molecule	Isoth	Progr	Isoth	Progr	Isoth	Progr	Isoth	Progr
Disulfoton sulfone	PS_3	1.38	1.18	1.87	1.25	2.23	1.43	2.43	1.36
Disulfoton O-analog	PS_2	0.46	0.63	0.47	0.66	0.39	0.55	0.33	0.65
Disulfoton O-analog sulfoxide	PS_2	1.06	1.02	[d]	[d]	[d]	[d]	[d]	[d]
Disulfoton O-analog sulfone	PS_2	1.04	1.01	1.48	1.16	2.41	1.46	[d]	[d]
Dition®	PS	10.9	2.34	14.6	2.23	11.6	2.40	11.2	2.16
Dursban®	PS	0.98	1.00	0.94	0.98	0.45	0.65	0.58	0.82
Dursban O-analog	P	0.94	0.97	0.97	1.00	0.88	0.95	0.79	0.93
Dyfonate®	PS_2	0.55	0.72	0.56	0.75	0.28	0.46	0.34	0.66
Dyfonate O-analog	PS	0.42	0.60	0.46	0.65	0.38	0.54	0.32	0.64
Endosulfan I	S	1.52	1.22	1.43	1.14	0.68	0.83	0.87	0.95
Endosulfan II	S	1.85	1.35	2.21	1.33	1.23	1.09	1.93	1.24
Endosulfan sulfate	S	2.30	1.48	3.03	1.45	2.63	1.50	3.20	1.46
EPN	PS	3.43	1.66	4.18	1.59	2.93	1.58	3.21	1.46
Ethion	P_2S_4	2.05	1.41	2.42	1.36	1.27	1.12	1.65	1.19
EXD	S_4	0.46	0.62	0.49	0.69	0.20	0.37	0.25	0.57
Famphur	PS_2	2.20	1.46	3.33	1.50	3.83	1.75	3.60	1.55
Fenson	S	0.95	0.98	1.15	1.04	0.82	0.91	1.00	1.00
Fenthion	PS_2	1.00	1.00	1.08	1.02	0.52	0.72	0.79	0.93
Fenthion sulfoxide	PS_2	1.92	1.36	3.12	1.47	3.00	1.60	3.06	1.44
Fenthion sulfone	PS_2	1.94	1.36	3.14	1.47	3.29	1.66	3.42	1.50
Fenthion O-analog	PS	0.80	0.89	0.95	0.99	0.78	0.88	0.85	0.95
Fenthion O-analog sulfoxide	PS	1.60	1.27	2.71	1.42	3.87	1.76	3.16	1.45
Fenthion O-analog sulfone	PS	1.60	1.27	2.68	1.42	4.26	1.80	3.55	1.54
Folpet	S	1.24	1.12	1.66	1.20	0.99	1.00	1.44	1.14
Gardona®	P	1.49	1.21	1.62	1.19	1.12	1.05	1.29	1.09
Geigy G-28029	PS_3	3.64	1.69	4.11	1.58	1.76	1.27	2.50	1.37
Genite 923®	S	1.47	1.20	1.77	1.23	1.23	1.09	1.49	1.15
Hempa	P	0.10	0.23	0.11	0.22	0.16	0.31	0.09	0.25
Imidan®	PS_2	3.30	1.64	4.97	1.68	3.01	1.60	4.17	1.60
Imidoxon	PS	2.44	1.51	4.19	1.59	3.54	1.70	[d]	[d]
Lethane A-70®	S_2	0.36	0.55	0.59	0.77	0.75	0.87	1.06	1.02
Lethane 384®	S	0.27	0.40	0.20	0.41	0.14	0.30	0.17	0.41
Malathion	PS_2	0.94	0.98	0.93	0.97	0.75	0.87	0.78	0.92
Malaoxon	PS	0.74	0.85	0.77	0.88	0.95	0.99	0.78	0.92
Menaxon	PS_2	3.28	1.63	5.84	1.75	2.05	1.39	[d]	[d]
Merphos	PS_3	1.25	1.13	0.93	0.97	0.33	0.51	0.36	0.68
		1.64	1.29	1.54	1.17	0.77	0.88	0.86	0.95
Mesurol®, phenol	S	0.28	0.42	0.24	0.44	0.09	0.21	0.26	0.58
Mesurol sulfoxide	S	0.73	0.85	1.02	1.00	0.92	0.96	[d]	[d]
Mesurol sulfone	S	0.80	0.90	1.14	1.04	1.07	1.02	[d]	[d]
Metepa	P	0.27	0.41	0.24	0.44	0.27	0.44	0.24	0.54
Methiotepa	PS	0.28	0.43	0.23	0.43	0.13	0.28	0.16	0.39
Methyl parathion	PS	0.74	0.85	0.84	0.93	0.79	0.90	0.92	0.97
Methyl trithion®	PS_3	1.88	1.36	2.40	1.36	1.00	1.00	1.80	1.21
Mevinphos	P	0.15	0.29	0.17	0.34	0.18	0.34	0.15	0.38
Mobam®, phenol	S	0.26	0.38	0.26	0.45	0.11	0.23	0.34	0.66
Molinate	S	0.27	0.42	0.23	0.43	0.10	0.22	0.13	0.35
Morestan®	S_2	1.40	1.18	1.60	1.18	0.61	0.78	1.05	1.01
Naled	P	0.35	0.55	0.39	0.61	0.25	0.43	0.25	0.57
Nemacide®	PS	0.72	0.84	0.65	0.80	0.38	0.54	0.37	0.69
Ovex	S	1.50	1.21	1.82	1.24	1.34	1.15	1.59	1.17

TABLE GC 65—(Continued)
PESTICIDES

Column Packing		P1		P2		P3		P4	
Temperature (°C)		T1		T1		T1		T1	
Gas: Flow Rate (ml/min)		160,N_2		160,N_2		160,N_2		160,N_2	
Column									
length, cm		240		240		240		240	
diameter, mm, O.D.		6		6		6		6	
mm, I.D.		4		4		4		4	
form		na		na		na		na	
material		G		G		G		G	
Detector		D1		D1		D1		D1	
		$r_{parathion}$							
Compound	Ps in molecule	Isoth	Progr	Isoth	Progr	Isoth	Progr	Isoth	Progr
Oxydemetonmethyl sulfone	PS_2	0.77	0.88	1.23	1.08	2.00	1.38	[d]	[d]
Parathion	PS	1.00	1.00	1.00	1.00	1.00	1.00	1.00	1.00
Paraoxon	P	0.81	0.90	0.87	0.95	1.31	1.14	1.02	1.00
Pebulate	S	0.24	0.36	0.16	0.28	0.08	0.17	0.07	0.21
Phenothiazine	S	1.18	1.08	1.53	1.17	0.70	0.84	1.74	1.20
Phorate	PS_3	0.41	0.60	0.36	0.60	0.19	0.35	0.24	0.53
Phorate sulfoxide	PS_3	0.91	0.96	1.17	1.05	1.15	1.05	1.52	1.16
Phorate sulfone	PS_3	0.93	0.97	1.17	1.05	1.33	1.14	1.52	1.16
Phorate O-analog	PS_2	0.33	0.50	0.31	0.54	0.25	0.43	0.23	0.51
Phorate O-analog sulfoxide	PS_2	0.71	0.83	0.93	0.97	1.43	1.17	1.41	1.14
Phorate O-analog sulfone	PS_2	0.71	0.83	0.93	0.97	1.46	1.18	1.42	1.14
Phosalone	PS_2	4.07	1.77	4.94	1.68	3.64	1.72	3.95	1.58
Phosfon®	P	0.58	0.75	0.37	0.60	0.64	0.81	0.32	0.64
Phosphamidon	P	0.75	0.85	0.78	0.89	1.26	1.12	0.92	0.97
Pirazinon®	PS	0.32	0.50	0.32	0.56	0.35	0.52	0.25	0.57
Plantvax®	S	2.64	1.55	4.71	1.65	[d]	[d]	[d]	[d]
Potasan®	PS	3.94	1.73	5.23	1.70	5.58	1.98	4.79	1.70
Prometryne	S	0.80	0.89	0.78	0.90	0.40	0.57	0.53	0.79
Propyl thiophrophosphate	P_2S_2	0.98	1.00	0.79	0.90	0.48	0.68	0.49	0.77
Ronnel	PS	0.83	0.93	0.77	0.88	0.42	0.60	0.48	0.76
Ruelene®	P	1.05	1.02	1.14	1.04	1.01	1.00	1.09	1.03
Schradan®	P_2	0.56	0.70	0.53	0.73	0.16	0.31	0.57	0.81
Shell SD-8280	P	0.98	1.00	1.14	1.04	0.78	0.89	0.92	0.97
Shell SD-8436	P	1.57	1.24	2.00	1.29	1.21	1.09	1.62	1.18
Shell SD-8448	P	1.70	1.33	1.85	1.25	1.31	1.14	1.33	1.11
Stauffer N-2788	PS_2	0.76	0.86	0.74	0.86	0.40	0.57	0.46	0.75
Sulfoxide	S	0.14	0.24	0.07	0.13	0.05	0.12	0.02	0.06
Sulphenone®	S	1.23	1.11	1.60	1.19	1.38	1.16	1.69	1.19
Tepa	P	0.21	0.33	0.27	0.46	0.24	0.40	0.26	0.58
Tepp	P_2	0.05	0.12	0.05	0.12	0.06	0.14	0.04	0.12
Tetrachlorothiophene	S	0.09	0.21	0.08	0.17	0.04	0.09	0.04	0.12
Tetradifon	S	3.90	1.73	4.75	1.66	3.25	1.65	3.38	1.49
Thanite®	S	0.66	0.78	0.60	0.78	0.58	0.76	0.56	0.79
Thiotepa	PS	0.28	0.41	0.28	0.48	0.14	0.29	0.21	0.48
Thiram	S_4	0.30	0.47	0.43	0.64	0.38	0.54	0.43	0.72
Union Carbide UC-8305	PS	0.54	0.68	0.64	0.79	0.50	0.70	0.67	0.87
Velsicol VCS-506	PS	4.27	1.79	4.76	1.66	1.86	1.32	2.79	1.40
Velsicol VCS-506 O-analog	P	3.38	1.66	4.18	1.59	2.10	1.40	2.75	1.39
Vitavax®	S	1.66	1.33	2.20	1.33	1.24	1.10	2.16	1.31
Zinophos®	PS	0.30	0.48	0.31	0.54	0.17	0.33	0.22	0.51
Zytron®	PS	0.94	0.97	0.86	0.94	0.49	0.68	0.67	0.88
Parathion—t_r (min)									
isothermally, 200 °C		4.75		7.25		3.65		10.3	
temp. programmed			7.50		10.3		7.10		10.8

P = phosphorus, S = sulfur, t_r = retention time, isoth = isothermal, progr = temperature programmed. Gas holdup (time for nonadsorbed substance to pass from injection port to detector) was not subtracted in determining relative retention times. Gas holdup was 0.30 min. in temperature-programmed analyses. [d] = no peak appeared.

TABLE GC 65—(Continued)
PESTICIDES

All columns conditioned at 320°C overnight.

Column Packing P1 = 5% (w/w) OV-101 (dimethyl silicone) on Gas Chrom Q (80/100 mesh).
P2 = 5% (w/w) OV-17 (phenyl methyl silicone-50% phenyl) on Gas Chrom Q (80/100 mesh).
P3 = 5% (w/w) OV-210 (trifluoropropyl silicone) on Gas Chrom Q (80/100 mesh).
P4 = 5% (w/w) OV-225 (cyanopropyl phenyl silicone) on Gas Chrom Q (80/100 mesh).

Temperature Programmed 150°C at beginning; 10°/min increase for 15 min. held at 300 °C; isothermal 150°–300°C to obtain best ratio.

Detector D1 = flame photometric is phosphorus or sulfur made.

Literature Bowman, M. C. and M. Beroza, *J. Ass. Offic. Agr. Chem.*, **53**, 499 (1970).

TABLE GC 66
PESTICIDES, TRIAZINES

Column Packing	P1	P1	P1
Temperature (°C)	T1	T2	T3
Gas; Flow Rate (ml/min)	He,115	He,115	He,115
Column			
length, m	2.5	2.5	2.5
diameter, mm, I.D.	3.5	3.5	3.5
form	tubing	tubing	tubing
material	Al	Al	Al
Detector	FI	FI	FI
Compound		t_r	
Ametryne®	24.6	16.0	10.87
Atratone	12.9	8.50	5.98
Atrazine	17.5	11.5	7.80
Chlorazine	5.67	4.09	2.91
Ipazine	8.97	6.30	4.41
Prometone	11.3	7.55	5.20
Prometryne®	21.1	13.7	9.37
Propazine	14.9	9.92	6.69
Simazine	20.4	13.2	9.05
Simetone	14.7	9.18	6.77
Simetryne®	28.3	18.45	—
Trietazine	10.6	7.24	5.04

Column Packing P1 = 2.5% Versamid 900, 60/80 mesh Diatoport S.

Temperature Injection 335 °C, T1 = 180 °C, T2 = 190 °C, T3 = 200 °C.

Literature Henkel, H. G. and W. Ebing, *J. Gas Chromatogr.*, **2**, 217 (1964).

TABLE GC 67
PESTICIDES, ORGANOPHOSPHATES

Column Packing	P1	P1	P2	P2	P1	P2
Temperature (°C)	200	220	200	220	180	180
Gas; Flow Rate (ml/min)	120,N_2	60,N_2	120,N_2	60,N_2	120,N_2	60,N_2
Column						
length, ft	6	6	6	6	6	6
diameter, mm, I.D.	4	4	4	4	4	4
form	coiled	coiled	coiled	coiled	coiled	coiled
material	glass	glass	glass	glass	glass	glass
Detector	D1	D1	D1	D1	D1	D1

	(a-parathion)				(b-diazinon)	
Compound	a	a	a	a	b	b
Trichlorfon	<0.1	—	<0.1	—	0.11	0.06,0.20
Dichlorvos	<0.1	—	<0.1	—	0.11	0.19
Mevinphos	0.13	0.15	0.16	0.18	0.24	—
Demeton thiono oxygen analog	0.23	0.26	0.22	0.24	0.39	—
Demeton thiono	0.26	0.31	0.19	—	—	—
Zinophos®	0.28	0.29	0.20	0.23	—	—
Phorate oxygen analog	0.29	0.31	0.25	0.27	0.50	—
Dimethoate oxygen analog	0.31	0.34	0.44	0.45	—	—
Naled	0.08,0.34	0.10,0.38	<0.1	—	0.12,0.38	0.20,0.84
Sulfotepp	0.34	0.37	0.25	0.28	0.62	0.81
Bidrin®	0.35	0.35	0.55	0.55	—	—
Azodrin®	0.37	0.37	0.57	0.56	—	—
Phorate	0.38	0.40	0.24	0.27	0.67	—
Dyfonate oxygen analog	0.39	0.41	0.34	0.36	—	—
Demeton thiol	0.41	0.44	0.36	—	—	—
Dimethoate	0.42	0.44	0.49	0.52	0.76	—
Diazinon oxygen analog	0.46	0.46	0.40	0.39	—	—
Dioxathion	0.47	0.52	0.36	0.38	—	—
Schradan	0.51	0.54	1.30	1.21	—	—
Dyfonate®	0.52	0.55	0.35	0.37	—	—
Diazinon®	0.53	0.53	0.31	0.31	1.00	1.00
Methyl paraoxon	0.54	0.56	0.87	0.88	—	—
Phosphamidon	0.54,0.68	0.55,0.59	1.10	1.06	—	—
Disulfoton	0.55	0.57	0.35	0.38	—	—
Ronnel oxygen analog	0.65	0.67	0.60	0.60	—	—
Phorate oxygen analog sulfone	0.67	0.70	—	1.21	—	—
Malathion oxygen analog	0.69	0.70	0.86	0.83	—	—
Methyl parathion	0.70	0.73	0.76	0.78	—	—
Demeton thiono sulfone	0.73	0.73	1.16	1.15	—	—
Parathion oxygen analog	0.76	0.77	1.15	1.13	—	—
Ronnel	0.82	0.84	0.53	0.55	—	—
Sumithion®	0.84	0.85	0.86	0.86	—	—
Malathion	0.90	0.80	0.79	0.77	—	—
Phorate sulfoxide	0.90	0.92	1.08	1.09,1.20	—	—
Phorate sulfone	0.91	0.91	1.24	1.20	—	—
Zytron®	0.94	0.94	0.62	0.64	—	—
Dursban oxygen analog	0.95	0.93	0.92	0.90	—	—
Fenthion	0.97	0.97	0.66	0.66	—	—
Parathion	1.00	1.00	1.00	1.00	—	—
Dursban®	1.02	1.00	0.64	0.63	—	—
Amino parathion	1.07	1.04	0.80	0.78	—	—
Disulfoton oxygen analog sulfone or demeton thiol sulfone	1.08	1.05	2.19	2.06	—	—
Ruelene®	1.10	1.06	1.05	1.01	—	—
Ciodrin®	1.36	1.31	1.38	1.25	—	—
Disulfoton sulfone	1.42	1.42	2.19	2.06	—	—
Merphos	1.89	1.80	1.14	1.09	—	—
DEF®	1.89	1.80	1.14	1.09	—	—
Dasanit sulfone	1.93	1.05	4.35	3.53	—	—
Dasanit oxygen analog sulfone	2.02	1.87	5.37	4.62	—	—
Carbophenothion oxygen analog	2.17	2.06	1.76	1.64	—	—

TABLE GC 67—(Continued)
PESTICIDES, ORGANOPHOSPHATES

Column Packing	P1	P1	P2	P2	P1	P2
Temperature (°C)	200	220	200	220	180	180
Gas; Flow Rate (ml/min)	120,N_2	60,N_2	120,N_2	60,N_2	120,N_2	60,N_2
Column						
length, ft	6	6	6	6	6	6
diameter, mm, I.D.	4	4	4	4	4	4
form	coiled	coiled	coiled	coiled	coiled	coiled
material	glass	glass	glass	glass	glass	glass
Detector	D1	D1	D1	D1	D1	D1

		(a-parathion)			(b-diazinon)	
Compound	a	a	a	a	b	b
Methyl trithion®	2.22	2.11	1.47	1.37	—	—
Dasanit®	2.43	2.19	3.27	2.88	—	—
Dasanit oxygen analog	2.53	2.32	4.29	3.72	—	—
Ethion	2.58	2.34	1.73	1.59	—	—
Imidan oxygen analog	2.91	2.69	4.28	3.72	—	—
Carbophenothion	2.94	2.74	1.79	1.67	—	—
Cythioate	3.20	2.90	5.49	4.68	—	—
Carbophenothion oxygen analog sulfone	3.80	3.44	—	—	—	—
Imidan®	4.02	3.62	3.96	3.49	—	—
Carbophenothion oxygen analog sulfoxide	4.16	3.69	—	—	—	—
EPN	4.47	3.92	3.93	3.48	—	—
Phenkapton	4.84	4.21	2.84	2.52	—	—
Azinphosmethyl	5.15	4.62	4.91	4.25	—	—
Carbophenothion sulfone	5.15	4.56	5.42	4.75	—	—
Carbophenothion sulfoxide	5.55	4.92	5.31	4.67	—	—
Phosalone	5.60	4.86	5.21	4.45	—	—
SD-7438	6.31	5.44	4.21	3.69	—	—
Hercules 14503	7.48	6.30	5.49	5.21	—	—
Coumaphos	9.48	7.85	12.02	9.90	—	—
Thiono demeton	—	—	—	—	0.50	0.56
Thiol demeton-disulfoton oxygen analog	—	—	—	0.37	0.75	1.16

Column Packing	P1 = 10% DC-200 on Gas Chrom Q (80/100 mesh).
	P2 = 15% QF-1 and 10% DC-200 on Gas Chrom Q (80/100 mesh).
Detector	D1 = Thermionic detector.
Literature	Watts, R. R. and Storherr, R. W., *J. Ass. Offic. Anal. Chem.*, **52**, 513 (1969).

TABLE GC 68
PESTICIDES

Column Packing	P1	P1	P1	P1
Temperature (°C)	170	180	190	200
Gas; Flow Rate (ml/min)	70	70	70	70
Column				
length, ft	6	6	6	6
diameter, in., O.D.	$\frac{1}{4}$	$\frac{1}{4}$	$\frac{1}{4}$	$\frac{1}{4}$
I.D.	$\frac{5}{32}$	$\frac{5}{32}$	$\frac{5}{32}$	$\frac{5}{32}$
form	na	na	na	na
material	glass	glass	glass	glass
Detector	EC	EC	EC	EC

Compound	r_{aldrin}			
Phosdrin®	0.27	0.29	0.30	0.32
2,4-D(ME)	0.39	0.40	0.42	0.44
Thimet®	0.43	0.44	0.46	0.47
α-BHC	0.42	0.44	0.47	0.49
CDEC	0.44	0.46	0.48	0.50
2,4-D(IPE)	0.54	0.55	0.55	0.55
Simazine	0.52	0.53	0.53	0.54
Atrazine	0.55	0.56	0.56	0.57
Diazinon®	0.60	0.59	0.59	0.58
Lindane	0.54	0.56	0.58	0.60
2,4,5-T(ME)	0.66	0.65	0.65	0.64
β-BHC	0.57	0.58	0.60	0.61
2,4-D(BE)I	0.85	0.82	0.80	0.78
δ-BHC	0.65	0.66	0.68	0.69
Heptachlor	0.80	0.81	0.82	0.83
2,4,5-T(IPE)	0.89	0.87	0.85	0.83
2,4-D(BE)II	0.98	0.94	0.91	0.87
Dichlone	0.82	0.82	0.83	0.83
Dimethoate	0.96	0.93	0.89	0.86
Aldrin (REFERENCE)	1.00	1.00	1.00	1.00
Ronnel	1.01	0.98	0.94	0.91
1-Hydroxychlordene	1.04	1.03	1.03	1.02
Methyl Parathion	1.49	1.44	1.39	1.34
Heptachlor epoxide	1.53	1.50	1.46	1.43
Malathion	1.70	1.61	1.51	1.42
Dacthal®	1.64	1.57	1.51	1.44
Dyrene	1.53	1.48	1.42	1.36
***o,p'*-DDE**	1.67	1.60	1.53	1.46
Chlorobenside	1.62	1.56	1.51	1.45
Ethyl Parathion	2.09	1.98	1.87	1.76
Thiodan I	1.99	1.93	1.86	1.79
***p,p'*-DDE**	2.16	2.05	1.93	1.82
DDA(ME)	2.27	2.13	1.98	1.84
Captan	2.27	2.16	2.05	1.94
Folpet	2.22	2.12	2.02	1.91
Dieldrin	2.43	2.33	2.22	2.12
Perthane®	2.43	2.27	2.11	1.94
***o,p'*-DDD**	2.34	2.22	2.10	1.98
***o,p'*-DDT**	3.02	2.80	2.60	2.39
Endrin	2.76	2.64	2.53	2.42
Kepone®	2.97	2.83	2.69	2.56
***p,p'*-DDD**	3.22	2.98	2.77	2.55
Thiodan II	3.19	3.04	2.88	2.72
Ethion	4.08	3.76	3.40	3.05
***p,p'*-DDT**	4.04	3.73	3.43	3.12
Trithion®	4.08	3.78	3.47	3.16
Dilan I	6.2	5.60	5.02	4.43
Mirex	6.1	5.68	5.24	4.79
Methoxychlor	6.7	5.98	5.29	4.60
Dilan II	7.4	6.6	5.87	5.11
Tedion®	11.6	10.3	9.0	7.8
Guthion®	12.5	11.1	9.7	8.4

Column Packing P1 = 4% SE-30 and 6% QF-1 on Chromosorb W.H.P. (80/100 mesh).

Literature Personal Communication, EPA Laboratories, Perrine, Florida.

TABLE GC 69
PHENOLS

Column Packing	P1	P2	P3	P2	P4
Temperature (°C)	165	165	165	160	160
Gas; Flow Rate (ml/min)	220	220	30	46.1	63.4
Column					
length, m^1, ft^2	11^1	11^1	11^1	4^2	6^2
diameter, mm^1, in^2, I.D.	7^1	7^1	7^1	$\frac{1}{4}^2$	$\frac{1}{4}^2$
form	—	—	—	na	na
material	glass	glass	glass	na	na
Detector	TC	TC	na	FI	FI
Literature	1	1	2	3	3

Compound			t_r (phenol)		
2,6-Di-*tert*-butylphenol	—	—	—	0.653	11.13
2-Chlorophenol	—	—	—	0.572	0.761
2,3-Dimethylphenol	2.62	3.38	4.2	1.49	2.80
2,4-Dimethylphenol	2.28	2.93	3.6	1.22	2.45
2,5-Dimethylphenol	2.28	2.93	3.5	1.18	2.45
2,6-Dimethylphenol	2.01	2.77	2.9	0.830	1.96
3,4-Dimethylphenol	2.77	3.56	4.5	1.96	3.33
3,5-Dimethylphenol	2.44	3.12	3.8	1.63	2.84
2-Ethyl-4-methylphenol	3.46	—	—	—	—
2-Ethyl-5-methylphenol	3.25	—	—	—	—
3-Ethyl-5-methylphenol	3.66	4.85	—	—	—
4-Ethyl-2-methylphenol	3.46	4.85	—	—	—
4-Ethyl-3-methylphenol	4.00	—	—	—	—
5-Ethyl-2-methylphenol	3.46	—	—	—	—
2-Ethylphenol	2.14	—	—	—	—
3-Ethylphenol	2.44	—	3.5	—	—
4-Ethylphenol	2.44	0.37	—	—	—
2-Isopropylphenol	2.77	3.38	—	—	—
4-Isopropylphenol	3.22	—	5.2	—	—
3-Methoxyphenol	—	—	—	4.29	4.35
4-Methoxyphenol	—	—	—	3.89	3.97
2-Methylphenol	1.46	1.68	1.15	1.00	1.43
3-Methylphenol	1.57	1.77	1.9	1.29	1.69
4-Methylphenol	1.57	1.77	2.14	1.29	1.69
Phenol	1.00	1.00	1.0	1.00	1.00
2,3,5-Trimethylphenol	4.00	—	—	1.81	4.68
2,4,5-Trimethylphenol	4.00	—	—	1.81	4.68
2,4,6-Trimethylphenol	3.10	4.85	—	1.00	3.60
3,4,5-Trimethylphenol	—	—	—	3.00	6.52

Column Packing P1 = 30% silicone oil on 350–510μ acid washed C-22 firebrick.
P2 = 30% Cyprezon L on 350–510μ acid washed C-22 firebrick.
P3 = 30% Apiezon L on Celite 535.
P4 = 25 wt. % polyethylene glycol succinate (DEGS) on 60–80 mesh acid-washed chromasorb.

Literature
1. Kreyenbuhl, A. and H. Weiss, *Bull. Soc. Chim.* (*France*), **1880** (1959).
2. Irvine, L. and T. J. Mitchell, *J. Appl. Chem.*, **8**, 425 (1958).
3. Crouse, R. H., Garner, J. W., and H. J. O'Neill, *J. Gas Chromatogr.*, **1**, 18 (1963).

TABLE GC 70
PHENOLS

Column Packing	P1	P2	P3	P4	P5	P6
Temperature (°C)	200	200	200	200	200	180
Gas; Flow Rate	N_2	N_2	N_2	N_2	N_2	N_2
Column						
length, m	1	1	1	1	1	1
diameter, mm, I.D.	2.2	2.2	2.2	2.2	2.2	2.2
form	na	na	na	na	na	na
material	glass	glass	glass	glass	glass	glass
Detector	FID	FID	FID	FID	FID	FID
Compound	$r_{2,4\text{-dinitrophenol}}$			$r_{3,5\text{-dinitroanisole}}$		
2-Nitrophenol	0.09	0.10	0.09	—	—	—
2-Nitroanisole	0.27	0.28	0.27	—	—	—
3-Nitrophenol	2.91	2.56	3.04	—	—	—
3-Nitroanisole	0.18	0.21	0.17	—	—	—
4-Nitrophenol	4.82	4.02	5.00	—	—	—
4-Nitroanisole	0.32	0.34	0.31	—	—	—
2,3-Dinitrophenol	2.53	2.38	2.90	—	—	—
2,3-Dinitroanisole	2.94	2.86	3.00	—	—	—
2,4-Dinitrophenol	1.00	1.00	1.00	—	—	—
2,4-Dinitroanisole	3.40	3.28	3.36	—	—	—
2,5-Dinitrophenol	0.80	0.79	0.78	—	—	—
2,5-Dinitroanisole	1.69	1.70	1.59	—	—	—
2,6-Dinitrophenol	1.53	1.53	1.56	—	—	—
2,6-Dinitroanisole	0.83	0.86	0.80	—	—	—
3,4-Dinitrophenol	>20[a]	>20[a]	>20[a]	—	—	—
3,4-Dinitroanisole	2.91	2.84	2.82	—	—	—
3,5-Dinitrophenol	>20[a]	>20[a]	>20[a]	—	—	—
3,5-Dinitroanisole	1.27	1.30	1.15	—	—	—
2-Nitroanisole	—	—	—	0.25	0.26	0.36
3-Nitroanisole	—	—	—	0.18	0.20	0.33
4-Nitroanisole	—	—	—	0.28	0.31	0.43
2,3-Dinitroanisole	—	—	—	2.59	2.07	1.80
2,4-Dinitroanisole	—	—	—	2.48	2.03	1.76
2,5-Dinitroanisole	—	—	—	1.21	1.04	1.05
2,6-Dinitroanisole	—	—	—	0.66	0.61	0.62
3,4-Dinitroanisole	—	—	—	2.34	1.89	1.67
3,5-Dinitroanisole	—	—	—	1.00	1.00	1.00

[a] Not eluted in 45 min.

Column Packing P1 = 5% PEGA and 1% H_3PO_4 on Chromosorb W (60/80 mesh).
P2 = 5% BDS and 1% H_3PO_4 on Chromosorb W (60/80 mesh).
P3 = 5% PEGS and 1% H_3PO_4 on Chromosorb W (68/80 mesh).
P4 = 5% XE-60 on Chromosorb W.
P5 = 5% QF-1 on Chromosorb W.
P6 = 5% SE-30 on Chromosorb W.

Literature Hrivnak, J., Barnoky, L., and E. Beska, *J. Chromatogr.*, **44**, 437 (1969).

TABLE GC 71
PHENOLS

Column Packing	P1
Temperature (°C)	215
Gas; Flow Rate	—
Column	
length, cm	140
diameter, mm, I.D.	1.5
form	na
material	glass
Detector	EC

Compound	r^a 1-naphthol derivative
Phenol	16
***o*-Cresol**	17
4-Fluorophenol	19
***m*-Cresol**	20
2,5-Xylenol	20
***p*-Cresol**	21
Thymol	23
2-Isopropoxyphenol	23
3,5-Xylenol	23
4-Ethylphenol	28
2-Methoxyphenol	29
3-Ethyl-5-methylphenol	29
2-Chlorophenol	29
4-Isopropylphenol	31
3,4-Xylenol	31
3-Chlorophenol	33
3,5-Di-*tert*-butylphenol	36
4-Chlorophenol	38
4-*tert*-Butylphenol	38
4-Allylphenol	38
4-*sec*-Butylphenol	40
4-*tert*-Pentylphenol	50
4-Bromophenol	56
2,4-Dichlorophenol	58
Eugenol	62
4-Iodophenol	92
3,5-Dimethyl-4-methyl-thiophenol	94
4-*tert*-Octylphenol	95
1-Naphthol	100 (8.2 min)
2,4,5-Trichlorophenol	102
2-Naphthol	141
4-Cyclohexylphenol	153
2-Nitrophenol	168
3-Nitrophenol	178
4-Nitrophenol	238
4-Benzylphenol	285

[a] Retention data for 2,4-dinitrophenyl derivatives of phenols, relative to 1-naphthol derivative.

Column Packing P1 = 1% G.E. XE-60 and 0.1% Epikote 1001 on acid washed, dimethylchlorosilane coated Chromosorb G (60/80 mesh).

Literature Cohen, I. C., Norcup, J., Ruzicka, J. H. A., and B. B. Wheals, *J. Chromatogr.*, **44**, 251 (1969).

TABLE GC 72
PHENOLS

Column Packing	P1	P2	P3
Temperature (°C)	200	200	200
Gas; Flow Rate (ml/min)	N_2 (20 mm Hg)	— (20 mm Hg)	— (20 mm Hg)
Column			
length, ft	8 or 12	8 or 12	8 or 12
diameter, in., I.D.	$\frac{3}{16}$	$\frac{3}{16}$	$\frac{3}{16}$
form	coil	—	—
material	Cu	Cu	Cu
Detector	FI[a]	FI[a]	FI[a]

Compound		r_{phenol}	
2-Methylphenol	1.5	0.9	0.9
2,6-Dimethylphenol	2.25	0.75	0.7
2-Methylphenol	1.6	1.35	1.3
3-Methylphenol	1.65	1.35	1.3
2-Ethylphenol	2.2	1.1	1.0
2,4-Dimethylphenol	2.4	1.3	1.1
2,5-Dimethylphenol	2.4	1.25	1.15
2-Isopropylphenol	2.7	1.2	—
2,3-Dimethylphenol	2.9	1.5	1.35
2,6-Diethylphenol	3.7	0.95	0.8
3-Ethylphenol	2.5	1.7	1.6
4-Ethylphenol	2.45	1.85	1.5
3,5-Dimethylphenol	2.6	1.8	1.65
2-*n*-Propylphenol	3.1	1.35	1.0
4-Ethyl-2-methylphenol	3.6	1.8	1.2
2,4,6-Trimethylphenol	3.5	1.15	0.85
5-Ethyl-2-methylphenol	3.6	1.7	1.3
2-Ethyl-5-methylphenol	3.2	1.4	1.15
3,4-Dimethylphenol	3.0	2.15	2.0
2-Methyl-6-*n*-propylphenol	5.1	1.05	—
2,3,6-Trimethylphenol	4.1	1.2	1.0
3-Ethyl-2-methylphenol	3.85	1.9	1.55
2-*n*-Propylphenol	3.6	2.5	1.8
3-Isopropylphenol	3.2	2.1	1.6
4-Methyl-2-isopropylphenol	3.85	1.5	0.9
4-Ethyl-3-methylphenol	4.2	2.9	2.2
4-Isopropylphenol	3.2	2.2	1.5
2,4,5-Trimethylphenol	4.5	2.0	1.6
4-*n*-Propylphenol	3.8	2.6	1.55
3-Ethyl-5-methylphenol	3.8	2.3	1.85
2,3,5-Trimethylphenol	4.35	1.9	1.7
5-Methyl-2-isopropylphenol	4.2	1.6	1.1
2,3,4-Trimethylphenol	5.2	2.4	2.2
2-Methyl-5-isopropylphenol	4.55	2.0	1.35
4-*tert*-Butylphenol	4.3	2.75	1.8
4-Indanol	6.0	3.4	3.2
2,3,5,6-Tetramethylphenol	7.3	1.9	1.6
3,5-Diethylphenol	5.2	3.35	2.15
3,4,5-Trimethylphenol	5.5	3.3	3.35
2,6-Di-*n*-propylphenol	7.8	1.4	—
2.6-Diisopropylphenol	5.7	1.0	0.55
5-Indanol	6.5	4.3	4.3
4-*tert*-Amylphenol	6.9	4.2	2.4
2,3,4,5-Tetramethylphenol	8.9	3.55	3.5
2,6-Di-*tert*-butyl-4-methylphenol	10.5	1.1	0.3
Pentamethylphenol	16.0	3.8	3.4
2-Phenylphenol	13.0	4.45	6.2

[a] Alternate: density balance.

TABLE GC 72—(Continued)
PHENOLS

Column Packing P1 = Apiezon L.
P2 = sodium dodecylbenzenesulfonate.
P3 = 4,4′-diaminodiphenylsulfone.

Literature Fitzgerald, J. S., *Aust. J. Appl. Sci.*, **10**, 169 (1959).

TABLE GC 73
PHENOLS

Column Packing	P1	P2
Temperature (°C)	150	150
Gas; Flow Rate (ml/min)	40–45,He	40–45,He
Column		
length, m	2	2
diameter, mm, I.D.	3	3
form	na	na
material	na	na
Detector	TC	TC

Compound	$r_{anisole}$	
Anisole	1.00	1.00
Phenol	1.66	1.06
***o*-Cresol**	2.00	1.50
***p*-Cresol**	2.40	1.74
***m*-Cresol**	2.46	1.76
2,6-Xylenol	2.77	2.44
***o*-Ethylphenol**	2.85	2.31
2,5-Xylenol	3.30	2.52
2,4-Xylenol	3.43	2.64
***m*-Ethylphenol**	3.85	2.68
3,5-Xylenol	3.90	2.68
2,3-Xylenol	3.90	3.10
3,4-Xylenol	4.55	3.20
2-Fluorobutylphenol	5.00	4.00
3-Methyl-5-ethylphenol	5.50	3.80
2,3,5-Trimethylphenol	5.95	4.60
***p-tert*-Butylphenol**	6.10	4.60
3,4,5-Trimethylphenol	8.30	5.65
2,6-Diisopropylphenol	6.80	6.25
2,6-Di-*tert*-butylphenol	8.20	9.40
2,4-Di-*tert*-butylphenol	10.30	9.80
2,4-Difluorobutylphenol	13.40	12.20
2,4,6-Tri-*tert*-butylphenol	16.90	—

Column Packing P1 = PPO (polyphenylene oxide), 20% on Chromosorb (69/80 mesh).
P2 = Apiezon L, 20% on Chromosorb (69/80 mesh).

Literature Zakupra, V. A. and V. S. Dobrov, *Khim. Teknol. Topliva i Masel*, **14**, 55 (1969).

TABLE GC 74
PHENOLS

Column Packing	P1	P2	P3	P4
Temperature (°C)	160	160	169	169
Gas; Flow Rate (ml/min)	60,N_2	60,N_2	75,N_2	75,N_2
Column				
length, m	85	85	85	85
diameter	na	na	na	na
form	na	na	na	na
material	na	na	na	na
Detector	FI	FI	FI	FI

Compound	r_{phenol}			
***o*-Chlorophenol**	0.68	0.61	0.63	0.57
Phenol	1.00	1.00	1.00	1.00
2,6-Dichlorophenol	1.63	1.59	1.13	1.24
2,3-Dichlorophenol	1.96	2.02	1.87	1.92
2,4-Dichlorophenol	1.97	2.04	1.73	1.79
2,5-Dichlorophenol	2.03	2.20	1.81	1.87
3,4-Dichlorophenol	2.85	2.78	1.73	1.86
2,4,6-Trichlorophenol	3.81	3.84	2.30	2.54
***p*-Chlorophenol**	4.30	4.45	4.41	4.45
***m*-Chlorophenol**	4.56	4.65	4.50	4.55
2,4,5-Trichlorophenol	7.03	7.30	5.85	5.90

Column Packing P1 = Ucon 50 B-550 X, 9.1% on Chromosorb W, AW, 1.2% H_3PO_4, 0.2–0.3 minutes.
P2 = Ucon 50 B-550 X, 16.7% on Chromosorb W, AW, 1.2% H_3PO_4, 0.2–0.3 minutes.
P3 = Carbowax 20M, 9.1% on Chromosorb W, AW, 1.2% H_3PO_4, 0.2–0.3 minutes.
P4 = Carbowax 20M, 16.7% on Chromosorb W, AW, 1.2% H_3PO_4, 0.2–0.3 minutes.

Literature Nosal, A., Pasternak, A., and S. Witek, *Chem. Anal.* (*Warsaw*), **14,** 1117 (1969).

TABLE GC 75
PHENOLS

Column Packing	P1	P1	P1	P2	P2
Temperature (°C)	110	130	150	110	150
Gas; Flow Rate (ml/min)	100–233, He	100–233, He	100–233, He	100–233, He	100–233, He
Column					
length, m	1	1	1	2	2
diameter	na	na	na	na	na
form	na	na	na	na	na
material	na	na	na	na	na
Detector	na	na	na	na	na

Compound			$r_{anisole}$		
Anisole	1.00	1.00	1.00	1.00	1.00
Phenol	—	—	3.20	—	—
2-Methylphenol	—	—	4.46	—	—
3-Methylphenol	—	—	5.69	—	—
4-Methylphenol	—	—	5.69	—	—
2-Ethylphenol	—	—	6.87	—	—
3-Ethylphenol	—	—	9.25	—	—
4-Ethylphenol	—	—	9.25	—	—
2,3-Dimethylphenol	—	—	8.51	—	—
2,4-Dimethylphenol	—	—	8.16	—	—
2,5-Dimethylphenol	—	—	7.16	—	—
2,6-Dimethylphenol	—	—	4.86	—	—
3,4-Dimethylphenol	—	—	10.42	—	—
3,5-Dimethylphenol	—	—	9.41	—	—
2-Methylanisole	1.73	1.62	1.54	1.95	1.62
3-Methylanisole	—	1.84	1.73	1.95	1.62
4-Methylanisole	—	1.84	1.73	1.95	1.62
2-Ethylanisole	—	—	2.26	—	2.31
3-Ethylanisole	—	—	2.80	3.17	2.60
4-Ethylanisole	—	—	2.89	—	2.62
2,3-Dimethylanisole	4.35	3.52	3.03	3.67	2.82
2,4-Dimethylanisole	—	—	—	—	—
2,5-Dimethylanisole	3.45	2.90	2.57	3.24	2.37
2,6-Dimethylanisole	2.52	2.16	2.00	2.48	2.02
3,4-Dimethylanisole	4.98	4.05	3.42	4.19	2.98
3,5-Dimethylanisole	4.19	3.51	3.03	3.52	2.56
Phenetole	—	—	1.39	—	—
2-Methylphenetole	—	—	2.05	—	—
3-Methylphenetole	—	—	2.39	—	—
4-Methylphenetole	—	—	2.43	—	—
2-Ethylphenetole	—	—	2.23	—	—
3-Ethylphenetole	—	—	3.89	—	—
4-Ethylphenetole	—	—	4.03	—	—
2,3-Dimethylphenetole	—	—	4.35	—	—
2,4-Dimethylphenetole	—	—	—	—	—
2,5-Dimethylphenetole	—	—	3.38	—	—
2,6-Dimethylphenetole	—	—	2.14	—	—
3,4-Dimethylphenetole	—	—	5.30	—	—
3,5-Dimethylphenetole	—	—	4.32	—	—

Column Packing P1 = *n*-Decyl phthalate on Celite.
P2 = Dow-Corning silicone oil on Celite.

Literature Bergmann, G. and D. Jentzsch, *Z. Anal. Chem.*, **164**, 10 (1958).

TABLE GC 76

POLYOLS AND METHYLGLYCOSIDES—TRIFLUOROACETYL DERIVATIVES

Column Packing	P1	P1	P2	P2	P3	P3
Temperature (°C)	200	150	100	150	100	150
Gas; Flow Rate, (ml/min)	20,He	20,He	20,He	20,He	20,He	20,He
Column						
length, m	2	2	2	2	2	2
diameter mm, I.D.	3	3	3	3	3	3
form	na	na	na	na	na	na
material	steel	steel	steel	steel	steel	steel
Detector	FI	FI	FI	FI	FI	FI
Trifluoroacetyl derivative of:				I		
Glycol	1090	1115	801	756	740	706
Glycerol	1374	1456	954	856	913	837
Meso-erythritol	1550	1670	1002	870	1006	923
DL-Threitol	1554	1672	1010	872	1000	918
D-Ribitol	1678	1824	1047	868	1092	990
D-Xylitol	1715	—	1080	905	1125	1017
D-Galactitol	1848	1994	1130	935	1195	1081
Butanediol-1,3	1246	1286	907	865	860	823
Butanediol-1,4	1350	1400	1014	964	942	906
Octanediol-1,8	1751	1807	1398	1326	1336	1292
Pentaerythritol	1880	1967	1123	1000	1115	1044
Cyclopentanol	1024	1026	844	830	785	770
Cyclohexanol	1131	1128	942	935	881	876
Cyclododecanol	1745	1750	1560	1535	1474	1468
Cyclohexanediol-1,2 ***trans***	1424	1468	1112	1086	1049	1030
Cyclohexanediol-1,3 ***cis***	1477	1531	1122	1079	1053	1030
trans	1540	1586	1187	1147	1097	1078
Cyclohexanediol-1,4 ***cis***	1474	1530	1132	1086	1042	1026
trans	1518	1570	1175	1127	1097	1058
Cyclohexanediol-1,2 ***trans***	1527	1576	1233	1182	1140	1121
α-D-Methylarabinopyranoside	1626	1722	1256	1160	1192	1128
β-D-Methylarabinopyranoside	1596	1683	1216	1120	1182	1127
β-D-Methylribopyranoside	1736	1819	1343	1248	1260	1213
α-D-Methylxylopyranoside	1590	1672	1234	1130	1169	1136
α-D-Methylglucopyranoside	1986	2029	1321	1200	1340	1266
β-D-Methylglucopyranoside	2026	2073	1371	1220	1360	1274
α-D-Methylmannopyranoside	1927	2000	1320	1206	1350	1265
α-L-Methylrhamnopyranoside	1515	1600	1126	1042	1163	1109

Column Packing P1 = 10% OV-17 on Chromosorb PAW-DCMS (60/80 mesh).
P2 = 10% FS-1265 on Haloport F (30/60 mesh).
P3 = 5% SE-30 on Chromosorb PAW (60/80 mesh).

Literature Jung, G., Pauschmann, H., Voelter, W., Breitmaier, E., and E. Bayer, *Chromatographia*, **3**, 26 (1970).

TABLE GC 77

PURINES, PYRIMIDINES, (TMS DERIVATIVES)

Column Packing	P1
Temperature (°C)	na
Gas; Flow Rate (ml/min)	80–90,He
Column	
length, ft	6
diameter, in., O.D.	0.25
form	U-shaped
material	Borosilicate glass
Detector	FI

	Retention as Methylene Unit Values for TMS-Nucleosides and Related Compounds
Ribonucleosides	
α-Pseudouridine	23.32
β-Pseudouridine	23.75
Uridine	24.66
MO-Cytidine	24.91
5-Methyluridine	25.02
Inosine	25.96
Adenosine	26.54
Xanthosine	27.05
Guanosine	28.02
Cytidine	28.11
Deoxyribonucleosides	
MO-Deoxycytidine	23.79
Thymidine	24.00
Deoxyinosine	25.44
Deoxyadenosine	25.91
Deoxyguanosine	28.00
Deoxycytidine	No peak
Purines and Pyrimidines	
Uracil	13.30
Cytosine	15.18
5-Methylcytosine	15.44
Orotic acid	17.50
6-Methyl adenine	17.59
Hypoxanthine	17.92
6-Dimethyladenine	17.95
Alloxanthine	18.43
Adenine	18.52
Theophylline	18.74
Xanthine	20.05
7-Methylxanthine	20.19
2-Dimethylguanine	20.79
Guanine	21.15
Uric acid	21.22
Kinetin	22.04
Nucleotides	
Uridine-2′(3′)-monophosphate	28.28
Uridine-5′-monophosphate	28.96
Adenosine-2′-monophosphate	30.06
Adenosine-3′-monophosphate	30.28
Inosine-5′-monophosphate	30.40
Deoxyadenosine-5′-monophosphate	30.92
Adenosine-5′-monophosphate	31.14
Xanthosine-5′-monophosphate	31.34
Guanosine-2′(3′)-monophosphate	31.59
Guanosine-5′-monophosphate	31.79
Deoxyguanosine-5′-monophosphate	32.24
Cytidine-2′(3′)-monophosphate	No peak

Column Packing P1 = 5% GC-grade SE-30 on Chromosorb W(HP) (80/100 mesh).

Literature Butts, W. C. *J. Chromatogr. Sci.*, **8**, 474 (1970).

TABLE GC 78
QUINOLINES AND PYRIDINES

Column Packing	P1	P2	P3	P4	P5
Temperature (°C)	200	200	200	180	180
Gas; Flow Rate (ml/min)	H_2	H_2	H_2	He,225	He,225
Column					
length, cm	80	80	80	190	160
diameter, mm[1], in.[2], I.D.	6–8[1]	6–8[1]	6–8[1]	3/8[2]	3/8[2]
form	na	na	na	tubing	tubing
material	na	na	na	Cu	Cu
Detector	na	na	na	TC	TC
Literature	1	1	1	2	2

Compound	$r_{\text{quinoliac}}$			$r_{\text{benzo[b]thiophene}}$	
Quinoline	1.00	1.00	1.00	1.19	1.93
Isoquinoline	1.18	1.14	1.15	1.32	1.75
2-Methylquinoline	1.03	1.27	1.31	1.62	1.95
4-Methylquinoline	—	—	—	2.17	3.77
5-Methylquinoline	—	—	—	1.86	3.52
6-Methylquinoline	1.36	1.54	1.53	—	—
7-Methylquinoline	1.41	1.68	1.60	1.86	2.97
8-Methylquinoline	0.98	1.39	1.40	1.66	2.14
1-Methylisoquinoline	1.39	1.66	—	1.89	2.50
3-Methylisoquinoline	1.19	1.57	—	—	—
2,4-Dimethylquinoline	1.81	2.37	2.33	2.72	4.75
2,6-Dimethylquinoline	1.50	2.14	2.06	2.44	3.47
2,8-Dimethylquinoline	—	—	—	2.02	2.13
4,6-Dimethylquinoline	—	—	—	3.35	7.33
4,8-Dimethylquinoline	—	—	—	2.77	3.36
2,4,6-Trimethylquinoline	2.63	4.02	3.70	—	—
3-Bromoquinoline	—	—	—	3.31	4.19
2-Chloro-4-methylquinoline	—	—	—	4.45	11.5
Benzo(b)thiophene	—	—	—	1.00[a]	1.00[b]
2,4,6-Trimethylpyridine	0.14	0.23	0.037	—	—
2,3,5-Trimethylpyridine	0.23	0.26	0.050	—	—
2,3,4-Trimethylpyridine	0.30	0.48	—	—	—
2,3,5,6-Tetramethylpyridine	0.38	0.51	—	—	—
2-Methyl-5-ethylpyridine	0.17	0.29	—	—	—
4-Methyl-3-ethylpyridine	0.26	0.46	—	—	—

[a] t_r = 3–4 minutes.
[b] t_r = 5 minutes.

Column Packing
P1 = Reoplex 400 (polyoxyalkylene adipate), 20–30% (w/w) substrate on Celite.
P2 = Apiezon L, 20–30% (w/w) substrate on Celite.
P3 = Silicon E-301, 20–30% (w/w) substrate on Celite.
P4 = 10% DC-710 on Chromosorb G (60/80).
P5 = 5% DC-710 on (4 : 5 w/w) Chromosorb G and W (60/80).

Literature
1. Janak, J. and M. Hrivnac, *J. Chromatogr.*, **3**, 297 (1960).
2. Klemm, L. H., Shabtai, J., and F. H. W. Lee, *J. Chromatogr.*, **51**, 433 (1970).

TABLE GC 79
STEROIDS

Column Packing	P1
Temperature (°C)	211
Gas; Flow Rate	na
Column	
length, ft	6
diameter, mm	4
form	U-tube
material	glass
Detector	—

Compound	$r_{\text{cholestane}}$
Androstan-17-one	0.18
Androstane-3,17-dione	0.39
4-Androstene-3,17-dione	0.48
4-Androsten-17β-ol-3-one	0.49
Androstane-3,16-dione	0.40
4-Androstene-3,16-dione	0.48
4-Androstene-3,11,17-trione	0.57
4-Androstene-11α-ol-3,17-dione	0.85
4-Androstene-11β-ol-3,17-dione	0.81
19-Nor-4-androstene-3,17-dione	0.38
Androstane-3β,17β-diol	0.38
Androstane-3α,17β-diol	0.37
Androstan-3β-ol-17-one	0.37
Androstan-3α-ol-17-one	0.36
5-Androstene-3β,17β-diol	0.37
5-Androsten-3β-ol-17-one	0.36
17α-Methyl-5-androstene-3β,17β-diol	0.41
Androstane-3β,16β-diol	0.37
Estrone	0.46
Estradiol	0.48
Equilenin	0.61
Pregnane-3,20-dione	0.60
Pregnan-21-ol-3,20-dione	1.10
5α-Pregnane-3,20-dione	0.66
5α-Pregnane-3,11,20-trione	0.89
5α-Pregnane-3β,20α-diol	0.66
5α-Pregnane-3β,20β-diol	0.62
5α-Pregnan-20β-ol-3-one	0.66
5α-Pregnan-3β-ol-20-one	0.62
5α-Pregnan-3β-ol-11,20-dione	0.86
4-Pregnene-3,20-dione	0.78
Cholestanol	2.00
Epicholestanol	1.98
Coprostanol	1.80
Cholestan-3-one	2.14
4-Cholesten-3-one	2.64
Cholestanyl acetate	3.12
Cholestanyl methyl ether	1.88
Cholestanyl trifluoracetate	1.66
Cholestanyl trimethylsilyl ether	2.61
Tigogenin	2.71
Solanidan-3β-ol	2.24
7-Cholesten-3β-ol	2.21
Desmosterol	2.12
Cholesterol	1.95

Column Packing P1 = 1% SE-30 on Gas Chrom P (100/120 mesh).
Literature Vanden Heuvel, W. J. A. and E. C. Horning, *Biochim. Biophys. Acta*, **64**, 416 (1962).

TABLE GC 80
STEROIDS

Column Packing	P1	P2
Temperature (°C)	200	200
Gas; Flow Rate (ml/min)	30,Ar	30,Ar
Column		
length, cm	134	134
diameter, mm, I.D.	3.5–4.0	3.5–4.0
form	tubing	tubing
material	glass	glass
Detector	EC	EC

Compound	$r_{cholestane}$	
Androstanes		
5α-Androstane	0.078	—
5α-Androstan-3-one	0.175	0.805
5α-Androstan-17-one	0.161	0.622
5α-Androstan-3β-ol	0.163	0.427
5α-Androstan-17α-ol	0.163	0.368
5α-Androstan-17β-ol	0.166	0.388
5α-Androstane-3,17-dione	0.366	4.649
3α-Hydroxy-5α-androstan-17-one	0.336	1.957
3β-Hydroxy-5α-androstan-17-one	0.338	2.249
17β-Hydroxy-5α-androstan-3-one	0.386	2.630
5α-Androstane-3β,17β-diol	0.353	1.344
5α-Androstane-3,11,17-trione	0.475	9.036
5β-Androstane	0.071	—
5β-Androstan-3α-ol	0.147	0.388
5β-Androstane-3,17-dione	0.328	4.306
3α-Hydroxy-5β-androstan-17-one	0.306	2.051
17β-Hydroxy-5β-androstan-3-one	0.344	2.422
5β-Androstane-3α,17β-diol	0.322	1.202
5β-Androstane-3,11,17-trione	0.410	8.043
3α-Hydroxy-5β-androstane-11,17-dione	0.392	3.984
5α-Androst-2-en-17-one	0.155	0.631
Androst-4-en-3-one	0.221	1.287
Androst-4-ene-3,17-dione	0.452	7.247
17α-Hydroxyandrost-4-en-3-one	0.478	3.893
17β-Hydroxyandrost-4-en-3-one	0.483	4.144
Androst-4-ene-3,6,17-trione	0.642	—
Androst-4-ene-3-11,17-trione	0.555	12.64
11β-Hydroxyandrost-4-ene-3,17-dione	0.822	—
Androst-5-en-3β-ol	0.163	0.396
3β-Hydroxyandrost-5-en-17-one	0.332	1.990
Androst-5-en-17β-ol	0.167	0.374
Androst-5-ene-3β,17β-diol	0.353	1.242
Pregnanes		
5α-Pregnane	0.148	0.206
5α-Pregnan-3-one	0.335	1.381
5α-Pregnan-11-one	0.201	0.536
5α-Pregnan-20-one	0.278	0.861
5α-Pregnan-3α-ol	0.304	0.647
5α-Pregnan-3β-ol	0.311	0.733
5α-Pregnan-20α-ol	0.305	0.642
5α-Pregnan-20β-ol	0.282	0.566
5α-Pregnane-3,20-dione	0.636	6.295
3β-Hydroxy-5α-pregnan-20-one	0.593	3.143
20β-Hydroxy-5α-pregnan-3-one	0.644	3.837
5α-Pregnane-3β,20α-diol	0.652	2.268
5α-Pregnane-3β,20β-diol	0.598	2.010
5α-Pregnane-3,11,20-trione	0.884	15.37
3β-Hydroxy-5α-pregnane-11,20-dione	0.856	8.396
5β-Pregnane	0.134	0.181
5β-Pregnan-20-one	0.251	0.779
5β-Pregnan-3α-ol	0.283	0.653
5β-Pregnan-3β-ol	0.279	0.574

TABLE GC 80—(Continued)
STEROIDS

Column Packing	P1	P2
Temperature (°C)	200	200
Gas; Flow Rate (ml/min)	30,Ar	30,Ar
Column		
length, cm	134	134
diameter, mm, I.D.	3.5–4.0	3.5–4.0
form	tubing	tubing
material	glass	glass
Detector	EC	EC

Compound	$r_{cholestane}$	
Pregnanes		
5β-Pregnane-3,20-dione	0.568	5.754
3α-Hydroxy-5β-pregnan-20-one	0.531	2.821
3β-Hydroxy-5β-pregnan-20-one	0.523	2.486
5β-Pregnane-3α,20α-diol	0.589	2.065
5β-Pregnane-3α,20β-diol	0.544	1.806
5β-Pregnane-3,11,20-trione	0.769	13.38
5α-Pregnan-2-en-20-one	0.272	0.876
5α-Pregn-2-ene-11,20-dione	0.396	2.330
11α-Hydroxy-5α-pregn-2-en-20-one	0.484	2.166
16α-Methyl-5α-pregn-2-en-20-one	0.275	0.832
16α-Methyl-5α-pregn-2-ene-11,20-dione	0.387	2.132
Pregn-4-en-3-one	0.420	2.191
Pregn-4-ene-3,20-dione	0.791	9.936
Pregn-4-ene-3,6,20-trione	1.173	—
11α-Hydroxypregn-4-ene-3,20-dione	1.510	—
11β-Hydroxypregn-4-ene-3,20-dione	1.469	—
3β-Hydroxypregn-5-en-20-one	0.585	2.831
Cholanes		
5α-Cholane	0.403	0.475
5α-Cholan-12-one	0.721	1.857
5α-Cholan-12α-ol	0.706	1.096
5β-Cholane	0.365	0.431
5β-Cholan-12-one	0.648	1.581
5β-Cholan-12α-ol	0.618	0.918
Cholestanes		
5α-Cholestane	1.000	1.000
5α-Cholestan-3-one	2.291	6.953
5α-Cholestan-6-one	2.049	—
5α-Cholestan-3α-ol	2.120	3.237
5α-Cholestan-3β-ol	2.135	3.662
5α-Cholestane-3,6-dione	4.223	—
5β-Cholestane	0.903	0.897
5β-Cholestan-3-one	2.055	6.277
5β-Cholestan-3α-ol	1.918	3.240
5β-Cholestan-3β-ol	1.892	2.829
5α-Cholest-2-ene	—	1.006
Cholest-4-en-3-one	2.891	11.25
Cholest-4-ene-3,6-dione	4.414	—
Cholest-5-ene	0.999	0.967
Cholest-5-en-3-one	—	11.18
Cholest-5-en-3β-ol	2.120	3.327

Column Packing P1 = 1% (w/v) SE-30 silicone polymer on acid-washed, treated with 5% (v/v) dichlorodimethylsilane in toluene, Gas Chrom P (80/100 mesh).

P2 = 1.1% QF-1 fluoroalkylsilicone polymer on acid-washed, treated with 5% (v/v) dichlorodimethylsilane in toluene, Gas Chrom P (80/100 mesh).

Literature Brooks, C. J. W. and L. Hanaineh, *J. Biochem.*, **87**, 151 (1963).

TABLE GC 81
STEROIDS

Column Packing	P1	P2	P3	P4	P5
Temperature (°C)	250	250	250	250	250
Gas; Flow Rate (ml/min)	70,Ar	70,Ar	70,Ar	70,Ar	70,Ar
Column					
length, ft	6	6	6	6	6
diameter, mm	4	4	4	4	4
form	coiled	coiled	coiled	coiled	coiled
material	glass	glass	glass	glass	glass
Detector	na	na	na	na	na

Compound			$r_{estrone} \times 100$		
5α-Androstane	9	13	18	23	28
5α-Androstane-3α-ol	22	28	34	42	50
5α-Androstane-3β-ol	24	30	36	42	50
5α-Androstane-17β-ol	22	29	36	42	50
5α-Androstane-3α-ol,17-one (Androsterone)	82	84	84	85	87
5α-Androstane-3,17-dione	178	156	136	113	94
5α-Androstane-3α-ol,11,17-dione	144	132	124	115	105
5α-Androstane-3β-ol,3,16-dione	249	218	189	157	127
5α-Androstane-3α,11β-diol,17-one	145	142	140	138	135
5β-Androstane	8	12	16	21	26
5β-Androstane-3α-ol	23	28	33	39	45
5β-Androstane-17β-ol,3-one	102	97	94	91	89
5β-Androstane-3α-ol,17-one	88	85	83	80	80
5β-Androstane-3α-ol,11β-ol,17-one	150	144	137	129	122
5β-Androstane-3β-ol,17-one	79	79	79	78	79
5β-Androstane-3,17-dione	167	146	126	105	86
5β-Androstane-3α-ol,11,17-dione	147	134	122	107	96
4-Androstene-3-one,17α-ol	152	143	133	122	113
4-Androstene-3-one,17β-ol (Testosterone)	159	147	136	124	115
4-Androstene-3,17-dione	253	217	182	143	109
4-Androstene-14α-ol,3,17-dione	483	408	333	250	171
4-Androstene-3,11,17-trione	385	320	258	188	128
5-Androstene-3β-ol,17-one (DHA)	85	84	84	84	85
5-Androstene-3β,17β-diol	55	63	71	80	88
5-Androstene-3β,17α-diol	54	62	71	79	88
5-Androstene-3β,17β-diol,16-one	117	116	115	114	114
5-Androstene-3β-ol,16-one	95	93	89	87	86
5-Androstene-3β,16α-diol,17-one	116	116	115	114	115
Pregnanes					
5α-Pregnane	13	21	29	38	46
5α-Pregnane-3β-ol	35	46	58	70	81
5α-Pregnane-3β,20β-diol	80	93	107	121	135
5α-Pregnane-3α,20α-diol	80	93	109	126	141
5α-Pregnane-3β-ol,20-one	106	112	119	125	131
5β-Pregnane	11	19	26	34	42
5β-Pregnane-3α-ol	33	42	52	63	74
5β-Pregnane-3-one	60	63	69	74	80
5β-Pregnane-3β-ol,20-one	98	101	111	115	121
5β-Pregnane-3,20-dione	202	183	168	148	130
5β-Pregnane-3α,20β-diol	75	86	99	111	124
5β-Pregnane-3α-ol,20-one	108	111	115	118	121
5β-Pregnane-3α,20α-diol	81	93	107	118	130
5β-Pregnane-3α,17α,20α-triol	137	152	166	180	191
4-Pregnene-3,20-dione (Progesterone)	315	281	245	207	168
6-Dehydro-progesterone	309	276	244	208	175
4-Pregnene-3,20-dione,17α-ol	452	397	344	286	230
5-Pregnene-3β-ol,20-one	106	112	119	125	131
5-Cholestene-3β-ol (Cholesterol)	309	275	244	208	174

TABLE GC 81—(Continued)
STEROIDS

Column Packing	P1	P2	P3	P4	P5
Temperature (°C)	250	250	250	250	250
Gas: Flow Rate (ml/min)	70,Ar	70,Ar	70,Ar	70,Ar	70,Ar
Column					
length, ft	6	6	6	6	6
diameter, mm	4	4	4	4	4
form	coiled	coiled	coiled	coiled	coiled
material	glass	glass	glass	glass	glass
Detector	na	na	na	na	na

Compound	$r_{estrone} \times 100$				
1,3,5(10)-Estratrienes					
-3-ol,17-one (Estrone)	100	100	100	100	100
-3-ol,16-one	112	109	107	104	106
-3,17β-diol (Estradiol-17β)	67	77	86	96	104
-3,17α-diol	64	74	85	93	103
-3,16α-diol	67	75	84	91	100
-3,16β-diol	66	76	84	93	101
-3,16α,17β-triol (Estriol)	138	146	154	164	171
-3,16α,17α-triol	139	149	158	166	177
-3,16β,17α-triol	135	147	156	169	177
-3,16β,17β-triol	143	153	160	169	176
-2,3,16α,17β-tetrol	280	291	309	316	329
-3-methoxy-16,17-diketo	204	185	169	147	132
-2-methoxy-3-ol,17-one	139	137	137	132	131
-3-methoxy-17-one	74	78	81	85	88
-3-methoxy-17β-ol	51	62	72	84	95
-3,17β-diol,16-one	102	106	111	116	120
-3-ol,16,17-dione	276	247	217	184	152
-2-methoxy-3,16α,17β-triol	192	201	209	219	225
-3-methoxy-16α,17β-diol	100	114	127	140	153
Cholestanes					
5α-Cholestane	43	80	119	158	196
5α-Cholestane-3-one	211	252	297	338	380
5α-Cholestane-3β-ol	116	175	235	294	350
5β-Cholestane	40	73	109	144	177
5β-Cholestane-3-one	195	232	272	309	351
5β-Cholestane-3α-ol	102	157	211	270	327
5β-Cholestane-3β-ol	92	149	207	266	324
5-Cholestene-3β-ol (Cholesterol)	117	176	238	294	358
5-Cholestene-(Cholestene)	41	79	119	157	198

Column Packing
P1 = 5% QF-1 (fluorosilicone polymer, Dow Corning) on Gas Chrom Q (80/100 mesh.)
P2 = 3.75% QF-1 (fluorosilicone polymer) and 1.25% L-45 (methylsilicone, General Electric) on Gas Chrom Q (80/100 mesh).
P3 = 2.5% QF-1 (fluorosilicone polymer) and 2.5% L-45 (methylsilicone) on Gas Chrom Q (80/100 mesh).
P4 = 1.25% QF-1 (fluorosilicone polymer) and 3.75% L-45 (methylsilicone) on Gas Chrom Q (80/100 mesh).
P5 = 5% L-45 (methylsilicone) on Gas Chrom Q (80/100 mesh).

Literature
Touchstone, J. C., Murawec, T., and A. Nikolski, *J. Chromatogr. Sci.*, **8**, 221 (1970).

TABLE GC 82

STEROIDS, ESTROGENS, TMS DERIVATIVES

Column Packing	P1	P2	P3	P4	P5	P6	P7
Temperature (°C)	195	213	184	198	210	195	295
Gas; Flow Rate	—	—	—	—	—	—	—
Column							
length, ft	6–12	6–12	6–12	6–12	6–12	6–12	6–12
diameter, mm, I.D.	4–6	4–6	4–6	4–6	4–6	4–6	4–6
form	—	—	—	—	—	—	—
material	—	—	—	—	—	—	—
Detector	EC or FI	EC or FI	EC or FI	EC or FI	EC or FI	EC or FI	EC or FI

TMS derivatives	$r_{cholestane}$						
3-Hydroxyestra-1,3,5(10)-triene-17-one (Estrone)	1.15	0.62	2.33	1.51	0.52	3.25	—
2-Methoxy-3-hydroxyestra-1,3,5(10)-triene-17-one	2.06	0.91	4.27	2.29	0.79	—	—
2,3-Dihydroxyestra-1,3,5(10)-triene-17-one	1.84	0.97	2.83	2.03	0.89	3.28	1.43
3,6β-Dihydroxyestra-1,3,5(10)-triene-17-one	1.47	0.82	2.46	1.91	0.73	3.14	1.23
3,7α-Dihydroxyestra-1,3,5(10)-triene-17-one	1.39	0.73	1.89	1.81	0.68	2.54	1.14
3,11β-Dihydroxyestra-1,3,5(10)-triene-17-one	—	0.95	1.99	2.81	—	2.70	1.21
3,16α-Dihydroxyestra-1,3,5(10)-triene-17-one	1.74	1.03	2.84	1.75	—	3.76	1.39
3-Hydroxyestra-1,3,5(10)-triene-6,17-dione	3.52	1.18	8.53	5.16	0.93	14.50	3.06
3-Hydroxyestra-1,3,5(10)-triene-16,17-dione	2.24	1.22	3.86	2.32	—	—	—
Estra-1,3,5(10)-triene-3,17β-diol-6-one	2.45	1.33	3.18	2.94	—	—	1.91
Estra-1,3,5(10)-triene-3,17β-diol-11-one	—	1.12	—	1.67	—	—	—
Estra-1,3,5(10)-triene-3,17β-diol-16-one	2.11	1.14	3.08	2.21	1.09	4.68	1.64
Estra-1,3,5(10)-triene-3,17β-diol (Estradiol)	0.83	0.70	1.01	0.81	0.66	—	0.79
Estra-1,3,5(10)-triene-3,17α-diol	0.70	0.63	0.87	0.71	—	—	0.70
Estra-1,3,5(10)11(12)-tetraene-3,17α-diol	0.62	0.53	0.84	—	—	—	—
2-Methoxyestra-1,3,5(10)-triene-3,17β-diol	1.34	1.05	1.78	1.21	—	—	1.22
3-Methoxyestra-1,3,5(10)11(12)-tetraene-3,17α-diol	—	0.45	—	—	—	—	—
3-Methoxyestra-1,3,5(10)-triene-3,17β-diol	—	0.58	—	—	—	—	—
Estra-1,3,5(10)-triene-2,3,17β-triol	1.10	1.05	1.28	1.08	1.13	1.21	1.18
Estra-1,3,5(10)-triene-3,6α,17β-triol	1.24	1.04	1.34	1.19	1.16	1.44	1.19
Estra-1,3,5(10)-triene-3,6β,17β-triol	1.01	0.92	1.15	1.04	1.00	1.12	1.04
Estra-1,3,5(10)-triene-3,7α,17β-triol	—	0.71	0.81	0.89	0.79	—	—
Estra-1,3,5(10)-triene-3,11β,17β-triol	0.87	0.77	2.42	0.91	—	—	0.87
Estra-1,3,5(10)-triene-3,15α,17β-triol	1.21	1.06	1.23	1.24	—	—	1.20
Estra-1,3,5(10)-triene-3,16α-17β-triol (Estriol)	1.41	1.29	1.73	1.43	1.44	1.77	1.43
Estra-1,3,5(10)-triene-3,16β-17β-triol	1.67	1.39	1.78	1.63	1.52	1.96	1.60
Estra-1,3,5(10)-triene-3,16α,17α-triol	1.39	1.23	1.63	1.39	1.29	1.63	1.42
Estra-1,3,5(10)-triene-3,16β,17α-triol	1.18	1.11	1.37	1.20	1.25	1.32	1.27
2-Methoxyestra-1,3,5(10)-triene-3,16α,17β-triol	2.25	1.91	3.06	2.04	—	—	2.23
Estra-1,3,5(10)-triene-2,3,16α,17β-tetrol	1.91	1.94	2.45	1.82	—	2.12	—
Estra-1,3,5(10)-triene-3,6α,16α,17β-tetrol	1.56	1.70	2.18	1.85	—	2.08	—
Estetrol	3.02	2.37	—	2.56	—	5.96	—

Column Packing
P1 = 1% XE-60 on Gas Chrom P (100/140 mesh) or Anakrom AS (90/100 mesh) or Anakrom ABS (160/170 mesh).
P2 = 3% F-60 on Gas Chrom P (100/140 mesh) or Anakrom AS (90/100 mesh) or Anakrom ABS (160/170 mesh).
P3 = 1% Z on Gas Chrom P (100/140 mesh) or Anakrom AS (90/100 mesh) or Anakrom ABS (160/170 mesh).
P4 = 2% QF-1 on Gas Chrom P (100/140 mesh) or Anakrom AS (90/100 mesh) or Anakrom ABS (160/170 mesh).
P5 = 0.8% JXR on Gas Chrom P (100/140 mesh) or Anakrom AS (90/100 mesh) or Anakrom ABS (160/170 mesh).
P6 = 1% NGS on Gas Chrom P (100/140 mesh) or Anakrom AS (90/100 mesh) or Anakrom ABS (160/170 mesh).
P7 = JXR-XE-60 on Gas Chrom P (100/140 mesh) or Anakrom AS (90/100 mesh) or Anakrom ABS (160/170 mesh).

Literature
Adlercreutz, H., Salokangas, A., and T. Luukkainen, "*Gas Liquid Chromatography of Steroids*", J. K. Grant, Ed., Cambridge University Press, p.89, 1967.

TABLE GC 83

STEROIDS, HEPTAFLUOROBUTYRATES

Column Packing	P1	P2	P2
Temperature (°C)	190	240	190
Gas; Flow Rate (ml/min)	50,Ar	50,Ar	50,Ar
Column			
length, ft	5	5	5
diameter	na	na	na
form	na	na	na
material	glass	glass	glass
Detector	FI	FI	EC

Compound	$r_a{}^1$ progesterone Free Compound	Ester	Free Compound	Ester	$r_a{}^2$ HFB ester progesterone Ester of 3-enol
Testosterone	0.63	0.41	0.49	0.18	0.53
Androstenedione	0.52	0.37	0.74	0.25	0.79
Δ^4-Androstene-3:11:17-trione	0.68	0.49	1.21	0.39	1.34
Progesterone	1.00	0.60	1.00	0.31	1.00
Epitestosterone	0.60	0.32	0.46	0.13	0.43
Cholestenone	3.10	2.10	0.99	0.29	1.18
Testosterone acetate	0.85	0.62	0.74	0.23	0.75
20α-Hydroxypregn-4-en-3-one	1.14	0.79	0.75	0.25	0.93
20β-Hydroxypregn-4-en-3-one	1.09	0.65	0.66	0.22	0.77
Epitestosterone acetate	0.78	0.56	0.70	0.24	0.71
Etiocholanolone	0.40	0.32	0.27	0.25	—
Androsterone	0.35	0.26	0.25	0.22	—
Epiandrosterone	0.42	0.44	0.30	0.30	—
Cholesterol	2.29	2.09	0.33	0.32	—
Dehydroepiandrosterone	0.35	0.35	0.25	0.27	—
Pregnenolone	0.62	0.59	0.32	0.32	—
3-Methyl estradiol 17β	0.50	0.44	0.16	0.16	—
Estrone	0.55	0.40	0.33	0.27	—
Estradiol 17β	0.55	0.37	0.23	0.19	—

$_a{}^1$ Relative to progesterone.
$_a{}^2$ Relative to progesterone HFB derivative.

Column Packing P1 = 1% SE-30 (silicone rubber) on siliconized Celite (80/100 mesh).
P2 = 3% QF-1 (trifluoropropylmethyl silicone) on siliconized Celite (80/100 mesh).

Literature Exley D. and J. Chamberlain in "*Gas Chromatographic Determination of Hormonal Steroids*," F. Polvani, M. Surace, and M. Luisi, Editors, Academic Press, N.Y. p. 215, 218 (1967).

TABLE GC 84

STEROIDS, PLANT STEROLS AND DERIVATIVES

Column Packing	P1	P2	P3	P4	P5	P6	P7
Temperature (°C)	225	225	225	225	225	225	200
Gas; Flow Rate	Ar	na	na	na	na	na	na
Column							
length, ft	4	4	4	4	4	4	4
diameter	na	na	na	na	na	na	na
form	na	na	na	na	na	na	na
material	na	na	na	na	na	na	na
Detector	EC	EC	EC	EC	EC	EC	EC

Compound	$r_{cholestane}$						
5α-Cholestan-3β-ol	2.04	2.22	2.84	3.20	6.40	6.34	7.00
Cholesterol	1.93	2.15	3.10	2.90	6.70	6.95	7.50
Δ^7-Cholesten-3β-ol	2.15	2.51	3.64	—	7.32	8.08	8.90
Campesterol	2.56	2.77	—	3.84	8.80	9.35	—
Ergosterol	2.44	2.70	—	—	8.60	10.70	—
β-Sitosterol	3.25	3.60	5.50	4.40	10.70	11.60	12.50
Stigmasterol	2.82	3.10	—	3.90	8.90	9.90	11.50
Fucosterol	3.32	3.76	5.81	4.35	11.80	12.70	14.50
29-*iso*Fucosterol-(Δ^5*Avena* sterol)	3.30	3.75	—	4.40	12.50	13.40	—
α-Spinasterol	3.22	3.82	5.64	4.40	10.20	11.60	13.00
Acetates							
5α-Cholestan-3β-ol	3.04	3.40	4.13	5.11	5.96	5.70	7.40
Cholesterol	2.90	3.17	4.27	4.40	5.90	5.75	7.50
Δ^7-Cholesten-3β-ol	3.36	3.86	5.54	5.50	6.90	6.78	10.00
Brassicasterol	3.16	3.63	—	4.80	6.22	6.52	—
Ergosterol	3.61	4.30	6.00	5.62	7.50	8.67	—
Stigmasterol	4.11	4.75	6.40	5.80	8.25	8.24	11.90
Fucosterol	5.12	5.95	8.73	—	10.50	10.70	16.05
29-*iso*Fucosterol	5.18	6.31	9.28	—	10.75	11.45	17.60
Trimethylsilyl ethers							
Cholesterol	2.45	2.52	—	1.87	2.07	2.60	—
Δ^7-Cholesten-3β-ol	—	3.06	—	2.15	2.48	3.06	—
Brassicasterol	—	2.94	—	2.05	2.12	—	—
Ergosterol	2.96	3.24	—	2.28	3.05	—	—
Stigmasterol	3.56	3.71	—	2.50	2.70	3.82	—
Fucosterol	4.16	4.42	—	3.05	3.80	5.55	4.88
29-*iso*Fucosterol	4.27	4.44	—	3.60	3.93	—	—
Cycloartenol	4.66	5.11	—	3.78	4.30	6.05	5.07
Citrostadienol	—	7.72	—	4.45	6.59	8.90	7.70

Column Packing
P1 = 1% (w/w) SE-30 on acid-washed, siliconized Gas Chrom P.
P2 = 1% (w/w) DC-560 on acid-washed, siliconized Gas Chrom P.
P3 = 1% (w/w) DC-710 on acid-washed, siliconized Gas Chrom P.
P4 = 1% (w/w) FS-1265 on acid-washed, siliconized Gas Chrom P.
P5 = 1% (w/w) Siliconized HiEFF-8B (CHDMS) on acid-washed, siliconized Gas Chrom P.
P6 = 2% (w/w) PVP (polyvinylpyrrolidone) and 1% (w/w) unsiliconized HiEFF-8B on Gas Chrom P.
P7 = 1% (w/w) EGSP-Z on acid-washed, siliconized Gas Chrom P.

Literature
Knights, B. A. in "*Gas Liquid Chromatography of Steroids*", J. K. Grant, Editor, Cambridge University Press (1967).

TABLE GC 85
S COMPOUNDS

Column Packing	P1
Temperature (°C)	140 (110 for lower boilers)
Gas; Flow Rate (ml/min)	50,N_2
Column	
length, ft	10
diameter, in.	$\frac{1}{8}$
form	na
material	na
Detector	FI, EC

Compound	$r_{\text{diallyl disulfide}}$
Mercaptans	
Methyl mercaptan	0.04
Ethyl mercaptan	0.05
Allyl mercaptan	0.07
n-Butyl mercaptan	0.09
n-Hexyl mercaptan	0.22
n-Octyl mercaptan	0.54
Monosulfides	
Dimethyl sulfide	0.06
Diethyl sulfide	0.07
Di-*n*-propyl sulfide	0.15
n-Propyl allyl sulfide	0.17
Di-allyl sulfide	0.18
sec-Butyl allyl sulfide	0.20
Methyl allyl sulfide	0.09 (tentative)
Disulfides (saturated)	
Dimethyl disulfide	0.16
Diethyl disulfide	0.38
Methyl *n*-propyl disulfide	0.40
Di-*n*-propyl disulfide	0.70
Di-*n*-butyl disulfide	1.52
Di-*n*-amyl disulfide	3.22
n-Hexyl *n*-amyl disulfide	4.76
Di-*n*-hexyl disulfide	7.18
Disulfides (unsaturated)	
Methyl allyl disulfide	0.46
Ethyl allyl disulfide	0.66
n-Propyl allyl disulfide	0.84
Di-allyl disulfide	1.00
n-Hexyl allyl disulfide	2.67
Trisulfides	
Di-methyl trisulfide	0.76
Methyl *n*-propyl trisulfide	1.33
Methyl allyl trisulfide	1.63
Di-*n*-propyl trisulfide	2.28
n-Propyl allyl trisulfide	2.83
Di-allyl trisulfide	3.42
Others	
Allyl isothiocyanate	0.57
Isobutyl sulfoxide	3.49
Sulfur dioxide	0.04
Hydrogen sulfide	0.04

Column Packing P1 = 10% Carbowax 20M on Chromosorb W (60/80 mesh).

Literature Oaks, D. M., Hartmann, H., and K. P. Dimick, *Anal. Chem.*, **36**, 1560 (1964).

TABLE GC 86
TERPENES

Column Packing	P1	P2	P3	P4	P5	P6
Temperature (°C)	110	110	100	150	150	110
Gas; Flow Rate	na	na	na	na	na	na
Column						
length, cm	320	320	320	320	320	240
diameter, cm I.D.	0.60	0.60	0.60	0.60	0.60	0.60
form	na	na	na	na	na	na
material	na	na	na	na	na	na

Compound	$r_{limonene}$					
Santene	0.34	0.34	0.33	0.37	0.44	0.43
α-Fenchene	0.48	0.53	0.49	0.58	0.66	0.63
α-Thujene	—	0.42	—	—	—	—
α-Pinene	0.42	0.44	0.39	0.50	0.60	0.61
Camphene	0.56	0.54	0.52	0.59	0.66	0.62
β-Pinene	0.65	0.68	0.64	0.62	0.69	0.66
Myrcene	0.78	0.80	0.82	0.72	0.67	0.75
Δ^3-Carene	0.77	0.79	0.74	0.83	0.88	0.90
Δ^4-Carene	0.63	—	—	—	—	—
Sabinene	0.71	0.69	0.74	0.92	0.88	0.89
α-Phellandrene	0.83	0.86	0.86	0.93	0.90	0.92
α-Terpinene	0.90	0.91	0.96	0.94	0.92	0.91
Limonene	1.00	1.00	1.00	1.00	1.00	1.00
β-Phellandrene	0.98	1.01	1.00	0.99	0.99	1.01
Ocimene	1.23	1.19	—	—	—	—
γ-Terpinene	1.25	1.26	1.28	1.23	1.14	1.18
p-Cymene	1.52	1.45	1.86	1.26	0.99	0.97
Terpinolene	1.49	1.49	1.49	1.41	1.31	1.35
Tricyclene	0.42	0.40	0.38	0.48	0.57	0.55
Bornylene	0.35	—	—	—	—	—
o-Cymene	1.77	1.72	2.25	1.40	0.99	1.03

Column Packing
P1 = 7% polyethylene glycol adipate.
P2 = 5% polyethylene glycol 4000.
P3 = 5% tetra(-*O*-cyanoethyl)methyl methane.
P4 = 7% tri-*o*-cresyl phosphate.
P5 = 7% Apiezon L.
P6 = 3% silicone oil—all on ground unglazed tile.

Literature
Lukes, V. unpublished data as in E. Heftmann, *Chromatography*, 2nd edition, Reinhold Publishing Corp., N.Y., 1967, p.520.

TABLE GC 87
TERPENES

Column Packing	P1	P1	P1	P2
Temperature (°C)	110	130	150	110
Gas; Flow Rate (ml/min)	He,45	He,45	He,45	He,45
Column				
length, ft	6.5	6.5	6.5	6.0
diameter, in., I.D.	$\frac{3}{16}$	$\frac{3}{16}$	$\frac{3}{16}$	$\frac{3}{16}$
form	na	na	na	na
material	SS	SS	SS	SS
Detector	TC	TC	TC	TC

Compound		$r_{\alpha\text{-pinene}}$		
Monocyclics				
α-Pyronene	0.99	1.00	0.98	1.02
β-Pyronene	1.25	1.24	—	1.35
3-*p*-Menthene	1.42	1.35	1.34	1.18
trans-*p*-Menthane	1.43	1.35	1.35	0.97
cis-*p*-Menthane	1.50	1.45	1.45	1.10
α-Phellandrene	1.84	1.78	1.75	1.96
β-Phellandrene	2.28	2.24	2.12	2.46
α-Terpinene	2.08	1.89	1.80	2.09
1-*p*-Menthene	1.90	1.83	1.77	1.72
4(8)-*p*-Menthene	2.04	1.90	1.85	1.71
Dipentene	2.16	2.04	1.90	2.29
γ-Terpinene	2.72	2.47	2.46	2.88
3,8-*p*-Menthadiene	3.06	2.79	2.56	3.12
2,4(8)-*p*-Menthadiene	3.66	3.26	2.92	—
Terpinolene	3.18	2.83	2.75	3.51
Aromatics				
Cumene	1.27	1.26	1.17	2.02
m-Cymene	2.45	2.25	2.12	3.14
p-Cymene	2.46	2.35	2.13	3.15
o-Cymene	2.94	2.38	2.40	—
Bicyclics				
Bornylene	0.80	0.83	0.86	0.86
α-Pinene	1.00	1.00	1.00	1.00
α-Fenchene	1.13	1.14	1.16	1.22
Camphene	1.16	1.20	1.21	1.28
β-Pinene	1.52	1.44	1.41	1.58
Isocamphane	1.45	1.40	1.40	1.26
trans-Pinene	1.31	1.34	1.37	1.19
cis-Pinane	1.39	1.41	1.40	1.25
3-Carene	1.77	1.74	1.66	1.82
Alcohols				
α-Fenchyl alcohol	—	6.10	5.13	—
β-Fenchyl alcohol	—	6.49	5.60	—
trans-Dihydro-α-terpineol	8.21	6.90	5.60	—
β-Terpineol	—	7.47	6.24	—
cis-Dihydro-α-terpineol	—	7.56	6.14	—
Borneol	—	—	7.10	—
Isoborneol	—	—	7.89	—
α-Terpineol	—	—	7.98	—
Ketones and ethers				
1,4-Cineole	2.12	1.94	1.82	—
1,8-Cineole	2.33	2.24	2.08	—
Fenchone	4.92	4.08	3.77	—
Estragole	—	—	7.82	—
trans-Anethole	—	—	14.4	—
Miscellaneous				
Cyclofenchene	0.64	0.67	0.68	0.65
Tricyclene	0.90	0.91	0.94	0.96
Myrcene	1.46	1.44	1.33	1.70

Column Packing P1 = 25% didecyl phthalate on 50–100 mesh Celite 545.
P2 = 28.5% Carbowax 4000 on 35/80 Chromosorb C-44857.

Literature Zubyk, W. J. and A. Z. Conner, *Anal. Chem.*, **32**, 912 (1960).

TABLE GC 88

VITAMINS

Column Packing	P1	P2
Temperatures (°C)	125	230
Gas; Flow Rate (ml/min)	75,He	Ar,45
Column		
length, ft	4	8
diameter, in.	¼	—
form	na	—
material	glass	glass
Detector	FI	EC
Literature	1	2

Compound	Time (min.)	Relative to pyridoxol acetate	Retention times for vitamin B_6 Acetyl derivatives (min.)
Pyridoxol	8.5	1.00	—
Pyridoxamine[a]	19.6	2.31	18.0
Pyridoxal	3.1	0.36	6.0
Pyridoxine	—	—	9.0
Pyridoxal ethyl acetal	2.4	0.27	—
4-Pyridoxic acid lactone	2.2	0.26	—
5-Pyridoxic acid lactone	2.8	0.33	—
4-Deoxypyridoxol	3.0	0.35	—

[a] At 165°C, the retention time was 2.3 minutes.

Column Packing P1 = 1.2% SE-30 (silicone gum rubber) on acid-washed silanized Chromosorb G (90/100 mesh). P2 = 6.67% SE-30 and 0.33% NPGS on 100/120 mesh silanized Gas Chrom P.

Literature

1. Korytnyk, W., Fricke, G., and B. Paul, *Anal. Biochem.*, **17**, 66 (1966).
2. Sheppard, A.J. and A. R. Prosser, in "Methods in Enzymology", Vol. XVIII, Pt. A, S. P. Colowick and N. O. Kaplan, Editors in Chief, Academic Press, N.Y., 1970.

TABLE GC 89
ACIDS, KREBS CYCLE

Column Packing	P1	P2
Temperature (°C)	T1	T1
Gas; Flow Rate (ml/min)	60,He	60,He
Column		
length, ft	6	6
diameter, mm, I.D.	4	4
form	tubing	tubing
material	glass	glass
Detector	FI	FI
Compound	t_r	r*
Methyl pyruvate	1.86	0.169
Methyl lactate	2.25	0.205
Methyl fumarate	7.31	0.664
Methyl succinate	9.17	0.835
Methyl laurate	11.00	1.000
Methyl malate	19.91	1.811
Methyl α-ketoglutarate	21.43	1.929
Methyl citrate	29.66	2.698

* Relative to methyl laurate.

Column Packing P1 = 12.5% DEGS on acid washed Chromosorb W (45/60 mesh).
P2 = 6% XE-60 on Gas Pack S (45/60 mesh).

Temperature T1 = programmed at 80 °C for one minute, 80–100 °C at 2.9 °C/min., 100–133 °C at 4.7 °C/min., 133–220 °C at 4.8 °C/min., 220 °C for 2.5 minutes.

Literature Zaura, D. and J. Metcoff, *Anal. Chem.*, **41**, 1781 (1969).

TABLE GC 90
ACIDS, PHENOLIC

Column Packing	P1	P1	P2
Temperature (°C)	T1	T1	T2
Gas; Flow Rate (ml/min)	42,He	42,He	42,He
Column,			
length, ft	5	5	5
diameter, in., I.D.	$\frac{1}{8}$	$\frac{1}{8}$	$\frac{1}{8}$
form	na	na	na
material	na	na	na
Detector	na	na	na
Compound		t_r	
***p*-Hydroxybenzoic**	2.0	2.0	7.4
Vanillic	4.1	3.7	6.7
***o*-Coumaric**	5.2	4.3	8.9
***p*-Coumaric**	7.4	5.3	9.5
Protocatechuic	7.4	5.3	14.8
Ferulic	11.3	6.1	9.5
Syringic	11.8	6.5	9.5
***m*-Hydroxybenzoic**	14.4	7.9	7.4

Column Packing P1 = 10% SE-30 on Diatoport S treated with dimethylchlorosilane (60/80 mesh).
P2 = 10% DEGS on Diatoport S (60/80 mesh) treated with dimethylchlorosilane.

Temperature T1 = programmed: 100 °C for 3 minutes, then to 275 °C at 25 °C/min.
T2 = programmed: 100 °C for 3 minutes, then to 200 °C at 25 °C/min.

Literature Mendez, J. and F. J. Stevenson, *J. Gas Chromatogr.*, **4**, 483 (1966).

TABLE GC 91
ACIDS, ALIPHATIC; UNSATURATED ESTERS

Column Packing	P1			P2			P3			P4		
Temperature (°C)	150			150			150			150		
Gas; Flow Rate	—			—			—			—		
Column												
length, ft	12			12			12			12		
diameter, in., O.D.	$\frac{1}{4}$			$\frac{1}{4}$			$\frac{1}{4}$			$\frac{1}{4}$		
form	na			na			na			na		
material	aluminum			aluminum			aluminum			aluminum		
Compound	$V_{R'}$[a]	r[b]	I	$V_{R'}$	r	I	$V_{R'}$	r	I	$V_{R'}$	r	I
Methyl acrylate	0.128	0.194	569	0.140	0.450	743	0.110	0.512	766	0.120	0.78	844
Ethyl acrylate	0.205	0.310	664	0.165	0.550	784	0.155	0.720	831	0.160	1.03	911
Propyl acrylate	—	—	—	0.280	0.930	887	0.260	1.21	939	0.232	1.55	1006
Butyl acrylate	0.595	0.903	878	0.470	1.57	987	0.445	2.02	1040	0.380	2.45	1111
Pentyl acrylate	0.960	1.455	978	0.870	2.80	1108	0.715	3.28	1142	0.580	3.75	1210
Hexyl acrylate	1.595	2.42	1068	1.415	4.55	1200	1.155	5.30	1240	0.895	5.78	1312
Isopropyl acrylate	—	—	—	0.195	0.65	816	0.180	0.835	865	0.165	1.06	914
2-Methylpropyl acrylate	—	—	—	0.410	1.32	953	0.330	1.51	986	0.300	1.94	1055
Methyl methacrylate	0.217	0.329	677	0.210	0.680	829	0.175	0.830	865	0.160	1.02	909
Ethyl methacrylate	0.320	0.482	756	0.249	0.830	863	0.245	1.12	920	0.210	1.40	980
Propyl methacrylate	0.528	0.798	856	0.402	1.34	957	0.375	1.79	1016	0.315	2.10	1085
Butyl methacrylate	0.884	1.34	962	0.690	2.30	1068	0.640	2.94	1118	0.485	3.24	1179
Pentyl methacrylate	1.46	2.212	1064	1.23	3.97	1176	1.035	4.75	1214	0.760	4.90	1272
Hexyl methacrylate	2.40	3.64	1165	2.07	6.68	1281	1.630	7.60	1312	1.115	7.43	1369
Octyl methacrylate	—	—	—	5.10	17.00	1463	3.905	18.45	1489	—	—	—
2-Methylpropyl methacrylate	0.750	1.135	928	0.620	1.94	1032	0.475	2.21	1060	0.410	2.64	1128
3-Methylbutyl methacrylate	—	—	—	4.05	13.51	1416	3.025	14.42	1439	—	—	—
Ethyl crotonate	0.441	0.667	820	0.430	1.39	965	0.395	1.80	1016	0.340	2.12	1086
Propyl crotonate	0.710	1.075	917	0.670	2.23	1065	0.625	2.87	1112	0.535	3.35	1188
Butyl crotonate	1.190	1.804	1021	1.10	3.67	1157	1.005	4.74	1210	0.760	4.90	1272
Isopropyl crotonate	0.540	0.818	863	0.435	1.50	978	0.385	1.77	1015	0.377	2.24	1090
Pentyl crotonate	1.905	2.89	1119	1.84	6.13	1265	1.515	7.22	1307	1.171	7.55	1372
3-Methylpropyl crotonate	0.978	1.481	983	0.900	2.80	1094	0.780	3.58	1156	0.630	4.06	1232
3-Methylbutyl crotonate	1.61	2.44	1082	1.413	4.71	1208	1.255	5.92	1257	0.980	6.32	1332
Vinyl acetate	0.115	0.173	548	0.096	0.320	679	0.070	0.341	688	0.086	0.571	770
Vinyl propionate	0.190	0.286	648	0.150	0.490	763	0.125	0.580	790	0.135	0.870	861
Vinyl butyrate	0.310	0.466	750	0.260	0.810	859	0.200	0.950	892	0.200	1.29	963
Vinyl pentanoate	—	—	—	0.380	1.29	950	0.340	1.56	990	0.280	2.00	1060
Vinyl hexanoate	—	—	—	0.635	2.08	1037	0.565	2.59	1095	0.430	3.07	1162

TABLE GC 91—(Continued)

ACIDS, ALIPHATIC; UNSATURATED ESTERS

Column Packing	P1			P2			P3			P4		
Temperature (°C)	150			150			150			150		
Gas; Flow Rate	—			—			—			—		
Column												
length, ft	12			12			12			12		
diameter, in., O.D.	$\frac{1}{4}$			$\frac{1}{4}$			$\frac{1}{4}$			$\frac{1}{4}$		
form	na			na			na			na		
material	aluminum			aluminum			aluminum			aluminum		
Compound	$V_{R'}$[a]	r[b]	I	$V_{R'}$	r	I	$V_{R'}$	r	I	$V_{R'}$	r	I
Vinyl isobutyrate	—	—	—	0.150	0.518	773	0.175	0.815	857	0.145	1.04	912
Vinyl 3-methylbutyrate	—	—	—	0.285	0.950	889	0.250	1.15	928	0.210	1.50	996
Vinyl 4-methylpentanoate	—	—	—	0.500	1.64	998	0.450	2.09	1050	0.350	2.50	1118
Allyl formate	0.130	0.200	571	0.120	0.380	714	0.100	0.455	743	0.120	0.780	844
Allyl acetate	0.210	0.310	670	0.190	0.600	801	0.155	0.720	831	0.144	0.960	892
Allyl propionate	0.340	0.521	770	0.310	0.970	895	0.260	1.18	933	0.230	1.48	995
Allyl butyrate	0.520	0.790	853	0.500	1.60	990	0.400	1.84	1020	0.355	2.29	1096
Allyl pentanoate	0.880	1.33	960	0.830	2.60	1090	0.660	3.07	1123	0.535	3.45	1192
Allyl hexanoate	1.430	2.16	1056	1.375	4.30	1192	1.075	4.89	1219	0.830	5.35	1294
Allyl octanoate	—	—	—	—	—	—	—	—	—	1.960	12.66	1490
Allyl isobutyrate	0.430	0.649	815	0.324	1.08	916	0.295	1.37	961	0.280	1.81	1041
Allyl 3-methylbutyrate	0.710	1.082	915	0.620	1.94	1032	0.500	2.29	1065	0.410	2.64	1130
Allyl 4-methylpentanoate				—	—	—	0.885	4.02	1181	0.640	4.27	1240
Isopropenyl acetate	0.185	0.274	644	0.140	0.460	744	0.130	0.605	782	0.095	0.680	815
Isopropenyl propionate	—	—	—	0.220	0.749	846	0.210	0.970	894	0.165	1.10	921
Isopropenyl butyrate	—	—	—	0.385	1.263	944	0.340	1.58	988	0.225	1.60	1012
Isopropenyl pentanoate	—	—	—	0.600	1.97	1036	0.570	2.60	1095	0.365	2.60	1022
Isopropenyl hexanoate	—	—	—	1.005	3.30	1138	0.895	4.16	1189	—	—	—
Isopropenyl isobutyrate	—	—	—	0.265	0.870	875	0.250	1.17	928	0.200	1.43	988
Isopropenyl 3-methylbutyrate	—	—	—	—	—	—	0.410	1.91	1029	0.330	2.20	1084
Isopropenyl 4-methylpentanoate	—	—	—	0.795	2.61	1092	0.715	3.33	1144	0.510	3.40	1184

[a] Corrected for dead volume.
[b] Relative to nonane.

Column Packing
P1 = 10% SE-30 on acid-washed silanized Celite 560 (60/80 mesh).
P2 = 10% OV-17 on acid-washed silanized Celite 560 (60/80 mesh).
P3 = 10% OV-25 on acid-washed silanized Celite 560 (60/80 mesh).
P4 = 10% XE-60 on acid-washed silanized Celite 560 (60/80 mesh).

Literature

Allen, I. D. and J. K. Haken, *J. Chromatogr.*, **51**, 415 (1970).

TABLE GC 92
ALCOHOLS AND DERIVATIVES

Column Packing	P1	P2	P3	P1	P2	P3	P1	P2	P3	P1	P2	P3	P1	P2	P3	P1	P2	P3	P1	P2	P3	P1	P2	P3
Temperature (°C)	—	—	—	—	—	—	—	—	—	—	—	—	—	—	—	—	—	—	—	—	—	—	—	—
Gas; Flow Rate (ml/min)	He	He	He	He	He	He	He	He	He	He	He	He	He	He	He	He	He	He	He	He	He	He	He	He
Column																								
length, m	2	2	2	2	2	2	2	2	2	2	2	2	2	2	2	2	2	2	2	2	2	2	2	2
diameter, mm, O.D.	4	4	4	4	4	4	4	4	4	4	4	4	4	4	4	4	4	4	4	4	4	4	4	4
form	—	—	—	—	—	—	—	—	—	—	—	—	—	—	—	—	—	—	—	—	—	—	—	—
material	—	—	—	—	—	—	—	—	—	—	—	—	—	—	—	—	—	—	—	—	—	—	—	—
Detector	TC	TC	TC	TC	TC	TC	TC	TC	TC	TC	TC	TC	TC	TC	TC	TC	TC	TC	TC	TC	TC	TC	TC	TC
	Retention Index																							
Compound	I			Silyl ethers			Acetates			Trifluoroacetates			Propionates			Pentafluoro-propionates			Butyrates			Heptafluoro-butyrates		
Temperature: 100°																								
Methanol	980	892	>400	—	—	—	970	836	505	—	—	—	1010	916	612	—	—	—	1085	997	704	—	—	—
Ethanol	1027	922	480	—	—	—	990	898	602	—	—	—	1063	970	695	—	—	—	1127	1045	784	—	—	—
2-Propanol	985	912	490	—	—	—	1000	912	610	—	—	—	1040	970	742	—	—	—	1100	1050	822	—	—	—
***n*-Propanol**	1120	1030	562	—	—	—	1084	986	695	—	—	—	1150	1050	792	—	—	—	1215	1135	882	—	—	—
2-Methyl-2-propanol	962	882	527	—	—	—	975	845	595	—	—	—	1030	998	691	—	—	—	1082	1048	799	—	—	—
***2-Butanol**	1115	1014	590	—	—	—	1112	990	740	—	—	—	1145	1065	847	—	—	—	1215	1140	925	—	—	—
***2-Methyl-1-propanol**	1180	1083	612	—	—	—	1133	1025	758	—	—	—	1155	1090	852	—	—	—	1205	1170	939	—	—	—
***n*-Butanol**	1212	1138	655	—	—	—	1177	1085	805	977	814	652	1232	1151	896	882	760	687	1295	1230	985	882	730	737
2-Methyl-2-butanol	1074	1002	622	—	—	—	1150	917	680	—	—	—	1175	1026	754	—	—	—	1190	1080	815	—	—	—
***3-Methyl-2-butanol**	1170	1090	672	—	—	—	1095	1067	812	—	—	—	1175	1130	908	—	—	—	1237	1210	994	—	—	—
3-Pentanol	1182	1097	685	—	—	—	1165	1077	835	—	—	—	1226	1142	930	—	—	—	1280	1220	1014	—	—	—
2-Pentanol	1195	1112	682	—	—	—	1170	1080	840	—	—	—	1233	1152	927	—	—	—	1280	1225	1010	—	—	—
2-Methyl-1-butanol	1270	1202	724	—	—	—	1220	1132	865	—	—	—	1242	1197	952	—	—	—	1305	1285	1044	—	—	—
3-Methyl-1-butanol	1270	1205	726	—	—	—	1222	1135	867	1068	877	755	1245	1200	954	952	820	784	1307	1289	1046	965	835	835
***n*-Pentanol**	1307	1245	752	948	910	887	1263	1182	899	1160	980	842	1316	1257	990	1060	919	882	1375	1326	1082	1062	923	920
***n*-Hexanol**	1400	1342	855	1050	1007	982	1350	1299	1002	1255	1070	945	1405	1354	1090	1156	1025	974	1460	1430	1177	1155	1025	1017
***n*-Heptanol**	1490	1450	945	1152	1104	1078	1443	1385	1095	1348	1185	1046	1485	1455	1192	1255	1120	1071	1545	1526	1270	1255	1126	1114
***n*-Octanol**	—	—	—	1248	1200	1177	—	—	—	—	—	—	—	—	—	—	—	—	—	—	—	—	—	—
***n*-Nonanol**	—	—	—	1355	1298	1270	—	—	—	—	—	—	—	—	—	—	—	—	—	—	—	—	—	—
***n*-Decanol**	—	—	—	1448	1395	1372	—	—	—	—	—	—	—	—	—	—	—	—	—	—	—	—	—	—

TABLE GC 92—(Continued)
ALCOHOLS AND DERIVATIVES

Column Packing	P1	P2	P3	P1	P2	P3	P1	P2	P3	P1	P2	P2	P1	P2	P3	P1	P2	P3	P1	P2	P3	P1	P2	P3
Temperature (°C)	—	—	—	—	—	—	—	—	—	—	—	—	—	—	—	—	—	—	—	—	—	—	—	—
Gas; Flow Rate (ml/min)	He	He	He	He	He	He	He	He	He	He	He	He	He	He	He	He	He	He	He	He	He	He	He	He
Column																								
length, m	2	2	2	2	2	2	2	2	2	2	2	2	2	2	2	2	2	2	2	2	2	2	2	2
diameter, mm, O.D.	4	4	4	4	4	4	4	4	4	4	4	4	4	4	4	4	4	4	4	4	4	4	4	4
form	—	—	—	—	—	—	—	—	—	—	—	—	—	—	—	—	—	—	—	—	—	—	—	—
material	—	—	—	—	—	—	—	—	—	—	—	—	—	—	—	—	—	—	—	—	—	—	—	—
Detector	TC	TC	TC	TC	TC	TC	TC	TC	TC	TC	TC	TC	TC	TC	TC	TC	TC	TC	TC	TC	TC	TC	TC	TC
	Retention Index																							
Compound	I			Silyl ethers			Acetates			Trifluoroacetates			Propionates			Pentafluoro-propionates			Butyrates			Heptafluoro-butyrates		
Temperature: 140°, 160°*																								
***n*-Hexanol**	1455	1350	856	—	—	—	1380	1301	999	1157	960	825	1415	1355	1085	—	907	859	1485	1426	1177	—	907	896
***n*-Heptanol**	1542	1452	955	1152	1087	1078	1490	1385	1095	1258	1073	940	1535	1455	1182	1107	1016	962	1580	1526	1273	1107	1016	1010
***n*-Octanol**	1648	1555	1061	1242	1185	1177	1597	1487	1198	1347	1172	1040	1648	1555	1284	1216	1114	1062	1668	1624	1377	1215	1114	1105
***n*-Nonanol**	1744	1655	1160	1355	1275	1265	1689	1582	1290	1449	1270	1137	1730	1650	1385	1328	1215	1160	1762	1728	1470	1325	1215	1201
2-Octanol	1530	1411	990	1182	1152	1111	1487	1398	1127	1150	1075	967	1590	1430	1231	1108	1001	987	1643	1507	1298	1110	1101	1032
***n*-Decanol**	—	—	—	1442	1365	1360	—	—	—	1543	1367	1232	—	—	—	1431	1311	1256	—	—	—	1430	1311	1298
***n*-Dodecanol**	—	—	—	1610	1568	1564	—	—	—	1736	1555	1421	—	—	—	1630	1505	1481	—	—	—	1632	1505	1480
***n*-Tetradecanol**	—	—	—	1797	1750	1754	—	—	—	—	—	—	—	—	—	1832	1701	1668	—	—	—	1836	1701	1668
***n*-Hexadecanol**	—	—	—	1978	1942	1938	—	—	—	2107*	1929*	1805*	—	—	—	2031*	1894*	1859*	—	—	—	2034*	1891*	1862*
***n*-Octadecanol**	—	—	—	2158	2135	2128	—	—	—	2287*	2121*	1901*	—	—	—	2239*	2097*	2054*	—	—	—	2241	2095	2063
Temperature 180°																								
***n*-Octanol**	1565	1552	1057	—	—	—	1648	1552	1286	—	1622	1370												
***n*-Nonanol**	1758	1655	1160	1720	1587	1295	1750	1650	1385	1830	1724	1470												
***n*-Decanol**	1863	1740	1263	1815	1674	1385	1864	1748	1476	1927	1822	1568												
***n*-Dodecanol**	2040	1925	1471	1996	1868	1587	2044	1995	1676	2110	2023	1768												
***n*-Tetradecanol**	2228	2112	1669	2172	2062	1790	2222	2178	1873	2288	2212	1964												
***n*-Hexadecanol**	2417	2297	1864	2356	2254	1948	2404	2370	2068	2470	2400	2154												
***n*-Ocadecanol**	2596	2488	2060	2540	2448	2178	2598	2560	2262	2662	2595	2352												

* Temperature 160

Packing
P1 = 20% DEGS on Chromosorb W.
P2 = 20% Carbowax 20M on Chromosorb W.
P3 = 20% SE-30 on Chromosorb W.

Literature
Zarazir, D., Chovin, P., and G. Guiochon, *Chromatographia*, 3, 180 (1970).

TABLE GC 93
ALIPHATICS, MISCELLANEOUS

Column Packing	P1	P2	P3	P4	P5	P6	P7	P8	P9	P10	P11	P12	P13	P14	P15	P16	P17	P18	P19	P20	P21
Temperature (°C)	100	100	100	100	100	100	100	100	100	100	100	100	100	100	100	100	100	100	100	100	100
Gas; Flow Rate	—	—	—	—	—	—	—	—	—	—	—	—	—	—	—	—	—	—	—	—	—
Column																					
length, ft	6	6	6	6	6	6	6	6	6	6	6	6	6	6	6	6	6	6	6	6	6
diameter, in.	$\frac{1}{4}$	$\frac{1}{4}$	$\frac{1}{4}$	$\frac{1}{4}$	$\frac{1}{4}$	$\frac{1}{4}$	$\frac{1}{4}$	$\frac{1}{4}$	$\frac{1}{4}$	$\frac{1}{4}$	$\frac{1}{4}$	$\frac{1}{4}$	$\frac{1}{4}$	$\frac{1}{4}$	$\frac{1}{4}$	$\frac{1}{4}$	$\frac{1}{4}$	$\frac{1}{4}$	$\frac{1}{4}$	$\frac{1}{4}$	$\frac{1}{4}$
form	tubing	tubing	tubing	tubing	tubing	tubing	tubing	tubing	tubing	tubing	tubing	tubing	tubing	tubing	tubing	tubing	tubing	tubing	tubing	tubing	tubing
material	SS	SS	SS	SS	SS	SS	SS	SS	SS	SS	SS	SS	SS	SS	SS	SS	SS	SS	SS	SS	SS
Detector	TC	TC	TC	TC	TC	TC	TC	TC	TC	TC	TC	TC	TC	TC	TC	TC	TC	TC	TC	TC	TC
Compound											r										
											(C_7H_6)										
n-Heptane	1.0	1.0	1.0	1.0	1.0	1.0	1.0	1.0	1.0	1.0	1.0	1.0	1.0	1.0	1.0	1.0	1.0	—	—	—	—
Cyclohexane	1.14	0.66	0.73	0.86	0.78	0.65	0.71	0.90	0.74	0.81	0.94	—	0.70	0.73	0.67	0.71	0.80	—	—	—	—
Benzene	1.02	1.90	1.45	2.32	0.83	1.01	0.86	1.11	1.34	1.53	2.78	2.69	1.18	0.96	1.00	1.40	1.14	—	—	—	—
Carbon tetrachloride	1.02	1.27	1.17	1.59	0.74	0.76	0.83	0.95	1.02	1.15	1.60	1.59	0.90	0.83	0.75	1.20	1.05	—	—	—	—
Chloroform	0.65	2.72	3.53	2.40	0.78	0.78	0.74	0.95	1.31	1.41	1.80	1.61	1.00	0.67	0.67	1.25	0.80	—	—	—	—
Ethylene dichloride	0.85	3.68	2.69	3.28	0.78	1.11	0.94	—	1.75	2.23	3.60	3.32	1.41	0.87	1.05	1.60	1.14	—	—	—	—
Methyl ethyl ketone	0.64	1.49	1.19	2.17	0.52	0.57	0.66	0.72	1.01	1.35	3.60	5.31	1.12	1.43	—	—	0.80	—	—	—	—
Ethanol	0.35	2.22	2.20	—	0.65	0.78	0.60	—	1.77	2.64	7.73	13.40	2.75	—	—	—	2.80	—	—	—	—
											($CHClCCl_2$)										
Methylene chloride	0.31	0.66	—	—	—	—	—	—	0.37	0.41	0.37	0.36	—	—	0.36	—	—	0.21	0.20	0.49	0.24
trans-1,2-Dichloroethene	0.38	0.46	—	—	—	—	—	—	0.31	0.32	0.28	0.31	—	—	0.30	—	—	0.30	0.38	0.36	0.28
cis-1,2-Dichloroethene	0.51	1.00	—	—	—	—	—	—	0.79	0.97	0.59	0.68	—	—	0.66	—	—	0.37	0.43	1.23	0.50
Chloroform	0.59	1.23	—	—	—	—	—	—	0.85	0.78	0.59	0.63	—	—	0.56	—	—	0.44	0.50	0.51	0.53
1,1,1-Trichloroethane	0.68	0.54	—	—	—	—	—	—	0.63	0.68	0.59	0.63	—	—	0.66	—	—	0.55	0.54	0.68	0.57
Carbon tetrachloride	0.78	0.54	—	—	—	—	—	—	0.63	0.59	0.59	0.57	—	—	0.66	—	—	0.69	0.69	0.68	0.66
1,2-Dichloroethane	0.70	1.64	—	—	—	—	—	—	1.05	1.12	1.29	1.17	—	—	0.84	—	—	0.58	0.60	2.31	0.72
1,2,2-Trichloroethene	1.0	1.0	—	—	—	—	—	—	1.0	1.0	1.0	1.0	—	—	1.0	—	—	1.0	1.0	1.0	1.0
											(C_6H_{12})										
Cyclopentane	0.52	0.73	—	—	—	—	—	—	0.50	0.56	0.58	0.57	—	—	0.50	—	—	0.59	0.49	0.31	0.55
Methylcyclopentane	0.78	0.88	—	—	—	—	—	—	0.75	0.78	0.79	0.75	—	—	0.74	—	—	0.81	0.73	0.49	0.78
Benzene	1.00	2.49	—	—	—	—	—	—	1.86	1.87	3.30	2.80	—	—	1.40	—	—	1.00	1.00	0.68	1.11
Cyclohexane	1.0	1.0	—	—	—	—	—	—	1.0	1.0	1.0	1.0	—	—	1.0	—	—	1.0	1.0	1.0	1.0
Methylcyclohexane	1.63	1.34	—	—	—	—	—	—	1.64	1.54	1.47	1.46	—	—	1.62	—	—	1.52	1.60	1.05	1.50

TABLE GC 93—(Continued)
ALIPHATICS, MISCELLANEOUS

Column Packing
P1 = 10% (w/w) Silicone grease on Chromosorb P.
P2 = 10% (w/w) Carbowax 20M on Chromosorb P.
P3 = 10% (w/w) (*n*-butyl O)$_3$ PO on Chromosorb P.
P4 = 10% (w/w) (ortho-$CH_3 \cdot C_6H_5O$)$_3$ PO on Chromosorb P.
P5 = 10% (w/w) (*n*-octyl)$_2$ PO(OH) on Chromosorb P.
P6 = 10% (w/w) (*n*-octyl O)$_2$ PO(OH) on Chromosorb P.
P7 = 10% (w/w) (2-ethylhexyl O) 2-ethylhexyl PO(OH) on Chromosorb P.
P8 = 10% (w/w) (2-ethylhexyl O)$_2$ PO(OH) on Chromosorb P.
P9 = 10% (w/w) (*n*-butyl O) *n*-butyl PO(OH) on Chromosorb P.
P10 = 10% (w/w) (*n*-butyl O)$_2$ PO(OH) on Chromosorb P.
P11 = 10% (w/w) (phenyl O) phenyl PO(OH) on Chromosorb P.
P12 = 10% (w/w) (phenyl O)$_2$ PO(OH) on Chromosorb P.
P13 = 10% (w/w) (2-ethylhexyl) hydrogen PO(OH) on Chromosorb P.
P14 = 10% (w/w) (2-ethylhexyl) PO(OH)$_2$ on Chromosorb P.
P15 = 10% (w/w) (2-ethylhexyl O) PO(OH)$_2$ on Chromosorb P.
P16 = 10% (w/w) (phenyl O) PO(OH)$_2$ on Chromosorb P.
P17 = 10% (w/w) (*n*-octyl O) PO(OH)$_2$ on Chromosorb P.
P18 = 10% (w/w) Apiezon L on Chromosorb P.
P19 = 10% (w/w) Apiezon L and Lexan on Chromosorb P.
P20 = 10% (w/w) Apiezon L and Benton 3,4 on Chromosorb P.
P21 = 10% (w/w) Apiezon L and (*n*-butyl O)$_2$ PO(OH) on Chromosorb P.

Literature
Szymanski, H. A., Annino, R., Roaldi, A., Galeza, J., and J. R. Ferraro, *J. Chromatogr. Sci.*, **7**, 439 (1969).

TABLE GC 94
ALCOHOLS

Column Packing	P1	P2	P3
Temperature (°C)	65	65	65
Gas; Flow Rate (ml/min)	He	He	He
Column			
length, ft	200	200	200
diameter, in., O.D.	0.0625	0.0625	0.0625
in., I.D.	0.010	0.010	0.010
form	capillary tubing	capillary tubing	capillary tubing
material	SS	SS	SS
Detector	na	na	na

Compound	t_r		
Methanol	4.0	3.3	3.8
Ethanol	5.0	3.9	4.9
2-Propanol	5.6	4.1	5.3
***tert*-Butanol**	6.0	4.1	5.7
***n*-Propanol**	8.6	6.3	8.5
2-Butanol	10.0	6.6	9.7
***tert*-Pentanol**	12.0	6.9	11.3
***iso*-Butanol**	12.9	8.5	12.5
2-Pentanol	15.5	10.2	15.1
3-Methyl-2-butanol	17.5	9.6	15.8
3-Pentanol	18.7	10.8	17.8
***n*-Butanol**	19.0	12.0	19.5
Neopentanol	21.0	12.0	20.0
2-Methyl-1-butanol	30.8	18.7	30.5
3-Methyl-1-butanol	32.7	19.3	32.3
***n*-Pentanol**	44.2	26.6	43.4

Column Packing P1 = 10% (v/v) *Igepal CO-210 in chloroform.
P2 = 10% (v/v) *Igepal CO-530 in chloroform.
P3 = 10% (v/v) 9 parts *Igepal CO-210 plus one part Igepal CO-530 in chloroform.

Literature Schwartz, R. D. and R. G. Mathews, *J. Chromatogr. Sci.*, **7**, 593 (1969).

* Available from General Aniline and Film Corp.

TABLE GC 95
ALKALOIDS, TOBACCO

Column Packing	P1	P2	P3
Temperature (°C)	190	180	190
Gas; Flow Rate (ml/min)	45,He	50,He	48,He
Column			
length, m	1	1	1
diameter, mm, O.D.	6	6	6
form	U-tubes	U-tubes	U-tubes
material	glass	glass	glass
Detector	na	na	na

Compound	t_r Retention time, min.		
3-Pyridyl methyl ketone	4.3	3.1	4.3
3-Pyridyl ethyl ketone	6.1	5.0	5.3
3-Pyridyl *n*-propyl ketone	8.1	7.0	6.6
Nicotine	8.6	8.2	5.2
Nornicotine	16.1	14.3	12.3
Myosmine	16.4	14.7	13.4
Anabasine	19.4	18.1	13.8
Nicotyrine	21.0	18.3	19.4
Metanicotine	23.5	20.9	16.5
Anatabine	25.2	22.5	21.1
2,3′-Dipyridyl	31	26	29
***N*-Methyl nicotinamide**	42	30	64
Nornicotyrine	73	55	101
Cotinine	79	63	85

Column Packing P1 = Polypropylene glycol, MWX 1025—20% w/w on Firebrick (Fisher "Columpak") base, and washed.
P2 = Polypropylene glycol, MWX 1500—20% w/w on Firebrick (Fisher "Columpak") base, and washed.
P3 = Polypropylene glycol, molecular weight 20,000—20% w/w on Firebrick (Fisher "Columpak") base, and washed.

Literature Quin, L. D., *J. Org. Chem.*, **24**, 911 (1959).

TABLE GC 96
ALKALOIDS

Column Packing	P1
Temperature (°C)	T1
Gas; Flow Rate (ml/min)	15 psi, Ar[a]
Column	
length, cm	180
diameter, cm, I.D.	0.4
form	U-tubes
material	glass
Detector	EC

Compound	t_r
Lupin alkaloids	
Cytisine	5.1
Methylcytisine	4.3
Methylcytisine *N*-oxide	5.8
Lupanine	5.5
13-Hydroxylupanine	11.6
Matrina	8.5
Lupinine	1.5[c]
Sparteine	5.9[c]
α-Isosparteine	5.2[c]
13-Hydroxysparteine	14.3[c]
Amaryllidaceae	
Galanthine	19.0
Acetylcaranine	10.5
Lycorenine	10.6
Galanthamine	7.8
Crinine	9.5
Powelline	15.8
Tasettine	15.2
Belladine	8.7
Papaveraceae	
Codeine	8.2
Gnoscopine	90.6
Laudanosine	21.0
Morphine	11.0
Neopine	9.1
Papaverine	35.3
Thebaine	13.2
Indole alkaloids	
Brucine	80.0[b]
Coronaridine	8.2[b]
Ibogamine	15.4
Ibogaine	35.1
Serpentine	16.8[b]
Strychnine	25.9[b]
Voacangine	40.3
Steroidal alkaloids	
Solanidine	40.6[a,b]
Solasodine	74.3[a,b]
Tomatidine	77.3[a,b]
Miscellaneous	
Atropine	5.0
Caffeine	1.6
Cinchonine	6.7[b]
Cocaine	4.8[b]
Corydaline	16.2[b]
Cryptopine	50.8
Himbacine	12.7[b]
Piperine	33.0
Protopine	44.7
Quinine	11.8[b]

[a] Except as noted, 10 psi.

Column Packing P1 = 2–3% SE-30 on Chromosorb W (80/100 mesh) washed with conc. HCl followed by 1% methanolic KOH dried at 100 °C.

Temperature T1 = 204 °C except [b] temperature 222 °C; [c] temperature 160 °C.

Literature Lloyd, H. A., Fales, H. M., Highet, P. F., Vanden Heuvel, W. J. A., and W. C. Wildman, *J. Amer. Chem. Soc.*, **82**, 3791 (1960).

TABLE GC 97
AMINES, ALIPHATIC

Column Packing	P1	P2	P3	P4	P5	P6	P7	P8	P9	P10	P11
Temperature (°C)	60	60	60	60	60	60	60	60	60	60	60
Gas; Flow Rate (ml/min)	25,N_2	25,N_2	25,N_2	25,N_2	25,N_2	25,N_2	25,N_2	25,N_2	25,N_2	25,N_2	25,N_2
Column											
length, m	1.6	1.6	1.6	1.6	1.6	1.6	1.6	1.6	1.6	1.6	1.6
diameter, mm, I.D.	4	4	4	4	4	4	4	4	4	4	4
form	—	—	—	—	—	—	—	—	—	—	—
material	glass	glass	glass	glass	glass	glass	glass	glass	glass	glass	glass
Detector	na	na	na	na	na	na	na	na	na	na	na
Compound						$V_{R'}$					
Methylamine	17.0	17.3	10.8	4.5	2.3	0.7	1.2	0.5	0.8	0.3	—
Ethylamine	17.9	11.5	8.9	4.9	2.8	0.9	1.6	0.6	0.4	0.3	—
Isopropylamine	13.7	9.6	6.5	3.9	2.7	0.8	1.7	0.7	0.4	0.4	1.0
tert-Butylamine	15.8	7.5	5.2	3.6	2.6	0.8	1.7	0.8	0.6	0.3	1.5
n-Propylamine	22.1	14.7	14.0	7.5	5.2	1.5	3.2	1.4	0.8	0.6	1.7
sec-Butylamine	19.7	14.6	10.3	5.8	5.3	1.6	3.5	1.7	1.0	0.6	3.0
Isobutylamine	21.7	16.5	11.4	8.0	6.5	—	4.3	2.1	1.2	0.9	3.5
tert-Pentylamine	19.9	15.2	9.5	7.2	6.2	1.8	3.9	1.9	1.2	0.9	4.7
n-Butylamine	39.0	24.9	25.1	14.9	11.1	3.1	—	3.2	1.7	1.3	4.7
Neopentylamine	12.5	12.4	10.2	7.2	6.9	1.8	4.6	2.6	1.6	1.1	—
1-Methyl-*n*-butylamine	30.2	34.4	17.1	13.1	10.8	3.2	6.9	3.4	1.8	2.0	6.9
3-Aminopentane	22.5	21.0	12.8	11.0	9.9	3.1	6.4	3.5	1.8	1.4	—
Isopentylamine	49.9	69.5	35.1	24.6	17.8	4.8	—	4.9	2.8	2.1	8.5
n-Pentylamine	62.5	80.2	51.0	32.7	24.9	6.5	—	6.8	3.7	3.0	12.5
n-Hexylamine	107.0	—	—	65.5	46.2	13.3	—	14.5	7.9	6.3	32.0
Dimethylamine	11.6	6.2	3.6	2.2	1.5	0.6	1.0	0.4	0.4	0.2	—
N-Methylisopropylamine	11.8	7.5	4.8	2.8	3.1	1.0	1.9	—	0.6	—	—
Diethylamine	14.4	9.6	5.6	3.5	2.8	1.0	1.9	0.9	0.6	0.4	2.8
N-Ethylisopropylamine	10.7	7.2	4.9	3.7	3.1	1.1	2.1	1.1	0.7	0.5	4.4
N-Methyl-*sec*-butylamine	17.5	10.4	7.1	5.7	4.9	1.4	3.3	—	—	—	6.3
N-Ethyl-*n*-propylamine	15.0	11.4	7.6	5.9	5.0	1.8	3.4	1.7	1.1	0.8	6.8
Diisopropylamine	5.9	5.0	3.7	3.0	2.9	1.1	1.9	1.2	0.8	0.5	6.2
N-Methyl-*n*-butylamine	25.9	19.1	12.6	9.5	7.5	2.6	6.3	2.9	1.6	1.2	7.7
Di-*n*-propylamine	15.4	15.2	10.6	8.5	8.6	3.1	5.8	3.5	2.0	1.5	15.8
Di-*sec*-butylamine	6.1	8.5	7.4	8.0	8.4	3.6	6.2	4.4	2.6	2.0	36.4
Diisobutylamine	4.6	8.0	7.7	8.5	9.4	4.3	6.9	5.2	3.1	2.5	43.0

TABLE GC 97—(Continued)
AMINES, ALIPHATIC

Column Packing	P1	P2	P3	P4	P5	P6	P7	P8	P9	P10	P11
Temperature (°C)	60	60	60	60	60	60	60	60	60	60	60
Gas; Flow Rate (ml/min)	25,N_2	25,N_2	25,N_2	25,N_2	25,N_2	25,N_2	25,N_2	25,N_2	25,N_2	25,N_2	25,N_2
Column											
length, m	1.6	1.6	1.6	1.6	1.6	1.6	1.6	1.6	1.6	1.6	1.6
diameter, mm, I.D.	4	4	4	4	4	4	4	4	4	4	4
form	—	—	—	—	—	—	—	—	—	—	—
material	glass	glass	glass	glass	glass	glass	glass	glass	glass	glass	glass
Detector	na	na	na	na	na	na	na	na	na	na	na
Compound	$V_{R'}$										
Di-*n*-butylamine	35.0	52.8	39.4	36.0	35.2	12.5	23.6	14.4	8.3	6.6	—
Trimethylamine	1.5	0.8	0.6	0.6	0.6	0.4	0.3	0.2	0.2	0.2	—
***N,N*-Dimethylethylamine**	—	1.5	1.0	0.7	0.9	0.4	—	0.5	0.3	0.3	1.9
***N,N*-Dimethyl-*n*-propylamine**	1.9	1.9	1.5	1.5	1.5	0.7	1.1	0.7	0.5	—	4.1
***N*-Methyldiethylamine**	2.5	2.1	1.7	1.5	1.7	0.8	1.2	0.8	0.6	0.4	4.2
***N,N*-Dimethylisopropylamine**	4.0	2.7	2.0	1.8	1.9	0.8	1.4	0.9	0.5	0.4	4.5
Triethylamine	2.7	2.8	2.3	2.2	2.5	1.1	1.8	1.2	0.8	0.7	8.3
***N,N*-Dimethyl-*tert*-butylamine**	5.9	4.4	3.5	3.3	3.7	1.4	2.3	1.5	0.7	0.6	—
***N,N*-Dimethyl-*sec*-butylamine**	3.3	3.2	2.7	2.8	3.1	1.4	2.1	1.5	1.0	0.7	10.3
***N,N*-Dimethyl-*n*-butylamine**	3.3	2.7	2.8	2.9	3.4	1.4	2.2	1.6	0.9	0.7	—
Tri-*n*-propylamine	2.0	5.6	5.3	8.0	9.0	4.4	6.5	5.4	3.6	2.7	—
Tri-*n*-butylamine	—	29.7	31.0	39.5	51.8	28.9	41.5	32.7	24.1	22.2	—

Column Packing

P1 = 20% (w/w) ethylene glycol on 5% PEI (polyethyleneimine) and acid washed Celite (100/120 mesh).
P2 = 20% (w/w) Carbowax 200 on 5% PEI and acid washed Celite (100/120 mesh).
P3 = 20% (w/w) Carbowax 400 on 5% PEI and acid washed Celite (100/120 mesh).
P4 = 20% (w/w) Carbowax 600 on 5% PEI and acid washed Celite (100/120 mesh).
P5 = 20% (w/w) Carbowax 1000 on 5% PEI and acid washed Celite (100/120 mesh).
P6 = 20% (w/w) Carbowax 20M on 5% PEI and acid washed Celite (100/120 mesh).
P7 = 20% (w/w) Carbowax 1000 on 5% KOH (potassium hydroxide) and acid washed Celite (100/120 mesh).
P8 = 20% (w/w) Carbowax 20M on 5% KOH (potassium hydroxide) and acid washed Celite (100/120 mesh).
P9 = 20% (w/w) Polyox WSR-35 on 5% KOH (potassium hydroxide) and acid washed Celite (100/120 mesh).
P10 = 20% (w/w) Polyox WSR-205 on 5% KOH (potassium hydroxide) and acid washed Celite (100/120 mesh).
P11 = 20% (w/w) Apiezon L on 5% KOH (potassium hydroxide) and acid washed Celite (100/120 mesh).

Literature

Smith, J. R. L. and D. J. Waddington, *J. Chromatogr.*, **42,** 183 (1969).

TABLE GC 98
AMINES

Column Packing	P1	P2	P3
Temperature (°C)	189	190	195
Gas; Flow Rate	na	na	na
Column			
length, ft	6	6	6
diameter, mm	4	4	4
form	U-tube	U-tube	U-tube
material	glass	glass	glass
Detector	EC	EC	EC

	$r^a_{n\text{-docosane}}$											
Derivative[b]	Cyclo-$C_{12}NH_2$	n-$C_{12}NH_2$	n-$C_{14}NH_2$	n-$C_{16}NH_2$	Cyclo-$C_{12}NH_2$	n-$C_{12}NH_2$	n-$C_{14}NH_2$	n-$C_{16}NH_2$	Cyclo-$C_{12}NH_2$	n-$C_{12}NH_2$	n-$C_{14}NH_2$	n-$C_{16}NH_2$
Free amine	0.097	0.07	0.13	0.32	0.084	0.052	0.12	0.26	0.11	0.071	0.16	0.36
TFAc	0.14	0.12	0.27	0.58	0.14	0.13	0.28	0.59	0.38	0.38	0.85	1.93
PFPr	0.13	0.12	0.25	0.55	0.13	0.11	0.25	0.56	0.30	0.32	0.74	1.65
HFBu	0.15	0.14	0.28	0.62	0.14	0.12	0.30	0.64	0.34	0.36	0.81	1.82
Ac	0.37	0.33	0.68	1.49	0.39	0.32	0.76	1.59	0.97	0.89	1.96	4.47
Pr	0.46	0.42	0.90	1.90	0.49	0.42	1.00	2.16	1.13	1.07	2.37	5.38
Bu	0.63	0.55	1.18	2.51	0.72	0.62	1.37	3.02	1.52	1.45	3.23	7.21
Methyl carbamate	0.33	0.27	0.59	1.26	0.34	0.27	0.60	1.33	0.56	0.48	1.07	2.39
Methyl thiolcarbamate	0.73	0.61	1.33	2.83	0.82	0.66	1.49	3.29	1.48	1.35	3.01	6.69
Methanesulfonamide	1.05	0.88	1.91	4.15	1.05	0.85	1.88	4.11	3.23	3.38	8.10	18.7
Acetone-SB	0.16	0.14	0.28	0.63	0.14	0.11	0.25	0.56	0.17	0.14	0.31	0.67
TF Acetone-SB	0.11	0.089	0.19	0.41	0.10	0.08	0.18	0.39	0.14	0.11	0.26	0.56
Cyclopent-SB	0.49	0.42	0.91	1.93	0.46	0.39	0.88	1.96	0.59	0.50	1.11	2.43
Cyclohex-SB	0.67	0.60	1.28	2.73	0.65	0.57	1.27	2.83	0.79	0.70	1.56	3.43
Cyclohept-SB	1.00	0.89	1.89	3.99	0.98	0.87	1.95	4.33	1.20	1.07	2.33	5.17

[a] t_r relative to n-docosane.
[b] Abbreviations: TFAc, trifluoroacetyl; PFPr, pentafluoropropionyl; HFBu, heptafluorobutyryl; Ac, acetyl; Pr, propionyl; Bu, butyryl; SB, eneamine or Schiff base; TF Acetone, 1,1,1-trifluoroacetone; Cyclopent, cyclopentanone; Cyclohex, cyclohexanone; Cyclohept, cycloheptanone.
[c] Sources of liquid phases: SE-52: General Electric; F-60-Z: Dow Corning; CNSI: General Electric (β-cyanoethyl methyl polysiloxane).

Column Packing[c] P1 = 5% SE-52 (w/w) on acid washed silanized Gas Chrom S (100/120 mesh).
P2 = 7% F-60-Z (w/w) on acid washed silanized Gas Chrom S (100/120 mesh).
P3 = 5.5% (w/w) CNSI on acid washed silanized Gas Chrom S (100/120 mesh).

Literature

Vanden Heuvel, W. J. A., Gardiner, W. L., and E. C. Horning, *Anal. Chem.*, **36**, 1550 (1964).

TABLE GC 99
AMINES, PENTAFLUOROPROPIONAMIDES

Column Packing	P1	P2	P3
Temperature (°C)	190	210	150
Gas; Flow Rate (ml/min)	35	40	6.5
Column			
length, m	2	2	50
diameter, mm, I.D.	6	6	0.3
form	na	na	na
material	glass	glass	glass
Detector	EC	EC	EC

Compound	t_r		
2-Aminobiphenyl	3.4	1.4	1.9
1-Naphthylamine	3.8	2.9	2.0
2-Naphthylamine	4.6	4.5	2.6
2-Methyl-1-naphthylamine	5.4	4.4	2.7
3-Methyl-2-naphthylamine	5.9	3.9	3.2
3-Aminobiphenyl	7.4	7.7	4.8
9-Aminofluorene	7.8	6.6	5.4
4-Aminobiphenyl	8.4	9.0	5.6
2-Aminostilbene (*trans*)	12.9	10.9	9.9
2-Aminodibenzofuran	14.8	18.2	11.3
2-Aminofluorene	18.1	22.0	15.8
4-Aminophenanthrene	21.3	24.0	20.0
3-Aminostilbene (*trans*)	24.5	33.3	24.0
9-Aminophenanthrene	26.8	30.0	25.6
1-Aminoanthracene	26.9	29.3	25.8
9-Aminoanthracene	27.0	37.7	22.4
4-Aminostilbene (*trans*)	27.8	39.1	28.4
1-Aminophenanthrene	29.2	35.3	28.9
3-Aminophenanthrene	31.7	47.4	35.6
2-Aminophenanthrene	35.5	55.0	41.6
2-Aminoanthracene	37.1	56.3	43.8
Benzidine	58.3	192.0	35.4

Column Packing P1 = 7.5% DC-200 and 5% QF-1 on Gas Chrom Q (80/100 mesh).
P2 = 2% HI-EFF 3 BP on Gas Chrom Q (80/100 mesh).
P3 = Apiezon L, capillary column.

Literature Masuda, Y. and D. Hoffmann, *J. Chromatogr. Sci.*, **7**, 694 (1969).

TABLE GC 100
ANILINES, SUBSTITUTED

Column Packing	P1
Temperature (°C)	200
Gas; Flow Rate (ml/min)	20,He
Column	
length, ft	8
diameter, in., O.D.	¼
form	tubing
material	copper
Detector	TC

Compound	t_r
2-Methylaniline	8
3-Methylaniline	8
4-Methylaniline	8
2-Ethylaniline	10.5
4-Ethylaniline	11
2,6-Dimethylaniline	11.5
2,4-Dimethylaniline	11.5
3,5-Dimethylaniline	11.5
2,5-Dimethylaniline	11.5
2,3-Dimethylaniline	13.5
3,4-Dimethylaniline	13.5
2,4,6-Trimethylaniline	15.5
2,3,6-Trimethylaniline	19.5
2,4,5-Trimethylaniline	19.5
2,3,5-Trimethylaniline	16.5
2,3,4-Trimethylaniline	23.5
3,4,5-Trimethylaniline	23
2-Methyl-5-ethylaniline	15
2-Ethyl-5-methylaniline	14
2-Methyl-6-ethylaniline	13.5
2-Methyl-4-ethylaniline	16
2-Ethyl-4-methylaniline	15
2-Methyl-3-ethylaniline	17
2-Ethyl-3-methylaniline	14
3-Methyl-4-ethylaniline	16
3-Ethyl-4-methylaniline	16
2,4,5,6-Tetramethylaniline	32
2,3,5,6-Tetramethylaniline	32

Column Packing P1 = Dow Corning 710 Silicone Fluid (40 : 100) on C-22 firebrick (30/60 mesh).

Literature Jones, J. H., Ritchie, C. D., and K. S. Heine, Jr., *J. Ass. Offic. Agr. Chem.*, **41**, 749 (1958).

TABLE GC 101
AMINO ACID ESTERS

Column Packing	P1	P2	P3
Temperature (°C)	T1	T2	T3
Gas; Flow Rate (ml/min)	N_2,128	N_2,30	Ar,60
Column			
length	2 m	106 cm	8 ft
diameter	0.63 cm, I.D.	3 mm, I.D.	0.5 cm, I.D.
form	coiled	na	dual U-shaped
material	SS	glass	glass
Detector	FI	FI	EC
Literature	1	2	3

Derivative	min.[a]	t_r min-sec.[b]	min.[c]
Alanine	10.9	4–48	—
α-Alanine	—	—	14
β-Alanine	—	—	26
Allohydroxyproline	—	—	75
Alloisoleucine	—	—	18
Allothreonine	—	—	34
α-Aminoadipic acid	—	—	40
α-Aminobutyric acid	—	—	17
β-Aminobutyric acid	—	—	28
γ-Aminobutyric acid	—	—	32
β-Aminoisobutyric acid	—	—	23
Arginine	32.7	—	—
Aspartic acid	24.3	12–06	84
Cysteine	26.0	13–48	44
Cystine	41.9	—	—
Ethionine	—	—	55
Glutamic acid	27.2	14–36	102
Glycine	15.7	6–51	24
Histidine	28.3	12–06	—
Hydroxyproline	—	14–24	69
Isoleucine	14.5	6–12	19
Leucine	16.4	6–24	22
Lysine	32.1	22.00	—
Methionine	23.6	13–00	51[d]
Methionine sulfone	—	23–00	49
Methionine sulfoxide	23.6	—	51[d]
S-Methylcysteine	—	—	47
Norleucine	—	—	23
Norvaline	—	—	20
Ornithine	30.8	21–06	33
Phenylalanine	24.7	14–12	60
Pipecolic acid	—	—	26
Proline	18.4	8–48	29
Sarcosine	11.7	—	—
Serine	19.1	11–00	39
Threonine	20.9	9–36	35
Tryptophan	38.8	28–45	—
Tyrosine	35.3	20–12	132
Valine	12.4	5–06	16

[a] *n*-Butyl-*N*-trifluoroacetyl esters.
[b] *n*-Amyl-*N*-acetyl esters.
[c] *n*-Propyl-*N*-acetyl esters.
[d] Methionione is converted to methionine sulfoxide during esterification.

Column Packing
P1 = Gas-Chrom A (60–80 mesh) coated with 1.0% neopentylglycol succinate polyester.
P2 = Chromasorb G.H.P. (80–100 mesh), 50% coated with 0.7% Carbowax 6000 and 50% coated with 0.7% Carbowax plus 0.05% tetracyanoethyl pentaerythritol.
P3 = Chromasorb W (60–80 mesh) acid washed, coated with 1% Carbowax 1540; conditioned 48 hrs at 148 °C with 150 ml/min argon.

Temperature
T1 = Initial temperature = 75 °C for 5 minutes then 5.6 °/min. for 21 minutes followed by 7.9 °/minute to 220 °C.
T2 = Initial temperature = 100–240 °C at 6 °/minute started immediately after solvent front.
T3 = Initial temperature = 125 °C for 23 minutes, then abruptly increased to 148 °C.

Literature
1. Zomzely, C., Marco, G., and E. Emery, *Anal. Chem.*, **34**, 1414 (1962).
2. Coulter, J. R. and C. S. Hann, *J. Chromatogr.*, **36**, 42 (1968).
3. Johnson, D. E., Scott, S. J., and A. Meister, *Anal. Chem.*, **33**, 669 (1961).

TABLE GC 102
ANILINES, SUBSTITUTED

Column Packing	P1	P2	P3
Temperature (°C)	137	137	137
Gas; Flow Rate (ml/min)	N_2	N_2	N_2
Column			
length, ft	4	4	4
diameter, mm, I.D.	3 to 4	3 to 4	3 to 4
form	tubing	tubing	tubing
material	glass	glass	glass
Detector	automatic titrator	automatic titrator	automatic titrator

Compound	V_R	V_R	V_R
2-Fluoroaniline	0.72	—	0.63
2,4-Difluoroaniline	0.74	0.75	—
2,5-Difluoroaniline	0.79	1.08	—
4-Fluoroaniline	0.96	1.28	1.1
3-Fluoroaniline	—	0.99	1.52
3-Fluoro-4-methylaniline	1.88	2.28	—
4-Methylaniline	1.95	1.63	1.75
4-Fluoro-2-methylaniline	2.0	2.28	—
3-Methylaniline	2.0	1.56	1.86
***N*-Methylaniline**	2.05	1.28	1.73
2-Methylaniline	2.05	1.53	1.8
***N,N*-Dimethyl-2-methylaniline**	2.56	0.71	1.07
***N,N*-Dimethylaniline**	2.60	1.05	1.68
2-Chloroaniline	2.8	2.34	2.6
***N*-Ethylaniline**	3.18	1.66	2.33
2-Methoxyaniline	3.4	2.7	3.7
***N*-Methyl-2-methylaniline**	3.4	—	3.05
***N*-Methyl-3-methylaniline**	3.4	—	3.28
4-Chloroaniline	3.7	5.25	4.4
***N*-Ethyl-*N*-methylaniline**	3.76	—	—
3-Chloroaniline	3.8	5.35	4.4
2,4-Dimethylaniline	3.84	2.60	3.36
***o*-Phenylenediamine**	3.92	7.95	6.6
2,5-Dimethylaniline	4.0	2.7	3.14
4-Methoxyaniline	4.1	5.5	—
2,6-Dimethylaniline	4.25	2.5	3.22
3,4-Dimethylaniline	4.27	3.27	3.8
2-Chloro-*N*-methylaniline	4.3	—	3.7
3-Methoxyaniline	4.4	5.8	6.6
2,3-Dimethylaniline	4.6	3.34	4.0
***N,N*-Dimethyl-2-chloroaniline**	4.8	—	—
2-Bromoaniline	4.9	4.5	5.0
4-Methoxyaniline	4.9	—	5.7
***N,N*-Dimethyl-4-methylaniline**	5.1	1.7	2.74
***N*-Ethyl-2-methylaniline**	5.21	2.34	3.66
***N,N*-Dimethyl-3-methylaniline**	5.4	1.8	2.9
***N,N*-Diethylaniline**	5.45	1.9	3.0
***N-n*-Propylaniline**	5.82	2.7	4.0
***N*-Ethyl-3-methylaniline**	5.9	3.02	4.4
***N*-Ethyl-4-methylaniline**	6.0	2.9	4.4
4-Ethoxyaniline	6.2	—	8.8
4-Bromoaniline	6.4	10.0	8.8
3-Bromoaniline	6.4	10.0	9.0
***N*-Methyl-4-chloroaniline**	7.35	—	6.9
3-Bromo-4-methylaniline	8.8	—	—
2-Nitroaniline	9.8	—	—
2,3-Dichloroaniline	10.0	—	—
2,4-Dichloroaniline	10.0	—	—
2-Iodoaniline	10.1	—	—
***N,N*-Diethyl-4-methylaniline**	10.5	3.2	5.0
***N,N*-Diethyl-3-methylaniline**	10.8	3.1	5.1
***N-n*-Butylaniline**	11.0	4.7	7.0
4-Iodoaniline	13.0	—	23.2
3-Iodoaniline	13.1	22.4	23.2
2,5-Dimethoxyaniline	14.0	—	—
3,4-Dichloroaniline	14.6	—	—
5-Amino-2-methoxyaniline	16.4	6.4	—
***N*-Isoamylaniline**	16.8	6.4	—
***N,N*-Dimethyl-4-bromoaniline**	17.3	6.8	13.3

Column Packing P1 = 40% paraffin wax on acid-washed Celite 545 (coarse graded).
P2 = 40% Lubrol MO (polyethylene oxide) on acid washed Celite 545 (coarse graded).
P3 = 40% benzyldiphenyl on acid washed Celite 545 (coarse graded).

Literature James, A. T., *Anal. Chem.*, **28**, 1564 (1956).

TABLE GC 103
DRUGS

Column Packing	P1
Temperature (°C)	T1
Gas; Flow Rate (ml/min)	200, N_2
Column	
length, ft	2
diameter, mm, I.D.	3
form	U-tube
material	all-glass system
Detector	micro-coulometer

Compound	$r_{phenothiazine}$*
Promazine	0.9
Phenothiazine	1.0
Chlorpromazine-*N*-oxide (CPNO)	1.5
Chlorpromazine (CP)	2.0
Demethylchlorpromazine (Nor_1CP)	3.0
2-Chlorophenothiazine (2-Cp)	4.0
Didemethylchlorpromazine (Nor_2CP)	4.1
Chlorpromazine-*N*-oxide sulfoxide (CPNSO)	4.8
Chlorpromazine sulfoxide (CPSO)	5.0
Acetyl-demethylchlorpromazine (acetyl-Nor_1CP)	6.0
Demethylchlorpromazine sulfoxide (Nor_1CPSO)	6.2
Acetyl-fluphenazine	6.2
Acetyl-didemethylchlorpromazine (acetyl-Nor_2CP)	6.5
Didemethylchlorpromazine sulfoxide (Nor_2CPSO)	6.7
2-Chlorophenothiazine sulfoxide (2-CpSO)	7.0
Acetyl-demethylchlorpromazine sulfoxide (acetyl-Nor_1CPSO)	7.9
Acetyl-didemethylchlorpromazine sulfoxide (acetyl-Nor_2CPSO)	8.0
Acetyl-7-hydroxychlorpromazine (acetyl-7-OH-CP)	8.1

* Phenothiazine = 1.0 min.

Column Packing P1 = 0.15% XE-60 (silicone polymer) + 0.1% Carbowax 20M on glass microbeads (40/60 mesh) mixed with 4% AR grade, filter aid, diatomaceous earth.

Temperature T1 = programmed 130 °C to 230 °C at 10°/min, linear.

Literature Johnson, D. E. and H. P. Burchfield in "*Lecture on Gas Chromatography*", Mattock, L. R., and H. A. Szymanski, Eds., Plenum Press, N.Y., 1965.

TABLE GC 104
DRUGS

Column Packing	P1	P2
Temperature (°C)	210	210
Gas; Flow Rate (ml/min)	50,He	50, He
Column		
length, ft	4	4
diameter, in., O.D.	¼	¼
form	—	—
material	glass	glass
Detector	FI	FI

Compound	t_r cm[a]	r	t_r(cm)	r*
Aspirin	0.48	0.09	0.72	0.09
			1.03	0.13
Phenylpropanolamine HCl	not detected		0.94	0.12
Pheniramine maleate	2.59	0.47	4.20	0.53
Diphenhydramine HCl	2.84	0.52	4.95	0.63
Methapyrilene HCl	5.49	1.00	7.35	0.93
Chlorpheniramine maleate*	5.49	1.00	7.87	1.00
Dextromethorphan HBr	6.77	1.23	12.15	1.54
Diphenylpyraline HCl	6.78	1.23	10.85	1.38
Salicylamide	7.69	1.40	1.20	0.15
Phenacetin	7.90	1.44	2.45	0.31
Brompheniramine maleate	8.27	1.51	10.85	1.38
Glyceryl guaiacolate	9.41	1.71	2.37	0.30
Caffeine	10.29	1.87	3.93	0.50
Phenindamine tartrate	10.30	1.88	12.59	1.60
			13.72	1.74
Chlorcylizine HCl	10.69	1.95	16.19	2.06
Pyrilamine maleate	15.08	2.75	16.19	2.06

[a] Recorder speed = 1 in/min.
* Relative to chlorpheniramine maleate.

Column Packing P1 = 2% SE-30 + 2% Carbowax 20M on Diatoport S (80/100 mesh).
P2 = 10% Silicone oil DC-200 on Diatoport S (60/80 mesh).

Literature Reiss, T. J., *J. Ass. Offic. Agr. Chem.*, **53**, 609 (1970).

TABLE GC 105
DRUGS

Column Packing	P1
Temperature (°C)	Col. 220
Gas; Flow Rate (ml/min)	N_2,30
Column	
length, m	1.83
diameter, mm, I.D.	3.2
form	na
material	S
Detector	HFI

Compound	$r_{tetraphenylethylene}$
***p*-Acetamidophenol**	0.50
Acetophenetidin	0.39
Acetylsalicylic acid	0.09
Allylphenylbarbituric acid (Alphenal)	0.94
Alphaprodine	0.29
Amobarbital	0.35
Amphetamine	0.08
Anileridine	>2.5
Aprobarbital	0.24
Barbital	0.21
Butabarbital	0.32
Caffeine	0.50
Carisoprodol	0.20
Chlorpromazine	[a]
Diallylbarbituric acid (Allobarbital)	0.26
Dibutylphthalate	0.50
Diphenylhydantoin	2.31
Di-*n*-propylbarbituric acid	0.25
Ephedrine	0.11
Ethinamate	0.12
Ethotoin	0.55
5-Ethyl-3-methyl-5-phenylhydantoin (Mephenytoin)	0.61
Glutethimide	0.59
Hexobarbital	0.58
***p*-Hydroxyphenobarbital**	2.00
Itobarbital (Butalbital)	0.29
Mebutamate	0.17
Mephobarbital	0.64
Meprobamate	0.16
Methadone	0.57
Methamphetamine	0.09
Methapyrilene	0.38
5-(Methylbutyl)barbituric acid	0.36
Methyprylon	0.27
Methyl salicylate	0.09
Morphine	>3.0
Nialamide	[a]
Nicotine	0.01
Pentobarbital	0.38
Pethidine (Meperidine)	0.23
***p*-Phenetidine**	0.10
Phenobarbital	0.82
2-Phenylglutaramide	0.71
Phenyl salicylate	0.25
Promazine	[a]
***d*-Propoxyphene**	0.28
Salicylamide	0.18
Salicylic acid	0.11
Secobarbital	0.42
Sulfosalicylic acid	0.10
Tetraphenylethylene	1.00[b]
Theobromine	0.80
Theophylline	1.18
Thiopental	0.43
Tybamate	0.24

[a] No peak observed.
[b] Retention time of tetraphenylethylene, 19.80 minutes.

Column Packing P1 = 100/120 mesh Chromosorb, W-HP 6% QF I and 3% SE-30 (7 : 3).

Literature Parker, K.D., Wright, J. A., Halpern, A. F., and C. H. Hine, *J. Forensic Sci.*, **8**, 125 (1968).

TABLE GC 106
DRUGS

Column Packing	P1	P1
Temperature (°C)	210	230
Gas; Flow Rate (cc/min)	30,N_2	30,N_2
Column		
length, ft	2½	2½
diameter, in., I.D.	0.093	0.093
form	na	na
material	SS	SS
Detector	FI	FI

Compound	$r_{probarbital}$	
Metharbital	0.14	0.27
Narconumal	0.20	0.27
Barbital	0.66	0.70
Hexobarbital	0.73	0.88
Butabarbital	0.96	1.04
Probarbital	1.00 (5.6 min.)	1.00 (2.3 min.)
Butethal	1.01	1.04
Allylbarbituric acid	1.02	1.00
Methallatal	1.09	1.17
Amobarbital	1.11	1.17
Aprobarbital	1.12	1.09
Diallylbarbituric acid	1.14	1.13
Mephobarbital	1.20	1.23
Pentobarbital	1.25	1.30
Butalbital	1.26	1.39
Thiopental	1.27	1.30
Secobarbital	1.55	1.56
Thiamylal	1.57	1.57
Vinbarbital	1.59	1.52
Hexethal	1.82	1.69
Cyclopal	3.26	2.91
Propallylonal	3.99	3.48
Cyclobarbital	4.14	3.70
Butallylonal	4.87	4.25
Kemithal	5.55	4.95
Heptabarbital	5.70	4.95
Phenobarbital	6.79	5.65
Alphenal	9.72	8.05
Amobarbital metabolite	6.27	4.74

TABLE GC 106—(Continued)
DRUGS

Salicylic acid	0.27	0.09
Acetylsalicylic acid	0.34	0.09
Phenacetin	0.46	0.65

Column Packing P1 = 1.5% (w/w) SE-30 and 2% (w/w) Carbowax 20M on acid-washed firebrick (100/120 mesh).

Literature Parker, K. D., Fontan, C. R., and P. L. Kirk, *Anal. Chem.*, **35**, 418 (1963).

TABLE GC 107
DRUGS

Column Packing	P1	P1	P1
Temperature (°C)	T1	T2	T3
Gas; Flow Rate (ml/min)	na,25	na,25	na,25
Column			
length, m	1.83	1.83	1.83
diameter, mm, I.D.	3	3	3
form	U-shaped	U-shaped	U-shaped
material	glass	glass	glass
Detector	D1	D1	D1

Compound	t_r		
Apoatropine	5.20	—	—
Apomorphine	—	13.0	4.60
Apomorphine HCl	—	—	4.60
Atropine	—	—	—
Atropine sulfate	5.20[a]	—	—
	8.30[a]	—	—
Brucine	—	—	36.8
Brucine sulfate	—	—	36.8
Caffeine	2.4	—	—
Caffeine citrated	2.4	—	—
Caffeine HBr	2.4	—	—
Caffeine HCl	2.4	—	—
Cinchonine	—	8.2	3.5

TABLE GC 107—(Continued)
DRUGS

Column Packing	P1	P1	P1
Temperature (°C)	T1	T2	T3
Gas; Flow Rate (ml/min)	na,25	na,25	na,25
Column			
length, m	1.83	1.83	1.83
diameter, mm, I.D.	3	3	3
form	U-shaped	U-shaped	U-shaped
material	glass	glass	glass
Detector	D1	D1	D1
Compound		t_r	
Cinchonine sulfate	—	—	3.5
Cinchonidine	—	8.4	3.6
Cocaine	8.6	2.6	—
Cocaine HCl	8.6	2.6	—
Codeine	16.4	4.2	1.7
Codeine HCl	—	—	1.7
Codeine phosphate	—	—	1.7
Cotarnine	8.0[b]	—	—
Cryptopine	—	26.0	8.6
Cryptopine oralate	—	—	8.6
Diacetylmorphine (heroin)	40.2	8.7	3.5
Diacetylmorphine HCl (heroin HCl)	—	—	3.5
Dihydrocodeinone (hydrocodone)	18.3	5.0	2.3
Dihydrocodeinone bitartrate (hydrocodone bitartrate)	—	—	2.3
Dihydromorphine	23.4	5.6	2.4
Dihydromorphinone (hydromorphone)	27.5	6.2	2.9
Dihydromorphinone HCl (hydromorphone HCl)	—	—	2.9
Emetine	—	—	75.0
Ethylmorphine	17.7	4.7	2.2
Ethylmorphine HCl (C dionin)	—	—	2.2
Eucaine	2.1	—	—
Eucaine HCl	2.1	—	—
Gelsemine	—	—	4.6
Gelsemine sulfate	—	—	4.6
Homatropine	5.0	—	—
Homatropine HBr	5.0	—	—
Hydrastine	—	—	10.10
Hydrastine HCl	—	—	10.10
Hydrastinine	3.7[b]	—	—
Hydrastinine HCl	3.7	—	—
Laudanosine	—	10.0	3.8
3,O-Monoacetylmorphine	24.0	6.0	2.7
6,O-Monoacetylmorphine	29.6	6.8	2.9
Morphine	27.2	6.0	2.8
Morphine sulfate	—	—	2.8
Nalorpine	41.0	8.9	3.6
Neopine	17.5	—	2.0
Neopine HBr	—	4.4	2.0
Noscapine	—	45.6	13.7
Papaverine	—	17.4	6.1
Papaverine HCl	—	—	6.1
Physostigmine	8.8[b]	—	—
Physostigmine salicylate	8.8[b]	—	—
Pilocarpine	6.5	—	—
Pilocarpine HCl	6.5	—	—
Quinidine	—	—	5.8
Quinidine sulfate	—	—	5.8
Quinine	—	15.7	5.8
Quinine ethylcarbonate	—	19.1	6.5
Quinine HCl	—	—	5.8
Scopolamine	13.7	—	—
Scopolamine HBr	13.7	—	—
Sparteine	2.1	—	—
Sparteine sulfate	2.1	—	—
Strychnine	—	34.0	11.7
Strychnine nitrate	—	—	11.7
Thebaine	23.6	5.7	2.5

[a] Multiple peaks.
[b] Major peak, decomposition resulted in additional minor peaks.

Column Packing P1 = 100/140 mesh Gas Chrom P, washed with concentrated HCl and methanolic potassium hydroxide and treated with hexamethyldisilazane, coated with 0.1% polyethylene glycol 9000 and after drying with 1.15% silicone rubber SE-30.

Temperature Injection temperature, 325 °C.
T1 = 175 °C.
T2 = 200 °C.
T3 = 225 °C.
Detector na.

Detector D1 = argon β-ionization.

Literature Brochmann-Hanssen, E. and A. Baerheim Svendsen, *J. Pharm. Sci.*, **51**, 1096 (1962).

TABLE GC 108
DRUGS

Column Packing	P1	P1
Temperature (°C)	220	250
Gas; Flow Rate (ml/min)	N_2,15.8	N_2,14.6
Column		
length, m	0.915	0.915
diameter, mm, O.D.	3.2	3.2
form	na	na
material	Pyrex glass	Pyrex glass
Detector	HFI	HFI

Compound	$r_{meprobamate}$	
Carisoprodol	0.53	0.60
Chlormethazonone (Chlormezonone)	2.30	2.17
Chlorpromazine	2.21	2.09
Ectylurea	0.09	[a]
Ethopropazine	1.80	1.73
Fluphenazine	1.35	1.35
Frenquel (Azacyclonol HCl)	1.26	1.59
Librium (Chlordiazepoxide)	6.88	5.91
Mepazine	2.65	2.55
Meprobamate	1.00[b]	1.00[c]
Methoxypromazine	3.47	3.18
Perphenazine	4.18	3.91
Phenaglycodol	0.18	[a]
Prochlorperazine	4.24	4.00
Promazine	1.24	1.68
Promethazine	1.73	0.97
Pyrathiazine	2.79	2.73
Reserpine	1.80	1.64
Thiopropazate	4.18	3.91
Thioridazine	20.00	17.00
Trifluoperazine	3.21	1.41
Triflupromazine	0.68	0.62
Trimeprazine	0.95	0.91

[a] With solvent.
[b] Retention time of meprobamate, 3.4 min.
[c] Retention time of meprobamate, 1.1 min.

Column Packing P1 = 100/120 mesh silanized Gas Chrom P, 1% Hi-Eff-8B.
Literature Jain, N. C. and P. L. Kirk, *Microchem. J.*, **12**, 242 (1967).

TABLE GC 109
DRUGS

Column Packing	P1	P1
Temperature (°C)	205	270
Gas; Flow Rate	—	—
Column		
length, ft	6	6
diameter, mm, I.D.	5	5
form	—	—
material	—	—
Detector	FI	FI

Generic name	$r_{chlorpromazine}$	
Trifluomeprazine	0.43	—
Triflupromazine	0.46	—
Promethazine	0.53	—
Trimeprazine	0.58	—
Promazine	0.60	—
Ethopropazine	0.70	—
Ethyl isobutrazine	0.91	—
Methdilazine	0.92	—
Chlorpromazine	1.00[a]	1.00[b]
Pyrathiazine	1.11	—
Mepazine	1.13	—
Methopromazine	1.20	—
Trifluoperazine	1.69	—
Propiopromazine	2.38	—
Prochlorperazine	—	2.72
Thioridazine	—	3.47
Thiopropazate	—	6.97
Pipamazine	—	8.83
Perphenazine	—	>23.90
Fluphenazine	—	>23.90

[a] Time: 24.9 min.
[b] Time: 3.80 min.

Column Packing P1 = 2% SE-30 on Gas Chrom S (80/100 mesh).
Literature Johnson, D. E. and H. P. Burchfield in "*Lecture on Gas Chromatography*", Mattock, L. R. and H. A. Szymanski, Eds., Plenum Press, N.Y., p. 109 1965.

TABLE GC 110
DRUGS

Column Packing	P1
Temperature (°C)	T1
Gas; Flow Rate (ml/min)	N_2,30
Column	
length, cm	1.525
diameter, mm, O.D.	3.2
form	na
material	SS preferred, or Copper or Aluminum
Detector	H_2,FI

Compound	t_r
Benoxinate HCl	2.5
Butacaine sulfate	3.2
Butethamine HCl	1.65
Chloroprocaine HCl	2.0
Dyclonine HCl	0.80
Isobucaine HCl	0.73
Lidocaine HCl	0.92
Mepivacaine HCl	1.3
Meprylcaine HCl	[a]
Metabutethamine HCl	2.2
Metabutoxycaine HCl	2.1
Nacpaine HCl	0.73
Parethoxycaine HCl	0.87
Piperocaine HCl	1.03
Procaine HCl	1.20
Propoxycaine HCl	[a]
Pyrrocaine HCl	1.08
Tetracaine HCl	1.45

[a] No detector response with up to 5 μl of 2% solution at column temperature from 100–225 °C.

Column Packing P1 = 100/120 mesh siliconized glass beads, 0.1% SE-30.
Temperature T1 = injection 265 °C, column 225 °C.
Literature Koehler, H. M. and J. J. Hefferren, *J. Pharm. Sci.*, **53**, 745 (1964).

TABLE GC 111
DRUGS

Column Packing	P1	P1
Temperature (°C)	220	250
Gas; Flow Rate (ml/min)	N_2,60	N_2,54.5
Column		
length, m	0.915	0.915
diameter, mm, O.D.	3.2	3.2
form	spiral	spiral
material	glass	glass
Detector	HFI	HFI

Compound	$r_{codeine}$	
Atropine	0.46	0.45
Caffeine	0.25	[a]
Cinchonine	3.47	2.67
Cocaine	0.31	0.33
Codeine	1.00[b]	1.00[c]
Colchicine	5.1	5.25
Dihydrocodeinone (Hydrocodone)	1.44	1.25
Dihydrohydroxycodeinone (Oxycodone)	1.84	1.67
Heroin	2.31	1.83
Hydrastine	[d]	13.80
Morphine	4.03	3.0
Narcotine (Noscapine)	[d]	11.8
Papaverine	8.59	6.5
Piperine	8.12	5.5
Quinine	8.06	5.42
Strychnine	19.4	12.8
Thebaine	1.75	1.42
Theobromine	0.56	0.55

[a] With solvent.
[b] Retention time of codeine, 3.2 min.
[c] Retention time of codeine, 1.2 min.
[d] Did not emerge.

Column Packing P1 = 100/120 mesh silanized Gas Chrom P, 1% Hi-Eff-8B.
Literature Jain, N. C. and P. L. Kirk, *Microchem. J.*, **12**, 229 (1967).

TABLE GC 112
GASES

Column Packing	P1	P2	P3	P4	P5	P6	P6	P7	P8	P9	P10	P11	P12	P11	P13	P14
Temperature (°C)	35	35	35	30	22	100	−78	ambient	−72	−196	23	T1	40	T2	35	na
Gas; Flow Rate (ml/min)	30,Ar	30,Ar	30,Ar	60,Ar	30,Ar	65,He	26.7,Ar	na	100,He	60,H_2	60,O^2	60,O_2	N_2	He	H_2	He
Column																
length, ft(a) m(b),	6(a)	6(a)	11(a)	4(b)	36(a)	1.2(b)	na	na	2.5(a)	10(a)	10(a)	10(a)	0.35(b)	2(a)	1½(a)	na
diam., mm(b), in.(a), I.D.	0.125(a)	0.125(a)	0.125(a)	4(b)	0.125(a)	5(b)	na	na	0.25(a)	0.25(a)	0.25(a)	0.25(a)	4(b)	na	0.125(a)	na
form	na	na	na	na	na	na	na	na	na	na	na	na	na	na	na	na
material	na	na	na	na	na	na	na	na	na	na	na	na	na	na	na	na
Detector	TC	TC	TC	TC	TC	EC	TC	TC	TC	TC	TC	TC	TC	EC	na	EC
Literature	1	1	1	2	3	4	3	3	5	6	6	6	7	8	1	9
Compound	t_r	t_r	t_r	t_r	t_r	t_r	t_r	t_r	t_r	V_R	V_R	V_R	t_r	t_r	t_r	t_r
Hydrogen	1.5	2.0	2.2	1.5	8.0	1.1	—	—	—	—	—	—	2	0.3	—	—
Oxygen	—	4.0	4.0	1.8	—	2.0	—	—	3.5	—	—	—	—	0.6	—	1.00
Nitrogen	4.5	4.0	5.0	1.8	—	3.4	—	4.0	—	—	—	—	—	1.5	—	1.96
Methane	7.0	11.0	13.2	2.8	—	5.0	—	7.5	—	—	—	660	3	2.5	2	2.5
Carbon monoxide	20.0	5.5	10.1	2.1	—	10.0	—	—	—	—	—	—	—	5	—	—
Ethane	—	—	—	10.5	—	—	—	—	—	—	—	—	17	—	—	—
Carbon dioxide	—	—	—	15.5	—	—	—	—	—	—	—	—	5	12	5	—
Ethylene	—	—	—	17.5	—	—	—	—	—	—	—	—	12	—	—	—
Helium	—	—	—	—	5.5	—	7.0	—	—	180	—	—	—	—	—	—
Neon	—	—	—	—	6.8	1.0	8.8	—	—	360	—	—	—	—	—	—
Argon	—	—	—	—	—	2.0	—	—	2.5	—	150	160	—	—	—	—
Krypton	—	—	—	—	—	4.0	—	5.5	—	—	240	540	—	—	—	—
Xenon	—	—	—	—	—	13.8	—	—	—	—	840	960	—	—	—	—

Column Packing

P1 = molecular sieve No. 5A (36/52 mesh) activated for 3 hours at 350 °C.
P2 = charcoal (36/52 mesh) activated for 2 hours at 250 °C.
P3 = molecular sieve and charcoal (36/52 mesh).
P4 = silica gel (40/60 mesh).
P5 = 109 g activated 5A Linde molecular sieve (calcium aluminum silicate).
P6 = molecular sieve No. 5A (36/60 mesh) activated at 400 °C.
P7 = 1 part activated charcoal with 10 parts molecular sieve No. 5A.
P8 = molecular sieve No. 5A (30/70 mesh) activated for one hour at 300 °C.
P9 = activated charcoal (20/40 mesh).
P10 = silica gel (20/40 mesh).
P11 = molecular sieve No. 5A (20/40 mesh).
P12 = activated charcoal (75/100 mesh).
P13 = activated charcoal (36/52 mesh).
P14 = molecular sieve No. 5A.

Temperature

T1 = 23 °C to elution of methane, then to 100 °C to elute xenon.
T2 = room temperature for 3 min., then programmed at 10 °/min.

Literature

1. *Gas Analysis by Gas Chromatography*, Warren Spring Lab, D.S.I.R., Stevenage, England, 1962.
2. Boreham, G. R. and F. A. Marhoff, Gas Council Res. Comm. GCS4, 1958.
3. Krejcí, M., Tesařík, K., and J. Janák, *Gas Chromatogr.*, Inst. Soc. Amer. Symp. 1959, H. J. Noebels, *et al.*, Ed., Academic Press, N.Y. (1961), p. 255.
4. Berry, R., *Proc. 4th Symp. Gas Chromatogr.*, Hamburg, 1962, M. Van Swaay, Ed., Butterworths, London (1963), p. 321.
5. Lard, E. W. and R. C. Horn, *Anal. Chem.*, **32**, 878 (1960).
6. Greene, S. A., *Anal. Chem.*, **31**, 480 (1959).
7. Ray, N. H., *J. Appl. Chem.*, **4**, 82 (1954).
8. Manufacturer's literature, W. G. Pye & Co., Ltd.
9. Manufacturer's literature, Gas Chromatography, Ltd.

TABLE GC 113
GASES

Column Packing	P1	P2	P3	P4	P5	P6	P7	P8	P9	P10	P11	P12	P12	P13	P14	P15	P16	P17	P18	P18	P19
Temperature (°C)	32	32	32	32	32	32	32	32	32	32	100	30	32	50	30	78	100	100	78	35	100
Gas; Flow Rate	—	—	—	—	—	—	—	—	—	—	—	—	—	—	—	—	—	—	—	—	—
Column																					
length, ft	10	10	10	10	10	10	10	10	10	10	6	6	3	10	6	6	6	8	8	16	8
diameter, in., O.D.	$\frac{1}{4}$	$\frac{1}{4}$	$\frac{1}{4}$	$\frac{1}{4}$	$\frac{1}{4}$	$\frac{1}{4}$	$\frac{1}{4}$	$\frac{1}{4}$	$\frac{1}{4}$	$\frac{1}{4}$	$\frac{1}{4}$	$\frac{1}{8}$	$\frac{1}{8}$	$\frac{1}{4}$	$\frac{1}{4}$	$\frac{1}{4}$	$\frac{1}{4}$	$\frac{1}{8}$	$\frac{1}{8}$	$\frac{1}{8}$	$\frac{1}{4}$
form	tubing	tubing	tubing	tubing	tubing	tubing	tubing	tubing	tubing	tubing	tubing	tubing	tubing	tubing	tubing	tubing	tubing	tubing	tubing	tubing	tubing
material	SS	SS	SS	SS	SS	SS	SS	SS	SS	SS	SS	SS	SS	SS	SS	SS	SS	SS	SS	SS	SS
Detector	TC	TC	TC	TC	TC	TC	TC	TC	TC	TC	TC	TC	TC	TC	TC	TC	TC	TC	TC	TC	TC
Compound	t_r																				
Air	2.22	0.73	2.25	0.65	2.14	0.65	3.04	1.10	3.04	1.08	0.96	1.11	0.69	0.67	1.20	0.40	0.62	0.81	0.64	1.60	0.49
CO_2	2.60	0.94	2.63	0.73	2.56	0.62	3.04	1.12	3.32	1.08	1.58	—	0.93	0.78	3.66	0.60	0.62[b]	1.07	0.76	1.65	0.73
N_2O	2.73	0.92	2.80	0.80	2.74	0.74	3.04	1.12	3.46	1.15	1.81	—	1.06	0.78	4.80[b]	0.84	0.62	1.07	0.79	4.57	0.76
NO	2.28	0.73	2.29	0.67	2.16	0.65	3.04	1.12	3.04	1.08	1.00	1.98	0.69	0.70	—	0.47	0.62	0.81	0.60	1.95	0.50
NO_2	NR	NR	NR	0.97	4.5[a,b] 5.1[a,b]	17.6[b]	3.42	1.43	NR	18.3[a,b,c]	1.10	2.42[b]	2–2.6[a,b]	4.8± 0.1	3.75[b]	1.70	NR	7.2[b]	1.25[b]	8.05[b]	2.9–3.1[b]
H_2S	3.25	1.47	4.09	1.07	2.86	0.88	3.04	1.12	3.50	1.23	3.65	—	2.46[b]	1.05	—	1.75	0.62	2.39	1.35	—	1.44
SO_2	4.13	9.30	11.02	1.90	5.37	1.97	3.18[c]	1.25	4.34	1.36	7.38	2.02	6.9[b]	1.92	3.60[b]	3.88	>20[b]	4.95[b]	4.26[b]	9.1[b]	4.32[b]
Cl_2	6.01	2.34	8.80	2.10	3.83	1.21	3.31[c]	1.25	5.62	1.65	3.71	—	3.05[a,b]	1.74[b]	—	1.49	>10[b]	5.13	5.37[b]	—	2.57[b]
HCl	NR	NR	10.8[a,b]	0.88	2.72	1.89[b]	3.04	1.16	3.32	1.37	3.65[b]	—	2.04[b]	2.26[b]	—	1.49	NR	2.55[b]	NR	—	2.55[b]
NH_3	2.83	1.99	2.33	1.14[b]	4.8[a,b]	0.87	3.67[b]	NR	NR	1.40[b]	1.09	—	1.60[b]	1.02[b]	—	0.89	NR	—	1.35	—	1.28[b]

[a] Peak exhibited considerable leading. [b] Peak exhibited considerable tailing. [c] Peak was considerably broader than the rest of the peaks on that column.

Column Packing

P1 = 10% Silicone oil SF-96 on Chromosorb T (40/60 mesh).
P2 = 10% Triacetin on Chromosorb T (40/60 mesh).
P3 = 10% Di-*n*-decyl phthalate on Chromosorb T (40/60 mesh).
P4 = 10% Arochlor 1232 on Chromosorb T (40/60 mesh).
P5 = 10% QF-1 (FS-1265) Fluoruon Chromosorb T (40/60 mesh).
P6 = 10% Silicone XE-60 (nitrile) on Chromosorb T (40/60 mesh).
P7 = 10% Fluorolube HG-1200 grease on Chromosorb W, AW, DMCS (60/80 mesh).
P8 = 10% Kel-F 90 grease on Chromosorb W, AW, DMCS (60/80 mesh).
P9 = 10% Halocarbon 11-14 on Chromosorb W, AW, DMCS (60/80 mesh).
P10 = 10% Silicone DC-200 oil on Chromosorb W, AW, DMCS (60/80 mesh).
P11 = 3% H_3PO_4 on Porapak Q (80/100 mesh).
P12 = Polypak 1 (80/120 mesh).
P13 = 10% Di-*n*-ethylhexyl adipate on Chromosorb W, AW (30/100 mesh).
P14 = Porapak Q (80/100 mesh).
P15 = Porapak Q5 (80/100 mesh).
P16 = Porosil B (80/100 mesh).
P17 = 5% Arochlor 1232 on Porapak Q (50/80 mesh).
P18 = 5% Arochlor 1232 on Porapak R (50/80 mesh).
P19 = Porapak Q (50/80 mesh).

Literature

Bethea, R. M. and M. C. Meador, *J. Chromatogr. Sci.*, 7, 655 (1969).

TABLE GC 114
GASES

Column Packing	P1	P2	P3
Temperature (°C)	60	48	48
Gas; Flow Rate	He,na	He,na	He,na
Column			
length	—	—	—
diameter	—	—	—
form	—	—	—
material	—	—	—
Detector	—	—	—
Literature	1	2	3

Compound	$r_{chlorine}$		
Chlorine monofluoride	0.48	0.60	—
Hydrogen fluoride	0.60	0.70	—
Chlorine	1.00	1.00	1.00
Chlorine trifluoride	1.45	1.40	—
Chlorine pentafluoride	0.85	—	—
Bromine	2.60	2.28	16.7
Bromine pentafluoride	—	3.05	—
Chlorine monoxide	1.76	—	—
Chlorine dioxide	1.45	—	—
Chlorine oxidifluoride	1.28	—	—
Fluorine monoxide	0.40	—	—
Uranium hexafluoride	6.00	5.25	—

Column Packing P1 = Kel F-40.
P2 = Silicone 702.
P3 = Silica Gel.

Literature

1. Hamlin, A. G., Iveson, G., and Phillips, T. R., *Anal. Chem.*, **35**, 2037 (1963).
2. Ellis, J. F. and Iveson, G., in "*Gas Chromatography*", R. P. W. Scott (Ed.), Butterworths, London, 1960, p. 308.
3. Janak, J., Nedorost, M., and V. Bubenikova, *Chem. Listy*, **51**, 890 (1957), *Collec. Czech. Chem. Commun.*, **22**, 1799 (1957).

TABLE GC 115
GASES: ISOTOPES OF HYDROGEN

Column Packing	P1
Temperature (°C)	−196
Gas; Flow Rate (ml/min)	Ne,200
Column	
length, mm	300
diameter mm, I.D.	4
form	coiled
material	Cu
Detector	dual; catharometer and ionization chamber

Compound	t_r
H_2	14.0
HD	17
HT	19
D_2	22
DT	25
T_2	30

Column Packing P1 = activated alumina sieved between 125–150 microns then covered with ferric hydroxide, dried and activated in vacuum for 24 hours at 220 °C, sieved a second time, then deactivated with carbon dioxide.

Literature
Genty, C. and R. Schott, *Anal. Chem.*, **42**, 7 (1970).

TABLE GC 116
GLYCOL ETHERS

Column Packing	P1	P1	P2	P2
Temperature (°C)	140	170	140	170
Gas; Flow Rate (ml/min)	22.5,He	22.5,He	22.5,He	22.5,He
Column				
length, m	2.4	2.4	2.4	2.4
diameter, mm, I.D.	3,6	3,6	3,6	3,6
form	na	na	na	na
material	na	na	na	na
Detector	FI	FI	FI	FI

Compound	I			
Methyl ether of				
Ethylene glycol	1098	1150	534	744
Diethylene glycol	1474	1398	826	907
Triethylene glycol	1833	1700	1114	1150
Tetraethylene glycol	—	1980	1413	1416
Ethyl ether of				
Ethylene glycol	1148	1219	622	718
Diethylene glycol	1492	1500	900	951
Triethylene glycol	1850	1771	1183	1198
Tetraethylene glycol	—	2048	1458	1463
Propyl ether of				
Ethylene glycol	1223	1272	723	819
Diethylene glycol	1573	1558	1002	1046
Triethylene glycol	1926	1823	1283	1299
Tetraethylene glycol	—	2099	1557	1555
Butyl ether of				
Ethylene glycol	1316	1326	826	887
Diethylene glycol	1668	1626	1109	1139
Triethylene glycol	2020	1892	1379	1396
Tetraethylene glycol	—	2168	1651	1655
Monoisopropyl ether of				
Ethylene glycol	1154	—	652	—
Monoisobutyl ether of				
Ethylene glycol	1234	—	789	—
Monoisoamyl ether of				
Ethylene glycol	1296	—	892	—
Ethylene glycol	1430	1432	634	652
Diethylene glycol	1782	1707	900	925
Triethylene glycol	2132	1983	1143	1177
Ethylene glycol (min)	3.80	1.60	—	—
Diethylene glycol (min)	20.8	6.80	—	—

Column Packing P1 = 10% NPGS on Chromosorb W, AW-DMSC (80/100 mesh).
P2 = 10% Apiezon L on Chromosorb W, AW-DMCS (80/100 mesh).

Literature Singliar, M. and J. Dykyj, *Collect. Czech. Chem. Commun.*, **34**, 773 (1969).

TABLE GC 117
KETONES

Column Packing	P1
Temperature (°C)	130
Gas; Flow Rate (ml/min)	$55,N_2$
Column	
length, ft	10
diameter, in., O.D.	$\frac{1}{4}$
form	—
material	Al
Detector	na

Compound	$r_{benzene}$
Acetone	0.306
Methyl ethyl ketone	0.568
Diethyl ketone	0.927
Methyl propyl ketone	0.970
Methyl butyl ketone	1.705
Ethyl butyl ketone	2.74
Methyl pentyl ketone	2.98
Propyl butyl ketone	4.41
Methyl hexyl ketone	5.04
Dibutyl ketone	7.46
Ethyl hexyl ketone	7.93
Methyl hexyl ketone	8.63
Ethyl heptyl ketone	13.38
Methyl octyl ketone	14.68
Dipentyl ketone	20.9
Ethyl octyl ketone	22.4
Methyl nonyl ketone	24.9
Ethyl nonyl ketone	37.6

Column Packing P1 = 10% SE-30 (methyl silicone polymer) on acid-washed Celite 560 (60/80 mesh).

Literature Haken, J. K. and P. Souter, *J. Gas Chromatogr.*, **3**, 348 (1965).

TABLE GC 118
HYDROCARBONS

Column Packing	P1	P1	P2	P2	P3	P3	P3	P4	P4	P4	P5	P5	P6	P7
Temperature (°C)	28	30	43	52	86	108	86	95	114	124	96	124	28	125
Gas; Flow Rate (ml/min)	70–130, He	70–130, He	50–75, He	50–75, He	28–35, He	28–35, He	28–35, He	40–80, He	40–80, He	40–80, He	50–100, He	50–100, He	20	60
Column														
length, ft	35	35	15	15	26	26	26	10	10	10	30	30	8	14
diameter, in., O.D.	$\frac{1}{4}$	$\frac{1}{4}$	$\frac{1}{4}$	$\frac{1}{4}$	$\frac{1}{4}$	$\frac{1}{4}$	$\frac{1}{4}$	$\frac{1}{4}$	$\frac{1}{4}$	$\frac{1}{4}$	$\frac{1}{4}$	$\frac{1}{4}$	$\frac{1}{4}$	$\frac{1}{4}$
form	spiral	spiral	spiral	spiral	spiral	spiral	spiral	spiral	spiral	spiral	spiral	spiral	spiral	spiral
material	copper	copper	copper	copper	copper	copper	copper	copper	copper	copper	copper	copper	copper	copper
Detector	na	na	na	na	na	na	na	na	na	na	na	na	na	na
Compound						r_{hexane}							r_{ethane}	r_{hexane}
Carbon monoxide	—	—	—	—	—	—	—	—	—	—	—	—	0.003	—
Methane	0.002	—	—	—	—	—	—	—	—	—	—	—	0.095	—
Ethane	0.016	0.017	0.023	—	—	—	0.024	—	—	—	—	—	1.000	—
Ethylene	0.019	0.02	0.028	0.03	—	—	—	—	—	—	—	—	2.10	—
Carbon dioxide	0.038	0.044	0.06	0.059	—	—	—	—	—	—	—	—	1.85	—
Propane	0.047	0.050	0.061	0.066	0.06	—	0.07	0.07	0.09	0.10	0.07	—	5.0	—
Propylene	0.077	0.086	0.102	0.125	0.75	—	0.08	0.09	—	0.12	0.09	—	—	—
Isobutane	0.091	0.093	0.11	0.12	—	—	0.13	—	0.16	0.18	—	—	—	—
Cyclopropane	—	0.121	—	—	—	—	—	—	—	—	—	—	—	—
n-Butane	0.138	0.142	0.164	0.176	0.19	—	0.19	0.20	0.215	0.247	0.19	—	—	—
Neopentane	0.147	0.15	0.17	0.188	—	—	—	—	—	—	—	—	—	—
Acetylene	—	0.19	—	—	—	—	—	—	—	—	—	—	—	—
1-Butene	0.210	0.233	0.256	0.28	—	—	0.19	0.20	0.21	—	—	—	—	—
Isobutylene	0.223	0.247	0.271	0.29	—	—	—	—	—	—	—	—	—	—
Allene	—	0.267	—	—	—	—	—	—	—	—	—	—	—	—
trans-2-Butene	0.270	0.305	0.325	0.36	0.28	—	0.234	0.25	0.26	0.306	0.26	—	—	—
Isopentane	0.290	0.295	0.32	0.36	0.33	—	0.35	0.38	0.40	0.43	—	—	—	—
cis-2-Butene	0.317	0.358	0.381	0.405	—	—	0.255	0.27	0.28	0.31	0.27	—	—	—
n-Pentane	0.378	0.388	0.41	0.455	0.43	0.46	0.44	0.46	0.47	0.506	0.45	0.55	—	0.55
1,1-Dimethylcyclopropane	0.366	0.39	—	0.476	0.35	—	0.347	—	—	—	—	—	—	—
3-Methyl-1-butene	0.375	0.40	0.426	0.48	0.35	—	0.335	—	—	—	—	—	—	—
2,2-Dimethylbutane	0.52	0.52	0.54	0.593	0.557	—	0.594	—	0.65	—	—	—	—	0.56
1,3-Butadiene	0.45	0.54	0.555	0.595	—	—	—	—	—	—	—	—	—	—
Methylacetylene	—	0.58	—	—	—	—	—	—	—	—	—	—	—	—
1-Pentene	0.545	0.595	0.616	0.655	—	—	—	—	0.485	—	—	—	—	—
3,3-Dimethyl-1-butene	0.615	—	0.66	0.69	—	—	—	—	—	—	—	—	—	—
2-Methyl-1-butene	0.635	0.69	0.705	0.760	0.48	—	0.48	—	—	—	—	—	—	—
2-Methylpentane	0.740	0.725	0.74	0.776	0.773	—	0.768	0.787	0.79	0.825	0.80	—	—	—
trans-2-Pentene	0.660	0.735	0.748	0.79	0.58	0.60	—	—	0.57	0.575	0.58	—	—	—

TABLE GC 118—(Continued)
HYDROCARBONS

Column Packing	P1	P1	P2	P2	P3	P3	P3	P4	P4	P4	P5	P5	P6	P7
Temperature (°C)	28	30	43	52	86	108	86	95	114	124	96	124	28	125
Gas; Flow Rate (ml/min)	70–130, He	70–130, He	50–75 He	50–75, He	28–35, He	28–35, He	28–35, He	40–80, He	40–80, He	40–80, He	50–100, He	50–100, He	20	60
Column														
length, ft	35	35	15	15	26	26	26	10	10	10	30	30	8	14
diameter, in., O.D.	$\frac{1}{4}$	$\frac{1}{4}$	$\frac{1}{4}$	$\frac{1}{4}$	$\frac{1}{4}$	$\frac{1}{4}$	$\frac{1}{4}$	$\frac{1}{4}$	$\frac{1}{4}$	$\frac{1}{4}$	$\frac{1}{4}$	$\frac{1}{4}$	$\frac{1}{4}$	$\frac{1}{4}$
form	spiral	spiral	spiral	spiral	spiral	spiral	spiral	spiral	spiral	spiral	spiral	spiral	spiral	spiral
material	copper	copper	copper	copper	copper	copper	copper	copper	copper	copper	copper	copper	copper	copper
Detector	na	na	na	na	na	na	na	na	na	na	na	na	na	na
Compound						r_{hexane}							r_{ethane}	r_{hexane}
1-Ethyl-1-*trans*-2-dimethyl-cyclopropane	—	2.6	2.30	—	2.1	—	—	—	—	—	—	—	—	—
2,4-Dimethyl-1-pentene	—	—	2.26	2.16	—	—	—	—	—	—	—	—	—	—
Cyclohexane	—	2.6	2.46	2.36	1.66	1.77	1.86	1.85	—	1.80	1.85	1.27	—	—
n-Heptane	—	2.5	2.38	2.25	2.23	2.10	2.22	2.14	2.02	1.97	2.15	1.43	—	—
2,3-Dimethyl-2-butene	2.42	2.70	2.59	2.47	—	—	1.47	—	—	—	—	—	—	1.71
Cyclopentadiene	—	2.92	2.82	2.68	—	—	—	—	—	—	—	—	—	—
2,5-Dimethylhexane	—	—	2.81	2.70	2.62	—	2.30	—	—	—	—	—	—	—
2,3-Dimethyl-1-pentene	—	2.95	2.60	—	—	—	—	—	—	—	—	—	—	—
3-Methyl-2-ethyl-1-butene	—	3.25	2.81	—	—	—	—	—	—	—	—	—	—	—
2,4-Dimethylhexane	—	—	—	2.82	2.73	—	—	—	—	—	—	—	—	—
2,4,4-Trimethyl-1-pentene	—	—	3.1	—	2.62	—	—	—	—	—	—	—	—	—
2,2,3-Trimethylpentane	—	—	—	3.24	2.82	—	—	—	—	—	—	—	—	2.0
1-Methylcyclopentene	—	3.76	—	3.3	1.99	—	1.84	—	—	—	—	—	—	—
2,3-Dimethylbutadiene	—	—	3.8	3.6	1.6	—	—	—	—	—	—	—	—	2.25
1-Heptene	—	3.6	3.48	—	—	—	—	—	—	—	—	—	—	—
3,4-Dimethyl-*cis*-2-pentene	—	3.77	3.34	—	2.2	—	—	—	—	—	—	—	—	—
3,4-Dimethyl-*trans*-2-pentene	—	3.94	3.43	—	2.2	—	—	—	—	—	—	—	—	—
cis-3-Heptene	—	3.9	3.60	—	—	—	—	—	—	—	—	—	—	—
2-Methylheptane	—	—	—	3.70	3.57	—	—	—	—	—	—	—	—	—
2,4,4-Trimethyl-2-pentene	—	—	3.8	—	2.84	—	—	—	—	—	—	—	—	—
trans-3-Heptene	—	4.25	3.84	—	—	—	—	—	—	—	—	—	—	—
2,3,4-Trimethylpentane	4.7	—	3.90	3.75	3.28	—	3.55	—	—	—	—	—	—	—
Methylcyclohexane	—	4.2	3.88	—	2.83	—	3.2	3.0	—	2.72	3.1	2.10	—	—
trans-2-Heptene	—	4.6	4.10	—	2.77	—	—	—	—	—	2.5	—	—	—
Ethylcyclopentane	—	4.76	4.38	4.0	3.10	—	3.2	3.1	—	2.77	3.2	—	—	—
2,3,3-Trimethylpentane	—	—	—	4.06	3.42	—	—	—	—	—	—	—	—	—
3-Ethylhexane	—	—	—	4.08	4.03	—	4.20	—	—	3.29	—	—	—	2.34
cis-2-Heptene	—	5.0	4.53	—	3.0	—	—	—	—	—	2.7	—	—	—
3,4,4-Trimethyl-1-pentene	—	—	—	4.53	—	—	—	—	—	—	—	—	—	—

TABLE GC 118—(Continued) HYDROCARBONS

Column Packing	P1	P1	P2	P2	P3	P3	P3	P4	P4	P4	P5	P5	P6	P7
Temperature (°C)	28	30	43	52	86	108	86	95	114	124	96	124	28	125
Gas; Flow Rate (ml/min)	70–130, He	70–130, He	50–75, He	50–75, He	28–35, He	28–35, He	28–35, He	40–80, He	40–80, He	40–80, He	50–100, He	50–100, He	20 He	60 He
Column														
length, ft	35	35	15	15	26	26	26	10	10	10	30	30	8	14
diameter, in., O.D.	1/4	1/4	1/4	1/4	1/4	1/4	1/4	1/4	1/4	1/4	1/4	1/4	1/4	1/4
form	spiral	spiral	spiral	spiral	spiral	spiral	spiral	spiral	spiral	spiral	spiral	spiral	spiral	spiral
material	copper	copper	copper	copper	copper	copper	copper	copper	copper	copper	copper	copper	copper	copper
Detector	na	na	na	na	na	na	na	na	na	na	na	na	na	na
Compound						r_{hexane}							r_{ethane}	r_{hexane}
Cyclohexene	—	5.40	4.9	4.63	2.9	—	—	—	—	—	—	—	—	—
2,3-Dimethyl-2-pentene	—	5.54	4.83	—	2.2	—	1.89	—	—	—	—	—	—	2.8
2-Methylpentadiene	—	5.6	5.0	4.73	3.26	—	—	—	—	—	—	—	—	2.6
3,3-Dimethyl-2-ethyl-1-butene	—	—	—	4.84	4.00	—	3.88	—	—	—	—	—	—	—
2,5-Dimethyl-1-hexene	—	—	5.9	5.5	2.3	—	1.96	—	—	—	—	—	—	—
2-Ethyl-1,3-butadiene	—	6.1	—	5.1	—	—	—	—	—	—	—	—	—	—
2,3,3-Trimethyl-1-pentene	—	—	—	5.15	5.0	4.4	5.1	4.99	4.17	3.8	4.68	2.8	—	—
n-Octane	—	6.3	5.62	5.05	2.4	—	2.02	—	—	—	—	—	—	—
2-Methylpentadiene	—	6.5	5.8	5.4	3.11	—	—	—	—	—	—	—	—	—
3,4,4-Trimethyl-2-pentene	—	—	—	5.63	3.66	—	—	—	—	—	—	—	—	2.7
1-Methyl-*trans*-2-ethylcyclopentane	—	—	6.55	5.85	4.9	—	4.87	—	—	4.08	—	—	—	—
3-Methyl-1,3-pentadiene	—	6.64	—	5.5	—	—	2.12	—	—	—	—	—	—	—
1-Methylcyclopentadiene	—	7.4	—	5.7	2.44	—	2.0	—	—	—	—	—	—	—
2,3,4-Trimethyl-2-pentene	—	—	7.28	—	4.41	—	—	—	—	—	—	—	—	—
2-Methyl-2-ethylcyclopentene	—	—	7.4	—	—	—	—	—	—	6.7	—	—	—	—
2,3,4-Trimethyl-1-pentene	—	—	7.50	—	4.02	—	—	—	—	—	—	—	—	—
2,3-Dimethylbutane	0.74	0.74	0.758	0.794	0.726	—	0.68	—	—	—	—	—	—	0.63
cis-2-Pentene	0.725	0.812	0.813	0.825	—	—	—	—	—	—	0.65	—	—	—
3-Methylpentane	0.875	0.873	0.976	0.910	0.84	—	0.87	0.886	—	0.91	—	—	—	0.91
1,4-Pentadiene	—	0.895	0.90	0.936	—	—	—	—	—	—	—	—	—	—
2-Methyl-2-butene	0.850	0.935	0.92	0.942	0.60	—	0.60	—	—	—	—	—	—	—
1,1,2-Trimethylcyclopropane	—	0.94	—	0.98	—	—	—	—	—	—	—	—	—	—
n-Hexane	1.00	1.00	1.00	1.00	1.00	1.00	1.00	1.00	1.00	1.00	1.00	1.00	—	1.00
3-Methyl-1-pentene	—	1.06	1.04	—	—	—	—	—	—	—	—	—	—	—
4-Methyl-1-pentene	1.01	1.06	1.04	—	—	—	—	—	—	—	—	—	—	—
Cyclopentane	1.02	1.14	1.14	1.15	—	—	—	—	—	—	—	—	—	—
4-Methyl-*cis*-2-pentene	—	1.14	1.14	—	—	—	—	—	—	—	—	—	—	—
4-Methyl-*trans*-2-pentene	1.08	1.18	1.16	—	—	—	—	—	—	—	—	—	—	—
2,2-Dimethylpentane	—	—	—	1.19	1.2	—	—	—	—	—	—	—	—	—

TABLE GC 118—(Continued) HYDROCARBONS

Column Packing	P1	P1	P2	P2	P3	P3	P3	P4	P4	P4	P5	P5	P6	P7
Temperature (°C)	28	30	43	52	86	108	86	95	114	124	96	124	28	125
Gas; Flow Rate (ml/min)	70–130, He	70–130, He	50–75, He	50–75, He	28–35, He	28–35, He	28–35, He	40–80, He	40–80, He	40–80, He	50–100, He	50–100, He	20 He	60 He
Column														
length, ft	35	35	15	15	26	26	10	10	10	10	30	30	8	14
diameter, in., O.D.	$\frac{1}{4}$	$\frac{1}{4}$	$\frac{1}{4}$	$\frac{1}{4}$	$\frac{1}{4}$	$\frac{1}{4}$	$\frac{1}{4}$	$\frac{1}{4}$	$\frac{1}{4}$	$\frac{1}{4}$	$\frac{1}{4}$	$\frac{1}{4}$	$\frac{1}{4}$	$\frac{1}{4}$
form	spiral	spiral	spiral	spiral	spiral	spiral	spiral	spiral	spiral	spiral	spiral	spiral	spiral	spiral
material	copper	copper	copper	copper	copper	copper	copper	copper	copper	copper	copper	copper	copper	copper
Detector	na	na	na	na	na	na	na	na	na	na	na	na	na	na
Compound						r_{hexane}							r_{ethane}	r_{hexane}
Isopropylcyclopropane	—	1.21	—	—	—	—	—	—	—	—	—	—	—	—
2,3-Dimethyl-1-butene	1.17	1.28	1.24	1.20	—	—	—	—	—	—	—	—	—	—
2,4-Dimethylpentane	1.30	1.32	1.21	1.20	1.21	—	1.25	—	—	—	—	—	—	—
Isoprene	1.29	1.51	1.46	1.41	0.88	—	—	—	—	—	—	—	—	—
2,2,3-Trimethylbutane	—	—	—	1.43	1.28	—	—	—	—	—	—	—	—	—
Cyclopentene	—	1.55	1.53	1.47	1.08	—	—	—	—	—	—	—	—	—
1-Hexene	1.44	1.57	1.51	1.47	—	—	—	—	—	—	—	1.15	—	—
2-Methyl-1-pentene	1.485	1.62	1.55	1.54	1.07	—	1.02	—	—	—	—	—	—	—
***trans*-3-Hexene**	1.47	1.65	1.575	1.54	1.13	—	—	1.13	1.1	1.07	1.2	—	—	—
4,4-Dimethyl-*trans*-2-pentene	—	1.67	1.58	1.57	—	—	—	—	—	—	—	—	—	—
***trans*-2-Hexene**	1.61	1.76	1.67	1.66	1.2	—	—	1.23	1.2	1.19	1.4	—	—	—
***cis*-3-Hexene**	1.61	1.76	1.67	1.66	1.2	—	—	1.23	1.2	1.19	1.4	—	—	—
3,3-Dimethylpentane	—	2.0	1.7	—	1.46	—	—	—	—	—	—	—	—	—
2-Ethyl-1-butene	1.69	1.84	1.80	1.75	1.23	—	—	—	—	—	—	—	—	—
Methylcyclopentane	1.74	1.78	1.74	1.71	1.315	1.34	1.39	1.39	1.37	1.38	1.39	—	—	—
2-Methyl-2-pentene	1.73	1.93	1.85	—	1.24	—	1.17	—	—	—	—	—	—	—
***cis*-2-Hexene**	1.83	2.01	1.9	1.82	1.23	—	—	1.23	1.2	1.19	1.43	—	—	—
3-Methylhexane	—	—	—	1.84	1.85	—	1.88	—	—	1.69	—	—	—	1.47
1-*trans*-3-Pentadiene	1.73	2.06	1.97	1.92	—	—	—	—	—	—	—	—	—	—
3,3-Dimethylpentene	—	2.25	2.02	—	—	—	—	—	—	—	—	—	—	—
2,2,4-Trimethylpentane	2.47	—	2.02	1.95	2.08	—	2.08	—	—	—	—	—	—	—
3-Methyl-*trans*-2-pentene	1.92	2.14	2.1	2.00	—	—	1.25	—	—	—	—	—	—	—
2,3-Dimethylpentane	—	2.2	1.92	—	1.67	—	1.81	—	—	—	—	—	—	—
1-*cis*-3-Pentadiene	1.9	2.31	2.19	2.1	—	—	—	—	—	—	—	—	—	—
4,4-Dimethyl-*cis*-2-pentene	—	—	—	2.08	—	—	—	—	—	—	—	—	—	—
3-Ethylpentane	—	—	—	2.09	1.91	—	2.06	—	—	1.82	—	—	—	1.49
3-Methylcyclopentene	—	2.32	—	2.12	1.31	—	—	—	—	—	—	—	—	—
3-Methyl-*cis*-2-pentene	2.12	2.35	2.2	2.12	1.35	—	1.34	—	—	—	—	—	—	—
2,3,3-Trimethyl-1-butene	—	2.54	2.22	2.18	1.45	—	—	—	—	—	—	—	—	—

TABLE GC 118—(Continued) HYDROCARBONS

Column Packing	P1	P1	P2	P2	P3	P3	P3	P4	P4	P4	P5	P5	P6	P7
Temperature (°C)	28	30	43	52	86	108	86	95	114	124	96	124	28	125
Gas; Flow Rate (ml/min)	70–130, He	70–130, He	50–75, He	50–75, He	28–35, He	28–35, He	28–35, He	40–80, He	40–80, He	40–80, He	50–100, He	50–100, He	20 He	60 He
Column														
length, ft	35	35	15	15	26	26	26	10	10	10	30	30	8	14
diameter, in., O.D.	1/4	1/4	1/4	1/4	1/4	1/4	1/4	1/4	1/4	1/4	1/4	1/4	1/4	1/4
form	spiral	spiral	spiral	spiral	spiral	spiral	spiral	spiral	spiral	spiral	spiral	spiral	spiral	spiral
material	copper	copper	copper	copper	copper	copper	copper	copper	copper	copper	copper	copper	copper	copper
Detector	na	na	na	na	na	na	na	na	na	na	na	na	na	na
Compound						r_{hexane}							r_{ethane}	r_{hexane}
1-Ethyl-1-*cis*-2-dimethylcyclopropane	—	2.5	2.20	—	2.1	—	—	—	—	—	—	—	—	—
3-Ethyl-3-hexene	—	—	7.6	6.3	5.18	—	—	—	—	3.62	—	—	—	3.19
1-*trans*-2-Dimethylcyclohexane	—	—	7.32	6.56	5.34	—	—	5.42	—	4.56	6.5	3.9	—	—
1-Methylcyclopentadiene	—	8.3	—	6.35	2.50	—	2.0	—	—	—	—	—	—	—
1-*trans*-4-Dimethylcyclohexane	—	—	5.82	5.34	4.5	—	—	—	—	—	—	—	—	—
1-*cis*-3-Dimethylcyclohexane	—	—	6.07	5.49	4.9	—	4.87	—	—	3.97	—	—	—	—
Cycloheptane	—	—	—	—	5.2	—	—	5.2	—	—	5.5	—	—	—
1-*cis*-4-Dimethylcyclohexane	—	—	7.85	7.03	5.28	—	—	—	—	—	—	—	—	—
1-*trans*-3-Dimethylcyclohexane	—	—	8.0	7.13	5.48	—	5.80	—	—	4.63	—	—	—	—
3-Ethyl-2-hexene	—	—	8.6	7.1	5.56	—	—	—	—	3.90	—	—	—	3.53
1-Octene	—	—	8.47	7.55	5.22	—	—	4.8	—	—	5.6	—	—	—
2,2,3,4-Tetramethylpentane	—	—	—	—	5.60	—	—	—	—	—	—	—	—	—
1-Methyl-*cis*-2-ethylcyclopentane	—	—	9.9	—	6.2	—	—	—	—	5.1	—	—	—	—
1-*cis*-2-Dimethylcyclohexane	—	—	10.3	9.11	6.5	—	7.15	6.78	5.9	5.63	7.1	4.5	—	—
Ethylcyclohexane	—	—	10.4	9.2	6.9	—	7.15	6.78	5.9	5.63	7.12	4.5	—	—
5-Methylcyclopentadiene	—	9.7	—	—	—	—	—	—	—	—	—	—	—	—
2,5-Dimethyl-1,5-hexadiene	—	—	12.8	11.2	5.8	—	4.9	—	—	—	—	—	—	—
2,2,3,4-Tetramethylpentene	—	—	—	—	6.95	—	—	—	—	—	—	—	—	—
Methylcycloheptane	—	—	13.7	12.0	8.6	—	9.1	8.52	7.4	6.9	—	5.6	—	—
Cyclooctane	—	—	—	—	4.1	—	—	—	—	10.2	14.9	8.5	—	—
Benzene	—	19	—	14.2	3.67	3.6	2.80	2.70	2.56	2.50	4.0	2.55	—	7.12
Toluene	—	—	—	30.7	8.66	7.9	6.45	6.10	5.48	5.10	9.2	5.37	—	11.8
Ethylbenzene	—	—	—	—	18.1	15.1	13.5	12.2	10.2	9.2	16.8	9.08	—	18.0
p-Xylene	—	—	—	—	19.3	16.2	14.5	13.6	—	—	21.0	11.0	—	—
m-Xylene	—	—	—	—	20.5	17.1	15.0	13.8	11.3	10.2	20.2	10.6	—	19.2
o-Xylene	—	—	—	—	25.9	21.2	18.6	16.7	13.65	12.2	27.0	13.6	—	24.2
Styrene	—	—	—	—	30.6	—	—	—	—	13.2	29.0	—	—	34.0
Ethyl ether	—	1.47	—	1.30	0.85	—	—	—	—	—	—	—	—	—
Acetone	—	—	—	7.4	6.0	—	0.95	—	—	0.9	—	—	—	4.0

TABLE GC 118—(Continued)
HYDROCARBONS

Column Packing	P1	P1	P2	P2	P3	P3	P3	P4	P4	P4	P5	P5	P6	P7
Temperature (°C)	28	30	43	52	86	108	86	95	114	124	96	124	28	125
Gas; Flow Rate (ml/min)	70–130, He	70–130, He	50–75, He	50–75, He	28–35, He	28–35, He	28–35, He	40–80, He	40o80, He	40–80, He	50–100, He	50–100, He	20 He	60 He
Column														
length, ft	35	35	15	15	26	26	26	10	10	10	30	30	8	14
diameter, in., O.D.	$\frac{1}{4}$	$\frac{1}{4}$	$\frac{1}{4}$	$\frac{1}{4}$	$\frac{1}{4}$	$\frac{1}{4}$	$\frac{1}{4}$	$\frac{1}{4}$	$\frac{1}{4}$	$\frac{1}{4}$	$\frac{1}{4}$	$\frac{1}{4}$	$\frac{1}{4}$	$\frac{1}{4}$
form	spiral	spiral	spiral	spiral	spiral	spiral	spiral	spiral	spiral	spiral	spiral	spiral	spiral	spiral
material	copper	copper	copper	copper	copper	copper	copper	copper	copper	copper	copper	copper	copper	copper
Detector	na	na	na	na	na	na	na	na	na	na	na	na	na	na
Compound	r_{hexane}												r_{ethane}	r_{hexane}
2-Methyl-2-propanol	—	—	—	—	—	—	—	—	—	—	—	—	—	4.5
1-Bromopropane	—	—	—	—	—	—	—	—	—	1.93	—	—	—	4.57
3-Pentanone	—	—	—	—	—	—	—	—	—	—	—	—	—	9.1
3,3-Dimethyl-2-butanone	—	—	—	—	—	—	—	—	—	—	—	—	—	8.4
2,2-Dimethyl-3-pentanone	—	—	—	—	—	—	—	—	—	—	—	—	—	11.0
2,2,3-Trimethyl-3-pentanone	—	—	—	—	—	—	—	—	—	—	—	—	—	—
2,2,3-Trimethyl-3-pentanol	—	—	—	—	—	—	—	—	—	—	—	—	—	30.5
3-Ethyl-3-hexanol	—	—	—	—	—	—	—	—	—	—	—	—	—	39.6
Water	—	—	—	—	—	—	—	—	—	—	—	—	—	2.2

Column Packing

P1 = 35% dimethylsulfolane and dipropylsulfone on firebrick (30/60 mesh).
P2 = 33% dimethylsulfolane on firebrick (100/120 mesh).
P3 = 5% 7,8-benquinoline on firebrick (100/120 mesh).
P4 = 20% dinonyl phthalate ester on Celite (30/60 mesh).
P5 = 7% di-*n*-propyl tetrachlorophthalate ester on Celite (30/60 mesh).
P6 = Davison's 950 silica gel (60/200 mesh).
P7 = 15% Carbowax 600 on Gas Chrom P (80/100 mesh).

Literature

Csicsery, S. M. and H. Pines, *J. Chromatogr.*, **9**, 34 (1962).

TABLE GC 119
HYDROCARBONS, AROMATIC

Column Packing	P1	P1
Temperature (°C)	150	200
Gas; Flow Rate (ml/min)	95,He	95,He
Column		
length, ft	15	15
diameter, in., O.D.	¼	¼
form	tubing	tubing
material	copper	copper
Detector	TC	TC

Compound	$r_{n\text{-propylbenzene}}$	
***n*-Propylbenzene (reference)**	(1.00)	(1.00)
Methylbenzene (toluene)	0.36	0.48
Ethylbenzene	0.62	0.64
1,4-Dimethylbenzene (*p*-xylene)	0.67	0.70
1,3-Dimethylbenzene	0.68	0.71
1,2-Dimethylbenzene	0.77	0.82
Isopropylbenzene	0.83	0.86
1-Methyl-3-ethylbenzene	1.06	—
1-Methyl-4-ethylbenzene	1.09	1.06
1-Methyl-2-ethylbenzene	1.21	1.17
1,3,5-Trimethylbenzene	1.19	1.13
1,2,4-Trimethylbenzene	1.38	1.29
1,2,3-Trimethylbenzene	1.67	1.52
Isobutylbenzene	1.33	1.26
***sec*-Butylbenzene**	1.34	1.27
***n*-Butylbenzene**	1.77	1.56
1-Methyl-3-isopropylbenzene	1.39	1.27
1-Methyl-4-isopropylbenzene	1.48	1.35
1-Methyl-2-isopropylbenzene	1.58	1.42
1-Methyl-3-*n*-propylbenzene	1.71	1.51
1-Methyl-4-*n*-propylbenzene	1.76	1.58
1-Methyl-2-*n*-propylbenzene	1.90	1.67
1,3-Diethylbenzene	1.62	1.45
1,4-Diethylbenzene	1.79	1.58
1,2-Diethylbenzene	1.81	1.60
1,3-Dimethyl-5-ethylbenzene	1.82	1.59
1,4-Dimethyl-2-ethylbenzene	2.04	1.73
1,2-Dimethyl-4-ethylbenzene	2.17	1.89
1,2-Dimethyl-3-ethylbenzene	2.50	2.06
1-Methyl-3,5-diethylbenzene	2.68	2.14
1,2,4,5-Tetramethylbenzene	2.77	2.28
1,2,3,5-Tetramethylbenzene	2.87	2.33
1-Methyl-3,4-diethylbenzene	3.00	—
1,2-Diisopropylbenzene	3.06	2.51
1,4-Diisopropylbenzene	3.26	2.49
1,2,3,4-Tetramethylbenzene	3.48	2.81
1,3,5-Triethylbenzene	3.87	1.13
Pentamethylbenzene	7.08	4.87
α-Methylstyrene	1.22	1.14
β-Methylstyrene (*trans*-)	1.64	1.47
Benzofuran	1.35	1.33
Indan	1.82	1.69
Indene	1.92	1.80
1-Methylindene	2.22	1.98
1-Methylindan	2.30	1.99
5-Methylindan	3.25	2.67
4-Methylindan	3.46	2.81
2-Methylindene	3.48	2.96
3-Methylindene	3.50	2.82
1,2,3,4-Tetrahydronaphthalene	3.84	3.13
Naphthalene	4.68	3.51

Column Packing P1 = 25% Apiezon L grease on firebrick (30/60 mesh).
Literature Chang, T. L. and C. Karr, Jr., *Anal. Chim. Acta*, **21**, 474 (1959).

TABLE GC 120
HYDROCARBONS, AROMATIC AND ALIPHATIC

Column Packing	P1	P1
Temperature (°C)	86	115
Gas; Flow Rate (Kp/cm²)	N_2,4	N_2,4
Column		
length, m	200	200
diameter, mm, I.D.	0.2	0.2
form	capillary	capillary
material	na	na
Detector	FI	FI

Compound	I	
***n*-Pentane**	500.0	500.0
1-Hexene	583.1	584.0
***trans*-3-Hexene**	591.4	590.6
***cis*-3-Hexene**	593.0	593.7
***trans*-2-Hexene**	596.7	596.4
***n*-Hexane**	600.0	600.0
***cis*-2-Hexene**	604.6	605.4
Benzene	646.6	653.6
1-Heptene	682.8	683.5
***trans*-3-Heptene**	687.5	687.4
***cis*-3-Heptene**	691.7	692.3
***trans*-2-Heptene**	698.7	698.7
***cis*-2-Heptene**	704.3	705.1
***n*-Heptane**	700.0	700.0
Toluene	754.2	760.7
1-Octene	782.3	782.9
***trans*-4-Octene**	784.2	784.1
***cis*-4-Octene**	787.9	788.6
***cis*-3-Octene**	789.5	790.2
***trans*-3-Octene**	788.4	788.0
***trans*-2-Octene**	797.7	797.3
***cis*-2-Octene**	802.8	803.6
***n*-Octane**	800.0	800.0
Ethylbenzene	844.3	851.3
***p*-Xylene**	858.1	864.9
***m*-Xylene**	860.3	866.3
***trans*-4-Nonene**	884.2	884.4
***cis*-4-Nonene**	884.8	886.0
***o*-Xylene**	880.1	888.1
1-Nonene	882.2	882.8
***trans*-3-Nonene**	886.6	886.5
***cis*-3-Nonene**	887.0	888.1
***cis*-2-Nonene**	901.5	902.6
***trans*-2-Nonene**	896.6	896.6
***n*-Nonane**	900.0	900.0
Isopropylbenzene	903.5	910.8
***n*-Propylbenzene**	932.6	940.2
***m*-Ethyltoluene**	945.6	952.1
***p*-Ethyltoluene**	948.0	955.1
***o*-Ethyltoluene**	961.2	968.8
1,3,5-Trimethylbenzene	965.3	971.2
1,2,4-Trimethylbenzene	982.2	990.4
***cis*-4-Decene**	982.2	983.4
***trans*-4-Decene**	982.2	982.7
1-Decene	982.2	982.7
***trans*-2-Decene**	996.7	996.6
***cis*-2-Decene**	1001.2	1002.2
***cis*-5-Decene**	981.0	982.1
***trans*-5-Decene**	984.0	984.7
***trans*-3-Decene**	985.4	985.5
***cis*-3-Decene**	985.4	986.4
***n*-Decane**	1000.0	1000.0
1,2,3-Trimethylbenzene	1007.5	—
1,3-Diethylbenzene	1025.0	—
1-Methyl-3-*n*-propylbenzene	1030.5	1037.1
***n*-Butylbenzene**	1033.0	1040.0
1,2-Diethylbenzene	1036.0	1043.6
1-Methyl-2-*n*-propylbenzene	1042.0	1050.5
1,3-Dimethyl-4-ethylbenzene	—	1070.8
***n*-Undecane**	1100.0	1100.0

Column Packing P1 = Squalane.
Literature Soják, L. and A. Bučinská, *J. Chromatogr.*, **51**, 75 (1970).

TABLE GC 121
HYDROCARBONS AND OXYGENATED COMPOUNDS

Column Packing	P1
Temperature (°C)	25
Gas; Flow Rate (ml/min)	81,He
Column	
length, cm	61
diameter, cm, O.D.	0.5
form	tubing
material	glass
Detector	FI

Compound	$r_{2,3\text{-dimethylbutane}}$
Hydrocarbons	
***n*-Butane**	0.16
Isopentane	0.38
***n*-Pentane**	0.49
2,2-Dimethylbutane	0.69
2,3-Dimethylbutane	1.00
2-Methylpentane	1.18
3-Methylpentane	1.38
2,4-Dimethylpentane	1.80
Methylcyclopentane	2.03
***n*-Hexane**	2.15
3,3-Dimethylpentane	2.48
2-Methylhexane	2.70
2,3-Dimethylpentane	2.87
2,2,4-Trimethylpentane	3.25
3-Ethylpentane	3.34
***trans*-1,3-Dimethylcyclopentane**	3.46
***cis*-1,3-Dimethylcyclopentane**	3.61
***trans*-1,2-Dimethylcyclopentane**	3.75
***n*-Heptane**	4.92
2,4-Dimethylhexane	5.17
Methylcyclohexane	5.46
3,3-Dimethylhexane	5.94
Benzene	7.45
3-Methylheptane	7.96
2,2,5-Trimethylhexane	7.96
3-Ethylhexane	7.97
3-Methyl-3-ethylpentane	8.13
3,4-Dimethylhexane	8.21
1,1-Dimethylcyclohexane	10.3
1-Methyl-1-ethylcyclopentane	11.2
***trans*-1,2-Dimethylcyclohexane**	11.9
Isopropylcyclopentane	13.7
1,1,3-Trimethylcyclohexane	17.5
Toluene	21.7
3,3-Diethylpentane	24.3
Isobutylcyclopentane	30.6
Isopropylcyclohexane	40.5
***p*-Xylene**	59.4
***n*-Decane**	87.6
Oxygenated Compounds	
Ethyl ether	0.93
Isopropyl ether	2.24
Ethyl formate	2.52
Methyl acetate	2.84
Acetone	3.06
Isobutyraldehyde	3.35
Isopropyl formate	4.05
Ethyl acetate	6.46
***n*-Propyl formate**	6.98
Methyl ethyl ketone	7.47
Isopropyl acetate	8.45
Ethyl alcohol	9.7
Isovaleraldehyde	10.5
Isobutyl formate	13.2
***n*-Propyl acetate**	15.4
***sec*-Butyl acetate**	22.2
Diethyl ketone	36.6

Column Packing P1 = di-*n*-butyl phthalate (25 : 75) on Sil-O-Cel C-22 firebrick (60/80 mesh).

Literature Farrington, P. S., Pecsok, R. L., Meeker, R. L., and T. J. Olson, *Anal. Chem.*, **31**, 1512 (1959).

TABLE GC 122
HYDROCARBONS, OLEFINS

Column Packing	P1	P2
Temperature (°C)	24	0
Gas; Flow Rate (cc/min)	0.6	H_2,3
Column		
length, ft	300	100m
diameter, in.	0.012	0.01
form	coiled capillary	capillary
material	glass	SS
Detector	FI	FI
Literature	1	2

Compound	$r_{n\text{-hexane}}$	$r_{pentane}$
Ethylene	0.003	—
Propylene	0.034	—
2-Methylpropene	0.11	—
1-Butene	0.11	—
trans-2-Butene	0.14	—
cis-2-Butene	0.18	—
3-Methyl-1-butene	0.22	—
1-Pentene	0.31	—
2-Methyl-1-butene	0.34	—
trans-2-Pentene	0.38	—
3,3-Dimethyl-1-butene	0.40	—
cis-2-Pentene	0.41	—
2-Methyl-2-butene	0.44	—
Cyclopentene	0.60	—
4-Methyl-1-pentene	0.63	1.89
3-Methyl-1-pentene	0.64	—
2,3-Dimethyl-1-butene	0.70	—
cis-4-Methyl-2-pentene	0.71	2.12
trans-4-Methyl-2-pentene	0.74	2.34
2-Methyl-1-pentene	0.88	2.94
1-Hexene	0.89	3.01
2-Ethyl-1-butene	0.99	—
n-Hexane	1.00	—
trans-3-Hexene	1.03	3.54
cis-3-Hexene	1.03	3.48
trans-2-Hexene	1.05	3.73
3-Methycyclopentene	1.08	—
2-Methyl-2-pentene	1.08	3.81
4,4-Dimethyl-1-pentene	1.11	—
trans-3-Methyl-2-pentene	1.11	—
4-Methylcyclopentene	1.12	—
cis-2-Hexene	1.16	4.05
cis-3-Methyl-2-pentene	1.23	—
trans-4,4-Dimethyl-2-pentene	1.31	—
3,3-Dimethyl-1-pentene	1.36	—
2,3-Dimethyl-2-butene	1.37	5.28
2,3,3-Trimethyl-1-butene	1.40	2.16
3,4-Dimethyl-1-pentene	1.52	—
cis-4,4-Dimethyl-2-pentene	1.53	—
2,4-Dimethyl-1-pentene	1.60	—
1-Methylcyclopentene	1.63	—
3-Methyl-1-hexene	1.68	—
3-Ethyl-1-pentene	1.73	—
2,4-Dimethyl-2-pentene	1.77	—
5-Methyl-1-hexene	1.80	—
2,3-Dimethyl-1-pentene	1.80	—
trans-2-Methyl-3-hexene	1.87	—
4-Methyl-1-hexene	1.92	—
cis-4-Methyl-2-hexene	1.97	—
2-Ethyl-3-methyl-1-butene	1.97	—
trans-4-Methyl-2-hexene	1.97	—
trans-5-Methyl-2-hexene	2.02	—
Cyclohexene	2.08	—
cis-3,4-Dimethyl-2-pentene	2.25	—
2-Methyl-1-hexene	2.42	—
trans-3,4-Dimethyl-2-pentene	2.46	—
1-Heptene	2.50	—
2-Ethyl-1-pentene	2.50	—
trans-3-Methyl-3-hexene	2.65	—
trans-3-Heptene	2.76	—
cis-3-Heptene	2.84	—
trans-3-Methyl-2-hexene	2.84	—
2-Methyl-2-hexene	2.87	—
cis-3-Methyl-3-hexene	2.90	—
trans-2-Heptene	2.95	—
3-Ethyl-2-pentene	2.97	—
cis-3-Methyl-2-hexene	3.07	—
2,3-Dimethyl-2-pentene	3.19	—
cis-2-Heptene	0.00	—

Column Packing P1 = GE Silicone fluid (SF-96) (from 10% in *n*-hexane).
P2 = Squalane.

Literature

1. Polgár, A. G., Holst, J. J. and S. Groennings, *Anal. Chem.*, **34**, 1226 (1962).
2. Dunning, R. W. and J. A. Leonard, *Chromatographia*, **2**, 293 (1969).

TABLE GC 123
NITRILES

Column Packing	P1	P2	P3
Temperature (°C)	70	70	100
Gas; Flow Rate (ml/CO sec)	60,H_2	100,He	100,He
Column			
length, m	5	3.67	7.32
diameter, mm, I.D.	4	6.3	6.3
form	na	coiled	coiled
material	na	Al	Al
Detector	na	TC	TC
Literature	1	2	2

Compound	$r_{acrylonitrile}$	t_r	t_r
Acetonitrile	0.67	10.1	11.5[a]
Acrylonitrile	1.00	8.2[a]	10.8[a]
Propionitrile	1.42	12.0	18.2
Methacrylonitrile	1.93	8.1[a]	15.7
Allyl cyanide	2.54	—	—
***cis*-Crotononitrile**	2.61	—	—
***trans*-Crotononitrile**	3.32	—	—

[a] not resolved.

Column Packing P1 = 10% (w/w) paraffin wax on Haloport F.
P2 = 20% Carbowax on Chromosorb W (30/60 mesh).
P3 = 15% Craig[b] polyester succinate on Chromosorb W (30/60 mesh).

[b] Varian-Aerograph.

Literature

1. Taramasso, M. and A. Guerra, *J. Gas Chromatogr.*, **3**(4), 138 (1965).
2. Lysyj, I., *Anal. Chem.*, **32**, 771 (1960).

TABLE GC 124
OLIGOSACCHARIDES, TRIMETHYLSILYL ETHER DERIVATIVES

Column Packing	P1	P1	P2	P2	P2	P3
Temperature (°C)	165	165	210	210	185	185
Gas; Inlet pressure	1.8hg/ cm^2,N_2	1.8hg/ cm^2,N_2	1.8hg/ cm^2,N_2	1.8hg/ cm^2,N_2	1.8hg/ cm^2,N_2	1.8hg/ cm^2,N_2
Column						
length, ft	6	6	6	6	6	6
diameter, mm, I.D.	4	4	4	4	4	4
form	na	na	na	na	na	na
material	na	na	na	na	na	na
Detector	FI	FI	FI	FI	FI	FI

	r*					
Derivative	Aldose[a]	Alditol[b]	Aldose[a]	Alditol[b]	Aldose[a]	Alditol[b]
Lactose	0.39	0.61	0.41	0.63	0.31	0.46
	0.56	—	0.66	—	0.50	—
Melibiose	0.75	1.00	0.80	1.00	0.81	1.00
	0.80	—	0.97	—	0.92	—
Maltose	0.45	0.74	0.45	0.75	0.36	0.55
	0.51	—	0.52	—	0.44	—
Gentiobiose	0.63	0.80	0.66	0.87	0.73	0.91
	0.80	—	0.96	—	1.09	—
Trehalose	0.54	—	0.54	—	0.37	—
LacNH$_2$[c]	0.75	0.89	2.72	2.22	1.65	1.23
	0.82	—	3.28	—	3.51	—
GalgalNH$_2$[d]	0.89	0.91	2.59	2.41	3.18	1.30
	1.16	—	5.82	—	4.39	—
Fucosyllactose	2.68	—	2.89	—	—	—
	3.20	—	3.48	—	—	—
Melibitol (min.)	—	9.9	—	9.4	—	15.7

[a] TMS ether of intact compound. [b] TMS ether of $NaBH_4$-reduced compound.
[c] *N*-Acetyllactosamine. [d] 3-β-Galactosyl-*N*-acetylgalactosamine.
* TMS ether of melibitol.

Column Packing P1 = SE-30, 2.2%, on Gas-Chrom P, AW-silanized (100/120 mesh).
P2 = QF-1, 3.0%, on Gas-Chrom P, AW-silanized (100/120 mesh).
P3 = NGS, 1.0%, on Gas-Chrom P, AW-silanized (100/120 mesh).

Literature

Huttunen, J. K. and T. A. Miettinen, *Anal. Biochem.*, **29**, 448 (1969).

TABLE GC 125
ORGANICS, VARIOUS COMPOUNDS

Packing	P1		P2		P3		P4		P5		P6		P7		P8	
Temperature (°C)	100		100		100		100		100		100		100		100	
Gas; Flow Rate (ml/min)	He,79–82		He,70		He,76–78		He,48–50		He,F1		He,52		He,F2		He,46	
Column																
length, m	4		4		4		4		4		4		4		4	
diameter	na		na		na		na		na		na		na		na	
form	na		na		na		na		na		na		na		na	
material	na		na		na		na		na		na		na		na	
Detector	na		na		na		na		na		na		na		na	
Component	t_R', min.[a]	Ratio[b], r	t_R', min.	Ratio, r	t_R', min.	Ratio, r	t_R', min.	Ratio, r	t_R', min.	Ratio, r	t_R', min.	Ratio, r	t_R', min.	Ratio, r	t_R', min.	Ratio, r
Methane	0.00	0.00	0.00	0.00	0.00	0.00	0.00	0.00	0.10	0.06	0.00	0.00	0.00	0.00	0.10	0.04
Ethane	0.10	0.04	0.15	0.07	0.00	0.00	0.30	0.08	0.25	0.14	0.00	0.00	0.20	0.07	0.30	0.11
Propane	0.50	0.21	0.40	0.19	0.10	0.25	0.70	0.19	0.45	0.25	0.00	0.00	0.50	0.17	0.60	0.21
Butane	1.20	0.50	1.00	0.48	0.20	0.50	1.80	0.49	0.95	0.53	0.10	0.40	1.35	0.46	1.35	0.48
Pentane	2.40	1.00	2.10	1.00	0.40	1.00	3.70	1.00	1.80	1.00	0.25	1.00	2.95	1.00	2.80	1.00
Hexane	4.90	2.04	4.80	2.29	0.80	2.00	8.40	2.27	3.50	1.94	0.45	1.80	6.35	2.15	5.70	2.04
Heptane	9.70	4.04	10.8	5.14	1.45	3.62	18.4	4.97	6.95	3.86	0.70	2.80	13.7	4.64	11.6	4.14
Octane	19.0	7.92	23.4	11.1	2.70	6.75	38.9	10.5	13.7	7.61	1.10	4.40	29.4	9.67	24.2	8.35
2-Methylpropane	0.90	0.38	0.80	0.38	0.25	0.63	1.35	0.37	0.70	0.39	0.10	0.40	0.95	0.32	0.95	0.34
2-Methylbutane	2.00	0.83	1.70	0.81	0.30	0.75	3.05	0.83	1.20	0.67	0.15	0.60	2.35	0.80	1.90	0.68
2-Methylpentane	3.80	1.58	3.75	1.78	0.60	1.50	6.60	1.78	2.70	1.50	0.25	1.00	4.95	1.68	4.15	1.48
Ethylene	0.15	0.06	0.15	0.07	0.05	0.12	0.20	0.05	0.15	0.08	0.00	0.00	0.20	0.07	0.20	0.07
Propylene	0.50	0.21	0.45	0.21	0.20	0.50	0.75	0.20	0.40	0.22	0.10	0.40	0.55	0.19	0.60	0.21
Isobutene	1.05	0.44	1.05	0.50	0.30	0.75	1.80	0.49	0.80	0.44	0.15	0.60	1.20	0.41	1.35	0.48
Butene-1	1.05	0.44	1.05	0.50	0.30	0.75	1.80	0.49	0.80	0.44	0.15	0.60	1.20	0.41	1.35	0.48
trans-Butene-2	1.20	0.50	1.30	0.62	0.35	0.88	2.20	0.60	0.95	0.53	0.20	0.80	1.45	0.49	1.65	0.59
cis-Butene-2	1.40	0.58	1.40	0.67	0.40	1.00	2.35	0.64	1.00	0.56	0.20	0.80	1.60	0.54	1.80	0.64
Pentene-1	2.30	0.96	2.05	0.98	0.60	1.50	3.95	1.04	1.40	0.78	0.25	1.00	2.50	0.85	2.80	1.00
Hexene-1	4.40	1.83	4.80	2.29	1.00	2.50	8.45	2.28	2.95	1.64	0.45	1.80	5.55	1.88	5.80	2.07
trans-Pentene-2	2.40	1.00	2.25	1.07	0.60	1.50	4.25	1.15	1.85	1.03	0.30	1.20	3.00	1.02	3.25	1.16
cis-Pentene-2	2.60	1.08	2.40	1.14	0.70	1.75	4.75	1.28	2.00	1.11	0.30	1.20	3.10	1.05	3.50	1.25
trans-Hexene-2	5.00	2.08	5.30	2.52	1.20	3.00	9.40	2.54	3.60	2.00	0.60	2.40	6.35	2.15	6.60	2.36
cis-Hexene-2	5.30	2.20	5.60	2.67	1.30	3.25	10.0	2.70	3.90	2.17	0.60	2.40	6.70	2.27	7.10	2.54
Propadiene	0.80	0.33	0.75	0.35	0.45	1.13	1.35	0.37	0.55	0.31	0.30	1.20	0.80	0.27	1.20	0.43
Butadiene-1,3	1.40	0.48	1.30	0.62	0.55	1.38	2.30	0.62	0.90	0.50	0.40	1.60	1.35	0.46	1.85	0.66
2-Methylbutadiene-1,3	2.20	0.92	2.80	1.33	1.00	2.50	5.00	1.35	2.00	1.11	0.60	2.40	3.10	1.05	3.75	1.34

TABLE GC 125—(Continued)
ORGANICS, VARIOUS COMPOUNDS

Packing	P1		P2		P3		P4		P5		P6		P7		P8	
Temperature (°C)	100		100		100		100		100		100		100		100	
Gas; Flow Rate (ml/min)	He,79–82		He,70		He,76–80		He,48–50		He,F1		He,52		He,F2		He,46	
Column																
length, m	4		4		4		4		4		4		4		4	
diameter	na		na		na		na		na		na		na		na	
form	na		na		na		na		qa		na		na		na	
material	na		na		na		na		na		na		na		na	
Detector	na		na		na		na		na		na		na		na	
Component	$t_{R'}$, min.[c]	Ratio[b], r	$t_{R'}$, min.	Ratio, r	$t_{R'}$, min.	Ratio, r	$t_{R'}$, min.	Ratio, r	$t_{R'}$, min.	Ratio, r	$t_{R'}$, min.	Ratio, r	$t_{R'}$, min.	Ratio, r	$t_{R'}$, min.	Ratio, r
Acetylene	0.20	0.08	0.25	0.12	0.30	0.75	0.40	0.11	0.15	0.08	0.20	0.80	0.15	0.05	0.50	0.18
Propyne	0.70	0.29	0.80	0.38	0.70	1.75	1.50	0.41	0.55	0.31	0.50	2.00	0.90	0.31	1.65	0.59
Butyne-1	1.30	0.54	1.55	0.74	1.10	2.75	2.60	0.70	0.95	0.53	0.60	2.40	1.35	0.46	2.50	0.89
Benzene	7.5	3.12	12.9	6.14	7.6	19.0	21.3	5.61	3.75	3.26	4.60	18.4	7.3	4.42	18.2	6.52
Toluene	14.9	6.21	29.9	14.2	13.2	33.0	46.9	12.3	7.50	6.52	7.60	30.4	16.5	10.0	37.4	13.4
Ethylbenzene	29.7	12.4	58.4	27.8	21.2	53.0	90.8	23.9	13.8	12.0	11.2	44.8	32.0	19.4	69.0	24.6
***p*-Xylene**	29.7	12.4	66.3	31.6	23.3	58.2	101.7	26.8	15.0	13.0	12.2	48.8	37.0	22.4	76.0	27.1
***m*-Xylene**	29.7	12.4	66.3	31.6	23.3	58.2	101.7	26.8	15.0	13.0	12.8	51.2	37.0	22.4	76.0	27.1
***o*-Xylene**	34.1	14.2	80.0	38.1	30.1	75.2	121.7	32.0	17.3	14.0	16.3	65.2	43.9	27.6	92.1	32.9
Isopropylbenzene	44.2	18.4	90.8	43.2	28.2	70.5	139.4	36.7	21.2	18.4	13.9	55.6	49.1	29.8	102.0	36.4
Cyclopentane	3.9	1.62	4.00	1.90	1.00	2.50	7.0	1.84	2.85	1.64	0.40	1.60	5.55	1.88	5.20	1.86
Methylcyclopentane	6.1	2.54	6.65	3.17	1.30	3.25	11.5	3.03	4.50	2.50	0.55	2.20	8.80	2.98	8.20	2.93
Cyclohexane	7.9	3.29	9.3	4.90	1.95	4.88	15.4	4.07	5.75	3.19	0.80	3.20	12.5	4.24	11.2	4.00
Methylcyclohexane	12.2	5.08	15.8	7.52	2.65	6.64	25.5	6.71	8.65	4.80	1.00	4.00	20.3	6.88	16.9	6.05
Methanol	2.0	0.83	3.6	1.38	7.6	19.0	4.5	1.18	2.0	1.11	7.00	28.0	2.2	0.79	8.9	3.18
Ethanol	2.0	0.83	4.8	1.85	9.0	22.5	6.2	1.63	2.0	1.11	7.90	31.6	3.7	1.32	8.9	3.18
Propanol	3.8	1.58	10.4	4.00	16.2	40.5	14.1	3.71	3.6	2.00	12.4	49.6	4.9	1.75	18.6	6.64
Butanol	8.0	3.33	24.0	9.24	29.8	74.5	32.5	8.55	8.4	4.66	20.6	82.4	12.8	4.57	40.0	14.3
2-Propanol	3.0	1.25	4.9	2.33	8.3	20.8	8.0	2.09	2.0	1.11	7.50	30.0	2.5	0.89	9.8	3.52
2-Butanol	5.8	2.42	11.0	5.24	14.6	36.5	18.8	4.95	4.5	2.50	11.7	46.8	7.0	2.50	20.1	7.18
***tert*-Butyl alcohol**	3.2	1.33	5.6	2.67	6.9	17.2	10.0	2.36	2.3	1.28	6.60	26.4	3.2	1.14	8.4	3.00
***tert*-Amyl alcohol**	7.1	2.96	14.1	6.71	13.4	33.5	24.8	6.53	5.0	2.78	11.4	45.6	5.5	1.96	20.2	7.21
Acetaldehyde	1.3	0.54	2.0	0.95	2.0	5.00	2.6	0.68	0.8	0.44	1.85	7.40	0.70	0.25	2.7	0.96
Propionaldehyde	2.1	0.88	3.9	1.86	3.4	8.50	5.4	1.42	1.6	0.89	3.05	12.2	2.4	0.86	5.0	1.80
Butyraldehyde	4.2	1.75	8.2	3.90	5.8	14.5	11.8	3.12	3.4	1.90	4.70	18.8	5.7	2.03	10.3	3.68
Valeraldehyde	9.2	3.83	18.2	8.67	10.7	26.8	26.4	6.95	7.6	4.22	7.80	31.2	14.2	5.62	21.6	7.71

TABLE GC 125—(Continued)
ORGANICS, VARIOUS COMPOUNDS

Packing	P1		P2		P3		P4		P5		P6		P7		P8	
Temperature (°C)	100		100		100		1000		100		100		100		100	
Gas; Flow Rate (ml/min)	He,79–82		He,70		He,76–78		He,48–50		He,F1		He,52		He,F2		He,46	
Column																
length, m	4		4		4		4		4		4		4		4	
diameter	na		na		na		na		na		na		na		na	
form	na		na		na		na		na		na		na		na	
material	na		na		na		na		na		na		na		na	
Detector	na		na		na		na		na		na		na		na	
Component	$t_{R'}$, min.[a]	Ratio[b], r	$t_{R'}$, min.	Ratio, r	$t_{R'}$, min.	Ratio, r	$t_{R'}$, min.	Ratio, r	$t_{R'}$, min.	Ratio, r	$t_{R'}$, min.	Ratio, r	$t_{R'}$, min.	Ratio, r	$t_{R'}$, min.	Ratio, r
Acetone	2.6	1.08	4.4	2.10	4.2	10.5	5.7	1.50	2.0	1.11	4.80	19.2	2.4	0.86	5.6	2.00
2-Butanone	4.8	2.00	9.0	4.29	6.9	17.2	12.4	3.26	3.8	2.11	6.90	27.6	6.3	2.25	11.4	4.07
2-Pentanone	9.7	4.04	17.3	8.24	10.8	27.0	24.6	6.47	7.0	3.89	9.60	38.4	12.6	4.50	21.1	7.54
2-Heptanone	32.7	13.6	76.0	13.6	33.7	84.3	—	—	25.7	14.3	24.8	99.2	31.6	11.3	86.1	30.8
Methyl acetate	3.7	1.54	4.5	2.14	3.9	9.75	6.4	1.67	2.2	1.22	4.00	16.0	3.0	1.07	6.8	2.41
Ethyl acetate	6.0	2.50	7.7	3.67	5.6	14.0	11.6	3.05	4.0	2.22	5.05	20.2	5.6	2.00	11.6	4.12
***n*-Propyl acetate**	11.7	4.88	15.9	7.57	9.4	23.5	24.4	6.42	7.8	4.33	7.60	30.4	11.3	4.04	21.8	7.79
***n*-Butyl acetate**	23.0	9.58	33.4	15.9	16.4	41.0	52.1	13.7	15.0	8.33	12.2	48.8	24.4	0.87	44.9	16.0
Diethyl ether	2.4	1.00	2.6	1.24	1.0	2.50	4.2	1.12	1.9	1.05	0.80	3.20	2.6	0.93	3.6	1.29
Diisopropyl ether	4.8	2.00	5.2	2.48	1.3	3.25	8.8	2.30	3.5	1.94	3.35	13.4	4.9	1.75	6.2	2.21
Di-*n*-butyl ether	31.3	13.0	44.0	20.9	8.2	20.5	75.9	20.0	23.0	12.8	—	—	44.4	15.8	50.7	18.1

[a] $t_R{}'$ = adjusted retention volume, air = 0.

[b] r = relative retention referred to *n*-pentane.

Flow Rate F1 = 55 ml/min except for mixture 9, 91 ml/min; F2 = 46 or 50 ml/min, except mixture 9, 94 ml/min.

Column Packing
P1 = 20% DC-200 on 60–80 mesh Chromosorb R.
P2 = 20% diisodecyl phthalate on 60–80 mesh Chromosorb R.
P3 = 20% Carbowax 1500 (polyethylene glycol) on 60–80 mesh Chromosorb R.
P4 = 20% di-2-ethylhexyl sebacate on 60–80 mesh Chromosorb R.
P5 = 20% DC high vacuum silicone grease on 60–80 mesh Chromosorb R.
P6 = 20% PEGS on 60–80 mesh Chromosorb R.
P7 = 20% Apiezon L grease on 60–80 mesh Chromosorb R.
P8 = 20% UCON LB-550-X (polypropylene glycol) on 60–80 mesh Chromosorb R.

Literature

1. Dal Nogare, S. and R. S. Juvet, Jr., "*Gas-Liquid Chromatography*," Interscience, N.Y., 1966, p 407.
2. Scholly, P. R. and N. Brenner, "*Gas Chromatography*," Noebels, H. J., Wall, R. F. and N. Brenner, Eds., Academic Press, N.Y., 1961, p 263.

TABLE GC 126
MISCELLANEOUS ORGANICS, ALIPHATIC AND AROMATIC

Column Packing	P1	P2	P3	P4	P5	P6	P7	P8	P9	P1	P2	P3	P5	P6	P7	P8	P9
Temperature (°C)	50	50	50	50	50	50	50	50	50	100	100	100	100	100	100	100	100
Gas; Flow Rate	He	He	He	He	He	He	He	He	He	He	He	He	He	He	He	He	He
Column																	
length, m	4	4	4	4	4	4	4	4	4	4	4	4	4	4	4	4	4
diameter	na	na	na	na	na	na	na	na	na	na	na	na	na	na	na	na	na
form	na	na	na	na	na	na	na	na	na	na	na	na	na	na	na	na	na
material	na	na	na	na	na	na	na	na	na	na	na	na	na	na	na	na	na
Detector	na	na	na	na	na	na	na	na	na	na	na	na	na	na	na	na	na
Compound	$r_{pentane}$																
***n*-Paraffins**																	
Methane	0.00	0.00	0.00	0.00	0.00	0.02	0.00	0.01	0.02	—	—	—	—	—	—	—	—
Ethane	0.04	0.04	0.05	0.04	0.07	0.06	0.00	0.05	0.05	0.07	0.08	0.04	0.00	0.14	0.00	0.07	0.11
Propane	0.12	0.12	0.14	0.14	0.21	0.15	0.16	0.12	0.14	0.19	0.19	0.21	0.25	0.25	0.00	0.17	0.21
Butane	0.35	0.36	0.35	0.40	0.48	0.41	0.44	0.37	0.39	0.48	0.49	0.50	0.50	0.53	0.40	0.46	0.48
Hexane	2.71	2.88	2.46	2.30	2.14	2.45	1.89	2.66	2.61	2.29	2.27	2.04	2.00	1.94	0.80	2.15	2.04
Heptane	7.10	7.42	5.95	5.17	4.76	5.98	4.00	7.20	6.55	5.14	4.97	4.04	3.62	3.86	2.80	4.64	4.14
Octane	—	19.1	14.25	11.9	10.6	14.8	9.22	—	—	11.1	10.5	7.92	6.75	7.61	4.40	9.67	8.35
Branched Paraffins																	
2-Methylpropane	0.25	0.25	0.29	0.28	0.41	0.29	0.44	0.24	0.27	0.38	0.37	0.38	0.63	0.39	0.40	0.32	0.34
2-Methylbutane	0.78	0.77	0.77	0.76	0.69	0.77	0.67	0.74	0.78	0.81	0.83	0.83	0.75	0.67	0.60	0.80	0.68
2-Methylpentane	2.09	2.04	1.91	1.73	1.52	1.85	1.44	1.95	1.97	1.78	1.78	1.58	1.50	1.50	1.00	1.68	1.48
Neopentane	0.41	0.39	0.45	—	0.34	0.68	0.44	0.36	0.41	0.52	0.53	0.58	0.50	1.00	0.40	0.44	0.46
Isooctane	7.01	6.73	5.69	4.50	4.00	5.62	4.00	6.27	0.59	4.81	4.89	4.00	3.00	3.78	3.00	4.28	3.80
Olefins																	
Ethylene	0.04	0.04	0.04	0.09	0.07	0.03	0.11	0.03	0.05	0.07	0.05	0.06	0.12	0.08	0.00	0.07	0.07
Propylene	0.14	0.13	0.14	0.26	0.21	0.13	0.44	0.11	0.16	0.21	0.20	0.21	0.50	0.22	0.40	0.19	0.21
Isobutene	0.38	0.37	0.36	0.66	0.62	0.36	0.78	0.31	0.44	0.50	0.49	0.44	0.75	0.44	0.60	0.41	0.48
Butene-1	0.38	0.37	0.36	0.66	0.62	0.36	0.78	0.31	0.44	0.50	0.49	0.44	0.75	0.44	0.60	0.41	0.48
trans-Butene-2	0.50	0.47	0.44	0.80	0.76	0.44	1.11	0.40	0.56	0.62	0.60	0.50	0.88	0.53	0.80	0.49	0.59
cis-Butene-2	0.55	0.52	0.50	0.92	0.97	0.50	1.11	0.45	0.62	0.67	0.64	0.58	1.00	0.56	0.80	0.54	0.64
Pentene-1	1.01	0.99	0.89	1.43	1.31	0.87	1.33	0.84	1.12	0.98	1.04	0.96	1.50	0.78	1.00	0.85	1.00
trans-Pentene-2	1.24	1.20	1.06	1.73	1 66	1.06	2.11	1.05	1.36	1.07	1.15	1.00	1.50	1.03	1.20	1.02	1.16
cis-Pentene-2	1.31	1.28	1.14	1.89	1.83	1.14	2.11	1.08	1.45	1.14	1.28	1.08	1.75	1.11	1.20	1.05	1.25
Hexene-1	2.87	2.75	2.23	3.43	3.03	2.21	2.78	2.32	2.92	2.29	2.28	1.83	2.50	1.64	1.80	1.88	2.07
trans-Hexene-2	3.17	3.12	2.54	3.78	3.41	2.54	3.56	2.66	3.30	2.52	2.54	2.08	3.00	2.00	2.40	2.15	2.36
cis-Hexene-2	3.52	3.40	2.80	4.28	4.00	2.79	3.56	2.88	3.66	2.67	2.70	2.20	3.25	2.17	2.40	2.27	2.54
Propadiene	0.25	0.23	0.20	0.74	0.90	0.20	1.11	0.17	0.35	0.35	0.37	0.33	1.13	0.31	1.20	0.27	0.43
Butadiene-1,3	0.50	0.48	0.39	1.33	1.31	0.38	1.67	0.34	0.62	0.62	0.62	0.48	1.38	0.50	1.60	0.46	0.66
2-Methylbutadiene-1,3	1.46	1.44	1.02	3.33	2.90	1.01	3.00	1.04	1.70	1.33	1.35	0.92	2.50	1.11	2.40	1.05	1.34

TABLE GC 126—(Continued)
MISCELLANEOUS ORGANICS, ALIPHATIC AND AROMATIC

Column Packing	P1	P2	P3	P4	P5	P6	P7	P8	P9	P1	P2	P3	P4	P5	P6	P7	P8
Temperature (°C)	50	50	50	50	50	50	50	50	50	100	100	100	100	100	100	100	100
Gas; Flow Rate	He	He	He	He	He	He	He	He	He	He	He	He	He	He	He	He	He
Column																	
length, m	4	4	4	4	4	4	4	4	4	4	4	4	4	4	4	4	4
diameter	na	na	na	na	na	na	na	na	na	na	na	na	na	na	na	na	na
form	na	na	na	na	na	na	na	na	na	na	n	na	na	na	na	na	na
material	na	na	na	na	na	na	na	na	na	na	na	na	na	na	na	na	na
Detector	na	na	na	na	na	na	na	na	na	na	na	na	na	na	na	na	na
Compound	$r_{pentane}$																
Alkynes																	
Acetylene	0.06	0.06	0.05	0.50	0.76	0.04	0.78	0.03	0.14	0.12	0.11	0.08	0.75	0.08	0.80	0.05	0.18
Propyne	0.28	0.20	0.23	1.41	1.72	0.20	2.22	0.10	0.45	0.38	0.41	0.29	1.75	0.31	2.00	0.31	0.59
Butyne-1	0.65	0.61	0.45	2.72	2.97	0.30	3.33	0.34	0.98	0.74	0.70	0.54	2.75	0.53	2.40	0.46	0.89
Inorganic Compounds																	
Carbon dioxide	0.04	0.04	0.03	0.15	0.17	0.03	0.44	0.03	0.04	0.07	0.05	0.08	0.25	0.06	0.40	0.05	0.09
Water	—	—	—	—	—	—	—	—	—	3.04	1.84	2.17	60.8	1.94	80.0	2.8	5.29
Aromatics																	
Benzene	—	—	—	—	—	—	—	—	—	6.14	5.61	3.12	19.0	3.26	18.4	4.42	6.52
Toluene	—	—	—	—	—	—	—	—	—	14.2	12.3	6.21	33.0	6.52	30.4	10.0	13.4
Ethylbenzene	—	—	—	—	—	—	—	—	—	27.8	23.9	12.4	53.0	12.0	44.8	19.4	24.6
p-Xylene	—	—	—	—	—	—	—	—	—	31.6	26.8	12.4	58.2	13.0	48.8	22.4	27.1
o-Xylene	—	—	—	—	—	—	—	—	—	38.1	32.0	14.2	75.2	15.0	65.2	27.6	32.9
m-Xylene	—	—	—	—	—	—	—	—	—	31.6	26.8	12.4	58.2	13.0	51.2	22.4	17.1
Isopropylbenzene	—	—	—	—	—	—	—	—	—	43.2	36.7	18.4	70.5	18.4	55.6	29.8	36.4
Alcohols																	
Methanol	—	—	—	—	—	—	—	—	—	1.38	1.18	0.83	19.0	1.11	28.0	0.79	3.18
Ethanol	—	—	—	—	—	—	—	—	—	1.85	1.63	0.83	22.5	1.11	31.6	1.32	3.18
Propanol	—	—	—	—	—	—	—	—	—	4.00	3.71	1.58	40.5	2.00	49.6	1.75	6.64
Butanol	—	—	—	—	—	—	—	—	—	9.24	8.55	3.33	74.5	4.66	82.4	4.57	14.3
2-Propanol	—	—	—	—	—	—	—	—	—	2.33	2.09	1.25	20.8	1.11	30.0	0.89	3.52
2-Butanol	—	—	—	—	—	—	—	—	—	5.24	4.95	2.42	36.5	2.50	46.8	2.50	7.18
tert-Butyl alcohol	—	—	—	—	—	—	—	—	—	2.67	2.36	1.33	17.2	1.28	26.4	1.14	3.00
tert-Amyl alcohol	—	—	—	—	—	—	—	—	—	6.71	6.53	2.96	33.5	2.78	45.6	1.96	7.21
Aldehydes																	
Acetaldehyde	—	—	—	—	—	—	—	—	—	0.95	0.68	0.54	5.00	0.44	7 40	0.25	0.96
Propionaldehyde	—	—	—	—	—	—	—	—	—	1.86	1.42	0.88	8.50	0.89	12.2	0.86	1.80
Butyraldehyde	—	—	—	—	—	—	—	—	—	3.90	3.12	1.75	14.5	1.90	18.8	2.03	3.68
Valeraldehyde	—	—	—	—	—	—	—	—	—	8.67	6.95	3.83	26.8	4.22	31.2	5.62	7.71

TABLE GC 126—(Continued)
MISCELLANEOUS ORGANICS, ALIPHATIC AND AROMATIC

Column Packing	P1	P2	P3	P4	P5	P6	P7	P8	P9	P1	P2	P3	P5	P6	P7	P8	P9
Temperature (°C)	50	50	50	50	50	50	50	50	50	100	100	100	100	100	100	100	100
Gas; Flow Rate	He	He	He	He	He	He	He	He	He	He	He	He	He	He	He	He	He
Column																	
length, m	4	4	4	4	4	4	4	4	4	4	4	4	4	4	4	4	4
diameter	na	na	na	na	na	na	na	na	na	na	na	na	na	na	na	na	na
form	na	na	na	na	na	na	na	na	na	na	na	na	na	na	na	na	na
material	na	na	na	na	na	na	na	na	na	na	na	na	na	na	na	na	na
Detector	na	na	na	na	na	na	na	na	na	na	na	na	na	na	na	na	na
Compound									$r_{pentane}$								
Ketones																	
Acetone	—	—	—	—	—	—	—	—	—	2.10	1.50	1.08	10.5	1.11	19.2	0.86	2.00
2-Butanone	—	—	—	—	—	—	—	—	—	4.29	3.26	2.00	17.2	2.11	27.6	2.25	4.07
2-Pentanone	—	—	—	—	—	—	—	—	—	8.24	6.47	4.04	27.0	3.89	38.4	4.50	7.54
2-Heptanone	—	—	—	—	—	—	—	—	—	13.6	—	13.6	84.3	14.3	99.2	11.3	30.8
Esters																	
Methyl acetate	—	—	—	—	—	—	—	—	—	2.14	1.67	1.54	9.75	1.22	16.0	1.07	2.41
Ethyl acetate	—	—	—	—	—	—	—	—	—	3.67	3.05	2.50	14.0	2.22	20.2	2.00	4.12
n-Propyl acetate	—	—	—	—	—	—	—	—	—	7.57	6.42	4.88	23.5	4.33	30.4	4.04	7.79
n-Butyl acetate	—	—	—	—	—	—	—	—	—	15.9	13.7	9.58	41.0	8.33	48.8	0.87	16.0
Ethers																	
Diethyl ether	—	—	—	—	—	—	—	—	—	1.24	1.12	1.00	2.50	1.05	3.20	0.93	1.29
Diisopropyl ether	—	—	—	—	—	—	—	—	—	2.48	2.30	2.00	3.25	1.94	13.4	1.75	2.21
Di-*n*-butyl ether	—	—	—	—	—	—	—	—	—	20.9	20.0	13.0	20.5	12.8	—	15.8	18.1
Cyclopentanes																	
Cyclopentane	—	—	—	—	—	—	—	—	—	1.90	1.84	1.62	2.50	1.64	1.60	1.88	1.86
Methylcyclopentane	—	—	—	—	—	—	—	—	—	3.17	3.03	2.54	3.25	2.50	2.20	2.98	2.93
Cyclohexanes																	
Cyclohexane	—	—	—	—	—	—	—	—	—	4.90	4.07	3.29	4.88	3.19	3.20	4.24	4.00
Methylcyclohexane	—	—	—	—	—	—	—	—	—	7.52	6.71	5.08	6.64	4.80	4.00	6.88	6.05

Column Packing
P1 = Diisodecyl phthalate, 20% on (60/80) Chromosorb.
P2 = Di-2-ethylhexyl sebacate, 20% on (60/80) Chromosorb.
P3 = DC-200, 20% on (60/80) Chromosorb.
P4 = Dimethylsulfolane, 33% on (60/80) Chromosorb.
P5 = Carbowax 1500, 20% on (60/80) Chromosorb.
P6 = DC HV grease, 20% on (60/80) Chromosorb.
P7 = Polydiethylene glycol succinate, 20% on (60/80) Chromosorb.
P8 = Apiezon L, 20% on (60/80) Chromosorb.
P9 = Polyethylene glycol (UCON LB-550-X).

Literature
Scholly, P. R. and N. Brenner, in *Gas Chromatography*, edited by Noebels, H. J., Wall, R. F., and N. Brenner, Academic Press, N.Y. 1961, p. 265.

TABLE GC 127
ORGANICS

Column Packing	P1	P2	P3	P4	P5	P1	P2	P3	P4	P5	P1	P2	P3	P4	P5
Temperature (°C)	157	157	157	157	157	100	100	100	100	100	32	32	32	32	32
Gas; Flow Rate (ml/min)	He	He	He	He	He	He	He	He	He	He	50,He	50,He	50,He	50,He	50,He
Column															
length, ft	6	6	6	6	6	6	6	6	6	6	6	6	6	6	6
diameter, in.	$\frac{3}{16}$	$\frac{3}{16}$	$\frac{3}{16}$	$\frac{3}{16}$	$\frac{3}{16}$	$\frac{3}{16}$	$\frac{3}{16}$	$\frac{3}{16}$	$\frac{3}{16}$	$\frac{3}{16}$	$\frac{3}{16}$	$\frac{3}{16}$	$\frac{3}{16}$	$\frac{3}{16}$	$\frac{3}{16}$
form	—	—	—	—	—	—	—	—	—	—	—	—	—	—	—
material	—	—	—	—	—	—	—	—	—	—	—	—	—	—	—
Detector	—	—	—	—	—	—	—	—	—	—	—	—	—	—	—
Compound									t_r						
Air	0.28	0.3	0.3	0.39	0.32	0.31	0.29	0.23	0.40	0.42	0.39	0.30	0.35	0.33	0.44
Methane	—	0.35	0.39	0.46	0.40	0.39	0.59	0.33	0.52	0.45	0.49	0.47	0.61	0.8	0.66
Carbon dioxide	—	—	0.39	0.62	0.50	—	—	0.56	0.78	0.85	0.83	1.1	1.72	1.7	3.6
Ethylene	—	0.5	0.52	0.70	0.53	0.54	0.88	0.75	1.1	0.95	1.27	2.3	3.2	4.0	3.8
Acetylene	—	—	0.58	0.62	0.64	—	0.88	0.95	1.2	1.4	1.46	2.3	4.8	5.0	11.0
Ethane	—	0.55	0.58	0.70	0.53	0.54	0.88	0.95	1.3	1.0	1.76	3.3	4.35	5.6	4.5
Water	0.48	0.65	1.0	1.4	1.7	1.00	1.2	3.5	3.2	7.1	4.88	5.9	50.0	33.2	110
Propylene	—	—	0.91	1.4	0.91	—	—	2.3	3.6	2.6	5.0	16	18.7	26	27
Propane	—	1.16	0.91	1.4	0.91	1.00	2.9	2.6	3.6	2.6	7.9	21	22.8	29	24
Methyl chloride	—	1.21	1.1	1.6	1.3	1.2	3.2	2.8	4.5	4.2	8.6	19	—	34	50
Vinyl chloride	—	1.7	1.6	2.2	1.6	1.6	5.6	4.7	7.1	6.8	—	—	—	—	—
Methyl alcohol	0.54	1.1	1.5	1.9	2.2	1.4	2.9	5.6	7.8	10.0	—	—	—	—	—
Ethylene oxide	—	—	1.35	—	2.1	—	—	4.2	7.8	9.5	—	—	—	—	—
Ethyl chloride	—	2.5	2.3	3.1	2.7	2.7	9.7	8.5	13.5	13	—	—	—	—	—
Acetone	1.00	3.5	3.4	4.5	5.8	5.3	17	15	36	41	—	—	—	—	—
Ethyl alcohol	0.75	2.2	2.8	3.4	4.1	2.8	8.2	14	19.5	26	—	—	—	—	—
Pentane	1.11	5.6	4.5	5.8	3.6	6.7	—	20	36	22	—	—	—	—	—
Cyclopentane	1.3	7.1	5.5	7.5	4.5	10.0	—	25	49	28	—	—	—	—	—
Butadiene	0.73	—	2.0	2.7	1.9	2.4	7.9	7.5	11	9.1	—	—	—	—	—
Isopropyl alcohol	0.94	3.8	4.5	5.7	6.7	5.2	18	27	40	54	—	—	—	—	—
Acetonitrile	1.09	3.2	3.2	4.2	6.2	4.9	12	13	20	41	—	—	—	—	—
Acrylonitrile	1.22	4.1	4.4	5.7	7.7	7.0	21	21	32	58	—	—	—	—	—
Diethyl ether	0.93	4.9	4.2	5.5	4.3	6.6	30	22	35	28	—	—	—	—	—

TABLE GC 127—(Continued)
ORGANICS

Column Packing	P1	P2	P3	P4	P5	P1	P2	P3	P4	P5	P1	P2	P3	P4	P5
Temperature (°C)	157	157	157	157	147	100	100	100	100	100	32	32	32	32	32
Gas: Flow Rate (ml/min)	He	He	He	He	He	He	He	He	He	He	50,He	50,He	50,He	50,He	50,He
Column															
length, ft	6	6	6	6	6	6	6	6	6	6	6	6	6	6	6
diameter, in.	$\frac{3}{16}$	$\frac{3}{16}$	$\frac{3}{16}$	$\frac{3}{16}$	$\frac{3}{16}$	$\frac{3}{16}$	$\frac{3}{16}$	$\frac{3}{16}$	$\frac{3}{16}$	$\frac{3}{16}$	$\frac{3}{16}$	$\frac{3}{16}$	$\frac{3}{16}$	$\frac{3}{16}$	$\frac{3}{16}$
form	—	—	—	—	—	—	—	—	—	—	—	—	—	—	—
material	—	—	—	—	—	—	—	—	—	—	—	—	—	—	—
Detector	—	—	—	—	—	—	—	—	—	—	—	—	—	—	—
Compound									t_r						
Methylene chloride	1.2	4.2	3.7	5.3	5.0	5.5	19	19	29	34	—	—	—	—	—
n-Propyl alcohol	1.2	5.0	6.5	7.8	9.5	7.7	29	47	65	85	—	—	—	—	—
tert-Butyl alcohol	1.25	6.1	7.2	8.2	10.5	7.7	39	55	74	98	—	—	—	—	—
Hexane	2.1	13.5	8.5	13	8.1	—	—	—	—	—	—	—	—	—	—
Cyclohexane	2.35	17	10.5	17	10.0	—	—	—	—	—	—	—	—	—	—
Benzene	2.65	14.5	10.0	16	13.0	—	—	—	—	—	—	—	—	—	—
Nitroethane	2.55	9.7	8.5	13	22	—	—	—	—	—	—	—	—	—	—
Ethylene glycol	3.1	10.5	23	28	53	—	—	—	—	—	—	—	—	—	—
Propylene glycol	4.0	18.5	34	42	69	—	—	—	—	—	—	—	—	—	—
Formic acid	—	2.05	—	26	13.5	—	—	—	—	—	—	—	—	—	—
Acetic acid	—	4.3	14.5	26	21	—	—	—	—	—	—	—	—	—	—
Propionic acid	—	10.5	32	—	46	—	—	—	—	—	—	—	—	—	—

Column Packing P1 = Porapak P.
P2 = Porapak Q.
P3 = Porapak R.
P4 = Porapak S.
P5 = Porapak T.

Literature Porapak Bulletin, Waters Assoc., Inc., 61 Fountain Street, Framingham, Mass., 1965.

TABLE GC 128
PESTICIDES

Column Packing	P1	P2	P3	P4	P5	P6	P7	P8
Temperature (°C)	200	180	180	195	180	200	190	200
Gas; Flow Rate (ml/min)	100	70	43	70	60	70	100	60
Column								
length, ft	6	6	6	6	6	6	6	6
diameter, in., O.D.	$\frac{1}{4}$	$\frac{1}{4}$	$\frac{1}{4}$	$\frac{1}{4}$	$\frac{1}{4}$	$\frac{1}{4}$	$\frac{1}{4}$	$\frac{1}{4}$
I.D.	$\frac{5}{32}$	$\frac{5}{32}$	$\frac{5}{32}$	$\frac{5}{32}$	$\frac{5}{32}$	$\frac{5}{32}$	$\frac{5}{32}$	$\frac{5}{32}$
form	—	—	—	—	—	—	—	—
material	glass	glass	glass	glass	glass	glass	glass	glass
Detector	EC	EC	EC	EC	EC	EC	EC	EC
Compound				r_{aldrin}				
CDAA	0.16	0.10	0.36	0.39	0.22	0.20	0.17	0.18
Phosdrin®	0.17	0.16	0.69	*0.74*, 0.88	*0.37*, 0.63	0.33	0.32	0.32
CDEC	0.38	0.36	0.68	0.86	0.51	0.50	0.47	0.57
Phorate	0.38	0.37	0.65	0.71	0.49	0.48	0.44	0.50
α-BHC	0.39	0.35	0.64	1.24	0.49	0.49	0.45	0.53
2,4-D(isopropyl ester)	0.41	0.40	0.93	0.96	0.59	0.56	0.54	0.54
Simazine	0.41	0.40	0.92	3.8	0.60	0.54	0.54	0.68
Atrazine	0.42	0.41	0.92	2.86	0.60	0.57	0.56	0.66
β-BHC	0.43	0.40	0.96	5.7	0.63	0.61	0.58	0.80
Dimethoate	0.44	0.46	2.36	5.9	1.06	0.88	0.94	1.02
Lindane	0.47	0.44	0.82	1.89	0.60	0.60	0.57	0.67
δ-BHC	0.49	0.47	1.12	5.2	0.70	0.69	0.66	0.92
Diazinon®	0.51	0.52	0.77	0.72	0.61	0.59	0.58	0.63
Daconil 2787	0.52	0.51	2.45	3.3	1.22	1.14	1.14	1.25
2,4-D(butyl ester)	0.60, *0.70*	0.060, *0.69*	1.29, *1.45*	1.33, *1.61*	0.87, *1.00*	0.79, *0.89*	0.80, *0.90*	0.79, *0.92*
2,4,5-T(isopropyl ester)	0.67	0.67	1.30	1.41	0.89	0.84	0.85	0.84
Methyl parathion	0.69	0.70	2.55	5.1	1.49	1.35	1.41	1.43
Heptachlor	0.79	0.78	0.86	1.00	0.82	0.82	0.81	0.83
Malathion	0.87	0.94	2.96	3.3	1.60	1.58	1.48	1.63
Aldrin (reference)	1.00	1.00	1.00	1.00	1.00	1.00	1.00	1.00
Ethyl parathion	1.00	1.00	3.92	4.3	1.96	1.79	1.89	1.81
Dyrene®	1.19	1.27	1.97	7.8	1.44	1.37	1.44	1.87
Heptachlor epoxide	1.24	1.28	1.88	2.68	1.47	1.43	1.47	1.54
Chlorbenside	1.35	1.41	1.86	3.8	1.54	1.47	1.51	1.93
2,4-D(butoxyethanol ester)	1.44, *1.70*	1.59, *1.89*	2.91, *3.41*	3.6, *4.7*	*1.97*, 2.30	1.75, *2.03*	1.89, *2.23*	1.92, *2.36*

TABLE GC 128—(Continued)
PESTICIDES

Column Packing	P1	P2	P3	P4	P5	P6	P7	P8
Temperature (°C)	200	180	180	195	180	200	190	200
Gas; Flow Rate (ml/min)	100	70	43	70	60	70	100	60
Column								
length, ft	6	6	6	6	6	6	6	6
diameter, in., O.D.	$\frac{1}{4}$	$\frac{1}{4}$	$\frac{1}{4}$	$\frac{1}{4}$	$\frac{1}{4}$	$\frac{1}{4}$	$\frac{1}{4}$	$\frac{1}{4}$
I.D.	$\frac{5}{32}$	$\frac{5}{32}$	$\frac{5}{32}$	$\frac{5}{32}$	$\frac{5}{32}$	$\frac{5}{32}$	$\frac{5}{32}$	$\frac{5}{32}$
form	—	—	—	—	—	—	—	—
material	glass	glass	glass	glass	glass	glass	glass	glass
Detector	EC	EC	EC	EC	EC	EC	EC	EC
Compound				r_{aldrin}				
2,4-D(P.G.B. ether ester)	1.47	1.61	2.78	2.61	*1.94*, 2.70	*1.75*, 4.18	1.89	*1.79*, 4.70
***o,p'*-DDE**	1.49	1.60	1.56	2.64	1.53	1.47	1.55	1.84
Thiodan®	*1.54*, 2.11	*1.62*, 2.25	*2.51*, 4.64	*3.1*, 10.0	*1.92*, 3.04	*1.81*, 2.75	*1.89*, 2.96	*1.97*, 3.62
DDA(methyl ester)	1.58	1.72	2.83	5.7	2.04	1.88	2.00	2.28
Dieldrin	1.85	1.93	2.95	4.0	2.27	2.15	2.23	2.40
***p,p'*-DDE**	1.85	2.00	2.04	3.4	1.92	1.83	1.96	2.23
***o,p'*-DDD**	1.89	2.04	2.49	5.9	2.10	1.99	2.12	2.66
2,4-D(isooctyl ester)	1.94, *2.32*	2.20, *2.64*	3.61, *4.34*	2.89, 3.3, *3.7*	2.53, *3.02*	2.24, *2.58*	2.45, *2.92*	2.30, *2.86*
Endrin	2.03	*2.18*, 2.50	3.56	4.4	2.60	2.45	2.58	2.93
Perthane®	2.19	2.48	1.90	3.9	2.10	1.96	2.16	2.76
2,4,5-T(butoxyethanol ester)	2.30, *2.69*	2.61, *3.11*, 3.34	4.33, *5.05*	5.7, *7.25*	3.04, *3.56*	2.69, *3.10*	2.99, *3.48*	3.04, *3.66*, 4.88
***p,p'*-DDD**	2.32	2.61	3.64	10.2	2.82	2.60	2.83	3.49
***o,p'*-DDT**	2.47	2.69	2.60	4.6	2.55	2.40	2.60	3.17
Ethion	2.50	2.84	5.43	6.0	3.38	3.08	3.46	4.44
Kepone®	2.72	2.91	2.65	—	2.68	2.58	2.75	2.85
Dilan®	2.74, *3.35*	3.16, *3.94*	10.4, *11.7*	—	5.36, *6.29*	4.63, *5.33*	5.16, *6.00*	5.68, *6.35*
Carbophenothion	2.80	3.22	4.80	8.1	3.43	3.20	3.49	4.66
2,4,5-T(isooctyl ester)	3.05, *3.63*	3.56, *4.31*	5.21, *6.23*	0.80, 4.3, *5.5*	3.89, *4.64*	3.37, *4.00*	3.92, *4.67*	0.23, 3.50, *4.32*
***p,p'*-DDT**	3.08	3.50	3.93	8.3	3.42	3.16	3.50	4.20
Methoxychlor	4.68	5.70	6.45	—	5.46	4.67	5.46	8.25
Guthion®	4.95	6.25	—	—	10.0	8.45	10.0	—
Tetradifon	5.03	6.05	18.5	—	9.84	8.1	9.3	—
Mirex	5.76	6.59	2.65, *3.77*	—	5.10	4.85	5.35	6.34
Multi-Peak Compound								
Fenthion	*0.91*, 1.85, 2.60	*0.98*, 2.12, 3.10	—	3.7, 7.7	*1.28*, 2.42, 3.53	0.36, *1.20*, *2.26*, 2.63, 3.24	*1.24*, 2.40, 3.56	—

TABLE GC 128—(Continued)
PESTICIDES

Column Packing	P1	P2	P3	P4	P5	P6	P7	P8
Temperature (°C)	200	180	180	195	180	200	190	200
Gas; Flow Rate (ml/min)	100	70	43	70	60	70	100	60
Column								
length, ft	6	6	6	6	6	6	6	6
diameter, in., O D	$\frac{1}{4}$	$\frac{1}{4}$	$\frac{1}{4}$	$\frac{1}{4}$	$\frac{1}{4}$	$\frac{1}{4}$	$\frac{1}{4}$	$\frac{1}{4}$
I D	$\frac{5}{32}$	$\frac{5}{32}$	$\frac{5}{32}$	$\frac{5}{32}$	$\frac{5}{32}$	$\frac{5}{32}$	$\frac{5}{32}$	$\frac{5}{32}$
form	—	—	—	—	—	—	—	—
material	glass	glass	glass	glass	glass	glass	glass	glass
Detector	EC	EC	EC	EC	EC	EC	EC	EC

Compound	r_{aldrin}							
Chlorobenzilate®	1.50, 1.85, *2.34*	1.58, 1.98, *2.56*	1.58, 2.08, *4.19*	—	1.53, 1.97, *3.02*	1.28, 1.47, 1.84, *2.70*	1.55, 1.97, *2.95*	1.48, 1.85, 2.25, *3.29*
Dicofol	0.97	0.83, *1.00*, 2.06	1.67, *1.96*	—	0.60, 1.14, 1.37, *4.31*	0.61, 1.08, 1.27, 4.17, *4.45*	0.58, *1.31*, 3.93, 5.03	0.65, 1.28, 1.51, 2.29, *6.6*
Chlordane	0.70, *0.80*, 1.44, 1.61	0.69, *0.79*, 1.17, 1.47, 1.67, 2.67	*0.87*, 1.29, 1.80, 1.98, 3.27	— —	0.78, *0.83*, 1.57, 1.73, 2.79	0.36, 0.77, *0.82*, 1.05, *1.51*, 1.65, 2.63	0.77, *0.82*, 1.08, 1.57, 1.77	0.76, *0.83*, 1.17, 1.70, 1.88, 3.45
Strobane®	1.78, *2.31*, 3.15, 3.65	1.79, *2.37*, 2.81, 3.21, 3.82	1.57, 2.62, 2.90, 3.08, *4.84*, 6.5	—	2.38, 2.62, 3.11, *3.72*	1.28, 1.64, 2.26, 2.44, *2.88*, 3.38	2.00, 2.56, *3.04*, 3.60	1.31, *2.34*, 2.64, 3.67, 4.43, 6.1
Toxaphene	1.61, 2.24, 2.61, *3.00*, 3.52, 4.93	1.82, 2.38, 2.85, *3.32*, 3.88, 5.64	2.72, 2.98, 3.2, *5.0*, 6.1, 6.7, 7.8	—	2.62, 3.16, *3.83*, 4.26	1.63, 2.06, 2.25, 2.87, *3.49*, 4.53, 5.5	3.10, *3.80*, 5.00	2.34, 3.25, *3.54*, 4.2, 5.8, 7.3, 7.9

Figures in *italics* represent major peaks.

Column Packing
P1 = 10% DC-200.
P2 = 3% OV-1.
P3 = 5% QF-1.
P4 = 3% DEGS.
P5 = 2% OV-1 and 3% QF-1.
P6 = 4% SE-30 and 6% QF-1.
P7 = 5% DC-200 and 7.5% QF-1.
P8 = 1.5% OV-17 and 1.95% QF-1.

Literature

Thompson, J. F., Walker, A. C., and R. F. Moseman, *J. Ass. Offic. Agr. Chem.*, **52**, 1263 (1969).

TABLE GC 129
PESTICIDES

Column Packing	P1		P2		P1		P2	
Temperature (°C)	200		200		210		210	
Gas; Flow Rate (ml/min)	N_2,120		N_2,120		N_2,120		N_2,120	
Column								
length, ft	6		6		6		6	
diameter, mm, I.D.	4		4		4.5		4.5	
form	na		na		—		—	
material	G		G		Al		Al	
Detector	EC		EC		D1		D1	
Literature	1		1		2		2	
Pesticide	r	ng[a]	r	ng	r	μg[b]	r	μg
CDAA	0.1	4	0.1	4	—	—	—	—
DDVP	0.1	10	—	—	—	—	0.11, *0.23*	1
Monuron	0.1	30	0.1	250	0.18	1.5	0.15	2
Diuron	0.11	10–15	*0.15*, 1.68	50–60	0.19	2	0.19	2
Neburon	0.11	10–15	*0.14*, 1.82	35–40	0.19	2	0.27	2
Hexachlorocyclopentadiene	0.12	0.4	0.11	0.5	—	—	—	—
Phosdrin®	0.13	18	0.31	25	—	—	—	—
Simazine	*0.18*, 0.40	100–300	0.26, *0.54*	300	0.45	4	0.58	5
Propazine	*0.18*, 0.42	100–300	0.26, 0.56	300	0.47	5	0.59	5
Altrazine	0.18, 0.42, 1.71, 2.18, 5.0	100–300	0.26, *0.55*	300	—	—	0.59	5
TEPP	*0.21*, 0.89, 2.95	3500–4500	1.69	5000	—	—	—	—
TCNB	0.28	0.50–0.6	0.38	0.5	0.36	1	0.45	1
Methyl ester 2,4-D	0.29	10	0.42	4–5	0.36	4	0.48	1
CIPC	0.31	1000–2000	0.41	2000–5000	0.37	3.5	0.48	1
Thimet®	0.35**	20–30**	0.45	10	—	—	—	—
Vegadex®	0.37	3	0.46	3	0.44	4	0.52	4
BHC	*0.38*, 0.47	1–2	—	—	—	—	—	—
BHC (alpha)	0.38	0.4	0.46, 0.60	0.3, 1.7	0.46	0.75	0.53	1
Dimethoate	0.40	4.5	0.97	8	—	—	—	—
Isopropyl ester 2,4-D	0.41	10	0.55	5	0.44	1.5	0.59	1
BHC (beta)	0.42	1.75	0.68	0.5	0.50	0.75	0.65	1
Hexachlorobenzene	0.43	0.3–0.4	0.42	0.3	0.49	0.4	0.50	0.5
Methoxychlor (tech)	0.46, 1.64, *4.5*	8–10	0.51, 3.23, *4.8*	10	0.50, 2.87, *3.90*	7.5	0.58, 3.03, 3.35, *4.20*	7
Chlordane	0.55	0.6	0.56	0.5	—	—	—	—
Lindane (gamma BHC)	0.47	0.5	0.58	0.3	0.52	0.75	0.63	1
PCNB®	0.:0	0.3–0.4	0.61	0.3	0.54	0.75	0.68	0.5

TABLE GC 129—(Continued) PESTICIDES

Column Packing	P1		P2		P1		P2	
Temperature (°C)	200		200		210		210	
Gas; Flow Rate (ml/min)	N_2,120		N_2,120		N_2,120		N_2,120	
Column								
length, ft	6		6		6		6	
diameter, mm, I.D.	4		4		4.5		4.5	
form	na		na		—		—	
material	G		G		Al		Al	
Detector	EC		EC		D1		D1	
Literature	1		1		2		2	
Pesticide	r	ng[a]	r	ng	r	μg[b]	r	μg
Diazinon®	0.50**	20–25**	0.06**	25**	—	—	—	—
Dichlone	0.54	3–4	*0.86*, 2.04	4	0.57	2.25	0.91	4
Tetraiodoethylene	0.54**	4**	0.51**	2**	0.63	8	0.59**	8**
Isobutyl ester 2,4-D	0.61	5	—	—	0.63	2	0.80	2
Endosulfan alcohol	0.63, 1.28, 1.64, *2.07*	14	0.1, 0.74, 1.25, 2.13, *2.68*	15	—	—	—	—
Isopropyl ester 2,4,5-T	0.66	2	0.84	2	0.65	2	0.85	3
Methyl parathion	0.68**	2–3**	1.42**	2.5**	—	—	—	—
***n*-Butyl ester 2,4-D**	0.71	40	0.95	10	0.69	2.5	0.92	2
Ronnel	0.78	1	0.97	1	0.77	2.5	0.97	3
Dimite®	0.78, *0.95*	15–20	0.82, 1.31	15–20	—	—	0.82, 1.29	5
Heptachlor	0.79	0.75–1	0.81	0.5	0.81	1	0.84	1
Malathion	0.87**	20–30**	1.48**	20–30**	—	—	—	—
1-Hydroxychlordene	0.97	1.3	1.11	1	—	—	—	—
Parathion	0.98**	15**	1.88 **	5**	—	—	—	—
Dacthal®	0.99	1.2	1.52	1.5	—	—	1.45	2
Aldrin	1.00*	0.75–1*	1.00***	0.6***	1.00**	1**	1.00****	1****
Kelthane®	1.01	2–3	—	—	—	—	1.27	5
Chlorothion	1.05	60–70	1.35, 1.82, *2.27*	2.5	0.99	2	—	—
Chloropropylate	1.06, 1.75, *2.32*	60–70	1.30, *2.79*	60–70	—	—	—	—
Chlorobenzilate®	1.06, 1.63, 2.28	60–70	1.30, 2.33, *2.73*	60–70	2.07	7	2.60	7
***n*-Butyl ester 2,4,5-T**	1.08	2	1.15, *1.30*	6	1.05	2	1.20, *1.33*, 2.97	2
Telodrin®	1.10	1	1.14	0.8	1.09	1	1.15	1
Chloranil	1.11, 1.28	50	*0.37*, 1.52, 1.73	5	—	—	0.44	8
Isodrin	1.15	1	1.25	0.8	—	—	1.20	2
***o,p′*-TDE olefin**	1.17	16	1.21	10	1.12	5	1.18	2
Folpet	1.20	4	2.03	4–5	—	—	1.97	8

TABLE GC 129—(Continued)
PESTICIDES

Column Packing	P1		P2		P1		P2	
Temperature (°C)	200		200		210		210	
Gas; Flow Rate (ml/min)	N_2,120		N_2,120		N_2,120		N_2,120	
Column								
length, ft	6		6		6		6	
diameter, mm, I.D.	4		4		4.5		4.5	
form	na		na		—		—	
material	G		G		Al		Al	
Detector	EC		EC		D1		D1	
Literature	1		1		2		2	
Pesticide	r	ng[a]	r	ng	r	μg[b]	r	μg
Dyrene	1.21	75	1.50, 1.80, 2.50	20	1.15, 1.38, 1.61, 1.87, 2.23	25	*1.33*, 1.77, 2.08, 2.37, 2.67	25
Captan	1.22	40–50	2.10	4	1.16	3.5	2.03	3
Sulphenone	1.24	4	2.62	3–4	1.19	8	2.43	8
Heptachlor epoxide	1.24	1	1.47	1	1.20	0.8	1.41	1
***o,p'*-Dichlorodiphenylmono-chloroethane**	1.34	—	1.53	—	—	—	1.43	—
Chlorbenside	1.36	2	1.51	1.8	1.25	3.5	1.45	5
***p,p'*-TDE olefin**	1.42	2–3	1.50	7	—	—	1.43	1
Chlordane (alpha)	1.43	1	1.57	—	—	—	1.47	1
Perthane olefin	1.47	40	1.31, 1.32	40	1.35	6	1.24	4
***o,p'*-DDE**	1.48	2	1.46	2	1.37	2	1.41	2
Propylene glycol butyl ether ester 2,4-D	1.51	20	1.66, *1.92*, 4.4	30	1.34, 3.12	6	*1.89*, 4.22	6
Endosulfan	1.58, 2.16	1–2	1.89, 2.92	2–3	*1.44*, 1.91	3	*1.79*, 2.73	2
Endosulfan I	1.58	1.5	1.89	1.5	—	—	—	—
Ovex	1.58	3	2.78	3–4	1.46	5	2.57	2
Chlordane (beta)	1.60	1	1.73	1.2	1.46	1	1.59	2
***p,p'*-Dichlorodiphenylmono-chloroethane**	1.62	—	1.94	—	—	—	1.81	—
Delta Trichloro "237"	1.67	2	1.70	1.5	—	—	—	—
Butoxy ethanol ester 2,4-D	1.78	12	1.98, *2.32*	30	1.50	5	1.91	5
DDT (tech)	1.81, 2.44, *3.03*	5	0.39, 1.88, 2.48, 2.70, *3.28*	3.5	0.35, 1.69, 2.21, *2.70*	4	0.45, 1.74, 2.25, *2.95*	5
***o,p'*-DDE**	1.81	1	1.88	1.5	—	—	1.74	1
Dieldrin	1.83	1–1.5	2.22	1–1.5	1.72	1.75	2.06	2
***p,p'*-TDE**	1.86	2	2.04	1.5	—	—	1.89	2
Iso-octyl ester 2,4-D	2.00, 2.37	30	—	—	1.70, *1.98*	9	2.15, *2.51*	9

TABLE GC 129—(Continued)
PESTICIDES

Column Packing	P1		P2		P1		P2	
Temperature (°C)	200		200		210		210	
Gas; Flow Rate (ml/min)	N_2,120		N_2,120		N_2,120		N_2,120	
Column								
length, ft	6		6		6		6	
diameter, mm, I.D.	4		4		4.5		4.5	
form	na		na		—		—	
material	G		G		Al		Al	
Detector	EC		EC		D1		D1	
Literature	1		1		2		2	
Pesticide	r	ng[a]	r	ng	r	μg[b]	r	μg
Endrin	2.05, 2.29	2	2.55	2.5–3	*1.90*, 2.09	3	2.40	2
Ethyl hexyl ester 2,4-D	2.07	8	1.66, 1.94, *2.18*	25	1.77	10	1.90, *2.14*	10
Perthane®	2.11	50–150	*2.02*, 2.60	100–150	1.91	4.5	1.85	5
Endosulfan II	2.16	2	2.92	2	—	—	—	—
***p,p′*-Methoxychlor, monochloroethylene analog**	2.22	35	2.34	35–40	—	—	—	—
Endrin aldehyde	2.30	4	3.98	4.5	—	—	3.59	3
***o,p′*-DDT**	2.44	4	2.48	3	2.21	3.5	2.25	5
Ethion	2.46**	10–12**	3.23	10	—	—	—	—
Endrin alcohol (d)	2.50	3.5	3.45	4	—	—	3.19	3
Endosulfan sulfate	2.67	5	5.40	6	—	—	—	—
Kepone®	2.70	4–5	2.67	7	2.35	3	2.45	1
***p,p′*-TDE**	2.70	1.5	2.70	1.5	—	—	2.45	2.5
Iso-octyl ester 2,4,5-T	2.73, 3.14, 3.73	10	1.82, 3.44, *3.90*, *4.6*	20–30	2.28, *2.62*, *3.08*, 3.60	10	2.78, *3.11*, 3.63	10
Carbophenothion	2.83	10	3.33	4	2.44	8	2.97	8
Prolan	2.83	5	4.90	10–15	2.40	11	0.29, 0.66, 0.75, 1.28, 3.30, *4.3*	10
Perthane, trichloroethane analog	2.83	6	2.37	5	—	—	—	—
Dilau®	2.83, *3.37*	5–10	4.40, 4.90, *5.70*	15	2.05, 2.40, 2.86	15	3.30, 3.70, 4.3, *4.9*	5
Butoxy ethanol ester 2,4,5-T	2.85	4	3.08, *3.58*	8	2.32	5	2.85	5
***p,p′*-Methoxychlor olefin**	2.91	8	3.00	10–12	2.52	8	—	—
***p,p′*-DDT**	3.03	4	3.28	3	2.70	3.5	2.95	5
Bulan®	3.35	4	5.70	5–10	2.40, *2.89*	10	4.3, *4.9*	15
***p,p′*-Methoxychlor, dichloroethane analog**	3.36	250	3.70	250	—	—	—	—

TABLE GC 129—(Continued)
PESTICIDES

Column Packing	P1		P2		P1		P2	
Temperature (°C)	200		200		210		210	
Gas; Flow Rate (ml/min)	N_2,120		N_2,120		N_2,120		N_2,120	
Column								
length, ft	6		6		6		6	
diameter, mm, I.D.	4		4		4.5		4.5	
form	na		na		—		—	
material	G		G		Al		Al	
Detector	EC		EC		D1		D1	
Literature	1		1		2		2	
Pesticide	r	ng[a]	r	ng	r	μg[b]	r	μg
Delta keto "153"	3.50	5	6.05	7	—	—	5.5	3
Imidan®	3.87**	15**	7.54	10	—	—	—	—
Methoxychlor	4.5	8–10	—	—	—	—	—	—
Guthion®	5.0	50	9.3**	50–100**	—	—	—	—
Tetradifon	5.4	5–7	8.75	8–10	4.22	8	7.5	8
BEP ester 2,4,5-T	0.16, 0.67, 1.06, *2.79*, 3.24, 5.2, 6.9	35	—	—	—	—	—	—
Chlordane (tech)	0.44, 0.62, 0.70, 0.79, 0.95, 1.14, *1.42*, 1.59, 2.56	12	0.72, *0.81*, 1.07, 1.55, *1.72*, 2.78	7	0.52, 0.66, 0.74, 0.82, 0.96, 1.14, *1.36*, 1.50, 2.3	10	0.53, *0.82*, 1.24, 1.34, *1.45*, *1.58*, 1.82, 1.92, 2.36	10
BEP ester 2,4-D (tech)	0.73, 1.84, 2.10, 3.33, 4.9	50	0.25, 0.82, 1.32, 1.66, 2.92, 3.28, *6.4*	40–50	0.71, 1.21, 1.53, 1.81, *3.51*, 8.5	18	0.90, 1.93, 3.12, *4.2*, 9.6	18
Strobane®	1.07, 1.29, 1.50, 1.76, 1.90, 2.05, *2.28*, 2.64, 3.04, 3.55	25	1.09, 1.38, 1.53, 1.69, 2.16, 2.52, 3.01, 3.56, 4.57	40	0.91, 1.07, 1.24, 1.44, 1.63, 2.07, 2.35, 2.66, 3.17, 3.77	18	1.08, 1.31, 1.47, 2.00, 2.24, 2.32, *2.74*, 3.19, 4.1	18
Toxaphene	1.18, 1.51, 1.76, *2.34*, 2.63, *3.06*, *3.61*, 4.3, 4.5, 5.0	20	1.70, 2.25, 2.37, 2.55, *3.03*, 3.66, 4.10, *4.70*, 5.8	25–30	1.29, 1.45, 1.55, 1.87, *2.17*, 2.45, *2.85*, *3.36*, 4.03, 4.5	22	1.08, 1.34, 1.59, 1.97, 2.17, 2.34, 2.63, 2.72, 3.07, *3.27*, *3.64*, 4.1, 5.1	22
Co-Ral	9.5	100–150	24.7	50–75	7.6	25	20.0	150
BHC (tech)	—	—	*0.46*, 0.58, 0.68	0.4	*0.44*, 0.52	1.25	*0.53*, 0.64, 0.74	1
Kelthane (tech)	—	—	0.58, 0.74, 1.09, *1.31*, 1.86, 2.21, 2.37	6	*0.99*, 3.82	15	0.28, *1.27*, 1.48, 2.06, 4.0, 4.3	6

TABLE GC 129—(Continued) PESTICIDES

Column Packing	P1		P2		P1		P2	
Temperature (°C)	200		200		210		210	
Gas; Flow Rate (ml/min)	N_2,120		N_2,120		N_2,120		N_2,120	
Column								
length, ft	6		6		6		6	
diameter, mm, I.D.	4		4		4.5		4.5	
form	na		na		—		—	
material	G		G		Al		Al	
Detector	EC		EC		D1		D1	
Literature	1		1		2		2	
Pesticide	r	ng[a]	r	ng	r	μg[b]	r	μg
Dioxathion	—	—	*0.66*, 1.13, 14.27	—	—	—	—	—
p,p′-Kelthane	—	—	1.31	5	—	—	—	—
Meta Systox® R	—	—	1.70	2000–3000	—	—	—	—
Paraoxon	—	—	2.25**	15–20**	—	—	—	—
p,p′-Methoxychlor	—	—	4.80	10	3.90	7	4.20	7
Mirex	—	—	5.15	6	—	—	4.33	3
Dichloral urea	—	—	—	—	0.15	1.5	0.12	0.2
Carbophenothion, thiolsulfoxide	—	—	—	—	0.15, 1.78, 1.89	150	—	—
Ethylene dichloride	—	—	—	—	0.16	1	0.10	0.4
Chlorpicrin	—	—	—	—	0.17	1	0.13	2
Carbophenothion, sulfoxide	—	—	—	—	0.17, 1.78, 2.46, *4.5*	60	—	—
Ethylene dibromide	—	—	—	—	0.18	0.75	0.11**	0.5**
p-Dichlorobenzene	—	—	—	—	0.18	0.75	0.12	0.5
Ethide	—	—	—	—	0.19	1.5	0.11	2
BHC (delta)	—	—	—	—	0.55	0.75	—	—
Pentachlorophenol	—	—	—	—	0.60	7.5	—	—
p,p′-DDD olefin	—	—	—	—	1.34	2	—	—
Chlordane (gamma)	—	—	—	—	1.35	1.5	—	—
Methyl ester DDA	—	—	—	—	1.48	7	—	—
DDE	—	—	—	—	1.69	2.5	—	—
o,p′-DDD	—	—	—	—	1.72	2	—	—
Aramite®	—	—	—	—	1.76, 1.84	15	2.34, 2.57	15
p,p′-DDD	—	—	—	—	2.12	2.5	—	—
Carbophenothion, sulfon	—	—	—	—	4.30	30	—	—

TABLE GC 129—(Continued)
PESTICIDES

Column Packing	P1		P2		P1		P2	
Temperature (°C)	200		200		210		210	
Gas; Flow Rate (ml/min)	N_2,120		N_2,120		N_2,120		N_2,120	
Column								
length, ft	6		6		6		6	
diameter, mm, I.D.	4		4		4.5		4.5	
form	na		na		—		—	
material	G		G		Al		Al	
Detector	EC		EC		D1		D1	
Literature	1		1		2		2	
Pesticide	r	ng[a]	r	ng	r	μg[b]	r	μg
BEP ester 2,4,5-T (tech)	—	—	—	—	0.66, 1.06, 2.32, 2.72, 3.85, *5.2*	30	0.26, 0.59, 0.85, 1.32, 2.82, 3.14, *6.1*	30

When more than one peak is present the major peak(s) is in italic.
* Aldrin reference. Retention time approximately 4.25 minutes.
** Not chlorinated.
*** Aldrin reference. Retention time approximately 3.5 minutes.
**** Aldrin reference. Retention time approximately 5.5 minutes.
[a] Response in ng for ½ full scale deflection (F.S.D.) at DC voltage at which 1 ng of heptachlor epoxide causes ½ F.S.D. at 10^{-9} amps-full-scale (A.F.S.) attenuator 1.
[b] Response in μg for ½ F.S.D., 6452.

Column Packing
P1 = 10% DC-200 on Gas Chrom Q (80/100) or Anakrom ABS (80/90).
P2 = Equal portions of 15% QF-1 on Gas Chrom Q (80/100) and 10% DC-200 on Gas Chrom Q (80/100).

Detector
D1 = Dohrmann coulometer, Model C-100, Halogen cell, Model T-200-S.

Literature
1. Pesticide Analytical Manual, FDA, Vol. 1, Tables 311.5A, B.
2. Pesticide Analytical Manual, FDA, Vol. 1, Tables 312.51-A, B.

TABLE GC 130
PESTICIDES, CARBAMATES AND ENOLS

Column Packing	P1			P2			P3		P4	—	—
Column Temperature (°C)	—			—			—		—	—	—
Gas; Flow Rate, (ml/min)	N_2,114			N_2,91			N_2,48		N_2,12.5	N_2,12.5	N_2,12.5
Column											
length, m	1.83			1.83			1.83		0.71	0.71	0.71
diameter, mm(a), in.(b)	5(a)			5(a)			5(a)		0.1(b)	0.1(b)	0.1(b)
form	U-shaped			U-shaped			U-shaped		na	na	na
material	G			G			G		na	na	na
Detector	FI			FI			FI		FI/EC	FI/EC	FI/EC
Literature	1			1			1		2	—	—
Compound	t_r										
Column Temperature	180°	200°	220°	180°	200°	220°	150°	170°	130°	150°	165°
4-Diallylamino-3,5-methylphenyl *N*-methylcarbamate	>107.0	73.5	32.0	76.0	37.0	19.0	19.2	8.9	—	—	—
Banol	18.7	9.6	5.5	14.3	8.6	5.3	3.2	2.0	—	—	—
Bayer 37344 (mesurol®)	> 80.0	80.0	36.0	>70.0	24.6	13.3	—	5.6	—	—	—
Baygon	7.0	4.0	2.7	8.1	5.1	2.7	2.9	1.8	—	—	—
Carbaryl	—	—	—	—	—	—	—	—	—	2.3	—
2-Chloro-3,4-dimethylphenyl *N*-methylcarbamate	17.0	8.8	5.1	14.7	8.7	5.4	3.2	2.0	—	—	—
3,4-Diisopropylphenyl *N*-methylcarbamate	—	—	—	22.2	—	—	—	—	—	—	—
3,5-Diisopropylphenyl *N*-methylcarbamate	49.0	21.6	11.2	—	12.2	6.9	6.5	3.4	—	—	—
3-Dimethylamino-4-methylphenyl *N*-methylcarbamate	64.0	27.5	13.7	18.8	10.4	6.2	5.5	3.1	—	—	—
3,5-Dimethylphenyl-*N*-methylcarbamate	24.7	11.8	6.6	7.7	4.8	3.2	19.0	—	—	—	—
3,5-Dimethylphenol	—	—	—	—	—	—	2.54	—	—	—	—
Dimetilan	—	—	—	—	—	—	—	—	—	—	8.5
Isolan®	—	—	—	—	—	—	—	—	5.6	—	—
1-Isopropyl-3-methylpyrazolon	—	—	—	—	—	—	—	—	2.0	—	—
Matacil	53.0	23.4	11.7	16.7	9.4	5.7	4.7	2.8	—	—	—
4-Dimethylamino-3,5-xylenol	—	—	—	—	—	—	—	—	1.0	—	—
3-Methylphenol	18.3	9.3	5.2	4.9	3.3	2.3	1.8	1.3	—	—	—
3-Methylphenyl-*N*-methylcarbamate	18.3	9.1	5.2	4.9	3.3	2.3	12.8	6.0	—	—	—
1-Naphthol	—	—	—	—	—	—	—	—	—	2.3	—
3-Methyl-5-hydroxypyrazole-5-dimethylcarbamate	—	—	—	—	—	—	—	—	—	—	4.2
Phenol	13.8	7.0	4.1	3.0	2.2	1.6	1.3	0.95	—	—	—
Phenyl-*N*-methylcarbamate	13.8	7.1	4.1	3.1	2.2	1.6	8.4	4.2	—	—	—
Pyrolan	—	—	—	—	—	—	—	—	—	6.8	—
1-Phenyl-3-methylpyrazolon	—	—	—	—	—	—	—	—	—	3.7	—
Zectran®	57.5	24.2	12.1	22.1	12.3	7.1	6.1	3.4	10.7	—	—

Column Packing
P1 = 60/80 mesh Chromosorb W (HMDS), 20% Carbowax 20M.
P2 = 60/80 mesh Chromosorb W (HMDS), 30% Apiezon N.
P3 = 60/80 mesh Chromosorb W (HMDS), 6% QF-1 and 4% SE-30.
P4 = 5% QF-1 on A.W. Chromosorb W (100/200).

Temperature
Injection 265 °C, Column listed above, Detector 266 °C.

Literature
1. Strother, Allen, *J. Gas Chromatogr.*, **6**, 111 (1968).
2. Eberle D. O. and F. A. Gunther, *J. Ass. Offic. Agr. Chem.*, **48**, 927 (1965).

TABLE GC 131
PYRIDINES

Column Packing	P1	P2	P3	P4	P5	P6	P7	P8	P9	P10	P11	P12
Temperature (°C)	130 ± 0.5	130 ± 0.5	130 ± 0.5	130 ± 0.5	130 ± 0.5	130 ± 0.5	130 ± 0.5	130 ± 0.5	130 ± 0.5	130 ± 0.5	130 ± 0.5	128
Gas; Flow Rate (ml/min)	He, 60 ± 0.1	He, 60 ± 0.1	He, 60 ± 0.1	He, 60 ± 0.1	He, 60 ± 0.1	He, 60 ± 0.1	He, 60 ± 0.1	He, 60 ± 0.1	He, 60 ± 0.1	He, 60 ± 0.1	He, 60 ± 0.1	He, 170
Column												
length, m	2	2	2	2	2	2	2	2	2	2	2	2
diameter, in.	$\frac{1}{4}$	$\frac{1}{4}$	$\frac{1}{4}$	$\frac{1}{4}$	$\frac{1}{4}$	$\frac{1}{4}$	$\frac{1}{4}$	$\frac{1}{4}$	$\frac{1}{4}$	$\frac{1}{4}$	$\frac{3}{4}$	$\frac{3}{8}$
form	tubing	tubing	tubing	tubing	tubing	tubing	tubing	tubing	tubing	tubing	tubing	tubing
material	aluminum	aluminum	aluminum	aluminum	aluminum	aluminum	aluminum	aluminum	aluminum	aluminum	aluminum	Cu
Detector	TC	TC	TC	TC	TC	TC	TC	TC	TC	TC	TC	TC
Literature	1	1	1	1	1	1	1	1	1	1	1	2

Compound	$r_{2,5\text{-dimethylpyridine}}$											r^d
Pyridine	0.30	0.29	0.33	0.40	0.35	0.34	0.34	0.40	0.44	0.40	0.45	0.08
2-Methylpyridine	0.47	0.46	0.51	0.53	0.50	0.48	0.50	0.53	0.51	0.52	0.53	0.15
2,6-Dimethylpyridine	0.73	0.72	0.73	0.66	0.70	0.67	0.67	0.68	0.69	0.65	0.55	—
3-Methylpyridine	0.64	0.64	0.69	0.74	0.71	0.71	0.71	0.81	0.86	0.82	0.88	0.14
4-Methylpyridine	0.64	0.64	0.71	0.71	0.71	0.74	0.69	0.83	0.86	0.86	0.95	0.16
2-Ethylpyridine	0.81	0.81	0.80	0.82	0.83	0.79	0.79	0.78	0.76	0.77	0.66	0.19
2,5-Dimethylpyridine	1.00	1.00	1.00	1.00	1.00	1.00	1.00	1.00	1.00	1.00	1.00	0.21
2,4-Dimethylpyridine	1.00	1.00	1.05	0.96	1.01	1.02	1.01	1.07	1.05	1.08	1.11	0.34
2,3-Dimethylpyridine	1.09	1.08	1.14	1.09	1.11	1.10	1.11	1.21	1.13	1.19	1.15	—
4-Ethylpyridine	1.19	1.18	1.26	1.22	1.30	1.33	1.27	1.43	1.47	1.46	1.51	0.27
2,4,6-Trimethylpyridine	1.50	1.51	1.49	1.30	1.36	1.35	1.42	1.42	1.35	1.33	1.11	0.30
2-Methyl-5-ethylpyridine	1.79	1.80	1.82	1.57	1.74	1.75	1.72	1.70	1.49	1.65	1.60	—
4-Isopropylpyridine	1.77	1.76	1.81	1.71	1.87	1.92	1.83	1.94	1.93	1.92	1.96	—
4-*n*-Propypyridine	2.07	2.09	2.15	2.07	2.15	2.23	2.19	2.40	2.28	2.32	2.31	—
3-Ethyl-4-methylpyridine	2.55	2.61	2.77	2.55	2.77	3.02	2.75	3.08	2.77	3.24	3.37	—
3,5-Dimethylpyridine	—	—	—	—	—	—	—	—	—	—	—	0.30
2-Chloropyridine	—	—	—	—	—	—	—	—	—	—	—	0.24
2,5-Dichloropyridine	—	—	—	—	—	—	—	—	—	—	—	0.46
2,6-Dichloropyridine	—	—	—	—	—	—	—	—	—	—	—	0.60
3,5-Dichloropyridine	—	—	—	—	—	—	—	—	—	—	—	0.34

TABLE GC 131—(Continued)
PYRIDINES

[a] Stationary liquid substrates.

Substrate	Character of Material	Source
Squalane	Branched-chain C_{30} alkane	Distillation Products Industries
Mineral oil	High molecular weight alkanes and cycloalkanes	E. R. Squibb and Sons
Apiezon-L	Saturated hydrocarbon lubricant	James G. Biddle Co.
Silicone high-vacuum grease	Silicone lubricant	Dow Corning Corp.
Silicone oil, D.C. 703	Phenylsilicones, mol. wt. 570	Distillation Products Industries
Octoil	Di-2-ethylhexyl phthalate	Consolidated Vacuum Corp.
Octoil-S	Di-2-ethylhexyl sebacate	Consolidated Vacuum Corp
Tri-*m*-cresyl phosphate	..	Distillation Products Industries
Diphenyl phthalate	..	Distillation Products Industries
Monohydroxyethyl-trihydroxypropyl-ethylenediamine[b]	..	Visco Products Co., Inc.
Tide[c]	Commercial detergent containing alkylaryl sulfonate	Procter and Gamble

[b] Abbreviated as MTE.

[c] Investigated by using the dried commercial detergent as a column packing (3).

[d] Reference compound: benzo(b)thiophene; t_r = 10–15 min.

Column Packing

P1 = 10% Squalane on Chromosorb (40/60 mesh).
P2 = 10% Mineral oil Chromosorb (40/60 mesh).
P3 = 10% Apiezon L Chromosorb (40/60 mesh).
P4 = 10% Silicone high vacuum grease on Chromosorb (40/60 mesh).
P5 = 10% Silicone oil DC-703 on Celite 545 (60/40 mesh).
P6 = 10% Octoil on Chromosorb (40/60 mesh).
P7 = 10% Octoil-S on Chromosorb (40/60 mesh).
P8 = 10% TCP (tri-*m*-cresyl phosphate [tri-*m*-tolyl phosphate]) on Chromosorb (40/60 mesh).
P9 = Tide.
P10 = 10% DDP (diphenyl phthalate) on Chromosorb (40/60 mesh).
P11 = 10% MTE (monohydroxyethyltrihydroxypropylethylenediamine) on Chromosorb (40/60 mesh).
P12 = 10% DC-710 on Chromosorb G (60/80 mesh).

Literature

1. Decora, A. W. and G. U. Dinneen, *Anal. Chem.*, **32**, 164 (1960).
2. Klemm, L. H., Shabtai, J., and F. H. W. Lee, *J. Chromatogr.*, **51**, 433 (1970).

TABLE GC 132
PLASTICIZERS

Column Packing	P1
Temperature (°C)	283
Gas; Flow Rate (ml/min)	50,He
Column	
length, cm	60
diameter, cm	0.5
form	helical
material	glass
Detector	TC

Compound	$r_{dibutyl\ phthalate}$
Triethyl phosphate	0.05
Dimethyl phthalate	0.24
Glycerol tripropionate	0.25
Diethyl phthalate	0.35
Diisobutyl phthalate	0.74
Diisobutyl azelate	0.91
Dibutyl phthalate	1.0
Bis(2-methoxyethyl) phthalate	1.3
Dibutyl isophthalate	1.4
Isobutyl palmitate	1.7
Dibutyl sebacate	1.7
2-Methoxyethyl palmitate	1.9
Bis(2-ethylbutyl) phthalate	2.2
Bis(2-ethylhexyl) adipate	3.1
Diisooctyl adipate	3.1
Triphenyl phosphate	5.1
Bis(2-ethylhexyl) phthalate	5.1
Diethylene glycol dibenzoate	5.4
Bis(3,5,5-trimethylhexyl) phthalate	7.6
Bis(2-ethylhexyl) isophthalate	9.1
Tris(*m*-tolyl) phosphate	9.5

Column Packing P1 = 2.0 g Apiezon oil K on 4.7 g Celite 545 (50 mesh).

Literature Lewis, J. S. and H. W. Patton, *Gas Chromatography, Symposium held under the auspices of Analysis Instrumentation Div. of Instrument Soc. of America* (1957). Coates, V. J., Hoebels, H. J., and I. S. Fagerson, Editors, Academic Press, Publishers, N.Y. (1958).

TABLE GC 133
PYRROLES

Column Packing	P1	P1
Temperature (°C)	190	190
Gas; Flow Rate (ml/min)	66,H_2	28,N_2
Column		
length, ft	4	4
diameter, in.	¼	¼
form	tubing	tubing
material	SS	SS
Detector	TC	FID

Compound	V_R'	
Pyrrole	1.61	—
2,5-Dimethylpyrrole	3.31	—
2,4-Dimethylpyrrole	3.44	—
1,2,5-Trimethylpyrrole	1.30	—
1-Ethyl-2,5-dimethylpyrrole	1.52	—
1-Allyl-2,5-dimethylpyrrole	1.25	—
1-*n*-Propyl-2,5-dimethylpyrrole	1.39	—
1-*n*-Butyl-2,5-dimethylpyrrole	1.34	—
2-Methyl-3-[2,5-dimethylpyrrolyl-(1)]-butane	1.90	—
3-[2,5-Dimethylpyrrolyl-(1)]-pentane	1.27	—
2,2-Dimethyl-4-[2,5-dimethylpyrrolyl-(1)]-propane (1-neopentyl-2,5-dimethylpyrrole)	1.78	—
1-*n*-Amyl-2,5-dimethylpyrrole	1.30	—
1-*n*-Hexyl-2,5-dimethylpyrrole	1.90	—
1-*n*-Heptyl-2,5-dimethylpyrrole	1.41	—
1-Methylpyrrole	—	1.60
2-Methylpyrrole	—	2.63
1-Ethylpyrrole	—	2.32
2-Ethylpyrrole	—	3.79
1-*n*-Propylpyrrole	—	3.19
2-*n*-Propylpyrrole	—	4.96
1-Isopropylpyrrole	—	2.85
2-Isopropylpyrrole	—	4.23
1-Allylpyrrole	—	3.04
2-Allylpyrrole	—	4.62
1-Propargylpyrrole	—	3.59
1-*prim*-Butylpyrrole	—	1.26
2-*prim*-Butylpyrrole	—	2.05
1-*sec*-Butylpyrrole	—	2.94
2-*sec*-Butylpyrrole	—	4.03
2-*tert*-Butylpyrrole	—	2.95

Column Packing P1 = DC-550 (% not given) on (42/60 mesh) C-22 firebrick.

Literature Kaluza, G. A. and F. Martin, *J. Gas Chromatogr.*, **5**, 562 (1967).

TABLE GC 134

PURINES, PYRIMIDINES AND RELATED COMPOUNDS

Column Packing	P1	P1	P1
Temperature (°C)	200	230	255
Gas; Flow Rate	na	na	na
Column			
length, ft	6	6	6
diameter, mm, I.D.	4	4	4
form	U-tube	U-tube	U-tube
Detector	FI	FI	FI
Compound		t_r (min.)	
3-Methyluridine	5.8	2.1	—
Uridine 2′,3′,5′-triacetate	30.4	5.4	—
3-Methyluridine 2′,3′,5′-triacetate	—	2.9	—
2′,3′-*O*-Isopropylideneuridine	6.5	1.5	—
2′-Deoxyuridine 3′,5′-diacetate	—	3.4	—
3-Methylthymidine	4.9	1.4	—
Thymidine 3′,5′-diacetate	—	3.6	—
3-Methylthymidine 3′,5′-diacetate	—	1.7	—
1-β-D-Glucopyranosyluracil 2′,3′,4′,6′-tetraacetate	—	10.4	—
1-β-D-Glucopyranosyl-3-methyluracil 2′,3′,4′,6′-tetraacetate	—	6.7	—
1-β-D-Glucopyranosyl-4-methoxy-2(1*H*)-pyrimidone 2′,3′,4′,6′-tetraacetate	54	9.1	—
1-β-D-Glucopyranosylthymine 2′,3′,4′,6′-tetraacetate	—	8.5	—
1-β-D-Glucopyranosyl-3-methylthymine 2′,3′,4′,6′-tetraacetate	28.7	5.3	—
Adenosine 2′,3′,5′-triacetate	—	5.7	1.4
***N*-Acetyladenosine 2′,3′,5′-triacetate**	—	15.7	3.8
2′,3′-*O*-Isopropylidineadenosine	4.0	1.0	—
2′-Deoxyadenosine 3′,5′-diacetate	—	2.6	—
***N*-Acetyl-2′-deoxyadenosine 3′,5′-diacetate**	—	8.7	—
1-Methylinosine 2′,3′,5′-triacetate	—	60.0	13.5
6-Methoxy-9-β-D-ribofuranosylpurine 2′,3′,5′-triacetate	26.3	4.8	1.3

Column Packing P1 = 1% QF-1 on siliconized Gas Chrom P (100/140 mesh).
Literature Miles, H. T. and H. M. Fales, *Anal. Chem.*, **34**, 860 (1962).

TABLE GC 135

SULFUR COMPOUNDS, DIALKYLSULFIDES

Column Packing	P1	P2	P3	P4	P5
Temperature (°C)	100	100	100	100	100
Gas; Flow Rate (ml/min)	25,N_2	25,N_2	25,N_2	25,N_2	25,N_2
Column					
length, cm	125	125	125	125	125
diameter, mm, I.D.	4	4	4	4	4
form	U-shaped	U-shaped	U-shaped	U-shaped	U-shaped
material	glass	glass	glass	glass	glass
Detector	TC	TC	TC	TC	TC
Compound			$r_{benzene}$		
Dimethyl sulfide	0.34	—	—	—	—
Methylethyl sulfide	0.66	—	—	—	—
Diethyl sulfide	1.17	—	—	—	—
Ethylisopropyl sulfide	1.62	1.74	1.34	1.53	1.91
Methylisobutyl sulfide	1.96	2.05	1.58	1.80	2.28
Methyl*sec*-butyl sulfide	1.96	2.19	1.70	1.89	2.34
Diisopropyl sulfide	1.96	2.19	1.58	1.89	—
Propylisopropyl sulfide	3.08	3.12	2.43	2.96	3.63
Ethylisobutyl sulfide	3.08	3.45	2.70	3.29	4.02
Ethyl*sec*-butyl sulfide	3.08	3.45	2.70	3.29	4.02
Dipropyl sulfide	4.35	4.98	3.88	4.64	5.91
Ethylbutyl sulfide	4.35	4.98	3.88	4.64	5.91
Isopropylisobutyl sulfide	4.35	4.01	3.10	3.96	4.79
Isobutyl*sec*-butyl sulfide	8.25	7.75	5.71	7.83	9.51
Propylbutyl sulfide	8.25	9.19	6.94	8.74	11.2
Dibutyl sulfide	15.2	—	—	—	—
V_g **Benzene**	37	46	44	65	39

Column Packing P1 = 20% Siliconol on Kieselguhr (50/60 mesh).
P2 = 20% 7,8-Benzoquinoline on Kieselguhr (50/60 mesh).
P3 = 20% 5,6-Benzoquinoline on Kieselguhr (50/60 mesh).
P4 = 20% 2,4-Dimethyl-7,8-benzoquinoline on Kieselguhr (50/60 mesh).
P5 = 20% Phenanthrene on Kieselguhr (50/60 mesh).
Literature Petranek, J., *J. Chromatogr.*, **5**, 254 (1961).

TABLE GC 136
STEROIDS AND DERIVATIVES

Column Packing		P1	P1	P1	P1	P1
Temperature (°C)		—	—	—	—	—
Gas; Flow Rate (ml/min)		100,Ar	100,Ar	100,Ar	100,Ar	100,Ar
Column						
length, ft		6	6	6	6	6
diameter, mm, I.D.		5	5	5	5	5
form		U-shaped	U-shaped	U-shaped	U-shaped	U-shaped
material		borosilicate glass	borosilicate glass	borosilicate glass	borosilicate glass	borosilicate glass
Detector		FI	FI	FI	FI	FI
				$r_{cholestane}$		
Chemical name	Trivial name	PC[b]	TMSi[a]	CMDSi[b]	HFB[c]	CDFA[d]
Diketone compounds						
4-Androstene-3,17-dione		1.44	—	—	—	—
1,4-Androstadiene-3,17-dione		1.72	—	—	—	—
4-Pregnene-3,20-dione	Progesterone	2.17	—	—	—	—
Monohydroxy compounds						
5α-Cholestan-3β-ol	Cholestanol	2.85	2.60	5.81	1.98	3.79
5-Cholesten-3β-ol	Cholesterol	2.85	2.55	5.18	1.87	3.62
5,7-Cholestadien-3β-ol	7-Dehydrocholesterol	D[e]	3.05	6.54	D	D
24β-Methyl-5,7,22(9β,10α)-cholestatrien-3β-ol	Lumisterol	2.54	1.67	3.72	D	D
24β-Methyl-5,7,22-cholestatrien-3β-ol	Ergosterol	3.80	3.38	7.27	D	D
24β-Ethyl-5-cholestaen-3β-ol	Sitosterol	4.72	4.56	9.27	3.30	6.21
Monohydroxymonoketone compounds						
5α-Androstan-3α-ol-17-one	Androsterone	0.97	0.46	1.43	0.36	0.84
5α-Androstan-3β-ol-17-one	Epiandrosterone	1.00	0.61	1.94	0.51	1.16
5α-Androstan-17β-ol-3-one	Androstanolone	1.19	0.60	2.22	D	1.28
5β-Androstan-3α-ol-17-one	Ethiocholanolone	0.88	0.53	1.61	0.41	0.98
5-Androstan-3β-ol-17-one	Dehydroepiandrosterone	1.01	0.61	1.89	0.47	1.05
4-Androsten-17α-ol-3-one	Epitestosterone	1.57	0.77	2.41	0.61	1.44
4-Androsten-17β-ol-3-one	Testosterone	1.64	0.96	3.01	0.83	1.78
5α-Pregnan-3β-ol-20-one	Allopregnanolone	1.49	1.03	2.82	0.80	1.74
5β-Pregnan-3α-ol-20-one	Epipregnanolone	1.33	0.88	2.40	0.65	1.51
4-Pregnen-20α-ol-3-one		2.65	1.95	5.33	1.61	3.22
4-Pregnen-20β-ol-3-one		2.33	1.75	4.79	1.37	2.85
5-Pregnen-3β-ol-20-one		1.54	1.02	2.84	0.78	1.67
1,3,5(10)-Estratrien-3-ol-17-one	Estrone	2.05	0.89	2.48	0.52	1.22
Monohydroxydiketone compounds						
5α-Androstan-3α-ol-11,17-dione	11-Ketoandrosterone	1.59	0.78	2.27	0.58	1.33
5β-Androstan-3α-ol-11,17-dione	11-Ketoethiocholanolone	1.51	0.93	2.40	0.69	1.70
4-Pregnen-17α-ol-3,20-dione	17-Hydroxyprogesterone	—	2.29	5.90	2.03	1.98

TABLE GC 136—(Continued)
STEROIDS AND DERIVATIVES

Column Packing		P1	P1	P1	P1	P1
Temperature (°C)		—	—	—	—	—
Gas; Flow Rate (ml/min)		100,Ar	100,Ar	100,Ar	100,Ar	100,Ar
Column						
length, ft		6	6	6	6	6
diameter, mm, I.D.		5	5	5	5	5
form		U-shaped	U-shaped	U-shaped	U-shaped	U-shaped
material		borosilicate glass	borosilicate glass	borosilicate glass	borosilicate glass	borosilicate glass
Detector		FI	FI	FI	FI	FI
				$r_{cholestane}$		
Chemical name	Trivial name	PC[b]	TMSi[a]	CMDSi[b]	HFB[c]	CDFA[d]
Dihydroxy compounds						
5α-Androstane-3α,17β-diol	Dihydroandrosterone	1.01	0.43	2.91	0.25	1.03
5-Androstene-5β,17α-diol		1.02	0.48	3.31	0.26	1.09
5-Androstene-3β,17β-diol		1.10	0.53	3.93	0.34	1.33
5α-Pregnane-3β,20β-diol		1.58	1.00	6.98	0.58	2.33
5β-Pregnane-3α,20α-diol	Pregnanediol	1.60	0.87	6.28	0.53	2.18
5-Pregnene-3β,20α-diol		—	1.09	7.22	0.65	2.46
1,3,5(10)-Estratriene-3,17β-diol	17β-Estradiol	2.34	0.70	4.90	0.36	2.58
Dihydroxymonoketone compounds						
5α-Androstane-3β,11β-diol-17-one		2.90	1.95	8.76	0.77	3.71
5β-Androstane-3α,11β-diol-17-one	11-Hydroxyethiocholanolone	2.42	1.48	6.84	0.57	2.95
1,3,5(10)-Estratriene-3,17β-diol-16-one		4.38	1.52	9.84	1.06	4.48
Trihydroxy compounds						
5-Androstene-3β,16α,17α-triol		—	1.06	15.22	0.42	3.06
5β-Pregnane-3α,17α,20α-triol	Pregnanetriol	—	1.52	12.10	D	D
1,3,5(10)-Estratriene-3,16α,17β-triol	Estriol	—	1.26	19.79	0.46	3.66

[a] Cholestane retention time: 14.6 min.
[b] Cholestane retention time: 5.2 min.
[c] Cholestane retention time: 21.5 min.
[d] Cholestane retention time: 11.2 min.
[e] D means that the compound decomposes.
PC, parent compound; TMSi, trimethylsilyl ethers; CMDSi, chloromethyldimethylsilyl ethers; HFB, heptafluorobutyrates; CDFA, chlorodifluoroacetates.

Column Packing P1 = 2% SE-30 (methylpolysiloxane) plus 1% TMCBA (tetramethylcyclobutanediol adipate) on acid-washed, silanized Gas Chrom P (100/120 mesh).

Literature Pinelli, A., *J. Chromatogr.*, **43**, 223 (1969).

TABLE GC 137

STEROIDS, ANDROSTANES AND DERIVATIVES

Column Packing	P1	P2	P3	P4
Temperature (°C)	260	260–262	232	238
Gas; Flow Rate (ml/min)	N_2	N_2	74,Ar	36,Ar
Column				
length, ft	5	6	6	5
diameter, in., I.D.	$\frac{1}{8}$	$\frac{1}{8}$	$\frac{1}{8}$	$\frac{1}{8}$
form	tube	tube	tube	tube
material	copper	SS	SS	copper
Detector	FI	FI	EC	EC

	$r_{cholestane}$															
Compound	Free	Acetate	TMSi	TFA	Free	Acetate	TMSi	TFA	Free	Acetate	TMSi	TFA	Free	Acetate	TMSi	TFA
5α-Androstane	0.21	—	—	—	0.27	—	—	—	0.09	—	—	—	0.13	—	—	—
5β-Androstane	0.20	—	—	—	0.25	—	—	—	0.08	—	—	—	0.12	—	—	—
5α-Androstane-17β-ol	0.32	0.39	0.33	—	0.60	0.71	0.38	—	0.47	0.44	0.20	0.16	0.71	0.58	0.22	0.16
5β-Androstane-17β-ol	0.41	0.52	0.49	—	0.89	1.04	0.52	—	0.84	0.80	—	0.32	1.48	1.27	—	—
5α-Androstane-17-one	0.32	—	—	—	0.65	—	—	—	0.39	—	—	—	0.61	—	—	—
5β-Androstane-17-one	0.42	—	—	—	0.95	—	—	—	0.71	—	—	—	1.15	—	—	—
5α-Androstane-3β-ol	0.31	0.39	0.34	0.24	0.65	0.70	0.39	—	0.47	0.47	0.21	0.18	0.75	0.71	0.24	0.20
5β-Androstane-3β-ol	0.29	0.35	0.31	0.22	0.53	0.59	0.32	—	0.38	0.37	—	0.13	0.61	0.57	—	0.15
5α-Androstane-3-one	0.33	—	—	—	0.77	—	—	—	0.49	—	—	—	0.73	—	—	—
5β-Androstane-3-one	0.32	—	—	—	0.68	—	—	—	0.42	—	—	—	0.67	—	—	—
5α-Androstane-3α,17β-diol	0.54	0.89	0.58	0.28	1.67	2.05	0.49	—	2.27	1.69	0.31	0.29	3.94	2.48	0.28	0.27
5β-Androstane-3α,17β-diol	0.51	0.87	0.61	0.32	1.68	1.95	0.50	—	2.08	1.89	0.35	0.34	3.85	2.82	0.37	0.30
5α-Androstane-3β,17β-diol	0.59	0.98	0.69	0.36	1.89	2.21	0.66	—	2.44	2.20	0.44	0.41	4.19	3.18	0.42	0.24
5β-Androstane-3β,17β-diol	0.49	0.86	0.56	—	1.56	1.90	0.51	—	1.93	1.66	0.28	0.28	3.50	2.42	0.28	0.23
5α-Androstane-3α-ol, 17-one	0.51	0.58	0.47	—	2.00	2.14	0.87	—	1.83	1.61	0.58	0.67	3.73	2.62	0.74	0.80
5β-Androstane-3α-ol, 17-one	0.47	0.56	0.46	0.28	1.90	2.15	0.94	—	1.80	1.76	0.72	0.73	3.58	2.86	1.03	0.61
5α-Androstane-3β-ol, 17-one	0.55	0.66	0.55	0.30	2.10	2.44	1.16	—	2.10	2.00	0.91	0.94	3.97	3.44	1.20	0.67
5β-Androstane-3β-ol, 17-one	—	—	—	—	1.98	2.98	0.86	—	1.88	—	—	—	—	—	—	—
5α-Androstane-3-one, 17β-ol	0.59	0.76	0.64	0.35	2.58	2.81	1.32	—	2.61	2.38	1.07	1.18	4.61	3.88	1.27	1.24
5β-Androstane-3-one, 17β-ol	0.54	0.69	0.59	0.33	2.44	2.53	1.18	—	2.20	1.98	0.91	0.97	4.18	3.44	1.13	1.14

TABLE GC 137—(Continued)
STEROIDS, ANDROSTANES AND DERIVATIVES

Column Packing	P1	P2	P3	P4
Temperature (°C)	260	260–262	232	238
Gas; Flow Rate (ml/min)	N_2	N_2	74,Ar	36,Ar
Column				
length, ft	5	6	6	5
diameter, in , I D	$\frac{1}{8}$	$\frac{1}{8}$	$\frac{1}{8}$	$\frac{1}{8}$
form	tube	tube	tube	tube
material	copper	SS	SS	copper
Detector	FI	FI	EC	EC

	$r_{cholestane}$															
Compound	Free	Acetate	TMSi	TFA	Free	Acetate	TMSi	TFA	Free	Acetate	TMSi	TFA	Free	Acetate	TMSi	TFA
5α-Androstane-3-one, 17α-ol	0.56	0.75	0.60	—	2.03	2.32	1.16	—	2.22	2.12	0.94	—	4.36	3.31	1.09	—
5β-Androstane-3-one, 17α-ol	0.53	0.61	0.47	0.33	2.14	2.14	0.91	—	2.05	1.76	0.62	0.50	3.97	2.91	0.81	0.78
5α-Androstane-3,17-dione	0.58	—	—	—	2.64	—	—	—	2.27	—	—	—	4.00	—	—	—
5β-Androstane-3,17 dione	0.50	—	—	—	2.48	—	—	—	1.98	—	—	—	3.64	—	—	—
Androstane-5-en-3β-ol, 17-one	0.61	0.64	0.55	0.33	2.01	2.38	1.10	—	2.09	1.99	0.90	0.84	4.00	3.38	1.16	1.02

Column Packing P1 = 3% SE-30 elastomer on Anakrom ABS (40/50 mesh).
P2 = 3% XE-60 on Diatoport S (80/100 mesh).
P3 = 3% NB Seb on Chromosorb W (30/60 mesh).
P4 = 3% Hi-Eff-8B on Anakrom ABS (40/50 mesh).

Literature Hartman, I. S. and H. H. Wotiz, *Biochim. Biophys. Acta*, **90**, 334 (1964).

TABLE GC 138
STEROIDS, BILE ACIDS AND TMS DERIVATIVES

Column Packing	P1	P2	P3	P1	P2	P3	P1	P2	P3	P1	P2	P3
Temperature (°C)	230	260	260	230	260	260	230	260	260	230	260	260
Gas; Flow Rate (ml/min)	40,He	60,He	60,He	40,He	60,He	60,He	40,He	60,He	60,He	40,He	60,He	60,He
Column												
length, ft	6	6	6	6	6	6	6	6	6	6	6	6
diameter, mm, I.D.	4	4	4	4	4	4	4	4	4	4	4	4
form	U-tube	U-tube	U-tube	U-tube	U-tube	U-tube	U-tube	U-tube	U-tube	U-tube	U-tube	U-tube
material	glass	glass	glass	glass	glass	glass	glass	glass	glass	glass	glass	glass
Detector	FI	FI	FI	FI	FI	FI	FI	FI	FI	FI	FI	FI
Compound	Methyl cholanoate[c] $r_{methyl\ deoxycholate}$						TMS derivatives[d] $r_{TMS\ derivatives\ of\ methyl\ deoxycholate}$					
Methyl cholanoate	0.15	0.35	0.20	0.18	0.40	0.24	—	—	—	—	—	—
12α	0.27	0.55	0.39	0.37	0.60	0.49	0.59	0.58	0.62	0.66	0.60	0.64
12β	0.31	0.56	0.41	—	—	—	0.55	0.54	0.56	—	—	—
7α	0.34	0.60	0.44	0.40	0.64	0.51	0.65	0.61	0.63	0.62	0.57	0.57
7β	0.36	0.59	0.46	0.43	0.67	0.52	0.71	0.71	0.75	0.81	0.78	0.83
6β	0.38	0.62	0.47	—	—	—	0.63	0.62	0.63	—	—	—
6α	0.45	0.65	0.52	—	—	—	0.67	0.65	0.72	—	—	—
3β	0.44	0.64	0.51	0.55	0.72	0.60	0.83	0.81	0.86	1.20	0.82	1.20
3α	0.49	0.62	0.52	0.50	0.72	0.58	0.91	0.85	0.93	0.89	0.72	0.92
12-Keto	0.49	0.57	0.43	0.60	0.63	0.51	—	—	—	—	—	—
7-Keto	0.57	0.58	0.46	0.72	0.66	0.57	—	—	—	—	—	—
6-Keto	0.78	0.64	0.58	—	—	—	—	—	—	—	—	—
3-Keto	1.00	0.71	0.61	1.06	0.76	0.71	—	—	—	—	—	—
7α,12α	0.70	0.91	0.84	0.90	0.99	0.99	0.72	0.69	0.60	0.63	0.63	0.53
6β,12α	0.77	0.94	0.91	—	—	—	0.79	0.80	0.69	—	—	—
3β,12α	0.86	1.00	1.00	1.16	1.16	1.20	1.02	1.00	0.90	1.35	1.23	1.19
3β,7α	0.94	1.06	1.10	1.29	1.18	1.34	0.96	0.96	0.87	1.14	1.08	1.00
3α,12α	1.00	1.00	1.00	1.07	1.12	1.19	1.00	1.00	1.00	0.93	1.18	1.22
3α,7α	1.18	1.08	1.14	1.22	1.18	1.27	1.09	1.04	1.00	1.00	0.84	0.92
3α,6β	1.20	1.11	1.21	—	—	—	1.08	1.10	0.98	—	—	—
3α,7β	1.27	1.06	1.13	—	—	—	1.24	1.21	1.14	—	—	—
3α,6α	1.50	1.20	1.32	—	—	—	1.20	1.12	1.08	—	—	—
3β,6β	—	—	—	1.27	1.20	1.29	—	—	—	1.40	1.33	1.31
7,12-Diketo	1.45	0.91	0.85	1.97	0.96	1.06	—	—	—	—	—	—
3,12-Diketo	2.86	1.05	1.29	3.52	1.15	1.51	—	—	—	—	—	—
3,7-Diketo	3.05	1.04	1.21	3.96	1.17	1.52	—	—	—	—	—	—
12-Keto-7α	—	—	—	1.40	1.03	1.12	—	—	—	2.43	0.90	1.18
7-Keto-12α	1.10	1.13	0.88	1.56	1.09	1.13	2.74	0.93	1.35	3.42	0.97	1.54
12-Keto-6β	1.19	0.99	1.11	—	—	—	2.13	0.95	1.27	—	—	—
12-Keto-3α	1.61	1.00	1.12	—	—	—	3.19	1.33	1.91	—	—	—

TABLE GC 138—(Continued)
STEROIDS, BILE ACIDS AND TMS DERIVATIVES

Column Packing	P1	P2	P3	P1	P2	P3	P1	P2	P3	P1	P2	P3
Temperature (°C)	230	260	260	230	260	260	230	260	260	230	260	260
Gas; Flow Rate (ml/min)	40,He	60,He	60,He	40,He	60,He	60,He	40,He	60,He	60,He	40,He	60,He	60,He
Column												
length, ft	6	6	6	6	6	6	6	6	6	6	6	6
diameter, mm, I.D.	4	4	4	4	4	4	4	4	4	4	4	4
form	U-tube	U-tube	U-tube	U-tube	U-tune	U-tube	U-tube	U-tube	U-tube	U-tube	U-tube	U-tube
material	glass	glass	glass	glass	glass	glass	glass	glass	glass	glass	glass	glass
Detector	FI	FI	FI	FI	FI	FI	FI	FI	FI	FI	FI	FI
Compound	Methyl cholanoate[c] $r_{\text{methyl deoxycholate}}$						TMS derivatives[d] $r_{\text{TMS derivatives of methyl deoxycholate}}$					
7-Keto-3α	1.80	1.03	1.20	1.97	1.23	1.42	3.65	1.34	1.98	4.16	1.41	2.30
7-Keto-3β	—	—	—	2.12	1.26	1.41	—	—	—	—	—	—
3-Keto-12α	1.82	1.07	1.17	2.23	1.19	1.41	4.10	1.09	1.67	3.04	1.17	1.85
3-Keto-7α	2.40	1.23	1.43	2.32	1.27	1.60	3.81	1.17	1.84	4.12	1.08	1.60
3-Keto-6α	2.93	1.32	1.59	—	—	—	5.04	1.32	1.99	—	—	—
3β,7α,12α	1.90	1.63	2.08	2.78	1.88	2.71	0.96	0.96	0.79	1.18	1.01	0.79
3α,7β,12α	2.20	1.52	2.07	—	—	—	1.05	1.09	0.90	1.00	1.05	0.87
3α,7α,12α	2.33	1.66	2.26	2.68	1.77	2.73	1.05	1.12	0.84	—	—	—
3α,6β,7β	2.33	1.68	2.30	2.10	1.78	2.34	1.38	1.15	1.11	—	—	—
3α,6β,7α	2.42	1.84	2.65	—	—	—	1.44	1.45	1.18	1.41	1.44	1.17
3α,6α,7β	2.67	1.69	2.25	—	—	—	1.88	1.92	1.71	—	—	—
3α,6α,7α	3.14	1.85	2.56	—	—	—	—	—	—	—	—	—
7-Keto-3α,6α	1.78	1.48	1.59	—	—	—	—	—	—	—	—	—
7-Keto-3α,12α	3.51	1.59	2.27	4.62	1.95	2.89	4.95	1.59	2.03	5.45	1.62	1.95
12-Keto-3α,7α	3.85	1.73	2.57	—	—	—	4.37	1.72	1.97	—	—	—
7,12-Diketo,3α	4.55	1.50	2.18	—	—	—	4.55	1.88	3.38	—	—	—
3-Keto-7α,12α	4.79	1.88	2.79	6.10	1.96	3.12	5.84	1.35	1.68	5.07	1.09	1.45
6-Keto-3α,7β	—	—	—	2.06	1.52	1.88	—	—	—	—	—	—
7-Keto-3β,12α	—	—	—	4.50	1.85	2.87	—	—	—	7.00	1.96	2.46
3,7-Diketo-12α	5.38[a]	1.54	2.31	7.14[b]	1.84	3.04	12.5	1.56	3.10	12.7	1.78	3.76
3,12-Diketo-7α	—[b]	1.91	3.30	—[b]	—[b]	—[b]	—	—	—	—	—	—
3,7,12-Triketo	6.40	1.39	2.25	9.24	1.64	2.98	—	—	—	—	—	—

[a] The peak was not well formed, so this value is tentative.
[b] This material is retained too long on the column to provide meaningful data.
[c] Absolute times of elution of methyl deoxycholate are: QF-1, 29.0 min; OV-1, 38.4 min; OV-17, 44.0 min.
[d] Bis(trimethylsilyl) ether of methyl deoxycholate = 1.00. Absolute times of elution of this derivative are: QF-1, 10.0 min; OV-1, 27.3 min; OV-17, 14.2 min.

Column Packing P1 = 3% QF-1 (fluorosilicone) on Gas Chrom Q (100/120 mesh).
P2 = 3% OV-1 (silicone) on Gas Chrom Q (100/120 mesh).
P3 = 3% OV-17 (silicone) on Gas Chrom Q (100/120 mesh).

Literature Elliott, W. H., Walsh, L. B., Mui, M. M., Thorne, M. A., and C. M. Siegfried, *J. Chromatogr.*, **44**, 452 (1969).

TABLE GC 139
SULFUR COMPOUNDS, SULFONES

Column Packing	P1	P1	P2	P2	P3	P3	P4	P4	P5	P5	P6
Temperature (°C)	200	250	200	250	200	250	200	225	200	250	200
Gas; Flow Rate (ml/min)	165,He	165,He	200,He	200,He	140,He	110,He	143,He	143,He	165,He	165,He	165,He
Column											
length, ft	3	3	3	3	3	3	3	3	3	3	3
diameter, in., O.D.	$\frac{1}{4}$	$\frac{1}{4}$	$\frac{1}{4}$	$\frac{1}{4}$	$\frac{1}{4}$	$\frac{1}{4}$	$\frac{1}{4}$	$\frac{1}{4}$	$\frac{1}{4}$	$\frac{1}{4}$	$\frac{1}{4}$
form	tubing	tubing	tubing	tubing	tubing	tubing	tubing	tubing	tubing	tubing	tubing
material	aluminum	aluminum	aluminum	aluminum	aluminum	aluminum	aluminum	aluminum	aluminum	aluminum	aluminum
Compound						$V_{R'}$					
Methyl sulfoxide	11	—	14	—	11	—	23	—	122	—	109
Ethyl ,,	25	—	36	—	19	—	49	—	141	—	122
n-Propyl ,,	46	—	74	—	43	—	91	—	193	—	162
Isopropyl ,,	38	—	58	—	36	—	69	—	123	—	105
n-Butyl ,,	95	—	156	—	92	—	192	—	298	—	286
Isobutyl ,,	58	—	94	—	61	—	105	—	153	—	135
sec-Butyl ,,	77	—	124	—	72	—	133	—	178	—	157
tert-Butyl ,,	13	—	16	—	14	—	26	—	122	—	111
n-Amyl ,,	211	—	346	—	189	—	393	—	525	—	495
Isoamyl ,,	147	—	242	—	134	—	287	—	359	—	325
n-Hexyl ,,	495	110	790	206	368	89	882	318	1000	195	891
2-Hydroxyethyl	61	26	76	34	47	38	109	60	541	122	417
Ethyl-2-hydroxyethyl sulfoxide	71	28	86	36	67	41	144	87	1210	198	1270
Methyl-*p*-nitrophenyl ,,	570	123	680	196	281	87	882	353	—	905	5400
Phenyl ,,	775	187	1250	328	478	114	1780	565	—	695	4630
3-Aminophenyl[a] ,,	—	—	—	—	—	—	—	—	—	—	—
4-Aminophenyl[a] ,,	—	—	—	—	—	—	—	—	—	—	—
4-Chlorophenyl ,,	—	520	5000	980	1010	197	—	1370	—	1630	—
4-Hydroxyphenyl[a] ,,	—	—	—	—	—	—	—	—	—	—	—
Sulfolane	44	—	70	—	46	—	109	—	370	—	300
3-Methylsulfolane	54	—	78	—	54	—	128	—	343	—	267
2,4-Dimethylsulfolane	58	—	86	—	58	—	123	—	297	—	215
Dibenzothiophene	2300	445	3730	670	1000	167	4300	1330	—	2040	—
Sulfolene[a]	—	—	—	—	—	—	—	—	—	—	—
3-Methylsulfolene[a]	—	—	—	—	—	—	—	—	—	—	—
2,4-Dimethyl-3-sulfolene[a]	—	—	—	—	—	—	—	—	—	—	—
Butadiene[a]	—	—	—	—	—	—	—	—	—	—	—
Methyl sulphonal	240	72	386	128	215	70	608	218	1210	215	900
sym.-Trimethylene trisulfone[b]	—	—	—	—	—	—	—	—	—	—	—

[a] Did not emerge in 30 min at 250°C on Silicone gum rubber column.

[b] A suitable solvent in which this compound is soluble could not be found.

Column Packing

P1 = 16% Apiezon M on acid-washed, silanized Gas Chrom Z.
P2 = 16% Apiezon L on acid-washed, silanized Gas Chrom Z.
P3 = 16% Silicone gum rubber on acid-washed, silanized Gas Chrom Z.
P4 = 16% Silicone oil 550 on acid-washed, silanized Gas Chrom Z.
P5 = 16% Carbowax 20M on acid-washed, silanized Gas Chrom Z.
P6 = 16% Carbowax 1500 on acid-washed, silanized Gas Chrom Z.

Literature

Cates, V. E. and C. E. Meloan, *J. Chromatogr.*, **11**, 472 (1963).

TABLE GC 140
TERPENES, SESQUITERPENES

Column Packing	P1	P2	P3	P4	P5	P6	P6	P6	P7
Temperature (°C)	155	170	130	165	132	132	165	205	160
Gas; Flow Rate (cc/min)	15–40,He	15–40,He	15–40,He	15–40,He	15–40,He	15–40,He	15–40,He	15–40,He	15–40,He
Column									
length, ft	16–50	16–50	16–50	16–50	16–50	16–50	16–50	16–50	16–50
diameter, in.	0.125	0.125	0.125	0.125	0.125	0.125	0.125	0.125	0.125
form	na	na	na	na	na	na	na	na	na
material	na	na	na	na	na	na	na	na	na
Detector	TC	TC	TC	TC	TC	TC	TC	TC	TC
Compound					I				
Cubebene	1368	—	—	—	—	—	—	—	—
α-Longipinene	—	—	1359.5	—	—	—	—	—	—
α-Ylangene	1401.5	1396.1	—	1454.5	—	—	1538.5	—	1653
β-Elemene	1410	—	—	—	—	—	—	—	—
α-Bourbonene	1410	—	—	—	—	—	—	—	—
α-Copaene	1410.2	1400.5	1378.5	1459	1447	1521.5	1551.3	1593	1665
Cyclosativene	1411.9	1399.7	—	—	—	—	1549	—	(1684)
Longicyclene	1417.1	1409.1	1371	1454	1465	1518.8	1554	1598.5	1684
Cyclocopacamphene	1417.8	—	—	—	1467.2	—	1555.4	—	1685.6
β-Bourbonene	1418.3	1411.7	1386	1477.3	1477.5	1547	1586.5	1618	1714
β-Farnesene	1429.2	—	—	—	1509	—	1668	—	1818.5
Sativene	1434.7	1420.7	—	—	—	—	1594.5	—	(1738)
Cyperene	1446.6	1432.5	1398	1501	1493	1562	1606	1650	1736.5
α-Gurjunene	—	1435.2	1413	1500.5	1471	1558	1591	1633	1712.5
Caryophyllene	1451.7	1445.3	1417.5	1523	(1587)	1618.5	1655.5	1695.5	1835.5
Longifolene	1464.0	1440.2	1404	1517.5	1520	1600	1643	1697	1802.5
Isosativene	1464.4	1440.9	—	—	—	—	1639	—	(1797)
Calarene	1466.0	1459.7	1435	1535.5	1513	1618	1655.5	1700	1806
β-Ylangene	—	—	~1417.5	—	—	—	—	—	—
β-Copaene	—	—	~1422.5	—	—	—	—	—	—
α-Cedrene	1473.4	1445.0	~1414	1516	1518	1597.5	1640	1689	1788.5
Thujopsene	1476.1	1458.3	1430.5	1542.3	1540	1643	1684.2	1732	1858.5
Aromadendrene	~1477	—	—	—	—	—	—	—	—
α-Maaliene	~1477	—	—	—	—	—	—	—	—
γ-Curcumene	1481.9	—	—	—	1532.5	—	—	—	—
β-Cedrene	1482.4	1454.8	1421	1533.5	1539.5	1624.5	1670	1714	1834.5
α-Curcumene	1483	1480.4	(1475)	1589	1557.5	—	1787.5	(1814)	1992.5
ε-Muurolene	1484.8	(1474.0)	(1445)	(1561.5)	1552.5	(1675.5)	1713.8	(1759.5)	1893.5
Humulene	1487.2	1476.6	1446.8	1561.5	1583.5	1681	1719	1765	1929.5
Santalene (minor)	—	1459.5	1441	1535	1522	1644.5	1671	1702	1830
Santalene (major)	—	1470.5	1454	1548	1533.5	1658	1683	1714.5	1843.5

TABLE GC 140—(Continued)
TERPENES, SESQUITERPENES

Column Packing	P1	P2	P3	P4	P5	P6	P6	P6	P7
Temperature (°C)	155	170	130	165	132	132	165	205	160
Gas; Flow Rate (cc/min)	15–40,He	15–40,He	15–40,He	15–40,He	15–40,He	15–40,He	15–40,He	15–40,He	15–40,He
Column									
length, ft	16–50	16–50	16–50	16–50	16–50	16–50	16–50	16–50	16–50
diameter, in	0.125	0.125	0.125	0.125	0.125	0.125	0.125	0.125	0.125
form	na	na	na	na	na	na	na	na	na
material	na	na	na	na	na	na	na	na	na
Detector	TC	TC	TC	TC	TC	TC	TC	TC	TC
Compound					1				
Selina-4(14),7-diene	1491.9	1475.7	—	—	1542	—	1694	—	1852.5
δ-Selinene	1504.5	—	—	—	—	—	1728.5	—	—
γ-Muurolene	1505.7	—	—	—	1545	—	1725	—	1889
γ-Amorphene	1506.4	—	—	—	1544.5	—	1724	—	1896.5
α-Himachalene	1508.0	—	1444	1561.5	1533.5	1662.5	1704.5	1755	1870
α-Amorphene	1509.5	1491.7	—	1582.5	1535	—	1724.5	—	1897
Zizaene	1511.6	1481.8	—	—	1562	—	1706.3	—	1879.5
β-Bisabolene	1512.9	1510.3	1496.5	1592.5	1548	1726	1745.5	1772	1909.5
β-Curcumene	1513.6	1510.4	—	—	1547.5	—	1756	—	1922.5
α-Zingaberene	—	(1479.6)	—	1583.5	—	—	1738	(1762)	—
Valencene	1525.6	1508.8	1457	~1600	1581	1725.5	1760	1801	1948
β-Himachalene	1529.7	—	1491	1607.5	1578	1717	1752.5	1799	—
β-Selinene	1530.2	1506.3	—	1595	1597.5	—	1766.5	1815.5	1958
γ-Bisabolene	1531.3	(1505)	—	1601	—	—	1765.5	1815	—
α-Muurolene	1531.3	1507.7	1495	~1600	1558.5	1726	1752.5	1792	1927.5
α-Pyrovetivene	1533.9	1521.5	—	—	—	—	1817	—	2026.0
α-Selinene	1534.5	—	—	—	—	—	—	—	—
e-Bulgarene	1538	—	—	—	—	—	—	—	—
δ-Cadinene	1546.4	(1526.4)	~1504	1628.5	—	—	1784	1818	1959
Calamenene	~1550	—	—	—	—	—	—	—	—
γ-Cadinene	1554.9	1523.5	1506.5	1623.5	1587	1762	1792.3	1835.5	1978.5
Selina-4(14),7(11)-diene	1572.0	—	—	—	1611.5	—	1816.3	—	(2018)
Selina-3,7(11)-diene	1580	—	—	—	—	—	—	—	—
β-Vetivenene	1583.0	1563.3	—	—	—	—	1885	—	2111

Values in parenthesis have been obtained from single trace.

Column Packing
P1 = 0.5–3% Apiezon L on Chromosorb G, silanized.
P2 = 0.5–3% SF-96 on Chromosorb G, silanized.
P3 = 0.5–3% SE-30 on Chromosorb G, silanized.
P4 = 0.5–3% DC-710 on Chromosorb G, silanized.
P5 = 0.5–3% QF-1 on Chromosorb G, silanized.
P6 = 0.5–3% Carbowax 20M on Chromosorb G, silanized.
P7 = 0.5–3% DEGS on Chromosorb G, silanized.

Literature

Andersen, N. H., and M. S. Falcone, *J. Chromatogr.*, **44**, 52 (1969).

TABLE GC 141
TERPENES, SESQUITERPENES

	P1	P2
Column Packing	P1	P2
Temperature (°C)	150	180
Gas; Flow Rate (ml/min)	60.6,N_2	68.0,N_2
Column		
length, m	2.25	1.87
diameter, mm	4	4
form	—	—
material	SS	glass
Detector	FI	FI

Compound	t_r	V_r	t_r	V_r
Kessane	5.8	351.5	1.5	102.0
α-Kessyl alcohol	10.7	648.4	7.2	489.6
Kessanol	12.5	757.5	12.4	843.2
Kessyl acetate	17.0	1030.2	7.4	503.2
Kessanyl acetate	17.8	1078.7	7.0	476.0
Kessoglycol	27.0	1636.2	74.1	5038.8
8-Acetoxykessan-2-ol	35.5	2151.3	40.3	2740.4
2-Acetoxykessan-8-ol	40.0	2424.0	73.3	4984.4
Kessoglycol diacetate	53.7	3254.2	38.2	2597.6
Maaliol	6.6	400.0	3.2	217.6
Valeranone	9.8	593.9	4.8	326.4
Cryptofauronol	10.3	624.2	9.4	639.2
Fauronyl acetate	27.8	1684.7	23.8	1618.4
Kanokonol	28.3	1715.0	65.5	4454.0
Kanokonyl acetate	42.3	2563.4	36.7	2495.6

Column Packing P1 = 1.5% SE-30 on Chromosorb W (60/80 mesh).
P2 = 12% DEGS on Gas Chrom P (80/100 mesh).

Literature Furuya, T. and H. Kojima, *J. Chromatogr.*, **29**, 341 (1967).

TABLE GC 142
VITAMINS

Column Packing	P1
Temperature (°C)	260
Gas; Flow Rate (ml/min)	60,Ar
Column	
diameter	—
form	—
material	—
Detector	—

Compound	$r_{n\text{-}C_{28}H_{58}}$
α-Tocopherol	2.34
ζ_2-Tocopherol	1.88
γ-Tocopherol	1.76
β-Tocopherol	1.68
5-Methyltocol	1.50
***n*-Tocopherol**	1.41
δ-Tocopherol	1.25
Tocol	1.10
***n*-$C_{28}H_{58}$**	1.00

Column Packing P1 = 5% Apiezon N high vacuum grease on Celite (100/120 mesh).

Literature Kofler, M., Sommer, P. F., Bolliger, H. R., Schmidli, B., and M. Vecchi, "*Vitamins and Hormones*", Vol. 20, R. S. Harris and I. G. Wool, Editors, Academic Press, N.Y., 1962, p. 432.

Volume One

SECTION II.II

CHROMATOGRAPHIC DATA
Liquid Chromatography Tables

Wherever possible, tables were arranged according to classes of chemical compounds. However, this was not always possible when diverse compounds were chromatographed under the same experimental conditions. The reader is referred to the master compound index for specific compounds which may appear in diverse tables.

Techniques

n.a. = not available from original source.
Solvent composition always v/v unless otherwise stated; if solid is dissolved in solvent, g/v.

TABLE LC 1
COLUMN CHROMATOGRAPHY OF PHENOLS, FLAVONOIDS AND DERIVATIVES ON SEPHADEX

Packing	P1	P1	P1	P1	P1
Column[a]					
length	35 cm	35 cm	35 cm	35 cm	35 cm
diameter	2.5 cm	2.5 cm	2.5 cm	2.5 cm	2.5 cm
material	G	G	G	G	G
Solvent	S1	S2	S3	S4	S5
Flow rate	25 ml/hr.	25 ml/hr.	25 ml/hr.	25 ml/hr.	25 ml/hr.
Temperature	rt	rt	rt	rt	rt
Detection	D1	D1	D1	D1	D1
Literature	1	1	1	1	1
Compound[b]	K_d	K_d	K_d	K_d	K_d
Phenol	1.8	2.0	1.1	1.95	—
Hydroquinone	1.05	2.4	1.7	2.4	—
Resorcinol	2.1	2.65	1.6	2.5	—
Catechol	2.05	1.9	1.05	2.5	2.2
Phloroglucinol	2.3	3.15	1.1	3.25	—
Pyrogallol	2.4	2.4	1.3	3.25	2.0
***o*-Hydroxybenzoic acid**	1.0	2.1	1.6	3.4	1.2
Arbutin	1.45	1.55	0.85	—	—
***o*-Nitrophenol**	1.45	2.45	1.8	2.1	—
Saligenin	1.85	2.1	1.55	1.9	—
Guaiacol	2.2	1.8	1.45	2.1	—
***o*-Hydroxybenzaldehyde**	2.3	2.3	1.55	2.3	—
***o*-Cresol**	2.6	2.55	1.85	3.0	—
Orcinol	3.0	2.6	1.5	—	—
***o*-Chlorophenol**	3.15	3.4	1.55	3.0	—
***p*-Hydroxydiphenyl**	4.8	5.4	2.9	—	—
***m*-Hydroxybenzoic acid**	1.0	1.8	0.85	—	—
***p*-Hydroxybenzoic acid**	1.0	1.45	0.8	—	—
2,3-Dihydroxybenzoic acid	2.6	1.6	0.4	—	2.0
2,4-Dihydroxybenzoic acid	0.85	2.7	1.1	—	1.4
2,5-Dihydroxybenzoic acid	1.0	3.0	1.7	—	1.4
2,6-Dihydroxybenzoic acid	0.7	1.45	0.4	—	2.95
3,4-Dihydroxybenzoic acid	1.7	2.1	0.7	—	—
3,5-Dihydroxybenzoic acid	2.2	1.9	0.7	—	—
3,4,5-Trihydroxybenzoic acid	1.05	2.5	1.05	—	1.9
2,3,4-Trihydroxybenzoic acid	2.05	2.65	1.1	—	2.0
Vanillic acid	0.85	1.95	0.85	—	—
Syringic acid	0.8	1.95	0.85	—	—
Chlorogenic acid	1.6	3.9	1.35	—	—
Chalcone	3.1	3.1	—	—	—
3-Hydroxyflavone	4.2	5.0	4.0	—	0.5
Rutin	5.6	6.5	3.9	—	0.26
Quercitrin	11.4	c	1.85	—	0.24
Robinetin	c	c	0.60	—	4.62
Quercetin	c	c	0.65	—	—
Hesperidin	c	c	0.8	—	1.95
Morin	c	c	1.2	—	2.16
Catechin	c	c	1.3	—	0.34
Dihydroquercetin	c	c	1.6	—	1.03
Naringenin	c	c	5.6	—	4.9
Hesperitin	c	c	6.25	—	1.3
Fisetin	c	c	—	—	3.33

[a] V_0 = 40 ml, V_i = 55 ml. [b] 1 ml samples containing 200 μg of compound in the eluant were used.
[c] Irreversibly adsorbed.

Packing P1 = Sephadex G-25, medium particle size grade.
Solvent S1 = H_2O.
S2 = 0.05*M* NaCl.
S3 = 0.1*M* NH_4OH.
S4 = 0.1*M* acetic acid.
S5 = 0.01*M* sodium molybdate.
Detection D1 = Phenols were analyzed by ultraviolet absorption at 280 nm or coupled with diazotized *p*-aminobenzoic acid and measured at 420 nm.
Literature 1. Woof, J. B., and J. S. Pierce, *J. Chromatogr.*, **28**, 94 (1967).

TABLE LC 2
ION EXCHANGE CHROMATOGRAPHY OF α- AND β- ASPARTYL OLIGO PEPTIDES

Packing	P1	P2	P3	P4
Column				
length	128 cm	128 cm	134 cm	134 cm
diameter	0.624 cm	0.624 cm	0.624 cm	0.624 cm
material	na	na	na	na
Solvent	S1	S1	S2	S3
Flow rate	0.50 ml/min.	0.50 ml/min.	0.50 ml/min.	0.50 ml/min.
Temperature	60° C	60 °C	ambient	ambient
Detection	D1	D1	D1	D1
Literature	1	1	1	1

Compound	t_r, min.	t_r, min.	t_r, min.	t_r, min.
β-Asp-Gly-Asn	161	167	375	785
β-Asp-Gly-Gln	161	—	335	690
β-Asp-Ser	168	176	410	930
β-Asp-Thr	174	—	315	690
β-Asp-Gln	176	—	350	745
β-Asp-Asn	178	—	375	825
β-Asp-Gly-Gly	190	196	315	690
β-Asp-Gly-Ala	204	—	270	560
β-Asp-Gly	215	222	345	775
α-Asp-Gly-Gln	230	—	—	—
α-Asp-Gly-Asn	—	235	—	—
α-Asp-Ala	237	242	295	635
β-Asp-Thr	—	260	—	—
α-Asp-Asn	—	261	—	—
Aspartic Acid	249	266	230	525
Serine	249	266	—	—
α-Asp-Ser	255	265	—	—
β-Asp-Gly-Pro	265	265	320	—
β-Asp-Gly-Val	265	270	300	—
Threonine	265	278	—	—
α-Asp-Gln	273	—	—	—
α-Asp-Gly-Gly	292	—	—	—
α-Asp-Gly-Ala	—	310	—	—
α-Asp-Gly	303	316	144	290
β-Asp-Val	335	—	280	—
Glutamic Acid	336	353	120	220
Glycine	336	360	70	75
α-Asp-Ala	362	380	—	—
α-Asp-Gly-Pro	—	395	—	—
α-Asp-Gly-Val	—	415	—	—
Alanine	407	424	—	—
NH_4Cl	407	436	—	—
Proline	429	446	—	—
β-Asp-Ile	503	—	315	—
β-Asp-Leu	548	—	355	750
α-Asp-Val	—	573	—	—
Valine	643	656	—	—
α-Asp-Ile	—	895	—	—
α-Asp-Leu	—	943	—	—

Packing P1 = Technicon Chromobeads, Type A.
P2 = Dowex® 50-X8, 25 ± 5 μ.
P3 = Dowex® 1-X4, 20–30 μ, formate form.
P4 = Dowex® 1-X4, 20–30 μ, acetate form.

Solvent S1 = Sodium phosphate buffer, pH 1.82, 0.5*M* in Na^+ (0.25 mole $Na_2HPO_4 \cdot 7H_2O$ and 0.75 mole H_3PO_4 per liter).
S2 = 0.1*M* formic acid.
S3 = 0.5*M* acetic acid.

Detection D1 = Automatic colorimetry using ninhydrin reagent.

Literature
1. Dorer, F. E., Haley, E. E., and D. L. Buchanan, *Anal. Biochem.*, **19**, 370 (1967).

TABLE LC 3
COLUMN CHROMATOGRAPHY OF NUCLEIC ACID COMPONENTS ON SEPHADEX GELS

Packing	P1	P2	P3	P3	P3
Column					
length	25 cm	25 cm	25 cm	90 cm	90 cm
diameter	1.5 cm	1.5 cm	1.5 cm	2.2 cm	2.2 cm
material	G	G	G	G	G
Solvent	S1	S1	S1	S2	S3
Flow rate	30 ml/hr.	30 ml/hr.	30 ml/hr.	1–2 ml/min.	1–2 ml/min.
Temperature	rt	rt	rt	rt	rt
Detection	UV	UV	UV	UV	UV
Literature	1	1	1	2	2
Compound[a]	K_d	K_d	K_d	K_d	K_d
Adenine	6.0	4.26	2.42	3.35	3.38
Adenosine	3.36	2.92	1.87	2.50	2.59
Deoxyadenosine	3.23	2.81	1.81	—	—
AMP	0.91	1.37	1.35	—	—
dAMP	0.94	1.39	1.38	—	—
ADP	0.41	0.74	1.03	—	—
ATP	0.29	0.50	0.80	—	—
Guanine	3.35	2.80	2.02	3.35	3.40
Guanosine	2.43	2.29	1.77	2.55	2.60
Deoxyguanosine	2.66	2.40	1.73	—	—
GMP	0.8	1.07	1.25	—	—
Hypoxanthine	2.17	1.89	1.56	—	—
Inosine	1.26	1.31	1.25	—	—
Deoxyinosine	1.26	1.30	1.14	—	—
IMP	0.45	0.71	0.85	—	—
Xanthine	3.43	2.94	2.14	—	—
Xanthosine	1.83	1.95	1.56	—	—
XMP	0.60	0.79	1.09	—	—
Uric acid	3.15	2.89	2.00	—	—
Cytidine	1.43	1.26	1.29	1.70	1.66
Cytosine	0.99	1.06	1.13	1.85	1.80
Deoxycytosine	1.09	1.16	1.20	—	—
CMP	0.4	0.54	0.87	—	—
Uracil	1.56	1.50	1.18	1.80	1.82
Uridine	1.07	1.21	1.12	1.70	1.66
Deoxyuridine	1.09	1.15	1.05	—	—
UMP	0.45	0.74	0.86	—	—
Thymine	1.86	1.63	1.24	—	—
Thymine riboside	1.00	1.15	1.09	—	—
Thymidine	1.24	1.29	1.04	—	—
TMP	0.52	0.73	0.92	—	—
TDP	0.28	0.44	0.60	—	—
TTP	0.19	0.32	0.56	—	—
Orotic acid	1.31	1.22	1.22	—	—
Orotidine	0.62	0.70	0.93	—	—

[a] Samples were 100 μl of a 2 mg/ml (or saturated) solution or mixture of compounds in Literature 1.

Packing
P1 = Sephadex G-10.
P2 = Sephadex G-15.
P3 = Sephadex G-25.

Solvent
S1 = 0.13*M* ammonium formate, pH 6.0.
S2 = 0.02*M* phosphate buffer, pH 7.6.
S3 = 0.005% ammonium carbonate, pH 7.6.

Literature
1. DeBersaques, J., *J. Chromatogr.*, **31**, 223 (1967).
2. Zadražil, S., Šormová, Z., and F. Šorm, *Coll. Czech. Chem. Commun.*, **26**, 2643 (1961).

TABLE LC 4
GEL FILTRATION OF AMINO ACIDS ON SEPHADEX G-10

Packing	P1	P1	P1	P1	P1	P1	P1	P1	P1	P1
Column[a]										
length	142 cm	142 cm	142 cm	142 cm	142 cm	142 cm	142 cm	142 cm	142 cm	142 cm
diameter	1 cm	1 cm	1 cm	1cm	1 cm	1 cm	1 cm	1 cm	1 cm	1 cm
material	M1	M1	M1	M1	M1	M1	M1	M1	M1	M1
Solvent	S1	S2	S3	S4	S5	S6	S7	S8	S9	S10
Flow rate	F1	F1	F1	F1	F1	F1	F1	F1	F1	F1
Temperature	25 °C	25 °C	25 °C	25 °C	25 °C	25 °C	25 °C	25 °C	25 °C	25 °C
Detection	D1	D1	D1	D1	D1	D1	D1	D1	D1	D1
Literature	1	1	1	1	1	1	1	1	1	1
Substance[b]	K_d	K_d	K_d	K_d	K_d	K_d	K_d	K_d	K_d	K_d
Ribonuclease	0	0	0	0	0	0	0	0	0	0
Bacitracin	0	0	0.26	—	0	—	—	—	—	—
Tetraglycine	0.33	0.10	0.38	0.42	—	0.18	—	—	—	—
Leu-Gly-Gly	0.38	—	—	—	—	—	—	—	—	—
Gly-Asp	—	0.27	—	—	—	—	—	—	—	—
Val-Try-Arg	—	0.42	—	—	—	—	—	—	—	—
Lysine	0.20	0.06	0.29	0.35	0.17	0.17	0.16	0.19	0.14	0.24
Histidine	0.31	0.07	0.34	0.39	—	—	0.32	—	—	—
Ornithine	—	0.07	—	0.37	—	—	—	—	—	—
Arginine	—	0.08	0.48	0.60	0.27	—	—	—	—	—
Threonine	—	0.28	—	0.51	—	—	—	—	—	—
Valine	0.38	0.28	0.50	0.69	—	—	—	—	—	—
Proline	—	0.29	0.38	0.47	—	—	—	—	—	—
Glycine	0.44	0.29	0.48	0.54	0.34	0.36	0.35	—	0.35	0.32
Serine	—	0.30	—	—	—	—	—	—	—	—
Isoleucine	—	0.30	0.57	0.86	—	—	—	—	—	—
Leucine	0.40	0.32	0.58	—	—	0.39	—	—	—	—
Norleucine	—	0.32	0.59	0.90	—	—	—	—	—	—
Methionine	0.40	—	—	0.58	—	—	—	—	—	—
Glutamic acid	0.44	0.41	0.49	0.61	—	—	0.29	0.14	—	0.13
Aspartic acid	—	0.53	—	0.54	—	—	—	—	—	0.15
Phenylalanine	0.86	0.64	1.11	1.83	0.71	0.81	0.65	0.56	0.74	0.61
Tyrosine	1.28	0.93	1.73	2.85	0.96	1.16	0.88	0.61	1.11	0.38
Tryptophan	—	2.74	—	—	—	—	—	—	—	2.88
Ammonia	0.56	0.13	0.68	0.79	0.46	0.39	0.43	0.67	0.38	0.75
Urea	1.09	1.06	1.09	1.17	1.03	1.07	0.95	0.94	1.02	1.06
Glucosamine	—	0.07	0.49	0.51	—	—	—	—	—	—
Glucose	—	0.40	0.49	0.51	—	—	—	—	—	—
Acetic acid	—	1.0	—	—	—	—	—	—	—	—

[a] V_0 = 41 ml, V_i = 43 ml.
[b] Samples were 0.5–8 μ moles in 1 ml, <40 μ bead size.

Packing	P1	= Sephadex G-10, W_r = 0.96.
Material	M1	= Perspex chromatographic tube.
Solvent	S1	= 0.2*M* acetic acid, pH 2.7, freshly poured column.
	S2	= 0.2*M* acetic acid, pH 2.7, after exposure of gel to all other media.
	S3	= 0.2*M* acetic acid with 0.5*M* NaCl.
	S4	= 0.2*M* acetic acid with 2.0*M* NaCl.
	S5	= 0.1 or 0.3*M* pyridine-acetic acid (equimolar), pH = 5.
	S6	= 0.35*M* sodium citrate, pH 5.28.
	S7	= 1.0*M* pyridine with 0.03*M* acetic acid, pH 6.7.
	S8	= 1.0*M* pyridine with 0.5*M* NaCl, pH 8.5
	S9	= 0.05*M* tris-H_2SO_4 buffer, pH 8.07.
	S10	= 0.01*M* NaOH.
Flow rate	F1	= 10 ml/hr.
Detection	D1	= Automatic ninhydrin colorimetry of the effluent. Glucose was determined by an orcinol reagent procedure and acetic acid by ultraviolet absorption.
Literature		1. Eaker, D., and J. Porath, *Separ. Sci.*, **2**, 518 (1967).

TABLE LC 5
SALTING-OUT CHROMATOGRAPHY OF ALCOHOLS

Packing	P1	P1	P1	P1	P1	P1
Column						
length	ca. 25 cm	ca. 25 cm	ca. 25 cm	ca. 25 cm	ca. 25 cm	ca. 25 cm
diameter[a]	ca. 2 cm	ca. 2 cm	ca. 2 cm	ca. 2 cm	ca. 2 cm	ca. 2 cm
material	G	G	G	G	G	G
Solvent	H_2O	S1	S2	S3	S4	S5
Flow rate	0.5 cm/min.	0.5 cm/min.	0.5 cm/min.	0.5 cm/min.	0.5 cm/min.	0.5 cm/min.
Temperature	rt	rt	rt	rt	rt	rt
Detection	D1	D1	D1	D1	D1	D1
Literature	1	1	1	1	1	1
Compound[b]	C	C	C	C	C	C
Glycerol	0.427	0.453	0.473	0.622	0.843	1.16
Propylene glycol	0.477	0.647	0.740	1.29	2.21	3.83
Methanol	0.507	0.670	0.804	1.15	1.68	2.33
Ethanol	0.673	0.843	1.12	2.00	3.72	—
Isopropanol	0.753	1.02	1.53	3.48	7.76	—
***tert*-Butanol**	0.883	1.19	1.96	5.12	13.4	—
***n*-Propanol**	1.28	1.71	2.57	5.45	12.3	—
***sec*-Butanol**	1.56	2.26	3.79	9.30	26.5	—
Isobutanol	2.47	3.34	5.17	12.3	—	—
***n*-Butanol**	2.37	3.52	5.67	13.8	—	—
***n*-Amyl alcohol**	6.63	11.0	—	—	—	—

Packing	P2	P2	P2	P2	P2
Column					
length	ca. 25 cm	ca. 25 cm	ca. 25 cm	ca. 25 cm	ca. 25 cm
diameter[a]	ca. 2 cm	ca. 2 cm	ca. 2 cm	ca. 2 cm	ca. 2 cm
material	G	G	G	G	G
Solvent	H_2O	S1	S2	S3	S4
Flow rate	0.5 cm/min.	0.5 cm/min.	0.5 cm/min.	0.5 cm/min.	0.5 cm/min.
Temperature	rt	rt	rt	rt	rt
Detection	D1	D1	D1	D1	D1
Literature	1	1	1	1	1
Compound[b]	C	C	C	C	C
Glycerol	0.586	0.635	0.688	0.901	1.20
Propylene glycol	0.644	0.856	1.08	1.77	3.07
Methanol	0.839	0.950	1.13	1.59	2.29
Ethanol	0.839	1.08	1.46	2.46	4.57
Isopropanol	0.744	1.21	1.61	3.58	8.34
***tert*-Butanol**	0.662	1.20	1.90	4.78	12.9
***n*-Propanol**	1.03	1.62	2.35	5.01	—
***sec*-Butanol**	1.07	1.73	2.88	7.48	—
Isobutanol	1.33	2.20	3.38	8.74	—
***n*-Butanol**	1.28	2.17	3.53	8.44	—
***n*-Amyl alcohol**	2.69	5.03	8.33	—	—

[a] Columns had cross-sectional areas of 3.90 cm^2.
[b] Samples were about 0.1 mmole alcohol in 1 ml of water.

Packing P1 = Dowex® 1-X8, $SO_4^=$ form, 200–400 mesh, 44.0% interstitial volume.
P2 = Dowex® 50-X8, ammonium form, 200–400 mesh, 39.1% bed volume.

Solvent S1 = 0.5*M* $(NH_4)_2SO_4$. S4 = 3.0*M* $(NH_4)_2SO_4$.
S2 = 1.0*M* $(NH_4)_2SO_4$. S5 = 4.0*M* $(NH_4)_2SO_4$.
S3 = 2.0*M* $(NH_4)_2SO_4$.

Detection D1 = 6 ml fractions were mixed with 0.1*N* sodium dichromate in conc. H_2SO_4 and the resulting Cr(III) is measured spectrophotometrically.

Literature 1. Sargent R., and W. Rieman III, *J. Phys. Chem.* **61**, 356 (1957).

TABLE LC 6
CHROMATOGRAPHY OF STEROIDS AND DERIVATIVES ON LIPOPHILIC SEPHADEX GELS

Packing	P1	P1	P1	P1	P2	P2	P2	P3	P4
Column									
length	na	na	na	na	na	na	na	na	na
diameter	na	na	na	na	na	na	na	na	na
material	na	na	na	na	na	na	na	na	na
Solvent	S1	S2	S3	S4	S1	S3	S4	S1	S1
Flow rate	F1	F1	F1	F1	F1	F1	F1	F1	F1
Temperature	18–23°C	18–23°C	18–23°C	18–23°C	18–23°C	18–23°C	18–23°C	18–23°C	18–23°C
Detection	D1	D1	D1	D1	D1	D1	D1	D1	D1
Literature	1	1	1	1	1	1	1	1	1
Compound[a]	V_e[b]	V_e[b]	V_e[b]	V_e[b]	V_e[b]	V_e[b]	V_e[b]	V_e[b]	V_e[b]
5α-Cholestane	45.5	69.0	70.0	—	54.5	75.5	—	47.0	57.0
5α-3-one	45.0	72.5	73.5	76.0	53.0	79.0	98.0	45.0	55.0
5α-3β-ol	61.5	69.0	70.0	76.0	67.0	71.0	91.0	54.0	64.0
5β-3-one	—	74.5	75.0	—	—	—	—	—	55.0
5β-3β-ol	59.0	70.5	72.0	—	64.5	74.0	93.0	—	—
5β-3,12-dione	—	77.5	—	—	—	82.0	—	—	—
5β-3α,12α-diol	73.5	72.5	72.0	—	80.0	72.5	84.5	59.0	68.5
5β-3,7,12-trione	44.0	90.0	85.0	73.5	52.0	100.5	94.0	44.0	53.0
5β-3α,7α,12α-triol	151.5	77.5	75.0	73.5	132.5	74.0	82.5	74.5	87.0
5-en-3β-OCH_3	42.0	68.0	69.0	76.0	53.0	75.5	100.5	45.0	55.0
5-en-3β-Acetate	43.5	70.5	71.5	76.0	—	82.5	102.5	43.0	54.0
5α-Pregnane	48.5	74.5	77.0	—	59.5	—	—	52.0	62.5
Progesterone	48.0	88.0	86.5	77.5	56.5	92.5	93.0	—	57.0
5α-Androstane	51.0	78.0	78.5	—	62.5	83.5	—	54.5	—
5α-Androstan-17-one	50.0	84.5	83.0	83.5	59.5	90.5	100.5	52.5	—
5α-Androstan-17β-ol	71.0	81.5	81.5	83.5	77.0	80.5	95.0	63.5	—
Estrone	239.5	115.5	114.0	99.0	202.5	133.0	92.0	138.0	—
Estradiol	—	113.5	109.5	98.0	—	113.0	91.0	—	—

[a] The notations refer to specific structural features of the cholestane nucleus unless otherwise indicated; sample size was usually 3–15 mg of each compound in 0.1–0.3 ml.
[b] Elution volume expressed as percent of total bed volume.

Packing P1 = Sephadex LH-20.
P2 = Sephadex G-25-36 (Sephadex G-25 with a relative methoxyl content of 36% w/w).
P3 = Sephadex G-15-38.
P4 = Sephadex G-25-40.
Solvent S1 = dichloromethane.
S2 = isopropanol.
S3 = *n*-propanol.
S4 = methanol.
Flow rate F1 = 0.2 ml/min. per cm^2.
Detection D1 = Fractions of 0.5–1.0% of the total bed volume were analyzed gravimetrically, by quantitative gas chromatography or by thin-layer chromatography.
Literature 1. Eneroth, P., and E. Nyström, *Biochim. Biophys. Acta*, **144**, 153 (1967).

TABLE LC 7
SOLUBILIZATION CHROMATOGRAPHY OF ETHERS, CARBOXYLIC ACIDS AND HYDROCARBONS

Packing	P1	P1	P1	P1	P1	P1	P2	P2
Column								
length	15 cm	15 cm	15 cm	15 cm	15 cm	15 cm	15 cm	15 cm
diameter[a]	2 cm	2 cm	2 cm	2 cm	2 cm	2 cm	2 cm	2 cm
material	G	G	G	G	G	G	G	G
Solvent	S1	S2	S3	S4	S5	S6	S5	S6
Flow rate	F1	F1	F1	F1	F1	F1	F1	F1
Temperature	rt	rt	rt	rt	rt	rt	rt	rt
Detection	D1	D1	D1	D1	D1	D1	D1	D1
Literature	1	1	1	1	1	1	1	1
Compound	C	C	C	C	C	C	C	C
Ethers								
Diisopropyl ether	3.51	3.11	2.88	2.13	1.30	—	—	—
Di-*n*-propyl ether	4.94	4.16	3.37	2.08	1.65	—	—	—
Ethyl *n*-butyl ether	5.25	4.19	3.60	2.48	1.69	—	—	—
Di-*n*-butyl ether	12.5	9.36	6.81	3.90	1.83	—	—	—
Anisole	15.2	11.6	9.09	5.36	3.08	—	—	—
Diphenyl ether	—	—	54.6	19.2	6.65	—	—	—
Diisoamyl ether	—	—	—	—	2.28	—	—	—
Di-*n*-amyl ether	—	—	—	—	3.03	—	—	—
Acids								
Butyric acid	2.17	—	1.81	1.45	1.24	—	—	—
***n*-Caproic acid**	4.99	—	3.35	2.01	1.40	—	—	—
***n*-Caprylic acid**	18.0	—	7.74	3.29	1.43	—	—	—
***n*-Capric acid**	—	—	24.4	6.22	1.63	—	—	—
Hydrocarbons								
Benzene	—	6.62	5.69	4.21	2.69	—	4.55	—
Toluene	—	10.2	8.21	5.36	3.00	—	5.53	—
***m*-Xylene**	—	—	14.2	7.76	3.75	—	7.41	—
Mesitylene	—	—	—	13.0	5.12	—	9.66	—
Naphthalene	—	—	—	—	—	2.92	15.8	7.48
β-Methylnaphthalene	—	—	—	—	—	3.37	22.4	9.40
1,4-Dimethyl-naphthalene	—	—	—	—	—	4.21	30.2	11.0

[a] Column had cross-sectional area of 3.9 cm^2.

Packing P1 = Dowex® 50-X4, H^+ form, 200–400 mesh.
P2 = Dowex® 1-X4, acetate form, 200–400 mesh.

Solvent S1 = H_2O.
S2 = 1.0*M* acetic acid.
S3 = 2.0*M* acetic acid.
S4 = 4.0*M* acetic acid.
S5 = 6.0*M* acetic acid.
S6 = 8.0*M* acetic acid.

Flow rate F1 = 0.4–0.5 cm/min.

Detection D1 = Fractions of 5 ml were mixed with 5 ml of 0.02*M* sodium dichromate in conc. H_2SO_4, diluted with 25 ml H_2O, and the resultant Cr (III) measured spectrophotometrically.

Literature 1. Sherma, J., and W. Rieman III, *Anal. Chim. Acta*, **20**, 357 (1959).

TABLE LC 8
SALTING-OUT CHROMATOGRAPHY OF ALDEHYDES AND KETONES

Packing	P1	P1	P1	P1	P1	P1
Column						
length	25 cm	25 cm	25 cm	25 cm	25 cm	25 cm
diameter[a]	2 cm	2 cm	2 cm	2 cm	2 cm	2 cm
material	G	G	G	G	G	G
Solvent	H_2O	S1	S2	S3	S4	S5
Flow rate	F1	F1	F1	F1	F1	F1
Temperature	rt	rt	rt	rt	rt	rt
Detection	D1	D1	D1	D1	D1	D1
Literature	1	1	1	1	1	1
Compound[b]	C	C	C	C	C	C
Formaldehyde	1.05	0.93	1.00	0.98	0.97	0.88
Acetaldehyde	0.80	0.98	1.22	1.52	2.18	2.92
Acetone	0.70	0.93	1.53	3.00	6.12	10.9
Acetoin	0.76	0.96	1.38	2.91	6.13	10.8
Diacetyl	1.25	2.06	2.20	4.60	8.50	13.9
2,5-Hexanedione	1.06	1.47	2.56	6.87	22.4	—
Diacetone alcohol	0.78	1.36	2.31	7.31	22.5	—
Propionaldehyde	1.37	1.86	2.50	4.58	7.70	—
Methyl ethyl ketone	1.28	2.00	3.02	8.21	21.1	—
Cyclopentanone	1.70	2.59	4.01	8.83	—	—
2,3-Pentanedione	1.92	2.59	4.49	12.8	—	—
2,4-Pentanedione	1.90	2.73	4.34	12.9	21.8	—
Methyl isopropyl ketone	2.12	3.59	5.76	16.5	—	—
Butyraldehyde	2.82	4.42	5.81	—	—	—
Methyl propyl ketone	2.51	4.56	7.52	19.6	—	—
Diethyl ketone	2.52	4.42	7.16	19.6	—	—
Cyclohexanone	3.09	5.17	8.61	—	—	—

[a] Columns had cross-sectional area of 3.90 cm^2.
[b] 0.050–0.100 mmole samples taken.

Packing P1 = Dowex® 1-X8, sulfate form, 200–400 mesh.

Solvent S1 = 0.50*M* $(NH_4)_2SO_4$.
S2 = 1.00*M* $(NH_4)_2SO_4$.
S3 = 2.00*M* $(NH_4)_2SO_4$.
S4 = 3.00*M* $(NH_4)_2SO_4$.
S5 = 4.00*M* $(NH_4)_2SO_4$.

Flow rate F1 = 0.2–0.8 cm per minute.

Detection D1 = Detection was by the differential pH method of Roe and Mitchell in which 5 ml of 0.1*M* hydroxylamine hydrochloride is added to each fraction and the pH is determined after 5 minutes; the amount of carbonyl compound is related to Δ pH. Alternatively, fractions were mixed with 5 ml of 0.1*N* sodium dichromate in conc. H_2SO_4, diluting with 25 ml H_2O and measuring the absorbance of the resulting Cr (III).

Literature
1. Breyer, A., and W. Rieman III, *Anal. Chim. Acta*, **18**, 207 (1958).

TABLE LC 9
GEL CHROMATOGRAPHY OF PROTECTED PEPTIDES ON METHYLATED SEPHADEX

Packing		P1	P1	P2	P2
Column					
length		L1	L1	L2	L2
diameter		2 cm	2 cm	2 cm	2 cm
material		G	G	G	G
Solvent		S1	S2	S1	S2
Flow rate		F1	F1	F1	F1
Detection		D1	D1	D1	D1
Literature		1	1	1	1
Protected Peptide [a,b]	MW	V_e[c]	V_e[c]	V_e[c]	V_e[c]
Z·Gly-Gly·OMe	280	1.06	1.55	1.00	1.22
Z·Gly-Gly-Gly·OMe	337	0.97	1.51	0.98	—
Z·Gly-Gly-Gly-Gly·OMe	394	0.93	1.53	1.00	—
Z·Val-Lys(BOC)·OH	479	0.68	0.72	0.78	—
Z·Lys(BOC)-Val-Tyr-Pro·O*t*Bu	795	0.69	0.98	0.86	0.96
H·Lys(BOC)-Val-Tyr-Pro·O*t*Bu	661	0.79	1.00	1.00	1.00
Z·Val-Lys(BOC)-Val-Tyr-Pro·O*t*Bu	894	0.66	0.93	0.83	0.94
H·Val-Lys(BOC)-Val-Tyr-Pro·O*t*Bu	760	0.71	0.97	—	0.94
H·Tyr(*t*Bu)-Ser(*t*Bu)·O*t*Bu	436	0.80	1.04	1.00	1.00
Z·Val-Tyr(*t*Bu)-Ser(*t*Bu)·O*t*Bu	669	0.74	0.90	0.89	0.97
Z·Leu-Val-Tyr(*t*Bu)-Ser(*t*Bu)·O*t*Bu	782	0.72	—	—	—
Z·Leu-Leu-Val-Tyr(*t*Bu)-Ser(*t*Bu)·O*t*Bu	761	0.71	0.91	1.00	0.95
Z·His-Leu-Leu-Val-Tyr(*t*Bu)-Ser(*t*Bu)·O*t*Bu	1032	0.71	0.84	0.86	0.92
H·His-Leu-Leu-Val-Tyr(*t*Bu)-Ser(*t*Bu)·O*t*Bu	898	0.72	0.78	0.83	0.89
Z·Val-Tyr-Val-His-Pro-Phe·OMe	908	0.76	1.18	1.07	1.06
Z·Arg(NO_2)-Val-Tyr-Val-His-Pro-Phe·OMe	1109	0.77	—	0.84	—
H·Arg(NO_2)-Val-Tyr-Val-His-Pro-Phe·OMe	975	0.85	—	0.88	—
Z·Asp(OBzl)-Arg(NO_2)-Val-Tyr-Val-His-Pro-Phe·OMe	1314	0.74	0.72	0.78	—
Z·Asp(O*t*Bu)-Glu(NH_2)-Leu-Ala-Glu(O*t*Bu)-Ala-Phe-Pro-Leu-Glu(O*t*Bu)-Phe·O*t*Bu	1636	0.60	—	—	—
Z·Glu(O*t*Bu)-Asp(O*t*Bu)-Glu(NH_2)-Leu-Ala-Glu(O*t*Bu)-Ala-Phe-Pro-Leu-Glu(O*t*Bu)-Phe·O*t*Bu	1821	0.50	—	—	—

[a] Abbreviations: Z = benzyloxycarbonyl; BOC = *tert*-butyloxycarbonyl; OMe = methyl ester; OBzl = benzyl ester; O*t*Bu = *tert*-butyl ester; *t*Bu = *tert*-butyl ether.
[b] Samples were 0.2–0.8 mg of compound in 0.5–1.0 ml.
[c] Elution volume relative to cholesterol-4-^{14}C, added to each sample.

Packing P1 = Methylated Sephadex G-25.
P2 = Methylated Sephadex G-50.
Length L1 = length resulting from placing 25 g of packing in the column.
L2 = length resulting from placing 12.5 g of packing in the column.
Solvent S1 = $CHCl_3$-methanol (1 : 1).
S2 = ethanol-H_2O (96.5 : 3.5).
Flow rate F1 = 0.5–1.0 ml/min.
Detection D1 = Fractions analyzed by ninhydrin colorimetry or by TLC using the chlorine-*o*-tolidine reaction. Cholesterol determined by gas-flow counting of radioactivity.
Literature 1. Nyström, E., and J. Sjövall, *J. Chromatogr.*, **24**, 209 (1966).

TABLE LC 10
CHROMATOGRAPHY OF HYDROXY- AND KETOPROGESTERONES ON LIPOPHILIC SEPHADEX

Packing	P1	P1	P2	P2
Column				
length	na	na	na	na
diameter	na	na	na	na
material	na	na	na	na
Solvent	S1	S2	S1	S2
Flow rate	na	na	na	na
Temperature	ambient	ambient	ambient	ambient
Detection	na	na	na	na
Literature	1,2	1,2	1,2	1,2

Hydroxyprogesterones[a]	V_e[b]	V_e[b]	V_e[b]	V_e[b]
Progesterone	1.00	1.00	1.00	1.00
11β-ol	1.21	1.46	—	1.81
17α-ol	1.19	1.47	—	1.78
21-ol	1.21	1.29	1.14	1.41
11β,21-ol	1.47	2.06	1.35	3.03
11β,17α-ol	1.52	2.33	—	3.89
17α,21-ol	1.41	2.05	—	3.20
11β,17α,21-ol	1.77	3.66	1.61	7.62
Total column volume[c]	1.34	1.78	1.81	2.18

Ketoprogesterones[d]	V_e[e]	V_e[e]		V_e[e]
P^4-3,20-one	1.00	1.00		1.00
P^4-20α-ol-3-one	1.07	1.30		1.53
P^4-20β-ol-3-one	1.03	1.24		1.44
P^4-11β-ol-3,20-one	1.21	1.46		1.81
6α-Me-P^4-11β-ol-3,20-one	1.16	1.39		1.68
6α-Me-P^4-3,11,20-one	1.08	1.08		1.09
P^4-17α,21-ol-3,20-one	1.41	2.05		3.20
6α-Me-P^4-17α,21-ol-3,20-one	1.36	1.94		2.76
P^4-11β,21-ol-3,20-one	1.47	2.06		3.03
P^4-21-ol-3,11,20-one	1.41	1.62		2.07
P^4-11β,17α-ol-3,20-one	1.52	2.33		3.89
P^4-17α-ol-3,11,20-one	1.34	1.64		2.22
P^4-11β,17α,21-ol-3,20-one	1.77	3.66		7.62
P^4-17α,21-ol-3,11,20-one	1.57	2.41		4.34

[a] Notations refer to substitute in progesterone.
[b] Elution volume relative to progesterone.
[c] Relative to progesterone elution volume.
[d] Notations as used by Bush.
[e] Elution volume relative to P^4-3,20-one.

Packing P1 = Sephadex G-25–36 (Sephadex G-25 with 36% w/w relative methoxyl content).
P2 = Sephadex LH-20.

Solvent S1 = heptane—$CHCl_3$—methanol (2 : 1 : 1).
S2 = heptane—$CHCl_3$—ethanol (5 : 5 : 1), saturated with H_2O by addition until incipient turbidity at 23 °C.

Literature
1. Sjövall, J., Nyström E., and E. Haahti, *Advances in Chromatogr.*, **6**, 119 (1968).
2. Bush, I. E., "*The Chromatography of Steroids*," Pergamon Press (1961).

TABLE LC 11
SOLUBILIZATION CHROMATOGRAPHY OF ALCOHOLS AND PHENOLS

Packing	P1	P1	P1	P1	P1	P2	P2	P2	P2	P2
Column										
length	20 cm	20 cm	20 cm	20 cm	20 cm	20 cm	20 cm	20 cm	20 cm	20 cm
diameter[a]	2 cm	2 cm	2 cm	2 cm	2 cm	2 cm	2 cm	2 cm	2 cm	2 cm
material	G	G	G	G	G	G	G	G	G	G
Solvent	H_2O	S1	S2	S3	S4	S5	S1	S2	S3	S4
Flow rate	F1	F1	F1	F1	F1	F1	F1	F1	F1	F1
Temperature	rt	rt	rt	rt	rt	rt	rt	rt	rt	rt
Detection	D1	D1	D1	D1	D1	D1	D1	D1	D1	D1
Literature	1	1	1	1	1	1	1	1	1	1
Compound[b]	C	C	C	C	C	C	C	C	C	C
Catechol	3.06	2.41	1.96	1.49	0.994	67.1	46.5	34.5	18.9	12.4
Phenol	4.66	3.65	3.03	2.03	1.39	56.1	36.5	26.8	13.6	9.02
***o*-Cresol**	7.85	5.66	4.38	2.56	1.57	115.9	61.2	42.1	19.9	10.9
***o*-Nitrophenol**	12.8	8.67	6.46	2.99	1.76	86.1	46.2	26.1	10.4	5.84
***tert*-Amyl alcohol**	2.19	1.86	—	—	—	—	—	—	—	—
***n*-Amyl alcohol**	4.61	3.85	3.30	2.21	1.43	—	2.91	2.29	1.50	1.02
***n*-Hexyl alcohol**	7.97	6.12	4.88	2.88	1.46	—	6.17	4.40	2.38	1.36
***n*-Heptyl alcohol**	14.9	10.8	8.28	3.81	1.65	—	14.1	9.79	4.12	2.12
***n*-Octyl alcohol**	27.5	20.7	13.8	5.39	1.85	—	—	—	—	—
***n*-Nonyl alcohol**	55.1	36.7	21.6	7.47	1.98	—	—	—	—	—
***n*-Decyl alcohol**	—	—	—	9.60	2.69	—	—	—	—	—
***n*-Undecyl alcohol**	—	—	—	—	3.55	—	—	—	—	—
***n*-Dodecyl alcohol**	—	—	—	—	4.77	—	—	—	—	—
Benzyl alcohol	5.71	4.30	3.76	2.53	1.55	—	—	—	—	—
Cyclohexanol	4.80	4.04	3.58	2.55	1.67	—	—	—	—	—
***o*-Chlorophenol**	13.5	8.48	6.80	3.70	2.03	—	—	—	—	—

[a] Columns had cross-sectional areas of 3.9 cm^2.
[b] Samples were a maximum of 0.2 mmole of each compound in 1.0 ml of up to 50% acetic acid.

Packing P1 = Dowex® 50-X8, H^+ form, 200–400 mesh.
P2 = Dowex® 1-X8, acetate form, 200–400 mesh.

Solvent S1 = 1.0*M* acetic acid. S3 = 4.0*M* acetic acid. S5 = 0.5*M* acetic acid.
S2 = 2.0*M* acetic acid. S4 = 6.0*M* acetic acid.

Flow rate F1 = 0.4–0.5 cm per minute.

Detection D1 = Six ml fractions were mixed with 0.02*M* sodium dichromate in conc. H_2SO_4, diluted with 25 ml H_2O, and the resultant Cr·III) measured spectrophotometrically.

Literature 1. Sherma, J., and W. Rieman III, *Amal. Chim. Acta.* **18**, 214 (1958).

TABLE LC 12
COLUMN CHROMATOGRAPHY OF AMINO ACIDS ON POLYSTYRENE

Packing	P1	P1	P1	P1	P1	P1	P1	P1	P1	P1
Column										
length	40 cm	40 cm	40 cm	43 cm	84 cm	84 cm	34.5 cm	34.5 cm	34.5 cm	34.5 cm
diameter	1.6 cm	1.6 cm	1.6 cm	1.5 cm	1.5 cm	1.5 cm	2.9 cm	2.9 cm	2.9 cm	2.9 cm
material	G	G	G	G	G	G	G	G	G	G
Solvent	S1	S2	S3	S4	S4	S5	S6	S7	S8	S9
Flow rate	1–2 ml/min	1–2 ml/min	1–2 ml/min	1–2 ml/min	1–2 ml/min	1–2 ml/min	1–2 ml/min	1–2 ml/min	1–2 ml/min	1–2 ml/min
Temperature	rt	rt	rt	rt	rt	rt	rt	rt	rt	rt
Detection	D1	D1	D1	D1	D1	D1	D1	D1	D1	D1
Literature	1	1	1	1	1	1	1	1	1	1
Compound[a]	V_e	V_e	V_e	V_e	V_e	V_e	V_e	V_e	V_e	V_e
Glycine	51	55	45	60	108	108	—	137	136	140
Glutamic acid	51	—	48	—	—	—	—	—	—	—
Histidine · HCl	51	—	45	—	—	—	—	—	—	—
Leucine	81	—	75	108	—	—	211	—	—	—
Tyrosine	81	—	75	—	—	—	—	248	—	—
Lysine · HCl	—	55	—	—	—	—	—	—	—	—
Taurine	—	55	—	—	—	—	—	—	—	—
Methionine	—	76	—	—	—	—	—	—	—	—
Isoleucine	—	—	—	—	156	—	—	—	—	—
Histidine	—	—	—	—	—	—	142	140	—	—

[a] Samples were 1 ml of sample medium containing 3–6 mg of each compound.

Packing P1 = Amberlite® XAD-2 polystyrene resin, 20–50 mesh.

Solvent
S1 = H_2O, sample medium was 0.4*N* HCl.
S2 = 0.001*N* NaCl, sample medium was 0.001*N* NaCl.
S3 = 0.001*N* NaCl, sample medium was 0.4*N* HCl.
S4 = H_2O, sample medium was HCl, pH 1.
S5 = H_2O, sample medium was NaOH, pH 8 or 13.
S6 = H_2O, sample medium was H_2O.
S7 = H_2O, sample medium was 0.05 or 0.1*N* HCl.
S8 = H_2O, sample medium was NaOH, pH 10.
S9 = 0.001*N* HCl for eluant and sample medium.

Detection D1 = Amino acids were estimated by ninhydrin colorimetry.

Literature
1. Zaika, L. L., *J. Chromatogr.*, **49**, 222 (1970).

TABLE LC 13
GEL FILTRATION OF PURINES AND PYRIMIDINES

Packing	P1	P1	P1	P1
Column				
length	35 cm	35 cm	35 cm	35 cm
diameter	3.5 cm	3.5 cm	3.5 cm	3.5 cm
material	G	G	G	G
Solvent	dist. H_2O	0.05*M* NaCl	S1	S2
Flow rate	F1	F1	F1	F1
Temperature	ambient	ambient	ambient	ambient
Detection	UV	UV	UV	UV
Literature	1	1	1	1
Compound[a]	K_d	K_d	K_d	K_d
Ribonucleic acid	0.0	—	—	—
Diphosphopyridine nucleotide	0.0	0.8	—	—
Adenosine triphosphate	0.0	0.6	—	—
Adenylic acid	0.1	1.2	—	0.1
Guanylic acid	0.4	1.3	0.9	0.1
Cytidylic acid	0.1	0.8	0.7	0.1
Uridylic acid	0.1	0.8	0.7	0.1
Adenosine	1.7	1.8	—	1.8
Guanosine	1.6	—	1.8	—
Cytidine	1.2	—	1.2	—
Uridine	1.0	—	1.0	—
Inosine	1.2	1.3	—	0.3
Adenine	2.2	2.4	—	1.2
Cytosine	1.6	—	1.4	1.3
Uracil	1.1	—	1.2	—
Hypoxanthine	1.6	1.6	—	—
Xanthine	1.8	—	—	—
Dimethylxanthine	1.3	—	—	—
Trimethylxanthine	1.1	—	—	—

[a] 0.5–142 mg samples tested.

Packing P1 = Sephadex G-25 with water regain = 2.9 g water/g dry substance and wet density = 1.099, 50–100 mesh. The Sephadex was swelled in 0.05*M* NaCl for 30 minutes and the fine particles removed by decantation before packing; the column was equilibrated with the solvent before addition of the samples.

Solvent S1 = Phosphate solution, $\mu = 0.05$, pH = 7.
S2 = 0.01*M* NH_4OH, pH = 10.06.

Flow rate F1 = 2 ml/in. with hydrostatic pressure of 60 cm.

Literature 1. Gelotte, B., *J. Chromatogr.*, **3**, 330 (1960).

TABLE LC 14
GEL CHROMATOGRAPHY OF PEPTIDES

Packing	P1	P1	P1	P1	P1	P1	P1
Column							
length	102 cm	102 cm	102 cm	102 cm	102 cm	102 cm	102 cm
diameter	1.2 cm	1.2 cm	1.2 cm	1.2 cm	1.2 cm	1.2 cm	1.2 cm
material	G	G	G	G	G	G	G
Solvent	S1	S2	S3	S4	S5	S6	S7
Flow rate	na	na	na	na	na	na	na
Temperature	rt	rt	rt	rt	rt	rt	rt
Detection	D1	D1	D1	D1	D1	D1	D1
Literature	1	1	1	1	1	1	1

Compound[a]	K_{av}[b]	K_{av}	K_{av}	K_{av}	K_{av}	K_{av}	K'_{av}
Glycine	0.30	0.32	0.36	0.28	0.17	0.24	0.25
Phenylalanine	0.54	0.76	0.98	0.54	0.26	0.53	0.46
Gly-Phe	0.55	1.10	1.43	0.55	0.19	0.43	0.33
Phe-Gly	0.43	0.82	1.03	0.42	0.22	0.47	0.28
Gly-Gly-Phe	0.49	0.89	1.10	0.48	0.15	0.33	0.27
Phe-Gly-Gly	0.29	0.36	0.39	0.26	0.12	0.25	0.25
Alanine	0.27	0.33	0.34	0.25	0.15	0.23	0.23
Gly-Ala	0.25	0.38	0.42	0.24	0.11	0.22	0.17
Ala-Gly	0.25	0.36	0.39	0.23	0.09	0.22	0.15
Leucine	0.29	0.36	0.43	0.29	0.15	0.28	0.25
Gly-Leu	0.31	0.57	0.65	0.31	0.11	0.23	0.18
Leu-Gly	0.25	0.37	0.45	0.26	0.11	0.26	0.17
Proline	0.24	0.27	0.30	0.25	0.16	0.26	0.18
Gly-Pro	0.23	0.35	0.37	0.24	0.09	0.21	0.17
Pro-Gly	0.23	0.31	0.35	0.23	0.11	0.23	0.18

[a] Sample size was 1 μ mole in 0.5 ml solvent.
[b] $K_{av} = V_e - V_0/V_t - V_0$; V_0 = 41 ml and V_t = 110 ml for this column.

Packing P1 = Sephadex G-10, expanded in 0.2*M* acetic acid.
Solvent S1 = 0.2*M* acetic acid.
S2 = 0.2*M* acetic acid plus 0.5*M* NaCl.
S3 = 0.2*M* acetic acid plus 1.0*M* NaCl.
S4 = 0.2*M* acetic acid plus 0.1*M* phenol.
S5 = 0.01*M* NaOH.
S6 = 0.01*M* NaOH plus 0.5*M* NaCl.
S7 = 0.2*M* acetic acid, after washing the column with 1*M* pyridine.
Detection D1 = One ml fractions are analyzed by spectroscopy at 420 nm using 2,4,6-trinitrobenzenesulfonic acid or ninhydrin reagents.
Literature 1. Ziska, P., *J. Chromatogr.*, **53**, 385 (1970).

TABLE LC 15
GEL CHROMATOGRAPHY OF NUCLEIC ACID COMPONENTS

Packing	P1	P2	P3
Column			
length	153 cm	153 cm	153 cm
diameter	1.3 cm	1.3 cm	1.3 cm
material	G	G	G
Solvent	S1	S1	S1
Flow rate	1 ml/min.	1 ml/min.	1 ml/min.
Temperature	rt	rt	rt
Detection	UV	UV	UV
Literature	1	1	1

Compound[a]	K_{av}[b]	K_d	K_{av}[b]	K_d	K_{av}[b]	K_d
Adenine	2.90	3.51	3.30	4.62	4.20	6.56
Adenosine	1.95	2.15	2.00	2.79	2.16	3.37
Adenosine-5′-monophosphate	0.60	0.80	0.42	0.58	0.16	0.25
Guanine	2.20	2.70	2.30	3.22	2.52	3.93
Guanosine	1.70	1.90	1.50	2.09	1.36	2.21
Guanosine-5′-monophosphate	0.60	0.80	0.30	0.43	0.12	0.18
Uracil	1.00	1.20	0.84	1.18	0.72	1.12
Uridine	0.80	1.00	0.61	0.85	0.48	0.75
Uridine-5′-monophosphate	0.50	0.70	0.07	0.10	0.08	0.12
Cytosine	1.35	1.55	1.07	1.50	0.96	1.50
Cytidine	1.05	1.25	0.84	1.18	0.68	1.06
Cytidine-3′(2′)-phosphate	0.55	0.75	0.27	0.37	0.08	0.12
Thymine	1.10	1.30	1.03	1.45	1.00	1.56
Thymidine	0.90	1.10	0.73	1.02	0.48	0.75
Thymidine-5′-monophosphate	0.55	0.75	0.18	0.26	0.20	0.31
Purine	—	—	2.00	2.79	2.16	3.37
Xanthine	—	—	1.57	2.20	1.44	2.31
Xanthosine	—	—	1.07	1.50	0.84	1.94
Hypoxanthine	—	—	1.57	2.20	1.48	2.31
Inosine	—	—	1.10	1.50	0.85	1.95
5-Methylcytosine	—	—	1.10	1.60	1.00	1.56
RNS	0.00	0.00	0.00	0.00	0.00	0.00
DNS	0.00	0.00	0.00	0.00	0.00	0.00

[a] Samples were 150 μl of solution containing 1–2 mg/ml of solute.

[b] $K_{av} = \frac{V_e - V_0}{V_t - V_0}$

Packing P1 = Sephadex G-25 fine.
P2 = Sephadex G-15.
P3 = Sephadex G-10.

Solvent S1 = $0.01M\ (NH_4)_2CO_3$ solution.

Literature 1. Gorbach, G., and J. Henke, *J. Chromatogr.*, **37**, 231 (1968).

TABLE LC 16
SOLUBILIZATION CHROMATOGRAPHY OF KETONES

Packing	P1	P1	P1	P1	P1	P1
Column						
length	20 cm	20 cm	20 cm	20 cm	20 cm	20 cm
diameter[a]	2 cm	2 cm	2 cm	2 cm	2 cm	2 cm
material	G	G	G	G	G	G
Solvent	S1	S2	S3	S4	S5	S6
Flow rate	0.4 cm/min.	0.4 cm/min.	0.4 cm/min.	0.4 cm/min.	0.4 cm/min.	0.4 cm/min.
Temperature	rt	rt	rt	rt	rt	rt
Detection	D1	D1	D1	D1	D2	D2
Literature	1	1	1	1	1	1
Compound[b]	C	C	C	C	C	C
Methyl-*n*-butyl ketone	4.14	3.04	4.33	3.58	4.08	3.32
Methyl-*n*-amyl ketone	5.95	4.00	6.13	4.96	6.00	4.41
Methyl *n*-hexyl ketone	10.1	6.94	11.3	8.16	10.4	6.46
Methyl isobutyl ketone	3.21	2.76	—	—	—	—
Methyl *n*-heptyl ketone	17.1	11.6	—	—	—	—
Methyl *n*-octyl ketone	33.6	21.3	—	—	—	—
Methyl *n*-nonyl ketone	63.1	39.6	—	—	—	—
Acetophenone	12.1	9.56	—	—	—	—

[a] Columns had cross-sectional areas of 3.9 cm^2.
[b] Samples were 0.2 mmole of compound dissolved in 1.0 ml of at least 50% of eluant.

Packing P1 = Dowex® 50-X8, H^+ form, 200–400 mesh.

Solvent
S1 = 2.0*M* methanol.
S2 = 4.0*M* methanol.
S3 = 1.0*M* ethanol.
S4 = 2.0*M* ethanol.
S5 = 1.0*M* acetic acid.
S6 = 2.0*M* acetic acid.

Detection
D1 = Fractions of 5 ml were mixed with 5 ml of 0.1*M* hydroxylamine hydrochloride and the pH determined after 5 minutes. The difference between this pH and that of a fraction containing no ketone is proportional to the amount of ketone present.
D2 = Fractions were mixed with 5.0 ml of 0.1*N* sodium dichromate in conc. H_2SO_4, diluted with 25 ml H_2O, and the resultant Cr (III) measured.

Literature
1. Sherma, J., and W. Rieman III, *Anal. Chim. Acta*, **19**, 138 (1958).

TABLE LC 17
COLUMN CHROMATOGRAPHY OF NUCLEOTIDES ON POLY-*N*-VINYL PYRROLIDONE

Packing	P1		P1	P1	P1
Column					
length	10.8 cm		39.5 cm	39.5 cm	39.5 cm
diameter	0.9 cm		1.0 cm	1.0 cm	1.0 cm
material	G		G	G	G
Solvent	H_2O, pH 6		S1	S2	S3
Flow rate	0.3 ml/min.		na	na	na
Temperature	25 °C		25 °C	25 °C	25 °C
Detection	UV		UV	UV	UV
Literature	1		2	2	2

Compound[a]	V_e, ml	width, ml[b]	V_e, ml	V_e, ml	V_e, ml
Vitamin B_{12}	6.0	—	—	—	—
AMP	7.4	0.9	—	—	—
UMP	7.9	0.9	—	—	—
2′-AMP	8.0	—	—	—	—
Riboflavin	8.1	—	—	—	—
5-Aminouracil	9.0	0.9	—	—	—
Thymine	9.3	—	36.2	34	28.8
Cytosine	9.9	1.0	31.2	34	32.5
6-Methyluracil	10.4	—	—	—	—
5-Nitrouracil	10.5	—	—	—	—
Riboflavin phosphate	10.7	—	—	—	—
Xanthine	12.6	1.6	—	—	—
Adenosine	13.8	1.3	—	—	—
Adenine	19.2	1.9	43.8	73.8	70.0
Guanine	—	—	50.0	67.5	37.5
1-Methylguanine	—	—	—	52.5	—

[a] Samples were 0.1–0.5 mg in 0.2 ml water for Literature 1 and 0.08–0.1 mg in 0.5 ml of solvent for Literature 2.
[b] Peak width at one-half maximum concentration.

Packing P1 = Polyclar AT poly-*n*-vinyl pyrrolidone powder (GAF Corporation), 53–250 μ particles.

Solvent S1 = 0.005*N* HCl adjusted to pH 3.5 with dilute NH_4OH.
S2 = 0.02*M* NH_4Cl + 0.2*M* NH_4OH, pH 7.
S3 = 0.02*M* NH_4Cl + 0.2*M* NH_4OH, pH 10.3.

Literature
1. Lerner, J., Dougherty, T. M., and A. I. Schepartz, *J. Chromatogr.*, **37**, 455 (1968).
2. Dougherty, T. M., and A. I. Schepartz, *J. Chromatogr.*, **43**, 398 (1969).

TABLE LC 18
GEL CHROMATOGRAPHY OF PHOSPHATE ESTERS

	P1	P2	P3
Packing	P1	P2	P3
Column			
length	231 cm	239 cm	239 cm
diameter	1.1 cm[a]	1.1 cm[b]	1.1 cm[c]
material	G	G	G
Solvent	H_2O	H_2O	H_2O
Flow rate	12–18 ml/hr.	12–18 ml/hr.	12–18 ml/hr.
Temperature	21 °C	21 °C	21 °C
Detection	UV	UV	UV
Literature	1	1	1

Compound	MW[d]	V_e	K_d	V_e	K_d	V_e	K_d
Blue Dextran 2000	2×10^6	78	0	93	0	93	0
Coenzyme-A	768	78	0	120	0.32	123	0.26
IDP	429	84	0.09	141	0.57	156	0.57
FMN	514	93	0.22	141	0.57	156	0.57
Pyridoxal-P	250	93	0.22	132	0.46	162	0.62
Vitamin B_{12}	1357	105	0.39	156	0.75	192	0.89
Pyridoxamine-P[e]	250	96	0.26	138	0.53	195	0.92
Acetone	58	147	1.00	177	1.00	204	1.00
dUMP	308	93	0.22	129	0.43	222	1.16
IMP	348	96	0.26	159	0.79	243	1.35
dGMP	347	102	0.35	153	0.71	294	1.81
dCMP	307	96	0.26	141	0.58	294	1.81
Riboflavin	376	273	2.82	312	2.61	306	1.92
dAMP	331	105	0.39	156	0.75	330	2.14

[a] Void volume was 78 ml, total volume 207 ml, internal volume 69 ml.
[b] Void volume was 93 ml, total volume 214 ml, internal volume 84 ml.
[c] Void volume was 93 ml, total volume 214 ml, internal volume 111 ml.
[d] Weights of free phosphoric acids.
[e] Pyridoxyl-P applied to G-25 column in the acid form.

Packing P1 = Sephadex G-10, 40–120 μ, swelled in distilled water, and fines decanted.
P2 = Sephadex G-15, 40–120 μ, swelled in distilled water, and fines decanted.
P3 = Sephadex G-25, 100–300 μ.

Literature 1. Lerner, J., and A. I. Schepartz, *J. Chromatogr.*, **39**, 135 (1969).

TABLE LC 19

COLUMN CHROMATOGRAPHY OF PROTEINS USED TO CALIBRATE SEPHADEX GELS FOR MOLECULAR WEIGHT DETERMINATIONS

Packing	P1	P2
Column		
length	70 cm	LI
diameter	1.5 cm	2 cm
material	G	na
Solvent	S1	S2
Flow rate	10–20 ml/hr.	10–20 ml/hr.
Temperature	rt	rt
Detection	UV	na
Literature	1	2

Compound[a]	MW	rad[b]	V_e	K_d	K_d	V_e/V_0
Cytochrome C (horse heart)	13,400	1.74	97.5	0.803	0.70	—
Myoglobin (horse)	16,900	2.07	93.5	0.747	—	—
Trypsin inhibitor (from soya bean)	21,500	2.26	90.5	0.707	—	1.93
Chymotrypsinogen	25,000	2.24	91.0	0.714	0.54	1.88
Ovalbumin	45,000	2.80	78.5	0.569	—	—
Serum albumin (bovine)	67,000	3.61	72.0	0.452	—	1.32
Transferrin (human)	88,000	4.00	70.0	0.424	—	—
Serum albumin dimer (bovine)	134,000	4.35	62.0	0.314	—	—
γ-Globulin (human)	158,000	5.22	57.5	0.252	—	—
Trypsin	24,000	2.41	—	—	0.54	1.84
Pepsin	35,500	—	—	—	—	1.62
Peroxidase-I	40,000	3.02	—	—	—	1.57
Enolase	80,000	—	—	—	—	1.29
Creatine phosphate kinase	81,000	—	—	—	—	1.25

[a] Samples applied in 1–2 ml buffer.
[b] rad = Stokes radius α (nm).

Packing	P1 = Sephadex G-200.
	P2 = Sephadex G-100.
Length	L1 = length not specified; bed volumes were 300–150 ml.
Solvent	S1 = 0.1*M* Tris, pH = 8.0.
	S2 = phosphate buffer pH 7.2 + 0.5*M* NaCl.
Literature	1. Summerell, J. M., and D. A. L. Davies, *Biochim. Biophys. Acta*, **207**, 95 (1970).
	2. Determann, H., and W. Michel, *J. Chromatogr.*, **25**, 306 (1966).

TABLE LC 20
GEL FILTRATION OF VITAMINS AND ALKALOIDS

Packing	P1	P1	P1	P1
Column				
length	35 cm	35 cm	35 cm	35 cm
diameter	3.5 cm	3.5 cm	3.5 cm	3.5 cm
material	G	G	G	G
Solvent	dist. H_2O	0.05*M* NaCl	S1	S2
Flow rate	F1	F1	F1	F1
Temperature	ambient	ambient	ambient	ambient
Detection	UV	UV	UV	UV
Literature	1	1	1	1

Compound[a]	K_d	K_d	K_d	K_d
Ascorbic acid	0.9	—	—	—
Riboflavin	1.6	—	—	0.8
Nicotinamide	1.4	—	—	1.3
Thiamine hydrochloride	>2.9	1.0	—	0.3
Pyridoxine hydrochloride	>3.3	1.0	—	0.7
Codeine hydrochloride	>3.5	1.2	—	—
Morphine hydrochloride	>3.3	1.3	—	—
Quinine sulfate	>2.9	1.6	—	—

[a] 0.5–142 mg samples tested.

Packing P1 = Sephadex G-25 with water regain = 2.9 g water/g dry substance and wet density = 1.099, 50–100 mesh. The Sephadex was swelled in 0.05*M* NaCl for 30 minutes and the fine particles removed by decantation before packing; the column was equilibrated with the solvent before addition of the sample.

Solvent S1 = phosphate solution, $\mu = 0.05$, pH = 7.
S2 = 0.01*M* NH_4OH, pH = 10.6.

Flow rate F1 = 2 ml/in. with a hydrostatic pressure of 60 cm.

Literature
1. Gelotte, B., *J. Chromatogr.*, **3**, 330 (1960).

TABLE LC 21
GEL PERMEATION CHROMATOGRAPHY OF HYDROCARBONS AND ETHERS

Packing	P1
Column	
length	L1
diameter	⅜ in., O.D.
material	SS
Solvent	benzene
Flow rate	1.0 ml/min
Temperature	24 °C
Detection	RI
Literature	1

Compound[a]	V_e	Width, ml[b]
P-4000	58.3	3.34
$(n\text{-}C_{18}H_{37})_2O$	62.8	3.60
$(n\text{-}C_{12}H_{25})_2O$	67.4	3.70
Dioctyl ether[c]	76.0	3.81
***n*-Dodecane[c]**	84.2	3.96
***n*-Decane**	88.9	3.82
***n*-Heptane[c]**	99.2	4.57
***n*-Hexane**	101.5	3.89
***n*-Pentane[c]**	105.4	3.97
Diethyl ether[c]	107.5	4.16
Dioctyl ether[c]	(53.3)	(4.24)
***n*-Dodecane[c]**	(58.3)	(4.25)
***n*-Nonane**	(63.0)	(4.20)
***n*-Heptane[c]**	(67.3)	(4.41)
***n*-Pentane[c]**	(72.4)	(4.48)
Diethyl ether[c]	(73.6)	(4.47)
Toluene	(81.0)	(—)
Ethylbenzene	(75.2)	(—)
***p*-Xylene**	(76.2)	(—)
Cumene	(71.0)	(—)
Naphthalene	(81.5)	(6.20)
Biphenyl	(75.2)	(5.92)
Diphenylmethane	102.7	5.08
Diphenoxyethane	96.0	4.94
Dicumyl peroxide	84.8	6.02
Anisole	114.6	4.56
Phenetole	107.6	4.43
Cyclohexane[c]	110.7	4.59
Cyclohexane[c]	(76.2)	(5.44)
Styrene	114.2	4.82
2-Methylpentene-1	91.8	4.85
4-Methylpentene-1	90.6	4.21
Heptene-1	(67.3)	(4.41)
Decene-1	82.7	4.56
Isooctane	(65.6)	(4.28)
Methyl methacrylate	98.2	4.95

[a] Sample sizes were typically 0.1 ml of 4% solute in benzene.
[b] Peak widths were obtained by drawing tangents to each side of the curves and reporting the number of ml of eluant at the base of the triangle.
[c] Compound run more than once.

Packing P1 = 40 Å stryenedivinylbenzene gel, permeable to alkanes with MW < 450.
Length L1 = 12 ft, except 8 ft for data in parentheses.
Literature 1. Hendrickson, J. G., *J. Chromatogr.*, **32**, 543 (1968).

TABLE LC 22
CATION EXCHANGE CHROMATOGRAPHY OF NUCLEOTIDES, NUCLEOSIDES AND NUCLEIC BASES

Packing	P1	P2
Column		
length	23 cm	50 cm
diameter	0.6 cm	0.9 cm
material	na	na
Solvent	S1	S2
Flow rate	0.25 ml/min.	60 ml/hr.
Temperature	48 °C	30 °C
Detection	UV	UV
Literature	1	2

Compound	V_e	Peak width (ml)	V_e[a]
Pseudouridine	5.5	0.67	—
Uridine	6.2	0.68	29.9
Ribothymidine	6.6	0.73	—
3-Methyluridine	6.8	0.68	—
Uracil	8.2	0.59	43.3
Inosine	9.2	0.55	59.9
1-Methylinosine	9.2	0.75	—
4-Thiouridine	10.5	0.83	—
4-Thiouracil	10.8	0.63	—
Thymine	10.5	0.63	—
Hypoxanthine	12.2	0.68	107.0
Guanosine	12.4	0.88	118.6
***N*²-Dimethylguanosine**	15.2	1.11	—
***N*²-Methylguanosine**	17.6	1.21	—
Adenosine	18	1.07	153.3
Bis(4-thiuridine) disulfide	20	2.12	—
***N*⁶-Methyladenosine**	22.3	1.30	—
Guanine	24.3	1.09	189.8
Cytidine	24.4	1.13	—
7-Methylguanine	35	1.43	—
***N*⁶-Dimethyladenosine**	42.5	2.36	—
Adenine	62	2.4	206.1
Cytosine	67	2.4	167.0
1-Methyladenosine	72	2.6	—
7-Methylinosine	73	—	—
3-Methylcytidine	72.5	2.5	—
7-Methylguanosine	104	3.9	—
1,7-Dimethylguanosine	125	4.5	—
5′ UMP	—	—	9.1
5′ GMP	—	—	14.7
5′ CMP	—	—	24.5
5′ AMP	—	—	36.0

[a] Samples were 1 ml aliquots of mixture in pH 3.0 citrate buffer.

Packings P1 = Bio-Rad A-6.
P2 = Aminex A-4 resin.

Solvent S1 = 0.4*M* ammonium formate, pH 4.65.
S2 = 0.20*M* sodium citrate buffer, pH 3.0, Na^+ = 0.25*M* adjusted with NaCl, for 40 minutes, followed by a 4-component gradient from a Technicon Autograd: Chamber 1 contained 50 ml of 0.25*M* sodium acetate, pH 6.4, Na^+ = 0.25*M*; chamber 2—50 ml of 0.50*M* sodium acetate, pH 6.4, Na^+ = 0.50*M*, and chambers 3 and 4—50 ml of 1.00*M* sodium acetate, pH 6.4, Na^+ = 1.00*M*.

Literature
1. Uziel, M., Koh, C. K., and W. E. Cohn, *Anal. Biochem.*, **25**, 96 (1968).
2. Murakami, F., Rokushika, S., and H. Hatano, *J. Chromatogr.*, **53**, 585 (1970).

TABLE LC 23
ANION EXCHANGE CHROMATOGRAPHY OF HYDROXY ACIDS

Packing	P1	P2
Column		
length	750 mm	1065 mm
diameter	6 mm	4 mm
material	na	na
Solvent	S1	S2
Flow rate	F1	F1
Temperature	T1	T1
Detection	D1	D1
Literature	1	1

Monohydroxy Acid	D_v	D_v
2-Hydroxyvaleric	30.3	27.7
4-Hydroxyvaleric	2.93	10.6
2-Hydroxybutyric	20.3	17.6
3-Hydroxybutyric	3.98	11.2
4-Hydroxybutyric	2.94	10.6
3-Hydroxypropionic	3.60	11.6
2-Hydroxy-3-methylvaleric	43.7	44.7
2-Hydroxyisovaleric	27.8	26.1
2-Hydroxy-2-methylbutyric	17.2	17.8
2-Hydroxyisobutyric	11.3	13.6
2-Hydroxymethylisobutyric	3.29	15.1
D-Allonic	9.20	8.32
D-Altronic	11.8	7.50
D-Gluconic	12.5	7.21
D-Idonic	12.6	6.82
2-Deoxy-D-arabino-hexonic	2.06	6.44
2-Deoxy-D-lyxo-hexonic	1.97	6.74
3-Deoxy-D-lyxo-hexonic	7.07	6.61
3-Deoxy-D-xylo-hexonic	8.82	6.85
6-Deoxy-D-galactonic	11.5	8.31
2,6-Dideoxy-D-ribo-hexonic	2.32	8.45
2-Deoxy-DL-erythro-pentonic	2.65	8.44
3-Deoxy-D-erythro-pentonic	9.5	8.3
2,5-Dihydroxyvaleric	13.6	10.8
DL-3,5-Dihydroxy-3-methylvaleric	4.21	11.8
DL-erythro-2,3-Dihydroxybutyric	22.2	11.5
3-Deoxy-2-C-hydroxymethyl-tetronic	15.8	8.5
D-2,4-Dihydroxy-3,3-dimethylbutyric	19.8	22.0
2-Tetrahydrofuroic	18.7	10.9
2,5-Anhydro-D-gluconic (chitaric)	33.2	12.9
2,5-Anhydro-D-mannonic (chitonic)	25.5	10.6
2,5-Anhydro-D-talonic	21.7	8.55
Arabinuronic	37.4	15.2
Lyxuronic	29.4	11.6
Riburonic	20.9	8.4
Xyluronic	29.3	14.0
6-Deoxy-gluco-hepturonic	3.16	7.52
Quinic	14.5	7.50
Shikimic	6.50	11.1
3-Hydroxy-2-methyl-1,4-pyron (maltol)	1.14	1.40

Packing P1 = Dowex® 1-X8, acetate form, 24–27 μ particle diameter.
P2 = Dowex® 1-X8, acetate form, 13–18 μ particle size.

Solvent S1 = 0.5*M* acetic acid.
S2 = 0.08*M* sodium acetate plus acetic acid to give pH 5.9.

Flow rate F1 = 2.3–5.3 ml cm^{-2} min^{-1}.

Temperature T1 = The eluant was boiled continuously under reflux before being fed into the column, the temperature of which was kept at 30 °C by circulating thermostated water in the column jacket.

Detection D1 = Automatic colorimetry of the effluent simultaneously by the carbazole method, the chromic acid method and the periodate-formaldehyde method.

Literature 1. Martinsson, E., and O. Samuelson, *Chromatographia*, **3**, 406 (1970).

TABLE LC 24
STEROL ACETATES

Packing	P1
Column	
length	100 cm
diameter	0.45 cm I.D.
material	G
Solvent	S1
Flow rate	15–20 ml/hr
Temperature	rt
Detection	D1
Literature	1, 2

Sterol acetate[a]	Retention value[b]
4,4-Dimethylcholest-5-ene-3β-ol	0.640
4,4-Dimethylcholestane-3β-ol	0.667
4,4,14α-Trimethylcholestane-3β-ol	0.709
4,4,14α-Trimethylcholest-8-ene-3β-ol	0.73
4β-Methylcholestane-3β-ol	0.750
Cholest-3,5-diene-3-β-ol	0.786
4β-Methylcholest-5-ene-3β-ol	0.794
4,4,14α-Trimethylcholest-7,9-diene-3β-ol	0.804
4α-Methylcholestane-3β-ol	0.804
4,4,14α-Trimethylcholest-7-ene-3β-ol	0.83
Cholestane-3β-ol	0.860
4,4-Dimethylcholest-8(14)-ene-3β-ol	0.870
4α-Methylcholest-5-ene-3β-ol	0.901
14α-Methylcholestane-3β-ol	0.918
14α-Methylcholest-8-ene-3β-ol	0.944
4,4-Dimethylcholest-7-ene-3β-ol	0.956
Cholest-5-ene-3β-ol	1.000
Sitosterol	1.04
14α-Methylcholest-7,9-diene-3β-ol	1.077
Stigmasterol	1.09
14α-Methylcholest-7-ene-3β-ol	1.105
Cholest-8(14)-ene-3β-ol	1.206
4α-Methylcholest-7-ene-3β-ol	1.22
4,4,14α-Trimethylcholest-8,24-diene-3β-ol	1.34
Cholest-7-ene-3β-ol	1.405
4α-Methylcholest-7,9-diene-3β-ol	1.435
4,4,14α-Trimethylcholest-7,24-diene-3β-ol	1.48
Cholest-7,9-diene-3β-ol	1.640
4,4,14α-Trimethylcholest-7,9,24-triene-3β-ol	1.66
Cholest-5,7-diene-3β-ol	1.77
Cholest-7,14-diene-3β-ol	1.955
Cholest-5,24-diene-3β-ol	1.98
Ergosterol	2.04
Cholest-8,24-diene-3β-ol	2.42

[a] Samples were 2–10 mg total sterol acetates in solvent.
[b] Elution volumes relative to cholesterol acetate = 1.00.

Packing P1 = Davison Code 12 silica gel, 200–325 mesh, surface area 805 m^2/g, average pore diameter 20 Å.

Solvent S1 = 16% benzene in pentane or hexane.

Detection D1 = 5–7 ml fractions of effluent were determined by the Liebermann–Burchardt reaction followed by colorimetry at 440 and 660 nm.

Literature
1. Klein, P. D., *Separ. Sci.*, **1**, 523 (1965).
2. Klein, P. D., and P. A. Szczepanik, *J. Lipid Res.*, **3**, 460 (1962).

TABLE LC 25
ANION EXCHANGE CHROMATOGRAPHY OF SACCHARIDES

Packing	P1	P2
Column		
length	100 cm	150 cm
diameter	0.6 cm	0.62 cm, I.D.
material	na	SS
Solvent	S1	S2
Flow rate	20 ml/hr.	68 ml/hr.
Temperature	T1	55 °C
Detection	D1	D2
Literature	1	2

Saccharide	Relative retention volumes Rhamnose = 100	Glucose = 100	t_r, hr[a]
Sucrose	12	—	1.22
Melezitose	14	—	—
Raffinose	30	—	1.44
Cellobiose	31	—	—
Maltose	52	—	3.06
Lactose	117	—	3.34
Rhamnose	100	—	—
Lyxose	—	22	—
Ribose	—	30	3.54
Mannose	—	34	—
Fucose	—	52	10.00
Arabinose	—	62	9.57
Fructose	—	64	9.13
Xylose	—	75	11.35
Sorbose	—	76	—
Galactose	—	77	10.53
Glucose	—	100	12.64
Mannoheptulose	—	—	12.08

[a] Mean elution time of 7 runs.

Packing P1 = Dowex® 1-X4, −400 mesh, total capacity 3.5 meg/dry g.
P2 = Dowex® 1-X8, borate form, 6–12 μ diameter particles.

Solvent S1 = (A) 0.15*M* boric acid, 0.5*M* 2,3-butanediol, 0.1% Brij®-35, 0.5 ml toluene per liter, pH = 7 at 23 °C;
(B) 0.8*M* boric acid, 1.0*M* 2,3-butanediol, 0.1% Brij®-35, 0.5 ml toluene per liter, pH = 7 at 23 °C. See T1 below for elution schedule.
S2 = 0.029*M* $Na_2B_4O_7$–0.057*M* H_3BO_3 increasing linearly in concentration with time (20 hours) to 0.147*M* $Na_2B_4O_7$–0.283*M* H_3BO_3.

Temperature T1 = temperature between 40 and 60 °C. Elution was begun with buffer A, column temperature 40°. At an effluent volume of 230 ml the temperature was increased to 60°. After 240 ml of the buffer had passed through the column, buffer A above the resin bed was removed and replaced by buffer B. Elution was continued up to the total effluent volume of 520 ml. The back pressure was 100–120 psi with buffer A and 50–60 psi with buffer B. Samples were applied in 1-ml aliquots of buffer A.

Detection D1 = heat 1 ml aliquots with aniline-acetic acid-orthophosphoric acid reagent.
D2 = sulfuric acid-phenol colorimetry at 480 and 490 nm.

Literature
1. Walborg, Jr., E. F., Ray, D. B., and L. E. Öhrberg, *Anal. Biochem.*, **29**, 437 (1969); from *J. Chromatogr.*, **49**, D31, (1970).
2. Jolley, R. L., and C. D. Scott, *J. Chromatogr.*, **47**, 275 (1970).

TABLE LC 26
SALTING-OUT CHROMATOGRAPHY OF AMINES

Packing	P1	P1	P1	P1	P1
Column					
length	9.7 cm	9.7 cm	9.7 cm	9.7 cm	9.7 cm
diameter[a]	ca. 2 cm	ca. 2 cm	ca. 2 cm	ca. 2 cm	ca. 2 cm
material	G	G	G	G	G
Solvent	S1	S2	S3	S4	S5
Flow rate	F1	F1	F1	F1	F1
Temperature	rt	rt	rt	rt	rt
Detection	D1	D1	D1	D1	D1
Literature	1	1	1	1	1
Compound[b]	C	C	C	C	C
Monoethanolamine	1.21	1.69	2.26	3.37	4.84
Diethanolamine	1.10	1.56	2.17	3.62	—
Triethanolamine	1.58	2.36	3.78	7.01	—
Ethylenediamine	1.36	2.08	3.39	7.96	—
Diethylenetriamine	1.36	2.31	2.26	13.7	—
Triethylenetetramine	1.18	2.29	—	19.2	—
Tetraethylenepentamine	1.09	2.34	—	31.7	—
Methylamine	1.65	2.23	3.57	6.35	10.34
Morpholine	1.69	2.90	5.57	12.4	—
Ethylamine	2.02	3.31	6.21	13.0	28.9
Isopropylamine	2.51	4.47	9.80	23.1	59.1
Tetramethylenediamine	2.84	5.21	12.0	33.0	—
***tert*-Butylamine**	2.82	5.55	12.8	—	—
***n*-Propylamine**	3.02	5.51	12.4	—	—
Diethylamine	3.02	6.09	15.2	—	—
***sec*-Butylamine**	3.85	8.04	20.5	—	—
***n*-Butylamine**	5.29	11.2	28.9	—	—
Triethylamine	5.07	12.7	37.5	—	—
Pyridine	10.0	14.3	29.9	—	—
***n*-Amylamine**	10.3	23.3	—	—	—
Aniline	20.1	29.9	—	—	—
Benzylamine	17.5	—	—	—	—

[a] Columns had cross-sectional areas of 3.90 cm^2.
[b] Sample size = 0.2 mmole each.

Packing P1 = Dowex 50-X4, K^+ form, 200–400 mesh.

Solvent S1 = 0.44*M* K_3PO_4.
S2 = 0.88*M* K_3PO_4.
S3 = 1.32*M* K_3PO_4.
S4 = 1.76*M* K_3PO_4.
S5 = 2.20*M* K_3PO_4.

Flow rate F1 = 0.50–0.75 cm per minute.

Detection D1 = Mono-, di-, and triethanolamines: six ml fractions are mixed with 0.1*N* $K_2Cr_2O_7$, and the resultant Cr(III) analyzed spectrophotometrically; aliphatic amine containing a single amine group and tetraethylenediamine are extracted with $CHCl_3$ and then titrated with 0.01*N* HCl; aromatic amines are determined by ultraviolet spectrophotometry at 260 nm; and ethylenediamine and its polymers are treated with HNO_2 and the alcohol formed determined by the dichromate method above.

Literature
1. Sargent R., and W. Rieman III, *Anal. Chim. Acta*, **17**, 411 (1957).

TABLE LC 27
GEL CHROMATOGRAPHY OF HYDROCARBONS

Packing	P1	P1	P1
Column			
length	60 cm	60 cm	60 cm
diameter	10 mm, I.D.	10 mm, I.D.	10 mm, 1.D.
material	G	G	G
Solvent	S1	S2	S3
Flow rate	0.2 mm/sec	0.3 mm/sec	0.1 mm/sec
Temperature	22 °C	22 °C	22 °C
Detector	RI	RI	RI
Literature	1	1	1

Compound	V_e	V_e	V_e
***n*-Hexatriacontane**	17.5	7.7	16.8
***n*-Hexadecane**	19.3	16.2	22.0
***n*-Dodecane**	19.9	17.6	22.9
***n*-Heptane**	20.8	21.6	25.0
Squalene	—	10.9	19.7
Diphenyl	20.9	21.4	24.6
***p*-Terphenyl**	20.5	19.0	22.0
Benzene	22.0	26.3	27.8
Picene	20.8	20.9	36.3
Chrysene	21.1	21.8	33.4
Phenanthrene	21.6	22.2	31.6
Naphthalene	22.8	23.3	29.6
Rubrene	19.0	15.8	—
Truxene	—	19.1	—
Pyrene	19.0	23.6	47.8
Coronene	29.9	32.8	51.8
Decacyclene	—	45.5	—
Cyclohexane	—	—	21.3
Decalin	20.7	21.1	22.8
Indan	21.4	—	—
Tetralin	21.1	22.9	22.0
Fluoranthrene	—	22.1	—
Dihydrophenanthrene	—	21.3	25.6
Octahydrophenanthrene	—	21.4	26.0
Dihydrotetracene	—	21.0	—
Dioctylbenzene	18.5	14.5	19.3
Tridecylbenzene	19.3	15.8	20.0
Diisopropylbenzene	20.6	19.3	—
Ethylbenzene	—	22.7	—
Dimethyl naphthalene	—	22.3	—
Trimethyl naphthalene	—	21.8	—
Methyl anthracene	—	22.0	—
Cholesteryl stearate	—	12.2	—
Cholesteryl palmitate	—	12.6	—
Cholesteryl laureate	—	13.2	—
Cholesterol	—	18.3	—

Packing P1 = Merckogel OR-500 vinyl acetate gel.
Solvent S1 = cyclohexane.
S2 = methylene chloride.
S3 = isopropanol.
Literature 1. Oelert, H. H., *J. Chromatogr.*, **53**, 243 (1970).

TABLE LC 28
GEL CHROMATOGRAPHY OF FLAVONOID COMPOUNDS

Packing	P1
Column	
length	45 cm
diameter	2.5 cm
material	G
Solvent	S1
Flow rate	3–5 ml/min
Temperature	rt
Detection	UV
Literature	1

Compound[a]	V_e/V_0[b]
Flavone	
Apigenin	5.3
Luteolin	6.3
Resokempferol	5.9
Fisetin	6.6
Robinetin	7.4
Kempferol	7.5
Quercitin	8.3
Morin	4.4
Myricitin	9.2
3-O-Methylquercitin	5.6
Azaleatin	5.9
Isorhamnetin	7.0[c]
Tamarixetin	7.3
Rhamnetin	7.7
Penta-O-methylquercitin	3.4
Hexa-O-methylmyricitin	3.1
Quercitin pent-acetate	2.7
Quercitrin	4.9
Rutin	4.0
Robinin	3.3
Flavanone	
Naringenin	5.4
Eriodictyol	5.8
Taxifolin	5.5
Hesperetin	5.2[c]
Naringin	3.6
Astilbin	4.2
Hesperidin	3.4[c]
Catechin	
***d*-Catechin**	5.2

[a] Samples: 2.5 mg/0.5 ml.
[b] Under these conditions, $V_e/V_0 = 2.2\ K_d + 1$.
[c] Sample dissolved in dioxane–methanol (1 : 1).

Packing P1 = Sephadex LH-20.
Solvent S1 = methanol.
Literature
1. Johnston, K. M., Stern, D. J., and A. C. Waiss, Jr., *J. Chromatogr.*, **33**, 540 (1968).

TABLE LC 29
AUTOMATIC ION EXCHANGE CHROMATOGRAPHY OF CARBOHYDRATES AND RELATED COMPOUNDS

Packing	P1
Column	
length	150 cm
diameter	0.9 cm
material	G
Solvent	S1
Flow rate	na
Temperature	rt
Detection	D1
Literature	1

Compound	V_e
Turanose	45
Lactose	45
Glucose	51–52
Galactose	54
Mannose	54
Sorbose	54–55
Rhamnose	57–58
Fructose	57–58
Glyceraldehyde (dimer)	56–58
Glyceraldehyde (monomer)	62–64
Ascorbic acid	65–66
Ribose	69
Methylglyoxal	68–69
Hydroxymethylfurfural	76–77
Dihydroxyacetone	78–79
Levulinic acid	79–80
Hydroxyproline	115–116
Kojic acid	137–139
Xylose	58–59
Arabinose	62–63
Lyxose	58–59
Xylulose[a]	57–59
	64–67
Acetaldehyde	70.5
Crotonaldehyde	113
Glycoaldehyde (dimer)	62.5–64.5
Glycoaldehyde (monomer)	64–66
Acrolein[a]	64.5
	79.5
	104

[a] More than one peak observed.

Packing P1 = Amberlite® IR-120
Solvent S1 = 0.2*N* sodium citrate.
Detection D1 = ninhydrin colorimetry at 440 and 570 nm.
Literature 1. Zacharius, R. M., and W. L. Porter, *J. Chromatogr.*, **30**, 192 (1967).

TABLE LC 30

ANION EXCHANGE CHROMATOGRAPHY OF BENZOIC ACID DERIVATIVES AND RELATED COMPOUNDS

Packing	P1
Column	
length	150 cm
diameter	0.62 cm
material	SS
Solvent	S1
Flow rate	F1
Temperature	25 °C
Detection	D1
Literature	1

Compound (Acid)	V_e, ml
Phenol	636
Hippuric	817
***p*-Cresol**	834
***p*-Aminobenzoic**	863
3-Methoxy-4-hydroxymandelic	871
Homovanillic	916
Syringic	975
***p*-Hydroxyphenyllactic**	1010
***p*-Hydroxymandelic**	1019
***p*-Hydroxyphenylacetic**	1020
Anthranilic	1064
Benzoic	1073
***m*-Hydroxyphenylacetic**	1125
Vanillic	1156
Phloretic	1194
Folic	1212
***o*-Hydroxyhippuric**	1253
Salicylacetic	1272
3-Hydroxyanthranilic	1287
***o*-Hydroxyphenylacetic**	1318
***m*-Hydroxybenzoic**	1343
***p*-Hydroxybenzoic**	1349
***o*-Hydroxybenzoic**	1350
3-Methoxy-4-hydroxycinnamic	1390
Homogentisic	1417
***a*-Resorcylic**	1430
***p*-Hydroxycinnamic**	1443

Packing P1 = Dowex® 1-X8, 5–10μ.
Solvent S1 = 0.015 *M* acetate buffer (pH 4.4) at the start varied up to 6.0 *M* acetate buffer (pH 4.4) at the end.
Flow rate F1 = 25 cc/hr.
Temperature 25 °C at the start and changed to 60 °C after 15 hours.
Detection D1 = Continuous-flow ultraviolet spectrophotometer.
Literature
1. Katz, S., and C. A. Burtis, *J. Chromatogr.*, **40**, 270 (1969).

TABLE LC 31
ANION EXCHANGE CHROMATOGRAPHY OF PURINES AND PYRIMIDINES

Packing	P1
Column	
length	150 cm
diameter	0.62 cm
material	SS
Solvent	S1
Flow rate	F1
Temperature	25 °C
Detection	D1
Literature	1

Compound	V_e, ml
Thiamine	14
Cytosine	20
Cytidine	24
Deoxycytidine	25
Pseudouridine	29
Uridine	38
Deoxyuridine	39
Uracil	42
Thymidine	45
Theobromine	39
Caffeine	50
Deoxyinosine	62
Inosine	66
Hypoxanthine	75
Deoxyadenosine	94
Adenosine	101
1-Methylguanine	118
7-Methylguanine	120
Theophylline	137
Adenine	152
Xanthosine	163
Xanthine	172
Guanosine	214
Deoxyguanosine	217
6-Methylaminopurine	240
Guanine	241
Uric acid	550

Packing P1 = Dowex® 1–X8, 5–10 μ.
Solvent S1 = 0.015*M* acetate buffer (pH 4.4) at the start varied up to 6.0*M* acetate buffer (pH 4.4) at the end.
Flow rate F1 = 25 cc/hr.
Temperature 25 °C at the start and changed to 60 °C after 15 hr.
Detection D1 = Continuous-flow ultraviolet spectrophotometer.
Literature 1. Katz, S., and C. A. Burtis, *J. Chromatogr.*, **40**, 270 (1969).

TABLE LC 32
ION EXCHANGE CHROMATOGRAPHY OF PENICILLIN AND RELATED COMPOUNDS

Packing	P1
Column	
length	150 cm
diameter	0.6 cm
material	G
Solvent	S1
Flow rate	35 ml/hr.
Temperature	60 °C
Detection	D1
Literature	1

Compound	t_r (min)
L-Cysteine	249
Glutathione (ox.)	275
DL + *meso*-Lanthionine	296, 319[a]
D-Penicillamine	331
L-Cysteine-glutathione disulfide	394
L-Cystine	473
L-Cysteine-D-penicillamine disulfide	504
D-Penicillamine disulfide	545
DL + *allo*-Cystathionine	548, 554[a]
L-Isoleucine	566
L-Cysteine trisulfide	595
L-Cysteine-D-penicillamine trisulfide	616
L-Cysteine-L-homocysteine disulfide	618
L-Homocysteine-D-penicillamine disulfide	640
D-Penicillamine trisulfide	655
L-Cysteine-L-homocysteine trisulfide	675
L-Homocystine	723
D-Penicillamine tetrasulfide	735
Cysteamine-glutathione disulfide	737
L-Homocysteine trisulfide	796
L-Cysteine-cysteamine disulfide	976

[a] Twin peaks.

Packing P1 = Technicon Chromabeads Type A cation exchange resin.

Solvent S1 = Autograd gradient elution device (Technicon Model NC-1 amino acid analyzer) was used with 75 ml of buffer in each chamber as follows:
1 to 4, pH 2.875; 5, pH 3.80; 6, pH 3.80 + pH 5.0 (1 : 2); 7 to 9, pH 5.0.

Detection D1 = ninhydrin colorimetry.

Literature
1. Purdie, J. W., Gravelle, R. A., and D. E. Hanafi, *J. Chromatogr.*, **38**, 347 (1968).

TABLE LC 33
AUTOMATIC ION EXCHANGE CHROMATOGRAPHY OF AMINO ACIDS

Packing	P1
Column	
length	140 cm
diameter	0.6 cm
material	G
Solvent	S1
Flow rate	0.5 ml/min.
Temperature	60 °C
Detection	D1
Literature	1

Compound[a]	t_r, min.	Time ratio[b]
Cysteic acid	48.0	0.0719
Hydroxyproline	181.0	0.2712
Aspartic acid	193.0	0.2891
Threonine	212.5	0.3184
Serine	224.0	0.3356
Glutamic acid	270.5	0.4052
Proline	300.5	0.4502
Glycine	357.0	0.5348
Alanine	384.0	0.5783
Valine	480.5	0.7199
Cystine	534.5	0.8007
Methionine	557.0	0.8345
Isoleucine	623.5	0.9341
Leucine	646.5	0.9685
Norleucine	667.5	1.0000
Tyrosine	707.0	1.0592
Phenylalanine	726.0	1.0876
Ammonia	789.0	1.1820
***n*-Hydroxylysine**	864.0	1.2951
***allo*-Hydroxylysine**	871.0	1.3049
Lysine	945.5	1.4165
Histidine	989.0	1.4816
Arginine	1168.5	1.7506

[a] Sample was 0.5 μ mole of each acid in 0.5 ml.
[b] Elution time relative to norleucine.

Packing P1 = Technicon Chromobeads Type A resin.
Solvent S1 = Gradient from pH 2.875 to 5.00 using the standard procedure with pH 2.875, 3.8 and 5.0 buffers and the Technicon Autograd 9-chamber device.
Detection D1 = ninhydrin colorimetry.
Literature
1. Exss, R. E., Hill, H. D., and G. K. Summer, *J. Chromatogr.*, **42**, 444 (1969).

TABLE LC 34
GEL FILTRATION OF BUFFER COMPONENTS

Packing	P1	P1
Column		
length	35 cm	35 cm
diameter	3.5 cm	3.5 cm
material	G	G
Solvent	dist. H_2O	0.05*M* NaCl
Flow rate	F1	F1
Temperature	ambient	ambient
Detection	D1	D1
Literature	1	1

Compound[a]	K_d	K_d
Sodium hydroxide	2.2	—
Potassium hydroxide	2.3	—
Ammonium hydroxide	0.8	—
Sodium bicarbonate	0.8	—
Sodium carbonate	0.8	—
Sodium phosphate	0.7	—
Sodium tetraborate	1.7	—
Sodium acetate	0.7	—
Sodium citrate	0.5	—
Hydrochloric acid	0.8	—
Formic acid	0.9	—
Acetic acid	0.9	—
Citric acid	0.9	—
Diethylbarbituric acid	1.2	—
Glycine	0.9	—
Triethylamine	1.0	—
Pyridine	1.2	—
γ-Collidine	> 5	0.9

[a] 0.5–142 mg samples tested.

Packing P1 = Sephadex G-25 with water regain = 2.9 g water/g dry substance and wet density = 1.099, 50–100 mesh. The Sephadex was swelled in 0.05*M* NaCl for 30 minutes and the fine particles removed by decantation before packing; the column was equilibrated with the solvent before addition of the sample.

Solvent S1 = phosphate solution, μ = 0.05, pH = 7.
S2 = 0.01*M* NH_4OH, pH = 10.6.

Flow rate F1 = 2 ml/in. with a hydrostatic pressure of 60 cm.

Detection D1 = Acids and bases by titration with dilute NaOH and HCl, respectively; salts by titration after passage through an anion or cation exchanger.

Literature
1. Gelotte, B., *J. Chromatogr.*, **3**, 330 (1960).

TABLE LC 35
CHROMATOGRAPHY OF PRIMARY AROMATIC AMINES ON ALGINIC ACID AND CARBOXYMETHYLCELLULOSE COLUMNS

Packing	P1	P2
Column		
length	L1	L2
diameter	na	na
material	na	na
Solvent	S1	H_2O
Flow rate	2 ml/min	F1
Temperature	ambient	ambient
Detection	D1	D1
Literature	1	1

Compound[a]	Volume range, ml	V_e, ml	Volume range, ml	V_e, ml
o-Nitroaniline	13–19	16.0	—	16.5
p-Nitroaniline	19–25	22.0	—	16.5
m-Nitroaniline	95–115	—	—	32.0
o-Aminobenzoic acid	54–68	60.5	12–20	16.5
p-Aminobenzoic acid	74–98	86.0	20–30	25.0
m-Aminobenzoic acid	120–150	—	45–70	57.5
o-Arsanilic acid	10–18	14.0	—	—
p-Arsanilic acid	25–40	32.5	—	11.0
m-Arsanilic acid	22–45	—	—	—
4-Aminosalicylic acid	42–60	52.0	—	16.5
5-Aminosalicylic acid	60–80	71.5	—	—
p-Aminoacetophenone	—	67.0	—	23.0
Sulfanilamide	—	62.5	—	19.0
p-Aminohippuric acid	—	—	—	19.0

[a] In most cases, 0.05 ml samples containing 1 g/l were used.

Packing P1 = alginic acid, 50–150 mesh.
P2 = carboxymethylcellulose, hydrogen form; sodium salt of the exchanger is treated with 1*M* HCl and washed with H_2O to absence of chloride.

Length L1 = Columns had a cross-sectional area of 0.94 cm^2 and were filled with 4 g of exchanger. Length and diameter not given.
L2 = Column with cross-sectional area of 0.94 cm^2 containing 2 g of exchanger.

Solvent S1 = 1*M* acetic acid.

Flow rate F1 = 1 ml/min initial rate decreasing to 0.5 ml/min.

Detection D1 = Colorimetry with *p*-dimethylaminobenzaldehyde.

Literature
1. Lepri, L., Desideri, P. G., Coas, V., and D. Cozzi, *J. Chromatogr.*, **49**, 239 (1970).

TABLE LC 36
GEL PERMEATION CHROMATOGRAPHY OF LOW MOLECULAR WEIGHT COMPOUNDS: MISCELLANEOUS ORGANICS

Packing	P1	P1	P1
Column			
length	L1	L2	L3
diameter	⅜ in., O.D.	⅜ in., O.D.	⅜ in., O.D.
material	na	na	na
Solvent	S1	S2	S1
Flow rate	1 ml/min.	1 ml/min.	1 ml/min.
Temperature	130 °C	25 °C	130 °C
Detection	RI	RI	RI
Literature	1	1	2

Compound[a]	V_e, cm^3	V_e, cm^3	V_e, cm^3
Benzene	186.3	—	—
Toluene	180.0	—	—
Ethylbenzene	175.5	—	—
***o*-Xylene**	175.4	—	—
***m*-Xylene**	174.6	—	—
***p*-Xylene**	174.6	—	—
Chlorobenzene	182.2	—	—
Iodobenzene	181.3	—	—
Cyclohexane	181.1	—	—
Bromocyclohexane	176.6	—	—
Cyclohexanol	181.3	—	—
Acetonitrile	195.0	142.5	—
Ethylene glycol	—	120.3	—
Aniline	—	126.6	—
Acetic acid	—	129.0	—
Chloroform	—	130.5	—
Ethyl formate	—	138.0	—
Acetone	—	142.0	—
Carbon tetrachloride	—	139.0	—
Styrene	—	—	171.2

[a] Samples are 2 ml aliquots of 0.25 to 1.0% solutions.

Packing P1 = a rigid, crosslined polystyrene gel characterized by Waters Associates.

Length L1 = four 4 ft columns in series having exclusion limits of 3×10^3, 250, 60 and 60 Å.

L2 = four 4 ft columns in series having exclusion limit of about 40 Å.

L3 = four 4 ft columns in series having exclusion limits of 3×10^3, 45, 45, and 45 Å.

Solvent S1 = *o*-dichlorobenzene.

S2 = tetrahydrofuran.

Literature
1. Cazes, J., and D. R. Gaskill, *Separ. Sci.*, **4**, 15 (1969).
2. Cazes, J., and D. R. Gaskill, *Separ. Sci.*, **2**, 426 (1967).

TABLE LC 37

GEL PERMEATION CHROMATOGRAPHY OF LOW MOLECULAR WEIGHT COMPOUNDS: ALCOHOLS AND GLYCOLS

Packing	P1	P1	P1
Column			
length	L1	L2	L3
diameter	⅜ in., O.D.	⅜ in., O.D.	⅜ in., O.D.
material	na	na	na
Solvent	S1	S2	S1
Flow rate	1 ml/min.	1 ml/min.	1 ml/min.
Temperature	130 °C	25 °C	130 °C
Detection	RI	RI	RI
Literature	1	1	2

Compound[a]	V_e, cm^3	V_e, cm^3	V_e, cm^3
Methanol	208.8	132.5	—
Ethanol	196.4	131.4	—
***n*-Propanol**	189.7	127.3	—
***iso*-Propanol**	186.8	125.8	—
***n*-Butanol**	183.6	—	—
***iso*-Butanol**	184.4	—	—
***sec*-Butanol**	182.5	—	—
***tert*-Butanol**	181.2	—	—
Allyl alcohol	194.4	—	—
***n*-Pentanol**	179.3	—	—
***n*-Hexanol**	174.6	—	—
***n*-Heptanol**	170.6	113.1	—
1,2-Ethanediol	—	—	202.5
1,2-Propanediol	—	—	186.7
1,4-Butanediol	—	—	182.9
1,3-Butanediol	—	—	179.7
2,2-Dimethyl-1,3-propanediol	—	—	175.8
2,2,4-Trimethyl-1,3-pentanediol	—	—	180

[a] Samples are 2 ml aliquots of 0.25 to 1.0% solutions.

Packing P1 = A rigid, crosslined polystyrene gel characterized by Waters Associates.

Length L1 = four 4 ft columns in series having exclusion limits of 3 × 10^3, 250, 60, and 60 Å.
L2 = four 4 ft columns in series having exclusion limits of about 40 Å.
L3 = four 4 ft columns in series having exclusion limits of 3 × 10^3, 45, 45 and 45 Å.

Solvent S1 = *o*-dichlorobenzene.
S2 = tetrahydrofuran.

Literature
1. Cazes, J., and D. R. Gaskill, *Separ. Sci.*, **4**, 15 (1969).
2. Cazes, J., and D. R. Gaskill, *Separ. Sci.*, **2**, 426 (1967).

TABLE LC 38

GEL PERMEATION CHROMATOGRAPHY OF LOW MOLECULAR WEIGHT COMPOUNDS: ALIPHATIC HYDROCARBONS

Packing	P1	P1	P1
Column			
length	L1	L2	L3
diameter	⅜ in., O.D.	⅜ in., O.D.	⅜ in., O.D.
material	na	na	na
Solvent	S1	S2	S1
Flow rate	1 ml/min	1 ml/min	1 ml/min
Temperature	130 °C	25 °C	130 °C
Detection	RI	RI	RI
Literature	1	1	2

Compound[a]	V_e, cm^3	V_e, cm^3	V_e, cm^3
***n*-Pentane**	176.0	133.8	—
***n*-Hexane**	172.2	—	—
***n*-Heptane**	168.0	124.6	—
***n*-Octane**	164.4	—	—
***n*-Nonane**	161.2	117.0	152.5
***n*-Decane**	158.4	—	—
***n*-Undecane**	155.7	—	—
***n*-Dodecane**	153.1	—	—
***n*-Tridecane**	150.7	—	—
***n*-Tetradecane**	148.2	—	—
***n*-Octadecane**	141.3	94.5	—
***n*-Eicosane**	138.3	91.8	—
***n*-Octacosane**	130.0	83.2	—
***n*-Pentatriacontane**	125.4	—	—
***n*-Hexatriacontane**	124.5	79.1	113.8
2,2,4-Trimethylpentane	166.4	119.0	—
3-Methylpentane	173.5	129.5	—
2-Methylpentene-1	174.5	131.4	—
4-Methylpentene-1	174.1	129.0	—

[a] Samples are 2 ml aliquots of 0.25 to 1.0% solutions.

Packing P1 = a rigid, crosslined polystyrene gel characterized by Waters Associates.

Length L1 = four 4 ft columns in series having exclusion limits of 3×10^3, 250, 60, and 60 Å.

L2 = four 4 ft columns in series having exclusion limit of about 40 Å.

L3 = four 4 ft columns in series having exclusion limits of 3×10^3, 45, 45, and 45 Å.

Solvent S1 = *o*-dichlorobenzene.

S2 = tetrahydrofuran.

Literature
1. Cazes, J., and D. R. Gaskill, *Separ. Sci.*, **4**, 15 (1969).
2. Cazes, J., and D. R. Gaskill, *Separ. Sci.*, **2**, 426 (1967).

TABLE LC 39
COLUMN PARTITION CHROMATOGRAPHY OF RARE EARTHS

Packing	P1	P1	P2	P3	P4
Column					
length	31 cm	31 cm	20.8 cm	20.8 cm	21 cm
diameter	0.9 cm	0.9 cm	0.9 cm	0.9 cm	0.9 cm
material	G	G	G	G	G
Solvent	S1	S2	S2	S3	S3
Flow rate[a]	F1	F1	F1	F2	F3
Temperature	rt	rt	rt	rt	rt
Detection	D1	D1	D1	D1	D1
Literature	1	1	1	1	1
Rare earth[b]	K_d	K_d	K_d	K_d	K_d
La	—	0.18	0.28	—	—
Pr	2.18	1.13	0.48	—	—
Nd	—	2.03	0.52	0.5	0.46
Sm	10.3	7.91	2.02	—	1.08
Eu	—	11.5	2.30	0.90	1.54
Gd	17.4	15.4	3.03	0.83	1.77
Tb	—	29.6	6.06	2.12	4.67
Dy	—	33.4	7.51	2.80	7.93
Ho	40.4	—	10.3	4.20	11.45
Er	66.8	56.4	11.6	7.70	37.6[c]
Lu	—	98.2	23.8	—	—
Tm	—	—	16.5	16.4	57.5[c]
Yb	—	—	26.9	25.1	93.1[c]

[a] The solvent reservoir was held under constant pressure of 2–15 psi, depending upon the desired flow rate.
[b] Samples were 50 μl of 0.1*M* solutions of each rare earth in 0.1*M* acid.
[c] Obtained on a 19 cm, more tightly-packed column.

Packing
P1 = 33% methylenebis(di-*n*-hexylphosphine oxide) (MHDPO) on Kel-F Grade 300, 60–80 mesh.
P2 = 10% MHDPO on Hyflo Super Cel, < 0.074 mm.
P3 = 10% methylenebis[di-(2-ethylhexyl)-phosphine oxide] (MEHDPO) on Hyflo Super Cel, < 0.074 mm.
P4 = 20% MEHDPO on Plaskon 3200 Powder, 200 mesh.

Solvent
S1 = 12*M* HNO_3.
S2 = 16*M* HNO_3.
S3 = 9.0*M* HNO_3.

Flow rate
F1 = 0.20 ml/cm²/min.
F2 = 0.27 ml/cm²/min.
F3 = 0.31 ml/cm²/min.

Detection
D1 = Fractions of effluent were diluted with a constant amount of H_2O, Chlorophosphonazo III reagent was added, and the rare earth complex determined spectrophotometrically.

Literature
1. O'Laughlin, J. W., and D. F. Jensen, *J. Chromatogr.*, **32**, 567 (1968).

TABLE LC 40
AUTOMATIC ION EXCHANGE CHROMATOGRAPHY OF ORGANIC ACIDS

Packing	P1	P1	P2	P2
Column				
length	1350 mm	1350 mm	900 mm	900 mm
diameter	6 mm	6 mm	6 mm	6 mm
material	G	G	G	G
Solvent	S1	S2	S3	S4
Flow rate	F1	F1	F2	F2
Temperature	30 °C	30 °C	28 °C	28 °C
Detection	D1	D1	D2	D2
Literature	1	1	2	2

Compound[a]	D_v	D_v	D_v	D_v
Aldonic acids:				
Glycolic	18.6	14.8	—	15.7
Glyceric	20.2	11.9	25.5	12.6
D-Erythronic	18.8	10.4	—	—
D-Threonic	19.5	10.6	—	—
D-Arabinonic	14.2	8.89	—	—
D-Lyxonic	19.5	10.3	—	—
D-Ribonic	9.17	9.24	—	—
D-Xylonic	15.7	8.19	—	—
D-Galactonic	11.3	7.51	—	—
D-Gluconic	12.5	7.21	15.1	—
D-Gulonic	13.5	7.70	—	—
D-Mannonic	17.5	9.49	—	—
D-Talonic	6.25	7.37	—	—
D-*glycero*-L-*manno*-Heptonic	14.2	7.70	—	—
D-*glycero*-D-*gulo*-Heptonic	10.8	6.63	—	—
6-Deoxy-D-mannonic	19.3	10.9	—	—
Aldobionic acids:				
Cellobionic	5.86	3.71	8.26	—
Lactobionic	5.09	3.15	7.35	—
Maltobionic	7.35	3.69	8.17	—
Melibionic	4.62	2.64	—	—
Methylated aldonic acids:				
2,3,5-Tri-O-methyl-D-galactonic	5.38	2.64	—	—
3,5,6-Tri-O-methyl-D-gluconic	7.12	3.90	—	—
2,3,4,6-Tetra-O-methyl-D-gluconic	4.76	2.18	—	—
Saccharinic acids:				
2-Hydroxypropionic (lactic)	15.1	13.8	—	14.3
D-*threo*-2,3-Dihydroxybutyric	16.9	12.0	—	—
2,4-Dihydroxybutyric	14.6	11.7	—	—
3,4-Dihydroxybutyric	3.39	9.31	—	—
2-Methyl-2,3-dihydroxypropionic	14.4	11.1	—	—
D-*threo*-2,4,5-Trihydroxyvaleric	11.8	9.17	—	—
α-D-Glucoisosaccharinic	6.09	6.06	—	—
β-D-Glucoisosaccharinic	14.8	6.46	—	—
α-D-Glucosaccharinic	5.41	6.72	—	—
α-D-Glucometasaccharinic	6.84	7.16	—	—
β-D-Glucometasaccharinic	9.56	7.59	—	—
Uronic acids:				
D-Galacturonic	21.4	8.40	—	—
D-Glucuronic	44.1	11.7	—	—
L-Guluronic	24.0	10.7	—	—
L-Iduronic	29.9	12.7	—	—
D-Mannuronic	36.5	12.9	—	—

TABLE LC 40 (Continued)
AUTOMATIC ION EXCHANGE CHROMATOGRAPHY OF ORGANIC ACIDS

Packing	P1	P1	P2	P2
Column				
length	1350 mm	1350 mm	900 mm	900 mm
diameter	6 mm	6 mm	6 mm	6 mm
material	G	G	G	G
Solvent	S1	S2	S3	S4
Flow rate	F1	F2	F2	F2
Temperature	30 °C	30 °C	28 °C	28 °C
Detection	D1	D1	D2	D2
Literature	1	1	2	2
Compound[a]	D_v	D_v	D_v	D_v
Aldehydo and keto acids:				
Glyoxylic	65.6	20.8	—	—
Levulinic	3.64	13.2	—	—
2-Keto-D-gluconic	95.0	13.7	—	—
5-Keto-D-gluconic	38.8	13.9	—	—
Heterocyclic acids:				
2-Furoic	>60	>60	—	—

[a] Samples were 1–4 mg; lactones were saponified, if present, before the sample was applied.

Packing P1 = Dowex 1-X8, acetate form, 26–32 μ particles.
P2 = Dowex 1-X8, acetate form, 18–23 μ particles.

Solvent S1 = 0.50*M* acetic acid.
S2 = 0.08*M* sodium acetate, pH 5.90.
S3 = 0.05*M* sodium acetate.
S4 = 0.10*M* sodium acetate.

Flow rate F1 = 2.8–7.7 ml cm^{-2} min^{-1}.
F2 = 2.5–3.3 ml cm^{-2} min^{-1}.

Detection D1 = Column effluent was divided into two streams, one of which was analyzed continuously by the carbazole colorimetric method and the other by the chromic acid colorimetric method.
D2 = Technicon Auto Analyzer colorimetric dichromate method.

Literature
1. Samuelson, O., and L. Thede, *J. Chromatogr.*, **30**, 558 (1967).
2. Larsson, U-B., Norstedt, I., and O. Samuelson, *J. Chromatogr.*, **22**, 104 (1966).

TABLE LC 41
AUTOMATIC ION EXCHANGE CHROMATOGRAPHY OF AMINO ACIDS

Packing	P1
Column	
length	120 cm
diameter	0.6 cm
material	G
Solvent	S1
Flow rate	F1
Temperature	T1
Detection	D1
Literature	1

Compound	t_r (min)[a]
Cysteic acid	45
Cysteine sulfinic acid	46
Homocysteic acid	46
Phosphothreonine	48
Taurine	72
Homotaurine	72
Phosphoethanolamine	81
2-Aminoethylphosphonic acid	96
Urea	98
Hypotaurine	100
Aspartic acid	190
Glutathione (reduced)	198
Hydroxyproline	201
Methionine sulfoxide[b]	226, 236
Threonine	234
Serine	249
Asparagine	275
Glutamic acid	288
Homoserine	300
Glutamine	305
Sarcosine	328
Proline	367
α-Aminoadipic acid	377
Glutathione (oxidized)	394
Glycine	403
Alanine	419
Citrulline	433
α-Amino-*n*-butyric acid	443
Valine	466
Cystine	493
Homocitrulline	513
Methionine	523
Alloisoleucine	540
Cystathionine[c]	548
3,4-Dihydroxyphenylalanine	552
Isoleucine	561
Selenomethionine	570
Glucosamine	571
Leucine	575
Cysteine-homocysteine mixed disulfide	590
Norleucine	593
Galactosamine	615
Tyrosine	615
Phenylalanine	645
Homocystine	686
β-Alanine	696
β-Aminoisobutyric acid	714
5-Hydroxytryptophan	753
Δ-Aminolevulinic acid	757

TABLE LC 41 (Continued)
AUTOMATIC ION EXCHANGE CHROMATOGRAPHY OF AMINO ACIDS

Packing	P1
Column	
length	120 cm
diameter	0.6 cm
material	G
Solvent	S1
Flow rate	F1
Temperature	T1
Detection	D1
Literature	1

Compound	t_r (min)[a]
Kynurenine	759
γ-Aminobutyric acid	795
Tryptophan	863
S-Adenosylhomocysteine	879
Ethanolamine	912
Ammonia	950
Hydroxylysine	957
Ornithine	1007
Lysine	1023
Histidine	1064
1-Methylhistidine	1085
Anserine	1098
3-Methylhistidine	1102
Carnosine	1111
Homocarnosine	1114
Arginine	1247

[a] The mean effluent volume in ml for each compound = $(t_r - 16)/2$.
[b] Two peaks.
[c] Authentic L-cystathionine.

Packing P1 = Technicon Chromobeads Type B resin, 17 μ particles, washed successively with acetone, H_2O, 6*N* HNO_3, H_2O, 2*M* LiOH, and H_2O before use. Columns regenerated at 70°C after each run by pumping 0.2*M* LiOH through for 30 minutes followed by pH 2.80, 0.2*M* lithium buffer for a further 90 minutes.

Solvent S1 = Autograd Chambers of Technicon Autoanalyzer (1965 version) were filled as follows with lithium citrate buffers: (1) 5 ml methanol + 70 ml pH 2.80, 0.2*M* Li^+ buffer; (2) 3 ml methanol +72 ml pH 2.80, 0.2*M* Li^+ buffer; (3) 75 ml pH 2.80, 0.2*M* Li^+ buffer; (4 and 5) 75 ml pH 3.80, 0.2*M* Li^+ buffer; (7, 8 and 9) 75 ml pH 6.10, 1.2*M* Li^+ buffer.

Flow Rate F1 = 30 ml/hr. for 21 hours.

Temperature T1 = 35°C for 6.5 hours (maximum operating pressure = 325–375 psi), and 70°C for the remainder (175–200 psi maximum pressure).

Detection D1 = ninhydrin colorimetry.

Literature
1. Perry, T. L., Stedman, D., and S. Hansen, *J. Chromatogr.*, **38**, 462 (1968).

TABLE LC 42
PARTITION CHROMATOGRAPHY OF HYDROCARBONS

Packing	P1	P1	P1	P1
Column				
length[a]	3 m	3 m	3 m	3 m
diameter	¼ in. O.D.	¼ in. O.D.	¼ in. O.D.	¼ in. O.D.
material	SS	SS	SS	SS
Solvent	S1	S1	S1	S1
Flow rate	na	na	na	na
Temperature	T1	T2	T3	T4
Detection	RI	RI	RI	RI
Literature	1	1	1	1

Compound[b]	V_g[c]	V_g	V_g	V_g
n-Butane	3.33	3.24	3.16	3.01
2-Methylbutane	4.62	4.38	4.12	3.70
n-Pentane	4.90	4.60	4.32	3.84
2,2-Dimethylbutane	5.24	4.84	4.48	3.86
2,3-Dimethylbutane	5.89	5.50	5.15	4.58
3-Methylpentane	6.12	5.79	5.44	4.88
2-Methylpentane	6.23	5.89	5.54	4.99
n-Hexane	6.87	6.36	5.90	5.11
2,2-Dimethylpentane	7.30	6.70	6.18	5.29
2,3-Dimethylpentane	—	—	6.82	—
2,4-Dimethylpentane	7.40	6.83	6.32	5.48
3-Ethylpentane	8.49	7.73	7.08	5.99
2-Methylhexane	8.76	7.95	7.21	6.00
n-Heptane	9.50	8.61	7.82	6.51
2,2,4-Trimethylpentane	8.65	8.11	7.62	6.75
2,4-Dimethylhexane	10.15	9.24	8.43	7.09
n-Octane	13.20	11.80	10.65	8.75
n-Nonane	17.75	15.90	14.35	11.65
2-Methylbutene-1	—	—	2.41	—
3-Methylpentene-1	2.98	2.82	2.67	2.40
2-Methylpentene-1	—	—	2.78	—
2,3-Dimethylbutene-1	—	—	2.74	—
Pentene-1	2.10	2.00	1.91	1.75
Hexene-1	3.10	2.94	2.79	2.53
1,5-Hexadiene	1.47	1.38	1.31	1.17
Heptene-1	4.23	3.98	3.73	3.32
Octene-1	5.78	5.45	5.13	4.61
2,4,4-Trimethylpentene-1	4.83	4.58	4.35	3.94
Cyclopentane	4.55	4.31	4.09	3.70
Methylcyclopentane	5.82	5.50	5.19	4.66
Cyclohexane	6.60	6.11	5.68	4.93
Methylcyclohexane	8.16	7.71	7.29	6.52
Ethylcyclohexane	11.30	10.49	9.78	8.50
Cyclopentene	2.45	2.36	2.28	2.14
Cyclohexene	3.26	3.08	2.92	2.64
Benzene	0.554	0.550	0.548	0.539
Toluene	0.804	0.786	0.770	0.738
Ethylbenzene	1.04	1.01	0.989	0.935
o-Xylene	1.06	1.03	1.00	0.950
m-Xylene	1.16	1.12	1.08	1.02
p-Xylene	1.17	1.14	1.10	1.03
n-Propylbenzene	1.38	1.33	1.26	1.13
n-Butylbenzene	1.98	1.90	1.84	1.71
n-Pentylbenzene	2.53	2.48	2.38	2.12
n-Hexylbenzene	3.89	3.58	3.30	2.83
n-Octylbenzene	6.84	6.37	5.92	5.18
n-Nonylbenzene	9.41	8.56	7.80	6.52
Isopropylbenzene	—	—	1.26	—
1,2,4-Trimethylbenzene	—	—	1.54	—
1,3,5-Trimethylbenzene	—	—	1.57	—
tert-Butylbenzene	—	—	1.45	—
sec-Butylbenzene	—	—	1.63	—
o-Diethylbenzene	—	—	1.71	—

TABLE LC 42 (Continued)
PARTITION CHROMATOGRAPHY OF HYDROCARBONS

Packing	P1	P1	P1	P1
Column				
length[a]	3 m	3 m	3 m	3 m
diameter	¼ in. O.D.	¼ in. O.D.	¼ in. O.D.	¼ in. O.D.
material	SS	SS	SS	SS
Solvent	S1	S1	S1	S1
Flow rate	na	na	na	na
Temperature	T1	T2	T3	T4
Detection	RI	RI	RI	RI
Literature	1	1	1	1

Compound[b]	V_g[c]	V_g	V_g	V_g
***m*-Diethylbenzene**	—	—	1.81	—
***p*-Diethylbenzene**	—	—	1.78	—
1,3,5-Triethylbenzene	—	—	2.88	—
1-Methyl-4-isopropylbenzene	—	—	1.84	—
Tetralin	—	—	1.83	—
Naphthalene	0.643	0.620	0.593	0.550
CH_2Cl_2	—	—	0.107	—
$CHCl_3$	—	—	0.232	—
CCl_4	1.52	1.49	1.45	1.39

[a] Units of specific retention volume (Vg) are cc/g of squalene.
[b] Samples were 0.01–1.0 μl of pure solute.
[c] Units of specific retention volume (V_g) are cc/g of squalene.

Packing P1 = 25% w/w squalene coated on silanized 100–200 mesh Chromosorb P.
Solvent S1 = Acetonitrile, deaerated.
Temperature T1 = 15.05 °C.
T2 = 20.00 °C.
T3 = 25.00 °C.
T4 = 35.10 °C.
Literature 1. Locke, D. C., *J. Chromatogr.*, **35**, 24 (1968).

TABLE LC 43
GEL CHROMATOGRAPHY OF ORGANOPHOSPHORUS PESTICIDES

Packing	P1	P1
Column		
length	L1	L1
diameter	2 cm	2 cm
material	G	G
Solvent	S1	S2
Flow rate	1.0 ml/min.	1.0 ml/min.
Temperature	rt	rt
Detection	D1	D1
Literature	1	1

Compound[a]	V_e[b]	V_e[b]
Aspon	82	66
Ethion	92	82
Sulfotep	93	75
Malathion	95	86
Mecarbam	96	85
Diazinon	98	77
Demeton-S	99	79
Pyrimithate	99	81
Disulfoton	100	87
Parathion	100 (Elution vol. 43.5 ml)	100 (75.5 ml)
Fenitrothion	101	114
Phorate	103	89
Demeton-O-methyl	103	93
Phorate sulfone	104	98
Malathion-O-analogue	104	76
Chlorthion	107	118
Fenthion	108	112
Thionazin	108	89
Thiometon	108	100
Parathion-methyl	109	114
Fenchlorphos	109	101
Parathion-O-analogue	112	86
Chlorfenvinphos	114	77
Bromophos	115	100
Mevinphos	116	79
Phosphamidon	117	72
Phorate-O-analogue	118	74
Trichlorphon	121	80
Dichlorvos	122	83
Morphothion	124	106
Demeton-S-methylsulfone	125	99
Demeton-S-methyl	130	86
Thionazin-O-analogue	133	79
Tepp	134	73
Bidrin	141	77
Phorate sulfoxide	147	82
Dimethoate	148	95
Dimefox	150	71
Demeton-S-methylsulfoxide	200	79
Azinphos-ethyl	—	113
Azinphos-methyl	—	133
Carbophenothion	—	98
Coumaphos	—	99
Crufomate	—	73
Dibrom	—	86
Dichlofenthion	—	89
Ethoate-methyl	—	87
Phenkapton	—	104
Haloxon	—	112

TABLE LC 43 (Continued)
GEL CHROMATOGRAPHY OF ORGANOPHOSPHORUS PESTICIDES

Packing	P1	P1
Column		
length	L1	L1
diameter	2 cm	2 cm
material	G	G
Solvent	S1	S2
Flow rate	1.0 ml/min.	1.0 ml/min.
Temperature	rt	rt
Detection	D1	D1
Literature	1	1
Compound[a]	V_e[b]	V_e[b]
Oxydementonmethyl	—	79
Phosalone	—	104
Schradan	—	59
Vamidothion	—	73

[a] Samples contained 20 μg of pesticide in 1.0 ml of solvent.
[b] Elution volume relative to parathion = 100.

Packing P1 = Sephadex LH-20, swelled for 24 hours in the solvent to be used.
Length L1 = the column length was not specified. Swollen slurry was poured into columns to give a packed bed volume of ca.75 ml.
Solvent S1 = acetone.
S2 = ethanol.
Detection D1 = Gas chromatograpic examination of fractions. Demeton-S-methylsulfoxide and—sulfone were determined by a colorimetric total phosphorus method.
Literature 1. Ruzicka, J. H., Thomson, J., Wheals, B.B., and N. F. Wood, *J. Chromatogr.*, **34**, 15 (1968).

TABLE LC 44
LIQUID CHROMATOGRAPHY OF INSECTICIDES

Packing	Corasil II (Waters Associates)
Column	
length	50 cm
diameter	2.3 mm, I.D.
material	na
Solvent	*n*-Hexane
Flow rate	1.5 ml/min. (280–320 psi)
Temperature	ambient
Detection	RI
Literature	1

Compound	Elution Volume, ml	Peak Width, ml
Aldrin	1.17	0.21
DDD	2.32	0.46
***p,p'*-DDT**	1.64	0.29
Endrin	10.20	2.04
Lindane	3.72	0.75

Literature 1. Bombaugh, K. J., Levangie, R. F., King, R. N., and L. Abrahams, *J. Chromatogr. Sci.*, **8**, 660 (1970).

TABLE LC 45
STEROIDS, GIBBERELLINS, PIGMENTS AND ESTERS

Packing	P1	P1	P2
Column			
length	100 cm	100 cm	100 cm
diameter	0.9 cm	0.9 cm	0.9 cm
material	G	G	G
Solvent	S1	S2	S3
Flow rate	6 ml/hr.	6 ml/hr.	6 ml/hr.
Temperature	18 °C	18 °C	18 °C
Detection	D1	D1	D1
Literature	1	1	1

Compound[a]	S.E.V.[b]	S.E.V.[b]	S.E.V.[b]
***n*-Hexatriacontane**	51.3	50.3	—
***n*-Tetradecane**	65.4	64.0	—
Methyl lignocerate	53.4	—	—
Methyl behenate	54.5	—	—
Methyl arachidate	55.6	—	—
Methyl stearate	56.8	—	—
Methyl myristate	59.5	—	—
Methyl laurate	60.8	—	—
Cholesteryl palmitate	49.0	—	—
Cholesteryl butyrate	54.5	—	—
Cholesteryl acetate	56.3	53.3	—
Cholesteryl benzoate	54.0	52.2	—
Cholestane	62.7	62.8	—
Coprostane	62.6	62.6	—
5α-Cholane	65.3	65.9	—
5α-Pregnane	67.7	68.2	—
5α-Androstane	68.5	69.3	—
α-Amyrin acetate	57.1	—	—
Stigmasteryl acetate	55.2	—	—
Cholesteryl TMSi	55.9	—	—
Cholestan-3-one	60.0	54.5	—
Cholest-4-ene-3,6-dione	58.6	—	—
Cholesteryl chloride	59.7	—	—
Progesterone	61.3	—	—
Isopetasin	56.9	50.3	—
Isopetasyl acetate	59.1	51.1	—
Isopetasone	61.3	53.1	—
S-Petasin	54.2	—	—
Tetralin	71.0	—	—
Decalin	73.8	—	—
Naphthalene	73.2	—	—
Ceryl alcohol	88.1	—	—
Lignoceryl alcohol	91.3	—	—
Behenyl alcohol	94.6	—	—
Arachidyl alcohol	99.0	—	—
Stearyl alcohol	103.3	—	—
Cetyl alcohol	108.2	—	—
Myristyl alcohol	113.2	62.2	—
Cholesterol	110.2	60.8	—
Lanosterol	82.3	58.5	—
Dihydrolanosterol	84.5	—	—
Cycloartenol	86.5	—	—
Stigmasterol	105.1	59.2	—
β-Sitosterol	107.4	—	—
Campesterol	109.2	—	—
Phytol	93.5	—	—
***trans,trans*-Farnesol**	97.7	—	—
***cis,trans*-Farnesol**	93.5	—	—
Nerolidol	77.7	—	—
Isopetasol	119.6	58.2	—
Methyl deoxycholate	156	—	—
Methyl hyodeoxycholate	236	—	—

TABLE LC 45 (Continued)
STEROIDS, GIBBERELLINS, PIGMENTS AND ESTERS

Packing	P1	P1	P2
Column			
length	100 cm	100 cm	100 cm
diameter	0.9 cm	0.9 cm	0.9 cm
material	G	G	G
Solvent	S1	S2	S3
Flow rate	6 ml/hr.	6 ml/hr.	6 ml/hr.
Temperature	18 °C	18 °C	18° C
Detection	D1	D1	D1
Literature	1	1	1

Compound[a]	S.E.V.[b]	S.E.V.[b]	S.E.V.[b]
Methyl ursodeoxycholate	134.5	—	—
Methyl chenodeoxycholate	156	—	—
Estrone	360	—	—
Octacosyl acetate	51.2	48.1	—
Myristyl acetate	58.9	55.2	—
Ceryl alcohol	78.1	53.9	—
Lignoceric acid	—	54.2	—
Lauric acid	—	63.0	—
Stigmasteryl acetate	55.2	52.1	—
5α-Cholan-24-ol	—	63.2	—
5α-Cholane-3α,24-diol	—	59.3	—
5α-Cholane-3α,12α,24-triol	—	59.9	—
5α-Cholane-3α,7α,12α,24-tetrol	—	59.2	—
Lithocholic acid(3α-OH)	—	58.6	—
Deoxycholic acid (3α,12α-OH)	—	59.9	—
Hyodeoxycholic acid (3α,6α-OH)	—	56.0	—
Ursodeoxycholic acid (3α,7β-OH)	—	56.0	—
Chenodeoxycholic acid (3α,7α-OH)	—	59.6	—
Cholic acid (3α,7α,12α-OH)	—	59.4	—
Gibberellin A_1	—	64.5	—
Gibberellin A_3	—	64.5	—
Gibberellin A_4	—	61 6	—
Gibberellin A_5	—	59.1	—
Gibberellin A_7	—	61.6	—
Gibberellin A_9	—	56.8	—
Triolein	42.2	—	52.0
Tripalmitin	43.2	39.8	—
Lecithin[a]	—	36.8	—
Phosphatidic acid[a]	—	43.8	—
Distearin (2-OH)	—	43.2	—
Distearin (3-OH)	—	44.0	—
Monostearin (1,2-diOH)	280	55.4	—
Chlorophyll *a*	45.0	—	73.5
Chlorophyll *b*	44.8	—	70.0
β-Carotene	47.4	—	70.0
Squalene	52.0	51.0	58.5
β-Sitosterol	107.4	59.4	69.0
Cholesterol	110.2	60.8	70.0
Lanosterol	82.3	58.5	68.5
Petasin	56.9	50.3	76.5
S-Petasin	54.2	—	80.0

[a] Sample mixtures commonly contained 0.5 mg of each component in 0.1 ml of solvent.
[b] SEV = measured elution volume/total column volume.

Packing P1 = N1114–50%—LH-20 (Sephadex LH-20 converted to hydroxyalkoxypropyl—Sephadex using the olefin oxide Nedox 1114; the product contains 50% by weight of hydroxyalkyl groups).
P2 = Sephadex LH-20 (unmodified).

Solvent S1 = benzene.
S2 = benzene-*iso*propanol (75 : 25).
S3 = benzene-ethanol (1 : 9).

Detection D1 = Fractions of 0.1 to 0.5 ml were determined by quantitative gas chromatography or thin-layer chromatography.

Literature 1. Brooks, C. J. W., and R. A. B. Keates, *J. Chromatogr.*, **44**, 509 (1969).

TABLE LC 46
CHROMATOGRAPHY OF PURINES AND RELATED COMPOUNDS ON COLUMNS OF SEPHADEX G-10

Packing	P1	P1
Column		
length	100 cm	235 cm
diameter	1 cm	1.3 cm
material	G	G
Solvent	S1	S2
Flow rate	F1	na
Temperature	rt	rt
Detection	UV	UV
Literature	1	2

Compound[a]	V_e, ml	V_e^{0} [b]
Purines		
6-Succinoaminopurine	52	—
6-Trimethylammoniumpurinide	65	—
1-Methyladenine	75	—
2-Hydroxypurine	82	—
6-Carboxypurine	82	—
9-Methylhypoxanthine	83	—
Caffeine	86	—
Adenine-1-*N*-oxide	91	—
1-Methylhypoxanthine	92	—
7-Methylhypoxanthine	93	—
Theobromine	95	—
6-Chloro-9-methoxymethylpurine	103	—
Purine	106	120
1,7-Dimethylxanthine	107	—
Theophylline	107	—
Hypoxanthine	113	136
6-Methylpurine	116	—
7-Methylxanthine	120	—
6,8-Dihydroxypurine	122	—
Isoguanine	135	—
1-Methylxanthine	139	—
N^2-Dimethylguanine	144	—
7-Methylguanine	153	—
Xanthine	156	152
6-Methoxypurine	159	—
1-Methylguanine	166	—
Uric acid	171	—
8-Hydroxy-7-methylguanine	174	—
N^2-Methylguanine	184	—
2-Aminopurine	187	—
Guanine	199	151
6-Chloropurine	206	—
8-Hydroxyguanine	225	—
6-Methylaminopurine	250	—
Adenine	251	135
6-Mercaptopurine	275	—
6-Bromopurine	278	—
6-Dimethylaminopurine	281	—
6-Cyanopurine	284	—
6-Methylmercaptopurine	319	—
Imuran (Azathioprine)	371	—
6-Iodopurine	400	—
2,6-Diaminopurine	439	—
6-Thioguanine	555	—
6-Furfurylaminopurine	788	—
6-Phenylaminopurine	1,754	—

TABLE LC 46 (Continued)
CHROMATOGRAPHY OF PURINES AND RELATED COMPOUNDS ON COLUMNS OF SEPHADEX G-10

Packing	P1	P1
Column		
length	100 cm	235 cm
diameter	1 cm	1.3 cm
material	G	G
Solvent	S1	S2
Flow rate	F1	na
Temperature	rt	rt
Detection	UV	UV
Literature	1	2
Compound[a]	V_e, ml	V_e^{0b}
Purine ribosides, deoxyribosides and ribonucleotides		
7-Methylinosine	53	—
5′-GMP	56	0.22
2′,3′-AMP	59	—
7-Methylguanosine	59	—
Inosine	77	268
2′-Deoxyinosine	78	—
Xanthosine	104	—
Guanosine	122	283
2′-Deoxyguanosine	126	267
Guanosine (ribose $\xrightarrow{IO_4}$ dialdehyde)	142	—
Adenosine	153	267
2′-Deoxyadenosine	153	251
Azapurines		
8-Azahypoxanthine	188	—
8-Azaguanine	258	152
8-Azaxanthine	298	153
8-Azaadenine	421	136
8-Aza-2,6-diaminopurine	441	—
Pyrazolo[3,4-*d*]*pyrimidines*		
4-Hydroxy	160	—
4,6-Dihydroxy	217	—
4-Amino	383	—
Pyrimidines, nucleosides and nucleotides		
5′-CMP	44	0.13
5′-UMP	45	0.11
Pseudouridine	64	—
Uridine	69	244
Cytidine	70	243
Pyrimidine	70	—
Thymidine	75	242
Cytosine	85	111
Uracil	87	112
5-Methylcytosine	87	125
Orotic acid	91	156
Thymine	96	126
Miscellaneous aromatic and heterocyclic compounds		
Creatinine	61	—
Urea	71	—
Imidazole	73	—
Allantoin	78	—
Cyclohexanone	80	—
Hippuric acid	84	—
Pyridine	105	—
Pyrazole	117	—

TABLE LC 46 (Continued)
CHROMATOGRAPHY OF PURINES AND RELATED COMPOUNDS ON COLUMNS OF SEPHADEX G-10

Packing	P1	P1
Column		
length	100 cm	235 cm
diameter	1 cm	1.3 cm
material	G	G
Solvent	S1	S2
Flow rate	F1	na
Temperature	rt	rt
Detection	UV	UV
Literature	1	2
Compound[a]	V_e, ml	V_e^{0} [b]
Miscellaneous aromatic and heterocyclic compounds		
4-Aminoimidazole-5-carboxamide	154	—
L-Tryptophan	158	—
Pyrrole	173	—
Chlorogenic acid	444	—
Benzimidazole	474	—

[a] Samples were approximately 1–2 mg of compound in 1.0 ml of 0.1*N* Li_2CO_3 for Literature 1, 50–300 μg for Literature 2.
[b] $V_e^0 = V_e - V_0/V_0$.

Packing P1 = Sephadex G-10, water regain 1.0 ± 0.1 g/g, particle size 40–120 μ.

Solvent S1 = phosphate buffer, 0.05*M*, pH = 7.0.
S2 = 0.01*M*$(NH_4)_2CO_3$, pH 9.0.

Flow rate F1 = 24.4 ml/cm²/hr (back pressure = 3.5 psi).

Literature
1. Sweetman, L., and W. L. Nyhan, *J. Chromatogr.*, **32**, 662 (1968).
2. Wasternack, C., and H. Reinbothe, *J. Chromatogr.*, **48**, 553 (1970).

TABLE LC 47
GEL FILTRATION CHROMATOGRAPHY OF PHOSPHATE ESTERS AND RELATED COMPOUNDS

Packing	P1	P1	P2	P2	P3	P3	P4	P5	P2
Column									
length	46 cm[a]	46 cm[a]	52 cm[b]	52 cm[b]	58 cm[c]	58 cm[c]	78 cm[d]	36 cm[e]	52 cm[b]
diameter	na	na	na	na	na	na	na	na	na
material	na	na	na	na	na	na	na	na	na
Solvent	S1	S2	S1	S3	S1	S4	S1	S1	S2
Flow rate	na	na	na	na	na	na	na	na	na
Temperature	20–27 °C	20–27 °C	20–27 °C	20–27 °C	20–27 °C	20–27 °C	20–27 °C	20–27 °C	20–27 °C
Detection	D1	D1	D1	D1	D1	D1	D1	D1	D1
Literature	1	1	1	1	1	1	1	1	1
Compound[f]	$K_d \pm 0.05$	$K_d \pm 0.05$	$K_d \pm 0.05$	$K_d \pm 0.05$	$K_d \pm 0.05$	$K_d \pm 0.05$	$K_d \pm 0.05$	$K_d \pm 0.05$	$K_d \pm 0.05$
Phosvitin	0.00	0.00	0.00	0.00	0.00	0.00	0.03	0.21	—
Myoinositol hexaphosphate	0.00	0.06	0.06	0.30	0.45	0.61	0.79	0.90	0.19
Pyrophosphate	0.04	0.25	0.31	0.67	0.81	0.83	0.97	—	—
Adenosine triphosphate	0.16	0.35	0.60	0.93	0.91	0.93	—	—	—
Orthophosphate	0.26	0.40	0.56	0.77	0.91	0.89	0.97	0.98	—
Fructose	0.63	0.65	0.75	0.81	0.89	0.93	0.97	0.98	—
Adenosine monophosphate	0.74	1.00	0.91	1.30	1.14	1.14	1.07	—	—
Myoinositol tripyrophosphate	—	—	0.06	—	0.48	0.61	—	—	0.19
Alkaline hydrolysis myoinositol pentaphosphate	—	—	0.06	—	0.56	0.62	—	—	0.19
Chicken blood myoinositol pentaphosphate	—	—	0.06	—	0.54	0.64	—	—	0.19
Myoinositol tetraphosphate	—	—	0.12	—	0.57	0.68	—	—	0.25
Myoinositol triphosphate	—	—	0.19	—	0.64	0.75	—	—	0.33
Myoinositol diphosphate	—	—	0.31	—	0.70	0.79	—	—	0.44
Myoinositol monophosphate	—	—	0.44	—	0.86	0.86	—	—	0.59
Glycerol phosphoryl myoinositol	—	—	0.53	—	0.97	0.97	—	—	0.56
Myoinositol	—	—	0.75	—	0.83	0.86	—	—	0.75

[a] V_t = 300 ml, V_0 = 120 ml. V_i = 130 ml.
[b] V_t = 500 ml, V_0 = 210 ml. V_i = 250 ml.
[c] V_t = 710 ml, V_0 = 270 ml, V_i = 410 ml.
[d] V_t = 770 ml, V_0 = 260 ml, V_i = 470 ml.
[e] V_t = 1020 ml, V_0 = 290 ml, V_i = 700 ml.
[f] Samples contained 200–300 μg P.

Packing
P1 = Sephadex G-15, 40–120 μ.
P2 = Sephadex G-25, 100–300 μ.
P3 = Sephadex G-50, 100–300 μ.
P4 = Sephadex G-100, 40–120 μ.
P5 = Sephadex G-200, 140–400 μ.

Solvent
S1 = 0.01*M* LiCl.
S2 = 0.20*M* LiCl.
S3 = 2.00*M* LiCl.
S4 = 0.10*M* LiCl.

Detection
D1 = Aliquots of each fraction were analyzed for total phosphorus. Inorganic phosphate was analyzed by the ascorbic acid method, fructose by the anthrone method, and inositol by periodate oxidation.

Literature
1. Steward, J. H., and M. E. Tate, *J. Chromatogr.*, **45**, 400 (1969).

TABLE LC 48
PARTITION CHROMATOGRAPHY OF SUGARS ON ION EXCHANGE RESINS

Packing	P1	P2	P2	P1	P2	P3	P4	P5	P5	P6	P6
Column											
length	55 cm	55 cm	55 cm	55 cm	55 cm	na	na	na	na	84–89 cm	84–89 cm
diameter	4 mm	4 mm	4 mm	4 mm	4 mm	na	na	na	na	6 mm	6 mm
material	na	na	na	na	na	na	na	na	na	G	G
Solvent	S1	S2	S3	S4	S5	S6	S2	S2	S2	S7	S7
Flow rate	F1	F1	F1	F1	F1	na	na	na	na	F2	F2
Temperature	75 °C	75 °C	75 °C	75 °C	75 °C	75 °C	75 °C	75 °C	100 °C	75 °C	90 °C
Detection	D1	D1	D1	D1	D1	na	na	na	na	D2	D2
Literature	1	1	1	1	1	2	2	2	2	3	3
Compound	D_v	D_v	D_v	D_v	D_v	D_v	D_v	D_v	D_v	D_v	D_v
Erythrose	3.1	1.9	2.3	—	—	—	—	—	—	—	—
Threose	3.8	1.4	1.8	—	—	—	—	—	—	—	—
Arabinose	10.1	3.5	4.2	—	—	7.6	11.2	14.2	7.9	10.5	8.9
Lyxose	8.9	2.7	3.2	—	—	6.6	6.5	8.1	5.5	9.7	7.9
Ribose	6.6	—	4.6	—	—	5.0	12.3	9.2	5.6	6.9	6.5
Xylose	12.5	2.7	3.2	2.3	1.9	9.2	7.3	8.6	5.7	14.0	11.5
Allose	16.6	6.0	8.3	—	—	—	—	—	—	—	—
Altrose	16.6	4.2	5.6	—	—	—	—	—	—	—	—
Galactose	23.4	—	7.6	—	—	16.1	22.0	27.5	15.2	23.0	18.1
Glucose	28.1	4.8	6.0	3.1	3.0	19.5	16.8	22.5	13.6	29.7	22.0
Gulose	20.4	4.9	6.3	—	—	—	—	—	—	—	—
Mannose	16.4	—	5.9	—	—	11.8	17.4	25.0	14.6	15.2	12.4
Talose	10.7	5.8	8.3	—	—	—	—	—	—	—	—
Fructose	13.5	—	6.8	—	—	10.3	18.7	20.1	11.3	14.6	12.1
3-Ribohexulose	9.8	4.7	6.6	—	—	—	—	—	—	—	—
6-Deoxy-D-glucose	7.3	0.9	1.1	—	—	—	—	—	—	—	—
Rhamnose	4.8	1.4	1.6	—	—	3.5	3.3	4.1	2.7	4.8	4.0
1,6-Anhydro-β-D-glucofuranose	6.3	1.4	1.6	—	—	—	—	—	—	—	—
1,6-Anhydro-β-D-glucopyranose	3.8	2.3	2.9	—	—	—	—	—	—	—	—
2,3-Di-O-methyl-6-deoxyallose (mycinose)	0.53	0.4	0.4	—	—	—	—	—	—	—	—
Mannobiose	41.6	—	—	2.9	7.8	—	—	—	—	—	—
Xylobiose	35.8	5.4	7.3	3.0	2.0	—	—	—	—	—	—
Cellobiose	—	—	—	4.4	4.9	—	—	—	—	—	—
Isomaltose	—	—	—	6.5	8.6	—	—	—	—	—	—
Trehalose	—	—	—	6.8	8.2	—	—	—	—	—	—
Cellotriose	—	—	—	5.8	8.7	—	—	—	—	—	—
Gentianose	—	—	—	9.6	12.3	—	—	—	—	—	—
1-Ketose	—	—	—	8.7	16.2	—	—	—	—	—	—

TABLE LC 48 (Continued)
PARTITION CHROMATOGRAPHY OF SUGARS ON ION EXCHANGE RESINS

Packing	P1	P2	P2	P1	P2	P3	P4	P5	P5	P6	P6
Column											
length	55 cm	55 cm	55 cm	55 cm	55 cm	na	na	na	na	84–89 cm	84–89 cm
diameter	4 mm	4 mm	4 mm	4 mm	4 mm	na	na	na	na	6 mm	6 mm
material	na	na	na	na	na	na	na	na	na	G	G
Solvent	S1	S2	S3	S4	S5	S6	S2	S2	S2	S7	S7
Flow rate	F1	F1	F1	F1	F1	na	na	na	na	F2	F2
Temperature	75 °C	75 °C	75 °C	75 °C	75 °C	75 °C	75 °C	75 °C	100 °C	75 °C	90 °C
Detection	D1	D1	D1	D1	D1	na	na	na	na	D2	D2
Literature	1	1	1	1	1	2	2	2	2	3	3
Compound	D_v	D_v	D_v	D_v	D_v	D_v	D_v	D_v	D_v	D_v	D_v
Maltotriose	—	—	—	6.6	7.7	—	—	—	—	—	—
Planteose	—	—	—	9.1	19.3	—	—	—	—	—	—
Cellotetraose	—	—	—	7.9	15.4	—	—	—	—	—	—
Nystose	—	—	—	14.9	27.8	—	—	—	—	—	—
Sachyose	—	—	—	18.5	64.7	—	—	—	—	—	—
Cellopentaose	—	—	—	10.9	27.4	—	—	—	—	—	—
Cellohexaose	—	—	—	15.2	44.4	—	—	—	—	—	—
Digitoxose	—	—	—	—	—	0.70	0.8	0.8	0.6	0.8	0.7
2-Deoxy-D-ribose	—	—	—	—	—	1.3	2.0	2.0	1.4	1.8	1.4
2-Deoxy-D-galactose	—	—	—	—	—	2.8	4.3	4.3	2.9	3.7	3.1
2-Deoxy-D-glucose	—	—	—	—	—	3.5	2.9	3.2	2.4	—	4.1
Fucose	—	—	—	—	—	4.2	7.1	8.9	5.1	4.4	3.9
Tagatose	—	—	—	—	—	10.0	11.3	10.9	7.4	16.7	13.1
Sorbose	—	—	—	—	—	11.0	13.0	13.4	8.5	16.9	13.7

Packing
P1 = strongly basic anion exchange resin T_5C, $SO_4^{=}$ form, 14–17 μ.
P2 = Dowex® 50-X8, Li^+ form, 17–21 μ.
P3 = Dowex® 1-X8, $SO_4^{=}$ form.
P4 = Amberlite® IR-120, Na^+ form.
P5 = Amberlite® IR-120, K^+ form.
P6 = Pharmacia cross-linked dextran quaternary ammonium exchanger, sulfate form, 5–30 μ particles.

Solvent
S1 = 88% ethanol–water.
S2 = 92.4% ethanol–water.
S3 = 96% ethanol–H_2O.
S4 = 70% ethanol–water.
S5 = 85% ethanol–water.
S6 = 86% ethanol–H_2O.
S7 = 95% ethanol–water.

Flow rate
F1 = 2.2–4.4 ml cm^{-2} min^{-1}.
F2 = 2.3–5.0 ml cm^{-2} min^{-1}.

Detection
D1 = automatic analysis of effluent by orcinol method.
D2 = automatic analysis of effluent by anthrone method.

Literature

1. Martinsson, E., and O. Samuelson, *J. Chromatogr.*, **50**, 429 (1970).
2. Samuelson, O., *Ion Exchange*, **Vol. 2**, p. 171, J. Markinsky, Ed., Marcel Dekker, New York, 1969.
3. Jonsson, P. and O. Samuelson, *J. Chromatogr.*, **26**, 197 (1967).

TABLE LC 49
GEL PERMEATION CHROMATOGRAPHY OF MONOMERS AND OLIGOMERS IN NYLON 6, NYLON 66, AND NYLON 12

Packing		P1	P1	P2	P2	P3	P4	P4	P5
Column									
length		L1	L2	L3	L4	L5	L6	L7	L8
diameter		15.6 mm	15.6 mm	15.6 mm	15.6 mm	15.6 mm	15.6 mm	15.6 mm	15.6 mm
material		G	G	G	G	G	G	G	G
Solvent		0.1*N* HCl	0.1*N* HCl	0.1*N* HCl	0.1*N* HCl	0.1*N* HCl	S1	S1	S2
Flow rate		30 ml/hr.	30 ml/hr.	30 ml/hr.	30 ml/hr.	20 ml/hr.	30 ml/hr.	30 ml/hr.	20 ml/hr.
Temperature		rt	rt	rt	rt	rt	rt	rt	rt
Detection		UV	UV	UV	UV	UV	UV[b]	UV[b]	UV[b]
Literature		1	1	1	1	1	2	2	2
Species	MW	V_e	V_e	V_e	V_e	V_e	V_e	V_e	V_e
Nylon 6									
Cyclic monomer	113	135	234	144	243	131	—	—	—
Cyclic dimer	226	122	211	133	223	115	—	—	—
Cyclic trimer	339	115	192	125	210	105	—	—	—
Cyclic tetramer	452	110	180	120	201	99	—	—	—
Cyclic pentamer	565	107	172	117	193	93	—	—	—
Cyclic hexamer	678	—	—	—	—	89	—	—	—
Linear monomer[a]	131	—	—	—	—	—	165	264	114
Linear dimer[a]	244	—	—	—	—	—	130	227	102
Linear trimer[a]	357	—	—	—	—	—	115	195	64
Linear tetramer[a]	470	—	—	—	—	—	103	176	59
Linear pentamer[a]	583	—	—	—	—	—	95	162	—
Linear hexamer[a]	696	—	—	—	—	—	88	150	—
Linear heptamer[a]	809	—	—	—	—	—	82	140	—
Linear octamer[a]	922	—	—	—	—	—	72	132	—

TABLE LC 49 (Continued)
GEL PERMEATION CHROMATOGRAPHY OF MONOMERS AND OLIGOMERS IN NYLON 6, NYLON 66, AND NYLON 12

Packing		P1	P1	P2	P2	P3	P4	P4	P5
Column									
length		L1	L2	L3	L4	L5	L6	L7	L8
diameter		15.6 mm	15.6 mm	15.6 mm	15.6 mm	15.6 mm	15.6 mm	15.6 mm	15.6 mm
material		G	G	G	G	G	G	G	G
Solvent		0.1*N* HCl	0.1*N* HCl	0.1*N* HCl	0.1*N* HCl	0.1*N* HCl	S1	S1	S2
Flow rate		30 ml/hr.	30 ml/hr.	30 ml/hr.	30 ml/hr	20 ml/hr.	30 ml/hr	30 ml/hr.	20 ml/hr.
Temperature		rt	rt	rt	rt	rt	rt	rt	rt
Detection		UV	UV	UV	UV	UV	UV[b]	UV[b]	UV[b]
Literature		1	1	1	1	1	2	2	2
Species	MW	V_e	V_e	V_e	V_e	V_e	V_e	V_e	V_e
Nylon 66									
Cyclic monomer	226	122	—	133	222	117	—	—	—
Cyclic dimer	452	114	—	123	205	104	—	—	—
Cyclic trimer	678	104	—	114	190	90	—	—	—
Linear monomer[a]	244	—	—	—	—	—	124	212	—
Linear dimer[a]	470	—	—	—	—	—	102	170	—
Linear trimer[a]	696	—	—	—	—	—	88	143	—
Linear tetramer[a]	922	—	—	—	—	—	78	127	—
Nylon 12									
Linear monomer[a]	215	—	—	—	—	—	153	242	—
Linear dimer[a]	410	—	—	—	—	—	116	193	—
Linear trimer[a]	605	—	—	—	—	—	102	180	—
Linear tetramer[a]	800	—	—	—	—	—	91	167	—
Linear pentamer[a]	995	—	—	—	—	—	83	153	—
Linear hexamer[a]	1190	—	—	—	—	—	78	137	—
Linear heptamer[a]	1385	—	—	—	—	—	—	127	—

[a] DNP derivative. [b] Wavelength in 360–450 nm range selected.

Packing
P1 = Sephadex G-15, 40–120 μ, W_r (water regain, g/g of dry gel) = 1.3.
P2 = Sephadex G-25, 20–80 μ, W_r = 2.3.
P3 = Bio-Gel P-4, 200–400 mesh, W_r = 2.0.
P4 = Sephadex LH-20, 25–100 μ, S_r (solvent regain, ml/g of dry gel) = 1.7.
P5 = Bio-Beads S-X2, 200–400 mesh, S_r = 0.95.

Length
1. Gel packed in 100 or 150 cm tubes (in all cases) to give a column with—V_t = 171 ml, V_0 = 67 ml, V_i = 70 ml.
2. V_t = 286 ml, V_0 = 111 ml, V_i = 117 ml.
3. V_t = 174 ml, V_0 = 69 ml, V_i = 83 ml.
4. V_t = 281 ml, V_0 = 112 ml, V_i = 133 ml.
5. V_t = 166 ml, V_0 = 52 ml, V_i = 80 ml.
6. V_t = 174 ml, V_0 = 72.5 ml, V_i = 71 ml.
7. V_t = 271 ml, V_0 = 113 ml, V_i = 111 ml.
8. V_t = 172 ml, V_0 = 50 ml, V_i = 62 ml.

Solvent
S1 = 0.05*N* HCl in methanol.
S2 = 0.05*N* HCl in methanol-benzene (1 : 1).

Literature
1. Mori, S., and T. Takeuchi, *J. Chromatogr.*, **49**, 230 (1970).
2. Mori, S., and T. Takeuchi, *J. Chromatogr.*, **50**, 419 (1970).

TABLE LC 50
PURINES AND PYRIMIDINES

Packing	P1	P1	P1	P1	P1	P1	P1	P1	P1	P1
Column										
length	40 cm	43 cm	84 cm	84 cm	84 cm	34.5 cm	34.5 cm	34.5 cm	34.5 cm	34.5 cm
diameter	1.6 cm	1.5 cm	1.5 cm	1.5 cm	1.5 cm	2.9 cm	2.9 cm	2.9 cm	2.9 cm	2.9 cm
material	G	G	G	G	G	G	G	G	G	G
Solvent	S1	S2	S2	S3	S4	S4	S5	S6	S7	S8
Flow rate	1–2 ml/min.	1–2 ml/min.	1–2 ml/min.	1–2 ml/min.	1–2 ml/min.	1–2 ml/min.	1–2 ml/min.	1–2 ml/min.	1–2 ml/min.	1–2 ml/min.
Temperature	rt	rt	rt	rt	rt	rt	rt	rt	rt	rt
Detection	UV	UV	UV	UV	UV	UV	UV	UV	UV	UV
Literature	1	1	1	1	1	1	1	1	1	1
Compound[a]	V_e	V_e	V_e	V_e	V_e	V_e	V_e	V_e	V_e	V_e
Guanosine	—	236	—	—	—	—	—	—	657	351
Inosinic acid	—	96	—	—	—	—	—	—	—	—
Hypoxanthine	—	96	—	180	188	—	248	243	256	—
Inosine	200	232	—	408	421	579	580	—	568	453
Uracil	—	—	163	164	—	—	—	—	—	—
Uridine	—	—	235	227	—	—	—	—	316	—
Adenosine	—	—	—	—	490	—	—	—	735	327
Uric acid	—	—	—	180	188	—	—	127	—	—
5'-Inosinic acid	—	—	—	—	—	137	169	—	191	—
Creatinine	—	—	—	—	—	294	153	—	170	—
5'-Adenylic acid	—	—	—	—	—	—	—	—	192	—
Adenine	—	—	—	—	—	—	—	—	204	—
Cytosine	—	—	—	—	—	—	—	—	148	—
Cytidine	—	—	—	—	—	—	—	—	181	—
Guanine	—	—	—	—	—	—	—	—	199	—

[a] Samples were 1 ml of sample medium containing 3–6 mg of each compound.

Packing P1 = Amberlite® XAD-2 polystyrene resin, 20–50 mesh.
Solvent S1 = 0.001*N* NaCl, sample medium was 0.001*N* NaCl.
S2 = H_2O, sample medium was HCl, pH 1.
S3 = H_2O, sample medium was NaOH, pH 8 or 13.
S4 = H_2O, sample medium was H_2O.
S5 = H_2O, sample medium was 0.05 or 0.1*N* HCl.
S6 = H_2O, sample medium was NaOH, pH 10.
S7 = 0.001*N* HCl for eluant and sample medium.
S8 = 0.10*N* HCl for eluant and sample medium.

Literature

1. Zaika, L. L., *J. Chromatogr.*, **49**, 222 (1970).

TABLE LC 51
COLUMN GEL CHROMATOGRAPHY OF PHENOLS ON POLYACRYLAMIDES

Packing	P1	P2
Column		
length	97 cm	97 cm
diameter	0.5 cm	0.5 cm
material	G	G
Solvent	S1	S1
Flow rate	1 ml/min.	1 ml/min.
Temperature	rt	rt
Detection	RI	RI
Literature	1	1

Compound[a]	K_d	K_d
Benzene	1.69	—
Phenol	2.20	1.43
Benzoic acid	1.04	—
2-Hydroxyphenol	2.81	—
3-Hydroxyphenol	2.90	—
4-Hydroxyphenol	2.58	1.60
2,3-Dihydroxyphenol	3.41	—
3,5-Dihydroxyphenol	3.50	—
2-Methylphenol	2.42	—
3-Methylphenol	2.44	—
2,4-Dimethylphenol	2.48	—
2,6-Dimethylphenol	2.44	—
2-Carboxyphenol	1.42	—
3-Carboxyphenol	1.40	—
4-Carboxyphenol	1.45	—
2-Chlorophenol	3.04	—
3-Chlorophenol	3.40	—
4-Chlorophenol	3.18	1.77
2,4-Dichlorophenol	4.48	—
2,4,6-Trichlorophenol	4.98	—
2-Nitrophenol	2.94	—
3-Nitrophenol	3.74	—
4-Nitrophenol	3.75	1.94
2,4-Dinitrophenol	2.37	1.76

[a] Samples were 50 μl.

Packing P1 = Bio-Gel, P-2, 100–200 mesh, polyamide gel, exclusion limit 200–2600.
P2 = Bio-Gel P-6, 100–200 mesh, exclusion limit 1000–5000.

Solvent S1 = 0.010*M* NaCl.

Literature 1. Streuli, C. A., *J. Chromatogr.*, **47**, 355 (1970).

TABLE LC 52
SEPHADEX GEL CHROMATOGRAPHY OF MONOSUBSTITUTED PHENOLS

Packing	P1	P1
Column		
length	12.5 cm	12.5 cm
diameter	1.5 cm	1.5 cm
material	G	G
Solvent	S1	S2
Flow rate	na	na
Temperature	rt	rt
Detection	UV	UV
Literature	1	1

Compound[a]	K_d[b]	K_d[b]
Phenol	7.2	0.58
***m*-Cresol**	10.2	0.72
***p*-Cresol**	10.1	0.76
***p*-Ethylphenol**	12.6	0.93
***p*-Hydroxybenzoic acid**	16.0	—
***m*-Hydroxybenzaldehyde**	7.8	1.05
***p*-Hydroxybenzaldehyde**	7.7	1.42
***p*-Methoxyphenol**	8.0	0.60
Resorcinol	12.7	—
Quinol	8.8	—
***m*-Nitrophenol**	16.2	2.96
***p*-Nitrophenol**	16.9	5.41
***m*-Fluorophenol**	11.0	1.10
***p*-Fluorophenol**	9.3	0.99
***m*-Chlorophenol**	25.6	2.38
***p*-Chlorophenol**	22.8	2.85
***m*-Bromophenol**	36.1	3.20
***p*-Bromophenol**	32.5	3.90
***p*-Iodophenol**	57.8	6.00
***m*-Aminophenol**	—	0.49
***p*-Aminophenol**	—	0.43

[a] Samples were 0.5–1.0 mg phenol in 0.5 ml solvent.
[b] V_0 = 7.5 ml, V_i = 8.6 ml for the column used.

Packing P1 = Sephadex G-10.

Solvent S1 = acetate buffer, pH = 4.0, μ = 0.1 (all phenols are unionized in this solvent).
S2 = 0.1*N*NaOH (all phenols are ionized).

Literature 1. Brook, A. J. W., and S. Housley, *J. Chromatogr.*, **41**, 202 (1969).

TABLE LC 53
GEL FILTRATION OF AROMATIC ACIDS AND OTHERS

Packing	P1	P1	P1	P1
Column				
length	35 cm	35 cm	35 cm	35 cm
diameter	3.5 cm	3.5 cm	3.5 cm	3.5 cm
material	G	G	G	G
Solvent	dist. H_2O	0.05 *M* NaCl	S1	S2
Flow rate	F1	F1	F1	F1
Temperature	ambient	ambient	ambient	ambient
Detection	D1	D1	D1	D1
Literature	1	1	1	1

Compound[a]	K_d	K_d	K_d	K_d
Benzoic acid	0.5	—	—	—
Salicylic acid	0.3	1.6	—	0.5
Anthranilic acid	0.6	—	—	—
Sulfanilic acid	0.3	1.1	—	—
Picric acid	0.4	2.5	2.7	0.4
Cinnamic acid	0.3	—	—	—
Phthalic acid	1.1	—	—	—
Phenol	0.7	1.7	—	1.0
Aniline	1.5	—	—	—
Benzyl alcohol	1.3	—	—	1.3
Salicyl alcohol	1.4	—	—	—

[a] 0.5–142 mg samples tested.

Packing P1 = Sephadex G-25 with water regain = 2.9 g water/g dry substance and wet density=1.099, 50–100 mesh. The Sephadex was swelled in 0.05 *M* NaCl for 30 minutes and the fine particles removed by decantation before packing; the column was equilibrated with the solvent before addition of the sample.

Solvent S1 = phosphate solution, $\mu = 0.05$, pH = 7.
S2 = 0.01 *M* NH_4OH, pH = 10.6.

Flow rate F1 = 2 ml/in. with a hydrostatic pressure of 60 cm.

Detection D1 = Acids and bases by titration with dilute NaOH and HCl, respectively; salts by titration after passage through an anion or cation exchanger; amino acids (except tryptophan and tyrosine) with ninhydrin reagent; tryptophan, tyrosine and other aromatic substances by ultraviolet spectrophotometry.

Literature 1. Gelotte, B., *J. Chromatogr.*, **3**, 330 (1960).

TABLE LC 54
GEL PERMEATION CHROMATOGRAPHY OF ALCOHOLS

Packing	P1
Column	
length	L1
diameter	$\frac{3}{8}$ in., O.D.
material	SS
Solvent	benzene
Flow rate	1.0 ml/min.
Temperature	24 °C
Detection	RI
Literature	1

Compound[a]	V_e	Width, ml[b]
Methanol	(92.0)	(5.47)
Ethanol	127.3	4.53
Isopropyl alcohol	117.6	4.52
***n*-Propanol**	(81.3)	(5.47)
Benzyl alcohol	(82.7)	(5.22)
***tert*-Butyl alcohol**	110.7	4.60
***n*-Butanol**	112.7	4.82
3-Heptanol	(68.7)	(4.65)
***n*-Heptanol[c]**	(69.8)	(4.77)
***n*-Heptanol[c]**	101.0	4.40
Diethylene glycol monomethyl ether	97.3	4.88
2,2,4-Trimethylpentanol	98.7	4.70
***n*-Decanol[c]**	93.5	4.70
***n*-Decanol[c]**	82.2	4.51
Ionol	(60.4)	(5.79)
Triphenylcarbinol	(65.8)	(—)
Dipropylene glycol	(68.1)	(7.16)
Tripropylene glycol	(61.8)	(5.36)
Tetraethylene glycol	(86.7)	(4.77)
Allyl alcohol	110.0	5.11
2-Butene-1-ol	103.8	5.30

[a] Sample sizes were typically 0.1 ml of 4% solute in benzene.
[b] Peak widths were obtained by drawing tangents to each side of the curves and reporting the number of ml of eluant at the base of the triangle.
[c] Compound run more than once.

Packing P1 = 40 Å styrenedivinylbenzene gel, permeable to alkanes with MW < 450.

Length L1 = 12 ft, except 8 ft for data in parentheses.

Literature 1. Hendrickson, J. G., *J. Chromatogr.*, **32**, 543 (1968).

TABLE LC 55
COLUMN CHROMATOGRAPHY OF ALKALOID DRUGS

Packing	P1	P1	P1	P1	P1
Column					
length[a]	12 cm	12 cm	12 cm	12 cm	12 cm
diameter	1.2 cm	1.2 cm	1.2 cm	1.2 cm	1.2 cm
material	G	G	G	G	G
Solvent	S1	S2	S3	S4	S5
Flow rate	F1	F1	F1	F1	F1
Temperature	rt	rt	rt	rt	rt
Detection	D1	D1	D1	D1	D1
Literature	1	1	1	1	1

Drug	t_r, min	t_r	t_r	t_r	t_r
2-(6-Carboxyhexylamino)-ethylguanidine	>180	105	—	50	42
Guanethidine N-oxide	>180	116	—	126	77
Guanethidine	>180	—	—	—	114
Cyclizine	—	—	69	—	—
Morphine	—	—	72	—	—
Pethidine	—	—	87	—	—
Strychnine	—	—	110	52	—
Methadone	—	>360	120	93	—
Ephedrine	—	—	123	69	—
Bethanidine	—	122	127	—	—
Amphetamine	—	—	136	94	—

[a] Two 6 cm columns connected in series.

Packing P1 = carboxymethyl cellulose.
Solvent S1 = citrate/phosphate buffer, pH 5.0, Na^+ 0.05*M*.
S2 = citrate/phosphate buffer, pH 7.0, Na^+ 0.05*M*.
S3 = borate buffer, pH 8.5, Na^+ 0.05*M*.
S4 = borate buffer, pH 10.0, Na^+ 0.05*M*.
S5 = borate/chloride buffer, pH 10.0, Na^+ 0.10*M*.
Flow rate F1 = 0.33 ml/min.
Detection D1 = Continuous drug-dye (bromocresol green) complex extraction and colorimetric assay system.
Literature
1. McMartin, C., Simpson, P., and N. Thorpe, *J. Chromatogr.*, **43**, 77 (1969).

TABLE LC 56
GEL FILTRATION OF AMINO ACIDS

Packing	P1	P1
Column		
length	35 cm	35 cm
diameter	3.5 cm	3.5 cm
material	G	G
Solvent	dist. H_2O	0.05*M* NaCl
Flow rate	F1	F1
Temperature	ambient	ambient
Detection	D1	D1
Literature	1	1

Compound[a]	K_d	K_d
Alanine	0.8	—
Serine	0.8	—
Leucine	0.8	—
Methionine	0.8	—
Proline	0.8	—
Hydroxyproline	0.8	—
Phenylalanine	1.0	—
Tyrosine	1.1	—
Tryptophan	1.9	—
Aspartic acid	0.2	0.8
Glutamic acid	0.2	0.8
Lysine hydrochloride	>3.3	1.0
Arginine hydrochloride	>13	1.0
Histidine hydrochloride	>3.6	0.9

[a] 0.5–142 mg samples were tested.

Packing P1 = Sephadex G-25 with water regain = 2.9 g water/g dry substance and wet density = 1.099, 50–100 mesh. The Sephadex was swelled in 0.05*M* NaCl for 30 minutes and the fine particles removed by decantation before packing; the column was equilibrated with the solvent before addition of the sample.
Solvent S1 = Phosphate solution, $\mu = 0.05$, pH = 7.
S2 = 0.01*M* NH_4OH, pH 10.6.
Flow rate F1 = 2 ml/in. with a hydrostatic pressure of 60 cm.
Detection D1 = Amino acids (except tryptophan and tyrosine) with ninhydrin reagent; tryptophan, tyrosine and other aromatic substances by ultraviolet spectrophotometry.
Literature
1. Gelotte, B., *J. Chromatogr.*; **3**, 330 (1960).

TABLE LC 57
GEL PERMEATION CHROMATOGRAPHY OF LOW MOLECULAR WEIGHT COMPOUNDS: CARBOXYLIC ACIDS

Packing	P1	P1	P1
Column			
length	L1	L2	L3
diameter	$\frac{3}{8}$ in, O.D.	$\frac{3}{8}$ in, O.D.	$\frac{3}{8}$ in, O.D.
material	na	na	na
Solvent	S1	S2	S1
Flow rate	1 ml/min	1 ml/min	1 ml/min
Temperature	130 °C	25 °C	130 °C
Detection	RI	RI	RI
Literature	1	1	2

Compound[a]	V_e, cm^3	V_e, cm^3	V_e, cm^3
Acetic acid	202.3	—	194.4
Propionic acid	191.6	—	184.6
***n*-Butyric acid**	186.9	—	178.4
***n*-Valeric acid**	180.7	—	174.7
***n*-Hexanoic acid**	176.6	—	168.9
***n*-Heptanoic acid**	172.0	—	165.8
***n*-Octanoic acid**	169.0	—	160.2
***n*-Nonanic acid**	—	—	157.1
***n*-Decanoic acid**	—	—	154.0
***n*-Undecanoic acid**	—	—	150.8
***n*-Dodecanoic acid**	—	—	148.0
Myristic acid	—	—	143.1
Palmitic acid	—	—	139.6
Stearic acid	—	—	132.8

[a] Samples are 2 ml aliquots of 0.25 to 1.0% solutions.

Packing P1 = A rigid, crosslined polystyrene gel characterized by Waters Associates.

Column L1 = four 4 ft columns in series having exclusion limits of 3×10^3, 250, 60, and 60 Å.
L2 = four 4 ft columns in series having exclusion limit of about 40 Å.
L3 = four 4 ft columns in series having exclusion limits of 3×10^3, 45, 45, and 45Å.

Solvent S1 = *o*-dichlorobenzene.
S2 = tetrahydrofuran.

Literature
1. Cazes, J., and D. R. Gaskill, *Separ. Sci.*, **4**, 15 (1969).
2. Cazes, J., and D. R. Gaskill, *Separ. Sci.*, **2**, 426 (1967).

TABLE LC 58
COLUMN GEL CHROMATOGRAPHY OF CARBOXYLIC ACIDS ON POLYACRYLAMIDES

Packing	P1	P2
Column		
length	97 cm	97 cm
diameter	0.5 cm	0.5 cm
material	G	G
Solvent	S1	S1
Flow rate	1 ml/min	1 ml/min
Temperature	rt	rt
Detection	RI	RI
Literature	1	1

Compound[a]	K_d	K_d
Hydrochloric acid	1.29	—
Trichloroacetic acid	1.17	—
Chloroacetic acid	0.96	—
Acetic acid	1.08	—
Lactic acid	0.79	0.98
Acrylic acid	0.97	—
Crotonic acid	1.28	1.00
Oxalic acid	1.04	—
Succinic acid	1.22	—
Malic acid	1.00	1.03
Tartaric acid	1.08	—
Maleic acid	1.26	1.11
Fumaric acid	0.93	0.96
Citric acid	1.19	1.04
Glycine	0.93	—
4-Aminobenzoic acid	2.75	—

[a] Samples of 50 μl.

Packing P1 = Bio-Gel P-2, 100–200 mesh polyamide gel, exclusion limit, 200–2600.
P2 = Bio-Gel P-6, 100–200 mesh, exclusion limit, 1000–5000.

Solvent S1 = 0.010 *M* NaCl.

Literature
1. Streuli, C. A., *J. Chromatogr.*, **47**, 355 (1970).

TABLE LC 59
CATION EXCHANGE CHROMATOGRAPHY OF UNCOMMON AMINO ACIDS

Packing	P1	P1
Column		
length	150 cm	50 cm
diameter	0.9 cm	0.9 cm
material	G	G
Solvent	S1	S1
Flow rate	0.5 ml/min	0.5 ml/min
Temperature	T1	T2
Detection	D1	D1
Literature	1	1

Compound	V_e[a]	V_e[a]
L-Canavanine	—	370
L-2,4-Diaminobutyric acid	—	105
S-Aminoethylcysteine	—	153
L-Cystine disulfoxide	122,155,190[b]	—
***allo*-Hydroxy-D-proline**	149	—
L-2-Thiolhistidine	168	—
DL-Homocysteic acid	40	—
S-Methylcysteine	247	—
S-Ethylcysteine	306	—
S-*n*-Butyl-L-cysteine	492	—
S-Carbamyl-L-cysteine	192	—
L-Ethionine	483	—
D-Penicillamine	440	—
L-Djenkolic acid	452	—
α,ε-Diaminopimelic acid	447	—
Argininosuccinic acid	551	—
***meso*-Cystine**	412	—
L-Homocystine	629	83
DL + *meso*-Homocystine	629	—

[a] Corrected back for pre-runs of 12 ml (150 cm column) and 40 ml (50 cm column).
[b] Three peaks obtained.

Packing P1 = Amberlite® IR-120, Na^+ form, particles < 400 mesh.
Solvent S1 = start with 0.2*N* sodium citrate buffer, pH 3.25, and change to 0.2*N* sodium citrate, pH 4.25, after 11.67 hours.
S2 = 0.2*N* sodium citrate, pH 4.25.
Temperature T1 = begin at 30°C, change to 50°C after 11.67 hours.
T2 = begin at 30°C and change to 50°C after 14 hours.
Detection D1 = Automatic ninhydrin colorimetry of effluent.
Literature 1. Frimpter, G. W., and A. Bass, *J. Chromatogr.*, 7, 429 (1962).

TABLE LC 60
GEL PERMEATION CHROMATOGRAPHY OF MISCELLANEOUS COMPOUNDS

Packing	P1
Column	
length	L1
diameter	⅜ in., O.D.
material	SS
Solvent	benzene
Flow rate	1.0 ml/min.
Temperature	24 °C
Detection	RI
Literature	1

Compound[a]	V_e	Width, ml[b]
Propylene oxide	119.7	4.41
Phenyl glycidyl ether	106.4	5.07
Butyl glycidyl ether	97.6	4.56
***p,p′*-Bisphenol diglycidyl ether**	82.3	5.69
Tetrahydrofuran	113.4	4.75
Furan	112.9	4.90
Propargyl chloride	108.8	4.90
Phenylacetylene	(79.2)	(5.67)
***n*-Propyl sulfide**	92.6	4.96
Carbon disulfide	134.7	4.72
Diethyl carbonate	102.8	4.23
Propylene carbonate	(76.5)	(5.47)
Nitroethane	(79.8)	(—)
Nitrobenzene[c]	(79.5)	(5.58)
Nitrobenzene[c]	116.0	4.75
Acetonitrile	(84.5)	(5.04)
Propionitrile	115.4	4.78
Hexanonitrile	100.4	4.20
Adiponitrile	102.8	4.23

[a] Sample sizes were typically 0.1 ml of 4% solute in benzene.
[b] Peak widths were obtained by drawing tangents to each side of the curves and reporting the number of ml of eluant at the base of the triangle.
[c] Compound run more than once.

Packing P1 = 40 Å styrenedivinylbenzene gel, permeable to alkanes with MW < 450.
Length L1 = 12 ft, except 8 ft for data in parentheses.
Literature 1. Hendrickson, J. G., *J. Chromatogr.*, **32**, 543 (1968).

TABLE LC 61
GEL PERMEATION CHROMATOGRAPHY OF PHENOLS AND ORGANIC ACIDS

Packing	P1
Column	
length	L1
diameter	$\frac{3}{8}$ in., O.D.
material	SS
Solvent	benzene
Flow rate	1.0 ml/min.
Temperature	24 °C
Detection	RI
Literature	1

Compound[a]	V_e	Width, ml[b]
Phenol	125.00	6.25
***p*-Cresol**	129.3	6.13
***p-tert*-Butylphenol**	101.9	6.16
Bisphenol "A" (*p, p′*)[d]	(76.2)	(12.95)
Bisphenol "A" (*p, p′*)[d]	117.7	8.37
***o,p′*-Bisphenol "A"**	110.0	6.97
2,6-Diisopropylphenol	95.5	—
4,4′,6,6′-Tetramethyl-2,2′-dihydroxybiphenyl	89.9	4.52
4,4′,6,6′-Tetra-*tert*-butyl-2,2′-dihydroxybiphenyl	64.9	6.23
f-Methoxyphenol	139.8	5.60
Acetic acid	131.7	20.5[c]
Caproic acid	115.6	23.4[c]
Heptanoic acid	112.2	18.6[c]
Palmitic acid	80.3	15.1[c]

[a] Sample sizes were typically 0.1 ml of 4% solute in benzene.
[b] Peak widths were obtained by drawing tangents to each side of the curves and reporting the number of ml of eluant at the base of the triangle.
[c] Tailing.
[d] Compound run more than once.

Packing P1 = 40 Å styrenedivinylbenzene gel, permeable to alkanes with MW < 450.
Length L1 = 12 ft, except 8 ft for data in parentheses.
Literature 1. Hendrickson, J. G., *J. Chromatogr.*, **32**, 543 (1968).

TABLE LC 62
COLUMN CHROMATOGRAPHY OF STEROIDS ON SEPHADEX LH-20

Packing	P1
Column	
length	60 cm
diameter	1 cm
material	G
Solvent	S1
Flow rate	F1
Temperature	rt
Detection	na
Literature	1

Compound[a]	V_e[b]
Cholest-5-en-3β-ol (cholesterol)	1.00
Cholest-5-en-3β-yl (cholesteryl) acetate	0.67[c]
Cholesta-3,5-dien-7-one	0.67
3β-Hydroxycholest-5-en-7-one	1.17[c]
Cholest-5-ene-3β,7α-diol	1.50
Cholest-5-ene-3β,7β-diol	1.57
Cholest-5-ene-3β,20α-diol	1.17
Cholest-5-ene-3β,25-diol	1.47
5α-Cholestane-3β,5,6β-triol	5.33
Cholest-5-ene-3β,20α,22R-triol	2.36
3β-Hydroxycholest-5-ene-17α-hydroperoxide	2.80
3β-Hydroxycholest-5-ene-20α-hydroperoxide	2.13
3β-Hydroxycholest-5-ene-25-hydroperoxide	2.90
3β-Hydroxy-5α-cholest-6-ene 5-hydroperoxide	3.60
3β-Hydroxychol-5-en-24-ic acid	>6.0
3β-Hydroxypregn-5-en-20-one	1.17
Pregn-5-ene-3β,20α-diol	1.90
Pregn-5-ene-3β,20β-diol	1.73
Pregn-5-en-3β-ol	1.23
3β-Hydroxyandrost-5-en-17-one	1.23
Androst-5-ene-3β,17β-diol	2.33

[a] Samples were 20–200 mg sterol in 0.5 ml methylene chloride.
[b] V_e relative to cholesterol = 1.00. The absolute elution volume of cholesterol was 30 ml.

Packing P1 = Sephadex LH-20.
Solvent S1 = methylene chloride.
Flow rate F1 = 0.5–1.0 ml/min.
Literature 1. Van Lier, J. E.,and L. L. Smith, *J. Chromatogr.*, **41**, 40 (1969).

TABLE LC 63
ION EXCHANGE CHROMATOGRAPHY OF COMPOUNDS RELATED TO UREA

Packing	P1	P2	P2	P2	P3	P3
Column						
length	15 cm	15 cm	15 cm	15 cm	15 cm	15 cm
diameter	0.8 cm	0.8 cm	0.8 cm	0.8 cm	0.8 cm	0.8 cm
material	G	G	G	G	G	G
Solvent	S1	S2	S3	S4	S1	S2
Flow rate	F1	F1	F1	F1	F1	F1
Temperature	35 °C	35 °C	35 °C	35 °C	35 °C	35 °C
Detection	D1	D1	D1	D1	D1	D1
Literature	1	1	1	1	1	1

Compound	V_e	V_e	V_e	V_e	V_e	V_e
Biuret	15.9	22.2	21.0	—	23.5	23.7
Thiourea	21.5	27.1	26.9	26.9	34.5	34.7
Dicyanodiamide	27.3	33.8	34.3	34.2	34.7	34.5
Nitroguanidine	—	40.7	37.5	—	—	—
Urea	106.7	113.1	108	58	18.4	18.2
Methylurea	136.6	154.1	145.1	—	18.9	18.4
Ethylurea	—	226	203	—	20.2	19.6
***tert*-Butylurea**	—	—	—	—	30.4	29.9

Packing P1 = Amberlite® CG-120, H^+ form, 325–400 mesh.
P2 = Dowex® 50-X8, H^+ form, 200–400 mesh.
P3 = Dowex® 1-X8, Cl^- form, 200–400 mesh.

Solvent S1 = H_2O. S3 = 0.1*N* HCl.
S2 = 0.01*N* HCl. S4 = 1.0*N* HCl.

Flow rate F1 = 0.46 ml/min.

Detection D1 = Automatic thermal reaction energy detection.

Literature
1. Nomura, N., Shiho, D. I., Ohsuga, K., and M. Yamada, *J. Chromatogr.*, **42**, 229 (1969).

TABLE LC 64
GEL CHROMATOGRAPHY OF NAD-GLYCOHYDROLASE AND OTHER PROTEINS

Packing	P1	P2	P2
Column			
length	60 cm	80 cm	80 cm
diameter	3 cm	2 cm	2 cm
material	G	G	G
Solvent	S1	S1	S2
Flow rate	9 ml/hr.[a]	5 ml/hr.[a]	5 ml/hr.[a]
Temperature	6°C	6°C	6°C
Detection	D1	D1	D1
Literature	1	1	1

Compound	V_e	V_e^b/V_0	V_e	V_e^c/V_0	V_e	V_e^d/V_0
γ-Globulin	286	1.46	117	1.13	118	1.13
Bovine serum albumin	320	1.63	134[e]	1.29	140	1.34
NADase I	329	1.68	133	1.27	142	1.36
NADase II	343	1.75	147	1.42	148	1.42
Cytochrome-C	420	2.14	192	1.85	194	1.85
D-Glucose	470	2.40	230	2.21	242	2.30
Ovalbumin	—	—	154	1.48	160	1.53
Myoglobin	—	—	188	1.81	191	1.83

[a] Ascending development.
[b] V_0 = 196.0 ml.
[c] V_0 = 103.7 ml.
[d] V_0 = 104.5 ml.
[e] BSA monomer.

Packing P1 = Sephadex G-200.
P2 = Sephadex G-100.

Solvent S1 = 0.05*M* phosphate buffer, 0.05*M* KCl, pH = 7.35.
S2 = 0.15*M* Tris-HCl buffer, 0.15*M* KCl, pH = 7.35.

Detection D1 = Fractions were analyzed by ultraviolet absorption spectrophotometry. Glucose was determined by the anthrone method (Literature 2), and NADase by the ADH method or the cyanide addition method (Literature 3).

Literature
1. Fehrenbach, F. J., *J. Chromatogr.*, **41**, 43 (1969).
2. Scott, T. A., Sr., and E. H. Melvin, *Anal. Chem.*, **25**, 1656 (1953); E. E. Morse, *Anal. Chem.*, **19**, 1012 (1947).
3. Kaplan, N. O., *J. Biol. Chem.*, **191**, 447 (1951).

TABLE LC 65

ION EXCHANGE CHROMATOGRAPHY OF CARBOXYLIC ACIDS

Packing	P1
Column	
length	12.0 cm
diameter[a]	ca. 2 cm
material	G
Solvent	0.1 *N* HCl
Flow Rate	0.7 cm/min
Temperature	rt
Detection	D1
Literature	1

Acid	C
Lactic	1.45
Tartaric	2.09
Malic	2.57
Citric	3.92
Fumaric	5.84
Maleic	7.90
Formic	2.19
Acetic	1.62
Propionic	3.34
***n*-Butyric**	6.39
Isobutyric	12.4
***n*-Valeric**	15.1
Oxalic	5.70
Malonic	4.86
Succinic	4.67
Glutaric	6.90
Adipic	11.4

[a] Cross-sectional area = 3.90 cm^2.

Packing P1 = Dowex 1-X8, Cl^- form, 200–400 mesh.

Detection D1 = Six ml fractions are mixed with dichromate in 50% H_2SO_4 with subsequent spectrophotometric measurement of the resultant Cr (III).

Literature

1. Sargent, R., and W. Rieman III, *Anal. Chim. Acta*, **18**, 203 (1958).

TABLE LC 66

ION EXCHANGE CHROMATOGRAPHY OF POLYAMINES AND RELATED COMPOUNDS

Packing	P1
Column	
length	7 cm
diameter	0.9 cm
material	na
Solvent	S1
Flow rate	2 ml/min
Temperature	80 °C
Detection	D1
Literature	1

Compound	t_r, min
Arginine	4
Putrescine	25
1,3-Diaminopropane	27
Cadaverine	32
Acetylspermidine B	34
Acetylspermidine A	42
Agmatine	59
Spermidine	64
Iminobispropylamine	64
Acetylspermine	68
Spermine	78

Packing P1 = Bio-Rex 70, minus 400 mesh.

Solvent S1 = 100 ml of 0.438*M* pyridinium acetate, pH 5.7, was passed through the column, followed by 0.5*M* pyridinium acetate, pH 4.4.

Detector D1 = Technicon A.A. detection system with adjustable range expander.

Literature

1. Morris, D. R., Koffron, K. L., and C. J. Okstein, *Anal. Biochem.*, **30**, 451 (1969).

TABLE LC 67

GEL PERMEATION CHROMATOGRAPHY OF RESIN AND FATTY ACID METHYL ESTERS

Packing	P1
Column	
length	200 cm
diameter	1 cm
material	G
Solvent	ethyl ether
Flow rate	16 ml/hr (gravity flow)
Temperature	13 °C
Detection	RI
Literature	1

Compound[a]		V_e, ml
Methyl ester		
Lignocerate	(24:0)	86
Behenate	(22:0)	87
Arachidate	(20:0)	89
Stearate	(18:0)	90
Palmitate	(16:0)	92
Oleate	(18:1)	93
Myristate	(14:0)	94
Linoleate	(18:2)	94
Laurate	(12:0)	96
Linolenate	(18:3)	96
Caprate	(10:0)	97
Nonanoate	(9:0)	98
Tetrahydroabietate (8β,9α,13α-H)		111
Pimarate		112
Tetrahydroabietate (8α,9α,13α-H)		112
Palustrate		113
Isopimarate		115
Abietate		116
Levopimarate		117
Dehydroabietate		119
Neoabietate		126

[a] Samples were <50 mg of total esters in a 100 μl total volume.

Packing P1 = Styragel, 40 Å porosity, particle size 37–70 μ.

Literature 1. Zinkel, D. F., and L. C. Zank, *Anal. Chem.*, **40**, 1145 (1968).

TABLE LC 68

SUPERCRITICAL FLUID CHROMATOGRAPHY OF FATTY ACIDS, LIPIDS, AND MISC. COMPOUNDS

Packing	P1	P2	P3
Column			
length	5 ft	6 ft	6 ft
diameter	4.5 mm, I.D.	4.5 mm, I.D.	4.5 mm, I.D.
material	SS	SS	SS
Solvent	pentane	pentane	pentane
Flow rate	F1	F1	F1
Temperature	200 °C	200 °C	200 °C
Detection	MA	MA	MA
Literature	1	1	1

Compound	r_t, min.	r_t, min.	r_t, min.
Butanoic acid	8.28	—	—
Hexanoic acid	8.85	—	—
Octanoic acid	9.69	—	—
Dodecanoic acid	13.5	—	—
Hexadecanoic acid	15.9	—	—
Octadecanoic acid	16.8	—	—
Lecithin	48.1	—	—
Cephalin	53.7	—	—
Tripalmitin	—	14.4	—
Squalene	—	14.4	—
Glycol-bis-cyanoethylether	—	75.5	—
Ethylene glycol	—	—	15.0
Triethylene glycol	—	—	18.1

Packing P1 = Porapak Q (150–200 mesh).
P2 = Durapak (octane/Porasil C, 120–150 mesh).
P3 = Durapak (Carbowax/Porasil C, 80–100 mesh).

Flow rate F1 = 1.5 ml/min, 680 psi.

Literature 1. Cashaw, J. L., Segura, R., and A. Zlatkis, *J. Chromatogr. Sci.*, **8**, 369 (1970).

TABLE LC 69
ION EXCHANGE CHROMATOGRAPHY OF POLYOLS

Packing	P1
Column	
length	50–55 cm
diameter	0.8 cm
material	G
Solvent	S1
Flow rate	25 ml/hr.
Temperature	35 °C
Detection	D1
Literature	1

Polyol[a]	Peak fraction number[b]
Glycerol	12
Threitol	26
Erythritol	39
Xylitol	46
Arabitol	71
Ribitol	79
Glucitol	101
Galactitol	114
Mannitol	117
Fucitol	99
Rhamnitol	67

[a] Sample dissolved in 1–2 ml of 0.1*M* borate, pH 9–10, buffer.
[b] Fractions = 5 ml.

Packing P1 = De-Acidite FF, borate form, 3–5 cross-linked, passing 200 mesh.
Solvent S1 = 0.18*M* boric acid, adjusted to pH 9 with triethylamine, for 80 fractions, followed by 0.36*M* boric acid also adjusted to pH 9.
Detection D1 = Fractions were oxidized with periodate and the resulting formaldehyde determined by spectrophotometry.
Literature 1. Spencer, N., *J. Chromatogr.*, **30**, 569 (1967).

TABLE LC 70
GEL PERMEATION CHROMATOGRAPHY OF CARBONYLS

Packing	P1
Column	
length	L1
diameter	⅜ in., O.D.
material	SS
Solvent	benzene
Flow rate	1.0 ml/min.
Temperature	24 °C
Detection	RI
Literature	1

Compound[a]	V_e	Width, ml[b]
Acetone[c]	(78.6)	(4.66)
Acetone[c]	115.3	4.34
Acetone[c]	101.5	4.52
***n*-Butyraldehyde**	110.4	4.46
2-Butanone	110.3	4.10
Ethyl acetate	108.6	4.34
Isobutyl methyl ketone	102.2	4.40
Dimethyl terephthalate	100.6	5.30
***n*-Heptaldehyde**	98.1	6.40
Dimethyl adipate	94.2	4.17
Dimethyl sebacate	73.8	4.62

[a] Sample sizes were typically 0.1 ml of 4% solute in benzene.
[b] Peak widths were obtained by drawing tangents to each side of the curves and reporting the number of ml of eluant at the base of the triangle.
[c] Compound run more than once.

Packing P1 = 40 Å styrenedivinylbenzene gel, permeable to alkanes with MW < 450.
Length L1 = 12 ft, except 8 ft for data in parentheses.
Literature 1. Hendrickson, J. G., *J. Chromatogr.*, **32**, 543 (1968).

TABLE LC 71
COLUMN POLYACRYLAMIDE GEL CHROMATOGRAPHY OF SUGARS AND ALCOHOLS

Packing	P1
Column	
length	60 cm
diameter	d1
material	G
Solvent	H_2O
Flow rate	2 ml/hr.
Temperature	25 °C
Detection	RI
Literature	1
Compound[a]	K_d
D-Glucose	0.921
L-Glucose	0.920
Mannose	0.930
Rhamnose	0.879
Glucuronic acid	0.415
Galactose	0.917
Fructose	0.929
Xylose	0.931
Sucrose	0.854
Raffinose	0.750
Stachyose	0.715
Formic acid	0.674
Formamide	0.939
Methanol	0.839
***n*-Propanol**	0.820
Isopropanol	0.777

[a] Samples were 0.1 ml containing 0.1 mg of solute.

Packing P1 = Polyacrylamide P-2, 200–400 mesh, slurred 24 hours with water and de-gassed under high vacuum.

Diameter d1 = Column diameter not given; cross-sectional area = 0.785 cm^2, V_0 = 17.1 ml, V_t = 47.7 ml, V_i = 28.7 ml.

Literature 1. Brown, W., *J. Chromatogr.*, **53**, 572 (1970).

TABLE LC 72
GEL FILTRATION OF AROMATIC AND HETEROCYCLIC ACIDS AND GLYCOLS

Packing	P1		P1
Column			
length	40 cm		40 cm
diameter	1.2 cm		1.2 cm
material	G		G
Solvent	H_2O		S1
Flow rate	35 ml/hr		35 ml/hr
Temperature	rt		rt
Detection	F		F
Literature	1		1
Compound[a]	V_e	K_{av}	V_e
Kynurenic acid	29	0.30	—
5-Hydroxyindoleacetic acid	29	0.30	250[b]
4-Hydroxyphenylacetic acid	30	0.33	—
3-Methoxy-4-hydroxymandelic acid	31	0.37	65
3-Methoxy-4-hydroxyphenylacetic acid	31	0.37	88[b]
3,4-Dihydroxyphenylacetic acid	31	0.37	—
Anthranilic acid	32	0.40	—
Indole-3-acetic acid	32	0.40	203[b]
Xanthurenic acid	32	0.40	—
Vanillic acid	35	0.50	—
Tyrosine	51	1.03	—
3,4-Dihydroxyphenylalanine	58	1.27	59
3-Methoxy-4-hydroxyphenethylglycol	71	1.70	68
3,4-Dihydroxyphenethylglycol	91	2.37	—

[a] 1–100 μg in 0.2 ml solvent.
[b] Tailed.

Packing P1 = Sephadex G-15, 40–120 μ, W_r 1.5 g/g.

Solvent S1 = 0.050 *M* NaCl adjusted to pH 4.0 with dilute HCl.

Literature 1. Demetriou, J. A., Macias R, F. M., McArthur, M. J., and J. M. Beattie, *J. Chromatogr.*, **34**, 342 (1968).

TABLE LC 73
COLUMN GEL CHROMATOGRAPHY OF AMIDES ON POLYACRYLAMIDES

Packing	P1	P2
Column		
length	97 cm	97 cm
diameter	0.5 cm	0.5 cm
material	G	G
Solvent	S1	S1
Flow rate	1 ml/min	1 ml/min
Temperature	rt	rt
Detection	RI	RI
Literature	1	1
Compound[a]	K_d	K_d
Urea	1.37	1.17
Biuret	1.88	1.29
Acetamide	1.06	—
***N*-Methylacetamide**	0.94	—
***N,N*-Dimethylacetamide**	0.86	0.89
Acrylamide	1.17	—
***N*-tert-Butylacrylamide**	0.97	—
***N*-Vinyl-2-pyrrolidone**	1.23	—

[a] Samples of 50 μl.

Packing P1 = Bio-Gel P-2, 100–200 mesh, polyamide gel; exclusion limit 200–2600.
P2 = Bio-Gel P-6, 100–200 mesh, exclusion limit, 1000–5000.

Solvent S1 = 0.010 *M* NaCl.

Literature 1. Streuli, C. A., *J. Chromatogr.*, **47**, 355 (1970).

TABLE LC 74
CHROMATOGRAPHY OF FOOD PRESERVATIVES ON POLYAMIDE COLUMNS

Packing	P1	P1
Column		
length	45 cm	45 cm
diameter	1.7 cm	1.7 cm
material	G	G
Solvent	S1	S2
Flow rate	0.25 ml/min.	0.25 ml/min.
Temperature	rt	rt
Detection	UV	UV
Literature	1	1
Compound[a]	V_e,ml	V_e,ml
Dehydroacetic acid	100	—
Sorbic acid	150	—
Benzoic acid	200	—
***p*-Hydroxybenzoic acid**	275	65
Salicylic acid	450	—
Ethyl *p*-hydroxybenzoate	—	115
***n*-Propyl *p*-hydroxybenzoate**	—	145
***n*-Butyl *p*-hydroxybenzoate**	—	220

[a] 10 μ mole samples dissolved in 5 ml of solvent.

Packing P1 = Woelm polyamide, washed with distilled H_2O, methanol and $CHCl_3$, successively, air dried, and suspended in the elution solvent before packing the column.

Solvent S1 = 350 ml H_2O-acetic acid-methanol (5.5 : 0.5 : 4) followed by the same solvent (3 : 1 : 6 v/v).
S2 = H_2O–NH_3 (99 : 1 w/w).

Literature 1. Nagasawa, K., Yoshidome, H., and R. Takeshita, *J. Chromatogr.*, **43**, 478 (1969).

TABLE LC 75

GEL PERMEATION CHROMATOGRAPHY OF ORGANIC HALIDES

Packing	P1
Column	
length	L1
diameter	$\frac{3}{8}$ in., O.D.
material	SS
Solvent	benzene
Flow rate	1.0 ml/min.
Temperature	24 °C
Detection	RI
Literature	1

Compound[a]	V_e	Width, ml[b]
Methyl iodide	(88.3)	(—)
Ethyl bromide	(81.9)	(5.18)
Chloroform	(82.3)	(5.70)
Carbon tetrachloride	(80.0)	(5.70)
Ethylene dichloride	(80.8)	(5.28)
Methylchloroform	(77.8)	(5.44)
Tetrachloroethane	(78.0)	(—)
Pentachloroethane	(77.0)	(—)
Perchloroethane	(77.6)	(—)
1,4-Dichlorobutane	(74.0)	(—)
Benzene hexachloride	(70.7)	(—)
Chlorobenzene	(80.6)	(5.28)
Bromobenzene	(81.7)	(5.45)
Iodobenzene	(81.7)	(5.80)
p-Dibromobenzene[c]	103.9	4.85
p-Dichlorobenzene[c]	102.0	4.93

[a] Sample sizes were typically 0.1 ml of 4% solute in benzene.
[b] Peak widths were obtained by drawing tangents to each of the curves and reporting the number of ml of eluant at the base of the triangle.
[c] Compound run more than once.

Packing P1 = 40 Å styrenedivinylbenzene gel, permeable to alkanes with MW < 450.
Length L1 = 12 ft, except 8 ft for data in parentheses.
Literature 1. Hendrickson, J. G., *J. Chromatogr.*, **32**, 543 (1968).

TABLE LC 76

COLUMN CHROMATOGRAPHY OF IODOTYROSINES, ON SEPHADEX AND DERIVATIVES

Packing	P1
Column	
length	18 cm
diameter	1.5 cm
material	G
Solvent	S1
Flow rate	F1
Temperature	rt
Detection	UV
Literature	1

Compound[a]	K_d
Tyrosine	0.32
3-Iodotyrosine	0.36
3,5-Diiodotyrosine	0.52
3,5-Diiodobenzaldehyde	3.10
Thyronine	0.52
3-Iodothyronine	0.93
3,5-Diiodothyronine	1.13
3,5,3′-Triiodothyronine	2.35
3,5,3′,5′-Tetraiodothyronine (thyroxine)	5.20
3′,5′-Diiodothyronine	1.95
3,3′,5′-Triiodothyronine	4.40
3,5-Diodothyroacetic acid	1.54
3,5,3′-Triiodothyroacetic acid	2.60

[a] 1.5 mg in 0.5 ml 0.02*N* NaOH.

Packing P1 = Sephadex G-25.
Solvent S1 = 0.02*N* NaOH.
Flow rate F1 = 0.55 ml/min.
Literature 1. Blasi, F. and R. V. DeMasi, *J. Chromatogr.*, **28**, 34 (1967).

TABLE LC 77
GEL FILTRATION OF HETEROCYCLIC AMINES

Packing	P1
Column	
length	40 cm
diameter	1.2 cm
material	G
Solvent	S1
Flow rate	35 ml/hr
Temperature	rt
Detection	F
Literature	1

Compound[a]	V_e	K_{av}
Tyramine	55	1.17
DL-Metanephrine	55	1.17
DL-Octopamine	58	1.27
3-Methoxy-4-hydroxyphenethylamine	60	1.33
DL-Normetanephrine	62	1.40
L-Epinephrine	62	1.40
DL-Norepinephrine	69	1.67
3,4-Dihydroxyphenethylamine	72	1.73
***N,N*-Dimethyltryptamine**	87	2.23
***N,N*-Dimethyl-5-hydroxytryptamine**	113	3.10
Tryptamine	115	3.17
***N*-Acetyl-5-methoxytryptamine**	130	3.67
5-Methoxytryptamine	148	4.27
Serotonin (5-hydroxytryptamine)	157	4.57
***N*-Acetyl-5-hydroxytryptamine**	249	7.63
Indole	328	10.27

[a] 4–100 μg in 0.2 ml solvent.

Packing P1 = Sephadex G-15, 40–120 μ, W_r 1.5 g/g.
Solvent S1 = 0.050*M* NaCl adjusted to pH 4.0 with dilute HCl.
Literature

1. Demetriou, J. A., Macias R. F. M., McArthur, M. J., and J. M. Beattie, *J. Chromatogr.*, **34**, 342 (1968).

TABLE LC 78
GEL PERMEATION CHROMATOGRAPHY OF TRIGLYCERIDES

Packing	P1
Column	
length	160 ft.
diameter	⅜ in.
material	na
Solvent	S1
Flow rate	F1
Temperature	na
Detection	RI
Literature	1

Compound	mol. wt.	V_e,ml	W,ml[a]
Polystyrene	867,000	873	14.50
Triarachidin	976	1132	14.30
Tristearin	892	1147	14.60
Tripalmatin	807	1163	13.35
Trimyristin	723	1183	14.30
Trilaurin	639	1203	13.10
Tricaprin	555	1228	13.55
O-Dichlorobenzene	147	1579	14.85

[a] W = peak width.

Packing P1 = Cross-linked polystyrene (Styragel), 35 μ, 500 Å.
Solvent S1 = Tetrahydrofuran.
Flow rate F1 = 0.4 ml/min., 300 psi.
Literature

1. Bombaugh, K. J., Dark, W. A., and R. F. Levangie, *J. Chromatogr. Sci.*, **7**, 44 (1969).

TABLE LC 79
METAL IONS

Resin AG Dowex® 1-X4, 100–200 mesh, lactate form.

Solvent Water-methanol mixtures containing lactic acid.

Method of determination An aliquot of a solution containing μg amounts of a rare earth mixture was added to an equilibrated column and eluted with the solvent. Elements were determined in effluent fractions by emission spectroscopy.

Distribution coefficients[a]
Aqueous methanol mixtures containing 0.05*M* lactic acid

Water	10%	20%	25%	30%	40%	50%	60%	70%	80%	100%
Element										
Ba	—	—	—	—	—	—	—	—	—	—
Sr	—	—	—	—	—	—	—	—	—	—
Sc		365	222	143	70	39.5	24	15	—	—
Y		636	348	210	88	42	20.5	—	—	—
Lu			940	565	240	113	59	32	12	—
Yb			780	465	195	92	48	25	9.5	—
Tm			650	385	158	74	38.5	19	—	—
Er			530	315	127	59	30	13	—	—
Ho			440	260	108	51	26	9.5	—	—
Dy		636	348	210	88	42	20.5	—	—	—
Tb		510	274	160	64	29.5	10	—	—	—
Gd		350	190	112	46	19	—	—	—	—
Eu		286	155	92	37	14	—	—	—	—
Sm	1130	216	120	72	30.5	10.5	—	—	—	—
Nd	500	115	69	44	19.5	—	—	—	—	—
Pr	370	88	53	34	12.5	—	—	—	—	—
Ce	285	69	41	26	—	—	—	—	—	—
La	172	46	28.5	18	—	—	—	—	—	—

(— = K_d < 10).

25% water-methanol mixtures with varying lactic acid concentration

Lactic acid	0.01*M*	0.02*M*	0.05*M*	0.1*M*	0.2*M*	0.3*M*	0.4*M*
Element							
Sc	150	180	222	250	267	260	240
Y	240	281	348	397	425		
Lu	680	790	940				
Yb	560	650	780	870			
Tm	460	538	650	735			
Er	370	417	530	600			
Ho	307	360	440	500			
Dy	240	281	348	397	425		
Tb	190	225	274	310	330	330	290
Gd	139	160	190	213	230	230	212
Eu	115	132	155	172	185	185	175
Sm	90	103	120	133	140	140	130
Nd	52	59	69	76	81	82	73
Pr	39.5	45	53	58.5	63	64	57.5
Ce	32	35	41	46.5	50	51	47.5
La	25	27	28.5	29.5	30	30	28.5

[a] $K_d = V_e - V_M$/Mass of resin in grams.

Literature Faris, J. P., *J. Chromatogr.*, **32**, 795 (1968).

TABLE LC 80
METAL IONS

Resin Dowex® 1-X10, NO_3^--form, 200–400 mesh.
Solvent Aqueous nitric acid.
Method of determination Column elution and batch equilibration methods.

Distribution coefficients

$[HNO_3]$	0.1*M*	0.2*M*	0.5*M*	1*M*	2*M*	4*M*	6*M*	8*M*	10*M*	12*M*	14*M*
Th(IV)	—	—	—	1.3	6.3	53	220	310	225	130	67
Pa(V)				1	3.2	15	30	41	34	23	12
U(VI)	—	—	—	1.3	3.3	8	13	14.5	11	8	6
Np(IV)			sl	27	120	900	2200	2800	2000	900	300
Np(V,VI)				~2	~4	~10	~12	~15	~20		
Pu(IV)			~30	100	400	2500	7200	7900	4400	1450	350
Zr(IV)	—	—	—	—	—	1.9	2.4	2.2	1.7	sl	—
Hf(IV)	—	—	—	—	—	sl	sl	sl	—	—	—
Mo(VI)	30	15	6	3	sl	—	—	—	—	—	—
Tc(VII)	~3620	1700	600	250	100	27	12	6	3.8	~3	~3
Re(VII)	~ 800	650	285	120	43	14	6	2.5	—	—	—
Rh(III)	—	—	—	—	sl	sl	sl	—	—	—	—
Pd(II)	~13	~24	55	67	62	50	32	19	11	6.5	4
Au(III)				1360	1220	980	700	440	225	92	30
Hg(II)	—	—	1.2	5	9	8.2	6.2	4	2.5	1.3	—
Tl(I)	sl	—	—	—	—	—	—	—	—	sl	sl
Sn(IV)	—	—	—	—	—	—	—	—	—	sl	sl
Pb(II)	—	—	—	1	2.1	1	sl	—	—	—	—
Bi(III)	sl	1	6	13.5	21	26	21	10	3.6	—	—
La(III)	—	—	—	—	sl	4.6	6.1	4.9	2.4	—	—
Ce(III)	—	—	—	—	—	3.7	6.0	3.7	—	—	—
Pr(III)	—	—	—	—	—	2.7	4.0	2.4	—	—	—
Nd(III)	—	—	—	—	—	—	sl	—	—	—	—
Ru	[	not reproducible, sl to ~50						]	sl	—	—
Tl(III)	[	not reproducible, ~5 to ~70									]

— = No adsorption, dist. coeff. < 1.
sl = Slight adsorption.

Elements not retained by resin—Li, Na, K, Rb, Cs, Be, Mg, Ca, Sr, Ba, Ra, Ti(IV), V(V), Cr(III), Mn(II), Fe(III), Co(II), Ni(II), Cu(II), Ag(I), Zn(II), Cd(II), B(III), Al(III), Ga(III), In(III), As(V), Sb(V), Te(IV), Ir(III), Sc(III), Y(III), Pm(III), Sm(III)-Lu(III), Ac, Am(III), Cm(III), Pu(III).

Literature Faris, J. P., and R. F. Buchanan, *Anal. Chem.*, **36**, 1158 (1964).

TABLE LC 81
METAL IONS

Resin Dowex® 1-X8, Br^- form, 100–200 mesh.
Solvent Mixtures of various organic solvents, HBr and water.
Method of determination Batch method employing 20 ml solvent, 5 mg element and 1 g resin.

Weight distribution coefficients

	Acetic acid concentration, %, constant 0.6*M* HBr					
Ion	0	20	40	60	80	90
UO_2(II)	<1	<1	<1	4	150	250
Th(IV)	<1	<1	<1	<1	5	18
Ce(III)	<1	<1	<1	<1	6	8
Mo(VI)	21	10	11	42	150	$>10^3$
V(IV)	<1	26	36	49	71	70
Fe(III)	<1	1	5	450	$>10^3$	500
Al(III)	<1	<1	<1	<1	6	12
In(III)	9	17	44	$>10^3$	$>10^3$	$>10^3$
Bi(III)	$>10^3$	$>10^3$	$>10^3$	$>10^3$	$>10^3$	$>10^3$
Mg(II)	<1	<1	<1	<1	4	7
Ca(II)	<1	<1	<1	<1	2	3
Sr(II)	<1	<1	<1	<1	1	1
Pb(II)	$>10^3$	$>10^3$	$>10^3$	$>10^3$	ppt	ppt
Zn(II)	25	160	900	880	850	850
Cd(II)	$>10^3$	$>10^3$	$>10^3$	$>10^3$	$>10^3$	$>10^3$
Cu(II)	2	50	50	350	$>10^3$	$>10^3$
Mn(II)	<1	<1	<1	13	17	$>10^3$
Co(II)	<1	<1	<1	2	130	$>10^3$
Ni(II)	<1	<1	<1	<1	9	11

	Acetone concentration, %, constant 0.6*M* HBr					
Ion	0	20	40	60	80	90
UO_2(II)	<1	<1	<1	<1	10	35
Th(IV)	<1	<1	<1	<1	13	34
Ce(III)	<1	<1	<1	<1	11	35
Mo(VI)	21	13	14	19	36	<1
V(IV)	<1	<1	<1	1	5	11
Fe(III)	<1	7	12	27	2	<1
Al(III)	<1	<1	<1	<1	<1	5
In(III)	9	19	36	24	<1	<1
Bi(III)	$>10^3$	$>10^3$	$>10^3$	$>10^3$	$>10^3$	900
Mg(II)	<1	<1	<1	<1	6	14
Ca(II)	<1	<1	<1	<1	6	14
Sr(II)	<1	<1	<1	1	9	14
Pb(II)	$>10^3$	$>10^3$	$>10^3$	$>10^3$	$>10^3$	$>10^3$
Zn(II)	25	80	32	150	80	110
Cd(II)	$>10^3$	$>10^3$	$>10^3$	950	240	160
Cu(II)	2	21	18	51	13	8
Mn(II)	<1	<1	<1	<1	6	57
Co(II)	<1	<1	<1	<1	12	100
Ni(II)	<1	<1	<1	<1	7	17

	Ethanol concentration, %, constant 0.6*M* HBr					
Ion	0	20	40	60	80	90
UO_2(II)	<1	<1	<1	<1	1	1.2
Th(IV)	<1	<1	<1	<1	2	3
Ce(III)	<1	<1	<1	<1	3	7
Mo(VI)	21	21	22	24	25	27
V(IV)	<1	<1	<1	<1	1	2
Fe(III)	<1	<1	2	3	20	112
Al(III)	<1	<1	<1	<1	<1	<1

TABLE LC 81 (Continued)
METAL IONS

Ion	Ethanol concentration, %, constant 0.6*M* HCl					
	0	20	40	60	80	90
In(III)	9	19	41	110	10^3	$>10^3$
Bi(III)	$>10^3$	$>10^3$	$>10^3$	$>10^3$	$>10^3$	$>10^3$
Mg(II)	<1	<1	<1	<1	2	2
Ca(II)	<1	<1	<1	<1	3	4
Sr(II)	<1	<1	<1	2	5	7
Pb(II)	$>10^3$	$>10^3$	$>10^3$	$>10^3$	$>10^3$	$>10^3$
Zn(II)	25	100	310	650	660	650
Cd(II)	$>10^3$	$>10^3$	$>10^3$	$>10^3$	$>10^3$	$>10^3$
Cu(II)	2	3	4	11	16	560
Mn(II)	<1	<1	<1	<1	1	3
Co(II)	<1	<1	<1	<1	1	5
Ni(II)	<1	<1	<1	<1	2	3

Ion	Methanol concentration, %, constant 0.6*M* HBr					
	0	20	40	60	80	90
UO_2(II)	<1	<1	<1	<1	<1	6
Th(IV)	<1	<1	<1	<1	<1	<1
Ce(III)	<1	<1	<1	<1	<1	<1
Mo(VI)	21	19	17	9	5	5
V(IV)	<1	<1	<1	<1	<1	<1
Fe(III)	<1	<1	2	5	19	133
Al(III)	<1	<1	<1	<1	<1	<1
In(III)	9	14	25	49	240	$>10^3$
Bi(III)	$>10^3$	$>10^3$	$>10^3$	$>10^3$	$>10^3$	$>10^3$
Mg(II)	<1	<1	<1	<1	<1	<1
Ca(II)	<1	<1	<1	<1	<1	<1
Sr(II)	<1	<1	<1	<1	<1	<1
Pb(II)	$>10^3$	$>10^3$	$>10^3$	$>10^3$	$>10^3$	$>10^3$
Zn(II)	25	57	150	290	290	340
Cd(II)	$>10^3$	$>10^3$	$>10^3$	$>10^3$	$>10^3$	$>10^3$
Cu(II)	2	3	4	8	116	620
Mn(II)	<1	<1	<1	<1	<1	<1
Co(II)	<1	<1	<1	<1	<1	1
Ni(II)	<1	<1	<1	<1	<1	<1

Ion	Methyl glycol concentration, %, constant 0.6*M* HBr					
	0	20	40	60	80	90
UO_2(II)	<1	<1	<1	<1	1	15
Th(IV)	<1	<1	<1	<1	<1	<1
Ce(III)	<1	<1	<1	<1	<1	<1
Mo(VI)	21	25	19	12	5	<1
V(IV)	<1	<1	<1	<1	<1	<1
Fe(III)	<1	1	2	4	13	5
Al(III)	<1	<1	<1	<1	<1	<1
In(III)	9	12	25	38	14	6
Bi(III)	$>10^3$	$>10^3$	$>10^3$	$>10^3$	$>10^3$	$>10^3$
Mg(II)	<1	<1	<1	<1	<1	<1
Ca(II)	<1	<1	<1	<1	<1	<1
Sr(II)	<1	<1	<1	<1	<1	<1
Pb(II)	$>10^3$	$>10^3$	$>10^3$	$>10^3$	$>10^3$	$>10^3$
Zn(II)	25	63	245	407	300	$>10^3$
Cd(II)	$>10^3$	$>10^3$	$>10^3$	$>10^3$	$>10^3$	$>10^3$
Cu(II)	2	60	500	185	75	70
Mn(II)	<1	<1	<1	<1	<1	3
Co(II)	<1	<1	<1	<1	<1	20
Ni(II)	<1	<1	<1	<1	<1	<1

TABLE LC 81 (Continued)
METAL IONS

Ion	*n*-Propanol concentration, %, constant 0.6*M* HBr					
	0	20	40	60	80	90
UO_2(II)	<1	<1	<1	<1	7	13
Th(IV)	<1	<1	<1	2	11	11
Ce(III)	<1	<1	<1	2	14	38
Me(VI)	21	7	8	10	13	21
V(IV)	<1	<1	<1	1	4	6
Fe(III)	<1	<1	<1	<1	<1	75
Al(III)	<1	<1	<1	<1	4	4
In(III)	9	17	32	75	130	600
Bi(III)	>10^3	<10^3	>10^3	>10^3	>10^3	>10^3
Mg(II)	<1	<1	<1	<1	6	8
Ca(II)	<1	<1	<1	1	8	16
Sr(II)	<1	<1	<1	2	15	42
Pb(II)	>10^3	>10^3	>10^3	>10^3	>10^3	>10^3
Zn(II)	25	69	110	163	200	>10^3
Cd(II)	>10^3	>10^3	>10^3	>10^3	>10^3	>10^3
Cu(II)	2	11	15	15	132	274
Mn(II)	<1	<1	<1	<1	5	12
Co(II)	<1	<1	<1	1	6	16
Ni(II)	<1	<1	<1	<1	6	11

Ion	Tetrahydrofuran concentration, %, constant 0.6*M* HBr					
	0	20	40	60	80	90
UO_2(II)	<1	<1	<1	<1	<1	3
Th(IV)	<1	<1	<1	<1	7	10
Ce(III)	<1	<1	<1	<1	8	24
Mo(VI)	21	35	22	12	7	<1
V(IV)	<1	<1	<1	2	3	5
Fe(III)	<1	<1	<1	<1	<1	<1
Al(III)	<1	<1	<1	<1	<1	<1
In(III)	9	19	11	<1	<1	<1
Bi(III)	>10^3	>10^3	>10^3	>10^3	>10^3	500
Mg(II)	<1	<1	<1	<1	3	8
Ca(II)	<1	<1	<1	<1	7	20
Sr(II)	<1	<1	<1	2	14	35
Pb(II)	>10^3	>10^3	>10^3	620	190	190
Zn(II)	25	88	167	55	20	28
Cd(II)	>10^3	>10^3	>10^3	796	200	180
Cu(II)	2	13	5	5	3	3
Mn(II)	<1	<1	<1	<1	2	4
Co(II)	<1	<1	<1	<1	6	26
Ni(II)	<1	<1	<1	<1	3	5

Ion	Final HBr concentration, *M*, constant 90% acetic acid			
	0.15	0.3	0.6	0.9
UO_2(II)	300	270	250	220
Th(IV)	15	21	18	13
Ce(III)	26	18	8	4
Mo(VI)	>10^3	>10^3	>10^3	>10^3
V(IV)	96	84	70	48
Fe(III)	>10^3	>10^3	500	180
Al(III)	20	18	12	8
In(III)	>10^3	>10^3	>10^3	>10^3
Bi(III)	>10^3	>10^3	>10^3	>10^3

Ion	Final HBr concentration, *M*, constant 90% acetone			
	0.15	0.3	0.6	0.9
UO_2(II)	33	34	35	46
Th(IV)	125	82	34	15
Ce(III)	91	73	35	16
Mo(VI)	>10^3	85	<1	<1
V(IV)	21	18	11	8
Fe(III)	<1	<1	<1	<1
Al(III)	22	19	5	8
In(III)	<1	<1	<1	<1
Bi(III)	>10^3	>10^3	>10^3	>10^3

TABLE LC 81 (Continued)
METAL IONS

	Final HBr concentration, *M*, constant 90% acetic acid					Final HBr concentration, *M*, constant 90% acetone			
Ion	0.15	0.3	0.6	0.9	Ion	0.15	0.3	0.6	0.9
Mg(II)	13	10	7	5	**Mg(II)**	17	16	14	8
Ca(II)	6	5	3	1	**Ca(II)**	24	23	14	6
Sr(II)	3	2	1	<1	**Sr(II)**	46	35	14	5
Pb(II)	ppt	ppt	ppt	ppt	**Pb(II)**	$>10^3$	$>10^3$	$>10^3$	$>10^3$
Zn(II)	850	850	850	870	**Zn(II)**	110	120	110	200
Cd(II)	$>10^3$	$>10^3$	$>10^3$	$>10^3$	**Cd(II)**	270	400	160	141
Cu(II)	$>10^3$	$>10^3$	$>10^3$	$>10^3$	**Cu(II)**	8	8	8	7
Mn(II)	$>10^3$	$>10^3$	$>10^3$	$>10^3$	**Mn(II)**	22	27	57	61
Co(II)	$>10^3$	$>10^3$	$>10^3$	$>10^3$	**Co(II)**	45	67	100	115
Ni(II)	20	15	11	11	**Ni(II)**	29	23	17	12

	Final HBr concentration, *M*, constant 90% Methanol					Final HBr concentration, *M*, constant 90% Methanol			
Ion	0.15	0.3	0.6	0.9	Ion	0.15	0.3	0.6	0.9
UO_2(II)	<1	<1	1.2	1.4	**UO_2(II)**	<1	<1	5.7	8.2
Th(IV)	<1	2	3	4	**Th(IV)**	<1	<1	<1	<1
Ce(III)	3	5	7	8	**Ce(III)**	<1	<1	<1	<1
Mo(VI)	$>10^3$	630	27	45	**Mo(VI)**	$>10^3$	11	5	16
V(IV)	<1	1	2	2	**V(IV)**	<1	<1	<1	<1
Fe(III)	35	81	112	124	**Fe(III)**	24	37	133	300
Al(III)	<1	<1	<1	<1	**Al(III)**	<1	<1	<1	<1
In(III)	$>10^3$	$>10^3$	$>10^3$	$>10^3$	**In(III)**	320	$>10^3$	$>10^3$	$>10^3$
Bi(III)	10^3	10^3	$>10^3$	$>10^3$	**Bi(III)**	$>10^3$	$>10^3$	$>10^3$	$>10^3$
Mg(II)	<1	<1	2	3	**Mg(II)**	<1	<1	<1	<1
Ca(II)	<1	2	3	4	**Ca(II)**	<1	<1	<1	<1
Sr(II)	3	4	7	10	**Sr(II)**	<1	<1	<1	<1
Pb(II)	$>10^3$	$>10^3$	$>10^3$	$>10^3$	**Pb(II)**	$>10^3$	$>10^3$	$>10^3$	$>10^3$
Zn(II)	620	620	650	650	**Zn(II)**	300	300	340	190
Cd(II)	$>10^3$	$>10^3$	$>10^3$	$>10^3$	**Cd(II)**	$>10^3$	$>10^3$	$>10^3$	$>10^3$
Cu(II)	380	450	560	$>10^3$	**Cu(II)**	320	600	620	$>10^3$
Mn(II)	<1	1	3	5	**Mn(II)**	<1	<1	<1	<1
Co(II)	1	2	5	16	**Co(II)**	<1	<1	<1	<1
Ni(II)	1	2	3	3	**Ni(II)**	<1	<1	<1	<1

	Final concentration, *M*, constant 90% methyl glycol					Final HBr concentration, *M*, in pure H_2O			
Ion	0.15	0.3	0.6	0.9	Ion	0.15	0.3	0.6	0.9
UO_2(II)	4	8	15	120	**UO_2(II)**	<1	<1	<1	<1
Th(IV)	<1	<1	<1	<1	**Th(IV)**	<1	<1	<1	<1
Ce(III)	<1	<1	<1	2	**Ce(III)**	<1	<1	<1	<1
Mo(VI)	7	2	<1	2	**Mo(VI)**	54	47	21	14
V(IV)	<1	<1	<1	<1	**V(IV)**	<1	<1	<1	<1
Fe(III)	12	9	5	5	**Fe(III)**	<1	<1	<1	<1
Al(III)	<1	<1	<1	<1	**Al(III)**	<1	<1	<1	<1
In(III)	16	9	6	4	**In(III)**	2	4	9	17
Bi(III)	$>10^3$	$>10^3$	$>10^3$	$>10^3$	**Bi(III)**	$>10^3$	$>10^3$	$>10^3$	$>10^3$
Mg(II)	<1	<1	<1	<1	**Mg(II)**	<1	<1	<1	<1
Ca(II)	<1	<1	<1	<1	**Ca(II)**	<1	<1	<1	<1
Sr(II)	<1	<1	<1	<1	**Sr(II)**	<1	<1	<1	<1
Pb(II)	$>10^3$	$>10^3$	$>10^3$	$>10^3$	**Pb(II)**	$>10^3$	$>10^3$	$>10^3$	$>10^3$
Zn(II)	$>10^3$	$>10^3$	$>10^3$	830	**Zn(II)**	3	8	25	80
Cd(II)	$>10^3$	$>10^3$	$>10^3$	$>10^3$	**Cd(II)**	$>10^3$	$>10^3$	$>10^3$	$>10^3$
Cu(II)	140	90	70	60	**Cu(II)**	<1	5	2	2
Mn(II)	<1	1	3	10	**Mn(II)**	<1	<1	<1	<1
Co(II)	3	6	20	68	**Co(II)**	<1	<1	<1	<1
Ni(II)	<1	<1	<1	1	**Ni(II)**	<1	<1	<1	<1

TABLE LC 81 (Continued)
METAL IONS

Ion	Final HBr concentration, *M*, constant 90% *n*-propanol			
	0.15	0.3	0.6	0.9
UO_2(II)	6	10	13	14
Th(IV)	9	10	11	12
Ce(III)	26	35	38	39
Mo(VI)	38	25	21	41
V(IV)	5	6	6	7
Fe(III)	84	77	75	60
Al(III)	4	4	4	3
In(III)	>10^3	>10^3	600	260
Bi(III)	>10^3	>10^3	>10^3	160
Mg(II)	6	7	8	9
Ca(II)	10	11	16	16
Sr(II)	25	33	42	43
Pb(II)	>10^3	>10^3	>10^3	>10^3
Zn(II)	>10^3	390	>10^3	407
Cd(II)	>10^3	>10^3	>10^3	>10^3
Cu(II)	340	234	274	320
Mn(II)	8	9	12	15
Co(II)	10	12	16	34
Ni(II)	8	10	11	11

Ion	Final HBr concentration, *M*, constant 90% THF			
	0.15	0.3	0.6	0.9
UO_2(II)	1	1.3	3	8
Th(IV)	42	30	10	6
Ce(III)	42	42	24	16
Mo(VI)	47	10	<1	<1
V(IV)	31	8	5	3
Fe(III)	<1	<1	<1	<1
Al(III)	<1	<1	<1	<1
In(III)	<1	<1	<1	<1
Bi(III)	>10^3	>10^3	500	150
Mg(II)	12	11	8	5
Ca(II)	24	23	20	15
Sr(II)	66	64	35	22
Pb(II)	190	190	190	190
Zn(II)	28	31	28	37
Cd(II)	295	226	180	143
Cu(II)	3	3	3	3
Mn(II)	10	10	4	17
Co(II)	17	17	26	56
Ni(II)	15	13	5	2

Literature Klakl, E. and J. Korkisch, *Talanta*, **16**, 1177 (1969).

TABLE LC 82
METAL IONS

Resin Dowex® 50–X8, H^+ form, 100–200 mesh.
Solvent Different solvents composed of 0–90% organic solvent, 10% HCl, and water up to 100%.
Method of determination Batch method employing 20 ml solvent, 5 mg of element and 1 g of resin. The mixture was equilibrated for 12 hours, the resin filtered off, and the element determined in the filtrate.

Weight distribution coefficients

Ion	Methanol concentration, %, constant 0.6*M* HCl					
	0	20	40	60	80	90
UO_2(II)	40	45	67	144	207	85
Th(IV)	>10^3	>10^3	>10^3	>10^3	>10^3	>10^3
Zr(IV)	>10^3	>10^3	>10^3	>10^3	>10^3	>10^3
Ce(III)	>10^3	>10^3	>10^3	>10^3	>10^3	>10^3
Mo(VI)	~2	~2	~2	~2	~2	~2
V(V)	5	5	6	~5	~4	~4
Fe(III)	84	96	89	57	34	6
Al(III)	186	300	>10^3	>10^3	>10^3	>10^3
Ga(III)	242	324	637	>10^3	242	2.4
In(III)	1	1	<1	<1	<1	<1
Bi(III)	<1	<1	<1	<1	<1	<1
Mg(II)	33	40	50	109	188	288
Ca(II)	74	106	192	447	>10^3	>10^3
Pb(II)	~1	9	12	18	35	50
Zn(II)	39	35	21	5	<1	<1
Cd(II)	6	6	5	4	2	2
Cu(II)	32	34	41	53	31	6.6
Mn(II)	80	100	140	180	190	200
Co(II)	44	50	65	161	197	209
Ni(II)	46	49	72	102	168	250

TABLE LC 82 (Continued)
METAL IONS

Ion	*n*-Propanol concentration, %, constant 0.6*M* HCl					
	0	20	40	60	80	90
UO_2(II)	40	61	41	102	159	266
Th(IV)	>10^3	>10^3	>10^3	>10^3	>10^3	>10^3
Zr(IV)	>10^3	>10^3	>10^3	>10^3	>10^3	>10^3
Ce(III)	>10^3	>10^3	>10^3	>10^3	>10^3	>10^3
Mo(VI)	~2	~2	~2	~2	~2	~2
V(V)	5	~10	~30	~40	~10	~5
Fe(III)	84	100	100	81	54	<1
Al(III)	186	315	680	>10	>10^3	>10^3
Ga(III)	242	247	786	>10^3	769	11
In(III)	1	<1	<1	<1	<1	<1
Bi(III)	<1	<1	<1	<1	2.3	22.7
Mg(II)	33	37	66	143	390	522
Ca(II)	74	115	175	400	>10^3	>10^3
Pb(II)	~1	5.3	7	21	70	110
Zn(II)	39	26	25	5	<1	<1
Cd(II)	6	6	6	6	6	6
Cu(II)	32	52	52	63	72	22
Mn(II)	80	118	147	189	241	248
Co(II)	44	55	86	143	376	376
Ni(II)	46	55	92	153	520	10^3

Ion	Isopropanol concentration, %, constant 0.6*M* HCl					
	0	20	40	60	80	90
UO_2(II)	40	60	120	224	400	500
Th(IV)	>10^3	>10^3	>10^3	>10^3	>10^3	>10^3
Zr(IV)	>10^3	>10^3	>10^3	>10^3	>10^3	>10^3
Ce(III)	>10^3	>10^3	>10^3	>10^3	>10^3	>10^3
Mo(VI)	~2	~2	~2	~2	~2	~2
V(V)	5	~5	~5	~5	~5	~5
Fe(III)	84	166	166	166	166	166
Al(III)	186	>10^3	>10^3	>10^3	>10^3	>10^3
Ga(III)	242	>10^3	>10^3	>10^3	3	<1
In(III)	1	<1	<1	<1	<1	<1
Bi(III)	<1	<1	<1	<1	<1	<1
Mg(II)	33	35	38	100	153	260
Ca(II)	74	68	75	182	405	>10^3
Pb(II)	~1	20	20	25	38	16
Zn(II)	39	38	37	38	36	33
Cd(II)	6	4	<1	<1	<1	<1
Cu(II)	32	40	42	44	46	53
Mn(II)	80	214	342	703	>10^3	>10^3
Co(II)	44	52	44	40	52	47
Ni(II)	46	55	55	55	62	76

Ion	Methyl glycol concentration, %, constant 0.6*M* HCl					
	0	20	40	60	80	90
UO_2(II)	40	45	51	56	61	2
Th(IV)	>10^3	>10^3	>10^3	>10^3	>10^3	>10^3
Zr(IV)	>10^3	>10^3	>10^3	>10^3	>10^3	>10^3
Ce(III)	>10^3	>10^3	>10^3	>10^3	>10^3	>10^3
Mo(VI)	~2	~2	~2	~2	~2	~2
V(V)	5	~5	~5	~5	~5	~5
Fe(III)	84	148	146	96	4	4
Al(III)	186	>10^3	>10^3	>10^3	>10^3	>10^3
Ga(III)	242	323	>10^3	>10^3	7	7
In(III)	1	1	<1	<1	<1	<1
Bi(III)	<1	104	137	118	176	370
Mg(II)	33	47	78	151	254	292

TABLE LC 82 (Continued)
METAL IONS

Ion	Methyl glycol concentration, %, constant 0.6M HCl					
	0	20	40	60	80	90
Ca(II)	74	144	183	491	10^3	$>10^3$
Pb(II)	~1	25	23	10	<1	<1
Zn(II)	39	36	10	<1	<1	<1
Cd(II)	6	<1	<1	<1	<1	<1
Cu(II)	32	48	60	58	8	<1
Mn(II)	80	227	304	475	567	215
Co(II)	44	187	329	1066	414	80
Ni(II)	46	51	107	196	208	138

Ion	Acetic acid concentration, %, constant 0.6M HCl					
	0	20	40	60	80	90
UO_2(II)	40	42	54	65	22	14
Th(IV)	$>10^3$	$>10^3$	$>10^3$	$>10^3$	$>10^3$	$>10^3$
Zr(IV)	$>10^3$	$>10^3$	$>10^3$	$>10^3$	$>10^3$	$>10^3$
Ce(III)	$>10^3$	$>10^3$	$>10^3$	$>10^3$	$>10^3$	$>10^3$
Mo(VI)	~2	~2	~2	~2	~2	~2
V(V)	5	~5	~5	~5	~5	~5
Fe(III)	84	89	78	36	2	1.5
Al(III)	186	$>10^3$	$>10^3$	$>10^3$	$>10^3$	$>10^3$
Ga(III)	242	308	324	180	<1	<1
In(III)	1	<1	<1	<1	<1	<1
Bi(III)	<1	<1	23	31	2.3	3.5
Mg(II)	33	57	81	102	174	184
Ca(II)	74	118	202	236	$>10^3$	$>10^3$
Pb(II)	~1	9	8	38	96	367
Zn(II)	39	31	7	<1	<1	<1
Cd(II)	6	<1	<1	<1	<1	<1
Cu(II)	32	49	42	17	2.6	1.5
Mn(II)	80	216	115	105	71	23
Co(II)	44	63	63	96	100	120
Ni(II)	46	$>10^3$	$>10^3$	$>10^3$	$>10^3$	$>10^3$

Ion	Tetrahydrofuran concentration, %, constant 0.6M HCl					
	0	20	40	60	80	90
UO_2(II)	40	77	76	77	11	<1
Th(IV)	$>10^3$	$>10^3$	$>10^3$	$>10^3$	$>10^3$	$>10^3$
Zr(IV)	$>10^3$	$>10^3$	$>10^3$	$>10^3$	$>10^3$	$>10^3$
Ce(III)	$>10^3$	$>10^3$	$>10^3$	$>10^3$	$>10^3$	$>10^3$
Mo(VI)	~2	~2	~2	~2	~2	~2
V(V)	5	~5	~5	~5	4	3
Fe(III)	84	166	134	17	~1	~1
Al(III)	186	$>10^3$	$>10^3$	$>10^3$	$>10^3$	$>10^3$
Ga(III)	242	268	704	39	8	~1
In(III)	1	<1	<1	<1	<1	<1
Bi(III)	<1	<1	<1	<1	<1	<1
Mg(II)	33	38	60	109	159	179
Ca(II)	74	188	207	281	422	655
Pb(II)	~1	16	10	4	<1	<1
Zn(II)	39	24	10	<1	<1	<1
Cd(II)	6	<1	<1	<1	<1	<1
Cu(II)	32	34	41	29	<1	<1
Mn(II)	80	184	274	292	355	176
Co(II)	44	52	69	138	108	6
Ni(II)	46	52	95	172	227	113

TABLE 82 LC (Continued)
METAL IONS

Ion	Final HCl concentration, M, constant 90% methanol				
	0.15	0.3	0.6	0.9	1.2
UO_2(II)	>10^3	880	85	9	<1
Th(IV)	>10^3	>10^3	>10^3	>10^3	>10^3
Zr(IV)	>10^3	>10^3	>10^3	>10^3	>10^3
Ce(III)	>10^3	>10^3	>10^3	>10^3	>10^3
Mo(VI)	>2	>2	~2	~2	~2
V(V)	>5	>5	5	~5	~5
Fe(III)	>10^3	123	6	<1	<1
Al(III)	—	>10^3	10^3	>10^3	>10^3
Ga(III)	>100	>10	2.4	<1	<1
In(III)	~1	~1	<1	<1	<1
Bi(III)	<1	147	<1	193	<1
Mg(II)	>10^3	843	288	78	66
Ca(II)	>10^3	>10^3	>10^3	>10^3	10^3
Pb(II)	—	73	50	25	13
Zn(II)	22	<1	<1	<1	<1
Cd(II)	5	2	2	2	<1
Cu(II)	300	58	6.6	<1	<1
Mn(II)	>10^3	703	200	44	25
Co(II)	>10^3	561	209	52	29

Ion	Final HCl concentration, M, constant 90% *n*-propanol				
	0.15	0.3	0.6	0.9	1.2
UO_2(II)	530	330	266	15	<1
Th(IV)	>10^3	>10^3	>10^3	>10^3	>10^3
Zr(IV)	>10^3	>10^3	>10^3	>10^3	>10^3
Ce(III)	>10^3	>10^3	>10^3	>10^3	>10^3
Mo(VI)	>2	2	~2	~1	<1
V(V)	>5	~5	~5	~3	<1
Fe(III)	>10^3	89	<1	<1	<1
Al(III)	—	>10^3	>10^3	>10^3	>10^3
Ga(III)	17	15	11	<10	>1
In(III)	<1	<1	<1	<1	<1
Bi(III)	<1	147	22	176	<1
Pb(II)	<1	147	22.7	17.6	<1
Mg(II)	>10^3	>10^3	522	462	340
Ca(II)	>10^3	>10^3	>10^3	>10^3	>10^3
Pb(II)	—	125	110	96	82
Zn(II)	<1	<1	<1	<1	<1
Cd(II)	6	6	6	2	<1
Cu(II)	518	118	22	<1	<1
Mn(II)	>10^3	>10^3	114	100	80
Co(II)	>10^3	601	376	147	19
Ni(II)	>10^3	>10^3	10^3	313	166

Ion	Final HCl concentration, M, constant 90% isopropanol				
	0.15	0.3	0.6	0.9	1.2
UO_2(II)	>10^3	>10^3	500	70	<1
Th(IV)	>10^3	>10^3	>10^3	>10^3	>10^3
Zr(IV)	>10^3	>10^3	>10^3	>10^3	>10^3
Ce(III)	>10^3	>10^3	>10^3	>10^3	>10^3
Mo(VI)	~2	~2	~2	~2	~2
V(V)	~5	~5	~5	~5	~5
Fe(III)	>10^3	170	166	~1	<1
Al(III)	—	>10^3	>10^3	>10^3	>10^3
Ga(III)	~1	<1	<1	<1	<1
In(III)	<1	<1	<1	<1	<1
Bi(III)	<1	<1	<1	<1	<1
Mg(II)	>10^3	>10^3	260	194	84

TABLE LC 82 (Continued)
METAL IONS

Ion	Final HCl concentration, *M*, constant 90% isopropanol				
	0.15	0.3	0.6	0.9	1.2
Ca(II)	$>10^3$	$>10^3$	$>10^3$	$>10^3$	$>10^3$
Pb(II)	—	191	116	90	80
Zn(II)	39	38	33	30	25
Cd(II)	<1	<1	<1	<1	<1
Cu(II)	300	128	53	5	<1
Mn(II)	$>10^3$	$>10^3$	$>10^3$	274	180
Co(II)	$>10^3$	851	47	20	12
Ni(II)	$>10^3$	$>10^3$	76	60	52

Ion	Final HCl concentration, *M*, constant 90% methyl glycol				
	0.15	0.3	0.6	0.9	1.2
UO_2(II)	370	200	2	<1	<1
Th(IV)	$>10^3$	$>10^3$	$>10^3$	$>10^3$	$>10^3$
Zr(IV)	$>10^3$	$>10^3$	$>10^3$	$>10^3$	$>10^3$
Ce(III)	$>10^3$	$>10^3$	$>10^3$	$>10^3$	$>10^3$
Mo(VI)	~2	~2	~2	~2	~2
V(V)	~5	~5	~5	~5	~5
Fe(III)	6	5	4	2.6	<1
Al(III)	$>10^3$	$>10^3$	$>10^3$	$>10^3$	$>10^3$
Ga(III)	~10	~8	7	~1	<1
In(III)	<1	<1	<1	<1	<1
Bi(III)	<1	136	370	193	<1
Mg(II)	813	292	292	116	50
Ca(II)	$>10^3$	$>10^3$	$>10^3$	$>10^3$	$>10^3$
Pb(II)	—	2	<1	<1	<1
Zn(II)	<1	<1	<1	<1	<1
Cd(II)	2	~1	<1	<1	<1
Cu(II)	44	5	<1	<1	<1
Mn(II)	703	365	215	12	3
Co(II)	$>10^3$	$>10^3$	414	16	3
Ni(II)	$>10^3$	523	138	39	27

Ion	Final HCl concentration, *M*, constant 90% acetic acid				
	0.15	0.3	0.6	0.9	1.2
UO_2(II)	140	40	14	<1	<1
Th(IV)	$>10^3$	$>10^3$	$>10^3$	$>10^3$	$>10^3$
Zr(IV)	$>10^3$	$>10^3$	$>10^3$	$>10^3$	$>10^3$
Ce(III)	$>10^3$	$>10^3$	$>10^3$	$>10^3$	$>10^3$
Mo(VI)	~2	~2	~2	~2	~2
V(V)	~5	~5	~5	~5	~5
Fe(III)	3	2	1.5	<1	<1
Al(III)	$>10^3$	$>10^3$	$>10^3$	$>10^3$	$>10^3$
Ga(III)	<1	<1	<1	<1	<1
In(III)	<1	<1	<1	<1	<1
Bi(III)	<1	147	~3	214	<1
Mg(II)	$>10^3$	548	184	135	68
Ca(II)	$>10^3$	$>10^3$	$>10^3$	$>10^3$	$>10^3$
Pb(II)	—	765	367	90	70
Zn(II)	<1	<1	<1	<1	<1
Cd(II)	<1	<1	<1	<1	<1
Cu(II)	4	3	1.4	<1	<1
Mn(II)	241	114	23	6	<1
Co(II)	$>10^3$	143	120	12	<1
Ni(II)	$>10^3$	$>10^3$	$>10^3$	25	20

TABLE LC 82 (Continued)
METAL IONS

Ion	Final HCl concentration, *M*, constant 90% tetrahydrofuran				
	0.15	0.3	0.6	0.9	1.2
UO_2(II)	42	<1	<1	<1	<1
Th(IV)	$>10^3$	$>10^3$	$>10^3$	$>10^3$	$>10^3$
Zr(IV)	$>10^3$	$>10^3$	$>10^3$	$>10^3$	$>10^3$
Ce(III)	$>10^3$	$>10^3$	$>10^3$	$>10^3$	$>10^3$
Mo(VI)	~2	~2	~2	~2	~2
V(V)	~3	~3	3	~3	~3
Fe(III)	~1	~1	~1	<1	<1
Al(III)	$>10^3$	$>10^3$	$>10^3$	$>10^3$	$>10^3$
Ga(III)	>1	>1	~1	<1	<1
In(III)	<1	<1	<1	<1	<1
Bi(III)	<1	<1	<1	<1	<1
Mg(II)	$>10^3$	522	179	57	41
Ca(II)	$>10^3$	$>10^3$	655	490	324
Pb(II)	—	<1	<1	<1	<1
Zn(II)	<1	<1	<1	<1	<1
Cd(II)	<1	<1	<1	<1	<1
Cu(II)	5	<1	<1	<1	<1
Mn(II)	700	189	176	10	<1
Co(II)	$>10^3$	$>10^3$	6	<1	<1
Ni(II)	$>10^3$	313	113	17	10

Literature Korkisch, J. and S. S. Ahluwalia, *Talanta*, **14**, 155 (1967).

TABLE LC 83
METAL IONS

Resin Bio Rad AG 50W–X8, H^+ form, 100–200 and 200–400 mesh.
Solvent Hydrochloric acid–ethanol mixtures.
Method of determination Batch method employing 250 ml solution, 2.50 g resin and 5 meq of element.

Element	WEIGHT DISTRIBUTION COEFFICIENTS Percentage ethanol, constant 0.10*M* HCl						
	0	20	40	60	80	90	95
Ga(III)	$\sim 10^4$	$>10^4$	$>10^4$	$>10^4$	$>10^4$	6730	153
Sn(IV)	$\sim 10^4$	ppt.	ppt.	ppt.	43.1	3.5	1.1
Fe(III)	9000	$>10^4$	$>10^4$	$>10^4$	$>10^4$	3340	176
U(VI)	758	1300	3200	$>10^4$	$>10^4$	4960	3330
Mn(II)	1360	1580	3020	$>10^4$	$>10^4$	$>10^4$	$>10^4$
Co(II)	1270	1450	2840	6500	$>10^4$	$>10^4$	$>10^4$
Ni(II)	1230	1470	2950	10^4	$>10^4$	$>10^4$	$>10^4$
Fe(II)	1220	1410	2830	6400	$>10^4$	9800	1010
Zn(II)	1030	1200	2150	2250	876	47.5	5.7
Cu(II)	1010	1200	2190	4410	4280	1250	430
Mg(II)	860	990	$>10^3$	$>10^3$	$>10^3$	$>10^4$	$>10^4$
In(III)	806	210	193	153	83	29.2	6.3
Cd(II)	410	367	332	317	123	81	20.8
Mo(VI)	10.9	35.0	43.0	39.8	36.1	38.8	40.9
Rh(III)	4.2	3.1	2.2	1.5	0.9	0.6	<0.5
Tl(III)	2.1	2.4	2.7	1.9	2.2	2.6	2.4
Hg(II)	1.6	1.1	0.8	0.6	<0.5	<0.5	<0.5
Pd(II)	1.6	0.8	1.0	0.8	0.9	0.7	0.5
Pt(IV)	1.4	1.5	1.6	2.1	2.7	3.1	3.4
As(V)	1.4	~2	~5	~3	<1	<1	<1
Ir(IV)	1.4	1.5	1.4	1.3	1.0	1.1	0.9
W(VI)[a]	1.1	1.3	1.7	1.8	8.1	3.0	—
Se(IV)	1.1	1.3	1.0	1.5	1.7	1.0	0.8

TABLE LC 83 (Continued)
METAL IONS

Element	Percentage ethanol, constant 0.10*M* HCl						
	0	20	40	60	80	90	95
Au(III)	0.8	1.1	0.9	1.2	1.3	0.8	1.0
Ge(III)	0.5	0.6	0.4	0.5	1.7	1.8	1.7
Mo(VI)[a]	<0.5	<0.5	<0.5	<0.5	<0.5	<0.5	<0.5

[a] H_2O_2 present.

Element	Percentage ethanol, constant 0.20*M* HCl						
	0	20	40	60	80	90	95
Fe(III)	3400	3600	5100	6000	1410	31.7	15.7
Ga(III)	3040	8680	$>10^4$	$>10^4$	3860	446	6.9
Mn(II)	510	690	1240	2280	5280	5410	4260
Co(II)	460	625	1120	2160	5010	$>10^4$	9330
Ni(II)	450	610	1110	2140	4960	8380	6320
Fe(II)	430	580	1020	2010	4370	1460	249
Cu(II)	380	424	706	1060	1040	308	69
Zn(II)	361	368	489	299	48.3	5.2	1.7
Mg(II)	350	361	369	1510	1990	6400	5100
U(VI)	252	284	469	847	1270	1430	1133
In(III)	110	53	48.8	38.1	17.9	6.0	2.9
Cd(II)	84	118	88	34.8	13.7	2.7	0.6
Sn(IV)	45	3400	1610	18.1	3.6	0.7	0.5
V(V)	7.0	19.7	39.1	84.0	202	ppt.	ppt.
V(V)[a]	6.5	6.6	9.0	57	488	529	336
Mo(VI)	4.5	17.7	25.9	25.7	23.1	12.6	7.4
Mo(VI)[a]	<0.5	<0.5	<0.5	<0.5	<0.5	<0.5	<0.5

[a] H_2O_2 present.

Element	Percentage ethanol, constant 0.50*M* HCl					
	0	20	40	60	80	90
Ga(III)	260	633	1470	4650	428	3.5
Fe(III)	225	226	304	361	153	6.6
Mn(II)	84	90	162	285	608	730
Mg(II)	74	78	126	234	601	1140
Cr(III)	73	93	118	170	279	421
Co(II)	72	93	159	305	671	1830
Ni(II)	70	96	165	315	689	880
Fe(II)	66	69	131	252	408	136
Cu(II)	65	88	119	176	195	35.9
U(VI)	58	67	111	182	264	220
Zn(II)	64	64	48.3	17.8	4.1	1.5
Cs(I)	44.2	63	108	229	852	—
V(IV)	44.1	53	84	157	286	508
Be(II)	42.3	47.3	69	114	170	—
Ti(IV)	39.1	121	265	634	1700	ppt.
Rb(I)	33.2	42.8	73	165	571	—
K(I)	29.1	47.3	89	201	838	—
Na(I)	13.5	19.1	34.1	79	254	—
Li(I)	8.1	10.8	17.1	28.8	50	—
In(III)	7.6	7.8	8.5	7.2	4.2	2.4
Cd(II)	6.5	6.0	4.6	1.4	<0.5	<0.5
Sn(IV)	6.2	5.9	3.6	2.2	1.3	0.6
V(V)	5.0	19.5	42.6	86	143	142
V(V)[a]	2.1	4.7	23.9	59	129	140
Mo(VI)	<0.5	7.6	11.0	12.8	11.0	7.3
Mo(VI)[a]	<0.5	<0.5	<0.5	<0.5	<0.5	<0.5
Bi(III)	<0.5	<0.5	<0.5	<0.5	<0.5	<0.5

[a] H_2O_2 present.

TABLE LC 83 (Continued)
METAL IONS

Element	Percentage ethanol, constant 1.0*M* HCl				
	0	20	40	60	80
Gd(III)	183	259	498	1460	~ 10^4
Yb(III)	153	178	331	841	4340
Ba(II)	128	193	615	1920	5140
Sc(III)	120	345	925	3680	>10^4
Al(III)	61	74	124	251	502
Sr(II)	60	76	162	543	3490
Ga(III)	42.6	102	204	347	5.8
Ca(II)	41.3	55	106	310	1570
Fe(III)	33.5	31.9	39.4	47.3	6.9
Cr(III)	26.7	35.9	53	83	114
Ni(II)	21.9	19.7	41.8	82	182
Co(II)	21.3	27.6	38.2	78	175
Mn(II)	20.2	28.2	42.4	78	157
Mg(II)	20.1	19.3	33.0	71	166
Fe(II)	19.8	25.8	46.9	81	184
U(VI)	19.2	22.1	34.2	58	70
Cs(I)	19.1	27.0	44.1	96	358
Cu(II)	17.5	17.8	21.9	23.2	24.7
Zn(II)	16.0	9.7	5.3	3.5	2.4
Rb(I)	15.4	21.4	36.8	81	326
K(I)	13.9	20.0	39.3	94	609
Be(II)	13.3	13.5	19.0	31.6	50.5
Ti(IV)	11.9	18.2	57	181	496
V(IV)	7.2	15.4	22.6	38.8	76
Na(I)	6.9	9.7	16.6	36.5	138
Li(I)	3.8	5.1	7.8	13.1	28.8
In(III)	1.8	1.8	1.9	1.3	0.7
Sn(IV)	1.6	1.4	0.8	<0.5	<0.5
Cd(II)	1.6	0.5	0.2	<0.2	<0.2
V(V)	1.1	4.7	10.3	23.1	43.0
Mo(VI)	0.8	4.2	6.9	7.3	5.8
Bi(III)	0.8	0.6	<0.5	<0.5	<0.5
Hg(II)	<0.5	<0.5	<0.5	<0.5	<0.5
Ge(IV)	<0.5	<0.5	<0.5	<0.5	<0.5

Element	Percentage ethanol, constant 2.0*M* HCl					
	0	20	40	60	70	80
Zr(IV)	489	925	2720	6400	>10^4	>10^4
Th(IV)	239	298	672	3170	>10^4	>10^4
La(III)	48.1	68	143	474	1200	~10^4
Gd(III)	36.2	49.8	116	313	610	1280
Ba(II)	36.0	74	197	951	3260	—
Y(III)	29.7	32.4	72	196	490	1800
Sc(III)	28.8	57	140	470	1140	—
Yb(III)	27.4	36.2	59	151	298	—
Sr(II)	17.8	22.3	49.3	192	542	—
Al(III)	12.5	12.7	16.9	44.8	69	118
Ca(II)	12.2	16.3	31.7	90	202	—
Cs(I)	10.4	10.4	17.1	32.6	60	—
Rb(I)	8.1	10.0	16.4	39.6	77	—
Cr(III)	7.9	7.5	12.3	27.2	33.3	—
Ga(III)	7.8	13.5	20.9	6.3	1.5	< 0.5
K(I)	7.4	10.0	21.0	54	118	—
Ni(II)	7.2	7.4	11.7	23.1	31.5	47.3
U(VI)	7.0	7.2	10.4	15.6	16.1	15.4
Co(II)	6.7	8.9	15.4	30.4	44.7	—
Mg(II)	6.2	7.2	12.1	23.8	36.2	55
Mn(II)	6.0	7.7	10.6	18.4	26.2	—
Be(II)	5.2	5.6	7.0	10.8	14.9	—

TABLE LC 83 (Continued)
METAL IONS

Element	Percentage ethanol, constant 2.0*M* HCl					
	0	20	40	60	70	80
Fe(III)	5.2	5.0	6.2	5.0	3.1	0.6
V(IV)	5.0	5.8	7.6	13.3	16.4	17.8
Cu(II)	4.2	4.2	4.8	5.1	2.8	0.9
Fe(II)	4.1	4.3	6.9	8.5	6.3	1.8
Na(I)	3.8	5.5	8.9	24.5	57	—
Ti(IV)	3.7	3.8	9.8	41.9	88	123
V(V)	<0.5	2.9	5.2	10.3	13.4	—
In(III), Sn(IV) Bi(III), Hg(II) Ge(IV), Mo(VI)[a]	<0.5	<0.5	<0.5	<0.5	<0.5	<0.5

[a] H_2O_2 present.

Element	Percentage ethanol, constant 3.0*M* HCl				
	0	20	40	60	70
Th(IV)	114	142	363	2360	—
Zr(IV)	61	152	343	2800	ppt.
La(III)	18.8	32.6	72	239	948
Ba(II)	18.5	42.6	143	1340	—
Gd(III)	15.3	22.8	51	178	495
Sc(III)	14.9	30.9	76	281	—
Y(III)	13.6	17.7	32.6	128	288
Yb(III)	12.2	15.2	25.2	76	—
Sr(III)	10.0	13.1	29.8	162	—
Ca(II)	7.3	8.4	17.2	69	—
Cs(I)	5.9	5.7	9.3	19.9	—
Rb(I)	5.3	6.5	10.6	28.7	—
K(I)	4.9	7.1	14.3	40.4	—
Cr(III)	4.8	3.9	4.9	11.4	—
Al(III)	4.7	4.3	7.3	13.2	—
Mg(II)	4.3	3.7	5.9	12.3	—
Co(II)	4.2	3.8	6.2	11.8	—
Mn(II)	3.9	3.5	5.5	8.6	—
Fe(III)	3.6	1.8	2.4	2.2	—
V(IV)	3.5	4.3	5.6	7.4	—
U(VI)	3.5	3.5	5.5	6.3	—
Fe(II)	2.9	3.0	4.1	4.3	—
Na(I)	2.7	3.8	6.8	20.5	—
Ti(IV)	2.4	3.1	5.3	22.1(6.8[a])	—
Ni(II)	2.0	2.1	4.5	6.7	—

[a] H_2O_2 present.

Literature

1. Strelow, F. W. E., Van Zyl, C. R., and C. J. C. Bothma, *Anal. Chim. Acta*, **45**, 81 (1969).

TABLE LC 84
METAL IONS

Resin AG1–X8, 100–200 and 200–400 mesh, sulfate form.
Solvent Aqueous H_2SO_4 solutions.
Method of determination Batch procedure employing 1mmole of element, 250 ml solvent and 2.500 g resin.

Weight distribution coefficients[b,c]

$[H_2SO_4]$	0.01N	0.03N	0.10N	0.20N	0.50N	1.00N	2.00N	3.00N	4.00N
Cr(VI)	25000	18000	12000	7800	4400	2100	800	435	302
Mo(VI)	60000	527	533	671	484	232	52	13.7	4.6
W(VI)[a]	—	—	528	457	337	222	127	96	110
Mo(VI)[a]	—	—	2560	1400	451	197	74	43.3	33.0
Ir(IV)	1010	690	450	310	220	180	160	160	170
Ir(III)	625	525	388	270	218	160	118	92	75
Ta(V)[a]	—	—	1860	1070	310	138	50	11.4	3.9
Zr(IV)	hydrolyzes	$>10^3$	1350	704	211	47.3	11.0	5.2	2.9
U(VI)	1160	1130	521	248	91	26.6	9.3	4.8	2.9
Hf(IV)	hydrolyzes	$>10^3$	4700	701	57	12.0	3.2	1.9	1.2
Th(IV)	116	82	34.6	21.4	8.3	3.7	2.0	1.1	0.6
Sc(III)	64	44.5	21.5	10.9	4.8	2.6	1.5	0.9	0.5
V(V)[a]	370	102	45.4	10.9	4.6	2.5	2.1	1.9	1.9
Fe(III)	54	39.9	15.6	9.1	3.6	1.4	0.9	0.6	<0.5
Nb(V)[a]	—	—	120	96	3.4	<0.5	<0.5	<0.5	<0.5
Bi(III)	—	—	17.7	4.7	2.1	0.9	0.5	<0.5	<0.5
Rh(III)	39.0	30.0	12.8	5.4	0.9	<0.5	<0.5	<0.5	<0.5
V(V)	1410	320	6.5	3.3	1.6	0.7	<0.5	<0.5	<0.5
In(III)	7.4	5.1	2.4	0.8	<0.5	<0.5	<0.5	<0.5	<0.5
Cr(III)	5.1	3.4	2.1	0.7	0.5	<0.5	<0.5	<0.5	<0.5
Se(IV)	8.1	5.3	1.1	<0.5	<0.5	<0.5	<0.5	<0.5	<0.5
V(IV)	3.4	1.7	0.9	<0.5	<0.5	<0.5	<0.5	<0.5	<0.5
As(III)	2.4	1.5	0.9	0.6	<0.5	<0.5	<0.5	<0.5	<0.5
As(V)	1.3	0.8	0.6	<0.5	<0.5	<0.5	<0.5	<0.5	<0.5
Ga(III)	1.2	0.8	0.6	<0.5	<0.5	<0.5	<0.5	<0.5	<0.5
Yb(III)	1.2	0.8	0.6	<0.5	<0.5	<0.5	<0.5	<0.5	<0.5
Ti(IV)[a]	hydrolyzes	hydrolyzes	hydrolyzes	0.5	<0.5	<0.5	<0.5	<0.5	<0.5

[a] H_2O_2 present.
[b] D values calculated on basis of resin weight in chloride form.
[c] The following elements were found to have distribution coefficients $D < 0.5$ in 0.01, 1.00, and 4.00N H_2SO_4 and are not included in the Table: Cd(II), Mn(II), Zn(II), Cu(II), Co(II), Ni(II), Be(II), Mg(II), Li(I), Na(I), K(I), Rb(I), Cs(I), and Tl(I). The distribution coefficients of Gd(III), Y(III), Ce(III), and La(III) are similar to those of Yb(III) and are not included in the Table. The following elements form sulfate precipitates or insoluble oxycompounds at low acid concentrations: Ca(II), Sr(II), Ba(II), Ra(II), Sn(IV), Sb(III), and Sb(V). Pd(II) and Hg(II) show considerable absorption at low sulfuric acid concentration, but the absorption partly seems to be due to a slow precipitation reaction within the resin particles. Solutions of Ag(I) sulfate are not stable in the presence of the resin and slowly form small amounts of precipitate. Halide-free solutions of Au(III) and Pt(IV) also are unstable.

Literature Strelow, F. W. E., and C. J. C. Bothma, *Anal. Chem.*, **39**, 595 (1967).

TABLE LC 85
METAL IONS

Resin AG 50W–X8, H^+ form, 100–200 mesh.
Solvent Aqueous solutions of HNO_3 or H_2SO_4.
Method of determination Batch procedure employing 5.00 meq of element, 2.50 g dry resin, and 250 ml solvent. This was reduced to 1.00 meq, 0.500 g dry resin and 50 ml solvent for rare earths and platinum metals.

Weight distribution coefficients

K_d Values in HNO_3

[HNO_3]	0.1*N*	0.2*N*	0.5*N*	1.0*N*	2.0*N*	3.0*N*	4.0*N*
Zr(IV)	$>10^4$	$>10^4$	$>10^4$	6500	652	112	30.7
Hf(IV)	$>10^4$	$>10^4$	$>10^4$	2400	166	61	20.8
Th(IV)	$>10^4$	$>10^4$	$>10^4$	1180	123	43.0	24.8
La(III)	$>10^4$	$>10^4$	1870	267	47.3	17.1	9.1
Ce(III)	$>10^4$	$>10^4$	1840	246	44.2	15.4	8.2
Yb(III)	$>10^4$	$>10^4$	1150	193	41.3	16.0	9.0
Er(III)	$>10^4$	$>10^4$	1100	182	38.2	14.9	8.0
Y(III)	$>10^4$	$>10^4$	1020	174	35.8	13.9	10.0
Sm(III)	$>10^4$	$>10^4$	1000	168	29.8	10.9	7.2
Gd(III)	$>10^4$	$>10^4$	1000	167	29.2	10.8	6.9
In(III)	$>10^4$	$>10^4$	680	118	23.0	10.1	5.8
Sc(III)	$>10^4$	3300	500	116	23.3	11.6	7.6
Cr(III)	5100	1620	418	112	27.8	19.2	10.9
Hg(I)	$>10^4$	7600	640	94	33.5	19.2	13.6
Ga(III)	$>10^4$	4200	445	94	20.0	9.0	5.8
Al(III)	$>10^4$	3900	392	79	16.5	8.0	5.4
Fe(III)	$>10^4$	4100	362	74	14.3	6.2	3.1
Ba(II)	5000	1560	271	68	13.0	6.0	3.6
Sr(II)	3100	775	146	39.2	8.8	6.1	4.7
Pb(II)	$>10^4$	1420	183	35.7	8.5	5.5	4.5
Ca(II)	1450	480	113	35.3	9.7	4.3	1.8
Cd(II)	1500	392	91	32.8	10.8	6.8	3.4
Co(II)	1260	392	91	28.8	10.1	6.1	4.7
Mn(II)	1240	389	89	28.4	11.4	7.1	3.0
Ni(II)	1140	384	91	28.1	10.3	8.6	7.3
Cu(II)	1080	356	84	26.8	8.6	4.8	3.1
Zn(II)	1020	352	83	25.2	7.5	4.6	3.6
Bi(III)	893	305	79	25.0	7.9	3.7	3.0
U(VI)	659	262	69	24.4	10.7	7.4	6.6
Mg(II)	794	295	71	22.9	9.1	5.8	4.1
Tl(I)	173	91	41.0	22.3	9.9	5.8	3.3
Ag(I)	156	86	36.0	18.1	7.9	5.4	4.0
Hg(II)	4700	1090	121	16.9	5.9	3.9	2.8
Cs(I)	148	81	34.8	16.8	7.6	4.7	3.4
Be(II)	553	183	52	14.8	6.6	4.5	3.1
Ti(IV)	1410	461	71	14.6	6.5	4.5	3.4
V(IV)	495	157	35.6	14.0	4.7	3.0	2.5
Rb(I)	118	68	29.1	13.4	6.6	4.1	2.9
K(I)	99	59	26.2	11.4	5.7	3.5	2.6
Te(IV)[a]	40.3	19.7	8.5	5.0	2.4	0.6	0.2
Pd(II)	97	62	23.5	9.1	3.4	2.7	2.5
Rh(III)	78	44.7	19.5	7.8	4.1	2.1	1.0
Na(I)	54	29.4	12.7	6.3	3.4	2.0	1.3
Li(I)	33.1	18.6	8.0	3.9	2.6	1.7	1.1
V(V)	20.0	10.9	4.9	2.0	1.2	0.8	0.5
Mo(VI)	Ppt.	5.2	2.9	1.6	1.0	0.8	0.6
Nb(V)	11.6	6.3	0.9	0.2	0.1	0.1	0.1
Se(IV)	<0.5	<0.5	<0.5	<0.5	<0.5	<0.5	<0.5
As(III)	<0.1	<0.1	<0.1	<0.1	<0.1	<0.1	<0.1

[a] Determined by taking 2.50 meq element.

TABLE LC 86
METAL IONS

Resin AG 50W–X8, H^+ form.
Solvent Acetone-hydrochloric acid solutions.
Method of determination Batch procedure.

Weight distribution coefficients

Constant 0.10*M* HCl with various amounts of acetone

Element	0%	20%	40%	60%	80%	90%	95%
Ga	$>10^4$	$>10^4$	$>10^4$	$>10^4$	1610	6.6	5.0
Fe(III)	$>10^4$	$>10^4$	$>10^4$	$>10^4$	28.6	2.4	0.5
Ca	1940	2890	5020	$>10^4$	$>10^4$	$>10^4$	$>10^4$
Mn(II)	1090	1670	3190	5700	6070	1490	67
Co(II)	1080	1500	2800	6500	$>10^4$	720	15
Zn	1010	939	848	380	7.6	2.4	1.2
Cu(II)	990	1510	2450	3210	578	16.7	0.9
Fe(II)	904	1440	3740	$\sim 10^4$	985	40.2	2.4
Mg	826	1100	2180	4450	5900	5900	5900
In(III)	815	795	452	106	2.6	0.5	<0.5
U(VI)	758	1090	2010	3280	824	53	3.9
Cr(III)	678	845	1480	2810	2920	1350	463
V(IV)	604	846	1920	3970	4270	859	178
Tl(I)	386	402	625	prec.	prec.	prec.	prec.
Cd	370	348	325	93	2.8	2.1	1.3
V(V)[a,b]	99	120	170	274	288	65	11.8
Te(IV)	39.3	127	236	prec.	prec.	prec.	prec.
Nb(V)[a]	27.2	30.8	prec.	prec.	prec.	prec.	prec.
V(V)[a,c]	15.0	5.4	5.3	5.3	11.7	11.2	10.1
Rh(III)	4.2	4.9	4.8	5.1	5.4	5.0	4.7
W(VI)[a]	1.7	1.6	1.8	1.9	1.7	1.6	1.3
Pd(II)	1.6	2.0	1.8	1.9	1.7	1.6	1.5
Hg(II)	1.6	1.7	1.4	1.2	1.1	0.9	0.9
Pt(IV)	1.4	1.6	1.4	1.7	1.6	1.4	1.3
As(V)	1.4	2.3	3.0	4.0	5.4	7.2	9.7
As(III)	1.4	1.4	1.5	1.4	1.7	1.4	1.6
Se(IV)	1.4	prec.	prec.	54	40.0	34.8	—
Sb(III)	prec.	prec.	prec.	prec.	prec.	5.9	13.7
Sn(IV)	prec.	prec.	prec.	1.6	1.1	1.3	1.0
Bi(III)	prec.	prec.	prec.	prec.	1.6	1.3	0.8
Mo(VI)[a]	0.9	0.8	0.7	0.8	0.6	0.7	0.6
Tl(III)	0.8	0.6	0.5	0.4	0.3	0.3	0.3
Au(III)	0.3	0.2	0.3	0.3	0.3	0.3	0.2

K_d Values in H_2SO_4

$[H_2SO_4]$	0.1*N*	0.2*N*	0.5*N*	1.0*N*	2.0*N*	3.0*N*	4.0*N*
La(III)	$>10^4$	$>10^4$	1860	329	68	24.3	12.1
Ce(III)	$>10^4$	$>10^4$	1800	318	66	23.8	11.8
Sm(III)	$>10^4$	$>10^4$	1460	269	56	20.1	10.0
Y(III)	$>10^4$	$>10^4$	1380	253	49.9	18.0	9.4
Yb(III)	$>10^4$	$>10^4$	1330	249	48.1	17.3	8.8
Gd(III)	$>10^4$	$>10^4$	1390	246	46.6	17.9	8.9
Er(III)	$>10^4$	$>10^4$	1300	242	48.6	16.7	8.5
Bi(III)[a]	$>10^4$	$>10^4$	6800	235	32.3	11.3	6.4
Ga(III)	$>10^4$	3500	618	137	26.7	10.0	4.9
Al(III)	$>10^4$	8300	540	126	27.9	10.6	4.7
Hg(II)	7900	1790	321	103	34.7	16.8	12.2
In(III)	$>10^4$	3190	376	87	17.2	6.5	3.8
Mn(II)	1590	610	165	59	17.4	8.9	5.5

TABLE LC 86 (Continued)
METAL IONS

K_d Values in H_2SO_4

(H_2SO_4)	0.1*N*	0.2*N*	0.5*N*	1.0*N*	2.0*N*	3.0*N*	4.0*N*
Fe(III)	$>10^4$	2050	255	58	13.5	4.6	1.8
Cr(III)	198	176	126	55	18.7	0.9	0.2
Th(IV)	$>10^4$	3900	263	52	9.0	3.0	1.8
Tl(I)	452	236	97	49.7	20.6	11.6	8.7
Tl(III)	6500	1490	205	47.4	12.0	7.2	5.2
V(IV)	1230	490	140	46.6	11.5	2.4	0.4
Ni(II)	1390	590	140	46.0	16.5	6.1	2.8
Fe(II)	1600	560	139	46.0	15.3	9.8	6.6
Cd(II)	1420	540	144	45.6	14.8	6.6	4.3
Zn(II)	1570	550	135	43.2	12.2	4.9	4.0
Co(II)	1170	433	126	42.9	14.2	6.2	5.4
Cu(II)	1310	505	128	41.5	13.2	5.7	3.7
Mg(II)	1300	484	124	41.5	13.0	5.6	3.4
Sc(III)	5600	1050	141	34.9	8.5	4.4	3.4
Be(II)	840	305	79	27.0	8.2	3.9	2.6
Cs(I)	175	108	52	24.7	9.1	4.8	3.5
Rb(I)	148	91	43.8	21.3	8.3	4.4	3.1
K(I)	138	86	41.1	19.4	7.4	3.7	2.9
Rh(III)	80	49.3	28.5	16.2	4.5	2.2	1.3
Pd(II)	109	71	32.5	13.9	6.0	3.8	2.7
Hf(IV)	2690	1240	160	12.1	1.7	1.0	0.7
U(VI)	596	118	29.2	9.6	3.2	2.3	1.8
Ti(IV)	395	225	45.8	9.0	2.5	1.0	0.4
Na(I)	81	47.7	20.1	8.9	3.7	2.6	1.7
Li(I)	48.0	28.2	11.7	5.8	3.0	1.6	1.1
Te(IV)	ppt.	30.8	9.8	5.2	2.6	0.6	0.3
Zr(IV)	546	474	98	4.6	1.4	1.2	1.0
V(V)	27.1	15.2	6.7	2.8	1.2	0.7	0.4
Nb(V)	14.2	7.4	4.0	1.9	0.7	0.5	0.3
Mo(VI)	ppt.	5.3	2.8	1.2	0.5	0.3	0.2
Se(IV)	<0.5	<0.5	<0.5	<0.5	<0.5	<0.5	<0.5
As(III)	<0.1	<0.1	<0.1	<0.1	<0.1	<0.1	<0.1

[a] Determined by taking 0.075 mmole element.

Literature Strelow, F. W. E., Rethemeyer, R., and C. J. C. Bothma, *Anal. Chem.*, **37**, 106 (1965).

Constant 0.2*M* HCl with various amounts of acetone

Element	0%	20%	40%	60%	80%	90%	95%
La	9120	$>10^4$	$>10^4$	$>10^4$	$>10^4$	$>10^4$	$>10^4$
Ga	3040	$>10^4$	$>10^4$	$>10^4$	22.9	5.3	4.2
Fe(III)	2700	3300	4650	2130	3.0	1.5	0.9
Sr	810	1037	1838	4210	$>10^4$	$>10^4$	$>10^4$
Ca	631	852	1340	2880	7300	$>10^4$	$>10^4$
Mn(II)	342	474	875	1550	1510	287	10.9
Co(II)	340	456	860	1480	1830	100	2.3
Ni(II)	340	423	798	1720	1780	1430	1260
Zn	336	396	384	49.8	2.7	2.1	1.4
Fe(II)	328	467	788	1840	320	3.2	1.2
Cu(II)	317	406	638	708	103	3.9	0.9
Mg	281	356	620	1190	1430	1600	1600
U(VI)	252	346	571	796	372	7.8	3.4
Cr(III)	242	354	498	742	750	420	219
V(IV)	197	244	398	756	770	248	58
Tl(I)	129	164	238	prec.	prec.	prec.	prec.
Be	117	307	621	1740	1860	1630	—

TABLE LC 86 (Continued)
METAL IONS

Constant 0.2*M* HCl with various amounts of acetone

Element	0%	20%	40%	60%	80%	90%	95%
In(III)	110	77	42.4	12.4	0.7	<0.5	<0.5
Sn(IV)	99	6070	10.9	0.8	0.6	<0.5	<0.5
Cs	98	131	183	308	678	1128	—
Cd	84	75	56	7.1	1.7	0.5	<0.5
Rb	71	95	162	334	763	1225	—
K	64	85	154	349	805	1380	—
V(V)[a,c]	10.6	9.7	8.5	5.5	17.8	14.1	4.9
Na	29.9	42.2	76	160	333	495	—
Li	17.1	24.5	38.0	68	127	190	—
Te(IV)	17.0	28.1	54.2	51.2	3.7	2.4	2.1
Nb(V)[a]	10.8	12.6	prec.	prec.	prec.	prec.	prec.
V(V)[a,b]	61	69	94	135	90	24.1	4.9
W(VI)[a]	1.8	2.0	1.9	1.7	1.8	1.6	1.3
Bi(III)	prec.	prec.	prec.	1.0	1.0	0.8	0.8
Mo(VI)[a]	0.8	0.9	1.3	1.1	0.9	0.8	0.6
Tl(III)	0.2	0.2	0.2	0.2	0.2	0.2	0.2
Au(III)	0.2	0.2	0.2	0.2	0.2	0.2	0.2

Constant 0.5*M* HCl with various amounts of acetone

Element	0%	20%	40%	60%	80%	90%	95%
La	1880	2810	6100	~10^4	>10^4	>10^4	>10^4
Ba	431	674	1411	3950	prec.	prec.	prec.
Al	278	464	751	1120	1180	1250	1310
Ga	245	1110	1770	168	7.2	1.9	1.1
Fe(III)	205	215	240	43.2	1.6	1.3	1.0
Sr	188	248	498	1050	2720	6180	—
Ca	131	178	283	632	1310	2140	—
Cr(III)	102	128	186	260	214	150	—
Mn(II)	74	92	151	243	233	14.3	2.8
Co(II)	72	89	145	256	127	4.0	1.2
Fe(II)	71	88	149	243	12.4	1.9	1.2
Ni(II)	70	85	149	257	304	249	—
Mg	67	71	112	173	381	399	432
Cu(II)	64	77	104	92	6.0	0.8	0.6
Zn	64	75	29.9	3.7	1.5	1.3	1.1
Pb(II)	62	71	27.4	14.1	7.6	4.7	2.9
U(VI)	58	77	117	138	15.1	1.4	0.9
V(IV)	42.7	52	79	121	106	38.2	—
Cs	42.4	51	77	142	299	358	—
Be	42.1	56	86	132	164	143	—
Ti(IV)	39.1	108	206	538	814	877	457
Tl(I)	37.7	51	43.9	prec.	prec.	prec.	—
Rb	32.2	44.0	76	153	352	490	—
K	29.5	41.1	73	176	388	632	—
V(V)[a,b]	25.1	28.3	37.3	62	23.9	6.1	—
Na	13.9	20.1	33.2	76	170	291	—
Li	7.6	10.7	16.4	30.4	58	91	—
In(III)	7.6	4.7	2.7	1.4	0.6	<0.5	<0.5
Cd	6.5	4.6	1.4	0.6	<0.5	<0.5	<0.5
Sn(IV)	6.3	3.6	2.0	1.3	0.7	<0.5	<0.5
Te(IV)	4.7	13.3	10.4	6.7	2.3	0.9	0.6
Nb(V)[a]	3.8	5.4	prec.	prec.	prec.	prec.	prec.
V(V)[a,c]	3.8	3.4	2.6	3.8	5.9	3.7	—
Bi(III), **Au(III)**, **Mo(VI)**[a], **Tl(III)**	<0.5	<0.5	<0.5	<0.5	<0.5	<0.5	<0.5

TABLE LC 86 (Continued)
METAL IONS

Constant 1.00M HCl with various amounts of acetone						
Element	0%	20%	40%	60%	80%	90%
Zr	9200	>10⁴	>10⁴	>10⁴	>10⁴	6300
Th	2220	3370	9500	>10⁴	>10⁴	>10⁴
Hf	3360	7350	>10⁴	>10⁴	>10⁴	6700
La	277	387	960	2340	>10⁴	>10⁴
Ce(III)	246	353	818	2080	5800	>10⁴
Y	175	238	448	1390	2810	—
Yb	157	224	454	1080	2060	—
Sc	231	575	1100	3270	975	—
Ba	127	191	470	940	prec.	prec.
Al	61	76	120	204	308	420
Sr	60	82	154	408	1640	—
Ga	42.6	115	165	5.5	3.3	1.1
Ca	40.8	46.7	102	207	663	—
Cr(III)	32.7	42.3	63	97	94	42.1
Fe(III)	29.2	30.1	35.0	10.1	1.4	0.6
Ni(II)	21.9	25.0	40.4	68	78	51
Co(II)	21.3	26.4	43.4	72	5.7	1.3
Mn(II)	20.2	25.7	40.0	66	28.2	2.4
Mg	20.2	23.4	34.8	58	106	191
Fe(II)	19.4	24.0	37.9	48.0	0.7	<0.5

Constant 2.00M HCl with various amounts of acetone						
Element	0%	20%	40%	60%	70%	80%
Zr	521	1120	3070	8700	659	3.1
Th	230	335	703	4900	>10⁴	>10⁴
Hf	169	358	1610	4280	695	8.9
La	57	79	155	470	1580	>10³
Ce(III)	51	75	140	438	1310	>10³
Ba	36	64	168	759	prec.	prec.
Y	28.9	41.3	77	199	401	—
Sc	31.6	60	158	753	1390	>10³
Yb	27.0	39.6	67	180	328	—
Sr	17.8	23.9	52	174	606	—
Ca	12.2	17.0	34.8	90	198	—
Al	12.2	12.5	19.5	43.7	75	134
Cs	10.4	12.0	18.3	33.6	45.6	—
Rb	8.1	11.5	18.4	38.0	56	—
Ga	7.8	11.0	3.4	<0.5	<0.5	<0.5
K	7.4	9.0	15.5	37.5	61	—
U(VI)	7.3	7.5	9.2	4.1	1.6	<0.5
Ni(II)	6.3	6.6	10.4	17.3	20.1	17.8
Co(II)	6.2	6.8	10.0	9.3	3.1	1.1
Mg	6.2	6.8	9.2	21.6	33.9	48.4

[a] 0.1% H_2O_2. [b] Partial reduction to V(IV). [c] 2 hr. equilibration time.

Literature Strelow, F. W. E., Victor, A. H., van Zyl, C. R., and C. Eloff, *Anal. Chem.*, **43**, 870 (1971).

TABLE LC 87
METAL IONS

Resin	Dowex® 50W–X8, H^+ form, 200–400 mesh.
Solvent	20% Dimethylsulfoxide-10% 6*M* HCl-10% 2.5*M* sodium acetate-60% H_2O.
Method of determination	Elution through 1 g columns of resin; distribution coefficients were calculated using the expression $K_d = (V_R - V_M)/m$, with m = 1 g and V_M = 0.7 ml.

Metal Ion (0.01*M*)	BTV	VEP	K_d	EV
UO_2(II)	50	65	64.3	85
Fe(III)	115	140	139.3	195
Cr(III)	10	11	10.3	13
Cu(II)	22	25	24.3	31
Co(II)	26	31	30.3	39
Ni(II)	29	34	33.3	40
Mo(VI)	5	9	8.3	17
Mo(IV/V)	1	2	1.3	2
V(V)	3	4	3.3	7
V(IV)	20	23	22.3	26
Cd(II)	3	3	2.3	5
Zn(II)	18	20	19.3	23
Mn(II)	24	30	29.3	40
Pb(II)	14	15	14.3	20
Th(IV)	>100	—	—	—
Sc(III)	>100	—	—	—
Al(III)	>100	—	—	—
Mg(II)	23	24	23.3	30
Ca(II)	60	75	74.3	90
Sr(II)	100	—	—	—
Ba(II)	>100	—	—	—
Be(II)	20	20	19.3	25
Pd(II)	1	2	1.3	3
Hg(II)	1	2	1.3	5
Bi(III)	1	3	2.3	10
Ga(III)	>100	—	—	—
In(III)	3	4	3.3	6
Zr(IV)	>100	—	—	—
Ti(IV)	20	24	23.3	31
Sn(II/IV)	6	8	7.3	12

BTV = breakthrough volume in ml.
VEP = volume of elution peak in ml (retention volume).
EV = elution volume in ml (volume for elution of entire peak).
K_d = distribution coefficient.

Literature Janauer, G. E., Korkisch, J., and S. A. Hubbard, *Talanta*, **18**, 767 (1971).

TABLE LC 88
METAL IONS

Resin	Dowex® 50W–X8, H^+ form, 100–200 mesh.
Solvent	95% tetrahydrofuran—5% 1*M* HNO_3 with various concentrations of dithizone.
Method of determination	Batch procedure.

Weight distribution coefficients

[dithizone]	0.000*M*	0.001*M*	0.01*M*
Ag	8×10^3	<1	<1
Cu	$>10^3$	<1	<1
Bi	31	<1	<1
Tl	$>10^3$	$>10^3$	$>10^3$
Hg	<1	<1	<1
Co	$>10^3$	12	<1
Sb	3.6	2.4	<1
Pb	75	107	200
Na	$>10^3$	$>10^3$	$>10^3$
Cs	1.2×10^4	$>10^4$	$>10^4$
Ca	$>10^3$	$>10^3$	$>10^3$
Sr	$>10^4$	$>10^4$	$>10^4$
Sc	$>10^4$	$>10^4$	$>10^4$
Yb	$>10^4$	$>10^4$	$>10^4$
Ce	$>10^4$	$>10^4$	$>10^4$
Am	$>10^5$	$>10^5$	$>10^5$
U	$>10^3$	$>10^3$	$>10^3$
Th	$>10^3$	$>10^3$	>700
Pu	8×10^3	8×10^3	$>10^3$
Ti	$>10^3$	$>10^3$	$>10^3$
Hf	2×10^4	$>10^4$	2×10^3
Mn	2×10^4	$>10^4$	$>10^4$
Fe	$>10^3$	$>10^3$	$>10^3$
Zn	$>10^3$	$>10^3$	$>10^3$
Cd	$>10^3$	$>10^3$	331
Ni	$>10^3$	$>10^3$	$>10^2$

Literature Orlandini, K. A., and J. Korkisch, *Anal. Chim. Acta*, **43**, 459 (1968).

TABLE LC 89
METAL IONS

Resin	Dowex® 50–X8, H^+ form, 100–200 mesh.
Solvent	Different percentages of tetrahydrofuran at constant overall acidity of 0.6*N* HNO_3.
Method of determination	Batch method employing 20 ml solvent, 5 mg of element and 1 g of resin. The mixture was equilibrated for 12 hours, the resin filtered off, and the element determined in the filtrate.

Weight distribution coefficients

	Percentage of tetrahydrofuran					
Metal ion	0	20	40	60	80	90
UO_2(II)	40	96	134	243	51	40
Mg(II)	18	21	36	72	357	$>10^3$
Ca(II)	38	50	102	145	760	$>10^3$
Sr(II)	68	109	228	676	$>10^3$	$>10^3$
Al(III)	132	137	258	479	878	$>10^3$

TABLE LC 89 (Continued)
METAL IONS

	Percentage of tetrahydrofuran					
Metal ion	0	20	40	60	80	90
In(III)	173	235	630	$>10^3$	$>10^3$	$>10^3$
Pb(II)	105	185	463	934	$>10^3$	$>10^3$
Bi(III)	135	280	630	592	195	47
Zn(II)	46	58	157	704	$>10^3$	$>10^4$
Cd(II)	46	49	178	10^3	10^4	$>10^4$
Y(III)	$>10^3$	$>10^3$	$>10^4$	$>10^4$	$>10^4$	$>10^4$
La → Lu(III)	$>10^3$	$>10^3$	$>10^4$	$>10^4$	$>10^4$	$>10^4$
Ti(IV)	70	127	300	10^3	$>10^3$	$>10^3$
Zr–Hf(IV)	$>10^4$	$>10^4$	$>10^4$	$>10^4$	$>10^4$	$>10^4$
Th(IV)	$>10^4$	$>10^4$	$>10^4$	$>10^4$	$>10^4$	1,420
Cr(III)	125	230	600	$>10^3$	$>10^3$	$>10^4$
Mn(II)	40	61	130	405	$>10^3$	$>10^3$
Fe(III)	171	200	287	550	$>10^3$	$>10^3$
Co(II)	32	42	119	306	$>10^3$	$>10^3$
Ni(II)	30	40	115	316	$>10^3$	$>10^3$

Literature Korkisch, J., *Separ. Sci.*, **1**, 163 (1966).

TABLE LC 90
RARE EARTH METAL IONS

Resin Dowex® 1–X4, αHIB form, 100–200 mesh.
Solvent Mixtures of αHIB (α-hydroxyisobutyric acid) and organic solvents.
Method of determination Elution of the element through a column of resin. Metals determined in fractions of effluent by copper spark emission spectroscopy.

Distribution coefficients[a]

25% Aqueous methanol mixtures with varying αHIB concentration

	Conc. αHIB					
Element	0.005*M*	0.0075*M*	0.010*M*	0.0125*M*	0.025*M*	0.050*M*
Sc	67	76	83	89	111	137
Y	223	250	272	290	250	424
Ho	275	308	331	352	425	520
Dy	223	250	272	290	350	425
Th	172	192	210	222	269	325
Gd	113	131	142	155	192	250
Eu	96	105	114	122	152	191
Sm	67	76	83	89	111	137
Nd	26	29	31	33	40	48
Pr	17	18.5	20	21	25.5	30.5
Ce	11	11.5	12	12	13	14
La	<10	<10	<10	<10	<10	<10

TABLE LC 90 (Continued)
RARE EARTH METAL IONS

Aqueous methanol mixtures, 0.0125M in αHIB concentration

Element	% Water 5	10	15	20	25	30	40	50	60	80	100
Ba	—	—	—	—	—	—	—	—	—	—	—
Sr	—	—	—	—	—	—	—	—	—	—	—
Sc		650	275	146	89	65	42	30	23	11	—
Y					290	188	88	48	28	—	—
Lu					850	540	250	131	75	30	11
Yb					690	440	208	110	64	25	—
Tm					560	365	170	90	51	19	—
Er					470	305	141	74	43	15	—
Ho					352	230	110	59	34	10	—
Dy					290	188	88	48	28	—	—
Tb					222	138	61	30	16	—	—
Gd				280	155	96	42	22	11	—	—
Eu			460	226	122	73	31	15	—	—	—
Sm		990	350	166	89	54	23	10	—	—	—
Nd		355	128	61	33	21	—	—	—	—	—
Pr	640	210	80	38	21	12	—	—	—	—	—
Ce	320	110	45	22	12	—	—	—	—	—	—
La	90	31	13	—	—	—	—	—	—	—	—
Cm		400	145	70	~40	~25	~12	—	—	—	—
Am		300	109	54	31	~19	—	—	—	—	—

(— = K_d < 10).

Aqueous ethanol mixtures, 0.0125M in αHIB concentration

Element	% Water 5	10	15	20	25	30	40	50	60	80
Ba	—	—	—	—	—	—	—	—	—	—
Sr	—	—	—	—	—	—	—	—	—	—
Sc			880	365	190	116	57	35	26	11
Y					565	305	118	54	29	10
Lu						830	325	154	84	32.5
Yb						720	280	130	69	26
Tm						600	233	108	58	21.5
Er						490	187	87	46	17
Ho						412	158	72	37.5	13
Dy					565	305	118	54	29	10
Tb				1000	445	240	88	39	20.5	—
Gd				790	350	185	64	29	15	—
Eu				630	280	148	52	23.5	12	—
Sm				480	215	110	38.5	17	9.5	—
Nd			600	200	86	43	14	—	—	—
Pr			335	122	56	28.5	—	—	—	—
Ce			225	83	37	18	—	—	—	—
La			82	29	13	—	—	—	—	—

(— = K_d < 10)

TABLE LC 90 (Continued)
RARE EARTH METAL IONS

Aqueous isopropanol mixtures, 0.125*M* in αHIB concentration

Element	% Water 5	10	15	20	25	30	40	50	60	80
Ba	—	—	—	—	—	—	—	—	—	—
Sr	—	—	—	—	—	—	—	—	—	—
Sc				480	252	150	70	41	29	13
Y					510	278	110	52	28.5	11
Lu						690	280	138	77	31
Yb						590	235	115	64	25.5
Tm						530	205	103	57	22
Er						420	165	80	44.5	18
Ho						340	143	66	37	15
Dy					510	278	110	52	28.5	11
Tb				840	385	205	77	35.5	19	—
Gd				635	280	145	51	23	12	—
Eu				510	222	110	38.5	17	—	—
Sm				405	173	88	29.5	12.5	—	—
Nd			590	185	76	36.5	11	—	—	—
Pr			420	128	52	25	—	—	—	—
Ce			310	92	36.5	17	—	—	—	—
La			150	41	15	—	—	—	—	—

(— = K_d < 10).
[a] $K_d = V_e/m$.

Literature Data supplied by J. P. Faris, Argonne National Laboratory.

TABLE LC 91
RARE EARTH METAL IONS

Resin Dowex® 1–X4, NO_3^- form, 200–400 mesh.
Solvent HNO_3-methanol mixtures.
Method of determination 5–200 μg of element was eluted through columns containing 1 g of resin. Metals determined in fractions by copper spark emission spectroscopy.

Distribution coefficients[a]

10% HNO_3–90% methanol with varying acid concentration

Element	Overall *M* HNO_3 0.01	0.02	0.05	0.10	0.30	0.70	1.00	1.20	1.57
La	365	660	1380	~2380	ads	ads	ads	ads	ads
Ce	205	390	800	1335	~3100	ads	ads	ads	ads
Pr	112	210	460	780	~1800	ads	ads	ads	ads
Nd	51	96	204	357	805	1510	ads	ads	ads
Sm	9	17	35	58	130	242	315	360	430
Eu	6	10	19	31	67	118	150	170	202
Gd	3	6	12.5	20	39	63	77	87	103
Tb	2.5	5	10	15	29	46	55	61	70
Dy } **Lu** }	~2	~4	~8	~12.5	25– 23	39– 29	44– 29	46– 28	47– 24
Y	1.5	3.3	6.5	11	21	29	31	31	30
Sc	—	—	—	1.1	3.2	5.2	5.2	4.9	2.4

ads = irreversibly adsorbed.

TABLE LC 91 (Continued)
RARE EARTH METAL IONS

Varying percentages of 7*M* HNO³ in methanol (constant acid molarity)

	7*M* HNO_3			
Element	5%	10%	20%	40%
Nd	ads	1510	260	41
Sm	~680	242	66	16
Eu	265	118	38	11
Gd	120	63	24	7.5
Tb	77	46	20	7
Dy	55	39	19	7
Ho	47	36	19	7
Er	42	34	19	7
Tm	39	32	19	7
Yb	36	30	19	6.5
Lu	34	29	19	6.5
Y	36	29	17	7
Sc	1.5	5.2	5.7	4.3

[a] $K_d = V_e/m$.
ads = irreversibly adsorbed.

Literature Data supplied by J. P. Faris, Argonne National Laboratory.

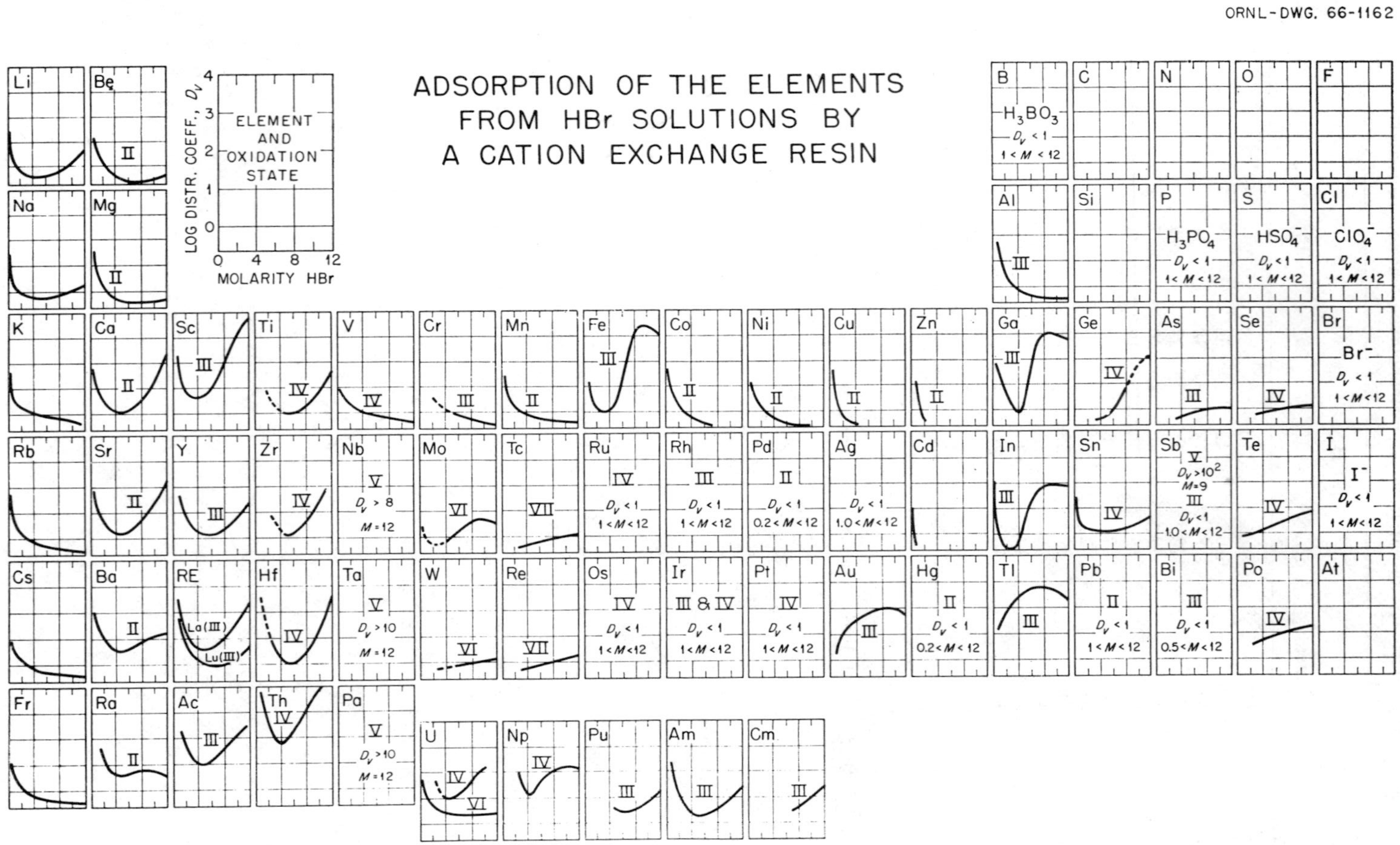

Figure LC-1. Resin used was Dowex® No. 1-X-10, Cl-form. Krause, K. A., and F. Nelson, *ASTM Special Technical Publication* **195**, p. 36 (1958); courtesy of F. Nelson.

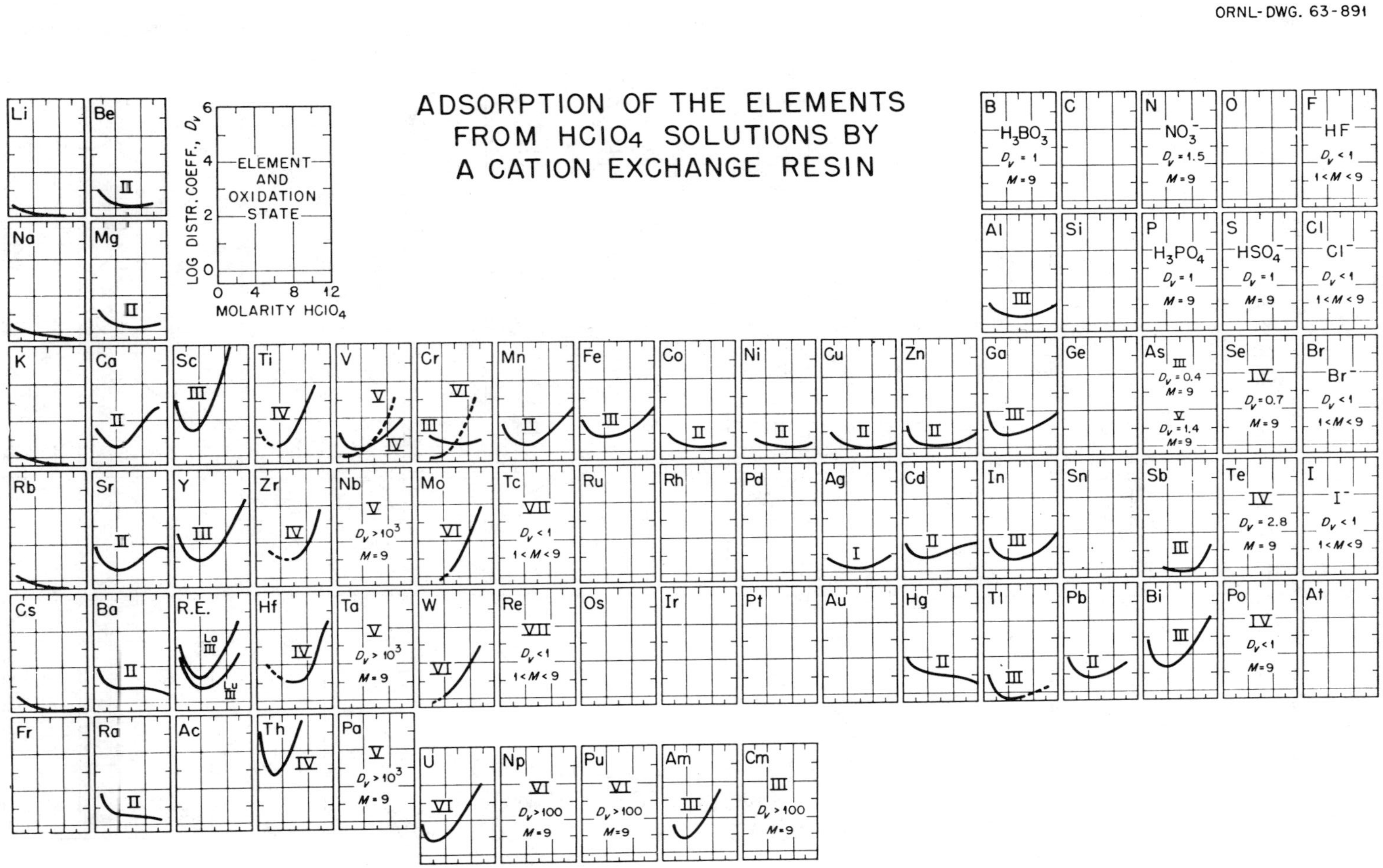

Figure LC-2. Resin used was Dowex® No. 50-X4, H^+ form. Nelson, F., Murase, T., and K. A. Krause, *J. Chromatogr.* **13**, 532 (1964); courtesy of F. Nelson.

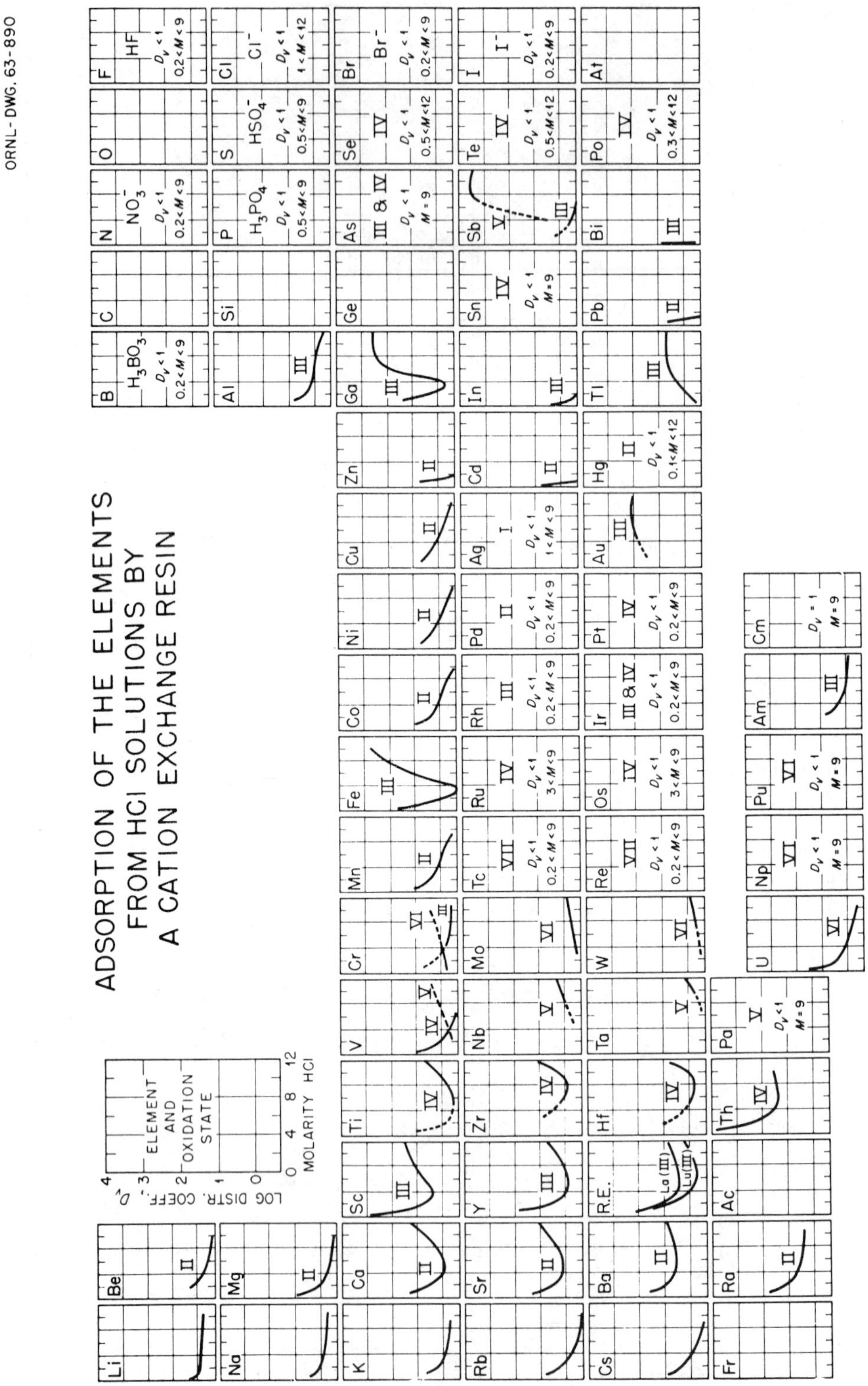

Figure LC-3. Resin used was Dowex® No. 50-X4, H⁺ form. Nelson, F., Murase, T., and K. A. Krause,, *J. Chromatogr.* **13**, 533 (1964); courtesy of F. Nelson.

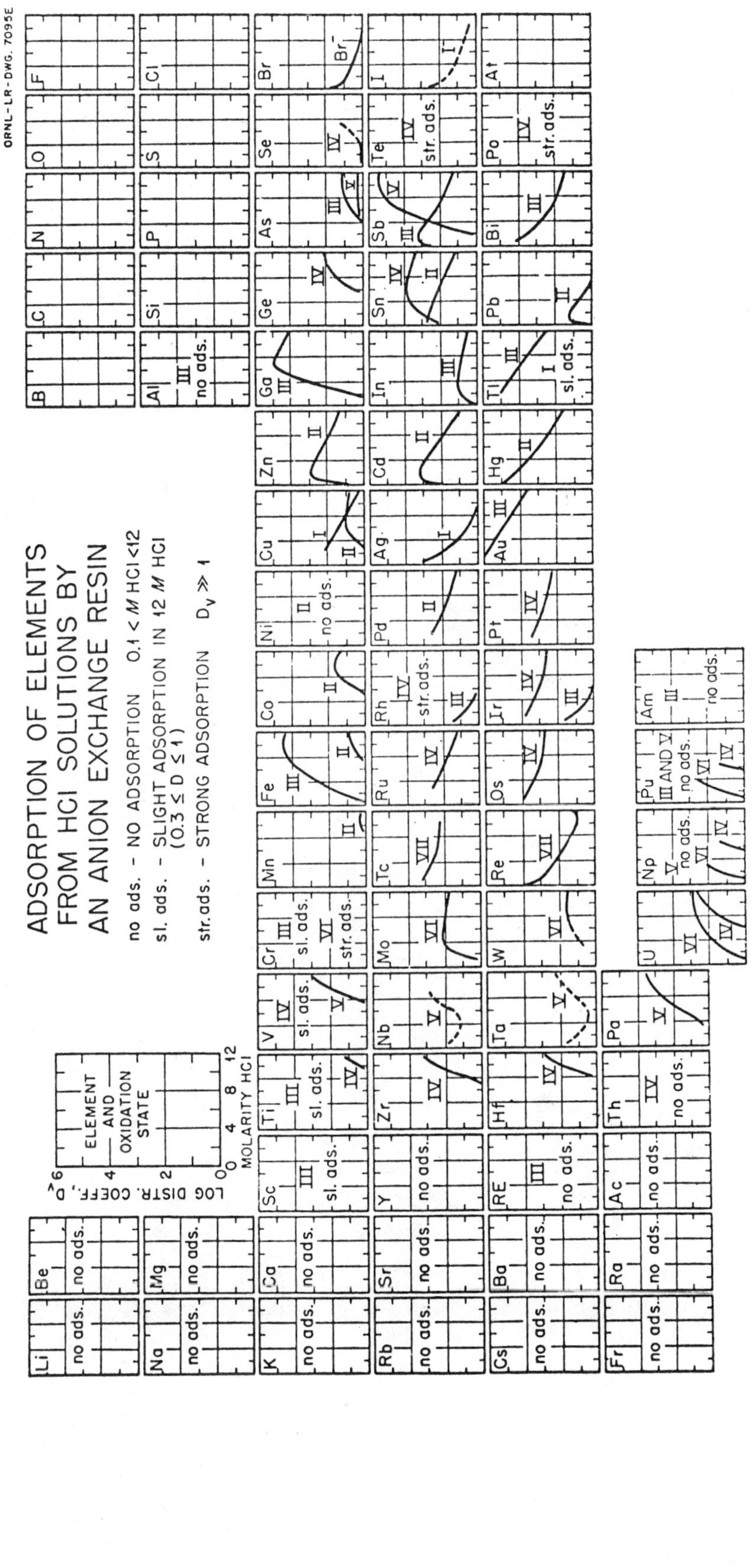

Figure LC-4. Resin used was Dowex® No. 50-X4, H⁺ form. Nelson, F., and D. C. Michelson, *J. Chromatogr.* **25**, 438 (1966); courtesy of F. Nelson.

VOLUME ONE
SECTION II.III

CHROMATOGRAPHIC DATA
Paper Chromatography Tables

Wherever possible, tables were arranged according to classes of chemical compounds. However, this was not always possible when diverse compounds were chromatographed under the same experimental conditions. The reader is referred to the master compound index for specific compounds which may appear in diverse tables.

Techniques

n.a. = not available from original source.
Solvent composition always v/v unless otherwise stated; if solid is dissolved in solvent, g/v.

TABLE PC 1
ALCOHOLS, POLYHYDRIC

Paper	na	na	na	na	na	na	na						
Solvent	S1	S2	S3	S4	S5	S6	S7						
Detection	—	—	—	—	—	—	—	D1	D2	D3	D4	D5	D6
Literature	1	1	1	1	1	1	1	2	2	2	2	2	2
Compound	$R_F \times 100$							Color				Sensitivity (μg)	
Arabitol	22	28	—	—	—	43	53	—	—	—	—	—	—
1,2,4-Butanetriol	—	—	—	20	—	—	—	—	—	—	—	—	—
1,3-Butyleneglycol	—	—	—	76	35	—	—	—	—	—	—	—	—
1,4-Butyleneglycol	—	—	—	74	26	—	—	—	—	—	—	—	—
2,3-Butyleneglycol	—	—	—	78	45	—	—	—	—	—	—	—	—
Diethyleneglycol	—	—	—	72	8	—	—	—	—	—	—	—	—
Dulcitol	16	17	18	—	—	31	48	—	gy	tan	y	25	20
Erythritol	35	31	34	—	—	49	63	gy	bk	lv	p	10	10
Ethylene glycol	64	—	—	47	8	—	83	—	—	—	—	—	—
Glycerol	48	—	—	18	—	58	75	—	gy	w	lv	—	10
Inositol	6	7	8	—	—	11	—	o	o	o	r	2	2
Mannitol	19	20	19	—	—	34	48	—	gy	du-y	y	10	5
1,2-Propylene glycol	—	—	—	67	24	—	—	—	—	—	—	—	—
1,3-Propylene glycol	—	—	—	55	14	—	—	—	—	—	—	—	—
Ribitol (Adonitol)	25	28	28	—	—	40	—	gy	bk	—	lv	—	10
Sorbitol	18	17	19	—	—	31	—	—	gy	tan	lt-br	20	25
Triethyleneglycol	—	—	—	83	8	—	—	—	—	—	—	—	—
Xylitol	20	24	—	—	—	—	—	—	—	—	—	—	—

Solvent
S1 = *n*-butanol-acetic acid-water (4 : 1 : 5 v/v).
S2 = *n*-butanol-ethanol-water (4 : 1 : 5 v/v).
S3 = ethyl acetate-acetic acid-water (3 : 1 : 3 v/v).
S4 = chloroform-ethanol (3 : 2 v/v).
S5 = water-saturated ethyl ether.
S6 = *n*-propanol-ethyl acetate-water (7 : 1 : 2 v/v).
S7 = *tert*-amyl alcohol-*n*-propyl alcohol-water (4 : 1 : 1.5 v/v).

Detection
D1 = modified Fleury, 10 minutes heating.
D2 = modified Fleury, 30 minutes heating.
D3 = *p*-anisidine phosphate over Fleury reagent.
D4 = *p*-anisidine phosphate over Fleury reagent, ultraviolet.
D5 = sensitivity (μg), daylight.
D6 = sensitivity (μg), ultraviolet.

Literature
1. Block, R. J., Durrum, E. L., and G. Zweig, *Paper Chromatography and Paper Electrophoresis*, Academic Press, N.Y., 1958, p. 205.
2. Lambou, M. G., *Anal. Chem.*, **28**, 1216 (1956).

TABLE PC 2
ALKALOIDS

Paper Solvent Literature	P1 S1 1	P1 S2 2	P1 S3 2	P1 S4 3	P1 S5 3	P1 S6 3	P1 S7 4
Compound				$R_F \times 100$			
3-Acetylpyridine	—	—	—	—	61	—	—
Ajmaline	—	—	—	—	—	—	76
Anabasine	—	—	—	—	21	—	—
Atropine	63	63	—	—	—	—	—
Berberine	30	35	—	—	—	—	—
Betaine	14	17	11	—	—	—	—
Boldine	49	59	—	—	—	—	—
Brucine	49	52	—	—	—	—	—
Caffeine	69	67	72	—	—	—	—
Chelidonine	68	67	—	—	—	—	—
Choline	19	23	10	—	—	—	—
Cinchonamine	81	84	—	—	—	—	—
Cinchonidine	78	81	—	—	—	—	—
Cinchonine	90	79	—	—	—	—	—
Cliradon	—	—	—	61	—	—	—
Codeine	46	51	—	—	—	—	—
Colarnine	35	47	—	—	—	—	—
Conessine	51	53	—	—	—	—	—
Corynantheine	91	85	—	—	—	—	—
Cryptopine	63	60	—	—	—	—	—
Delphenine	84	88	—	—	—	—	—
Deserpine	—	—	—	—	—	—	41
Dicodid	—	—	—	39	—	—	—
Dihydronicotyrine	—	—	—	—	36	—	—
Dilaudid	—	—	—	32	—	—	—
2,3′-Dipyridyl	—	—	—	—	95	—	—
Dolantin	—	—	—	70	—	—	—
Dromoran	—	—	—	71	—	—	—
Ephedrine	70	66	90	—	—	—	—
Eserine	78	72	—	—	—	—	—
"Eukodal"	—	—	—	34	—	—	—
Gelsemine	65	69	—	—	—	—	—
Hordenine	53	52	88	—	—	—	—
Hydrastinine	24	40	—	—	—	—	—
Isonicotinic aid	—	—	—	—	51	65	—
Metanicotine	—	—	—	—	25	—	—
4-Methylamino-1-(3-pyridyl)-1-butanol	—	—	—	—	12	—	—
"*N*-Methylanabasine"	—	—	—	—	36	—	—
"*N*-Methylmyosmine"	—	—	—	—	13	—	—
6-Methylpyridine-2-aldehyde	—	—	—	—	—	64 (90)	—
2-Methylpyridine-5-carboxylic acid	—	—	—	—	—	69	—
6-Methylpyridine-2-carboxylic acid	—	—	—	—	—	52	—
2-Methyl-6-(3-pyridyl)-tetrahydro-1,2-oxazine·HCl	—	—	—	—	96	—	—
Methyl reserpate	—	—	—	—	—	—	52; 76
Morphine	4	7	—	31	—	—	—
Myosmine	—	—	—	—	92	—	—
Narcotine	81	80	—	—	—	—	—
Nicotinamide	71	65	65	—	86	—	—
Nicotine	44	50	—	23	33	—	—
Nicotinic acid	—	—	—	—	62	71	—
Nicotyrine	—	—	—	—	98	—	—
Nornicotine	—	—	—	—	17	—	—
Nornicotyrine	—	—	—	—	97	—	—
Oxynicotine	—	—	—	—	30	—	—
Papaverine	80	72	—	—	—	—	—
Pervitin	—	—	—	65	—	—	—
β-Picoline	—	—	—	—	87	—	—
α-Picolinic acid	—	—	—	—	—	59	—
Pilocarpine	40	47	—	—	—	—	—

TABLE PC2—(Continued)
ALKALOIDS

Paper	P1	P1	P1	P1	P1	P1	P1
Solvent	S1	S2	S3	S4	S5	S6	S7
Literature	1	2	2	3	3	3	4
Compound	$R_F \times 100$						
Piperine	95	90	—	—	—	—	—
Polamidon	—	—	—	80	—	—	—
α-Pyridinealdehyde	—	—	—	—	—	68[a] (86)	—
β-Pyridinealdehyde	—	—	—	—	—	67 (90)	—
γ-Pyridinealdehyde	—	—	—	—	—	59 (86)	—
2,6-Pyridinedialdehyde	—	—	—	—	—	83 (93)	—
2,3-Pyridinedicarboxylic acid	—	—	—	—	—	57	—
2,4-Pyridinedicarboxylic acid	—	—	—	—	—	60	—
2,5-Pyridinedicarboxylic acid	—	—	—	—	—	71	—
2,6-Pyridinedicarboxylic acid	—	—	—	—	—	49	—
3,4-Pyridinedicarboxylic acid	—	—	—	—	—	50	—
Quinidine	91	80	—	—	—	—	—
Quinine	87	80	—	—	—	—	—
Rauwolscine	—	—	—	—	—	—	63
Rescinnamine	—	—	—	—	—	—	26
Reserpine	—	—	—	—	—	—	34
Serpentine	—	—	—	—	—	—	5
Sparteine	63	69	—	—	—	—	—
Strychnine	73	69	—	—	—	—	—
Thebaine	68	67	—	—	—	—	—
Theobromine	41	42	45	—	—	—	—
Theophylline	64	59	57	—	—	—	—
Trigonelline	8	13	9	—	—	—	—
Veratrine	89	85	—	—	—	—	—
Yohimbine	67	72	—	—	—	—	—

[a] Dimedate derivative.

Paper P1 = W1, SS2043b, or equivalent.

Solvent S1 = four ml of glacial acetic acid is added to 100 ml of *n*-butanol, and the mixture is saturated with water at 20 °C.

S2 = ten ml of glacial acetic acid is added to 100 ml of *n*-butanol, and the mixture is saturated with water.

S3 = *n*-butanol-ethanol-NH_4OH-H_2O (90 : 10 : 1 : 97 v/v) are mixed, and the upper phase is used.

S4 = *n*-butanol-formic acid-water (12 : 1 : 7 v/v).

S5 = *tert*-amyl alcohol-0.2*M* acetate buffer, pH 5.6 (1 : 1 v/v).

S6 = *sec*-butanol-formic acid-water (75 : 15 : 10 v/v).

S7 = 10% acetic acid in 5% sodium acetate, saturated with *n*-butyl ether.

Literature

1. Munier, R. and M. Macheboeuf, *Bull. Soc. Chim. Biol.*, **31**, 1144–1162 (1949).
2. Tso, T. C. and R. N. Jeffery, *Arch. Biochem. Biophys.*, **43**, 269–285 (1953).
3. Jatzkewitz, H., *Z. Physiol. Chem.*, **292**, 94–100 (1953).
4. Boscott, R. J. and A. B. Kar, *Nature*, **176**, 1077–1078 (1955).

TABLE PC 3
ALKALOIDS

Paper	P1	P1	P2	P3	P4	P4	P5	P6	P7	P8	P9
Solvent	S1	S2	S3	S4	S5	S6	S7	S8	S9	S10	S11
Technique	na	na	T1	T1	na	na	na	na	na	na	T2
Literature	1	1	2	2	3,4	3	3,4	5	5	5	6
Compound	$R_F \times 100$										
Genista Alkaloids											
Anagyrine	31	24	—	—	—	—	—	—	—	—	—
Calycotomine	36	26	—	—	—	—	—	—	—	—	—
Cytisine	17	12	—	—	—	—	—	—	—	—	—
Epilupinine	76	36	—	—	—	—	—	—	—	—	—
Hydroxylupanine	50	22	—	—	—	—	—	—	—	—	—
Hydroxysparteine	43	35	—	—	—	—	—	—	—	—	—
***l*-α-Isosparteine**	52	56	—	—	—	—	—	—	—	—	—
Lupanine	66	37	—	—	—	—	—	—	—	—	—
Lupinine	72	33	—	—	—	—	—	—	—	—	—
Methylcytisine	23	10	—	—	—	—	—	—	—	—	—
17-Oxosparteine	82	37	—	—	—	—	—	—	—	—	—
Retamine	45	14	—	—	—	—	—	—	—	—	—
Sparteine	59	45	—	—	—	—	—	—	—	—	—
Papaveraceae Alkaloids											
Allocryptopine	—	—	—	—	72	—	42	—	—	—	—
Aporheine	—	—	—	—	63	—	30	—	—	—	—
Armepavine	—	—	—	—	65	—	40	—	—	—	—
Berberine	—	—	—	—	63	20	—	—	—	—	—
Chelerythrine	—	—	—	—	57	—	—	—	—	—	—
Chelidonine	—	—	—	—	75	—	61	—	—	—	—
Chelilutine	—	—	—	—	67	—	—	—	—	—	—
Chelirubine	—	—	—	—	57	—	—	—	—	—	—
Coptisine	—	—	—	—	51	11	—	—	—	—	—
Corysamine	—	—	—	—	77	51	—	—	—	—	—
Cryptopine	—	—	—	—	73	—	34	—	—	—	—
Dehydrocorydaline	—	—	—	—	79	72	—	—	—	—	—
Glaucamine	—	—	—	—	60	—	62	—	—	—	—
(−)-β-Hydrastine	—	—	—	—	76	—	64	—	—	—	—
Hydrastinine	—	—	—	—	51	—	18	—	—	—	15
Jatrorrhizine	—	—	—	—	60	34	—	—	—	—	—
Macarpine	—	—	—	—	51	—	—	—	—	—	—
Mecambrine	—	—	—	—	58	—	55	—	—	—	—
Morphine	—	—	—	—	36	—	13	—	—	—	36
Muramine	—	—	—	—	77	—	48	—	—	—	—
Narcotine	—	—	—	—	69	—	82	—	—	—	—
Palmatine	—	—	—	—	59	32	—	—	—	—	—
Papaverine	—	—	—	—	71	—	81	—	—	—	27
Protopine	—	—	—	—	69	—	40	—	—	—	—
Rhoeadine	—	—	—	—	73	—	52	—	—	—	—
Rhoeagenine	—	—	—	—	64	—	44	—	—	—	—
Roemeridine	—	—	—	—	33	—	8	—	—	—	—
Sanguilutine	—	—	—	—	62	—	—	—	—	—	—
Sanguinarine	—	—	—	—	46	—	—	—	—	—	—
Sanguirubine	—	—	—	—	42	—	—	—	—	—	—
Stylophylline	—	—	—	—	70	—	83	—	—	—	—
Stylopine	—	—	—	—	75	—	76	—	—	—	—
Tetrahydroberberine	—	—	—	—	80	—	76	—	—	—	—
Thebaine	—	—	—	—	63	—	34	—	—	—	—
Solonaceous Alkaloids											
Apoatropine	—	—	81	82	—	—	—	—	—	—	—
Belladonnine	—	—	20	38	—	—	—	—	—	—	—
Hyoscyamine	—	—	57	65	—	—	—	—	—	—	42
Meteloidine	—	—	39	66	—	—	—	—	—	—	—
Norhyoscyamine	—	—	60	56	—	—	—	—	—	—	—
Oscine	—	—	45	7	—	—	—	—	—	—	—
Scopolamine	—	—	29	85	—	—	—	—	—	—	—
Tigloidine	—	—	71	83	—	—	—	—	—	—	—
Tropine	—	—	8	28	—	—	—	—	—	—	55
Valeroidine	—	—	62	71	—	—	—	—	—	—	—

TABLE PC 3—(Continued)
ALKALOIDS

Paper	P1	P1	P2	P3	P4	P4	P5	P6	P7	P8	P9
Solvent	S1	S2	S3	S4	S5	S6	S7	S8	S9	S10	S11
Technique	na	na	T1	T1	na	na	na	na	na	na	T2
Literature	1	1	2	2	3, 4	3	3, 4	5	5	5	6
Compound	$R_F \times 100$										
Miscellaneous Alkaloids											
Aconitine	—	—	—	—	—	—	—	84	70	54	—
Brucine	—	—	—	—	—	—	—	27	73	—	—
Caffeine	—	—	—	—	—	—	—	74	—	—	73
Cevadine	—	—	—	—	—	—	—	87	—	84	—
Cinchonine	—	—	—	—	—	—	—	—	—	51	3
Cocaine	—	—	—	—	—	—	—	—	52	71	58
Colchicine	—	—	—	—	—	—	—	—	>80	—	83
Demecolcine	—	—	—	—	—	—	—	49	>80	—	—
Emetine	—	—	—	—	—	—	—	30	21	37	—
Ephedrine	—	—	—	—	—	—	—	—	0	47	63
Eserine	—	—	—	—	—	—	—	57	—	—	41
Heroin	—	—	—	—	—	—	—	—	76	—	33
Hordenine	—	—	—	—	—	—	—	—	—	27	—
Hydrastinine	—	—	—	—	—	—	—	—		—	—
Lobeline	—	—	—	—	—	—	—	87		6	—
Physostigmine	—	—	—	—	—	—	—	—		8	—
Pilocarpine	—	—	—	—	—	—	—	16	29	—	—
Quinidine	—	—	—	—	—	—	—	72	—	47	—
Quinine	—	—	—	—	—	—	—	70	51	40	3
Sparteine	—	—	—	—	—	—	—	16	—	>90	—
Strychnine	—	—	—	—	—	—	—	40	48	9	—
Theobromine	—	—	—	—	—	—	—	43	—	—	—
Theophylline	—	—	—	—	—	—	—	65	—	—	—
Veratridine	—	—	—	—	—	—	—	82	24	—	—
Vincamine	—	—	—	—	—	—	—	—	72	—	—

Paper
P1 = SS2045bM.
P2 = W1 saturated with pH 3.8 buffer.
P3 = W1 saturated with pH 6.8 buffer.
P4 = W1.
P5 = W1 impregnated with 0.15*M* KH_2PO_4.
P6 = paper impregnated with citrate buffer (pH 3.8).
P7 = paper impregnated with formamide plus 1% acetic acid.
P8 = paper impregnated with formamide.
P9 = W1 impregnated with 5% zirconium phosphate.

Solvent
S1 = isopropanol-HCl-H_2O (8 : 1 : 1 v/v).
S2 = *n*-butanol-acetic acid-H_2O (50 : 2 : 17 v/v).
S3 = butanol saturated with McIlvaine's buffer (pH 3.8).
S4 = butanol saturated with McIlvaine's buffer (pH 6.8).
S5 = *n*-butanol-acetic acid-H_2O (10 : 1 : 3 v/v).
S6 = ethanol-H_2O (3 : 2 v/v).
S7 = *n*-butanol saturated with water.
S8 = *n*-butanol-H_2O-citric acid (50 : 50 : 1 v/v/w).
S9 = chloroform.
S10 = cyclohexane-diethylamine (9 : 1 v/v).
S11 = 5% acetic acid.

Technique
T1 = descending development.
T2 = ascending development after 24 hours equilibration at 28 ± 1°C.

Literature

1. Barnasconi, R., Gill, S., and E. Steinigger, *Pharm. Acta Helv.*, **40**, 253 (1965).
2. Karawaya, M. S. and S. I. Balboa, *J. Pharm. Sci. U. Anals. Rep.*, **4**, 57 (1963); *J. Chromatogr.*, **22**, D11 (1966).
3. Slavik, J., *Collect. Czech. Chem. Commun.*, **32**, 4434–4435 (1967).
4. Slavik, J. and J. Appelt, *Collect. Czech. Chem. Commun.*, **30**, 3690 (1965).
5. Macek, K., in *Chromatography* (E. Heftmann, Ed.) Reinhold, N.Y. 1967, p. 622.
6. Coussio, I. D., Marini-Bettolo, G. B., and J. Moscatelli, *J. Chromatogr.*, **11**, 238 (1963).

TABLE PC 4
ALKALOIDS

Paper	P1	P1	P2	P2							
Solvent	S1	S2	S3	S4							
Detection	—	—	—	—	D1	D2	D3	D4	D5	D6	D7
Literature	1	1	1	1							
Compound	$R_F \times 100$				Colors						
Aconitine	80	70	72	55	ft. b	bn → y	p	y	b	b	r
Apomorphine	73	46	26	12	dk b	bn → g	b/bk → p	g-y	g	gy	—
Atropine	75	89	28	14	ft. b	ft. y	b/bk → p	—	b	b	—
Brucine	68	73	14	29	—	ft. y	b	y	b	b	—
Cocaine	80	84	34	20	—	ft. y	p	—	b	g	—
Codeine	65	83	14	6	ft. y	ft. y	p	y	gb	b	r
Diamorphine	74	89	32	16	—	ft. y	p	y	gb	b	—
Emetine	79	87	17	12	b-y	y	rp	gy	b	vt to b	—
Homatropine	68	82	23	—	—	bn → y	b/bk → p	—	gb	b	—
Hyoscine	65	90	16	5	ft. b	y	b/bk → p	y	ft. b	g	—
Hyoscyamine	75	91	24	16	—	y	b/bk → p	—	b	b	—
Morphine	62	82	11	3	—	bn → y	b/bk	y	g → b	b	r
Nicotine	64	97	8	4	—	bn → y	b/bk → p	—	—	—	—
Papaverine	88	85	54	39	y	bn → y	p	y	b	b	—
Physostigmine	81	93	38	19	ft. b	bn → y	b/bk → p	y	ft. b	b	—
Pilocarpine	50	92	13	5	ft. lt b	ft. y	b/bk → p	—	b	ft. g	—
Quinidine	92	91	57	40	b	ft. y	p	y	ft. b	b	—
Quinine	87	90	56	38	b	ft. y	p	y	ft. b	g → b	—
Sparteine	74	95	20	5	—	bn	b/bk	y	b	g → b	—
Strychnine	74	76	23	12	—	ft. y	p	—	ft. b	g → b	—
Thebaine	78	86	33	14	—	bn → y	b/bk → p	y	g	b	—
Yohimbine	67	67	15	30	y-g	y	p	y	g	g → b	—
Alphaprodine	92	90	56	33	—	ft. y	b/bk-p	—	b	b-g	—
Dextromethorphan	92	84	60	36	—	y	b/bk	—	b	b-g	—
Dextromoramide	94	87	70	45	—	y	r-p	—	b	ft. g	—
Dihydromorphinone	64	92	12	3	—	bn-y	ft. p	y	ft. b	—	—
Dihydrocodeinone	66	90	15	5	—	bn-y	b/bk	y	ft. b	ft. bg	—
Dionyl	84	86	48	16	b	ft. y	b/bk	y	b	b-g	r
Dipipanone	92	80	82	65	—	y	r-p	r-bn	ft. b	b	—
Ethylmorphine	64	89	32	10	—	bn-y	b/bk	y	ft. b	b-g	—
Levorphanol	84	84	55	20	—	ft. y	p	y	b	—	—
Methadone	86	87	70	57	—	bn-y	ft. p	—	b	b-g	—
Nalorphine	65	84	33	10	—	bn	b/bk	y	b	ft. b	—
Pholcodeine	43	91	5	—	—	y	—	y-bn	b	b	—
Pethidine	76	89	0	0	—	y	p	—	b	—	—
Amethocaine	58	64	47	45	ft. b	y	p	y	b	ft. g	y
Amydricaine	75	85	50	46	—	bn-y	rp	r	b	ft. g-b	—
Amylocaine	77	92	60	75	—	y	p	—	b	ft. g	—
Cinchocaine	84	67	82	70	ft. b	bn-y	p	y	b	b-g	—
Lignocaine	68	90	60	37	—	bn-y	b-y to w	ft. y	b	b	—
Neostigmine	55	87	30	7	—	bn-y	p-r to y	—	b	g	—
Mepacrine	75	69	30	4	y	y	bn-gy	bn-y	g	g	y

Paper P1 = W1.
P2 = W1 buffered with 5% sodium dihydrogen citrate.

Solvent S1 = *n*-butanol-glacial acetic acid-water (120 : 30 : 50).
S2 = *n*-butanol-1*N* sodium acetate-1*N* HCl (7 : 120 : 60).
S3 = *n*-butanol-5% w/v citric acid (180 : 20).
S4 = *iso*-amyl alcohol-5% citric acid (180 : 20).

Detection D1 = UV light.
D2 = Iodine.
D3 = Iodoplatinate.
D4 = KMnO4.
D5 = Cobalt thiocyanate.
D6 = Bromocresol green.
D7 = Ehrlich.

Technique Ascending; solvent S2 can also be used for short descending runs.

Literature
1. Jackson, J. V. and M. S. Moss, in *Chromatographic and Electrophoretic Techniques*, Vol. 1, Smith, I., Ed., Interscience Publishers, Division of John Wiley & Sons, New York, N.Y., 1969, pp 520–521.

TABLE PC 5

AMINES, ADRENALINE AND RELATED COMPOUNDS

Paper	P1	P1	P1	P2	P1
Solvent	S1	S2	S3	S3	S4
Technique	T1	T1	T2	T2	T2
Literature	1	1	2	3	2
Compound			$R_F \times 100$		
Adrenaline	—	—	44	50	36
2-Aminoheptane	—	76	—	—	—
2-Amino-2′-hydroxydiethylamine	—	42	—	—	—
Benzylamine	70	66	—	—	—
Butanefrin·HCl	—	50	—	—	—
Cobefrin·HCl	—	42	—	—	—
Corbasil	—	—	41	—	—
***n*-Decylamine**	87	—	—	—	—
Dexedrine	—	74	—	—	—
Dihydroxyphenylalanine	—	—	23	35	19
Epinine	—	—	76	—	—
Ethylenediamine	—	39	—	—	—
***n*-Heptylamine**	83	—	—	—	—
Hexamethylenediamine	—	48	—	—	—
***n*-Hexylamine**	80	—	—	—	—
2-Hydroxyethylamine	—	48	—	—	—
***m*-Hydroxyphenylethanolamine**	—	—	—	—	56
Hydroxytyramine	—	—	32	45	39
Methyl adrenaline	—	—	80	—	—
***N*-Methyl corbasil**	—	—	71	—	—
Noradrenaline	—	—	22	30	28
Noradrenaline·HCl	—	34	—	—	—
Norephedrine	—	53	—	—	—
***n*-Octylamine**	85	—	—	—	—
Paredrin	—	64	—	—	—

Paper P1 = W1.
P2 = SS597.

Solvent S1 = *n*-butanol-acetic acid-water (5 : 1 : 4 v/v).
S2 = *n*-butanol-acetic acid-water (2 : 1 : 1 v/v).
S3 = phenol-H_2O (80 : 20).
S4 = *n*-butanol-acetic acid-H_2O (4 : 1 : 5).

Technique T1 = development at room temperature overnight.
T2 = ascending or descending, tank flushed with CO_2 30 minutes before development, dry in N_2.

Literature

1. Baker, W., Harborne, J. B., and W. D. Ollis, *J. Chem. Soc.*, 1952, 3215–3217.
2. James, W. O., *Nature*, **161**, 851–852 (1948).
3. Goldenberg, M., Faber, M., Alston, E. J., and E. C. Chargaff, *Science*, **109**, 534–535 (1949).

TABLE PC 6
AMINES, AROMATIC

Paper	P1	P1	P2	P2	P2	P2	P2
Solvent	S1	S2	S3	S4	S5	S6	S7
Literature	1	1	2	3	3	3	4
Compound	$R_F \times 100$						
1-Amino-3-acetaminobenzene	—	—	—	—	—	33	—
1-Amino-4-acetaminobenzene	—	—	—	—	—	24	—
2-Amino-4-chlorophenol	—	—	—	—	—	71	—
2-Amino-4,6-dinitrophenol	—	—	—	—	—	1	—
2-Amino-5-nitrophenol	—	—	—	—	—	53	—
2-Aminophenol	—	—	—	—	—	50	—
3-Aminophenol	—	—	—	—	—	36	—
4-Aminophenol	—	—	—	—	—	29	—
2-Aminophenol-5-sulfonamide	—	—	—	—	—	7	—
2-Aminostilbene	—	—	—	16	24	—	—
Aniline	90	90	8	—	—	—	30
Benzidine	87	89	—	—	—	—	—
4-Bromoaniline	66	83	—	—	—	—	—
2-Chloroaniline	87	—	—	23	31	—	—
3-Chloroaniline	65	81	—	—	—	—	—
4-Chloroaniline	80	84	—	—	—	—	38
3-Chloro-4-methoxyaniline	84	85	—	—	—	—	—
3-Chloro-6-methoxyaniline	35	66	—	—	—	—	—
3-Chloro-2-methylaniline	45	77	—	—	—	—	—
3-Chloro-6-methylaniline	35	66	—	—	—	—	—
4-Chloro-2-methylaniline	55	79	—	28	23	—	64
1,2-Diamino-4-nitrobenzene	—	—	—	—	—	46	—
1,3-Diamino-4-nitrobenzene	—	—	—	—	—	41	—
2,4-Dichloroaniline	2	20	—	—	—	—	—
2,5-Dichloroaniline	2	10	—	33	34	—	—
3,3′-Dichlorobenzidine	1	19	—	—	—	—	—
5,5′-Dichloro-3,3′-dimethoxybenzidine	3	56	—	—	—	—	—
3,3′-Dimethoxybenzidine	82	89	—	—	—	—	—
2,3-Dimethylaniline	—	—	—	29	30	—	—
2,4-Dimethylaniline	—	—	—	35	45	—	—
2,5-Dimethylaniline	—	—	—	33	42	—	77
2,6-Dimethylaniline	—	—	—	42	54	—	—
N,N-Dimethylaniline	—	—	72	—	—	—	—
2-Ethoxyaniline	—	—	—	34	34	—	69
2-Methoxyaniline	—	—	—	—	—	—	44
2-Methoxy-5-chloroaniline	—	—	—	16	21	—	61
2-Methoxy-5-methylaniline	—	—	—	31	31	—	—
2-Methylaniline	—	—	—	—	—	—	57
N-Methylaniline	—	—	31	—	—	—	—
2-Methyl-5-chloroaniline	—	—	—	27	25	—	—
1-Naphthylamine	—	—	—	—	—	—	48
2-Naphthylamine	—	—	—	—	—	—	42
2-Nitroaniline	20	63	—	—	—	—	—
3-Nitroaniline	51	70	—	—	—	—	—
4-Nitroaniline	42	64	—	—	—	—	—
2-Nitro-4-chloroaniline	4	19	—	—	—	—	—
4-Nitro-2-chloroaniline	9	31	—	—	—	—	—
2-Nitro-4-methoxyaniline	10	37	—	—	—	—	—
4-Nitro-2-methoxyaniline	10	38	—	—	—	—	—
2-Nitro-4-methylaniline	6	24	—	—	—	—	—
3-Nitro-6-methylaniline	19	50	—	—	—	—	—

TABLE PC 6—(Continued)
AMINES, AROMATIC

Paper	P1	P1	P2	P2	P2	P2	P2
Solvent	S1	S2	S3	S4	S5	S6	S7
Literature	1	1	2	3	3	3	4
Compound	$R_F \times 100$						
4-Nitro-2-methylaniline	14	46	—	—	—	—	—
***o*-Tolidine**	—	—	—	—	—	—	6
2,4,5-Trichloroaniline	0	3	—	—	—	—	—

Paper P1 = Niederschlag FN14 (German).
P2 = W3.

Solvent S1 = 5% acetic acid, paper impregnated with 0.17 weight (compared to original paper) 5–20% 1-bromonaphthalene in hexane.
S2 = 30% acetic acid, paper impregnated with 0.18 weight as S1.
S3 = cyclohexane, paper impregnated with 50% dimethylformamide in ethanol.
S4 = heptane, paper impregnated with 25% dimethylformamide in benzene.
S5 = heptane, paper impregnated with 20% ethylene glycol monoacetate in ethanol.
S6 = chloroform-ethyl acetate (1 : 1 v/v), paper impregnated with 20% formamide in ethanol.
S7 = heptane, paper impregnated with 20% formamide in ethanol.

Detection Diazotized *p*-nitraniline.

Technique Not specified.

Literature

1. Bassl, A., Heckemann, H. J., and E. Baumann,, *J. Prakt. Chem.*, **36**, 276 (1967); *J. Chromatogr.*, **36**, D81 (1968).
2. Nosek, J. and J. Gasparič, *Collect. Czech. Chem. Commun.*, **32**, 2030 (1967); *J. Chromatogr.*, **32**, D11 (1968).
3. Gemzová, I. and J. Gasparič, *Mikrochim. Acta*, 1966, pp 313, 318; *J. Chromatogr.*, **26**, D5 (1967).
4. Gasparič, J. and B. Kloucek, *Collect. Czech. Chem. Commun.*, **31**, 106 (1966); *J. Chromatogr.*, **22**, D8 (1966).

TABLE PC 7

AMINES, ALIPHATIC POLYAMINES AND AMINO ALCOHOLS

Paper	P1	P1	P1	P1	P2	
Solvent	S1	S2	S3	S4	S5	
Detection	—	—	—	—	—	D1
Literature	1	1	1	1	1	
Compound	$R_F \times 100$					Color
3-Aminopropanol	82	78	56	35	19	bl
***n*-Butylamine**	88	86	85	73	—	gy-bl-vt
1-Chloro-2-aminoethane	80	74	59	45	—	bl-br
Diethanolamine	—	—	—	—	38	—
Diethylenetriamine	45	36	12	2	—	br-bl
***N,N*-Diethylethylenediamine**	78	77	49	21	—	gr-br
Diisopropanolamine	—	—	—	—	72	—
Dipropylenetriamine	57	48	15	2	—	d-gy-bl
Ethylamine	82	78	63	47	—	gy-bl-vt
Ethylenediamine	51	44	20	5	—	gy-bl
Hexamethylenediamine	76	73	42	13	—	d-bl-vt
***N*-(2-Hydroxyethyl)-ethylenediamine**	64	60	26	8	5	br-gy-bl
Isopropanolamine	82	73	55	35	—	bl
Methylamine	75	72	50	31	—	gy-bl-vt
***N*-Methyldiethanolamine**	—	—	—	—	70	—
Monoethanolamine	75	67	45	25	17	bl
Monoisopropanolamine	—	—	—	—	31	—
Pentamethylenediamine	72	66	36	10	—	bl-vt
***n*-Propylamine**	85	83	75	61	—	gy-bl-vt
Tetramethylenediamine	66	59	29	8	—	bl-vt
Triethanolamine	—	—	—	—	67	—
Triethylenetetramine	42	32	6	1	—	br-bl
Triisopropanolamine	—	—	—	—	at front	—
Trimethylenediamine	58	49	23	6	—	br-vt

Paper P1 = W3.
P2 = E 208 (slow).

Solvent S1 = methanol-concentrated HCl (1 : 1 v/v).
S2 = ethanol-concentrated HCl (1 : 1 v/v).
S3 = *n*-propanol-concentrated HCl (2 : 1 v/v).
S4 = *n*-butanol-concentrated HCl (4 : 1 v/v).
S5 = ethyl acetate-dimethylformamide (2 : 1 v/v).

Detection D1 = Ninhydrin-pyridine.

Literature
1. Borecky, J., *Mikrochim. Acta*, pp 283, 284 (1966); *J. Chromatogr.*, **28**, D1 (1967).

TABLE PC 8
AMINES, CHOLINE ESTERS

Paper	P1	P1	P1	P1	P1	P2	P1	P1
Solvent	S1	S2	S3	S4	S5	S6	S7	S8
Technique	T1	T1	T1	T1	T1	T2	na	na
Literature	1	1	1	1	1	2	3	3
Compound*	$R_F \times 100$							
Acetylcholine	14	17	24	46	33	46	18	36
Acetyl-β-methylcholine	19	23	35	56	37	55	—	—
Acetylsalicylcholine	—	—	—	—	—	67	—	—
Acetylthiamine	—	—	—	—	—	33	—	—
Acetylthiocholine iodide	—	—	—	—	—	58	—	—
Benzoylcholine	28	30	43	65	49	71	—	—
Betaine	—	—	—	—	—	46	—	—
Butyrylcholine	28	29	43	66	46	66	—	—
Carbaminoylcholine	—	—	—	—	—	30	7	14
Choline	9	13	24	38	25	38	14	31
Lactylcholine	18	—	—	—	—	—	—	—
Propionylcholine	22	27	35	59	—	57	—	—
Salicylcholine	—	—	—	—	—	64	82	97
							60	69
Succinyldicholine iodide	—	—	—	—	—	12	—	—
Valerylcholine	—	55	—	—	—	—	—	—

* Chlorides unless stated otherwise.

Paper P1 = W4.
P2 = Munktell OB.

Solvent S1 = *n*-butanol saturated with water.
S2 = *n*-butanol-1-propanol-water (4 : 2 : 1 v/v).
S3 = *n*-propanol-water (9 : 1 v/v).
S4 = *n*-propanol-formic acid-water (8 : 1 : 1 v/v).
S5 = *n*-propanol-benzyl alcohol-water (5 : 2 : 2 v/v).
S6 = *n*-butanol-ethanol-acetic acid-water (8 : 2 : 1 : 3 v/v).
S7 = ethyl acetate-pyridine-water (50 : 35 : 15 v/v).
S8 = ethyl acetate-pyridine-water (50 : 30 : 20 v/v).

Detection Dipicrylamine.

Technique T1 = either ascending or descending development.
T2 = saturated chamber, 20 ± 0.5°C, ascending development, 40 cm in 16 hours.

Literature
1. Whitaker, V. P. and S. Wijesundra, *Biochem. J.*, **51**, 348–351 (1952) and **52**, 475–479 (1952).
2. Augustinsson, K. B. and M. Grahn, *Acta Chem. Scand.*, **7**, 906–912 (1953).
3. Malyoth, G. and H. W. Stein, *Biochem. Z.*, **322**, 165–167 (1951).

TABLE PC 9
AMINES, GUANIDINES

Paper	1	1	1	1
Solvent	S1	S2	S3	S4
Literature	1	1	1	1
Compound	$R_F \times 100$			
Alkyl guanidines				
H	49	49	34	60
Methyl	56	52	46	66
Ethyl	69	65	60	74
n-Propyl	80	72	71	79
iso-Propyl	80	72	73	79
Allyl	70	66	61	76
n-Butyl	83	78	77	84
iso-Butyl	83	78	77	84
n-Amyl	89	83	81	85
iso-Amyl	88	83	84	85
iso-Amylene (galegine)	86	79	77	86
Benzyl	81	78	68	82
Hydroxy-n-alkyl guanidines				
2-Hydroxyethyl	44	51	32	61
3-Hydroxypropyl	51	55	39	65
4-Hydroxybutyl	56	58	41	66
5-Hydroxyamyl	63	68	58	75
2-Hydroxybutyl	64	67	58	74
Carboxyalkyl guanidines (guanidino-acids)				
α-G-Acetic (glycocyamine)	33	31	4	38
α-G-Propionic	47	42	14	47
β-G-Propionic	57	33	7	43
α-G-*n*-Butyric	59	53	22S	55
γ-G-*n*-Butyric	61	34	9	47
α-G-*n*-Valeric	72	61	38S	65
α-G-*iso*-Valeric	69	57	36S	62
δ-G-*n*-Valeric	69	37	11	50
α-G-*n*-Caproic	77	71	47S	72
α-G-*iso*-Caproic	82	65	40S	69
α-G-β-Methylvaleric	80	65	40S	68
ε-G-*n*-Caproic	72	56	11	57
guanidino-alkyl guanidines (α, ω-diguanidino-alkanes)				
G—$(CH_2)_2$—G^{++}	16	0–20S	0	0–38S
G—$(CH_2)_3$—G	16	0–34S	0	0–50S
G—$(CH_2)_4$—G (Arcaine)	36	0–45S	0	56
G—$(CH_2)_5$—G	43	47	0–15S	57
G—$(CH_2)_6$—G	51	49	0–20S	59
G—$(CH_2)_{10}$—G	79	63	30	74
1-alkyl-3-nitroguanidines[a]				
H	58	76	55	78
Methyl	68	80	65	81
Ethyl	82	88	81	86
n-Propyl	91	93	88	88
iso-Propyl	86	93	91	89
Allyl	80	89	85	88
n-Butyl	92	95	93	92
tert-Butyl	97	96	93	92
Benzyl	97	97	87	93
Tolyl	97	97	87	98
Nitroarginine	22	45	45	54

TABLE PC 9—(Continued)
AMINES, GUANIDINES

Paper	1	1	1	1
Solvent	S1	S2	S3	S4
Literature	1	1	1	1
Compound	$R_F \times 100$			
Miscellaneous guanidines				
Agmatine	16	8S	12	14S
Aminoguanidine	—	4	0	5S
Arginine	18	13	0	20S
Argininosuccinic acid	20	18	0	27
Argininosuccinic anhydride	15	17	0	79
Bornyl guanidine	—	93	98	89
Canavanine	18	19	0	27
Creatine	37	36	8	41
Creatinine	45	56	36	60
Creatine phosphate	37	31	12	39
Cyanamide	53	75	61	75
α : δ-Diguanidino-valeric acid	49	52	31	58
α : ε-Diguanidino-caproic acid	49	52	31	58
Homoarginine	19	17	0	20S
α-G-Glutaric acid	32	27	0	38
α-G-β-Hydroxy butyric acid	25	35	6	45
α-G-β-Hydroxy propionic acid	39	39	11	47
α-G-β-Hydroxyphenylpropionic acid	—	—	—	—
α-G-β-Imidazolyl propionic acid	27	56	14	—
α-G-β-Indolyl propionic acid	62	70	27S	73
α-G-β-Phenylpropionic acid	75	66	45	73
α-G-γ-Methiol-butyric acid	60	63	30	67
di(α-G-β-Thiol propionic acid)	23	24	0	27
***N*-Formamidino-proline**	49	44	18	43
***N*-Formamidino-hydroxyproline**	26	28	3	40
α-G-Succinic acid	28	25	0	36
α-G-Succinic anhydride	11	29	0	36
α-Keto-δ-G-valeric acid	33	18S	8	44
Octopine	19	22	0	36
Taurocyamine	25	47	19	55

[a] Negative results with Sakaguchi and diacetyl reagents. Positive detection with PCF reagent. (See Volume II, Section II.I.)

Paper P1 = W1.
Solvent S1 = *n*-butanol-glacial acetic acid-water (120 : 30 : 50).
S2 = *n*-butanol-pyridine-water (65 : 65 : 65).
S3 = isopropanol-concentrated NH_4OH-water (200 : 10 : 20).
S4 = amyl alcohol-pyridine-water (42 : 84 : 74).
Technique Ascending, no equilibration.
Literature 1. Smith, I. and H. G. Horsewell, in *Chromatographic and Electrophoretic Techniques*, Vol. 1, Smith, I., Ed., Interscience Publishers, Division of John Wiley & Sons, Inc., New York, N.Y. 1969, pp 288–289.

TABLE PC 10
AMINES, GUANIDINES

Paper	na	na	na	na	na	na	na	na
Solvent	S1	S2	S3	S4	S5	S6	S7	S8
Literature	1	1	1	1	1	1	1	1
Compound	$R_F \times 100$							
Agmatine·H_2SO_4 (guanidobutylamine)	26	—	97	97	14	9	10	12
Aminoguanidine·H_2SO_4	—	97	97	91	11	—	30	42
Arcaine·H_2SO_4 (L-4-diguanidobutane)	32	—	97	95	0	2	10	16
Arginine	17	—	86	68	6	4	16	20
Canavanine·H_2SO_4	11	51	59	37	2	—	—	16
Creatine	29	—	—	68	16	9	55	40
Creatine phosphate	—	—	—	—	11	—	—	—
Creatinine	41	—	92	87	37	30	61	57
***N,N'*-Dibutylguanidine**	90	—	—	—	91	—	—	—
***N,N'*-Diethylguanidine**	78	97	97	—	—	—	90	86
***N,N*-Dimethylguanidine**	53	—	97	97	51	18	78	57
***N,N'*-Dimethylguanidine**	54	95	95	95	65	—	80	70
Guanidine·HCl	30	—	89	78	37	18	66	44
Guanidoacetic acid	30	—	71	47	10	11	40	31
γ-Guanidoalanine	37	75	75	61	17	—	—	42
γ-Guanidobutyric acid	53	—	91	74	16	11	51	40
γ-Guanidohistidine picrate	—	92	92	77	20	—	—	57
γ-Guanidohistidine·H_2SO_4	21	84	84	70	24	—	—	60
Methylguanidine	47	—	95	92	46	17	72	55
Octopine	15	—	—	47	2	—	40	22
Taurocyamine	19	—	61	40	16	20	52	48

Solvent
S1 = *n*-butanol-acetic acid-water (60 : 15 : 25).
S2 = phenol-water (80 : 20). Stable indefinitely.
S3 = Add 1 ml of conc. NH_4OH to 200 ml of S2 solvent just before use.
S4 = phenol-ethanol-water (120 : 40 : 40) plus NH_3 as above. This solvent is usually prepared by adding to the S2 (150 ml), water (10 ml), and ethanol (40 ml). It is stable.
S5 = ethanol-water-conc. NH_4OH (90 : 5 : 5).
S6 = *tert*-butanol-methyl ethyl ketone-water (2 : 2 : 1). To 200 ml add 8 ml of diethylamine.
S7 = methanol-water (80 : 20). To 200 ml add 8 ml of pyridine.
S8 = *n*-butanol-pyridine-water (1 : 1 : 1).

Detection Sakaguchi reaction.

Literature
1. Jepson, J. B. and I. Smith, *Nature*, 172, 1100–1101 (1953).

TABLE PC 11
AMINO ACIDS

Paper	P1	P2	P2	P2	P2	P1	P2	na	na	P3	P3	P2
Solvent	S1	S2	S3	S4	S5	S6	S7	S8	S9	S10	S11	S12
Technique	T1	T2	na	na	na	na	na	na	na	na	na	na
Literature	1	2	2	2	2	3–6	7	8	8	9	3,9	6
Compound	$R_F \times 100$											
Adenine	—	—	—	—	—	56	—	—	—	—	—	58
Alanine	47	22	30	19	23	30	27	—	—	—	52	—
Arginine	24	—	—	—	—	15	—	—	—	—	16	—
Asparagine	22	—	—	—	—	12	—	—	—	—	21	—
Aspartic acid	22	—	—	—	—	23	17	—	—	—	40	—
***n*-Butyric acid**	—	31	39	22	29	—	—	—	—	—	—	—
Cystathionine	—	—	—	—	—	10	—	—	—	46	19	—
Cysteic acid	—	—	—	—	—	10	—	32	11	71	39	42
Cysteine	—	—	—	—	—	10	—	—	—	—	18	58,23
Cysteine methyl ester	—	—	—	—	—	16	—	—	—	—	—	—
Cystine	13	—	—	—	—	9	—	—	—	30	20	33
DL-β,β-Dimethylaspartic acid	—	—	—	—	—	32	—	—	—	—	—	—
***S*-Ethylcysteine**	—	—	—	—	—	54	—	—	—	—	—	—
Glutamic acid	34	—	—	—	—	28	19	—	—	—	48	—
Glutamine	34	—	—	—	—	17	—	—	—	—	31	—
Glutathione	—	—	—	—	—	24	—	—	—	—	—	—
Glycine	30	16	20	15	17	23	26	—	—	—	37	—
Heptanoic acid	—	66	68	49	58	—	—	—	—	—	—	—
Histidine	16	—	—	—	—	12	23	—	—	—	28	35
Homocysteic acid	—	—	—	—	—	12	—	32	16	27	46	41
Homocystine	—	—	—	—	—	50	—	—	—	39	31	62,34
Homocystine thiolacetone	—	—	—	—	—	—	—	—	—	85	81	—
Homoserine	—	—	—	—	—	30	—	—	—	—	—	58
Hydroxyproline	43	—	—	—	—	22	—	—	—	—	45	—
Isoleucine	80	—	—	35	—	67	—	—	—	—	68	—
Leucine	80	—	—	38	—	70	—	—	—	—	70	—
Lysine	19	—	—	—	—	12	13	—	—	—	15	—
Methionine	68	—	—	—	—	56	—	—	—	—	60	66
Methionine sulfone	—	—	—	—	—	24	—	52	36	—	43	56
Methionine sulfoxide	—	—	—	—	—	22	—	40	33	—	45	55
DL-*threo*-β-Methylaspartic acid	—	—	—	—	—	30	—	—	—	—	—	—
***N*-Methylcysteine**	—	—	—	—	—	12	—	—	—	—	—	—
***S*-Methylcysteine**	—	—	—	—	—	37	—	—	—	—	—	—
***S*-Methylcysteine *S*-oxide**	—	—	—	—	—	17	—	—	—	—	—	—
***S*-Methylcysteine sulfoxide**	—	—	—	—	—	—	—	50	17	—	—	—
***S*-Methylglutathione**	—	—	—	—	—	28	—	—	—	—	—	—
Nonanoic acid	—	—	74	58	68	—	—	—	—	—	—	—
Norleucine	—	58	61	41	50	—	—	—	—	—	—	—
Norvaline	—	46	52	31	40	—	—	—	—	—	—	—
Octanoic acid	—	—	72	55	64	—	—	—	—	—	—	—
Phenylalanine	72	—	—	—	—	60	—	—	—	—	60	—
Proline	59	—	—	—	—	34	—	—	—	—	55	—
Serine	27	—	—	—	—	22	—	—	—	—	42	—
Taurine	—	—	—	—	—	16	—	66	17	—	31	60
Threonine	39	—	—	—	—	26	—	—	—	—	50	—
Tryptophan	62	—	—	—	—	50	—	—	—	—	42	—
Tyrosine	51	—	—	—	—	45	60	—	—	—	50	—
Valine	62	—	—	27	—	51	—	—	—	—	65	—

Paper P1 = W2.
P2 = W1.
P3 = W3MM.

Solvent S1 = phenol-*n*-butanol-methyl ethyl ketone-propionic acid-acetic acid-H_2O (20 : 20 : 50 : 10 : 10 : 20 v/v).
S2 = *n*-butanol-acetic acid-H_2O (4 : 1 : 5 v/v).
S3 = *n*-butyl acetate-*n*-butanol-acetic acid-H_2O (47 : 9 : 28 : 16 v/v).

TABLE PC 11—(Continued)
AMINO ACIDS

Solvent
S4 = *tert*-amyl alcohol-dioxane-H_2O (5 : 1 : 2 v/v).
S5 = *tert*-amyl alcohol-pyridine-H_2O (2 : 1 : 1 v/v).
S6 = *n*-butanol-acetic acid-H_2O (12 : 3 : 5 v/v).
S7 = *n*-butanol-pyridine-H_2O (1 : 1 : 1 v/v).
S8 = pyridine-acetone-NH_3-H_2O (45 : 30 : 5 : 20 v/v).
S9 = isopropanol-formic acid-H_2O (75 : 12.5 : 12.5 v/v).
S10 = acetone-0.5% aqueous urea (6 : 4 v/v).
S11 = methanol-pyridine-H_2O (80 : 4 : 20 v/v).
S12 = ethanol-acetic acid-H_2O (65 : 1 : 34 v/v).

Detection Ninhydrin.

Technique
T1 = circular chromatography.
T2 = ascending.

Literature

1. Barbiroli, G., *Mikrochim. Ichnoanal. Acta*, p 655 (1965).
2. Trzaska, J. and G. N. Kowkabany, *J. Chromatogr.*, **26**, 141 (1967).
3. Smith, I., Ed., *Chromatographic and Electrophoretic Techniques*, Vol. I, Interscience Publishers, Division of John Wiley & Sons, Inc., New York, pp 112–113 (1969).
4. Schoental, R. and D. J. Rive, *Biochem. J.* **97**, 466 (1965).
5. Burrows, I. E., Shaw, G., and D. V. Wilson, *J. Chem. Soc.*, **C**, p 42 (1968).
6. Gaitonde, M. K. and G. E. Gaull, *Biochem. J.*, 102, 959 (1967).
7. Zamorani, A., Pifferi, P. G., Lanzarini, G., and A. Riva, *Gazz. Chim. Ital.*, **98**, 472 (1968).
8. Perry, T. L., Dixon, G. H., and S. Hansen, *Nature*, **206**, 895 (1965).
9. Bessman, S. P., Koppanyi, Z. H., and R. A. Wapnir, *Anal. Biochem.*, **18**, 215 (1967).

TABLE PC 12
CARBOHYDRATES, SUGAR PHOSPHATES

Paper	P1	P1	P1	P1	P1	P1
Solvent	S1	S2	S3	S4	S5	S6
Literature	1	1	2	3	3	3
Compound			$R_F \times 100$			
2,3-Diphosphoglycerate	—	—	—	11	7	15
D-Fructose-1,6-diphosphate	40	24	37	8	8	13
D-Fructose-6-phosphate	34	44	37	17	36	54
D-Glucose-1-phosphate	27	60	31	14	36	44
D-Glucose-6-phosphate	38	48	—	12	29	50
D-Glyceraldehyde-3-phosphate	—	—	—	7	19	—
Glycerol-1-phosphate	—	—	—	26	39	54
Orthophosphate	63	28	67	33	21	50
Phosphoenolpyruvate	52	46	—	—	—	—
2-Phosphoglycerate	46	18	—	27	41	23
3-Phosphoglycerate	50	35	—	23	22	28
Pyrophosphate	46	8	40	—	—	—
D-Ribose-1-phosphate	—	—	—	15	40	50
D-Ribose-5-phosphate	—	—	46	—	—	—

Paper P1 = W1.

Solvent
S1 = methanol-formic acid-water (80 : 15 : 5 v/v).
S2 = methanol-ammonium hydroxide-water (60 : 10 : 30 v/v).
S3 = acetone-35% chloroacetic acid (60 : 40 v/v).
S4 = ethyl acetate-acetic acid-water (3 : 3 : 1 v/v).
S5 = methyl cellosolve-methyl ethyl ketone-3*N* NH_4OH (7 : 2 : 3 v/v).
S6 = ethyl acetate-formamide-pyridine (6 : 4 : 1 v/v).

Detection Phosphomolybdic acid.

Technique Ascending and descending.

Literature

1. Bandursky, R. S. and B. Axelrod, *J. Biol. Chem.*, **193**, 405–410 (1951).
2. Burrows, S., Grylls, F. S. M., and J. S. Harrison, *Nature*, **170**, 800–801 (1952).
3. Mortimer, D. C., *Can. J. Chem.*, **30**, 653–660 (1952).

TABLE PC 13
CARBOXYLIC ACIDS

Paper	P1	P1	P1	P1	P1	P1	P1	P1	P1	P1[a]	P1
Solvent	S1	S2	S3	S4	S5	S6	S7	S8	S9	S10	S11
Literature	1	1	1	1	1	1	1	1	1	1	1
Compound						$R_F \times 100$					
Aconitic	36	81	78	—	70	73	2	9	—	—	—
Adipic	86	—	—	—	—	79	20	38	—	57.5	—
Ascorbic	—	—	—	—	44	—	—	—	32	—	—
Azelaic	—	—	—	—	—	—	—	—	—	94.0	—
Caffeic	—	—	—	—	—	—	—	—	—	—	63
Chlorogenic	—	—	—	—	—	—	—	—	—	—	29
Citric	26	31	37	15	38	23	2	6	46	5.5	22
Fumaric	63	—	86	90	—	82	17	35	—	48.5	80
Glutaric	78	—	80	—	—	—	—	—	—	46.5	66
Glyceric	—	—	—	—	—	—	—	—	49	—	—
Glycerate (Ca)	—	—	—	—	—	—	—	—	48	—	—
Glycolic	59	51	—	45	—	42	34	37	—	—	—
Glycolate (K)	—	—	—	—	—	—	—	—	64	—	—
Isocitric	—	—	—	—	—	—	—	—	46	—	—
Itaconic	—	—	—	—	—	—	—	—	82	—	—
α-Ketoglutaric	—	63	—	36	—	63	8	28	69	—	25
Lactic	72	66	77	72	77	59	48	61	80	46.0, 60.0	—
Levulinic	91	—	—	—	94	—	—	—	—	—	—
Maleic	—	—	46	—	46	38	13	25	78	33.5	—
Malic	42	39	44	25	49	32	9	23	55	11.5	30
Malonic	48	63	60	52	61	51	7	17	74	25.0	—
Oxalic	18	—	5	12	—	2	2	5	18	13.5	11
Pimelic	—	—	—	—	—	—	—	—	—	73.0	—
Pyruvate (K)	—	—	—	—	—	—	—	—	83	—	—
Sebacic	—	—	—	—	—	—	—	—	—	100.0	—
Suberic	—	—	—	—	—	—	—	—	—	84.5	—
Succinic	66	70	72	76	78	66	14	30	79	34.0	60
Syringic	95	—	—	—	—	—	—	—	—	—	—
Tartaric	19	22	23	9	26	12	6	15	33	4.5	10
Tricarballylic	52	63	67	—	—	—	—	—	—	—	—

[a] R_F relative to sebacic acid.

Paper P1 = W1.

Solvent
S1 = phenol-water-formic acid (75 : 25 : 1).
S2 = mesityl oxide-water-formic acid (75 : 75 : 36).
S3 = butanol-formic acid-water (10 : 2 : 15).
S4 = ether-acetic acid-water (13 : 3 : 1).
S5 = ethyl acetate-acetic acid: water (3 : 1 : 1).
S6 = ethyl cellosolve-NH_4OH-water (80 : 5 : 15).
S7 = ethanol-NH_4OH-water (80 : 5 : 15).
S8 = CCl_4-acetic acid-water (4 : 1 : 1).
S9 = ether-formic acid-water (5 : 2 : 1).
S10 = *n*-amyl formate-water-formic acid (7 : 1 : 2).
S11 = 1-pentanol-water-formic acid (20 : 12 : 1).

Detection Acid-base indicators; ammoniacal silver nitrate.

Literature
1. Block, R. J., Durrum, E. L., and G. Zweig, *Paper Chromatography and Paper Electrophoresis*, Academic Press, N.Y., 1958, p 233.

TABLE PC 14
INORGANICS, ANIONS

Paper	P1	P1	P1	P1	P1	P1	P1	P1	P1	P1	P1	P1	P1
Solvent	S1	S2	S3	S4	S5	S6	S7	S8	S9	S10	S11	S12	S13
Literature	1	1	1	1	1	1	1	1	1	1	1	1	1
Anion							$R_F \times 100$						
Fluoride	4	6	24	40	46	48	56	63	71	77	78	88	92
Chloride	6	13	32	47	55	56	68	74	77	76	88	94	96
Bromide	9	16	36	46	55	60	67	71	75	69	85	92	94
Iodide	18	28	48	62	60	70	71	76	77	80	81	85	87
Thiocyanate	26	39	59	68	77	77	79	76	80	81	81	85	88
Chlorate	21	28	47	60	67	74	77	74	81	80	85	89	92
Bromate	5	9	27	38	50	56	62	64	73	72	81	87	90
Iodate	0	0	4	14	23	30	39	47	60	65	76	83	86
Perchlorate	30	42	60	68	70	75	75	77	77	84	89	93	94
Periodate				0	0	0	0	0	0	0	0	0	0
	0	0	4	14	26	36	45	49	62	65	77	83	86
										82			
Sulfite	0	0	5	21	20	32	40	47	63	70	77	85	89
Selenite	0	1	5	14	26	32	43	51	65	68	79	91	94
Tellurite	0	0	2	4	15	24	36	44	58	61	73	82	84
Sulfate	0	0	4	13	32	44	54	60	71	75	84	97	100
Selenate	0	0	8	26	38	44	51	60	68	78	84	93	96
Thiosulfate	0	0	6	21	37	44	48	50	64	77	90	95	97
Persulate	0	1	7	20	33	38	49	46	67	80	91	97	100
Nitrite	16	26	43	55	69	59	62	71	76	79	85	88	93
Phosphite	0	0	5	21	32	42	48	52	65	74	80	92	92
Arsenite	7	8	11	20	25	28	33	47	46	60	70	77	80
Nitrate	16	27	45	59	56	67	67	72	78	84	83	90	95
Orthophosphate	0	0	2	10	9	29	29	43	63	66	80	91	96
Arsenate	0	0	1	16	10	33	34	48	62	72	83	89	96
Metaphosphate	0	0	0	0	0	0	0	0	57	72	82	87	90
				8	17	22	30	41	68	82	96	100	100
Pyrophosphate	0	0	0	4	8	20	26	45	60	74	80	83	91
				16	23	30	37	—	—	—	92	90	100
Perrhenate	30	45	67	76	80	78	82	82	81	81	79	81	80
Chromate	0	0	3	17	25	30	41	38	54	73	83	88	93
Molybdate	0	0	8	25	36	43	55	52	64	72	86	90	93
Tungstate	0	0	3	20	38	37	52	50	67	82	85	94	100
Vanadate	0	1	3	6	15	19	25	22	37	50	70	81	87
Borate	7	12	13	15	16	20	24	24	37	48	56	70	72
Ferricyanide	0	1	15	19	39	33	48	62	74	88	93	100	100
									67	82			
Ferrocyanide	0	0	2	1	20	11	37	51	66	82	94	100	100
Formate	6	17	31	53	61	63	63	67	73	82	88	94	96
Oxalate	0	0	3	17	26	36	44	45	65	82	89	94	100
Acetate	7	18	31	54	61	62	63	67	77	82	89	94	100

Paper P1 = W1.

Solvent Solvents are mixtures of isopropanol-1.5*N* ammonia:

S1 = 95 : 5.
S2 = 90 : 10.
S3 = 80 : 20.
S4 = 70 : 30.
S5 = 65 : 35.
S6 = 60 : 40.
S7 = 55 : 45.
S8 = 50 : 50.
S9 = 40 : 60.
S10 = 30 : 70.
S11 = 20 : 80.
S12 = 10 : 90.
S13 = 5 : 95.

Technique Ascending.

Literature
1. Grassini, G. and L. Ossicini, *J. Chromatogr.*, 7, 351 (1962).

TABLE PC 15
INORGANICS, CATIONS AND ANIONS

Paper	P1	P1	P1	P1	P1
Solvent	S1	S2	S3	S4	S5
Literature	1	1	1	1	1
Ions			$R_F \times 100$		
K	4	11	—	—	—
Rb	4	11	—	—	—
Cs	4	11	—	—	—
Be	33	46	52	61	66
Mg	8	16	24	37	45
Ca	4	8	12	19	25
Sr	1	4	7	13	15
Ba	1	2	4	6	8
Zn	74	100	100	100	100
Cd	80	100	100	100	100
Hg(II)	100	100	100	100	100
Cu(II)	10	100	100	100	100
Ag	0	10t[a]	10–86t	55–100t	55–100t
Sn(II)	83	100	100	100	100
Pb	40	60	60	64	65
As(III)	54	78	100	—	—
Sb	90	100	100	100	100
Bi	80	100	92	85	78
Cr	3	6	9	23	30
Al	6	16	23	36	40
Fe(III)	11	58	100	100	100
Ni	7	13	20	32	33
Co	6	14	24	38	55
Rh	59	94	100	100	100
Pd	58	93	100	100	100
Pt	20t	95	100	100t	100t
Au	67	97	100	100	100
Mn(II)	9	19	25	32	33
Sc	2	8	15	30	38
Y	2	7	13	24	30
La	0	3	8	17	24
Ti	7	19	26	43	49
V	12	20	27	43	47
Zr	0	1	3	11	14
Th	0	3	9	15	18
U(VI)	11	21	30	43	47
U(IV)	0	2	6	17	25
Ga	4	10	26	65	62
In	78	100	100	100	100
Tl(III)	93	100	100	100	100
Ce(III)	2	4	8	15	23
$B_4O_7^{2-}$	60	74	77	80	83
$Cr_2O_7^{2-}$	11t	17	22	36	40
SeO_3^{2-}	63	78	100	100	100

[a] t = Tailing.

Paper P1 = W1.

Solvent S1 = butanol 100 ml + hydrobromic acid 10 ml + water 90 ml (two phases), butanol layer was used as solvent.
S2 = butanol fraction of solvent 1 + hydrobromic acid 10 ml (one phase).
S3 = butanol fraction of solvent 1 + hydrobromic acid 20 ml (one phase).
S4 = butanol fraction of solvent 1 + hydrobromic acid 40 ml (one phase).
S5 = butanol fraction of solvent 1 + hydrobromic acid 60 ml (one phase).
The solvents were used after 24 hr standing in the chromatography jars.

Technique Ascending, 22 °± 2 °C.

Literature Kertes, S. and M. Lederer, *Anal Chim. Acta.*, **15**, 543 (1956).

TABLE PC 16
NITROGEN HETEROCYCLICS, AMINOHYDROXYPYRIMIDINES AND IMIDAZOLES

Paper	P1	P1	P1	P1	P1	P1	P1	
Solvent	S1	S2	S3	S4	S5	S6	S7	
Detection	—	—	—	—	—	—	—	D1
Literature	1	1	1	1	1	1	1	
Compound				$R_F \times 100$				
4-Amino-5-imidazolecarboxamide	23	34	66	57	58	58	70	a[a]
4-Amino-5-imidazolecarboxamidine	16	22	55	42	35	41	62	a
4,5-Diamino-2,6-dihydroxypyrimidine	0	5	30	6	5	6	—	F
3,4-Diamino-6-hydroxypyrimidine	6	20	60	32	40	30	60	F
4,5-Diamino-2-hydroxypyrimidine	7	31	45	27	7	6	4	F
2,6-Dihydroxypurine (Xanthine)	22	60	22	32	37	43	45	a
2,8-Dihydroxypurine	14	27	26	30	33	40	60	F
6,8-Dihydroxypurine	30	33	31	26	35	38	40	a
2-Hydroxypurine	16	37	48	37	40	41	80	a
6-Hydroxypurine (Hypoxanthine)	25	54	59	50	41	41	60	a
8-Hydroxypurine	33	63	50	54	51	46	56	a
Purine	60	75	75	60	66	60	42	a
2,4,5-Triaminopyrimidine	10	33	58	23	36	30	55	F
4,5,6-Triaminopyrimidine	11	35	—	37	53	38	64	F
2,6,8-Trihydroxypurine (Uric acid)	5	17	12	14	23	22	40	—

[a] a = absorbs UV light; F = fluorescence.

Paper	P1 = W1 (10 × 50 cm strips).
Solvent	S1 = butanol-water (86 : 14).
	S2 = propanol-water (77 : 23).
	S3 = butanol-propionic acid-water (43 : 27 : 30).
	S4 = *tert*-butanol-ethyl methyl ketone-formic acid-water (40 : 30 : 15 : 15).
	S5 = *tert*-butanol-ethyl methyl ketone-ammonia (15*N*)-water (40 : 30 : 10 : 20).
	S6 = isopropanol-hydrochloric acid (10*N*)-water (65 : 18.4 : 16.6).
	S7 = ammonium bicarbonate (16%).
Detection	D1 = UV light except for uric acid which was detected by spraying with Folin-Ciocalteau reagent.
Technique	Uni-dimensional descending technique at room temperature (20 to 25 °C).
Literature	1. Khattak, M. N., Barker, N. T., and J. H. Grun, *Analyst*, **91**, 526 (1966).

TABLE PC 17
NITROGEN HETEROCYCLICS, IMIDAZOLES

Paper	P1	P1	P1		
Solvent	S1	S2	S3		
Detection	—	—	—	D1	D2
Literature	1	1	1		
Compound[a]	$R_F \times 100$			Color[b]	
4-Amino-5-carboxamide imidazole	42	55	28	dp. b	dp. b
Anserine	15	24	4	—	—
Carnosine	16	24	4	r	r
Ergothioneine	21	41	4	o	o
1-Guanidino-2(4-imidazolyl)-propionic acid	27	56	14	r	r
Histamine	20	47	56	r	r
Histamine β-*N*-acetyl	53	75	80	r	r
Histamine 1-methyl	23	41	50	—	—
Histamine β-*N*-ethyl	34	50	84	r	r
Histamine β-*N*-diethyl	36	52	95	r	r
Histamine β-*N*-dimethyl	28	43	89	r	r
Histidine	16	25	6	r	r
Histidinol	17	45	41	r	r
Hydroxymethyl imidazole	42	70	58	r	r
Imidazole	48	79	77	o-bn	o
Imidazole acetic acid	36	32	7	r	r
Imidazole acetol	35	45	36–50	r	r
Imidazole acrylic acid	52	47	9	r	r
Imidazole aldehyde	61	79	56	(o)	(o)
Imidazole carboxylic acid	26	32	8	(y)	(bn)
Imidazole glycerol	27	50	25	r	r
Imidazole lactic acid	28	34	8	r	r
Imidazole propionic acid	38	38	12	r	r
Imidazole pyruvic acid	25	34	4	r	r
Methyl histidine (from anserine)	16	26	6	—	—
1-Methyl imidazole	44	78	74	—	—
2-Thiol histidine	15	33	4	r	o-y
2-Thiol imidazole	64	80	58	r	o-y
2-Thiol imidazole propionic acid	59	43	9	r	o-y
2-Thiol-4-methyl imidazole	77	85	67	r	o-y
4-Ureido imidazole-5-carboxylic acid	5–30	42	10	v	v

[a] All substituents are in the 4 position unless otherwise specified.
[b] Faint colors are shown in brackets.

Paper	P1 = W1.
Solvent	S1 = *n*-butanol-glacial acetic acid-water (120 : 30 : 50).
	S2 = *n*-butanol-pyridine-water (65 : 65 : 65).
	S3 = isopropanol-conc. NH_4OH-water (160 : 8 : 16).
Detection	D1 = anisidine.
	D2 = sulfanilic acid.
Technique	Ascending.
Literature	1. Smith, I., in *Chromatographic and Electrophoretic Techniques*, Vol. 1, Smith, I., Ed., Interscience Publishers, Division of John Wiley & Sons, Inc., New York, N.Y. 1969, p 277.

TABLE PC 18
NITROGEN HETEROCYCLICS, NICOTINIC ACID AND RELATED COMPOUNDS

Paper	P1	P1	P1	P1	na	na	na
Solvent	S1	S2	S3	S4	S5	S6	S7
Technique	T1	na	na	na	T2	T3	T3
Literature	1	2	2	2	3	4	4
Compound				$R_F \times 100$			
Carbamyl-*N*′,2′-hydroxyethylpyridinium bromide	—	—	74	36	—	—	—
3-Carbamyl-*N*′-ribosidylpyridinium bromide	—	—	62	33	—	—	—
3-Carbamyltetraacetyl-*N*′-D-glucosidylpyridinium bromide	—	—	94	84	—	—	—
Cinchomeronic acid	—	—	—	—	—	29	—
Cinchomeronic acid hydrazine	—	—	—	—	—	—	13
Diethylnicotinamide	—	—	90	—	—	—	—
DPN	—	—	43	10	—	—	—
Hydrazine	—	—	—	—	—	—	15
Isonicotinic acid	—	—	—	—	—	54	—
Isonicotinic acid hydrazide	—	—	—	—	—	—	37[a]
***N*′-Methylnicotinamide chloride**	5	—	71	47	—	—	—
***N*′-Methylnicotinuric acid betaine**	—	—	57	26	—	—	—
***N*-Methyl-2-pyridone-3-carboxylic acid**	—	—	—	—	79	—	—
***N*-Methyl-2-pyridone-5-carboxylic acid**	—	—	—	—	70	—	—
***N*-Methyl-2-pyridone-5-carboxylic amide**	—	—	—	—	57	—	—
Nicotinamide	67	69	—	—	55	—	—
Nicotinic acid	30	21	—	—	68	66	—
Nicotinic acid hydrazide	—	—	—	—	—	—	49
Nicotinuric acid	13	11	—	—	—	—	—
Picolinic acid	—	—	—	—	—	47	—
Picolinic acid hydrazide	—	—	—	—	—	—	64
2,4-Pyridine dicarboxylic acid	—	—	—	—	—	17	—
2,4-Pyridine dicarboxylic acid hydrazide	—	—	—	—	—	—	0
Pyridoxine	—	—	—	—	74	—	—
Quinolinic acid	—	—	—	—	—	28	—
Quinolinic acid hydrazide	—	—	—	—	—	—	7
TPN	—	—	30	0	—	—	—
Trigonelline	7	—	—	—	16	—	—

[a] Isonicotinic acid hydrazide is decomposed to the amide.

Paper P1 = W1.

Solvent
S1 = *n*-butanol-acetone-water (45 : 5 : 50).
S2 = *n*-butanol-saturated with water.
S3 = 60% acetone.
S4 = 80% acetone.
S5 = isobutanol-98% formic acid-water (100 : 20 : 45).
S6 = *n*-butanol-acetic acid-water (4 : 1 : 2).
S7 = 10% NH_3-*n*-butanol (2 : 10).

Technique
T1 = paper strip, descending, saturated tank, 8 hours.
T2 = descending.
T3 = circular.

Literature

1. Johnson, B. C. and P. H. Lin, *J. Amer. Chem. Soc.*, **75**, 2971–73 (1953).
2. Kodicek, E. and K. K. Reddi, *Nature*, **168**, 475–77 (1951).
3. Allouf, R. and R. Munier, *Bull. Soc. Chim. Biol.*, **34**, 196–203 (1952); *Chem. Abstr.*, **47**, 447e (1953).
4. Itai, T., Oba, T. and S. Kamiya, *Eisei Shikenjo Hôkoku*, **72**, 87–90 (1954); *Chem. Abstr.*, **49**, 6350h (1955).

TABLE PC 19
NUCLEIC ACID DERIVATIVES

Paper	P1	P1	P1	P1	P1	P1	P1	P1	P1
Solvent	S1	S2	S3	S4	S5	S6	S7	S8	S9
Literature	1	1	1	1	1	1	1	1	1
Compound					$R_F \times 100$				
Thymidine	57	66	65	61	35	83	74	73	70
Deoxycytidine	34	63	59	57	30	67	74	53	49
Deoxyadenosine	37	40	38	38	13	67	49	48	48
Deoxyguanosine	25	35	28	30	9	69	48	45	42
Thymidylic acid	8	7	7	5.5	0	43	23	16	16
Deoxycytidylic acid	0	5	5	3.5	0	35	21	8	11
Deoxyadenylic acid	0	6	6	3.5	0	34	19	7	9
Deoxyguanylic acid	0	3	1	2.8	0	31	11	7	8
***TrThymidine**	96	—	—	—	—	—	—	—	—
MMTrThymidine	100	—	—	—	—	—	—	—	—
DMTrThymidine	100	—	—	—	—	—	—	—	—
T-OAc	93	—	—	—	77	—	—	93	92
pT-OAc	8	—	—	—	—	—	—	—	50
Cyanoethyl pT	—	—	—	—	22	—	—	83	52
Trichloroethyl pT	42	—	—	—	—	—	—	—	61
Trichloroethyl pT-OAc	57	—	—	—	—	—	—	—	—
TrTpT	62	—	—	—	—	—	—	—	—
MMTrTpT-OAc	57	—	—	—	—	—	—	—	—
d-pC^{An}-OAc	23	—	—	—	—	—	—	—	73
d-pC^{An}	—	—	—	—	—	—	—	—	43
d-pA^{Bz}-OAc	10	—	—	—	—	—	—	63	—
d-pA^{Bz}	—	—	—	—	—	—	—	25	—
d-pA^{Bz}-OBz	—	—	—	—	—	—	—	90	—
d-pG^{Ac}-OAc	—	—	—	—	—	—	—	—	55
d-pGpGpG	—	—	—	—	—	—	—	—	18
MMTrTpC^{An}-OAc	67	—	—	—	—	—	—	—	—
d-TpC^{An}-OAc	41	—	—	—	—	—	—	—	—
dTpA	—	—	—	—	—	—	—	15	—
DMTrTpA^{Bz}-OAc	75	—	—	—	—	—	—	—	—
DMTrTpA^{Bz}	60	—	—	—	—	—	—	—	77
DMTrTpA^{Bz}pG^{Ac}-OAc	50	—	—	—	—	—	—	—	—
MMTrTpG^{Ac}-OAc	56	—	—	—	—	—	—	—	—

* The abbreviations used in this table are derived from those used by Khorana *et al.* See: Ohtsuka, E., Moom, M. W., and H. G. Khorana, *J. Amer. Chem. Soc.*, **87**, 2956 (1965). Jacob, T. M. and H. G. Khorana, *J. Amer. Chem. Soc.*, **87**, 2971 (1965). The basic system of abbreviations used for polynucleotides and their protected derivatives is as has been used in previous papers in this series and is in current use in *J. Biol. Chem.* Thus, the single letters A, T, C, and G represent the nucleosides of respectively adenine, thymine, cytosine, and guanine. The letter "p" to the left of the nucleoside initial indicates a 5′-phosphomonoester group and the same letter to the right indicates a 3′-phosphomonoester group. Thus, in going from the left to the right the polynucleotide chain is specified in the $C_{3'}$–$C_{5'}$ direction. The protecting groups on the purine or pyrimidine rings are designated by two-letter abbreviations added as superscripts after the nucleoside initial: thus A^{Bz} for *N*-benzoyldeoxyadenosine, C^{An} for *N*-anisoyldeoxycytidine, G^{Ac} for *N*-acetyldeoxyguanosine. The acetyl group at the 3′-hydroxyl group of a nucleoside is shown by -OAc added after the nucleoside initial: thus d-pTpG^{Ac}-OAc for the dinucleotide, 5′-O-phosphorylthymidylyl-(3′→5′)-*N*3′-O-diacetyldeoxyguanosine.

Paper P1 = W3MM.

Solvent S1 = acetonitrile-ammonium acetate (80 : 20) (all Solvents prepared just before use).
S2 = acetonitrile-ammonium acetate-28% ammonium hydroxide (70 : 20 : 10).
S3 = acetonitrile-ammonium acetate-28% ammonium hydroxide-isopropanol (60 : 20 : 10 : 10).
S4 = acetonitrile-ammonium acetate-28% ammonium hydroxide-*n*-butanol (60 : 20 : 10 : 10).
S5 = acetonitrile-ammonium acetate-28% ammonium hydroxide (80 : 10 : 10).
S6 = acetonitrile-ammonium chloride, pH 5 (60 : 40).
S7 = acetonitrile-ammonium acetate-28% ammonium hydroxide (60 : 30 : 10).
S8 = acetonitrile-ammonium acetate (70 : 30).
S9 = acetonitrile-ammonium chloride, pH 5 (70 : 30).

Detection Ultraviolet light (254 nm).

Technique Ascending technique (with no prior equilibration).

Literature 1. Gabriel, T. F., *J. Chromatogr.*, **36**, 518 (1968).

TABLE PC 20
OXO COMPOUNDS, ALDEHYDES, KETONES AND DERIVATIVES

Paper	na	na	P1	P2	na	P3	P3	P1	P1
Solvent	S1	S2	S3	S4	S5	S6	S6	S7	S8
Literature	1	2	3	4	5	6	6	7	7
	$R_F \times 100$								
Compound	2,4-Dinitrophenylhydrazone					Girard T	Girard P	Free aldehyde and ketone	
Acetaldehyde	56	18	—	—	67	—	19	—	—
Acetone	42	30	85	—	—	—	19	—	—
Acetovanillone	—	—	—	—	—	—	—	60	85
Acrolein	—	21	—	—	—	—	—	—	—
***p*-Anisaldehyde**	—	—	—	—	—	41	—	—	—
Benzaldehyde	—	—	68	—	81	45	—	—	—
Biacetyl	—	—	0	—	—	—	—	—	—
***n*-Butyraldehyde**	—	42	85	—	—	36	38	—	—
***n*-Caproaldehyde**	—	57	—	—	—	—	—	—	—
5-Carboxyvanillin	—	—	—	—	—	—	—	12	37
Cinnamaldehyde	—	—	48	—	—	—	—	—	—
Crotonaldehyde	—	22	—	—	90	—	—	—	—
Cyclohexanone	23	—	—	—	—	—	—	—	—
Cyclopentanone	29	—	—	—	—	—	—	—	—
Decyl aldehyde	—	—	86	—	—	—	—	—	—
Dehydrovaniillin	—	—	—	—	—	—	—	31	streak
2,2-Dimethylcyclohexanone	15	—	—	—	—	—	—	—	—
2,4-Dinitrophenylhydrazine	—	—	5	—	32	—	—	—	—
Di-*n*-propyl ketone	—	—	—	—	—	51	51	—	—
Formaldehyde	60	9	75	—	6	0	0	—	—
5-Formylvanillin	—	—	—	—	—	—	—	20	53
Furfural (Furfuraldehyde)	—	6	57	—	70	29	28	—	—
Girard Reagent	—	—	—	—	—	9	10	—	—
Glycolaldehyde	—	—	0	—	—	—	—	—	—
Glyoxal (bis)	—	—	—	—	23	—	—	—	—
***p*-Hydroxybenzaldehyde**	—	—	—	—	—	36	—	—	—
Isobutyraldehyde	32	—	—	—	—	37	38	—	—
Isovalerylaldehyde	—	—	84	—	—	—	—	—	—
Methyl-*n*-butyl ketone	—	65	—	—	—	—	—	—	—
Methyl ethyl ketone	31	43	90	38	—	—	29	—	—
5-Methyl-2-furfural	—	—	—	—	—	39	35	—	—
Methyl glyoxal	—	—	—	—	20	—	—	—	—
Methyl heptyl ketone	—	—	—	—	—	59	62	—	—
Methyl isopropyl ketone	—	—	91	48	—	—	—	—	—
Methyl-*n*-propyl ketone	—	53	90	54	—	—	—	—	—
Nonyl aldehyde	—	—	87	—	—	—	—	—	—
***n*-Propionaldehyde**	—	32	57	—	91	—	29	—	—
Salicylaldehyde	—	—	30	—	—	48	—	—	—
Syringaldehyde	—	—	—	—	—	—	—	37	79
***n*-Valeraldehyde**	—	54	0	—	—	—	—	—	—
Vanillin	—	—	—	—	—	—	—	44	87
Vanillovanillone	—	—	—	—	—	—	—	47	86

Paper P1 = W1.
P2 = heavy filter paper soaked in sodium silicate and 6*N* HCl, dried at 110 °C.
P3 = W1 impregnated with olive oil.

Solvent S1 = ethanol-petroleum ether (80 : 20).
S2 = anhydrous methanol-heptane (1 : 1).
S3 = 5% ether in petroleum ether (65–110 °C).
S4 = ethyl ether-petroleum ether (110 °C) (5 : 95).
S5 = dibutyl ether-dimethylformamide-tetrahydrofuran (85 : 15 : 4) (fresh).
S6 = *n*-butanol-ethanol-water (27 : 3 : 10).
S7 = *n*-butanol saturated with 2% aqueous NH_4OH.
S8 = *n*-butanol-pyridine-water (10 : 3 : 3).

Detection Derivatives viewed in daylight; free compounds.

TABLE PC 20—(Continued)
OXO COMPOUNDS, ALDEHYDES, KETONES AND DERIVATIVES

Technique Ascending or descending development.

Literature
1. Sykora, V. and Z. Procházka, *Chem. Listy*, **47**, 1674–1675 (1953); from *Chem. Abstr.*, **48**, 3852e (1954).
2. Meigh, D. F., *Nature*, **170**, 579 (1952).
3. Rice, R. G., Keller, G. J., and J. G. Kirchner, *Anal. Chem.*, **23**, 194–195 (1951).
4. Kirchner, J. G. and G. J. Keller, *J. Amer. Chem. Soc.*, **72**, 1867–1868 (1950).
5. Schmidt, F. and H. U. Bauriedl, *Naturwissenschaften*, **43**, 470 (1956).
6. Seligman, R. B., Edmonds, M. D., and L. A. Lee, *Chem. Ind.* (London), 1195 (1954).
7. Pearl, I. A. and D. L. Beyer, *J. Amer. Chem. Soc.*, **76**, 2224 (1954) and **76**, 6106–6108 (1954).

TABLE PC 21
OXO COMPOUNDS, KETONES

Paper	P1	P1	P1	P1	P1	P1	P1	P1	P1
Solvent	S1	S2	S3	S4	S5	S6	S7	S8	S9
Literature	1	1	1	1	1	1	1	1	1
Compound	$R_F \times 100$								
Hexanophenone	0	2	3	5	24	44	52	69	86
4′-Methylacetophenone	14	25	35	37	54	66	71	82	88
4′-Phenylacetophenone	0	2	3	3	6	13	33	46	64
1-Phenyl-1,3-butanedione	4	5	7	14	22	44	52	61	71
***trans*-4-Phenyl-3-buten-2-one**	8	15	23	29	42	55	70	75	84
Phenyl-2-propanone	29	39	47	45	59	75	80	84	90
Phenyl-2-thienyl ketone	2	2	3	6	20	33	44	51	62
2-Tridecanone	0	1	1	1	3	10	18	34	73

Paper P1 = Reeve Angel SB-2 ion exchange paper in chloride form, 18 × 39 cm sheets.

Solvent Methanol molarity in water.
S1 = water. S4 = 10.0*M*. S7 = 16.0*M*.
S2 = 6.0*M*. S5 = 12.0*M*. S8 = 18.0*M*.
S3 = 8.0*M*. S6 = 14.0*M*. S9 = 20.0*M*.

Detection 3,5-dinitrobenzoic acid.

Technique Descending development, saturated chamber, 34 cm development distance, 1.5–3.0 hours.

Literature 1. Sherma, J. and L. H. Pignolet, *Anal. Chim. Acta.*, **34**, 185–192 (1966).

TABLE PC 22
OXO COMPOUNDS, 4-NITROPHENYLHYDRAZONES OF AROMATIC CARBONYL COMPOUNDS

Paper	P1	P1	P1	P1	
Solvent	S1	S2	S3	S4	
Detection	—	—	—	—	D1
Literature	1	1	1	1	
Compound			$R_F \times 100$		Color
Acetophenone	56	45	45	43	p-r-p
Anisaldehyde	27	26	—	—	pk
Benzalacetophenone	—	54	—	—	—
Benzaldehyde	43	44	41	37	rt
Benzophenone	94	58	—	94	br
Butyrophenone	84	60	50	46	rt
Cinnamaldehyde	60	43	—	7	—
2,4-Dimethylacetophenone	86	50	20	41[a]	—
2,5-Dimethylacetophenone	81	50	—	37[a]	—
3,4-Dimethylacetophenone	72	53	50	29[a]	d-pk
Furfural	8	33	—	—	—
2-Hydroxyacetophenone	5	35	38	7	p
3-Hydroxyacetophenone	5	24	15	2	—
4-Hydroxyacetophenone	5	3	13	2	lv
***m*-Hydroxybenzaldehyde**	4	24	12	2	—
***p*-Hydroxybenzaldehyde**	3	23	—	2	d-p
***m*-Hydroxybenzophenone**	93	54	—	83	—
***o*-Hydroxybenzophenone**	89	14	—	15	—
	22	47	—	72	—
***p*-Hydroxybenzophenone**	6	14	27	3	d-p
2-Hydroxy-3-methoxybenzaldehyde	4	20	—	—	p
2-Hydroxy-5-methylbenzophenone	9	34	—	—	—
***o*-Methoxybenzaldehyde**	29	30	—	—	lt-p
5-Methyl-2-furfural	14	40	—	—	—
2-Naphthaldehyde	2	8	—	—	br
Piperonal	17	24	—	—	o
Propiophenone	76	44	46	26[a]	rt
Salicylaldehyde	7	34	27	6	rt
Vanillin	3	3	—	—	—
Veratraldehyde	8	11	—	—	p

[a] Smears.

Paper P1 = SS2043 B gl.

Solvent S1 = cyclohexane-benzene-dipropylene glycol (70 : 30 : 3), paper impregnated with 35% solution of formamide in ethanol.

S2 = dibutyl ether-*N*,*N*-dimethylformamide-tetrahydrofuran (85 : 15 : 4), paper impregnated with 50% solution of dimethylformamide.

S3 = dibutyl ether-*N*,*N*-dimethylformamide-tetrahydrofuran (85 : 15 : 4), paper impregnated with 25% solution of dimethylformamide.

S4 = cyclohexane-benzene-formamide (15 : 12 : 7), paper impregnated with 20% solution of formamide.

Detection D1 = potassium hydroxide.

Technique Ascending development

Literature

1. Barker, E. D. and E. Sawicki, *Anal. Chem.*, **40**, 985 (1968); *J. Chromatogr.*, **39**, D10 (1969).

TABLE PC 23
OXO COMPOUNDS, AZINES OF AROMATIC CARBONYLS

Paper	P1	P1	P1	P1	P1	P1	P1
Solvent	S1	S2	S3	S4	S5	S6	S7
Literature	1	1	1	1	1	1	1
Compound	$R_F \times 100$						
4-Acetoaminobenzalazine	0	—	—	1	11	1	—
Acetophenone azine	62	—	—	80	—	92	52
4-Aminoacetophenone	—	—	—	—	4	—	—
4-Aminoacetophenone azine hydrochloride	3	—	—	2	—	—	—
Anisaldazine	22	21	22	37	—	49	17
Benzalazine	54	65	56	75	—	83	42
Benzophenone azine	57	70	55	85	71	95	42
Butyrophenone azine	86	—	—	92	—	98	73
4-Carboxybenzalazine	0	—	—	—	—	—	—
2,4-Dihydroxyacetophenone azine	—	—	—	3	8	—	—
2,5-Dihydroxyacetophenone azine	2	—	—	5	31	1	—
2,6-Dihydroxyacetophenone azine	—	—	—	5	5	—	—
2,4-Dihydroxybenzophenone azine	—	—	—	—	5	—	—
2,3-Dimethoxybenzalazine	20	—	—	64	61	68	—
2,4-Dimethylacetophenone azine	84	—	4	97	—	—	58
2,5-Dimethylacetophenone azine	86	—	—	96	—	95	67
3,4-Dimethylacetophenone azine	79	—	—	92	—	95	57
4-Dimethylaminobenzalazine	8	—	—	22	44	79	—
2-Hydroxyacetophenone azine	33	44	38	64	59	68	—
4-Hydroxyacetophenone azine	0	—	—	0	17	2	—
2-Hydroxybenzalazine	2	2	3	53	21	2	—
4-Hydroxybenzophenone azine	0	—	—	43	27	7	—
2-Hydroxy-3-methoxybenzalazine	3	—	4	41	32	12	—
2-Hydroxy-4-methoxybenzophenone azine	12	—	—	92	42	71	—
2-Hydroxy-4-methoxybenzophenone hydrazone	19	—	—	—	—	20 49	—
2-Hydroxy-5-methylbenzophenone azine	41	—	—	94	70	93	—
2-Methoxybenzalazine	31	36	41	67	55	69	28
Piperonal azine	9	9	8	41	37	32	—
Propiophenone azine	78	84	75	91	—	—	60
Salicylaldazine	31	23	24	45	—	56	—
Terephthaladazine	30	—	—	—	14 74	—	—
3,4,5-Trimethoxybenzalazine	4	—	—	30	17	16	—
Vanillin azine	0	—	—	0	11	2	—
Veratraldazine	4	—	—	4	13	13	—

Paper P1 = Schleicher and Schuell a gl.

Solvent S1 = *N,N*-dimethylformamide-cyclohexane (20 : 80), paper impregnated with 75% *N,N*-dimethylformamide-25% acetone.

S2 = *N,N*-dimethylformamide-cyclohexane (20 : 80), paper impregnated with 50% *N,N*-dimethylformamide-50% acetone.

S3 = decalin-*N,N*-dimethylformamide (80 : 20), paper impregnated with 50% *N,N*-dimethylformamide-50% acetone.

S4 = cyclohexane-*N,N*-dimethylformamide-ethyl acetate-water (120 : 120 : 20 : 40), paper impregnated with 75% *N,N*-dimethylformamide-25% acetone.

S5 = dibutyl ether-*N,N*-dimethylformamide-tetrahydrofuran (85 : 15 : 4), paper impregnated with 75% *N,N*-dimethylformamide-25% acetone.

S6 = cyclohexane-*N,N*-dimethylformamide-carbon tetrachloride (20 : 1 : 4), paper impregnated with 25% *N,N*-dimethylformamide-75% ethyl alcohol (95%).

S7 = cyclohexane-methylformanilide (4 : 1), paper impregnated with 10% methylformanilide-90% ethyl alcohol (95%).

Detection UV light.

Technique Ascending development at 20 °C.

Literature 1. Barker, E. D., *J. Chromatogr.*, **27**, 398 (1967).

TABLE PC 24
OXYGEN HETEROCYCLICS, ANTHOCYANINS

Paper		P1	na	P1	P1
Solvent		S1	S2	S3	S4
Technique		T1	T1	T1	T1
Literature		1	1	1	1
Compound	No. sugar residues	$R_F \times 100$			
Cyanidin glycosides					
3-Diglucoside	—	33	22	34	61
3,5-Diglucoside	—	28	6	16	40
3-Monogalactoside	—	37	24	7	26
3-Monoglucoside	—	38	25	7	26
3-Rhamnoglucoside	—	37	25	19	43
3-Rhamnoglucosido-5-glucoside	—	25	8	36	59
3-Xyloglucoside	—	36	24	24	51
Delphinidin glycosides					
3,5-Diglucoside	—	15	3	8	32
3,7-Diglucoside	2	30	10	38	70
3-Gentiobioside	2	30	26	21	41
3-Gentiotrioside	3	25	10	35	52
5-Glucoside-3-rutinoside	3	29	13	40	58
5-Glucoside-3-sambubioside	3	24	18	43	—
5-Glucoside-3-sophoroside	3	25	10	60	68
7-Glucoside-3-sophoroside	3	18	4	73	73
3-Monoglucoside	—	26	11	3	18
3-Rhamnoglucoside	—	30	15	11	37
Malvidin glycosides					
3,5-Diglucoside	—	31	3	13	42
3-Monoglucoside	—	38	15	6	29
3-Rhamnoglucosido-5-glucoside	—	30	5	40	63
Pelargonidin glycosides					
Aglycone	0	80	80	0	—
3,5-Diglucoside	2	31	14	23	45
3,7-Diglucoside	2	30	10	38	70
3-Galactoside	1	39	37	13	33
3-Gentiobioside	2	30	26	21	47
3-Gentiotrioside	3	25	10	35	52
3-Glucoside	1	44	38	14	35
5-Glucoside	1	51	49	18	57
7-Glucoside	1	46	51	15	—
5-Glucoside-3-rutinoside	3	29	13	40	58
5-Glucoside-3-sambubioside	3	24	18	43	—
5-Glucoside-3-sophoroside	3	25	10	60	68
7-Glucoside-3-sophoroside	3	18	4	73	73
3-(2^G-Glucosylrutinoside)	3	33	15	63	73
3-Lathyroside	2	35	31	34	60
3-Rhamnoside	1	71	64	22	53
3-Rhamnoside-5-glucoside	2	46	24	39	70
3-Rutinoside	2	37	30	22	44
3-Sambubioside	2	37	34	31	60
3-Sophoroside	2	36	30	38	65
Peonidin glycosides					
3,5-Diglucoside	—	31	10	17	44
3-Monoglucoside	—	41	30	9	33
3-Rhamnoglucosido-5-glucoside	—	29	12	37	60
Petunidin glycosides					
3,5-Diglucoside	—	24	4	8	32
3-Monoglucoside	—	35	14	4	22
3-Rhamnoglucoside	—	35	16	13	42
3-Rhamnoglucosido-5-glucoside	—	23	6	37	61

TABLE PC 24—(Continued)
OXYGEN HETEROCYCLICS, ANTHOCYANINS

Paper	P1 = W1.
Solvent	S1 = *n*-butanol-acetic acid-water (4 : 1 : 5, upper layer).
	S2 = *n*-butanol-2*N* acetic acid (1 : 1, upper layer).
	S3 = water-12*N* HCl (97 : 3).
	S4 = water-acetic acid-12*N* HCl (82 : 15 : 3).
Detection	UV and visible light.
Technique	T1 = descending.
Literature	1. Harborne, J. B., *J. Chromatogr.*, **1**, 473 (1958).

TABLE PC 25
OXYGEN HETEROCYCLICS, ANTHOCYANIDINS

Paper	P1	P1	P2	P2	P1		
Solvent	S1	S2	S3	S4	S5		
Detection	—	—	—	—	—	D1	D1
Technique	na	na	na	na	na		
Literature	1	1	2	2	1	1	1
Compound	$R_F \times 100$					Colors	
Apigeninidin	75	74	—	—	44	y	y-br
Capensinidin	88	79	—	—	—	p	—
Cyanidin	49	68	69	69	22	ma	pk
Delphinidin	32	42	35	52	13	p	mv
Hirsuitidin	78	66	—	—	36	p	mv
Luteolinidin	61	56	—	—	35	o	r-br
Malvidin	60	58	53	90	27	p	mv
Pelargonidin	68	80	80	82	33	r	o-r
Peonidin	63	71	72	87	30	ma	pk
Petunidin	46	52	45	75	20	p	mv
Pulchellidin	50	48	—	—	24	p	—
Rosinidin	76	77	—	—	39	ma	pk
Tricetinidin	46	38	—	—	28	or	—

Paper	P1 = SS2043b.
	P2 = W1.
Solvent	S1 = acetic acid-conc. HCl-water (30 : 3 : 10).
	S2 = *n*-butanol-acetic acid-water (4 : 1 : 5, upper layer).
	S3 = *n*-butanol-2*N* HCl.
	S4 = *m*-cresol-5.5*N* HCl-acetic acid (1 : 1 : 1).
	S5 = formic acid-conc. HCl-water (5 : 2 : 3).
Detection	D1 = UV or visible light.
Literature	1. Harborne, J. B., *J. Chromatogr.*, **1**, 473 (1958).
	2. Bate-Smith, E. C., *Biochem. J.*, **58**, 123 (1954).

TABLE PC 26
OXYGEN HETEROCYCLICS, FLAVONOIDS

Paper Solvent Literature	P1 S1 1,2	P1 S2 1,2	P1 S3 1,2	P1 S4 3	P1 S5 4	P1 S6 4	P1 S7 4	P1 S8 4	P1 S9 2	P1 S10 2	P1 S11 2
Compound						$R_F \times 100$					
Aurone aglycones											
Aureusidin	57	—	29	—	—	—	—	—	1	10	57
Leptosidin	76	—	80	—	—	—	—	—	1	19	76
Maritimetin	53	—	13	—	—	—	—	—	—	5	53
Sulfuretin	80	—	70	—	—	—	—	—	1	19	80
Aurone glucosides											
Aureusin	—	—	35	—	—	—	—	—	1	16	28
Cernuoside	—	—	45	—	—	—	—	—	2	25	49
Leptosin	—	—	73	—	—	—	—	—	3	33	51
Maritimein	—	—	—	—	—	—	—	—	—	21	43
Sulfurein	28	—	69	—	—	—	—	—	3	41	49
Chalcone aglycones											
Butein	78	51	66	—	—	—	—	—	1	19	83
Okanin	56	38	—	—	—	—	—	—	—	8	56
Stillopsidin	—	—	—	—	—	—	—	—	1	—	65
Chalcone glucosides											
Coreopsin	56	—	64	—	—	—	—	—	5	43	56
Marein	45	—	—	—	—	—	—	—	—	22	38
Stillopsin	—	—	—	—	—	—	—	—	2	—	47
Flavanone aglycones											
Butin	91	—	—	—	98	—	—	55	20	64	94
Eriodictyol	—	—	—	29	—	—	—	—	—	56	85
Hesperitin	89	—	96	60	97	94	—	50	11	67	—
Homoeriodictyol	88	—	—	61	—	—	—	—	14	67	—
Naringenin	89	—	—	49	—	—	—	—	16	66	69
Pinocembrin	97	—	98	80	99	97	—	58	—	—	—
Sakuranetin	90	—	—	—	—	—	—	—	9	65	—
Flavone aglycones											
Acacetin	91	91	88	75	94	80	0	0	—	—	—
Apigenin	89	83	88	33	87	89	0	0	—	—	—
Auranetin	91	—	93	—	92	99	34	63	—	—	—
Baicalein	—	—	—	61	97	83	—	—	—	—	—
Chrysin	—	—	—	78	86	—	0	0	—	—	—
Chrysoeriol	—	—	—	47	—	—	—	—	—	—	—
Diosmetin	85	80	86	47	—	—	—	—	—	—	—
Luteolin	78	66	66	17	—	—	—	—	—	—	—
Norwogonin	—	—	—	51	85	83	0	0	—	—	—
Saponaretin	56	86	80	—	—	—	—	—	—	—	—
Tectochrysin	—	—	—	93	98	94	—	27	—	—	—
Tricin	73	72	87	—	—	—	—	—	—	—	—
Vitexin	43	82	63	—	—	—	—	—	—	—	—
Flavone glycosides											
Apiin	57	—	75	—	—	—	—	42	6	—	—
Cosmetin	65	—	78	—	—	—	—	25	4	—	—
Galuteolin	82	—	65	—	—	—	—	7	0	—	—
Linarin	61	—	55	—	—	—	—	60	14	—	—
Naringin	70	—	75	—	51	86	77	80	—	—	—
Neo-hesperidin	67	—	75	—	38	88	72	81	—	—	—
Rhoifolin	58	—	74	—	—	—	—	46	9	—	—
Saponarin	42	—	65	—	—	—	—	64	31	—	—
Flavonol aglycones											
Azaleatin	48	49	50	—	—	—	—	—	—	—	—
Fisetin	73	58	32	9	—	—	—	—	—	—	—
Gossypetin	21	—	9	—	59	51	7	12	—	—	—
Isorhamnetin	74	53	66	—	—	—	—	—	—	—	—
Kampferol	83	55	58	30	90	77	4	10	—	—	—
Morin	79	73	34	—	71	58	13	27	—	—	—
Myricetin	43	28	13	—	78	45	—	12	—	—	—
Nortangeretin	76	—	92	—	79	60	4	10	—	—	—

TABLE PC 26—(Continued)
OXYGEN HETEROCYCLICS, FLAVONOIDS

Paper	P1	P1	P1	P1	P1	P1	P1	P1	P1	P1	P1
Solvent	S1	S2	S3	S4	S5	S6	S7	S8	S9	S10	S11
Literature	1,2	1,2	1,2	3	4	4	4	4	2	2	2
Compound						$R_F \times 100$					
Flavonol aglycones											
Patuletin	78	—	13	—	81	76	6	10	—	—	—
Quercetagetin	22	—	6	0	17	61	13	19	—	—	—
Quercetin	64	41	29	10	81	67	4	7	—	—	—
Rhamnetin	72	53	66	—	92	73	3	8	—	—	—
Robinetin	40	36	18	—	41	58	3	8	—	—	—
Flavonol glycosides											
Astralagin	70	—	74	—	—	—	—	43	13	—	—
Avicularin	70	—	61	—	—	—	—	31	7	—	—
Cannabiscitrin	47	—	32	—	—	—	—	25	5	—	—
Dactylin	38	—	63	—	—	—	—	62	27	—	—
Gossypin	87	—	81	—	2	—	—	—	—	—	—
Gossypitrin	49	—	39	—	11	54	17	14	—	—	—
Hyperin	55	—	56	—	—	—	—	35	9	—	—
Isoquercitrin	58	—	54	—	40	79	24	37	8	—	—
Myricitrin	60	—	39	—	35	87	—	44	15	—	—
Nicotiflorin	54	—	64	—	—	—	—	54	23	—	—
Quercetrin	72	—	58	—	50	79	45	49	19	—	—
Quercimeritrin	37	—	40	—	50	80	27	7	0	—	—
Reynoutrin	65	—	—	—	—	—	—	32	6	—	—
Robinin	40	—	73	—	—	—	—	75	54	—	—
Rutin	45	—	46	—	15	83	45	51	23	—	—
Spiraeoside	48	—	33	—	—	—	—	13	0	—	—
Isoflavone aglycones											
Biochanin A	—	—	—	77	—	—	—	—	—	—	—
Daidzein	92	—	—	35	—	—	—	—	8	62	—
Formononetin	94	—	—	72	—	—	—	—	7	67	—
Genistein	94	—	—	41	—	—	—	—	4	59	—
Irigenin	91	—	—	—	—	—	—	—	16	79	—
Irigenol	54	—	—	—	—	—	—	—	2	33	—
Orobol	—	—	—	20	—	—	—	—	—	—	—
Podospicatin	—	—	—	86	—	—	—	—	—	—	—
Pratensein	—	—	—	54	—	—	—	—	—	—	—
Prunetin	94	—	—	—	—	—	—	—	3	67	—
Tectirigenin	90	—	—	—	—	—	—	—	9	70	—
Isoflavone glucosides											
Genisten	67	—	—	—	—	—	—	—	25	75	—
Iridin	66	—	—	—	—	—	—	—	48	87	—
Tectiridin	67	—	—	—	—	—	—	—	38	83	—

Paper P1 = W2.

Solvent
S1 = *n*-butanol-acetic acid-water (4 : 1 : 5, upper phase).
S2 = acetic acid-concentrated HCl-water (30 : 3 : 10).
S3 = phenol-water (saturated).
S4 = benzene-acetic acid-water (125 : 72 : 3).
S5 = ethyl acetate-water (saturated).
S6 = *iso*-propanol-water (6 : 4).
S7 = *n*-heptane-*n*-butanol-water (29 : 14 : 57).
S8 = acetic acid-water (15 : 85).
S9 = water.
S10 = acetic acid-water (30 : 70).
S11 = *n*-butanol-acetic acid (73 : 27).

Technique One-dimensional descending chromatography.

Literature

1. Bate-Smith, E. C., "*Biochemical Symposium N. 3*", Cambridge, Cambridge University Press, 1950.
2. Harborne, J. B., *J. Chromatogr.*, **2**, 581 (1959).
3. Wong, E. and A. O. Taylor, *J. Chromatogr.*, **9**, 449 (1962).
4. Gage, T. B., Douglass, C. D., and S. H. Wender, *Anal. Chem.*, **23**, 1582 (1951).

TABLE PC 27
PEROXIDES, ORGANIC

Paper	P1	P2	P2	P3	P4
Solvent	S1	S2	S3	S4	S5
Literature	1	1	1	1	2
Compound			$R_F \times 100$		
tert-Amyl hydroperoxide	—	—	43	—	—
Benzoyl peroxide	—	—	—	—	19
n-,iso-,sec-Butyl hydroperoxide	—	—	31	—	—
tert-Butyl hydroperoxide	spreads across front	35	22	—	—
tert-Butyl 2-hydroxyethylperoxide	—	—	55	80	—
tert-Butylperbenzoate	—	—	100	—	—
Cumene hydroperoxide	—	—	53	82	—
Cumyl,2-hydroxy-2-phenylethylperoxide	—	—	76	—	—
Cyclohexane hydroperoxide	—	—	30	—	—
Cyclohexene hydroperoxide	—	—	14	—	46
Diacetylperoxide	37	18	—	—	—
Di-_n_-heptylperoxide	—	—	95	18	—
Di-2-hydroxyethylperoxide	—	—	—	88	—
Dihydroxymethylperoxide	25	—	—	—	—
2,5-Dimethylhexane-2,5-dihydroperoxide	—	—	20	—	—
Ethyl hydroperoxide	38	0	0	—	—
Ethylidenebis(_tert_-butyl)peroxide $(Me_3COO)_2CHMe$	—	—	95	49	—
_n_-Heptan-3,3′-dihydroperoxide	—	—	7	—	—
_n_-Heptyl hydroperoxide,3-methylhexyl-3-hydroperoxide	—	—	80	—	—
_n_-Heptyl,2-hydroxyethylperoxide	—	—	92	—	—
Hydrogen peroxide	5	0	0	68	83
2-Hydroxyethylhydroperoxide	22	0	0	—	—
Hydroxymethylhydroperoxide	9	0	0	—	—
Isochroman hydroperoxide	—	—	—	—	50
Isochroman peroxide	—	—	—	—	22
Methyl hydroperoxide	36	0	0	—	—
Peracetic acid	26	5	0	—	—
Per-_n_-butyric acid, per-isobutyric acid	—	37	30	—	—
Percrotonic acid	—	23	—	—	—
Per-_n_-hexoic acid	—	71	—	—	—
Perpropionic acid	54	16	11	—	—
Per-_n_-valeric acid	—	57	50	—	—
Succinyl peroxide	—	—	—	—	82
Tetrahydrofuran hydroperoxide	—	—	—	—	76
Tetralin hydroperoxide	—	—	—	—	36

Paper P1 = W3MM paper impregnated with 5% ethylene glycol solution in acetone.
P2 = Paper impregnated with 20% glycol solution.
P3 = Paper impregnated with 5% silicone oil solution in petroleum ether (80–100 °C).
P4 = Acetylated paper.

Solvent S1 = _n_-butanol-petroleum ether (80–100 °C) (1 : 9).
S2 = chloroform-petroleum ether (80–100 °C) (1 : 1).
S3 = diethyl ether-petroleum ether (80–100 °C) (5 : 95).
S4 = water-ethanol-chloroform (20 : 17 : 2).
S5 = ethyl acetate-dioxane-water (20 : 45 : 46)

Detection Ferrous thiocyanate.

Technique Paper held between two glass plates clamped together because of volatility of compounds; saturated chamber.

Literature 1. Cartlidge, J. and C. F. H. Tipper., _Anal. Chim. Acta._, **22**, 106 (1960).
2. Rieche, A. and M. Schulz, _Angew. Chem._, **70**, 694 (1958).

TABLE PC 28
PESTICIDES, INSECTICIDES, ORGANOPHOSPHORUS

Paper	P1	P1	P1	P1	P1	P2	P3	P3
Solvent	S1	S2	S3	S4	S5	S6	S7	S8
Literature	1	1	1	1	2	2	3	3
Compound	$R_F \times 100$							
Co-Ral	—	—	—	—	81	23	—	—
Delnav	80[a] 95	18–39[b]	—	—	—	—	—	—
Diazinon	48 98	24 96	—	—	77	38	—	—
Dibrom	—	—	—	—	78	—	—	—
Dipterex	69	34	—	—	78	25	—	—
Di-Syston	—	—	—	—	61	23	—	—
EPN	69	34	—	—	68	25	—	—
Ethion	—	—	—	—	63	22	—	—
Guthion	10	89	51	30	77	30	—	—
Korlan (Trolene)	—	—	—	—	51	21	—	—
Malathion	62	81	—	90	81	34	—	—
Menazon	—	—	—	—	—	—	9	4
Methylparathion	35	81	—	72	—	—	—	—
Nemacide (VC-13)	—	—	—	—	53	22	—	—
Parathion	75	51	—	—	75	24	—	—
Phosdrin	11 4	98	15 4	7 1	84	56 63	—	—
Rogor (Dimethoate)	—	—	—	—	79	54	78	65
Rogor-oxon	—	—	—	—	—	—	18	6
Ronnel	96	12	—	—	—	—	—	—
Systox (Demeton)	96 59	34 90	—	62	73 84	32 46	—	—
Systox-S-methylsulfoxide	—	—	—	—	—	—	41	14
Systox-S-methylsulfone	—	—	—	—	—	—	61	36
TEPP	—	—	—	—	76	71	—	—
Thimet	—	—	—	—	61	25	—	—
Trithion	96	6	—	—	51	20	—	—
Vamidothion	—	—	—	—	—	—	70	40

[a] Two R_F values indicate two isomers.
[b] Indicates R_F range.

Paper P1 = W1.
P2 = SS acetylated.
P3 = W20.

Solvent S1 = dimethylformamide (20%) (stationary phase); 2,2,4-trimethylpentane (mobile solvent).
S2 = heavy mineral oil (10%) (stationary phase), 50% dimethylformamide in water (mobile solvent).
S3 = 10% formamide in acetone (stationary phase), 2,2,4-trimethylpentane (mobile solvent).
S4 = 30% formamide (stationary phase), 2,2,4-trimethylpentane (mobile solvent).
S5 = mineral oil (8%, BDH) (stationary phase), methanol-H_2O-0.5% NH_4OH (mobile solvent).
S6 = mineral oil (2%, BDH) (stationary phase), 70% aqueous acetone (mobile solvent).
S7 = 20% formamide in acetone (stationary phase), benzene-chloroform (6 : 4) (mobile solvent).
S8 = 30% dimethylformamide (stationary phase), *n*-hexane (mobile solvent).

Technique One-and two-dimensional chromatography.

Literature
1. Mitchell, L. C., *J. Ass. Offic. Agr. Chem.*, **43**, 810 (1960).
2. MacRae, H. F. and W. P. McKinley, *J. Assoc. Offic. Agr. Chem.*, **44**, No. 2, 207 (1961).
3. Smart, N. A. and A. R. C. Hill, *J. Chromatogr.*, **30**, 626 (1967).

TABLE PC 29
PHENOLS, CHLORINATED, AND RELATED COMPOUNDS

Paper	P1	P1	P2	P2	P2
Solvent	S1	S2	S3	S4	S5
Technique	T1	T1	T2	T2	T2
Literature	1,2	1,2	1,2	1,2	1,2
Compound	$R_F \times 100$				
N-Acetyl-*S*-(2-nitrophenyl)-L-cysteine	—	—	0	—	39
N-Acetyl-*S*-(4-nitrophenyl)-L-cysteine	—	—	0	—	42
2-Amino-3-chlorophenol	—	—	—	95	—
2-Amino-4-chlorophenol	—	—	—	62	—
2-Amino-5-chlorophenol	—	—	—	51	—
2-Amino-6-chlorophenol	—	—	—	93	—
3-Amino-2-chlorophenol	—	—	—	91	—
3-Amino-4-chlorophenol	—	—	—	83	—
3-Amino-5-chlorophenol	—	—	—	51	—
4-Amino-2-chlorophenol	—	—	—	45	—
4-Amino-3-chlorophenol	—	—	—	67	—
5-Amino-2-chlorophenol	—	—	—	70	—
m-Aminophenol	—	—	—	40	—
o-Aminophenol	—	—	—	17	—
p-Aminophenol	—	—	—	20	—
m-Chloroaniline	—	—	—	100	—
o-Chloroaniline	—	—	—	100	—
p-Chloroaniline	—	—	—	100	—
3-Chlorocatechol	41	—	—	—	—
4-Chlorocatechol	33	—	—	—	—
2-Chloro-3-nitrophenol	—	—	75	—	69
2-Chloro-4-nitrophenol	—	—	53	—	66
2-Chloro-5-nitrophenol	—	—	50	—	69
2-Chloro-6-nitrophenol	—	—	100	—	60
3-Chloro-2-nitrophenol	—	—	100	—	69
3-Chloro-4-nitrophenol	—	—	7	—	69
3-Chloro-5-nitrophenol	—	—	16	—	84
4-Chloro-2-nitrophenol	—	—	100	—	66
4-Chloro-3-nitrophenol	—	—	14	—	87
5-Chloro-2-nitrophenol	—	—	100	—	66
L-2-Chlorophenylmercapturic acid	—	56	—	—	—
L-3-Chlorophenylmercapturic acid	—	61	—	—	—
L-4-Chlorophenylmercapturic acid	—	60	—	—	—
Chloroquinol	17	—	—	—	—
4-Chlororesorcinol	19	—	—	—	—
2-Chlorothiophenol	—	50	—	—	—
3-Chlorothiophenol	—	61	—	—	—
4-Chlorothiophenol	—	59	—	—	—
3,4-Dichlorocatechol	54	62	—	—	—
3,5-Dichlorocatechol	51	—	—	—	—
4,5-Dichlorocatechol	42	52	—	—	—
2,3-Dichlorophenol	—	84	—	—	—
2,4-Dichlorophenol	—	85	—	—	—
2,5-Dichlorophenol	93	—	—	—	—
2,6-Dichlorophenol	—	73	—	—	—
3,4-Dichlorophenol	—	95	—	—	—
3,5-Dichlorophenol	—	94	—	—	—
2,3-Dichlorophenylmercapturic acid	—	70	—	—	—
L-2,4-Dichlorophenylmercapturic acid	—	61	—	—	—
L-2,6-Dichlorophenylmercapturic acid	—	53	—	—	—
3,4-Dichlorophenylmercapturic acid	—	75	—	—	—
L-3,5-Dichlorophenylmercapturic acid	—	65	—	—	—
2,3-Dichloroquinol	38	—	—	—	—
2,5-Dichloroquinol	33	—	—	—	—
2,6-Dichloroquinol	39	—	—	—	—
2,4-Dichlorothiophenol	—	2	—	—	—
2,6-Dichlorothiophenol	—	56	—	—	—
3,5-Dichlorothiophenol	—	39	—	—	—
m-Nitrophenol	—	—	9	—	81

TABLE PC 29—(Continued)
PHENOLS, CHLORINATED, AND RELATED COMPOUNDS

Paper	P1	P1	P2	P2	P2
Solvent	S1	S2	S3	S4	S5
Technique	T1	T1	T2	T2	T2
Literature	1,2	1,2	1,2	1,2	1,2
Compound	$R_F \times 100$				
o-Nitrophenol	—	—	100	—	57
p-Nitrophenol	—	—	2	—	59

Paper
P1 = W1.
P2 = W4.

Solvent
S1 = benzene-acetic acid-water (1 : 1 : 2).
S2 = *n*-butanol-ethanol-3*N* $(NH_4)_2$ CO_2 buffer (10 : 11 : 19).
S3 = petroleum ether (90 °C) saturated with 98% formic acid.
S4 = chloroform-acetic acid-water (2 : 1 : 1).
S5 = *n*-butanol-3*N* $(NH_4)_2CO_3$-3*N* NH_3 solution (4 : 3 : 3).

Technique
T1 = descending.
T2 = see Literature 1.

Literature
1. Bray, H. G., Thorpe, W. V., and K. White, *Biochem. J.*, **46**, 271 (1950).
2. Azouz, W. M., Parke, D. V., and R. T. Williams, *Biochem. J.*, **59**, 410 (1955).

TABLE PC 30
PHENOLS, POLYPHENOLS

Paper	P1	P1	na	na
Solvent	S1	S2	S3	S4
Technique	T1	T1	na	na
Literature	1	1	2	2
Compound	$R_F \times 100$			
Alectoronic acid	—	—	45	—
Atranorin	—	—	63	—
Baeomycesic acid	—	—	50	64
Barbatic acid	—	—	65	87
Catechin	5	75	—	—
L-*epi*-Catechin	5	68	—	—
L-*epi*-Catechin gallate	5	9	—	—
α-Collatolic acid	—	—	66	—
Diffractaic acid	—	—	80	83
Divaric acid	—	—	41	—
Divaricatic acid	—	—	83	—
Divaricatinic acid	—	—	77	—
Evernic acid	—	—	61	—
Everninic acid	—	—	53	—
Fumarprotocetraric acid	—	—	—	38
Gallic acid	15	7	—	—
DL-Gallocatechin	3	6	—	—
L-*epi*-Gallocatechin	3	45	—	—
Gallocatechin-α-gallate	2	7	—	—
L-*epi*-Gallocatechin gallate	3	7	—	—
Lecanoric acid	—	—	33	—
Lobaric acid	—	—	54	—
Methyl orsellate	—	—	62	—
Microphyllic acid	—	—	91	—
Norstictic acid	—	—	—	62

TABLE PC 30—(Continued)
PHENOLS, POLYPHENOLS

Paper	P1	P1	na	na
Solvent	S1	S2	S3	S4
Technique	T1	T1	na	na
Literature	1	1	2	2
Compound	$R_F \times 100$			
Obtusatic acid	—	—	60	—
Olivetol-carboxylic acid	—	—	55	—
Olivetoric acid	—	—	74	—
β-Orcinol-carboxylic acid	—	—	33	—
β-Orcinol-carboxylic acid dimethyl ether	—	—	66	—
Orcinol-dicarboxylic acid monomethyl ether	—	—	4	—
Orsellic acid	—	—	15	—
Perlatolic acid	—	—	93	—
Physodalic acid	—	—	—	68
Protocetraric acid	—	—	—	45
Psoromic acid	—	—	—	62
Rhizonic acid	—	—	64	—
Salazinic acid	—	—	—	50
Sphaerophorin	—	—	88	—
Sphaerophorol-carboxylic acid	—	—	80	—
Squamatic acid	—	—	28	28
Stictic acid	—	—	—	52
Thamnolic acid	—	—	—	37
Usnic acid	—	—	92	—

Paper P1 = W1.

Solvent S1 = phenol saturated with water.
S2 = *n*-butanol-acetic acid (4 : 1) plus water short of saturation.
S3 = *n*-butanol saturated with concentrated NH_4OH.
S4 = butanol-acetone-water (5 : 1 : 2).

Detection $FeCl_3$.

Technique T1 = aluminum tank, saturated, ascending.

Literature
1. Roberts, E. A. H. and D. J. Wood, *Biochem. J.* **49**, 414–422 (1951).
2. Mitsuno, M., *Pharm. Bull.* (Tokyo) **1**, 170–173 (1953); *Chem. Abstr.* **48**, 7716i (1954).

TABLE PC 31
PHENOLS

Paper	P1	P1	P1	P1
Solvent	S1	S2	S3	S4
Technique	T1	T1	na	T2
Literature	1–3	1–3	1–3	1–3
Compound	$R_F \times 100$			
Alphanaphthol	—	—	—	100
Betanaphthol	—	—	—	100
***o*-Bromophenol**	47	88	—	—
***p*-Bromophenol**	32	81	—	—
4-Chloro-*m*-cresol	—	—	55	—
4-Chloro-*o*-cresol	—	—	66	—
6-Chloro-*m*-cresol	—	—	73	—
6-Chloro-*o*-cresol	—	—	89	—
4-Chloro-2,3-dimethylphenol	—	—	73	—
4-Chloro-2,5-dimethylphenol	—	—	76	—
4-Chloro-3,5-dimethylphenol	—	—	72	—
6-Chloro-3,4-dimethylphenol	—	—	79	—
4-Chloro-5-ethyl-3-methylphenol	—	—	81	—
***o*-Chlorophenol**	69	78	—	—
***p*-Chlorophenol**	12	77	—	—
***m*-Cresol**	—	—	34	100
***o*-Cresol**	—	—	50	—
***m*-Cyanophenol**	—	—	—	61
***p*-Cyanophenol**	—	—	—	20
4,6-Dichloro-*m*-cresol	—	—	84	—
4,6-Dichloro-*o*-cresol	—	—	91	—
2,4-Dichloro-3,5-dimethylphenol	—	—	91	—
2,6-Dichloro-3,4-dimethylphenol	—	—	92	—
2,4-Dichloro-5-ethyl-3-methylphenol	—	—	92	—
2,2′-Dihydroxydiphenyl	—	—	—	77
2,3-Dimethylphenol	—	—	62	—
2,5-Dimethylphenol	—	—	60	—
3,4-Dimethylphenol	—	—	44	—
3,5-Dimethylphenol	—	—	59	—
5-Ethyl-3-methylphenol	—	—	70	—
***m*-Hydroxybenzoic acid**	—	—	—	0
***o*-Hydroxybenzoic acid**	—	—	—	27
***p*-Hydroxybenzoic acid**	—	—	—	0
***o*-Hydroxydiphenyl**	—	—	—	100
***o*-Iodophenol**	93	89	—	—
***p*-Iodophenol**	92	56	—	—
***m*-Nitrophenol**	—	—	—	65
***o*-Nitrophenol**	—	—	—	44
***p*-Nitrophenol**	—	—	—	10
Phenol	—	—	—	100
2,4,6-Trichloro-*m*-cresol	—	—	90	—
2,4,6-Trinitrophenol	—	—	—	89

Paper P1 = W1.

Solvent S1 = benzene-acetic acid-water (2 : 2 : 1).
S2 = hydrochloric acid, 20% w/v.
S3 = xylene saturated with formamide.
S4 = benzene-diethylamine-water (3 : 2 : 5).

Technique T1 = two-dimensional chromatography was performed, first with solvent S1, then at right angles with solvent S2.
T2 = ascending.

Literature
1. Chogvill, H. S. and D. E. Bissing, *Anal. Chem.*, **32**, 440 (1960).
2. Husain, S., *J. Chromatogr.*, **18**, 197 (1965).
3. Matthews, R. W., *J. Chromatogr.*, **20**, 190 (1965).

TABLE PC 32
PHOSPHORUS COMPOUNDS, PHOSPHATE ESTERS

Paper	P1	P1	P1	P1	P1	P1
Solvent	S1	S2	S3	S4	S5	S6
Literature	1	1	2	2	2	2
Compound & abbreviation	R_a[a] × 100	R_F × 100	R_b[b] × 100	R_a[a] × 100	R_F × 100	R_F × 100
Adenine (Ad)	140	—	160	30	72	60
Adenosine (Ao)	100	—	157	16	84	49
Adenosine diphosphate (ADP)	24	—	82	0	18	5
Adenosine-3′-phosphate (AMP-3′)	80	—	122	0	30	22
Adenosine-5′-phosphate (AMP-5′)	70	—	100	0	26	20
Adenosine triphosphate (ATP)	12	—	75	0	15	4
Creatine phosphate (Cr-P)	98[c]	—	86[c]	62[c]	17	decomp
Cytidine triphosphate (CTP)	7	—	52	0	11	3
Dihydroxyacetone phosphate (DHAP)	71	65	68	23	decomp	5
2,3-Diphosphoglyceric acid (2,3-PPGA)	54	—	38	1	4	12
Diphosphopyridine nucleotide (DPN)	10	—	102	0	decomp	4
Erythrose-4-phosphate	57	64	—	—	—	—
Fructose (F)	68	—	77	21	98	26
Fructose-1,6-diphosphate (F-1,6-PP)	35	—	37	1	9	8
Fructose-1-phosphate (F-1-P)	52	—	40	1	27	12
Fructose-6-phosphate (F-6-P)	50	—	48	3	31	19
Glucose (G)	59	—	69	16	77	22
Glucose-1-phosphate (G-1-P)	41	—	48	2	30	20
Glucose-6-phosphate (G-6-P)	41	40	46	2	26	12
Glyceraldehyde-3-phosphate (GAP)	73	60	40	23	decomp	15
Glycerol-1-phosphate (Gl-1-P)	76	—	65	52	35	14
Glycerol-2-phosphate (Gl-2-P)	83	—	65	52	35	14
Guanosine diphosphate (GDP)	5	—	39	0	7	5
Guanosine-3′-phosphate (GMP-3′)	56	—	70	0	12	18
Guanosine-5′-phosphate (GMP-5′)	51	—	56	0	10	18
Guanosine triphosphate (GTP)	2	—	26	0	6	2
Hypoxanthine (Hy)	115	—	110	31	66	48
Inosine (I)	75	—	96	16	65	36
Inosine diphosphate (IDP)	19	—	43	0	13	9
Inosine monophosphate (IMP)	49	—	58	0	17	17
Inosine triphosphate (ITP)	7	—	35	0	10	5
Orthophosphate (PO_4)	100	72	68	100	19	28
Phospho-enol-pyruvate (PEP)	106	—	62	73	26	27
6-Phosphogluconic acid (6-PG)	49	53	42	2	22	12
2-Phosphoglycerate (2-PGA)	74	—	45	20	26	16
3-Phosphoglycerate (3-PGA)	79	—	45	32	23	18
Phosphoribulose pyrophosphates	40	36[d]	—	—	—	—
Pyrophosphate (P_2O_7)	52	—	52	2	11	11
Ribose (R)	86	—	84	40	78	30
Ribose-5-phosphate (R-5-P)	56	57	53	4	25	18
Ribulose-1,5-diphosphate (Ru-1,5-PP)	46	54	35	0	6	8
Ribulose-5-phosphate (Ru-5-P)	55	61	—	—	—	—
Sedoheptulose-7-phosphate (S-7-P)	41	43	46	0	28	12
Triphosphopyridine nucleotide (TPN)	5	—	71	0	decomp	11
Uridine diphosphate glucose (UDPG)	13	—	25	0	60	5
Xylulose-5-phosphate	55	52	—	—	—	—

[a] $R_a = R_F$ relative to $PO_4^{\equiv}$.
[b] $R_b = R_F$ relative to AMP-5′.
[c] Partial decomposition.
[d] Streaked zone.

Paper P1 = W541 for separation of phosphates and W1 for free sugars.

Solvent S1 = *n*-butanol-*n*-propanol-acetone-80% (w/v) formic acid-30% (w/v) trichloroacetic acid (40 : 20 : 25 : 25 : 1.5 run twice in same direction), 0.05 gm tetrasodium salt of EDTA/100 ml solvent.

TABLE PC 32—(Continued)
PHOSPHORUS COMPOUNDS, PHOSPHATE ESTERS

S2 = methanol-90% formic acid-H_2O (80 : 15 : 5) and 0.05 gm of tetrasodium salt of EDTA/100 ml solvent.
S3 = isobutyric acid-*N* NH_4OH (10 : 6).
S4 = isopropyl ether-*n*-butanol-90% (w/w) formic acid (3 : 31 : 2 run twice).
S5 = *n*-propanol-concentrated NH_3-H_2O (6 : 3 : 1).
S6 = *n*-butanol-acetic acid-water (60 : 15 : 25).

Detection Hanes-Isherwood reagent.
Technique Ascending development in tanks or jars.
Literature

1. Wood, T., *J. Chromatogr.*, **35**, 352 (1968).
2. Wood, T., *J. Chromatogr.*, **6**, 142 (1961).

TABLE PC 33
PIGMENTS, PORPHYRINS

Paper	P1	P1	P1	P1	P1	P1	P1
Solvent	S1	S2	S3	S4	S5	S6	S7
Literature	1	2–4	2–4	5	1	6	7
Compound	$R_F \times 100$						
Aetioporphyrin	—	—	—	—	100	—	—
All porphyrin esters	100	—	—	—	100	100	—
Chlorin a_2 (methyl ester)	—	55	32	77	—	—	—
Cobalt protoporphyrin	60	—	—	—	—	—	—
Coproporphyrin	54	—	—	—	54	47	25
Deuteroporphyrin	88	—	—	—	88	79	—
Dioxymonovinylhydroxyethyldeuteroporphyrin	—	26	0	0	—	—	—
Dioxyprotoporphyrin	—	34	0	0	—	—	—
Etioporphyrin	100	—	—	—	—	—	—
Haemin	—	—	—	—	70	63	—
Hematoporphyrin	87	30	5	5	87	77	7
Heminporphyrin	70	—	—	—	—	—	—
Hexacarboxylic porphyrin	—	—	—	—	—	—	50
Mesochlorin	—	88	89	96	—	—	—
Mesoporphyrin	86	82	82	96	86	81	—
Monovinylhydroxyethyldeuteroporphyrin	—	40	28	0	—	—	—
Protoporphyrin	84	75	80	96	84	75	—
Pyrrochlorin	—	96	97	96	—	—	—
Pyrroporphyrin	—	92	96	96	—	—	—
Rhodochlorin	—	90	91	96	—	—	—
Rhodoporphyrin	—	77	82	96	—	—	—
Uroporphyrin	26	—	—	—	26	19	90

Paper P1 = W1.
Solvent S1 = 2,5-dimethylpyridine-water (1 : 1).
S2 = kerosene-chloroform (4 : 2.6.
S3 = kerosene-*n*-propanol (6 : 1).
S4 = trichloroethylene.
S5 = 2,6-lutidine-H_2O, NH_3 vapor.
S6 = 2,4-lutidine H_2O, NH_3 vapor.
S7 = 0.1*M* LiCl, NH_3 atmosphere.

Detection UV or daylight.
Technique Ascending.
Literature

1. Kehl, R. and W. Stich, *Z. Physiol. Chem.*, **289**, 6–10 (1951).
2. Chu, T. C., Green, A. A., and E. J. H. Chu, *J. Biol. Chem.*, **190**, 643 (1951).
3. Chu, T. C. and E. J. H. Chu, *J. Amer. Chem. Soc.*, **75**, 3021–3023 (1953).
4. Lucas, J., Vassilaras, L., and L. Petraitis, *Fed. Proc.*, **13**, 256 (1954).
5. Barrett, J., *Biochemistry*, **64**, 626–639 (1956).
6. Kehl, R. and W. Stich, *Z. Physiol. Chem.*, **290**, 151 (1952).
7. With, T. K., *Scand. J. Clin. Lab. Invest.*, **9**, 395 (1957).

TABLE PC 34
PHOSPHORUS COMPOUNDS, ALKYL PHOSPHATES

Paper	P1
Solvent	S1
Literature	1
Compound	$R_F \times 100$
Ammonium *n*-butyl *n*-propyl phosphate	82
Ammonium dibenzyl hydrogen phosphate	86
Ammonium diethyl phosphate	84
Ammonium ethyl *n*-propyl phosphate	80
***tert*-Butyl ethyl hydrogen phosphate**	88
***tert*-Butyl phosphate**	28
Cyclohexylammonium benzyl *n*-butyl phosphate	90
Cyclohexylammonium isopropyl *n*-propyl phosphate	82
Cyclohexylammonium isopropyl *n*-propyl pyrophosphate	63
Diammonium P^1,P^2-di-*n*-butyl pyrophosphate	82
Diammonium P^1,P^2-diethyl pyrophosphate	48
P^1,P^2-Di-*n*-propyl pyrophosphate	77
Ethyl *n*-propyl hydrogen phosphate	84
***n*-Propyl dihydrogen phosphate**	28

Paper P1 = W30.
Solvent S1 = 2-propanol-ammonia-water (7 : 2 : 1).
Detection Acid-molybdate or isatin reagent.
Literature 1. Blackburn, G. M., Cohen, J. S., and Lord Todd, *J. Chem. Soc. C*, p 239 (1966).

TABLE PC 35
PHOSPHORUS COMPOUNDS, ETHYL ESTERS OF CONDENSED PHOSPHORIC ACIDS

Paper	P1
Solvent	S1
Literature	1
Compound	$R_F \times 100$
Orthophosphoric acid (H_3PO_4)	26
Pyrophosphoric acid ($H_4P_2O_7$)	13
Diethyl orthophosphate	77
Diethyl pyrophosphate	59
Diethyl triphosphate	48
Ethyl orthophosphate	39
Ethyl pyrophosphate	29
Tetraethyl pyrophosphate	93
Tetraethyl triphosphate	91
Triethyl pyrophosphate	86
Triethyl triphosphate	71

Paper P1 = SS2043b Mgl.
Solvent S1 = *iso*-butanol-*iso*-propanol-ethanol-H_2O-25% NH_4OH (25 : 20 : 20 : 34 : 0.7).
Detection Acid molybdate.
Technique Development at room temperature.
Literature 1. Wicker, W., Walter, H. P., and E. Thilo, *Chem. Ber.*, **97**, 2385 (1964).

TABLE PC 36
PURINES, PYRIMIDINES AND DERIVATIVES

Paper	P1	P1	P1	P1	P1	P1	P1	P1	P1
Solvent	S1	S2	S3	S4	S5	S6	S7	S8	S9
Literature	1	1	1	1	1	1	1	1	1
Compound					$R_F \times 100$				
5-Acetylamino-6-amino-3-methyluracil	44	26	59	48	78	39	40	0	25
5-Acetyluracil	57	54	66	70	71	64	52	36	63
Adenine	48	40	65	71	33	63	37	6	33
Adenine-*N*′-oxide	43	29	61	41	32	32	37	5	8
Adenosine	32	64	56	63	33	51	20	3	25
Adenosine-5′-diphosphate	2	4	6	12	40	0	0	0	0
Adenosine-5′-phosphate	8	8	27	25	42	5	3	0	0
Adenosine-5′-triphosphate	0	4	6	13	39	0	0	0	0
***S*-Adenosylmethionine**	5	35	17	14	7	8	1	0	0
5-Amino-6-carboxyuracil	29	16	43	25	24	10	27	11	10
6-Amino-2,8-dihydroxypurine	15	9	31	23	33	—	4,17	0	5
6-Amino-1,3-dimethyl-5-formylaminouracil	50	58	61	51	51	43	48	8	27
2-Amino-4,6-dimethylpyrimidine	74	92	82	89	78	86	70	31	67
6-Amino-2,4-dimethylpyrimidine	73	87	72	69	75	70+	69	4	16, 30
6-Amino-1,3-dimethyluracil	72	83	75	76	pk str.	75	74	38	61
2-Amino-4,6-dioxy-5-methylpyrimidine	38	17	56	45	68	33	38	5	14
2-Amino-4,6-dioxypyrimidine	35	21	51	48	pk str.	39	33	6	22
6-Amino-5-formylamino-4-hydroxypyrimidine	21	22	40	30	19	18	21	0	7
6-Amino-5-formylamino-1-methyl-4-oxypyrimidine	39	30	55	42	39	33	41	6	17
6-Amino-5-formylamino-3-methyl-4-oxypyrimidine	38	45	53	35	30	25	40	6	13
6-Amino-5-formylamino-1-methyluracil	22	27	42	35	27	19	22	1	7
6-Amino-5-formylamino-3-methyluracil	38	26	54	50	50	38	31	5	25
6-Amino-5-formylaminouracil	15	16	31	26	22	16	12	0	6
2-Amino-4-hydroxy-6-methylpyrimidine	58	38	72	70	63	63	44	6	36
2-Amino-8-hydroxypurine	19	24	28	37	21	24	8	0	6
6-Amino-2-hydroxypurine	30	22	52	34	32	24	19	0	6
6-Amino-8-hydroxypurine	31	27	51	55	44	52	10	0	19
2-Amino-4-hydroxypyrimidine	50	35	66	66	55	57	40	4	33
4-Amino-5-imidazolecarboxamide	45	58	58	56	50	48	36	1	25
4-Amino-5-imidazolecarboxamide ribonucleoside	32	57	50	48	48	36	21	0	19
6-Amino-3-methyl-5-(*N*-formylmethylamino)uracil	59	35	64	64	83	58	52	3	52
2-Amino-5-methyl-4-oxypyrimidine	61	47	72	71	67	64	48	7	38
2-Amino-6-methylpurine	42	43	62	64	27	59	32	2	26
2-Amino-4-methylpyrimidine	69	88	82	87	63	83	65	39	72
5-Amino-6-methyluracil	32	31	54	46	25	40	16	1	28
6-Amino-1-methyluracil	49	57	58	53	pk str.	44	51	11	27
2-Aminopurine	32	37	57	60	17	48	23	1	23
2-Aminopyrimidine	62	80	78	77	52	72	58	35	60
6-Amino-2-thiouracil	50	38	57	str.	65	57	45	18	53
5-Aminouracil	25	31	41	46	21	36	12	5	22
6-Aminouracil	34	22	47	46	pk str.	38	32	6	22
5-Aminouridine	14	23	35	42	24	25	7	0	14
6-Azathymine	71	46	73	82	78	83	70	66	77
6-Azauridine	36	42	49	53	63	42	27	9	39
Barbituric acid	43	37	45	28	51	12	52	31	21
5-Bromodeoxyuridine	64	53	68	81	85	75	56	32	77
5-Bromouridine	46	42	58	69	72	61	39	17	65
5-Butoxymethyluracil	88	85	86	96	95	93	89	87	87
6-Carboxypurine	30	16	43	22	35	10	27	4	4
Cytidine	32	47	47	45	45	29	19	0	12
Cytidine-5′-phosphate	7	6	22	20	50	4	2	0	0
Cytosine	45	49	53	48	45	37	32	0	16
Cytosine-5-carboxylic acid	9	23	—	22	26	8	19	0	3
Deoxyadenosine	43	74	66	70	35	62	31	5	35
Deoxycytidine	48	61	57	55	58	40	29	0	18
Deoxycytidine-5′-phosphate	15	7	29	22	64	4	6	0	0
Deoxyguanosine	33	36	50	55	24	39	18	1	20
Deoxyinosine	38	42	49	58	31	45	26	4	24
Deoxyuridine	50	45	58	67	80	59	41	13	52

TABLE PC 36—(Continued)
PURINES, PYRIMIDINES AND DERIVATIVES

Paper	P1	P1	P1	P1	P1	P1	P1	P1	P1
Solvent	S1	S2	S3	S4	S5	S6	S7	S8	S9
Literature	1	1	1	1	1	1	1	1	1
Compound	$R_F \times 100$								
Deoxyuridine-5′-phosphate	21	4	38	17	83	6	12	0	0
2,5-Diamino-4,6-deoxypyrimidine	3	3	8	14	—	2	0	0	0
2,6-Diamino-5-formylamino-4-hydroxypyrimidine	12	10	30	20	19	11	9	0	1
2,6-Diamino-5-formylamino-4-methoxypyrimidine	50	52	61	47	51	26	40	0	5
4,6-Diamino-5-formylamino-2-methylpyrimidine	38	48	57	37	40	20	31	0	0
2,6-Diamino-7-methylpurine	28	30	42, 52	33	14	16	15	4, 24	0
4,5-Diamino-6-methylpyrimidine	63	69	64	60	49	55	52	0	13
2,4-Diamino-6-oxypyrimidine	38	28	54	36	51, pk	23	29	3	9
4,6-Diamino-2-oxypyrimidine	43	29	42	19	52	13	49	35	10
2,6-Diaminopurine hemisulfate	34	31	56	39	16	28	29	0	11
4,6-Diaminopyrimidine	53	60	63	56	47	46	49	13	15
4,6-Diamino-2-thiopyrimidine	59	80	62	68	56	52	44	2	25
6,8-Dihydroxy-2-methylpurine	29	23	54	50	43	36	29	5	18
6,8-Dihydroxypurine	23	18	38	42	41	36	20	2	19
2,4-Dihydroxypyrimidine-6-methylsulfone	58	60	61	39	66	23	62	50	61
2,4-Dimethoxy-5-methylpyrimidine	90	95	93	93	89	92	91	97	93
2-Dimethylamino-6-hydroxypurine	51	36	68	61	37	52	42	4	28
6-Dimethylaminopurine	63	71	77	88	52	82	55	12	65
6-Dimethylaminopurine ribonucleoside	53	83	74	80	58	74	36	13	61
1,7-Dimethylguanine	29	60	49	39	25	23	17	0	7
1,7-Dimethylhypoxanthine	66	70	73	64	52	57	60	29	40
4,6-Dimethylpyrimidine	84	—	89	—	74	—	81	80	77
1,3-Dimethyluracil	86	91	85	87	97	84	88	82	83
5,6-Dimethyluracil	69	70	75	80	85	77	69	40	66
1,3-Dimethyluric acid	47	43	64	60	67	49	42	16	42
1,7-Dimethyluric acid	47	35±	65	63	82	48	40	17	43
3,7-Dimethyluric acid	39	38	60	50	57	40±	41	9	34
1,3-Dimethylxanthine	70	55	78	80	61	75	67	50	65
1,7-Dimethylxanthine	66	42	76	77	58	68	70	42	56
3,7-Dimethylxanthine	58	46	70	64	52	56	62	30	42
3,8-Dimethylxanthine	56	38	70	69	48	64	56	14	45
Dithiouracil	76	73	77	88	88	86	83	90	89
5-Ethoxymethyluracil	73	59	78	86	89	79	70	49	75
5-Ethoxymethyluridine	56	51	66	70	85	63	—	—	—
5-Ethoxymethyl-6-methyluracil	77	69	81	88	81	84	78	53	76
5-Fluorodeoxyuridine	59	42	65	72	86	66	50	29	73
5-Fluorouracil	54	42	62	70	75	63	55	38	64
5-Formylamino-3-methyl-6-methyl-aminouracil	48	30	63	52	54	39	49	7	22
5-Formylcytosine	42	48	55	52	35	40	33	7	26
5-Formyldeoxyuridine	42	63	55	64	73	53	26	5	44
5-Formyluracil	44	56	50	56	58	47	39	24	54
5-Formyluridine	30	43	38	45	56	34	20	4	31
6-Furfurylaminopurine	78	85	87	94	79	89	77	58	85
Guanine	23	19	45	42	22	36	13	0	11
Guanosine	20	27	36	44	29	32	11	0	11
Guanosine-5′-phosphate	5	4	17	22	39	2	0	0	0
5-Hydroxydeoxyuridine	32	18	47	57	68	46	25	6	40
6-Hydroxy-2-methylaminopurine	47	34	58	60±	49	50±	30	0	25
5-Hydroxymethylcytosine	38	42	49	40	44	30	23	0	10
5-Hydroxymethyldeoxycytidine	26	56	53	44	54	34	—	—	—
5-Hydroxymethyldeoxyuridine	36	39	49	59	73	47	25	4	36
8-Hydroxy-7-methylguanine	22	15	46	43	43	35	23	4	11
4-Hydroxy-2-methylmercaptopyrimidine	83	59	82	88	76	85	86	78	81
5-Hydroxymethyl-6-methyluracil	44	39	55	55	75	49	40	9	34
5-Hydroxymethylorotic acid	26	22	32	30	53	15	19	0	5
5-Hydroxymethylorotic acid lactone	45	37	49	46	49	38	47	37	49
2-Hydroxy-6-methylpurine	36	30	55	45	40	33	27	2	11
2-Hydroxy-4-methylpyrimidine	54	48	66	60	49	55	41	7	38

TABLE PC 36—(Continued)

PURINES, PYRIMIDINES AND DERIVATIVES

Paper	P1	P1	P1	P1	P1	P1	P1	P1	P1
Solvent	S1	S2	S3	S4	S5	S6	S7	S8	S9
Literature	1	1	1	1	1	1	1	1	1
Compound	$R_F \times 100$								
4-Hydroxy-6-methylpyrimidine	66	47	76	75	54	72	58	33	59
5-Hydroxymethyluracil	37	36	48	56	65	46	29	8	33
5-Hydroxymethyluridine	24	31	38	48	59	32	13	0	19
8-Hydroxypurine	42	40	63	76	31	66	26	13	51
4-Hydroxypyrimidine	58	46	68	70	45	65	54	25	51
Hypoxanthine	38	35	53	57	28	49	25	6	26
Inosine	24	36	39	48	30	35	14	0	14
Isobarbituric acid	34	4, 14	44	53	57	42	31	9	36
8-Mercapto-6-methylpurine	66	49	73	77	49	77	58	45	74
5-Methoxymethyluracil	61	50	68	72	81	65	57	29	58
5-Methoxymethyluridine	43	37	55	58	74	47	35	7	39
6-Methoxypurine	72	54	82	84	58	84	68	52	69
6-Methoxypurine ribonucleoside	62	81	78	77	59	70	60	29	58
1-Methyladenine	46	47	59	35	31	18	41	1	8
2-Methyladenine	41	46	65	71	41	64	33	0	12
3-Methyladenine	53	55	61	48	46	43	45	4	12
7-Methyladenine	45	57	60	58	30	42	35	0	15
9-Methyladenine	57	67	71	66	40	59	47	12	30
2-Methyladenine sulfate	57	51	69	76	44	68	46	3	30
1-Methyladenosine	32	51	53	28	42	15	22	0	5
6-Methylamino-5-(*N*-formylmethylamino)uracil	38	29	57	42	60	31	45	7	19
6-Methylaminopurine	61	61	77	86	49	80	47	8	53
6-Methylaminopurine deoxyribonucleoside	30	52	53	26	38	12	18	0	6
6-Methylaminopurine ribonucleoside	42	77	68	70	51	62	25	7	43
5-Methylaminouracil	34	40	61	61	35	53	10, 21	11	45
5-Methylbarbituric acid	60	33	54	24±	63	10	65	55	46±
5-Methylcytidine	36	47	53	41	49	30	20	0	12
1-Methylcytosine	57	61	63	55	55	45	44	2	18
5-Methylcytosine	53	55	60	55	52	44	38	0	15
5-Methyldeoxycytidine	56	67	65	59	63	46	35	1	21
5-Methyldeoxycytidine-5′-phosphate	19	7	36	27	65	8	10	0	0
1-Methylguanine	40	25	50	45	32	30	23	0	7
3-Methylguanine	36	22	53	34	33	25	21	2	7
7-Methylguanine	35	24	55	46	26	35	19	2	11
1-Methylguanine sulfate	30	36	54	48	20	35	18	2	14
1-Methylguanosine	30	49	50	47	38	32	16	1	13
7-Methylguanosine	28	unstable	45	28	40	16	15	0	5
5-Methyl-3-hydroxybarbituric acid	43	25	53	51	63	45	43	25	52
1-Methylhypoxanthine	51	37	63	62	36	51	42	13	30
2-Methylhypoxanthine	45	32	62	60	39	50	35	8	28
3-Methylhypoxanthine	38	30	52	40	30	31	27	4	11
7-Methylhypoxanthine	49	35	62	64	36	50	42	12	28
8-Methylhypoxanthine	47	33	63	62	46	57	30	5	33
9-Methylhypoxanthine	50	37	59	57	36	47	42	11	25
1-Methylinosine	37	56	54	45	49	37	30	3	19
7-Methylinosine	26	unstable	38	29	31	15	13	0	4
6-Methylmercaptopurine	79	68	82	90	55	82	78	71	77
1-Methyl-6-methylamino-5-(*N*-formylmethylamino)uracil	50	46	64	52	75	44	49	12	32
3-Methyl-6-methylamino-5-(*N*-formylmethylamino)uracil	67	41	76	64	89	56	70	27	51
3-Methyl-6-methylaminouracil	76	43	78	84	pk str.	77	73	38	67
5-Methylorotic acid	57	27	44	—	78	18	51	8	16
6-Methylpurine	58	52	76	81	46	77	46	22	57
6-Methyl-2-thiouracil	72	56	76	70	78	73	77	64	75
1-Methylthymine	75	67	78	72	85	69	75	56	68
2-Methyl-4,5,6-triaminopyrimidine sulfate	45	55	64	43	35	32	36	0	16±
1-Methyluracil	66	50	69	70	81	60	64	38	57

TABLE PC 36—(Continued)
PURINES, PYRIMIDINES AND DERIVATIVES

Paper	P1	P1	P1	P1	P1	P1	P1	P1	P1
Solvent	S1	S2	S3	S4	S5	S6	S7	S8	S9
Literature	1	1	1	1	1	1	1	1	1
Compound	$R_F \times 100$								
3-Methyluracil	77	61	78	81	91	77	75	58	70
6-Methyluracil	59	50	68	71	77	66	59	29	55
3-Methyluridine	62	83	68	70	92	65	50	18	59
5-Methyluridine	45	46	57	64	74	55	35	8	47
1-Methyluric acid	25	28	42±	42	—	27	20	3	30
3-Methyluric acid	22	28	41	38	36	24	19	3	18
7-Methyluric acid	26	25	49	37	54	22	20	3	22
1-Methylxanthine	52	35	67	69	47	65	49	25	51
3-Methylxanthine	38	43	55	60	35	51	38	16	31
7-Methylxanthine	40	32	56	57	32	47	42	15	28
Orotic acid	38	23	39	31	57	15	34	9	6
Orotidine	17	17	30	20	66	7	12	0	0
Orotidine monophosphate	8	2	12	12	74	1	1	0	0
Propyl-2-thiouracil	88	76	88	93	95	95	93	92	92
Purine	58	53	73	73	35	69	45	22	52
5-Ribosyluracil	21	25	35	43	53	34	12	2	0
Spongothymidine	55	63	61	79	81	66	42	12	59
6-Succinoaminopurine	47	7	61	17	55	6	27	4	45
1,3,7,9-Tetramethyluric acid	79	87	80	80	93	71	79	58	65
Theophylline-7-acetic acid	77	45	74	36	88	17	72	54	54
2-Thiobarbituric acid	54	50	46	41	72	21	58	39	15
2-Thiocytosine	58	34, 93	59	89	55	87	45	5	33
5′-Thiomethyladenosine	57	87	75	82	54	77	45	14	54
5′-Thiomethylinosine	48	53	61	62	45	53	38	11	44
2-Thio-orotic acid	47	25	44	—	69	—	44	21	13
2-Thiothymine	77	63	77	92	87	83	75	70	84
Thymidine	62	59	68	77	87	71	48	19	61
Thymidine-5′-phosphate	24	5	36	27	88	6	13	0	0
Thymine	62	56	68	77	80	70	61	34	61
4,5,6-Triaminopyrimidine sulfate	37	48	57	40	22	32	29	0	2, 14
1,3,7-Trimethylxanthine	82	88	84	81	79	79	85	74	73
1,3,8-Trimethylxanthine	80	61	84	86	75	83	80	63	76
Uracil	50	41	56	65	69	55	46	21	47
Uracil-6-acetic acid	43	17	51	26	65	15	34	14	33
Uracil-5-carboxylic acid	44	9	46	23	54	10	38	26	44
Uric acid	14	18	30	27	22	5	7	4	12
Uridine	35	33	46	55	67	42	25	6	35
Uridine-5′-diphosphate	0	0	7	12	66	0	0	0	0
Uridine diphosphate glucose	0	6	8	16	69	2	0	0	0
Uridine-5′-phosphate	12	4	23	24	70	6	5	0	0
Uridine-5′-triphosphate	3	3	5	15	63	0	0	0	0
Xanthine	24	27	45	46	25	41	21	0	22
Xanthosine	18	32	34	28	26	13	12	0	10

± following a figure indicates an abnormally high variability in replicate determinations of the R_F values.

Paper P1 = W1.
Solvent S1 = *tert*-butanol-methyl ethyl ketone-formic acid-water (40 : 30 : 15 : 15).
S2 = *tert*-butanol-methyl ethyl ketone-water-ammonium hydroxide (40 : 30 : 20 : 10).
S3 = *n*-butanol-glacial acetic acid-water (50 : 25 : 25).
S4 = upper phase from a mixture of water-*sec*-butanol-*tert*-butanol (48.4 : 43 : 8.6).
S5 = isopropyl alcohol-water-conc. HCl (56 : 18.4 : 16.6).
S6 = upper phase from a mixture of *sec*-butanol and water.
S7 = ethyl acetate-formic acid-water (70 : 20 : 10).
S8 = upper phase from a mixture of ethyl acetate-water-formic acid (60 : 35 : 5).
S9 = *tert*-butanol-methyl ethyl ketone-water-formic acid (44 : 44 : 11 : 0.26).
Detection UV light.
Technique Ascending development at 28 °C; tank equilibrated with solvent or aqueous phase for two-phase solvents.
Literature 1. Fink, K. and W. S. Adams, *J. Chromatogr.*, **22**, 118 (1966).

TABLE PC 37
PURINES, PYRIMIDINES, NUCLEOTIDES AND RELATED COMPOUNDS

Paper	P1	P1	P1	P1	P1	P1	P1	P1	P1
Solvent	S1	S2	S3	S4	S5	S6	S7	S8	S7[a]
Technique	T1	T2	T3	T4	T5	T6	T6	T6	T6
Literature	1	1	1	1	1	2	2	2	2
Compound	$R_F \times 100$								
Adenine	34	46	74	55	48	35	24	42	16
Adenosine	17	54	59	54	55	—	—	43	12
Allantoin	27	32	28	62	—	—	—	—	—
5-Aminouracil	—	—	—	—	—	70	—	8	2
6-Aminouracil	—	—	—	—	—	14	11	27	12
AMP	17	16	38	17	81	—	—	—	—
Caffeine	—	—	—	—	—	66	—	73	54
CMP	15	13	36	22	88	—	—	—	—
Cytidine	14	46	58	65	75	—	—	—	—
Cytosine	34	51	73	64	—	14	18	45	14
Deoxyadenosine	31	63	66	64	58	—	—	44	12
Deoxy-AMP	26	18	50	22	—	—	—	—	—
Deoxy-CMP	22	15	48	28	—	—	—	—	—
Deoxycytidine	22	56	66	73	77	—	—	—	—
Deoxycytidylic acid	—	—	—	—	—	—	—	6	2
Deoxy-GMP	17	9	36	13	—	—	—	—	—
Deoxyguanosine	18	41	48	55	79	—	—	36	16
Deoxyinosine	20	49	47	69	82	—	—	—	—
Deoxyuridine	29	45	51	73	88	—	—	51	22
1,3-Dimethyluracil	—	—	—	—	—	75	100	—	100
Dithiouracil	—	—	—	—	—	64	—	58	77
GMP	12	8	26	10	—	—	—	—	—
Guanine	12	29	45	43	—	32	—	23	10
Guanosine	10	30	41	53	76	—	—	—	—
Hypoxanthine	23	43	52	62	—	18	—	37	15
IMP	14	10	24	10	89	—	—	—	—
Inosine	12	42	36	66	79	—	—	—	—
Isocytosine	—	—	—	—	—	20	23	43	23
6-Mercaptopurine	—	—	—	—	—	29	—	34	31
6-Methylthiouracil	—	—	—	—	—	59	—	69	66
6-Methyluracil	—	—	—	—	—	46	48	66	42
5-Nitrouracil	—	—	—	—	—	10	17	47	48
Orotic acid	22	31	54	49	88	—	—	—	—
Purine	—	—	—	—	—	45	33	54	19
Theobromine	—	—	—	—	—	30	—	73	48
Theophylline	—	—	—	—	—	45	72	71	48
Thiouracil	—	—	—	—	—	47	—	65	59
Thymidine	42	54	67	77	85	—	—	73	44
Thymine	52	56	71	72	—	49	48	70	44
TMP	36	17	50	32	—	—	—	—	—
UMP	22	15	36	26	92	—	—	—	—
Uracil	38	43	60	70	80	30	40	65	35
Uric acid	19	*dec.*	19	*dec.*	—	25	—	9	3
Uridine	18	40	46	72	86	—	—	40	12
Xanthine	19	33	38	51	64	—	100	20	19
Xanthosine	35	31	33	62	75	—	—	—	—

[a] = The paper could not be pretreated as suggested in S7 as the second solvent for two-dimensional chromatography.
dec. = partial decomposition.

Paper P1 = W1 (see below for impregnation).

Solvent S1 = isopropyl ether-*n*-butanol-98% formic acid (30 : 30 : 20).

S2 = *n*-butanol-*n*-propanol-ethanol-25% NH_4OH-H_2O (40 : 40 : 10 : 45 : 15).

S3 = 80% (v/v) formic acid-*n*-butanol-*n*-propanol-acetone-30% (w/v) trichloroacetic acid (25 : 40 : 20 : 25 : 15).

S4 = methanol-isopropanol-25% NH_4OH-H_2O (45 : 30 : 15 : 10).

S5 = H_2O-*n*-propanol-25% NH_4OH (90 : 10 : 1).

S6 = *n*-butanol equilibrated with 0.4*M* sodium phosphate buffer, pH 6.86 (prepared by dissolving 0.2 mole of sodium dihydrogen phosphate and 0.2 mole of disodium hydrogen phosphate in distilled water and diluting to 1 liter).

TABLE PC 37—(Continued)
PURINES, PYRIMIDINES, NUCLEOTIDES AND RELATED COMPOUNDS

Solvent S7 = 8 volumes of acetonitrile were mixed with 2 volumes 0.4*M* sodium phosphate buffer, pH 6.86; paper is sprayed with pH 6.86 sodium phosphate buffer and dried before spotting the compounds.
S8 = 70% acetonitrile-H_2O (v/v), adjusted to pH 10 (apparent pH) with concentrated NH_4OH, paper is pretreated by spraying with water adjusted to pH 10 (ammoniacal), and drying in air.

Detection UV light (254 nm).

Technique T1 = descending (20–24 °C, 5 hours).
T2 = descending (26 °C, 12 hours in solvent-saturated atmosphere).
T3 = ascending (20–24 °C, 18 hours in solvent-saturated atmosphere).
T4 = descending (20–24 °C, 12 hours in partially water-saturated atmosphere).
T5 = ascending (20–24 °C, 3–4 hours).
T6 = two-dimensional chromatography.

Literature
1. Gerlach, E., Driesbach, R. H., and B. Deuticke, *J. Chromatogr.*, **18**, 81 (1965).
2. Berger, B. A. and C. E. Hedrick, *Anal. Biochem.*, **16**, 260 (1966).

TABLE PC 38
SILICONE COMPOUNDS, METHYL PHENYL POLYSILOXANES

Paper	P1	P1
Solvent	S1	S2
Literature	1	1
Compound	$R_F \times 100$	
$[(CH_3)_2(C_6H_5)Si]_2O$	71	77
$[(CH_3)_2SiO]_3OSi(C_6H_5)_2$	80	84
$[(CH_3)_2SiO]_3OSi(CH_3)(C_6H_5)$	17	14
$[(CH_3)_3SiO]_2Si(C_6H_5)_2$	49	52
$[(CH_3)_3SiO]_2Si(CH_3)(C_6H_5)OSi(CH_3)_2$	17	13
$(C_6H_5)(CH_3)_2SiOSi(C_6H_5)_2Cl$	82	88
cis-$[(CH_3)(C_6H_5)SiO]_3$	88	91
trans-$[(CH_3)(C_6H_5)SiO]_3$	97	92
$[(CH_3)(C_6H_5)_2Si]_2O$	93	94
$[(CH_3)_2SiO]_3[(C_6H_5)_2SiO]$	43	29
$[(CH_3)_2SiO]_2[(CH_3)(C_6H_5)SiO]_2$	63	57
$(CH_3)_3SiOSi(C_6H_5)_2OSi(CH_3)_2OSi(CH_3)_3$	36	23
$[(CH_3)_3SiOSi(CH_3)(C_6H_5)]_2O$	39	33
$[(C_6H_5)_2(OC_2H_5)Si]_2O$	76	92
$[(CH_3)_3SiOSi(C_6H_5)(CH_3)]_2OSi(CH_3)_2$	25	17
$[(CH_3)(C_6H_5)SiO]_4$	80	85
$[(CH_3)_3SiOSi(C_6H_5)_2]_2O$	75	70
$(CH_3)_3SiO[Si(CH_3)(C_6H_5)O]_3Si(CH_3)_3$	36	28
$[(CH_3)_3SiOSi(CH_3)(C_6H_5)OSi(CH_3)(C_6H_5)]_2O$	41	29
$[(CH_3)_3SiOSi(CH_3)(C_6H_5)OSi(CH_3)(C_6H_5)]_2OSi(CH_3)(C_6H_5)$	49	33

Paper P1 = W1.

Solvent S1 = water-dimethylformamide (1 : 8), paper impregnated with 5% solution of Apiezon L (Carlo Erba) in toluene.
S2 = water-dimethylformamide (1 : 8), paper impregnated with 5% solution of silicone elastomer (VChZ-kolin) in toluene.

Detection Fuming nitric acid.

Technique After D1 the chromatograms were immersed in distilled water and after drying were ironed with a gently warmed flat iron. While still moist they were placed between two glass plates and suitably compressed by weights.

Literature
1. Franc, J. and J. Cecová, *Collect. Czech. Chem. Commun.*, **33**, 1572 (1968); *J. Chromatogr.*, **40**, D20 (1969).

TABLE PC 39
STEROIDS, BILE ACIDS

Paper	P1	P1	P1	P1	P1	P1
Solvent	S1	S2	S3	S4	S5	S6
Literature	2	1	3	4	4	5
Compound	$R_F \times 100$					
Chenodeoxycholic	78	—	—	58	—	—
Cholic	36	71	—	28	79	—
Dehydrocholic	—	65	—	—	—	—
Deoxycholic	75	92	—	58	80	—
Glycochenodeoxycholic	—	—	—	13	68	—
Glycocholic	5	—	—	3	51	—
Glycodeoxycholic	—	—	—	17	68	—
Hyodeoxycholic	67	—	—	44	—	—
Methyl-3,12-diketocholanate	—	—	70	—	—	—
Methyl-7,12-diketocholanate	—	—	42	—	—	—
Methyl-3-hydroxy-7,12-diketocholanate	—	—	23	—	—	—
Methyl-12-hydroxy-3,7-diketocholanate	—	—	53	—	—	—
Methyl-12-keto-3-acetoxycholanate	—	—	76	—	—	—
Methyl-12-ketocholanate	—	—	83	—	—	—
Methyl-7-keto-3,12-dihydroxycholanate	—	—	58	—	—	—
Methyl-12-keto-3-hydroxycholanate	—	—	74	—	—	—
Methyl-3,7,12-triketocholanate	—	—	9	—	—	—
Norcholic	—	69	—	—	—	—
Taurochenodeoxycholic	—	—	—	0	—	29
Taurocholic	0	—	—	0	6	18
Taurodeoxycholic	—	—	—	0	16	32
Triformylnorcholic	—	92	—	—	—	—

Paper P1 = W1.

Solvent S1 = isopropyl ether-heptane (7 : 3), paper saturated with 70% acetic acid.
S2 = *n*-propyl alcohol-ethanolamine-water (9 : 5 : 5).
S3 = *n*-butanol-*n*-propanol-water (6 : 2 : 1).
S4 = amyl acetate-heptane (5 : 5), paper saturated with 70% acetic acid.
S5 = amyl acetate-heptane (8 : 2).
S6 = amyl acetate-heptane (85 : 15).

Technique Ascending.

Literature

1. Kritchevsky, D. and M. R. Kirk, *Arch. Biochem. Biophys.*, **35**, 346–351 (1952).
2. Sjövall, J., *Acta Chem. Scand.*, **6**, 1552–1553 (1952).
3. Tanaka, K. and K. Takeda, *J. Biochem.* (Tokyo), **39**, 333–337 (1952).
4. Block, R. J., Durrum, E. L., and G. Zweig, *Paper Chromatography and Paper Electrophoresis*, Academic Press, N.Y., 1958, pp 276–280.
5. Sjövall, J., *Arkiv. Kemi*, **8**, 299–301 (1955).

TABLE PC 40
STEROIDS, $C_{19\text{-}20}$(17-OXYSTEROIDS)

Paper	P1	P1	P1	P2	P3	P3	P4	P4	P4	P1
Solvent	S1	S2	S3	S4	S2	S5	S6	S7	S8	S4
Temperature	—	—	—	—	—	—	38 °C	22 °C (38 °)	22 °C	—
Literature	1	1	1	1	1	1	1	1	1	1
Compound	$R_{cortexone}$									
Group I[a]										
Androsta-4,9(11)-diene-3,17-dione	—	—	—	1.92	—	—	1.90	—	—	—
Androsta-4,11-diene-3,17-dione	—	—	1.25	—	1.66	1.15	2.37	(1.40)	—	—
Androsta-1,4-diene-3,17-dione	1.38	1.10	1.18	1.16	1.33	1.10	1.17	1.45	—	—
5α-Androst-1-ene-3,17-dione	4.40	—	—	—	—	—	—	(1.14)	—	—
Androst-4-ene-3,17-dione	3.80	1.32	1.26	1.92	1.62	1.17	1.90	1.45	1.11	—
Androst-4-ene-3,6,17-trione	1.00	1.38	—	0.84	—	1.07	0.56	1.00	—	—
Androsta-1,4-diene-3,11,17-trione	—	—	—	—	—	—	—	0.71	—	—
5α-Androst-1-ene-3,11,17-trione	—	—	—	—	—	—	—	1.04	—	—
Androst-4-ene-3,11,17-trione	0.80	1.20	1.07	0.79	0.89	1.01	0.50	0.80	0.90	—
6α-Hydroxy-androst-4-ene-3,17-dione	—	0.06	—	—	—	—	—	—	—	—
6β-Hydroxy-androst-4-ene-3,17-dione	0.18	0.20	0.36	—	—	0.49	—	0.35	0.46	—
11α-Hydroxy-androst-4-ene-3,17-done	—	—	0.22	—	—	0.32	—	(0.18)	—	—
11β-Hydroxy-androst-4-ene-3,17-dione	0.22	0.35	0.56	0.14	0.35	0.68	—	0.50 (0.53)	0.69	—
12α-Hydroxy-androst-4-ene-3,17-dione	—	—	0.44	—	—	0.46	—	(0.38)	—	—
14α-Hydroxy-androst-4-ene-3,17-dione	—	—	0.27	—	—	0.39	—	(0.21)	—	—
15α-Hydroxy-androst-4-ene-3,17-dione	—	—	0.19	—	—	0.28	—	(0.14)	—	—
19-Hydroxy-androst-4-ene-3,17-dione	—	—	0.18	—	—	0.25	—	0.17	—	—
Group II[b]										
17β-Hydroxy-androsta-1,4-dien-3-one	0.27	0.34	0.40	0.60	0.69	0.83	0.48	0.65 (0.57)	—	—
17α-Hydroxy-androst-4-en-3-one	—	—	0.81	—	1.13	—	1.27	1.07	—	—
17β-Hydroxy-androst-4-en-3-one	0.73	0.50	0.70	1.03	1.01	0.96	1.05	0.95 (0.92)	—	—
1-Dehydro-testololactone	—	—	0.60	—	0.25	0.69	0.11	(0.28)	—	—
Testololactone	—	—	0.84	—	0.44	0.81	0.24	(0.50)	—	—
11β-Hydroxy-androst-4-ene-3,11-dione	—	0.15	0.35	0.10	0.21	0.43	—	0.25	—	—
17α-Hydroxy-androst-4-ene-3,11-dione	—	—	0.41	—	0.33	—	—	0.43	—	—
6β,17β-Dihydroxy-androst-4-en-3-one	—	0.03	0.07	—	—	0.11	—	—	—	—
11β,17β-Dihydroxy-androst-4-en-3-one	0.05	0.05	0.11	0.07	0.09	0.19	—	—	—	—
17α-Methyl-17β-hydroxy-androst-4-en-3-one	—	0.80	0.77	1.34	—	1.10	1.44	1.23	—	—
17α-Methyl-17β-hydroxy-androst-4-ene-3,11-dione	—	0.33	0.45	—	—	0.67	—	0.40	—	—
Group III[c]										
5α-Androst-2-ene-17-one	20.4	—	—	—	—	—	—	—	—	6.45
3β-Chloro-androst-5-ene-17-one	20.4	—	—	—	—	—	—	—	—	6.45
5α-Androstane-3,17-dione	5.5	—	—	—	—	—	—	—	—	3.14
5β-Androstane-3,17-dione	5.5	—	—	—	—	—	—	—	—	3.14
6-Hydroxy-1-androstan-17-one	3.3	—	—	—	—	—	—	—	—	—
3β-Hydroxy-androst-5-ene-17-one	—	0.87	0.84	—	1.50	—	1.75	(1.18)	—	1.37
3α-Hydroxy-5α-androst-9-ene-17-one	2.22	—	—	—	—	—	—	—	—	—
3α-Hydroxy-5β-androst-9-ene-17-one	1.56	—	—	—	—	—	—	—	—	—
3α-Hydroxy-5α-androstan-17-one	2.22	1.36	1.00	—	1.60	—	2.63	(1.40)	—	2.58
3β-Hydroxy-5α-androstan-17-one	—	0.83	0.84	—	1.45	—	2.20	(1.26)	—	1.76
3α-Hydroxy-5β-androstan-17-one	1.56	0.93	0.90	—	1.50	—	2.28	(1.26)	—	—
3β-Hydroxy-5β-androstan-17-one	—	—	1.00	—	1.70	—	2.58	(1.35)	—	—
5α-Androstane-3,7,17-trione	—	1.29	—	—	—	—	—	—	—	1.76
5α-Androstane-3,11,17-trione	1.67	—	1.22	—	1.32	1.14	—	(1.13)	—	—
5β-Androstane-3,11,17-trione	1.67	—	1.22	—	1.32	1.14	—	(1.13)	—	—
3β-Hydroxy-androst-5-ene-11,17-dione	—	—	0.50	—	0.45	0.76	—	(0.32)	—	—
3β-Hydroxy-5α-androstane-6,17-dione	—	0.11	—	—	—	0.74	—	—	—	—
3α-Hydroxy-5α-androstane-11,17-dione	0.67	—	—	—	—	0.74	—	(0.60)	—	—
3β-Hydroxy-5α-androstane-11,17-dione	—	—	0.50	—	0.44	0.75	—	(0.34)	—	—
3α-Hydroxy-5β-androstane-11,17-dione	0.58	0.38	0.65	—	0.66	0.77	0.35	(0.47)	—	0.31
3β-Hydroxy-5β-androstane-11,17-dione	—	—	0.67	—	0.75	0.90	—	(0.50)	—	—
11β-Hydroxy-5α-androstane-3,17-dione	0.38	—	—	—	—	—	—	—	—	—

TABLE PC 40—(Continued)
STEROIDS, $C_{19\text{-}20}$(17-OXYSTEROIDS)

Paper	P1	P1	P1	P2	P3	P3	P4	P4	P4	P1
Solvent	S1	S2	S3	S4	S2	S5	S6	S7	S8	S4
Temperature	—	—	—	—	—	—	38 °C	22 °C (38 °C)	22 °C	—
Literature	1	1	1	1	1	1	1	1	1	1
Compound					$R_{cortexone}$					
Group III[c]										
3β,11β-Dihydroxy-androst-5-en-17-one	—	—	0.12	—	—	0.36	—	—	—	—
3α,11β-Dihydroxy-5α-androstan-17-one	0.22	—	0.29	—	—	0.61	—	—	—	—
3β,11β-Dihydroxy-5α-androstan-17-one	—	—	0.14	—	0.20	—	—	(0.08)	—	—
3α,11β-Dihydroxy-5β-androstan-17-one	0.11	—	0.20	—	—	0.49	—	—	—	0.05

Paper P1 = W1 impregnated with propylene glycol.
P2 = W1 impregnated with carbitol.
P3 = W1 impregnated with formamide.
P4 = W1.

Solvent S1 = ligroine.
S2 = cyclohexane-benzene (1 : 1).
S3 = toluene.
S4 = methyl cyclohexane.
S5 = benzene.
S6 = petroleum ether-methanol-water (10 : 8 : 2).
S7 = petroleum ether-benzene-methanol-water (33 : 17 : 40 : 10).
S8 = petroleum ether-toluene-methanol-water (25 : 25 : 35 : 15).

Detection UV light and Zimmermann or Raymond (*m*-dinitrobenzene) reaction.

	UV	*DNB*
[a] Group I	+	+
[b] Group II	+	−
[c] Group III	−	+

Literature 1. Neher, R., *Steroid Chromatography*, Elsevier, 1964, pp 169–173.

TABLE PC 41
STEROIDS, 17-KETOSTEROIDS

Paper	P1	P2	P2
Solvent	S1	S2	S2
Literature	1	1	1
Compound	R_{AND}[a]	R_{DOC}[b] × 100	R_{CORT}[c]
5β-Androstane-3,11,17-trione	75	75	—
4-Androstene-3,17-dione	170	205	—
3α,11β-Dihydroxy-5α-androstan-17-one	10	12	350
3β,11β-Dihydroxy-5α-androstan-17-one	—	7.5	215
3β,11β-Dihydroxy-5β-androstan-17-one	4.8	9	225
3α-Hydroxy-5β-androstane-11,17-dione	26	20	670
11β-Hydroxy-5α-androstane-3,17-dione	17	27	1030
3α-Hydroxy-5α-androstan-17-one (androsterone)	100	228	—
3α-Hydroxy-5β-androstan-17-one	70	175	—
3β-Hydroxy-5α-androstan-17-one (epiandrosterone)	70	185	—
3β-Hydroxy-5β-androstan-17-one	100	225	—
3α-Hydroxy-5β-androst-9-en-17-one	70	180	—
3β-Hydroxy-5-androsten-17-one (dehydroepiandrosterone)	70	135	—

[a] Mobility relative to androsterone as reference steroid.
[b] Mobility relative to deoxycorticosterone as reference steroid.
[c] Mobility relative to cortisone as reference steroid.

Paper P1 = W1 impregnated with propylene glycol.
P2 = W1 impregnated with DMSO.

Solvent S1 = petroleum ether, saturated with propylene glycol.
S2 = heptane, saturated with DMSO.

Technique Zaffaroni type paper partition chromatography.

Literature 1. Katz, S. and J. R. Broich, *J. Chromatogr.* **6**, 514 (1961).

TABLE PC 42
STEROIDS AND STEROID ESTERS

Paper	P1		P1				P1	
Solvent	S1		S2				S3	
Literature	1		1				1	
	Free	Ft	Free	Ft	Ac	Pr	Free	Ft
Compound[a]	$R_F \times 100$							
Cholestenone	96	—	98	—	—	—	—	—
Oestrone	—	—	—	—	—	—	—	22
Androstane derivatives								
Androsta-1,4-diene-3,17-dione	—	—	—	—	—	—	10	—
Androsta-1,4-dien-17β-ol-3-one	—	—	—	—	—	—	—	29
5α-Androstane-3α,17β-diol	—	77	—	84	—	—	—	87
5β-Androstane-3α,17β-diol	—	78	—	88	—	—	—	89
5β-Androstane-3α,11β-diol-17-one	—	—	—	3	—	—	—	27
5α-Androstane-3,17-dione	5	—	13	—	—	—	46	—
5β-Androstane-3,17-dione	5	—	13	—	—	—	45	—
5α-Androstan-3α-ol-17-one	7.5	32	11	51	59	78	55	68
5α-Androstan-3β-ol-17-one	—	25	—	54	62	77	—	68
5β-Androstan-3α-ol-17-one	5.5	29	8	52	57	73	31	74
5β-Androstan-3β-ol-17-one	6	30	—	—	—	—	47	82
5β-Androstan-17β-ol-3-one	5	25	9	43	—	—	—	75
5α-Androstan-17-one	95	—	97	—	—	—	—	—
Androst-5-ene-3β,17α-diol	—	76	—	86	—	—	—	89
Androst-5-ene-3β,17β-diol	—	78	—	85	—	—	—	89
Androst-4-ene-11β,17β-diol-3-one	—	—	—	2	—	—	—	13
Androst-4-ene-3,17-dione	3	—	3	—	—	—	20	—
Androst-4-ene-3,11,17-trione	—	—	2	—	—	—	24	—
Androst-5-en-3β-ol	—	93	—	93	—	—	—	—
Androst-4-en-11β-ol-3,17-dione	—	—	0	—	—	—	2	—
Androst-4-en-17α-ol-3-one	—	8	3	19	20	39	16	49
Androst-4-en-17β-ol-3-one	—	8	3	20	23	42	16	49
Androst-5-en-3β-ol-17-one	5	28	9	51	56	74	29	72
5β-Androst-9(11)-en-3α-ol-17-one	7.5	32	11	58	64	81	51	76
17α-Methyl-5α-androstane-3β,17β-diol	11	48	—	66	—	—	—	82
17α-Methylandrost-5-ene-3β,17β-diol	—	48	—	55	—	—	—	79
17α-Methylandrost-4-en-17β-ol-3-one	0	—	6	—	32	—	25	—
19-Norandrost-4-en-17β-ol-3-one	—	8	—	16	—	—	—	48
Pregnane derivatives								
16α-Methylpregn-4-ene-3,20-dione	—	—	—	—	—	—	62	—
Pregna-4,16-diene-3,20-dione	—	—	—	—	—	—	49	—
Pregna-5,16-dien-3β-ol-20-one	—	—	—	—	—	—	58	86
5α-Pregnane-3α,20α-diol	—	86	—	88	—	—	—	93
5β-Pregnane-3α,20α-diol	—	86	—	88	—	—	—	94
5α-Pregnane-3,20-dione	19	—	36	—	—	—	68	—
5β-Pregnane-3,20-dione	19	—	36	—	—	—	68	—
5α-Pregnane-3α,17α,20α-triol	—	23	—	36	—	—	—	70
5β-Pregnane-3β,16β,20α-triol	—	22	—	—	—	—	—	73
5α-Pregnan-3β-ol	—	91	—	93	—	—	—	—
5β-Pregnan-3β-ol	—	83	—	93	—	—	—	—
5α-Pregnan-3α-ol-20-one	—	70	35	76	—	—	—	86
5β-Pregnan-3α-ol-20-one	—	65	28	77	—	—	—	89
Pregn-5-ene-3β,20α-diol	—	85	—	88	—	—	—	92
Pregn-5-ene-3β,20β-diol	—	84	—	89	—	—	—	93
Pregn-4-ene-3,20-dione	4	—	11	—	—	—	48	—
Pregn-4-ene-3,11,20-trione	—	—	0	—	—	—	8	—
Pregn-4-en-20β-ol-3-one	—	19	—	28	33	52	—	72
Pregn-5-en-3β-ol-20-one	—	61	23	76	80	86	60	86

[a] Ft = formate; Ac = acetate; Pr = propionate; mono-, di-, or tri-ester as appropriate.

Paper P1 = W1, 2, 4, and 3MM.

Solvent S1 = formic acid-petroleum ether (80–100 °C) (1 : 1).
S2 = formic acid-petroleum ether-methanol (10 : 9 : 1).
S3 = formic acid-petroleum ether-benzene (2 : 1 : 1).

Technique The top layer of the solvent mixture is the mobile phase. The paper is equilibrated for four hours in the vapors of the bottom layer before development.

Literature
1. Edwards, R. W. H., *J. Chromatogr.*, **12**, 212 (1963).

TABLE PC 43
STEROIDS, C_{20-21}

Paper	P1	P2	P2	P2	P3	P3	P4
Solvent	S1	S2	S3	S4	S3	S5	S6
Temperature	—	—	—	—	—	—	38 °C
Literature	1	1	1	1	1	1	1
Compound	$R_{cortexone}$						
Pregna-4,16-diene-3,20-dione	3.14	5.10	—	—	—	—	1.70
Pregna-4,9(11)-diene-3,20-dione	3.14	5.10	—	—	—	—	1.70
Pregna-1,4-diene-3,20-dione	1.94	—	—	—	1.84	—	1.55
Pregn-4-ene-3,20-dione	3.14	5.10	1.50	1.28	1.86	1.21	1.70
20β-Hydroxy-pregn-4-en-3-one	—	—	—	0.99	1.59	—	1.35
20α-Hydroxy-pregn-4-en-3-one	—	0.53	—	0.88	1.50	—	1.25
3-Oxo-18-hydroxy-eti-4-enic acid-lactone (20 → 18)	—	—	—	1.14	0.91	—	0.90
Pregn-4-ene-3,6,20-trione	—	2.58	—	—	—	—	—
Pregn-4-ene-3,11,20-trione	1.16	1.86	1.00	1.22	1.35	—	1.15
Pregn-4-ene-3,18,20-trione	—	—	—	—	1.40	—	1.30
17-Hydroxy-pregn-4-ene-3,20-dione	0.92	0.62	0.64	0.86	0.95	1.01	0.93
11β-Hydroxy-pregn-4-ene-3,20-dione	0.70	0.44	0.52	0.81	0.74	0.95	0.80
6β-Hydroxy-pregn-4-ene-3,20-dione	0.73	0.31	0.40	0.59	0.48	0.75	0.60
14α-Hydroxy-pregn-4-ene-3,20-dione	—	—	—	0.71	0.64	0.87	0.60
16α-Hydroxy-pregn-4-ene-3,20-dione	0.47	0.10	0.23	—	—	0.49	0.30
18-Hydroxy-pregn-4-ene-3,20-dione	—	—	—	0.73	0.90	0.95	1.78[a]
Pregn-4-ene-3,6,11,20-tetrone	—	0.56	—	—	—	—	—
3-Oxo-11β,18-epoxy-eti-4-enic acid-lactone	—	—	—	—	—	—	—
(20 → 18) (aldosterone lactone)	—	—	—	0.99	0.44	0.91	0.52
3,20-Dioxo-11β-hydroxy-eti-4-enic acid-lactone (18 → 11)	—	—	—	—	0.32	—	—
17-Hydroxy-pregn-4-ene-3,11,20-trione	0.46	0.14	0.25	—	—	0.59	0.38
6β-Hydroxy-pregn-4-ene-3,11,20-trione	—	—	0.14	—	—	0.39	—
11β,17-Dihydroxy-pregn-4-ene-3,20-dione	—	—	0.07	—	—	0.26	—
6β,17-Dihydroxy-pregn-4-ene-3,20-dione	—	—	0.05	—	—	0.14	—

[a] Decomposition product.

Paper W1 impregnated with:
P1 = carbitol.
P2 = propylene glycol.
P3 = formamide.
P4 = W1.

Solvent S1 = methyl cyclohexane.
S2 = ligroine.
S3 = cyclohexane-benzene (1 : 1).
S4 = toluene.
S5 = benzene.
S6 = petroleum ether-benzene-methanol-water (33 : 17 : 40 : 10).

Detection UV light (all positive); tetrazolium blue (all negative).

Technique Not specified.

Literature
1. Neher, R., *Steroid Chromatography*, Elsevier, 1964, p 187.

TABLE PC 44
STEROIDS, C_{21}

Paper	P1	P2	P2	P3	P3
Solvent	S1	S2	S3	S4	S5
Temperature	—	—	—	22 °C	22 °C
Literature	1	1	1	1	1
Compound			$R_{cortexone}$		
5α-Pregnane-3,20-dione	4.07	6.7	—	—	—
5β-Pregnane-3,20-dione	4.07	6.7	—	—	—
3β-Hydroxy-pregn-5-en-20-one	1.92	2.80	1.04	1.65	2.65
3α-Hydroxy-5α-pregnan-20-one	—	3.11	—	1.87	3.57
3β-Hydroxy-5α-pregnan-20-one	—	—	1.05	1.79	2.98
3α-Hydroxy-5β-pregnan-20-one	2.27	2.89	—	1.79	3.14
3β-Hydroxy-5β-pregnan-20-one	1.92	—	1.04	1.71	3.14
5α-Pregnane-3,11,20-trione	1.45	—	—	1.50	1.81
5β-Pregnane-3,11,20-trione	1.45	—	—	1.50	1.81
3α-Hydroxy-5α-pregnane-11,20-dione	—	—	0.88	—	—
3β-Hydroxy-5α-pregnane-11,20-dione	0.74	—	—	0.65	0.46
3α-Hydroxy-5β-pregnane-11,20-dione	0.92	—	—	0.83	0.59
3β-Hydroxy-5β-pregnane-11,20-dione	0.92	—	—	0.83	0.69
11β-Hydroxy-5β-pregnane-3,20-dione	1.16	—	—	1.25	0.86
3α,17-Dihydroxy-5β-pregnan-20-one	—	0.20	—	—	—
3β,21-Dihydroxy-pregn-5-en-20-one (BT+)	—	—	0.47	0.68	—
17-Hydroxy-5β-pregnane-3,11,20-trione	0.74	—	—	0.71	0.29
3α,17-Dihydroxy-5β-pregnane-11,20-dione	0.41	—	—	0.19	—

Paper	W1 impregnated with: P1 = carbitol. P2 = propylene glycol. P3 = W1.
Solvent	S1 = methyl cyclohexane. S2 = ligroine. S3 = toluene. S4 = petroleum ether-benzene-methanol-water (33 : 17 : 40 : 10). S5 = petroleum ether-methanol-water (100 : 80 : 20).
Detection	UV light (all negative); tetrazolium blue (all negative); $SbCl_3$ (positive).
Technique	Not specified.
Literature	1. Neher, R., *Steroid Chromatography*, Elsevier, 1964, p 19.

TABLE PC 45
STEROIDS, C_{21}

Paper	P1	P2	P3	P3	P3	P3
Solvent	S1	S2	S3	S4	S5	S6
Temperature	—	—	22 °C	38 °C	38 °C	38 °C
Literature	1	1	1	1	1	1
Compound			$R_{cortisol}$			
3β,21-Dihydroxy-5α-pregnane-11,20-dione	> 6.0	3.71	3.19	1.74	—	1.52
3α,21-Dihydroxy-5β-pregnane-11,20-dione	> 6.0	3.75	3.71	1.92	—	2.18
3β,11β,21-Trihydroxy-5α-pregnan-20-one	4.00	2.63	2.64	1.67	—	1.61
3α,11β,21-Trihydroxy-5β-pregnan-20-one[a]	5.43	2.60	3.24	2.01	2.32	1.62
3β,17,21-Trihydroxy-5α-pregnan-20-one	2.34	2.50	2.87	1.63	—	2.51
3α,17,21-Trihydroxy-5α-pregnan-20-one	3.98	2.78	2.87	1.70	—	2.56
3α,17,21-Trihydroxy-5β-pregnan-20-one	2.80	2.79	2.81	1.81	—	2.71
17,21-Dihydroxy-5α-pregnane-3,11,20-trione	5.47	3.11	2.80	1.71	—	1.60
17,21-Dihydroxy-5β-pregnane-3,11,20-trione	5.31	2.84	2.83	1.73	1.83	1.66
3β,17,21-Trihydroxy-5α-pregnane-11,20-dione	0.40	0.79	0.70	0.77	—	1.78
3α,17,21-Trihydroxy-5β-pregnane-11,20-dione	0.42	1.01	0.89	0.88	1.55	1.92
3α,11β,21-Trihydroxy-5β-pregnan-18-al-20-one[b]	0.52	1.01	0.76	0.65	—	1.54
11β,17,21-Trihydroxy-5α-pregnane-3,20-dione	2.67	1.61	1.86	1.43	—	1.40
11β,17,21-Trihydroxy-5β-pregnane-3,20-dione	2.70	1.43	1.99	1.51	1.71	1.49
3α,11β,17,21-Tetrahydroxy-5α-pregnan-20-one	0.40	0.40	0.55	0.82	—	1.37
3β,11β,17,21-Tetrahydroxy-5α-pregnan-20-one	0.20	0.22	0.33	0.57	—	1.12
3α,11β,17,21-Tetrahydroxy-5β-pregnan-20-one	0.29	0.28	0.38	0.63	1.11	1.51

[a] The corresponding 5α-pregnane isomer ("allo-TH-corticosterone") was not available; it is supposed to run slightly faster than the 5β-isomer.
[b] "Tetrahydroaldosterone".

Paper W1 impregnated with:
P1 = propylene glycol.
P2 = formamide.
P3 = W1.

Solvent S1 = toluene.
S2 = chloroform.
S3 = benzene-methanol-water (100 : 50 : 50).
S4 = toluene-ethyl acetate-methanol-water (90 : 10 : 50 : 50).
S5 = hexane-*tert* butanol-methanol-water (50 : 22.5 : 22.5 : 20).
S6 = isooctate-*tert* butanol-water (10 : 5 : 9).

Detection UV light (all negative); tetrazolium blue (all positive).

Technique Not specified.

Literature 1. Neher, R., *Steroid Chromatography*, Elsevier, 1964, p 194.

TABLE PC 46
SULFUR COMPOUNDS, SULFONIC ACIDS

Paper	P1	P2	P3	P2	P2	P2
Solvent	S1	S2	S2	S3	S4	S5
Technique	T1	T1	T1	T2	T2	T2
Literature	1	1	1	2	2	2
Compound			$R_F \times 100$			
o-Aminonaphtholsulfonic acid isomer						
1,2,0	93	—	—	—	—	—
1,2,4	61	—	—	—	—	—
1,2,5	39	—	—	—	—	—
1,2,6	44	—	—	—	—	—
1,2,7	47	—	—	—	—	—
1,2,8	27	—	—	—	—	—
2,1,0	93	—	—	—	—	—
2,1,4	70	—	—	—	—	—
2,1,5	41	—	—	—	—	—
Naphthalene derivative						
-1,5-disulfonic acid	—	—	—	86	70	46
-1,6-disulfonic acid	—	—	—	81	65	47
-1,7-disulfonic acid	—	—	—	86	70	51
-2,6-disulfonic acid	—	—	—	76	58	46
-2,7-disulfonic acid	—	—	—	76	60	46
-1-sulfonic acid	—	—	—	71	57	78
-2-sulfonic acid	—	—	—	64	49	78
-1,3,5,7-tetrasulfonic acid	—	—	—	98	86	15
-1,3,5-trisulfonic acid	—	—	—	96	76	27
-1,3,6-trisulfonic acid	—	—	—	88	72	24
-1,3,7-trisulfonic acid	—	—	—	96	76	27
1,2-Naphthoquinone derivative						
1-diazide	—	90	98	—	—	—
2-diazide	—	92	97	—	—	—
4-sulfonic acid 1-diazide	—	51	65	—	—	—
4-sulfonic acid 2-diazide	—	45	71	—	—	—
5-sulfonic acid 1-diazide	—	41	52	—	—	—
5-sulfonic acid 2-diazide	—	40	62	—	—	—
6-sulfonic acid 1-diazide	—	36	47	—	—	—
7-sulfonic acid 1-diazide	—	41	56	—	—	—
8-sulfonic acid 1-diazide	—	55	73	—	—	—

Paper P1 = W1.
P2 = W4.
P3 = W4, impregnated with 5% solution of $NaHCO_3$.

Solvent S1 = *n*-propanol-5% $NaHCO_3$ (2 : 1).
S2 = butanol-pyridine-H_2O (3 : 1 : 4).
S3 = HCL-H_2O (1 : 3).
S4 = saturated aqueous K_2SO_4.
S5 = *n*-propanol-NH_3 (2 : 1).

Technique T1 = 3 mm spots on 10 cm strips, developing 8–72 hours at 20 °C, damp atmosphere.
T2 = descending development in chamber.

Literature
1. Volodarskii, I. L. and V. V. Kozlov, *Zh. Anal. Khim.*, **23**, 763 and 764 (1968); *J. Chromatogr.*, 38, **D97** and **D98** (1968).
2. Latinák, J., *Mikrochim. Acta*, p 350 (1966).

TABLE PC 47
SULFUR COMPOUNDS, THIOUREAS

Paper	P1
Solvent	S1
Literature	1
Compound	R_{Ph}[a] × 100
N-**Allyl-4-morpholine thiocarboxamide**	131
N-**Allylthiourea**	26
N-**Benzylthiourea**	90
N-p-**Bromophenylthiourea**	100
N-**3-Butenylthiourea**	61
N-n-**Butylthiourea**	84
N-DL-*sec*-**Butylthiourea**	74
N-tert-**Butylthiourea**	100
N-trans-**Crotylthiourea**	78
5,5-Diallylthiobarbituric acid	137
N′,N″-**Dibutylthiourea**	141
5,5-Diethylthiobarbituric acid	130
N′,N″-**Diethylthiourea**	124
N′-**Dimethylthiourea**	60
N′,N″-**Dimethylthiourea**	42
N-**Ethylenethiourea**	26
N-**Ethyl-4-morpholine thiocarboxamide**	127
N′,N″-**Ethyl phenyl thiourea**	141
N-**Ethyl-1-pyrrolidinethiocarboxamide**	137
5-Ethylthiobarbituric acid	0
N-**Ethylthiourea**	15
N-p-**Iodophenylthiourea**	106
N-**Isobutylthiourea**	76
N-**Isopentylthiourea**	107
N-**Isopropylthiourea**	41
N-α-**Methallylthiourea**	62
5-Methyl-5-allylthiobarbituric acid	119
N′,N″-**Methyl butyl thiourea**	135
N′,N″-**Methyl ethyl thiourea**	89
N-**Methyl-4-morpholine thiocarboxamide**	117
N′,N″-**Methyl phenyl thiourea**	140
N-**Methyl-1-piperidine thiocarboxamide**	137
N-**Methyl-1-pyrrolidinethiocarboxamide**	135
4-Methylthiosemicarbazide	20
N-**Methylthiourea**	40
N-β-**Phenylethylthiourea**	110
1-Phenylthiosemicarbazide	55
N-n-**Propylthiourea**	42
Tetramethylthiourea	143
Thioacetamide	25
Thiobarbituric acid	0
Thiobenzamide	79
Thiosemicarbazide	0
Thiourea	0
N-m-**Tolylthiourea**	125
N-o-**Tolylthiourea**	119
N-p-**Tolylthiourea**	122
1,1,4-Trimethylsemicarbazide	132
Trimethylthiourea	121

[a] R_{Ph} = Migration relative to *N*-phenylthiourea.

Paper	P1 = W1.
Solvent	S1 = chloroform saturated with water.
Detection	Grote's Reagent.
Technique	Saturated chamber, ascending development time = 110–120 minutes.
Literature	1. Kjaer, A. and A. Rubenstein, *Acta. Chem. Scand.*, **7**, 528–536 (1953); *Nature*, **171**, 840–841 (1953).

TABLE PC 48
VITAMINS AND RELATED COMPOUNDS

Paper	P1	P1	P1	P1	P1	P1	P1	P1	P1	P1	P1	P1	P1
Solvent	S1	S2	S2	S3	S4	S5	S5	S6	S7	S8	S9	S10	S10
Technique	T1	T2	T3	T2	T1	T2	T3	T1	T1	T1	T1	T2	T3
Literature	1	1	1	1	1	1	1	1	1	1	1	1	1
Compound	$R_F \times 100$												
β-Carotene (C)	51	81	98	84	92	87	95	94	13	32	79	86	90
Chlorophylls (Chl) *a* + *b*	—	—	—	—	—	4	4	8	—	—	—	—	—
Coenzyme Q_{10} (Q_{10})	5	9	16	47	70	83	90	91	—	—	13	55	85
Pheophytin (Ph)	—	—	—	—	17	47	58	77	—	—	—	9	74
Plastoquinone A (PQ_9)	8	18	23	77	82	85	94	91	—	5	18	78	87
α-Tocopherol (T)	—	5	13	26	46	48	64	82	—	—	10	32	78
α-Tocopherylquinone (TQ)	—	—	5	5	14	41	50	73	—	—	5	9	65
Vitamin A (A)	—	—	—	4	11	29	31	51	—	—	—	9	38
Vitamin D_2 (D_2)	—	—	5	8	13	40	41	59	—	—	—	9	52
Vitamin K_1 (K_1)	20	50	90	78	87	85	94	91	8	10	53	79	87
Xanthophylls (X)	—	—	—	—	—	—	—	8	—	—	—	—	—

Paper P1 = SS No. 288.
Solvent S1 = cyclohexane.
S2 = cyclohexane-benzene (7 : 3).
S3 = cyclohexane-benzene (3 : 7).
S4 = benzene.
S5 = benzene-$CHCl_3$ (1 : 1).
S6 = $CHCl_3$.
S7 = isooctane.
S8 = petroleum ether (b.p. 30–60 °C).
S9 = CCl_4.
S10 = toluene.
Technique T1 = ascending development, room temperature in dark.
T2 = walls of tank lined with solvent-wetted filter paper and T1.
T3 = unlined tank walls and T1.
Literature 1. Lichtenthaler, H. J., *J. Chromatogr.*, **13**, 166 (1964).

TABLE PC 49
VITAMINS A_1, A_2 AND THEIR DERIVATIVES

		P1	
Paper		P1	
Solvent		S1	
Literature		1	
Compound	λ_{max} in petroleum ether (nm)	$R_F{}^a \times 100$	UV
Anhydrovitamin A_1	350, 370, 390	20 ± 1	yF
Anhydrovitamin A_2	350, 370, 390	33 ± 1	yF
β-Carotene (see Chapter XI)	—	0	o
Isoanhydrovitamin A_1	330, 350, 370	29 ± 3	bl F
Vitamin A_1 acetate	325	35 ± 3	bl F
Vitamin A_1 acid	350	65 ± 2	r a
Vitamin A_1 alcohol	325	82 ± 3	bl F
Vitamin A_1 aldehyde	370	55 ± 1	r a
Vitamin A_1 palmitate	325	0	bl F
Vitamin A_2 acetate	350	39 ± 2	gr F
Vitamin A_2 alcohol	350	93 ± 1	gr F
Vitamin A_2 aldehyde	385	67 ± 3	r a

[a] Mean values of 10 observations.

TABLE PC 49—(Continued)
VITAMINS A_1, A_2 AND THEIR DERIVATIVES

Paper P1 = W1.
Solvent S1 = 90% methanol.
Detection UV light.
Technique Reverse-phase circular, development time 4–5 hours.
Literature 1. Jungalwala, F. B. and H. R. Cama, *J. Chromatogr.*, **8**, 535 (1962).

TABLE PC 50
VITAMINS AND RELATED COMPOUNDS

Paper	P1	P2	P2	P3	P4	P5	P2
Solvent	S1	S2	S3	S4	S5	S6	S7
Technique	T1	T1	T1	T1	T1	T2	T1
Literature	1	2	2	3	3	4	2
Compound	$R_F \times 100$						
Coenzyme Q_6	—	54	—	—	—	—	31
Coenzyme Q_7	—	49	—	—	—	—	23
Coenzyme Q_8	—	42	—	—	—	—	18
Coenzyme Q_9	—	36	—	—	—	—	11
Coenzyme Q_{10}	—	27	—	—	—	ca. 30	6
Coenzyme Q_6 hydroquinone	—	—	—	—	—	—	66
Coenzyme Q_7 hydroquinone	—	—	—	—	—	—	60
Coenzyme Q_8 hydroquinone	—	—	—	—	—	—	55
Coenzyme Q_9 hydroquinone	—	—	—	—	—	—	41
Coenzyme Q_{10} hydroquinone	—	—	—	—	—	—	26
Hydrogenated ε-tocopherol	—	—	—	22	—	—	—
Tocol	—	—	—	65	50	—	—
α-Tocopherol	50	—	33[a]	10	31	40–50	—
β-Tocopherol	72	—	—	22	37	—	—
γ-Tocopherol	72	—	49[a]	22	37	—	—
δ-Tocopherol	84	—	62[a]	42	45	—	—
ε-Tocopherol	—	—	—	53	—	—	—
α-Tocopherylquinone	—	84	58	—	—	—	—
Vitamin K_1	—	59	—	—	—	—	—
Vitamin K_2	—	29	—	—	—	—	—

[a] = *d*-tocopherols.

Paper P1 = W1 coated with vaseline.
P2 = paper impregnated with silicone.
P3 = paper impregnated with a 5% solution of liquid paraffin in petroleum ether.
P4 = W4 impregnated with a 5% solution of olive oil in petroleum ether.
P5 = silicone-coated glass fiber paper with alcohol fluorescein.

Solvent S1 = 75% ethanol.
S2 = *n*-propanol-H_2O (4 : 1).
S3 = ethanol-H_2O (3 : 1).
S4 = 65% ethanol in H_2O.
S5 = 90% ethanol in H_2O.
S6 = 2,2,4-trimethylpentane-*iso*-propylacetate (200 : 0.5).
S7 = *n*-propanol-H_2O (7 : 3).

Technique T1 = reversed phase chromatography.
T2 = glass fiber chromatography, ascending in tanks containing isooctane, 200 ml and isopropylacetate, 0.5 ml, development time, approximately 20 minutes.

Literature
1. Brown, F., *Biochem. J.*, **51**, 237 (1952); **52**, 523 (1952).
2. Lester, R. L. and T. Ramasarma, *J. Biol. Chem.*, **234**, 672 (1959).
3. Green, J., Marcinkiewics, S., and D. McHale, *J. Chromatogr.*, **10**, 158 (1963).
4. Atlas, S. J. and K. G. Pinter, *Anal. Biochem.*, **17**, 258 (1966).

TABLE PC 51
VITAMINS AND RELATED COMPOUNDS

Paper	P1	P1										
Solvent	S1	S2										
Detection	—	—		D1	D2	D3	D4	D5	D6	D7	D8	D9
Literature	1	1										
Compound	$R_F \times 100$		Quantity used μg	Color								
p-Aminobenzoic acid	80	69	250	—	—	—	—	—	—	br	br	—
Biotin	79	78	250	—	—	—	br	—	—	—	—	—
Choline	87	52	250	—	—	—	br	—	—	—	—	—
Folic acid	34	30	12.5	—	—	—	—	—	—	—	—	F
Inositol	21	12	50	br	—	—	—	—	—	—	—	—
Menadione	94	—	250	—	—	—	—	—	—	—	—	y[b]
Niacin	83	68	250	—	—	—	br	pk	o	—	—	—
Nicotinamide	85	69	250	—	y	—	br	—	y	—	—	—
Pantothenic acid	66	38	250	—	—	p	—	—	—	—	—	—
Pyridoxine	87	60	250	—	—	—	—	—	—	br	—	F
Riboflavin	91	32	20	—	—	—	—	—	—	—	—	y,F
Thiamine	93	55	250	—	F	y	—	—	—	—	—	—
α-Tocopherol	89	77	250[a]	bk	—	—	—	—	—	—	—	—
Vitamin A	90	—	250	—	—	—	—	—	—	—	—	y[b]
Vitamin B_{12}	92	31	50	—	—	—	—	—	—	—	—	pk
Vitamin C	34	34	250	br	—	—	—	—	—	—	—	—

[a] In mg.
[b] After phenol run only.

Paper P1 = W1.

Solvent S1 = phenol-buffer (100 : 25; aqueous buffer-6.3% sodium citrate, 3.7% KH_2PO_4).
S2 = *n*-butanol-propionic acid-water (freshly prepared from equal volumes of solution A (1246 ml *n*-butanol + 84 ml water) and solution B (620 ml propionic acid + 790 ml water)).

Detection D1 = ammoniacal silver nitrate.
D2 = ferricyanide-nitroprusside.
D3 = ninhydrin.
D4 = iodine vapor.
D5 = 2,6-dichlorophenolindophenol.
D6 = cyanogen bromide.
D7 = ferric chloride.
D8 = phenol-hypochlorite reagent.
D9 = UV. Order for multiple detection: D9 (UV), D3, D4, D7, or D1.

Technique Two-dimensional paper chromatography descending techniques. Time of run: 18–22 hours (S1); 14–16 hours (S2). Temperature of run: 24 ° ± 0.5 °C.

Literature 1. Gadsden, E. L., Edwards, C. H., and G. A. Edwards, *Anal. Chem.*, **32**, 1415 (1960); *J. Chromatogr.*, **6**, D32 (1961).

TABLE PC 52

ALCOHOLS, ALCOHOL DERIVATIVES

Paper	P1	P2	P3	P3
Solvent	S1	S2	S3	S3
Detection	D1	D1	D2	D2
Literature	1	2	3	3
	$R_F \times 100$			
Compound	3,5-Dinitro-benzoate	3,5-Dinitro-benzoate	3-Nitro-phthalate	Diphenate
Alcohol				
Allyl alcohol	—	—	32	41
Benzyl alcohol	—	—	50	49
1-Butanol	57	64	55	56
2-Butanol	61	—	48	54
Cinnamic alcohol	—	—	50	53
Cyclohexanol	—	—	57	62
Cyclopentanol	—	—	49	55
Ethanol	39	40	26	37
1-Heptanol	—	—	70	69
1-Hexanol	72	79	67	65
Methanol	24	21	15	24
2-Methyl-1-butanol	—	—	61	59
3-Methyl-1-butanol	65	—	63	60
2-Methyl-1-propanol	55	—	53	55
1-Pentanol	66	72	64	61
2-Pentanol	—	—	59	58
2-Phenylethanol	—	—	52	56
3-Phenyl-1-propanol	—	—	55	60
1-Propanol	46	50	42	48
2-Propanol	51	52	35	47

Paper P1 = W1.
P2 = SS2043b impregnated with 50% *N,N*-dimethylformamide in acetone.
P3 = W1 or W4.

Solvent S1 = heptane saturated with methanol.
S2 = decalin saturated with dimethylformamide.
S3 = isoamyl alcohol-concentrated ammonia-H_2O (30 : 15 : 5 v/v).

Detection D1 = Rhodamine 6 GBN-500-UV light
D2 = 4-methylumbelliferone-UV light

Literature

1. Meigh, D. F. *Nature*, **169**, 706–707 (1952).
2. Sundt, E. and M. Winter, *Anal. Chem.*, **29**, 851–852 (1957).
3. Siegel, A. and K. Schlögl, *Mikrochemie ver Mikrochim. Acta*, **40**, 383–389 (1953).

TABLE PC 53

ALCOHOLS

Paper	P1	P1	P1	P1	—
Solvent	S1	S2	S3	S4	—
Literature	1	1	1	1	—
Compound		$R_F \times 100$			Smallest amount detected μg
Aliphatic alcohols					
Primary					
Cinnamyl alcohol	25	—	—	—	0.6
***o*-Hydroxyethylguaiacol**	—	11	—	—	0.2
***p*-Methoxybenzyl alcohol**	13	—	—	—	0.6
1-Octanol	64	—	—	—	17
2-Phenoxyethanol	—	—	86	—	5.6
3-Phenyl-1-propanol	31	—	—	—	1.9
Secondary					
Benzhydrol	34	—	—	—	0.6
Benzoin	23	—	—	—	1.9
Cholesterol	90	—	—	—	50
Tertiary					
Triphenylmethanol	57	—	—	—	1.9
Glycols					
Diethylene glycol	—	—	62	—	1.9
Ethylene glycol	—	—	52	—	5.6
Triethylene glycol	—	—	63	—	1.9
Polyhydroxy compounds					
Glycerol	—	—	35	—	1.9
Pentaerythritol	—	—	36	—	0.6
Sugar alcohols					
Dulcitol	—	—	9	23	0.6
Mannitol	—	—	8	21	0.6
Sorbitol	—	—	8	19	0.6

Paper P1 = 6 × 22.5 in. W1.

Solvent S1 = heptane saturated with methanol.
S2 = petroleum ether saturated with water.
S3 = *n*-butanol saturated with H_2O.
S4 = *n*-butanol-acetic-acid-H_2O (4 : 1 : 5 v/v).

Detection 3,5-Dinitrobenzoyl chloride-rhodamine.

Technique Development in conventional solvent system, steam two minutes prior to spraying.

Literature

1. Brownell, H. H., *J. Chromatogr.*, **19**, 556 (1965).

TABLE PC 54
ALKALOIDS

Paper	P1	P1	P1	P1	P1	P1													
Solvent	S1	S2	S3	S4	S5	S6													
Detection	—	—	—	—	—	—	D1	D2	D3	D4	D5	D6	D7	D8	D9	D10	D11	D12	D13
Literature	1	1	1	1	1	1	1	1	1	1	1	1	1	1	1	1	1	1	1
Compound			$R_F \times 100$										Color[a]						
Aconitine[b]	84	94	40	92	2	1	—	—	—	—	—	+	—	—	—	—	—	—	—
Ajmaline	79	90	35	67	8	2	—	22	—	62	62	—	—	17	—	+	33	40	—
(+)-Biotin	74	0	28	27	3	0	—	—	—	—	—	—	—	—	—	+	33	5	15, 20
Boldine	55	89	2	9	0	0	—	23, 52	52, 54	24, 35	60	23, 69	—	—	68	+	—	—	—
Brazilin[b]	86	2	77	9	3	0	6	—	—	—	—	+	17	—	68	+	—	17	—
Caffeine[b]	66	58	47	91	61	55	—	—	—	—	—	—	—	—	—	—	—	—	—
Cocaine[b]	77	90	13	62	41	14	—	—	—	—	—	—	—	—	—	+	—	—	—
Codeine[b]	39	82	1	22	0	0	—	—	—	—	—	59	—	—	—	+	—	—	—
Corynanthine[b]	84	95	14	19	0	0	40	—	—	—	—	+	—	—	—	+	—	—	—
Cyanidin chloride	93	22	89	4	6	0	3	59	55, 65	63	63	52	—	51	59, 43	++	—	58	—
***d*-Cycloserine(*d*-4-amino-3-isoxalidone)**[c]	9	2	0	0	0	0	—	—	—	—	—	+	—	—	39	++	33	6	17
***dl*-Desthiobiotin(5-methyl-2-imidazolidone**[c] **4-caproic acid)**[d]	78	0	54	59	8	1	—	—	—	—	—	—	—	—	—	—	33	5	22, 20
Heroin[b]	61	90	4	78	2	1	—	—	—	—	—	—	—	—	—	+	—	—	—
Ibogaine[b]	80	94	28	66	5	0	4	—	—	—	—	70	—	—	40	+	33	44	27
Laudanosine[b]	68	92	10	25	4	0	33	—	—	—	—	—	—	—	—	+	33	—	—
Nicotine[b]	23	84	0	19	0	0	—	—	—	+	9	70	—	—	—	+	—	24	70
Papaverine[b]	64	90	38	96	42	37	—	—	—	—	—	+	+	—	—	+	—	—	—
Pyocyanine chloride[b]	48	23	4	55	0	0	17	59	62	23	62	—	—	—	41	+	—	—	—
Quinidine[b]	60	91	8	82	0	0	33	+	—	—	+	—	—	—	—	+	33	—	—
Quinine[b]	55	90	5	87	1	0	38	—	—	—	—	—	—	—	—	+	—	—	—
3-Quinuclidinol[b]	7	26	0	0	0	0	—	—	—	—	—	—	—	—	—	—	33	—	—
Vulpinic acid[b]	94	90	94	92	89	88	2	—	—	—	—	—	—	—	—	+	—	—	—

[a] For interpretation, see Table PC 120.
[b] Dragendorff reagent, positive.
[c] Ninhydrin reagent, violet.
[d] *p*-Dimethylaminobenzaldehyde reagent, pale yellow.

TABLE PC 54—(Continued)
ALKALOIDS

Paper	P1	= W1.
Solvent	S1	= methyl ethyl ketone-acetone-formic acid-water (40 : 2 : 1 : 6).
	S2	= methyl ethyl ketone-diethylamine-water (921 : 2 : 77).
	S3	= methyl isobutyl ketone-formic acid-water (10 parts ketone saturated with 1 part 4% formic acid).
	S4	= chloroform-methanol-formic acid-water (10 parts chloroform saturated with a mixture of 1 part methanol and 1 part 4% formic acid).
	S5	= benzene-methyl ethyl ketone-formic acid-water (a mixture of 9 parts benzene and 1 part ketone saturated with 1 part 2% formic acid).
	S6	= benzene-formic acid-water (10 parts benzene saturated with 1 part 2% formic acid).
Detection	D1	= ultraviolet light.
	D2	= diazotized sulfanilic acid (0.3% solution in dioxane-water 1 : 2).
	D3	= diazotized 4-benzoylamino-2,5-dimethoxyaniline (0.6% solution in dioxane-water 1 : 2).
	D4	= diazotized *o*-dianisidine (0.6% solution in dioxane-water 1 : 2).
	D5	= *p*-nitrobenzenediazonium fluoborate (0.4% solution in dioxane-water 1 : 2).
	D6	= 2,5-dibromoquinone-4-chloroimide (0.5% solution in dioxane-acetone 4 : 1).
	D7	= 2,4-dinitrophenylhydrazine (*ca.* 0.1% solution in 1*N* HCl).
	D8	= ferric chloride (2% aqueous solution).
	D9	= phosphomolybdic acid (2% aqueous solution).
	D10	= potassium permanganate (1% aqueous solution).
	D11	= bromophenol blue (*ca.* 0.05% solution in ethanol).
	D12	= Ehrlich reagent (1% *p*-dimethylaminobenzaldehyde in 1*N* HCl).
	D13	= *p*-dimethylaminocinnamaldehyde (0.1% solution in 1*N* HCl).
Technique	Descending development, 40 cm.	
Literature	1. Reio, L., *J. Chromatogr.*, **47**, 60–85 (1970).	

TABLE PC 55

AMINES, ALIPHATIC, 3,5-DINITROBENZOYL DERIVATIVES

Paper	P1	P1	P1	P1	P1
Solvent	S1	S2	S3	S4	S5
Literature	1,2	1,2	1,2	1,2	1,2
Compound	$R_F \times 100$				
Ammonia	0	0	0	4	70
***iso*-Amylamine**	9	8	26	—	62
***n*-Amylamine**	8	9	29	85	—
***iso*-Butylamine**	4	3	17	—	66
***n*-Butylamine**	4	3	18	77	—
***sec*-Butylamine**	5	4	15	—	—
***tert*-Butylamine**	10	10	30	—	—
***n*-Decylamine**	41	tailing	—	—	32
Di-*n*-butylamine	25	85	—	—	—
Diethylamine	3	30	57	—	60
Dimethylamine	0	6	18	82	74
Di-*n*-propylamine	8	58	77	—	49
***n*-Dodecylamine**	60	tailing	—	—	19
Ethylamine	1	0	4	49	71
***n*-Hexadecylamine**	82	—	—	—	6
***n*-Hexylamine**	12	19	47	—	56
Methylamine	0	0	0	28	74
***n*-Octadecylamine**	89	—	—	—	3
***n*-Octylamine**	23	39	69	—	44
***n*-Propylamine**	2	0	9	65	68
***n*-Tetradecylamine**	—	—	—	—	11

Paper P1 = W1, W3MM or W4.

Solvent S1 = hexane, paper impregnated with 50% dimethylformamide.
S2 = hexane-benzene (4 : 1 v/v), paper impregnated with formamide.
S3 = hexane-benzene (3 : 2 v/v), paper impregnated with formamide.
S4 = benzene-$CHCl_3$ (1 : 1 v/v), paper impregnated with formamide.
S5 = 90% aqueous acetic acid, paper impregnated with 1-bromonaphthalene.

Detection Stannous chloride, followed by *p*-dimethylaminobenzaldehyde.

Technique Not specified.

Literature
1. Gasparič, J., in *Paper Chromatography*, Hais, I. M. and K. Macek, Editors, Academic Press, New York, 1963, pp 414–431.
2. Gasparič, J. and J. Borecký, *J. Chromatogr.*, **5**, 466 (1961).

TABLE PC 56

AMINES, ALIPHATIC

Paper	P1
Solvent	S1
Literature	1
Compound	$R_F \times 100$
Adrenaline	45
Agmatine	5
Allylamine	50
***n*-Amylamine**	77
Benzylamine	68
***n*-Butylamine**	70
1,4-Diaminobutane (putrescine)	16
1,2-Diaminoethane	14
1,6-Diaminohexane	20
1,5-Diaminopentane (cadaverine)	17
1,3-Diaminopropane	15
Dimethylamine	43
β,β-Diphenylethylamine	72
Ephedrine	75
Ethanolamine	33
Ethylamine	45
Glucosamine	24
***n*-Heptylamine**	85
Histamine	19
Isoamylamine	77
Isopropylamine	57
Methylamine	37
β-Phenylethylamine	72
β-Phenyl-β-hydroxyethylamine	65
***n*-Propylamine**	58
Spermine	7
Tryptamine	67
Tyramine	62

Paper P1 = W4.

Solvent S1 = *n*-butanol (40%), acetic acid (10%), water (50%).

Literature
1. Bremner, J. M. and R. H. Kenten, *Biochem. J.*, **49**, 651–655 (1951).

TABLE PC 57

AMINES, AMINO ALCOHOLS

Paper	P1	P1	P1
Solvent	S1	S2	S3
Literature	1	1	1

Compound	$R_F \times 100$		
Alaninol	25	23	83
Argininol	8	7	—
Aspartidiol	19	23	82
Ethanolamine	18	15	74
Glutamidiol	21	23	85
Histidinol	8	12	—
Isoleucinol	50	63	100
Leucinol	53	63	100
Lysinol	8	7	78
Phenylalaninol	54	70	100
Prolinol	28	26	100
Serinol	16	15	69
Threoninol	17	31	—
Tyrosinol	40	54	81
Valinol	40	49	100

Paper P1 = W1.

Solvent S1 = *n*-butanol-acetic acid-water (77 : 6 : 17 v/v).
S2 = *n*-butanol saturated with 0.1% NH_4OH.
S3 = phenol saturated with 0.1% NH_4OH.

Detection Ninhydrin.

Literature 1. Fromageot, C., Jutisz, M., Meyer, D., and L. Pénasse, *Compt. Rend.*, **230**, 1905–1906 (1950); *Biochim. Biophys. Acta*, **6**, 283–289.

TABLE PC 58

AMINES, CATECHOLAMINES

Paper	P1	
Solvent	S1	
Literature	1	

Compound	$R_F \times 100$	Color
Adrenaline	26	gr-bl
***m*-Adrenaline**	50	p
3,4-Dihydroxyphenylacetic acid	74	y
DOPA	14	pk
Dopamine	38	pk
Ferulic acid	68	bl-y
Homovanillic acid	84	bl
3-Hydroxy-4-methoxycinnamic acid	81	p
3-Hydroxy-4-methoxymandelic acid	66	p
***o*-Hydroxyphenylacetic acid**	84	p
α-Methyl-DOPA	32	pk
Noradrenaline	19	gr-bl
Nor-*m*-adrenaline	44	p
L-Phenylephrine	49	pk
Tyramine	49	pk
Vanillic acid	84	p
Vanilmandelic acid	68	p

Paper P1 = W1.

Solvent S1 = *n*-butanol-acetic acid-H_2O (14 : 4 : 50 v/v).

Detection Diazotized nitraniline.

Literature 1. Floch, H., *Biol. Med. (Paris)*, **53**, 36 (1964); *J. Chromatogr.*, **19**, D4 (1965).

TABLE PC 59
AMINES, AMINOBENZENE DERIVATIVES

Paper	P1	P1	P1	P1	P1	P1													
Solvent	S1	S2	S3	S4	S5	S6													
Detection	—	—	—	—	—	—	D1	D2	D3	D4	D5	D6	D7	D8	D9	D10	D11	D12	D13
Literature	1	1	1	1	1	1	1	1	1	1	1	1	1	1	1	1	1	1	1
Compound	$R_F \times 100$						Color[a]												
2-Aminobenzophenone[c,d]	96	95	93	95	94	87	48	—	—	—	8	59	5	—	—	+	—	9	23
2-Amino-5-chlorotoluene[b]	93	92	94	92	93	93	—	—	—	62	3	—	—	—	—	+	—	1	22
1-Aminonaphthalene[d]	91	92	90	93	87	84	33	19	14	24	23	41	—	26	68	+	—	7	22
								15	63	25		23							23
2-Aminonaphthalene[c,d]	87	92	87	92	84	78	26	62	12	25	19	70	—	59	68	++	—	9	22
								9			23								
Aniline	69	93	56	92	45	37	—	8	—	11	10	—	—	—	—	++	—	7	22
Benzidine[b,d]	56	92	19	80	3	2	—	51	58	63	51	—	—	43	54	++	—	14	23
														68	68				25
p-Bromoaniline	91	89	93	91	90	87	—	—	—	—	3	—	—	—	—	+	—	7	22
2,7-Diaminofluorene[c,d]	33	87	7	46	0	0	26	38	45	45	51	50	—	52	60	++	—	15	65
								57	52	52	59			53	68				25
1,5-Diaminonaphthalene[b]	68	90	54	78	0	0	26	53	23	68	69	63	—	25	69	++	—	9	23
								23	65	53		23		69					65
2,7-Diaminonaphthalene	52	87	23	58	1	0	26	52	64	65	63	52	—	57	60	++	—	10	23
								62			65								65
2,4-Dichloroaniline	95	95	89	93	92	91	8	—	—	—	—	+	—	—	—	—	—	7	21
N,N-Dimethylaniline	79	93	76	97	78	74	—	8	—	65	60	39	—	—	70	+	—	—	—
3,5-Dinitro-4-methylaniline	94	90	95	93	95	92	59	—	—	—	—	—	—	—	—	—	—	3	63
N-Methylaniline	77	92	75	97	77	73	—	9	+	62	9	27	—	—	33	+	—	4	15
												41							
m-Nitroaniline	94	95	91	92	91	86	57	—	—	—	—	—	—	—	—	—	—	6	22
o-Nitroaniline	93	92	89	91	90	87	53	—	—	—	—	—	—	—	—	+	—	—	65
p-Nitroaniline	93	92	93	88	86	79	53	—	—	—	—	—	—	—	—	+	—	12	65
																		64	25
2-Nitro-4-methylaniline	93	96	90	96	95	92	59	—	—	—	—	—	—	—	+	—	—	64	24
2-Nitro-6-methylaniline	91	92	93	95	96	92	57	—	—	—	—	—	—	—	—	—	—	4	26
3-Nitro-5-methylaniline	94	95	92	94	93	91	57	—	—	—	—	—	—	—	—	—	—	7	23
3-Nitro-6-methylaniline	94	92	93	95	90	87	+	—	—	—	—	—	—	—	—	—	—	1	22
4-Nitro-6-methylaniline	92	94	92	95	90	85	51	—	—	—	—	—	—	—	—	—	—	8	49
																			54
4-Nitro-*o*-phenylenediamine[d]	94	92	87	71	90	23	56	—	—	—	—	+	—	—	51	+	—	8	25
m-Phenylenediamine	15	77	0	9	0	0	—	10	23	19	10	35	—	—	69	++	—	8	19
									11	65									23

TABLE PC 59—(Continued)

AMINES, AMINOBENZENE DERIVATIVES

Paper	P1	P1	P1	P1	P1	P1													
Solvent	S1	S2	S3	S4	S5	S6													
Detection	—	—	—	—	—	—	D1	D2	D3	D4	D5	D6	D7	D8	D9	D10	D11	D12	D13
Literature	1	1	1	1	1	1	1	1	1	1	1	1	1	1	1	1	1	1	1
Compound	$R_F \times 100$						Color[a]												
***o*-Phenylenediamine**[b,c]	74	83	0	57	13	5	+	—	—	62	63	+	+	64	68	++	—	9	23
***p*-Phenylenediamine**[b,d]	28	59	0	5	0	0	—	50	51	52	9	49	—	43	40	++	—	15	23
														25	68				35
***o*-Tolidine**	73	90	63	92	42	34	—	45	48	45	49	42	—	68	43	++	—	12	23
								62	57	62		70		54					25
***m*-Toluidine**[d]	75	93	62	92	64	52	—	8	—	65	—	23	—	48	71	++	—	6	15
										56									21
***o*-Toluidine**[b]	79	93	76	92	71	63	—	10	6	64	14	38	—	+	69	++	—	6	17
										65									20
***p*-Toluidine**[b]	66	92	48	88	39	30	—	—	6	13	6	—	—	+	71	+	—	7	17
										65									20

[a] For interpretation, see Table PC 120.
[b] Ninhydrin reagent, pale violet.
[c] Dragendorff reagent, positive.
[d] *p*-Dimethylaminobenzaldehyde, pale yellow.

Paper P1 = W1.

Solvent
S1 = methyl ethyl ketone-acetone-formic acid-water (40 : 2 : 1 : 6).
S2 = methyl ethyl ketone-diethylamine-water (921 : 2 : 77).
S3 = methyl isobutyl ketone-formic acid-water (10 parts ketone saturated with 1 part 4% formic acid).
S4 = chloroform-methanol-formic acid-water (10 parts chloroform saturated with a mixture of 1 part methanol and 1 part 4% formic acid).
S5 = benzene-methyl ethyl ketone-formic acid-water (a mixture of 9 parts benzene and 1 part ketone saturated with 1 part 2% formic acid).
S6 = benzene-formic acid-water (10 parts benzene saturated with 1 part 2% formic acid).

Detection
D1 = UV.
D2 = diazotized sulfanilic acid (0.3% solution in dioxane-water, 1 : 2).
D3 = diazotized 4-benzoylamino-2,5-dimethoxyaniline (0.6% solution in dioxane-water, 1 : 2).
D4 = diazotized *o*-dianisidine (0.6% solution in dioxane-water, 1 : 2).
D5 = *p*-nitrobenzenediazonium fluoborate (0.4% solution in dioxane water 1 : 2).
D6 = 2,6-dibromoquinone-4-chloroimide (0.5% solution in dioxane-acetone 4 : 1).
D7 = 2,4-dinitrophenylhydrazine (*ca.* 0.1% solution in 1*N* HCl).
D8 = ferric chloride (2% aqueous solution).
D9 = phosphomolybdic acid (2% aqueous solution).
D10 = potassium permanganate (10% aqueous solution).
D11 = bromphenol blue (*ca.* 0.05% solution in ethanol).
D12 = Ehrlich's reagent (1% *p*-dimethylaminobenzaldehyde in 1*N* HCl).
D13 = *p*-dimethylaminocinnamaldehyde (0.1% solution in 1*N* HCl).

Technique Descending development, 40 cm.

Literature
1. Reio, L., *J. Chromatogr.*, **47**, 60–85 (1970).

TABLE PC 60
AMINES, NITROGEN COMPOUNDS, BIOLOGICALLY ACTIVE

Paper	P1	P1	P1	P1	P1	P1													
Solvent	S1	S2	S3	S4	S5	S6													
Detection	—	—	—	—	—	—	D1	D2	D3	D4	D5	D6	D7	D8	D9	D10	D11	D12	D13
Literature	1	1	1	1	1	1													
Compound	$R_F \times 100$						Color[a]												
Acetylcholine[c]	25	3	0	0	0	0	—	—	—	—	—	—	—	—	—	—	33	—	—
***N*-Acetylhistamine**	6	15	0	0	0	0	—	15 8	62	64 56	7 24	—	—	—	—	—	—	—	—
***N*-Acetyltyramine**	85	82	73	52	21	5	—	7	60	64	23 8	—	—	—	68	+	—	—	—
Acetylurea	59	51	27	27	4	2	—	—	—	—	—	—	—	—	—	—	—	4	22
2,2′-Aminomethyl-1,3-propanediol[b]	10	8	0	0	0	0	—	—	—	—	—	—	—	57	—	+	—	—	—
Benzylamine	48	88	2	1	0	0	—	—	—	8	6	—	53	—	—	+	33	—	—
Biuret = carbamylurea	35	25	8	3	0	0	—	—	—	—	—	—	—	—	—	—	—	5	17
Celocurine[c]	5	22	0	0	0	0	+	—	—	—	—	58	—	—	—	+	—	—	—
Cinnamylephedrine[d]	87	94	63	94	47	30	—	—	—	—	—	—	—	—	—	+	—	—	—
Creatine	21	0	0	0	0	0	—	—	—	—	—	—	27	—	—	—	—	—	—
Cyanuric acid	37	22	7	1	0	0	—	—	—	—	—	—	—	—	—	—	—	6	17
***dl*-Desoxyephedrine**[d]	52	82	4	25	0	0	—	72	—	—	—	—	—	—	—	—	—	—	—
2,3-Diaminopyridine	36	14	0	0	0	0	33	8 59	56	56	19 63	23	—	60	69	+	—	6	23
2-Diethylaminoethanethiol[c]	12	92	0	0	0	0	—	—	—	—	—	51	—	—	45	++	—	—	—
Diliuric acid (5-nitrobarbituric acid)	5	3	0	0	0	0	57	—	—	—	—	—	—	+	—	—	+	—	—
β-(3,4-Dimethoxyphenyl)-ethylamine	23	5	2	3	0	0	—	6	—	58 60	6 60	23 56	—	57	47	+	—	5	57 8
2-Dimethylaminomethylpyrrole[b]	47	10	1	0	0	0	+	63	62	64	8	70	17	—	—	—	33	17 22	69
1,3-Dimethylurea	66	51	24	28	4	2	—	—	—	—	—	—	—	—	—	—	33	5	22
Ethanolamine[b]	4	0	0	0	0	0	—	—	—	—	—	+	23	—	—	—	33	5	17
3-Hydroxy-*N*-methylpiperidine	11	2	0	0	0	0	—	—	—	—	—	—	—	—	—	+	—	—	17
1-Hydroxypyridine-2-thione[c]	92	89	88	95	94	83	—	—	—	—	—	7	—	—	—	+	—	—	—
Hydroxyurea	27	12	3	0	0	0	—	—	72	72	—	+	—	27	30	++	—	7	22
Isomescaline[b]	54	87	3	5	0	0	—	—	—	—	—	—	25	—	—	—	33	—	26
***N*-Isopropylnoradrenaline**	40	53	0	0	0	0	—	62	63	23	8 65	23 52	27	—	28 68	++	33	—	—
***dl*-Metanephrine**	27	4	0	0	0	0	—	59 9	12	60 64	6 23	35	—	71	30	++	—	—	—
***p*-Methoxybenzylamine**[b]	50	79	1	.1	0	0	—	—	—	+	—	—	53	—	—	+	33	5	—

TABLE PC 60—(Continued)

AMINES, NITROGEN COMPOUNDS, BIOLOGICALLY ACTIVE

Paper Solvent Detection Literature	P1 S1 — 1	P1 S2 — 1	P1 S3 — 1	P1 S4 — 1	P1 S5 — 1	P1 S6 — 1	D1	D2	D3	D4	D5	D6	D7	D8	D9	D10	D11	D12	D13
Compound			R$_F$ × 100										Color[a]						
β-(3-Methoxy-4-hydroxyphenyl)-ethylamine	12	69	0	0	0	0	—	17	59	59	58 25	—	—	—	68	++	—	—	—
3-Methoxy-4,5-methylene-dioxyamphetamine	59	84	1	5	0	0	+	—	—	—	—	—	24	—	—	—	33	—	—
β-(*p*-Methoxyphenyl)-ethylamine	42	85	7	11	5	2	34	58	60	7 64	8 63	—	—	58	58 71	+	33	20	60 14
1-Methylhistamine	0	2	0	0	0	0	34	—	57	56	+	—	—	—	—	—	—	—	—
N-Methylmescaline[c]	78	79	3	16	0	0	—	—	—	—	—	+	—	—	—	+	33	—	—
N-Methyl-2-pyridone-5-carboxamide	39	19	7	12	1	0	—	—	—	—	—	—	—	8	—	+	—	+	17
Noradrenalone	21	7	0	0	0	0	—	57 18	57 18	60	57	52	—	43	68	++	—	5	62
dl-Normetanephrine	15	71	0	0	0	0	—	59	10	56 23	58 23	35	—	—	30	+	—	—	—
β-Phenyl-β-propyl-thioethylamine[b,c]	68	92	3	32	0	0	—	—	—	—	—	60	60	—	—	+	—	+	+
O-Phosphoethanolamine	0	0	0	0	0	0	—	—	—	—	—	—	23	—	—	—	+	4	—
Thiourea	49	33	13	2	0	0	—	—	—	60	58	24	—	—	71	++	—	5	—
Uramil (5-aminobarbituric acid)	6	1	0	0	0	0	+	—	—	—	—	—	—	—	29	—	—	5	59
p-Urazine (diurea)[c]	5	1	0	0	0	0	—	+	60	43 51	+	—	—	—	39	+	—	—	—

[a] For interpretation, see Table PC 120.
[b] *p*-Dimethylaminobenzaldehyde (2% solution in acetic anhydride), pale yellow.
[c] Dragendorff reagent, positive.
[d] Detected by KI-BI$_3$ reagent.

TABLE PC 60—(Continued)
AMINES, NITROGEN COMPOUNDS, BIOLOGICALLY ACTIVE

Paper P1 = W1.

Solvent
- S1 = ethyl methyl ketone-acetone-formic acid-water (40 : 2 : 1 : 6 parts by volume).
- S2 = ethyl methyl ketone-diethylamine-water (921 : 2 : 77 parts by volume).
- S3 = methyl isobutyl ketone-formic acid-water (10 parts ketone saturated with 1 part 4% formic acid).
- S4 = chloroform-methanol-formic acid-water (10 parts of chloroform saturated with a mixture of 1 part methanol and 1 part 4% formic acid).
- S5 = benzene-ethyl methyl ketone-formic acid-water (a mixture of 9 parts benzene and 1 part ketone saturated with 1 part 2% formic acid).
- S6 = benzene-formic acid-water (10 parts benzene saturated with 1 part 2% formic acid).

Detection
- D1 = ultraviolet light.
- D2 = diazotized sulfanilic acid (0.3% solution in dioxane-water, 1 : 2).
- D3 = diazotized 4-benzoylamino-2,5-dimethoxyaniline (0.6% solution in dioxane-water, 1 : 2).
- D4 = diazotized *o*-dianisidine (0.6% solution in dioxane-water 1 : 2).
- D5 = *p*-nitrobenzenediazonium fluoborate (0.4% solution in dioxane-water, 1 : 2).
- D6 = 2,6-dibromoquinone-4-chloroimide (0.5% solution in dioxane-acetone, 4 : 1).
- D7 = ninhydrin reagent (2% solution in butanol saturated with water).
- D8 = ferric chloride (2% aqueous solution).
- D9 = phosphomolybdic acid (2% aqueous solution).
- D10 = potassium permanganate (1% aqueous solution).
- D11 = bromophenol blue (*ca.* 0.05% solution in ethanol).
- D12 = Ehrlich's reagent (1% *p*-dimethylaminobenzaldehyde in 1 *N* HCl).
- D13 = *p*-dimethylaminocinnamaldehyde (0.1% solution in 1 *N* HCl).

Technique Descending 40 cm.

Literature

1. Reio, L., *J. Chromatogr.*, **47**, 60–85 (1970); *J. Chromatogr.*, **13**, 475–496 (1964).

TABLE PC 61
AMINES, HYDRAZINES

Paper	P1	P1	P1
Solvent	S1	S2	S2
Technique	T1	T1	T2
Literature	1	1	1
Compound	$R_F \times 100$		
1-Acetyl-2,2-dimethylhydrazine	76	—	—
1-Benzoyl-2,2-dimethylhydrazine	88	—	—
1,2-Diacetyl-1,2-dimethylhydrazine	74	—	—
1,2-Diacetylhydrazine	48	—	—
1,2-Dibutyrylhydrazine	91	—	—
1,2-Diethylhydrazine·2HCl	30	55	25
1,2-Diformyl-1,2-dimethylhydrazine	60	—	—
1,2-Diformylhydrazine	26	—	—
1,1-Dimethylhydrazine·HCl	29	53	20
1,2-Dimethylhydrazine·2HCl	27	57	23
1,2-Dipropionylhydrazine	81	—	—
1-Formyl-2,2-dimethylhydrazine	72	—	—
Hydrazine·2HCl	0	39	16
Methylhydrazine sulfate	13	47	17
Trimethylhydrazine·HCl	—	42	23

Paper P1 = W1 in sheets and size 9 disks.
Solvent S1 = *iso*-amyl alcohol-acetic acid-water (10 : 1.5 : 10).
S2 = *n*-butanol-acetic acid-water (4 : 1 : 5).
Detection Ehrlich's reagent.
Technique T1 = circular chromatography.
T2 = ascending chromatography.
Literature 1. Hinman, R. L., *Anal. Chim. Acta*, **15**, 125–128 (1956).

TABLE PC 62
AMINO ACIDS, DERIVATIVES

Paper	P1	P1	P1	P1	P1
Solvent	S1	S2	S3	S4	S5
Technique	T1	T1	T1	T1	T2
Literature	1	1	1	1	2
Amino acid	$R_F \times 100$				
Alanine	25	32	52	48	51
Arginine	33	30	82	52	43
Asparagine	7	10	23	48	—
Aspartic acid	0	2	2	66	4
Cysteic acid	0	0	0	68	3
Cystine	—	—	—	—	19 (di-)
Glutamic acid	0	2	2	62	4
Glutamine	7	13	29	48	—
Glycine	17	19	34	43	41
Histidine	13	21	54	57	40
Isoleucine	68	56	76	44	70
Leucine	68	56	81	45	70
Lysine	76	56	82	2	83 (di-)
Methionine	—	—	—	—	64
Phenylalanine	64	51	84	32	60
Proline	20	30	59	55	42
Serine	12	17	30	53	32
Threonine	15	23	35	57	42
Tryptophan	56	46	74	14	—
Tyrisine	—	—	—	—	46 (*o*-mono-)
Tyrosine	—	—	—	—	69 (di-)
Valine	47	45	73	48	65

Paper P1 = W1.
Solvent S1 = toluene-pyridine-ethylene chlorohydrin-0.8*M* NH_3 (5 : 1 : 3 : 3 v/v).
S2 = *n*-butanol-3% NH_3 (1 : 1 v/v).
S3 = *iso*-amyl alcohol-phenol-water (1 : 1 : 1 v/v).
S4 = 1.5*M* phosphate buffer, pH 6.
S5 = *n*-butanol-1*N* NH_3 (5 : 3 v/v, upper layer).
Detection Cl_2, then *o*-tolidine for pipsyl derivatives.
Technique T1 = dinitropyridyl derivatives.
T2 = pipsyl derivatives.
Literature
1. Signor, A., Biondi, L., Terbojevich, M., and P. Pajetta, *Gazz. Chim. Ital.*, **94**, 619 (1964); *J. Chromatogr.* **17**, D6 (1965).
2. Fletcher, J. C., *Biochem. J.*, **102**, 815 (1967).

TABLE PC 63
AMINO ACIDS

Paper	P1	P1	P1	P1	P1	P1													
Solvent	S1	S2	S3	S4	S5	S6													
Detection	—	—	—	—	—	—	D1	D2	D3	D4	D5	D6	D7	D8	D9	D10	D11	D12	D13
Literature	1	1	1	1	1	1													
Compound	$R_F \times 100$						Color[a]												
N-Acetyl-*dl*-alanine	73	1	24	16	1	0	—	—	—	—	—	—	—	—	—	—	+	—	—
N-Acetyl-L-glutamic acid	59	0	7	4	0	0	—	—	—	—	—	—	—	—	—	—	+	—	—
N-Acetyl-L-histidine	5	0	0	0	0	0	—	15,8	62	68,56	7,24	—	—	—	—	—	—	—	—
N[α]-Acetylkynurenine	87	55	67	60	10	1	39	—	—	—	—	—	—	—	70	+	+	8	23
O-Acetylserine	7	0	0	0	0	0	—	—	—	—	—	—	23	—	—	—	33	—	—
N-Acetyltryptophanamide	78	65	54	33	4	1	27	—	—	+	9	70	—	—	—	++	—	23,27	25
N-Acetyl-*l*-tyrosine	85	80	76	57	18	4	—	12	—	+	7	—	—	—	71	++	+	—	—
Alanine	6	0	0	0	0	0	—	—	—	—	—	—	24	—	—	—	33	4	—
β-Alanine[b]	6	0	0	0	0	0	—	—	—	—	—	—	25	17	—	—	33	+	—
dl-α-Amino-*n*-butyric acid[b]	8	0	0	0	0	0	—	—	—	—	—	—	24	—	—	—	33	4	—
DL-γ-Aminobutyric acid	8	0	0	0	0	0	—	—	—	—	—	—	—	—	—	—	30	—	—
ε-Aminocaproic acid	18	0	0	0	0	0	—	—	—	—	—	—	23	—	—	—	33	—	—
Aminohydroxyacetic acid[b]	17	7	0	0	0	0	—	—	—	—	—	—	—	—	—	—	—	5	—
Cysteic acid	0	0	0	0	0	0	—	—	—	—	—	—	—	—	—	—	+	4	—
l-Cysteinesulfinic acid[b]	3	0	0	0	0	0	—	—	—	—	—	—	23,64	—	—	++	++	—	—
3,5-Diiodothyronine	75	10	18	0	0	0	—	62	—	—	—	—	23	—	38	+	—	4	—
3,5-Dinitrotyrosine	44	0	0	0	0	0	53	—	—	—	—	—	60	—	—	—	—	—	—
dl-Djenkolic acid	0	0	0	0	0	0	—	—	—	—	—	—	23	—	—	+	33	5	17
DL-Ethionine	20	1	0	0	0	0	—	—	—	—	—	—	—	—	—	++	—	2	—
N-Formyl-*dl*-tryptophan[b]	86	5	79	25	9	1	—	—	—	+	8	52	—	—	—	++	+	23,70	25
N-Formyl-*l*-tyrosine[b]	81	2	58	3	0	0	—	10	+	63	23	—	23	—	39	++	+	—	—
Glycine	3	0	0	0	0	0	—	—	—	—	—	—	24	—	—	—	33	4	—
Glycylglycine	8	0	0	0	0	0	—	—	—	—	—	—	23	—	—	—	33	—	—
dl-Homocysteine	2	3	0	0	0	0	—	—	—	—	—	59	23	—	30	++	33	4	—
l-Homocysteinethiolactone	22	4	0	5	0	0	—	—	—	—	—	—	23	—	32	+	33	4	—
dl-Homocystine	2	0	0	0	0	0	—	—	—	—	—	—	23	—	—	+	33	4	—
Hydantoin-5-acetic acid	38	0	3	1	0	0	—	—	—	—	—	—	—	—	—	—	30	—	—
5-Hydroxykynurenine	58	5	19	0	0	0	40	62	61	54	17	38	57	—	43,38	+	—	9	22
3-Iodotyrosine	42	1	0	0	0	0	—	11	—	—	63	—	23	—	38	++	—	—	—
dl-Methionine	18	1	0	0	0	0	—	—	—	—	—	—	24	—	—	+	33	—	—
dl-Methionine methylsulfonium chloride	27	0	0	0	0	0	—	—	—	—	—	—	23	—	—	++	—	4	—
DL-Methionine sulfoximine	2	0	0	0	0	0	—	—	—	—	—	—	—	—	—	—	33	5	—
dl-Methionine sulfone[b]	7	0	0	0	0	0	—	—	—	—	—	—	23	—	—	—	33	—	—
dl-Methionine sulfoxide	2	0	0	0	0	0	—	—	—	—	—	—	23	—	—	++	—	—	—
dl-5-Methoxytryptophan	39	1	1	0	0	0	—	—	—	+	10	26	23	—	—	+	—	27	29
3-Methoxy-*dl*-tyrosine	25	0	0	0	0	0	—	64	+	56	23,52	+	26	—	68	++	33	—	—

TABLE PC 63—(Continued)
AMINO ACIDS

Paper	P1	P1	P1	P1	P1	P1													
Solvent	S1	S2	S3	S4	S5	S6													
Detection	—	—	—	—	—	—	D1	D2	D3	D4	D5	D6	D7	D8	D9	D10	D11	D12	D13
Literature	1	1	1	1	1	1													
Compound	$R_F \times 100$						Color[a]												
S-Methyl-*l*-cysteine[b]	10	0	0	0	0	0	—	—	—	—	—	—	23	—	—	++	33	6	14
N-Methyl-2-piperidinecarboxylic acid[b]	4	0	0	0	0	0	—	—	—	—	—	—	—	—	—	—	33	—	—
N-Methyltaurine[b]	3	0	0	0	0	0	—	72	—	—	72	—	—	—	—	—	33	—	12
6-Methyltryptophan	49	1	1	0	0	0	34	60	—	61	60	—	26	—	—	+	33	25,48	25,28
dl-α-Methyl-*m*-tyrosine	49	2	0	0	0	0	—	10	11	65	64	29,23	23	—	44	++	33	—	—
Mimosine	2	0	0	0	0	0	34	—	—	—	—	—	23	—	69	—	33	—	—
3-Nitrotyrosine	35	0	0	0	0	0	53	—	—	—	—	—	23	—	71	+	—	—	—
dl-Norleucine	36	0	0	0	0	0	34	—	—	—	—	—	23	—	—	—	33	4	—
dl-Norvaline[b]	24	0	0	0	0	0	—	—	—	—	—	—	23	—	—	—	33	5	—
dl-Penicillamine (β-mercaptovaline)[b]	73	19	0	0	0	0	—	72	—	—	72	6	6	+	45	++	+	5	17
l-Phenylalanine methyl ester	53	86	1	59	0	0	—	—	—	—	+	17	18	—	—	—	33	+	—
l-Phenylalaninol	35	87	0	2	0	0	—	—	—	—	7	—	23	—	—	—	33	—	—
dl-*O*-Phosphoserine	0	0	0	0	0	0	—	—	—	—	—	—	23	—	—	—	+	—	—
dl-*O*-Phosphothreonine	0	0	0	0	0	0	—	—	—	—	—	—	23	—	—	—	+	—	—
4-Piperidinecarboxylic acid (isonipecotic acid)	1	0	0	0	0	0	—	—	—	—	—	—	23	—	—	—	—	—	—
dl-Serine	3	0	0	0	0	0	—	—	—	—	—	—	24	—	—	+	33	—	—
Taurine	7	0	0	0	0	0	—	—	—	—	—	—	23	—	—	—	33	—	—
l-2-Thiolhistidine	2	0	0	0	0	0	—	9	64	17	62	62	23	—	38	++	—	5	17
Thioproline (thiazolidine-4-carboxylic acid)[b]	9	2	0	0	0	0	—	—	—	—	72	—	26	—	—	++	+	—	+
dl-Threonine	5	3	0	0	0	0	—	—	—	—	—	—	24	—	—	+	33	—	—
dl-Tryptophanol	39	90	2	0	0	0	+	—	—	—	—	—	27	—	71	++	33	23,58	24,25
l-Tyrosine methyl ester[b]	20	80	0	0	0	0	—	10	60	62	59	+	24	—	33	++	33	4	64
l-Tyrosinol	35	0	0	0	0	0	—	—	—	—	—	—	26	—	27	++	—	—	—

[a] For interpretation, see Table PC 120.

[b] *p*-Dimethylaminobenzaldehyde (2% solution in acetic anhydride) reagent, pale yellow.

TABLE PC 63—(Continued)
AMINO ACIDS

Paper P1 = W1.

Solvent
- S1 = ethyl methyl ketone-acetone-formic acid-water (40 : 2 : 1 : 6 parts by volume).
- S2 = ethyl methyl ketone-diethylamine-water (921 : 2 : 77 parts by volume).
- S3 = methyl isobutyl ketone-formic acid-water (10 parts ketone saturated with 1 part 4% formic acid).
- S4 = chloroform-methanol-formic acid-water (10 parts of chloroform saturated with a mixture of 1 part methanol and 1 part 4% formic acid).
- S5 = benzene-ethyl methyl ketone-formic acid-water (a mixture of 9 parts benzene and 1 part ketone saturated with 1 part 2% formic acid).
- S6 = benzene-formic acid-water (10 parts benzene saturated with 1 part 2% formic acid).

Detection
- D1 = ultraviolet light.
- D2 = diazotized sulfanilic acid (0.3% solution in dioxane-water 1 : 2).
- D3 = diazotized 4-benzoylamino-2,5-dimethoxyaniline (0.6% solution in dioxane-water 1 : 2).
- D4 = diazotized *o*-dianisidine (0.6% solution in dioxane-water 1 : 2).
- D5 = *p*-nitrobenzenediazonium fluoborate (0.4% solution in dioxane-water 1 : 2).
- D6 = 2,6-dibromoquinone-4-chloroimide (0.5% solution in dioxane-acetone 4 : 1).
- D7 = ninhydrin reagent (2% solution in butanol saturated with water).
- D8 = ferric chloride (2% aqueous solution).
- D9 = phosphomolybdic acid (2% aqueous solution).
- D10 = potassium permanganate (1% aqueous solution).
- D11 = bromophenol blue (0.05% solution in ethanol).
- D12 = Ehrlich reagent (1% *p*-dimethylaminobenzaldehyde in 1*N* HCl).
- D13 = *p*-dimethylaminocinnamaldehyde (0.1% solution in 1*N* HCl).

Technique One dimensional disc., 40 cm development distance.

Literature
1. Reio, L., *J. Chromatogr.*, **13**, 475–496 (1964) and *J. Chromatogr.*, **47**, 60–85 (1970).

TABLE PC 64

AMINO ACIDS AND RELATED COMPOUNDS (INDOLES, IMIDAZOLES, GUANIDINES)

Paper	P1	P1	P1	P1	P1	P1	P1	P1
Solvent	S1	S2	S3	S4	S5	S6	S7	S8
Detection								
Literature	1	1	1	1	1	1	1	1
	$R_F^a \times 100$							
Aspartic acid	23	20	17	6	2	6	40	20
Glutamic acid	28	33	26	12	5	6	48	20
Serine	22	35	35	22	20	30	42	33
Glycine	23	42	42	26	18	18	37	29
Threonine	26	48	48	33	37	50	50	36
Alanine	30	58	58	41	33	20	52	37
Valine	51	78	78	62	48	35	65	48
iso-Leucine	67	85	85	71	52	45	68	56
Leucine	70	85	85	73	55	52	70	60
Histidine	11	65s	72s	50	21	18	28	24
Lysine	12	42s	77s	57	16	10	15s	13
Arginine	15	83s	86s	66	9	5	16s	15
Phenylalanine	60	84	84	72	45s	52	60	63
Tyrosine	45	60	60	43	31	30	50	60
Tryptophan	50	77	77	58	36	50	42s	62
Proline	34	90	90	75	38	25	55	34
Hydroxyproline	22	67	67	44	24	15	45	30
Cysteine	8	22	11s	5s	2	5	18s	14
Cystine	5	15	20s	16	4	6	18s	15s
Cysteic acid	7	6	7	5	6	19	41	26
Methionine	50	82	82	62	41	39	60	53s
Methionine sulfoxide	27	78	78	45	18	30	45	39
Methionine sulfone	22	60	64	55	19	17	43	31
β-Alanine	33	64	64	45	22	22	50	28
α-Amino adipic acid	30	40	30	16	2	8	33	18
α-Amino-*n*-butyric acid	40	68	72	53	36	33	60	41
γ-Amino-*n*-butyric acid	40	78	78	56	24	20	50	26
β-Aminoisobutyric acid	49	76	76	52	27	30	57	32
δ-Amino laevulinic acid	37	—	d	—	—	—	—	23
Asparagine	12	40	40	22	10	12	21	20
Argininosuccinic acid	20	—	—	—	—	—	—	18
Argininosuccinic anhydride	15	—	—	—	—	—	—	17
Citrulline	18	63	63	40	11	12	36	23
Cystathionine	10	16	31	18	0	7	19	14
Ethanolamine	38	72	94	83	57	65	67	43

TABLE PC 64—(Continued)
AMINO ACIDS AND RELATED COMPOUNDS (INDOLES, IMIDAZOLES, GUANIDINES)

Paper	P1	P1	P1	P1	P1	P1	P1	P1
Solvent	S1	S2	S3	S4	S5	S6	S7	S8
Detection								
Literature	1	1	1	1	1	1	1	1
	$R_F{}^a \times 100$							
Glutamine	17	27	57	29	10	10	31	23
Homocystine	18	19	41	21	3	11	30s	24
Ornithine	12	17	70s	45	12	11	12s	11
Phosphoethanolamine	12	28	30	13	0	3	29	10
N-Methylhistidine	16	74	91	70	18	21	35s	27
Taurine	20	35	40	28	32	35	50	38
Urea	50	75	75	60	41	50	57	51
β-Amino-*n*-butyric acid	42	84	84	65	36	30	61	36
α-Aminoisobutyric acid	44	70	70	57	41	31	70	43
α-Amino-*n*-caproic acid	67	80	85	73	54	60	70	60
ε-Amino-*n*-caproic acid	53	80	90	73	26	34	74	45
α-Amino-*n*-caprylic acid	79	92	92	85	89	80	70	72
ω-Aminocaprylic acid	66	—	90	—	—	—	—	48
1 : 1 Aminocyclopropane carboxylic acid	35	—	59	—	—	—	—	35
α-Amino-pimelic acid	35	45	34	19	5	11	50	25
α-Amino-*n*-valeric acid	56	77	77	65	47	46	66	50
δ-Amino-*n*-valeric acid	50	84	84	63	25	27	57s	31
Arginosuccinic acid	20	—	—	—	—	—	—	18
Azetidine-2-carboxylic acid	24	—	82	—	—	—	—	29
Baikaiin	38	90	90	75	40	41	61	42
Canavanosuccinic acid	—	—	—	—	—	—	—	—
Canaline	50	—	76s	—	—	—	—	19
α, γ-Diamino-*n*-butyric acid	13	17	75s	60s	30	22	18s	20s
β, ε-Diamino-*n*-caproic acid	17	47s	89	68	12	18	15s	11
α, γ-Diaminoglutaric acid	8	—	21	—	—	—	—	17
α, β-Diaminopimelic acid	10	17	20	10	1	4	13	11
α, β-Diaminopropionic acid	15	10	55s	42	25	22	10	11
Diethoxyalanine	59	93	93	81	53	50	73	57
2 : 3 DOPA	29	d	d	d	21	d	34	59
2 : 4 DOPA	27	d	d	d	24	d	36	54
2 : 5 DOPA	26	d	d	d	d	d	35	50
2 : 6 DOPA	30	d	d	d	33	62	36	53
3 : 4 DOPA	23	25	33	21	13	15	41	52
3 : 5 DOPA	26	d	d	d	21	d	34	53

TABLE PC 64—(Continued)

AMINO ACIDS AND RELATED COMPOUNDS (INDOLES, IMIDAZOLES, GUANIDINES)

Paper	P1	P1	P1	P1	P1	P1	P1	P1
Solvent	S1	S2	S3	S4	S5	S6	S7	S8
Detection								
Literature	1	1	1	1	1	1	1	1
	$R_F{}^a \times 100$							
Homoarginine	19	—	—	—	—	—	—	17
Homoserine	26	47	55	35	18	33	50	30
Hydroxy-α-aminopimelic acid	25	75	78	61	22	—	48	39
Hydroxyglutamic acid	19	13	11	8	1	11	41	19
Hydroxyglutamic acid allo	19	13	11	8	1	11	41	19
Hydroxyleucine	48	68	68	54	48	38	64	49
Hydroxy-nor-leucine allo	50	68	68	54	48	32	62	51
Hydroxy-nor-leucine threo	48	68	68	54	44	32	62	51
Hydroxylysine	11	28	58	39	5	—	12s	9
Hydroxymethylproline	25	—	85	—	—	—	—	33
Hydroxyvaline	35	58	59	46	42	32	62	38
Hydroxy-nor-valine threo	35	58	59	48	38	26	57	40
Hydroxy-nor-valine allo	35	58	59	46	38	28	61	41
5-Hydroxypipecolinic acid	29	75	71	54	25	32	56	41
Hydroxyproline allo	24	71	71	50	20	29	48	34
Hypoglycin A	61	—	77	—	—	—	—	52
iso-Asparagine	18	35	37	22	8	12	28	20
iso-Glutamine	23	50	53	28	10	—	35	23
Kynurenine	46/40	70	77	64	30	60/53	35/50	47/42
Methoxinine	39	87	90	71	38	36	53	40
α-Methylserine	26	51	53	36	29	35	46	36
α-Methyl*iso*serine	27	51	53	36	30	37	40	33
γ-Methylglutamic acid	33	28/35	31	18	7	15	50	25/32
γ-Methylglutamine	26	67	68	45	15	28	43	36
γ-Methylene glutamic acid	22	26	30	17	4	4	48	22
γ-Methylene glutamine	23	67	69	44	13	28	35	32
β-1-(naphth-2-ol) alanine	50	71	72	54	29	—	—	62
Phenylglycine	48	72	81	65	35	33	61	57
Phenylserine allo	50	68	72	60	38	73	62	58
Phenylserine threo	44	66	72	60	38	86	48	58
Phenylserine 4-NO_2 allo	44	61	72	59	33	83	56	67
Phenylserine 4-NO_2 threo	41	64	74	60	35	92	50	64
Phenylserine 4-Cl allo	55	67	76	65	47s	84	60	71
Phenylserine 4-Cl threo	50	67	76	65	47s	94	57	67

TABLE PC 64—(Continued)
AMINO ACIDS AND RELATED COMPOUNDS (INDOLES, IMIDAZOLES, GUANIDINES)

Paper	P1	P1	P1	P1	P1	P1	P1	P1
Solvent	S1	S2	S3	S4	S5	S6	S7	S8
Detection								
Literature	1	1	1	1	1	1	1	1
				$R_F{}^a \times 100$				
Piperidine-2-carboxylic acid	44	93	93	80	41	39	64	45
Piperidine-3-carboxylic acid	45	96	95	75	30	13	57	33
Piperidine-4-carboxylic acid	45	90	90	70	27	22	58	28
Piperidine-2, 6-dicarboxylic acid	38	47	—	—	46	—	—	31
Sarcosine	27	77	77	57	33	29	51	32
Thiolhistidine	19	17	26	18	8	25s	26	38
Threonine allo	22	48	48	32	27	47	55	31
Cystathionine	10	16	31	18	0	7	19	14
Cysteamine	37	80	80	80	55	—	64	66
Cysteine	8	22	11s	15	2	6	18s	16s
Cysteic acid	7	6	7	5	6	19	41	26
Cysteine sulfinic acid	15	12	13	8	5	14	40	24
meso Cystine	8	12	17	15	5	5	16	14
Cystine	5	15	20s	16	4	6	18s	15
Djenkolic acid	9	22	38	18	3	50	17	17
Ethionine	58	84	84	70	45	57	61	60
Felinine	44	85	85	70	46	54	49	44
Homo-cysteine	18	32	40	21	3	11	30s	24
Homo-cystine	18	19	41	21	3	11	30s	24
Homo-cysteic acid	11	14	10	5	7	12	53	31
Lanthionine	6	9	23	10	2	6	10	11
β-Methyl lanthionine	9	15	38	18	3	7	24	15
Penicillamine	19	50	55	33	11	22	42	25
Penicillamine disulfide	15	50	60s	33	12	25	40s	23
S-Benzyl penicillamine	82	95	95	89	70	90	62	71
S-Methyl cysteine	33	72	78	58	35	41	41	45
S-Ethyl cysteine	48	80	88	66	41	59	48	55
S-Propyl cysteine	60	87	89	72	55	71	54	65
S-Butyl cysteine	72	91	88	78	60	82	57	70
S-Amyl cysteine	78	92	95	80	69	86	60	75
S-Benzyl cysteine	70	91	92	79	54	80	50	75

TABLE PC 64—(Continued)
AMINO ACIDS AND RELATED COMPOUNDS (INDOLES, IMIDAZOLES, GUANIDINES)

Paper	P1	P1	P1	P1	P1	P1	P1	P1
Solvent	S1	S2	S3	S4	S5	S6	S7	S8
Detection								
Literature	1	1	1	1	1	1	1	1
	$R_F{}^a \times 100$							
Methionine	50	85	85	62	41	39	60	53s
Methionine sulfoxide	27	78	78	55	19	30	45	39
Methionine sulfone	22	60	64	45	18	17	43	31
α-Methyl cysteine	15	—	56	—	—	—	—	23
S-Methyl methionine	60	—	87	—	—	—	—	59
Adrenaline	40	67	d	d	d	47	69	60s
Agmatine	26	81	97	97	14	9	10	12
Allantoin	22	50	50	26	6	13	47	52
p-Amino benzoic acid (PABA)	88	80	64	40	26	38	75	75
p-Amino hippuric acid	48	75	75	45	21	40	64	52
Anthranilic acid	95	80	65	47	41	74	77	85
Cadaverine	17	33s	95	96	43	74	18s	16
Chondrosamine (galactosamine)	20	18	68	50	29	28	37s	43
Dopamine 2 : 5	28	d	d	d	83	d	55	74
Dopamine 3 : 4	28	d	d	d	45	d	40	60
Ephedrine	78	96	96	76	83	—	88	81
Ethanolamine	38	72	94	83	57	65	67	43
Galactosamine	20	18	68	50	29	28	37s	43
Glucosamine	21	18	62	42	30	29	50s	42
Glucosaminic acid	10	16	20	18	8	19	17	11
Histamine	15	91	96	96	52	73	32s	47
Histidinol	17	—	—	—	—	—	—	45
3-Hydroxy anthranilic acid	84	55	35	27	24	—	70	80
5-Hydroxy tryptamine	46	73	—	—	59	95	32	71
3-Hydroxytyramine	28	—	—	—	45	—	40	60
Kynurenic acid	58	72	72	72	78	30	67	65
N-Methyl glucamine	21	53	89	70	44	12	43s	37
β-Phenylethylamine	76	90	98	95	79	99	77	80
Putrescine	15s	19	95	95	36	62	15s	13

TABLE PC 64—(Continued)
AMINO ACIDS AND RELATED COMPOUNDS (INDOLES, IMIDAZOLES, GUANIDINES)

Paper	P1	P1	P1	P1	P1	P1	P1	P1
Solvent	S1	S2	S3	S4	S5	S6	S7	S8
Detection								
Literature	1	1	1	1	1	1	1	1
	R_F[a] × 100							
Tyramine	65	75s	98	96	72	97	77	75
Urea	50	75	75	60	41	50	57	51
Metanephrine	53	—	98	—	—	—	—	66
Nor-Metanephrine	45	—	91	—	—	—	—	63
Cyanoacetic acid	55	—	41	—	—	—	—	48

[a] Abbreviations: s = streaks, d = decomposes.

Paper P1 = W1.

Solvent S1 = *n*-butanol-glacial acetic acid-water (120 : 30 : 50).
S2 = phenol-water (160 g : 40 ml).
S3 = solvent S2-conc. NH_4OH (200 : 1).
S4 = solvent S2-ethanol-water (150 : 40 : 10).
S5 = ethanol-water-conc. NH_4OH (180 : 10 : 10).
S6 = *tert*-butanol-water-methyl ethyl ketone-diethylamine (80 : 80 : 40 : 8).
S7 = methanol-water-pyridine (160 : 40 : 8).
S8 = *n*-butanol-pyridine-water (60 : 60 : 60).
(Solvents require 4–12 hours for 8 in. rise from origin).

Technique Ascending, without equilibration.

Literature
1. Smith, I., in *Chromatographic and Electrophoretic Techniques*, Smith, I., Ed., Vol. I, Interscience Publishers, Division of John Wiley & Sons, Inc., 1969, pp. 112–118.

TABLE PC 65
AMINO ACIDS, DINITROPHENYL DERIVATIVES

Paper	P1	P1	P1	P1	P1	P1
Solvent	S1	S2	S3	S4	S5	S6
Literature	1	1	1	1	1	1
DNP-Amino acid	R_F[a] × 100					
Aspartic acid	1	11	8	3	67	—
Glutamic acid	1	9	10	6	63	—
Glycine	33	16	36	14	39	—
Serine	20	16	29	13	53	—
Threonine	30	21	41	17	59	—
Alanine	46	28	52	25	49	—
Histidine (*bis*-DNP)	70	23	54	32	18	—
Dinitrophenol	52	42	62	45	38	—
Lysine (*bis*-DNP)	83	38	65	49	14	—
Tyrosine (*bis*-DNP)	83	38	75	55	3	—
Methionine	60	42	63	50	54	—
Phenylalanine	69	60	76	58	42	—
Valine	60	48	69	68	62	—
Leucine	63	45	78	70	61	—
Proline	45	34	53	50	65	—
Tryptophan	61	45	75	69	57	—
Cystine	47	40	24	28	18	—
Cysteic acid	—	—	—	—	—	25
Lysine (ε-DNP)	—	—	—	—	—	55
Arginine	—	—	—	—	—	64
Tyrosine (O-DNP)	—	—	—	—	—	70
Histidine (*im*-DNP)	—	—	—	—	—	33

[a] Dinitraniline runs very fast in the organic solvents and very slowly in phosphate buffer. It is consequently easily recognized on chromatograms. The R_F values of the DNP-amino acids are very variable and those quoted here should be taken only as an indication of position.

Paper P1 = W1.

Solvent S1 = toluene-chloroethanol-pyridine-5*N* NH_4OH (5 : 3 : 1.5 : 0.5), freshly prepared.

S2 = *sec*-butanol saturated with pH 6 buffer (1.021 g potassium hydrogen phthalate in 45.5 ml 0.1*N* NaOH made up to 100 ml); papers dipped in buffer and dried at room temperature before use.

S3 = *n*-butanol saturated with 0.1% (w/v) NH_4OH; use organic phase, prepare fresh.

S4 = same as S2 with *tert*-amyl alcohol in place of butanol.

S5 = 78 g $NaH_2PO_4{\cdot}2H_2O$ and 44.5 g $Na_2HPO_4{\cdot}2H_2O$ per liter water, pH = 6.

S6 = *n*-butanol-acetic acid-water (120 : 30 : 50); prepare fresh.

Technique Ascending, 20 °C, run in dark.

Literature

1. Mills, G. L. and D. Beale, in *Chromatographic and Electrophoretic Techniques*, Vol. I, Smith, I., Ed., Interscience Publishers, Division of John Wiley & Sons, Inc., 1969, p.181.

TABLE PC 66
ANTIBIOTICS, MACROLIDE

Paper	na	P1	P2	P2
Solvent	S1	S2	S3	S4
Technique	na	T1	na	na
Literature	1	2	3	3
Compound	$R_F \times 100$			
Acumycin	—	—	69	80
Angolamycin	—	—	16	20
Carbomycin	71	—	68	79
1,2-Diacetylol eandomycin (DAO)	—	90	—	—
1,3-Diacetylol eandomycin (DAO)	—	65	—	—
2,3-Diacetylol eandomycin (DAO)	—	35	—	—
Erythromycin	27	—	4	11
Foromacidin A	—	—	22	43
Foromacidin B	—	—	66	78
Foromacidin C	—	—	74	80
Methymycin	56	—	—	—
1-Monoacetyloleandomycin (MAO)	—	50	—	—
2-Monoacetyloleandomycin	—	25	—	—
3-Monoacetyloleandomycin	—	10	—	—
Narbomycin	—	—	85	91
Oleandomycin	—	5	—	—
Picromycin	—	—	47	62
Triacetyloleandomycin	—	95	—	—
Tylosin	—	—	7	16

Paper P1 = W4 strips impregnated with 50% formamide in methanol.
P2 = formamide pretreated paper.

Solvent S1 = nonyl alcohol-CCl_4-*n*-propanol (75 : 75 : 2 v/v).
S2 = benzene-cyclohexane (2 : 1 v/v, formamide saturated).
S3 = benzene.
S4 = benzene-$CHCl_3$ (3 : 1 v/v).

Detection Bioautography.

Technique T1 = descending in cylindrical jar.

Literature

1. Sokolski, W. T., Ullmans, S., Koffler, H., and P. A. Tetrault, *Antibiot. Chemother.*, **4**, 1057 (1954).
2. Lees, T. M., de Muria, J., and W. H. Boegmann, *J. Chromatogr.*, **5**, 126 (1961).
3. Bickel, H., Gäumann, E., Hütter, R., Sackmann, W., Vischer, E., Voser, W., Wettstein, A., and H. Zähner, *Helv. Chim. Acta*, **45**, 1396 (1962).

TABLE PC 67
ANTIBIOTICS, STREPTOMYCINS, NEOMYCINS, KANAMYCINS

Paper	P1	P1	P1	P1
Solvent	S1	S2	S3	S4
Technique	na	na	T1	na
Literature	1	1	2	1
Compound	$R_F \times 100$			
Catenulin	—	29	52	—
Dihydrostreptomycin	40	70	—	34
Hydroxystreptomycin	41	—	—	—
Kanamycin	—	—	57	15
Kanamycin A	—	35	—	—
Kanamycin B	—	60	—	—
Mannosidodihydrostreptomycin	—	25	—	—
Mannosidostreptomycin	10	40	—	—
Neomycin A (Neamine)	—	85	43	—
Neomycin B	41	75–85	26	—
Neomycin C	53	26	30	—
Neomycin complex	—	3–45	—	5
Paromomycin	—	—	—	23
Streptomycin	49	100	—	38
Streptothricin	—	—	—	27
Streptothricin I	27	—	—	—
Streptothricin II	16	—	—	—
Streptothricin III	22	—	—	—
Vancomycin	—	—	—	40
Viomycin	—	—	—	16

Paper P1 = W1.

Solvent S1 = *n*-butanol-methanol-H_2O-*p*-toluenesulfonic acid (10 : 40 : 20 : 1 v/v/v/w).
S2 = *n*-butanol (H_2O saturated)-2% *p*-toluenesulfonic acid.
S3 = *n*-propanol-acetic acid-H_2O (9 : 1 : 10 v/v/v) (compounds as sulfate salts).
S4 = methanol-2% aqueous NaCl (2 : 1 v/v), paper impregnated with 5% Na_2SO_4.

Detection Bioautography; diacetyl-α-naphthol.

Technique T1 = descending.

Literature

1. Vondráček, M., and I. M. Hais in *Paper Chromatography* (Hais, I. M. and Macek, K., Eds.), Academic Press, New York, 1963, p 686.
2. Rinehart, K. L., Argoudelis, A. D., Gross, W. A., Sohler, A., and D. P. Schaffner, *J. Amer. Chem. Soc.*, **82**, 3938 (1960).

TABLE PC 68

ANTIBIOTICS, TETRACYCLINE

Paper	P1	P1	P1	P1	P1
Solvent	S1	S2	S3	S4	S5
Literature	1	2	3	4	5
Compound			$R_F \times 100$		
Anhydrodemethylchlortetracycline	—	—	—	71	—
Anhydrotetracycline	—	—	—	62	—
Chlortetracycline	50	58	70–73	48	62
***iso*-Chlortetracycline**	—	—	—	48	—
Dehydrochlortetracycline	—	—	—	—	39
Demethylchlortetracycline	—	51	56–60	25	39
Demethyltetracycline	—	—	38–41	—	22
Epianhydrodemethychlortetracycline	—	—	—	44	—
Epianhydrotetracycline	—	—	—	40	—
Epichlortetracycline	8	—	—	—	33
Epidemethylchlortetracycline	—	—	—	17	20
Epidemethyltetracycline	—	—	—	—	10
Epioxytetracycline	—	—	—	—	12
Epitetracycline	5	—	—	15	14
Oxytetracycline	13	36	—	27	24
Tetracycline	28	29	47–50	39	20
***iso*-Tetracycline**	—	—	—	22	—

Paper P1 = W1 or equivalent.

Solvent
S1 = $CHCl_3$-nitromethane-pyridine (10 : 20 : 30 v/v); paper wetted with pH 4.5 McIlvaine's buffer and blotted dry.
S2 = methyl butyl ketone-ethyl acetate-butanol (80 : 80 : 35 v/v); paper impregnated with a pH 4.6 buffer, 0.2*M* Na_2HPO_4-0.1*M* citric acid (9.35 : 10.65 v/v).
S3 = $CHCl_3$-*n*-butanol (4 : 1 v/v) saturated with McIlvaine's buffer, pH 4.5; paper impregnated with the same buffer saturated with the organic phase; circular development.
S4 = *n*-butanol-acetic acid-H_2O (4 : 1 : 5 v/v, top layer).
S5 = $CHCl_3$-*n*-butanol (9 : 1 v/v); paper impregnated with 0.3*N* H_3PO_4 + 0.1% trichloroacetic acid.

Detection Bioautography.

Technique Modification of the method of Selzer and Wright Literature 1.

Literature
1. Selzer, G. B., and W. W. Wright, *Antibiot. Chemother.*, **7**, 292 (1957).
2. Canterelli, G., *Farm. (Pavia), Ed. Pract.*, **17**, 728 (1963).
3. Urx, M., Vondráckova, J., Kovarik, L., Horský, O., and M. Herold, *J. Chromatogr.*, **11**, 62 (1963).
4. Kelly, R. G. and D. A. Buyske, *Antibiot. Chemother.*, **10**, 604 (1960).
5. McCormick, J. R. D., Sjoalander, N. O., Hirsch, U., Jensen, E. R., and A. P. Doerschuk, *J. Amer. Chem. Soc.*, **79**, 4561 (1957).

TABLE PC 69

ANTIOXIDANTS

Paper	P1	P1	P2	—
Solvent	S1	S2	S3	—
Technique	T1	T1	T1	—
Literature	1	1	2	1
Compound	$R_F \times 100$		R_{PG}* × 100	Sensitivity μg
BHA (Butylated hydroxyanisole)	56	0	65	1–2
BHT (Butylated hydroxytoluene)	9	0	38	20
DPPD (*N,N'*-diphenyl-*p*-phenylenediamine)	—	—	34	—
DTBH (2,5-di-*tert*-butylhydroquinone)	—	—	60	—
NDGA (Nordiydroguaiaretic acid)	75	5	81	1–2
PG (Propyl gallate)	75	36	100	1–2
Santoquin (6-ethoxy-2,2,4-trimethyl-1,2-dihydroquinoline)	—	—	65	—
α-Tocopherol	—	—	19	—

* R_{PG} = distance moved relative to PG.

Paper
P1 = 8 × 8″ sheets of paper.
P2 = acetylated W1.

Solvent
S1 = immobile phase-soybean oil, mobile phase-80% methanol.
S2 = immobile phase-soybean oil, mobile phase-water.
S3 = 60% acetone.

Detection Phosphomolybdic acid.

Technique T1 = Ascending development.

Literature
1. Mitchell, L. C., *J. Ass. Offic. Agr. Chem.*, **40**, 909–915 (1957).
2. Dehority, B. A., *J. Chromatogr.*, **2**, 384–387 (1959).

TABLE PC 70
CARBOHYDRATES, SUGARS

Paper	P1	P1	P1	P1	P1	P1	P1	P1			
Solvent	S1	S2	S3	S4	S5	S6	S7	S8			
Detection									D1	D2	D3
Literature	1	1	1	1	1	1	1	1			
Compound	$R_{glucose} \times 100$								Color		
3-O-MeGlucose	160	181	147	152	136	156	163	—	y-bn	bn	—
Xylulose	155	214	128	160	130	158	136	149	gy-bn	r-bn	gn-bn
Rhamnose	152	172	130	—	133	178	162	151	gy-bn	r-bn	—
Deoxyribose	150	202	133	—	140	194	176	186	bn	r-bn	—
Deoxyglucose	146	—	—	—	—	—	—	—	r	—	gn
Ribulose	142	200	125	—	125	152	142	162	g	—	gn-bn
Ribose	137	189	122	—	122	142	138	154	gy-bn	r-bn	—
Fucose	135	146	113	—	117	128	132	161	gy-bn	r-bn	—
Xylose	130	158	117	124	117	128	130	122	gy-bn	r-bn	—
Arabinose	112	132	110	114	108	117	116	138	gy-bn	r-bn	—
Mannose	105	120	108	—	110	106	108	118	gy-bn	bn	—
Fructose	105	118	108	112	108	109	109	133	gy	bn	r
Sorbose	102	120	102	—	103	103	97	107	gy	—	r
Glucose	100	100	100	100	100	100	100	100	y-bn	bn	—
Galactose	90	86	89	96	92	91	90	116	y-bn	bn	—
Sucrose	80	80	89	77	91	77	75	107	y-bn	—	r
Palatinose	64	71	86	67	84	67	65	—	y	bn	r
Lactulose	64	57	74	64	74	60	56	—	—	bn	r
Maltose	47	63	80	53	73	56	50	93	gy	bn	—
Cellobiose	45	61	61	—	68	50	48	84	gy-bn	bn	—
Isomaltose	35	48	62	53	65	46	50	—	y	bn	—
Lactose	32	47	64	53	58	41	38	98	y-bn	bn	—
Melibiose	35	42	64	—	58	39	38	80	gr-bn	bn	—
Raffinose	30	37	53	41	56	33	32	71	y-bn	—	r
Maltotriose	27	40	64	34	54	31	31	—	gy	bn	—
Isomaltotriose	19	23	38	34	38	21	24	—	y-bn	bn	—
Glucuronic A	20	14	23	85	28	23	26	29	r-bn	y-b	b
Galacturonic A	15	14	35	—	—	—	—	33	gn	y-b	b
Dulcitol	105	86	97	—	95	102	106	118	—	—	—
Mannitol	110	92	100	—	100	102	106	118	—	—	—
Sorbitol	105	86	93	—	97	98	100	124	—	—	—
Inositol	35	32	41	—	38	34	36	58	—	—	—

TABLE PC 70—(Continued)
CARBOHYDRATES, SUGARS

Paper P1 = W3MM.

Solvent
S1 = isopropanol-water (160 : 40); R_F glucose = 37.
S2 = ethyl acetate-pyridine-water (120 : 50 : 40); R_F glucose = 28.
S3 = *n*-butanol-pyridine-water (80 : 80 : 40); R_F glucose = 45.
S4 = *n*-butanol-acetic acid-water (120 : 30 : 50); R_F glucose = 25.
S5 = isopropanol-pyridine-water (120 : 40 : 40); R_F glucose = 50.
S6 = *n*-propanol-ethyl acetate-water (140 : 20 : 40); R_F glucose=36
S7 = isopropanol-*n*-butanol-water (140 : 20 : 40); R_F glucose = 36.
S8 = phenol-water-conc. NH_4OH (160 g : 40 ml : 1 ml).

Detection
D1 = aniline-diphenylamine phosphate.
D2 = aniline phosphate.
D3 = naphthoresorcinol.

Technique Descending chromatogram, 47 cm, run 16 hours or longer, solvent drips off serrated paper.

Literature
1. Menzies, I. S. and J. W. T. Seakins, in *Chromatographic and Electrophoretic Techniques*, Vol. I, Smith, I., Ed., Interscience Publishers, Division of John Wiley & Sons, Inc., 1969, p. 313.

TABLE PC 71
CARBOHYDRATES, SUGAR ESTERS

Paper	P1	P1	P1	P1
Solvent	S1	S2	S3	S4
Literature	1,2	1,2	1,2	1,2
Compound	$R_F \times 100$			
Caffeyl-glucose	—	31	32	—
1-*p*-Cumaroyl-β-D-glucose	66	61	65	—
2-*p*-Cumaroyl-glucose	69	—	64	68
3-*p*-Cumaroyl-glucose	66	63	68	67
6-*p*-Cumaroyl-glucose	59	56	55	59
4-*p*-Cumaroyl-rhamnose	80	78	68	79
Ferulic acid	84	87	27	80
1-Feruloyl-β-D-glucose	60	—	57	—
2-Feruloyl-glucose	63	60	57	59
3-Feruloyl-glucose	62	57	65	58
4-Feruloyl-glucose	60	55	67	54
6-Feruloyl-glucose	55	47	49	48
Glucose	22	—	87	10
Glucosyl-(1 → 2)-6-*p*-cumaroyl-β-D-glucose from *Raphanus*	42	26	73	—
Glucosyl-(1 → 2)-6-feruloyl-β-D-glucose from *Raphanus*	39	20	67	—

Paper P1 = SS2043b (Mgl.) wedge strip form according to Matthias.

Solvent
S1 = *n*-butanol-acetic acid-water (4 : 1 : 2 v/v).
S2 = *n*-pentanol-acetic acid-water (4 : 1 : 5 v/v).
S3 = water-concentrated HCl (97 : 3 v/v).
S4 = *n*-butanol-ethanol-water (4 : 1 : 5 v/v).

Detection Aniline phthalate.

Literature
1. Birkofer, L., Kaiser, C., Donike, M., and W. Koch, *Z. Naturforsch.*, **20b**, 426 (1965); *J. Chromatogr.*, **20**, D5 (1965).
2. Matthias, W., *Naturwissenschaften*, **41**, 17 (1954).

TABLE PC 72

CARBOHYDRATES, LACTONES, GLYCOSIDES, AND SUGAR ACIDS

Paper	na	na	na	na
Solvent	S1	S2	S3	S4
Literature	1	1	2,3	4
Compound	$R_F \times 100$			R_G* × 100
Arabitol pentaacetate	—	—	85	—
L-Arabono-γ-lactone	—	39,59	—	—
D-Galactone-γ-lactone	—	—	35	—
Galacturonic acid	—	—	—	18
D-Glucoheptono-γ-lactone	—	23	13	—
D-Glucono-δ-lactone	—	—	22	—
D-Glucono-γ-lactone	—	56	32	—
D-Glucose	42	39	—	100
Glucuronic acid	—	—	—	27
Guluronic acid	—	—	—	28
2-Ketogluconic acid	—	—	—	21
5-Ketogluconic acid	—	—	—	47
Mannitol hexaacetate	—	—	85	—
D-Mannono-γ-lactone	—	39	25	—
Mannuronic acid	—	—	—	35
Methyl α-L-arabinoside	60	65	—	—
Methyl β-L-arabinoside	68	—	—	—
Methyl α-L-fucoside	76	—	—	—
Methyl α-D-galactoside	54	—	—	—
Methyl β-D-galactoside	51	—	—	—
Methyl β-D-glucofuranoside	—	—	30	—
Methyl α-D-glucoside	58	—	—	—
Methyl β-D-glucoside	63	—	—	—
Methyl β-lactoside	33	25	—	—
Methyl β-maltoside	37	33	—	—
Methyl α-D-mannofuranoside	74	—	—	—
Methyl α-D-mannoside	68	—	—	—
Methyl α-L-rhamnoside	86	87	—	—
Methyl α-D-xyloside	75	76	—	—
L-Rhamnono-γ-lactone	—	37,63	—	—
L-Rhamnono-γ-lactone	—	—	50	—
D-Xylono-γ-lactone	—	—	41	—

* R_F with respect to glucose.

Solvent

S1 = *tert*-pentyl alcohol-*n*-propanol-water (4 : 1 : 1.5 v/v).
S2 = *tert*-pentyl alcohol-*n*-propanol-water (4 : 1 : 1 v/v).
S3 = upper phase of *n*-butanol-ethanol-water (5 : 1 : 4 v/v).
S4 = pyridine-ethyl acetate-acetic acid-water (5 : 5 : 1 : 3 v/v), tank saturated with pyridine-ethyl acetate-water (11 : 40 : 6 v/v).

Literature

1. Cifonelli, J. A. and F. Smith, *J. Biol. Chem.*, **208**, 537–541 (1954).
2. Abdel-Akher, M. and F. Smith, *J. Amer. Chem. Soc.*, **73**, 5859–5860 (1951).
3. Hough, L., *Nature*, **165**, 400 (1950).
4. Fischer, F. G. and H. Dörfel, *Z. Physiol. Chem.*, **301**, 224–234 (1955).

TABLE PC 73

CARBOHYDRATES, METHYLATED SUGARS

Paper	P1
Solvent	S1
Literature	1,2
Compound	R_g* × 100
2,3-Dimethylarabinose	61
2,4-Dimethylarabinose	62
2,5-Dimethylarabinose	84
2,3-Dimethylgalactose	55
2,4-Dimethylgalactose	50
2,5-Dimethylgalactose	54
3,4-Dimethylgalactose	43
Dimethylmannose	66
3,4-Dimethylrhamnose	86
2,3-Dimethylxylose	73
Galactose	14
2-Methylarabinose	55
2-Methylgalactose	30
4-Methylmannose	40
3-Methylrhamnose	66
2-Methylxylose	38
Oligosaccharides	4
Rhamnose	35
Tetramethylgalactose	91
Tetramethylmannose	100
2,3,4-Trimethylarabinose	84
2,3,5-Trimethylarabinose	98
2,3,4-Trimethylgalactose	79
2,3,5-Trimethylgalactose	73
2,3,6-Trimethylgalactose	79
2,4,6-Trimethylgalactose	79
2,3,4-Trimethylglucose	86
3,4,6-Trimethylmannose	77
2,3,4-Trimethylxylose	94

* R_g = R_F relative to that of 2,3,4,6-tetramethyl D-glucose.

Paper P1 = W1 or W3.
Solvent S1 = *n*-butanol-ethanol-H_2O (4 : 1 : 5 v/v).
Detection Analine phthalate.
Literature

1. Alam, M. and R. J. Mc Ilroy, *J. Chem. Soc.* (C), 1967, 1577.
2. Aspinall, G. O., Bhavanandan, V. P., and T. B. Christensen, *J. Chem. Soc.*, 1965, 2682.

TABLE PC 74

CARBOHYDRATES, OLIGOSACCHARIDES AND DERIVATIVES

Paper	P1	P1	P1
Solvent	S1	S2	S3
Detection	D1	D1	D2
Technique	T1	T2	T3
Literature	1	2	3
Compound	R_G[a] × 100	R_F[b] × 100	R_a[c] × 100
Glucose	100.0	100	89
Isomaltoheptaose	7.0	—	—
Isomaltohexaose	11.2	—	—
Isomaltopentaose	17.1	—	—
Isomaltose	54.5	45	—
Isomaltotetraose	26.4	—	—
Isomaltotriose	36.8	29	—
Maltoheptaose	9.3	8	—
Maltohexaose	13.8	11	9
Maltopentaose	21.0	15	19
Maltose	70.4	65	66
Maltotetraose	33.3	21	26
Maltotriose	50.2	38	41

[a] R_F relative to glucose.
[b] R_F relative to free amine.
[c] R_a values are of *N*-(1-naphthyl)-ethylenediamines of sugars.

Paper P1 = W1.
Solvent S1 = *n*-butanol-pyridine-water (6 : 4 : 3).
S2 = ethyl acetate-pyridine-water (10 : 4 : 3).
S3 = *n*-butanol-ethanol-water-NH_4OH (40 : 12 : 16 : 1).
Detection D1 = Modified Seliwanoff's reagent.
D2 = Ninhydrin or UV.
Technique T1 = descending 20–30 hr, 22 °C.
T2 = descending 96 hr.
T3 = descending 18 hr, tank equilibration not necessary.
Literature

1. Myrbäck, K. and E. Willstaedt, *Arkiv. Kemi*, **6**, 417–425 (1953); *Chem. Abstr.*, **48**, 6490 (1954).
2. Whistler, R. L. and J. L. Hickson, *Anal. Chem.*, **27**, 1514–1517 (1955).
3. Wadman, W. H., Thomas, G. J., and A. B. Pardee, *Anal. Chem.*, **27**, 1193 (1954).

TABLE PC 75
CARBOHYDRATES, SUGARS AND GLUCOSIDES

Paper	P1	P1	P1	P1
Solvent	S1	S2	S3	S4
Literature	1	1	2	2
Compound	$R_F \times 100$			
L-Arabinose	—	—	18	13
6-Deoxy-L-talose	—	60	—	—
L-Fucose	—	43	—	—
Methyl-β-L-arabinopyranoside	—	—	40	31
2-C-Methyl-L-arabinose	—	—	37	34
Methyl-α-L-fucopyranoside	61	—	—	—
Methyl-2-C-methyl-β-L-arabinopyranoside	—	—	58	52
Methyl-2-C-methyl-β-L-ribopyranoside	—	—	74	92
Methyl-α-L-rhamnopyranoside	76	65	—	—
Methyl-β-D-ribopyranoside	—	—	55	86
2-C-Methyl-L-ribose	—	—	39	71
L-Rhamnose	—	48	—	—
D-Ribose	—	—	27	54

Paper P1 = W1.
Solvent S1 = ethyl acetate-*n*-propanol-H_2O (5 : 3 : 2 v/v).
S2 = ethyl acetate-pyridine-H_2O (2 : 1 : 2 v/v).
S3 = *n*-butanol-ethanol-H_2O (4 : 1 : 5 v/v).
S4 = S3, 100 ml, plus 5 g phenylboronic acid and 2.5 ml ethanol.
Detection Aniline hydrogen phthalate.
Technique Descending.
Literature
1. Collins, P. M. and W. G. Overend, *J. Chem. Soc.*, 1912 (1965).
2. Ferrier, R. J., Overend, W. G., Rafferty, G. A., Wall, H. M., and N. R. Williams, *J. Chem. Soc.* (C), 1092 (1968).

TABLE PC 76
CARBOHYDRATES, SUGAR ACIDS AND LACTONES

Paper	P1	P1	P1
Solvent	S1	S2	S3
Literature	1	1	1
Compound	R_g* × 100		
Allonolactone	112	102	109
Altronolactone	112	109	132
Fructuronic acid	0	64	62
Galactonolactone	100	100	100
Galacturonic acid	0	33	85
Gluconolactone	177	108	122
Glucuronic acid	0	42	46
Glucuronolactone	—	110	125
Gulonolactone	77	89	84
Idonolactone	112	103	111
Mannonolactone	85	91	112
Tagaturonic acid	0	59	62
Talonolactone	103	98	107

* Relative to galactonolactone.

Paper P1 = W1.
Solvent S1 = pyridine-ethyl acetate-water (20 : 72 : 23 v/v).
S2 = pyridine-ethyl acetate-acetic acid-water (5 : 5 : 1 : 3 v/v).
S3 = ethyl acetate-acetic acid-water (3 : 1 : 3 v/v).
Detection Alkaline silver nitrate.
Technique Descending chromatography.
Literature
1. Hickman, J. and G. Ashwell, *J. Biol. Chem.*, **241**, 1426 (1966).

TABLE PC 77
CARBOXYLIC ACIDS AND DERIVATIVES

Paper	P1	P2	P3	P4	P5	P5	P6	P2	P2	P2	P7	P7	P7	P8	P9	P10	P11
Solvent	S1	S2	S3	S4	S5	S6	S7	S8	S9	S10	S11	S12	S13	S14	S15	S16	S17
Technique	T1	T1	T1	T1	T1	T1	T1	T1	T1	T1	T3	T3	T3	T3	T3	T3	T3
Literature	1	2	3	4, 5	6	7	7	8	8	8	9	9	9	9	9	9	9
Derivatives[a]:	H	DNPH	Free acid					$R_F \times 100$ Hg			Free acid						
Elaidic acid (18)[b]	—	—	34	—	—	—	—	—	—	—	—	—	—	—	—	—	—
β-Eleostearic acid (18)	—	—	—	—	—	—	—	77	86	81	—	—	—	—	—	—	—
Enanthic acid (7)	64	—	—	—	—	—	—	—	—	—	—	—	—	—	—	—	—
Erucic acid (22)	22	—	14	—	—	—	—	7	10	40	—	—	—	—	—	—	—
2-Hexadecenoic acid (16)	—	—	—	—	—	—	—	29	30	74	—	—	—	—	—	—	—
9-Hexadecenoic acid (16)	—	—	—	—	—	—	—	27	29	74	—	—	—	—	—	—	—
2-Hydroxylignoceric acid	—	—	—	—	92	0	—	—	—	—	—	—	—	—	—	—	—
2-Hydroxymyristic acid	—	—	—	—	94	89	—	—	—	—	—	—	—	—	—	—	—
2-Hydroxypalmitic acid	—	—	—	—	93	82	—	—	—	—	—	—	—	—	—	—	—
2-Hydroxystearic acid	—	—	—	—	93	64	—	—	—	—	—	—	—	—	—	—	—
Lauric acid (12)	38	43	70	63	—	—	62	—	—	—	85	76	81	70	85	87	96
Lignoceric acid (24)	—	—	—	11	21	—	1	—	—	—	—	—	—	—	—	—	—
Linoleic acid (18)	—	—	—	—	—	—	—	57	62	82	83	—	80	69	84	85	94
Linolenic acid (18)	—	—	45	—	—	—	—	78	87	84	83	—	80	69	84	85	94
Myristic acid (14)	34	33	56	48	80	—	51	—	—	—	76	65	73	57	73	72	81
2-Octadecenoic acid (18)	—	—	—	—	—	—	—	17	21	62	—	—	—	—	—	—	—
Oleic acid (18)	30	—	46	—	72	—	—	17	20	62	65	—	63	46	51	60	60
Palmitic acid (16)	30	25	39	35	72	—	33	—	—	—	66	55	62	45	52	63	60
Pelargonic acid (9)	51	—	—	—	—	—	—	—	—	—	—	—	—	—	—	—	—
Ricinoleic acid (18)	—	—	—	—	—	—	—	64	53	81	—	—	—	—	—	—	—
Sebacic acid (10)	—	—	93	—	—	—	—	—	—	—	—	—	—	—	—	—	—
Stearic acid (18)	24	17	29	26	59	—	19	—	—	—	55	48	50	38	41	53	48
Stearolic acid (18)	—	—	—	—	—	—	—	75	84	82	—	—	—	—	—	—	—
2-Tetradecenoic acid (14)	—	—	—	—	—	—	—	43	47	81	—	—	—	—	—	—	—
9-Tetradecenoic acid (14)	—	—	—	—	—	—	—	42	46	80	—	—	—	—	—	—	—
10-Undecenoic acid (11)	—	—	—	—	—	—	—	59	71	84	—	—	—	—	—	—	—
Undecylic acid (11)	40	—	80	77	—	—	—	—	—	—	—	—	—	—	—	—	—
Valeric acid (5)	84	—	—	—	—	—	—	—	—	—	—	—	—	—	—	—	—

[a] H = hydroxamates. DNPH = 2,4-dinitrophenylhydrazides. Hg = mercury acetate addition compounds.
[b] Number in parenthesis shows number of carbon atoms.

TABLE PC 77—(Continued)
CARBOXYLIC ACIDS AND DERIVATIVES

Paper	P2	P2	P2	P2	P2	P2	P2	P2	P2	P2	P2	P2	P2	P2
Solvent	S18	S19	S20	S21	S22	S23	S24	S25	S26	S27	S28	S29	S30	S31
Technique	T4	T4	T4	T5	T5	T5	T5	T5	T5	T5	T5	T5	T5	T5
Literature	7	7	8	10	10	10	10	10	10	11	11	12	7	13
Derivative[a]	DNPH $R_F \times 100$			Hydroxamates $R_F \times 100$										
Acetic acid	85	93	77	50	34	30	11	0	1	57	70	—	49	4
Acetyl phosphate	—	—	—	—	—	—	—	—	—	—	—	52	—	—
Adipic acid	—	—	—	—	—	—	—	—	—	54, 67	44, 60	—	—	—
Arachidic acid	13	27	9	—	—	—	—	—	—	—	—	—	—	—
Azelaic acid	—	—	—	—	—	—	—	—	—	63, 74	66 streaked	—	—	—
Behenic acid	8	18	5	—	—	—	—	—	—	—	—	—	—	—
Butyric acid	88	97	82	72	67	68	45	4	13	74	80	—	89	18
Butyryl phosphate	—	—	—	—	—	—	—	—	—	—	—	76	—	—
Capric acid	64	78	64	—	—	—	—	—	—	—	—	—	—	—
Caproic acid	83	93	81	84	86	83	81	26	61	87	90	—	84	52
Caproyl phosphate	—	—	—	—	—	—	—	—	—	—	—	86	—	—
Caprylic acid	74	87	74	—	—	—	—	—	—	—	—	—	86	90
Citric acid	—	—	—	—	—	—	—	—	—	90, 23	20, 29	—	—	—
Decanoic acid	—	—	—	90	90	—	92	92	97	—	—	—	—	—
Formic acid	—	—	—	42	26	—	8	0	0	45	57	—	23	2
Glutaric acid	—	—	—	—	—	—	—	—	—	47	36, 52	—	—	—
Heptadecanoic acid	25	41	19	—	—	—	—	—	—	—	—	—	—	—
Heptanoic acid	78	90	78	86	89	85	88	51	87	—	—	—	—	—
Iso-valeric acid	—	—	—	—	—	—	—	—	—	—	—	—	83	23
Lauric acid	54	70	51	—	—	—	—	—	—	—	—	—	—	—
Malonic acid	—	—	—	—	—	—	—	—	—	11, 23	19, 32	—	—	—
Myristic acid	42	60	38	—	—	—	—	—	—	—	—	—	—	—
Nonadecanoic acid	16	30	12	—	—	—	—	—	—	—	—	—	—	—
Nonanoic acid	69	81	68	89	89	90	90	88	97	—	—	—	—	—
Octanoic acid	—	—	—	88	89	—	90	77	95	—	—	—	—	—
Oxalic acid	—	—	—	—	—	—	—	—	—	14, 40	22, 28, 32	—	—	—
Palmitic acid	30	49	25	—	—	—	—	—	—	—	—	—	—	—
Pelargonic acid	—	—	—	—	—	—	—	—	—	—	—	—	87	94
Pentadecanoic acid	34	52	30	—	—	—	—	—	—	—	—	—	—	—
Pimelic acid	—	—	—	—	—	—	—	—	—	60, 73	52, 69	—	—	—
Propionic acid	—	—	—	62	51	50	25	1	5	68	78	—	70	10
Propionyl phosphate	—	—	—	—	—	—	—	—	—	—	—	65	—	—

TABLE PC 77—(Continued)
CARBOXYLIC ACIDS AND DERIVATIVES

Paper	P2	P2	P2	P2	P2	P2	P2	P2	P2	P2	P2	P2	P2	P2
Solvent	S18	S19	S20	S21	S22	S23	S24	S25	S26	S27	S28	S29	S30	S31
Technique	T4	T4	T4	T5	T5	T5	T5	T5	T5	T5	T5	T5	T5	T5
Literature	7	7	8	10	10	10	10	10	10	11	11	12	7	13
Derivative[a]	DNPH $R_F \times 100$			Hydroxamates $R_F \times 100$										
Pyruvic acid	—	—	—	—	—	—	—	—	—	59, 86	54, 62, 73	—	—	—
Sebacic acid	—	—	—	—	—	—	—	—	—	89	74, 89	—	—	—
Stearic acid	21	38	15	—	—	—	—	—	—	—	—	—	—	—
Succinic acid	—	—	—	—	—	—	—	—	—	40, 72	45, 52	—	—	—
Tartaric acid	—	—	—	—	—	—	—	—	—	10	19	—	—	—
Tridecanoic acid	45	62	43	—	—	—	—	—	—	—	—	—	—	—
Undecanoic acid	57	72	56	—	—	—	—	—	—	—	—	—	—	—
Valeric acid	—	—	—	79	78	81	67	11	32	83	84	—	83	35
Valeryl phosphate	—	—	—	—	—	—	—	—	—	—	—	81	—	—

Paper

P1 = acetylated W1.
P2 = W1.
P3 = paper impregnated with 12% olive oil in toluene.
P4 = paper saturated with paraffin (190–220°C).
P5 = developed at 40°C using W1 impregnated with paraffin oil (12–15%).
P6 = W1 drawn through 7% mineral oil solution.
P7 = W3 impregnated with 10% liquid paraffin in benzene.
P8 = W3 impregnated with 10% Dow-Corning stopcock grease.
P9 = W3 impregnated with 5% Dow-Corning high vacuum grease.
P10 = W3 impregnated with 10% vaseline.
P11 = W3 impregnated with 5% vaseline.

TABLE PC 77—(Continued)
CARBOXYLIC ACIDS AND DERIVATIVES

Solvent

S1 = ethyl acetate-tetrahydrofuran-water (0.6 : 3.5 : 4.7).
S2 = methanol-decaline (8 : 1).
S3 = 75% ethanol.
S4 = acetic acid-water (9 : 1).
S5 = 95% acetic acid.
S6 = 65% acetic acid.
S7 = acetic acid.
S8 = 90% methanol.
S9 = diethylene glycol-acetic acid acid-tetralin (60 : 20 : 11).
S10 = methanol.
S11 = 90% acetic acid.
S12 = glacial acetic acid-formic acid (88%)-hydrogen peroxide (30%), (6 : 1 : 1).
S13 = aqueous methanol 95%.
S14 = aqueous methanol 90%.
S15 = aqueous methanol 65%.
S16 = aqueous methanol 90%.
S17 = aqueous methanol 80%.
S18 = 90% methanol-acetic acid-tetralin (10 : 2 : 1 v/v).
S19 = 80% ethanol-acetic acid-tetralin (10 : 2 v/v).
S20 = 90% methanol-tetralin (10 : 1 v/v).
S21 = 1–butanol-acetic acid-water (4 : 1 : 5 v/v).
S22 = amyl alcohol-acetic acid-water (4 : 1 : 5 v/v).
S23 = amyl alcohol-formic acid-water (75 : 25 : 75 v/v).
S24 = octyl alcohol-formic acid-water (75 : 25 : 75 v/v).
S25 = benzene-formic acid-water (75 : 75 : 75 v/v).
S26 = benzene-acetic acid-water (100 : 75 : 100 v/v).
S27 = water-saturated isobutyric acid.
S28 = water-saturated phenol.
S29 = water-saturated butanol.
S30 = butanol-acetic acid-water (4 : 1 : 1 v/v).
S31 = benzene-formic acid (1 : 1 v/v).

Technique

T1 = reversed phase partition system.
T2 = circular chromatography.
T3 = reverse phase circular paper chromatography.
T4 = ascending technique at 30°C.
T5 = one dimensional chromatography.

Literature

1. Micheel, F. and H. Schweppe, *Angew. Chem.*, **66**, 136–137 (1954).
2. Hirayama, O. and M. Noda, *Abura Kagaku*, **5**, 16–18 (1956).
3. Kabrle, V. and R. Zahradnik, *Chem. Listy.*, **48**, 1189–1196 (1954).
4. Wagner, H., Abisch, L., and K. Bernhard, *Helv. Chim. Acta.*, **38**, 1536–1541 (1955).
5. Perila, O., *Acta Chem. Scand.*, **9**, 864–865 (1955).
6. Ballance, P. E. and W. M. Crombie, *Biochem. J.*, **69**, 632 (1958).
7. Block, R. J., Durrum, E. L., and G. Zweig, *Paper Chromatography and Paper Electrophoresis*, Academic Press, N.Y. (1958).
8. Inoue, Yoshiyuki, Noda, M., and O. Hirayama, *J. Amer. Oil Chem. Soc.*, **32**, 132–135 (1955).
9. Viswanathan, C. V. and B. Meera Bai, *J. Chromatogr.*, **7**, 507 (1962).
10. Thompson, A. R., *Australian J. Sci. Research*, **B4**, 180 (1951).
11. Fink, K. and R. M. Fink, *Proc. Soc. Exptl. Biol. Med.*, **70**, 654 (1949).
12. Stadtman, E. R. and H. A. Barker, *J. Biol. Chem.*, **184**, 769 (1950).
13. Ueno, Y., *J. Biochem.* (Japan), **48**, 161 (1960).

TABLE PC 78

CARBOXYLIC ACIDS, BENZOIC ACID DERIVATIVES

Paper Solvent Detection Literature	P1 S1 — 1	P1 S2 — 1	P1 S3 — 1	P1 S4 — 1	P1 S5 — 1	P1 S6 — 1	D1	D2	D3	D4	D5	D6	D7	D8	D9	D10	D11	D12	D13
Compound			$R_F \times 100$										Color[a]						
***m*-Nitrobenzyl alcohol**	95	90	85	84	88	83	27	—	—	—	—	—	—	—	—	+	—	+	—
***o*-Aminoacetophenone**	94	93	93	93	93	93	39	—	—	—	—	—	+	—	34	+	—	8	24
***m*-Aminoacetophenone**	90	89	81	91	74	66	57	8	—	64	62	—	7	—	+	+	—	6	19
***m*-Aminobenzaldehyde**	93	92	85	92	78	70	—	—	—	—	—	—	6	—	—	+	—	6	22
***o*-Nitrobenzaldehyde**	96	94	93	95	93	92	+	—	—	—	—	—	6	—	—	+	—	4	23
Thiobenzoic acid	94	58	92	88	87	86	—	+	60	+	+	64	—	65, 72	69	++	+	—	—
2,6-Dichlorobenzoic acid	91	39	82	84	84	65	—	—	—	—	—	—	—	—	—	—	++	—	—
***p*-Hydrazinobenzoic acid**	89	6	80	49	24	5	39	—	59	63, 56	—	—	—	—	—	+	—	5	17
***N*-Methylanthranilic acid**	94	31	86	93	86	84	33	6	—	12, 62	8	38	—	26, 51	45, 50	++	+	7, 3	21, 15
2-Amino-4-methylbenzoic acid	94	14	92	87	78	50	33	8	—	64	12	45	—	—	+	+	—	7	24
2-Amino-5-methylbenzoic acid	94	18	91	84	78	52	33	6	—	62	8	60	—	58	30	++	+	9	19, 65
3-Amino-4-methylbenzoic acid	90	4	80	66	33	7	34	8	—	64	12	43	—	—	—	+	—	5	22
4-Amino-3-methylbenzoic acid	91	6	88	68	46	15	—	8	—	64	10	—	—	—	—	+	—	6	23
3,5-Dimethylanthranilic acid	93	15	89	92	85	58	39	—	—	64	6	—	—	—	+	+	—	—	+
3,4-Diaminobenzoic acid	88	86	33	6	3	0	38	+	58	23, 56	59	56	—	63, 56	69	++	—	9	23, 25
2,5-Dinitrobenzoic acid	92	48	89	67	79	24	26	—	—	—	—	—	—	—	—	—	++	58	—
3-Nitrophthalic acid	85	2	83	45	7	0	26	—	—	—	—	—	—	—	—	—	+	—	—

[a] For interpretation of color readings, see Table PC 120.

Paper P1 = W1.

Solvent
S1 = ethyl methyl ketone-acetone-formic acid-water (40 : 2 : 1 : 6 parts by volume).
S2 = ethyl methyl ketone-diethylamine-water (921 : 2 : 77 parts by volume).
S3 = methyl isobutyl ketone-formic acid-water (10 parts ketone saturated with 1 part 4% formic acid).
S4 = chloroform-methanol-formic acid-water (10 parts of chloroform saturated with a mixture of 1 part methanol and 1 part 4% formic acid).
S5 = benzene-ethyl methyl ketone-formic acid-water (a mixture of 9 parts benzene and 1 part ketone saturated with 1 part 2% formic acid).
S6 = benzene-formic acid-water (10 parts benzene saturated with 1 part 2% formic acid).

Detection
D1 = ultraviolet light.
D2 = diazotized sulfanilic acid (0.3% solution in dioxane-water 1 : 2).
D3 = diazotized 4-benzoylamino-2,5-dimethoxyaniline (0.6% solution in dioxane-water 1 : 2).
D4 = diazotized *o*-dianisidine (0.6% solution in dioxane-water 1 : 2).
D5 = *p*-nitrobenzenediazonium fluoborate (0.4% solution in dioxane-water 1 : 2).
D6 = 2,6-dibromoquinone-4-chloroimide (0.5% solution in dioxane-acetone 4 : 1).
D7 = 2,4-dinitrophenylhydrazine (*ca.* 0.1% solution in 1*N* HCl).
D8 = ferric chloride (2% aqueous solution).
D9 = phosphomolybdic acid (2% aqueous solution).
D10 = potassium permanganate (1% aqueous solution).
D11 = bromophenol blue (*ca.* 0.05% solution in ethanol).
D12 = Ehrlich reagent (1% *p*-dimethylaminobenzaldehyde in 1*N* HCl).
D13 = *p*-dimethylaminocinnamaldehyde (0.1% solution in 1*N* HCl).

Technique Descending, four-five hours, 40 cm.

Literature
1. Reio, L., *J. Chromatogr.*, **13**, 475–496 (1964).

TABLE PC 79

CARBOXYLIC ACIDS, LOWER FATTY ACIDS

Paper	P1	P1	P1	P1	P1
Solvent	S1	S2	S3	S4	S5
Technique	T1	T2	na	T3	T4
Literature	1	2	3	4	5
Compound	$R_F \times 100$				
Acetic	11	33	37	28	8
Bromoacetic	—	—	45	—	2
Bromobutyric	—	—	60	—	—
2-Bromopropionic	—	—	52	—	25
3-Bromopropionic	—	—	—	—	20
***n*-Butyric**	29	54	57	51	—
***n*-Caproic**	53	68	69	—	—
Chloroacetic	—	—	—	—	14
2-Chloropropionic	—	—	—	—	23
3-Chloropropionic	—	—	—	—	19
Dibromoacetic	—	—	—	—	10
2,3-Dibromopropionic	—	—	—	—	27
Dichloroacetic	—	—	—	—	27
Formic	10	31	37	—	—
***n*-Heptanoic**	62	72	—	—	—
Iodoacetic	—	—	50	—	—
Iodobutyric	—	—	62	—	—
Iodopropionic	—	—	55	—	—
***n*-Nonanoic**	67	—	—	—	—
***n*-Octanoic**	65	76	78	—	—
Propionic	19	44	48	41	8
Tribromoacetic	—	—	—	—	31
Trichloroacetic	—	—	—	—	45
***n*-Valeric**	41	60	—	56	—

Paper P1 = W1 or W4.

Solvent S1 = *n*-butanol-1.5*N* NH_4OH (1 : 1 v/v).
S2 = 95% ethanol-concentrated NH_4OH (100 : 1 v/v).
S3 = *n*-propanol-concentrated NH_4OH (7 : 3 v/v).
S4 = *n*-butanol-H_2O-diethylamine (100 : 15 : 1 v/v).
S5 = *n*-butanol-1.5*N* aqueous NH_3 (1 : 1 v/v).

Technique T1 = ascending, 20–24 hours.
T2 = ascending, 6–8 hours, 20–23 °C.
T3 = ascending, 20–25 cm, cylindrical glass jars, 25 °C.
T4 = saturated chamber, 25 °C.

Literature

1. Reid, R. L. and M. Lederer, *Biochem. J.*, **50**, 60–67 (1951).
2. Kennedy, E. P. and H. A. Barker, *Anal. Chem.*, **23**, 1033–1034 (1951).
3. Hashmi, M. H. and C. F. Cullis, *Anal. Chim. Acta*, **14**, 336–338 (1956).
4. Jones, A. R., Dowling, E. J., and W. J. Skraba, *Anal. Chem.*, **25**, 394 (1954).
5. Chittum, J. W., Gustin, T. A., McGuire, R. L., and J. T. Sweeney *Anal. Chem.*, **30**, 1213 (1958).

TABLE PC 80

CARBOXYLIC ACIDS, HIPPURIC ACID DERIVATIVES

Paper	P1	P1	P1	P1	P1	P1													
Solvent	S1	S2	S3	S4	S5	S6													
Detection	—	—	—	—	—	—	D1	D2	D3	D4	D5	D6	D7	D8	D9	D10	D11	D12	D13
Literature	1	1	1	1	1	1													
Compound	$R_F \times 100$						Color[a]												
o-Methoxyphenaceturic acid	86	6	76	83	34	5	—	—	—	—	—	—	5	—	—	—	+	—	—
2,3-Dihydroxyhippuric acid	85	5	76	10	5	0	30	8	62	62	10	28	13	28, 70	68	++	+	—	—
2,5-Dihydroxyhippuric acid	84	4	62	2	1	0	33	—	—	—	—	—	12, 13	70	30	+	+	—	—
Vanilloylglycine	76	2	43	9	1	0	—	8	—	65	8, 65	28	10	71	70	++	+	—	—
Caffeoylglycine	69	2	24	0	0	0	33	—	63	58	8, 63	+	17	3, 51	69	++	+	—	—
Feruoylglycine	76	2	45	13	1	0	33	6, 24	23, 13	23, 43	13	25	17, 15	—	69	++	+	—	—
Anthranylglycine	85	5	65	34	8	1	33	—	—	56	8	—	13, 17	—	—	++	+	7	22
o-Aminohippuric acid methyl ester	92	89	87	92	77	67	33	—	—	—	6	—	6	—	—	+	—	6	22
3-Hydroxyanthranylglycine	91	8	84	18	1	0	38	—	—	17	58	43	6	56	43, 69	++	+	5	22
Indole-3-acetylglycine	85	5	73	26	7	0	—	—	—	—	7	—	—	—	—	++	+	28	27
5-Hydroxyindole-3-acetylglycine	77	3	28	1	0	0	—	63	24	25	63, 65	71, 27	—	59	70	++	+	71, 28	30, 33
Indole-3-acrylglycine	91	6	74	14	3	0	34	6	63	23	58	—	17	—	70	++	+	13	23
Quinaldylglycine	90	7	87	82	58	10	34	—	—	—	—	—	22	—	—	+	+	—	—

[a] For interpretation, see Table PC 120.

Paper P1 = W1.

Solvent
S1 = ethyl methyl ketone-acetone-formic acid-water (40 : 2 : 1 : 6 parts by volume).
S2 = ethyl methyl ketone-diethylamine-water (921 : 2 : 77 parts by volume).
S3 = methyl isobutyl ketone-formic acid-water (10 parts ketone saturated with 1 part 4% formic acid).
S4 = chloroform-methanol-formic acid-water (10 parts of chloroform saturated with a mixture of 1 part methanol and 1 part 4% formic acid).
S5 = benzene-ethyl methyl ketone-formic acid-water (a mixture of 9 parts benzene and 1 part ketone saturated with 1 part 2% formic acid).
S6 = benzene-formic acid-water (10 parts benzene saturated with 1 part 2% formic acid).

Detection
D1 = ultraviolet light.
D2 = diazotized sulfanilic acid (0.3% solution in dioxane-water 1 : 2).
D3 = diazotized 4-benzoylamino-2,5-dimethoxyaniline (0.6% solution in dioxane-water 1 : 2).
D4 = diazotized *o*-dianisidine (0.6% solution in dioxane-water 1 : 2).
D5 = *p*-nitrobenzenediazonium fluoborate (0.4% solution in dioxane-water 1 : 2).
D6 = 2,6-dibromoquinone-4-chloroimide (0.5% solution in dioxane-acetone 4 : 1).
D7 = *p*-dimethylaminobenzaldehyde (2% solution in acetic anhydride).
D8 = ferric chloride (2% aqueous solution).
D9 = phosphomolybdic acid (2% aqueous solution).
D10 = potassium permanganate (1% aqueous solution).
D11 = bromophenol blue (*ca.* 0.05% solution in ethanol).
D12 = Ehrlich reagent (1% *p*-dimethylaminobenzaldehyde in 1*N* HCl).
D13 = *p*-dimethylaminocinnamaldehyde (0.1% solution in 1*N* HCl).

Technique Descending development, 40 cm.

Literature
1. Reio, L., *J. Chromatogr.*, **13**, 475–496 (1964).

TABLE PC 81
CARBOXYLIC ACIDS, α-KETO ACIDS AND DERIVATIVES

Paper	P1	P1	P1	P1	P1	P1	P1	P1	P1	P1	P1
Solvent	S1	S2	S3	S4	S5	S6	S7	S8	S9	S10	S11
Literature	1	1	2	3	4	5	6	7	7	8	8
	$R_F \times 100$										
Compound	Free acids		2,4-Dinitrophenylhydrazones							Nitro-2-R-[a] quinoxalinols	
Acetoacetic acid	—	—	—	50	55	—	—	—	—	—	—
Acetone	—	—	—	—	93	—	—	—	—	—	—
Cinnamoylformic acid	75	95	—	—	—	—	—	—	—	—	—
Dimethylpyruvic acid	60	78	—	—	—	—	—	—	—	86	85
Glyoxylic acid	—	—	—	—	—	—	—	39	71,77	—	—
o-Hydroxyphenylpyruvic acid	—	—	—	81	—	—	—	—	—	—	—
p-Hydroxyphenylpyruvic acid	—	—	—	70	—	—	—	75	83	—	—
β-Indolpyruvic acid	—	—	—	—	—	—	—	87	83	—	—
α-Keto-ε-aminocaproic acid	—	—	—	—	—	—	—	72	95	—	—
α-Ketobutyric acid	—	—	76	—	—	—	—	63	80	73	74
α-Ketoglutaric acid	21	7	51	16	12	4	69	33	62	14	—
α-Ketoisocaproic acid	—	—	—	—	—	—	—	83	88	—	—
α-Ketoisovaleric acid	—	—	83	—	—	—	—	80	86	—	—
α-Keto-γ-methiolbutyric acid	—	—	—	—	—	—	—	43	82	—	—
D-α-Keto-β-methylvaleric acid	—	—	—	—	—	—	—	77	86	—	—
β-Mercaptopyruvic acid	—	—	—	—	—	—	—	47	77	—	—
Oxalacetic acid	16	4	8	8	24	6	78	32,47	63,76	12	13
Phenylpyruvic acid	—	—	—	78	—	91	—	87	86	—	—
Pyruvic acid	29	46	64	36	41	62,80	45	51	75,81	65	57
β-Sulfopyruvic acid	—	—	—	—	—	—	—	18,22	62,83	—	—

[a] R = alkyl portion of corresponding acid.

Paper
P1 = W4.

Solvent
S1 = *n*-butanol-water-propionic acid (10 : 7 : 5).
S2 = α-*n*-dimethylaminoisobutyric nitrile-*n*-butanol-water (4 : 6 : 3).
S3 = water-saturated *n*-butanol-formic acid (95 : 5).
S4 = *tert*-amyl alcohol-ethanol-water (5 : 1 : 4).
S5 = *n*-butanol-ethanol-0.5*N* NH_4OH (7 : 1 : 2).
S6 = *n*-butanol-1*N* $NaHCO_3$ (1 : 2).
S7 = glycine-NaOH buffer (0.1*M*) pH 8.4.
S8 = *n*-butanol-water-ethanol (5 : 1 : 1).
S9 = methanol-benzene-*n*-butanol-water (4 : 2 : 2 : 2).
S10 = *n*-butanol-5*N* NH_4OH (2 : 1).
S11 = benzyl alcohol-5*N* NH_4OH (2 : 1).

Literature

1. Wieland, T. and E. Fischer, *Naturwissenschaften*, **36**, 219 (1949).
2. Magasanik, B. and H. E. Umbarger, *J. Amer. Chem. Soc.*, **72**, 2308–2309 (1950).
3. Altmann, S. M., Crook, E. M., and S. P. Datta, *Biochem. J.*, **49**, lxiii (1951).
4. El Hawary, M. F. S. and R. H. S. Thompson, *Biochem. J.*, **53**, 340–347 (1953).
5. Seligson, D. and B. Shapiro, *Anal. Chem.*, **24**, 754–755 (1952).
6. Virtanen, A. I., Miettinen, J. K. and H. Kunttu, *Acta. Chem. Scand.*, **7**, 38–44 (1953).
7. Meister, A. and P. A. Abendschein, *Anal. Chem.*, **28**, 171–173 (1956).
8. Hockenhull, D. J. D., *Nature*, **171**, 982 (1953).

TABLE PC 82
CARBOXYLIC ACIDS, PHENOLS, AND RELATED COMPOUNDS

Paper	P1	P1	P1	P1	P1	P1	P1	P1	P1	P1	P1
Solvent	S1	S2	S3	S4	S5	S6	S7	S8	S9	S10	S11
Literature	1	1	1	1	1	2	3	4,5,7	4	6	7
Compound	$R_F \times 100$										
I. Aromatic Acids											
p-Aminobenzoic acid	—	—	—	—	—	—	9	—	—	—	—
Anthranilic acid	—	—	—	—	—	—	26	—	—	—	—
Benzoic acid	—	—	—	—	—	—	41	—	—	—	—
Caffeic acid	—	19	36	78	53	—	—	—	—	—	—
Cinnamic acid	—	—	—	—	—	—	62	—	—	—	—
o-Coumaric acid	—	—	—	—	—	—	27	—	—	—	—
Gallic acid	dec.	0	streaks	60	dec.	43	—	68	8	—	—
Gentisic acid	68	26	53	89	34	51	—	—	—	—	—
Hippuric acid	—	—	—	—	—	—	31	—	—	—	—
Homogentisic acid	—	9	90	75	34	—	—	—	—	—	—
3-Hydroxyanthranilic acid	—	38	streaks	85	65	—	—	90	—	—	—
m-Hydroxybenzoic acid	39	53	70	88	50	—	—	—	—	—	—
o-Hydroxybenzoic (salicylic acid)	78	88	66	90	43	65	—	95	84	—	—
p-Hydroxybenzoic acid	23	55	54	87	68	—	—	—	—	—	—
m-Hydroxycinnamic acid	44	58	39	93	60	—	—	—	—	—	—
o-Hydroxycinnamic acid	35	70	44	93	75	—	—	—	—	—	—
p-Hydroxycinnamic acid	28	58	31	93	67	—	—	—	—	—	
m-Hydroxyhippuric acid	33	9	75	78	19	—	—	—	—	—	—
o-Hydroxyhippuric acid	22	43	68	86	30	—	—	—	—	—	—
p-Hydroxyhippuric acid	15	7	72	78	18	—	—	—	—	—	—
5-Hydroxyindoleacetic acid	32	16	58	76	45	—	—	—	—	—	—
m-Hydroxymandelic acid	37	7	87	68	19	—	—	—	—	—	—
o-Hydroxymandelic acid	58	13	82	54	30	—	—	—	—	—	—
	(68)	—	(88)	(68)	—	—	—	—	—	—	—
p-Hydroxymandelic acid	33	7	87	68	16	—	—	—	—	—	—
m-Hydroxyphenylacetic acid	46	49	83	92	46	—	—	—	—	—	—
o-Hydroxyphenylacetic acid	76	57	86	91	49	—	—	—	—	—	—
p-Hydroxyphenylacetic acid	42	49	82	92	50	—	—	—	—	—	—
m-Hydroxyphenyllactic acid	51	12	86	82	21	—	—	—	—	—	—
o-Hydroxyphenyllactic acid	59	29	84	85	28	—	—	—	—	—	—
p-Hydroxyphenyllactic acid	45	13	85	82	19	—	—	—	—	—	—
m-Hydroxyphenylpropionic acid	54	61	77	92	68	—	—	—	—	—	—
o-Hydroxyphenylpropionic acid	70	74	77	91	72	—	—	—	—	—	—
p-Hydroxyphenylpropionic acid	51	61	73	93	69	—	—	—	—	—	—
m-Hydroxyphenylpyruvic acid	dec.	13	streaks	74	dec.	—	—	—	—	—	—
o-Hydroxyphenylpyruvic acid	dec.	21	85	70	43	—	—	—	—	—	—

TABLE PC 82—(Continued)
CARBOXYLIC ACIDS, PHENOLS, AND RELATED COMPOUNDS

Paper	P1	P1	P1	P1	P1	P1	P1	P1	P1	P1	P1
Solvent	S1	S2	S3	S4	S4	S6	S7	S8	S9	S10	S11
Literature	1	1	1	1	1	2	3	4,5,7,	4	6	7
Compound	$R_F \times 100$										
o-Hydroxyphenylpyruvic lactone	54	88	44	85	87	—	—	—	—	—	—
Kynurenic acid	52	16	37	55	21	—	—	64	—	—	—
Mandelic acid	—	—	—	—	—	—	28	—	—	—	—
3-Methoxy-4-hydroxybenzoic (vanillic acid)	22	80	66	89	60	—	—	92	81	—	—
3-Methoxy-4-hydroxycinnamic acid	27	80	35	88	63	—	—	—	—	—	—
3-Methoxy-4-hydroxyhippuric acid	16	18	66	74	16	—	—	—	—	—	—
3-Methoxy-4-hydroxymandelic acid	28	16	84	71	20	—	—	—	—	—	—
3-Methoxy-4-hydroxyphenylacetic acid	39	66	82	88	streaks	—	—	—	—	—	—
3-Methoxy-4-hydroxyphenylpropionic acid	43	81	74	91	65	—	—	—	—	—	—
Phenylacetic acid	—	—	—	—	—	—	44	—	—	—	—
Phenylpropionic acid	—	—	—	—	—	—	57	—	—	—	—
Protocatechuic acid	6	16	73	83	streaks	—	—	85	35	—	—
Pyrogallol carboxylic acid	—	—	—	—	—	71	—	—	—	—	—
Quinolinic acid	—	—	—	—	—	—	—	—	50	—	—
α-Resorcylic acid	24	9	64	78	39	—	—	—	—	—	—
β-Resorcylic acid	39	38	48	92	34	55	—	93	54	—	—
γ-Resorcylic acid	77	11	50	54	48	—	—	—	—	—	—
Sulfanilic acid	—	—	—	—	—	—	11	—	—	—	—
Syringic acid	18	79	57	87	dec.	—	—	—	—	—	—
Tannic acid	—	—	—	—	—	99	—	—	—	—	—
Xanthurenic acid	8	3	24	53	23	—	—	69	—	—	—
II. Phenols											
Catechol	—	—	—	—	—	96	—	91	74	—	—
m-Cresol	—	—	—	—	—	—	—	—	—	94	—
o-Cresol	—	—	—	—	—	—	—	—	—	93	—
p-Cresol	—	—	—	—	—	—	—	—	—	92	—
p-Cresolmonomethylol	—	—	—	—	—	—	—	—	—	89	—
2,4-Dimethylol-*o*-cresol	—	—	—	—	—	—	—	—	—	79	—
2,6-Dimethylol-*p*-cresol	—	—	—	—	—	—	—	—	—	82	—
2,4-Dimethylolphenol	—	—	—	—	—	—	—	—	—	61	—
2,6-Dimethylolphenol	—	—	—	—	—	—	—	—	—	67	—
Hydroquinone	—	—	—	—	—	96	—	88	69	—	—
m-Hydroxybenzyl alcohol	—	—	—	—	—	—	—	—	—	81	—
o-Hydroxybenzyl alcohol	—	—	—	—	—	—	—	—	—	84	—
p-Hydroxybenzyl alcohol	—	—	—	—	—	—	—	—	—	79	—
Hydroxyquinol	—	—	—	—	—	48	—	—	—	—	—
Orcinol	—	—	—	—	—	—	—	91	75	—	—

TABLE PC 82—(Continued)
CARBOXYLIC ACIDS, PHENOLS, AND RELATED COMPOUNDS

Paper	P1	P1	P1	P1	P1	P1	P1	P1	P1	P1	P1
Solvent	S1	S2	S3	S4	S5	S6	S7	S8	S9	S10	S11
Literature	1	1	1	1	1	2	3	4,5,7	4	6	7
Compound						$R_F \times 100$					
Phenol	—	—	—	—	—	97	—	—	—	93	—
Phloroglucinol	—	—	—	—	—	96	—	76	16	—	—
Pyrogallol	—	—	—	—	—	94	—	77	34	—	—
Resorcinol	—	—	—	—	—	97	—	91	63	—	—
2,4,6-Trimethylolphenol	—	—	—	—	—	—	—	—	—	34	—
III. Conjugated Phenols											
m-Aminophenol	—	—	—	—	—	—	—	75	—	—	90
p-Aminophenol	—	—	—	—	—	—	—	64	—	—	81
m-Aminophenyl glucoside	—	—	—	—	—	—	—	33	—	—	23
m-Aminophenyl sulfuric acid	—	—	—	—	—	—	—	33	—	—	35
p-Aminophenyl sulfuric acid	—	—	—	—	—	—	—	33	—	—	10
7-Coumarinylglucoside	—	—	—	—	—	—	—	67	—	—	46
7-Coumarinyl sulfuric acid	—	—	—	—	—	—	—	53	—	—	78
p-(2,4-Dihydroxyphenyl)-azophenol	—	—	—	—	—	—	—	90	—	—	25
p-(2,4-Dihydroxyphenyl)-azophenyl sulfuric acid	—	—	—	—	—	—	—	78	—	—	11
7-Hydroxycoumarin (umbelliferone)	—	—	—	—	—	—	—	95	—	—	64
7-Hydroxy-4-methylcoumarin	—	—	—	—	—	—	—	90	—	—	77
m-Hydroxyphenylsulfamic acid	—	—	—	—	—	—	—	22	—	—	15
3-Hydroxyquinoline	—	—	—	—	—	—	—	88	—	—	70
6-Hydroxyquinoline	—	—	—	—	—	—	—	77	—	—	90
7-Hydroxyquinoline	—	—	—	—	—	—	—	74	—	—	90
8-Hydroxyquinoline	—	—	—	—	—	—	—	70	—	—	90
4-Methylcoumarin-7-yl sulfuric acid	—	—	—	—	—	—	—	66	—	—	90
8-Quinolyl glucoside	—	—	—	—	—	—	—	55	—	—	50
3-Quinolyl sulfuric acid	—	—	—	—	—	—	—	74	—	—	82
6-Quinolyl sulfuric acid	—	—	—	—	—	—	—	54	—	—	62
7-Quinolyl sulfuric acid	—	—	—	—	—	—	—	58	—	—	62
8-Quinolyl sulfuric acid	—	—	—	—	—	—	—	56	—	—	61

TABLE PC 82—(Continued)
CARBOXYLIC ACIDS, PHENOLS, AND RELATED COMPOUNDS

Paper P1 = W1 or W4.

Solvent S1 = *iso* propanol-conc. NH_4OH-H_2O (8 : 1 : 1 v/v).
S2 = benzene-propionic acid-water (2 : 2 : 1 v/v).
S3 = 20% KCl.
S4 = *n*-butanol-acetic acid-water (8 : 2 : 2 v/v).
S5 = *n*-butanol-pyridine-dioxane-water (70 : 20 : 5 : 5 v/v).
S6 = *n*-butanol-pyridine-saturated NaCl (1 : 1 : 2 v/v).
S7 = benzene-acetic acid-water (2 : 2 : 1 v/v).
S8 = *n*-butanol-acetic acid-water (4 : 1 : 5 v/v).
S9 = *m*-cresol-acetic acid-water (50 : 2 : 48 v/v).
S10 = *n*-butanol-NH_4OH (4 : 1 v/v).
S11 = ethyl methyl ketone-2*N* NH_4OH (2 : 1 v/v).

Detection See Section II.I, Volume Two.

Technique Ascending or descending development.

Literature

1. Armstrong, M. D., Shaw, K. N. F., and P. E. Wall, *J. Biol. Chem.*, **218**, 293 (1956).
2. Evans, R. A., Parr, W. H., and W. C. Evans, *Nature*, **164**, 674 (1949).
3. Robinson, D., Smith, J. N., and R. T. Williams, *Biochem. J.*, **50**, 221 (1951).
4. Bate-Smith, E. C., *Biochem. Soc. Symp.* (Cambridge, England), **3**, 62 (1949).
5. Gaffney, G. W., Schreier, K., DiFerrante, N., and K. I. Altman, *J. Biol. Chem.*, **206**, 695 (1954).
6. Freeman, J. H., *Anal. Chem.* **24**, 955 and 2001 (1952).
7. Smith, J. N., *Biochem. J.*, **60**, 436 (1955).

TABLE PC 83

CARBOXYLIC ACIDS, MISCELLANEOUS

Paper Solvent Detection Literature	P1 S1 — 1	P1 S2 — 1	P1 S3 — 1	P1 S4 — 1	P1 S5 — 1	P1 S6 — 1	D1	D2	D3	D4	D5	D6	D7	D8	D9	D10	D11	D12	D13
Compound	$R_F \times 100$						Color[a]												
4-Chlorocinnamic acid	94	19	93	90	88	82	—	—	—	—	—	—	—	—	—	+	+	—	—
4-Bromomandelic acid	93	42	91	75	67	12	—	—	—	—	—	—	—	—	—	+	+	—	—
3-Methyl-3-phenylglutaric acid	88	6	90	79	61	18	—	—	—	—	—	—	—	—	—	—	+	—	—
3-Pyridylacetic acid	34	1	7	34	0	0	—	—	—	—	—	—	—	—	—	—	+	—	—
4,5-Imidazole-dicarboxylic acid	0	3	0	0	0	0	—	—	—	—	+	—	—	—	—	—	+	—	—
Imidazole-4-acetic acid	5	0	0	0	0	0	34	15,60	62	64	8,65	—	—	—	—	—	+	—	—
1-Methylimidazole-4-acetic acid	6	0	0	0	0	0	—	—	+	56	+	—	—	—	—	—	+	—	—
Dihydrourocanic acid	8	0	0	0	0	0	—	15	60	65	7	—	—	—	—	—	+	—	—
Pyrrole-2-carboxylic acid	87	73	82	44	34	9	—	7,11	64	23	12,64	—	—	62	—	+	+	22,23	39,69
3-Indolylacetic acid methyl ester	93	94	84	88	87	84	+	—	60	60	8	—	—	—	—	++	—	18,69	24,25
3-Indolylglyoxylic acid	88	8	86	18	24	5	—	—	—	56	—	+	3	64	3,72	—	++	—	—
3-Indolylpyruvic acid	91	7	0	23	28	4	3	58	62	63	+	63	24	24	69	+	+	17	23,65

[a] For interpretation of readings, see Table PC 120.

Paper P1 = W1.

Solvent S1 = ethyl methyl ketone-acetone-formic acid-water (40 : 2 : 1 : 6 parts by volume).
S2 = ethyl methyl ketone-diethylamine-water (921 : 2 : 77 parts by volume).
S3 = methyl isobutyl ketone-formic acid-water (10 parts ketone saturated with 1 part 4% formic acid).
S4 = chloroform-methanol-formic acid-water (10 parts of chloroform saturated with a mixture of 1 part methanol and 1 part 4% formic acid).
S5 = benzene-ethyl methyl ketone-formic acid-water (a mixture of 9 parts benzene and 1 part ketone saturated with 1 part 2% formic acid).
S6 = benzene-formic acid-water (10 parts benzene saturated with 1 part 2% formic acid).

Detection D1 = ultraviolet light.
D2 = diazotized sulfanilic acid (0.3% solution in dioxane-water 1 : 2).
D3 = diazotized 4-benzoylamino-2,5-dimethoxyaniline (0.6% solution in dioxane-water 1 : 2).
D4 = diazotized *o*-dianisidine (0.6% solution in dioxane-water 1 : 2).
D5 = *p*-nitrobenzenediazonium fluoborate (0.4% solution in dioxane-water 1 : 2).
D6 = 2,6-dibromoquinone-4-chloroimide (0.5% solution in dioxane-acetone 4 : 1).
D7 = 2,4-dinitrophenylhydrazine (*ca.* 0.1% solution in 1*N* HCl).
D8 = ferric chloride (2% aqueous solution).
D9 = phosphomolybdic acid (2% aqueous solution).
D10 = potassium permanganate (1% aqueous solution).
D11 = bromophenol blue (*ca.* 0.05% solution in ethanol).
D12 = Ehrlich reagent (1% *p*-dimethylaminobenzaldehyde in 1*N* HCl).
D13 = *p*-dimethylaminocinnamaldehyde (0.1% solution in 1*N* HCl).

Technique Descending development, 40 cm, 4–5 hours.

Literature
1. Reio, L., *J. Chromatogr.*, **13**, 475–496 (1964).

TABLE PC 84
DRUGS AND MISCELLANEOUS COMPOUNDS

Paper Solvent Detection Literature	P1 S1 — 1	P1 S2 — 1	P1 S3 — 1	P1 S4 — 1	P1 S5 — 1	P1 S6 — 1	D1	D2	D3	D4	D5	D6	D7	D8	D9	D10	D11	D12	D13
Compound	$R_F \times 100$						Color[a]												
Acridine	76	90	70	95	79	73	7	58	—	—	+	—	—	3	3 71	—	—	58	—
Aminophenylpyridone[b]	68	72	42	93	20	9	—	—	—	8	8	62	—	58	—	+	—	8	20
2-Amino-4-picoline	28	77	4	23	0	0	34	—	—	62	—	—	—	57	8 72	—	33	—	22
2-Amino-6-picoline	38	80	4	22	0	0	34	6	58	62	9	+	—	58	9 72	—	33	+	17
2-Aminopyridine[b,c]	27	78	2	8	0	0	34	—	—	—	—	70	—	—	—	—	33	5	22
3-Aminopyridine[b]	24	69	2	15	0	0	—	—	—	—	—	—	—	—	44	—	33	7	22
4-Aminopyridine	17	65	0	1	0	0	—	—	—	60	59	—	—	59	7 22	—	33	6	17 23
Aminopyrine (pyramidon)[b,c]	55	93	35	95	35	31	+	—	—	—	—	—	—	26	68	++	—	+	17
Atebrin (quinacrine)	47	92	1	7	0	0	3	—	—	—	64	—	—	—	—	+	—	—	60
Bilirubin	95	3	0	94	0	0	58	64	41 25	44 68	—	48	—	—	71	+	—	+	7
Biliverdin	95	0	0	95	0	0	34	—	—	—	—	71	—	—	71	—	—	—	—
Chlorpromazine	81	64	42	88	22	15	39	—	—	—	60	52	—	17	18 49	++	—	—	—
***N'*,*N'*-Diethylsulfanilamide**[c]	95	92	90	96	91	81	—	—	—	—	—	60	—	—	—	—	—	8	65
Dromoran	70	91	17	29	46	0	—	9	—	—	—	—	—	—	71	+	—	—	—
2-Ethoxy-6,9 diaminoacridine (rivanol)	79	42	3	0	0	0	48	9	64	63	62	—	—	—	—	++	—	12	65
Folic acid	10	0	0	0	0	0	44	—	—	60	—	—	—	—	—	+	+	—	17
7-Hydroxycoumarin	92	90	87	74	63	11	39	11	18 23	23 63	59	—	—	—	44 71	++	—	—	—
Isoniazid (isonicotinoylhydrazine)	64	63	60	23	14	0	5	—	59	65	3 64	55	—	60	45 68	++	33	58	64
Levomepromazine-5-sulfoxide[b]	68	30	3	58	0	0	—	—	—	—	—	—	—	57	+	+	—	29	27 70
Methopromazine[b]	77	95	62	92	39	24	—	—	—	—	+	70	+	24 25	28 68	++	33	26 23	23 25

TABLE PC 84—(Continued)
DRUGS AND MISCELLANEOUS COMPOUNDS

Paper	P1	P1	P1	P1	P1	P1													
Solvent	S1	S2	S3	S4	S5	S6													
Detection	—	—	—	—	—	—	D1	D2	D3	D4	D5	D6	D7	D8	D9	D10	D11	D12	D13
Literature	1	1	1	1	1	1													
Compound	$R_F \times 100$						Color[a]												
Perphenazine[b]	62	87	10	83	4	2	18	—	—	—	—	+	+	17	62 / 39	+	33	—	—
Phenothiazine	95	94	91	92	93	89	34	—	—	—	—	45	+	62	38	+	—	17 / 40	17 / 23
Phenothiazine-5-oxide	88	80	73	92	53	33	—	—	—	—	—	—	42	41	41 / 30	+	—	59 / 39	57 / 56
Sulfanilamide[c]	84	82	67	8	4	0	—	—	—	64	6	+	—	—	—	+	—	9	19

[a] For interpretation, see Table PC 120.
[b] Dragendorff reagent, positive.
[c] *p*-Dimethylaminobenzaldehyde, pale yellow.

Paper P1 = W1.

Solvent
S1 = ethyl methyl ketone-acetone-formic acid-water (40 : 2 : 1 : 6 parts by volume).
S2 = ethyl methyl ketone-diethylamine-water (921 : 2 : 77 parts by volume).
S3 = methyl isobutyl ketone-formic acid-water (10 parts ketone saturated with 1 part 4% formic acid).
S4 = chloroform-methanol-formic acid-water (10 parts of chloroform saturated with a mixture of 1 part methanol and 1 part 4% formic acid).
S5 = benzene-ethyl methyl ketone-formic acid-water (a mixture of 9 parts benzene and 1 part ketone saturated with 1 part 2% formic acid).
S6 = benzene-formic acid-water (10 parts benzene saturated with 1 part 2% formic acid).

Detection
D1 = ultraviolet light.
D2 = diazotized sulfanilic acid (0.3% solution in dioxane-water 1 : 2).
D3 = diazotized 4-benzoylamino-2,5-dimethoxyaniline (0.6% solution in dioxane-water 1 : 2).
D4 = diazotized *o*-dianisidine (0.6% solution in dioxane-water 1 : 2).
D5 = *p*-nitrobenzenediazonium fluoborate (0.4% solution in dioxane-water 1 : 2).
D6 = 2,6-dibromoquinone-4-chloroimide (0.5% solution in dioxane-acetone 4 : 1).
D7 = 2,4-dinitrophenylhydrazine (*ca.* 0.1% solution in 1*N* HCl).
D8 = ferric chloride (2% aqueous solution).
D9 = phosphomolybdic acid (2% aqueous solution).
D10 = potassium permanganate (1% aqueous solution).
D11 = bromophenol blue (*ca.* 0.05% solution in ethanol).
D12 = Ehrlich reagent (1% *p*-dimethylaminobenzaldehyde in 1*N* HCl).
D13 = *p*-dimethylaminocinnamaldehyde (0.1% solution in 1*N* HCl).

Technique Descending development, 40 cm.

Literature
1. Reio, L., *J. Chromatogr.*, **47**, 60–85 (1970) and *J. Chromatogr.*, **13**, 475–496 (1964).

TABLE PC 85
DRUGS, ANTIHISTAMINES AND TRANQUILIZERS

Paper	P1	P1	P2	P2	P3							
Solvent	S1	S2	S3	S4	S4							
Detection	—	—	—	—	—	D1	D2	D3	D4	D5	D6[a]	D7
Literature	1	1	1	1	1							
Compound	$R_F \times 100$					Color						
Acetylpromazine maleate	92	65	—	45	22	—	bn-y	b/bk	y	g	—	—
Antazoline	92	84	72	70	25	—	bn-y	b to p	y	b	—	b
Benactyzine	88	86	—	78	42	—	bn-y	p	y	ft. b	−b	—
Benzhexol	95	76	85	88	77	—	bn-y	p	ft. y	ft. b	−b	—
Captodiame	95	0	—	67	77	—	bn	y	y	b	—	—
Chlorpromazine	95	52	60	53	40	b	bn	b/bk	g	g-b	bn to b/bk	Violet
Chlorcyclizine	91	87	73	60	62	—	y	b/bk to p	y	ft. b	−b	—
Diphenhydramine	88	85	65	45	34	—	y	r-p	ft. y	b	−b	—
Hydrallazine	—	82	—	3	57	—	ft. y	p-yg	—	ft. g	—	—
Hydroxyzine	90	89	—	56	52	—	y	b	g-y	b	—	—
Meclozine	97	75	92	86	95	—	y	b-rp	g-y	g	—	—
Mepyramine maleate	87	92	40	24	40	b	bn-y	b/bk	ft. y	b	—	—
Phenindamine tartrate	95	70	68	44	42	—	y	p	y	ft. b	—	—
Prochlorperazine	96	54	—	17	0	—	ft. y	ft. p	y	ft. b	—	—
Promazine	90	66	60	41	38	—	bn-y	b/bk	y	b	−b	r
Perphenazine	88	64	22	12	55	gb	bn	b/bk	gy	b	b	—
Pecazine	94	62	67	37	37	—	bn-y	b/bk	bn	b	−b	—
Promethazine-8-chlorotheophyllinate	93	71	66	46	47	—	bn-g	b/bk to p	gy	b	−b	—
Triflorperazine	94	—	—	—	55	y	y	b/bk	gy	gb	−b	—
Triflorpromazine	93	—	80	—	62	y	y	b/bk	y	gb	−b	—
Thiopropazate	—	63	—	37	—	—	bn-y	ft. p	bn	gb	−b	—

[a] A negative sign before the color indicates reaction after exposure to NH_3 vapor.

Paper P1 = W1.
P2 = W1 buffered with 5% sodium dihydrogen citrate.
P3 = W1 buffered with 1% potassium dihydrogen phosphate.

Solvent S1 = *n*-butanol-glacial acetic acid-water (120 : 30 : 50).
S2 = *n*-butanol-1*N* sodium acetate-1*N* HCl (7 : 120 : 60).
S3 = *n*-butanol-5% w/v citric acid (180 : 20).
S4 = *iso*-amyl alcohol-5% w/v citric acid (180 : 20).

Detection D1 = UV light.
D2 = Iodine.
D3 = Iodoplatinate.
D4 = $KMnO_4$.
D5 = Cobalt thiocyanate.
D6 = Bromocresol green.
D7 = Ehrlich.

Technique Ascending; solvent S2 can be used for short descending runs.

Literature
1. Jackson, J. V. and M. S. Moss in *Chromatographic and Electrophoretic Techniques*, Vol. 1, Smith, I., Ed., Interscience Publishers, Division John Wiley & Sons, Inc., New York, N.Y., 1969, p 522.

TABLE PC 86
DRUGS, BARBITURATES

Paper	P1	P1	P1	P1	P1
Solvent	S1	S2	S3	S4	S5
Literature	1	1	1	1	1
Compound	$R_{pentothal} \times 100$				
Allobarbitone	83	72	58	48	51
Amylobarbitone	96	95	93	95	15
Barbitone	75	65	50	25	70
Butobarbitone	85	76	70	80	27
Cyclobarbitone	90	81	66	68	20
Hexobarbitone	91	89	80	105	14
Methylphenobarbitone	76	88	71	96	10
Pentobarbitone	99	94	89	97	16
Phenobarbitone	77	62	33	2	48
Quinalbarbitone	100	98	98	95	10
Thialbarbitone	97	96	93	97	5
Thiopentone	100	100	100	100	3
Aprobarbitone	—	—	—	—	24
Heptabarbitone	—	—	—	—	18
Methobarbitone	—	—	—	—	3
Nealbarbitone	—	—	—	—	16
Secbutobarbitone	—	—	—	—	29
Talbutal	—	—	—	—	24

Paper P1 = W1.
Solvent S1 = *n*-butanol-water-concentrated NH_4OH (340 : 57 : 3).
S2 = *n*-butanol-concentrated NH_4OH (150 : 50).
S3 = amyl alcohol-concentrated NH_4OH (180 : 20).
S4 = solvent tray : chloroform; tank : concentrated NH_4OH.
S5 = phosphate buffer, pH 7.4 (1.5 g KH_2PO_4, 7.9 g Na_2HPO_4, water to one liter).
Technique Ascending.
Literature
1. Jackson, J. V., in *Chromatographic and Electrophoretic Techniques*, Vol. I, Smith, I., Ed., Interscience Publishers, Division of John Wiley & Sons, Inc., New York, N.Y., 1969, pp 500–502.

TABLE PC 87
DRUGS, SULFA

Paper	na	na	P1
Solvent	S1	S2	S3
Literature	1	1	2
Compound	$R_F \times 100$		
Sodium sulfadiazine	—	13	—
Sodium sulfamerazine	—	20	—
Sodium sulfamethazine	—	30	—
Sodium sulfapyridine	—	48	—
Sodium sulfathiazole	—	34	—
Sulfacetamide	71	44	—
Sulfadiazine	24	50	24
Sulfaguanidine	43	36	30
Sulfamerazine	34	50	42–46
Sulfamethazine	48	48	65
Sulfanilamide	56	42	55
Sulfapyridine	70	51	65
Sulfathiazole	48	46	47

Paper P1 = W2.
Solvent S1 = organic phase-*n*-butanol-acetic acid-H_2O (50 : 15 : 60 v/v); *p*-dimethylaminobenzaldehyde is added to a final concentration of 0.5%.
S2 = organic phase-*n*-butanol-NH_4OH-H_2O (4 : 1 : 3 v/v).
S3 = *n*-butanol saturated with water-pyridine (8 : 1 v/v).
Detection Modified Ehrlich's reagent.
Technique One- or two-dimensional on strips; ascending.
Literature
1. Robinson, R., *Nature*, **168**, 512–513 (1951).
2. deReeder, P. L., *Anal. Chim. Acta*, **8**, 325–326 (1953).

TABLE PC 88
DYES, INDICATOR

Paper	P1	P1	P1	na	na	P2	P3
Solvent	S1	S2	S3	S4	S5	S6	S7
Technique	T1	T1	T1	na	T2	T3	T2
Literature	1	1	1	2	3	4	3
Compound	$R_F \times 100$						
Bromcresol green	84	47	24	—	32	79	—
Bromcresol purple	68	43	10	58	19	76	—
Bromphenol blue	—	55	19	57	27	74	—
Bromphenol red	—	—	—	—	11	68	—
Bromthymol blue	93	79	63	37	50	88	—
Chlorophenol red	—	17	1	—	9	67	—
Congo red	—	0	0	—	—	—	—
Cresol red	—	41	12	—	12	67	—
Dibromofluorescein	—	—	—	—	—	—	26(24)
Fluorescein	—	—	—	—	—	—	18
Indigo carmine	0	0	0	32	—	—	—
Metacresol purple	—	—	—	—	16	66	—
Methyl orange	77	55	26	—	—	—	—
Methyl red	73	59	33	41	—	—	—
Methyl violet	95	88	86	1	—	—	—
Neutral red	—	66	53	—	—	—	—
Phenolphthalein	100	92	89	10	—	—	—
Phenol red	—	18	1	73	6	54	—
Tetrabromofluorescein	—	—	—	—	—	—	45(34)
Tetrachlorotetrabromofluorescein	—	—	—	—	—	—	64
Tetraiodofluorescein	—	—	—	—	—	—	52
Thymol blue	100	90	75	37	70	85	—
Thymolphthalein	100	92	92	—	—	—	—

Paper
P1 = W2.
P2 = W3MM equilibrated overnight with solvent mixture.
P3 = W1.

Solvent
S1 = isopropanol (NH_3 atmosphere).
S2 = *n*-butanol (NH_3 atmosphere).
S3 = *n*-amyl alcohol (NH_3 atmosphere).
S4 = 3% aqueous NaCl.
S5 = *tert*-amyl alcohol-NH_4OH (20 : 5).
S6 = *tert*-amyl alcohol-acetic acid-water (20 : 5 : 5).
S7 = *n*-butanol-water-NH_4OH-ethanol (200 : 88 : 2 : 40).

Detection
Visual.

Technique
T1 = strips in cylinder, ascending saturated tank.
T2 = one- or two-dimensional; S5 used in 2nd dimension ascending.
T3 = descending, 26 hours.

Literature
1. Lederer, M., *Science*, **112**, 504–505 (1950).
2. Taira, T., Yamatodani S., and S. Fujii, *J. Agr. Chem. Soc. Jap.* **25**, 121–123, 1953–52 from *Chem. Abstr.*, **47**, 6302 (1953).
3. Graichen, C., *J. Ass. Offic. Agr. Chem.*, **34**, 795–799 (1951).
4. Franglen, G. T., *Nature*, **175**, 134 (1955).

TABLE PC 89
HYDROCARBONS, AROMATIC

Paper	P1	P1	P1	P1
Solvent	S1	S2	S3	S4
Literature	1	1	1	1
Compound	$R_F \times 100$			
Anthanthrene	31	16	30	17
Anthracene	30	21	22	16
2,3-Benzofluorene	17	6	5	0
3,4-Benzopyrene	26	14	23	15
Chrysene	0	0	0	0
1,2 : 5,6-Dibenzanthracene	12	5	12	15
1,2 : 3,4-Dibenzopyrene	19	9	17	11
1,2 : 4,5-Dibenzopyrene	20	11	20	12
1,2 : 6,7-Dibenzopyrene	0	0	0	0
3,4 : 9,10-Dibenzopyrene	17	7	14	9
Fluoranthene	38	26	36	24
20-Methylcholanthrene	11	8	10	6
Pyrene	45	31	37	28

Paper P1 = Impregnated paper.
Solvent S1 = DHPTBr[a] 0.50*M*; Di-*n*-butyl ether, stationary phase.
S2 = TMUA[a] 0.05*M*; Di-*n*-butyl ether, stationary phase.
S3 = DHTPBr 0.50*M*; Decahydronaphthalene, stationary phase.
S4 = TMUA 0.05*M*; Decahydronaphthalene, stationary phase.
Detection UV light.
Technique Ascending, 16–18 hours, nitrogen atmosphere.
Literature 1. Stuchlik, M., Krasnec, L., and L. Csiba, *J. Chromatogr.*, **30**, 543 (1967).

[a] DHPTBr = 1-(2,3-dihydroxypropyl)-theobromine.
TMUA = 1,3,7,9-tetramethyluric acid.

TABLE PC 90
HYDROCARBONS, AROMATIC

Paper	P1	P1
Solvent	S1	S2
Literature	1	1
Compound	$R_F{}^a \times 100$	
1,2-Benzanthracene	265	444
Benzo(c)phenanthrene	296	586
1′-Methyl-1,2-benzanthracene	204	250
2′-Methyl-1,2-benzanthracene	298	525
3-Methyl-1,2-benzanthracene	330	590
3′-Methyl-1,2-benzanthracene	320	570
4-Methyl-1,2-benzanthracene	282	500
4′-Methyl-1,2-benzanthracene	250	334
5-Methyl-1,2-benzanthracene	274	380
6-Methyl-1,2-benzanthracene	300	452
7-Methyl-1,2-benzanthracene	345	595
9-Methyl-1,2-benzanthracene	193	294
10-Methyl-1,2-benzanthracene	280	364
1-Methylbenzo(c)phenanthrene	264	520
2-Methylbenzo(c)phenanthrene	338	695
3-Methylbenzo(c)phenanthrene	290	576
4-Methylbenzo(c)phenanthrene	240	347
5-Methylbenzo(c)phenanthrene	325	584
6-Methylbenzo(c)phenanthrene	366	744

[a] Relative to 3,4-benzopyrene.

Paper P1 = Acetylated W1.
Solvent S1 = ethanol-toluene-water (17 : 4 : 1 v/v).
S2 = methanol-ether-water (4 : 4 : 1 v/v).
Detection UV light, tetrachlorophthalic anhydride.
Technique Descending development.
Literature 1. Spotswood, T. M., *J. Chromatogr.*, **3**, 101 (1960).

TABLE PC 91
INORGANICS, ANIONS

Paper	P1	P1	P1	P2
Solvent	S1	S2	S3	S3
Literature	1	1	1	1
Anion	$R_F \times 100$			
Chloride	22	51	66	100
Bromide	10	33	43	100
Iodide	3	7	12	84
Thiocyanate	2	6	10	81
Chlorate	12	34	49	100
Bromate	28	60	75	100
Iodate	60	87	88	100
Periodate	0	5	7	100
Nitrite	21	49	63	100
Selenite	64	86	89	86
Selenate	19	80	90	100
Tellurite	0–25	0–25	3–27	34–86
Thiosulfate	8	43	83	100
Chromate	3	6	0–17	100
Molybdate	0–11	0–22	18	72–100
Vanadate	0	0	5	4–100
Arsenite	75	78	79	80
Arsenate	61	84	88	91
Orthophosphate	66	88	92	100
Ferrocyanide	0	3	43	100
Ferricyanide	0	0	0	100

Paper P1 = Amberlite SB-2 stongly basic anion exchange paper.
P2 = W1.
Solvent S1 = 0.1*N* KNO_3.
S2 = 0.5*N* KNO_3.
S3 = 1.0*N* KNO_3.
Technique Ascending.
Literature 1. Ossicini, L., *J. Chromatogr.*, **9**, 114 (1962).

TABLE PC 92
INORGANICS, CATIONS

Paper	P1	P1	P1
Solvent	S1	S2	S3
Literature	1	1	1
Cation	$R_F \times 100$		
Ag^+	50	36	17
Tl^+	48	46	25
Bi^{3+}	88	60	5
Cd^{2+}	72	44	8
Pb^{2+}	53	27	4
Hg^{2+}	78	56	9
Cu^{2+}	63	34	4
Fe^{3+}	elongated comet	—	3
Co^{2+}	70	52	10
Ni^{2+}	68	52	10
Mn^{2+}	75	52	13
Zn^{2+}	72	45	9
Al^{3+}	78	36	4
Sc^{3+}	75	44	4
Y^{3+}	79	50	4
La^{3+}	66	32	2
Ce^{3+}	69	29	3
Zr^{4+}	comet → 100	comet → 39	comet → 8
Be^{2+}	85	71	12
Mg^{2+}	84	62	17
Ca^{2+}	65	45	8
Sr^{2+}	62	48	11
Ba^{2+}	52	35	5
$Co(NH_3)_6^{3+}$	55	27	2
$Co(en)_3^{3+}$	66	29	3
ReO_4^-	72	76	94

Paper P1 = W CM-50 carboxymethyl-cellulose ion-exchange paper.
Solvent S1 = 0.1*N* sodium Acetate—1*N* acetic acid.
S2 = 0.5*N* sodium Acetate—0.5*N* acetic acid.
S3 = 0.1*N* sodium Acetate—0.1*N* acetic acid.
Technique Ascending.
Literature 1. Lederer, M., *J. Chromatogr.*, **29**, 306 (1967).

TABLE PC 93
INORGANICS, ANIONS

Paper	P1	na	na	na
Solvent	S1	S2	S3	S4
Technique	T1	na	T2	na
Literature	1	2	3	4

Anion	$R_F \times 100$			
Arsenate	—	5	0	—
Arsenite	—	19	21	—
Bromate	—	25	13	—
Bromide	47	36	16	—
Carbonate	—	6	—	—
Chlorate	—	42	—	98
Chloride	23	24	10	—
Chromate	—	0	0	—
Dichromate	—	—	—	17
Ferricyanide	—	—	0	91
Ferrocyanide	—	—	0	100
Fluoride	0	—	0	—
Iodate	—	9	3	—
Iodide	71	47	30	99
Molybdate	—	—	—	38
Nitrate	—	40	24	—
Nitrite	—	25	20	—
Oxalate	—	—	0	—
Periodate	—	—	0	—
Perrhenate	—	—	—	94
Phosphate	—	4	0	82
Picrate	—	—	7	—
Selenite	—	—	—	78
Sulfate	—	7	0	—
Sulfide	—	—	0	—
Tellurite	—	—	—	91
Tetraborate	—	—	—	74
Thiocyanate	—	56	45	—
Thiosulfate	—	—	0	—

Paper P1 = W1 or W3.
Solvent S1 = pyridine-10% H_2O.
S2 = butanol-pyridine-1.5*N* NH_3 (2 : 1 : 2 v/v).
S3 = butanol-1.5*N* NH_4OH.
S4 = *n*-butanol-40% HBr-water (10 : 1 : 9)—organic phase + 10 ml. 40% HBr.

TABLE PC 93—(Continued)

Detection Silver nitrate and others (see Section II).
Technique T1 = paper strip chromatography, saturated tank.
T2 = paper cylindrical chromatography, ascending; saturated tank, 25 cm in twenty-four hours.
Literature
1. Burstall, F. H., Davies, G. R., Linstead, R. P., and R. A. Wells, *J. Chem. Soc.*, **1950**, 516–528.
2. Pollard, F. H. and J. F. W. McOmie, *Endeavour* **10**, 213–221 (1951).
3. Lederer, M., *Science*, **110**, 115–116 (1949); *Austral. J. Sci.*, **11**, 174 (1949).
4. Kertes, S. and M. Lederer, *Anal. Chim. Acta.*, **15**, 543–547 (1956).

TABLE PC 94
INORGANICS, CATIONIC COMPLEXES

Paper	P1
Solvent	S1
Literature	1

Compound[a]	$R_F \times 100$	Sensitivity µg	Color
$[Co(en)_2Cl_2]Cl$	33	8	d-vt-br
$[Co(en)_3]Cl_3$	21	8	d-pk-vt
$[Cr(en)_3]_2(SO_4)_3$	16.8	8	d-vt
$[Co(NH_3)_6]Cl_3$	16.5	15	y
$[Ni(en)_3]Cl_2$	21	5	d-vt
$[Co(NH_3)_3(H_2O)]Br_3$	18.4	15	lt-y-pk
$[Co_4(en)_6(OH)_6]Cl_6$[b]	20	8	pk-br
$[Cr_4(en)_6(OH)_6]Cl_6$[b]	23	8	d-vt-pk
$CoCl_2$[c]	77	40	pk-y
$NiCl_2$[c]	78	40	lt-y
$Cr_2(SO_4)_3 \cdot K_2SO_4$	20	60	lt-du-gr-gy

[a] (en) = ethylenediamine.
[b] Streaked zones.
[c] Detected spots became more prominent on storage.

Paper P1 = Paper strips.
Solvent S1 = *n*-butanol-acetic acid-pyridine-water (30 : 6 : 24 : 20 v/v).
Detection 0.2% ninhydrin in acetone.
Technique Descending development.
Literature 1. Singh, C., *J. Chromatogr.*, **18**, 194 (1965).

TABLE PC 95
INORGANICS, METAL IONS

Paper	P1	P1	P1	P1	P1	P1	P1	P1	P1	P1	P1	P1	P1
Solvent	S1	S2	S3	S4	S5	S6	S7	S8	S9	S10	S11	S12	S13
Technique	T1	T1	T1	T1	T1	T1	T1	T2	T2	T3	T4	na	T5
Literature	1	1	1	1	1	1	1	2	2	3	4	5	6
Cation	$R_F \times 100$												
Ag(I)	2	0	0	26–57	0	18	12	—	—	10	0–68	T	0–100
Al(III)	37	4	21	0– 3	42	10	15	—	—	—	21–45	53	88–100
As(III)	50	46	83	83–88	65	45	15	13	27	—	79–90	33	—
Au(III)	—	—	—	—	—	—	—	—	—	100	38–55	—	—
Ba(II)	4	0	9	0	8	6	17	—	—	—	0–31	51	74–87
Bi(III)	94	41	76	41–60	81	20	34	30	T	—	44–62	30	88–100
Ca(II)	—	—	18	1– 4	27	12	20	—	—	—	23–38	50	86–100
Cd(II)	100	37	97	82–93	82	15	28	22	32	100	64–76	45	86–100
Co(II)	32	3	22	65–82	38	12	24	—	—	—	58–76	51	87–96
Cr(III)	47	4	22	0– 3	42	10	65	—	—	—	21–45	50	93–100
Cu(II)	47	5	31	65–76	46	15	63	6	20	74	56–68	40	82–100
Fe(II)	—	—	—	—	—	—	—	—	—	—	75–100	—	—
Fe(III)	56	5	42	89–98*	73	13	73	38	80	92	80–100	62	81–100
Hg(II)	100	6	100	—	90	50	83	74	36	—	61–76	60	—
Hg_2(II)	8	—	—	84–90*	0	50	9	—	—	—	0–72	50*	82–94
K(I)	8	6	19	0– 3	14	11	23	—	—	—	—	—	—
Mg(II)	33	7	24	1– 5	40	13	20	—	—	—	0–21	50	86–96
Mn(II)	36	6	24	15–22	48	13	25	—	—	—	46–66	42	71–100
MoO_4^{-2}	—	—	—	—	—	—	—	—	—	69	—	—	64–91
Na(I)	—	—	24	0– 6	24	11	25	—	—	—	—	—	—
Ni(II)	34	3	22	1– 6	36	12	24	—	—	—	21–39	46	81–100
Pb(II)	16	0	39	12–23	33	9	18	—	—	—	21–62	38	83–94
Pd(II)	—	—	—	—	—	—	—	—	—	57	43–60	—	75–90
Pt(IV)	—	—	—	—	—	—	—	—	—	77	—	—	—
Sb(III)	85	T	99	88–95*	81	0	13	21	68	—	58–100*	63	81–98
Sn(II)	97	55	100	86–99	84	70	13	T	83	—	68–75	60	68–87
Sn(IV)	—	—	100	0	84	65	15	—	—	—	61–78	—	71–87
Sr(II)	11	0	9	0	15	8	18	—	—	—	23–32	42	75–87
Th(IV)	—	—	—	—	—	—	—	—	—	11	16–44	T	87–100
Tl(I)	—	—	—	—	—	—	—	—	—	—	0–59	71	31–100
UO_2(II)	—	—	—	—	—	—	—	10	30	28	54–72	53	81–93
VO(II)	—	—	—	—	—	—	—	—	—	13	—	T	87–96
WO_4^{-2}	—	—	—	—	—	—	—	—	—	0	—	—	0–100
Zn(II)	93	51	98	80–93	84	12	30	—	—	—	64–75	57	81–96
Zr(IV)	—	—	—	—	—	—	—	—	—	0	0–68	60T	0

T = Trailing zone; * = Trace at origin.

TABLE PC 95—(Continued)
INORGANICS, METAL IONS

Paper
P1 = W1 (untreated paper).

Solvent
S1 = ethanol-5*M* HCl (9 : 1 v/v).
S2 = 100 ml *n*-butanol saturated by shaking with 100 ml 1*M* HCl.
S3 = 100 ml *n*-butanol saturated by shaking with 100 ml 4*M* HCl.
S4 = di-*n*-propyl ketone-10*M* HCl (85 : 15 v/v).
S5 = diethyl ether-methanol-H_2O-conc. HCl (50 : 30 : 15 : 4 v/v).
S6 = 95 ml *n*-butanol + 5 ml acetoacetic ester saturated by shaking with 100 ml 0.1*M* HNO_3.
S7 = *n*-butanol-H_2O-glacial acetic acid-acetoacetic ester (50 : 35 : 10 : 5 v/v).
S8 = methyl ethyl ketone-10% aqueous HCl (9 : 1 v/v).
S9 = cyclohexanone-30% aqueous HCl (9 : 1 v/v).
S10 = chloroform-acetone-amyl alcohol-conc. HCl (47.6 : 23.8 : 23.8 : 4.5) (radial development).
S11 = formic acid-conc. HCl-acetone (3 : 3 : 4 v/v).
S12 = cyclohexanol-formic acid (3 : 7 v/v).
S13 = HCl-H_3PO_4-H_2O (10 : 1 : 9 v/v).

Detection
See Section II.I, Volume Two.

Technique
T1 = ascending, saturated chamber.
T2 = descending.
T3 = radial.
T4 = ascending, glass jars.
T5 = ascending, glass jars, development 55 minutes at 31–32 °C.

Literature

1. Pollard, F. H. and G. Nickless, "*Handbook of Analytical Chemistry*", McGraw-Hill, N.Y., 1963, pp. 10–48.
2. Halmekoski, J. and F. Sundholm, *Suomen Kemistilehti*, B36, **63**, (1963); from *J. Chromatogr.*, **15**, D1 (1964).
3. Bhatnagur, R. P. and N. S. Poonia, *Anal. Chim. Acta.*, **30**, 211 (1964); from *J. Chromatogr.*, **15**, D2 (1964).
4. Qureshi, M. and F. Khan, *J. Chromatogr.*, **34**, 222 (1968).
5. Qureshi, M. and F. Khan, *J. Indian Chem. Soc.*, **41**, 673 (1964).
6. Qureshi, M. and F. Khan, *Anal. Chem.*, **39**, 1329 (1967).

TABLE PC 96
INORGANICS, METAL IONS

Paper	P1	P1	P1	P1	P1	P1	P2	P1	P1	P1	P3	P4	P4
Solvent	S1	S2	S3	S4	S5	S6	S7	S8	S9	S10	S11	S12	S13
Technique	na	na	na	na	na	na	T1	T2	T2	T3	T4	T5	T5
Literature	1	1	1	1	2	2	3	4	4	5	6	7	8
Metal	$R_F \times 100$												
Ag(I)	7	97	13	98	0	31	40	T	—	0– 9	17	—	—
Al(III)	98	23–87	98	52–91	97	98	27	100	100	0–20	0,52	—	—
As(III)	90	99	90	99	92	20	30	—	—	0–66	—	—	—
Au(III)	0	8	—	—	0	0	—	—	—	94–100	12[a]	10	—
Ba(II)	—	—	—	—	—	—	14	—	—	0	32	6	—
Be(II)	5–90	0–18	0–90	—	98	98	37	—	—	26–52	18,46	—	—
Bi(III)	0	9	1	23	1	0	10	95	—	45–80	2	—	T
Ca(II)	—	—	—	—	—	—	13	—	—	0– 8	49	—	—
Cd(II)	0	23	2	56	1	1	63	100	2	85–100	21	—	70
Ce(III)	—	—	—	—	—	—	64	—	—	0– 8	16	3	—
Co(II)	0	0	0	1	94	55	64	95	92	54–75	58	39	55
Cr(III)	98	98	99	99	98	98	66	100	100	0–20	—	—	—
Cu(II)	0	63	3	84	82	3	74	100	76	60–76	21	61	55
Eu(III)	—	—	—	—	—	—	—	92	—	—	—	—	—
Fe(III)	0	0	0	1	50–99	1	88	100	8	87–100	0	51	88
Ga(III)	3	1	2	2	95	0– 9	—	—	—	90–100	—	—	—
Ge(IV)	96	98	96	98	96	99	—	—	—	—	—	—	—
Hg(II)	0	12	1	22	0	0	83	10	0	0–90	5	—	—
Hg_2(II)	—	—	—	—	—	—	70	—	—	95–100	—	—	—
In(II)	2	0	2	1	10	2	23	100	2	80–92	35	4–51	—
In(III)	0	0	0	0	0	0	—	—	0	68–92	—	—	—
K(I)	—	—	—	—	—	—	5	—	—	0– 7	—	—	—
La(III)	97	98	98	98	94	95	—	100	—	0– 7	14	6	—
Li(I)	—	—	—	—	—	—	19	—	—	—	—	—	—
Mg(II)	—	—	—	—	—	—	19	—	—	0–20	79	—	—
Mn(II)	8	12	10	30	98	95	23	—	90	21–36	64	—	—
Mo(VI)	0	0	0	0	—	—	61	—	0	83–97	—	—	—
Na(I)	—	—	—	—	—	—	8	—	—	—	—	—	—
Ni(II)	79	99	80	99	98	95	58	100	96	0–20	57	39	0
Pb(II)	21	98	45	99	0	30	48	100	—	4–47	15	—	T
Pd(II)	0	9	0	10	0	0	—	100	—	50–85	51[a]	—	—
Pt(IV)	1	1	—	—	1	1	—	—	2	77–94	28[a]	—	—
Sb(III)	1–28	8	0–30	16–40	3	0	0	—	—	88–100	48[a]	—	—
Sc(III)	0	0	10	1	99	93	—	69	5	—	4	—	—
Se(IV)	91	91	91	93	88	12	—	—	—	0–16	—	—	—
Sn(II)	0	0	0	0	1	1	90	—	0	—	—	—	—
Sn(IV)	—	—	—	—	—	—	0	—	—	—	—	—	—
Sr(II)	—	—	—	—	—	—	13	—	—	0	48	16	—

TABLE PC 96—(Continued)
INORGANICS, METAL IONS

Paper	P1	P1	P1	P1	P1	P1	P1	P1	P1	P1	P3	P4	P4
Solvent	S1	S2	S3	S4	S5	S6	S7	S8	S9	S10	S11	S12	S13
Technique	na	na	na	na	na	na	T1	T2	T2	T3	T4	T5	T5
Literature	1	1	1	1	2	2	3	4	4	5	6	7	8
Metal	$R_F \times 100$												
Te(IV)	0–82	2–82	0–80	30–80	20	5	—	—	—	85–100	—	—	—
Th(IV)	96	30	98	45	95	93	59	6	0	0– 7	—	—	—
Ti(IV)	0	0	0	0	95	95	0	—	11	0–37	—	—	—
Tl(I)	64	98	66	98	0	0	26	—	—	0–22	33	—	—
U(VI)	0	0	0	1	95	13	94	5	0	68–90	70[a]	—	—
V(IV)	—	—	—	—	—	—	—	—	0	22–40	—	—	—
V(V)	3	3	—	—	—	—	76	—	—	—	—	—	5
W(VI)	0	0	0	0	6	0	—	—	—	—	—	—	—
Y(III)	98	98	98	98	—	—	—	90	79	0–100	23	—	—
Zr(IV)	1	0	2	0	94	96	—	19	2	0–10	3	—	—

[a] Solvent is 6*N* HCl. T = Trailing zone.

Paper
- P1 = W1.
- P2 = Carboxymethyl-cellulose paper.
- P3 = Amberlite WA-2 hydrogen-form carboxylic acid paper.
- P4 = Amberlite sulfonic resin paper, hydrogen-form.

Solvent
- S1 = 1*M* aqueous NH_4SCN, paper impregnated with 0.2*M* tri-*n*-butylamine in benzene.
- S2 = 7*M* aqueous NH_4SCN, paper impregnated with 0.2*M* tri-*n*-butylamine in benzene.
- S3 = 1*M* aqueous NH_4SCN, paper impregnated with 0.2*M* tri-*n*-octylamine in benzene.
- S4 = 7*M* aqueous NH_4SCN, paper impregnated with 0.2*M* tri-*n*-octylamine in benzene.
- S5 = 0.5*M* HBr, paper impregnated with 0.2*M* tri-*n*-octylamine hydrobromide in benzene.
- S6 = 5*M* HBr, paper impregnated with 0.2*M* tri-*n*-octylamine hydrobromide in benzene.
- S7 = *n*-butanol-15% aqueous trichloroacetic acid-1-phenyl-1,3-butanedione (100 : 100 : 1 v/v/w).
- S8 = 3*M* HNO_3, paper impregnated with 20% by volume tri-*n*-butyl phosphate in CCl_4.
- S9 = 0.5*M* HCl, paper impregnated with 0.1*M* bis(di-*n*-hexylphosphinyl) methane in CCl_4.
- S10 = *n*-butanol-dioxane-10*M* HCl (6 : 2 : 2 v/v), paper impregnated with stannic tungstate.
- S11 = 3*N* sodium acetate-3*N* acetic acid, or 6*N* HCl.
- S12 = acetic acid-2.4*N* HCl (30 : 70 v/v).
- S13 = acetone-6*N* HCl (9 : 1 v/v).

Detection
See Section II.I.

Technique
- T1 = descending upper phase of mixed solvent in trough, lower in bottom of chromatographic tank.
- T2 = ascending or radial development.
- T3 = ascending development 110 mm, saturation chamber.
- T4 = ascending development, spots placed 4 cm behind the liquid front to avoid double spotting.
- T5 = ascending development.

Literature

1. Przeszlakowski, S., *Chem. Anal.* (Warsaw), **12**, 60 (1967).
2. Przeszlakowski, S. and E. Soczewiuski, *Chem. Anal.* (Warsaw), **11**, 898 (1966).
3. Elbeih, I. I. M. and M. A. Abou-Elnaga, *Chemist-Analyst*, **56**, 99 (1967).
4. O'Laughlin, J. W. and C. V. Banks, *Anal. Chem.*, **36**, 1223 (1964).
5. Qureshi, M., Mathur, K. N., and A. H. Israili, *Talanta*, **16**, 503 (1969).
6. Lederer, M., *J. Chromatogr.*, **29**, 306 (1967); M. Lederer and R. Marini-Bettolo, *ibid*, **43**, 149 (1969).
7. Ossicini, L. and M. Lederer, *J. Chromatogr.*, **34**, 292 (1968) and previous papers in series.
8. Sherma, J., *Anal. Chem.*, **39**, 1497 (1967) and *J. Chromatogr.*, **35**, D53 (1968).

TABLE PC 97
INORGANICS, CATIONS

Paper	P1	P1	P1
Solvent	S1	S2	S3
Literature	1	1	1
Cation	$R_F \times 100$		
Ag^+	17–77 (comet)	8	46–59 (comet)
Hg^+	100	1	92
Pb^{2+}	44	—	48
Cu^{2+}	53	0	56
Cd^{2+}	92	7 (slight comet)	77
Bi^{3+}	75	0	85
Hg^{2+}	100	1	100
As^{3+}	92	—	65
AsO_4^{3-}	86	44	59
Sb^{3+}	88	22	82
Sn^{2+}	91	—	82
Al^{3+}	42	0	2
Cr^{3+}	38	—	27
Fe^{3+}	79	0	86
Zn^{2+}	93	0–24	90
Mn^{2+}	48	19	25
Co^{2+}	45	0	45
Ni^{2+}	52	0	3
Ca^{2+}	40 (comet)	—	0
Sr^{2+}	30	—	0
Ba^{2+}	22	—	0
Mg^{2+}	47	—	3
Li^+	46	—	4
K^+	41	—	5
Rb^+	43	—	5
Cs^+	46	—	3
Th^{4+}	0	—	0
UO_2^{2+}	46	8	64
Tl^+	41 (comet)	4	100
Tl^{3+}	41–100	0	100
MoO_4^{2-}	73	41	92 (comet)
Be^{2+}	60	0	22
In^{3+}	57	5	82
Ga^{3+}	100	5	68
La^{3+}	0	0	0
Ce^{3+}	0	0	0
Y^{3+}	0	0	0
Sc^{3+}	0	0	0
Ti^{4+}	64	29	64

TABLE PC 97—(Continued)

Paper P1 = W1.

Solvent S1 = butanol-conc. HCl-conc. HF-water (100 : 50 : 2 : 48).
S2 = butanol-conc. H_2SO_4-water-conc. HF (100 : 9.5 : 85.5 : 5).
S3 = acetone-conc. HCl-conc. HF-water (90 : 5 : 1 : 4).
(All solvents are one phase.)

Technique Ascending.

Literature 1. Grand-Clement, A., Jakovac, Z., Lederer, M., and E. Pluchet, Proc. Intern Symposium Microchem., Birmingham University, Great Britain, August 1958. p 231–242 (published 1959). Reprinted in *Chromatogr. Rev.*, **12**, 284, 1970.

TABLE PC 98
INORGANIC PHOSPHATES

Paper	P1	P1	P1
Solvent	S1	S2	S3
Literature	1	1	1
Compound	$R_F \times 100$		
Monophosphate $(PO_4)^{-3}$	69	73	79
Diphosphate $(P_2O_7)^{-4}$	44	53	68
Triphosphate $(P_3O_{10})^{-5}$	29	39	58
Tetraphosphate $(P_4O_{13})^{-6}$	17	29	47
Pentaphosphate $(P_5O_{16})^{-7}$	11	22	36
Hexaphosphate $(P_6O_{19})^{-8}$	7	16	25
Heptaphosphate $(P_7O_{22})^{-9}$	4	11	15
Octaphosphate $(P_8O_{25})^{-10}$	—	8	—
Trimetaphosphate $(P_3O_9)^{-3}$	20	32	39
Tetrametaphosphate $(P_4O_{12})^{-4}$	8	18	22

Paper P1 = SS2040a.

Solvent S1 = *iso*-propanol-water-trichloracetic acid-20% NH_3 (75 : 25 : 5 : 0.3 v/v/w/v).
S2 = *iso*-propanol-water-20% trichloracetic acid-25% NH_3 (70 : 10 : 20 : 0.3 v/v).
S3 = methanol-trichloracetic acid (to 100 ml TCA, diluted to 500 ml, add 22.7 ml NH_3)-diluted acetic acid (20 ml 96% acetic acid plus 80 ml water) (60 : 10.3 : 5 v/v).

Detection Phosphomolybdic acid reagent.

Technique Ascending development, 16 °C, 14 hours.

Literature 1. Hettler, H., *J. Chromatogr.*, **1**, 389 (1958).

TABLE PC 99
LIPIDS AND RELATED COMPOUNDS

Paper	P1	P1	P1	P1
Solvent	S1	S2	S3	S4
Literature	1	1	1	1
Compound	$R_F \times 100$			
Choline	11	72	30	40
α,β-Dipalmitolein	93	88	91	92
β-Glycerophosphate	0	46	3	7
Glycerylphosphorylcholine	0	70	25	76
α-Monopalmitin	93	92	91	90
Phosphatidic acid	Streak	Streak	Streak	Streak
Phosphorylcholine	0	45	0	8
Oleic acid	95	92	94	94
Saturated lecithin	87	89	78	86
Saturated lysolecithin	67	89	61	85
Unsaturated lecithin (dipalmitoleyl)	81	89	78	86
Unsaturated lysolecithin	70	87	80	80

Paper P1 = W1.

Solvent S1 = *n*-butanol saturated with H_2O.
S2 = ethanol: H_2O (8 : 1).
S3 = *n*-propanol: H_2O (8 : 1).
S4 = *n*-propanol: acetic acid:H_2O (8 : 1 : 1).

Detection See Section II.I, Volume Two.

Technique Descending development, room temperature, saturated chamber.

Literature
1. Huennekens, F. M., Hanahan, D. J., and M. Uziel, *J. Biol. Chem.*, **206**, 443–447 (1954).

TABLE PC 100
NITRO COMPOUNDS

Paper	P1	P1	P1	P1	P1
Solvent	S1	S2	S3	S4	S5
Literature	1	1	1	1	1
Compound	$R_F \times 100$				
***p*-Nitrobenzoic acid**	1	71	43	81	32
***p*-Nitroethylbenzene**	91	92	91	70	90
***p*-Nitrophenylacetic acid**	2	60	42	81	34
1-(*p*-Nitrophenyl)ethanol	20	92	93	92	30
2-(*p*-Nitrophenyl)ethanol	15	91	90	90	25
***p*-Nitrotoluene**	92	93	90	72	91

Paper P1 = W4.

Solvent S1 = *n*-hexane-diisopropyl ether (5 : 1 v/v) run on paper treated with formamide-saturated ether and dried.
S2 = 2-butanone-2*N* NH_3 (1 : 1).
S3 = *n*-butanol-ammonia-water (4 : 1 : 5).
S4 = ethanol-water (7 : 3 v/v) on paper treated with 5% v/v olive oil in ether and dried.
S5 = *n*-hexane-diisopropyl ether (5 : 1 v/v) on paper treated with a saturated solution of Carbowax 4000 in ether-ethanol (50 : 1) and dried.

Detection Aqueous HCl.

Technique Descending development, for 12 cm.

Literature
1. Chakraborty, J. and J. N. Smith, *Biochem. J.*, **93**, 389 (1964).

TABLE PC 101
NITROGEN HETEROCYCLICS, INDOLES

Paper	P1	P1	P1	P1	P1	P1													
Solvent	S1	S2	S3	S4	S5	S6													
Detection	—	—	—	—	—	—	D1	D2	D3	D4	D5	D6	D7	D8	D9	D10	D11	D12	D13
Literature	1	1	1	1	1	1													
Compound	$R_F \times 100$						Color[a]												
5-Bromoindole	95	96	93	94	94	93	—	—	—	—	—	—	—	—	+	+	—	23, 22	32, 68
6-Methoxyindole[b,c]	93	93	93	94	93	94	57	60	—	+	64	25	—	43	68	++	—	25, 64	52, 35
5-Nitroindole[c]	94	95	90	94	92	84	60	—	—	—	—	+	—	—	—	+	—	62, 23	40, 28
5-Methylisatin	92	85	87	89	78	42	60	—	—	—	—	—	—	—	—	—	—	59	+
5-Hydroxy-2-methylindole[b,c]	93	91	88	78	82	44	—	7	56	23, 56	10, 64	27, 69	18	27, 51	39, 69	++	—	22	43, 53
2-Methyl-3-ethyl-5-aminoindole	90	88	80	58	62	39	57	62	64	68	63	70	—	+	50, 37	+	—	7, 59	62, 23
2-Methyl-3-ethyl-5-dimethylaminoindole[d]	84	92	36	92	18	12	3	64	52, 60	52, 63	23, 52	38, 69	—	+	23, 28	++	33	40	62, 23
N-1-Methyl-2-methyl-3-ethyl-5-dimethylaminoindole	95	94	53	92	73	62	27	64	63	24	23, 56	38, 42	—	—	23, 28	++	33	40	23
3-Indoleacetaldehyde	91	85	87	93	90	75	—	—	—	—	—	—	+	60	39	+	—	23	25
7-Methyl-3-indolealdehyde	93	90	92	93	88	62	34	—	—	—	—	62	17	—	—	—	—	18	—
Indole-2-carboxylic acid	93	16	90	76	71	28	33	+	53, 25	25	+	—	—	—	—	+	+	17, 23	34, 27
5-Methoxyindole-2-carboxylic acid ethyl ester	93	93	92	94	92	91	33	—	—	—	—	62	—	—	—	—	—	—	—
2-Methyl-5-methoxy-3-indolylacetamide	86	79	77	79	37	16	59	+	7	60	8	28	—	—	—	+	—	40	52, 24
3-Indolylacetic acid hydrazide[c]	92	90	89	47	82	4	—	+	+	62	8	—	—	—	30	++	—	6, 58	22, 23
Indolon-2	85	82	79	88	68	55	—	—	—	—	—	62	—	—	68	+	—	6	59
4-Hydroxyindole[b,c]	91	92	89	64	74	34	—	58, 17	59, 15	23, 25	64, 25	69, 28	70	43, 68	33, 27	++	—	29, 38	35, 65
4-Hydroxyskatole	91	92	91	75	91	67	—	60	52	56	62	33	—	51	39	++	—	29, 71	33, 27
6-Hydroxyskatole	94	90	91	68	82	47	+	12, 63	12, 64	54, 66	63, 65	35	51	60, 3	45, 42	++	—	35	43, 28
N-Formyl-*dl*-tryptophan[c]	86	5	79	25	9	1	—	—	—	+	8	52	—	—	—	++	+	23, 70	25
N-Acetyltryptophanamide	78	65	54	33	4	1	27	—	—	59	9	70	—	—	—	++	—	23, 27	25

TABLE PC 101—(Continued)
NITROGEN HETEROCYCLICS, INDOLES

Paper Solvent Detection Literature	P1 S1 — 1	P1 S2 — 1	P1 S3 — 1	P1 S4 — 1	P1 S5 — 1	P1 S6 — 1	D1	D2	D3	D4	D5	D6	D7	D8	D9	D10	D11	D12	D13
Compound	$R_F \times 100$						Color[a]												
6-Methyltryptophan[b]	49	1	1	0	0	0	34	60	—	61	60	—	—	—	—	+	33	25, 48	25, 28
dl-5-Methoxytryptophan[b]	39	1	1	1	0	0	—	—	—	+	10	26	—	—	—	+	—	27	27, 29
dl-Tryptophanol[b]	39	90	2	0	0	0	+	—	—	—	—	—	—	—	71	++	33	23, 58	24, 25
α-Methyltryptamine	59	85	2	0	0	0	—	—	—	—	—	—	—	—	44	+	—	22, 30	23
5-Methyltryptamine	66	86	3	0	0	0	—	—	—	—	+	+	—	—	40	+	—	23	25
7-Methyltryptamine[b]	67	91	4	0	0	0	—	—	—	—	+	—	—	—	—	+	—	23	30
2-Methyl-5-nitrotryptamine[b,c]	72	92	5	0	0	0	56	—	—	—	—	—	—	—	—	+	—	3	22
6-Methoxytryptamine[b,c]	58	85	3	0	0	0	39	—	—	23	15	71	40	8	51, 44	++	—	30, 38	27, 29
N-ω-Methyl-5-hydroxytryptamine	52	64	2	0	0	0	—	59	61	65	60	69	—	—	70	+	—	25, 70	33, 23
2-Methyl-*N*,*N*-dimethyltryptamine	81	92	6	28	3	2	—	—	—	—	8	—	57	—	44	+	—	57, 70	26, 70
d-Lysergic acid	59	2	4	0	0	0	27	—	—	—	17	—	—	—	40	+	—	24, 48	25, 68
d-Bromolysergic acid diethylamide[d]	82	89	9	93	7	2	57	—	—	—	—	—	—	—	71	+	—	—	—
Norharman[d]	55	86	7	31	1	0	33	—	—	—	—	—	—	—	—	—	33	—	—
Harmol[b]	64	77	2	9	0	0	33	—	—	—	—	57	—	—	69	++	33	+	—
Tetrahydroharman[d]	57	88	5	4	0	0	39	—	—	—	8	52	—	—	39	++	—	—	—
6-Hydroxytetrahydroharman	51	21	3	0	0	0	48	60	+	+	64	70	—	—	69	+	33	70	27
6-Methoxytetrahydroharman[c]	55	84	4	6	0	0	3	+	+	56	59	71	—	—	40	+	33	—	70
Indican	37	35	0	0	0	0	—	—	—	55	8	25	—	—	—	+	—	63	26, 23
5-Hydroxyindole	92	91	87	43	59	22	—	64	55, 5	24, 25	56, 63	28	16	55	43, 68	++	—	23	43, 28
Gramine-*N*-oxide	59	8	7	7	0	0	—	—	—	64	7	+	17	58	59, 71	+	33	58	22, 24
5-Methoxygramine	48	86	2	1	0	0	—	—	—	—	8	23	—	—	—	+	33	4	26, 71
N-Acetyltryptamine	90	88	87	85	67	45	33	57	—	—	10	—	6	—	30	++	—	8, 24	24, 25
N-Methyltryptamine	51	80	3	0	0	0	—	—	—	13	8	+	—	—	+	+	33	17, 71	23, 25

TABLE PC 101—(Continued)
NITROGEN HETEROCYCLICS, INDOLES

Paper Solvent Detection Literature	P1 S1 — 1	P1 S2 — 1	P1 S3 — 1	P1 S4 — 1	P1 S5 — 1	P1 S6 — 1	D1	D2	D3	D4	D5	D6	D7	D8	D9	D10	D11	D12	D13
Compound			$R_F \times 100$										Color[a]						
N,N-Dimethyltryptamine	48	92	5	5	2	1	—	—	—	58	58	—	—	58	52	++	—	58, 23	23
N,N-Diethyltryptamine	68	93	4	6	0	0	—	—	—	+	8	+	—	—	70	+	33	17, 23	23
5-Fluorotryptamine	67	87	4	0	0	0	—	—	—	—	—	—	—	58	—	—	—	17	24, 27
Psylocybin	8	0	0	0	0	0	—	—	—	—	—	—	—	—	70	—	—	16, 26	27
5-Methoxytryptamine	52	83	2	0	0	0	—	6	—	60	8	—	—	—	70	++	—	26	27, 25
N-Acetyl-5-hydroxytryptamine	83	82	60	18	1	0	—	17, 60	24	25	18	+	—	—	30	+	—	69	30, 43
N-Acetyl-5-methoxytryptamine = melantonin	90	88	78	84	48	28	—	6	—	—	8	—	—	—	70	++	—	26	25, 28
Indole-3-carbinol	94	93	93	93	92	92	57	—	—	64	9	—	58	17	57, 20	++	—	58, 17	24
Tryptophol	90	88	90	81	72	59	34	58	—	—	8	24	70	—	71	++	—	7, 63	23, 25
L-Tryptophan methyl ester	51	84	1	1	0	0	—	—	—	—	8	23	—	—	—	++	33	3	23, 25
5-Chloroindole-2-carboxylic acid	93	19	92	72	68	15	34	6	24	25	9, 63	—	—	57	—	—	+	—	26
Indole-5-carboxylic acid	91	9	90	63	51	15	30	6	59	64	10	—	—	57	+	++	+	23	30
3-Indolylacetic acid methyl ester	93	94	84	88	87	84	+	—	60	60	8	—	—	—	—	++	—	8, 69	24, 25
3-Indolylacetamide	85	77	65	50	19	7	—	6	—	58	9	—	—	—	44, 71	++	—	26, 69	23, 25
3-Indolylglyoxylic acid	88	8	86	18	24	5	—	—	—	56	—	+	3	64	3, 72	—	++	—	—
3-Indolylglyoxylamide	94	93	92	93	92	88	—	—	—	—	—	—	7	—	—	+	—	—	17
3-Indolylglyoxyldimethylamide	95	93	93	94	92	92	—	—	—	—	—	—	7	—	—	+	33	—	+
3-Indolypyruvic acid	91	7	0	23	28	4	3	58	62	63	+	63	—	24	69	+	+	17	23, 65
3-Indolylacrylic acid	94	8	91	66	50	14	34	63, 23	—	24, 64	64	23	58	63	70	++	+	40	23

TABLE PC 101—(Continued)
NITROGEN HETEROCYCLICS, INDOLES

Paper Solvent Detection Literature	P1 S1 — 1	P1 S2 — 1	P1 S3 — 1	P1 S4 — 1	P1 S5 — 1	P1 S6 — 1	D1	D2	D3	D4	D5	D6	D7	D8	D9	D10	D11	D12	D13
Compound	$R_F \times 100$						Color[a]												
5-Hydroxy-3-indolylacetic acid methyl ester	93	91	89	75	63	25	—	17	+	+	+	—	—	—	—	—	—	34	30, 27
5-Hydroxy-3-indolylacetamide	68	57	20	1	0	0	+	64	23	23, 65	64	68	—	70	71	++	—	51	25, 28
5-Methoxy-3-indolylacetic acid	91	10	91	71	53	11	—	3	—	+	7	—	60	26, 23	44, 71	++	+	26, 27	27, 30
5-Methoxy-3-indolylacetamide	86	74	63	54	15	5	—	—	—	—	7	—	—	—	71	+	—	26, 27	27, 24
Harmalol	42	45	0	0	0	0	39	17, 64	64	71, 67	64	43, 71	—	—	71	—	—	—	
Harmine	52	88	2	12	0	0	32	—	—	—	—	—	—	57	3, 71	—	—	—	+
Harmane	62	88	5	15	0	0	33	—	—	—	—	—	—	—	—	—	—	—	—
Ergotamine	80	91	16	47	0	0	34	—	—	—	—	+	—	—	—	+	—	26, 70	+

[a] For interpretation, see Table PC 120. [b] Ninhydrin reagent, pale pink. [c] *p*-Dimethylaminobenzaldehyde, pale brown. [d] Dragendorff reagent, positive.

Paper P1 = W1.

Solvent
S1 = ethyl methyl ketone-acetone-formic acid-water (40 : 2 : 1 : 6).
S2 = ethyl methyl ketone-diethylamine-water (921 : 2 : 77).
S3 = methyl isobutyl ketone-formic acid-water (10 parts ketone saturated with 1 part 4% formic acid).
S4 = chloroform-methanol-formic acid-water (10 parts of chloroform saturated with a mixture of 1 part methanol and 1 part 4% formic acid).
S5 = benzene-ethyl methyl ketone-formic acid-water (a mixture of 9 parts benzene and 1 part ketone saturated with 1 part 2% formic acid).
S6 = benzene-formic acid-water (10 parts benzene saturated with 1 part 2% formic acid).

Detection
D1 = ultraviolet light.
D2 = diazotized sulfanilic acid (0.3% solution in dioxane-water 1 : 2).
D3 = diazotized 4-benzoylamino-2,5-dimethoxyaniline (0.6% solution in dioxane-water 1 : 2).
D4 = diazotized *o*-dianisidine (0.6% solution in dioxane-water 1 : 2).
D5 = *p*-nitrobenzenediazonium fluoborate (0.4% solution in dioxane-water 1 : 2).
D6 = 2,6-dibromoquinone-4-chloroimide (0.5% solution in dioxane-acetone 4 : 1).
D7 = 2,4-dinitrophenylhydrazine (*ca.* 0.1% solution in 1*N* HCl).
D8 = ferric chloride (2% aqueous solution).
D9 = phosphomolybdic acid (2% aqueous solution).
D10 = potassium permanganate (1% aqueous solution).
D11 = bromophenol blue (*ca.* 0.05% solution in ethanol).
D12 = Ehrlich reagent (1% *p*-dimethylaminobenzaldehyde in 1*N* HCl).
D13 = *p*-dimethylaminocinnamaldehyde (0.1% solution in 1*N* HCl).

Technique Descending development, 40 cm.

Literature
1. Reio, L., *J. Chromatogr.*, **47**, 60–85 (1970) and *J. Chromatogr.*, **13**, 475–496 (1964).

TABLE PC 102
NITROGEN HETEROCYCLICS, INDOLES AND RELATED COMPOUNDS

Paper	P1	P1	P1	P1	P1	P1					
Solvent	S1	S2	S3	S4	S5	S6					
Detection							D1	D2	D3	D4	D5
Literature	1	1	1	1	1	1					
Compound			$R_F \times 100$						Color[a]		
N-Acetylindoxyl	82	85	90	F	65	66	—	sl o	—	—	y fd
N-Acetyltryptophan	42	88	70	80	74	70	—	r-p	—	—	—
Anthranilic acid	34	90	82	67	73	68	y [dk b]	imm. y → o	—	—	lt y
Bufothionine	35s	47	72	—	43	49	—	wk b-p	—	—	—
Citrulline	6	18	23	63	F	96	—	sl y	—	p	—
N,*N*-Diethyltryptamine (as HCl)	98	85	83	—	62	63	—	r-p	—	—	—
5,6-Dimethoxytryptamine (as HCl)	65	50	62	—	24	27	—	imm. b-g → b	sl bn-o	p-gy	r-o
N,*N*-Dimethyl-5-hydroxytryptamine (as acetate)	90	60	79	—	43	47	—	rp b-p → b	—	—	r-bn
N,*N*-Dimethyltryptamine (as acetate)	92	74	82	—	60	62	—	r-p	—	—	—
α,α-Dimethyltryptamine (as acetate)	90	82	87	—	60	60	—	r-p	wk pk	—	—
β,β-Dimethyltryptamine (as oxalate)	92	81	79	—	58	63	—	neg	—	b-p	—
Ethyl indolylacetate	95	95	98	F	41	50	—	pk-p	sl pk-p	—	—
N-Ethyltryptamine (as HCl)	93	83	79	—	60	60	—	r-p	—	wk bn	—
α-Ethyltryptamine (as acetate)	92	84	86	—	52	57	—	r-p	—	wk p	—
Gramine	95	72	70	F	66	67	—	wk sl bn	—	—	—
Hippuric acid	48	85	66	80	85	89	—	—	—	—	—
3-Hydroxyanthranilic acid	20s	84	73s	41s	59s	56	y-bn [b]	sl bn-y	—	—	bn
5-Hydroxyanthranilic acid	21	60	67	33	65	72	[b]	y-o	—	—	wk o-bn
5-Hydroxyindolylacetic acid	18	76	73	46	48	42	—	rp b → gy → b	sl pk-p	—	r-bn
5-Hydroxyindolylacetic acid 5-sulfate ester	5	28	—	—	70	65	—	rp → sl b	sl pk	—	—
4-Hydroxytryptamine (as creatinine sulfate)	54	48	72	—	33	34	—	rp b-p → gy → b	—	gy-p	r-bn
5-Hydroxytryptamine (as creatinine sulfate)	56	48	70	—	34	34	—	rp b-p → gy → b	—	gy-p	r-bn
6-Hydroxytryptamine (as creatinine sulfate)	53	44	72	—	33	34	—	imm b-g → b	—	gy-p	p-r
5-Hydroxytryptophan	15	27	40	46	35s	38s	—	b → g-b	—	gy-p	r-bn
Indole	90	95	F	F	—	—	—	imm pk-p → p	sl pk	—	—
Indole-3-aldehyde	92	92	95	98	33s	38	—	sl bn → pk-p	sl pk	—	—
Indole-2-carboxylic acid	42	93	73	54	34	25	—	sl pk fd	—	—	lt o
Indole-3-carboxylic acid	25	90	93	60	40	23	—	sl b-pk [r]	—	—	o-y
Indolylacetamide	80	88	90	95	53	57	—	sl pk-p → gy-p	sl pk	—	—
3-Indolylacetic acid	35	90	84	75	60	55	—	pk-p → gy → bn	str pk	—	—
3-Indolylacetonitrile	87	95	93	F	33	36	—	wk r-p → bn	sl gy-v	—	—
Indolylacetyl-asparagine	22	70	60	78	70	70	—	pk-p	sl v	—	—
Indolylacetyl-aspartic acid	7	80	55	48	77	75	—	pk-p	sl v	—	—
Indolylacetyl-glutamic acid	7	85	43	51	78	72	—	pk-p → b-gy	sl v	—	—
Indolylacetyl-glutamine	25	72	58	82	70	70	—	pk-p → bg-y	sl v	—	—
Indolylacetyl-glycine	36	82	65	82	65	65	—	pk-p → b-gy	sl v	—	—

TABLE PC 102—(Continued)
NITROGEN HETEROCYCLICS, INDOLES AND RELATED COMPOUNDS

Paper Solvent Detection Literature	P1 S1 1	P1 S2 1	P1 S3 1	P1 S4 1	P1 S5 1	P1 S6 1	D1	D2	D3	D4	D5
Compound	$R_F \times 100$						Color[a]				
3-Indolylacrylic acid	36	90	94	70	14	12	—	y-o → bn [g]	rp pk → o [y]	—	o
γ-(3-Indolyl)butyric acid	48	95	93	82	47	36	—	b-p [r]	sl y-bn	—	—
C-(3-Indolyl)-glycine	13	43	46	73	67	68	—	neg	—	o-y	—
3-Indolyl-glycolic acid	28	80 dec.	56	50	78	73	—	str pk	str pk	—	—
3-Indolyl-glyoxylic acid	41 dec.	77	72	60 dec.	40	48	[wk bn]	sl gy-p	sl g-y	—	—
3-Indolyl-lactic acid	42	86	71	63	65	58	—	p → b-gy	sl r-p	—	—
Indolyl-propane 1,2-diol	82	82	90	92	60	64	—	p → gy	sl y-bn	—	—
β-(3-indolyl)propionic acid	40	92	89	78	47	39	—	b-p[r]	sl y-bn	—	—
Indoxyl O-acetate	—	91	95	—	50	53	—	pk → bn	b	—	sl o
Indoxyl O-phosphate	3	35	50	42	85	82	—	rp p → bn	—	gy	y
Indoxyl sulfate	60	46	83	42	59	66	[lt b]	rp o-bn	b [dk g]	—	—
Isatin	81	87	95	95	62	62	—	—	—	—	—
Kynurenic acid	37	58	65	70	32	46	[g-b]	—	—	[b]	—
Kynurenine	20	43	48	80	66	66	[lt b]	imm o	[g]	p	—
5-Methoxyindolyl acetic acid	30	85	82	78	48	42	—	rp b → gy	pk-p	—	wk o
5-Methoxytryptamine (as HCl)	76	64	72	—	30	35	—	rp b-p → b-gy	—	bn-p	—
6-Methoxytryptamine (as HCl)	75	65	72	—	30	35	—	rb b	—	wk p	wk o
5-Methoxytryptophan	20	42	52	76	41	41	—	bp	—	bn-p	—
N-Methyl-5-hydroxytryptamine (as oxalate)	80	56	72	—	38	42	—	gy-p	—	wk bn	r-bn
Methyl indolyl-acetate	95	95	98	F	53	57	—	pk-p	sl pk-p	—	—
7-Methyl-3-indolyl-acetonitrile	88	98	95	F	25	28	—	wk r-p → bn	sl gy-v	—	—
1-Methyltryptamine (as HCl)	84	70	75	—	—	—	—	sl bn-p	—	p	—
2-Methyltryptamine (as creatinine sulfate)	87	75	80	—	39	45	—	v sl r-p	sl g-b	b-gy	—
N-Methyltryptamine (as HCl)	88	73	76	—	58	59	—	r-p	—	wk bn	—
α-Methyltryptamine (as acetate)	88	78	82	—	57	57	—	r-p	—	wk p	—
N-Methyltryptophan	37	60	65	95	70	74	—	pk-p	—	sl p	—
Porphobilinogen	3	59	19	—	85	—	—	rp b-r	—	gy-y	pk fd
Pyrrole-2-carboxylic acid	25	84	71	50	78	66	—	pk-r → r-bn	wk bn	—	r
Skatole	92	98	F	F	—	—	—	imm b-p → b	sl y-bn	—	—
Tryptamine	83	72	76	—	48	51	—	r-p	—	r-p	—
Tryptophan	25	50	56	77	55	55	—	p	—	gy-p	—
Tryptophan-4-carboxylic acid	8	41	30s	41	54	45	[b acid lt b alk]	v sl y	—	r-p	—

TABLE PC 102—(Continued)
NITROGEN HETEROCYCLICS, INDOLES AND RELATED COMPOUNDS

Paper	P1	P1	P1	P1	P1	P1					
Solvent	S1	S2	S3	S4	S5	S6					
Detection							D1	D2	D3	D4	D5
Literature	1	1	1	1	1	1					
Compound	$R_F \times 100$						Color[a]				
Tryptophanamide	64	57	77	F	25	25	—	pk-p	—	p	—
Tryptophanol	80	66	78	95	54	58	—	b-r	—	wk b-gy	—
Urea	42	50	52	73	83	83	—	imm y	—	—	—
Xanthurenic acid	10	55	64	75	22	32	[b]	—	—	—	r-bn

s = Streaking zone.

[a] For colors, see master abbreviation list; imm = immediate; [ ] indicate fluorescence; neg = negative; sl = slow.

Paper P1 = W1.

Solvent S1 = isopropanol-conc. NH_4OH-water (200 : 10 : 20).
S2 = *n*-butanol-glacial acetic acid-water (120 : 30 : 50).
S3 = *n*-butanol-pyridine-water (60 : 60 : 60).
S4 = 80% phenol-conc. NH_4OH (200 : 1).
S5 = 20% w/v aqueous KCl.
S6 = 8% w/v aqueous NaCl-glacial acetic acid (200 : 2).

Detection D1 = visible or [UV] light.
D2 = Ehrlich.
D3 = acid oxidation.
D4 = ninhydrin-pyridine.
D5 = sulfanilic acid.

Technique Ascending.

Literature
1. Jepson, J. B., in *Chromatographic and Electrophoretic Techniques*, Vol. I, Smith, I., Ed., Interscience Publishers, Division of John Wiley and Sons, Inc., New York, N.Y., 1969, pp 250–252.

TABLE PC 103
NITROGEN HETEROCYCLICS, INDOLES

Paper	P1	P1	P1	P1	P1	P1	P1	P2	P2	P2	P2	P2	P2	P2
Solvent	S1	S2	S3	S4	S5	S6	S7	S8	S9	S8	S10	S11	S12	S13
Technique	T1	T1	T1	T1	T1	T1	T1	T1	T1	T1	T1	T1	T1	T1
Literature	1	2	1	1	1	3	4	5	5	5	5	5	5	5
Compound	$R_F \times 100$													
N-Acetyltryptophan	88	—	68	58	46	—	—	—	—	—	—	—	—	—
Anthranilic acid	—	90	—	—	—	—	—	—	—	—	—	—	88	66
5-Benzyloxyindole-3-acetic acid	—	—	—	—	—	—	—	52	87	—	—	—	—	—
Bufotenin	—	—	—	—	—	—	—	—	—	90	60	51	—	—
N,*N*-Diethyltryptamine	—	—	—	—	—	—	—	—	—	—	83	66	—	—
5,6-Dimethoxytryptamine	50	—	66	—	63	—	—	—	—	—	—	—	—	—
α,α-Dimethyltryptamine	83	—	87	90	89	—	—	—	—	—	—	—	—	—
N,*N*-Dimethyltryptamine	76	—	80	92	—	—	—	—	—	—	75	64	—	—
α-Ethyltryptamine	84	—	86	89	—	79	—	—	—	—	84	58	—	—
Gramine	74	—	76	—	—	80	—	—	—	—	—	—	—	—
3-Hydroxyanthranilic acid	—	84	—	—	—	—	—	—	—	—	—	—	85	60
5-Hydroxyanthranilic acid	—	54	—	—	—	—	—	—	—	—	—	—	—	—
5-Hydroxyindoleacetic acid	77	—	70	31	20	17	—	24	70	25	—	—	—	—
7-Hydroxyindoleacetic acid	—	—	—	—	—	9	—	—	—	—	—	—	—	—
3-Hydroxykynruenine	—	—	—	—	—	—	—	—	—	—	—	—	30	56
D-3-Hydroxykynurenine	—	24	—	—	—	—	—	—	—	—	—	—	—	—
L-3-Hydroxykynurenine	—	29	—	—	—	—	—	—	—	—	—	—	—	—
D-5-Hydroxykynurenine	—	14	—	—	—	—	—	—	—	—	—	—	—	—
L-5-Hydroxykynurenine	—	17	—	—	—	—	—	—	—	—	—	—	—	—
5-Hydroxytryptamine	47	38	70	65	57	52	—	—	—	—	—	—	—	—
6-Hydroxytryptamine	—	—	—	—	—	—	—	—	—	—	46	35	—	—
5-Hydroxytryptophan	—	—	—	—	—	11	—	18	15	25	36	42	—	—
Indican	—	34	—	—	—	—	—	—	—	—	—	—	—	—
Indole	97	—	97	93	93	79	78	91	92	—	—	—	—	—
Indole-3-acetic acid	90	—	80	52	38	25	43	—	—	—	—	—	—	—
Indole-3-acetonitrile	—	—	—	—	—	—	—	90	92	—	—	—	—	—
Indole-3-acrylic acid	—	—	—	—	—	—	80	43	86	—	—	—	—	—
Indole-3-aldehyde	90	—	—	—	—	72	46	90	86	—	—	—	—	—
Indole-3-butyric acid	95	—	91	65	55	44	53	—	—	—	—	—	—	—
Indole-3-butyronitrile	—	—	—	—	—	—	79	—	—	—	—	—	—	—
Indole-3-carboxylic acid	—	—	—	—	—	20	62	35	90	—	—	—	—	—
Indole-3-lactic acid	—	—	—	—	—	30	33	—	—	—	—	—	—	—
Indole-3-propionic acid	92	—	86	59	45	35	49	—	—	—	—	—	—	—
Indolylacetonitrile	93	—	—	92	—	75	—	—	—	—	—	—	—	—

TABLE PC 103—(Continued)
NITROGEN HETEROCYCLICS, INDOLES

Paper	P1	P1	P1	P1	P1	P1	P1	P2	P2	P2	P2	P2	P2	P2
Solvent	S1	S2	S3	S4	S5	S6	S7	S8	S9	S8	S10	S11	S12	S13
Technique	T1	T1	T1	T1	T1	T1	T1	T1	T1	T1	T1	T1	T1	T1
Literature	1	2	1	1	1	3	4	5	5	5	5	5	5	5
Compound	$R_F \times 100$													
Indolyl-3-pyruvic acid	—	—	—	—	—	12	—	—	—	—	—	—	—	—
Indoxyl sulfate	45	25	78	71	61	—	—	—	—	—	—	—	—	—
Kynuramine	—	—	—	—	—	—	—	—	—	—	66	69	—	—
Kynurenic acid	—	—	—	—	—	—	—	—	—	—	—	—	56	55
Kynurenine	—	—	—	—	—	—	—	—	—	—	—	—	37	67
D-Kynurenine	—	31	—	—	—	—	—	—	—	—	—	—	—	—
L-Kynurenine	—	37	—	—	—	—	—	—	—	—	—	—	—	—
5-Methoxy-*N*-dimethyltryptamine	—	—	—	—	—	—	—	—	—	—	70	47	—	—
5-Methoxytryptamine	60	—	73	—	76	—	—	—	—	—	64	34	—	—
6-Methoxytryptamine	62	—	75	—	76	—	—	—	—	—	—	—	—	—
Melatonin	—	—	—	—	—	—	—	—	—	—	87	39	—	—
N-Methylserotonin	—	—	—	—	—	—	—	—	—	85	50	42	—	—
N-Methyltryptamine	—	—	—	—	—	—	—	—	—	—	73	58	—	—
α-Methyltryptamine	78	—	82	89	87	—	—	—	—	—	—	—	—	—
1-Methyltryptamine	70	—	75	85	—	—	—	—	—	—	—	—	—	—
2-Methyltryptamine	72	—	78	85	—	—	—	—	—	—	—	—	—	—
N-Methyltryptophan	56	—	66	—	39	—	—	—	—	—	—	—	—	—
Psilocin	—	—	—	—	—	—	—	—	—	—	75	52	—	—
Serotonin	—	—	—	—	—	—	—	—	—	70	49	37	—	—
Skatole	97	—	97	95	95	90	—	92	92	—	—	—	—	—
Na-acetylkynurenine	—	—	—	—	—	—	—	—	—	—	—	—	82	81
Tryptamine	71	—	78	82	82	65	72	—	—	—	70	54	—	—
Tryptophan	50	44	62	42	24	18	34	—	—	—	50	62	—	—
Xanthurenic acid	—	—	—	—	—	—	—	—	—	—	—	—	56	45
Xanthurenic acid-8-methyl ester	—	—	—	—	—	—	—	—	—	—	—	—	55	51

TABLE PC 103—(Continued)
NITROGEN HETEROCYCLICS, INDOLES

Literature	5	6	7	7		7		7		7	
	Color reactions										
				D4		D5		D6		D7	
Compound	D1	D2	D3	Color	(μg)	Color	(μg)	Color	(μg)	Color	(μg)
Acetyl tryptophan			lt-bl	lt-br		ah					
***N*-Acetyl indoxyl**			bl	lt-br	10	o	3	lt-br	10	br-lt	3
***N*-Acetyl isatin**			br			lt-y		y	50	lt-y	5
5-Benzyloxyindole-3-acetic acid	bl					bl					
5-Bufotenin	bl	r				bl-p					
***N,N*-Diethyltryptamine**	p					r-p					
Dihydroxyindole			br			y					
***N,N*-Dimethyltryptamine**	p					r-p					
***N*-Ethyltryptamine**	p					r-p					
Gramine			gy	pk-ish-bl	5	y	1				
5-Hydroxyindoleacetic acid	bl	p-r				p					
6-Hydroxytryptamine	bl-gr					gr					
5-Hydroxytryptophan	bl	p-r				p					
Indican			lt-vt	bl	3	br-ish-o	3	y	10	y-ish-pk	1
Indican glucoside			y			pl-y					
Indigo disulfonate			lt-bl	bl		bk					
Indigo tetrasulfonate			lt-bl	bl		bk					
Indigotin			lt-bl	bl		bk					
Indole	gr		pl-gr	o-r	3	lt-r	3	r	3	pk-ish-br	3
Indole-3-acetamide			y-ish-br	pk	1	pk-ish-br	1	pk-ish-br	1	y-ish-br	3
Indole-3-acetic acid			ah	pk	1	bl-ish-pk	1	r	3	y-ish-br	1
Indole-3-acetonitrile	p-r		gr-ish-bl	gr	3	y-pk	1	ah-br	1	y	5
Indole-3-acrylic acid	bl					gr					
Indole-3-aldehyde			pl-y	lt-br	3	lt-br	3	y	10	y	3
Indole-3-butyric acid			lt-bl	br	1	bl-ish-gr	1	y	3	lt-br	1
Indole-3-butyronitrile			vt	o-br	3	bl-ish-lt-br	1	o	10	y-ish-br	10
Indole-3-carboxylic acid			bl	o	1	pk	1	r	3	o	3
Indole-3-propionic acid			lt-bl	lt-br	1	bl-ish-gr	1	y	3	lt-br	1
Indoxyl-acetate			gr	bl	1	gy-ish-br	3	lt-ah	10	br	1
Isatin			br	y	5	y	3	y	3	lt-y	3
Kynuramine	r-p					r					
5-Methoxy-*N*-dimethyltryptamine	bl					bl-p					
5-Methoxytryptamine	bl					bl-p					
Melatonin	bl-gr					bl					
2-Methylindole			gr	pk-ish-y		sc					
3-Methylindole			lt-bl	pk-ish-br	3	bl	1	y	3	lt-br	3
***N*-Methylserotonin**	bl	r				bl-p					

TABLE PC 103—(Continued)
NITROGEN HETEROCYCLICS, INDOLES

Literature	5	6	7	7		7		7		7	
				Color reactions							
				D4		D5		D6		D7	
Compound	D1	D2	D3	Color	(μg)	Color	(μg)	Color	(μg)	Color	(μg)
***N*-Methyltryptamine**	p					r-p					
Psilocin	gr					gr-bl					
Serotonin	bl	r				bl-p					
Skatole	p-bl					p					
Tryptamine	p			du-br		r-p		y	3		
Tryptophan	p		y-ish-gr*	lt-br	1	pk	1	y	1	lt-br	3

* After treatment with $HClO_4$

Paper
P1 = SS2071 or 2230.
P2 = W1.

Solvent
S1 = *n*-butanol-acetic acid-water (60 : 15 : 25).
S2 = *n*-butanol-acetic acid-water (4 : 1 : 5).
S3 = *n*-butanol-pyridine-water (1 : 1 : 1).
S4 = *iso*-propanol-NH_4OH-water (8 : 1 : 1).
S5 = *iso*-propanol-NH_4OH-water (20 : 1 : 2).
S6 = *iso*-propanol-NH_4OH-water (10 : 1 : 1).
S7 = *iso*-propanol-NH_4OH-water (80 : 5 : 15).
S8 = isopropanol-ammonia-water (8 : 1 : 1).
S9 = *n*-butanol-acetic acid-water (4 : 1 : 1).
S10 = *n*-butanol-acetic acid-water (12 : 3 : 5), 1st direction.
S11 = 20% aqueous potassium chloride, 2nd direction.
S12 = *n*-butanol-acetic acid-water (4 : 1 : 5), 1st direction.
S13 = distilled water, 2nd direction.

Detection
D1 = *p*-dimethylaminocinnamaldehyde.
D2 = *p*-nitroaniline.
D3 = fluorescence in ultraviolet light.
D4 = $FeCl_3$-$HClO_4$.
D5 = *p*-dimethylaminobenzaldehyde (Ehrlich reagent).
D6 = KNO_2-HNO_3.
D7 = cinnamaldehyde HCl.

Technique
T1 = both ascending and descending, chromatography.

Literature
1. Jepson, J. B., Smith, I., and B. J. Stevens, Private Communication (1954).
2. Dalgliesh, C. E., *Arch. Biochem. Biophys.*, **58**, 214 (1955).
3. Stowe, B. B. and K. V. Thimann, *Arch. Biochem. Biophys.*, **51**, 499 (1954).
4. Vlitos, A. J. and W. Meudt, *Contribs. Boyce Thompson Inst.*, **17**, 197 (1953).
5. Durkee, A. B. and J. C. Sirois, *J. Chromatogr.*, **13**, 173 (1964).
6. Cowles, E. J. and A. C. Hilding, *J. Chromatogr.*, **19**, 326 (1965).
7. Sen, S. P. and A. C. Leopold, *Physiol. Planat.*, **7**, 98 (1954).

TABLE PC 104
NITROGEN HETEROCYCLICS, QUINOLINE DERIVATIVES

Paper	P1	P1	P1	P1	P1	P1													
Solvent	S1	S2	S3	S4	S5	S6													
Detection	—	—	—	—	—	—	D1	D2	D3	D4	D5	D6	D7	D8	D9	D10	D11	D12	D13
Literature	1	1	1	1	1	1													
Compound	$R_F \times 100$						Color[a]												
4-Amino-7-chloroquinoline[b]	64	76	2	12	0	0	34	—	—	—	—	—	—	—	—	—	—	—	—
8-Amino-6-methoxyquinoline[b]	84	93	87	92	86	83	—	63	28, 63	23	63, 24	41	+	52	69	++	—	9, 62	63, 65
8-Aminoquinoline	87	93	84	95	87	80	40	20, 11	12, 62	23, 53	63	42, 23	—	+	3, 70	+	—	8	19, 65
1,3-Dihydroxyisoquinoline	83	89	78	86	80	59	—	6	15, 62	19, 65	8	23, 60	—	—	42, 68	++	—	+	—
4,8-Dimethoxyxanthurenic acid	92	85	88	95	90	78	38	—	—	—	—	—	—	—	—	—	—	—	—
4-Hydroxy-6-methoxyquinoline[b]	72	62	45	48	5	0	27	—	—	—	—	—	—	—	39	—	—	—	—
4-Hydroxy-6-methoxy-3-quinoline-carboxylic acid	87	11	89	48	4	0	34	—	—	—	—	—	—	58	—	—	+	—	—
4-Hydroxy-7-methoxy-3-quinoline-carboxylic acid	82	18	74	65	10	0	34	—	—	—	—	—	—	57	—	—	+	—	—
4-Hydroxy-8-methoxyquinoline-carboxylic acid	84	17	79	82	21	5	27	—	—	—	—	—	—	58	—	—	+	—	—
4-Hydroxy-*N*-methyl-2-quinoline	86	16	84	88	49	0	—	7	63	14	7	25	—	+	—	+	—	8	—
4-Hydroxyquinazoline	90	4	82	78	29	2	34	—	—	—	—	—	—	—	—	—	—	—	—
8-Mercaptoquinoline[b]	91	95	93	91	92	86	—	—	—	—	—	64	—	—	45, 70	+	—	—	—
6-Methoxy-4-quinoline-carboxylic acid (quininic acid)	82	9	61	63	24	3	33	—	—	—	—	—	—	—	—	—	+	—	—
4-Methoxyxanthurenic acid	74	35	44	75	8	4	37	—	—	—	—	—	—	—	—	—	+	—	—
8-Methoxyxanthurenic acid	62	5	27	9	3	0	33	—	—	—	—	—	—	—	—	—	+	—	—

[a] For interpretation, see Table PC 120. [b] Dragendorff reagent, positive.

Paper P1 = W1.

Solvent
- S1 = ethyl methyl ketone-acetone-formic acid-water (40 : 2 : 1 : 6).
- S2 = ethyl methyl ketone-diethylamine-water (921 : 2 : 77).
- S3 = methyl isobutyl ketone-formic acid-water (10 parts ketone saturated with 1 part 4% formic acid).
- S4 = chloroform-methanol-formic acid-water (10 parts of chloroform saturated with a mixture of 1 part methanol and 1 part 4% formic acid).
- S5 = benzene-ethyl methyl ketone-formic acid-water (a mixture of 9 parts benzene and 1 part ketone saturated with 1 part 2% formic acid).
- S6 = benzene-formic acid-water (10 parts benzene saturated with 1 part 2% formic acid).

Detection
- D1 = ultraviolet light.
- D2 = diazotized sulfanilic acid (0.3% solution in dioxane-water 1 : 2).
- D3 = diazotized 4-benzoylamino-2,5-dimethoxyaniline (0.6% solution in dioxane-water 1 : 2).
- D4 = diazotized *o*-dianisidine (0.6% solution in dioxane-water 1 : 2).
- D5 = *p*-nitrobenzenediazonium fluoborate (0.4% solution in dioxane-water 1 : 2).
- D6 = 2,6-dibromoquinone-4-chloroimide (0.5% solution in dioxane-acetone 4 : 1).
- D7 = 2,4-dinitrophenylhydrazine (*ca.* 0.1% solution in 1*N* HCl).
- D8 = ferric chloride (2% aqueous solution).
- D9 = phosphomolybdic acid (2% aqueous solution).
- D10 = potassium permanganate (1% aqueous solution).
- D11 = bromophenol blue (*ca.* 0.05% solution in ethanol).
- D12 = Ehrlich reagent (1% *p*-dimethylaminobenzaldehyde in 1*N* HCl).
- D13 = *p*-dimethylaminocinnamaldehyde (0.1% solution in 1*N* HCl).

Technique Descending development, 40 cm.

Literature
1. Reio, L., *J. Chromatogr.*, **47**, 60–85 (1970).

TABLE PC 105
NITROGEN HETEROCYCLICS, NAPHTHALENE AND QUINOLINE DERIVATIVES

Paper Solvent Detection Literature	P1 S1 — 1	P1 S2 — 1	P1 S3 — 1	P1 S4 — 1	P1 S5 — 1	P1 S6 — 1	D1	D2	D3	D4	D5	D6	D7	D8	D9	D10	D11	D12	D13
Compound			$R_F \times 100$										Color[a]						
1-Amino-2-naphthol	95	94	83	77	6	4	57	60	56	56	56	—	23	56	53	++	—	6	64
2-Amino-3-naphthol	88	91	80	55	36	9	33	12, 23	23, 65	65	65, 64	63, 28	—	56	68	++	—	8	30, 24
2-Hydroxy-6-naphthoic acid	93	9	90	53	37	1	33	13	24, 65	23, 65	12, 15	43, 63	—	60	30	++	+	—	—
3,5-Dihydroxy-2-naphthoic acid	94	36	92	33	36	1	6	15, 23	24	25, 28	64, 68	63, 29	—	28, 69	43, 68	++	+	5	—
5-Hydroxyisoquinoline	59	85	29	56	7	1	34	12, 15	20	23, 65	64, 55	35, 28	—	71	68	++	—	—	—
1,5-Dihydroxyisoquinoline	84	68	68	25	6	0	34	17, 15	63, 25	25, 28	15, 65	39	—	26	68	++	—	—	—
2-Hydroxyquinoline-4-carboxylic acid	79	0	37	8	5	0	34	—	—	—	—	—	—	—	—	—	+	—	—
5-Hydroxy-1-tetralone	94	92	92	92	87	68	17	9	60	—	—	8, 38	—	—	—	+	—	—	—
5-Hydroxy-1,4-naphthoquinone[b]	92	91	90	93	93	92	53	—	—	—	—	—	62	—	23, 71	+	—	—	—
1-Amino-5-naphthol	92	92	87	72	35	8	70	—	51	24, 70	63	—	—	+	70	+	—	8, 60	24
2,4-Dinitro-1-naphthol	92	87	92	93	94	93	3	—	—	—	—	—	—	72	—	++	—	—	—
2-Hydroxy-3-naphthoic acid hydrazide	92	42	92	93	80	37	57	16, 15	45, 25	24, 25	15	50, 33	—	—	50, 33	+	—	5	17
2,8-Dihydroxy-3-naphthoic acid	94	43	93	28	56	3	3	15, 23	56, 23	25	63, 35	28	—	51	43, 68	++	+	—	—
1,8-Dihydroxyanthracene (chrysazol)	95	95	93	96	93	94	—	—	—	—	—	—	—	—	69	—	—	—	—
1,5-Dihydroxyanthraquinone (anthrarufin)	92	94	93	95	93	92	57	—	—	—	—	—	—	—	—	—	—	—	—
1,8-Dihydroxyanthraquinone (chrysazin)	93	94	89	95	92	92	60	—	—	—	—	+	—	—	26	—	—	—	—
1,2,5,8-Tetrahydroxyanthraquinone (quinalizarin)	92	0	88	29	19	0	57	—	—	—	—	—	—	—	27	—	—	—	—
Carminic acid	38	0	0	0	0	0	17	—	—	—	—	—	—	—	—	—	33	—	—
1-Aminoanthraquinone	93	93	96	94	92	90	17	10	+	—	—	—	—	—	—	—	—	17	26
2-Aminoanthraquinone	91	90	95	92	87	82	18	—	—	—	—	—	—	—	—	—	—	17	27

[a] For interpretation of readings, see Table PC 120.
[b] Ninhydrin reagent, pale yellow.

TABLE PC 105—(Continued)

NITROGEN HETEROCYCLICS, NAPHTHALENE AND QUINOLINE DERIVATIVES

Paper P1 = W1.

Solvent
S1 = ethyl methyl ketone-acetone-formic acid-water (40 : 2 : 1 : 6).
S2 = ethyl methyl ketone-diethylamine-water (921 : 2 : 77).
S3 = methyl isobutyl ketone-formic acid-water (10 parts ketone saturated with 1 part 4% formic acid).
S4 = chloroform-methanol-formic acid-water (10 parts of chloroform saturated with a mixture of 1 part methanol and 1 part 4% formic acid).
S5 = benzene-ethyl methyl ketone-formic acid-water (a mixture of 9 parts benzene and 1 part ketone saturated with 1 part 2% formic acid).
S6 = benzene-formic acid-water (10 parts benzene saturated with 1 part 2% formic acid).

Detection
D1 = ultraviolet light.
D2 = diazotized sulfanilic acid (0.3% solution in dioxane-water 1 : 2).
D3 = diazotized 4-benzoylamino-2,5-dimethoxyaniline (0.6% solution in dioxane-water 1 : 2).
D4 = diazotized *o*-dianisidine (0.6% solution in dioxane-water 1 : 2).
D5 = *p*-nitrobenzenediazonium fluoborate (0.4% solution in dioxane-water 1 : 2).
D6 = 2,6-dibromoquinone-4-chloroimide (0.5% solution in dioxane-acetone 4 : 1).
D7 = 2,4-dinitrophenylhydrazine (*ca.* 0.1% solution in 1*N* HCl).
D8 = ferric chloride (2% aqueous solution).
D9 = phosphomolybdic acid (2% aqueous solution).
D10 = potassium permanganate (1% aqueous solution).
D11 = bromophenol blue (*ca.* 0.05% solution in ethanol).
D12 = Ehrlich reagent (1% *p*-dimethylaminobenzaldehyde in 1*N* HCl).
D13 = *p*-dimethylaminocinnamaldehyde (0.1% solution in 1*N* HCl).

Technique Descending four-five hours, 40 cm.

Literature
1. Reio, L., *J. Chromatogr.*, **13**, 475–496 (1964), and *J. Chromatogr.*, **47**, 60–85 (1970).

TABLE PC 106
NITROGEN HETEROCYCLICS, QUINOLINE DERIVATIVES

Paper	P1	P1	P1
Solvent	S1	S2	S3
Literature	1	1	1
Compound	$R_F \times 100$		
2,3-Dihydroxyquinoline	—	—	18
2,4-Dihydroxyquinoline	83	—	—
2,6-Dihydroxyquinoline	78	59	—
4,6-Dihydroxyquinoline	78	41	—
3-Hydroxyquinoline	—	—	30
6-Hydroxyquinoline	86	88	—
6-Hydroxyquinolyl-5-sulfuric acid	53	—	—
2-Quinolonyl-6-glucuronide	38	—	—
4-Quinolonyl-6-glucuronide	34	—	—
Quinolyl-6-glucuronide	25	—	—

Paper P1 = W1 or W4.

Solvent S1 = *n*-butanol-acetic acid-water (4 : 1 : 5).
S2 = benzene-*n*-butanol-concentrated NH_4OH (2 : 5 : 2).
S3 = benzene-ethanol-water (5 : 1 : 4).

Detection UV after exposure to HCl or NH_3.

Technique Development distance = 12–15 in.

Literature
1. Smith, J. N. and R. T. Williams, *Biochem. J.*, **60**, 284–290 (1955).

TABLE PC 107
PEROXIDES, ISOMERIC MONOSUBSTITUTED PEROXYBENZOIC ACIDS

Paper	P1	P2
Solvent	S1	S2
Literature	1	1
Compound	R_b[a] × 100	
***m*-Bromo**	123	120
***o*-Bromo**	66	106
***p*-Bromo**	114	142
***m*-Chloro**	91	130
***o*-Chloro**	51	117
***p*-Chloro**	80	142
***o*-Fluoro**	57	89
***p*-Fluoro**	10	106
***m*-Methyl**	157	153
***p*-Methyl**	131	136
***o*-Nitro**	11	33

[a] R_F relative to peroxybenzoic acid.

Paper P1 = 25% acetylated, Binzer.
P2 = impregnated with *N*-methylformamide.

Solvent S1 = hexane saturated with dioxane and dimethylformamide.
S2 = benzene.

Detection *p*-aminodimethylaniline hydrochloride.

Technique Equilibrate 30 minutes in tank, immerse in mobile phase for ascending development, 25 °C.

Literature
1. Kavcic, R., Plesnicar, B., and B. Hristovski, *J. Chromatogr.*, **22**, 381 (1966).

TABLE PC 108
NUCLEOTIDES, NUCLEOSIDES AND BASES

Paper Solvent Literature	na S1 1	na S2 1	na S3 1	na S4 1	na S5 1	na S6 1	na S7 1	na S8 1	na S9 1	na S10 1	na S11 1	na S12 1	na S13 1	na S14 1
Compound							$R_F \times 100$							
Adenosine 3′ : 5′-cyclic phosphate	—	74	67	—	—	—	27	—	—	65	—	—	—	—
Adenosine 5′-diphosphate	—	45	—	78	47	—	—	43	< 5	22	51	11	—	—
Adenosine 2′-monophosphate	—	70	77	—	—	—	52	—	—	39	65	—	—	—
Adenosine 3′-monophosphate	—	—	—	—	—	—	52	—	—	39	65	—	—	—
Adenosine 5′-monophosphate	—	60	85	72	50	—	52	56	< 5	26	65	27	—	29
Adenosine 5′-triphosphate	—	40	—	85	44	—	—	29	< 5	13	42	7	—	—
Cytidine 5′-diphosphate	—	40	—	—	54	—	—	33	< 5	—	47	—	—	—
Cytidine 2′ : 3′-monophosphate	63	60	—	—	—	—	72	—	—	66	—	—	—	—
Cytidine 2′ and 3′-monophosphates	86	—	—	—	—	—	84	—	—	42	—	—	—	—
Cytidine 5′-monophosphate	84	55	94	90	57	—	87	45	< 5	25	59	—	—	82
Cytidine 5′-triphosphate	—	32	—	—	50	—	—	21	< 5	—	33	—	—	—
Deoxyadenosine 5′-diphosphate	—	67	—	—	—	—	—	52	45	37	55	—	—	—
Deoxyadenosine 5′-monophosphate	—	73	—	70	—	—	—	60	53	40	72	—	—	—
Deoxyadenosine 5′-triphosphate	—	51	—	—	—	—	—	35	24	33	42	—	—	—
Deoxycytidine 5′-monophosphate	—	—	—	90	—	—	—	59	21	43	—	—	—	—
Deoxyguanosine 5′-diphosphate	—	38	—	—	—	—	—	—	6	—	—	—	—	—
Deoxyguanosine 5′-monophosphate	—	45	—	80	—	—	—	42	8	35	—	—	—	—
Deoxyguanosine 5′-triphosphate	—	28	—	—	—	—	—	—	5	—	—	—	—	—
Guanosine 5′-diphosphate	—	23	—	—	44	—	—	16	< 5	8	26	5	—	—
Guanosine 2′ and 3′-monophosphates	—	43	—	—	—	—	62	22	—	42	—	—	—	—
Guanosine 5′-monophosphate	—	38	—	82	47	—	73	30	< 5	8	36	13	—	53
Guanosine 5′-triphosphate	—	21	—	—	40	—	—	11	< 5	—	19	—	—	—
Inosine 5′-diphosphate	—	—	—	—	—	—	—	17	< 5	—	—	10	—	—
Inosine 5′-monophosphate	—	39	—	—	—	—	—	31	< 5	—	—	25	—	65
Inosine 5′-triphosphate	—	—	—	—	—	—	—	10	< 5	—	—	6	—	—
Thymidine 5′-diphosphate	—	39	—	—	—	—	—	37	6	—	—	—	—	—
Thymidine 5′-monophosphate	—	51	—	—	83	—	—	59	20	44	—	—	—	—
Thymidine 5′-triphosphate	—	30	—	—	—	—	—	25	6	—	—	—	—	—
Uridine 5′-diphosphate	—	28	—	—	72	—	—	18	< 5	—	26	15	—	—
Uridine 2′ : 3′-monophosphate	53	—	—	—	—	—	—	—	—	—	—	—	—	—
Uridine 2′ and 3′-monophosphates	78	—	—	—	—	—	—	—	—	36	—	—	—	—
Uridine 5′-monophosphate	76	43	—	90	74	—	—	32	< 5	24	34	33	—	80
Uridine 5′-triphosphate	—	19	—	—	70	—	—	16	< 5	—	20	5	—	—

TABLE PC 108—(Continued)
NUCLEOTIDES, NUCLEOSIDES AND BASES

Paper	na	na	na	na	na	na	na	na	na	na	na	na	na	na
Solvent	S1	S2	S3	S4	S5	S6	S7	S8	S9	S10	S11	S12	S13	S14
Literature	1	1	1	1	1	1	1	1	1	1	1	1	1	1
Compound	$R_F \times 100$													
Deoxynucleosides														
Deoxyadenosine	—	—	—	—	41	40	—	94	71	76	—	—	—	—
Deoxycytidine	—	—	—	—	—	26	71	82	67	75	—	—	—	—
Deoxyguanosine	—	—	66	—	30	14	50	—	45	57	—	—	—	—
Deoxyinosine	—	69	68	—	34	9	—	—	—	—	—	—	—	—
Deoxyuridine	—	—	—	—	79	—	—	—	25	—	—	—	—	—
Thymidine	—	—	—	—	87	49	—	83	—	—	—	—	36	—
Ribonucleosides														
Adenosine	—	87	59	59	40	34	36	89	59	65	90	—	—	14
Cytidine	—	80	89	83	51	13	76	76	40	62	76	—	—	70
Guanosine	—	50	—	47	29	13	36	52	23	42	—	46	—	43
Inosine	—	—	73	—	34	7	—	61	—	—	—	—	—	—
Uridine	—	60	90	84	75	14	—	60	—	—	—	70	—	70
Xanthosine	—	51	—	—	26	< 2	—	—	—	—	—	—	—	—
Purine and Pyrimidine Bases														
Adenine	—	95	44	43	40	40	—	94	55	65	94	—	—	11
Cytosine	—	95	83	75	51	27	74	85	58	66	88	—	—	72
Guanine	—	65	—	—	29	13	—	—	—	—	—	46	—	5
Hypoxanthine	—	70	60	—	28	18	—	—	—	—	—	—	—	—
Thymine	—	—	74	—	85	50	—	83	—	—	—	—	50	—
Uracil	—	70	80	75	75	33	—	72	—	—	—	70	—	63
Xanthine	—	54	—	—	25	—	—	—	—	—	—	—	—	—

Solvent

S1 = water-saturated $(NH_4)_2SO_4$[70g $(NH_4)_2SO_4$ plus 100 g H_2O] -0.1*M* pH 6 phosphate buffer-isopropanol (79 : 19 : 2).
S2 = isobutyric acid-2.0*N* NH_4OH (66 : 34).
S3 = isoamyl alcohol-5% aqueous $Na_2HPO_4 \cdot H_2O$ (1 : 1).
S4 = isoamyl alcohol-5% aqueous sodium citrate$\cdot 2H_2O$ (1 : 1).
S5 = isopropanol-concentrated HCl (65 : 16.7), dilute to 100 ml with H_2O.
S6 = water-saturated *n*-butanol-15*N* NH_4OH (100 : 1).
S7 = saturated $(NH_4)_2SO_4$ in H_2O-1*M* sodium acetate-isopropanol (80 : 18 : 2).
S8 = isobutyric acid-15*N* NH_4OH-H_2O (66 : 1 : 33).
S9 = isopropanol-15*N* NH_4OH-0.1*M* H_3BO_3 (7 : 1 : 2).
S10 = isopropanol-15*N* NH_4OH-0.1*M* H_3BO_3 (6 : 1 : 3).
S11 = isobutyric acid-1*N* NH_4OH-0.2*M* EDTA (100 : 60 : 0.8).
S12 = 95% ethanol-1*M* ammonium acetate, pH 5.0 (adjusted with acetic acid) (7 : 3).
S13 = ethyl acetate-H_2O-formic acid (60 : 35 : 5). Equilibrate for 16½ hours in lower phase; run for 3 hours in upper phase.
S14 = 13.8 g $NaH_2PO_4 \cdot H_2O$ in 900 ml H_2O, adjust pH to 6.8 with H_3PO_4 and dilute to 1 liter with H_2O, then add 600 g $(NH_4)_2SO_4$ and 20 ml *n*-propanol.

Detection UV light.

Technique Paper fastened into a cylinder, taped to the inner wall of a chamber containing 20 ml of solvent to equilibrate 3 hours at 30 °C, and then lowered into the solvent to develop 16 hours at 30 °C.

Literature

1. Schwarz, BioResearch Catalogue, Orangeburg, N.Y., 1969, pp 6–7, 48, 52.

TABLE PC 109
OXO COMPOUNDS, ANTHRAQUINONES

Paper	P1	
Solvent	S1	
Detection	—	D1
Literature	1	

Compound	$R_F \times 100$	Color
Alizarin	4	vt
Aloe-emodin	15	o
Anthragallol	0	gy
Chrysophanol	92	o
Dihydroxymethylanthraquinone	92	o
Emodin	52	pk
Endocrocin	0	pk
2-Methylquinizarin	92	p
Physcion	89	o
Purpurin	3	p
Quinizarin	89	p
Rhein	0	o
Rhodocladonic acid	0	o
Rubiadin	49	o-y
Rufigallic acid	0	gy-vt
Tetrahydroxyanthraquinone	2	o

Paper P1 = Toyo-Filter paper No. 2.
Solvent S1 = petroleum ether (45–70 °C), saturated with 97% methanol.
Detection D1 = spray with solution of 0.5% magnesium acetate in methanol.
Technique Development distance 30 cm, 5–6 hours, 24–25 °C, closed chamber.
Literature
1. Shibata, S., Takito, M., and O. Tanaka, *J. Amer. Chem. Soc.*, **72**, 2789–2790 (1950).

TABLE PC 110
OXO COMPOUNDS, ANTHRAQUINONE DERIVATIVES

Paper	P1
Solvent	S1
Literature	1

Compound	$R_F \times 100$
Anthraquinone	57
1-Bromoanthraquinone	64
2-Bromoanthraquinone	55
1-Bromo-2-chloroanthraquinone	33
1-Chloroanthraquinone	64
2-Chloroanthraquinone	50
1-Chloro-2-bromoanthraquinone	31
1-Chloro-5-bromoanthraquinone	70
2-Chloro-3-bromoanthraquinone	19
1-Chloro-2-fluoroanthraquinone	65
1-Chloro-5-fluoroanthraquinone	75
2-Chloro-3-fluoroanthraquinone	44
1-Chloro-2-iodoanthraquinone	28
1-Chloro-5-iodoanthraquinone	68
2-Chloro-3-iodoanthraquinone	17
1-Fluoroanthraquinone	81
2-Fluoroanthraquinone	75
1-Fluoro-2-chloroanthraquinone	71
1-Iodoanthraquinone	59
2-Iodoanthraquinone	50
1-Iodo-2-chloroanthraquinone	30

Paper P1 = Soviet paper M impregnated with 5% paraffin oil in petroleum ether.
Solvent S1 = 60% acetic acid saturated with paraffin oil.
Detection UV light.
Literature
1. Dokunikhin, N. S., Kolokolov, B. N., and O. A. Egorova, *Zh. Analit. Khim.*, **21**, 889 (1966).

TABLE PC 111
OXO COMPOUNDS, NAPHTHOQUINONES, SUBSTITUTED

Paper	P1	P2	
Solvent	S1	S2	
Technique	T1	T2	
Literature	1	2	
Compound	$R_F \times 100$		Color
Benzoquinones			
***p*-Benzoquinone**	87	—	pl br
2,5-Dichloro-*p*-benzoquinone	28	—	pl y
1,2-Naphthoquinone-3-sulfonic acid	—	85	p
3,4,5,6-Tetrachloro-*o*-benzoquinone	90	—	pl y-br
2,3,5,6-Tetrachloro-*p*-benzoquinone	22	—	pl y
***p*-Toluquinone**	85	73	y-br
***p*-Xyloquinone**	97	—	pl br
1,4-Naphthoquinones			
Unsubstituted	85	46	o-br
2-(*N*-Acetanilido)-3-chloro-	89	—	pl o-br
2-Acetylamino-	83	—	pk-o-br
2-Amino-	82	—	y-br
2-Amino-3-chloro-	83	—	o
2-Chloro-3-*n*-declyamino-	91	—	pl y
2-Chloro-3-dimethylamino-	87	—	o-pk
2-Chloro-3-ethylamino-	88	—	—
2,3-Dichloro-	41	—	y
2,3-Dimethoxy-	84	—	p-pk
2-Dimethylamino-	83	—	o
Ferric 2-chloro-3-hydroxy-	42	—	o-pk
2-Hydroxy-	47	—	o
2-Methoxy-	81	—	o
2-Methoxy-3-hydroxy-	37	—	p-pk
2-Methyl-	87	34	o-br
2-Methyl-3-acetyl-	83	—	pl br
2-Methylmercapto-	84	—	y
3-Methyl-6-succino-	53	—	p

Paper P1 = W1.
P2 = W2.

Solvent S1 = fusel oil-pyridine-water (3 : 2 : 1.5).
S2 = concentrated NH_4OH.

Detection Spray with 5% aqueous solution of NaOH.

Technique T1 = descending development, 39–40 cm, 15–18 hours, 18 °C.
T2 = ascending development, saturated chamber.

Literature

1. Sprosten, T. and E. G. Bassett, *Anal. Chem.*, **26**, 552–553 (1954).
2. Canady, W. J. and J. H. Roe, *J. Biol. Chem.*, **220**, 563–570 (1956).

TABLE PC 112
PHOSPHORUS COMPOUNDS, AROMATIC AND PYRIDYL PHOSPHATE ESTERS, AND RELATED COMPOUNDS

Paper	P1	P2
Solvent	S1	S2
Technique	T1	T2
Literature	1	2
Compound	$R_F \times 100$	
Benzylphosphate	—	11
Benzyl-α-pyridyl phosphate	—	64
4-Chlorophenyl phosphate	—	16
Di(*p*-chlorophenyl) pyrophosphate	84	—
Diphenyl pyrophosphate	70–73	—
Di-α-pyridyl phosphate	—	53
Ditolyl pyrophosphate	78	—
Inorganic phosphate	15–16	—
Inorganic pyrophosphate	10–11	—
Inorganic trimetaphosphate	25–26	—
Methylphenyl phosphate	—	57
Monethyl phosphate	26	—
Mono(*p*-chlorophenyl) phosphate	54–55	—
Monophenyl phosphate	39–41	8
Monophenyl pyrophosphate	30–32	—
Monotolyl phosphate	46–48	—
β-Naphthylphosphate	—	11
β-Naphthyl-α-pyridyl phosphate	—	64
Phenylbenzyl phosphate	—	72
Phenyl-α-pyridyl phosphate	—	62
Propylphenyl phosphate	—	70
Thymylphosphate	—	32
Thymyl-α-pyridyl phosphate	—	76

Paper P1 = MN 2261 (washed).
P2 = SS2043b (washed).

Solvent S1 = *n*-propanol-NH_3-H_2O (6 : 3 : 1).
S2 = *iso*-propanol-NH_3-H_2O (8 : 1 : 1).

Detection Molybdate acid spray reagent or UV light.

Technique T1 = ascending development, 15 hours at 22 °C.
T2 = descending development.

Literature

1. Hennrich, N. and F. Cramer, *J. Amer. Chem. Soc.* **87**, 1121 (1965).
2. Kampe, W., *Chem. Ber.*, **98**, 1038 (1965).

TABLE PC 113
OXYGEN HETEROCYCLICS, COUMARINS AND RELATED COMPOUNDS

Paper	P1	P1	P1	P1	P1	P1	P2	P3	P1	P2	P3	P1	P1	P1	P1	P1	P1	P1	P1	P1	na
Solvent	S1	S2	S3	S4	S5	S6	S6	S6	S7	S7	S7	S8	S9	S10	S11	S12	S13	S14	S15	S16	S17
Literature	1	1	1	1	1	1	1	1	1	1	1	1	1	1	1	1	1	1	1	1	2
Compound	$R_F \times 100$																				
Aesculetin	72	81	79	65	80	32	11	29	44	0	14	75	0	31	63	65	34	48	54	65	73
Aesculin	80	64	49	21	44	30	3	12	4	0	2	8	0	34	78	80	81	69	76	77	—
Bergapten	98	97	88	88	97	88	91	88	94	96	95	94	96	86	65	59	19	45	38	81	—
Caffeic acid	33	46	78	73	75	28	0	5	21	0	3	80	0	19	60	57	78	38	78	69	—
Chlorogenic acid	49	33	61	31	54	5	0	4	0	0	0	31	0	10	78	76	89	64	81	75	—
o-Coumaric acid	66	81	90	90	92	72	20	23	78	1	18	93	0	39	74	69	86	58	81	81	—
p-Coumaric acid	67	81	88	88	90	62	16	18	77	1	12	90	0	32	69	65	81	49	79	78	—
Coumarin	98	97	92	86	98	87	87	88	92	94	94	93	95	87	84	81	67	76	74	87	—
Daphnetin	84	83	81	68	83	60	10	40	74	0	58	78	2	38	69	70	61	54	62	72	—
Ferulic acid	77	89	84	81	90	54	11	13	55	1	13	90	0	26	68	63	78	48	77	74	—
Fraxetin	—	—	—	—	—	—	—	—	—	—	—	—	—	—	—	—	—	—	—	—	73
Fraxin	—	—	—	—	—	—	—	—	—	—	—	—	—	—	—	—	—	—	—	—	44
Herniarin	98	98	91	87	98	85	87	87	92	94	94	94	95	87	79	76	50	67	63	84	—
Imperatorin	98	97	91	92	97	91	91	92	91	96	95	96	97	90	65	50	22	47	59	90	—
Limettin	97	97	92	89	98	88	89	88	94	95	95	96	96	88	67	40	11	39	44	87	—
Osthol	98	98	92	91	97	92	89	90	96	96	94	89	96	89	72	71	17	53	66	91	—
Ostruthin	96	98	93	95	97	95	92	93	95	97	95	97	96	88	36	7	5	10	18	93	—
Phloroglucinol	32	16	69	56	52	70	60	65	62	50	68	68	18	35	72	69	52	61	68	80	—
Scopoletin	96	96	83	73	91	76	40	71	85	36	79	83	6	59	70	70	50	51	60	78	77
Scopolin	—	—	—	—	—	—	—	—	—	—	—	—	—	—	—	—	—	—	—	—	33
Skimmin	90	81	52	22	54	33	20	31	9	4	10	12	7	61	84	85	72	82	73	80	—
Umbelliferone	95	94	89	83	92	84	65	83	90	73	89	90	28	66	75	73	57	60	66	84	—
Xanthotoxin	96	97	87	88	96	87	85	86	93	95	94	93	95	84	76	67	33	58	57	83	—

Paper
P1 = W1.
P2 = W1 buffered with 0.1*M* Na_3BO_3.
P3 = W1 buffered with 0.1*M* Na_2HPO_4.

Solvent
S1 = phenol-water.
S2 = *m*-cresol-acetic acid-water (50 : 2 : 48).
S3 = *n*-butanol-acetic acid-water (4 : 1 : 5).
S4 = amyl alcohol-acetic acid-water (4 : 1 : 5).
S5 = benzyl alcohol-acetic acid-water (4 : 1 : 5).
S6 = *n*-butanol, saturated with water.
S7 = ethyl acetate, saturated with water.
S8 = ethyl acetate-2*N* HCl (1 : 1).
S9 = ethyl acetate-2*N* NH_3 (1 : 1).
S10 = 1-propanol-concentrated NH_4OH (7 : 3).
S11 = aqueous phase-*n*-butanol-acetic acid-H_2O (4 : 1 : 5).
S12 = aqueous phase-1-benzyl alcohol-acetic acid-H_2O (4 : 1 : 5).
S13 = water.
S14 = 10% acetic acid.
S15 = isopropanol-water (1 : 4).
S16 = isopropanol-water (2 : 3).
S17 = *n*-butanol-acetic acid-water (60 : 15 : 30).

Detection UV fluorescence.

Technique 35 cm ascending development, 21 ± 0.5 °C.

Literature
1. Swain, T., *Biochem. J.*, **53**, 200–208 (1953).
2. Krebs, K. and A. Wankmüller, *Naturwissenschaften*, **40**, 623–624 (1953).

TABLE PC 114

PEPTIDES, TYROSINE, DIIODOTYROSINE AND THYROXINE

Paper	P1	P1
Solvent	S1	S2
Literature	1	1
Compound	$R_F \times 100$	
3,5-Diiodo-L-tyrosyl-glycine	9	32
Glycyl-L-3,5-diiodotyrosine	13	28
Glycyl-3,5-diiodo-L-tyrosyl-glycine	5	27
Glycyl-L-thyroxine	27	57
Glycyl-L-thyroxyl-glycine	33	51
Glycyl-L-tyrosine	12	11
Glycyl-L-tyrosyl-glycine	9	12
L-Thyroxyl-glycine	37	58
L-Tyrosyl-glycine	14	12

Paper P1 = W3MM.

Solvent S1 = *n*-butanol-dioxane-2*N* NH_3 (4 : 1 : 5).
S2 = *n*-butanol-acetic acid-H_2O (78 : 10 : 12).

Literature 1. Shiba, T. and H. J. Cahnmann, *J. Org. Chem.*, **29**, 3063 (1964).

TABLE PC 115

PESTICIDES, ESTERS OF 2,4-DICHLORO-PHENOXYACETIC ACID

Paper	P1	P1	P1
Solvent	S1	S2	S3
Literature	1	1	1
Compound	$R_F \times 100$		
Butoxyethanol ester	53	28	—
***p*-Butyl ester**	44	—	60
Iso-octyl ester	32	—	—
Isopropyl ester	54	43	—
Propylene glycol butyl ether	42	—	76

Paper P1 = W1.

Solvent S1 = immobile phase—10% mineral oil in ether, mobile phase—65% acetonitrile in H_2O.
S2 = immobile phase—10% phenoxyethanol in ether, mobile phase—isooctane.
S3 = immobile phase—50% dimethylformamide in ether, mobile phase—isooctane.

Detection Silver nitrate, UV light.

Technique Paper washed three times with water by the ascending technique and then dried, prior to spotting and development.

Literature 1. Yip, G., *J. Ass. Offic. Agr. Chem.*, **47**, 343 (1964).

TABLE PC 116
PESTICIDES, CHLORINATED

Paper	na	na	
Solvent	S1	S2	
Literature	1	1	
Compound	$R_F \times 100$		Approximate minimum detectability μg
Aldrin	37	100	0.2
BHC	79, 83, 78, 93	16, 26, 68, 77	0.2–2.0
Chlordane	streak	96	0.2
Chlorobenzilate	100	50	0.2–0.5
2,4-D	95	0	0.05–0.2
DDE	48	98	0.1
DDT	60	87	0.1
Dieldrin	60	89	0.1
Endrin	62	83	0.1
Heptachlor	48	97	0.1
Heptachlor epoxide	72	87	0.1–0.5
Kelthane	76	37	0.1
Lindane	83	65	0.1
Methoxychlor	89	51	0.2
Mitox	62	78	1.0
PCNB	48	80	0.2–1.0
Perthane	59	not determ.	0.5
Rothane	75	streak	0.2
Strobane	streak	streak	0.2
Thiodan	55, 78	48, 95	0.5
Toxaphene	streak	streak	0.2

Solvent S1 = mineral oil (stationary phase); 75% acetone (mobile phase).
S2 = phenoxyethanol (stationary phase); 2,2,4-trimethylphetane (mobile phase).

Detection Silver nitrate, UV.

Literature
1. Mitchell, L. C., *J. Ass. Offic. Agr. Chem.*, **46**(6), 988 (1963).

TABLE PC 117
PESTICIDES, DITHIOCARBAMATE FUNGICIDES

Paper		P1	P2
Solvent		S1	S2
Literature		1	2
Trade or common name	Sensitivity μg	$R_F \times 100$	$R_X{}^a \times 100$
Copper oxychloride	5	22	—
Ferbam	0.1	81	100
Ceresan®	25	80	—
Urbazit	0.1	83	—
Cerenox-Special	0.1	78	—
Polyram	2.0	83	—
Cerenox	10	78	—
Thiram	0.1	82	100
Camatan	2.0	89	—
Brestan®	2.5	91	—
Ziram	0.1	80	100
Zineb	5	streaks	37
Maneb	—	—	37
Nabam	—	—	37

[a] Position relative to that of Ferbam.

Paper P1 = SS2043b.
P2 = glass fiber paper.

Solvent S1 = *n*-butanol-acetic acid-water (4 : 1 : 1).
S2 = 5% formamide in acetone (stationary), chloroform (mobile).

Detection Bioautographic detection.

Technique Ascending development.

Literature
1. Weltzien, H. C., *Naturwissenschaften*, **45**, 228 (1958).
2. McKinley, W. P. and S. A. Magarvey, *J. Ass. Offic. Agr. Chem.*, **43**, (3) 717 (1963).

TABLE PC 118
PESTICIDES, HERBICIDES

Paper	P1	P2	P3
Solvent	S1	S2	S3
Literature	1	1	2
Compound	$R_F \times 100$		
2,4-D	64[a]	39	58
Dalapon	—	—	45
2,4-DB	34	73	79
MCPA	50	58	68
MCPB	—	—	81
2,4,5-T	48	50	66
2,4,5-TP	31	82	—

[a] Separated as methyl esters.

Paper P1 = W1 impregnated with 10% mineral oil in ethyl ether.
P2 = W1 impregnated with 35% DMF (dimethylformamide).
P3 = Untreated paper, W1.

Solvent S1 = 50% acetonitrile in water.
S2 = 2,2,4-trimethylpentane.
S3 = liquid paraffin-benzene-acetic acid-cyclohexane (1 : 3 : 2 : 14).

Detection $AgNO_3$ UV.

Literature
1. Yip, G., *J. Ass. Offic. Agr. Chem.*, **47**, 343 (1964).
2. Abbott, D. C., Egan, H., Hammond, E. W., and J. Thompson, *Analyst*, **89**, 480 (1964).

TABLE PC 119
PESTICIDES, PYRETHRINS

Paper	P1
Solvent	S1
Literature	1
Compound	$R_F \times 100$
α-DL-*trans*-Allethrin	82
Allethrolone	9
***cis*-Chrysanthemic acid**	56
***trans*-Chrysanthemic acid**	54
Chrysanthemum dicarboxylic acid	0
Methyl pyrethrate	73
Pyrethrin I	85
Pyrethrin II	70

Paper P1 = W1 equilibrated with S1 overnight in the chromatographic chamber under reduced pressure.

Solvent S1 = petroleum ether (80–100 °C), saturated with CH_3OH.

Detection $KMnO_4$, benzidine.

Technique Ascending.

Literature
1. Quayle, J. R., *Nature*, **178**, 375–376 (1956).

TABLE PC 120
PHENOLS AND DERIVATIVES

Paper / Solvent / Detection / Literature	P1 S1 — 1	P1 S2 — 1	P1 S3 — 1	P1 S4 — 1	P1 S5 — 1	P1 S6 — 1	D1	D2	D3	D4	D5	D6	D7	D8	D9	D10	D11	D12	D13
Compound	$R_F \times 100$						Color												
o-tert-Butylphenol	95	93	95	95	95	94	—	7	8	63, 56	8, 24	43, 35	—	—	43, 68	++	—	—	—
2,6-Di-*tert*-butylphenol	95	93	94	95	95	94	—	17	17	17	9, 28	28, 25	—	—	43	+	—	—	—
o-Acetaminophenol	89	85	84	75	64	30	—	6, 10	9, 15	19, 65	7, 23	41, 37	—	26, 70	69	43, 66	—	—	—
m-Acetaminophenol	87	84	79	20	17	2	—	6, 9	15	19, 15	9, 15	38, 35	—	70	71	++	—	3	+
p-Acetaminophenol	83	81	72	9	7	0	—	14	12, 64	63, 65	62, 65	—	—	27	30	++	—	—	60
p-Methylaminophenol	34	85	13	16	0	0	56	—	—	+	63	53	—	+	68	++	—	52	21
4-Amino-3-methylphenol	40	82	10	12	3	0	55	—	—	64	7	—	—	—	69	++	—	7	21
2-Amino-4-nitrophenol	92	65	92	56	57	8	55	+	58	55	58	55	—	60	43, 68	++	—	6	23
2-Amino-5-nitrophenol	94	89	92	56	57	8	59	63	63	55, 65	55	21, 63	—	55	55, 69	++	—	15, 12	25, 28
4-Amino-2-nitrophenol	91	82	87	86	78	70	55	56	62	63	64	—	—	—	63, 68	++	—	9	19
p-Anisidine	49	86	27	79	24	17	70	8	59	65, 63	59	—	—	24	50, 53	++	—	9	19
o-Dianisidine	72	88	48	93	53	41	56	63	—	—	56	50	—	43, 68	53	++	—	15, 20	23, 65
4-Nitro-*o*-anisidine	92	90	94	90	92	86	65	—	—	—	—	—	—	—	—	+	—	12	65, 25
5-Nitro-*o*-anisidine	92	91	95	91	92	86	65	—	—	—	—	—	—	—	—	+	—	7	19
2-Nitro-*p*-anisidine	93	91	92	92	89	86	64	—	—	—	—	57	—	—	—	+	—	11	24, 25
5-Hydroxyindole	92	91	87	43	59	22	—	64	55, 65	24, 25	56, 63	28	16	55	43, 68	++	—	23	43, 28
2-Nitro-3-hydroxyacetophenone	96	53	94	93	89	88	57	—	—	—	+	—	+	—	—	—	—	—	—
Salicylaldoxime	95	91	93	89	88	68	—	—	—	63	—	3, 41	—	54	68	++	—	—	—
Salicylamide	92	79	86	76	69	44	33	7	8	65	10	41	—	23	—	++	—	6	21
5-Bromosalicylic acid	95	65	93	88	87	66	34	—	—	—	—	—	—	24	—	+	++	—	—
5-Nitrosalicylic acid	93	67	87	76	77	24	26	—	—	—	—	—	—	64	—	—	++	—	—
3-Hydroxybenzoic acid methyl ester	89	92	91	84	87	71	—	8	62	63, 65	12	43	—	+	—	++	—	—	—

TABLE PC 120—(Continued)
PHENOLS AND DERIVATIVES

Paper Solvent Detection Literature	P1 S1 — 1	P1 S2 — 1	P1 S3 — 1	P1 S4 — 1	P1 S5 — 1	P1 S6 — 1	D1	D2	D3	D4	D5	D6	D7	D8	D9	D10	D11	D12	D13
Compound	$R_F \times 100$						Color												
3-Hydroxy-4-nitrobenzoic acid	92	6	92	83	73	26	59	+	8	56	63	48	—	+	8, 34	+	+	—	57
3,5-Dichloro-4-hydroxybenzoic acid	94	93	93	74	76	15	—	—	—	—	8, 63	40	—	57	—	++	+	—	—
5-Hydroxyanthranilic acid	81	0	48	0	0	0	34	7	—	24, 65	3, 64	—	—	—	34	+	+	—	6
o-Methoxyphenylacetic acid	93	14	89	88	87	84	—	—	—	—	—	—	—	—	—	—	+	—	—
m-Hydroxyphenylacetic acid	91	5	87	19	22	1	—	6, 9	14	15, 65	8, 65	37, 25	—	—	71	++	+	—	—
m-Hydroxyphenylpropionic acid	93	10	88	37	41	3	—	6, 10	12	65	9, 65	38, 25	—	—	30, 71	++	+	—	—
o-Methoxycinnamic acid	95	16	94	91	90	76	33	—	—	—	—	—	—	—	—	++	+	—	—
m-Methoxycinnamic acid	95	14	94	90	84	69	33	—	—	—	—	—	—	—	—	++	+	—	—
dl-*p*-Hydroxymandelic acid	85	4	57	2	2	0	—	8	12, 64	65	6, 65	38, 25	—	3	71	++	+	—	—
o-Methoxymandelic acid	87	62	78	76	39	8	—	—	—	—	—	—	—	58	—	+	++	—	—
m-Methoxymandelic acid	90	18	78	68	46	5	—	—	—	—	—	—	—	+	—	+	++	—	—
p-Methoxymandelic acid	92	9	76	54	35	4	—	—	—	—	—	—	—	—	—	+	+	—	—
5-Hydroxy-3-indolylacetic acid methyl ester	93	91	89	75	63	25	—	17	+	+	+	—	—	—	—	—	—	34	30, 27
5-Methoxy-3-indolylacetic acid	91	10	91	71	53	11	—	3	—	+	7	—	60	26, 23	44, 71	++	+	26, 27	27, 30
5-Hydroxy-3-indolylacetamide	68	57	20	1	0	0	+	64	23	23, 65	64	68	—	70	71	++	—	51	25, 28
5-Hydroxyisophthalic acid	88	0	83	0	0	0	—	9	—	63	60	38, 25	—	+	—	+	+	—	—
2-Chloro-6-hydroxypyridine	91	76	88	88	88	87	—	—	—	—	—	—	—	57	—	—	—	—	—
2-Mercaptopyridine	72	64	58	68	24	15	63	—	—	62	—	8, 63	—	—	51, 69	++	—	—	—
4-Mercaptopyridine	45	15	12	14	0	0	3	5	9	11	6	59	—	—	45, 68	++	—	—	—
4-Isopropylpyrocatechol	94	94	92	86	86	66	—	59, 9	24	56, 24	64, 65	28	+	+	43, 68	++	—	17	17
3-Nitropyrocatechol	93	69	93	88	83	75	56	7, 19	63	65	63, 68	68	—	45	69, 9	+	—	3	58
3-Methoxy-4-hydroxyphenylacetic acid (homovanillic acid)	90	4	82	59	52	12	—	59, 17	—	—	63	33	—	—	69	+	+	—	—
3-Hydroxy-4-methoxycinnamic acid	91	4	79	59	28	5	30	9, 56	15, 8	63, 65	11, 55	38, 70	—	57	35, 69	++	+	—	—

TABLE PC 120—(Continued)
PHENOLS AND DERIVATIVES

Paper Solvent Detection Literature	P1 S1 — 1	P1 S2 — 1	P1 S3 — 1	P1 S4 — 1	P1 S5 — 1	P1 S6 — 1	D1	D2	D3	D4	D5	D6	D7	D8	D9	D10	D11	D12	D13
Compound	$R_F \times 100$						Color												
3,4-Dihydroxyphenylpropionic acid (hydrocaffeic acid)	87	1	76	13	5	0	—	6, 64	23, 19	23, 19	9, 65	25	—	53	43, 68	++	+	—	—
***dl*-3,4-Dihydroxymandelic acid**	60	0	25	0	0	0	—	58, 17	17	17	8, 65	28	—	43	68	++	+	—	—
3-Hydroxy-4-methoxymandelic acid	77	3	47	4	2	0	—	7, 10	8, 15	11, 25	13, 65	38, 35	—	58	71	+	+	—	—
4-Hydroxy-3-methoxymandelic acid	79	3	48	5	4	0	—	63	60	+	24, 3	38, 28	—	—	69	+	+	—	—
4-Nitrosoresorcinol	87	32	67	16	11	0	56	8, 56	23, 62	65	10, 65	+	—	—	69	++	—	—	—
2,4-Dinitrosoresorcinol	64	5	32	17	4	0	55	58	58	59	+	52	+	43	43	++	—	—	—
2,4-Dihydroxypropiophenone	89	80	90	88	87	68	34	8, 56	56, 64	23, 65	9, 65	6, 28	6	23, 65	34, 71	++	—	—	—
5-Bromo-2,4-dihydroxybenzoic acid	94	20	92	36	49	3	—	6, 59	19, 63	24, 65	12	—	—	24, 27	—	++	+	—	—
4-Ethoxy-2-hydroxybenzoic acid	94	35	94	92	90	71	—	56, 8	62	64	9	—	—	23	+	+	+	—	—
2,4-Dimethoxybenzoic acid	92	24	89	90	85	74	—	—	—	—	—	—	—	—	—	—	+	—	—
2,6-Dihydroxypyridine	58	6	26	81	4	4	40, 56	6, 10	65, 63	66	60, 55	43, 28	—	55	52, 68	++	—	18, 21	5, 52
2,6-Dihydroxyisonicotinic acid	62	0	10	0	0	0	40	6, 8	14	14, 62	6	25	—	—	+	+	+	—	—
2-Bromohydroquinone	94	85	92	46	61	12	—	—	56	60	59	—	—	—	41, 69	+	—	—	—
2,6-Dimethoxy-1,4-benzoquinone	87	84	81	93	83	78	57	—	—	—	—	—	—	—	—	—	—	—	—
2,5-Dimethoxybenzoic acid	92	88	89	88	85	83	+	—	—	—	—	—	—	—	—	—	++	—	—
3,6-Dihydroxyphthalimide	73	5	42	8	4	2	40	7	11	11	60	56	—	51	69, 3	++	—	—	—
1,2,4-Trihydroxybenzene	78	8	62	0	2	0	56	+	+	56	56	56	55	54	68, 55	++	—	+	27, 44
3,4-Dihydroxy-5-methoxybenzaldehyde	89	39	81	56	26	6	—	57	59	60	59	43, 70	9	70	43, 68	++	—	—	—
2,4,6-Trihydroxybenzoic acid	85	41	64	3	1	0	—	10	60, 63	23, 25	12, 65	26, 70	—	25, 26	39, 70	++	++	18, 23	30, 25
3,5-Dimethoxy-4-hydroxycinnamic acid (sinapic acid)	88	3	77	75	28	6	38	15, 23	63	25, 65	14, 62	25	—	21	52, 28	++	+	—	—
3,4,5-Trimethoxyaniline	75	83	52	94	51	35	+	58	58	65, 64	64	71	—	—	52, 70	++	—	7	19
Tetrahydroxy-*p*-benzoquinone	8	0	0	0	0	0	+	—	58	—	—	—	—	—	70	—	—	—	—

TABLE PC 120—(Continued)
PHENOLS AND DERIVATIVES

Paper P1 = W1.

Solvent
S1 = ethyl methyl ketone-acetone-formic acid-water (40 : 2 : 1 : 6).
S2 = ethyl methyl ketone-diethylamine-water (921 : 2 : 77).
S3 = methyl isobutyl ketone-formic acid-water (10 parts ketone saturated with 1 part 4% formic acid).
S4 = chloroform-methanol-formic acid-water (10 parts of chloroform saturated with a mixture of 1 part methanol and 1 part 4% formic acid).
S5 = benzene-ethyl methyl ketone-formic acid-water (a mixture of 9 parts benzene and 1 part ketone saturated with 1 part 2% formic acid).
S6 = benzene-formic acid-water (10 parts benzene saturated with 1 part 2% formic acid).

Color Index

The colors produced by the reagent are recorded as numbers and the corresponding shades can be found in the color index. The — sign indicates an uncertain reaction, which was too weak to deserve color estimation. Reactions with the reagents D10 and D11 are only indicated by the signs: —, + or + +. The + + sign means that a positive reaction was obtained immediately. In the few cases where the colors are recorded by a number placed on top of another number, *e.g.* $\frac{24}{57}$, this indicates that immediately upon spraying a violet spot (24 in the color index) appears which, within a few seconds, turns to light brown (57 in the color index). Usually most colors are unstable and after some time take on a brownish tone; this is to some extent caused by the chemical influence of other reagents used in the vicinity. This change in color is not recorded in the tables, nor is there any column for those compounds that, at this low concentration, are visible on the chromatograms because of their own color.

1 Zinc Yellow	25 Dark Violet	49 Sap Green
2 Lemon Cadmium	26 Light Violet	50 Cedar Green
3 Gold	27 Blue Violet Lake	51 Olive Green
4 Primrose Yellow	28 Delft Blue	52 Bronze
5 Straw Yellow	29 Ultramarine	53 Sepia
6 Deep Cadmium	30 Smalt Blue	54 Burnt Umber
7 Naples Yellow	31 Cobalt Blue	55 Vandyke Brown
8 Middle Chrome	32 Spectrum Blue	56 Raw Umber
9 Deep Chrome	33 Light Blue	57 Brown Ochre
10 Orange Chrome	34 Sky Blue	58 Raw Sienna
11 Spectrum Orange	35 Prussian Blue	59 Golden Brown
12 Scarlet Lake	36 Indigo	60 Burnt Yellow Ochre
13 Pale Vermilion	37 Oriental Blue	61 Copper Beech
14 Deep Vermilion	38 Kingfisher Blue	62 Burnt Sienna
15 Geranium Lake	39 Turquoise Blue	63 Venetian Red
16 Flesh Pink	40 Turquoise Green	64 Terra Cotta
17 Pink Madder Lake	41 Jade Green	65 Burnt Carmine
18 Rose Pink	42 Juniper Green	66 Chocolate
19 Madder Carmine	43 Bottle Green	67 Ivory Black
20 Crimson Lake	44 Water Green	68 Blue Grey
21 Rose Madder Lake	45 Mineral Green	69 Gunmetal
22 Magenta	46 Emerald Green	70 French Grey
23 Imperial Purple	47 Grass Green	71 Silver Grey
24 Red Violet Lake	48 May Green	72 White = colorless

Detection
D1 = ultraviolet light.
D2 = diazotized sulfanilic acid (0.3% solution in dioxane-water 1:2).
D3 = diazotized 4-benzoylamino-2,5-dimethoxyaniline (0.6% solution in dioxane-water 1 : 2).
D4 = diazotized *o*-dianisidine (0.6% solution in dioxane-water 1 : 2).
D5 = *p*-nitrobenzenediazonium fluoborate (0.4% solution in dioxane-water 1 : 2).
D6 = 2,6-dibromoquinone-4-chloroimide (0.5% solution in dioxane-acetone 4 : 1).
D7 = 2,4-dinitrophenylhydrazine (*ca.* 0.1% solution in 1*N* HCl).
D8 = ferric chloride (2% aqueous solution).
D9 = phosphomolybdic acid (2% aqueous solution).
D10 = potassium permanganate (1% aqueous solution).
D11 = bromophenol blue (*ca.* 0.05% solution in ethanol).
D12 = Ehrlich reagent (1% *p*-dimethylaminobenzaldehyde in 1*N* HCl).
D13 = *p*-dimethylaminocinnamaldehyde (0.1% solution in 1*N* HCl).

Technique One dimensional descending in rectangular glass jars 20 × 30 × 60 cm; equilibrated for one hour before use. Four to five hours developing time, 40 cm developing distribution.

Literature
1. Reio, L., *J. Chromatogr.*, **13**, 475–496 (1964).

TABLE PC 121
PHENOLS AND DERIVATIVES

Paper	P1	P1	P1	P1	P1	P1													
Solvent	S1	S2	S3	S4	S5	S6													
Detection	—	—	—	—	—	—	D1	D2	D3	D4	D5	D6	D7	D8	D9	D10	D11	D12	D13
Literature	1	1	1	1	1	1	1	1	1	1	1	1	1	1	1	1	1	1	1
Compound	$R_F \times 100$							Color[a]											
2-Amino-6-nitrophenol	94	92	95	94	88	87	+	59, 62	53	+	+	56	—	—	51	+	—	3	23, 65
***o*-Aminothiophenol**[b]	94	95	94	97	92	91	53	6	—	+	9	+	—	—	—	++	—	8	65, 63
***p*-Aminothiophenol**[b]	95	95	95	97	89	82	53	57	—	62	3	70	—	—	70	+	—	9	65, 63
***m*-Anisidine**	78	92	73	93	64	57	+	11	65, 11	23, 65	62	41	—	—	40	+	—	8	22, 64
***o*-Anisidine**[c]	78	93	77	92	70	61	54	12	+	62	11	38, 70	—	—	40	++	—	8	14, 64
9,10-Anthradiol	94	94	92	96	96	93	57	—	—	—	—	—	—	—	—	—	—	—	—
Arbutin	33	16	2	0	0	0	—	11, 17	64, 62	25, 56	25, 23	33, 25	—	—	38	++	—	—	—
***p*-Cyanophenol**	91	82	89	86	83	52	—	—	—	—	—	—	—	—	—	+	—	—	—
3,4-Dihydroxybenzoic acid ethyl ester	92	83	89	85	80	28	—	+	63	63	63	50, 70	—	51, 70	3, 68	++	—	—	—
3,5-Dihydroxy-4-methylbenzoic acid	91	5	87	4	4	0	—	7	62	63, 23	6	25	—	—	69	+	+	—	—
6,11-Dihydroxynaphthacenequinone	92	94	92	96	95	94	59	—	—	—	—	—	—	—	—	—	—	—	—
2,3-Dihydroxyphenylacetic acid	84	4	91	5	3	0	—	60	64	+	7	25	—	—	68	+	+	—	—
3,4-Dihydroxyphenylglycol	54	32	15	0	0	0	—	62, 60	19, 64	21, 17	59, 24	28	—	—	28, 68	++	—	—	—
3,4-Dihydroxypropiophenone	92	69	91	67	52	8	+	60	+	+	64, 56	60, 69	58	—	52, 68	++	—	—	—
2,3-Dihydroxypyridine	60	24	29	22	6	2	—	8, 13	56, 25	25, 35	9, 65	29, 24	—	28, 46	25, 33	+	—	—	—
2,5-Dihydroxypyridine	46	6	8	1	0	0	26	19, 9	49, 28	25, 28	14	38, 41	—	+	33	+	—	—	—
3,4-Dihydroxypyridine	34	4	3	1	0	0	—	8	—	—	14, 19	—	—	18, 48	33	+	—	—	—
2,5-Dimethoxyaniline	79	93	82	93	84	76	—	15, 63	25, 15	25, 19	63, 23	43, 28	—	43, 51	68	++	—	9, 60	63, 23
Diphenolic acid	93	22	91	48	76	3	—	10	10, 62	62, 56	8, 65	62, 35	—	—	68	++	+	—	17
***p*-Ethylphenol**	95	92	91	95	92	87	—	17	59	64	64	—	—	—	—	++	—	—	—
Gallic acid methyl ester	87	62	76	15	6	0	59	60	24, 65	65	56	60, 52	—	51, 66	50, 68	++	—	—	—

TABLE PC 121—(Continued)
PHENOLS AND DERIVATIVES

Paper	P1	P1	P1	P1	P1	P1													
Solvent	S1	S2	S3	S4	S5	S6													
Detection	—	—	—	—	—	—	D1	D2	D3	D4	D5	D6	D7	D8	D9	D10	D11	D12	D13
Literature	1	1	1	1	1	1	1	1	1	1	1	1	1	1	1	1	1	1	1
Compound	$R_F \times 100$							Color[a]											
9-Hydroxyanthracene	92	93	94	95	94	94	4	—	—	—	—	—	—	—	—	++	—	—	—
3-Hydroxy-2-hydroxymethylpyridine	29	45	1	4	0	0	33	7	63	24	17	38	—	—	—	++	—	—	—
3-Hydroxy-4-methoxybenzyl alcohol (isovanillyl alcohol)	84	78	75	67	34	18	27	10	19	65	8	28, 64	—	—	68	++	—	—	—
3-Hydroxy-4-methoxycinnamic acid	89	3	82	66	38	7	33	60	64	62, 65	9, 23	37	—	—	68	+	+	—	—
2-Hydroxy-5-methylaniline	92	87	46	48	8	3	—	20, 65	63	65, 63	62, 65	51	—	65	68	++	—	8	22, 63
3-Hydroxy-4-methylaniline	91	86	32	27	4	2	+	64	64, 65	65	64, 65	25	—	—	51	++	—	8	23, 61
o-Methoxybenzaldehyde	94	92	96	93	94	93	—	—	—	—	—	30, 68	8	—	+	+	—	—	—
4-Methoxy-2-methylaniline	93	93	46	88	28	23	—	—	—	64	3, 52	—	—	56	68	+	—	2	22, 62
2-Methoxy-4-nitro-5-methylaniline	93	93	90	93	95	92	52	—	—	—	—	—	—	—	—	+	—	9	52, 25
3-Methoxyprocatechol	88	87	84	78	64	42	57	57	63	59	10	25	—	59, 69	52, 68	++	—	—	—
4-Methyl-6-hydroxycoumarin	90	90	87	86	73	0	33	8, 62	60	65, 64	11	—	—	—	—	++	—	—	—
p-(Methylthio)-phenol	94	92	92	92	92	84	—	57, 63	8, 64	60, 66	62, 25	33	—	+	33	++	—	—	—
2-Nitro-*p*-cresol	95	84	91	93	90	85	3	—	—	—	—	+	—	—	—	+	—	—	—
4-Nitro-*m*-cresol	94	82	92	92	92	77	59	—	—	—	—	51	—	—	—	+	—	—	—
p-Nitrophenolsulfate	63	69	8	0	0	0	56	—	—	—	—	—	—	—	—	—	—	—	—
4-Nitropyrocatechol[d]	93	51	89	47	46	4	52	8	7	—	52, 60	12	—	52	6, 69	++	—	7	—
4-Nitropyrocatecholsulfate	79	21	8	0	0	0	58	—	—	—	—	—	—	—	—	+	—	—	—
4-Nitrosalicylic acid	92	57	92	52	85	32	52	—	—	—	—	—	—	58	—	—	++	6	22
5-Nitrovanillin	92	52	93	90	93	90	53	—	—	—	—	—	6	—	—	+	—	—	—
β-Piperonylacrylic acid	92	8	90	91	85	57	33	—	—	—	—	—	—	—	+	+	+	—	—
4-*n*-Propylresorcinol	95	93	95	73	83	37	—	9, 62	52, 19	65, 25	9, 12	24	—	+	43, 33	++	—	18	39
Salicylic acid hydrazide[c]	91	41	90	67	73	19	—	—	61	12, 65	6	43, 68	4	—	68	++	+	4	17

TABLE PC 121—(Continued)
PHENOLS AND DERIVATIVES

Paper	P1	P1	P1	P1	P1	P1													
Solvent	S1	S2	S3	S4	S5	S6													
Detection	—	—	—	—	—	—	D1	D2	D3	D4	D5	D6	D7	D8	D9	D10	D11	D12	D13
Literature	1	1	1	1	1	1	1	1	1	1	1	1	1	1	1	1	1	1	1
Compound	$R_F \times 100$						Color[a]												
Syringin	24	8	0	2	0	0	—	—	—	—	—	—	—	—	—	++	—	—	—
2,4,6-Tribromophenol	94	86	93	93	93	91	26	—	—	—	—	58	—	—	39	++	—	—	—
3,4,5-Trimethoxybenzyl alcohol	86	84	85	90	79	73	—	—	—	—	—	—	—	—	—	+	—	—	—

[a] For interpretation, see Table PC 120.
[b] Dragendorff reagent, positive.
[c] *p*-Dimethylaminobenzaldehyde, positive.
[d] Ninhydrin reagent, pale yellow.

Paper P1 = W1.

Solvent
S1 = ethyl methyl ketone-acetone-formic acid-water (40 : 2 : 1 : 6).
S2 = ethyl methyl ketone-diethylamine-water (921 : 2 : 77).
S3 = methyl isobutyl ketone-formic acid-water (10 parts ketone saturated with 1 part 4% formic acid).
S4 = chloroform-methanol-formic acid-water (10 parts of chloroform saturated with a mixture of 1 part methanol and 1 part 4% formic acid).
S5 = benzene-ethyl methyl ketone-formic acid-water (a mixture of 9 parts benzene and 1 part ketone saturated with 1 part 2% formic acid).
S6 = benzene-formic acid-water (10 parts benzene saturated with 1 part 2% formic acid).

Detection
D1 = ultraviolet light.
D2 = diazotized sulfanilic acid (0.3% solution in dioxane-water 1 : 2).
D3 = diazotized 4-benzoylamino-2,5-dimethoxyaniline (0.6% solution in dioxane-water 1 : 2).
D4 = diazotized *o*-dianisidine (0.6% solution in dioxane-water 1 : 2).
D5 = *p*-nitrobenzenediazonium fluoborate (0.4% solution in dioxane-water 1 : 2).
D6 = 2,6-dibromoquinone-4-chloroimide (0.5% solution in dioxane-acetone 4 : 1).
D7 = 2,4-dinitrophenylhydrazine (*ca.* 0.1% solution in 1*N* HCl).
D8 = ferric chloride (2% aqueous solution).
D9 = phosphomolybdic acid (2% aqueous solution).
D10 = potassium permanganate (1% aqueous solution).
D11 = bromophenol blue (*ca.* 0.05% solution in ethanol).
D12 = Ehrlich reagent (1% *p*-dimethylaminobenzaldehyde in 1*N* HCl).
D13 = *p*-dimethylaminocinnamaldehyde (0.1% solution in 1*N* HCl).

Technique Descending development, 40 cm.

Literature Reio, L., *J. Chromatogr.*, **47**, 60–85 (1970).

TABLE PC 122
PIGMENTS, ALGAL

Paper	P1	P1
Solvent	S1	S2
Literature	1	1
	$R_F \times 100$	
Pigment	1st dimension	2nd dimension
Carotenes	96	96
Chlorophyll *a*	84	29
Chlorophyll *b*	65	10
Chlorophyll *c*	20	0
Diadinoxanthin	54	44
Diatoxanthin	57	60
Dinoxanthin	54	44
Fucoxanthin	49	28
Lutein	80	86
Neodinoxanthin and neodiadinoxanthin	52	30
Neofucoxanthin A and B	49	8
Neoperidinin	41	23
Neoxanthin	32	5
Peridinin	41	23
Siphonaxanthin	30	5
Siphonein	58	38
Violaxanthin	64	50

Paper P1 = W3MM.
Solvent S1 = 4% *n*-propanol in petroleum ether (60–80 °C).
S2 = 30% chloroform in petroleum ether (60–80 °C).
Technique Two-dimensional chromatography.
Literature
1. Jeffrey, S. W., *Biol. Bull.*, **135**, 141 (1968); Jeffrey, S. W. and F. T. Haxo, *ibid.*, **135**, 149 (1968).

TABLE PC 123
PIGMENTS, PORPHYRIN METHYL ESTERS

Paper	P1	P1	P1
Solvent	S1	S2	S3
Literature	1	2	3
Compound	$R_F \times 100$		
Coproporphyrin I	47	37	—
Coproporphyrin III	67	34	—
Deuteroporphyrin IX	88	—	59
Haematoporphyrin IX	—	—	3
Haematoporphyrin IX dimethyl ester	—	—	59
Mesoporphyrin IX	89	—	59
Monovinylmonohydroxyethyldeuteroporphyrin IX	—	—	20
Protoporphyrin I	—	29	—
Protoporphyrin III	—	22	—
Protoporphyrin IX	84	—	51
Uroporphyrin I	17	37	—
Uroporphyrin III	—	40,54	—

Paper P1 = W1 and SS2043b.
Solvent S1 = kerosene-chloroform (4.0 : 2.6), then same direction kerosene-*n*-propanol (5 : 1).
S2 = 80% ethanol v/v.
S3 = kerosene-tetrahydropyran-methyl benzoate (5 : 1.4 : 0.35).
Detection Daylight or UV.
Technique Ascending.
Literature
1. Chu, T. C., Green, A. A., and E. J. H. Chu, *J. Biol. Chem.*, **190**, 643 (1951).
2. Serchi, G., *Ann. Chim.* (*Rome*), **43**, 253 (1953).
3. Chu, T. C. and E. J. H. Chu, *J. Biol. Chem.*, **208**, 537 (1959).

TABLE PC 124
STEROIDS

Paper	na	na	na	na	na	na	na	na	na	
Solvent	S1	S2	S3	S4	S5	S6	S7	S8	S9	
Literature	1	1	1	1	1	1	1	1	1	
Compound	R[a]	R[b]	R[b]	R[a]			$R_F \times 100$			Det.[e]
A. C_{19} steroids										
3β-Acetoxy-5-androsten-17-one	550	—	—	—	85	99	99	—	—	D4
1,4-Androstadiene-3,17-dione	100	—	—	—	27	—	78	—	—	D3
4,9(11)-Androstadiene-3,17-dione	165	—	—	—	50	—	78	—	—	D2
Androstane-3,17-dione	—	—	—	—	—	—	—	—	100	—
5α-Androstane-3,17-dione	270	—[c]	—	—	76	—	—	—	—	D1
4-Androstene-3,17-dione	165	—	530	—	50	89	78	—	—	D2
4-Androstene-3,6,17-trione	72	—	550	161	13	80	52	—	—	D2
4-Androstene-3,11,17-trione	68	—	480	151	11	78	43	—	—	D2
6β-Hydroxy-4-androstene-3,17-dione	—	100	89	74	—	43	19	90	—	D2
11α-Hydroxy-4-androstene-3,17-dione	—	65	49	45	—	25	6	82	—	D2
11β-Hydroxy-4-androstene-3,17-dione	—	150	140	100	—	53	26	—	—	D2
14α-Hydroxy-4-androstene-3,17-dione	—	90	79	55	—	33	12	87	—	D2
15α-Hydroxy-4-androstene-3,17-dione	—	—	40	41	—	22	—	82	—	D2
3β-Hydroxy-5-androsten-17-one	115	—	360	—	38	83	65	—	—	D4
17β-Hydroxy-4-androsten-3-one	89	150	200	144	22	74	52	—	—	D2
Testololactone	40	280	270	123	4.3	63	28	—	—	D2
B. C_{19} 17β-hydroxysteroids										
17β-Acetoxy-4-androsten-3-one	350	—	—	—	83	—	—	—	—	D2
5-Androstene-3β,17β-diol	—	38	57	—	—	62	—	—	—	D4
17β-Cyclopentylpropionoxy-4-androsten-3-one	585	—	—	—	99	—	—	—	—	D2
6β,17β-Dihydroxy-4-androsten-3-one	—	—	13	16	—	15	—	67	—	D2
11α,17β-Dihydroxy-4-androsten-3-one	16	7.4	6.5	9	—	7	—	47	—	D2
11β,17β-Dihydroxy-4-androsten-3-one	—	—	21	28	—	21	—	74	—	D2
14α,17β-Dihydroxy-4-androsten-3-one	—	—	18	21	—	12	—	67	—	D2
17α-Ethinyl-17β-hydroxy-4-androsten-3-one	74[d]	180[d]	255[d]	155[d]	16[d]	74[d]	52[d]	—	—	D2
17β-Hydroxy-1,4-androstadien-3-one	60	100	135	124	11	69	34	—	—	D3
17β-Hydroxy-4-androstene-3,11-dione	—	82	64	65	—	31	13	—	—	D2
17β-Hydroxy-4-androsten-3-one	89	150	200	144	22	74	52	—	—	D2
17α-Methyl-11α,17β-dihydroxy-4-androsten-3-one	—	—	13	20	—	13	—	58	—	D2
17α-Methyl-17β-hydroxy-4-androstene-3,11-dione	—	105	130	100	—	56	21	—	—	D2
17α-Methyl-17β-hydroxy-4-androsten-3-one	115	180	310	166	33	82	64	—	—	D2
17β-Propionoxy-4-androsten-3-one	425	—	—	—	85	—	—	—	—	D2
C. C_{18} steroids										
10ξ,17β-Dihydroxy-4-estren-3-one	—	—	—	—	—	11	—	66	—	D2
11α,17β-Dihydroxy-4-estren-3-one	—	—	—	—	—	4	—	39	—	D2

TABLE PC 124—(Continued)
STEROIDS

Paper	na	na	na	na	na	na	na	na	na	
Solvent	S1	S2	S3	S4	S5	S6	S7	S8	S9	
Literature	1	1	1	1	1	1	1	1	1	
Compound	R[a]	R[b]	R[b]	R[a]	$R_F \times 100$					Det.[e]
C. C_{18} steroids										
Estradiol	—	—	40	68	—	47	21	—	—	D4
Estradiol-17α	—	—	—	—	—	—	—	—	45	—
Estradiol-17β	—	—	—	—	—	—	—	—	27	—
4-Estrene-3,17-dione	132	—	—	—	40	—	74	—	—	D2
Estrone	66	130	160	142	14[d]	74	52	—	78	D4
11α-Hydroxy-4-estrene-3,17-dione	—	—	35	32	—	19	5	76	—	D2
17β-Hydroxy-4-estren-3-one	78	130	186	132	14	68	43	—	—	D2
D. C_{21} steroids										
21-Acetoxy-11β-hydroxy-4,17(20)-pregnadien-3-one	115	—	480	—	50	—	78	—	—	D2
11α-Acetoxy-4-pregnene-3,20-dione	125	—	530	—	25	—	71	—	—	D2
11β,21-Dihydroxy-4,17(20)-pregnadien-3-one	—	29	29	47	—	31	—	87	—	D2
3α,17α-Dihydroxy-5β-pregnane-11,20-dione	35	—	27	48	—	32	10	87	—	D1
6β,11α-Dihydroxy-5β-pregnane-3,20-dione	—	—	—	6.7	—	5	—	47	—	D1
3α,11α-Dihydroxy-5β-pregnan-20-one	40	—	43	74	—	41	—	87	—	D1
3β,11α-Dihydroxy-5α-pregnan-20-one	24	—	29	63	—	29	—	87	—	D1
6β,11α-Dihydroxy-4-pregnene-3,20-dione	—	7.4	4	5	—	4	—	38	—	D2
6β,17α-Dihydroxy-4-pregnene-3,20-dione	—	—	19	21	—	19	—	75	—	D2
11α,17α-Dihydroxy-4-pregnene-3,20-dione	—	14	9	13	—	9	—	61	—	D2
11β,17α-Dihydroxy-4-pregnene-3,20-dione	—	38	29	39	—	23	—	83	—	D2
11α-Hydroxy-4,16-pregnadiene-3,20-dione	40	100	100	104	4	47	19	—	—	D2
3β-Hydroxy-5,16-pregnadien-20-one	165	—	—	—	61	—	78	—	—	D4
3α-Hydroxy-5β-pregnane-11,20-dione	79	160	210	—	14	—	43	—	—	D1
3β-Hydroxy-5α-pregnane-11,20-dione	64	160	190	—	11	—	34	—	—	D1
3β-Hydroxy-5β-pregnane-11,20-dione	79	160	210	—	16	—	43	—	—	D1
11α-Hydroxy-5α-pregnane-3,20-dione	64	160	190	—	8	—	34	—	—	D1
11α-Hydroxy-5β-pregnane-3,20-dione	79	160	210	—	12	—	43	—	—	D1
11β-Hydroxy-5β-pregnane-3,20-dione	100	—	330	—	20	—	65	—	—	D1
17α-Hydroxy-5β-pregnane-3,11,20-trione	64	150	180	—	7	65	37	—	—	D2
3α-Hydroxy-5β-pregnan-20-one	195	—	—	—	72	—	93	—	—	D5
3β-Hydroxy-5α-pregnan-20-one	165	—	—	—	72	—	89	—	—	D5
3β-Hydroxy-5-pregnen-20-one	165	—	—	—	64	—	78	—	—	D4
6β-Hydroxy-4-pregnene-3,20-dione	63	150	160	112	9.7	69	34	—	—	D2
11α-Hydroxy-4-pregnene-3,20-dione	40	100	100	100	3.4	47	19	—	—	D2
11β-Hydroxy-4-pregnene-3,20-dione	60	180	210	142	11	71	43	—	—	D2
14α-Hydroxy-4-pregnene-3,20-dione	60	180	210	132	7.8	69	34	—	—	D2

TABLE PC 124—(Continued)
STEROIDS

Paper	na	na	na	na	na	na	na	na	na	
Solvent	S1	S2	S3	S4	S5	S6	S7	S8	S9	
Literature	1	1	1	1	1	1	1	1	1	
Compound	R[a]	R[b]	R[b]	R[a]	$R_F \times 100$					Det.[e]
D. C_{21} steroids										
15α-Hydroxy-4-pregnene-3,20-dione	37	100	77	82	1.7	40	13	—	—	D2
15β-Hydroxy-4-pregnene-3,20-dione	40	110	120	100	3.6	49	20	—	—	D2
16α-Hydroxy-4-pregnene-3,20-dione	40	80	92	74	2.4	43	15	—	—	D2
17α-Hydroxy-4-pregnene-3,20-dione	78	180	255	151	16	78	52	—	—	D2
21-Hydroxy-4-pregnene-3,20-dione	86	—	400	150	23	81	52	—	—	D6
6β-Hydroxy-4-pregnene-3,11,20-trione	—	91	57	58	—	28	—	—	—	D2
17α-Hydroxy-4-pregnene-3,11,20-trione	40	100	100	89	30.4	47	19	—	—	D2
Methyl-3,11-diketo-3,17(20)-(*cis*)-pregnadien-21-oate	150	—	580	—	57	—	78	—	—	D2
1,4-Pregnadiene-3,20-dione	165	—	—	—	55	—	78	—	—	D2
4,6-Pregnadiene-3,20-dione	270	—	—	—	68	—	96	—	—	D2
4,9(11)-Pregnadiene-3,20-dione	270	—	—	—	68	—	96	—	—	D2
4,16-Pregnadiene-3,20-dione	270	—	—	—	76	—	96	—	—	D2
5α-Pregnane-3,20-dione	350	—	—	—	83	—	—	—	—	D1
5β-Pregnane-3,20-dione	350	—	—	—	83	—	—	—	—	D1
5α-Pregnane-3,11,20-trione	125	—	—	—	42	89	78	—	—	D1
5β-Pregnane-3,11,20-trione	125	—	—	—	42	89	78	—	—	D1
4-Pregnene-3,20-dione	270	—	600	—	76	95	96	—	—	D2
4-Pregnene-3,11,20-trione	100	—	400	—	25	82	65	—	—	D2
E. C_{21} 21-hydroxysteroids										
21-Acetoxy-4β-bromo-17α-hydroxy-5β-pregnane-3,11,20-trione	40	—	—	142	—	71	39	—	—	D7
21-Acetoxy-11β,17α-dihydroxy-1,4-pregnadiene-3,20-dione	—	—	15[d]	25[d]	—	27[d]	7	82[d]	—	D8
21-Acetoxy-3α,17α-dihydroxy-5β-pregnane-11,20-dione	—	—	11	31	—	28	—	87	—	D9
21-Acetoxy-3β,17α-dihydroxy-5α-pregnane-11,20-dione	—	—	8	23	—	21	—	87	—	D9
21-Acetoxy-11β,17α-dihydroxy-4-pregnene-3,20-dione	17	—	29[d]	39[d]	—	34[d]	9	90	—	D6
21-Acetoxy-9α-fluoro-11β,17α-dihydroxy-4-pregnene-3,20-dione	—	—	—	23	—	27[d]	—	90	—	D6
21-Acetoxy-17α-hydroxy-1,4-pregnadiene-3,11,20-trione	—	—	39[d]	56[d]	—	40	12	90	—	D8
21-Acetoxy-17α-hydroxy-5β-pregnane-3,11,20-trione	40	150[d]	—	100	—	65[d]	28[d]	—	—	D7
21-Acetoxy-11β-hydroxy-4-pregnene-3,20-dione	40	260	210	136	—	71	38	—	—	D6
21-Acetoxy-15α-hydroxy-4-pregnene-3,20-dione	—	92	73	79	—	34	10	—	—	D6
21-Acetoxy-17α-hydroxy-4-pregnene-3,20-dione	49	180	170	150	—	73	43	—	—	D6
21-Acetoxy-17α-hydroxy-4-pregnene-3,11,20-trione	22	91[d]	64[d]	65	—	44	15	95	—	D6
21-Acetoxy-4,16-pregnadiene-3,20-dione	115	—	—	—	50	—	89	—	—	D6
21-Acetoxy-4-pregnene-3,20-dione	140	—	—	—	55	—	89	—	—	D6

TABLE PC 124—(Continued)
STEROIDS

Paper / Solvent / Literature	na / S1 / 1	na / S2 / 1	na / S3 / 1	na / S4 / 1	na / S5 / 1	na / S6 / 1	na / S7 / 1	na / S8 / 1	na / S9 / 1	
Compound	R[a]	R[b]	R[b]	R[a]	$R_F \times 100$					Det.[e]
E. C_{21} 21-hydroxysteroids										
21-Acetoxy-4-pregnene-3,11,20-trione	47	—	470	160	—	75	43	—	—	D6
21-Acetoxy-11β,17α,20α-trihydroxy-4-pregnen-3-one	—	—	—	5.5	—	9	—	61	—	D2
17α,21-Dihydroxy-4-pregnadiene-3,11,20-trione	—	—	—	—	—	—	—	57	—	D6
17α,21-Dihydroxy-4-pregnane-3,20-dione	28	42	36	35	—	33	—	87	—	D6
17α,21-Dihydroxy-5β-pregnane-3,11,20-trione	21	—	11	11	—	21	—	75	—	D7
17α,21-Dihydroxy-5β-pregnane-3,11,20-trione-3,20-ethylene glycol diketal	50	—	170	126	—	60	37	—	—	D1
6β,21-Dihydroxy-4-pregnene-3,20-dione	—	—	28	19	—	21	—	75	—	D6
11α,21-Dihydroxy-4-pregnene-3,20-dione	—	23	16	13	—	10	—	61	—	D6
11β,21-Dihydroxy-4-pregnene-3,20-dione	22	63	49	39	—	33	—	87	—	D6
14α,21-Dihydroxy-4-pregnene-3,20-dione	17	42	36	25	—	21	—	77	—	D6
15α,21-Dihydroxy-4-pregnene-3,20-dione	—	—	12	8.6	—	6	—	53	—	D6
17α,21-Dihydroxy-4-pregnene-3,11,20-trione	16	—	8.5	5.5	—	9	—	61	—	D6
17α,21-Dihydroxy-5-pregnene-3,11,20-trione-3,20-ethylene glycol diketal	36	—	140	100	—	47	23	91	—	D1
17α-Hydroxymethyl-17aα-hydroxy-D-homo-4-androstene-3,17-dione-(D-homo compound S)	—	71	33	35	—	30	—	87	—	D1
21-Hydroxy-4-pregnene-3,20-dione	86	—	400	150	23	81	52	—	—	D6
21-Hydroxy-4-pregnene-3,11,20-trione	28	—	140	69	—	43	—	92	—	D6
2-Methyl-21-acetoxy-11β,17α-dihydroxy-4-pregnene-3,20-dione	—	—	64	92	—	51	20	96	—	D6
2-Methyl-21-acetoxy-9α-fluoro-11β,17α-dihydroxy-4-pregnene-3,20-dione	—	—	49	56[d]	—	41[d]	15	95	—	D6
2-Methyl-9α-fluoro-11β,17α,21-trihydroxy-4-pregnene-3,20-dione	—	—	—	4	—	5.7	—	61	—	D6
2-Methyl-11β,17α,21-trihydroxy-4-pregnene-3,20-dione	—	—	—	9	—	10	—	69	—	D6
11β,17α,20α,21-Tetrahydroxy-4-pregnen-3-one	—	—	—	—	—	—	—	9	—	D2
11β,17α,20β,21-Tetrahydroxy-4-pregnen-3-one	—	—	—	—	—	—	—	12	—	D2
11β,17α,21-Trihydroxy-1,4-pregnadiene-3,20-dione	—	—	—	1.8	—	2.5	—	27	—	D8
6β,17α,21-Trihydroxy-4-pregnene-3,20-dione	—	—	—	—	—	24	—	22[d]	—	D6
11α,17α,21-Trihydroxy-4-pregnene-3,20-dione	—	—	1	1	—	1.3	—	12	—	D6
11β,17α,21-Trihydroxy-4-pregnene-3,20-dione	—	—	3.3	2.9	—	4	—	37	—	D6
11β,17α,21-Trihydroxy-5-pregnene-3,20-dione-3,20-ethylene glycol diketal	28	—	66	50	—	31	13	61	—	D1

TABLE PC 124—(Continued)
STEROIDS

Paper	na	na	na	na	na	na	na	na	na	
Solvent	S1	S2	S3	S4	S5	S6	S7	S8	S9	
Literature	1	1	1	1	1	1	1	1	1	
Compound	R[a]	R[b]	R[b]	R[a]	$R_F \times 100$					Det.[e]
F. Miscellaneous										
Cortisone	—	—	—	—	—	—	—	—	0	—
Dehydroisoandrosterone	—	—	—	—	—	—	—	—	94	—
Dienestrol	—	—	—	—	—	—	—	—	48	—
α,β-Diethylstilbestrol	—	—	—	—	—	—	—	—	50	—
Equilenin	—	—	—	—	—	—	—	—	40	—
Equilin	—	—	—	—	—	—	—	—	65	—
Estriol	—	—	—	—	—	—	—	—	0	—
17-Ethinylestradiol	—	—	—	—	—	—	—	—	52	—
Hexoestrol	—	—	—	—	—	—	—	—	48	—
17-Methyltestosterone	—	—	—	—	—	—	—	—	87	—
Progesterone	—	—	—	—	—	—	—	—	100	—
***cis*-Testosterone**	—	—	—	—	—	—	—	—	90	—
***trans*-Testosterone**	—	—	—	—	—	—	—	—	77	—

[a] Mobility relative to 4-pregnene-3,11,20-trione.
[b] Mobility relative to 11α-hydroxy-4-pregnene-3,20-dione.
[c] Solvent not suitable.
[d] Streaked.
[e] Preferred detection method.

Solvent

S1 = methyl cyclohexane saturated with diethylene glycol monoethyl ether.
S2 = toluene saturated with propylene glycol-MeOH (v/v).
S3 = cyclohexane-benzene (v/v), saturated with propylene glycol-MeOH (v/v).
S4 = benzene, saturated with formamide-MeOH (v/v).
S5 = petroleum ether (86–100)-MeOH-water (5 : 4 : 1).
S6 = toluene-petroleum ether-MeOH-water (5 : 5 : 7 : 3).
S7 = benzene-petroleum ether-MeOH-water (33.3 : 66.7 : 80 : 20).
S8 = benzene-methanol-water (2 : 1 : 1).
S9 = toluene, saturated with aqueous sodium *p*-toluenesulfonate.

Detection

D1 = 2,4-dinitrophenylhydrazine reagent.
D2 = ultraviolet absorption and 2,4-dinitrophenylhydrazine.
D3 = ultraviolet absorption.
D4 = phosphomolybdic acid reagent.
D5 = 2,4-dinitrophenylhydrazine and phosphomolybdic acid.
D6 = ultraviolet absorption, 2,4-dinitrophenylhydrazine and Tollens reagent.
D7 = 2,4-dinitrophenylhydrazine and Tollens reagent.
D8 = ultraviolet absorption and Tollens reagent.
D9 = Tollens reagent.

Literature

1. Reineke, L. M., *Anal. Chem.*, **28**, 1853–1858 (1956).

TABLE PC 125
STEROIDS, C_{21}

Paper	P1	P2	P2	P2	P2	P3	P3	P3	P3
Solvent	S1	S2	S3	S4	S5	S6	S7	S8	S9
Temperature						(38 °C) 22 °C	38 °C	38 °C	38 °C
Literature	1	1	1	1	1	1	1	1	1
Compound					$R_{cortisol}$				
21-Hydroxy-pregn-4-ene-3,20-dione	—	—	—	—	1.29	—	—	1.25	—
6β,21-Dihydroxy-pregn-4-ene-3,20-dione	—	—	3.24	—	0.96	—	1.80	—	1.43
11α,21-Dihydroxy-pregn-4-ene-3,20-dione	5.07	—	3.17	—	0.84	2.15	1.37	1.36	—
11β,21-Dihydroxy-pregn-4-ene-3,20-dione	—	—	3.51	—	1.31	—	1.95	1.43	1.78
16α,21-Dihydroxy-pregn-4-ene-3,20-dione	—	—	2.75	—	0.65	—	1.28	—	0.91
17,21-Dihydroxy-pregn-4-ene-3,20-dione	—	—	3.55	—	1.52	—	—	—	—
21-Hydroxy-pregn-4-en-18-al-11,20-dione[a]	~4.50	2.65	2.78	0.77	0.65	(1.39)	1.18	0.64	0.94
11β,21-Dihydroxy-pregn-4-ene-3,20-dione-18-acid-lactone (18 → 11)	—	—	—	—	0.47	—	—	—	—
19,21-Dihydroxy-pregn-4-ene-3,20-dione	4.11	—	2.79	—	0.78	2.34	1.40	1.24	1.05
20,20,21-Trihydroxy-pregn-4-en-3-one-18-acid-lactone (18 → 20)	2.46	—	2.75	—	1.33	2.76	1.67	1.39	—
11β,21-Dihydroxy-pregn-4-en-18-al-3,20-dione (aldosterone)	2.55	1.20	2.25	0.40	0.41	(1.22)	0.97	0.77	1.03
11β,21-Dihydroxy-17-iso-pregn-4-en-18-al-3,20-dione (17-iso-aldosterone)	1.83	1.26	2.22	0.41	0.50	(1.09)	0.92	0.73	0.73
18,21-Dihydroxy-pregn-4-ene-3,11,20-trione[b]	2.95	1.82	2.20	0.74	0.54	(1.20)	1.08	0.74	0.91
6β,21-Dihydroxy-pregn-4-ene-3,11,20-trione	1.87	—	1.62	—	0.51	0.91	0.75	0.50	1.08
17,21-Dihydroxy-pregn-4-ene-3,11,20-trione (cortisone)	2.51	1.95	2.39	0.95	1.02	1.91	1.38	1.10	1.08
6β,11β,21-Trihydroxy-pregn-4-en-3,20-dione	0.66	0.30	0.50	—	0.40	0.31	0.40	0.40	0.47
6β,17,21-Trihydroxy-pregn-4-ene-3,20-dione	0.49	0.31	0.58	0.60	0.66	0.44	0.59	0.82	0.70
11α,17,21-Trihydroxy-pregn-4-ene-3,20-dione	0.20	—	0.46	—	—	0.24	0.31	0.81	—
11β,18,21-Trihydroxy-pregn-4-ene-3,20-dione[b]	0.57	0.36	0.88	0.70	0.45	—	0.54	—	0.76
11β,19,21-Trihydroxy-pregn-4-ene-3,20-dione	0.69	0.60	1.37	—	0.44	0.80	0.77	0.79	0.57
17,19,21-Trihydroxy-pregn-4-ene-3,20-dione	0.19	0.13	0.36	—	0.46	0.20	0.35	0.53	0.49
11β,17,21-Trihydroxy-pregn-4-en-18-al-3,20-dione (17α-hydroxyaldosterone)	0.33	0.22	0.72	—	0.23	—	0.29	0.43	—
2α,11β,17,21-Tetrahydroxy-pregn-4-ene-3,20-dione	—	—	0.17	0.36	—	—	—	—	—
6β,11β,17,21-Tetrahydroxy-pregn-4-ene-3,20-dione	—	—	0.04	0.19	—	—	—	—	—

[a] Runs mostly in tailing spots, possibly due to tautomerism.
[b] BT reaction only very weak and slow, due to cyclohemiketal formation.

Paper
P1 = W1 impregnated with propylene glycol.
P2 = W1 impregnated with formamide.
P3 = W1.

Solvent
S1 = toluene.
S2 = benzene.
S3 = chloroform.
S4 = *n*-butanol-butyl acetate-water (15 : 85 : 5).
S5 = ethyl acetate-butyl acetate-water (15 : 85 : 5).
S6 = benzene-methanol-water (100 : 50 : 50).
S7 = toluene-ethyl acetate-methanol-water (90 : 10 : 50 : 50).
S8 = isooctane-*tert*-butanol-water (10 : 5 : 9).
S9 = cyclohexane-dioxane-methanol-water (100 : 100 : 50 : 25).

Detection UV light (all positive); tetrazolium blue (all positive).
Technique Not specified.
Literature
1. Neher, R., *Steroid Chromatography*, Elsevier, 1964, pp 181–182.

TABLE PC 126
STEROIDS, STEROLS AND VITAMINS

Paper	na	na	na	na	na	na	na	na	na	na	na	na	na	na	na	na	na
Solvent	S1	S2	S3	S4	S5	S6	S7	S8	S9	S10	S11	S12	S13	S14	S15	S16	S17
Literature	1	1	1	1	1	1	1	1	1	1	1	1	1	1	1	1	1
Compound	$R_F \times 100$																
β-Cholestanol	56	63	—	—	—	—	—	—	—	—	27	—	—	—	—	—	—
Cholesterol	52	56	49	73	56	30	77	0	—	63	33	—	29	49	50	35	42
Cholesterol acetate	—	—	—	—	—	11	—	—	—	—	9	—	—	—	—	—	—
7-Dehydrocholesterol	94	88	48	67	—	—	—	90	—	—	40	—	39	57	56	—	—
Ergosterol	95	84	43	68	—	47	60	0	—	37	42	—	0	56	0	—	89
Lumisterol	—	—	—	—	—	—	—	—	—	—	33	—	27	59	52	—	—
Phytosterol from rapeseed or olive oil	—	—	—	—	33–41	—	—	—	—	—	—	—	—	—	—	33–40	42
β-Sitosterol	54	65	43	65	47–58	—	80	—	—	—	25	—	0	51	42	26	38–33
Stigmasterol	53	52	—	—	—	—	80	—	—	30	27	—	0	—	—	—	38
Suprasterol II	—	—	—	—	—	—	—	—	—	—	—	—	36	52	57	—	—
Tachysterol	—	—	—	—	—	—	—	—	—	—	—	—	0–37, 98–93	0–61, 98	0–58, 98	—	—
α-Tocopherol	—	—	—	—	—	—	—	—	—	—	13	7	15	58	—	—	—
Vitamin A	—	—	—	—	—	—	—	—	90	—	30	—	98	—	—	—	—
Vitamin A acetate	—	—	—	—	—	—	—	—	50	—	1	—	32	—	—	—	—
Vitamin D_2	—	—	80	91	—	—	—	—	—	—	29	—	44	66	69	—	—
Vitamin D_3	—	—	76	91	—	—	—	—	—	—	36	—	46	68	68	—	—
Zymosterol	—	—	—	—	—	—	—	—	—	—	43	—	41	61	48	—	—

Solvent
S1 = ethanol-H_2O (8 : 2), paper impregnated with quilon.
S2 = methanol, paper impregnated with quilon.
S3 = methanol-H_2O-ethylene glycol monomethyl ether (65 : 20 : 20), paper impregnated with quilon.
S4 = methanol-H_2O (95 : 5), paper impregnated with quilon.
S5 = ethanol-H_2O (8 : 2), paper impregnated with quilon.
S6 = methanol-carbon tetrachloride-H_2O (18 : 5 : 2), paper impregnated with sodium stearate or palmitate.
S7 = hexane-ether (3 : 1), paper impregnated with aluminum oxide.
S8 = phenol-methanol-H_2O (13.5 : 30 : 56.5), paper impregnated with H_2O.
S9 = isopropanol-H_2O (1 : 1), paper impregnated with H_2O.
S10 = heptane, paper impregnated with phenylcellosolve.
S11 = acetic acid-H_2O (84 : 16), paper impregnated with paraffin.
S12 = acetonitrile-H_2O (6 : 4), paper impregnated with silicone grease.
S13 = ethylene glycol monoethyl ether-*n*-propanol-methanol-H_2O (35 : 10 : 30 : 25), paper impregnated with paraffin.
S14 = *n*-propanol-methanol-H_2O (15 : 82 : 3), paper impregnated with paraffin.
S15 = methanol-H_2O (95 : 5), paper impregnated with paraffin.
S16 = pyridine-H_2O (85 : 15), paper impregnated with petroleum (b.p. 220–240 °C).
S17 = acetic acid-H_2O (84 : 16), paper impregnated with paraffin.

Detection See Section II.I, Volume Two.
Technique Reversed-phase chromatography.
Literature
1. Peereboom, J. W. C., Roos, J. B., and H. W. Beekes, *J. Chromatogr.*, **5**, 500 (1961).

TABLE PC 127
STEROID GLYCOSIDES, CARDIAC GLYCOSIDES, AGLYCONES, AND THEIR ACETATES

Paper	P3	P1	P1	P5	P1		P1		P1	P1	P3	P4	na	P1, P2	P1, P2	P1, P2	P1, P2	P1, P2	P1, P2	P1, P2	P1, P2	P1, P2				
Solvent	S1	S2	S3	S4	S1		S5		S6	S7	S8	S9	S10	S11	S11	S11	S12	S13	S14	S15	S16	S17				
Technique	T1	T1	T1	na	T2		T2		T1	T1	T1	T1	T1	T3	T3	T3	T3	T3	T3	T3	T3	T3				
Literature	7	7	3, 7	8	1		1		2, 7	4, 7	5, 7	1, 7	6	9	9	9	9	9	9	9	9	8				
Compound	Migration in cm		R_F ×100		Migration cm hours 24	48	24	48	R_F × 100														D1	D2	D3	D4
Acetyldigitoxin-α and β	—	—	—	—	—	—	—	—	—	—	—	—	—	—	—	—	—	2	21	—	90–95	—	—	—	—	—
Acetyldigoxin-α	—	—	—	—	—	—	—	—	—	—	—	—	—	—	—	—	—	—	—	—	39	—	—	—	—	—
Acetyldigoxin-β	—	—	—	—	—	—	—	—	—	—	—	—	—	—	—	—	—	—	—	—	45	—	—	—	—	—
Acetylgitoxin-α	—	—	—	—	—	—	—	—	—	—	—	—	—	—	—	—	—	8	41	—	59	—	—	—	—	—
Acetylgitoxin-β	—	—	—	—	—	—	—	—	—	—	—	—	—	—	—	—	—	—	—	—	64	—	—	—	—	—
Allocymarin	—	—	—	—	—	—	—	—	—	—	—	—	—	—	—	—	—	—	—	—	48	—	—	—	—	—
Alloemicymarin	—	—	—	—	1.0	1.4	1.0	2.5	—	—	—	—	—	—	—	—	—	—	—	—	7	—	—	—	—	—
Alloperiplocymarin	—	—	—	—	11.8	28.9	2.5	7.1	—	—	—	—	—	—	—	—	—	—	—	—	—	—	—	—	—	—
Alloperiplogenin acetate	—	—	—	—	15.0	32.0	6.6	—	—	—	—	—	—	—	—	—	—	—	—	—	—	—	—	—	—	—
Allostrophanthidin	—	—	—	—	—	—	—	—	—	—	—	—	—	—	—	—	—	—	—	—	32	—	—	—	—	—
"α"-Anhydrodigitoxigenin	—	—	—	—	—	—	—	—	—	—	—	—	91	—	—	—	—	—	—	—	—	—	—	—	—	—
"β"-Anhydrodigitoxigenin	—	—	—	—	—	—	—	—	—	—	—	—	63	—	—	—	—	—	—	—	—	—	—	—	—	—
"β"-Anhydrodigoxigenin	—	—	—	—	—	—	—	—	—	—	—	—	73	—	—	—	—	—	—	—	—	—	—	—	—	—
"β"-Anhydrodigoxigenone	—	—	—	—	—	—	—	—	—	—	—	—	59	—	—	—	—	—	—	—	—	—	—	—	—	—
"β"-Anhydrosarmentogenin	—	—	—	—	—	—	—	—	—	—	—	—	69	—	—	—	—	—	—	—	—	—	—	—	—	—
"β"-Anhydrouzarigenin	—	—	—	—	—	—	—	—	—	—	—	—	94	—	—	—	—	—	—	—	—	—	—	—	—	—
Cymarin	5.5	11	—	—	4.0	6.2	2.0	5.1	—	—	—	—	—	—	—	—	—	—	—	—	55	—	—	—	—	—
Cymarin acetate	—	—	—	—	—	—	5.1	—	—	—	—	—	—	—	—	—	—	—	—	—	—	—	—	—	—	—
Cymarol	3.7	8	—	—	2.5	3.6	1.0	3.0	—	—	—	—	—	—	—	—	—	—	—	—	36	—	—	—	—	—
Deacetyldigilanid A	—	—	—	—	—	—	—	—	—	—	—	—	—	7	10	15	—	39	F	—	—	50	—	—	—	—
Deacetyldigilanid B	—	—	—	—	—	—	—	—	—	—	—	—	—	2	3	5	—	73	F	—	—	19	—	—	—	—
Deacetyldigilanid C	—	—	—	—	—	—	—	—	—	—	—	—	—	1	1	2	—	—	—	—	—	13	—	—	—	—
Digilanid A	—	—	—	—	—	—	—	—	—	—	—	—	—	37	46	49	—	28	F	3	—	85	—	—	—	—
Digilanid B	—	—	—	—	—	—	—	—	—	—	—	—	—	14	18	25	—	66	F	10	—	47	—	—	—	—
Digilanid C	—	—	—	—	—	—	—	—	—	—	—	—	—	8	12	15	—	86	F	25	—	33	—	—	—	—
Diginin	—	—	—	24	—	—	—	—	—	—	—	—	—	—	—	—	—	—	—	—	—	—	gr-bl	d-bl	bl	r
Digitalinum verum	—	—	—	—	—	—	—	—	17	—	—	—	—	—	—	—	—	95	F	—	—	16	—	—	—	—
Digitoxigenin	—	—	—	—	—	—	—	—	—	81	50	90	76	92	93	92	91	3	—	—	81	—	—	—	—	—
Digitoxigenin glycoside	—	—	—	—	—	—	—	—	—	—	42	—	—	—	—	—	—	—	—	—	—	—	—	—	—	—
Digitoxigenin acetate	—	—	—	—	—	—	—	—	—	—	91	—	89	—	—	—	—	—	—	—	—	—	—	—	—	—
Digitoxin	—	—	—	51	—	—	—	—	—	69	22	95	—	88	91	90	68	3	27	—	69	—	gr-bl	d-bl	bl	r
Digoxigenin	—	—	—	—	—	—	—	—	—	—	—	—	75	58	64	61	—	—	—	—	32	—	—	—	—	—

TABLE PC 127—(Continued)

STEROID GLYCOSIDES, CARDIAC GLYCOSIDES, AGLYCONES, AND THEIR ACETATES

Paper	P3	P1	P1	P5	P1		P1		P1	P1	P3	P4	na	P1, P2	P1, P2	P1, P2	P1, P2	P1, P2	P1, P2	P1, P2	P1, P2	P1, P2				
Solvent	S1	S2	S3	S4	S1		S5		S6	S7	S8	S9	S10	S11	S11	S11	S12	S13	S14	S15	S16	S17				
Technique	T1	T1	T1	na	T2		T2		T1	T1	T1	T1	T1	T3	T3	T3	T3	T3	T3	T3	T3	T3				
Literature	7	7	3, 7	8	1		1		2, 7	4, 7	5, 7	1, 7	6	9	9	9	9	9	9	9	9	8				
Compound	Migration in cm			R_F ×100	Migration cm hours 24	48	24	48	R_F × 100														D1	D2	D3	D4
Digoxigenin diacetate	—	—	—	—	—	—	—	—	—	—	—	—	70	—	—	—	—	—	—	—	—	—	—	—	—	—
Digoxigenone	—	—	—	—	—	—	—	—	—	—	—	—	68	—	—	—	—	—	—	—	—	—	—	—	—	—
Digoxin	—	—	—	—	—	—	—	—	—	—	—	—	—	68	75	76	—	29	72	—	16	—	—	—	—	—
Emicymarin	1.5	1.5	—	—	0.8	1.1	0.7	3.0	—	—	—	—	—	—	—	—	—	—	—	—	11	—	—	—	—	—
Emicymarin acetate	—	—	—	—	—	—	13.6	23.4	—	—	—	—	—	—	—	—	—	—	—	—	—	—	—	—	—	—
Gitaloxin	—	—	—	—	—	—	—	—	—	55	—	—	—	—	—	—	—	—	—	—	—	—	—	—	—	—
Gitorin	—	—	—	—	—	—	—	—	—	—	—	—	—	—	—	—	—	78	F	—	—	—	—	—	—	—
Gitoxigenin	—	—	—	84	—	—	—	—	—	42	—	55	26	82	84	84	80	17	52	—	42	—	—	lt-bl	bl	r
Gitoxigenin diacetate	—	—	—	—	—	—	—	—	—	—	—	—	88	—	—	—	—	—	—	—	—	—	—	—	—	—
Gitoxigenin glycoside	—	—	—	—	—	—	—	—	24	—	—	—	—	—	—	—	—	—	—	—	—	—	—	—	—	—
Gitoxin	—	—	—	70	—	—	—	—	29	29	—	80	—	76	81	82	40	18	55	—	29	—	gr-bl	lt-bl	bl	r
Glucogitaloxin	—	—	—	—	—	—	—	—	53	—	—	—	—	—	—	—	—	—	—	—	—	—	—	—	—	—
Isostrophanthidin	—	—	—	—	—	—	—	—	—	—	—	—	—	—	—	—	—	—	—	—	76	—	—	—	—	—
Nervifolin	—	—	—	—	—	—	—	—	—	—	—	—	—	—	—	—	—	—	—	—	55	—	—	—	—	—
Oleandrigenin	—	—	—	—	—	—	—	—	—	—	—	—	—	—	—	—	—	—	—	—	72	—	—	—	—	—
Oleandrin	—	—	—	—	—	—	—	—	—	—	—	—	—	—	—	—	—	—	—	—	93	—	—	—	—	—
Periplocymarin	—	—	—	—	16.2	38.9	3.8	10.0	—	—	—	—	—	—	—	—	—	—	—	—	—	—	—	—	—	—
Periplogenin	—	—	—	—	2.8	4.2	0.7	2.7	—	—	—	—	43	—	—	—	—	—	—	—	—	—	—	—	—	—
Periplogenin acetate	—	—	—	—	21.6	43.6	6.9	16.9	—	—	—	—	—	—	—	—	—	—	—	—	—	—	—	—	—	—
Purpurea glycoside A	—	—	—	84	—	—	—	—	—	—	—	—	—	—	—	—	—	—	—	—	—	—	gr-bl	lt-bl[a]	bl	r
Purpurea glycoside B	—	—	—	93	—	—	—	—	—	—	—	—	—	—	—	—	—	—	—	—	—	—	gr-bl	lt-bl	bl	r
Saponin	—	—	—	77	—	—	—	—	—	—	—	—	—	—	—	—	—	—	—	—	—	—	y	d-vt	—	—
Sarmentocymarin	4.1	7.5	>37	—	3.3	4.1	0.8	3.1	—	—	—	—	—	—	—	—	—	—	—	—	—	—	—	—	—	—
Sarmentogenin	—	—	8.0	—	2.5	2.5	1.0	2.0	—	—	—	—	10	—	—	—	—	—	—	—	—	—	—	—	—	—
Sarmentogenin acetate	—	—	—	—	—	—	15.5	25.3	—	—	—	—	—	—	—	—	—	—	—	—	—	—	—	—	—	—
Sarmentogenin diacetate	—	—	—	—	—	—	—	—	—	—	—	—	85	—	—	—	—	—	—	—	—	—	—	—	—	—
Sarmentoside A	—	—	—	—	2.3	2.3	0.7	1.3	—	—	—	—	—	—	—	—	—	—	—	—	—	—	—	—	—	—
Sarnovide mono(?) acetate	—	—	—	—	—	—	—	—	—	—	—	—	72	—	—	—	—	—	—	—	—	—	—	—	—	—
Sarverogenin	—	—	32.4	—	—	—	—	—	—	—	—	—	40	—	—	—	—	—	—	—	—	—	—	—	—	—
Sarverosid	8.5	17	—	—	—	—	—	—	—	—	—	—	—	—	—	—	—	—	—	—	—	—	—	—	—	—
Strophanthidin	—	—	13.6	—	—	—	—	—	—	—	—	—	16	—	—	—	—	—	—	—	40	—	—	—	—	—
Strophanthidin acetate	—	—	—	—	4.1	6.1	2.8	7.1	—	—	—	—	—	—	—	—	—	—	—	—	—	—	—	—	—	—

TABLE PC 127—(Continued)

STEROID GLYCOSIDES, CARDIAC GLYCOSIDES, AGLYCONES, AND THEIR ACETATES

Compound																							D1	D2	D3	D4
Paper	P3	P1	P1	P5	P1		P1		P1	P1	P3	P4	na	P1, P2	P1, P2	P1, P2	P1, P2	P1, P2	P1, P2	P1, P2	P1, P2	P1, P2				
Solvent	S1	S2	S3	S4	S1		S5		S6	S7	S8	S9	S10	S11	S11	S11	S12	S13	S14	S15	S16	S17				
Technique	T1	T1	T1	na	T2		T2		T1	T1	T1	T1	T1	T3	T3	T3	T3	T3	T3	T3	T3	T3				
Literature	7	7	3,7	8	1		1		2,7	4,7	5,7	1,7	6	9	9	9	9	9	9	9	9	8				
	Migration in cm			R_F ×100	Migration cm hours				R_F × 100																	
					24	48	24	48																		
Strophanthidol	—	—	14.0	—	2.0	2.6	0.6	1.4	—	—	—	—	5	—	—	—	—	—	—	—	26	—	—	—	—	—
Strophanthidol acetate	—	—	—	—	8.3	15.8	4.7	12.0	—	—	—	—	—	—	—	—	—	—	—	—	—	—	—	—	—	—
Strospeside	—	—	—	—	—	—	—	—	—	—	—	—	—	—	—	—	—	—	—	—	75	61	—	—	—	—
Uzarigenin	—	—	—	—	—	—	—	—	—	—	—	—	83	—	—	—	—	—	—	—	—	—	—	—	—	—
Verodoxin	—	—	—	—	—	—	—	—	81	—	—	—	—	—	—	—	—	—	—	—	—	—	—	—	—	—

F = Solvent front.

[a] The dark blue fluorescence of purpurea glycoside A is masked by the light blue fluorescence of gitoxigenin, which has the same R_F value.

Paper
P1 = W1.
P2 = SS2043b.
P3 = W1 (formamide treated).
P4 = W3MM.

Solvent
S1 = benzene, saturated with formamide.
S2 = benzene-chloroform (9 : 1), saturated with formamide.
S3 = chloroform saturated with formamide.
S4 = methanol-chloroform-water (10 : 10 : 6).
S5 = toluene, paper treated with propylene glycol.
S6 = chloroform-tetrahydrofuran-methyl glycol-formamide (30 : 45 : 1.5 : 9).
S7 = xylene-methyl ethyl ketone, v/v, saturated with formamide (ascending development).
S8 = benzene-chloroform (75 : 25).
S9 = chloroform, saturated with ethylene glycol.
S10 = benzene-methanol-H_2O (10 : 5 : 54).
S11 = mixtures of $CHCl_3$-methanol-H_2O.
S12 = benzene-$CHCl_3$ (2 : 3), paper treated with formamide.
S13 = octanol-pentanol-water-formamide (5 : 6 : 8 : 2), paper impregnated with the organic phase and developed with the aqueous phase.
S14 = octanol-pentanol-water-formamide (6 : 2 : 1 : 4).
S15 = octanol-pentanol-water-formamide (5 : 1 : 1 : 2).
S16 = isobutanol-H_2O.
S17 = butanol-pyridine-H_2O (3 : 1 : 3).

Detection
(See Section II.I for details).
D1 = anisaldehyde (daylight).
D2 = anisaldehyde (UV).
D3 = Zimmerman reagent.
D4 = Legal reagent (daylight) = nitroprusside.

Technique
T1 = descending technique.
T2 = descending partition chromatography.
T3 = partition chromatography.

Literature

1. Heftmann, E. and A. J. Levant, *J. Biol. Chem.*, **194**, 703–709 (1952).
2. Dutta, N. L., *Nature*, **175**, 85 (1955).
3. Okada, M., Yamada, A., and K. Kometani, *J. Pharm. Soc. Japan*, **72**, 930 (1952).
4. Eriksson, S. and J. Sjövall, *Acta Chem. Scand.*, **8**, 1099 (1954).
5. Wichtl, J., *Naturwissenschaften*, **43**, 58 (1956).
6. Bush, I. E. and D. A. H. Taylor, *Biochem. J.*, **52**, 643 (1952).
7. Block, R. J., Durrum, E. L., and G. Zweig, *Paper Chromatography and Paper Electrophoresis*, Academic Press, N.Y., 1958.
8. Günzel, C. and F. Weiss, *Z. Anal. Chem.*, **148**, 250 (1955); *Pharmazie*, **10**, 725 (1955).
9. Neher, R., *Steroid Chromatography*, Elsevier, Amsterdam (1964).

Volume One

SECTION II.IV

CHROMATOGRAPHIC DATA
Thin-Layer Chromatography Tables

Wherever possible, tables were arranged according to classes of chemical compounds. However, this was not always possible when diverse compounds were chromatographed under the same experimental conditions. The reader is referred to the master compound index for specific compounds which may appear in diverse tables.

Techniques

n.a. = not available from original source.
In TLC, technique is ascending development, unless otherwise stated.
Solvent composition always v/v unless otherwise stated; if solid is dissolved in solvent, g/v.

TABLE TLC 1
ALCOHOLS

Layer	L1	L1	L1	L2	L3	L4	L5	L6	L6	L6	L7	L7
Solvent	S1	S2	S3	S4	S5	S6	S6	S7	S8	S9	S10	S11
Technique	T1	T1	T1	T2	na	T3	T3	na	na	na	na	na
Literature	1	1	1	2	3	4	4	5	5	5	6	6
Compound						$R_F \times 100$						
Allyl alcohol	—	—	—	—	—	—	—	—	—	—	38^c	—
sec-Amyl alcohol	—	—	—	—	—	—	—	—	—	—	41^c	—
Benzyl alcohol	20	95	17	—	63^a	—	—	—	—	—	—	—
n-Butanol	25	92	25	18	65^a	54^a	50^a	66^b	68^b	70^b	43^c	—
sec-Butyl alcohol	—	—	—	—	—	—	—	—	—	—	40^c	—
Cyclohexanol	—	—	—	—	68^a	—	—	—	—	—	53	—
n-Decanol (*n*-Decyl alcohol)	—	—	—	26	81^a	89^a	80^a	—	—	—	—	80^c
Diacetone alcohol	—	86	36	—	—	—	—	—	—	—	—	—
Dodecyl alcohol	—	—	—	—	—	—	—	—	—	—	—	80^c
Ethanol	—	—	—	—	49^a	32^a	35^a	48^b	48^b	53^b	29^c	—
2-Ethyhexyl alcohol	—	—	—	—	81^a	83^a	77^a	—	—	—	—	—
Glycerol	—	—	—	—	5^a	—	—	—	—	—	—	—
1-Heptanol	—	—	—	—	—	77^a	66^a	—	—	—	63^c	—
n-Hexanol	35	95	27	20	73^a	71^a	60^a	78^b	83^b	85^b	56^c	—
3-Hexen-1-ol	—	—	—	25	—	—	—	—	—	—	—	—
Isobutanol	—	—	—	19	—	56^a	52^a	—	—	—	41^c	—
Isopentanol	—	—	—	21	—	—	—	—	—	—	54^c	—
Isopropyl alcohol	—	—	—	—	—	—	—	—	—	—	37^c	—
Methanol	—	—	—	—	40^a	15^a	24^a	38^b	38^b	—	19^c	—
Methyl heptenol	—	—	—	26	—	—	—	—	—	—	—	—
n-Nonanol	—	—	—	24	—	86^a	77^a	—	—	—	—	76^c
n-Octanol	43	96	30	23	80^a	82^a	72^a	—	—	—	—	72^c
2-Octanol	—	—	—	—	—	—	—	82^b	86^b	90^b	—	49^c
n-Pentanol	30	93	26	20	—	62^a	55^a	74^b	78^b	78^b	49^c	—
1-Propanol	—	—	—	—	—	44^a	43^a	58^b	58^b	60^b	34^c	—

[a] 3,5-Dinitrobenzoates of listed alcohols.
[b] Xanthogenates of listed alcohols.
[c] β-Alkoxypropionitriles.

Layer L1 = Aluminum oxide (loose layer).
L2 = Silica gel G.
L3 = Kieselguhr G (Merck).
L4 = Kieselguhr G impregnated with polyethylene glycol M600.
L5 = Kieselguhr G impregnated with phenoxyethanol.
L6 = cellulose powder Schleucher and Schüll No. 143.
L7 = Silica gel KSK.

Solvent S1 = hexane–acetone (3 : 1).
S2 = ether–ethanol (99 : 1), 100 ml shaken with 10 ml of water and upper layer used.
S3 = hexane–acetone (4 : 1).
S4 = methylene chloride.
S5 = benzene–methyl acetate (150 : 1).
S6 = petroleum (b.p. 100–140 °C).
S7 = *n*-propanol-petroleum (b.p. 100–150 °C)—concentrated NH_3 (6 : 3 : 2).
S8 = *n*-propanol-petroleum (b.p. 100–150 °C)—concentrated NH_3 (14 : 7 : 4).
S9 = *n*-propanol-petroleum (b.p. 100–150 °C)—concentrated NH_3 (6 : 2 : 2).
S10 = cyclohexane-ether-heptane (10 : 7 : 3).
S11 = carbon tetrachloride-ethyl acetate (20 : 5).

Detection See Section II.I, Volume Two.

Technique T1 = 20 cm distance.
T2 = 15 cm distance.
T3 = develop chromatogram three times.

Literature

1. Kučera, J., *Collect. Czech. Chem. Commun.*, **28**, 1341 (1963).
2. Attaway, J. A. and R. W. Wolford, *5th Intern. Symp. Gas Chromatogr.*, Brighton, England, Sept. 1964.
3. Braun, D., *Chimia* (Aarau), **19**, 82 (1965).
4. Diemair, W., Pfeilsticker, K., and I. Hölscher, *Z. Anal. Chem.*, **234**, 421 (1968).
5. Röhr, M. and D. Chiari, *Mikrochim. Acta*, **1967**, 138.
6. Buzlanova, M. M., Ul'yanova, V., and S. I. Obtemperanskaya, *Zh. Anal. Khim.*, **23**, 1426 (1968).

TABLE TLC 2
ALKALOIDS

Layer	L1	L2	L1	L2	L3				
Solvent	S1	S1	S2	S2	S3				
Detection						D1	D2	D3	D4
Technique	na	na	na	na	na				
Literature	1	1	1	1	2	2	2	2	2
Compound	$R_F \times 100$					Color			
Ajmaline	—	—	—	—	—	31	vt	—	—
Apoatropine	24	19	36	43	—	—	—	—	—
Atropine	9	8	18	20	17	lt bl	—	—	—
Brucine	15	10	22	25	46	bl	—	lt o	—
Cinchonine	—	—	—	—	34	br	bl	—	—
Cocaine	69	78	72	80	63	vt	—	—	—
Codeine	28	18	36	24	—	—	—	—	—
Colchicine	16	15	22	20	—	—	—	—	—
Dihydroergocornine	47	47	56	48	—	—	—	—	—
Dihydroergocristine	47	47	56	48	—	—	—	—	—
Dihydroergocryptine	47	47	56	48	—	—	—	—	—
Dihydroergotamine	34	28	39	29	—	—	—	—	—
Ephedrine	—	—	—	—	33	w	—	—	—
Ergometrine	18	14	27	17	—	—	—	—	—
Ergotamine	36	29	42	32	—	—	—	—	—
Hydrocodone	25	18	32	25	—	—	—	—	—
Hydromorphone	13	9	15	13	—	—	—	—	—
Morphine	13	9	14	9	—	—	—	—	—
Narcotine	72	78	74	81	76	p-ish br	—	—	bl
Nicotine	—	—	—	—	66	gy	—	—	—
Oxycodone	61	64	63	69	—	—	—	—	—
Papaverine	62	64	65	69	72	vt	—	—	—
Pethidine	52	48	61	60	—	—	—	—	—
Physostigmine	49	47	59	60	—	—	—	—	—
Pilocarpine	28	25	37	37	63	br	—	—	—
Quinidine	37	28	42	43	—	—	—	—	—
Quinine	31	23	35	35	53	br	bl[a]	—	—
Reserpine	63	72	67	75	—	—	—	—	—
Scopolamine	38	37	46	52	—	—	—	—	—
Sparteine	14	11	49	32	—	—	—	—	—
Strychnine	28	20	37	40	—	bl-vt	—	—	—
Thebaine	49	46	57	51	—	—	—	—	—

[a] Under UV.

Layer
L1 = Silica gel VEB Laborchemie Apolda.
L2 = Silica gel G Merck.
L3 = Kieselgel (D-O without binder, Caemag).

Solvent
S1 = benzene-acetone-ether-10% aqueous NH_3 (4 : 6 : 1 : 0.3).
S2 = benzene-acetone-ether-25% NH_3 (4 : 6 : 1 : 0.3).
S3 = chloroform-ethanol (9 : 1).

Detection
D1 = $K_2(PtI_6)$.
D2 = H_2SO_4.
D3 = $Ce(SO_4)_2$.
D4 = $Fe[Fe(CN)_6]$.

Literature
1. Zarnack, J. and S. Pfeifer, *Pharmazie*, **19**, 216 (1964).
2. Manzur-ul-Hague Hashmi, Parveen Shanaz, and N. A. Chughtai, *Mikrochim. Acta*, 1969, 452.

TABLE TLC 3
ALKALOIDS

Layer	L1	L1	L1	L1	L1	L2	L2	L3		
Solvent	S1	S2	S3	S4	S5	S6	S7	S8		
Detection									D1	D2
Literature	1	1	1	1	1	1	1	1		
Compound	$R_F \times 100$								Color	
Aconitine	68	>90	35	3	49	36	60	65	—	r-br
Ajmaline	47	42	12	3	30	6	13	56	bl	be
Apoatropine	54	67	40!	20	26	15	40	16	—	bl-vt
Arecoline	66	>90	56	34	48	0	0	0	—	w (on pk)
Aspidospermine	65	>90	54	20	49	50	60	65	—	w (on pk)
Atropine	38	40	16	5	12	0	10	17	—	bl-vt
Boldine	16	16	3	0	5	24	6	58	vt	be
Brucine	42	63	18	0	19	50	54	12	—	vt-br
Bulbocapnine	65	>90	35	7	54	78	70	48	bl	w (on pk)
Cephaeline	56	63	19	2	23	25	17	37	bl-vt	w (on pk)
Cinchonine	38	44	17	7	27	0	22	40	—	be-br
Cocaine	73	>90	65	36	58	84	77	62	—	vt
Codeine	38	53	16	4	26	12	27	35	—	—
Colchicine	47	41	4	0	4	11	0	57	—	lt gy
Cotarnine	60	>90	43	31	45	0	25	0	gr-y	vt
Cupreine	3	0	0	0	0	0	0	0	br-y	r-br
Dihydrocodeine	38	54	18	6	28	10	30	25	bl	bl-vt
Dihydrocodeinone	51	65	21	4	30	48	43	18	—	vt
Dihydroergocristine	42	30	3	0	7	15	7	69	bl-vt	br
Dihydroergotamine	21	12	0	0	3	7	0	61	vt	br
Dihydromorphinone	24	23	8	1	11	5	8	16	—	br-y
Emetine	67	>90	40	6	45	38	58	50	bl	r-br
Ephedrine	—	—	—	—	—	—	—	—	—	lt-br
Ergocristine	51	38	14	5	13	46	15	70	bl-vt	be/lt br
Ergocristinine	61	57	13	0	20	0	27	70	bl-vt	lt-br
Ergometrine	14	6	0	0	2	3	0	64	bl-vt	w (on pk)
Ergometrinine	42	25	3	0	8	12	10	62	bl-vt	bl-vt
Ergotamine	24	16	0	0	3	10	5	59	bl-vt	pk
Ergotaminine	24	51	0	0	14	42	15	68	bl-vt	pk
Homatropine	37	45	15	5	23	4	24	15	—	bl-vt
Hordenine	33	36	14	5	28	0	15	35	—	w (on pk)
Hydrastinine	66	>90	58	41	50	0	25	0	steel-bl	bl-vt
Lobeline	68	>90	48	14	48	55	60	55	—	r-br
Morphine	10	8	0	0	3	3	0	34	—	dp bl
Narceine	3	0	0	0	0	0	0	0	—	dp bl
Narcotine	72	>90	51	10	57	81	79	72	bl	lt y
Papaverine	67	>90	42	3	47	85	84	70	y	y
Physostygmine	65	>90	32	4	44	59	50	46	—	pk
Pilocarpine	41	52	9	0	13	32	25	55	—	lt br
Psicaine new	66	>90	60	35	53	83	82	59	—	y
Quinidine	33	40	15	0	25	12	18	50	bl	lt y
Quinine	19	26	7	0	17	9	18	43	bl	y-w
Rauwolscine	55	63	18	4	36	36	15	68	y-gr	pl be
Reserpine	72	80	20	0	46	63	35	69	gr-y	w (on pk)
Sarpagine	12	4	0	0	0	0	0	0	—	be
Scopolamine	56	60	19	3	34	30	0	52	—	vt
Scopoline	60	90	44	20	44	46	50	37	—	w (on pk)
Serpentine	24	15	0	0	4	0	0	0	dk br	r-br
Serpentinine	53	56	8	0	10	0	3	12	y-gr	y-br
Sparteine	70	90	68	68	55	0	55	5	—	vt
Strychnine	53	76	28	5	38	57	60	22	—	y
Thebaine	65	90	51	16	50	71	76	40	—	r-br
Tropacocaine	65	90	56	34	45	58	78	35	—	vt
Yohimbine	63	62	18	3	37	33	15	60	gr-bl	lt-y

TABLE TLC 3—(Continued)
ALKALOIDS

Layer	L1 = Silica gel G.
	L2 = Alumina G.
	L3 = Silica gel G pretreated with 0.1*N* NaOH.
Solvent	S1 = chloroform-acetone-diethylamine (50 : 40 : 10).
	S2 = chloroform-diethylamine (90 : 10).
	S3 = cyclohexane-chloroform-diethylamine (50 : 40 : 10).
	S4 = cyclohexane-diethylamine (90 : 10).
	S5 = benzene-ethyl acetate-diethylamine (70 : 20 : 10).
	S6 = chloroform.
	S7 = cyclohexane-chloroform (30 : 70) plus 0.05% of diethylamine (3 drops).
	S8 = methanol.
Detection	D1 = fluorescence color in UV light, 365 mμ.
	D2 = iodoplatinate reagent.
Technique	Ascending, development distance = 10 cm.
Literature	1. Waldi, D., Schnackerz, K., and F. Munter, *J. Chromatogr.*, **6**, 66–69 (1961).

TABLE TLC 4
ALKALOIDS, OPIUM ALKALOIDS

Layer	L1	L2	L3	L4	L4	L5	L3	L6	L3	L3	L3		
Solvent	S1	S2	S3	S4	S5	S6	S7	S7	S8	S9	S10		
Detection												D1	D1
Technique	T1	T1	T1	T1	T1	T1	T1	na	T2	na	T3		
Literature	1	2	3	4	4	4	5	6	7	7	8	8	8
Compound						$R_F \times 100$						Color	
Codeine	62	45	21	16	64	85	26	33	35	15	41	—	o
Codethyline	—	—	—	—	—	—	—	37	—	—	—	—	—
Cryptopine	—	—	34	—	—	—	—	—	—	—	—	—	—
Dextromethorphan	—	—	—	—	—	—	—	—	—	—	23	—	r
Dextromoramide	—	—	—	—	—	—	—	—	—	—	87	—	o
Didhydrocodeine	—	—	—	—	—	—	—	22	—	—	—	—	—
Ethylmorphine	—	—	—	—	—	—	—	—	—	—	41	dk	r
Hydrocodone	—	—	—	—	—	—	—	28	—	—	29	—	o
Hydromorphone	—	—	—	—	—	—	—	—	—	—	28	dk	y

TABLE TLC 4—(Continued)
ALKALOIDS, OPIUM ALKALOIDS

Layer	L1	L2	L3	L4	L4	L5	L3	L6	L3	L3	L3		
Solvent	S1	S2	S3	S4	S5	S6	S7	S7	S8	S9	S10		
Detection												D1	D1
Technique	T1	T1	T1	T1	T1	T1	T1	na	T2	na	T3		
Literature	1	2	3	4	4	4	5	6	7	7	8	8	8
Compound	$R_F \times 100$											Color	
Hydromorphone HCl	—	—	—	—	—	—	—	13	—	—	—	—	—
10-Hydroxycodeine	—	—	—	—	—	—	—	—	17	15	—	—	—
Ketobemidone	—	—	—	—	—	—	—	—	—	—	56	dk	o
Kryptopine	—	—	—	—	—	—	—	—	48	40	—	—	—
Laervorphanol	—	—	—	—	—	—	—	—	—	—	28	—	r
Laudanidine	—	—	—	—	—	—	—	—	47	28	—	—	—
Laudanine	—	—	26	—	—	—	—	—	47	28	—	—	—
Laudanosine	—	—	42	—	—	—	—	—	—	—	—	—	—
Methadone	—	—	—	—	—	—	—	—	—	—	48	—	r
Morphine	39	50	11	3	14	28	12	10	12	8	40	dk	o
Narceine	0	20	—	0	0	0	—	—	9	3	23	bl	vt
Narcotine (Noscapine)	92	—	68	100	92	100	74	—	97	87	81	bt lt bl	o
Narcotoline	—	—	58	—	—	—	—	—	88	71	—	—	—
Neopine	—	—	—	—	—	—	—	—	38	12	—	—	—
Oxycodone HCl	—	—	—	—	—	—	—	70	—	—	—	—	—
Papaverine	88	100	63	60	85	96	59	—	97	77	82	bt y	o
Porphyroxine	—	—	—	—	—	—	—	—	93	65	—	—	—
Protopine	—	—	38	—	—	—	—	—	46	42	—	—	—
Reticuline	—	—	—	—	—	—	—	—	28	18	—	—	—
Thebacone	—	—	—	—	—	—	—	59	—	—	31	y-ish	o
Thebaine	—	—	40	38	85	100	45	—	65	38	41	dk	y

Layer L1 = Silicic acid-heavy magnesium oxide (1 : 1) with $CaSO_4$ binder.
L2 = Aluminum oxide buffered to pH5.
L3 = Silica gel G.
L4 = heavy hydrated magnesium oxide prepared with 2.5% $CaCl_2 \cdot 6H_2O$ (5 : 9) (w/v).
L5 = heavy hydrated magnesium oxide prepared with 2% $MgSO_4 \cdot 7H_2O$ (5 : 9) (w/v).
L6 = Silica gel G+0.5 KOH.

Solvent S1 = dioxane.
S2 = ethanol-*n*-butanol-water (1 : 9 : 1).
S3 = benzene-methanol (8 : 2).
S4 = *n*-hexane-acetone (3 : 1).
S5 = chloroform-ethyl acetate (3 : 1).
S6 = chloroform-ethyl acetate (1 : 1).
S7 = xylene-methyl ethyl ketone-methanol-diethylamine (20 : 20 : 3 : 1).
S8 = methanol-chloroform (1 : 9).
S9 = ethanol-benzene (1 : 4).
S10 = methanol-acetone-triethanolamine (1 : 1 : 0.03).

Detection D1 = UV light and Dragendorff.

Technique T1 = 13.5 cm development distance.
T2 = 10 cm development distance, 45 minutes.
T3 = 30 minutes development time.

Literature

1. Borke, M. L. and E. R. Kirch, *J. Amer. Pharm. Assoc. Sci. Ed.*, **42**, 627 (1953).
2. Mariani, A. and O. Mariani-Marelli, *Rend. Ist. Super. Sanita.*, **22**, 759 (1959); *Chem. Abstr.* **54**, 11374 (1960).
3. Neubauer, D. and K. Mothes, *Planta Med.*, **9**, 466 (1961).
4. Ragazzi, E., Veronese, G. and C. Giacobazzi, "*Thin-Layer Chromatography*", Marini-Bettòlo, G.B., Ed., Elsevier, Amsterdam, 1964, 149.
5. Bayer, I., *J. Chromatogr.*, **16**, 237 (1964).
6. Teichert, K., Mutschler, E., and H. Rochelmeyer, *Deut. Apotheker-Ztg.*, **100**, 477 (1960).
7. Brochmann-Hanssen, E. and T. Furuya, *J. Pharm. Sci.*, **53**, 1550 (1964).
8. Bäumler, J. and S. Rippstein. *Pharm. Acta Helv.*, **36**, 382 (1961).

TABLE TLC 5
AMINES

Layer	L1	L1	L1	L2	L3	L4	L5
Solvent	S1	S2	S3	S4	S3	S5	S6
Literature	1	1	1	1	1	1	1
Compound	$R_F \times 100$						
Benzylamine	70	49	36	50	42	—	—
***n*-Butylamine**	—	—	—	—	—	43	—
Cadaverine	3	6	2	1	1	—	—
Diethylamine	—	—	—	—	—	—	68[a]
Dimethylamine	—	—	—	—	—	—	47[a]
Ethanolamine	30	18	11	10	10	7	—
Ethylamine	—	15	13	20	14	19	22[a]
Histamine	41	26	2	3	3	2	—
Isoamylamine	—	49	39	45	44	58	50[a]
Isobutylamine	—	—	—	—	—	—	42[a]
Methylamine	—	11	10	10	7	12	14[a]
Phenylethylamine	66	56	37	55	42	—	—
***n*-Propylamine**	—	30	19	30	23	31	35[a]
Putrescine	3	4	2	1	1	—	—
Tryptamine	60	54	43	90	45	—	—
Tyramine	56	44	38	55	40	28	—

[a] 3,5-Dinitrobenzamide derivative.

Layer L1 = Neutral Silica gel G plates.
L2 = Phosphate buffered (pH = 6.8) Silica gel G.
L3 = Sodium acetate buffered Silica gel G.
L4 = Cellulose layers.
L5 = Silica gel G.

Solvent S1 = 95% ethanol-25% NH_3 (4 : 1).
S2 = phenol-water (8 : 3).
S3 = butanol-acetic acid-water (4 : 1 : 5).
S4 = 70% alcohol.
S5 = *n*-amyl alcohol-acetic acid-water (4 : 1 : 5).
S6 = chloroform-96% EtOH (99 : 1).

Detection Ninhydrin or see Section II.I, Volume Two.

Literature 1. Teichert, K., Mutschler, E., and H. Rochelmeyer, *Deut. Apotheker-Ztg.*, **100**, 283 (1960).

TABLE TLC 6
AMINES, ARYLAMINE METABOLITES

Layer	L1	L1							
Solvent	S1	S2							
Detection			D1	D2	D3	D4	D5	D6	D7
Literature	1	1							
Compound	$R_F \times 100$					Color[a]			
4-Aminobiphenyl	45	41	gy	—	y	r	br	—	—
4-Acetamido-4′-hydroxybiphenyl	12	40	gy	br[b]	—	—	—	—	—
4-Amino-4′-hydroxybiphenyl	21	17	gr	gr	o	mv	—	—	—
4-Acetamido-3-hydroxybiphenyl	32	52	gy	br[b]	—	—	bl	—	—
4-Amino-3-hydroxybiphenyl	33	17	gy	bk	o	mv	bl	y	gr
4-(*N*-Hydroxyacetamido)-biphenyl	10–35	25–50	gy	bk[b]	pk	pk	—	—	p
4-(Hydroxyamino)-biphenyl	39	[c]	gy	bk	y	r	gr	mv[b]	—
2-Aminofluorene	45	26	gy	br[b]	y	r	—	—	—
2-Acetamido-7-hydroxyfluorene	11	44	gy	bk[b]	—	—	bl	—	gy
2-Amino-7-hydroxyfluorene	19	9	gr	gr	o	mv	bl	mv[b]	—
2-Acetamido-1-hydroxyfluorene	34	52	gy	br[b]	—	—	bl	—	gy
2-Amino-1-hydroxyfluorene	37	10	gy	bk	o	bl	bl	bl	—
2-(*N*-Hydroxyacetamido)-fluorene	10–30	25–50	gy	bk[b]	pk	pk	—	—	p
2-(Hydroxyamino)-fluorene	41	[c]	gy	bk	y	r	y	mv[b]	—
2-(*N*-Hydroxyacetamido)-naphthalene	10–30	25–48	gy	bk[b]	pk	pk	—	—	p
2-Acetamido-1-hydroxy-naphthalene	36	51	mv	mv	mv	mv	mv	pk	mv
N-Hydroxyacetanilide	10–30	20–45	gy	bk[b]	pk	pk	—	—	p
2-Hydroxyacetanilide	29	49	gy	bk[b]	—	—	bl	—	—

[a] Indicates no color.
[b] Visible after 10 minutes.
[c] Decomposed.

Layer L1 = Silica gel G.
Solvents S1 = petroleum ether (b.p. 40–60 °C) –acetone (7 : 3).
S2 = chloroform–ethyl acetate–acetic acid (6 : 3 : 1).
Detection D1 = Folin and Ciocalteau's reagent followed by 10% Na_2CO_3.
D2 = 2% ammoniacal $AgNO_3$.
D3 = *p*-dimethylaminobenzaldehyde (0.5% in ethanol containing 1 ml conc. HCl/100 ml).
D4 = *p*-dimethylaminocinnamaldehyde (1% in 50% ethanolic 3*N* HCl).
D5 = 2,6-dichloroquinonechlorimide (0.5% in ethanol) followed by 10% Na_2CO_3.
D6 = 0.5% sodium aminoprusside.
D7 = $FeCl_3 \cdot 6H_2O$ (5% in 95% ethanolic 0.1*N* HCl).
Technique Saturated chamber, 14 cm development.
Literature 1. Booth J. and E. Boyland, *Biochem. J.*, **91**, 364 (1964); *J. Chromatogr.*, **19**, D3 (1965).

TABLE TLC 7
AMINES AND HYDROCARBONS

Layer	L1	L1	L1	L1	
Solvent	S1	S2	S3	S4	
Detection					D1
Literature	1	1	1	1	
Compound		$R_F \times 100$			Color[a]
Aniline	—	14	30	60	vt → gy or bk
***o*-Toluidine**	—	—	45	—	r-vt → bl
***p*-Toluidine**	—	—	—	65	vt → br
2,4-Dimethylaniline	1	7	30	70	vt → br
2,6-Diethylaniline	—	20	—	—	vt → bl
4-Chloro-2-methylaniline	—	20	40	—	vt → gr
2-Chloro-6-methylaniline	—	40	65	—	vt → bl-vt
2-Chloro-4-methylaniline	—	30	60	—	vt → r-br
2,6-Dichloroaniline	14	60	60	—	gr-vt
2,4-Dichloroaniline	—	40	60	—	gr-vt
2,4,6-Trichloroaniline	—	60	—	—	r br
***N,N*-Dimethyl-*p*-phenylenediamine**	—	—	15	—	dk r-vt
Sulfanilamide	—	—	—	—	pl vt-gy
***N*-Methylaniline**	—	25	60	85	bl-vt → bl
***N*-Ethylaniline**	—	—	65	—	bl-vt → bl
***N*-Cyclohexylaniline**	—	30	—	—	vt → bl
Diphenylamine	4	60	—	—	vt → bl
***N*-Methyl-*o*-toluidine**	—	—	70	—	bl
***N,N*-Dimethylaniline**	—	35	70	85	bl
***N,N*-Diethylaniline**	—	40	75	—	bl
***N*-Methyldiphenylamine**	15	70	—	—	bl
Triphenylamine	20	75	—	—	bl
***N,N*-Dimethyl-*o*-toluidine**	—	—	70	—	pl gy-bl
***N,N*-Dimethyl-*p*-toluidine**	—	—	70	—	bl[b]
***N,N*,2,6-Tetramethylaniline**	25	60	80	—	bl-gy
***N,N*,2,4-Tetramethylaniline**	3	18	—	—	gy → gy-g
***N,N*-Diethyl-2,4-dimethylaniline**	—	—	75	—	gy[b]
1-Naphthylamine	—	—	—	70	vt → bl
***N,N*-Dimethyl-1-naphthylamine**	—	—	70	—	bl
4,4′-Bis(dimethylamino)benzophenone	—	2	35	70	gy
***N,N,N′,N′*-Tetramethylbenzidine**	—	4	45	85	gr
Pyridine	—	4	20	70	gr[c]
2,4,6-Trimethylpyridine	—	—	—	65	ft gy
Quinoline	—	—	30	65	none
Isoquinoline	—	—	—	60	ft gy
Indole	—	—	60	—	r-vt
Carbazole	2	50	60	85	vt
***N*-Methylcarbazole**	10	70	80	95	vt
***N*-Phenylcarbazole**	6	—	—	—	vt
3-(*m*-Chlorobenzoyl)-*N*-methylcarbazole	—	—	60	85	none
2-Aminopyridine	—	—	—	45	gy
2-Aminothiazole	—	—	—	45	r-br
Indoline	—	8	40	—	bl → vt
1,2,3,4-Tetrahydroquinoline	—	20	60	—	bl → bl-gy
6-Ethoxy-1,2-dihydro-2,2,4-trimethylquinoline	—	—	—	80	bl-gy
Iminodibenzyl	—	60	—	—	bl
Julolidine	—	35	—	—	gr
2,6-Dimethylpiperidine	—	—	—	—	vt → gy
2,2,6,6-Tetramethylpiperidine	—	—	—	10[d]	vt-br
***N*-Cyclohexylpiperidine**	—	4	—	20[d]	vt-br
Piperazine	—	—	—	5	dk br
***N*-Ethylmorpholine**	—	—	—	—	ft gy
Benzylamine	—	—	—	—	br-vt
Dibenzylamine	—	—	25	50	gy-vt
Tribenzylamine	—	—	—	—	none
Dicyclohexylamine	—	—	—	15[d]	vt-gy
***n*-Octylamine**	—	—	—	10[d]	vt-dk br
1-Methylheptylamine	—	—	—	—	br

TABLE TLC 7—(Continued)
AMINES AND HYDROCARBONS

Layer	L1	L1	L1	L1	
Solvent	S1	S2	S3	S4	
Detection					D1
Literature	1	1	1	1	
Compound		$R_F \times 100$			Color[a]
2-Ethylhexylamine	—	—	—	10[d]	r → br-vt
tert-Octylamine	—	—	—	—	vt-br
Diethylamine	—	—	—	5	vt-br
Dibutylamine	—	—	—	15	r → gy
Dimethyloctylamine	—	—	3	20[d]	ft
Tributylamine	—	—	—	—	pale p-pale gy
Diethylenetriamine	—	—	—	4	br
Acetanilide	—	—	6	—	none
N-Methyl-*N*-nitrosoaniline	—	—	65	—	pale bl-gr
Nitrosobenzene	—	45	—	—	none
4-Nitrotoluene	—	55	—	—	none
Azoxybenzene	—	50	—	—	ft y
N,N-Diethylhydroxylamine	—	—	20	55	olive gr
Trimethylamine oxide	—	—	—	3	none
N,N-Dimethylhexadecylamine oxide	—	—	—	16	ft gy
N,N-Dimethylaniline oxide	—	—	1	11	ft gy
N-Methyldiphenylamine oxide	—	—	—	35	ft bl
Pyridine oxide	—	—	3	35	ft gy
Trimethylanilinium benzenesulfonate	—	—	—	2	pale br
19-Methylacridinium chloride	—	—	—	[d]	remained y
Benzophenone	—	35	—	—	none
Benzyl alcohol	—	5	—	—	none
2,4-Pentanedione	—	10	50	—	pale br
Chrysene	15	—	—	—	ft br[c]
Pyrene	20	75	75	—	intense vt
Anthracene	20	—	—	—	vt → bl
Phenanthrene	25	—	—	—	o
Naphthalene	30	75	—	—	y → y-o
Fluoranthene	20	75	—	90	o
Fluorene	—	—	—	—	o
p-Terphenyl	10	—	—	—	ft
Biphenyl	—	—	—	—	y
Bibenzyl	12	—	—	—	ft y
Hexamethylbenzene	—	—	—	—	o → r-br
Mesitylene	35	—	—	—	ft y
p-Cymene	45	—	—	—	ft y
Diphenylfulvene	12	—	—	—	pale o
Azobenzene	7	66	—	—	—
4-Phenylazodimethylaniline	—	32	—	—	—
1-Phenylazo-2-naphthol (Sudan Yellow)	—	20	80	85	—
Sudan Red	—	11	—	—	—
Indophenol	—	3	60	85	—
Oracet Orange (Ciba)	—	8	40	69	—
Oracet Red (Ciba)	—	—	4	53	—
Orasol Scarlet 2B (Ciba)	—	—	—	17	—
Rhodamine B	—	—	—	7	—

[a] For color abbreviations see inside cover.
[b] Two closely placed spots of different shade.
[c] Very slowly.
[d] Highly elongated spot.

Layer L1 = Silica gel G or silica gel $HF_{254+366}$ (Merck), activated one hour at 110 °C.
Solvent S1 = cyclohexane.
S2 = benzene.
S3 = benzene-methanol (19 : 1).
S4 = chloroform-methanol (6 : 1).
Detection D1 = 2,6-Dichloroquinone-4-chlorimide.
Technique Ascending, saturated chamber.
Literature 1. Ross, J. H., *Anal. Chem.*, **60**, 2140–2141 (1968).

TABLE TLC 8

AMINES, EPINEPHRINE (ADRENALINE) AND RELATED COMPOUNDS

Layer	L1	L1	L1	L2	L2	L2
Solvent	S1	S2	S3	S1	S2	S3
Literature	1	1	1	1	1	1
Compound	$R_F \times 100$					
Epinephrine	46	85	91	8	22	11
Norepinephrine	29	76	87	34	39	6
Normetanephrine	56	85	93	10	59	69
Metanephrine	72	88	96	15	47	45
Dihydroxymandelic acid	9	75	44	15	32	13
Dopamine	39	82	89	—	49	55
Vanilmandelic acid	31	80	64	—	—	—
3-Methoxy-4-hydroxyphenylglycol	60	81	87	—	—	—
Dihydroxyphenylalanine	18	—	—	—	—	—
Adrenochrome	36	77	32	—	—	—
Tyramine	74	71	100	—	—	—
Serotonin	38	76	92	—	—	—
Dihydroxyphenylacetic acid	23	85	66	—	—	—
Homogentisic acid	55	79	80	—	—	—
5-Hydroxytryptophan	20	86	91	—	—	—
Triiodotyrosine	15	92	85	—	—	—
Phenylpyruvic acid	24	82	66	—	—	—
Tryptophan	42	67	11	—	—	—
Tyrosine	48	—	12	—	—	—

Layer L1 = Polyamide (Woelm). L2 = Silica gel.
Solvent S1 = isobutanol-acetic acid-cyclohexane (80 : 7 : 10). S2 = isopropanol-ammonia (4 : 1). (S1 and S2 are useful for two-dimensional TLC). S3 = *n*-butanol-acetic acid-water (4 : 1 : 1).
Detection Diazotized *p*-nitraniline or sulfanilic acid.
Literature 1. Sapira, J. D., *J. Chromatogr.*, **42**, 134 (1969).

TABLE TLC 9

AMINES, ADRENALINE DERIVATIVES

Layer	L1	L2	L3	L4	L1	L2	L3	L4	L1	L2	L3	L4	L1	L2	L3	L4
Solvent	S1	S1	S1	S1	S2	S2	S2	S2	S3	S3	S3	S3	S4	S4	S4	S4
Literature	1	1	1	1	1	1	1	1	1	1	1	1	1	1	1	1
Compound	$R_F \times 100$															
Adrenaline	27	6	5	15	33	10	18	27	27	5	9	13	42	22	24	28
Oxedrine	35	40	48	38	42	40	45	56	30	35	36	34	45	47	50	55
Noradrenaline	41	12	6	14	52	13	17	30	34	14	17	20	52	25	30	36
Adrenone	38	6	5	7	41	10	8	17	27	6	8	16	45	16	19	24
Carbadrine	57	14	12	18	56	21	27	37	38	19	22	22	60	31	36	42
Oxamphetamine	69	71	72	51	64	66	71	73	48	56	60	60	67	64	64	73
Isoprenaline	45	11	14	20	48	21	34	40	31	17	20	20	53	24	31	46
Metaoxedrin	44	47	53	40	48	50	53	61	34	39	40	36	51	47	52	61
3-Hydroxytyramine	53	17	12	28	54	25	29	46	37	19	21	23	56	24	31	42

Layer L1 = Kieselgel G buffered to pH 4. L2 = Kieselgel G treated with Na_2MoO_4.
L3 = Kieselgel G treated with Na_2WO_4. L4 = Kieselgel G treatedwith borax.
Solvent S1 = *n*-butanol saturated with aqueous sulfur dioxide solution (H_2SO_3).
S2 = *n*-butanol-acetic acid-H_2SO_3 (4 : 1 : 5).
S3 = *n*-amyl alcohol-acetic acid-ethanol-H_2SO_3 (4 : 1 : 1 : 5).
S4 = *n*-butanol-*n*-propanol-acetic acid-H_2SO_3 (4 : 1 : 1 : 5).
Detection Folin-Denis reagent.
Technique 19–21 °C temperature.
Literature 1. Halmekoski, J., *Suomen Kemistilehti*, **B36**, 58 (1963); from *J. Chromatogr.*, **15**, D9 (1963).

TABLE TLC 10
AMINES, 4-(PHENYLAZO) BENZENE SULFONAMIDE DERIVATIVES

Layer	L1
Solvent	S1
Literature	1

Compound	$R_F{}^a \times 100$
Primary amines	
4-PABSA	
Methylamine	47
Ethylamine	71
Propylamine	87
Isopropylamine	92
Butylamine	100
***sec*-Butylamine**	104
Isobutylamine	106
***tert*-Butylamine**	105
Amylamine	108
Isoamylamine	113
Hexylamine	115
Octylamine	120
Decylamine	127
Allylamine	81
Cyclopentylamine	100
Cyclohexylamine	105
Benzylamine	87
α-Phenethylamine	86
β-Phenethylamine	83
Amphetamine	94
Mescaline	15
Secondary amines	
4-PABSA	R_X
Dimethylamine	116
Diethylamine	143
Dipropylamine	150
Dibutylamine	>150
Diisobutylamine	>150
Diamylamine	>150
Diallylamine	147
Dipropargylamine	128
Dibenzylamine	141
β-Phenylisopropylmethylamine	126
Ephedrine	57
Pyrazole	138
Imidazole	94
Morpholine	99

[a] Relative to the derivative of butylamine.

Layer L1 = Microchromatoplates coated with alumina (Fluka, D5).
Solvent S1 = ethyl acetate-petroleum ether (62–82 °C, from Shell) (1 : 4), saturated with water.
Detection Visible colored spots, improved by exposure to iodine vapor.
Technique Development in a beaker covered with cork, 10–15 minutes, ascending.
Literature 1. Jart, A. and A. J. Bigler, *J. Chromatogr.*, **29**, 258–262 (1967).

TABLE TLC 11
AMINO ACIDS

Layer	L1	L1	L1	L1						
Solvent	S1	S2	S3	S4						
Detection						D1	D2	D3	D4	D5
Technique	T1	T2	T2	T2						
Literature	1	1	1	1						
Compound		$R_F \times 100$			μg^a			Color		
α-Alanine	33–38	34	22	61	0.05	vt	vt		–	
β-Alanine	33–39	37	14	66	0.05	gr	l		+	
Alloisoleucine	58–64	65	62	88	0.1	vt	pk		(+)	
Allothreonine	54–60	31	55	59	0.5	vt	pk		(+)	
α-Amino-*n*-butyric acid	42–47	42	31	78	0.05	vt	l		(+)	
β-Aminobutyric acid	41–46	43	21	87	0.05	l	y		+	
γ-Aminobutyric acid	42–48	46	13	78	0.2	vt	l		+	
ε-Amino-*n*-caproic acid	52–57	55	20	87	1.0	vt	pk		+	
***p*-Aminohippuric acid**	44–49	65	74	89	10.0	vt	y		+	
α-Aminoisobutyric acid	41–47	50	25	78	0.05	vt	l		+	
β-Aminoisobutyric acid	42–48	47	22	78	0.05	vt	l		+	
Arginine	3– 8	13	2	90	0.2	vt	pk		+	+
Asparagine	19–25	12	21	47	2.0	y	pk		+	
Aspartic acid	8–14	21	20	20	0.05	gr	vt		(+)	
Canavanine sulfate	12–17	7	11	66	0.2	vt	br-ish		+	+
Citrulline	19–24	16	15	68	0.1	vt	pk		+	
Creatine	2– 9	34	8	86	—	—	y	(+)	+	+
Creatinine	33–39	36	25	97	—	—	y		+	+
Cysteic acid	23–27	6	41	9	0.05	vt	y		(+)	
Cystine	17–22	4	26	35	1.0	br	pk		(y)	
α,γ-Diaminobutyric acid	36–42	8	20	73	2.0	vt	pk		+	
α,α-Diaminopimelic acid	2–10	5	6	36	2.0	vt	r		gy	
Dihydroxyphenylalanine	10–22	16	10	13	10.0	gy	l	+	(y)	
Diiodotyrosine	62–69	50	73	71	5.0	vt	l	(+)	+	
Dimethylcysteine	40–45	10	53	73	0.2	vt	pk		–	
Glutamic acid	15–20	30	21	25	0.05	vt	l		–	
Glutamine	25–29	14	18	59	0.5	vt	pk		(+)	
Glycine	28–33	22	19	46	0.05	br	pk		+	
Glycocyamine	0– 7	34	8	71	—	–	pk		(+)	+
Histidine	36–41	11	34	87	0.3	gy	l	+	(gy)	
Hydroxyglutamic acid	16–21	30	15	25	0.05	vt	pk		(+)	
Hydroxylysine	29–34	8	11	71	0.5	vt	l		+	
Hydroxyproline	26–30	22	19	66	1.0	y	bl		(+)	
β-Hydroxyvaline	62–69	41	81	65	0.5	vt	pk		(+)	
Isoleucine	58–64	65	62	88	0.1	vt	pk		(+)	
Kynurenine	54–60	39	47	80	0.5	br	pk		(y)	
Lanthionine	9–17	8	22	35	1.0	vt	o		pk	
Leucine	59–66	65	60	88	0.5	vt	pk		–	
Lysine	32–37	10	6	82	0.5	vt	r		+	
Methionine	56–62	50	54	88	0.5	vt	pk		–	
Methionine sulfone	37–42	20	36	73	0.2	vt	pk		(+)	
Methionine sulfoxide	24–28	19	26	82	0.2	vt	pk		(br)	
Norleucine	59–66	65	60	88	0.5	vt	l		–	
Norvaline	52–58	55	45	87	0.05	vt	r		(+)	
Ornithine	28–33	8	5	82	0.2	vt	r		+	
α-Phenylalanine	61–67	54	73	88	0.2	vt	l		(y)	
α-Phenylglycine	63–69	76	81	77	5.0	y	y	(+)	(br)	
Proline	41–46	35	19	87	0.5	y	bl		(+)	
Sarcosine	36–41	29	17	80	0.1	gy	y		+	
Serine	42–49	21	38	41	0.2	vt	o		y	
Taurine	46–52	21	44	51	0.2	vt	y		+	
Threonine	60–67	31	67	56	0.5	vt	pk		(+)	
Thyronine	63–69	62	82	88	2.0	br	br		–	
Thyroxine	69–76	73	84	86	2.0	br	y	–	(+)	
Tryptophan	56–62	41	60	88	0.5	vt	l		(br)	
Tyrosine	45–51	38	56	71	0.5	br	r	+	(pk)	
Valine	49–55	55	44	87	0.05	vt	pk		(+)	

[a] Detection limitation for ninhydrin reaction on two-dimensional chromatogram using combination of S1 and S2. The actual limit of detection is lower but varies somewhat with the drying conditions after applying the ninhydrin reagent.

TABLE TLC 11—(Continued)
AMINO ACIDS

Layer L1 = Cellulose layers.
Solvent S1 = *n*-butanol-acetone-diethylamine-water (10 : 10 : 2 : 5).
S2 = isopropanol-formic acid-water (20 : 1 : 5).
S3 = *sec*-butanol-methyl ethyl ketone-dicyclohexylamine-water (10 : 10 : 2 : 5).
S4 = phenol-water (75 : 25) plus 7.5 mg of NaCN; chamber is saturated with NH_3.
Detection D1 = ninhydrin-collidine reagent.
D2 = isatin.
D3 = Pauly's diazo reagent; + yellow to reddish.
D4 = Modified Reindel-Hoppe reagent; + bl.-bk color.
D5 = Na nitroprusside-K ferricyanide reagent; + red color.
Technique T1 = 1st-dimensional development.
T2 = S2, S3, or S4, unsaturated chamber; solvent migration, 16 cm.
Literature 1. von Arx, E. and R. Neher, *J. Chromatogr.*, **12**, 329 (1963).

TABLE TLC 12
AMINO ACIDS AND PEPTIDES, CARBOBENZOXY DERIVATIVES

Layer	L1	L1	L1	L1
Solvent	S1	S2	S3	S4
Literature	1	1	1	1
Compound	$R_F \times 100$			
Cbo[a]-Glycine	81	76	74	64
Cbo-Diglycine	66	65	56	56
Cbo-Triglycine	57	56	41	50
Cbo-Tetraglycine	51	43	26	45
Cbo-Diglycine ethyl ester	82	73	75	74
Cbo-Triglycine ethyl ester	76	67	68	69
Cbo-Alanine-Glycine ethyl ester	81	75	79	77
Cbo-DL-Alanine	77	77	74	61
Cbo-DL-Dialanine	72	72	68	59
Cbo-Glycine-DL-Alanine	68	68	63	56
Cbo-DL-Alanine-Glycine	68	68	61	57
Cbo-Glycine-DL-Alanine-Glycine	61	59	50	54
Cbo-DL-Phenylalanine-Glycine ethyl ester	86	83	80	76
Cbo-Glycine-DL-Phenylalanine	84	79	77	76
Cbo-Glycine-Glycine-DL-Phenylalanine ethyl ester	81	75	69	76
Cbo-Glycine-DL-Phenylalanine-Glycine ethyl ester	83	78	76	74
Cbo-DL-Alanine-DL-Phenylalanine	78	78	72	61
Cbo-Glycine-DL-Phenylalanine	72	76	69	65
Cbo-Glycine-L-Isoleucine	75	73	74	67
Cbo-Glycine-L-Leucine	74	74	69	65
Cbo-Glycine-Glycine-L-Leucine	71	69	63	65
Cbo-Glycine-L-Glutamic acid	72	71	65	62
Cbo-DL-Phenylalanine	74	76	77	75
Glycine-Glycine	17	25	4	15
DL-Alanine-Glycine	21	30	9	22
Glycine-L-Leucine	43	44	24	41
Glycine	22	29	10	21
DL-Alanine	27	34	14	27
DL-Phenylalanine	45	42	31	46
HCl-DL-Phenylalanine ethyl ester	59	52	44	65
HCl-Glycine ethyl ester	43	42	24	53

[a] Cbo = carbobenzoxy.

Layer L1 = Silica gel G (air-dried).
Solvent S1 = *n*-butanol-acetone-acetic acid-5% ammonium hydroxide-water (4.5 : 1.5 : 1 : 1 : 2).
S2 = *n*-butanol-acetic acid-5% ammonium hydroxide (5.5 : 3 : 1.5).
S3 = *n*-butanol-acetic acid-5% ammonium hydroxide-water (6 : 1 : 1 : 2).
S4 = *n*-butanol-acetic acid-pyridine-water (15 : 3 : 10 : 12).
Detection Ninhydrin and potassium dichromate.
Technique Ascending development, 2 hours for 8–10 cm.
Literature 1. Ehrhardt, E. and F. Cramer, *J. Chromatogr.*, **7**, 405 (1962).

TABLE TLC 13
AMINO ACIDS AND RELATED COMPOUNDS

Layer	L1	L1
Solvent	S1	S2
Literature	1	1
Compound	$R_F \times 100$	
Alanine	57	23
β-Alanine	46	19
α-Aminoadipic acid	62	3
α-Amino-*n*-butyric acid	65	29
γ-Amino-*n*-butyric acid	51	23
ε-Aminocaproic acid	76	23
p-Aminohippuric acid	82	25
4-Amino-5-imidazole carboxylic acid	43	63
α-Aminoisobutyric acid	74	30
β-Aminoisobutyric acid	60	26
δ-Aminolaevulinic acid	41	2
2-Amino-octanoic acid	93	81
Arginine	19	6
Argininosuccinic acid	31	0
Asparagine	21	14
Aspartic acid	48	1
Cadaverine	14	70
Citrulline	34	12
Cysteic acid	53	8
Cysteine	12	5
Cystine	6	3
2,4-Diaminobutyric acid	10	25
2,6-Diaminopimelic acid	16	1
Djenkolic acid	9	5
3,4-DOPA	43	6
Epinephrine	51	52
Ethanolamine	39	81
Ethionine	83	66
Formiminoglycine	45	16
Glucosamine	20	38
Glutamic acid	56	1
Glutamine	31	13
Glycine	37	16
Histamine	15	85
Histidine	11	26
Homoarginine	23	12
Homoserine	45	27
Hydroxylysine	10	17
Hydroxyproline	48	17
Isoleucine	90	63
Kynurenine	38	67
Leucine	90	69
Lysine	16	17
Methionine	78	51
Methionine sulfone	50	27
Methionine sulfoxide	39	20
1-Methylhistidine	14	18
3-Methylhistidine	12	14
Norleucine	92	73
Ornithine	11	15
Penicillamine	34	15
Phenylalanine	82	67
Phosphoethanolamine	43	0
Pipecolic acid	66	36
Proline	58	30
Putrescine	16	60
Sarcosine	48	24
Serine	39	27
Taurine	41	33
Threonine	51	61
Tryptophan	70	54

TABLE TLC 13—(Continued)
AMINO ACIDS AND RELATED COMPOUNDS

Layer	L1	L1
Solvent	S1	S2
Literature	1	1
Compound	$R_F \times 100$	
Tyrosine	72	35
Valine	79	44

Layer L1 = Cellulose powder MN-300, without binder and washed before use.

Solvent S1 = 2-propanol-butanone-1*N* HCl (60 : 15 : 25 v/v).
S2 = 2-methylpropanol-2-butanone-propanone-methanol-water-conc. ammonia (40 : 20 : 20 : 1 : 14 : 5 v/v).

Detection Ninhydrin-cadmium acetate reagent; Ehrlich's reagent for imidazole.

Technique S1 for first dimension, S2 for second dimension; saturated tank, 13 cm solvent migration, about 2.5 hours, 22°C.

Literature 1. Haworth, C. and J. G. Heathcote, *J. Chromatogr.*, **41**, 380–385 (1969).

TABLE TLC 14
DERIVATIVES OF AMINO ACIDS
N-2,4-DINITRO-5-AMINOPHENYL (DNAP)

Layer	L1	L1	L1
Solvent	S1	S2	S3
Literature	1	1	1
DNAP of	$R_F \times 100$		
Alanine	33	58	64
β-Alanine	57	73	67
Arginine	0	0	12
Aspartic acid	2	9	53
Glutamic acid	9	39	40
Glycine	17	36	49
N-α DNAP-Histidine	0	0	10
Isoleucine	70	87	86
Leucine	65	84	83
Methionine	47	69	74
Norleucine	67	86	84
Norvaline	59	82	92
Phenylalanine	50	76	88
Proline	38	60	83
Serine	4	12	39
Threonine	9	20	37
Tryptophan	46	69	52
Tyrosine	29	36	75
O-Mono-DNAP-Tyrosine	0	0	15
Valine	60	84	92
OH(3-Amino-4,6-dinitrophenol)	80	92	100
NH_2(4,6-Dinitro-*m*-phenylenediamine)	streaks	streaks	86

Layer L1 = Silica gel G activated at 110 °C for 30 minutes.

Solvent S1 = chloroform-*n*-propanol-acetic acid (70 : 30 : 1).
S2 = chloroform-*tert*-amyl alcohol-acetic acid (50 : 50 : 3).
S3 = benzene-pyridine-acetic acid (60 : 13 : 27).

Detection NO_2 followed by β-naphthol.

Technique 10 cm solvent migration.

Literature 1. Ratney, R. S., Godshalk, M. F., Joice, J., and K. W. James, *Anal. Biochem.*, **19**, 362 (1967).

TABLE TLC 15
AMINO ACIDS

Layer	L1	L2	L1	L2	L1	L2	L1	L2	L1	L2	L1	L2
Solvent	S1	S1	S2	S2	S3	S3	S4	S4	S5	S5	S6	S6
Technique	T1	T2	T1	T2	T1	T3	T1	T3	T1	T3	T4	T5
Literature	1	1	1	1	1	1	1	1	1	1	1	1
Compound	$R_F \times 100$											
Alanine	43	36	41	34	16	13	40	33	35	25	37	27
β-Alanine	45	39	33	27	9	6	30	24	31	21	37	29
α-Amino-*n*-butyric acid	53	45	48	41	24	20	47	41	45	34	47	37
γ-Amino-*n*-butyric acid	53	45	34	27	11	8	32	25	45	30	44	34
α-Amino-*n*-caprylic acid	89	89	80	78	73	66	90	87	88	91	76	73
α-Aminoisobutyric acid	—	47	49	42	25	21	49	43	48	37	46	34
β-Aminoisobutyric aicd	56	48	42	33	18	13	42	33	47	33	47	38
α-Amino-*n*-octanoic acid	—	—	—	—	—	—	—	—	—	—	—	—
α-Amino-*n*-valeric acid (Norvaline)	63	57	56	49	35	31	58	60	60	53	57	49
Arginine	22	18	7	4	4	3	21	14	19	9	5	3
Asparagine	22	16	22	16	5	5	25	18	14	11	22	19
Aspartic acid	27	21	31	21	1	0	17	15	18	14	14	9
Citrulline	28	23	32	23	4	2	27	20	20	16	20	12
Cysteic acid	14	12	36	31	2	1	21	15	13	9	26	19
Cystine	13	9	18	13	1	1	17	13	6	4	17	10
3,5-Diiodotyrosine	73	68	80	78	4	4	—	—	69	88	—	—
Ethanolamine	50	37	12	7	64	57	58	46	44	29	64	—
Glutamic acid	41	28	40	33	2	2	22	16	37	22	15	9
Glutamine	33	22	33	22	6	4	31	20	21	14	23	15
Glycine	33	24	30	22	8	7	29	24	22	15	27	23
Histidine	23	18	29	23	7	6	32	28	13	6	38	33
3-Iodotyrosine	65	58	74	74	8	8	—	—	65	75	—	—
Isoleucine	71	68	52	58	44	36	65	69	65	60	62	56
Leucine	73	71	64	61	50	40	68	74	69	64	64	59
Lysine	18	17	5	1	4	2	18	14	12	7	32	28
Methionine	59	50	57	62	24	18	55	55	53	46	56	48
Methionine sulfone	32	25	41	35	9	8	35	36	26	18	37	33
Methionine sulfoxide	30	23	31	26	7	6	31	31	23	16	22	20
1-Methyl histidine	21	17	29	25	9	8	36	33	12	7	27	22
3-Methyl histidine	29	21	36	33	8	6	37	32	20	10	31	21
Ornithine	15	15	6	2	3	5	16	15	10	5	26	24
Phenylalanine	67	62	64	59	36	32	64	67	59	60	65	59
Proline	45	38	42	33	20	17	41	39	39	29	39	33
Sarcosine	38	31	36	29	15	15	35	32	30	23	35	31
Serine	32	25	32	25	7	7	30	28	23	15	43	40
Taurine	28	22	42	36	17	15	42	40	29	22	50	42
Threonine	38	33	38	36	14	8	41	41	32	24	64	54
Tryptophan	58	50	64	61	22	19	55	51	49	42	63	54
Tyrosine	53	47	62	59	15	12	47	45	49	44	48	38
Urea	56	55	55	54	48	45	54	54	54	46	48	35
Valine	61	55	53	50	32	23	57	49	58	49	56	44

Layer L1 = Cellulose MN-300.
L2 = Avicel.

Solvent S1 = *n*-butanol-acetic acid-water (60 : 15 : 25).
S2 = *n*-butanol-pyridine-water (60 : 60 : 60).
S3 = isopropanol-water-conc. ammonia (200 : 20 : 10).
S4 = *n*-propanol-0.2*N* ammonia (3 : 1).
S5 = *n*-propanol-1*N* acetic acid (3 : 1).
S6 = *n*-butanol-acetone-diethylamine-water (70 : 70 : 14 : 35).

Detection Ninhydrin, Ehrlich's reagent; diazotized sulfanilic acid.

Technique T1 = ascending development, 25 °C, 15 cm, 5 hours.
T2 = ascending development, 25 °C, 12 cm, 4.5 hours.
T3 = ascending development, 25 °C, 12 cm, 4.3 hours.
T4 = ascending development, 25 °C, 15 cm, 2.5 hours.
T5 = ascending development, 25 °C, 12 cm, 3 hours.

Literature
1. Smith, I., Rider, L. J., and R. P. Lerner, *J. Chromatogr.*, **26**, 449–455 (1967).

TABLE TLC 16
AMINO ACIDS

Layer	L1	L1	L1	L1	L1	L1	L1	L1	L1
Solvent	S1	S2	S3	S4	S5	S6	S7	S3	S8
Literature	1	1	1	1	1	2	2	2	2
Compound	$R_F \times 100$								
α-Alanine	47	37	27	39	40	49	25	32	29
β-Alanine	33	26	27	30	29	49	20	33	30
α-Aminobutyric acid	—	—	—	—	—	54	25	32	—
β-Aminobutyric acid	—	—	—	—	—	48	32	36	—
γ-Aminobutyric acid	—	—	—	—	—	38	30	39	—
α-Aminocaprylic acid	66	65	60	58	60	—	—	—	69
α-Aminoisobutyric acid	—	—	—	—	—	57	26	35	—
β-Aminoisobutyric acid	—	—	—	—	—	46	29	38	—
Arginine-HCl	4	2	8	10	6	6	14	13	19
Asparagine	—	—	—	—	—	46	19	22	—
Aspartic acid	55	33	21	9	7	56	5	26	6
Citrulline	—	—	—	—	—	46	29	26	—
Cysteic acid	69	50	14	17	21	61	5	20	4
Cystine-HCl	39	32	16	27	22	8	9	7	12
Dihydroxyphenylalanine	—	—	45	—	—	—	—	—	34
Glutamic acid	63	35	27	14	15	55	7	32	10
Glutamine	—	—	—	—	—	55	28	24	—
Glycine	43	32	22	29	34	50	18	28	24
Histidine-HCl	33	20	6	38	42	33	24	10	32
Hydroxyproline	44	34	20	28	31	63	33	26	38
Isoleucine	60	53	46	52	58	60	36	47	49
Leucine	61	55	47	53	58	63	37	53	48
Lysine-HCl	3	2	5	18	11	5	8	10	9
Methionine	59	51	40	51	60	62	36	43	49
Norleucine	61	57	49	53	59	66	54	55	52
Norvaline	56	50	38	49	57	67	56	54	42
Ornithine	—	—	—	—	—	6	5	8	—
Phenylalanine	63	58	49	54	60	63	41	54	55
Proline	35	26	19	37	30	48	45	24	50
Sarcosine	31	22	17	34	31	—	—	—	37
Serine	48	35	22	27	31	52	19	29	20
Taurine	—	—	—	—	—	59	22	29	—
Threonine	50	37	25	37	40	66	18	28	26
Tryptophan	65	62	56	55	58	69	45	56	63
Tyrosine	65	62	56	55	58	65	36	50	47
Valine	55	45	35	48	56	56	29	38	40

Layer L1 = Air-dried Silica gel G plates.

Solvent
S1 = 96% ethanol-water (7 : 3).
S2 = *n*-propanol-water (7 : 3).
S3 = *n*-butanol-acetic acid-water (3 : 1 : 1).
S4 = *n*-propanol-34% ammonium hydroxide (7 : 3).
S5 = 96% ethanol-34% ammonium hydroxide (7 : 3).
S6 = *n*-propanol-water (1 : 1).
S7 = phenol-water (3 : 1).
S8 = phenol-water (75 : 25 w/w), 20 mg NaCN added per 100 g mixture.

Detection Ninhydrin (Moffat and Lytle modification).

Technique Ascending chromatography in saturated chamber, development distance 10 cm.

Literature
1. Brenner, M. and A. Niederwieser, *Experientia* (*Basel*), **16**, 378 (1960).
2. Opienska-Blauth, J., Kraczkowski, H., and H. Brzuszkiewicz, "*Thin Layer Chromatography*", Marini-Bettolo, G. B., Ed., Elsevier Publishing Co., Amsterdam 1964, p. 165.

TABLE TLC 17
AMINO ACIDS, DINITROPHENYL DERIVATIVES

Layer	L1	L1	L1	L1	L1	L1	L1	L1	L1	L2	L2
Solvent	S1	S2	S3	S4	S5	S6	S7	S8	S9	S10	S11
Literature	1	1	1	1	1	1	1	1	1	1	1
Compound						$R_F \times 100$					
Group I											
DNP-L-Arginine	0	0	0	0	0	4	0	14	78	45	—
DNP-L-Asparagine	0	0	0	2	32	58	0	24	71	—	14
DNP-L-Aspartic acid	0	0	4	4	42	72	0	0	2	—	14
Di-*S,N*-DNP-L-Cysteine	0	2	6	16	49	83	40	96	95	—	—
N,N′-Di-DNP-L-Cystine	0	0	0	0	16	53	0	14	72	28	—
DNP-DL-Glutamic acid	0	4	10	15	58	85	0	0	3	—	25
DNP-L-Glutamine	0	0	0	3	35	58	1	17	82	—	—
Di-DNP-L-Histidine	0	0	0	0	4	7	11	38	94	62	12
ε-DNP-L-Lysine	0	0	0	0	0	4	0	3	37	40	—
DNP-DL-Methionine sulfoxide	0	0	0	0	14	36	2	28	76	—	—
DNP-L-Serine	0	0	2	24	55	85	7	36	92	—	13
DNP-L-Threonine	0	2	8	28	65	90	13	49	97	—	18
Group II											
DNP-Glycine	6	25	31	51	83	Front	16	58	Front	—	33
Di-*N,N′*-DNP-L-Lysine	2	17	29	35	77	Front	26	92	Front	—	57
DNP-L-Tryptophan	9	27	40	79	97	Front	29	86	Front	—	58
Di-*O,N*-DNP-L-Tyrosine	2	20	33	40	80	Front	37	99	Front	47	60
Group III											
DNP-L-Alanine	24	50	54	79	93	Front	17	68	Front	—	55
DNP-DL-Methionine	26	54	54	77	95	Front	36	95	Front	—	72
DNP-L-Phenylalanine	26	54	59	77	95	Front	39	97	Front	—	—
DNP-L-Proline	28	55	54	74	96	Front	26	77	Front	—	66
Group IV											
DNP-L-Isoleucine	50	75	74	97	98	Front	49	98	Front	—	85
DNP-L-Leucine	49	73	71	95	97	Front	56	97	Front	—	84
DNP-L-Valine	41	70	67	95	98	Front	39	96	Front	—	80

Layer L1 = ITLC-SG (Gelman).
L2 = ITLC-SAF (Gelman).
Solvent S1 = isooctane-chloroform-acetic acid (100 : 50 : 5).
S2 = isooctane-chloroform-acetic acid (100 : 100 : 10).
S3 = isooctane-chloroform-acetic acid (100 : 150 : 5).
S4 = diethyl ether-acetic acid-water (100 : 1 : 1).
S5 = diethyl ether-acetic acid-water (100 : 3 : 3).
S6 = diethyl ether-acetic acid-water (100 : 5 : 5).
S7 = diethyl ether-methanol-7*M* ammonium hydroxide (100 : 6 : 2).
S8 = diethyl ether-methanol-7*M* ammonium hydroxide (100 : 8 : 3).
S9 = diethyl ether-methanol-7*M* ammonium hydroxide (100 : 10 : 4).
S10 = *n*-propanol-conc. ammonia (70 : 30).
S11 = chloroform-benzyl alcohol-glacial acetic acid (70 : 30 : 3).
Detection UV (360 nm and 254 nm).
Technique Gelman chamber, saturated chamber, development distance = 10–15 cm.
Literature
1. Haer, F. C., *An Introduction to Chromatography on Impregnated Glass Fiber*, Ann Arbor Science Publishers, Inc., Ann Arbor, Michigan, 1969, p. 101.

TABLE TLC 18
AMINO ACIDS, DINITROPHENYL DERIVATIVES

Layer	L1	L1	L1	L1	L1	L1	L1
Solvent	S1	S2	S3	S4	S4	S5	S5
Technique	T1	T1	T1	T1	T2	T2	T2
Literature	1	1	1	1	1	1	1
Compound	R_F[a] × 100			$R_{DNP-Leucine}$[b] × 100			
2,4-Dinitroaniline	90	90	72	115	128	131	101
2,4-Dinitrophenol	41	100	83	22	21	148	102
DNP-Alanine	34	54	60	32	33	59	66
DNP-β-Alanine	27	71	73	89	98	99	95
DNP-α-Aminobutyric acid	46	72	73	52	52	79	85
DNP-α-Aminocaprylic acid	79	92	83	105	108	108	101
DNP-Aspartate	2	13	9	6	5	7	6
Di-DNP-Cystine	—	3	1	0	0	0	2
DNP-Glutamate	1	26	31	12	12	12	12
DNP-Glycine	27	32	40	17	18	31	38
DNP-Isoleucine	64	83	81	107	107	100	101
Di-DNP-Histidine	53	11	8	5	4	12	16
DNP-Leucine	66	82	80	100	100	100	100
Di-DNP-Lysine	74	56	60	12	13	66	73
DNP-Methionine	55	70	69	43	43	72	81
DNP-Methionine sulfone	17	—	—	3	3	10	10
DNP-Norleucine	69	82	80	86	90	101	100
DNP-Norvaline	56	77	76	65	70	86	95
DNP-Ornithine	70	34	40	6	6	39	46
DNP-Phenylalanine	67	75	74	44	46	81	86
DNP-Proline	29	65	67	58	59	78	84
DNP-Sarcosine	23	56	57	34	35	59	65
DNP-Serine	15	11	11	9	10	7	8
DNP-Threonine	20	17	15	12	14	9	11
DNP-Tryptophan	65	69	69	23	25	54	61
Di-DNP-Tyrosine	76	58	60	17	16	57	65
DNP-Valine	53	79	77	76	81	91	98

[a] Average values for 6 single measurements.
[b] Development distance of DNP-leucine = 10 cm.

Layer L1 = Silica gel G layers.

Solvent S1 = toluene-pyridine-ethylene chlorohydrin-0.8*N* ammonium hydroxide (10 : 3 : 6 : 6), the upper phase is used for development, the lower for pretreatment of the layer.
S2 = chloroform-benzyl alcohol-acetic acid (70 : 30 : 3).
S3 = chloroform-*tert*-amyl alcohol-glacial acetic acid (70 : 30 : 3).
S4 = benzene-pyridine-acetic acid (40 : 10 : 1).
S5 = chloroform-methanol-acetic acid (95 : 5 : 1).

Technique T1 = development distance = 15 cm, ascending.
T2 = development distance of DNP-leucine = 10 cm, horizontal.

Literature
1. Brenner, M., Niederwieser, A. and G. Pataki, *Experientia* (Basel), **17**, 145 (1961).

TABLE TLC 19
AMINO ACIDS, IODOAMINO ACIDS AND DERIVATIVES

Layer	L1	L1	L1	L1	L1	L1
Solvent	S1	S2	S3	S4	S5	S6
Literature	1	1	1	1	1	1
Compounds			$R_F \times 100$			
Iodo Compounds						
Thyroxine	13	45	27	38	29	50
3,5,3′-Triiodothyronine	22	37	19	48	33	41
3,3′,5′-Triiodothyronine	9	49	31	35	—	—
3,5-Diiodothyronine	19	34	15	43	—	34
3,3′-Diiodothyronine	16	29	21	40	—	—
3-Monoiodotyrosine	6	25	10	16	13	25
3,5-Diiodotyrosine	3	35	13	10	7	37
3-Monoiodohistidine	5	7	2	9	9	5
3,5,3′,5′-Tetraiodothyroacetic acid	24	89	73	47	34	87
3,5,3′-Triiodothyroacetic acid	33	85	63	57	39	81
3,5-Diiodothyroacetic acid	28	87	64	58	—	77
3,5,3′5′-Tetraiodothyropyruvic acid	21	—	67	49	—	—
3,5,3′-Triiodothyropyruvic acid	30	—	59	58	—	—
3,5,3′,5′-Tetraiodothyrolactic acid	19	—	61	43	—	—
3,5,3′-Triiodothyrolactic acid	27	—	55	53	—	—
3,5,3′-Triiodothyronamine	58	—	23	65	—	—
Iodide	33	26	10	55	37	30
Acyl Derivatives						
N-Acetylthyroxine	21	82	52	45	—	—
N-Acetyl-3,5,3′-triiodothyronine	32	77	48	55	—	—
N-Acetyl-3,3′,5′-triiodothyronine	14	85	55	41	—	—
N-Acetyl-3,5-diiodothyronine	31	77	45	53	—	—
N-Acetyl-3,3′-diiodothyronine	27	79	43	47	—	—
N-Acetyl-3-monoiodotyrosine	13	72	—	—	21	—
N-Acetyl-3,5-diiodotyrosine	8	77	—	—	13	—
N-Acetyl-3-monoiodohistidine	11	33	—	—	15	—
O,N-Diacetylthyroxine	54	85	63	77	—	—
O,N-Diacetyl-3,5,3′-triiodothyronine	70	82	59	89	—	—
O,N-Diacetyl-3,3′,5′-triiodothyronine	40	86	63	67	—	—
O,N-Diacetyl-3,5-diiodothyronine	68	83	57	85	—	—
O,N-Diacetyl-3,3′-diiodothyronine	63	82	58	80	—	—
N-Carbobenzoxythyroxine	35	92	72	60	—	—
N-Carbobenzoxy-3,5,3′-triiodothyronine	49	88	63	67	—	—
N-Carbobenzoxy-3,3′,5′-triiodothyronine	32	92	75	57	—	—
N-Carbobenzoxy-3,5-diiodothyronine	45	87	61	66	—	—
N-Carbobenzoxy-3,3′-diiodothyronine	43	89	66	64	—	—
N-Carbobenzoxy-3-monoiodotyrosine	33	83	—	—	38	—
N-Carbobenzoxy-3,5-diiodotyrosine	26	89	—	—	30	—
N-Carbobenzoxy-3-monoiodohistidine	30	57	—	—	32	—
O,N-Dicarbobenzoxy-3,5-diiodotyrosine	66	—	—	—	56	—
O,N-Dicarbobenzoxythyroxine	65	—	80	82	—	—
O,N-Dicarbobenzoxy-3,5,3′-triiodothyronine	68	—	75	83	—	—
O,N-Dicarbobenzoxy-3,3′,5′-triiodothyronine	63	—	80	81	—	—
O,N-Dicarbobenzoxy-3,5-diiodothyronine	67	—	75	83	—	—
O,N-Dicarbobenzoxy-3,3′-diiodothyronine	67	—	76	83	—	—
N-Acetyl-O-methylthyroxine	33	—	—	57	—	—
N-Acetyl-O-methyl-3,5,3′-triiodothyronine	40	—	—	63	—	—
N-Acetyl-O-methyl-3,3′,5′-triiodothyronine	30	—	—	54	—	—
N-Acetyl-O-methyl-3,5-diiodothyronine	37	—	—	60	—	—
N-Acetyl-O-methyl-3,3′-diiodothyronine	35	—	—	58	—	—
O-Acetyl-tetraiodothyropyruvic acid	49	—	77	62	—	—
O-Acetyl-3,5,3′-triiodothyropyruvic acid	50	—	69	69	—	—
O-Acetyl-tetraiodothyrolactic acid	25	—	75	45	—	—
O-Acetyl-3,5,3′-triiodothyrolactic acid	32	—	32	54	—	—
O,O-Diacetyl-tetraiodothyrolactic acid	50	—	81	60	—	—
O,O-Diacetyl-3,5,3′-triiodothyrolactic acid	51	—	51	66	—	—
N-Acetyl-3,5,3′-triiodothyronamine	85	—	63	—	—	—
O,N-Diacetylthyroxamine	90	—	78	—	—	—

TABLE TLC 19—(Continued)
AMINO ACIDS, IODOAMINO ACIDS, AND DERIVATIVES

Layer	L1	L1	L1	L1	L1	L1
Solvent	S1	S2	S3	S4	S5	S6
Literature	1	1	1	1	1	1
Compound			$R_F \times 100$			
Methyl esters and methyl ethers						
Thyroxine methyl ester	77	—	40	—	—	—
3,5,3′-Triiodothyronine methyl ester	86	—	30	—	—	—
3,3′,5′-Triiodothyronine methyl ester	70	—	46	—	—	—
3,5-Diiodothyronine methyl ester	82	—	26	—	—	—
3-3′-Diiodothyronine methyl ester	79	—	36	—	—	—
3-Monoiodotyrosine methyl ester	75	—	24	—	—	—
3,5-Diiodotyrosine methyl ester	57	—	27	—	—	—
3-Monoiodohistidine methyl ester	60	—	7	—	—	—
Tetraiodothyroacetic acid methyl ester	88	—	85	—	—	—
3,5-3′-Triiodothyroacetic acid methyl ester	91	—	79	—	—	—
3,5-Diiodothyroacetic acid methyl ester	90	—	78	—	—	—
Tetraiodothyropyruvic acid methyl ester	90	—	82	—	—	—
3,5,3′-Triiodothyropyruvic acid methyl ester	93	—	75	—	—	—
Tetraiodothyrolactic acid methyl ester	85	—	77	—	—	—
3,5,3′-Triiodothyrolactic acid methyl ester	90	—	72	—	—	—
O-Methylthyroxine	32	—	32	47	—	—
O-Methyl-3,5,3′-triiodothyronine	33	—	27	50	—	—
O-Methyl-3,5-diiodothyronine	31	—	28	49	—	—
O-Methyl-tetraiodothyroacetic acid	51	—	87	58	—	—
O-Methyl-3,5,3′-triiodothyroacetic acid	48	—	85	63	—	—
O-Methyl-3,5-diiodothyroacetic acid	49	—	85	63	—	—

Layer L1 = Plates spread at 250μ thickness from slurry of silica gel (Machery-Nagel, N/HR UV_{254}; 30 g) and freshly prepared rice starch solution (11.5% : 65 ml); air-dried before use.

Solvent S1 = chloroform-methanol-*conc. ammonia (50 : 25 : 2.5 v/v).
S2 = acetic acid-methanol-*conc. ammonia, (40 : 20 : 3 v/v).
S3 = formic acid-methanol-chloroform (5 : 15 : 80 v/v).
S4 = ethyl acetate-methanol-*conc. ammonia (diluted 1 : 5) (50 : 20 : 10 v/v).
S5 = ethyl acetate-methanol-*conc. ammonia, (diluted 1 : 5) (50 : 13 : 10 v/v).
S6 = formic acid-methanol-chloroform (15 : 15 : 70 v/v).

Detection UV light.

Literature 1. Osborn, R. H., and T. H. Simpson, *J. Chromatogr.*, **40**, 219 (1969).

* sp. gr. 0.880 = conc.

TABLE TLC 20
ANTIBIOTICS, DRUGS

Layer	L1	L1	L1	L1	L1	L1	L2	L2	L2	L2
Solvent	S1	S2	S3	S4	S5	S6	S7	S7	S7	S7
Detection									D1	D2
Literature	1	1	1	1	1	1	2	2	2	2
Compound				$R_F \times 100$				R_G*	Color	
Acidomycin	74	—	—	—	—	—	—	—	—	—
Actinomycin C	68	—	—	—	—	—	—	—	—	—
Actinomycin J	73	—	—	—	—	—	—	—	—	—
Althiomycin	—	—	—	—	78	—	—	—	—	—
Amidinomycin[b]	—	—	—	—	—	—	53	108	—	r-vt
Ammarvomycin	38	—	—	—	—	—	—	—	—	—
Amminosidin	—	68	40	—	—	—	—	—	—	—
Amphomycin	53	—	—	—	—	—	—	—	—	—
Amphotericin	—	—	—	33	—	—	—	—	—	—
Amphotericin A	33	—	—	—	—	—	—	—	—	—
Antimycin A	—	—	—	72	81	—	—	—	—	—
Aureothricin	58	—	—	—	57	63	—	—	—	—
Bacitracin	—	—	—	58	13	—	—	—	—	—
Blasticidin S	—	70	—	—	—	—	—	—	—	—
Blastmycin	—	78	—	82	—	—	—	—	—	—
Carbomycin	55	—	—	—	—	—	—	—	—	—
Catenulin	—	60	54	—	—	—	—	—	—	—
Catenulin[a]	—	—	—	—	—	—	15	31	—	vt
Chromomycin A_3	—	—	—	—	—	65	—	—	—	—
Dihydrostreptomycin[b]	—	—	—	—	—	—	44	90	r	—
Enteromycin	73	—	—	—	—	—	—	—	—	—
Erythromycin	39	—	—	—	—	—	—	—	—	—
Etamycin	66	—	—	—	80	—	—	—	—	—
Fradiomycin[b]	—	—	—	—	—	—	10	20	—	vt
Gentamycin[b]	—	—	—	—	—	—	35	72	—	—
							28	57	—	vt
							20	41	—	—
Glebomycin[a]	—	—	—	—	—	—	41	84	r	—
Glucosamine[a]	—	—	—	—	—	—	49	100	—	r-vt
Homomycin	—	—	—	42	—	—	—	—	—	—
Hydroxystreptomycin[a]	—	—	—	—	—	—	32	66	r	—
Kanamycin	—	65	55	—	—	—	—	—	—	—
Kanamycin A[b]	—	—	—	—	—	—	17	35	—	vt
Kanamycin B[c]	—	—	—	—	—	—	15	31	—	vt
Kanamycin C[c]	—	—	—	—	—	—	23	47	—	vt
Leucomycin	58	—	—	—	—	—	—	—	—	—
Mitomycin C	54	—	—	—	45	38	—	—	—	—
Neomycin[b]	—	—	—	—	—	—	10	20	—	vt
Neomycin B	—	51	46	—	—	—	—	—	—	—
Netropsin[a]	—	—	—	—	—	—	51	105	r	—
Novobiocin	—	—	—	82	—	—	—	—	—	—
Nystatin	18	—	—	18	—	—	—	—	—	—
Olendomycin	29	—	—	—	—	—	—	—	—	—
Paromomycin	—	68	40	—	—	—	—	—	—	—
Paromomycin[b]	—	—	—	—	—	—	15	31	—	vt
Pentamycin	67	—	—	67	—	—	—	—	—	—
Picromycin	41	—	—	—	—	—	—	—	—	—
Pimaricin	34	—	—	34	—	—	—	—	—	—
Porfiromycin	50	—	—	—	50	48	—	—	—	—
Pyridomycin	38	—	—	—	18	—	—	—	—	—
Spiramycin	8	—	—	—	—	—	—	—	—	—
Streptomycin[b]	—	—	—	—	—	—	44	90	r	—
Streptothricin	—	26	52	—	—	—	—	—	—	—
Streptothricin[b]	—	—	—	—	—	—	26	53	—	vt
Telomycin	44	—	—	—	—	—	—	—	—	—
Tertiomycin A	86	—	—	—	—	—	—	—	—	—
Tertiomycin B	63	—	—	—	—	—	—	—	—	—
Thiolutin	65	—	—	—	64	70	—	—	—	—
Trichomycin	17	—	—	45	—	—	—	—	—	—

TABLE TLC 20—(Continued)
ANTIBIOTICS, DRUGS

Layer	L1	L1	L1	L1	L1	L1	L2	L2	L2	L2
Solvent	S1	S2	S3	S4	S5	S6	S7	S7	S7	S7
Detection									D1	D2
Literature	1	1	1	1	1	1	2	2	2	2
Compound				$R_F \times 100$				R_G*	Color	
Tylosin	59	—	—	—	—	—	—	—	—	—
Unamycin A	36	—	—	—	—	—	—	—	—	—
Viomycin	—	11	—	—	—	—	—	—	—	—
Viomycin[b]	—	—	—	—	—	—	21	43	—	br-vt
Zygomycin[b]	—	—	—	—	—	—	15	31	—	vt
Zygomycin A	—	62	56	—	—	—	—	—	—	—

[a] Hydrochloride.
[b] Sulfate.
[c] Free base.
* $R_G = R_F$ relative to glucosamine · HCl.

Layer L1 = Silica gel G.
L2 = MN cellulose 300.

Solvent S1 = butanol-acetic acid-water (3 : 1 : 1).
S2 = chloroform-methanol-17% ammonium hydroxide (2 : 1 : 1), upper phase.
S3 = *n*-propanol-pyridine-acetic acid-water (15 : 10 : 3 : 10).
S4 = ethanol-ammonium hydroxide-water (8 : 1 : 1).
S5 = ethanol-water (4 : 1).
S6 = ethyl acetate-methanol (20 : 3).
S7 = propanol-pyridine-acetic acid-water (15 : 10 : 3 : 12).

Detection Bioautography or detection reagents.
D1 = oxidized nitroprusside.
D2 = ninhydrin.

Literature
1. Ikekawa, T., Iwami, F., Akita, E., and H. Umezawa, *J. Antibiot.* (Tokyo) *Ser. A.*, **16**, 56 (1963); Akita, E. and T. Ikekawa, *J. Chromatogr.*, **12**, 250 (1963).
2. Ito, Y., Namba, M., Nagahane, N., Yamaguchi , T., and T. Okuda, *J. Antibiot.* (Tokyo) *Ser. A.*, **17**, 218 (1964).

TABLE TLC 21
ANTIBIOTICS, DRUGS

Layer	L1	L1	L2	L2
Solvent	S1	S2	S3	S4
Technique	T1	T1	T2	T2
Literature	1	1	1	1
Compound		$R_F \times 100$		
Amphotericin B	0	20	0	0
Ampicillin Na	97	98 (91)[a]	12	15
Bacitracin	no spot	no spot	no spot	no spot
Benzylpenicillin Na	90	90	61 (28)[a]	58
Cephaloridine	97[b]	90[b]	[b]	79[b]
Chlorotetracycline HCl	54[b]	69	0	0
Cloxacillin Na	65	38	64	77
Demethylchlorotetracycline HCl	52[b]	67	0	0
Dicloxacillin Na	47	22	65	77
Gramicidin	0	0	41	74[b]
Griseofulvin	23	60	59	83
Hetacillin K	96	98	30 (12)[a]	64 (15)[a]
Lincomycin HCl	96	95	0	13 (3)
Methicillin Na	93	93	52 (18)[a]	59
Nafcillin Na	47 (37)[a]	22[b]	64	77
Neomycin sulfate	no spot	no spot	no spot	no spot
Novobiocin	0	0	59	89
Oxacillin Na	74	49	65 (37)[a]	63
Oxytetracycline HCl	66[b]	75	0	0
Phenethicillin K	84	73 (55)[a]	66	77
Phenoxymethylpenicillin K	82	76 (56)[a]	52 (18)[a]	75
Polymyxin B sulfate	no spot	no spot	no spot	no spot
Pyrrolidinomethyltetracycline	59[b]	74	0	0
Ristocetin	74	97	0	0
Tetracycline HCl	60[b]	76	0	0
Tyrocidine HCl	0	0	1 (6)	0
Tyrothricin	0	0	23	0
Vancomycin HCl	91[b]	95	0	0

[a] Secondary spot.
[b] Streaking.

Layer L1 = Cellulose MN-300.
L2 = Silica gel G (Merck).
Solvent S1 = 0.1*M* NaCl solution.
S2 = 0.3*M* citric acid solution saturated with *n*-butanol.
S3 = isoamyl acetate-methanol-formic acid -water (65 : 20 : 5 : 10).
S4 = acetone-acetic acid (95 : 5).
Detection Bioautography or See Section II.I, Volume Two.
Technique T1 = plates activated at 110 °C for three minutes, ascending; 15 cm.
T2 = plates dried overnight at RT, ascending; 15 cm.
Literature 1. McGilveray, I. J. and Strickland, R. D., *J. Pharm. Sci.*, **56**, 78 (1967).

TABLE TLC 22
ANTIOXIDANTS AND ACCELERATORS IN VULCANIZED RUBBERS

Layer	L1	L1		
Solvent	S1	S2		
Detection			D1	D2
Literature	1, 2	1, 2		
Compound	$R_F \times 100$		Color	
1,1′-Bis-(4-hydroxyphenyl)-cyclohexane	0	93	—	pl-r-ish br
Copper dimethyldithiocarbamate	74	90	ctl	ct-y
N-Cyclohexyl-2-benzothiazole-sulfenamide	28	95	lm-y	w
N,N′-Diaryl-*p*-phenylenediamine	68	88	pl-bl	d-y-br
Dibenzothiazolyl disulfide	62	88	y-w	w-ish-gr
Dilauryl thiodipropionate	29	95	—	gy
N,N′-Di-β-naphthyl-*p*-phenylenediamine	72	86	du-ol	pl-y-ish br
Diphenylguanidine	0	87	—	lt-y-br
N,N′-Diphenyl-*p*-phenylenediamine	56	87	lt-bl gy	du-y
Di-*o*-tolylguanidine	0	19	—	br-ish-gy
6-Ethoxy-2,2,4-trimethyl-1,2-dihydroquinoline	35	91	pk	gy-ish-r br
Hexamethylenetetramine	0	89	—	ct-y
High temperature reaction product of diphenylamine and acetone	89	89	ct-gr	gr-ish-bl
2-Mercaptobenzimidazole	3	87	—	br-ish-gy
2-Mercaptobenzothiazole	33	90	—	ctl
2-Mercaptoimidazoline	0	75	—	pl-y-ish w
Nickel dibutyldithiocarbamate	90	86	—	w
Nickel isopropyl xanthate	93	83	—	w
N-Oxydiethylene-2-benzothiazole-sulfenamide	23	82	y-w	y-ish-w
N-Phenyl-*N′*-isopropyl-*p*-phenylenediamine	13	98	sky bl	pl-y-ish br
N-Phenyl-α-naphthylamine	86	90	gr-ish-w	du-ol
N-Phenyl-β-naphthylamine	84	89	pl-br-r pk	br-ish-gy
Piperidine pentamethylenedithiocarbamate	0	5	—	w
Polymer of 2,2,4-trimethyl-1,2-dihydroquinoline	74	95	—	br-ish w
Reaction product of phenyl-β-naphthylamine and acetone	83	92	ct-gr	ms
Selenium diethyldithiocarbamate	77	80	pl-br-r pk	o-pk
Sodium dibutyldithiocarbamate	0	75	—	sl-w
Sodium dimethyldithiocarbamate	50	82	—	pl-gy-ish w
Styrenated phenol	91	0, 92	—	d-br
Tellurium diethyldithiocarbamate	18	75	lm-y	w
Tetrabutylthiuram disulfide	94	94	bt-y-gr	lt-br-r
Tetraethylthiuram disulfide	56	89	—	w
Tetramethylthiuram disulfide	21	43	—	y-ish-w
Tetramethylthiuram monosulfide	32	82	—	w
4,4′-Thiobis-(6-*tert*-butyl-3-methylphenol)	48	0, 93	ct-y	d-br
o-Tolylbiguanidine	0	4	—	br-ol
Tri (nonylphenyl) phosphite	36	72	—	d-br
Zinc di-*n*-butyldithiocarbamate	92	90	—	w
Zinc diethyldithiocarbamate	78	11	ct-y	w
Zinc dimethyldithiocarbamate	59	0	—	w
Zinc ethylphenyldithiocarbamate	69	0	cr-y	w
Zinc salt of 2-mercaptobenzimidazole	5	86	—	gy
Zinc salt of 2-mercaptobenzothiazole	13	0	—	y-ish w

Layer L1 = Silica gel containing 5% calcinated gypsum.
Solvent S1 = chloroform-benzene (10 : 9).
S2 = ethanol.
Detection D1 = formaldehyde-sulfuric acid (1 : 4).
D2 = formaldehyde-sulfuric acid (1 : 4), but recorded after heating at 120 °C for one hour.
Technique A sample solvent extracted from rubber is spotted on the thin layer.
Literature
1. Nagasawa, K. and K. Ohta, *Jap. Anal.*, 16, 1286–1288 (1967); from *J. Chromatogr.*, **37**, D86–87 (1968).
2. Nagasawa, K. and K. Ohta, *Chem. Abstr.*, **68**, 79281c (1968).

TABLE TLC 23
CARBOHYDRATES

Layer	L1	L1	L1	L1	L1	L1	L1	L1	L1	L1	L1	L1	L1	L2
Solvent	S1	S2	S3	S4	S5	S6	S7	S8	S9	S10	S11	S12	S13	S2
Literature	1	1	1	1	1	1	1	1	1	1	1	1	1	1
Compound							$R_F \times 100$							
Arabinose	7	4	25	27	5	37	13	41	43	35	29	3	9	9
Dihydroxyacetone	22	21	43	42	37	54	22	55	57	47	51	15	30	30
Erythrose	16	16	23	45	19	34	30	56	54	45	51	9	23	29
Fucose	10	7	28	35	8	46	18	48	50	42	36	3	14	13
Galactose	3	26	23	24	46	35	9	37	40	34	20	0	7	4
Glucose	5	3	25	30	4	37	10	44	43	37	25	0	9	6
Glyceraldehyde	8	19	48	52	33	38	29	54	67	41	58	25	47	49
Lactose	0	0	15	14	0	23	3	18	25	22	7	0	2	1
Levulose (fructose)	4	3	24	28	7	37	13	39	42	37	25	2	9	8
Lyxose	13	8	31	37	9	44	21	52	53	44	40	4	17	12
Maltose	1	1	19	20	0	30	7	25	29	29	11	0	4	2
Mannoheptulose	4	2	25	29	3	36	10	42	41	36	20	0	9	6
Mannose	20	16	39	46	17	5	30	58	60	51	49	7	24	26
Melibiose	0	0	15	14	0	22	2	23	23	20	7	0	3	0
α-Methyl-D-mannoside	8	5	27	34	4	37	13	47	42	43	26	3	11	9
α-Methyl-D-glucoside	15	12	34	41	9	44	19	51	50	51	33	3	16	18
Raffinose	0	0	11	9	0	17	2	6	18	21	3	0	0	0
Rhamnose	21	16	40	47	18	54	33	5	63	53	5	9	27	28
Ribose	8	7	29	30	12	42	20	42	44	38	31	4	14	14
Sedoheptulose	5	3	23	25	4	34	10	37	48	31	23	3	8	7
Sorbose	7	4	26	31	5	38	12	46	43	41	29	3	11	9
Sucrose	2	2	21	24	2	33	8	35	35	36	16	0	5	3
Trehalose	2	2	19	20	1	26	17	34	26	30	10	0	4	1
Turanose	1	1	20	22	1	29	6	34	30	30	13	0	5	3
Xylose	13	9	32	39	7	45	21	52	52	46	43	4	19	14

Layer	L2	L2	L2	L2	L2	L2	L2	L2	L2	L2	L2	L2	L2	L2
Solvent	S3	S4	S14	S5	S6	S7	S8	S9	S10	S11	S15	S12	S13	S16
Literature	1	1	1	1	1	1	1	1	1	1	1	1	1	1
Compound							$R_F \times 100$							
Arabinose	15	26	51	7	22	12	42	24	41	33	30	13	13	12
Dihydroxyacetone	47	44	70	49	57	60	54	55	54	52	68	14	44	48
Erythrose	44	47	71	36	55	47	53	47	56	65	61	13	41	42
Fucose	26	37	69	11	31	20	51	35	52	49	42	9	24	23
Galactose	9	19	47	3	13	8	31	16	30	16	15	5	10	7
Glucose	13	22	50	4	18	10	40	20	35	24	20	8	10	9
Glyceraldehyde	52	58	79	38	52	55	64	55	62	55	59	21	62	54
Lactose	3	12	33	0	4	3	20	9	16	6	5	1	3	2
Levulose (fructose)	16	27	58	7	20	13	40	24	41	31	25	7	16	14
Lyxose	27	36	64	14	35	23	49	36	47	52	43	10	26	24
Maltose	5	17	48	2	6	4	30	9	29	11	8	3	5	3
Mannoheptulose	14	25	53	5	18	10	42	21	37	27	21	4	13	11
Mannose	43	47	74	26	49	46	55	47	59	63	52	13	41	40
Melibiose	2	8	24	0	3	2	15	8	11	22	2	0	1	1
α-Methyl-D-mannoside	18	30	70	6	20	14	41	20	46	31	24	14	13	12
α-Methyl-D-glucoside	34	45	74	12	37	22	46	36	60	57	43	13	24	26
Raffinose	2	7	22	0	1	2	13	8	9	2	3	0	2	0
Rhamnose	45	49	78	32	52	50	60	51	62	68	57	11	45	47
Ribose	28	36	69	16	35	26	47	36	49	49	43	9	28	27
Sedoheptulose	14	26	51	5	19	10	42	21	40	32	26	13	10	10
Sorbose	18	31	51	6	23	13	42	26	44	35	29	14	13	12
Sucrose	8	22	59	2	10	5	35	13	42	19	13	5	8	6
Trehalose	4	15	38	0	6	3	38	8	30	10	5	10	2	2
Turanose	7	21	55	2	9	5	36	10	36	20	13	5	6	5
Xylose	29	36	67	13	35	24	52	37	51	67	43	10	25	25

TABLE TLC 23—(Continued)
CARBOHYDRATES

Layer L1 = Silica gel, TLC grade, with binder (Fluka G) impregnated with sodium acetate (0.2*M*).
L2 = Silica gel, TLC grade, with binder (Fluka G) impregnated with monosodium phosphate (0.2*M*); both dried 24 hours at room temperature and activated 110 °C for one hour.

Solvent S1 = *n*-butanol-boric acid, 0.3*M* (9 : 1).
S2 = *n*-butanol-ethyl acetate-boric acid, 0.03*M* (70 : 15 : 15).
S3 = *n*-butanol-ethyl acetate-isopropanol-acetic acid-water (35 : 100 : 60 : 35 : 30).
S4 = *n*-propanol-water (85 : 15).
S5 = ethyl acetate-isopropanol-water (65 : 22 : 11).
S6 = ethyl acetate-isopropanol-acetic acid-water (100 : 60 : 35 : 30).
S7 = ethyl acetate-isopropanol-water (100 : 60 : 30).
S8 = *n*-butanol-methanol-boric acid, 0.03*M* (5 : 3 : 1).
S9 = ethyl acetate-acetic acid-methanol-water (60 : 15 : 05 : 10).
S10 = isopropanol-water (4 : 1).
S11 = acetone-water (9 : 1).
S12 = *n*-butanol-ethyl ether-water (4 : 5 : 1).
S13 = *n*-butanol-ethyl acetate-isopropanol-water (200 : 100 : 70 : 35).
S14 = *n*-propanol-pyridine-water (5 : 3 : 2).
S15 = acetone-water-chloroform-methanol (8 : 0.5 : 1 : 1).
S16 = *n*-butanol-ethyl acetate-isopropanol-water (35 : 100 : 60 : 30).

Detection Naphthoresorcinol-H_2SO_4 reagent.

Technique Monodimensional, upright plates, ascending, 22 °C.

Literature 1. Lato, M., Brunelli, B., Ciuffini, G., and T. Mezzetti, *J. Chromatogr.*, **39**, 407–417 (1969).

TABLE TLC 24
CARBOHYDRATES—DERIVATIVES

Layer	L1	L1	L2	L3	L3
Solvent	S1	S2	S3	S4	S5
Technique	T1	T1	T2	na	na
Literature	1	1	2	3	3
Compound	$R_F \times 100$				
Derivative	Phenylhydrazone			Phenylosazones	
Arabinose	75	63	91	65	—
Cellobiose	—	—	—	—	57
Fructose	—	—	39	—	—
Galactose	69	61	52	29	—
Glucose	68	64	39	22	—
Lactose	—	—	2	—	50
Maltose	—	—	12	—	57
Mannose	71	62	—	—	—
Rhamnose	83	76	—	78	—
Ribose	—	—	91	—	—
Sorbose	—	—	21	—	—
Xylose	77	68	72	—	—

Layer L1 = Acidic aluminum oxide.
L2 = Kieselguhr G prepared with 0.05*M* sodium tetraborate-dried at 80 °C for 30 minutes.
L3 = Silene EF (hydrated calcium silicate), air-dried and activated 24 hours at 110 °C.

Solvent S1 = butanol-acetone-H_2O (4 : 5 : 1).
S2 = butanol-acetone-H_2O (7 : 2 : 1).
S3 = chloroform-dioxane-tetrahydrofuran-0.1*M* sodium tetraborate (40 : 20 : 20 : 1.5).
S4 = chloroform-acetone-95% ethanol (5 : 3 : 3).
S5 = chloroform-acetone-ethanol-H_2O (10 : 10 : 6 : 1).

Detection *p*-methoxybenzaldehyde.

Technique T1 = development distance = 10 cm.
T2 = for preparation of derivatives from urine see original article, development distance = 10 cm.

Literature
1. Stroh, H. H. and W. Schüler, *Z. Chem.*, **4**, 188 (1964).
2. Rink, M. and S. Herrmann, *J. Chromatogr.*, **12**, 415 (1963).
3. Tore, J. P. *Anal. Biochem.*, **7**, 123 (1964).

TABLE TLC 25
CARBOHYDRATES, ACETYL DERIVATIVES

Layer	L1	L1	L2	L2	L3	L3
Solvent	S1	S2	S3	S4	S5	S6
Technique	T1	T1	T2	T2	T3	T3
Literature	1	1	2	2	3	3
Compound			$R_F \times 100$			
2,2′,3,3′,4′,6,6′-Hepta-*O*-acetyl-α-maltose	—	—	—	—	8	22
2,2′,3,3′,4′,6,6′-Hepta-*O*-acetyl-β-maltose	—	—	—	—	30	55
2′,3,3′,4′,6,6′-Hexaacetyl-β-maltose	—	—	—	—	10	27
Hexa-*O*-acetylxylobiose	—	—	72	84	—	—
Octa-*O*-acetyl-β-cellobiose	—	—	64	84	—	—
Octa-*O*-acetyl-gentiobiose	27	22	—	—	—	—
Octa-*O*-acetyl-β-laminaribiose	—	—	62	81	—	—
Octa-*O*-acetyl-α-maltose	—	—	79	85	—	—
Octa-*O*-acetyl-β-maltose	—	—	—	—	61	81
Octa-*O*-acetyl-sucrose	—	—	63	77	—	—
Penta-*O*-acetyl-α-D-galactofuranose	52	—	—	—	—	—
Penta-*O*-acetyl-β-D-galactofuranose	49	—	—	—	—	—
Penta-*O*-acetyl-α-D-galactopyranose	65	—	—	—	—	—
Penta-*O* acetyl-β-D-galactopyranose	56	—	81	77	—	—
Penta-*O*-acetyl-α-D-glucopyranose	66	—	82	79	—	—
Penta-*O*-acetyl-β-D-glucopyranose	68	63	81	79	—	—
Penta-*O*-acetyl-β-D-mannopyranose	57	52	—	—	—	—
Tetra-*O*-acetyl-α-D-xylopyranose	78	74	—	—	—	—
Tetra-*O*-acetyl-β-D-xylopyranose	71	69	84	84	—	—

Layer L1 = Mallinkrodt silicic acid + 10% starch binder (dried at 110 °C for 2 hours).
L2 = Silica gel G, dried overnight at 135 °C.
L3 = Silica gel G (Merck).

Solvent S1 = ethyl acetate-benzene (3 : 7).
S2 = methanol-benzene (2 : 98).
S3 = upper phase of benzene-ethanol-H_2O-ammonium hydroxide (200 : 47 : 15 : 1).
S4 = *n*-butanol-acetic acid-ethyl ether-water (9 : 6 : 3 : 1).
S5 = ethyl acetate-benzene (1 : 1).
S6 = ethyl acetate-benzene (7 : 3).

Detection See Section II.I, Volume Two.

Technique T1 = ascending development distance = 13 cm.
T2 = ascending.
T3 = 25 minutes development time.

Literature

1. Deferrari, J. O., deLederkremer, R. M., Matsuhiro, B., and J. F. Sproviero, *J. Chromatogr.*, **9**, 283 (1962).
2. Hay, G. W., Lewis, B. A., and F. Smith, *J. Chromatogr.*, **11**, 479 (1963).
3. Wolfram, M. L. and R. M. deLederkremer, *J. Org. Chem.*, **30**, 1560 (1965).

TABLE TLC 26
CARBOHYDRATES, METHYL DERIVATIVES

Layer	L1	L1	L1	L2	L2	L1	L1	L3
Solvent	S1	S2	S3	S3	S4	S5	S6	S7
Technique	T1	T1	T2	T3	T3	T2	T2	T1
Literature	1	2	3	3	3	3	3	4
Compound				$R_F \times 100$				
1,6-Anhydro-2,3,4-tri-*O*-methyl-β-D-glucopyranose	—	—	—	—	—	—	—	58
1,2,4,5-Diisopropylidenefructopyranose	—	—	—	—	—	5	47	—
2,3,4,5-Diisopropylidenefructopyranose	—	—	—	—	—	5	54	—
1,2,5,6-Diisopropylideneglucofuranose	—	—	—	—	—	5	52	—
1,2,4,5-Diisopropylidene-3-*O*-methylfructopyranose	—	—	—	—	—	17	91	—
2,3,4,5-Diisopropylidene-1-*O*-methylfructopyranose	—	—	—	—	—	17	91	—
1,2,5,6-Diisopropylidene-3-*O*-methylglucofuranose	—	—	—	—	—	17	87	—
2,4-Di-*O*-methyl-D-glucose	5	—	—	—	—	—	—	—
2,6-Di-*O*-methyl-D-glucose	5	—	—	—	—	—	—	—
4,6-Di-*O*-methyl-D-glucose	—	11	—	—	—	—	—	—
2,3-Di-*O*-methyl-D-xylose	15	—	—	—	—	—	—	—
Fructose	—	—	—	29	35	—	—	—
Glucose	—	—	55	—	—	—	—	—
1-*O*-Methylfructose	—	—	—	33	44	—	—	—
3-*O*-Methylfructose	—	—	—	54	61	—	—	—
6-*O*-Methylfructose	—	—	—	70	—	—	—	—
2-*O*-Methylglucose	—	—	72	—	—	—	—	—
3-*O*-Methylglucose	—	—	72	—	—	—	—	—
6-*O*-Methylglucose	—	—	64	—	—	—	—	—
α-Methylglucoside	—	—	70	—	—	—	—	—
Methyl 2,3,4-tri-*O*-ethyl-α-D-glucopyranoside	—	—	—	—	—	—	—	53
Methyl 2,3,4-tri-*O*-ethyl-β-D-glucopyranoside	—	—	—	—	—	—	—	57
Methyl 2,3,4-tri-*O*-methyl-α-D-glucopyranoside	—	—	—	—	—	—	—	46
Methyl 2,3,4-tri-*O*-methyl-β-D-glucopyranoside	—	—	—	—	—	—	—	51
2,3,4,6-Tetra-*O*-methyl-D-galactose	—	33	—	—	—	—	—	—
2,3,4,6-Tetra-*O*-methyl-D-glucose	38	45	—	—	—	—	—	—
2,3,4-Tri-*O*-methyl-D-galactose	—	16	—	—	—	—	—	—
2,3,6-Tri-*O*-methyl-D-galactose	—	21	—	—	—	—	—	—
2,4,6-Tri-*O*-methyl-D-galactose	17	20	—	—	—	—	—	—
2,3,4-Tri-*O*-methyl-D-glucose	—	34	—	—	—	—	—	—
2,3,6-Tri-*O*-methyl-D-glucose	18	25	—	—	—	—	—	—
2,4,6-Tri-*O*-methyl-D-glucose	13	23	—	—	—	—	—	—
3,4,6-Tri-*O*-methyl-D-glucose	—	27	—	—	—	—	—	—
2,3,4-Tri-*O*-methyl-D-xylose	28	50	—	—	—	—	—	—

Layer
L1 = Silica gel G.
L2 = Silica gel G prepared with 0.1*N* boric acid.
L3 = Kieselguhr G.

Solvent
S1 = upper phase of benzene-ethanol-H_2O-conc. NH_4OH. (200 : 47 : 15 : 1).
S2 = isopropyl ether-methanol (5 : 1).
S3 = *n*-butanol-acetone-H_2O (4 : 5 : 1).
S4 = *n*-propanol-acetone-H_2O (4 : 5 : 1).
S5 = petroleum ether.
S6 = *n*-butanol-petroleum ether (1 : 9).
S7 = benzene-methanol-ammonium hydroxide (80 : 20 : 0.3).

Detection
See Section II.I, Volume Two.

Technique
T1 = ascending.
T2 = 20 μg applied, 22–24 °C.
T3 = 20 μg applied, 22–24 °C; wedge-strip technique.

Literature

1. Hay, G. W., Lewis, B. A., and F. Smith, *J. Chromatogr.*, **11**, 483 (1963).
2. Tschesche, R. and G. Wulff, *Tetrahedron* **19**, 621 (1963).
3. Prey, V., Berbalk, H., and M. Kausz, *Mikrochim. Acta*, 1962, p. 449.
4. Ruckel, E. R. and C. Schuerch, *J. Org. Chem.* **31**, 2233 (1966).

TABLE TLC 27
CARBOXYLIC ACID ESTERS, PLASTICIZERS AND "SOFTENERS"

Layer	L1	L1	L1	L1	L1	L2	L1
Solvent	S1	S2	S3	S2	S4	S1	S1
Technique	T1	T1	T2	T2	T2	na	na
Literature	1	2	3	3	3	4	5
Compound	$R_F \times 100$		$R_{DS}{}^a \times 100$			$R_F \times 100$	
Adipic acid esters							
Benzyl octyl adipate	50	—	—	—	—	—	48
Diisobutyl adipate	—	—	83	86	85	—	—
Dinonyl adipate	50	—	—	—	—	44	50
Dioctyl adipate	49	—	—	—	—	42	49
2-Ethylhexyl adipate	50	—	—	—	—	—	—
Polyester adipate	0	—	—	—	—	2	—
Citric acid esters							
Acetyl tributyl citrate	32	62	53	85	70	24	—
Acetyl triethyl citrate	19	29	26	51	29	15	—
Acetyl tri-2-ethyl hexyl citrate	52	90, 96	—	—	—	46	—
Tributyl citrate	20	35	—	—	—	14	—
Triethyl citrate	15	17	—	—	—	12	—
Phosphoric acid esters							
Diphenylcresyl phosphate	43	—	—	—	—	51	45
Diphenyloctyl phosphate	38	—	—	—	—	42	36
2-Ethylhexyl diphenyl phosphate	—	—	46	77	58	—	—
Tributyl phosphate	—	—	—	—	—	—	12
Trichlorethyl phosphate	13	—	—	—	—	—	11
Tricresyl phosphate	49	—	42	86	69	53	51
Trioctyl phosphate	24	—	—	—	—	23	24
Triphenyl phosphate	42	—	33	80	50	47	44
Phthalic acid esters							
Benzylbutyl phthalate	53	—	—	—	—	—	53
Dibutyl phthalate	52	—	74	103	84	69	51
Dicyclohexyl phthalate	53	73	—	—	—	—	—
Didecyl phthalate	63	—	—	—	—	85	65
Di-(2-ethylhexyl)-phthalate	58	89	114	116	115	85	60
Diethyl phthalate	—	58	51	79	60	—	38
Dihexyl phthalate	57	—	—	—	—	80	—
Diisobutyl phthalate	—	74	—	—	—	—	—
Diisodecyl phthalate	57	—	—	—	—	84	—
Diisononyl phthalate	56	—	—	—	—	—	57
Diisooctyl phthalate	—	—	—	—	—	—	62
Dimethoxy ethyl phthalate	10	19	—	—	—	—	—
Dimethylcyclohexyl phthalate	48	—	—	—	—	—	—
Dimethyl phthalate	38	48	—	—	—	51	37
Dinonyl phthalate	60	—	101	118	114	—	60
Dioctyl phthalate	59	88	—	—	—	86	—
Sebacic acid esters							
Dibutyl sebacate	35	—	100	100	100	41	—
Dioctyl sebacate	47	—	—	—	—	61	58
Ethylhexyl sebacate	47	—	—	—	—	—	—
Sebacic acid polyester	0	—	—	—	—	2	—
Miscellaneous							
Benzenesulfonic acid *N*-butylamide	—	—	—	—	—	—	37
Butyl acetylricinoleate	40	—	—	—	—	—	—
***n*-Butylbenzene sulfonamide**	38	—	—	—	—	—	—
Butyl phthalyl butyl glycolate	—	—	43	90	65	—	—
Butyl stearate	—	—	161	123	128	—	—
Di-2-ethylhexyl adipate	—	—	—	—	—	44	50
Di-2-ethylhexyl sebacate	—	—	—	—	—	61	—
Di-2-ethylhexyl thiobutyrate	52	—	—	—	—	—	—

TABLE TLC 27—(Continued)
CARBOXYLIC ACID ESTERS, PLASTICIZERS AND "SOFTENERS"

Layer	L1	L1	L1	L1	L1	L2	L1
Solvent	S1	S2	S3	S2	S4	S1	S1
Technique	T1	T1	T2	T2	T2	na	na
Literature	1	2	3	3	3	4	5
Compound	$R_F \times 100$		$R_{DS}{}^a \times 100$			$R_F \times 100$	
Dioctyl azelate	—	—	—	—	—	—	54
2-Ethylhexyl *p*-hydroxybenzoate	16	—	—	—	—	—	—
Ethyl phthalyl ethyl glycolate	—	22	66	30	—	—	—
Glycerol triacetate	—	—	—	—	—	13	—
Methyl acetoricinoleate	33	—	—	—	—	—	—
Thiodibutyric acid di-(2-ethylhexyl ester)	—	—	—	—	—	—	48
Triacetin	18	—	18	34	17	—	—

[a] R_{DS} = R_F relative to dibutyl sebacate.

Layer L1 = Silica gel G activated at 120 °C for 30 minutes.
L2 = Kieselgel G (Merck).
Solvent S1 = methylene chloride.
S2 = ethyl acetate-benzene (5 : 95) for phthalates, (1 : 19) for citrates.
S3 = ethyl acetate-isooctane (1 : 9).
S4 = diethyl ether-hexane (4 : 1).
Detection See Section II.I, Volume Two.
Technique T1 = development distance 10 cm, time 20–30 minutes.
T2 = micro-adsorption (Stahl).
Literature
1. Braun, D., *Kunstoffe-Plastics*, **52**, 2 (1962).
2. Jaminet, F., *Farmaco (Pavia) Ed. Pract.*, **18**, 633 (1963).
3. Peereboom, J. W. C., *J. Chromatogr.*, **4**, 323 (1960).
4. Braun, D., *Chimia (Aarau)*, **19**, 79 (1965).
5. Rau, J. H. and H. Haase, *Melliand Textilber.*, **1965**, 1318.

TABLE TLC 28
CARBOXYLIC ACIDS, LICHEN SUBSTANCES

Layer	L1	L1	L1	L1		
Solvent	S1	S2	S3	S4		
Detection					D1	D2
Compound	$R_F \times 100$				Color	
Calycin	73	60	81	56	o-r	dk br-r
Epanorin	83	80	84	57	y	o
Pinastric acid	58	40	86	19	y	o
Pulvic acid	0	0	27	0	y	y
Pulvic dilactone	90	83	92	75	y	y
Rhizocarpic acid	82	78	84	53	y	o
Usnic acid	78	62	82	51	ft y	dk
Vulpinic acid	47	29	85	17	y	y

Layer L1 = Eastman Chromagram sheets, type K 301 R2.
Solvent S1 = chloroform-acetone (4 : 1).
S2 = hexane-chloroform-acetone (2 : 3 : 1).
S3 = toluene-glacial acetic acid (4 : 1).
S4 = hexane-chloroform-acetone (10 : 8 : 1).
Detection D1 = Daylight.
D2 = UV.
Technique Ascending, 5 cm.
Literature Santesson, J., *Acta Chem. Scand.*, **21**, 1167 (1967).

TABLE TLC 29
CARBOXYLIC ACIDS, ALIPHATIC ACIDS

Layer	L1
Solvent	S1
Literature	1

Compound	$R_F \times 100$
Unsubstituted monocarboxylic acids	
Formic	41
Acetic	45
Propionic	58
Butyric	69
Valeric	79
Caproic	86
Enanthic	92
Caprylic	95
2-Methylpropionic	71
2-Methylbutyric	80
2,2-Dimethylpropionic	81
2,2-Dimethylbutyric	88
2-Ethylbutyric	87
2-Methyl-2-ethylbutyric	92
2,2-Dimethylvaleric	92
2-Ethylpropionic	95
3,3-Dimethylbutyric	78
4-Methylvaleric	82
2,2,4,4-Tetramethylcaproic	97
Acrylic	53
Crotonic	64
Hydroxy acids	
Glycolic	25
Lactic	31
3-Hydroxypropionic	27
2-Hydroxybutyric	42
3-Hydroxybutyric	38
4-Hydroxybutyric	28
2-Hydroxyvaleric	55
2-Hydroxycaproic	66
2-Hydroxyhexanoic	77
Glyceric	12
Amino acids	
Glycine	7
Alanine	13
2-Aminobutyric	19
3-Aminobutyric	17
4-Aminobutyric	14
2-Aminoisobutyric	21
Norvaline	27
Valine	22
Leucine	32
Isoleucine	31
6-Aminocaproic	14
Halogen acids	
Difluoroacetic	67
Trifluoroacetic	78
Chloroacetic	52
2-Chloropropionic	64
2-Chlorobutyric	75
Dichloroacetic	64
Trichloroacetic	74
Trichlorolactic	69
Bromoacetic	54
Tribromoacetic	75
Iodoacetic	54

TABLE TLC 29—(Continued)
CARBOXYLIC ACIDS, ALIPHATIC ACIDS

Layer	L1
Solvent	S1
Literature	1
Compound	$R_F \times 100$
Other acids	
Glyoxylic	30
Pyruvic	49
Levulinic	57
Ethoxyacetic	55
Thioglycolic	52
Butane-1-sulfonic	64

Layer L1 = Cellulose powder HR_{300} (Machery-Nagel), 0.25 mm thickness, air-dried at room temperature overnight.
Solvent S1 = *n*-butanol-diethylamine-water (85 : 1 : 14).
Detection Ninhydrin.
Technique Almost all acids spotted as their diethylamine salts, dissolved in methanol; plates equilibrated four hours at 23 ± 1 °C, ascending development, 10 cm, in saturated tank, 85 minutes; after development the solvent is evaporated at 60 °C.
Literature 1. Gütlbauer, F., *J. Chromatogr.*, **45**, 104–112 (1969).

TABLE TLC 30
CARBOXYLIC ACIDS, CYCLIC DICARBOXYLIC ANHYDRIDE DERIVATIVES

Layer	L1
Solvent	S1
Literature	1
Compound	$R_F \times 100$
p-Anisidide of	
Pyromellitic anhydride	14
Citraconic anhydride	23
Maleic anhydride	26
Succinic anhydride	37
Trimellitic anhydride	47
Endomethylenetetrahydrophthalic anhydride	52
Phthalic anhydride	58
Tetrahydrophthalic anhydride	61
Hexahydrophthalic anhydride	65
Methylendomethylenetetrahydrophthalic anhydride	74
Tetrachlorophthalic anhydride	83
Hexachloroendomethylenetetrahydrophthalic anhydride	90

Layer L1 = MN-Kieselgel N-HR.
Solvent S1 = benzene-tetrahydrofuran-propionic acid (80 : 30 : 5).
Detection *p*-Dimethylaminobenzaldehyde.
Technique Preparation of derivatives: 100 mg of the investigated anhydride dissolved in 10 ml acetone are refluxed with 5 ml of a 1% solution of *p*-anisidine in acetone for 30 minutes. The excess of *p*-anisidine is removed by means of a strong acid cation exchanger.
Literature 1. Knappe, E. and I. Rohdewald, *Z. Anal. Chem.*, **232**, 341 (1967); from *J. Chromatogr.*, **35**, D57 (1968).

TABLE TLC 31
CARBOXYLIC ACIDS, DICARBOXYLIC ACIDS

Layer	L1	L2	L2	L2	L3	L3	L3	L4	
Solvent	S1	S2	S3	S4	S5	S6	S7	S8	
Detection									D1
Technique	T1	T2	T3	T3	T4	T4	T4	T5	
Literature	1	2	3	3	4	4	4	5	5
Compound				$R_F \times 100$					Color
Oxalic	14	5	0	0	6	10	16	7	bl-streaks
Malonic	21	14	13	5	13	20	55	69	ft bl
Succinic	28	30	28	23	20	30	74	86[a]	ft bl
Glutaric	36	39	35	28	27	38	77	91	no reaction
Adipic	43	43	42	34	32	47	80	—	—
Pimelic	55	53	47	36	—	—	—	—	—
Suberic	67	54	50	40	—	—	—	—	—
Azelaic	82	56	53	43	—	—	—	—	—
Sebacic	92	67	55	47	—	—	—	—	—
Malic	20	—	13	6	8	20	54	53	bl
Citric	—	5	2	2	4	12	49	37	bl
Tartaric	—	8	—	—	4	14	42	23	bl

[a] Sodium salt applied.

Layer
L1 = Silica gel G impregnated with polyethylene glycol M 1000 (1.0 : 0.5).
L2 = Silica gel G.
L3 = Silica gel G—kieselguhr G (1 : 1).
L4 = Silica gel H.

Solvent
S1 = diisopropyl ether-formic acid-water (90 : 7 : 3), saturated with polyethylene glycol M 1000.
S2 = ethyl alcohol 96%-water-25% ammonium hydroxide (100 : 12 : 16).
S3 = benzene-methyl alcohol-acetic acid (45 : 8 : 4).
S4 = benzene-dioxane-acetic acid (90 : 25 : 4).
S5 = benzene-ethanol-ammonium hydroxide (10 : 20 : 5).
S6 = butyl acetate-methanol-ammonium hydroxide (15 : 20 : 5).
S7 = butyl acetate-acetic acid-water (30 : 20 : 10).
S8 = water-saturated diethyl ether-formic acid (7 : 1).

Detection
D1 = acid ammonium molybdate solution.

Technique
T1 = saturated atmosphere, 12 cm development distance.
T2 = acids as the ammonium salts, unsaturated atmosphere, 10 cm, 2 hours.
T3 = 10 cm development distance, 45 minutes.
T4 = 6 cm development distance.
T5 = saturated tank, 14 cm ascending development for 1 hour, room temperature.

Literature

1. Knappe, E. and D. Peteri, *Z. Anal. Chem.*, **188**, 184 (1962).
2. Braun, D. and H. Greenen, *J. Chromatogr.*, **7**, 56 (1962).
3. Petrowitz, H-J. and G. Pastuska, *J. Chromatogr.*, **7**, 128 (1962).
4. Bancher, E., Scherz, H., and V. Prey, *Mikrochim. Ichnoanal. Acta*, 1963, 712 from Kirchner, J. G. "*Thin-Layer Chromatography*", Tech. of Org. Chem. XII, Perry, E. S., and A. Weissberger, Eds., Interscience Publishers, Division of John Wiley & Sons, Inc., New York, 1967, p. 247.
5. Ting, I. P. and W. M. Dugger, Jr., *Anal. Biochem.*, **12**, 574 (1965).

TABLE TLC 32
CARBOXYLIC ACIDS, 2,4-DINITROPHENYLHYDRAZONES OF KETO ACIDS

Layer	L1	L2	L3	L4	L4	L4	
Solvent	S1	S2	S2	S3	S4	S5	
Detection							D1
Literature	1	2	2	3	3	3	3
Compound	$R_{2,4\text{-DNPH}}$[a]			$R_F \times 100$			Color
2,4-DNPH of							
Acetoacetic	12	—	—	—	—	—	—
2,4-Dinitrophenylhydrazine	100	100	100	—	—	—	—
Glyoxylic acid	—	—	—	56	51	45	r
				70	66	63	o-br
Hydroxypyruvic acid	—	—	—	55	50	44	r
				60	57	53	gr
α-Ketobutyric	—	48	29	—	—	—	—
α-Ketocaproic	—	56	48	86	83	90	—
α-Ketoglutaric	0	5	2	60	20	24	gr
α-Ketoisocaproic acid	—	—	—	87	90	91	—
α-Ketoisovaleric	—	59	39	87	83	90	—
α-Keto-β-methylvaleric	—	58	48	—	—	—	—
Levulinic	—	56	49	—	—	—	—
Oxaloacetic	0	2	2	30	16	23	br
				40	30	26	gr
Phenylpyruvic acid	—	—	—	90	90	90	—
Phenylpyruvic (Isomer 1)	30	—	—	—	—	—	—
Phenylpyruvic (Isomer 2)	63	—	—	—	—	—	—
Pyruvic acid	—	—	—	63	60	57	br-vt
				75	77	73	gr-ish y
Pyruvic (Isomer 1)	15	28	20	—	—	—	—
Pyruvic (Isomer 2)	38	—	—	—	—	—	—

[a] R_F relative to 2,4-dinitrophenylhydrazine.

Layer L1 = Silica gel G, 0.3 mm layer, dried at 110 °C overnight.
L2 = Silica gel G, 0.1 mm layer, dried at 110–120 °C for 30 minutes.
L3 = Acidic silica gel G prepared from 30 g silica gel G + 5 ml propionic acid + 60 ml water.
L4 = Cellulose powder No. 143 (S & S).

Solvent S1 = isoamyl alcohol-0.25*N* ammonium hydroxide (20 : 1).
S2 = petroleum ether (60–80 °C)-ethyl formate (13 : 7) + 0.104*M* propionic acid/100 ml.
S3 = *n*-butanol-*n*-propanol-benzene-6% ammonia (15 : 50 : 15 : 20).
S4 = *n*-butanol-*n*-propanol-benzene-6% ammonia (15 : 55 : 15 : 15).
S5 = *n*-butanol-*n*-propanol-benzene-6% ammonia (15 : 15 : 20 : 15).

Detection D1 = Alcoholic potassium hydroxide.

Technique T1 = 10–11.5 cm development distance in saturated atmosphere, 40 minutes development.

Literature

1. Dancis, J., Hutzler, J., and M. Levitz, *Biochim. Biophys. Acta*, **78**, 85 (1963).
2. Ronkainen, P., *J. Chromatogr.*, **11**, 228 (1963).
3. Chiari, D. and M. Röhr, *Mikrochim. Acta*, 1967, 141; from *J. Chromatogr.*, **30**, D9 (1967); *Chem. Abstr.*, **66**, 52018 (1967).

TABLE TLC 33
CARBOXYLIC ACIDS, FATTY ACIDS, ESTERS AND ALCOHOLS

Layer			L1	L2	L1	L2	L1	L2	L2	L2	L2
Solvent			S1	S2	S3	S2	S4	S2	S5	S6	S7
Technique			T1	T2	T1	T2	T1	T2	T2	T2	T2
Literature			1	1	1	1	1	1	1	1	1
Compound			$R_F \times 100$								
Chain length	Place of unsaturation	Common name of acid	Acids		Methyl esters		Alcohols				
			Acetylenic								
22	13,14	Behenolic	43	40	60	14	64	21	7	15	28
18	9,10	Stearolic	34	64	58	39	61	38	18	26	43
18	6,7	Tariric	34	62	58	40	61	38	18	26	43
11	10,11	Undecynoic	19	87	39	81	39	71	44	55	61
			Ethylenic								
22	13,14	Erucic	52	20	74	7	58	15	3	9	18
18	9,10	Oleic	41	50	73	29	54	28	12	18	34
18	6,7	Petroselinic	41	52	73	30	54	28	12	18	34
11	10,11	Undecenoic	26	75	57	70	34	55	34	44	51
18	9,10;12,13	Linoleic	—	62	—	40	54	38	18	26	43
18	9,10;12,13;15,16	Linolenic	—	71	—	50	54	48	23	35	53
			Saturated								
22	nil	Behenic	52	3	74	0	58	0	0	0	4
18	nil	Stearic	45	40	71	14	54	21	7	15	28
11	nil	Undecanoic	28	66	57	60	34	38	26	32	38
20	nil	Arachidic	52	20	71	7	—	—	—	—	—
16	nil	Palmitic	44	50	73	29	—	28	12	18	34
14	nil	Myristic	44	64	73	39	—	38	18	26	43
9	nil	Pelargonic	28	75	59	70	—	—	—	—	—
12	nil	Lauric	—	71	—	50	—	48	23	35	53

Layer L1 = Silica gel G layers.
L2 = same impregnated with Silicone oil (Dow Corning Silicone Fluid, 200).

Solvent S1 = 5% ether-petroleum ether.
S2 = acetonitrile-acetic acid-water (70 : 10 : 20).
S3 = 2% ether-petroleum ether.
S4 = 20% ether-petroleum ether.
S5 = acetic acid-water (70 : 30).
S6 = acetic acid-water (80 : 20).
S7 = acetic acid-water (90 : 10).

Detection Sulfuric Acid.

Technique T1 = direct.
T2 = reversed phase.

Literature Roomi, M. W., Subbaram, M. R., and K. T. Achaya, *J. Chromatogr.*, **16**, 106 (1964) and *J. Chromatogr.*, **9**, 295 (1965).

TABLE TLC 34
CARBOXYLIC ACIDS, LICHEN SUBSTANCES

Layer	L1	L1	L1	L1	L2	L2	L2			
Solvent	S1	S2	S3	S4	S5	S6	S7			
Detection								D1	D2	
Literature	1	1	1	1	2	2	2	2	2	
								Color		
Compound	$R_F \times 100$								Daylight	UV_{365}
Atranorin	82	94	85	60	72	83	71	o	y	r-br
Bacomycesic acid	52	83	71	[a]	41	53	—	y(r)	y	o
Barbatolic acid	51	60	36	[a]	25	63	—	(dk)r	y	dk
Chloroatranorin	81	94	85	39	72	83	35	o	y	r-br
Decarboxythamnolic acid	—	—	—	—	21	52	—	dk	dirty y	dk
Fumarprotocetraric acid	10	46	25	[a]	9	24	—	dk	y	r-br
Norstictic acid	34	81	66	[a]	41	43	—	dk	y	(dk)r
Pannarin	81	93	85	49	70	75	45	y(r)	o	(dk)r
Physodalic acid	34	62	46	[a]	30	49	—	dk	y	r-br
Protocetraric acid	10	52	33	[a]	9	44	—	dk	y	r-br
Psoromic acid	52	82	73	[a]	41	44	—	gr-y	y	o-r
Salazinic acid	6	49	29	[a]	16	40	—	dk	y	r-br
Stictic acid	21	51	26	[a]	35	45	—	dk	o-r	dk
Thamnolic acid	21	45	22	[a]	12	32	—	dk	dirty y	dk

[a] R_F does not exceed 10.

Layer L1 = Eastman Chromagram sheets, type $K301R_2$.
L2 = Silica gel $HF_{254+366}$.

Solvent S1 = toluene-glacial acetic acid (9 : 1).
S2 = toluene-diethyl ether-glacial acetic acid (3 : 6 : 1).
S3 = toluene-diethyl ether-glacial acetic acid (7 : 12 : 1).
S4 = cyclohexane-chloroform-methyl ethyl ketone (30 : 15 : 2).
S5 = benzene-dioxane-glacial acetic acid (90 : 25 : 4).
S6 = butanol-acetone-water (5 : 1 : 2).
S7 = chloroform-acetone (1 : 1).

Detection D1 = *o*-Dianisidine.
D2 = *p*-Phenylenediamine.

Literature
1. Santesson, J., *Acta Chem. Scand.*, **21**, 1163 (1967).
2. Santesson, J., *Acta Chem. Scand.*, **19**, 2254–55 (1965); from *J. Chromatogr.*, **39**, D8 (1969) and **26**, D6 (1967).

TABLE TLC 35
CARBOXYLIC ACIDS, LICHEN SUBSTANCES

Solvent	L1	L1	L1	L1	L1		
Layer	S1	S2	S3	S4	S5		
Detection						D1	D2
Literature	1	1	1	1	1		
Compound	$R_F \times 100$					Color	
Depsides and depsidones							
Barbatic acid	40	54	34	[a]	[a]	r-br	—
Boninic acid	22	42	35	51	[a]	y-y-br	—
Confluentinic acid	24	43	28	[a]	[a]	r-br	—
Diffractaic acid	38	53	32	[a]	[a]	r-br	—
Divaricatic acid	32	49	27	[a]	[a]	r-br	—
Evernic acid	27	45	20	[a]	[a]	r-br	—
Grayanic acid	20	38	24	[a]	[a]	y-y-br	—
Homosekikaic acid	46	67	45	[a]	[a]	y-y-br	—
Hypothamnolic acid	3	6	1	25	34	y-y-br	—
Sekikaic acid	43	56	37	[a]	[a]	r-br	—
Sphaerophorin	43	60	38	[a]	[a]	r-br	—
Squamatic acid	5	7	3	30	56	y-y-br	—
Variolaric acid	2	4	2	21	42	r-br	—
Secondary front	58	70	55	57	82	—	—
Depsides							
Cryptochlorophaeic acid	—	—	—	45	65	—	br
Gyrophoric acid	—	—	—	26	45	—	r-vt
Hiascic acid	—	—	—	8	23	—	r-br
Remalinolic acid	—	—	—	33	56	—	r-vt
Sekikaic acid	—	—	—	56	81	—	r-br

[a] Travels with the secondary front.

Layer L1 = Eastman Chromagram sheets, Type K 301 R2.
Solvent S1 = toluene-butyric acid (19 : 1).
S2 = toluene-butyric acid (9 : 1).
S3 = dichloromethane-hexane-butyric acid (10 : 9 : 1).
S4 = toluene-glacial acetic acid (9 : 1).
S5 = dichloromethane-glacial acetic acid (9 : 1).
Detection D1 = Bis-diazotized benzidine.
D2 = first spray with 0.1% aq. soln. of Echtblausalz B, then with 1% KOH and heat to 100 °C for 30–45 sec.
Technique Ascending, 5 cm.
Literature 1. Santesson, J., *Acta Chem. Scand.*, **21**, 1165 and 1168 (1967).

TABLE TLC 36
CARBOXYLIC ACIDS, ORGANIC ACIDS

Layer	L1	L1	L2	L2	L3	L3
Solvent	S1	S2	S3	S4	S5	S5
Detection	—	—	—	—	D1	D1
Technique	T1	T1	T2	T2	T3	T3
Literature	1	1	2	2	3	3
Acid			$R_F \times 100$			Color
Citrate	5	31	—	—	—	—
***cis*-Aconitate**	7, 10	29, 82	—	—	61	ft bl
***dl*-Isocitrate**	8, 39	28, 82	—	—	42[a]	bl[a]
Glyoxalate	14	70	17	55	—	—
Oxalacetate	15	20	—	—	—	—
Malonate	17	64	—	—	—	—
***l*-Aspartate**	18	8	—	—	—	—
***l*-Malate**	21	46	—	—	—	—
***l*-Glutamate**	24	14	—	—	—	—
Succinate	27	80	80	65	—	—
α-Ketoglutarate	31	51	45	35	56	bl
Fumarate	32	89	97	70	96[a]	lt bl[a]
***l*-Lactate**	56	80	—	—	—	—
Pyruvate	60	78	—	—	—	—
Fluoroacetate	60	86	—	—	—	—
β-Hydroxybutyrate	66	88	—	—	81	ft bl
Glycolate	—	—	59	50	—	—
α-Hydroxyglutarate	—	—	25	17	—	—
α-Hydroxyglutaryl lactone	—	—	33	73	—	—
Citramalate	—	—	41	21	64	bl
Malate	—	—	17	11	—	—
Aspartate	—	—	—	—	4	ft bl
Glutamate	—	—	—	—	5	ft bl
Lactate	—	—	—	—	65	bl
Sudan Red G	—	—	—	—	88	—

[a] Sodium salt applied.

Layer L1 = pure cellulose.
L2 = ITLC-SG.
L3 = Silica gel H, dried for 3–4 hours.

Solvent S1 = ethanol-ammonia-water (8 : 2 : 1).
S2 = isobutanol-5*M* formic acid (2 : 3).
S3 = petroleum ether (b.p. 30 to 60 °C)-anhydrous diethyl ether-formic acid (28 : 12 : 1).
S4 = chloroform-methanol-formic acid (80 : 1 : 1).
S5 = water-saturated diethyl ether-formic acid (7 : 1).

Detection D1 = ammonium molybdate.

Technique T1 = Solvents can be used for 2-dimensional technique.
T2 = 15 cm ascending development.
T3 = 14 cm development, one hour, saturated tank, room temperature.

Literature

1. Goebell, H. and M. Klingenberg, *Chromatographic Symposium II*, 1962, published by Soc. Belge des Sciences Pharmaceutiques, Brussels, 1963, p. 153; *Chem. Abstr.*, **60**, 4444g (1964) and *J. Chromatogr.*, **15**, D5 (1964).
2. Bleiweis, A. S., Reeves, H. C., and S. J. Ajl, *Anal. Biochem.*, **20**, 335 (1967).
3. Ting, I. P. and W. M. Dugger, Jr., *Anal Biochem.*, **12**, 574 (1965).

TABLE TLC 37
CARBOXYLIC ACIDS, SUBSTITUTED BENZOIC ACIDS AND PHENOLIC COMPOUNDS

Layer	L1	L2	L3	L3		
Solvent	S1	S2	S3	S3		
Detection					D1	D2
Technique	T1	T2	na	na		
Literature	1	1	2	2	2	2
Compound		$R_F \times 100$			Color	
Benzoic acid	51	70	—	—	—	—
***o*-Hydroxybenzoic acid**	54	71	94	49	pl y-br	pl y-br
***m*-Hydroxybenzoic acid**	15	37	35	45	o	y
***p*-Hydroxybenzoic acid**	6	37	35	46	o-br	y o
***o*-Methoxybenzoic acid**	42	33	—	—	—	—
***m*-Methoxybenzoic acid**	45	62	—	—	—	—
***p*-Methoxybenzoic acid**	43	58	—	—	—	—
***o*-Aminobenzoic acid**	29	26	—	—	—	—
***m*-Aminobenzoic acid**	14	1	—	—	—	—
***p*-Aminobenzoic acid**	10	11	—	—	—	—
***o*-Nitrobenzoic acid**	43	38	—	—	—	—
***m*-Nitrobenzoic acid**	50	53	—	—	—	—
***p*-Nitrobenzoic acid**	49	53	—	—	—	—
***o*-Chlorobenzoic acid**	53	63	—	—	—	—
***m*-Chlorobenzoic acid**	65	76	—	—	—	—
***p*-Chlorobenzoic acid**	63	76	—	—	—	—
***o*-Bromobenzoic acid**	55	63	—	—	—	—
***m*-Bromobenzoic acid**	66	75	—	—	—	—
***p*-Bromobenzoic acid**	65	75	—	—	—	—
***o*-Methylbenzoic acid**	59	76	—	—	—	—
***m*-Methylbenzoic acid**	57	75	—	—	—	—
***p*-Methylbenzoic acid**	55	73	—	—	—	—
2,4-Dichlorobenzoic acid	72	74	—	—	—	—
3,4-Dichlorobenzoic acid	78	82	—	—	—	—
3,5-Dichlorobenzoic acid	83	86	—	—	—	—
1-Naphthoic acid	65	69	—	—	—	—
2-Naphthoic acid	55	69	—	—	—	—
***p*-*tert*-Butylbenzoic acid**	90	90	—	—	—	—
2,3-Dihydroxybenzoic acid	—	—	35	42	—	—
2,4-Dihydroxybenzoic acid	—	—	24	31	vt	y
2,5-Dihydroxybenzoic acid	—	—	16	43	w gy	p tan
2,6-Dihydroxybenzoic acid	—	—	18	38	r br	y
3,4-Dihydroxybenzoic acid	—	—	6	34	w tan	p tan
3,5-Dihydroxybenzoic acid	—	—	2	33	r p	gy y
2,3,4-Trihydroxybenzoic acid	—	—	5	28	pl br	pl br
3,4,5-Trihydroxybenzoic acid	—	—	0	28	pl br	gr br
2,4,6-Trihydroxybenzoic acid	—	—	2	21	p	y br
3-Hydroxy-4-methoxybenzoic acid	—	—	56	33	pk br	o
4-Hydroxy-3-methoxybenzoic acid	—	—	79	37	pk br	o
3,4-Dihydroxy-5-methoxybenzoic acid	—	—	10	29	w gy	pk gy
3,5-Dihydroxy-4-methoxybenzoic acid	—	—	7	40	r p	br y
3,5-Dimethoxy-4-hydroxybenzoic acid	—	—	75	31	pk br	r
Phenol	—	—	88	69	r o	y o
1,2-Dihydroxybenzene	—	—	30	62	pk gy	pk gy
1,3-Dihydroxybenzene	—	—	11	56	r p	gy y
1,4-Dihydroxybenzene	—	—	10	63	gy br	br
1,2,3-Trihydroxybenzene	—	—	4	55	pk br	br
1,3,5-Trihydroxybenzene	—	—	0	42	p	y br
2-Methoxyphenol	—	—	98	—	vt	o
1,2-Dihydroxy-3-methoxybenzene	—	—	54	53	w gy	pk gy
1,3-Dihydroxy-2-methoxybenzene	—	—	43	63	vt	gy y
2,6-Dimethoxyphenol	—	—	96	54	p	r

TABLE TLC 37—(Continued)

CARBOXYLIC ACIDS, SUBSTITUTED BENZOIC ACIDS AND PHENOLIC COMPOUNDS

Layer L1 = MN-Cellulose powder 300.
L2 = Kieselguhr G impregnated with 10% adipic acid triethylene glycol polyester.
L3 = MN cellulose.
Solvent S1 = isobutanol-25% ammonia-water (7 : 1.5 : 3) (upper layer).
S2 = diisopropyl ether-petroleum ether-undecane-carbon tetrachloride-formic acid-water (40 : 20 : 20 : 10 : 8 : 1).
S3 = 20% aqueous potassium chloride-glacial acetic acid (100 : 1).
Detection D1 = fast blue B salt.
D2 = diazotized sulfanilamide.
Technique T1 = development time 2 hours.
T2 = development time 30 minutes.
Literature 1. Knappe, E. and J. -I Stuck, *Z. Anal. Chem.*, **227**, 354 (1967); from *J. Chromatogr.*, **32**, D1 (1968).
2. Scheline, R. R., *J. Pharm. Pharmacol.*, **18**, 665 (1966); from *J. Chromatogr.*, **28**, D11 (1966).

TABLE TLC 38
DRUGS, MEDICINAL CARBAMATES

Layer	L1	L1	L2	L2	L2						
Solvent	S1	S1	S2	S2	S3						
Detection						D1[a]	D2[b]	D3[c]	D4[d]	D4[e]	D4[f]
Technique	T1	T2	T3	T4	T5						
Literature	1	1	1	1	1						
Compound	$R_F \times 100$					Color					
Urea	13	12	0	0	3	—	+	+	r	nc	nc
Methocarbamol	22	22	31	33	0	+	+	+	y	pk	r–p
Chlorphenesin carbamate	27	26	31	29	0	+	+	+	y	nc	nc
Styramate	29	28	19	21	0	+	+	+	y	gy–br	gy–br
Mephenesin carbamate	34	32	44	43	0	+	+	o	y	p	pk–br
Meprobamate	37	37	35	34	0	—	+	+	y	bl	bl
Mebutamate	39	38	41	40	0	—	+	+	y	bl	bl-gy
Hydroxyphenamate	42	41	50	50	4	+	+	+	y	gy-p	p-gy
Salicylamide	51	54	40	41	10	+ Fl	+	—	—	ft-br	ft-br
Urethane	60	61	57	61	11	—	+	+	y	ft-y	ft-y
Carisoprodol	64	65	79	81	25	—	+	+	y	p	bl
Tybamate	65	70	78	84	37	—	+	+	y	bl	dk-br
Methylpentynol carbamate	70	72	70	71	25	—	+	o	p	p	bl-gy
Ethinamate	72	75	74	75	37	—	+	o	y	bl	br-gy
Emylcamate	80	82	77	81	66	—	+	p-r	p	p	p

[a] + Indicates absorption, + Fl, absorption, blue fluorescence at center of spot.
[b] + Indicates blue-black spots.
[c] + Indicates blue-black spots, except as noted. The background darkens rapidly and blue-black spots merge into it.
[d] Color immediately after spraying.
[e] Color on heating at 110 °C.
[f] Color on standing in air for 15 min. after heating.

Layer L1 = Silica gel G and fluorescent indicator green.
L2 = Kieselguhr G impregnated with formamide.
Solvent S1 = lower layer obtained by shaking together acetic acid-carbon tetrachloride-chloroform-water (100 : 60 : 90 : 50).
S2 = benzene-chloroform (30 : 120), saturated with formamide.
S3 = carbon tetrachloride saturated with formamide.
Detection D1 = UV (254 mμ).
D2 = furfural-hydrochloric acid.
D3 = furfural-conc. sulfuric acid.
D4 = vanillin-sulfuric acid.
Technique T1 = 10 cm development distance, 26 min. development time.
T2 = 15 cm, 78 min.
T3 = 10 cm, 12 min., 12 min. drying time.
T4 = 15 cm, 21 min., 15 min. drying time.
T5 = 10 cm, 16 min., 12 min. drying time.
Literature 1. Davis, T. W. Mc.C., *J. Chromatogr.*, **29**, 287–292 (1967).

TABLE TLC 39
DRUGS

Layer	L1	L1	L1	L2	L2	L3	L3	L3	L4			
Solvent	S1	S2	S3	S2	S4	S5	S6	S7	S8			
Detection										D1	D1	D1
Technique	T1	T1	T1	T1	T1	T2	T2	T2	T2			
Literature	1	1	1	1	1	2	2	2	2	3	3	3
Compound				R_F[a] × 100							Color	
										Cold	85°[c]	UV F
Adiphenine	65	65	70	46	25	—	—	—	—	—	—	—
2-Aminoheptane	53	22	36	60	56	—	—	—	—	—	—	—
Amitriptyline	72	50	34	41	28	—	—	—	—	—	—	—
Amolanone	64	53	75	51	35	—	—	—	—	—	—	—
Amphetamine	34	28	33	59	53	—	—	—	—	—	—	—
Antazoline (Antistine)	8	11	6	61	38	31	72	40	38	cr	cr	—
Atropine	9	11	1	31	14	—	—	—	—	—	—	—
Azacyclonol	2	8	1	60	44	—	—	—	—	—	—	—
Benactyzine	63	69	72	57	28	—	—	—	—	—	—	—
Benzocaine	6	68	73	63	57	—	—	—	—	—	—	—
Benzphetamine	79	70	85	54	49	—	—	—	—	—	—	—
Betazole	19	20	47	38	17	—	—	—	—	—	—	—
Bromodiphenhydramine (Ambodryl®)	50	37	26	48	33	86	80	33	63	y	br	r-br
Bromphenlramine (Dimetan®)	—	—	—	—	—	42	47[b]	16	48	—	—	—
Buclizine	73	72	84	75	64	—	—	—	—	—	—	—
Butacaine	8	63	71	59	51	—	—	—	—	—	—	—
Butamben	7	68	76	63	60	—	—	—	—	—	—	—
Butethamine	5	58	49	58	47	—	—	—	—	—	—	—
Caffeine	4	55	42	48	34	—	—	—	—	—	—	—
Carbetapentane	63	37	30	44	18	—	—	—	—	—	—	—
Carbinoxamine (Clistin®)	29	21	6	5	0	33	40[b]	17	63	—	—	lt bl
Chlorcyclizine (Di-Paralene®)	49	44	30	43	19	90	68	40	49	—	gy br	
2-Chloroprocaine	3	53	52	46	26	—	—	—	—	—	—	—
Chlorothen (Tagathen®)	43	37	28	15	2	70	56[b]	29	66	pk	p	—
Chlorpheniramine (Chlor-Trimeton®)	38	19	6	8	1	40	41[b]	20	52	—	—	lt bl
Cinchonine	10	37	11	22	9	—	—	—	—	—	—	—
Clemizole	33	67	61	47	27	—	—	—	—	—	—	—
Cocaine	58	57	64	26	10	—	—	—	—	—	—	—
Codeine	7	28	6	24	10	—	—	—	—	—	—	—
Covatin	58	54	52	52	29	—	—	—	—	w	bl	—
Cyclizine	55	46	27	41	16	—	—	—	—	—	y	—
Cyclomethycaine	66	45	42	46	29	—	—	—	—	—	—	—
Cyclopentamine	52	10	2	52	40	—	—	—	—	—	—	—
Desimpramine	26	18	3	62	45	—	—	—	—	—	—	—
Diacetylmorphine (Heroin)	22	39	20	24	9	—	—	—	—	—	—	—
Diazepam	33	65	62	65	54	—	—	—	—	—	—	—
Diethylpropion	76	69	78	44	27	—	—	—	—	—	—	—
Dihydrocodeinone (Hydrocodone)	6	17	4	11	3	—	—	—	—	—	—	—
Dihydromorphinone (Hydromorphone)	5	15	3	13	5	—	—	—	—	—	—	—
Diphenhydramine (Benadryl®)	52	37	26	45	25	91	61	25	53	y	y	—
Diphenylpyraline	42	25	11	44	23	—	—	—	—	y	y	—
Doxylamine (Decapryn®)	—	—	—	—	—	52	38[b]	17	60	—	—	—
Dyclonine	54	51	46	46	31	—	—	—	—	—	—	—
Ephedrine	8	18	2	54	42	—	—	—	—	—	—	—
Epinephrine	1	7	11	62	40	—	—	—	—	—	—	—
Ethoheptazine	59	29	11	39	19	—	—	—	—	—	—	—
Guanethidine	0	3	0	15	5	—	—	—	—	—	—	—
Hydroxyamphetamine	3	29	36	59	21	—	—	—	—	—	—	—
Hydroxyzine (Atarax®)	8	59	39	56	25	27	70	62	84	—	br	—
Imipramine	61	35	18	39	25	—	—	—	—	—	—	—
Iproniazid	4	64	34	34	22	—	—	—	—	—	—	—
Isoamylhydrocupreine	12	60	22	34	25	—	—	—	—	—	—	—
Isocarboxazid	28	71	70	65	61	—	—	—	—	—	—	—
Levallorphan	31	60	53	63	47	—	—	—	—	—	—	—

TABLE TLC 39—(Continued)
DRUGS

Layer	L1	L1	L1	L2	L2	L3	L3	L3	L4			
Solvent	S1	S2	S3	S2	S4	S5	S6	S7	S8			
Detection										D1	D1	D1
Technique	T1	T1	T1	T1	T1	T2	T2	T2	T2			
Literature	1	1	1	1	1	2	2	2	2	3	3	3
Compound	$R_F{}^a \times 100$									Color		
										Cold	85°[c]	UV F
Lidocaine	39	70	69	47	23	—	—	—	—	—	—	—
Meclizine	69	71	84	74	60	—	—	—	—	—	br	—
Meperidine	55	48	21	49	24	—	—	—	—	—	—	—
Mephentermine	50	15	3	52	39	—	—	—	—	—	—	—
Mepivacaine	37	62	57	43	30	—	—	—	—	—	—	—
Metabutoxycaine	30	60	67	43	23	—	—	—	—	—	—	—
Metaraminol	1	33	44	68	21	—	—	—	—	—	—	—
Methadone	76	37	43	51	25	—	—	—	—	—	—	—
Methamphetamine	46	18	4	50	39	—	—	—	—	—	—	—
Methaphenilene	55	46	40	44	24	—	—	—	—	r-o	p	—
Methenamine	2	16	0	14	4	—	—	—	—	—	—	—
Methoxamine	13	28	53	63	56	—	—	—	—	—	—	—
Methoxyphenamine	42	21	4	53	42	—	—	—	—	—	—	—
Methylhexaneamine	38	25	39	59	57	—	—	—	—	—	—	—
Methylphenethylhydrazine	56	72	73	60	65	—	—	—	—	—	—	—
Methylphenidate	55	56	35	54	45	—	—	—	—	—	—	—
Morphine	2	28	4	23	10	—	—	—	—	—	—	—
Naphazoline	5	8	1	43	30	—	—	—	—	—	—	—
Nialamide	0	55	9	35	23	—	—	—	—	—	—	—
Nicotine	53	52	29	8	2	—	—	—	—	—	—	—
Nikethamide	24	64	44	38	25	—	—	—	—	—	—	—
Norepinephrine	2	21	15	66	50	—	—	—	—	—	—	—
Nylidrin	5	68	64	66	68	—	—	—	—	—	—	—
Orphenadrine	60	48	32	45	22	—	—	—	—	—	—	—
Papaverine	11	62	53	62	21	—	—	—	—	—	—	—
Pentylenetetrazol	13	63	67	53	51	—	—	—	—	—	—	—
Phenacaine	12	63	76	61	55	—	—	—	—	—	—	—
Phendimetrazine	54	61	41	28	15	—	—	—	—	—	—	—
Phenelzine	51	72	75	62	49	—	—	—	—	—	—	—
Phenindamine (Thephorin®)	55	53	35	41	25	90	67	48	65	w	lt br	y
Pheniramine (Trimeton®)	40	18	5	6	1	68	38[a]	19	63	—	—	lt bl
Phenmetrazine	24	44	11	49	37	—	—	—	—	—	—	—
Phenylephrine	5	21	33	60	45	—	—	—	—	—	—	—
Phenylpropanolamine	9	35	50	58	56	—	—	—	—	—	—	—
Phenyltoloxamine	46	53	33	42	18	—	—	—	—	o	o	—
Phenyramidol	14	68	66	57	40	—	—	—	—	—	—	—
Piperidolate	70	65	65	42	25	—	—	—	—	—	—	—
Piperocaine	63	45	42	47	29	—	—	—	—	—	—	—
Pipradol	74	44	66	59	51	—	—	—	—	—	—	—
Pivalylbenzhydrazine	29	72	67	69	59	—	—	—	—	—	—	—
Pramoxine	52	61	55	48	28	—	—	—	—	—	—	—
Procaine	5	52	47	39	18	—	—	—	—	—	—	—
Procyclidine	78	36	39	52	41	—	—	—	—	—	—	—
Propoxyphene	72	62	60	54	30	—	—	—	—	—	—	—
Propylhexedrine	54	15	3	54	46	—	—	—	—	—	—	—
3-Pyridinemethanol	3	60	40	30	5	—	—	—	—	—	—	—
Pyrilamine (Neo-Antergan®)	42	33	24	12	1	82	43[b]	27	56	pk	p	—
Pyrrobutamine (Pyronil®)	62	39	34	59	40	84	82	32	78	—	y	—
Quinine	5	47	11	37	14	—	—	—	—	—	—	—
Scopolamine	9	54	33	34	13	—	—	—	—	—	—	—
Strychnine	13	17	3	17	7	—	—	—	—	—	—	—
Tetracaine	18	50	28	38	14	—	—	—	—	—	—	—
Thenyldiamine	47	32	25	12	1	—	—	—	—	br	p	—
Thenylpyramine (Methapyrilene)	47	36	27	14	2	83	52[b]	30	56	r	gr	—
Thonzylamine (Neohetramine®)	41	38	27	29	12	65	57	32	59	w	p	—
Tranylcypromine	55	57	58	58	56	—	—	—	—	—	—	—

TABLE TLC 39—(Continued)
DRUGS

Layer	L1	L1	L1	L2	L2	L3	L3	L3	L4			
Solvent	S1	S2	S3	S2	S4	S5	S6	S7	S8			
Detection										D1	D1	D1
Technique	T1	T1	T1	T1	T1	T2	T2	T2	T2			
Literature	1	1	1	1	1	2	2	2	2	3	3	3
Compound	R_F[a] × 100									Color		
										Cold	85°[c]	UV F
Tripelennamine	50	35	27	12	3	92	48[b]	23	62	w	br	—
Triprolidine (Actidil®)	41	45	13	18	2	40	48[b]	26	81	—	—	bt bl
Yohimbine	9	68	61	55	11	—	—	—	—	—	—	—
Zoxazolamine	4	64	64	66	59	—	—	—	—	—	—	—

[a] Average of 3 or more different plates.
[b] Tailing occurred.
[c] Color after heating at 85 °C.

Layer L1 = 20 × 20 cm glass plates coated with silica gel G and 0.1*M* KOH, air-dried, then conditioned for 16 hours at 52% humidity and 22 °C.
L2 = 20 × 20 cm glass plates coated with silica gel G and 0.1*M* $NaHSO_4$, air-dried, then conditioned for 16 hours at 52% humidity and 22 °C.
L3 = silica gel (with or without $CaSO_4$).
L4 = aluminum oxide (with or without $CaSO_4$).

Solvent S1 = cyclohexane-benzene-diethylamine (75 : 15 : 10 v/v).
S2 = methanol.
S3 = acetone.
S4 = 95% ethanol.
S5 = benzene-dioxane-aqueous NH_3 (60 : 35 : 5).
S6 = ethanol-acetic acid-water (50 : 30 : 20).
S7 = methanol-butanol (60 : 40).
S8 = butanol-butyl ether-acetic acid (40 : 80 : 10).

Detection D1 = Mandelin's reagent.

Technique T1 = ascending, saturated chamber, development distance = 10 to 12 cm, in room at 52% humidity and 22 °C.
T2 = development distance 10 cm.

Literature
1. Fike, W. W., *Anal. Chem.*, **38**, 1698–99 (1966).
2. Cochin, J. and J. W. Daly, *J. Pharmacol. Exp. Ther.*, **139**, 160 (1963).
3. Fike, W. W. and I. Sunshine, *Anal. Chem.*, **37**, 128 (1965).

TABLE TLC 40
DRUGS (ANALGESICS AND ANTIPYRETICS)

Layer	L1	L2	L1	L2	L2	L2	L3	L3	L3	L3	L4
Solvent	S1	S1	S2	S2	S3	S4	S5	S6	S7	S8	S9
Literature	1	1	1	1	2	2	3	3	3	3	4
Compound	$R_F \times 100$										
Amidopyrine	4	4	69	75	16	21	53	95	73	85	64
Sodium noramidopyrine-methane-sulfonate	10	10	5	3	—	—	—	—	—	—	—
Phenazone (Antipyrine)	22	21	58	60	3	9	70	76	80	78	50
Paraacetamole	39	46	66	67	—	—	—	—	—	—	—
Phenacetin	52	57	74	81	25	57	63	32	70	60	—
Salicylamide	58	61	57	59	—	—	32	24	33	37	—
Acetosalicylic acid (Aspirin)	70	83	16	19	0	26	44	32	47	61	—
Phenylbutazone	76	90	34	40	—	—	—	—	—	—	—
Quinine	2	2	53	55	—	—	—	—	—	—	—
Codeine	4	3	33	38	—	—	—	—	—	—	—
Papaverine	12	11	72	80	—	—	—	—	—	—	—
Caffeine	25	25	52	53	—	—	—	—	—	—	—
Isopropylantipyrine	—	—	—	—	59	72	—	—	—	—	—
Acetanilide	—	—	—	—	44	66	50	38	57	48	—
Ethoxybenzamide	—	—	—	—	32	57	—	—	—	—	—
***N*-Acetyl-*p*-aminophenol**	—	—	—	—	16	48	—	—	—	—	—
Caffeine	—	—	—	—	4	13	—	—	—	—	—
Sulpyrin	—	—	—	—	0	0	—	—	—	—	—
Aminopropiron	—	—	—	—	2	3	—	—	—	—	—
Neophyllin M	—	—	—	—	30	3	—	—	—	—	—
Noscapine	—	—	—	—	49	71	—	—	—	—	—
Barbital	—	—	—	—	88	91	—	—	—	—	—
Allylisopropylacetylurea	—	—	—	—	61	87	—	—	—	—	—
Bromovalerylurea	—	—	—	—	61	85	—	—	—	—	—
Ethylquinine carbonate	—	—	—	—	9	—	—	—	—	—	—
Salicylic acid	—	—	—	—	—	—	24	8	24	41	—
Oxyphenbutazone	—	—	—	—	—	—	42	7	50	46	—
Indomethacin	—	—	—	—	—	—	68	1	71	77	—
Novaminsulfone	—	—	—	—	—	—	—	—	—	—	0
***N*-Acetyl-*p*-phenetidine**	—	—	—	—	—	—	—	—	—	—	70

Layer L1 = Silica gel VEB Laborchemie Apolda.
L2 = Silica gel G.
L3 = Polyamide.
L4 = Silica gel.

Solvent S1 = butyl acetate-chloroform-85% formic acid (6 : 4 : 2).
S2 = butyl acetate-acetone-*n*-butanol-10% aqueous ammonia (5 : 4 : 3 : 1).
S3 = ether.
S4 = ethyl acetate-ether (4 : 1).
S5 = chloroform-benzene-90% formic acid (5 : 1 : 0.1).
S6 = isopropanol-water-90% formic acid (1.5 : 6 : 0.1).
S7 = chloroform-90% formic acid (20 : 0.1).
S8 = cyclohexane-chloroform-glacial acetic acid (4 : 5 : 1).
S9 = chloroform-acetone-water (2: 9 : 0.5).

Detection Kieffer reagent.

Literature

1. Zarnack, J. and S. Pfeifer, *Pharmazie*, **19**, 216 (1964); from *J. Chromatogr.*, **18**, D18 (1965).
2. Tsuji, A. and A. Wada, *Eisei Shikenjo Hokoku*, **N.82**, 73–75 (1964); *Chem. Abstr.*, **65**, 8668g (1966).
3. Hsui, H. C., Shih, T. B., and K.-T. Wang, *J. Chromatogr.*, **41**, 491–493 (1969).
4. Kutlu, H., *Istanbul. Univ. Eczacjlik Fa. Mecmuasi*, **1**, 128–136 (1965); *Chem. Abstr.*, **65**, 8676f (1966).

TABLE TLC 41
DRUGS, DIURETICS AND ANTIHYPERTENSIVES

Layer	L1	L2	L2	L2	L2	L2	L2	L2	L2	L2
Solvent	S1	S2	S3	S4	S5	S6	S2	S7	S8	S9
Technique	T1	na	T2	T2	T2	T2	T2	T2	T2	T2
Literature	1	1	2	2	2	2	2	2	2	2
Compound	$R_F \times 100$									
Bendrofluazide	71	—	—	—	—	—	—	—	—	—
Benzflumethiazide	—	66	88	93	70	54	91	98	52, 84	98
Benzthiazide	58	55	85	100	60	69	33	61	43	80
Chlorothiazide	27	17	70	85	26	64	8	streak	42	50
Clopamide	36	—	—	—	—	—	—	—	—	—
Cyclopenthiazide	63	58	—	—	—	—	—	—	—	—
Cyclothiazide	59–64	—	—	—	—	—	—	—	—	—
Epithiazide	67	59	—	—	—	—	—	—	—	—
Flumethiazide	44	33	—	—	—	—	—	—	—	—
Hydrochlorothiazide	39	29	77	75	42	68	28, 55	46, 72	42	54
Hydroflumethiazide	53	41	—	—	—	—	—	—	—	—
Hygroton	46	—	—	—	—	—	—	—	—	—
Methyclothiazide	52	47	95	95	80	53	65	71	79	82
Polythiazide	58	47	93	95	86	55	72	70	80	82
Quinethazone	23	—	—	—	—	—	—	—	—	—
Spironolactone	54	—	—	—	—	—	—	—	—	—
Thiabutazide	62	57	—	—	—	—	—	—	—	—
Triamterene	0	—	—	—	—	—	—	—	—	—
Trichloromethiazide	67	57	100	90	70	62	79	22, 52	70	85

Layer L1 = Silica gel.
L2 = Silica gel G.

Solvent S1 = ethyl acetate containing 1.5% water.
S2 = ethyl acetate-benzene (8 : 2).
S3 = propanol-2-12*N* ammonia (8 : 2).
S4 = propanol-1-12*N* ammonia (8 : 2).
S5 = butanol-1-12*N* ammonia (8 : 2).
S6 = pentanol-1-12*N* ammonia (8 : 2).
S7 = ethyl acetate-12*N* ammonia (8 : 2).
S8 = chloroform-methanol (8 : 2).
S9 = cyclohexane-glacial acetic acid-acetone (4 : 1 : 5).

Detection Short UV light; modified Folin reagent.

Technique T1 = temperature 22 °C.
T2 = ascending, development distance = 5 cm.

Literature
1. Adam, R. and C. L. Lapiere, *J. Pharm. Belg.* **19**, 82 (1964); *Chem. Abstr.* **64**, 7970d (1966).
2. Smith, P. J. and T. S. Hermann, *Anal. Biochem.*, **22**, 134–144 (1968).

TABLE TLC 42
DRUGS, PSYCHOTROPIC

Layer	L1	L1		
Solvent	S1	S2		
Literature	1	1		
Compound	$R_F \times 100$		Color	Detection Reagent
Triperidol	90	—	mv	D1, D2
Haloperidol	79	—	bl-vt	D1
Phenergan®	53	—	bl-gr	D2
Parsidol	74	—	gr	D2
Diparcol®	69	—	sm	D2
Prazine®	41	—	r-br	D2
Esucos®	67	—	sm-pk	D2
Largactil®	55	—	r-vt	D2
Stemetil®	48	—	bl-pk	D2
Trilafon®	56	—	m-r	D2
Mapazine	45	—	dp-bl	D2
Nozinan	67	—	pl-bl	D2
Plegicil®	40	—	r-o	D2
Melleril	48	—	bl	D2
Thioproperazine	44	—	r	D2
Moditen or Sevinol	66	—	r-o	D2
Trifluperazine	50	—	r-o	D2
Atarax®	64	—	—	—
Azacyclonol	13	—	—	—
Pipradrol	56	—	—	—
Mephenesin	—	54	bt-pk	D4
Meprobamate	—	33	y	D4
Apesan	—	64	y-gr	D4
N Oblivon	—	73	gy-br	D4
Sinaxar®	—	20	pl-y	D4
Taractan®	54	—	o	D3
Librium®	78	—	pl-y	D3
Rimifon®	59	—	r	D5
Iproniazid	65	—	r-vt	D5
Niamide	43	—	ro	D5
Marplan®	75	—	vt	D5
Sursum	70	—	vt	D5
Amphetamine	57	—	—	—
Pevitin	43	—	—	—
Rilatin	80	—	—	—
Preludin®	70	—	—	—
Tofranil®	36	—	dp-bl	D6
Surmontil®	54	—	dp-bl	D6
Insidon	50	—	go y	D6

Layer L1 = Kieselgel G.

Solvent S1 = acetone-methanol-ammonia (50 : 50 : 1).
S2 = benzene-methanol (9 : 1).

Detection D1 = Dragendorff's reagent.
D2 = Procaine in chloroform, followed by alc. HNO_3.
D3 = Marquis reagent.
D4 = *p*-Dimethylaminobenzaldehyde in alc. H_2SO_4.
D5 = Triphenyltetrazolium chloride (2% in chloroform + alc. NaOH).
D6 = Alc. HNO_3.

Literature
1. Thomas, J. J. and L. Dryon, *J. Pharm. Belg.*, **19**, 481 (1964); *Chem. Abstr.*, **63**, 1660c (1965); from *J. Chromatogr.*, **21**, D7 (1966).

TABLE TLC 43
DRUGS, LOCAL ANESTHETICS

Layer	L1	L1	L1	L1	L1	L1	L1	L1	L1	L1	L1
Solvent	S1	S2	S3	S4	S5	S6	S7	S8	S9	S10	S11
Literature	1	1	1	1	1	1	1	1	1	1	1
Compound (common name)						$R_F \times 100$					
Cinchocaine hydrochloride (Dibucaine HCl)	10	19	46	65	75	42	51	47	75	61	20
Cocaine hydrochloride	17	34	65	75	82	57	67	60	80	70	30
Dextrocaine (Psicaine)	8	18	46	65	78	55	64	57	78	63	25
Dimethocaine (Larocaine®)	12	18	37	54	70	40	50	45	79	68	24
Diocaine	10	17	55	67	80	35	51	52	81	80	24
Ethoform (Benzocaine)	18	25	48	—	—	53	52	62	70	75	31
β-Eucaine	14	16	28	53	72	41	44	45	51	30	21
Hostocaine	10	17	26	42	68	27	—	—	—	—	5
Lidocaine (Xylocaine®)	10	17	50	56	78	52	62	56	81	65	22
Orthoform (Orthocaine®)	0	0	0	0	0	0	0	0	0	0	0
Phenacaine (Holocaine)	6	24	52	60	77	37	37	55	77	70	27
Procaine hydrochloride	0	14	31	52	65	27	34	43	60	54	11
Tetracaine hydrochloride	8	15	40	63	72	40	43	56	59	58	20
Trimecaine (Mesocaine)	5	16	35	57	74	50	64	55	80	62	23
Tutocaine	10	16	43	62	71	45	41	57	60	55	22

Layer L1 = Aluminum oxide, loose layers (activity III).
Solvent S1 = benzene.
S2 = benzene-ethanol (98 : 2).
S3 = benzene-ethanol (95 : 5).
S4 = benzene-ethanol (90 : 10).
S5 = benzene-ethanol (80 : 20).
S6 = chloroform.
S7 = chloroform-ethanol (99 : 1).
S8 = chloroform-butanol (98 : 2).
S9 = chloroform-acetone (1 : 1).
S10 = ether.
S11 = ether-petroleum ether (1 : 1).
Detection Munier modified Dragendorff reagent.
Literature 1. Šarsúnová, M., *Pharmazie*, **18**, 748 (1963).

TABLE TLC 44
DRUGS, SULFONAMIDES

Layer	L1	L2	L2	L3	L3	L3
Solvent	S1	S2	S3	S4	S5	S6
Literature	1	1	1	1	1	2
Compound			$R_F \times 100$			
N^1-Acetyl-2-sulfanilamido-3-methoxypyrazine	*T*	76	54	97	*T*	—
N^1Acetyl-3-sulfanilamido-6-methoxypyridazine	*T*	70	54	97	*T*	—
N^1-Benzoyl-sulfanilamide	50	85	59	10	7	—
N^4-Carbethoxysulfoethylthiodiazole	63	88	61	24	23	—
Cinnamein sulfanilamide	96	97	79	85	74	—
***N*,*N*-Dimethyl-acroylsulfanilamide**	38	74	54	22	12	—
Dimethylacrylsulfanilamide	—	—	—	—	—	82
N'-3,4-Dimethylbenzoyl sulfanilamide	85	86	61	10	8	40
2,4-Dimethyl-6-sulfanilamido-1,3-diazine	45	81	57	57	25	—
4-Homosulfanilamide salt of 1-sulfanilyl-2-thiourea	28	60	49	10	5	—
N^1-Isopropoxysulfanilamide	58	92	61	18	14	—
Maleylsulfanilamide	34	81	61	13	8	—
5-Methyl-3-sulfanilamidoisoxazole	—	—	—	—	—	80
Phthalylsulfacetamide	6	26	33	0	0	—
Phthalylsulfathiazole	6	50	44	1	0	—

TABLE TLC 44—(Continued)
DRUGS, SULFONAMIDES

Layer	L1	L2	L2	L3	L3	L3
Solvent	S1	S2	S3	S4	S5	S6
Literature	1	1	1	1	1	2
Compound			$R_F \times 100$			
Succinylsulfanilamide (sodium salt)	12	6	39	4	6	—
Sulfacetamide	19	—	38	12	7	31
Sulfadiazine	19	49	38	30	13	39
Sulfadimethoxine	—	—	—	—	—	67
Sulfa-4,6-dimethoxy-pyrimidine	46	85	57	58	45	72
Sulfaethylthiodiazole	41	80	61	14	13	—
Sulfaguanidine	78	50	64	47	33	15
Sulfamerazine	25	59	45	49	25	44
Sulfamethazine	35	69	51	67	46	52
Sulfa-5-methylthiodiazol	—	—	—	—	—	25
4′-Sulfamyl-2,4-diaminoazobenzene · HCl	—	—	—	—	—	47
Sulfanilamide	96	97	79	84	73	43
Sulfanilamido-4,5-dimethyloxazole	—	47 + T^a	58 + T^a	25	17	—
				6	5	
2-Sulfanilamido-5-methoxydiazine	25	60	45	35	35	56
2-Sulfanilamido-3-methoxypyrazine	34	63	50	25	17	70
3-Sulfanilamido-6-methoxypyridazine	36	73	50	47	30	61
2-Sulfanilamido-5-methylpyrimidine	26	59	45	50	26	46
Sulfanilamido-2-phenylpyrazole	56	86	57	13	9	77
Sulfaproxyline	—	—	—	—	—	33
Sulfapyridine	28	77	55	84	69	37
Sulfathiazole	42	80	57	18	12	41
Sulfathiazole-formaldehyde	—	—	—	0	6	—
Sulfathiourea	28	60	49	10	5	34
Sulfisomidine	39	74	48	20	17	19
Sulfisoxazole	34	77	53	8	7	51
Sulfonamidodimethoxytriazine	—	—	—	—	—	24

[a] T = tailing.

Layer L1 = Aluminum oxide G.
L2 = Silica gel G.
L3 = Polyamide.

Solvent S1 = ethyl acetate-methanol-25% ammonium hydroxide (17 : 3 : 3).
S2 = ethyl acetate-methanol-25% ammonium hydroxide (17 : 6 : 5).
S3 = methyl isobutyl ketone-acetone-25% ammonium hydroxide (1 : 4 : 1).
S4 = methyl isobutyl ketone-acetone-25% ammonium hydroxide (5 : 20 : 1).
S5 = ethyl acetate-methanol-25% ammonium hydroxide (17 : 3 : 1).
S6 = petroleum ether (60–80 °C)-chloroform-butanol (1 : 1 : 1).

Detection *p*-Dimethylaminobenzaldehyde.

Literature
1. Van der Venne, M. Th., and J. B. T'Siobbel, *Chromatogr. Symp., 2nd*, Brussels, 1962, p. 196.
2. Karpitschka, N., *Mikrochim. Acta*, 1963, 157.

TABLE TLC 45
DRUGS

Layer	L1	L1	L1	L1
Solvent	S1	S2	S3	S4
Literature	1	1	1	1
Compound		$R_F \times 100$		
Ephedrine	4	15	12	16
Nicotine	18	55	41	58
Brucine	3	10	25	32
Strychnine	5	13	24	30
Atropine	2	12	6	9
Quinine	2	41	25	31
Theobromine	11	50	38	56
Theophylline	19	53	50	64
Caffeine	35	64	61	74
Morphine	6	15	6	10
Codeine	4	18	20	23
Hydromorphone (Dilaudid®)	1	21	15	19
Dionin® (Ethylmorphine)	3	31	34	38
Papaverine	40	78	82	93
Narcotine	42	65	83	83
Cocaine	35	60	70	76
Prothipendyl (Dominal)	4	18	20	25
Sordinol	8	36	39	47
Captodiamine (Covatix)	18	51	59	79
Methaqualone (Revonal)	70	89	79	79
Octamylamine (Octinum® D)	5	23	32	37
Pheniramine (Avil®)	3	11	12	15
Tolazoline (Priscol®)	4	16	10	12
Synopen (Halopyramine)	4	19	21	25
Diphenhydramine (Benadryl®)	8	22	31	33
Antazoline (Antistine®)	4	17	20	23
Neo-Antergan® (Pyrilamine)	10	32	47	57
Diphenylpyraline (Mepiben)	9	35	42	49
Thenalidine (Sandostene®)	11	40	51	65
Tripelennamine (Pyribenzamine®)	10	43	50	64
Clemizole (Allercur®)	45	80	81	91
Tetracaine (Pantocaine)	15	47	40	48
Nupercaine (Dibucaine)	15	59	46	57
Phenacaine (Holocaine)	17	49	46	55
Lidocaine (Xylocaine®)	37	71	68	86
Novesine (Dorsacaine®)	17	64	58	73
Procaine (Novocaine)	10	53	37	39
Tutocaine (Butamine)	12	56	48	58
Propylhexedrine (Eventin®)	3	13	11	14
Methamphetamine (Pervitin®)	4	17	14	18
Reactivan	6	25	19	23
Methadone (Heptadon®)	16	29	35	55
Tepanil (Regenon®)	20	38	53	66
Perphenazine (Decentan)	6	48	29	33
Pethidine (Dolantin)	20	50	60	69
Phenmetrazine (Preludin®)	12	57	43	51
Hydroxyzine (Atarax®)	15	52	56	71
Ritalin (Rilatin)	12	54	55	68
Prethcamide (Micoren®)	84	32	61/66	70/79
Tryptizol	36	27	55	36
Valium	89	54	83	82
Librium (Methaminodiazepoxide)	9	76	79	89
Taractan (Truxal®)	20	45	57	72
Levomepromazine (Nozinan)	9	44	45	54
Promethazine (Phenergan®)	9	38	49	58
Imipramine (Tofranil®)	15	37	44	52
Thioridazine (Melleril)	9	24	38	44
Chlorpromazine (Largactil®)	10	24	37	39

TABLE TLC 45—(Continued)
DRUGS

Layer	L1 = Kieselgel G with fluorescence indicator.
Solvent	S1 = chloroform with 10% acetone.
	S2 = methanol.
	S3 = chloroform with 10% ethanol.
	S4 = chloroform with 20% ethanol.
Detection	UV light or potassium iodoplatinate.
Technique	45 minutes development time.
Literature	1. El Gendi, S., Kisser, W., and G. Machata, *Mikrochim. Acta*, **1965**, 125; from *J. Chromatogr.*, **23**, D22–23 (1966).

TABLE TLC 46
DYES, FAT-SOLUBLE

Layer	L1	L1	L1	L1	L1	L2	
Solvent	S1	S2	S3	S4	S5	S6	
Technique	T1	T1	T1	T1	T1	T2	
Literature	1	1	1	1	1	2	
Compound			$R_F \times 100$				Color
Butter yellow	4	23	57	59	66	40[a]	o-y
Oil Red OS	0	0	10	3	4	—	—
Orange SS	1	8	41	29	38	—	—
Sudan I	2	11	40	36	41	—	—
Sudan II	1	4	32	19	28	—	—
Sudan III	0	3	27	10	24	12[a]	carmine r
Sudan IV	0	2	18	4	10	—	—
Sudan Black B	—	—	—	—	—	28[a]	steel bl
Sudan Blue G	—	—	—	—	—	—	—
Sudan GN	—	48	63	64	74	—	—
	—	75	95	94	98	—	—
Sudan Orange	—	—	—	—	—	4[a]	y-o
Sudan Orange RR	—	—	—	—	—	29[a]	bt-r
Sudan R	—	—	—	—	—	13[a]	carmine r
Sudan Red G	4	26	54	53	58	—	—
Sudan Violet BR	—	—	—	—	—	53[a]	vt-bl
Sudan Yellow 3G	2	14	48	33	48	—	—
Yellow AB	6	43	61	71	71	—	—
Yellow OB	4	29	59	57	61	—	—

[a] Main spot.

Layer	L1 = Suspension of 10g starch and 10% of solution paraffin oil in petroleum ether, dried at room temperature 20 minutes.
	L2 = Silica gel G.
Solvent	S1 = 50% methanol.
	S2 = 70% methanol.
	S3 = 100% methanol.
	S4 = methanol-water-glacial acetic acid (10 : 3 : 1).
	S5 = methanol-water-glacial acetic acid (8 : 1 : 1).
	S6 = benzene.
Technique	T1 = reverse-phase chromatography.
	T2 = migration 5–6 cm.
Literature	1. Davidek, J. and G. Janiček, *J. Chromatogr.*, **15**, 542 (1964).
	2. Gänshirt, H., Waldi, D., and E. Stahl, "*Thin-Layer Chromatography*" (Eng. ed.) E. Stahl, Ed., Academic Press, New York, 1965, p. 345.

TABLE TLC 47
DYES, AZO DYESTUFFS

Layer	L1	L1	L1
Solvent	S1	S2	S3
Literature	1	1	1
Compound (abbreviation)		$R_F \times 100$	
4-(2′-Thiazolylazo)-resorcinol (TAR)	50	80	80
4-(2′-Benzothiazolylazo)-resorcinol (BTAR)	20	90	95
1-(2′Thiazolylazo)-2-naphthol (TAN)	70	85	90
1-(2′-Benzothiazolylazo)-2-naphthol (BTAN)	75	90	95
2-(2′-Thiazolylazo)-*p*-cresol (TAC)	80	85	90
4-(2′-Thiazolylazo)-phenol (TAP)	55	75	85
2-(2′-Thiazolylazo)-phloroglucinol (TAPh)	0	50	60(l) 80(vs)
2-(2′-Thiazolylazo)-pyrogallol (TAPyg)	0	0–60 (str)	80–95 (str)
4-(2′-Pyridylazo)-resorcinol (PAR)	25 (str)[a]	70	80
1-(2′-Pyridylazo)-2-naphthol (PAN)	68	95	99
4-(Phenylazo)-resorcinol (BAR)	70	65(l) 95(s)	90(l) 97(s)
Phenylazo-*m*-cresol (BAmC)	75	90	90
Benzene-*p*-sulfonic acid-azo-resorcinol (BAR-S)	0	40	60(vs) 70(l)
Benzene-*p*-sulfonic acid-azo-*m*-cresol (BAC-S)	0	65	70
2-(Salicylideneamino)-pyridine (SAMPy)	95	85	100
2,6-Bis-(salicylideneamino)-pyridine (diSAMPy)	0	50–95 (str)	75–100 (str)
2-(2′-Hydroxyphenyliminomethyl)-pyridine (POAP)	60	80–95 (str)	95
2,6-Bis-(2′hydroxyphenyliminomethyl)-pyridine (BPOP)	15	95 (str)	95
1-Hydroxy-(2′-naphthylideneamino)-pyridine (NAMPy)	70	95	98
2-(2′-Thiazolylazo)-phenol-4-sulfonic acid (TAP-S)	0	90	90
2-(2′-Thiazolylazo)-1-naphthol-4-sulfonic acid (TAN α4S)	0	95(l) 45(s)	90(l) 60(s) 50(vs)
(2′-Thiazolylazo)-1-naphthol-5-sulfonic acid (TAN α5S)	0	55	65
1-(2′-Thiazolylazo)-2-naphthol-6-sulfonic acid (TAN β6S)	0	55(l) 70(s) 80(vs)	70(l) 90(s)
1-(2′-Thiazolylazo)-2-naphthol-7-sulfonic acid (TAN β7S)	0	65	70
1-(2′-Thiazolylazo)-2-naphthol-8-sulfonic acid (TAN β8S)	0	55–75 (str)	70
(2′-Thiazolylazo)-1-naphthol-3,6-disulfonic acid (TAN α36S)	0	25(l) 40(s) 37(vs)	40(s) 50(l) 70(s)
1-(2′-Thiazolylazo)-2-naphthol-3,6-disulfonic acid (TAN β36S)	0	40	50
1-(2′-Thiazolylazo)-2-naphthol-6,8-disulfonic acid (TAN β68S)	0	0–40 (str)	40(l) 55(s) 70(vs)
Eriochrome Black T	0	45–60 (str)	80

[a] Terms in parentheses refer to type of spot; l = large; s = small; vs = very small; str = streaky.

Layer L1 = Silica gel G (Merck) with starch binder, activated at 110 °C for 20 minutes.
Solvent S1 = petroleum ether (40–60 °C)-diethyl ether-absolute ethanol (50 : 50 : 5).
S2 = *n*-butanol-absolute ethanol-2*N* aqueous ammonium hydroxide (60 : 20 : 20).
S3 = isopropanol-methyl ethyl ketone-conc. ammonia (40 : 30 : 30).
Detection Ethanolic solution of cupric acetate.
Literature
1. Pollard, F. H., Nickless, G., Samuelson, T. J., and R. G. Anderson, *J. Chromatogr.*, **16**, 231–233 (1964).

TABLE TLC 48
DYES, OIL-SOLUBLE

Layer		L1	L1	L1	L1	L1	L2	L2	L2	L3	L4
Solvent		S1	S2	S3	S4	S5	S6	S7	S8	S9	S10
Technique		na	na	na	na	T1	T1	T1	T1	T1	T1
Literature		1	1	1	1	1	2	2	2	2	2
Compound	C.I.[a]	$R_F \times 100$									
Azobenzene	—	64	63	45	88	65	—	—	—	—	—
Butter yellow (Dimethylazobenzene)	11,020	49	27	10	70	47	68	68	57	59	85
Ceres Red G	12,150	—	—	—	—	—	18	30	36	19	16
Ceres Yellow	12,700	—	—	—	—	—	54	74	75	56	54
Martius Yellow	10,315	—	—	—	—	—	0	28	0	0	0
***p*-Methoxyazobenzene**	—	54	40	22	70	55	—	—	—	—	—
Oil Orange SS	12,100	42	20	16	64	41	—	—	—	—	—
Oil Red OS	26,125	35	10	7	58	34	—	—	—	—	—
Quinoline Yellow SS	47,007	3	0	0	18	5	—	—	—	—	—
Solvent Yellow	11,000	28	6	3	37	18	—	—	—	—	—
Solvent Yellow 7	11,800	12	4	1	11	8	—	—	—	—	—
Sudan I	12,055	42	23	16	64	41	68	77	70	56	63
Sudan II	12,140	42	20	16	64	41	72	78	67	62	44
Sudan III	26,100	41	140	7	59	34	56	68	61	41	15
Sudan IV	26,105	42	145	10	63	38	56	68	61	38	15
Sudan G (Ceres Orange GN)	11,920	5	2	0	5	4	14	36	37	0	0
Yellow AB	11,380	62	50	34	82	64	25	46	41	22	88
Yellow OB	11,390	62	55	40	82	64	27	50	49	27	87
Yellow XP	12,740	—	—	—	—	—	60	81	80	68	40

[a] Color Index for the Society of Dyers and Colourists, Bradford, 1956.

Layer
L1 = Silica gel.
L2 = Silica gel G.
L3 = Aluminum oxide G.
L4 = Kieselguhr G.

Solvent
S1 = 1,2-dichloroethane.
S2 = xylene.
S3 = pentachloroethane.
S4 = chloroform.
S5 = 1,1,2-trichloroethane.
S6 = hexane-ethyl acetate (9 : 1).
S7 = petroleum ether-ether-acetic acid (70 : 30 : 1).
S8 = petroleum ether-ether-ammonia (70 : 30 : 1).
S9 = hexane-ethyl acetate (98 : 2).
S10 = cyclohexane.

Detection Visible light.

Technique T1 = Development distance, 10 cm.

Literature
1. Fujii, S. and M. Kamikura, *Shokuhin Eiseigaku Zasshi*, **4**, 96, 135 (1963); *Chem. Abstr.*, **59**, 11691 (1963).
2. Peereboom, J. W. C., *Chem. Weekbl.*, **57**, 625 (1961).

TABLE TLC 49
DYES, SYNTHETIC FOOD DYES

Layer	L1	L1	L2	L2	L3	L3	L4	L4
Solvent	S1	S1	S1	S1	S1	S1	S1	S1
Technique	T1	T2	T1	T2	T1	T2	T1	T2
Literature	1	1	1	1	1	1	1	1
Compound	$R_F \times 100$							
Chocolate Brown FB	0	0	0	0	0	0	0	0
Chocolate Brown HF	0	0	0	0	0	0	0	0
Brown FK	18	19	10	18				
	33	42	21	31	42	53	39	
	47	58	33	58	53	80	56	88
Black PN	16	15	4	11	7	10	9	25
Violet BNP	32	39	21	30	33	63	41	81
Indigo Carmine	—	—	—	—	—	—	—	—
Blue VRS	31	40	19	31	31	49	25	80
Green S	12	17	6	13	8	17	8	28
Yellow 2G	19	19	8	14	15	20	20	—
Naphthol Yellow	34	52	23	35	41	68	48	85
Tartrazine	15	15	5	11	5	8	10	21
Yellow RFS	27	44	19	27	22	38	36	70
	47	66	35	61			59	
Sunset Yellow	30	38	18	29	33	49	28	86
Yellow RY	3	4	—	0	6	7		
	1	1	—	2	12	17	15	8
Orange G	28	44	18	26	27	51	36	72
Orange RN	51	70	35	68	58	86	59	95
Erythrosine BS	53	67	35	69	56	84	58	94
Ponceau MX	25	33	14	26	31	41	27	
	53	65	36	68		87	64	88
Ponceau SX	25	37	15	23	27	38	35	67
	33	44	22	29				
	38	54	27	39				
Red 2G	30	33	16	25	30	43	30	73
Ponceau 4R	14	17	3	8	12	15	11	42
Fast Red E	36	44	23	36	41	63	37	83
Ponceau 3R	25	39	15	29	21	53	32	72
Amaranth	16	17	78	14	8	14	15	23
Red 10B	25	39	16	24	25	44	33	78
Red FB	36	53	23	28	32	48	52	80
Red 6B	20	32	14	22	16	29	23	47
Carmoisene	25	34	15	26	13	32	29	44

Layer L1 = Laboratory prepared silica gel.
L2 = Precoated silica gel plates (Merck), both L1 and L2 activated 1.5 hours at 160 °C, cooled.
L3 = Precoated layers (MN) with starch binder.
L4 = Precoated layers (MN) without starch binder.

Solvent S1 = *n*-butanol-ethyl methyl ketone-conc. ammonia-water (5 : 3 : 1 : 1 v/v).

Detection Visible light.

Technique T1 = sandwich chamber.
T2 = tank, saturated atmosphere, 12.5 cm development in both cases.

Literature
1. Graham, R. J. T. and A. E. Nya, *Fifth International Symposium on Chromatography and Electrophoresis*, Belgian Society of Pharmaceutical Sciences, Brussels, 1968, Ann Arbor-Humphrey Science Publ. Inc., Ann Arbor, Mich., 1969, p. 489.

TABLE TLC 50
HYDROCARBONS, ANTHRACENE DERIVATIVES

Layer	L1	L1	L1		
Solvent	S1	S2	S3		
Detection	—	—	—	D1	D2
Literature	1	1	1		
Compound	$R_F \times 100$			Color	
7,12-Dimethylbenz[*a*]anthracene	95[a]	98	98	vt	vt
7-Hydroxymethyl-12-methylbenz[*a*]anthracene	22	62	72	vt	vt
12-Hydroxymethyl-7-methylbenz[*a*]anthracene	40	78	85	vt	vt
7,12-Dihydroxymethylbenz[*a*]anthracene	0	28	45	vt	vt
12-Methylbenz[*a*]anthracene-7-carboxaldehyde	62	98	98	y gr	y gr
12-Methylbenz[*a*]anthracene-7-carboxylic acid	0	0	5	vt	vt
Methyl 12-methylbenz[*a*]anthracene-7-carboxylate	69	98	98	bl	bl
***cis*-5,6-Dihydro-5,6-dihydroxy-7,12-dimethylbenz[*a*]-anthracene**	0	23	38	dk vt	dk vt
***trans*-5,6-Dihydro-5,6-dihydroxy-7,12-dimethylbenz[*a*]-anthracene**	0	20	32	dk vt	dk vt
***cis*-5,6-Dihydro-5,6-dihydroxy-7-hydroxymethyl-12-methylbenz[*a*]anthracene**	0	5	25	dk vt	dk vt
***cis*-7-Acetoxymethyl-5,6-dihydro-5,6-dihydroxy-12-methylbenz[*a*]anthracene**	0	25	38	dk vt	dk vt
8,9-Dihydro-8,9-dihydroxy-7,12-dimethylbenz[*a*]anthracene	0	16	30	dk vt	dk vt
8,9-Dihydro-8,9-dihydroxy-7-hydroxymethyl-12-methyl-benz[*a*]anthracene[b]	0	5	21	dk vt	dk vt
3-Hydroxy-7,12-dimethylbenz[*a*]anthracene	28	70	79	vt	y gr
4-Hydroxy-7,12-dimethylbenz[*a*]anthracene	38	76	82	vt	pk
3-Hydroxy-7-hydroxymethyl-12-methylbenz[*a*]anthracene	0	22	39	vt	y gr
4-Hydroxy-7-hydroxymethyl-12-methylbenz[*a*]anthracene	0	26	42	vt	pk

[a] R_F 54 in petroleum ether (b.p. 80–100 °C)-benzene (19 : 1).
[b] The isomeric 12-hydroxymethyl-7-methyl derivatives had identical properties.

Layer L1 = Silica gel G (Merck).
Solvent S1 = benzene.
S2 = benzene-ethanol (19 : 1).
S3 = benzene-ethanol (9 : 1).
Detection D1 = Fluorescence in UV light.
D2 = Fluorescence in UV light after exposure to ammonia.
Literature 1. Boyland, E. and P. Sims, *Biochem. J.*, **104**, 398 (1967).

TABLE TLC 51
HYDROCARBONS, AROMATIC

Layer	L1	L1	L1	L1	L1	L2	L2	L3	L4	L5
Solvent	S1	S2	S3	S1	S4	S1	S4	S5	S6	S7
Technique	T1	T1	T1	T2	T2	T2	T2	T3	T3	T4
Literature	1	1	1	2	2	2	2	3	3	4
Compound	$R_F \times 100$							R_β[a] ×100	R_β[a] ×100	R_F ×100
Naphthalene	50	40	94	59	76	—	85	—	—	35
1-Methylnaphthalene	40	46	94	54	74	55	78	—	—	—
2-Methylnaphthalene	46	47	93	50	87	52	85	—	—	49
1,2-Dimethylnaphthalene	48	56	94	—	—	—	—	—	—	—
1,3-Dimethylnaphthalene	42	53	95	—	—	—	—	—	—	—
1,4-Dimethylnaphthalene	55	54	95	—	—	—	—	—	—	—
1,5-Dimethylnaphthalene	44	47	93	50	72	46	85	—	—	—
1,6-Dimethylnaphthalene	47	47	94	54	75	46	83	—	—	—
1,7-Dimethylnaphthalene	50	52	95	—	—	—	—	—	—	—
2,3-Dimethylnaphthalene	36	40	95	48	75	39	81	—	—	—
2,6-Dimethylnaphthalene	46	44	93	50	72	52	85	—	—	—
2,7-Dimethylnaphthalene	45	44	94	53	—	42	80	—	—	—
2-Ethylnaphthalene	55	52	93	—	—	—	—	—	—	—
2,3,5-Trimethylnaphthalene	50	47	93	—	—	—	—	—	—	—
2-Phenylnaphthalene	34	38	92	—	—	—	—	—	—	—
Indene	—	—	—	55	84	67	81	—	—	—
Biphenyl	45	45	95	45	74	58	79	—	—	30
3,5-Dimethylbiphenyl	40	42	93	—	—	—	—	—	—	—
4,4′-Dimethylbiphenyl	39	44	94	—	—	—	—	—	—	—
4,5-Benzindane	50	50	96	—	—	—	—	—	—	—
Acenaphthene	52	53	95	44	83	44	70	—	—	—
Fluorene	40	40	95	32	66	35	73	—	—	—
Phenanthrene	36	40	93	33	65	20	66	199	113	—
1-Methylphenanthrene	29	32	94	—	—	—	—	—	—	—
Anthracene	40	42	95	—	—	—	—	199	114	34
2-Methylanthracene	40	40	94	—	—	—	—	—	—	—
Pyrene	35	38	93	32	64	10	65	172	125	27
Fluoranthene	30	25	92	29	66	10	55	189	109	—
Chrysene	18	18	94	0	11	0	0	175	110	—
3,4-Benzo[*a*]pyrene	—	—	—	17	65	0	10	—	—	—
Tetrahydronaphthalene	—	—	—	—	—	—	75	—	—	—
1,2,4,5-Tetramethylbenzene	—	—	—	62	76	—	85	—	—	—
Triphenylene	—	—	—	—	—	—	—	149	107	—
Benz[*a*]anthracene	—	—	—	—	—	—	—	147	103	—
11 H-Benz[*b*]fluorene	—	—	—	—	—	—	—	133	108	—
Benz[*e*]pyrene	—	—	—	—	—	—	—	116	104	—
Perylene	—	—	—	—	—	—	—	114	91	—
Benz[*k*]fluoranthene	—	—	—	—	—	—	—	103	98	—
Benz[*a*]pyrene	—	—	—	—	—	—	—	100	100	—
Anthanthrene	—	—	—	—	—	—	—	70	71	—
sec-Butylbenzene	—	—	—	—	—	—	—	—	—	61
tert-Butylbenzene	—	—	—	—	—	—	—	—	—	70
Camphene	—	—	—	—	—	—	—	—	—	74
Cedrene	—	—	—	—	—	—	—	—	—	82
Cumene	—	—	—	—	—	—	—	—	—	62
p-Cymene	—	—	—	—	—	—	—	—	—	38
Dicyclopentadiene	—	—	—	—	—	—	—	—	—	72
Heptene	—	—	—	—	—	—	—	—	—	87
Hexyne	—	—	—	—	—	—	—	—		43
Limonene		—	—	—	—	—	—	—	—	41
Mesitylene	—	—	—	—	—	—	—	—	—	64
Myrcene	—	—	—	—	—	—	—	—	—	56
4-Nonene		—	—	—	—	—	—	—	—	94
Octadecene	—	—	—	—	—	—	—	—	—	88
α-Pinene	—	—	—	—	—	—	—	—	—	83
β-Pinene	—	—	—	—	—	—	—	—	—	80
Stilbene	—	—	—	—	—	—	—	—	—	20
Styrene	—	—	—	—	—	—	—	—	—	55
Terpinolene	—	—	—	—	—	—	—	—	—	64
Xylene	—	—	—	—	—	—	—	—	—	65

[a] R_F relative to benz[*a*]pyrene.

TABLE TLC 51—(Continued)
HYDROCARBONS, AROMATIC

Layer L1 = Silica gel PHH (loose-layer).
L2 = Neutral Al_2O_3 (loose-layer) (Act. I–II).
L3 = Cellulose powder MN-300.
L4 = Aluminum oxide G.
L5 = Silicic acid chromostrips using 5% starch binder, dried 105 °C for 15 minutes, then ½ hour at 3mm Hg over phosphorus pentoxide.

Solvent S1 = *n*-hexane.
S2 = cyclohexane.
S3 = benzene.
S4 = CCl_4.
S5 = dimethylformamide-water (1 : 1).
S6 = pentane-ether (19 : 1).
S7 = hexane (benzene free).

Detection See Section II.I, Volume Two.

Technique T1 = layer thickness 0.6–0.9 mm.
T2 = 20–45 minutes development time.
T3 = saturated chamber, ascending development for 15 cm (1½ hours) in dark.
T4 = 10 cm development distance, ascending.

Literature
1. Janak, J., *J. Chromatogr.*, **15**, 15 (1964).
2. Kucharczyk, N., Fohl, J., and J. Vymetal, *J. Chromatogr.*, **11**, 55 (1963).
3. Sawicki, E., Stanley, T. W., Elbert, W. C., and D. Pfaff, *Anal. Chem.*, **36**, 497 (1964).
4. Kirchner, J. G. and J. M. Miller, *Ind. Eng. Chem.*, **44**, 318 (1952).

TABLE TLC 52
INORGANICS, CATIONS

Layer	L1	L1			
Solvent	S1	S2			
Detection	—	—	D1	D2	—
Literature	1	1			
Ion	$R_F \times 100$		Color		Sensitivity of $TCNQ^-$, μg
Li	98	—	y	—	—
Na	89	39	o	bt bl	0.5
K	60	20	r	bt-bl	0.5
Rb	53	17	r-vt	bl	2
Cs	38	10	r-vt	gy	2
Be	99	100[a]	y-gr	y	20
Mg	95	96[a]	y	ol-gr	20
Ca	94	93[a]	o	gy-gr	20
Sr	93	15	r-pk	pl-gr	50
Ba	90	0	pk	bl	50
Cu^{++}	80[b]	95	—	gr	25
Ag	89[b]	38[b]	—	bt-bl	0.1
Zn	90	100[b]	—	bl	10
Cd	100	100	—	bl	10
Hg^+	98	0	—	bt-y	10
Hg^{++}	100	97	—	bt-y	10
Tl^+	30[c]	8	r	gy	40
Pb^{++}	90	2	o	bl	75

[a] Very narrow band. [b] Tailing. [c] Comet or long trail.

Layer L1 = deionized silica gel.

Solvent S1 = acetic acid-95% ethanol (5 : 95).
S2 = 2*N* HCl-*tert*-butanol (5 : 95).

Detection D1 = violuric acid.
D2 = 0.5% lithium tetracyanoquinodimethanide in 1 : 1 ethanol-water.

Technique Development distance 10 cm.

Literature
1. Druding, L. F., *Anal. Chem.*, **35**, 1583 (1963).

TABLE TLC 53
HYDROCARBONS, AROMATIC

Layer	L1
Solvent	S1
Literature	1
Compound	$R_F \times 100$
Hexadecylbenzene	74
Dodecylbenzene	70
Octylbenzene	69
Hexylbenzene	65
1,3,5-Triethylbenzene	65
Phenylcyclohexane	61
1,2,4-Trimethylbenzene	60
1-Propyl-2,4,6-trimethylbenzene	57
Pentamethylbenzene	53
1,2,4,5-Tetramethylbenzene	51
Tetrahydronaphthalene	49
2-Octadecylnaphthalene	65
Naphthalene	50
Acenaphthene	48
1,4-Dimethylnaphthalene	48
1,5-Dimethylnaphthalene	48
1,7-Dimethylnaphthalene	47
1,3-Dimethylnaphthalene	47
1,6-Dimethylnaphthalene	45
1-Methylnaphthalene	45
1-Ethylnaphthalene	45
2-Methylnaphthalene	44
2,4,6-Trimethylnaphthalene	43
4-Methyldiphenyl	43
2,6-Dimethylnaphthalene	42
3-Methyldiphenyl	42
4,4′-Dimethyldiphenyl	42
2,3-Dimethylnaphthalene	40
Diphenyl	40
Fluorene	38
Diphenylmethane	35
1-Phenylnaphthalene	39
9-Methylanthracene	38
2-Methylanthracene	38
9,10-Dimethylanthracene	38
Anthracene	34
Phenanthrene	34
2-Phenylnaphthalene	33
2,3-Benzofluorene	29
1,2-Benzofluorene	29
1,4-Diphenylbenzene	29
Pyrene	33
3-Methylpyrene	33
Chrysene	32
2,2′-Dinaphthyl	28
1,2-Dihydronaphthalene	46
9,10-Dihydrophenanthrene	35
Fluoranthene	31
Perylene	29

Layer L1 — Silica gel G.
Solvent S1 = diisobutylene
Detection Iodine vapors; UV light.
Literature 1. Pawlowska-Marzec, A., *Chem. Anal.* (Warsaw), **13**, 475–76 (1968).

TABLE TLC 54
HYDROCARBONS, TERPENE

Layer	L1	L1	L1	L1	L1	L2	L2
Solvent	S1	S2	S3	S4	S5	S6	S1
Technique	T1	T1	T1	T1	T1	T2	T2
Literature	1	1	1	1	1	2	2
Compound	$R_F \times 100$					R_{st}[a]	R_A[b]
β-Bisabolene	—	—	—	—	—	93.6	—
Camphene	84	76	79	80	79	—	—
α-Caryophyllene	50	35	47	33	27	18.9	—
β-Caryophyllene	62	60	52	65	75	42.2	—
γ-Caryophyllene	80	82	87	90	83	—	—
Cedrene	82	80	83	85	78	—	—
Copaene	—	—	—	—	—	116.1	—
Cuparene	—	—	—	—	—	—	539.1
p-Cymene	38	41	62	69	57	—	—
β-Elemene	—	—	—	—	—	33.1	—
α-Gurjunene	—	—	—	—	—	116.8	—
α-Himachalene	—	—	—	—	—	—	519
β-Himachalene	—	—	—	—	—	—	315.5
ar-Himachalene	—	—	—	—	—	116.4	—
Isolongifolene	—	—	—	—	—	—	744.7
Limonene	41	55	54	59	74	—	—
Longicyclene	—	—	—	—	—	—	839.9
Longifolene	—	—	—	—	—	—	535.7
α-Pinene	83	85	84	90	80	—	—
β-Pinene	80	75	80	85	75	—	—
β-Selinene	—	—	—	—	—	80.5	—
Terpinolene	64	67	60	65	89	—	—
Thujopsene	—	—	—	—	—	114.9	—

[a] $R_{st} = R_F$ relative to Sudan III.
[b] $R_A = R_F$ relative to azobenzene.

Layer L1 = Silicic acid chromostrips.
L2 = Silica gel impregnated with silver nitrate.

Solvent S1 = hexane.
S2 = 2,2-dimethylbutane.
S3 = cyclohexane.
S4 = methylcyclohexane.
S5 = isopentane.
S6 = benzene-acetone (95 : 5).

Detection See Section II.I, Volume Two.

Technique T1 = 10 cm development distance.
T2 = saturated chamber, ascending, 10 cm, 20–25 minutes.

Literature
1. Miller, J. M. and J. G. Kirchner, *Anal. Chem.*, **25**, 1107 (1953).
2. Gupta, A. S. and S. Dev, *J. Chromatogr.*, **12**, 189 (1963).

TABLE TLC 55
INORGANICS, NOBLE METALS

Layer	L1	L1	L2	L2	—	—	—	L1	L1	L1	
Solvent	S1	S2	S1	S2	—	—	—	S3	S4	S5	
Detection											D1
Technique	T1	T1	T1	T1	—	—	—	T2	T2	T2	
Literature	1	1	1	1	—	—	—	2	2	2	
Metal	$R_F \times 100$				Detection limit (μg)		Color	$R_F{}^a \times 100$			Color
Rhodium	00	—	—	—	D1	1.3	o	—	—	—	—
Ruthenium	28	79	62	68	D2	2.0	bl	—	—	—	—
Platinum	30	93	66	80	D2	0.2	p-r	56	30	10	y
Gold	51	—	—	—	D2	1.0	gr-ish bk	0	0	0	y
Osmium	62	—	—	—	D3	0.3	pk	—	—	—	—
Palladium	90	76	83	82	D2	0.1	pl-o	83	50	20	y
Iridium	95	—	—	—	in S1	3.0	pl-y	—	—	—	—
Copper	100	100	89	100	D2	0.1	bk	—	—	—	—
Tellurium	—	—	—	—	—	—	—	95	70	50	bk
Selenium	—	—	—	—	—	—	—	86	60	33	o

[a] Metals in form of their diethyldithiocarbamate complexes.

Layer
L1 = Alumina S.
L2 = Silica gel G.

Solvent
S1 = acetone-acetylacetone-2*M* hydrochloric acid (100 : 10 : 3).
S2 = acetone-water-conc. nitric acid (5 : 5 : 1).
S3 = benzene.
S4 = toluene.
S5 = xylene.

Detection
D1 = stannous chloride, KI.
D2 = Rubeanic acid.
D3 = Thiourea.

Technique
T1 = circular.
T2 = metals in form of diethyldithiocarbamate complexes, development time, 15 to 20 min.

Literature

1. Hashmi, M. H. and A. S. Adil, *Z. Anal. Chem.*, **227**, 172 (1967); from *J. Chromatogr.*, **30**, D15 (1967).
2. Rai, J. and V. P. Kukreja, *Chromatografia*, **3**, 501 (1970).

TABLE TLC 56
LIPIDS

Layer	L1	L1	L1	L1	L1	L1
Solvent	S1	S2	S3	S4	S5	S6
Technique	T1	T1	T1	T1	T1	T1
Literature	1	1	2	1	2	3
Compound	$R_F{}^a \times 100$					
Squalene	92	94	—	90	100	—
Hexadecyl palmitate	61	93	—	91	100	—
Vitamin A palmitate	0	46	—	87	100	—
Cholesteryl palmitate	63	90	93	93	100	—
Cholesteryl acetate	—	—	96	—	100	—
α-Tocopheryl acetate	32	83	—	91	100	—
α-Tocopherol	42	79	—	91	100	—
Tripalmitin	—	—	93	—	100	—
Triglycerides (unsaturated)	0	65	89	89	100	—
Hexadecanol	16	65	—	64	100	—
Vitamin D_3	7	55	—	74	100	—
Vitamin A alcohol	0	0	—	0	100	—
Stearic acid, linoleic acid	—	—	—	—	100	95
Cholesterol	8	44	67	69	100	—
7-Dehydrocholesterol	4	0, 38	—	48	100	—
1,3-Dipalmitin	—	—	53	—	100	—
1,3-Diolein	—	—	48	—	100	—
1-Monopalmitin	0	—	11	—	100	—
1-Monoölein	0	—	11	—	100	—
Phosphatidylethanolamines	0	0	0	—	94	85
Lecithins (from egg)	0	0	0	0	—	66
Lecithin (dipalmitoleyl-)	0	0	0	0	50	—
Lecithin (distearyl-)	0	0	0	0	46	—
Lecithin (dimyristyl-)	0	0	0	0	—	65
Sphingomyelins	0	0	0	0	40	50
Lysolecithins	0	0	0	0	—	31
Lysolecithin (2-palmitoleyl-)	0	0	0	0	23	—
Lysolecithin (2-stearyl)	0	0	0	0	18	—

[a] R_F values differ slightly on different batches of impregnated paper. Mixtures of compounds with close R_F values (*e.g.* the lysolecithins) have not been resolved.

Layer L1 = Silicic acid-impregnated glass fiber paper.

Solvent S1 = cyclohexane.
S2 = 2% ether in isooctane.
S3 = 5% ether in isooctane.
S4 = 1% methanol in cyclohexane.
S5 = 50% methanol in ether.
S6 = phenol (200 g) + ether (400 ml) + acetone (250 ml) + water (50 ml).

Detection H_2SO_4 or Dragendorff's reagent.

Technique T1 = ascending, saturated tank.

Literature
1. Sorrels, M. F. and R. Reiser, *J. Amer. Oil Chem. Soc.*, **34**, 131 (1957).
2. Dieckert, J. W. and R. Reiser, *J. Amer. Oil Chem. Soc.*, **33**, 123 and 535 (1956).
3. Brown, M., Yeadon, D. A., Goldblatt, L. A., and J. W. Dieckert, *Anal. Chem.*, **29**, 30 (1957).

TABLE TLC 57
LIPIDS

Layer	L1	L1	L1	L1	L1
Solvent	S1	S2	S3	S4	S5
Literature	1	1	1	1	1
Compound			$R_F \times 100$		
Cholesterol oleate	15	45	87	91	—
Cholesterol palmitate	20	47	83	96	—
Methyl stearate	17	40	72	99	—
Tripalmitin	0	15	62	99	—
Stearic acid	3	13	43	99	—
Myristic acid	1	10	41	99	—
Cholesterol acetate	—	—	65	63	—
β-Sitosterol acetate	—	—	69	55	—
Triolein	—	10	57	43[a]	95
Cholesterol	—	—	—	28	95
β-Sitosterol	—	—	—	26	92
Oleic acid	—	5	35	26[a]	92
Ergosterol	—	—	—	27	86
Phosphatidyl-ethanolamine	—	—	—	0	35
Erythro-sphingosine	—	—	—	0	34
Erythro-dihydro-sphingosine	—	—	—	0	30
Dipalmitoyl-phosphatidylcholine	—	—	—	0	29
Sphingomyelin	—	—	—	0	25

[a] These unexpected values were obtained repeatedly on purified samples from different sources.

Layer L1 = Test tubes (13 mm × 150 mm) lined with silicic acid.
Solvent S1 = *n*-hexane.
S2 = ethyl ether-*n*-hexane (3 : 97 v/v).
S3 = ethyl ether-*n*-hexane (15 : 85 v/v).
S4 = chloroform.
S5 = methyl alcohol-water-28% concentrated ammonium hydroxide (93 : 5 : 2).
Detection Iodine staining.
Technique 5 μg in 1 μl suitable solvent applied, ascending development, 25 °C, 10 cm.
Literature
1. Lie, K. B. and J. F. Nye, *J. Chromatogr.*, **8**, 75 (1962).

TABLE TLC 58
NITROGEN HETEROCYCLICS

Layer	L1	L1	L1	L1	L2	L2	L1
Solvent	S1	S2	S3	S4	S1	S2	S5
Technique	T1	T1	T1	T1	T2	T2	na
Literature	1	2	2	2	3	3	4
Compound	$R_F \times 100$						
Acridine	19	—	—	—	—	0	—
3,4-Benzacridine	75	—	—	—	—	—	—
3,4-Benzoquinoline	19	—	—	—	—	—	—
5,6-Benzoquinoline	13	—	—	—	—	—	—
7,8-Benzoquinoline	47	—	—	—	—	1	—
Carbazole	—	—	—	—	75	74	—
1,3-Dimethylisoquinoline	—	—	—	—	—	—	40
2,3-Dimethylquinoline	—	—	—	—	—	—	47
2,4-Dimethylquinoline	8	—	—	—	—	—	—
2,6-Dimethylquinoline	11	—	—	—	—	4	—
2,8-Dimethylquinoline	47	—	—	—	—	—	—
Indole	—	—	—	—	63	65	—
Isoquinoline	12	7	22	—	—	3	39
2-Methylcarbazole	—	—	—	—	70	77	—
2-Methylindole	—	—	—	—	60	65	—
3-Methylindole	—	—	—	—	70	65	—
5-Methylindole	—	—	—	—	65	65	—
7-Methylindole	—	—	—	—	65	64	—
1-Methylisoquinoline	—	—	—	—	29[a]	2	37
3-Methylisoquinoline	11	—	—	—	—	2	37
1-Methyl-3-phenylisoquinoline	—	—	—	—	—	—	75
2-Methyl-3-phenylquinoline	—	—	—	—	—	—	63
2-Methylquinoline	16	4	27	89	40[a]	4	48
3-Methylquinoline	—	—	—	—	—	—	46
4-Methylquinoline	13	2	22	82	40[a]	4	—
6-Methylquinoline	18	5	26	86	—	—	—
7-Methylquinoline	16	—	—	—	—	3	—
8-Methylquinoline	31	17	53	94	38[a]	3	—
Quinoline	11	9	30	—	—	2	47

[a] Tailing from origin.

Layer L1 = Silica gel G (Merck).
L2 = Loose-layer silica gel PHH (Spolana N.E.).

Solvent S1 = chloroform.
S2 = benzene.
S3 = benzene-methanol (95 : 5).
S4 = methanol.
S5 = benzene-ethyl acetate (1 : 1).

Detection See Section II.I, Volume Two.

Technique T1 = development distance = 10 cm.
T2 = mirror glass plates 100 × 200 mm and 200 mm × 20 mm used in glass chromatographic chamber; slope of coated plates was 20–30°.

Literature
1. Petrowitz, H-J., Pastuska, G., and S. Wagner, *Chem. Ztg.*, **89**, 7 (1965).
2. Petrowitz, H-J., *Chem. Ztg.*, **85**, 143 (1961); Petrowitz, H-J., *Materialpruefung*, **2**, 309 (1960).
3. Janak, J., *J. Chromatogr.*, **15**, 15 (1964).
4. Troszkiewicz, C., Suwinski, J., and W. Zielinski, *Chem. Anal.* (Warsaw), **13**, 4 (1968).

TABLE TLC 59
NITROGEN HETEROCYCLICS, INDOLES, 2,4-DINITROPHENYLTHIO DERIVATIVES

Layer	L1	L1	L1	L1	L1	L1	L1	L1	L1	L1	L1
Solvent	S1	S2	S3	S4	S5	S6	S7	S8	S9	S10	S11
Literature	1	1	1	1	1	1	1	1	1	1	1
Reaction product of DNPCl with	R_F[a] × 100										
3-Indoleacetamide	56	22	49	27	86	77	45	9	—	—	—
Ethyl 3-indoleacetate	89	75	85	73	87	86	90	73	—	—	—
3-Indoleacetonitrile	78	55	80	63	86	87	80	57	—	—	—
Tryptophol	75	32	85	73	85	84	76	74	—	—	—
3-Indoleacetone	90	62	83	60	90	85	85	64	—	—	—
Melatonin	70	28	50	25	90	78	53	7	—	—	—
3-Indoleacetic acid	20	7	13	8	76	66	9	~0	—	—	—
3-Indole-γ-butyric acid	44	16	36	~0	78	73	33	8	—	—	—
3-Indole-β-propionic acid	40T	15	~0	~0	81	72	29T	7	—	—	—
3-Indoleacetylaspartic acid	5	~0	~0	~0	31T	14T	~0	~0	86	61T	60T
3-Indole-L-lactic acid	7	~0	~0	~0	81	34	~0	~0	93	70	77
5-Hydroxy-3-indoleacetic acid	8	~0	7	~0	66	26	~0	~0	94	74	80
Tryptamine	9	7	37T	30T	68	42	10T	~0	65	57	63
Serotonin	8	~0	7	~0	63	24	~0	~0	57	26	46
***N*-Methyltryptamine**	8	~0	~0	~0	55	44	4	~0	59	62	65
Bufotenine	45	18	6	~0	58	47	16	~0	43T	53T	61T
3-Indoleacetyl-ε-L-lysine	~0	~0	~0	~0	45	10	~0	~0	24	11	15
2,4-Dinitrophenylsulfenyl chloride[b]	95	90	88	81	84	80	89	86	95	95	90
Ethyl 2,4-dinitrophenylsulfenate	97	92	88	81	84	81	91	86	94	94	91

[a] ~0 = spot remains on starting line or R_F value <3 usually with tailing from origin; T = Tailing.
[b] Streaking or multiple spot formation is observed in most solvents. The R_F value for the most intensely colored spot is reported in the Table.

Layer L1 = Eastman Chromagram silica thin layer sheets (6061).
Solvent S1 = diisopropyl ether-dimethylformamide (80 : 20).
S2 = diisopropyl ether-dimethylformamide (90 : 10).
S3 = hexane-acetone (50 : 50).
S4 = hexane-acetone (60 : 40).
S5 = chloroform-methanol (50 : 50).
S6 = benzene-methanol (75 : 25).
S7 = chloroform-methanol-carbon tetrachloride (10 : 1 : 9).
S8 = chloroform.
S9 = ethyl acetate-isopropanol-water-formic acid (65 : 25 : 5 : 2).
S10 = chloroform-methanol-acetic acid (80 : 15 : 2).
S11 = chloroform-methanol-acetic acid (70 : 30 : 0.5).
Detection Spots are yellow; sensitivity ~2 μg.
Technique Solution of indole derivative in CH_2Cl_2 and 99% HCO_2H is added to 2,4-dinitrophenylsulfenyl chloride in CH_2Cl_2, kept at 25 °C, 1–2 hours, and applied to silica sheets.
Literature 1. Raj, R. K. and O. Hutzinger, *J. Chromatogr.*, **44**, 199–201 (1969).

TABLE TLC 60
NITROGEN HETEROCYCLICS, PYRIDINES

Layer	L1	L1	L1
Solvent	S1	S2	S3
Literature	1	1	1
Compound		$R_F \times 100$	
2-Acetylpyridine	20	57	69
2-Aminopyridine	>0	27	50
3-Aminopyridine	>0	18	45
4-Aminopyridine	>0	5	14
2-Benzoylpyridine	9	62	71
2-Bromopyridine	26	63	72
3-Bromopyridine	13	57	67
2-Chloropyridine	28	61	70
3-Chloropyridine	16	56	65
2,4-Dimethylpyridine	4	28	49
2,6-Dimethylpyridine	6	36	59
2-Ethylpyridine	3	42	62
2-Fluoropyridine	24	62	69
2-Hydroxypyridine	0	6	20
3-Hydroxypyridine	0	23	53
4-Hydroxypyridine	0	0	2
3-Iodopyridine	13	58	70
2-Methylpyridine	3	30	54
3-Methylpyridine	6	35	55
4-Methylpyridine	4	27	48
2-*n*-Propylpyridine	12	47	64
Pyridine	4	29	54
Pyridine-2-aldehyde	11	51	67
Pyridine-3-aldehyde	4	33	58
Pyridine-4-aldehyde	5	36	56
Pyridine-2-carbinol	>0	18	45
Pyridine-3-carbinol	>0	13	39
Pyridine-4-carbinol	>0	4	39
Pyridine-2-carboxylic acid	0	2	4
Pyridine-3-carboxylic acid	0	6	6
Pyridine-4-carboxylic acid	0	5	5
Pyridine-2,6-dicarboxylic acid	4	3	5
2,4,6-Trimethylpyridine	2	26	51

Layer L1 = Silica gel (Merck).
Solvent S1 = chloroform.
S2 = ethyl acetate.
S3 = acetone.
Detection See Section II.I, Volume Two.
Technique Ascending development, 10 cm.
Literature 1. Petrowitz, H-J., Pastuska, G., and S. Wagner, *Chem. Ztg.*, **89**, 8 (1965).

TABLE TLC 61
NITROGEN HETEROCYCLICS, PYRIDINE DERIVATIVES

Layer	L1	L1	L1
Solvent	S1	S2	S3
Literature	1	1	1
Compound		$R_F \times 100$	
4-Amino-5-ethyl-2-methylpyridine *N*-oxide	30	14	13
4-Benzyloxy-5-ethyl-2-methylpyridine *N*-oxide	43	22	—
4-Butylmercapto-5-ethyl-2-methylpyridine *N*-oxide	32	14	9
4-Chloro-5-ethyl-2-methylpyridine	81	42	81
4-Chloro-5-ethyl-2-methylpyridine *N*-oxide	54	23	18
4-Chloro-3-methylpyridine	89	—	—
4-Chloro-3-methylpyridine *N*-oxide	43	30	—
2-Methyl-5-ethylpyridine	82	40	75
2-Methyl-5-ethylpyridine *N*-oxide	48	17	7
2-Methylpyridine	82	40	71
2-Methylpyridine *N*-oxide	33	11	4
3-Methylpyridine *N*-oxide	44	21	—
4-Nitro-5-ethyl-2-methylpyridine	81	71	80
4-Nitro-5-ethyl-2-methylpyridine *N*-oxide	73	55	53
4-Nitro-2-methylpyridine	82	70	80
4-Nitro-2-methylpyridine *N*-oxide	57	45	40
4-Nitro-3-methylpyridine *N*-oxide	62	51	—
4-Nitropyridine *N*-oxide	64	56	—
Pyridine	83	45	78
Pyridine *N*-oxide	24	9	3

Layer L1 = Aluminum oxide, activity II, 50–200 mesh.
Solvent S1 = benzene-methanol (9 : 1).
S2 = benzene-methanol (25 : 1).
S3 = benzene-dioxane (9 : 1).
Detection UV light; I_2 vapor.
Literature
1. Kost, A. N., Golovleva, L. A., Terent'ev, P. B., and M. Islam, *Zh. Anal. Khim.*, **21**, 860 (1966); from *J. Chromatogr.*, **26**, D4 (1967).

TABLE TLC 62
NUCLEOSIDES, NUCLEOTIDES AND OLIGONUCLEOTIDES

Layer	L1	L1	L1
Solvent	S1	S2	S3
Literature	1	1	1
Compound		$R_F \times 100$	
Adenosine	51	—	—
Adenosine-5′-phosphate	—	29	35
Adenylyl-(3′-5′)-uridine	—	60	—
Cytidine	24	—	—
Cytidine-5′-diphosphate	—	7	—
Cytidine-5′-phosphate	—	21	24
Deoxyadenosine-5′-phosphoric acid trichloroethyl ester	—	—	74
Deoxythymidine	69	—	—
Deoxythymidine-5′-phosphoric acid benzhydryl ester	37	—	—
Deoxythymidine-5′-phosphate	—	29	—
Deoxythymidine-5′-phosphoric acid trichloroethyl ester	37	—	—
Deoxythymidylyl-(3′-5′)-deoxythymidine	8	75	61
2′-O-Ethoxyethyleneuridylyl-(3′-5′)-2′,3′-O-ethoxymethyleneuridine	21	71	—
Guanosine	37	—	—
Guanosine-5′-phosphate	—	18	—
5-Hydroxymethyluridine	55	—	—
Inosine	36	72	—
Inosine-5′-phosphoric acid morpholidate	19	—	—
2′,3′-O-Isopropylideneadenosine	69	—	—
2′,3′-O-Isopropylidenecytidine	51	—	—
2′,3′-O-Isopropylideneguanosine	68	—	—
2′,3′-O-Isopropylideneinosine	66	—	—
2′,3′-O-Isopropylidene-4-thiouridine	92	—	—
2′,3′-O-Isopropylideneuridine	78	—	—
6-Methoxypurine riboside	64	—	—
1-Methylinosine	48	—	—
3-Methyl-4-thiouridine	80	—	—
3-Methylthymidine	71	—	—
5-Methyluridine (thymidine)	62	—	—
5-Methyluridine-5′-phosphate (thymidine-5′-phosphate)	—	19	37
5-Methyluridylyl-(3′-5′)-uridine	—	58	—
5′-O-Phosphoryldeoxythymidylyl-(3′-5′)-deoxythymidylyl-(3′-5′)-deoxythymidine	—	24	19
4-Thiomethylpyrimidin-2-one-1-(β-D-ribofuranoside)	64	—	—
4-Thiouridine	75	—	—
5′-O-Trichloroethylphosphoryldeoxyadenylyl-(3′-5′)-deoxyadenosine	6	76	68
5′-O-Trichloroethylphosphoryldeoxythymidylyl-(3′-5′)-deoxyguanylyl-(3′-5′)-deoxyguanosine	—	59	—
5′-O-Trichloroethylphosphoryldeoxythymidylyl-(3′-5′)-deoxythymidylyl-(3′-5′)-deoxyadenosine	—	73	—
5′-O-Trichloroethylphosphoryldeoxythymidylyl-(3′-5′)-deoxythymidylyl-(3′-5′)-deoxyguanosine	—	61	—
5′-O-Trichloroethylphosphoryldeoxythymidylyl-(3′-5′)-deoxythymidylyl-(3′-5′)-deoxythymidine	—	68	—
5′-O-Trichloroethylphosphoryldeoxythymidylyl-(3′-5′)-thymidine	13	—	—
Uridine	51	—	—
Uridine-5′-phosphate	—	21	30
Uridylyl-(3′-5′)-6-methoxypurine riboside	—	61	—
Uridylyl-(3′-5′)-1-methylinosine	—	58	—
Uridylyl-(3′-5′)-uridine	5	56	—

Layer L1 = Silica gel F_{254} (Merck).
Solvent S1 = water-saturated *n*-butanol.
S2 = isopropanol-ammonia-water (6 : 3 : 1).
S3 = *n*-butanol-acetic acid-water (5 : 2 : 3).
Detection Fluorescent indicator and UV light.
Technique Development distance = 6 cm.
Literature 1. Scheit, K. H., *Biochim. Biophys. Acta*, **134**, 217 (1967).

TABLE TLC 63
NUCLEOTIDES

Layer	L1	L1	L1	L1	L1	L1	L1	L1	L1	L1
Solvent	S1	S2	S3	S4	S5	S6	S7	S8	S9	S10
Technique	T1	T1	T1	T2	T2	T2	T2	T3	T3	T3
Literature	1	1	1	1	1	1	1	1	1	1
Compound*	$R_F \times 100$									
5′-AMP	11	52	65	68[a]	>80	>80	>80	>80	>80	>80
5′-IMP	13	59	74	40	60	73	>80	19	53	78
5 -GMP	6	40	51	28	45	57	65[b]	41	50[a]	72[a]
5′-CMP	15	64	75	70[a]	>80	—[c]	—	>80	>80	>80
5′-UMP	20	75	80	51	72	>80	—	20	64	>80
ADP	0	26	54	3	10	32	75[b]	3	29	70
IDP	0	30	63	0	4	14	49	0	8	55
GDP	0	17	45	0	2	9	34	0	13	61
CDP	0	33	64	8	20	45	>80	4	35	73
UDP	0	41	71	2	7	24	60	0	11	60
ATP	0	6	34	0	0	4	24	0	4	33
ITP	0	9	39	0	0	2	11	0	2	17
GTP	0	5	25	0	0	0	7	0	2	24
CTP	0	11	41	0	2	5	29	0	4	37
UTP	0	14	49	0	0	2	17	0	2	20
d-AMP	11	52	—							
d-GMP	6	41	—							
d-CMP	18	65	—							
TMP	24	74	—							
d-ATP	0	—	35							
d-GTP	0	—	26							
d-CTP	0	—	43							
TTP	0	—	52							
DPN	64	80	>80	68[a]	>80	>80	—	>80	>80	>80
DPNH	20	71	—							
TPN	3	60	—	14	43	>80	—	10	44[a]	80
TPNH	0	34	—							
ADPG	22	77	>80	30	57	>80	—	9	51[a]	>80
GDPM	12	72	>80	7	18	51	—	1	34	73
CDPG	27	>80	>80	45	67	>80	—	13	60	>80
UDPG	34	>80	>80	17	39[b]	77	—	1	27[b]	>80
UDPAG	44	>80	>80	32	55[b]	>80	—	2	38[b]	—
UDPGA	4	73	—	—	3	—	—	0	9	66

* For abbreviations, see Table TLC 160.
[a] Spot in second front.
[b] Elongated spot.
[c] — = not investigated.

Layer L1 = Poly(ethylenimine)-cellulose; preliminary ascending development with distilled H_2O, dry overnight.

Solvent
S1 = LiCl in water (0.25*M*).
S2 = LiCl in water (1.0*M*).
S3 = LiCl in water (1.6*M*).
S4 = formic acid-sodium formate pH 3.4 (0.5*M*).
S5 = formic acid-sodium formate pH 3.4 (1.0*M*).
S6 = formic acid-sodium formate pH 3.4 (2.0*M*).
S7 = formic acid-sodium formate pH 3.4 (4.0*M*).
S8 = 1.0*N* formic acid.
S9 = 2.0*N* formic acid-0.5*M* LiCl (1 : 1).
S10 = 2.0*N* formic acid-2.0*M* LiCl (1 : 1).

Detection Short wave UV light.

Technique
T1 = 0.5 mm thick layers; 10 cm development in saturated or open chamber; ascending, neutral pH.
T2 = 0.5 mm thick layers; 10 cm development in saturated or open chamber; ascending, at pH 3.4.
T3 = 0.5 mm thick layers; 10 cm development in saturated or open chamber; ascending, at pH 2.

Literature
1. Randerath, K. and E. Randerath, *J. Chromatogr.*, **16**, 111 (1964).

TABLE TLC 64
ORGANIC IODINE COMPOUNDS

Layer		L1	L1	L1	L1	L1	L1
Solvent		S1	S2	S3	S4	S5	S6
Literature		1	1	1	1	1	1
Compound	Trade name	$R_F \times 100$					
***N,N′*-Adipic-di-(3-amino-2,4,6-triiodo-benzoic acid)**	Biligrafin	9	27	33	0	0	33
***N,N′*-Diacetyl-3,5-diamino-2,4,6-triiodo-benzoic acid**	Gastrografin®	12	30	37	40	0	37
α-Ethyl-β-(2,4,6-triiodo-3-hydroxy-phenyl)-propionic acid	Teridax®	19	30	36	0	33	37
3-Acetylamino-2,4,6-triiodobenzoic acid	Triopac	26	40	47	0	0	44
Monoiodomethanesulfonic acid	Abrodil	28	43	48	0	0	44
3,5-Diiodo-4-pyridone-*N*-acetic acid	Ioduron	29	43	50	33	0	43
***N*-(3-Amino-2,4,6-triiodobenzoyl)-*N*-phenyl-β-aminopropionic acid**	Osbil	31	42	49	10	15	44
β-(3-Dimethylamino-methyleneamino-2,4,6-triiodophenyl)-propionic acid	Biloptin	35	43	50	40	28	50
α-(2,4,6-Triiodophenoxy)-butyric acid	Baygnostil	37	43	50	27	7	57
3,5-Diiodo-4-pyridone-*N*-acetic acid *n*-propyl ester	Propyliodan	87	63	71	0	60	67
3-Iodo-L-tyrosine		7	20	23	0	0	8
Iodogorgoic acid (3,5-diiodo-L-tyrosine)		7	20	23	0	0	8
3,5-Diiodo-L-thyronine		31	42	38	0	0	13
3,3′,5-Triiodo-L-thyronine		30	39	33	0	0	17
DL-Thyroxine		22	35	30	0	0	12
Potassium iodide		33	48	47	0	0	40
Potassium iodate		0	0	0	0	0	0

Layer L1 = Silica gel HF_{254} (Merck) + 2% starch.

Solvent S1 = ethyl acetate-isopropyl alcohol-25% NH_3 (55 : 35 : 20).
S2 = acetone-isopropyl alcohol-25% NH_3 (40 : 40 : 20).
S3 = isopropyl alcohol-25% NH_3 (80 : 20).
S4 = acetic acid-chloroform (5 : 95).
S5 = chloroform-methanol-pyridine (85 : 5 : 10).
S6 = ethyl acetate-methanol-diethylamine (50 : 40 : 20).

Detection 50% acetic acid and irradiation with UV light (2537 Å).

Literature 1. Stahl, E. and J. Pfeifle, *Z. Anal. Chem.* **200**, 382 (1964); from *J. Chromatogr.*, **18**, D16 (1965).

TABLE TLC 65
ORGANIC PEROXIDES

Layer	L1	L1	L1	L1	L1
Solvent	S1	S2	S3	S4	S5
Technique	T1	T1	T1	na	na
Literature	1	1	1	2	3
Compound	$R_F \times 100$				
Benzoyl peroxide	55	70	—	92	—
2,2-Bis(*tert*-butylperoxy)-butane	10	35	—	—	—
***sec*-Butylhydroperoxide**	—	—	—	66	—
***tert*-Butylhydroperoxide**	5	30	—	62	—
***tert*-Butylperacetate**	12	32	18	—	—
***tert*-Butylperbenzoate**	24	47	—	—	—
***tert*-Butylperoctoate**	28	55	—	—	—
***tert*-Butylperoxyacetate**	—	—	—	51	—
***tert*-Butylperoxybenzoate**	—	—	—	89	—
4-Chlorobenzoyl peroxide	74	94	—	—	—
Cumenehydroperoxide	11	33	9	—	—
Cyclohexanone peroxide					
Component A	21	38	—	—	—
Component B	0	12	—	—	—
Component C	0	10	—	—	—
Decahydro-9-naphthyl hydroperoxide	—	—	—	79	—
Di-*tert*-butylperoxide	0	39	—	—	—
2,4-Dichlorobenzoyl peroxide	81	88	—	—	—
1,1′-Dihydroperoxy-dicyclohexyl peroxide	—	—	—	—	74
1,1′-Dihydroperoxy-2,2′-dimethyldicyclohexyl peroxide	—	—	—	—	75
1,1′-Dihydroperoxy-3,3′-dimethyldicyclohexyl peroxide	—	—	—	—	75
1,1′-Dihydroperoxy-4,4′-dimethyldicyclohexyl peroxide	—	—	—	—	75
1,1′-Dihydroxydicyclohexyl peroxide	—	—	—	—	5
1,1′-Dihydroxy-2,2′-dimethyldicyclohexyl peroxide	—	—	—	—	5
1,1′-Dihydroxy-3,3′-dimethyldicyclohexyl peroxide	—	—	—	—	5
1,1′-Dihydroxy-4,4′-dimethyldicyclohexyl peroxide	—	—	—	—	5
Hydrogen peroxide	0	0	0	13	5
1-Hydroxy-1′-hydroperoxy-dicyclohexyl peroxide	—	—	—	—	12
1-Hydroxy-1′-hydroperoxy-2,2′-dimethyldicyclohexyl peroxide	—	—	—	—	14
1-Hydroxy-1′-hydroperoxy-3,3′-dimethyldicyclohexyl peroxide	—	—	—	—	14
1-Hydroxy-1′-hydroperoxy-4,4′-dimethyldicyclohexyl peroxide	—	—	—	—	14
Lauroyl peroxide	85	95	—	—	—
Methyl ethyl ketone peroxide					
Component A	16	42	—	—	—
Component B	10	10	—	—	—
Methyl isobutyl ketone peroxide					
Component A	25	55	—	—	—
Component B	0	12	—	—	—
1-Methyl-1-phenylethyl hydroperoxide	—	—	—	72	—
1-Methyl-1-phenylethyl xanthydryl peroxide	—	—	—	97	—
1-Methyl-1-phenylpropyl hydroperoxide	—	—	—	78	—
1-Phenylethyl hydroperoxide	—	—	—	76	—
1,2,3,4-Tetrahydro-1-naphthyl hydroperoxide	—	—	—	72	—

Layer L1 = Silica gel G.

Solvent S1 = toluene-carbon tetrachloride (2 : 1).
S2 = toluene-acetic acid (19 : 1).
S3 = petroleum ether (50–70 °C)-ethyl acetate (49 : 1).
S4 = chloroform-ether (3 : 1).
S5 = toluene-methanol (20 : 3).

Detection See Section II.I, Volume Two.

Technique T1 = 15 minutes development time.

Literature
1. Knappe, E. and D. Peteri, *Z. Anal. Chem.*, **190**, 386 (1962).
2. Coffee, E. C. J. and A. G. Davies, *J. Chem. Soc. C.*, **17**, 1493 (1966).
3. Buzlanova, M. M., Stepanovskaya, V. F., Nesterov, A. F., and V. L. Antonovsky, *Zh. Anal. Khim.*, **21**, 507 (1966); from *J. Chromatogr.*, **26**, D1 (1967).

TABLE TLC 66
ORGANIC SULFUR COMPOUNDS, ACCELERATORS

Layer		L1		
Solvent		S1		
Detection		—	D1	D2
Literature		1	1	1
Compound	Trade name	$R_F \times 100$	Color	
Tetramethylthiuram monosulfide	Unads	27	r	—
Tetrabutylthiuram monosulfide	Pentex	53	r-br	—
Tetramethylthiuram disulfide	Methyl Tuads	41	r	—
Tetraethylthiuram disulfide	Ethyl Tuads	57	r	—
Dipentamethylenethiuram tetrasulfide	Tetrone A	61, 56	p	—
Cyclic thiuram	Conac T	17, 12, 8	y-br	—
Piperidinium pentamethylenedithiocarbamate	Accelerator 552	54	p	—
Zinc dimethyldithiocarbamate	Methyl Zimate	49	r	—
Zinc diethyldithiocarbamate	Ethyl Zimate	64	r	—
Zinc dibutyldithiocarbamate	Butyl Zimate	77	r-br	—
Copper dimethyldithiocarbamate	Cumate	63	y-gr	—
Bismuth dimethyldithiocarbamate	Bismate	32	r	—
Selenium dimethyldithiocarbamate	Methyl Selenac	56, 51	r	—
Tellurium diethyldithiocarbamate	Tellurac	60, 48	r	—
Cadmium diethyldithiocarbamate	Cadmate	47	r	—
Lead diethyldithiocarbamate	Ethyl Ledate	54	r	—
2-Benzothiazyl-*N,N*-diethylthiocarbamyl sulfide	Ethylac	29	r	—
2-Mercaptobenzothiazole	MBT	25	y-br	—
Benzothiazyl disulfide	MBTS	44	—	y-br
Zinc benzothiazyl sulfide	Zetax	24, 0	—	y-br
2-Mercaptothiazoline	2-MT	11	—	y
***N-tert*-Butyl-2-benzothiazole sulfenamide**	Santocure NS	45	cy-y	cy-y
***N,N*-Diisopropyl-2-benzothiazole sulfenamide**	DIBS	56	y-br	y-br
***N*-Cyclohexyl-2-benzothiazole sulfenamide**	Santocure	48	o-br	o-br
***N*-Oxydiethylene-2-benzothiazole sulfenamide**	Amax	24	r-vt	r-vt
2-(2,6-Dimethyl-4-morpholinothio) benzothiazole	Santocure 26	31, 23	—	tan
***N,N*-Dimethylcyclohexylamine salt of dibutyldithiocarbamic acid**	RZ-50A	72	r-br	r-br
1,1-Methylene-dipiperidine-carbon disulfide reaction product	R-2 Crystals	streaks	p	p

Layer L1 = Silica gel G.
Solvent S1 = benzene-ethyl acetate-acetone (100 : 5 : 1).
Detection D1 = ninhydrin.
D2 = bismuth nitrate.
Technique Saturated tank, 15 cm development distance.
Literature 1. Kreiner, J. G. and W. C. Warner, *J. Chromatogr.*, **44**, 315–330 (1969).

TABLE TLC 67
ORGANO-METALLIC TIN COMPOUNDS

Layer	L1	L1	L1	L1	L1	L1
Solvent	S1	S2	S3	S4	S5	S6
Literature	1	1	1	1	1	1
Compound			$R_F \times 100$			
Dimethyltin chloride	0	0	0	0	2	0
Diethyltin chloride	0	0	0	0	9	0
Triethyltin chloride	2	7	48	17	—	—
Tetraethyltin	100	100	100	100	—	100
Tripropyltin acetate	9	34	71	29	—	—
Dibutylin dichloride	0	0	0	0	38	—
Tributyltin chloride	21	45	85	41	—	83
Tetrabutyltin	100	100	100	100	—	100
Dihexyltin dichloride	0	0	0	0	57	—
Trihexyltin chloride	45	54	92	60	—	—
Tetrahexyltin	100	100	100	100	—	100
Di-2-ethylhexyltin dichloride	0	0	0	0	—	—
Dioctyltin dichloride	0	0	0	0	68	—
Tri-2-ethylhexyltin chloride	78	69	100	82	—	—
Tetra-2-ethylhexyltin	100	100	100	100	—	—
Diphenyltin dichloride	0	0	0	0	29	—
Triphenyltin acetate	55	55	84	60	—	—
Tetraphenyltin	100	100	100	100	—	—
Butyltin trichloride	—	—	—	—	—	0
Butylthiostannous acid	0	0	0	0	—	0

Layer L1 = Silica gel.

Solvent S1 = butanol-pyridine (15 : 7) saturated with water.
S2 = butanol-ethanol (3 : 1) saturated with water.
S3 = butanol saturated with 25% ammonia.
S4 = upper phase of butanol + 2.5% ammonia.
S5 = isopropanol-(1 vol 1*N* sodium acetate + 1 vol 1*N* acetic acid) (2 : 1).
S6 = isopropanol-(2 vol 10% ammonium carbonate + 1 vol 5*N* ammonia) (2 : 1).

Detection Alcoholic pyrocatechol violet.

Technique Development distance, 10 cm.

Literature 1. Buerger, K., *Z. Anal. Chem.*, **192**, 280 (1962).

TABLE TLC 68
OXO COMPOUNDS, AROMATIC ALDEHYDES

Layer	L1	L1	L1	L1						
Solvent	S1	S2	S3	S4						
Detection	—	—	—	—	D1		D2		D3	
Literature	1	1	1	1						
					Color					
Compound	$R_F \times 100$				Daylight	UV	Daylight	UV	Daylight	
Benzylvanillin	61	46	67	64	y	y	—	—	—	
Coumarin	50	38	—	—	—	—	—	gr	o	
Dihydrocoumarin	62	50	—	—	—	—	—	—	o	
2,4-Dimethoxybenzaldehyde	50	44	56	58	y	bt y	—	—	—	
3-Ethylcoumarin	81	75	—	—	—	—	—	bl	o-ro	
Ethylvanillin (4-hydroxy-3-ethoxy-benzaldehyde)	46	42	62	55	y	y	—	—	—	
***m*-Hydroxybenzaldehyde**	46	37	15	23	—	—	—	y-gr	y[b]	
***p*-Hydroxybenzaldehyde**	32	27	11	20	y	y	—	—	—	
***o*-Methoxybenzaldehyde**	70	62	77	63	y	y	—	—	—	
***p*-Methoxybenzaldehyde**	64	53	70	59	y	y-gr	—	—	—	
3-Methylcoumarin	68	58	—	—	—	—	—	bl	o-ro	
6-Methylcoumarin	55	43	—	—	—	—	—	y	vt	
Methylvanillin (3,4-dimethoxybenzaldehyde)	41	38	62	55	y	o	—	—	—	
Piperonal (heliotropin)	66	60	72	63	y	dk spot	—	—	—	
Propenylguaethol (vanitrope, 1-ethoxy-2-hydroxy-4-propenylbenzene)	81	74	78	66	—	—	—	—	r-o	
Vanillin (4-hydroxy-3-methoxy-benzaldehyde)	30	27	34	29	y	o	—	—	ro	
***o*-Vanillin (2-hydroxy-3-methoxy-benzaldehyde)**	39	27	55	45	y	r	y[a]	y	y-o	

[a] weak.
[b] very weak.

Layer L1 = Silica gel G.
Solvent S1 = petroleum ether (50–70 °C)-ethyl acetate (5 : 2.5).
S2 = hexane-ethyl acetate (5 : 2).
S3 = chloroform-ethyl acetate (98 : 2).
S4 = decaline-methylene chloride-methanol (5 : 4 : 1).
Detection See Section II.I, Volume Two, for details.
D1 = hydrazine.
D2 = alc. KOH.
D3 = diazotized sulfanilic acid.
Technique 11.5 cm development distance.
Literature
1. Sundt, E. and A. Saccardi, *Food Technol.*, **16**, 89 (1962).

TABLE TLC 69
OXO COMPOUNDS, AMINOANTHRAQUINONES

Layer	L1		
Solvent	S1		
Detection	—	D1	D2[a]
Literature	1		

Anthraquinone derivative:	$R_F \times 100$	Color	
1-Amino-	61	brick-r	+
2-Amino-	38	o-y	+
1,2-Diamino-	40	vt-bl	vt-bl
1,3-Diamino-	22	r	o-br
1,4-Diamino-	26	vt	gy-gr
1,5-Diamino-	46	r	br-carmine
1,8-Diamino-	42	carmine	carmine-br
1,6-Diamino-	22	o-r	r-br
1,7-Diamino-	14	o-y	o-br
2,3-Diamino-	12	br	dk br
2,7-Diamino-	11	ochre-y	o-y
1,2,3-Triamino-	23	vt	vt
2,6-Diamino-	10	br	br
1,2,4-Triamino-	20	vt	gr
1,4,5,8-Tetraamino-	19	dk bl	dk bl
2-Methyl-1-amino-	69	brick-r	+
2-Methyl-4-amino-	63	brick-r	+
2-Methyl-5(8)-amino-	61	brick-r	+
2-Methyl-3-amino-	44	o-y	+
2-Methyl-6(7)-amino-	41	o-y	+
2-Methyl-1,3-diamino-	29	r	o-br
2-Methyl-1,4-diamino-	38	vt	gy-gr
2-Methyl-1,5-diamino-	51	r	carmine-r
2-Methyl-1,8-diamino-	46	carmine-r	br-carmine
2-Methyl-1,6-diamino-	28	o-r	br
2-Methyl-1,7-diamino-	18	o-y	br
2-Methyl-3,6(7)-diamino-	8	ochre-y	lt br
2-Methyl-1,4,5(8)-triamino-	22	vt	gy-gr
4-Methyl-1-amino-	66	brick-r	+
4-Methyl-2-amino-	38	o-y	+
4-Methyl-1,3-diamino-	21	r	r
4-Methyl-1,5-diamino-	53	r	br
4-Methyl-1,8-diamino-	47	carmine-r	carmine-br
4-Methyl-1,6-diamino-	29	o-r	br-carmine
4-Methyl-1,7-diamino-	17	o-y	lt br
4-Methyl-1,5,8-triamino-	23	vt	vt
1,3-Dimethyl-2,4-diamino-	32	r	o-r
1-Amino-4-benzoylamino-	49	vt	vt
2-Methyl-1-amino-4-benzoyl-amino-	57	vt	vt
2-Ethyl-1-amino-4-benzoylamino	62	vt	vt
1-Amino-4-chloro-	56	brick-r	+
1-Amino-5-chloro-	58	brick-r	+
1-Amino-8-chloro-	64	brick-r	+
1-Acetamino-4-chloro-	72	y	y
2-Amino-3-chloro-	55	y	y
1-Amino-2-bromo-	85	o-r	o-r
1-Amino-3-bromo-	74	r	+
1-Amino-4-bromo-	62	r	+
2-Methyl-1-amino-3-chloro-	77	brick-r	+
2-Methyl-1-amino-3-bromo-	79	brick-r	+
2-Amino-3-bromo-	55	o-y	+
2-Methyl-1-amino-4-bromo-	68	brick-r	+
1,4-Diamino-2-bromo-	59	vt	gy-gr
2-Amino-1,3-dibromo-	71	o-y	+
1-Amino-2,4-dibromo-	82	brick-r	+
1-Amino-2-chloro-4-bromo-	82	brick-r	+
2-Methoxy-1-amino-	63	brick-r	+
2-Methoxy-3-amino-	47	ochre-y	o-y

TABLE TLC 69—(Continued)
OXO COMPOUNDS, AMINOANTHRAQUINONES

Layer	L1		
Solvent	S1		
Detection	—	D1	D2[a]
Literature	1		
Anthraquinone derivative:	$R_F \times 100$	Color	
1-Methylamino-	86	carmine-vt	carmine-vt
1-Hydroxy-2,4-diamino-	<5	vt	gy-gr
1-(*p*-Tosylamino)-3-bromo-	70	y	y
1-Anilino-	90	carmine-vt	carmine-vt
1-(*p*-Nitroanilino)-	75	lt pk	lt pk
1-(*o*-Chloroanilino)-	88	carmine-vt	carmine-vt

[a] + Indicates that after spraying with Ehrlich's reagent the original color of the spots becomes deeper.

Layer L1 = Aluminum oxide (Reanal), without binder, loose layer.
Solvent S1 = hexane-acetone (3 : 1).
Detection D1 = original spot color.
D2 = Ehrlich's reagent.
Literature 1. Obruba, K., Šlosar, J., and J. Arient, *Mikrochim. Acta*, 1966, 262; from *J. Chromatogr.*, **35**, D66–D67 (1968).

TABLE TLC 70
OXO COMPOUNDS, KETONES

Layer	L1	L1	L1	L1	L1	L1	L1
Solvent	S1	S2	S3	S4	S5	S6	S7
Literature	1	1	1	1	1	1	1
Compound	$R_F \times 100$						
Phenyl-2-propanone[a]	42	53	68	72	76	66	75
Phenyl-2-propanone[b]	32	45	59	62	68	48	61
4′-Methylacetophenone[a]	37	48	70	74	81	58	73
4′-Methylacetophenone[b]	27	40	60	64	75	42	63
***trans*-4-Phenyl-3-buten-2-one**[a]	25	35	59	63	69	57	57
***trans*-4-Phenyl-3-buten-2-one**[b]	15	26	47	52	61	37	4
1-Phenyl-1,3-butanedione[a]	13	21	38	42	60	37	55
1-Phenyl-1,3-butanedione[b]	6	16	31	33	52	33	39
Hexanophenone[a]	15	27	58	64	83	18	57
Hexanophenone[b]	9	20	50	54	76	0	30
Phenyl-2-thienyl-ketone[a]	11	17	35	38	58	26	56
Phenyl-2-thienyl-ketone[b]	4	11	30	31	50	14	36
4′-Phenylacetophenone[a]	5	15	34	38	59	21	46
4′-Phenylacetophenone[b]	0	6	25	29	51	13	33
2-Tridecanone[a]	7	20	70	85	100	16	77
2-Tridecanone[b]	1	12	57	73	90	0	8

[a] R_F of front. [b] R_F of rear.

Layer L1 = Amberlite CG-400 anion-exchange layers, 200 μ thickness.
Solvent S1 = 12.0*M* methanol in water.
S2 = 14.0*M* methanol in water.
S3 = 17.0*M* methanol in water.
S4 = 19.0*M* methanol in water.
S5 = 22.0*M* methanol in water.
S6 = 6.0*M* acetone in water.
S7 = 8.0*M* acetone in water.
Detection 3,5-dinitrobenzoic acid.
Technique Ascending development, saturated tank, 15 cm, 60 minutes.
Literature 1. Sherma, J., Goldstein, D. A., and R. Lucek., *J. Chromatogr.*, **38**, 54–60 (1968).

TABLE TLC 71
OXO COMPOUNDS, BENZALDEHYDE AND RELATED COMPOUNDS

Layer	L1	L1	L1	L1	L2	L2	L2	L2
Solvent	S1	S2	S3	S4	S5	S6	S7	S2
Literature	1	1	1	2	3	3	3	3
Compound	R_F[a] × 100			R_F × 100				
Benzaldehyde	—	—	—	80	84	—	—	—
2,4-Dihydroxybenzaldehyde	66	20	79	—	—	—	—	—
2,5-Dihydroxybenzaldehyde	59	20	76	—	—	—	—	—
3,4-Dihydroxybenzaldehyde	17	2	32	6	—	—	—	—
3,4-Dihydroxy-5-methoxybenzaldehyde	21	10	18	—	—	—	—	—
2,3-Dimethoxybenzaldehyde	142	155	136	—	—	—	—	—
2,4-Dimethoxybenzaldehyde	116	130	109	—	—	—	—	—
2,5-Dimethoxybenzaldehyde	160	154	142	—	—	—	—	—
3,4-Dimethoxybenzaldehyde	98	143	113	45	—	—	—	—
3,5-Dimethoxybenzaldehyde	173	168	153	—	—	—	—	—
***m*-Hydroxybenzaldehyde**	66	24	90	—	3	30	47	11
***o*-Hydroxybenzaldehyde**	150	152	142	19	32	44	52	26
***p*-Hydroxybenzaldehyde**	51	16	78	17	2	15	24	5
4-Hydroxy-3,5-dimethoxybenzaldehyde	56	50	43	14	—	—	—	—
2-Hydroxy-3-ethoxybenzaldehyde	154	148	119	—	—	—	—	—
3-Hydroxy-4-ethoxybenzaldehyde	80	77	89	—	—	—	—	—
4-Hydroxy-3-ethoxybenzaldehyde	100	100	100	—	—	—	—	—
2-Hydroxy-3-methoxybenzaldehyde	133	125	87	53	3	4	4	6
3-Hydroxy-4-methoxybenzaldehyde	62	55	71	—	4	21	32	26
4-Hydroxy-3-methoxybenzaldehyde	81	71	80	27	4	13	16	13
***m*-Methoxybenzaldehyde**	172	158	152	—	—	—	—	—
***o*-Methoxybenzaldehyde**	158	149	138	—	—	—	—	—
***p*-Methoxybenzaldehyde**	147	147	141	65	19	39	—	—
3,4-Methylene-dihydroxybenzaldehyde	151	156	140	—	—	—	—	—
3,4,5-Trimethoxybenzaldehyde	110	136	123	—	—	—	—	—

[a] R_F values relative to 4-hydroxy-3-ethoxybenzaldehyde.

Layer L1 = Silica gel G.
L2 = Loose-layer aluminum oxide (Brockmann activity III).

Solvent S1 = benzene-ethyl acetate-acetic acid (90 : 5 : 5).
S2 = chloroform.
S3 = benzene-pyridine (9 : 1).
S4 = benzene-methanol (95 : 5).
S5 = benzene-ethanol (98 : 2).
S6 = benzene-ethanol (95 : 5).
S7 = benzene-ethanol (90 : 10).

Detection $KMnO_4$.

Technique 12 cm development distance, 60 minutes.

Literature
1. Klouwen, M. H., R. ter Heide, and J. G. J. Kok, *Fette Seifen Anstrichm.*, **65**, 414 (1963); *Chem. Abstr.*, **59**, 4525g (1963).
2. Petrowitz, H-J., *Z. Anal. Chem.*, **183**, 432 (1961).
3. Hěrmánek, S., Schwarz, V., and Z. Čekan, *Pharmazie*, **16**, 566 (1961).

TABLE TLC 72
OXO COMPOUNDS, 2,4-DINITROPHENYLHYDRAZONES

Layer	L1	L1	
Solvent	S1	S2	
Detection	—	—	D1
Literature	1	1	
Compound	$R_F{}^a \times 100$		Color
2,4-DNPH of:			
Acetone	61	25	gr-ish bl
Acetophenone	90	—	color deepening
Acrolein	62	42	gr
Anisaldehyde	50	21	color deepening
Benzaldehyde	76	—	color deepening
***n*-Butanal**	74	56	soft gr
***d*-Carvone**	115	122	—
***l*-Carvone**	113	125	color deepening
Citral	98	79	color deepening
	105	96	
Crotonaldehyde	67	45	color deepening
Cyclohexanone	75	61	gr
Decanal	103	89	soft gr
2-Decen-l-al	106	97	color deepening
Diacetyl	26	—	no color change
Dodecanal	107	91	soft gr
2-Dodecen-l-al	110	104	very weak gr
Ethanal	46	26	gr
Furfural	43	21	color deepening
Heptanal	93	79	soft gr
Hexanal	88	75	soft gr
***trans*-Hexen-2-al-l**	88	75	color deepening
Hexene-2-al-l	89	76	color deepening; after 5 min. very weak coloring
α-Ionone	120	132	dirty gr
β-Ionone	111	123	dirty gr
Isobutanal	82	65	soft gr
Isopentanal	82	69	soft gr
Methanal	40	23	gr
Methyl ethyl ketone	78	69	gr-ish bl
Methyl propyl ketone	89	81	gr-ish bl
Nonanal	99	86	soft gr
Octanal	98	84	soft gr
***n*-Pentanal**	81	68	soft gr
Propanal	64	31	gr

[a] R_F relative to butter-yellow.

Layer	L1 = Silica gel G.
Solvent	S1 = carbon tetrachloride-hexane-ethyl acetate (10 : 2 : 1). S2 = petroleum ether (40–60 °C)-diisopropyl ether (22 : 3).
Detection	Spray reagent: 0.2% $[K_3Fe(CN)_6]$ + 0.1% $FeCl_3 \cdot 6H_2O$ in 2*N* HCl; colors change with time.
Literature	1. Mehlitz, A., Gierschner, K., and T. Minas, *Chemiker Ztg.*, **87**, 573 (1963) from Kirchner, J. G., "*Thin Layer Chromatography*," *Techniques of Organic Chemistry*, Vol. XII, Perry, E. S., and A. Weissberger, Eds., Interscience Publishers, Inc., Division of John Wiley & Sons, Inc., New York, N.Y., 1967, pp. 380–381.

TABLE TLC 73
OXO COMPOUNDS, 2,4-DINITROPHENYLHYDRAZONES

Layer	L1	L1	L1	L2	L2	L3	L4
Solvent	S1[a]	S2	S3	S4[b]	S5	S6	S7[a]
Literature	1	1	1	1	1	1	1
Compound	R_F[c] × 100						
2,4-DNPH of:							
p-Toluylaldehyde	100	100	100	100	100	100	100
Anisaldehyde	51	75	44	44	51	58	streak
Acetophenone	streak	102	115	106	102	streak	112
Benzophenone	streak	121	180	122	158	115	135
Benzaldehyde	93	93	78	96	104	75	streak
Benzyl methyl ketone	152	112	162	87	117	114	110
Cinnamaldehyde	82	90	70	86	86	74	75
α-Pentylcinnamaldehyde	112	135	206	128	135	115	120
Cuminaldehyde	streak	120	154	117	120	110	115
p-Isopropyl-α-methylhydrocinnamaldehyde	168	129	216	98	123	130	150
Ethyl vanillin	3	61	20	5	11	1	7
Vanillin	2	54	13	4	6	1	0
Ethyl styryl ketone	streak	114	150	112	136	122	123
Furfural	60	87 96	45 90	46 72	42	94 67	88 97
5-Methylfurfural	65	93 113	70 125	52 82	43	110 84	53
5-Hydroxymethylfurfural	72	68 48	24 80	0	0	29 14	0
Piperonal	78	63	30	50	60	51	45
Hydratropic aldehyde	142	113	145	90	103	123	111
Isojasmon	185	135	230	130	140	165	170
Methyl α-naphthyl ketone	72	95	streak	98	100	98	streak
Methyl β-naphthyl ketone	72	95	streak	102	102	89	streak
Salicylaldehyde	45	79	33	45	41	1	21
Veratraldehyde	3	69	25	4	8	35	1
Phenylacetaldehyde	123	108	117	82	87	92	88
Phenylpropionaldehyde	132	107	125	71	75	80	80
p-Hydroxybenzaldehyde	3	66	14	10	9	2	0
Syringaldehyde	2	35	8	2	1	2	0

[a] Number of irrigations: 2.
[b] Number of irrigations: 5.
[c] R_F relative to 2,4-DNPH of *p*-toluylaldehyde.

Layer L1 = Acetylcellulose (Machery-Nagel).
L2 = Silica gel G (Merck).
L3 = Zinc carbonate (BDH) and plaster of Paris (30 : 2).
L4 = Anasil B (Analabs Inc.).

Solvents S1 = petroleum ether-methanol-ethyl acetate (9.5 : 1.5 : 1.0).
S2 = petroleum ether-dibutyl ether-ethyl acetate-methanol (6.5 : 1.0 : 1.5 : 1.0).
S3 = petroleum ether-dibutyl ether-ethyl acetate-methanol (8.5 : 0.5 : 0.5 : 0.5).
S4 = petroleum ether-benzene (1 : 1).
S5 = petroleum ether-benzene-dibutyl ether (2 : 1 : 1).
S6 = petroleum ether-pyridine (8 : 1).
S7 = petroleum ether-benzene-pyridine (8 : 1 : 0.25).

Detection Alc. KOH.

Literature 1. Dhont, J. H. and G. J. C. Mulders-Dijkman, *Analyst*, **92**, 431–435 (1967).

TABLE TLC 74
OXO COMPOUNDS, 4-NITROPHENYLHYDRAZONES

Layer	L1	L1	L1	L1	
Solvent	S1	S2	S3	S4	
Detection					D1
Literature	1	1	1	1	
Compound		$R_F \times 100$			Color
4-NPH of:					
Acetophenone	67	67	27	65	p-r
Anisaldehyde	53	72	21	57	pk
Benzaldehyde	58	60	30	63	r-br
Benzophenone	81	88	47	—	bt-r
Butyrophenone	73	77	—	—	r
Cinnamaldehyde	56, 72	59, 67	11	—	—
2,4-Dimethylacetophenone	70	78	34	—	r
2,5-Dimethylacetophenone	85	81	—	—	r
3,4-Dimethylacetophenone	72	86	—	80	y-r
Furfural	63	—	—	—	—
2-Hydroxyacetophenone	59	46	9	41	p-br
3-Hydroxyacetophenone	14	25	2	41	—
4-Hydroxyacetophenone	26	31	2	42	—
3-Hydroxybenzaldehyde	38	28	6	39	br
4-Hydroxybenzaldehyde	7	24	2	47	d p
***m*-Hydroxybenzophenone**	59, 64	91	26	—	p
***o*-Hydroxybenzophenone**	49, 89	53, 85	24	—	bt-r
***p*-Hydroxybenzophenone**	24, 47	50	—	—	r
2-Hydroxy-3-methoxybenzaldehyde	33	39	—	—	r-br
2-Hydroxy-5-methylbenzophenone	47	63	5	—	—
2-Methoxybenzaldehyde	81	63	—	—	—
5-Methyl-2-furfural	64	—	—	—	—
2-Naphthaldehyde	19	33	2	—	—
Piperonal	55	71	12	—	r-p
Propiophenone	67	67	—	—	o
Salicylaldehyde	48	43	23	51	r-br
Vanillin	36	34	—	—	r-br
Veratraldehyde	33	54	7	—	—

Layer L1 = Silica gel G, plates activated at 110–120 °C for one hour.
Solvent S1 = dichloromethane.
S2 = benzene-methanol (95 : 5).
S3 = benzene.
S4 = ethyl acetate-hexane (1 : 1).
Detection Potassium hydroxide.
Technique Ascending, 20 °C.
Literature 1. Barker, E. D. and E. Sawicki, *Anal. Chem.*, **40**, 985 (1968).

TABLE TLC 75
OXO COMPOUNDS, POLYNUCLEAR RING-CARBONYL COMPOUNDS

Layer	L1	L1	L2		
Solvent	S1	S2	S3		
Detection				D1[a]	D2[a]
Literature	1	1	1		
Compound		$R_F \times 100$		Color	
10,11-Dihydro-5H-dibenzo[*a,d*]cycloheptene-5-one	64	76	67	—	lt y
Fluoren-9-one	60	72	11[b]	dull	d r
2-Benzoylpyrene	50	79	08	lt y	—
9-Acetylphenanthrene	49	66	64	—	lt r
6-Benzoylchrysene	48	76	40	lt y	—
Thioxanthen-9-one	48	66	47	bl	y
11H-Benzo[*b*]fluorene-11-one	44	64	34	lt o	—
Anthrone	43	78	63	—	d r
3-Acetylpyrene	43	64	37	bl	—
9-Acetylanthracene	42	74	34	—	lt bl
Xanthen-9-one	38	53	63	—	bl
6H-Benzo[*c,d*]pyrene-6-one	31	49	19[b]	lt y-o	r-o
5,12-Dihydronaphthacen-5-one	30	64	24	y	r-o
Benzanthrone	29	58	39	gr-y	o
1-Indanone	29	55	93	—	bl-gr
Coumarin	13	47	92	—	gr-y
Phenalen-1-one	11	24	81	y	emerald gr
7H-Dibenzo[*c,h*]xanthen-7-one	8	29	7	bl	gr
7:8-Benzoflavone	0	10	34	bl	gr
9-Acridanone	0	8	83	lt bl	bl
6(5H)-Phenanthridinone	0	0	79	pl bl	lt bl
11,12-Dihydroindeno[2,1-*a*]fluorene-11,12-dione	0	0	0	dk r	—

[a] Failure to detect a spot (—).
[b] Tails.

Layer L1 = Aluminum oxide G.
L2 = Cellulose.
Solvent S1 = pentane-ether (19 : 1).
S2 = toluene.
S3 = dimethylformamide-water (35 : 65).
Detection D1 = fluorescence (on neutral plates).
D2 = fluorescence.
Technique Ascending, 15 cm.
Literature 1. Sawicki, E., Stanley, T. W., Elbert, W. C., and M. Morgan, *Talanta*, **12**, 609 (1965).

TABLE TLC 76

OXO COMPOUNDS, ALIPHATIC AND TERPENE 2,4-DINITROPHENYLHYDRAZONES

Layer	L1	L2	L3
Solvent	S1	S2	S1
Literature	1	1	1
Compound	$R_F \times 100$		
2,4-DNPH of:			
Butanal	39	10	24
Butan-2-one	37	13	28
Camphor	<5	0	0
Carvone	62	32	54
Carvotane acetone	67	49	67
Citral	38	23	44
Citronellal	49	32	52
Cycloheptanone	39	24	47
Cyclohexanone	35	17	37
Cyclooctanone	30	32	53
Cyclopentanone	25	11	28
Decanal	57	57	73
Dihydrocarvone	57	34	57
Dodecanal	60	76	84
Ethanal	22	5	11
Fenchone	<5	0	0
Heptanal	51	27	47
Hexanal	48	21	38
Heptan-2-one	50	34	53
Hexan-2-one	46	26	45
α-Ionone	60	54	73
β-Ionone	63	60	76
Isomenthone	66	54	73
Isopulegone	52	41	56
Methanal	23	2	6
Menthone	51	55	73
Methylheptenone	44	28	48
α-Methylionone	67	68	79
β-Methylionone	66	73	82
Nonanal	55	46	65
Nonan-2-one	55	55	70
Octanal	53	37	56
Octan-2-one	53	44	63
Pentanal	44	15	31
Pentan-2-one	41	19	35
Perilla-aldehyde	44	20	39
α-Phellandral	64	50	68
Piperitone	39	43	64
Propanal	33	7	17
Propan-2-one	22	7	18
Pseudoionone	51	40	60
Pulegone	47	57	64
α-Thujone	53	46	66
β-Thujone	63	46	66
Undecanal	59	66	78
Undecan-2-one	60	73	82

Layer L1 = Silica gel G (Merck) equilibrated with the atmosphere.
L2 = L1 impregnated with nitromethane by vapor phase adsorption.
L3 = L1 impregnated with acetonitrile by vapor phase adsorption.

Solvent S1 = benzene-hexane (1 : 1), five developments.
S2 = hexane, three developments.

Technique Ascending, tank saturation, 11.5 cm development distance, five or three times; best separation is 2-dimensional, S1 for first dimension followed by S2.

Detection See Section II.I, Volume Two.

Literature 1. Dhont, J. H. and G. J. C. Mulders-Dijkman, *Analyst*, **94**, 1093 (1969).

TABLE TLC 77
OXO COMPOUNDS, TERPENE ESSENTIAL OILS, CARBONYLS

Layer	L1 or L2	L1	L1 or L2	L1	L1	L1	L1 or L2	L1	L1	L1	L1 or L2	L1 or L2 or L3	L3
Solvent	S1	S2[a]	S3	S4[a]	S5	S6	S7	S8[a]	S9	S10	S11[b]	S12	S13
Literature	1	1	1	1	1	1	1	1	1	1	2	3	3
Compound	$R_F \times 100$												
Acetaldehyde	—	—	27[c]	—	—	—	39[c]	—	—	—	28[c]	20[c]	—
Anisaldehyde	—	—	22[c]	—	—	—	57[c]	—	—	—	33[c]	22[c]	41[d]
Benzaldehyde	—	—	46[c]	—	—	—	70[c]	—	—	—	50[c]	38[c]	—
Camphor	56	39	55	43	56	59	—	78	47	50	28[a]	18[d]	31[d]
Caprylaldehyde	—	—	—	—	—	—	—	—	—	—	—	21[d]	42[d]
Carvone	45	72	62	70	76	79	74[c]	75	62	65	37[a]	28[c]	—
Cinnamaldehyde	31	70	68	70	68	70	63[c]	78	52	45	9[a]	24[c]	—
Citral	45	64	57	47	62	62	46[c]	69	56	51	15[a]	19[c]	—
Citronellal	49	—	—	—	—	—	81[c]	—	—	—	33[c]	46[c]	—
Cuminaldehyde	—	—	—	—	—	—	—	—	—	—	—	29[d]	51[d]
Fenchone	51[c]	—	—	—	—	—	62[c]	—	—	—	33[c]	30[c]	—
Furfural	21	44	41	41	44	41	66	38	37	17	6[a]	14[c]	—
Isovanillin	54[c]	—	—	—	—	—	32[c]	—	—	—	—	2[c]	—
Lauric aldehyde	58	72	67	62	91	76	—	84	75	83	50[a]	—	—
Methylheptenone	48	74	62	58	69	70	—	75	57	60	35[a]	—	—
Perrillaldehyde	41[c]	—	—	—	—	—	66[c]	—	—	—	38[c]	26[c]	—
Piperonal	30[c]	—	—	—	—	—	54[c]	—	—	—	40[c]	22[c]	—
Propionaldehyde	58[c]	—	—	—	—	—	—	—	—	—	—	—	—
Pseudoionone	34[c]	—	—	—	—	—	68[c]	—	—	—	20[c]	20[c]	—
Pulegone	58	65	76	71	79	83	—	91	72	79	60[a]	—	—
Vanillin	7[c]	—	—	—	—	—	36[c]	—	—	—	9[c]	7[c]	—

[a] Alcohol free.
[b] Traces of alcohol.
[c] L2.
[d] L3.

Layer L1 = Silicic acid bound with starch.
L2 = Silica gel, starch bound.
L3 = Silica gel G.

Solvent S1 = ethyl acetate-*n*-hexane (3 : 17).
S2 = ethyl acetate-chloroform (1 : 9).
S3 = ethyl acetate-benzene (3 : 17).
S4 = ethyl acetate-chloroform (5 : 95).
S5 = 1-nitropropane-hexane (1 : 1).
S6 = ethyl acetate-hexane (3 : 7).
S7 = ethyl acetate-hexane (2: 3).
S8 = ethyl carbonate-chloroform (3: 17).
S9 = isopropyl formate-hexane (3 : 7).
S10 = isopropyl ether-hexane (1 : 1).
S11 = chloroform.
S12 = benzene.
S13 = ethyl acetate-benzene (5 : 95).

Detection See Section II.I, Volume Two.

Technique 10 cm development distance.

Literature
1. Miller, J. M. and J. G. Kirchner, *Anal. Chem.*, **25**, 1107 (1953).
2. Katayama, T., *Nippon Suisan Gakkaishi*, **26**, 814 (1960).
3. Schantz, M. von, Lopmeri, A., Stroemer, E., Salonen, R., and S. Brunni, *Farm. Aikakanslehti*, **71**, 52 (1962).

TABLE TLC 78
OXYGEN HETEROCYCLICS, ANTHOCYANINS AND ANTHOCYANIDINS

Layer	L1	L2	L2	L3	L3	L4	L4	L5	L5
Solvent	S1	S2	S3	S4	S5	S6	S7	S6	S7
Technique	T1	T1	na	T2	T2	T1	T1	T1	T1
Literature	1	1	1	1	1	1	1	1	1
Compound	$R_F \times 100$								
Anthocyanidins									
Cyanidin-3,5-diglucoside	—	36	52	—	—	—	—	—	—
Cyanidin-3-monoglucoside	46	—	—	—	—	—	—	—	—
Cyanidin-3-triglucoside	42	—	—	—	—	—	—	—	—
Delphinidin-3,5-diglucoside	—	35	26	—	—	—	—	—	—
Delphinidin-3-monoglucoside	36	—	—	—	—	—	—	—	—
Malvidin-3,5-diglucoside	54	44	33	—	—	—	—	—	—
Malvidin-3-monoglucoside	57	60	61	—	—	—	—	—	—
Paeonidin-3,5-diglucoside	—	56	37	—	—	—	—	—	—
Petunidin-3-monoglucoside	38	—	—	—	—	—	—	—	—
Anthocyanins									
Cyanidin	31	83	92	47	53	70	75	57	75
Delphinidin	21	—	—	41	34	50	60	50	72
Malvidin	37	95	94	62	61	87	84	84	91
Paeonidin	39	—	—	55	65	90	86	83	85
Pelargonidin	41	—	—	51	71	95	92	69	78
Petunidin	27	—	—	54	46	70	75	72	79

Layer L1 = Polyacrylonitrile-polyamide (7 : 2) mixed with 40 ml of 0.05*M* primary potassium phosphate.
L2 = Silica gel D–O (Camag.).
L3 = Cellulose MN-300.
L4 = Cellulose.
L5 = Polyamide.

Solvent S1 = pentanol-1-propanol-1-acetic acid-water (3 : 2 : 2 : 1) two parts heptanol-1 may be added to decrease streaking.
S2 = *n*-butanol-acetic acid-water (4 : 1 : 2).
S3 = ethyl formate-methyl ethyl ketone-formic acid-water (3 : 4 : 1 : 2).
S4 = formic acid-hydrochloric acid-water (10 : 1 : 3).
S5 = amyl alcohol-acetic acid-water (2 : 1 : 1).
S6 = acetic acid-hydrochloric acid-water (10 : 1 : 3); development distance, 12 cm.
S7 = acetic acid-water (3 : 2).

Technique Development distance 12 cm.

Detection UV light, after spraying with 10% solution of oxalic acid in acetone-water (1 : 1).

Literature (After) Kirchner, J. G., Thin Layer Chromatography, in Technique of Organic Chemistry, Vol. XII, Perry and Weissberger, Editors, Interscience Publishers, New York, p. 559, 1967.

TABLE TLC 79
OXYGEN HETEROCYCLICS, COUMARINS

Layer	L1	L1	L1	L1	L1	L1		
Solvent	S1	S1	S2	S2	S3	S3		
Detection							D1	D2
Literature	1	1	1	1	1	1	1	1
Compound	$R_F \times 100$	$R_S{}^a \times 100$	$R_F \times 100$	$R_S{}^a \times 100$	$R_F \times 100$	$R_S{}^a \times 100$		
2-Benzohydryl-1,3-indanedione	20	25	0	— —	9	12	— —	+
3,3′-Benzylidene-bis-(4-hydroxycoumarin)	44	55	66	73	54	74	++	— —
3,3′-Butylidene-bis-(4-hydroxycoumarin)	48	60	68	76	42	58	++	+
Coumarin	80	100	90	100	73	100	— —	++
β,β-Di-(4-hydroxy-3-coumarin)-ethyl chloride	15	19	16	18	5	7	++	+++
1-(1,3-Diketo-2-indanyl)-1-phenyl-2-nitropropane	24	30	87	97	9	12	— —	+
3,3′-Ethylidene-bis-(4-hydroxycoumarin)	44	55	63	70	44	60	++	++
4-Hydroxycoumarin	26	33	31	34	12	16	++	++
7-Hydroxycoumarin	74	93	77	86	71	97	++	++
4-Hydroxycoumarin-3-carboxylic acid ethyl ester	14	18	15	17	4	6	+++	+++
3,3′-Methylene-bis-(4-hydroxycoumarin)	28	35	46	51	15	21	++	++
3,3′-Methylene-bis-(4-hydroxy-7-methoxy-coumarin)	24	30	37	41	14	19	+	— —
3,3′-Methylene-bis-(4-propionoxycoumarin)	80	100	85	94	81	111	+	— —
3,3′-Propylidene-bis-(4-hydroxycoumarin)	44	55	56	62	44	60	++	+
3,3′-Thio-bis-(4-hydroxycoumarin)	17	21	32	36	12	16	+	+

[a] $R_S = R_F$ relative to coumarin.

Layer L1 = Silica gel G.
Solvent S1 = acetone-2,2,4-trimethylpentane-water (100 : 40 : 1).
S2 = acetone-ethyl acetate-petroleum ether-water (100 : 100 : 33 : 3.5).
S3 = 4-methyl-2-pentanone-acetone-petroleum ether (60 : 20 : 20).
Detection D1 = UV.
D2 = $KMnO_4$ (2537Å).
Technique Ascending, saturated tank.
Literature 1. Copenhaver, J. H. and M. J. Carver, *J. Chromatogr.*, **16**, 229 (1964).

TABLE TLC 80
PESTICIDES

Layer Solvent Literature	L1 S1 1	L1 S2 1	L1 S3 1	L1 S4 1	L1 S5 1	L1 S6 1	L1 S7 1	L1 S8 1	L1 S9 1	L1 S10 1
Compound					$R_F \times 100$					
Aldrin[a]	80	78	83	94	73	78	80	76	69	56
Aramite®	62[a]	68	79	62 84	59	71	56[a]	52[c] 56[a]	40[a]	28
Bayer 25141[b]	12	18	30	62	2	9[a] 13[c]	18	7	8	2
Binapacryl (Morocide®)[a]	55	69	70	85	66	68	60	41	42	29
Captan[a]	44	56	67	73	37	57	45	37	15	13
Carbophenothion (Trithion®)	73	75	83	92	73	76	78	60	52	38
Chlordane[a]	79	76	83	93	73	78	78	X[d]	65	41 53
Chlorthion®	0[a]	0[a]	0[a]	0[a] 85[a]	0 14 59	0 28[a] 67[a]	23[a] 61[a]	0 15[a]	7[a]	2[a] 22[a]
Co-Ral®	46	61	67	78	53	61	46	23	19	14
DDE,*p*,*p′*[a]	69	70	74	90	74	72	68	64	64	54
DDT,*o*,*p*[a]	68	69	72	90	72	70	68	58	58	48
DDT,*p*,*p′*[a]	66	70	72	90	72	70	68	58	58	46
Delnav®	0[a] 64	0[a] 72	0 83	4[a] 66 78[a] 86[c] 90[a]	0 58	0[a] 73	0[a] 44[a] 53	0 43	6[a] 18[a] 33	0 26
Demeton	25 66[a]	33 38[a]	54 81[a] 84[c]	65 85[a]	8 64	30 73	35[a] 60	31 61[a]	14 52[a]	14 34
Demeton, thiol isomer	26	33[c] 37[a]	54	68	8	31	34[a]	31	16	13
Demeton, thiol isomer sulfone	8	13[a]	24[a]	49[a]	0	6	X	7	5[a]	0[a]
Demeton, thiol isomer sulfoxide	0[a]	X	5	23[a]	0	0	X	0	7[a]	0 11[a]
Demeton, thiono isomer	26	38	0[a] 53 81	0[a] 67[a] 84	0 9 66	28 74	0[a] 33[a] 60	32	8 18 51	32
Diazinon®	0[a]	0[c]	0 48 56	2[a] 67[a]	0 5[a] 10[a]	0 22[a] 33[a]	0 28[a] 35[a]	0 28[a]	6[a] 13[a]	0 10[a]
Dicapthon	66	72	77	84	62	70	65	38	24	45
Dieldrin[a]	64	76	75	79	65	73	71	X	52	37
Dimethoate	8	11[a]	22	47	2	8	12	7	7	0
Dimethoate, oxygen analog	X	0[a]	5[a]	17[a]	0	0	X	0	6	0[a]
Di-syston	63	80	80	83	69	76	72	61	54	41
Di-syston, sulfone	29	49	60	64[a] 68[c]	22	45	40	23	8	5
Di-syston, sulfoxide	9	13	25	59 83[a]	0 69[a]	6 75[a]	19 71[a]	5	7 54[a]	3 40[a]
Dylox® (Dipterex)[a]	3	4	10	39	0	3	17	4	7	4
Endosulfan (Thiodan®)[a]	73	76	82	91	59 71	69 77	65 76	X X	32 58	26 40
Endosulfan, low mp[a]	72	75	82	91	69	76	75	X	56	40
Endosulfan, high mp[a]	59	64	76	85	57	67	64	X	32	27
Endrin[a]	66	78	78	83	68	76	73	X	57	39
EPN	60	78	78	81	68	73	70	48	41	29
Eradex	66	71	79	90	63	73	64	59[c] 63[a]	38 56	35
Ethion	66	80	80	85	71	77	74	57	50	37 44[a]
Genite®[a]	63	76	76	81	67	73	70	X	40	30
Guthion®	33	55	63	65	33	54	45	27	13	7
Heptachlor[a]	76	80	79	86	75	78	80	X	65	57
Heptachlor epoxide[a]	68	78	78	84	69	74	75	X	56	41
Karathane[a]®	66	78	79	85	73	76	75	58	49	X

TABLE TLC 80—(Continued)
PESTICIDES

Layer	L1	L1	L1	L1	L1	L1	L1	L1	L1	L1
Solvent	S1	S2	S3	S4	S5	S6	S7	S8	S9	S10
Literature	1	1	1	1	1	1	1	1	1	1
Compound					$R_F \times 100$					
Kelthane®[a]	68	74	76	88	68	74	70	51	35	33
Kepone®[a]	X	15	25	37	15	X	20	17	X	6
Lindane[a]	73	74	76	89	68	74	73	X	45	36
Malathion	40	63	72	77	46	63	44	35	25	22
Menazon[e]	0	0	0	35	0	0	8	0	4	0
Methoxychlor[a]	68	73	78	88	66	73	68	45	49	33
Methyl parathion	60	70	75	84	59	67	61	34	29	22
Mirex[a]	81	77	82	92	75	79	81	X	70	62
Naled (Dibrom)[a]	33	45	58	66	0	0	39	25	5	0
					18	36			12	11
Parathion	64	73	78	86	65	71	67	44	38	30
									41[a]	
Phorate	71	75	79	89	68	75	72	60	53	42
Phosdrin®[a]	X	23	31	57	3	9	11	7	6	2
			41	63		16	18	11		
Phosphamidon[a]	4	9	17	48	0	5	7	4	4	0
Phostex	74	77	81	91	71	77	76	57	4	42
									52	
Schradan	0[a]	0[a]	4[a]	X	0[a]	0[a]	X	X	4	X
Sevin®	20[a]	36[a]	51[a]	65[a]	19[a]	34[a]	35[a]	20[a]	6	6[a]
Sulfotepp	56	68	72	85	64	66	57	42	44	32
Sulphenone[a]	X	64	71	80	X	X	51	X	X	18
TDE[a]	X	78	82	92	75	79	83	X	X	41
Tetradifon (Tedion®)[a]	66	75	80	89	70	77	75	X	45	31
Toxaphene[a]	76	76	83	93	74	78	81	X	59	40
										53
VC-13®	63	72	73	90	69	68	65	52	55	38
Zectran®	28[a]	43[a]	60[a]	12[a]	23	44[a]	2	38[a]	10	0

[a] Spot appears after exposing sprayed plate to ultraviolet.
[b] *O,O*-Diethyl *O-p*-methylsulfinylphenyl phosphorothioate.
[c] Spot disappears after exposing sprayed plate to ultraviolet.
[d] X = No visible spot, too weak, or streaking.
[e] *S*-[(4,6-Diamino-*s*-triazin-2-yl)methyl]*O,O*-dimethyl phosphorodithioate.

Layer L1 = Silica gel G.
Solvent S1 = chloroform.
S2 = chloroform-ether (9 : 1).
S3 = chloroform-acetone (9 : 1).
S4 = chloroform-acetic acid (9 : 1).
S5 = benzene-ethyl acetate (9 : 1).
S6 = benzene-acetone (9 : 1).
S7 = benzene-acetic acid (9 : 1).
S8 = hexane-acetone (8 : 2).
S9 = hexane-methanol (9 : 1).
S10 = hexane-acetic acid (9 : 1).
Detection Bromine, fluorescein; $AgNO_3$/phenoxyethanol; UV irradiation.
Technique Development distance at least 10 cm.
Literature 1. Walker, K. C. and M. Beroza, *J. Ass. Offic. Agr. Chem.*, **46**, 250 (1963).

TABLE TLC 81
PESTICIDES

Layer	L1	L1	L1	L1	L1
Solvent	S1	S2	S3	S4	S5
Literature	1	1	1	1	1
Compound	$R_F \times 100$				
Hexachlorobenzene	70	76	80	90	90
Aldrin	61	71	77	90	90
Chlordane	52, 45, 37, 28, 20	69, 48, 22	75, 64, 57	90	90
DDE	51	69	80	87	88
Isobenzan	48	71	78	90	90
1-Chloro-2,2-bis(4-chlorophenyl)ethylene	48	69	78	88	89
Quintozene	43	68	82	87	89
DDT	40	58	67	87	88
α-BHC	35	61	68	85	86
γ-BHC	26	51	61	86	87
TDE	24	46	63	83	86
Trifluralin	19	74	76	88	88
Pentachlorophenyl acetate	19	55	65	80	87
Benfluralin	12	63	74	83	90
Bromophos-ethyl	13	57	64	83	84
Dichlofenthion	7	57	65	78	89
Dursban®	3	55	65	77	89
Fenoprop butyl	3	50	60	74	87
4,4′-Dichlorobenzophenone	2	47	58	63	87
Endrin	9	39	57	74	85
Dieldrin	7	39	53	75	85
Ethion	2	33	50	79	89
Dicofol	4	(50) 32	(60) 48	63	87
Fenoprop methyl	4	34	48	62	84
2,4,5-T *n*-butyl	2	31	50	57	86
Dinocap	0	30	50	61	88
2,4,5-T isobutyl	0	32	50	52	87
2,4,5-T isopropyl	0	32	50	53	87
Tetradifon	2	30	47	65	87
Bromoxynil octanoate	0	28	46	57	86
2,4,5-T *n*-propyl	0	27	45	49	87
Nitrofen	2	26	43	78	85
N,*N*-Dimethyl-*p*-phenylazoaniline	0	26	41	42	84
2,4,5-T ethyl	0	25	41	43	86
2,4-D *sec*-butyl	0	23	44	51	83
Parathion	4	23	38	51	85
2,4-D isobutyl	0	21	39	43	86
Dicloran	2	21	25	45	73
2,4-D *n*-butyl	0	20	37	47	84
2,4-D isopropyl	0	19	38	46	84
2,4,5-T methyl	1	18	33	43	82
2,4-D ethyl	0	17	31	43	82
Phenothiazine	1	16	28	65	75
2,4-Dichlorophenol	4	18	27	37	59
δ-BHC	4	14	28	80	84
Diazinon®	2	14	28	30	84
2,4-D methyl	0	13	21	32	79
2,4,5-T butoxyethyl	0	11	23	26	83
α-Naphthol	2	11	23	29	63
3,4-Dichloroaniline	4	7	13	42	84
Mercaptodimethur	0	7	22	23	65
Dioxathion	3	8	17	32	86
Malathion	2	5	14	22	79
2,4-D butoxyethyl	0	4	13	14	84
Folpet	0	0	12	27	80
p-Phenylazoaniline	0	4	10	24	68
Trichlorophenol	2	3	4	13	86
Captan	0	0	6	13	70
Linuron	0	2	5	11	68
Dithianone	0	0	0	13	81

TABLE TLC 81—(Continued)
PESTICIDES

Layer	L1	L1	L1	L1	L1
Solvent	S1	S2	S3	S4	S5
Literature	1	1	1	1	1
Compound	$R_F \times 100$				
Imidan®	0	0	3	10	73
Carbaryl	0	(11) 1	(24) 4	(30) 11	(64) 53
Ametryne®	1	4	6	10	59
2,4-Dichlorophenoxyethanol	0	2	3	8	52
Propazine	0	0	5	6	60
Azinphos-ethyl	0	0	4	5	73
Azinphos-methyl	0	0	2	4	68
Atrazine	0	2	4	5	60
Thiram	0	2	5	8	45
Dazomet	0	0	1	8	45
Simazine	0	0	0	1	43
Cyolane®	0	2	3	3	41
4-Nitrophenol	1	3	4	5	35
Crotoxyphos	0	0	0	0	41
Demeton-methyl	0	0	1	3	39
Diuron	0	0	0	0	41
Bromacil	0	0	0	0	41
Monuron	0	0	0	0	35
Fluometuron	0	0	2	3	33
1,1-Bis-(4-hydroxyphenyl)-2,2,2-trichloroethane	0	0	0	0	26
Methomyl	0	0	0	6	25
Fenuron	0	0	0	3	25
Haloxon	0	0	0	0	25
Coumaphos	0	0	5	6	14
Dimethoate	0	0	0	2	17
Thiabendazole	0	0	0	0	11
Warfarin	0	0	0	0	11
Trichlorfon	0	0	0	0	7
Pentachlorophenol	0	0	0	0	7
Amitrole	0	0	0	0	0

Layer L1 = Florisil.
Solvent S1 = hexane.
S2 = diethyl ether-hexane (6 : 94).
S3 = diethyl ether-hexane (15 : 85).
S4 = toluene.
S5 = acetone-toluene (1 : 9).
Detection See Section II.I, Volume Two.
Technique Plates equilibrated 40% relative humidity at 25 °C for 24 hours after spotting, saturated tank, over running technique, 15 cm, 15-20 minutes.
Literature 1. Hamilton, D. J. and B. W. Simpson, *J. Chromatogr.*, **39**, 186–194 (1969).

TABLE TLC 82
PESTICIDES, CHLORINATED

Layer	L1	L5	L1	L1	L2	L3	L3	L3	L3	L4	L3
Solvent	S1	S7	S1	S2	S3	S1	S4	S1	S2	S5	S6
Technique	T1	T1	T2	T2	T1	T1	T1	T2	T2	T2	T2
Literature	5	5	1	1	2	2	2	3	3	3	4
Compound	R_F	R_F	R_{DDD}	R_{TDE}	R_F	R_F	R_F	R_{TDE}	R_{TDE}	R_{TDE}	R_{TDE}
						×100					
Aldrin	86	81	206	191	67	72	77	21	14	43	50
BHC	50[c]	21[c]	9[a]	23[a]	—	—	—	—	—	—	—
	41	16	27[a]	52[a]	35[c]	40[c]	25[c]	—	—	—	114
	11	11	98[a]	83[a]	23[a]	31	18	—	—	—	—
	3	5	113[a]	112[a]	8[a]	8	13	—	—	—	—
	—	—	—	—	—	3	7	—	—	—	—
α-BHC	—	—	—	—	—	—	—	13	11	—	—
β-BHC	—	—	—	—	—	—	—	27	72	—	—
γ-BHC	—	—	—	—	—	—	—	10	25	—	—
BHC (T)	—	—	—	—	—	—	—	10	25	30	—
	—	—	—	—	—	—	—	27	72	75	—
	—	—	—	—	—	—	—	11	11	13	—
	—	—	—	—	—	—	—	13[e]	92	18	—
Captan	—	—	—	—	—	—	—	0	9	25	80[a]
Carbophenothion	—	—	—	—	—	—	—	42	101	19	131[a]
Chlorbenside	48	23	—	—	43[c]	43[c]	28[c]	—	—	0	—
	—	—	36	107	0[a]	0	3	13	12	20	—
Chlorbenzilate	—	—	—	—	5	0	8[c]	0	5	25	124
Chlordane	[a]	[a]	109[a]	122[a]	62[a]	65	63	—	—	—	—
	—	—	134[a]	148[b]	55[a]	58	53	—	—	—	—
	—	—	—	—	50[a]	42	45	—	—	—	—
	—	—	—	—	40[a]	32	23	—	—	—	—
Chlordane (T)	—	—	—	—	—	—	—	120[c,d]	110[c,d]	—	—
	—	—	—	—	—	—	—	140	120	29	—
	—	—	—	—	—	—	—	180	130	34	—
	—	—	—	—	—	—	—	200[e]	140[e]	37	—
CIPC	1	9	7	40	7	2	10	—	—	—	—
Dacthal	—	—	20	93	—	—	—	—	—	—	—
DDD(TDE)	—	—	—	—	32	30	15	—	—	—	100
p,p'-DDD(*p,p'*-TDE)	—	—	100	100	—	—	—	100	100	100	—
p,p'-DDE	80	50	201	168	65	65	53	200	140	340	50
o,p'-DDT	—	—	191	155	53	62	45	190	130	300	65
p,p'-DDT	63	25	165	144	48	50	30	160	120	220	68
Delta Keto "153"	—	—	—	—	—	—	—	9	23	56	—
Dichlone	—	—	—	—	8	7	20	16	0	0	—
	—	—	—	—	—	—	—	—	72[e]	—	—
Dicofol (Kelthane)	1	22	7	26	10	—	8–25[d]	6	28	0	109
Dieldrin	18	37	49	93	18	15	40	52	90	280	94
Diuron	—	—	—	—	—	—	—	0	0	0	—
Dyrene	—	—	—	—	—	—	—	15	51	20	—
Endosulfan (Thiodan)	—	—	—	—	32[c]	25[c]	45[c]	7	24	31[e]	—
	—	—	—	—	5[a]	2	17	88[e]	92[e]	0	88[a]
	—	—	—	—	—	—	—	—	—	—	111[a]
Endosulfan I	—	—	8[a]	22[a]	—	—	—	—	—	—	—
	—	—	80[a]	132[a]	—	—	—	—	—	—	—
Endosulfan II	—	—	8	22	—	—	—	—	—	—	—
Endrin	25	42	67	125	23	22	45	71	100	290	90
Endrin alcohol	—	—	—	—	—	—	—	0	0	0,14[e]	—
Endrin aldehyde	—	—	—	—	—	—	—	0	0	0	—
	—	—	—	—	—	—	—	—	—	23[e]	—
Heptachlor	78	60	196	168	60	63	60	200	140	370	74
Heptachlor epoxide	23	30	68	107	27	20	35	71	100	240	102
Hexachlorobenzene	89	90	238	198	80	82	85	270	170	—	27
Lindane	40	15	103	90	22	32	18	110	92	130	122
Methoxychlor	12	10	28	69	12	8	13	27	79	85	108
	—	—	—	—	—	—	—	33	—	—	—
Monuron	—	—	—	—	—	—	—	0	0	0	—
Ovex	—	—	15	74	13	5	8	18	76	61	128

TABLE TLC 82—(Continued)
PESTICIDES, CHLORINATED

Layer	L1	L5	L1	L1	L2	L3	L3	L3	L3	L4	L3
Solvent	S1	S7	S1	S2	S3	S1	S4	S1	S2	S5	S6
Technique	T1	T1	T2	T2	T1	T1	T1	T2	T2	T2	T2
Literature	5	5	1	1	2	2	2	3	3	3	4
Compound	R_F	R_F	R_{DDD}	R_{TDE}	R_F	R_F × 100	R_F	R_{TDE}	R_{TDE}	R_{TDE}	R_{TDE}
PCNB	—	—	191	175	80[a]	78	60	180	140	370	66[a]
	—	—	—	—	60[c]	—	—	—	—	—	94[a]
	—	—	—	—	60[a]	61[a]	—	—	—	—	—
Perthane	—	—	116	149	40	40	34	130	120	250	66
Perthane olefin	—	—	—	—	—	—	—	180	140	440	—
Ronnel	—	—	—	—	—	—	—	85	110	220	102
Strobane	—	—	—	—	—	—	—	120[c,d]	120[c,d]	230[c,d]	—
	—	—	—	—	—	—	—	170	130	300	—
Sulfenone	—	—	—	—	—	—	—	0	31	0	—
TCNB	—	—	180	164	50	60	47	170	130	320	—
***o,p′*-TDE**	—	—	—	—	—	—	—	—	—	110	—
	—	—	—	—	—	—	—	110	95	220	—
***o,p′*-TDE olefin**	—	—	—	—	—	—	—	160	130	290	—
	—	—	—	—	—	—	—	—	140	—	—
***p,p′*-TDE olefin**	—	—	—	—	—	—	—	180	140	320	—
Tedron (Tetradifon)	—	—	—	—	15	3	17	11	82	90	112
Telodrin (Isobenzan)	—	—	172	167	57	58	52	170	140	350	—
Toxaphene	—	—	—	—	—	—	—	120[c,d]	—	—	—
	—	—	streak	streak	—	—	—	170	120[c,d]	230[c,d]	—
	—	—	—	—	—	—	—	—	130	300	—
Vegadex	—	—	—	—	22[a]	18	30	—	—	—	—

[a] Multiple spots.
[b] Produced 2 distinct spots and several overlapping spots.
[c] Major peaks.
[d] Streak.
[e] Most intense spot.

Layer
L1 = Aluminum oxide G, the substrate was prepared by mixing 20 g aluminum oxide G with 36 ml of water + 7 ml of 1% silver nitrate.
L2 = MN-silica gel-HR.
L3 = Aluminum oxide G.
L4 = Aluminum oxide, 15% calcium sulfate.
L5 = Aluminum oxide G dipped into ethereal *N,N*-dimethylformamide.

Solvent
S1 = *n*-heptane.
S2 = 2% acetone in *n*-heptane.
S3 = *n*-heptane + 1% acetone.
S4 = isooctane-25% DMF.
S5 = immobile, 25% dimethylformamide in ethyl ether, mobile, 2,2,4-trimethylpentane.
S6 = immobile, 2.5% USP mineral oil absorbed on alumina, mobile, methanol-water (85 : 15).
S7 = isooctane.

Detection
See Section II.I, Volume Two.

Technique
T1 = Micro slides, ascending saturated chamber, development time, 5–10 minutes.
T2 = Ascending development.

Literature

1. Meeth, E. and G. L. Roach, *Internal Communication, FDA*, Kansas City, Mo. (1966).
2. Kovacs, Jr., M. F., *J. Ass. Offic. Agr. Chem.*, **49**, 368 (1966).
3. U.S. Department of Health, Ed. and Wel. *Pesticide Analytical Manual*, **Vol. 1**, Table 410.3A,B,C, Jan. (1968).
4. Wells, Clyde E. and H. Moya, Seattle District U.S. Food and Drug Administration, Office of Legislative and Governmental Services, **No. 170**, *Pesticides*, February 23 (1968).
5. Kovacs, Jr., M. F., *J. Ass. Offic. Agric. Chem.*, **48**, 1021 (1965).

TABLE TLC 83
PESTICIDES, INSECTICIDES, CARBAMATE

Layer	L1	L1	L1	L1	L1	L1
Solvent	S1	S2	S3	S4	S5	S6
Literature	1	1	1	1	1	1
Compound	$R_F \times 100$					
Carbamate and its enol						
Carbaryl	50	—	—	—	—	—
α-Naphthol	75	—	—	—	—	—
Dimetilan	—	70	—	—	—	—
3-Methyl-5-hydroxypyrazol-5-dimethylcarbamate	—	30	—	—	—	—
Isolan®	—	—	89	—	—	—
1-Isopropyl-3-methylpyrazolon	—	—	45	—	—	—
Pyrolan	—	—	74	—	—	—
1-Phenyl-3-methylpyrazolon	—	—	65	—	—	—
Zectran®	—	—	—	58	—	—
4-Dimethylamino-3,5-xylenol	—	—	—	75	—	—
Nitrosated enols						
1-Isopropyl-3-methyl-4-oximino-2-pyrazolin-5-one[a]	—	—	—	—	73, 75[d]	—
1-Phenyl-3-methyl-4-oximino-2-pyrazolin-5-one[b]	—	—	—	—	75, 78[d]	—
2-Nitroso-1-naphthol[c]	—	—	—	—	—	58[d]

[a] From isolan.
[b] From pyrolan.
[c] From carbaryl.
[d] Comparative R_F.

Layer L1 = Silica gel_{254}.
Solvent S1 = ethyl ether-hexane (60 : 40).
S2 = benzene-acetone (70 : 30).
S3 = acetone-methanol-benzene (20 : 10 : 80).
S4 = benzene-ethyl acetate (80 : 20).
S5 = benzene-methanol-acetone (80 : 20 : 20).
S6 = hexane-ethyl ether (1 : 1).
Technique Method of Stahl as used by Walker-Beroza.
Literature 1. Eberle, D. O. and F. A. Gunther, *J. Ass. Offic. Anal. Chem.*, **48**, 927 (1965).

TABLE TLC 84
PESTICIDES, ORGANOPHOSPHORUS

Layer	L1	L2	L3	L4	L5	L6	L7	L7	L7
Solvent	S1	S2	S3	S4	S5	S6	S7	S8	S9
Technique	T1	T2	T2	T3	T4	T5	T6	T6	na
Literature	1[a]	2[a]	2[a]	3	4	5	6	7	8
Compound	R_F	$R_{Co\text{-}Ral}$		$R_F \times 100$		R_{par}[d]		R_F	
Abate	—	131	112	—	—	—	—	—	—
Azinphosethyl	68	—	—	—	—	—	—	—	—
Azinphosmethyl (Guthion®), (Gusathion®)	57	67	69	18	4	52	32	70	—
(Benzamid-residue)[b]	—	—	—	—	—	—	24	60	—
Bay 25141 (Dasanit)	33	—	—	—	—	—	—	—	—
Bayer 29493 (Fenthion)	90	181	120	—	—	—	—	—	—
Bidrin®	3	32	0	—	—	—	—	—	—
Bomyl®	13	—	—	—	—	—	—	—	—
Carbophenothion (Trithion®)	96	284	144	65	88	147	—	—	—
Chlorthion®	—	—	—	34	—	—	31	82	43–45
(Chloro-nitrophenol)[b]	—	—	—	—	—	—	67	90	—
Ciodrin®	31	43[c]	31	—	—	—	—	—	—
Co-Ral® (Coumaphos)	72	100	100	—	—	—	—	—	—
Coumaphos *O*-analog	26	—	—	—	—	—	—	—	—
DDVP	—	7	62	—	—	—	—	—	—
Delnav (Dioxathion)	88	155	125	—	—	—	42,32, 100	94	—
Demeton (Systox)	94	394	145	—	44	—	43,58	86	—
Demeton-*O*-methyl	—	—	—	50	—	—	—	—	—
Demeton-*S*-methyl	—	—	—	13	—	—	—	—	—
Demeton thiol sulfone	10	—	—	—	—	—	—	—	—
Desethylphosphamidon	—	15	5	—	—	—	—	—	—
Diazinon®	82	367	125	49	—	111	44	86	—
(Diazine residue)[b]	—	—	—	—	—	—	100	100	—
Diazinoxon	—	365	126	—	—	—	—	—	—
Dibrom (Naled)	53	9[c]	71	—	—	—	—	—	—
Dicapthon	85	—	—	—	—	—	—	—	—
Dichlorvos	42	—	—	—	—	—	—	—	—
Dimethoate (Rogor®)	19	10	7	3	—	—	—	—	4–7
Dimethoate *O*-analog	0	—	—	—	—	—	—	—	—
Dimethoxon	—	0	0	—	—	—	—	—	—
Di-syston (Disulfoton)	100	406	143	65	—	—	—	—	—
Dursban®	—	350	140	—	—	—	—	—	—
Dylox® (Trichlorfon)	9	6	4	—	—	—	—	—	—
EPN	90	—	—	—	—	—	—	—	—
Ethion	92	300	147	—	122	—	—	—	—
Fac	—	—	—	—	—	—	—	—	20–26
Imidan®	57	55	79	—	9	—	—	—	—
Lebaycid	—	—	—	—	—	30,50	54,80	—	—
Malaoxon	20	0	30	—	—	—	—	—	—
Malathion	74	168	116	33	—	76	32,44, 58,66	99–100	52–54
Menazon	—	—	—	0	—	—	—	—	—
Meta-Systox® (Methyl demeton)	—	5	0	—	—	—	30,40, 48	87	62–64, 17,21
Methyl parathion	77	98	117	—	—	—	31	82	—
(Nitrophenol)[b]	—	—	—	—	—	—	81	89	—
Methyl trithion	90	189	140	—	—	—	—	—	—
Oxydemeton methyl	0	—	—	0	—	—	—	—	—
Paraoxon (Diethyl-4-nitrophenyl phosphate)	—	1	43	11	—	—	—	—	—
Parathion	81	177	131	45	53	100	42	84	65–68
(Nitrophenol)[b]	—	—	—	—	—	—	81	90	—
Phenkapton (Phencapton)	—	—	—	62	—	—	40,68	98–100	—
Phorate oxygen-analog	—	—	—	2	—	—	—	—	—
Phorate oxygen-analog sulfone	—	—	—	1	—	—	—	—	—
Phosdrin® (Mevinphos)	23	28	66	—	—	32	—	—	—
Phosphamidon	13	50	6	37	—	—	—	—	—

TABLE TLC 84—(Continued)
PESTICIDES, ORGANOPHOSPHORUS

Layer Solvent Technique Literature	L1 S1 T1 1[a]	L2 S2 T2 2[a]	L3 S3 T2 2[a]	L4 S4 T3 3	L5 S5 T4 4	L6 S6 T5 5	L7 S7 T6 6	L7 S8 T6 7	L7 S9 na 8
Compound	R_F	$R_{Co\text{-}Ral}$		R_F × 100		R_{par}[d]	R_F		
Potasan®	—	—	—	—	—	—	42,56	88	—
(Methylumbelliferone)[b]	—	—	—	—	—	—	80	92	—
Ronnel (Fenchlorphos)	93	325	139	60	—	—	—	—	—
Ruelene®	23	—	—	—	—	—	—	—	—
Thimet® (Phorate)	100	338	146	65	73	—	—	—	—
Thimet sulfone	—	0	12	—	—	—	—	—	—
Thimet sulfoxide	—	84	20	—	—	—	—	—	—
Thiometon	—	—	—	—	—	—	34,28, 58	90	—
Zinophos® (Thionaxin)	92	165	102	—	—	—	—	—	—
Zytron®	79	—	—	—	—	—	—	—	—

[a] Additional data in cited reference.
[b] Cleavage products of pesticide directly above.
[c] Streaks.
[d] $R_{par} = R_F$ relative to parathion migration.

Layer
L1 = SilicAR® 4 (Mallinkrodt).
L2 = Aluminum oxide G.
L3 = Eastman silica sheet.
L4 = Silica gel G, layer activated at 120 °C for two hours.
L5 = Silica gel H.
L6 = MN-Silica gel G-HR.
L7 = Silica gel.

Solvent
S1 = cyclohexane-acetone-chloroform (70 : 25 : 50).
S2 = immobile–20% dimethylformamide in acetone, mobile-methylcyclohexane.
S3 = *n*-hexane-ethyl acetate (3 : 1).
S4 = hexane-acetone (5 : 1).
S5 = 10% hexane in toluene.
S6 = acetone-hexane (15 : 85).
S7 = methanol-methylene chloride-10% ammonia (20 : 80 : 3).
S8 = methanol-methylene chloride-10% ammonia (7 : 13 : 1).
S9 = *n*-hexane-acetone (4 : 1).

Detection See Section II.I, Volume Two.

Technique
T1 = develop 23–29 °C, ascending 10 cm.
T2 = development distance 10 cm, ascending.
T3 = ascending development time—7 minutes.
T4 = thin strip, ascending, 16 cm.
T5 = development distance 15 cm, 25 °C.
T6 = 12 cm development distance, saturated chamber, 30–31 °C.

Literature

1. Getz, M. E. and H. G. Wheeler, *J. Ass. Offic. Agr. Chem.*, **51**, 1101 (1968).
2. Ragab, M. T. H., *Bull. Environ. Contam. Toxicol.*, **2**, 285 (1967).
3. Abbott, D. C., Burbridge, A. S., Thomson, J., and K. S. Webb, *Analyst*, **92**, 171 (1967).
4. Beckman, H. and W. Winterlin, *Bull. Environ. Contam. Toxicol.*, **1**, 78–85 (1966).
5. Mendoza, C. E., Wales, P. J., and D. F. Bray, *Analyst*, **93**, 689 (1968).

6–7. Fischer, R. and W. Klingelhoeller, *Pflanzenschutz Ber.*, **27**, 165 (1961); and *Arch. Toxikol.*, **19**, 119 (1961).

8. Baeumler, J. and S. Rippstein, *Helv. Chim. Acta*, **44**, 1162 (1961).

TABLE TLC 85
PESTICIDES, PYRETHRINS

Layer	L1	L1				
Solvent	S1	S1				
Detection	—	—	D1	D2	D3	D4
Literature	1	1				
Compound	R_F[b] × 100		Color			
	1st direction	2nd direction				
Pyrethrin I	50	62	+	+	+	—
Pyrethrin II	30	41	+	+	+	—
Isopyrethrin I	57	72	+	+	+	—
Isopyrethrin II	35	49	+	+	+	—
Pyrethrin I peroxide	—	23	(+)	(+)	+	+
Pyrethrin II peroxide	—	11	(+)	(+)	+	+
Isopyrethrin I peroxide	—	37 (23)	(+)	(+)	+	+
Isopyrethrin II peroxide	—	17	(+)	(+)	+	+
Lumipyrethrin	—	0	+	+	+	+ (+)
Allethrin	48	—	+[a]	+		
Piperonyl butoxide	35	—	+	+		
Bucarpolate	23	—	+[a]	+		
S 421	67	—	—	—	chlorine detection	
Butter Yellow	42	53	red	red	red	red
Indophenol	35	45	yellowish	yellowish	yellowish	yellowish
Sudan Red G	27	36	red	red	red	red

[a] $SbCl_5$. [b] Irradiation products.

Layer L1 = Silica gel G.
Solvent S1 = hexane-ethyl acetate (75 : 25).
Detection D1 = antimony trichloride or antimony pentachloride.
D2 = phosphmolybdic acid.
D3 = 2,4-dinitrophenylhydrazine.
D4 = potassium-iodide-starch reaction.
Technique Developing in saturated atmosphere; then UV irradiation; 2nd direction with same solvent.
Literature
1. Stahl, E., *Arch. Pharm.*, **293**, 531 (1960) from "*Thin-Layer Chromatography*," E. Stahl, Ed., Academic Press, Inc., New York, 1965, p. 364.

TABLE TLC 86
PESTICIDES, TRIAZINE HERBICIDES

Layer	L1	L1	L1	L1	L2	L2	L3	L3
Solvent	S1	S2	S3	S4	S5	S6	S7	S8
Detection	D1	D1	D1	D1	D1	D1	D1,D2,D3	D2
Technique	T1	T1	T1	T1	T2	T2	na	na
Literature	1	1	1	1	2	2	3	3
Compound	$R_F \times 100$							
Ametryne®	—	—	—	—	59	—	71	76
Atratone	—	—	51	—	34	—	36	50
Atrazine	78	30	84	—	—	37	68	57
Chlorazine	—	—	—	—	—	80	—	—
Ipazine	—	—	—	—	—	66	—	—
Prometone	—	—	56	59	42	—	45	63
Prometryne®	—	—	—	66	68	—	81	76
Propazine	83	37	—	80	—	48	75	69
Simazine	72	21	—	—	—	28	58	44
Simetone	—	—	46	—	26	—	—	—
Simetryne®	—	—	—	—	50	—	—	—
Trietazine	—	—	—	—	—	60	88	69
G 34360	—	—	—	—	—	—	58	50

Layer
L1 = Silica gel G dried at 100 °C.
L2 = Silica gel G dried at room temperature.
L3 = Silica gel G/UV_{254} (Machery-Nagel & Co.) for D2 or Silica gel G (Merck) for D1 and D3.

Solvent
S1 = ethyl acetate-petroleum ether (110–140 °C) (7 : 3).
S2 = ethyl acetate-petroleum ether (100–140 °C) (3 : 7).
S3 = chloroform-absolute ethanol-ethyl acetate (18 : 1 : 1).
S4 = toluene-acetic acid-water (10 : 10 : 1).
S5 = chloroform-nitromethane (1 : 1).
S6 = chloroform-nitromethane (5 : 1).
S7 = toluene-acetone (85 : 15).
S8 = carbon tetrachloride-absolute diethylamine (95 : 15), suitable for separation of simazine and G34360.

Detection
D1 = spray the dried plates with 0.02–0.1*N* silver nitrate solution, then expose to UV short-wave lamp or daylight.
D2 = UV (254 nm).
D3 = chlorine gas; after 1 minute spray with potassium iodide-starch reagent.

Technique
T1 = ascending, 15 cm development.
T2 = ascending, 10 cm development.

Literature

1. Stammbach, K., Kilchher, H., Friedrich, K., Larsen, M., and G. Székely, *Weed Res.*, **4**, 64 (1964).
2. Henkel, H. G., *Chimia* (Aaran), **18**, 252 (1964).
3. Delley, R., Friedrich, K., Karlhuber, B., Székely, G., and K. Stammbach, *Z. Anal. Chem.*, **228**, 32 (1967); from *J. Chromatogr.*, **34**, D44 (1968).

TABLE TLC 87
PHENOLS

Layer	L1	L1	L1	L1	L1	L2	L2	L2	L2	L2	L2
Solvent	S1	S2	S3	S4	S5	S6	S7	S8	S9	S10	S11
Literature	1	1	1	1	1	1	1	1	1	1	1
Compound						$R_F \times 100$					
Phenol	28	66	68	32	35	64	72	66	64	22	68
4-Methylphenol	30	—	—	33	38	68	74	71	70	26	71
4-Ethylphenol	32	—	—	34	40	71	77	76	76	31	74
4-Propylphenol	24	69	71	35	42	75	79	81	83	37	77
Guaiacol	63	68	68	53	49	—	83	86	78	75	85
4-Methylguaiacol	60	—	—	—	49	—	86	—	80	81	87
4-Ethylguaiacol	59	—	—	—	50	—	88	—	82	82	88
4-Propylguaiacol	57	68	66	53	52	—	89	88	85	84	90
Catechol	7	62	64	8	17	35	42	37	37	4	41
4-Methylcatechol	—	—	—	9	19	36	47	42	42	5	45
4-Ethylcatechol	—	—	—	10	20	38	54	47	49	6	49
4-Propylcatechol	9	64	59	13	22	42	59	53	57	7	54
Vanillin	37	58	53	—	36	66	81	—	53	47	67
4-Acetoguaiacone	33	56	53	—	34	68	84	—	54	50	69
4-Propioguaiacone	38	61	59	—	40	70	87	—	63	61	74
Protocatechuic aldehyde	—	51	54	3	7	14	40	14	14	—	27
4-Acetocatechol	—	50	56	5	7	18	46	18	15	—	28
4-Propiocatechol	—	52	57	6	10	22	50	22	21	—	32

Layer L1 = Silica gel (E. Merck AG), activated at 100–105 °C for 30 minutes.
L2 = Polyamide (Woelm AG).

Solvent S1 = chloroform.
S2 = ethyl acetate
S3 = *n*-propanol.
S4 = chloroform-methanol (99 : 1).
S5 = benzene-methanol (95 : 5).
S6 = benzene-ethyl acetate (3 : 7).
S7 = benzene-methanol-acetic acid (45 : 8 : 4).
S8 = benzene-dioxane-acetic acid (90 : 25 : 4).
S9 = *n*-butyl ether (saturated with water)-acetic acid (10 : 1).
S10 = *n*-butanol-benzene-water (1 : 40 : 40), organic phase.
S11 = xylene-methyl ethyl ketone-formamide (25 : 25 : 1).

Detection Folin-Denis reagent, followed by 25% ammonia.

Literature
1. Halmekoski, J. and H. Hannikainen, *Suomen Kemistilehti*, **B36**, 24 (1963); from *J. Chromatogr.*, **15**, D6–D7 (1964).

TABLE TLC 88
PHENOLS

Layer	L1	L1	L1	L1	L1	L2	L3	L3	L2	L2
Solvent	S1	S2	S3	S4	S5	S6	S7	S8	S4	S5
Literature	1	1	1	1	1	1	1	1	1	1
Compound	$R_F \times 100$									
Phenol	63	80	23	53	64	62	81	—	—	—
Hydroquinone	32	50	2	6	8	23	36	79	—	—
Resorcinol	38	58	2	8	—	22	33	77	—	—
Catechol	48	70	6	18	22	35	52	82	—	—
1,2,3-Trihydroxybenzene	27	44	2	3	5	18	27	75	—	—
1,2,4-Trihydroxybenzene	20	30	0	1	0	14	22	—	—	—
1,3,5-Trihydroxybenzene	17	27	0	0	0	10	17	63	—	—
2-Hydroxydiphenyl ether	77	91	65	80	93	83	95	^{t}	—	—
3-Hydroxydiphenyl ether	64	86	49	58	69	75	89	^{t}	—	—
4-Hydroxydiphenyl ether	59	83	40	55	65	78	92	^{t}	—	—
o-Aminophenol	24	40	—	—	—	45	67	—	—	—
m-Aminophenol	23	44	—	—	—	43	62	—	—	—
p-Aminophenol	9	12	—	—	—	23^{t}	38	—	—	—
o-Cresol	62	90	38	55	76	63	87	—	—	—
m-Cresol	59	88	30	50	65	57	85	—	—	—
p-Cresol	60	86	26	46	61	58	85	—	—	—
2,3-Dihydroxytoluene	51	78	11	25	—	44	65	75	—	—
2,5-Dihydroxytoluene	35	67	4	12	—	27	42	76	—	—
2,6-Dihydroxytoluene	44	—	7	18	—	26	42	—	—	—
3,4-Dihydroxytoluene	46	78	7	17	—	35	59	74	—	—
3,5-Dihydroxytoluene	40	66	3	8	—	23	42	74	—	—
2,2′-Dihydroxybiphenyl	58	86	14	34	43	56	79	69	—	—
2,4′-Dihydroxybiphenyl	50	78	9	24	—	32	50	54	—	—
2,5-Dihydroxybiphenyl	49	77	11	24	35	37	52	62	—	—
3,3′-Dihydroxybiphenyl	47	73	5	13	17	19	35	33	—	—
4,4′-Dihydroxybiphenyl	42	69	5^{t}	13^{t}	19	17	30	22^{t}	—	—
2,3,2′,3′-Tetrahydroxybiphenyl	30^{t}	50^{t}	1	3	4	19	33	61	—	—
3,4,3′,4′-Tetrahydroxybiphenyl	18	28	0	0	—	3	7	22	—	—
2,5,2′,5′-Tetrahydroxybiphenyl	19	28	0	0	—	4	9	67	—	—
2,6,2′,6′-Tetrahydroxybiphenyl	30	44	0	0	—	15	30	49	—	—
2,4,2′,4′-Tetrahydroxybiphenyl	22	30	0	0	—	6	10	60	—	—
4,4′-Dihydroxystilbene	45	65	0	0	—	19	34	0	—	—
4,4′-Dihydroxydiphenylmethane	48	66	—	13	20	36	58	^{t}	—	—
2,2′-Dihydroxydiphenylmethane	64	78	—	30	51	53	75	—	—	—
4-Hydroxybenzyl alcohol	35	48	—	5	7	37	57	—	—	—
p-Benzoquinone	—	—	56	77	—	—	99	—	—	—
Diphenoquinone	—	97	—	—	—	—	99	—	—	—
Tetrahydroxy-*p*-benzoquinone	—	—	0	0	—	—	17^{t}	—	—	—
2,5-Dihydroxy-*p*-benzoquinone	—	—	0	2^{t}	—	—	40^{t}	—	—	—
2,2′-Dihydroxy-5,5′-dimethylbiphenyl	60	83	22	38	55	67	85	^{t}	—	—
2-Hydroxy-4′,5-dimethyldiphenyl ether	82	92	53	78	95	89	97	0	—	—
4,4′-Dihydroxy-3,3′-dimethylbiphenyl	56	70	13	30	—	28	45		—	—
4-Hydroxy-2′,3-dimethyldiphenyl ether	—	—	—	85	—	90	97	—	—	—
2-Bromophenol	70	48	—	—	—	—	—	—	63	80
3-Bromophenol	65	32	—	—	—	—	—	—	53	72
4-Bromophenol	62	26	—	—	—	—	—	—	54	73
2,3-Dihydroxybiphenyl	57	—	39	—	—	—	—	—	53	77
2,4-Dihydroxybiphenyl	47	—	19	—	—	—	—	—	30	52
2,5-Dihydroxybiphenyl	49	—	26	—	—	—	—	—	35	53
2,6-Dihydroxybiphenyl	60	—	49	—	—	—	—	—	47	67
3,4-Dihydroxybiphenyl	52	—	25	—	—	—	—	—	37	58
Diphenylene oxide	92	78	89	—	—	—	—	—	99	99

^{t} = tailing.

TABLE TLC 88—(Continued)
PHENOLS

Layer	L1 = Silica gel G7731.
	L2 = Polyamide MN.
	L3 = Cellulose powder 300 GF_{254}.
Solvent	S1 = benzene-methanol-acetic acid (70 : 8 : 4).
	S2 = benzene-methanol-acetic acid (70 : 8 : 4) (2 developments).
	S3 = benzene-methanol-acetic acid (80 : 1 : 1).
	S4 = benzene-methanol-acetic acid (80 : 1 : 1) (2 developments).
	S5 = benzene-methanol-acetic acid (80 : 1 : 1) (3 developments).
	S6 = benzene-methanol-acetic acid (70 : 8 : 4).
	S7 = benzene-methanol-acetic acid (70 : 8 : 4) (2 developments).
	S8 = benzene-*n*-butyl alcohol-water (9 : 1 : 10) (lower phase).
Detection	Diazotized benzidine or diazotized sulfanilic acid.
Technique	Ascending development, 15 cm.
Literature	1. Joschek, H. I. and S. I. Muller, *J. Amer. Chem. Soc.*, **88**, 3276 (1966).

TABLE TLC 89
PHENOLS, ACYL

Layer	L1	L1	L1	L1
Solvent	S1	S1	S2	S2
Literature	1	1	1	1
	$R_{2,\text{propionylphenol}} \times 100$		$R_{4,\text{propionylphenol}} \times 100$	
Compound	Phenol	2,4-DNPH[a]	Phenol	2,4-DNPH[a]
2-Acetylphenol	76	85	—	—
4-Acetylphenol	—	—	74	93
2-Bromo-4-chloro-6-propionylphenol	133	126	—	—
2-Bromo-3,4-dimethyl-6-propionylphenol	103	109	—	—
2-Bromo-4-fluoro-6-propionylphenol	128	117	—	—
2-Bromo-4-methyl-6-propionylphenol	109	109	—	—
2-Bromo-4-propionylphenol	—	—	200	213
2-Bromo-6-propionylphenol	114	112	—	—
3-Bromo-4-propionylphenol	—	—	137	—
3-Bromo-6-propionylphenol	130	129	—	—
4-Bromo-6-propionylphenol	131	129	—	—
2-Butyrylphenol	111	109	—	—
4-Butyrylphenol	—	—	115	106
2-Caproylphenol	124	120	—	—
4-Caproylphenol	—	—	125	113
2-Caprylphenol	150	146	—	—
4-Caprylphenol	—	—	157	116
2-Caprylylphenol	129	135	—	—
4-Caprylylphenol	—	—	138	110
2-Chloro-4-bromo-6-propionylphenol	133	125	—	—
2-Chloro-6-bromo-4-propionylphenol	—	—	315	—
3-Chloro-2,4-dibromo-6-propionylphenol	137	127	—	—
3-Chloro-2,6-dibromo-4-propionylphenol	—	—	383	—
2-Chloro-4-propionylphenol	—	—	200	199
2-Chloro-6-propionylphenol	110	107	—	—
3-Chloro-4-propionylphenol	—	—	130	—
3-Chloro-6-propionylphenol	121	126	—	—
4-Chloro-6-propionylphenol	123	127	—	—
2,4-Dibromo-6-propionylphenol	145	140	—	—
2,6-Dibromo-4-propionylphenol	—	—	371	329
2,4-Dichloro-6-propionylphenol	132	123	—	—
2,3-Dimethyl-4-bromo-6-propionylphenol	146	140	—	—
2,3-Dimethyl-6-bromo-4-propionylphenol	—	—	395	—

TABLE TLC 89—(Continued)
PHENOLS, ACYL

Layer	L1	L1	L1	L1
Solvent	S1	S1	S2	S2
Literature	1	1	1	1
	$R_{2,\text{propionylphenol}} \times 100$		$R_{4,\text{propionylphenol}} \times 100$	
Compound	Phenol	2,4-DNPH[a]	Phenol	2,4-DNPH[a]
2,5-Dimethyl-6-bromo-4-propionylphenol	—	—	395	—
3,5-Dimethyl-4-bromo-6-propionylphenol	25	—	—	—
3,5-Dimethyl-2,4-dibromo-6-propionylphenol	57	—	—	—
2,3-Dimethyl-4-propionylphenol	—	—	169	141
2,3-Dimethyl-6-propionylphenol	113	124	—	—
2,4-Dimethyl-6-propionylphenol	115	122	—	—
2,5-Dimethyl-4-propionylphenol	—	—	165	149
2,6-Dimethyl-4-propionylphenol	—	—	186	228
3,4-Dimethyl-6-propionylphenol	87	98	—	—
3,5-Dimethyl-6-propionylphenol	59	—	—	—
4-Fluoro-6-propionylphenol	114	106	—	—
2-Formylphenol	94	85	—	—
4-Formylphenol	—	—	85	84
2-Methyl-4-bromo-6-propionylphenol	138	138	—	—
2-Methyl-6-bromo-4-propionylphenol	—	—	330	—
3-Methyl-2,4-dibromo-6-propionylphenol	137	133	—	—
3-Methyl-2,6-dibromo-4-propionylphenol	—	—	430	—
2-Methyl-4-propionylphenol	—	—	133	140
2-Methyl-6-propionylphenol	114	126	—	—
3-Methyl-4-propionylphenol	—	—	130	100
3-Methyl-6-propionylphenol	96	103	—	—
4-Methyl-6-propionylphenol	98	100	—	—
2-Propionylphenol	100	100	—	—
4-Propionylphenol	—	—	100	100
2,3,4-Tribromo-6-propionylphenol	148	127	—	—
2,3,6-Tribromo-4-propionylphenol	—	—	400	—
2-Valerylphenol	121	116	—	—
4-Valerylphenol	—	—	122	110

[a] DNPH = Dinitrophenylhydrazones.

Layer L1 = Silica gel.
Solvent S1 = benzene.
S2 = methylene chloride-acetic acid (96 : 4).
Literature 1. Martin, R. and J. M. Betoux, *Chim. Anal. (Paris)*, **49**, 128–138 (1967).

TABLE TLC 90
PHENOLS AND PHENOL ETHERS

Layer	L1	L1	L1				
Solvent	S1	S2	S3				
Detection	—	—	—	D1		D2[a]	
Literature	1	1	1				
				Color[b]			
Compound	R_{Thymol} × 100			20 °C	110 °C	20 °C	110 °C
Anethol	166	493	532	bl-vt	vt	lt y	bt y
Carvacrol	94	110	110	br	br	pk	br
Chavicol methyl ether	166	500	425	y	y-br	—	—
Cresol	107	102	131	br	y-br	y-br	ro
Dihydroeugenol	120	198	138	—	y-br	—	wk pk
Eugenol	124	137	104	bl-vt	gy-br	—	y-br
Eugenol benzyl ether	153	237	282	r-br	br	—	—
Eugenol methyl ether	148	128	235	bl-vt	gy	—	—
Guaiacol	113	145	138	—	gy	—	—
	124	177	177	gy-bl	gy-bl	—	br
Hydroquinone	2	0	0	—	y-br	—	br bl (UV)
Hydroquinone dimethyl ether	156	336	396	—	y-br	—	—
***p*-Hydroxyanisole**	37	14	35	—	—	—	vt
Isochavibetol	118	133	93	—	r-br	—	y
Isoeugenol	120	123	97	bl-vt	gy-br	bt-y	y-br
Isoeugenol benzyl ether	165	208	244	r-br	br	—	wk y
Isoeugenol methyl ether	152	130	223	bl-vt	gy-vt	bt y	y
Isomyristicin	158	191	339	vt	vt	—	y
Isosafrol	166	506	550	bl-vt	vt	bt y	bt y
Isothymol	67	63	81	—	—	—	r-br
3,4-Methylenedioxyphenol	42	18	39	gr	gy-gr	o	o-br
Myristicin	154	187	393	br	br	—	—
α-Naphthol	67	43	50	gy	gy	o	o-br
β-Naphthol	52	32	41	y-br	gy-br	pk	o-r
β-Naphthol ethyl ether	164	400	637	gy-br	gy	—	wk pk
β-Naphthol isobutyl ether	176	449	800	gy-br	gy	—	ro
β-Naphthol methyl ether	168	341	518	y-br	gy	—	wk pk
Orcin	3	0	0	—	br	o	o
Phenol	49	47	60	—	gy	—	br
Phloroglucinol	0	0	0	—	br	o-y	r-br
Propenylguaethol	138	146	165	y-br	br	—	wk pk
Pyrocatechol	10	0	10	bl	gy-bl	—	r-br
Pyrogallol	0	0	0	—	br	y	y-br
Resorcinol	0	0	0	gr	br	o-y	o
Safrol	164	545	573	gy	bl-bk	—	—
Thymol	100	100	100	r-br	r-br	pk	br

[a] Most colors can be intensified by spraying with sodium carbonate solution.
[b] Temperatures refer to heating TLC plates.

Layer L1 = Silica gel G layers (dried at 105 °C).
Solvent S1 = chloroform.
S2 = petroleum ether (80–100 °C)–acetic acid (95 : 5).
S3 = petroleum ether (80–100 °C)–pyridine (95 : 5).
Detection D1 = $SbCl_5$-CCl_4.
D2 = *p*-nitraniline-$NaNO_2$.
Technique Pure compounds applied in 1% acetone solution using 10–20 μg per spot.
Literature 1. Klouwen, M. H. and R. ter Heide, *Parfuem. Kosmet.*, **43**, 195 (1962).

TABLE TLC 91
PHENOLS, 4-ACETYL-2-NITROPHENOL DERIVATIVES (ANP)

Layer	L1	L2	L1	L2
Solvent	S1	S1	S2	S2
Detection	D1	D2	D1	D2
Literature	1	1	1	1
Compound	$R_F \times 100$			
ANP ether of:				
Phenol	62	60	82	65
***o*-Cresol**	65	65	81	86
***m*-Cresol**	60	52	71	70
***p*-Cresol**	65	65	82	85
2,3-Dimethylphenol	65	46	69	70
2,4-Dimethylphenol	69	62	82	80
2,5-Dimethylphenol	70	65	83	80
3,4-Dimethylphenol	62	64	83	81
Thymol	72	74	76	69
Guaiacol	44	34	61	60
3-Methoxyphenol	50	60	73	70
2,3-Dimethoxyphenol	50	28	54	20
2,6-Dimethoxyphenol	46	29	53	25
***o*-Chlorophenol**	53	52	65	49
2,4,6-Trichlorophenol	69	75	70	56
***o*-Nitrophenol**	34	19	45	69
***m*-Nitrophenol**	49	38	50	63
***p*-Nitrophenol**	55	40	53	64
***p*-Hydroxybenzaldehyde**	32	22	54	51
Vanillin	29	23	65	36
4-Hydroxy-3-methoxyacetophenone	34	13	41	15
4-Hydroxy-2-methoxyacetophenone	42	21	36	14
2-Hydroxy-4-methoxyacetophenone	40	19	38	24
α-Naphthol	58	53	73	81
β-Naphthol	69	62	65	51
***p*-Hydroxybenzoic acid**	0	7	0	0
Vanillic acid	0	4	0	0
Syringic acid	0	3	0	8
***p*-Hydrocinnamic acid**	0	4	0	13
Ferulic acid	0	4	5	0
***N*-ANP-*m*-Aminophenol**	28	27	36	10
***N*-ANP-*p*-Aminophenol**	45	25	44	44
Di-ANP-ether of:				
Pyrocatechol	20	10	39	11
Resorcinol	16	14	54	29
Hydroquinone	34	20	45	45
Orcinol	34	17	48	29
2-Methylresorcinol	22	23	49	19
Methylhydroquinone	26	28	45	16
Tri-ANP-ether of:				
Phloroglucinol	12	7	22	43
2,4,6-Trihydroxyacetophenone	5	3	13	29
N-ANP-derivative of:				
***o*-Aminophenol**	56	38	11	8

Layer L1 = Silica gel G.
L2 = Polygram SIL S-HR/UV_{254}.
Solvent S1 = cyclohexane-cyclohexanone (3 : 1).
S2 = methylene chloride-cyclohexane (5 : 1).
Detection D1 = Rhodamine B, UV.
D2 = UV.
Technique Ascending; 13 cm.
Literature 1. Wildenhain, W., Henske, G., and G. Bienert, *J. Chromatogr.*, **45**, 158–160 (1969).

TABLE TLC 92
PHENOLS, POLYNUCLEAR

Layer	L1	L2	L3
Solvent	S1	S2	S3
Literature	1	1	1
Compound		$R_F \times 100$	
Acacetin (methoxyflavone)	0	55	0
1-Anthrol	68	91	0
Apigenin	0	47	0
Astragalin	0	30	11
D-Catechin	—	8	28
Chrysin	0	61	0
1,4-Dihydroxyanthraquinone	4	90	0
6,11-Dihydroxynaphthacenequinone	0	93	0
Ellagic acid	0	13	0
Esculin	0	0	65
Fisetin	0	—	—
Flavanone	0	54	0
1-Hydroxyanthraquinone	12	90	0
4-Hydroxyxanthrone	28	59	0
4-Hydroxybenzylideneacetophenone	—	63	—
2-Hydroxycarbazole	30	49	0
6-Hydroxychrysene	53	90	0
4-Hydroxycoumarin	—	0	0
7-Hydroxycoumarin	0	43	14
2-Hydroxydibenzofuran	60	83	0
o-Hydroxydiphenyl	83	93	0
p-Hydroxydiphenyl	73	83	0
3-Hydroxyflavone	17	87	0
2-Hydroxyfluorene	71	87	0
3-Hydroxyfluorene	75	86	0
5-Hydroxyisoquinoline	19	9	81
7-Hydroxy-4-methylcoumarin	0	39	10
1-Hydroxypyrene	48	89	0
5-Hydroxyquinoline	30	8	85
4-Hydroxyquinoline	0	0	65
8-Hydroxyquinoline	3	10	81
Malvin	0	0	90
4′-Methoxy-3′,5,7-trihydroxyflavanone	0	57	5
3-Methylchromone	0	0	64
4-Methyldaphnetine	—	29	17
Morin	—	16	0
Myricetin	0	12	0
9-Naphthacenone	82	94	0
1-Naphthol	80	90	0
2-Naphthol	74	89	0
2-Quinolinol	—	53	—
Robinin	0	0	—
Rutin	0	47	23
Techtochrysin	3	80	—
4′,5,7-Trihydroxyflavanone	0	58	—

Layer L1 = Silica gel, Analtech.
L2 = Silica gel, Analtech., saturated with 0.3*M* sodium acetate.
T3 = polyamide.

Solvent S1 = triethylamine.
S2 = toluene-ethyl formate-formic acid (5 : 4 : 1).
S3 = water-ethanol-2-butanone-acetylacetone (65 : 15 : 15 : 5).

Detection See Section II.I, Volume Two.

Literature
1. Sawicki, E., Guyer, M., Schumacher, R., Elbert, W. C., and C. R. Engel, *Mikrochim. Acta*, **1027** (1968).

TABLE TLC 93
PIGMENTS, CAROTENOIDS

Layer	L1	L1	L1	L1	L1	L1	L1
Solvent	S1	S2	S3	S4	S5	S6	S7
Technique	T1	T1	T1	T1	T1	T1	T1
Literature	1	1	1	1	1	1	1
Compound	$R_F \times 100$						
β-Apo-8′-carotenic acid	28	38	30	15	5	0	0
Astacin	34	50	69	72	42	25	12
Canthaxanthin	58	65	79	80	55	37	20
Capsanthin	24	42	79	81	62	42	25
Capsorubin	19	37	74	76	60	42	25
β-Carotene	100	100	100	80	25	10	5
Cryptoxanthin	62	70	76	74	39	21	9
Dehydroadonirubin	47	59	74	76	48	30	15
Dihydrocanthaxanthin	55	61	70	65	35	17	8
Dihydrocanthaxanthin, red	22	30	37	34	18	8	4
Dihydrorhodoxanthin	81	85	87	87	69	47	27
3,4-Dihydroxy-β-carotene	50	58	72	81	56	37	20
Equinenone	91	92	90	72	35	18	6
Escholtzxanthin	12	22	25	22	8	1	0
Euglenanone	62	68	81	80	54	34	17
Fucoxanthin	27	44	84	98	95	85	75
Isozeaxanthin	34	56	92	91	57	36	19
Lutein	35	57	93	95	68	45	24
Lycopene	95	90	75	60	5	0	0
Lycophyll	8	20	22	20	7	0	0
Lycoxanthin	29	37	40	32	8	0	0
Neoxanthin	22	42	82	96	90	78	60
Physaliene	100	100	95	30	0	0	0
Rhodoxanthin	28	42	43	40	14	7	2
Rubixanthin	45	60	64	45	15	4	0
Tareoxanthin	28	42	84	92	78	62	43
3,4,3′,4′-Tetrahydroxy-β-carotene	18	33	86	95	81	62	44
Torularhodin	6	10	9	2	1	0	0
3,4,4′-Trihydroxy-β-carotene	31	50	88	92	76	53	38
Violaxanthin	30	52	76	93	88	73	55
Zeaxanthin	30	54	78	82	55	35	18

Layer L1 = Polyamide (Merck) + 15% cellulose MN-300.

Solvent S1 = petroleum ether (b.p. 100–120 °C)-methanol-methyl ethyl ketone (8 : 1 : 1).
S2 = petroleum ether-methanol-methyl ethyl ketone (4 : 1 : 1).
S3 = petroleum ether-methanol-methyl ethyl ketone (2 : 1 : 1).
S4 = methanol-methyl ethyl ketone (1 : 1).
S5 = water-methanol-methyl ethyl ketone (1 : 5 : 5).
S6 = water-methanol-methyl ethyl ketone (1 : 3 : 3).
S7 = water-methanol-methyl ethyl ketone (1 : 2 : 2).

Technique T1 = Ascending.

Literature
1. Egger, K. and H. Voigt, *Z. Pflanzenphysiol.*, **53**, 70 (1965); from *J. Chromatogr.*, **23**, D6 (1966).

TABLE TLC 94
PIGMENTS, CAROTENOIDS

Layer	L1	L1	L1	L1	L1	L1	L2	L3	L4	L4	L4
Solvent	S1	S2	S3	S4	S5	S6	S6	S7	S8	S9	S10
Technique	T1	T2	na	na	na	T3	T4	T5	na	na	na
Literature	1	2	3	3	4	5	6	7	8	8	8
Compound	$R_F \times 100$										
β-Apo-8′-carotenal	—	34	—	—	—	—	—	—	—	—	—
Bixin	51	8	—	—	—	—	—	—	—	—	—
Canthaxin	38	9	2	28	—	—	—	—	—	—	—
Capaxanthin	—	0	—	—	—	—	—	—	—	—	—
α-Carotene	—	—	—	—	—	—	75	0	26	66	80
β-Carotene	96	97	75	83	—	—	64	0	11	49	74
γ-Carotene	—	—	—	—	—	—	—	—	0	11	41
δ-Carotene	—	—	—	—	—	—	—	—	0	20	55
ε-Carotene	—	—	—	—	—	—	—	—	47	70	84
Crocetin	—	16	—	—	—	—	—	—	—	—	—
Cryptocapsin	—	—	—	—	47	—	—	—	—	—	—
Cryptoxanthin	—	—	—	—	—	—	—	7	—	—	—
Ethyl β-apo-8′-carotenoate	—	54	—	—	—	—	—	—	—	—	—
Isozeaxanthin	63	—	—	—	—	—	—	—	—	—	—
Lutein	—	—	—	—	—	—	—	56	—	—	—
Luteinepoxide	—	—	—	—	—	—	—	72	—	—	—
Lycopene	—	96	75	81	—	—	—	—	0	0–2	13
Lycopersene	—	—	—	—	—	30	—	—	—	—	—
Methyl β-apo-8′-carotenoate	—	47	—	—	—	—	—	—	—	—	—
Neoxanthin	—	—	—	—	—	—	—	95	—	—	—
Phytoene	—	—	—	—	—	21	—	—	—	—	—
Phytofluene	—	—	—	—	—	12	—	—	—	—	—
Rhodoxanthin	—	—	—	—	—	—	—	26	—	—	—
Violaxanthin	—	—	—	—	—	—	—	84	—	—	—
Zeaxanthin	17	—	—	—	—	—	—	55	—	—	—

Layer
L1 = Silica gel.
L2 = Aluminum oxide.
L3 = Kieselguhr impregnated with vegetable oil.
L4 = "Darlington" light magnesium oxide, activated 1 hour at 120 °C.

Solvent
S1 = *n*-hexane-ether (3 : 7).
S2 = petroleum ether-benzene-acetone-acetic acid (80 : 20 : 2 : 1).
S3 = petroleum ether (60–70 °C)-benzene-acetone (160 : 40 : 4).
S4 = petroleum ether (60–70 °C)-ethyl acetate-chloroform (160 : 40 : 4).
S5 = petroleum ether (60–80 °C)-acetone (98 : 2).
S6 = petroleum ether.
S7 = methanol-acetone-water (20 : 4 : 3) saturated with the impregnating oil.
S8 = petroleum ether (90–110 °C)-benzene (9 : 1).
S9 = petroleum ether (90–110 °C)-benzene (5 : 5).
S10 = petroleum ether (90–110 °C)-benzene (1 : 9).

Technique
T1 = development distance, 7.6 cm.
T2 = development distance, 10 cm.
T3 = circular development, quantitative method.
T4 = development distance, 30 cm, ascending.
T5 = reverse-phase partition chromatography.

Literature

1. Demole, E., *J. Chromatogr.*, **1**, 24 (1958).
2. Benk, E., Wolff, I., and H. Treiker, *Deut. Lebensm-Rundsch.*, **59**, 39 (1963).
3. Rispolo, G. and A. D. Giacomo, *Boll. Lab. Chim. Provincal. (Bologna)*, **13**, 587 (1962); *Chem. Abstr.*, **59**, 8044 (1963).
4. Cholnoky, L., Szabolcs, J., Cooper, R. D. G., and B. C. L. Weedon, *Tetrahedron Lett.*, 1963, 1257.
5. Lagoni, H. and A. Wortmann, *Milchwissenschaft*, **11**, 206 (1956).
6. Davídek, J. and J. Blattná, *J. Chromatogr.*, **7**, 204 (1962).
7. Egger, K., *Planta*, **58**, 664 (1962).
8. Bolliger, H. R., König, A., and V. Schweiter, *Chimia* (Aarau), **18**, 136 (1964); from *J. Chromatogr.*, **17**, D8 (1965).

TABLE TLC 95
STEROIDS

Layer	L1	L1	L1	L2	L2	L2
Solvent	S1	S2	S3	S4	S5	S6
Literature	1	1	1	1	1	1
Compound			$R_F \times 100$			
Mestranol	35	75	—	—	—	68
Methylandrostenolone acetate	48	—	—	—	—	—
Ethynodiol diacetate	—	50	—	—	—	—
Lynestrenol	—	20	58	—	—	87
Androstenediol dipropionate	—	—	31	—	—	—
Ethylestrenol	—	—	22	—	—	—
Nortestosterone acetate	72	95	—	—	—	70
Nortestosterone phenylpropionate	—	65	—	—	—	—
Nortestosterone benzoate	—	63	—	—	—	—
Nortestosterone decanoate	—	6	57	—	—	—
Testosterone acetate	62	92	—	—	—	75
cis-Testosterone acetate	67	—	—	—	—	73
Chlortestosterone acetate	34	—	—	—	—	80
Methyltestosterone acetate	43	—	—	—	—	—
Estrone acetate	—	86	—	—	—	—
Estrone benzoate	—	45	—	—	—	—
Ethynylestradiol acetate	—	64	—	—	—	—
Ethynylestradiol benzoate	11	57	—	—	—	—
Nortestosterone	—	—	—	21	—	—
Testosterone	—	—	—	27	—	—
Chlorotestosterone	—	—	—	31	—	—
cis-Testosterone	—	—	—	36	—	—
Methandienone	—	—	—	20	—	—
Norethisterone	—	—	—	29	—	—
Ethisterone	—	—	—	34[a]	—	—
Methylestrenolone	—	—	—	40	—	—
Dehydroandrosterone (iso)	—	—	—	46	—	—
Methyltestosterone	—	—	—	48	—	—
Androstanolone	—	—	—	56	—	—
Norethynodrel	—	—	—	61	—	—
Epiestriol (Estriol)	—	—	—	—	0	—
Fluoxymestrone	—	—	—	—	16	—
β-Estradiol	—	—	—	—	24	—
α-Estradiol	—	—	—	—	31	—
Ethinylestradiol	—	—	—	—	37	—
Androstenediol	—	—	—	—	41	—
Methandriol	—	—	—	—	50	—
Estrone	—	—	—	—	73	—
Testosterone	—	—	—	—	81	—
Desoxycortone acetate	—	—	—	—	—	38
Isopregnenone	—	—	—	—	—	50
Megestrol acetate	—	—	—	—	—	48
Progesterone	—	—	—	—	—	60
Stanazolol	—	—	—	—	—	70
Testosterone propionate	—	—	—	—	—	85
Testosterone isobutyrate	—	—	—	—	—	95

[a] Tailing.

Layer L1 = Kieselguhr G impregnated with 10% paraffin oil in light petroleum.
L2 = L1 impregnated with 10% propylene glycol in acetone.

Solvent S1 = 30% acetic acid.
S2 = 50% acetic acid.
S3 = 70% acetic acid.
S4 = cyclohexane-toluene (4 : 1).
S5 = toluene.
S6 = petroleum ether.

Detection *p*-Toluenesulfonic acid.

Technique Partition chromatography.

Literature 1. Sonanini, D. and L. Anker, *Pharm. Acta Helv.*, **42**, 56–59 (1967); from *J. Chromatogr.*, **32**, D46-D48 (1968).

TABLE TLC 96
STEROIDS

Layer	L1	L1	L1	L1
Solvent	S1	S1	S2	S2
Compound, systematic name	$R_F \times 100$	R_S[a] $\times 100$	$R_F \times 100$	R_S[a] $\times 100$
17α-Methyl-3,5-seco-4-nor-androstan-17β-ol-5-on-3-oic acid	48	109	—	—
N-Benzyl-3,5-seco-4-norcholestan-5β-ol-3-amide	50	125	—	—
4-Oxa-5α-cholestan-3-one	69	157	—	—
4-Benzyl-4-aza-5-cholesten-3-one	68	155	—	—
4,17α-Dimethyl-4-aza-5-androsten-17β-ol-3-one	27	61	—	—
4-(β-Hydroxyethyl)-4-aza-5-cholesten-3-one	44	100	—	—
4-Phenyl-4-aza-5-cholesten-3-one	66	150	—	—
4-Methyl-4-aza-5-cholesten-3-one	71	161	—	—
4-Oxa-5α-pregnane-3,20-dione	57	129	—	—
17α-Methyl-4-aza-5-androsten-17β-ol-3-one	16	36	—	—
4-Aza-5-cholesten-3-one	52	118	—	—
4-Methyl-4-aza-5-pregnene-3,20-dione	45	102	—	—
4-Methyl-4-aza-5-cholestene	81	184	—	—
4β,5β-Epoxypregnane-3,20-dione	68	149	—	—
4α,5α-Epoxypregnane-3,20-dione	71	161	—	—
4-Aza-5-pregnene-3,20-dione	42	95	—	—
3,5-Seco-4-norpregnan-20β-ol-5-on-3-oic acid	47	107	—	—
17α-Methyl-4-oxa-5-androsten-17β-ol-3-one acetate	64	145	—	—
17α-Methyl-4-aza-5α-androstan-17β-ol-3-one	11	25	—	—
3-Aza-A-homo-5α-cholestan-4-one	29	66	—	—
4-Methyl-4-aza-5-pregnen-20β-ol-3-one	29	66	—	—
3,5-Seco-4-norcholestane-3,5β-diol	26	59	—	—
3-Aza-A-homo-5β-cholestan-4-one	37	84	—	—
2′-Aminothiazolo(*d*-3,2)-5α-cholest-2-ene	43	98	—	—
4-(β-Hydroxyethyl)-4-aza-5-pregnen-20β-ol-3-one	14	32	—	—
3β-Acetoxy-6-aza-B-homo-5α-cholestan-7-one	50	113	—	—
N-(β-Hydroxyethyl)-3,5-seco-4-norcholestan-5β-ol-3-amide	59	66	—	—
4-(Dimethylaminoethyl)-17α-methyl-4-aza-5-androsten-17β-ol-3-one	16	36	—	—
Thiazolo(*d*-3,2)-5α-cholest-2-ene	61	139	—	—
4-Oxa-5α-cholestane	73	166	—	—
6-Acetyl-6-aza-B-homo-5α-cholestan-3β-ol acetate	60	136	—	—
Indolo(*b*-3,4)-5β-cholest-3-ene	70	159	—	—
3,5-Cholestadien-7-one	33	75	—	—
Quinoxalino(*b*-2,3)-5α-cholestane	67	152	—	—
2′-Aminothiazolo(*d*-3,2)-17α-methyl-5α-androst-2-en-17β-ol	28	64	—	—
1′-Nitrosoindolo(*b*-3,2)-5α-cholest-2-ene	67	152	—	—
g-Benzoquinoxalino(*b*-2,3)-5α-cholestane	71	161	—	—
1′-Nitrosoindolo(*b*-3,4)-5β-cholest-3-ene	63	143	—	—
1′-Aminoindolo(*b*-3,2)-5α-cholest-2-ene	64	145	—	—
6-Methyl-6-aza-B-homo-5α-cholestan-3β-ol	41	93	—	—
6-Methyl-6-aza-B-homo-5α-cholestan-3β-ol methiodide	46	114	—	—
3,5-Seco-4-norandrostan-17β-ol-5-on-3-oic acid	23	52	—	—
1′-Methylindolo(*b*-3,2)-5α-cholest-2-ene	51	116	—	—
1′-Methylindolo(*b*-3,4)-5β-cholest-3-ene	62	141	—	—
7a-Aza-B-homo-5α-cholestan-3α-ol-7-one acetate	54	123	—	—
4-Methyl-4-aza-5α-cholestan-3-one	57	130	—	—
4-Methyl-3-phenyl-4-aza-2,5-cholestadiene	62	141	—	—
3-Ethyl-4-methyl-4-aza-2,5-cholestadiene	65	148	—	—
2,3-Seco-5α-cholestane-2,3-diol	33	75	—	—
B-Nor-3,5-cholestadiene	66	150	—	—
3α-Amino-5α-cholestane hydrochloride	—	—	93	116
N-Benzyl-3,5-seco-4-norcholestan-5β-ol-3-amide	—	—	84	105
4-(β-Hydroxyethyl)-4-aza-5-cholesten-3-one	—	—	77	96
17α-Methyl-4-aza-5-androsten-17β-ol-3-one	—	—	71	89
4α,5α-Epoxypregnane-3,20-dione	—	—	90	112
4-Benzyl-17α-methyl-4-aza-5-androsten-17β-ol	—	—	79	99
17β-Hydroxy-4-(β-hydroxyethyl)-17α-methyl-4-aza-5-androstene	—	—	51	64
17α-Methyl-4-aza-5α-androstan-17β-ol-3-one	—	—	67	84

TABLE TLC 96—(Continued)
STEROIDS

Layer Solvent	L1 S1	L1 S1	L1 S2	L1 S2
Compound, systematic name	$R_F \times 100$	$R_S{}^a \times 100$	$R_F \times 100$	$R_S{}^a \times 100$
N,N-Bis-(β-chloroethyl)-3β-amino-5α-cholestane hydrochloride	—	—	32	40
N-(β-Hydroxyethyl)-3,5-seco-4-norcholestan-5β-ol-3-amide	—	—	62	77
6-Methyl-6-aza-B-homo-5α-cholestan-3β-ol methiodide	—	—	20	25
4-(Dimethylaminoethyl)-4-aza-5-pregnene-3,20-dione methiodide	—	—	20	25

[a] R_F relative to 5-androsten-3β-ol-17-one.

Layer L1 = Silica gel G (Merck), activated 110 °C one hour, stored over calcium chloride.
Solvent S1 = benzene-methanol-ethyl acetate (85 : 10 : 5).
S2 = chloroform-methanol-ammonia (85 : 14 : 1).
Detection Iodine vapors.
Technique Saturated atmosphere; room temperature.
Literature Doronboos, N. J. and R. K. Sharma, *J. Chromatogr.*, **29**, 393–395 (1967).

TABLE TLC 97
STEROID 2,4-DINITROPHENYLHYDRAZONES

Layer Solvent	L1 S1	L1 S2	L1 S3	L1 S4	L1 S5	L1 S6	L1 S7	L1 S8
Compound	$R_F \times 100$							
2,4-DNPH Derivative of:								
Testosterone	22	30	34	16	25	17	9	11
Epitestosterone	47	53	53	33	39	39	23	28
Androsterone	65	69	66	50	52	55	40	53
Epiandrosterone	40	43	44	26	39	26	14	20
Etiocholanolone	71	77	72	62	61	68	54	72
Epietiocholanolone	66	71	67	54	56	60	43	56
3α-OH-4-Androstenone	53	67	64	49	57	67	39	57
3β-OH-4-Androstenone	67	56	55	37	54	51	23	38
11β-OH-Androsterone	64	66	65	53	60	67	44	59
11β-OH-Etiocholanolone	71	72	68	60	65	71	54	73
Dehydroepiandrosterone	38	41	42	24	36	26	14	19
3β-Chloro-dehydroepiandrosterone	66	68	65	48	57	66	38	58

Layer L1 = Polyamide.
Solvent S1 = methylene chloride-methanol (2 : 8).
S2 = 1,2-dichloroethane-methanol (2 : 8).
S3 = trichloroethylene-methanol (2 : 8).
S4 = methylene chloride-methanol-water (5 : 7 : 3).
S5 = trichloroethylene-methanol-water (20 : 64 : 16).
S6 = acetone-methanol-acetic acid (20 : 80 : 0.5).
S7 = carbon tetrachloride-acetone-methanol-water (5 : 10 : 40 : 10).
S8 = acetone-ethanol-water (5 : 16 : 4).
Technique Ascending, reversed phase chromatography, paper-lined chamber, filled with mobile phase.
Literature Penzes, L., Menzel, P., and G. W. Oertel, *J. Chromatogr.*, **44**, 189–190 (1969).

TABLE TLC 98
STEROIDS

Layer	L1	L1	L1	L1	L1	L1		
Solvent	S1	S2	S3	S4	S5	S6		
Detection							D1	D2
Literature	1	1	1	1	1	1		
Compound			R_F[a] × 100				Color	
3β-Acetoxy-5α-androstane	51	—	—	—	—	—	br	o
16β-Acetoxy-5α-androstane	45	—	—	—	—	—	br	o
17β-Acetoxy-5α-androstane	55	—	—	—	—	—	br	o
3β-Acetoxy-5α-cholestane	52	—	—	—	—	—	—	y
3β-Acetoxycholest-5-ene	52	—	—	—	—	—	r-br	y
3β-Acetoxy-B-norcholest-5-ene	61	—	—	—	—	—	br	br
5α-Pregnan-3-one	—	56	—	—	—	—	br	w
5α-Cholestan-3-one	—	51–65	—	—	—	—	br	w
5β-Cholestan-3-one	—	54–67	—	—	—	—	br	br
Cholest-4-en-3-one	—	34	—	—	—	—	br	y
5α-Androstan-3α-ol	—	—	70	—	—	—	br	y
5α-Androstan-3β-ol	—	—	50	—	—	—	br	y
5α-Androstan-17β-ol	—	—	61	—	—	—	br	o
5α-Pregnan-3β-ol	—	—	49	—	—	—	br	y
5α-Cholestan-3α-ol	—	—	77	—	—	—	br	y
5α-Cholestan-3β-ol	—	—	50	—	—	—	br	y
5β-Cholestan-3α-ol	—	—	69	—	—	—	br	y
5β-Cholestan-3β-ol	—	—	65	—	—	—	br	y
Androst-5-en-3β-ol	—	—	52	—	—	—	br	o
Cholest-5-en-3β-ol	—	—	50	—	—	—	r	r
3α-Acetoxy-5α-androstan-17-one	—	28–43	—	—	—	—	r	o
3β-Acetoxy-5α-androstan-17-one	—	36	—	—	—	—	r	o
3β-Acetoxy-5α-pregnan-20-one	—	45	80–94	—	—	—	r, r	br, r
3β-Acetoxyandrost-5-en-17-one	—	—	75–90	—	—	—	r	vt
3β-Acetoxypregn-5-en-20-one	—	41–56	78–91	—	—	—	br, r	y, vt
Androst-4-ene-3,16-dione	—	—	37–48	—	—	—	br	w
Androst-4-ene-3,17-dione	—	—	39–51	—	—	—	vt	y-gr
Pregn-4-ene-3,20-dione	—	—	54	—	—	—	br	w
5α-Androstan-3α-ol-17-one	—	—	—	71	—	—	r	o
5α-Androstan-3β-ol-17-one	—	—	—	64	—	—	r	o
5β-Androstan-3α-ol-17-one	—	—	—	45	—	—	br	y
5β-Androstan-3β-ol-17-one	—	—	—	75	—	—	r	o
5α-Androstan-3α-ol-16-one	—	—	—	65	—	—	vt	y-gr
5α-Androstan-3β-ol-16-one	—	—	—	60	—	—	y	y-gr
5β-Androstan-17β-ol-3-one	—	—	—	55	—	—	br	o
Androst-5-en-3β-ol-17-one	—	—	—	60	—	—	r-vt	r-vt
Androst-4-en-17β-ol-3-one	—	—	—	48	—	—	br	y
B-Norandrost-5-en-3β-ol-17-one	—	—	—	61	—	—	r-br	br
Pregn-5-en-3β-ol-20-one	—	—	—	60–76	—	—	r	bl-vt
17α-Methylandrost-4-en-17β-ol-3-one	—	—	—	40–60	—	—	y	y
Pregna-5,16-dien-3β-ol-20-one	—	—	—	60–76	—	—	br	br
17α-Ethinylpregn-4-en-17β-ol-3-one	—	—	—	61–74	—	—	br	y
Oestrone	—	—	—	65–80	—	—	y	y
5α-Androstane-3β,17α-diol	—	—	—	—	58–71	—	br	br
5α-Androstane-3β,17β-diol	—	—	—	—	54–71	—	br	br
5β-Androstane-3β,17β-diol	—	—	—	—	68–82	—	br	br
Androst-5-ene-3β,17α-diol	—	—	—	—	61–73	—	br	br
Androst-5-ene-3β,17β-diol	—	—	—	38–52	56–71	—	br, br	br, br
Tigogenin acetate	—	55	—	—	—	—	br	br-y
Sarsasapogenin acetate	—	48	—	—	—	—	br	br-y
Hecogenin acetate		3–16	—	—	—	—	br	y
Tigogenin	—	—	38–50	—	—	—	br-y	y
Smilagenin	—	—	50–63	—	—	—	br-y	y
Hecogenin	—	—	—	53–68	—	—	br	y
Gitogenin	—	—	—	0–2	—	30–52	br, br	w, w
Yuccagenin	—	—	—	—	—	26–54	br	br
Chlorogenin	—	—	—	4	—	25–41	br, br	br, br
Mexogenin	—	—	—	2	—	28–42	br, br	w, w
Diosgenin	—	—	44	—	—	—	br	br
Yamogenin	—	—	38–50	—	—	—	br	br

TABLE TLC 98—(Continued)
STEROIDS

Layer	L1	L1	L1	L1	L1	L1		
Solvent	S1	S2	S3	S4	S5	S6		
Detection							D1	D2
Literature	1	1	1	1	1	1	1	1
Compound			$R_F{}^a \times 100$				Color	
Botogenin	—	—	16–28	53–68	—	—	br, br	br, br
Kryptogenin	—	—	—	3–16	—	55–70	br, br	br, br
3α-Dimethylamino-5α-cholestane	—	—	70–86	—	—	—	y[b]	—
3β-Dimethylamino-5α-cholestane	—	—	10–25	—	—	—	y	—
3α-Dimethylamino-5β-cholestane	—	—	17–36	—	—	—	y	—
3β-Dimethylamino-5β-cholestane	—	—	59	—	—	—	y	—
3α-Dimethylamino-cholest-5-ene	—	—	67	—	—	—	y	—
3β-Dimethylamino-cholest-5-ene	—	—	32	—	—	—	y	—
3β-Acetoxy-5α-conanine	—	—	—	—	100	—	y	—
Conanin-3-one	—	—	—	—	82	—	y	—
3β-Methylamino-5α-conanine	—	—	—	—	11	—	y	—
3β-Methylamino-con-5-enine	—	—	—	—	9–22	—	y	—
Conessine	—	—	—	—	30–50	—	y	—

[a] R_F value measured from spot center if spot width/length ≥ 1 : 2, otherwise R_F of upper and lower end of spot given.
[b] Bases detected by Dragendorff reagent without drying chromatograms.

Layer L1 = Alumina, neutral, Brockmann activity V.

Solvent S1 = hexane-benzene (7 : 3).
S2 = benzene.
S3 = benzene-ether (7 : 3).
S4 = ether.
S5 = ether-ethanol (98 : 2).
S6 = ether-ethanol (96 : 4).
(Hexane for chromatography, Merck; benzene dried over Na, distilled.)

Detection D1 = Conc. H_2SO_4.
D2 = D1, but examined in UV light.

Technique 100 μg applied; time of run = 10–20 min.

Literature 1. Cĕrný, V., Joska, J., and L. Läbler, *Collect. Czech. Chem. Commun.*, **26**, 1658 (1961); from *J. Chromatogr.*, 7, D51–52 (1962).

TABLE TLC 99
STEROIDS, STEROLS

Layer	L1	L1	L2	L1	L1	L2	L1	L3	L4	L5	L5	L6
Solvent	S1	S2	S2	S3	S4	S4	S5	S5	S6	S7	S8	S9
Technique	T1	T1	T1	T1	T1	T1	T2	T2	T3	T4	T4	T1
Literature	1	1	1	1	1	1	1	1	2	2	2	1
Compound				$R_F \times 100$					$R_{cholesterol}$		R_{ch}[a]	R_F
Campesterol	—	—	—	—	—	69[b]	25	23	—	—	—	50
Campestanol	—	—	—	—	—	51	25	26	—	—	—	51
Cholesterol	81[b]	78[b]	63[b]	31[b]	44[b]	68[b]	25	23	100	100	100	50
Allocholesterol	95[b]	92[b]	77[b]	43[b]	58[b]	82[b]	29	28	—	—	—	53
Desmosterol	82[b]	79[b]	—	—	47[b]	69[b]	25	14	—	—	—	50
Cholestanol	62	60	44	19	27	51	25	26	—	—	—	—
Coprostanol	76	66	—	—	—	69	35	42	—	—	—	69
β-Sitosterol	79[b]	80[b]	63[b]	32[b]	43[b]	68[b]	25	23	100	86	83	50
Stigmasterol	80[b]	76[b]	62[b]	—	44[b]	68[b]	25	23	100	93	91	52
β-Sitostanol	60	59	44	18	27	50	25	32	—	—	—	49
Lanosterol	57[b]	77[b]	—	—	59[b]	81[b]	42	41	137	84	97	78
Dihydrolanosterol	50[b]	55[b]	—	—	—	65[b]	42	45	138	—	—	77
Agnosterol	100	100	—	—	—	94	92	89	135	76	86	95
Dihydroagnosterol	—	—	—	—	—	—	92	95	134	—	—	—
Dihydrocholesterol	—	—	—	—	—	—	—	—	93	90	89	51
Ergosterol	—	—	—	—	—	—	—	—	89	116	122	52
7-Dehydrocholesterol	—	—	—	—	—	—	—	—	93	112	126	52
Zymosterol	—	—	—	—	—	—	—	—	102	—	—	—
Brassicasterol	—	—	—	—	—	—	—	—	—	100	100	—
Epicholesterol	—	—	—	—	—	—	—	—	—	90	116	—
21-Norcholesterol	—	—	—	—	—	—	—	—	—	—	—	50

[a] R_{ch} relative to cholesterol, except sterol acetates relative to cholesterol acetate.
[b] As brominated derivatives.

Layer
L1 = Silica gel G.
L2 = Alumina.
L3 = $AgNO_3$-treated silica gel G.
L4 = Kieselguhr G.
L5 = Kieselguhr G impregnated with 10% decane in petroleum ether.
L6 = Aluminum oxide G.

Solvent
S1 = benzene-ethyl acetate (2 : 1).
S2 = benzene-ethyl acetate (4 : 1).
S3 = benzene-ethanol (19 : 0.2).
S4 = benzene-ethanol (19 : 0.4).
S5 = chloroform.
S6 = cyclohexane-ethyl acetate (95.5 : 0.5).
S7 = acetic acid-water (9 : 1) saturated with undecane at 22–24 °C.
S8 = acetic acid-water (92 : 8) saturated with undecane at 22–24 °C.
S9 = benzene-ethanol (19 : 0.4), development distance 12 cm.

Detection $SbCl_3$, $SbCl_5$ or phosphomolybdic acid.

Technique
T1 = saturated tank, development distance = 12 cm.
T2 = saturated tank, development distance = 15 cm.
T3 = 20 cm development distance, 2–3 hours, 23 °C.
T4 = reversed phase, 20 cm development distance, 5 hours, 23 °C.

Literature
1. Ikan, R., Harel, S., Kashman, J., and E. D. Bergmann, *J. Chromatogr.*, **14**, 504 (1964); and Ikan, R. and M. Cudzinovski, *J. Chromatogr.*, **18**, 422 (1965).
2. Peereboom, J. W. C. and H. W. Beekes, *J. Chromatogr.*, **9**, 316 (1962).

TABLE TLC 100
STEROIDS

Layer	L1	L1	L1	L1	L1	L2	L2	L2
Solvent	S1	S2	S3	S4	S5	S6	S7	S8
Technique	T1	T1	T1	T1	T1	T1	T1	T1
Literature	1	1	1	1	1	2	2	2
Compound				$R_F \times 100$				
5α-Androstan-3α-ol-17-one	46	32	64	46	77	—	—	—
5α-Androstan-3β-ol-17-one	38	29	55	41	67	—	—	—
5α-Androstan-17β-ol-3-one	—	—	—	—	—	4	32	63
5α-Androstane-17α-methyl-17β-ol-3-one	—	—	—	—	—	3	32	63
5β-Androstane-3α,12α-diol-17-one	37	14	65	38	37	—	—	—
5α-Androstan-3α-ol-11,17-dione	28	20	33	36	50	—	—	—
5β-Androstan-3α-ol-12,17-dione	25	16	20	30	41	—	—	—
5β-Androstan-12α-ol-3,17-dione	20	21	29	25	50	—	—	—
5α-Androstane-3,17-dione	—	—	—	—	—	—	42	72
5β-Androstane-3,12,17-trione	30	20	14	48	75	—	—	—
Δ^4-Androsten-17β-ol-3-one	35	23	49	41	63	1	15	47
Δ^4-Androstene-17α-methyl-17β-ol-3-one	41	27	49	48	73	1	17	48
Δ^4-Androsten-17α-ethynyl-17β-ol-3-one	—	—	—	—	—	3	28	70
Δ^4-Androstene-3,17-dione	58	53	56	55	80	2	20	55
5α-1-Androstene-3,17-dione	—	—	—	—	—	—	32	69
1,4-Androstadiene-3,17-dione	—	—		—	—	0	14	44
Δ^4-Androstene-3,11,17-trione	36	40	27	52	82	—	—	—
3β-Acetoxy-5α-androstane	77	74	87	66	97	—	—	—

Layer L1 = Silica gel with 5% calcium sulfate binder.
L2 = Silica gel with 5% starch binder.

Solvent S1 = benzene-acetone (4 : 1).
S2 = benzene-methanol (9 : 1).
S3 = ether.
S4 = chloroform-acetone (3 : 1).
S5 = chloroform-methanol (97 : 3).
S6 = hexane-ethyl acetate (4 : 1).
S7 = hexane-ethyl acetate (1 : 1).
S8 = ethyl acetate.

Detection 10% molybdophosphoric acid in EtOH.

Technique T1 = saturated system.

Literature
1. Hara, S. and M. Takeuchi, *Endocrinol. Jap.*, **10**, 202 (1963).
2. Smith, L. L. and T. Foell, *J. Chromatogr.*, **9**, 339 (1962).

TABLE TLC 101
STEROIDS

Layer	L1	L1	L1	L1	L1	L1	L1	L1	L1	L2
Solvent	S1	S2	S3	S4	S5	S6	S7	S8	S9	S10
Technique	T1	T1	T1	T1	T1	T1	T1	T1	T1	na
Literature	1	1	1	2	2	2	2	2	2	3
Compound	R_F cortisone × 100									
3α,17α,21-Trihydroxypregnane-11,20-dione	53	38	59	—	—	—	—	—	—	—
3α,11β,17α,21-Tetrahydroxypregnan-20-one	33	22	55	—	—	—	—	—	—	—
Pregn-4-ene-17α,21-diol-3,20-dione	116	131	230	—	—	—	—	—	—	22
17α-Hydroxyprogesterone	145	152	—	470	300	235	206	<400	<533	—
Pregn-4-ene-21-ol-3,11,20-trione	125	160	100	80	85	78	58	—	—	—
Pregn-4-ene-21-ol,3-20-dione	140	200	300	—	—	—	—	—	—	38
3β,11,21-Trihydroxyallopregnan-20-one	73	75	100	—	—	—	—	—	—	—
3α,21-Dihydroxypregnane-11,20-dione	79	91	40	—	—	—	—	—	—	—
11β,17α,21-Trihydroxypregnane-3,20-dione	70	68	130	—	—	—	—	—	—	—
17α,21-Dihydroxypregnane-3,11,20-trione	108	92	140	—	—	—	—	—	—	—
17α,21-Dihydroxypregnane-3,20-dione	115	128	360	—	—	—	—	—	—	—
3α,17α,21-Trihydroxypregnan-20-one	80	59	170	—	—	—	—	—	—	—
Prednisone	94	82	80	80	85	78	58	80	80	—
Prednisilone	62	45	75	—	20	22	45	40	53	—
Dexamethasone	72	51	140	60	55	35	52	50	67	—
Triamcinolone	0	23	10	—	—	—	—	—	—	—
Cortisone	100	100	100	100	100	100	100	100	100	14
Hydrocortisone	66	50	75	30	30	35	74	40	73	9
Corticosterone	102	119	100	—	—	—	—	—	—	15
Aldosterone	85	85	35	—	30	109	37	100	146	—
Pregnenolone	—	—	—	520	290	287	213	<400	<533	—
Pregn-4-en-3α-ol-20-one	—	—	—	490	300	283	216	<400	<533	—
Pregnane-3,20-dion-21-ol	—	—	—	530	295	270	219	<400	<533	—
Allopregnane-3α,20α-diol	—	—	—	390	220	200	200	355	<533	—
Pregnane-3α,20α-diol	—	—	—	240	170	152	197	325	500	—
16-Methylene-1-dehydro-11β,17α-dihydroxy progesterone	—	—	—	180	170	196	136	215	233	—
16-Methylene prednisolone	—	—	—	—	15	52	55	40	73	—
Pregnanediol	—	—	—	—	—	—	—	—	—	—
5α-Pregnane-3,20-dione	—	—	—	—	—	—	—	—	—	67
5α-Pregnane-3β,20α-diol	—	—	—	—	—	—	—	—	—	30
5α-Pregnane-3β,20β-diol	—	—	—	—	—	—	—	—	—	32
Progesterone	—	—	—	—	—	—	—	—	—	65
Dexamethasone	—	—	—	—	—	—	—	—	—	—
11-Epicortisol	—	—	—	—	—	—	—	—	—	—
6α-Methylprednisolone	—	—	—	—	—	—	—	—	—	—
5α-Pregnane-3β,16α-diol-20-one	—	—	—	—	—	—	—	—	—	—
Pregna-5,16-dien-3β-ol-20-one	—	—	—	—	—	—	—	—	—	—
3β-Acetoxy-5,16-pregnadien-20-one	—	—	—	—	—	—	—	—	—	—
5α-Pregn-16-en-3β-ol-20-one	—	—	—	—	—	—	—	—	—	—
Pregn-5-ene-3β,17α-diol-20-one	—	—	—	—	—	—	—	—	—	—
Pregn-5-ene-3β,17α,21-triol-20-one	—	—	—	—	—	—	—	—	—	—
12α-Acetoxy-3α,7α-dihydroxy-5β-pregnan-20-one	—	—	—	—	—	—	—	—	—	4
Pregn-4-en-21-ol-3,20-dione acetate	—	—	—	—	—	—	—	—	—	—
Hydrocortisone acetate	—	—	—	—	—	—	—	—	—	—
11-Epicortisol-21-acetate	—	—	—	—	—	—	—	—	—	—
11-Epicortisol-11,21-diacetate	—	—	—	—	—	—	—	—	—	—
Pregn-4-ene-17α,21-diol-3,20-dione-21-acetate	—	—	—	—	—	—	—	—	—	—
Pregn-4-ene-17α,21-diol-3,20-dione-17,21-diacetate	—	—	—	—	—	—	—	—	—	—
17α-Hydroxyprogesterone caproate	—	—	—	—	—	—	—	—	—	—
Pregn-4-ene-17α,21-diol-3,11,20-trione-21-acetate	—	—	—	—	—	—	—	—	—	—
Prednisone acetate	—	—	—	—	—	—	—	—	—	—

TABLE TLC 101—(Continued)
STEROIDS

Layer	L2	L2	L2	L2	L3	L3	L3	L3	L3	L3	L3
Solvent	S11	S12	S13	S14	S15	S16	S17	S18	S19	S20	S21
Technique	na	na	na	na	na	na	na	na	na	na	na
Literature	3	3	3	3	3	4	4	4	4	4	4
Compound	$R_F \times 100$										
3α,17α,21-Trihydroxypregnane-11,20-dione	—	—	—	—	—	—	—	—	—	—	—
3α,11β,17α,21-Tetrahydroxypregnan-20-one	—	—	—	—	—	—	—	—	—	—	—
Pregn-4-ene-17α,21-diol-3,20-dione	13	29	30	31	49	63	75	49	70	40	48
17α-Hydroxyprogesterone	—	—	—	—	—	—	—	—	—	—	—
Pregn-4-ene-21-ol-3,11,20-trione	—	—	—	—	—	—	—	—	—	—	—
Pregn-4-ene-21-ol,3-20-dione	37	34	44	84	70	—	—	72	—	60	62
3β,11,21-Trihydroxyallopregnan-20-one	—	—	—	—	—	—	—	—	—	—	—
3α,21-Dihydroxypregnane-11,20-dione	—	—	—	—	—	—	—	—	—	—	—
11β,17α,21-Trihydroxypregnane-3,20-dione	—	—	—	—	—	—	—	—	—	—	—
17α,21-Dihydroxypregnane-3,11,20-trione	—	—	—	—	—	—	—	—	—	—	—
17α,21-Dihydroxypregnane-3,20-dione	—	—	—	—	—	—	—	—	—	—	—
3α,17α,21-Trihydroxypregnan-20-one	—	—	—	—	—	—	—	—	—	—	—
Prednisone	—	—	—	—	26	37	62	—	52	—	—
Prednisilone	—	—	—	—	—	30	50	—	29	—	—
Dexamethasone	—	—	—	—	—	—	—	—	—	—	—
Triamcinolone	—	—	—	—	—	20	16	—	6	—	—
Cortisone	11	13	18	17	32	44	67	—	57	—	—
Hydrocortisone	10	13	9	8	25	42	53	14	35	18	25
Corticosterone	12	14	19	25	—	—	—	—	—	—	—
Aldosterone	—	—	—	—	—	—	—	—	—	—	—
Pregnenolone	—	—	—	—	—	—	—	—	—	—	—
Pregn-4-en-3α-ol-20-one	—	—	—	—	—	—	—	—	—	—	—
Pregnane-3,20-dion-21-ol	—	—	—	—	—	—	—	—	—	—	—
Alleopregane-3α,20α-diol	—	—	—	—	—	—	—	—	—	—	—
Pregnane-3α,20α-diol	—	—	—	—	—	—	—	—	—	—	—
16-Methylene-1-dehydro-11β,17α-dihydroxy progesterone	—	—	—	—	—	—	—	—	—	—	—
16-Methylene prednisolone	—	—	—	—	—	—	—	—	—	—	—
Pregnanediol	—	—	—	—	—	—	—	—	—	—	—
5α-Pregnane-3,20-dione	70	89	63	97	—	—	—	—	—	—	—
5α-Pregnane-3β,20α-diol	18	55	38	48	—	—	—	—	—	—	—
5α-Pregnane-3β,20β-diol	18	59	38	48	—	—	—	—	—	—	—
Progesterone	57	66	60	97	—	—	—	—	—	—	—
Dexamethasone	—	—	—	—	—	37	48	—	33	—	—
11-Epicortisol	—	—	—	—	12	21	38	—	22	—	—
6α-Methylprednisolone	—	—	—	—	—	35	46	—	30	—	—
5α-Pregnane-3β,16α-diol-20-one	—	—	—	—	57	—	—	51	—	41	63
Pregna-5,16-dien-3β-ol-20-one	—	—	—	—	—	—	—	—	—	—	—
3β-Acetoxy-5,16-pregnadien-20-one	—	—	—	—	—	—	—	—	—	—	—
5α-Pregn-16-en-3β-ol-20-one	—	—	—	—	—	—	—	—	—	—	—
Pregn-5-ene-3β,17α-diol-20-one	—	—	—	—	—	—	—	—	—	—	—
Pregn-5-ene-3β,17α,21-triol-20-one	—	—	—	—	—	—	—	—	—	—	—
12α-Acetoxy-3α,7α-dihydroxy-5β-pregnan-20-one	12	4	7	13	—	—	—	—	—	—	—
Pregn-4-en-21-ol-3,20-dione acetate	—	—	—	—	90	—	—	92	—	79	84
Hydrocortisone acetate	—	—	—	—	67	—	—	62	—	50	73
11-Epicortisol-21-acetate	—	—	—	—	51	—	—	45	—	38	62
11-Epicortisol-11,21-diacetate	—	—	—	—	78	81	—	76	—	58	76
Pregn-4-ene-17α,21-diol-3,20-dione-21-acetate	—	—	—	—	75	81	—	73	—	58	80
Pregn-4-ene-17α,21-diol-3,20-dione-17,21-diacetate	—	—	—	—	85	—	—	—	—	—	—
17α-Hydroxyprogesterone caproate	—	—	—	—	—	—	—	—	—	—	—
Pregn-4-ene-17α,21-diol-3,11,20-trione-21-acetate	—	—	—	—	68	—	—	62	—	53	70
Prednisone acetate	—	—	—	—	64	—	—	63	—	49	67

TABLE TLC 101—(Continued)
STEROIDS

Layer	L4	L4	L4	L4	L5	L5	L5	L5	L5	L5	L5
Solvent	S22	S23	S24	S25	S26	S27	S18	S28	S19	S15	S29
Technique	T2	T2	T2	T2	na	na	na	na	na	na	na
Literature	5	5	5	5	6	6	6	6	6	6	6
Compound	$R_F \times 100$										
3α,17α,21-Trihydroxypregnane-11,20-dione	—	—	—	—	—	—	—	—	—	—	—
3α,11β,17α,21-Tetrahydroxypregnan-20-one	—	—	—	—	—	—	—	—	—	—	—
Pregn-4-ene-17α,21-diol-3,20-dione	—	15	51	—	20	35	10	28	35	47	51
17α-Hydroxyprogesterone	—	—	—	—	25	53	20	32	46	60	60
Pregn-4-ene-21-ol-3,11,20-trione	—	—	—	—	—	—	—	—	—	—	—
Pregn-4-ene-21-ol,3-20-dione	—	22	56	96	21	43	13	31	40	33	13
3β,11,21-Trihydroxyallopregnan-20-one	—	—	—	—	—	—	—	—	—	—	—
3α,21-Dihydroxypregnane-11,20-dione	—	—	—	—	—	—	—	—	—	—	—
11β,17α,21-Trihydroxypregnane-3,20-dione	—	—	—	—	—	—	—	—	—	—	—
17α,21-Dihydroxypregnane-3,11,20-trione	—	—	—	—	—	—	—	—	—	—	—
17α,21-Dihydroxypregnane-3,20-dione	—	—	—	—	—	—	—	—	—	—	—
3α,17α,21-Trihydroxypregnan-20-one	—	—	—	—	—	—	—	—	—	—	—
Prednisone	—	6	38	69	12	30	4	16	21	24	25
Prednisilone	—	5	35	46	11	20	2	9	13	29	42
Dexamethasone	—	—	—	—	—	—	—	—	—	—	—
Triamcinolone	—	—	—	—	5	20	0	3	5	18	—
Cortisone	—	7	38	75	17	31	5	16	24	34	31
Hydrocortisone	—	6	38	55	14	33	6	11	16	41	51
Corticosterone	—	—	—	—	—	—	—	—	—	—	—
Aldosterone	6	38	—	—	27	48	36	—	—	—	—
Pregnenolone	—	—	—	—	—	—	—	—	—	—	—
Pregn-4-en-3α-ol-20-one	—	—	—	—	—	—	—	—	—	—	—
Pregnane-3,20-dion-21-ol	—	—	—	—	—	—	—	—	—	—	—
Allopregnane-3α,20α-diol	—	—	—	—	—	—	—	—	—	—	—
Pregnane-3α,20α-diol	—	—	—	—	—	—	—	—	—	—	—
16-Methylene-1-dehydro-11β,17α-dihydroxy progesterone	—	—	—	—	—	—	—	—	—	—	—
16-Methylene prednisolone	—	—	—	—	—	—	—	—	—	—	—
Pregnanediol	—	—	—	—	32	52	30	45	49	76	93
5α-Pregnane-3,20-dione	—	—	—	—	—	—	—	—	—	—	—
5α-Pregnane-3β,20α-diol	—	—	—	—	—	—	—	—	—	—	—
5α-Pregnane-3β,20β-diol	—	—	—	—	—	—	—	—	—	—	—
Progesterone	5	36	—	—	37	58	50	—	—	—	—
Dexamethasone	—	—	—	—	13	39	5	10	17	35	52
11-Epicortisol	—	—	—	—	6	20	3	4	7	19	15
6α-Methylprednisolone	—	2	20	33	13	36	3	7	13	32	45
5α-Pregnane-3β,16α-diol-20-one	—	—	—	—	16	44	6	16	24	44	48
Pregna-5,16-dien-3β-ol-20-one	5	37	—	—	27	50	41	—	—	—	—
3β-Acetoxy-5,16-pregnadien-20-one	34	75	—	—	39	58	64	—	—	—	—
5α-Pregn-16-en-3β-ol-20-one	9	44	—	—	—	—	—	—	—	—	—
Pregn-5-ene-3β,17α-diol-20-one	—	—	—	—	25	46	35	—	—	—	—
Pregn-5-ene-3β,17α,21-triol-20-one	—	—	—	—	—	40	15	—	—	54	83
12α-Acetoxy-3α,7α-dihydroxy-5β-pregnan-20-one	—	—	—	—	—	—	—	—	—	—	—
Pregn-4-en-21-ol-3,20-dione acetate	—	—	—	—	35	60	22	50	57	67	50
Hydrocortisone acetate	—	—	—	—	29	52	12	35	41	62	81
11-Epicortisol-21-acetate	—	—	—	—	—	—	—	—	—	—	—
11-Epicortisol-11,21-diacetate	—	—	—	—	—	—	—	—	—	—	—
Pregn-4-ene-17α,21-diol-3,20-dione-21-acetate	—	—	—	—	34	52	23	54	60	74	—
Pregn-4-ene-17α,21-diol-3,20-dione-17,21-diacetate	—	—	—	—	—	—	—	—	—	—	—
17α-Hydroxyprogesterone caproate	—	—	—	—	35	61	41	53	60	71	83
Pregn-4-ene-17α,21-diol-3,11,20-trione-21-acetate	—	—	—	—	—	—	—	—	—	—	—
Prednisone acetate	—	—	—	—	—	—	—	—	—	—	—

TABLE TLC 101—(Continued)
STEROIDS

Layer	L1 = Silica gel G. L2 = Silica gel. L3 = Aluminum oxide, deactivated by shaking with 2.5% acetic acid for one hour. L4 = Silica gel with 5% rice starch binder. L5 = Magnesium silicate (without binder).
Detection	See Section II.I, Volume Two.
Solvent	S1 = chloroform-ethanol (9 : 1). S2 = chloroform-90% methanol. S3 = cyclohexane-ethyl acetate (1 : 1). S4 = chloroform-acetone (9 : 1). S5 = chloroform-acetone (4 : 1). S6 = cyclohexane-chloroform-acetic acid (7 : 2 : 1). S7 = methylene chloride-acetone (4 : 1). S8 = chloroform-acetic acid (9 : 1). S9 = methylene chloride-acetic acid (9 : 1). S10 = benzene-acetone (4 : 1). S11 = benzene-methanol (9 : 1). S12 = ether. S13 = chloroform-acetone (3 : 1). S14 = chloroform-methanol (97 : 3). S15 = benzene-dioxane (2 : 1). S16 = benzene-dioxane (1 : 1). S17 = benzene-dimethylformamide (9 : 1). S18 = chloroform-ethanol (99 : 1). S19 = chloroform-ethanol (96 : 4). S20 = benzene-ethanol (95 : 5). S21 = diisopropyl ether. S22 = hexane-ethyl acetate (4 : 1). S23 = hexane-ethyl acetate (1 : 1). S24 = ethyl acetate. S25 = benzene-(2-propanol) (4 : 1). S26 = benzene-ethanol (98 : 2). S27 = benzene-ethanol (9 : 1). S28 = chloroform-ethanol (98 : 2). S29 = ether-ethanol (98 : 2).
Technique	T1 = saturated atmosphere. T2 = saturated tank, ascending development 15 cm in 60 minutes.

Literature

1. Cavina, G. and C. Vicari, "Qualitative and Quantitative Analysis of Natural and Synthetic Corticosteroids by Thin-Layer Chromatography", in *Thin-Layer Chromatography*, G. B. Marini-Bettòlo, Ed., Elsevier, Amsterdam, 1964, p. 180.
2. Waldi, D., "Steroids", in *Thin-Layer Chromatography*, E. Stahl, Ed., Academic Press, N.Y., 1965, p. 262.
3. Hara, S. and M. Takeuchi, *Endocrinol. Jap.*, **10**, 202 (1963).
4. Schwarz, V. and K. Syhora, *Collect. Czech. Chem. Commun.*, **27**, 2567 (1962).
5. Smith, L. L. and T. Foell, *J. Chromatogr.*, **9**, 339 (1962).
6. Schwarz, V., *Pharmazie*, **18**, 122 (1963).

TABLE TLC 102
STEROIDS, CARDIAC GLYCOSIDES

Layer	L1	L1	L1	L1
Solvent	S1	S2	S3	S4
Literature	1	1	1	1
Compound	$R_F \times 100$			
Digitoxigenin (Dg)				
(3β,14β-Dihydroxy-5β-cardenolide)	62	68	74	68
-peracetate	76	84	82	80
Dg-β,D-glucoside	18	23	28	18
-peracetate	73	86	83	80
Dg-β,D-allomethyloside	35	47	49	39
-peracetate	73	79	83	78
Dg-2-desoxy-β,D-alloside	32	44	48	41
-peracetate	73	81	81	78
Dg-β,D-glucomethyloside	31	43	45	35
-peracetate	78	84	84	81
Dg-2-desoxy-β,D-glucoside	28	35	43	31
-peracetate	76	82	85	80
Dg-β,D-rhamnoside	25	34	35	28
-peracetate	78	84	85	80
Dg-α,D-rhamnoside	39	47	49	41
-peracetate	78	84	83	81
Digiproside	24	37	40	27
-peracetate	78	85	85	83
Glucodigifucoside	—	—	—	—
-peracetate	70	79	83	78
Neo-gluco-digifucoside	—	—	—	—
-peracetate	75	83	86	79
Vallaroside	51	62	68	56
-peracetate	76	85	81	80
Honghelin	51	67	69	57
-peracetate	70	79	84	77
Neriifolin	51	59	66	54
-peracetate	75	82	81	78
Veneniferin	70	78	78	75
Thevebioside	11	19	20	11
-peracetate	73	82	83	80
Odoroside H	31	45	55	35
-peracetate	72	78	83	76
Odorobioside G	5	8	11	5
-peracetate	67	74	83	74
Neo-odorobioside G	3	6	8	3
-peracetate	14	83	83	78
Evatromonoside	53	64	67	61
-peracetate	74	84	81	81
Dg-α,D-digitoxoside	56	70	70	64
-peracetate	79	82	82	80
Gluco-evatromonoside	10	18	13	13
-peracetate	76	87	83	82
Dg-bisdigitoxoside	47	59	60	51
-peracetate	78	83	82	81
Gluco-Dg-bisdigitoxoside	8	14	14	6
-peracetate	75	86	85	82
Digitoxin	39	53	57	47
-peracetate	79	—	—	83
Purpureaglycoside A	5	11	10	7
Lanatoside A	10	22	20	13
-peracetate	80	87	87	82
Somalin	66	72	76	71
-peracetate	77	85	82	79
Wallichoside	66	76	75	70
-peracetate	76	82	84	78
Odoroside A	54	67	72	60
-peracetate	72	78	83	78
Dg-3-tetrahydropyranylether	78	86	85	82

TABLE TLC 102—(Continued)
STEROIDS, CARDIAC GLYCOSIDES

Layer	L1	L1	L1	L1
Solvent	S1	S2	S3	S4
Literature	1	1	1	1
Compound	$R_F \times 100$			
Digitoxigenon				
(14β-Hydroxy-3-oxo-5β-cardenolide)	67	77	80	70
Menabegenin				
(3β,14β-Dihydroxy-5β,17βH-cardenolide)	56	67	73	64
-peracetate	69	78	79	76
Bufalin				
(3β,14β-Dihydroxy-5β-bufadienolide)	68	75	76	72
-peracetate	75	85	81	79
Uzarigenin (Uzg)				
(3β,14β-Dihydroxy-5α-cardenolide)	57	70	71	66
-peracetate	78	83	84	82
Desgluco-uzarin	18	21	24	16
-peracetate	75	83	84	80
Uzg-β,D-digitoxoside	47	62	62	57
-peracetate	76	84	85	81
Uzg-canaroside	49	52	56	50
-peracetate	77	84	80	81
Odoroside B	50	62	66	55
-peracetate	69	78	82	78
Acovenosigenin				
(1β,3β,14β-Trihydroxy-5β-cardenolide)	—	—	—	—
Acovenoside A	25	39	43	27
Periplogenin				
(3β,5β,14β-Trihydroxy-cardenolide)	41	55	62	48
-peracetate	54	67	72	60
Periplorhamnoside	14	21	22	13
-peracetate	64	75	77	72
Telocinobufagin				
(3β,5β,14β-Trihydroxy-bufadienolide)	49	59	64	51
-peracetate	57	70	74	63
Periplogenon				
(5β,14β-Dihydroxy-3-oxo-cardenolide)	49	60	68	51
7β-Hydroxy-digitoxigenin				
(3β,7β,14β-Trihydroxy-5β-cardenolide)	52	56	73	61
-peracetate	78	85	83	78
7β-Hydroxy-digitoxin	38	52	53	45
-peracetate	80	87	87	80
7-Oxo-digitoxigenon				
(14β-Hydroxy-3,7-dioxo-5β-cardenolide)	63	73	78	71
Sarmentogenin				
(3β,11α,14β-Trihydroxy-5β-cardenolide)	38	53	62	46
-peracetate	71	78	80	77
Rhodexin A	22	34	39	24
-peracetate	72	80	82	77
Sarnovide	18	32	42	20
-peracetate	68	75	83	76
Gamabufotalin				
(3β,11α,14β-Trihydroxy-5β-bufadienolide)	47	60	65	51
-peracetate	72	80	81	78
Mallogenin				
(3β,11β,14β-Trihydroxy-5α-cardenolide)	—	—	—	—
Malloside	25	36	38	30
-peracetate	71	79	84	76
Digoxigenin (Dxg)				
(3β,12β,14β-Trihydroxy-5β-cardenolide)	42	55	68	50
-peracetate	72	83	82	78
Dxg-monodigitoxoside	41	55	59	47
-peracetate	78	82	83	80
Dxg-bisdigitoxoside	32	44	51	38
-peracetate	76	83	82	78

TABLE TLC 102—(Continued)
STEROIDS, CARDIAC GLYCOSIDES

Layer	L1	L1	L1	L1
Solvent	S1	S2	S3	S4
Literature	1	1	1	1
Compound	$R_F \times 100$			
Digoxigenin (Dxg)				
Digoxin	26	42	51	36
-peracetate	76	83	85	80
Lanatoside C	7	15	17	9
-peracetate	79	88	88	80
Neo-digoxin	23	37	37	28
-peracetate	76	83	85	82
Digoxoside	26	41	49	26
-peracetate	76	83	87	80
12β-Hydroxy-bufalin				
(3β,12β,14β-Trihydroxy-5β-bufadienolide)	45	58	66	57
-peracetate	75	86	81	81
15α-Hydroxy-digitoxigenin				
(3β,14β,15α-Trihydroxy-5β-cardenolide)	50	62	73	59
-peracetate	79	84	83	82
15β-Hydroxy-digitoxigenin				
(3β,14β,15β-Trihydroxy-5β-cardenolide)	53	57	70	56
-peracetate	76	86	80	81
15-Oxo-digitoxigenon				
(14β-Hydroxy-3,15-dioxo-5β-cardenolide)	68	78	80	71
Gitoxigenin				
(3β,14β,16β-Trihydroxy-5β-cardenolide)	45	55	66	55
-peracetate	75	—	—	80
Desacetyl-rhodexin B	27	35	38	27
Rhodexin C	—	—	—	—
-peracetate	75	84	85	78
Strospeside	22	32	49	22
-peracetate	72	81	85	78
Gitoside	46	57	65	50
-peracetate	77	84	84	79
Gluco-gitoroside	5	11	9	7
-peracetate	77	84	83	81
Gitoxin	29	44	49	38
-peracetate	82	90	88	84
Purpureaglycoside B	3	8	8	4
Lanatoside B	6	15	15	9
-peracetate	78	86	86	81
Desacetyl-bufotalin				
(3β,14β,16β-Trihydroxy-5β-bufadienolide)	48	60	65	51
Gitaloxigenin				
(3β,14β-Dihydroxy-16β-formyloxy-5β-cardenolide)	62	72	76	66
-peracetate	66	75	78	72
Verodoxin	34	44	56	38
-peracetate	72	81	85	77
Gluco-verodoxin	—	—	—	—
-peracetate	70	79	86	74
Lanadoxin	59	68	72	62
-peracetate	78	85	81	82
Gluco-lanadoxin	12	17	19	13
-peracetate	77	84	84	80
Gitaloxin	45	56	66	50
-peracetate	78	87	82	82
Acetyl-gitaloxin	55	66	75	60
Gluco-gitaloxin	7	13	11	7
-peracetate	78	87	84	81
Oleandrigenin (Og)				
(3β,14β-Dihydroxy-16β-acetoxy-5β-cardenolide)	64	—	—	68
Rhodexin B	37	48	50	42
-peracetate	78	86	83	81

TABLE TLC 102—(Continued)
STEROIDS, CARDIAC GLYCOSIDES

Layer	L1	L1	L1	L1
Solvent	S1	S2	S3	S4
Literature	1	1	1	1
Compound	$R_F \times 100$			
Oleandrigenin (Og)				
Acoschimperoside P	46	60	60	51
-peracetate	76	84	85	80
Og-tridigitoxoside	43	55	64	51
Oleandrin	68	77	80	74
-peracetate	77	83	82	80
Cryptograndoside A	67	75	78	73
-peracetate	77	85	79	81
Bufotalin				
(3β,14β-Dihydroxy-16β-acetoxy-5β-bufadienolide)	56	70	73	65
-peracetate	74	80	80	78
16-epi-Gitoxigenin				
(3β,14β,16α-Trihydroxy-5β-cardenolide)	43	49	59	43
16-*epi*-Gitoxin	29	40	50	31
1β, 7β-Dihydroxy-digitoxigenin				
(1β,3β,7β,14β-Tetrahydroxy-5β-cardenolide)	42	50	68	49
-peracetate	69	78	78	75
1β,7β-Dihydroxy-digitoxigenon				
(1β,7β,14β-Trihydroxy-3-oxo-5β-cardenolide)	48	59	68	55
-peracetate	65	75	77	72
Bipindogenin				
(3β,5β,11α,14β-Tetrahydroxy-cardenolide)	—	—	—	—
Lokundjoside	—	—	—	—
-peracetate	59	72	78	66
Panogenin				
(3β,11β,14β,19-Tetrahydroxy-5α-cardenolide)	—	—	—	—
Panoside	10	18	22	12
-peracetate	65	74	79	71
Diginatigenin				
(3β,12β,14β,16β-Tetrahydroxy-5β-cardenolide)	27	36	55	33
-peracetate	78	81	84	76
Diginatin	17	33	41	22
-peracetate	77	87	84	82
16-Acetyl-diginatigenin				
(3β,12β,14β-Trihydroxy-16β-acetoxy-5β cardenolide)	46	54	65	49
Cannogenin				
(3β,14β-Dihydroxy-19-oxo-5β-cardenolide)	49	57	71	56
-peracetate	70	78	80	73
Apocannoside	59	68	72	65
-peracetate	68	78	81	74
Carpogenin				
(3α,14β-Dihydroxy-19-oxo-5β-cardenolide)	57	63	74	65
-peracetate	76	83	81	80
Cannogenol				
(3β,14β,19-Trihydroxy-5β-cardenolide)	29	41	47	34
-peracetate	72	80	81	78
Desgluco-ericordin	15	21	24	14
-peracetate	68	77	81	76
Corotoxigenin				
(3β,14β-Dihydroxy-19-oxo-5α-cardenolide)	45	54	65	55
-peracetate	68	77	79	75
Bovogenin A				
(3β,14β-Dihydroxy-19-oxo-5α-bufadienolide)	—	—	—	—
Bovoside A	46	—	60	53
-peracetate	73	79	81	78
Coroglaucigenin				
(3β,14β,19-Trihydroxy-5α-cardenolide)	32	43	49	34
-peracetate	71	78	78	78

TABLE TLC 102—(Continued)
STEROIDS, CARDIAC GLYCOSIDES

Layer	L1	L1	L1	L1
Solvent	S1	S2	S3	S4
Literature	1	1	1	1
Compound	$R_F \times 100$			
Coroglaucigenin				
Frugoside	13	22	19	14
-peracetate	74	81	85	80
Strophanthidin (Strdn)				
(3β,5β,14β-Trihydroxy-19-oxo-cardenolide)	38	51	67	41
-peracetate	54	65	73	59
Strdn-β,D-glucoside	—	—	—	—
-peracetate	50	63	76	57
Strdn-β,D-xyloside	7	14	18	9
-peracetate	55	67	78	60
Strdn-β,D-rhamnoside	4	10	13	4
-peracetate	54	66	76	60
Strdn-α,D-rhamnoside	14	26	32	15
Convallatoxin	12	20	25	12
-peracetate	59	69	79	64
Desgluco-cheirotoxin	12	21	47	13
-peracetate	51	61	78	54
Strophanthojavoside	14	23	28	13
-peracetate	54	65	76	57
Strophothevoside	18	31	39	18
-peracetate	51	62	75	53
Mansonin	33	51	60	36
-peracetate	52	65	73	56
Helveticoside	22	31	47	25
-peracetate	56	68	78	60
Erysimoside	—	—	—	—
-peracetate	58	70	76	71
Erycanoside	9	14	20	8
-peracetate	54	73	78	66
Corchoroside	26	33	48	26
-peracetate	52	66	78	57
Cymarin	32	49	61	36
-peracetate	51	65	74	57
Strdn-3-tetrahydropyranylether	57	67	78	62
Hellebrigenin				
(3β,5β,14β-Trihydroxy-19-oxo-bufadienolide)	41	53	67	47
-peracetate	54	67	75	61
Desglucohellebrin	12	—	29	17
-peracetate	60	72	78	66
Strophanthidol				
(3β,5β,14β,19-Tetrahydroxy-cardenolide)	27	43	48	32
-peracetate	57	67	74	58
Convallatoxol	8	13	11	6
-peracetate	59	71	77	65
Helveticosol	18	27	28	16
-peracetate	55	68	76	61
Cymarol	21	35	41	22
-peracetate	56	69	75	58
Hellebrigenol				
(3β,5β,14β,19-Tetrahydroxy-bufadienolide)	33	46	51	33
-peracetate	54	69	72	60
Strophanthidin acid				
(3β,5β,14β-Trihydroxy-cardenolide-19-acid)	11	13	19	10
Cymarin acid	12	20	23	8
-peracetate	64	45	51	72
Strophanthidin acid methylester				
(3β,5β,14β-Trihydroxycardenolide-19-acid methylester)	—	—	—	—
Cymarin acid methylester	35	52	62	35
-peracetate	54	70	74	60

TABLE TLC 102—(Continued)
STEROIDS, CARDIAC GLYCOSIDES

Layer	L1	L1	L1	L
Solvent	S1	S2	S3	S4
Literature	1	1	1	1
Compound	**$R_F \times 100$**			
Nigrescigenin (= *Sarmentosigenin A*)				
(3β,5β,11α,14β-Tetrahydroxy-19-oxo-cardenolide)	27	38	54	29
-peracetate	54	66	75	57
Sarmentoside A	6	15	20	7
-peracetate	55	68	77	59
Adonitoxigenin				
(3β,14β,16β-Trihydroxy-19-oxo-5β-cardenolide)	—	—	—	—
Adonitoxin	17	24	32	16
Alloglaucotoxigenin				
(3β,14β,15β-Trihydroxy-19-oxo-5α-cardenolide)	31	41	56	38
-peracetate	67	76	81	75
Antiarigenin				
(3β,5β,12β,14β-Tetrahydroxy-19-oxo-cardenolide)	20	31	53	26
β-Antiarin	—	—	—	—
-peracetate	62	71	80	67
Strophadogenin				
(3β,5β,14β,16β-Tetrahydroxy-19-oxo-cardenolide)	23	30	53	25
-peracetate	57	67	74	61
16-Formyl-strophadogenin				
(3β,5β,14β-Trihydroxy-16β-formyloxy-19-oxo-cardenolide)	24	40	57	26
-peracetate	46	56	72	45
Canarigenin				
(3β,14β-Dihydroxy-cardadien-4,20(22)-olide)	63	71	76	71
-peracetate	75	81	82	80
Canaridigitoxoside	51	64	68	58
-peracetate	76	84	83	80
Gluco-canaridigitoxoside	9	13	14	9
-peracetate	76	86	83	82
Scillarenin A				
(3β,14β-Dihydroxy-bufatrien-4,20,22-olide)	63	71	73	71
-peracetate	78	84	80	81
Scillarenin-β,D-glucoside	13	22	25	17
-peracetate	76	87	83	80
Scillaren A	6	11	8	6
-peracetate	72	83	82	75
Scilliglaucosidin				
(3β,14β-Dihydroxy-19-oxo-bufatrien-4,20,22-olide)	54	61	76	61
Scillirosidin				
(3β,8β,14β-Trihydroxy-6β-acetoxy-bufatrien-4,20,22-olide)	—	—	—	—
Scillirosid	11	19	16	11
-peracetate	71	80	80	75
Xysmalogenin				
(3β,14β-Dihydroxy-cardadien-5,20(22)-olide)	62	70	72	68
-peracetate	77	84	83	80
Tanghinigenin				
(3β,19β-Dihydroxy-7β,8β-epoxy-5β-cardenolide)	—	—	—	—
Desacetyl-tanghinin	45	59	62	47
-peracetate	76	83	85	78
Desacetyl-protanghinin	10	15	14	7
Protanghinin	23	33	33	25
-peracetate	75	82	84	80

TABLE TLC 102—(Continued)
STEROIDS, CARDIAC GLYCOSIDES

Layer	L1	L1	L1	L
Solvent	S1	S2	S3	S4
Literature	1	1	1	1
Compound	$R_F \times 100$			
Cerberigenin				
(3β,14β-Dihydroxy-11β,12β-epoxy-5β-cardenolide)	55	65	74	64
-peracetate	74	77	85	77
Desacetyl-cerbertin	43	59	62	48
-peracetate	69	80	80	77
Cerbertin	66	74	78	72
Adynerigenin				
(3β-Hydroxy-8β,14β-epoxy-5β-cardenolide)	—	—	—	—
Adynerin	54	64	70	57
-peracetate	71	78	82	76
10β-Hydroxy-19-nor-periplogenin				
(3β,5β,10β,14β-Tetrahydroxy-19-nor-cardenolide)	29	48	51	34
-peracetate	45	60	61	46
Resibufogenin				
(3β-Hydroxy-14β,15β-epoxy-5β-bufadienolide)	67	72	76	70
-peracetate	75	84	82	80
3β-Acetoxy-14β,15β-epoxy-5β-cardenolide	77	84	82	78
3β-Acetoxy-14α,15α-epoxy-5β-cardenolide	79	86	83	83
3-Oxo-14α,15α-epoxy-5β-cardenolide	72	80	79	76
3β-Hydroxy-15-oxo-5β,14α-cardenolide	61	67	78	69
-peracetate	75	82	82	79
3β-Acetoxy-15-oxo-5β,14β-cardenolide	77	86	85	80
3β-Acetoxy-15-oxo-5β,14β,17βH-cardenolide	77	84	82	80
3β-Hydroxy-5α,14α,17βH-cardenolide	73	77	78	77
-peracetate	82	88	86	86
Monoanhydro-gitoxigenin				
(3β,16β-Dihydroxy-5β-cardadien-14,20(22)-olide)	64	68	75	71
Dianhydro-gitoxigenin				
(3β-Hydroxy-5β-cardatrien-14,16,20(22)-olide)	71	77	78	77
14-Desoxy-14αH-digitoxigenin				
(3β-Hydroxy-5β,14α-cardenolide)	72	78	78	78
14-Desoxy-17βH-digitoxigenin				
(3β-Hydroxy-5β,14β,17βH-cardenolide)	72	78	76	77

Layer L1 = Silica gel G (Merck).
Solvent S1 = *n*-hexane-ethyl acetate-ethanol (15 : 75 : 10) + 2.2% water.
S2 = ethyl acetate-pyridine (90 : 10) + 2.7% water.
S3 = methyl ethyl ketone + 5.4% water.
S4 = toluene-ethyl acetate-*n*-propanol (20 : 66 : 14) + 2.2% water.
Technique Ascending development, 15 cm, 22 ± 2 °C.
Literature

1. Nover, L., Jüttner, G., Noack, S., Baumgarten, G., and M. Luekner, *J. Chromatogr.*, **39**, 419 (1969).

TABLE TLC 103
STEROID SAPOGENINS

Layer	L1	L1	L1	L1	L1	L1	L1	L1	L2	L2	L2
Solvent	S1	S2	S3	S4	S5	S6	S7	S8	S9	S5	S10
Literature	1	1	1	1	1	1	1	1	2	2	2
Compound						$R_F \times 100$					
Luvigenin	87	76	80	85	80	78	93	72	—	—	—
Neometeogenin	81	52	63	66	64	48	85	48	—	—	—
Meteogenin	81	52	63	66	64	48	85	48	—	—	—
Sarsasapogenin	65	39	46	51	51	42	67	31	—	—	—
Diosgenin	59	35	41	42	46	34	57	25	18	67	—
Tigogenin	59	35	39	39	46	29	53	23	18	67	—
Pennogenin	53	26	30	42	35	22	50	27	4	48	68
Gentrogenin	49	24	25	32	26	16	42	24	—	—	—
Hecogenin	46	22	22	37	21	16	50	25	2	31	58
Convallamarogenin	39	21	24	30	28	22	42	22	—	—	—
Isorhodeasapogenin	39	18	25	38	28	22	47	24	—	—	—
Rhodeasapogenin	37	18	25	32	28	22	47	22	—	—	—
Nogiragenin	27	11	15	20	18	15	37	18	—	—	—
Heloniogenin	26	7	13	16	16	11	30	18	—	—	—
Yonogenin	21	11	11	17	13	9	30	18	—	—	—
Gitogenin	19	7	11	16	9	11	35	15	—	—	—
Tokorogenin	9	2	2	12	3	4	27	10	—	—	—
Metagenin	6	1	1	9	2	1	23	7	—	—	—
Kitigenin	4	1	1	10	0	1	20	0	—	—	—
Kogagenin	3	1	1	8	0	0	23	5	—	—	—
Diosgenin 3-acetate	—	—	—	—	—	—	—	—	73	—	—
Tigogenin 3-acetate	—	—	—	—	—	—	—	—	74	—	—
Smilagenin 3-acetate	—	—	—	—	—	—	—	—	73	—	—
Hecogenin 3-acetate	—	—	—	—	—	—	—	—	26	89	—
Gentrogenin 3-acetate	—	—	—	—	—	—	—	—	25	89	—
Sarsapogenin 3-acetate	—	—	—	—	—	—	—	—	68	—	—
Chlorogenin	—	—	—	—	—	—	—	—	0	4	18
Kryptogenin	—	—	—	—	—	—	—	—	0	22	25
Tomatidine	0	—	—	—	2	3	—	—	0	2	3

Layer L1 = Silica gel.
L2 = Starch-bound layers of silica gel.

Solvent S1 = chloroform-ethanol (95 : 5).
S2 = chloroform-acetone (9 : 1).
S3 = benzene-acetone (85 : 15).
S4 = benzene-methanol (92 : 8).
S5 = *n*-hexane-ethyl acetate (1 : 1).
S6 = *n*-hexane-acetone (4 : 1).
S7 = benzene-ethanol (85 : 15).
S8 = benzene-ethanol (92 : 8).
S9 = hexane-ethyl acetate (4 : 1).
S10 = ethyl acetate.

Detection H_2SO_4; $SbCl_3$; phosphomolybdic acid.

Technique 15 cm development distance.

Literature
1. Matsumoto, N., *Chem. Pharm. Bull. (Japan)*, **11**, 1189 (1963); *Chem. Abstr.*, **59**, 15559 (1963).
2. Smith, L. L. and T. Foell, *J. Chromatogr.*, **9**, 339 (1962).

TABLE TLC 104
SULFUR COMPOUNDS, THIOPHENE DERIVATIVES

Layer	L1	L2	L2	
Solvent	S1	S2	S3	
Detection	—	—	—	D1
Literature	1	1	1	

Compound	$R_F \times 100$			Color	
				UV	Isatin
2-Methylthiophene[a]	76	—	—	—	y→gr bl
3-Methylthiophene	82	—	—	—	d bl
2-Ethylthiophene[a]	92	—	—	—	d y
2,5-Dimethylthiophene[a]	95	—	—	—	pk
2,3,5-Trimethylthiophene[a]	87	—	—	—	vt-pk
2,3-Dimethyl-4-ethylthiophene[a]	84	—	—	—	br
Tetramethylthiophene[a]	89	—	—	—	vt-pk
Methyl 3-thienyl sulfide[a]	92	—	—	—	o
n-Decyl 3-thienyl sulfide[a]	96	—	—	—	pl vt
Bithienyl	80	—	—	vt	bl
α-Terthienyl	57	—	—	y	wine r→bl gr
α-Quaterthienyl	26	—	—	lime gr	pl gr
5,5′-Dichlorobithienyl	89	—	—	lime gr	vt
5,5′-Dimethylbithienyl	70	—	—	vt	y
α-Phenyl-α-bithienyl	50	—	—	y	gr-bl
5,5′-Diphenylbithienyl	16	—	—	bl	wine r→mauve
5,5″-Dimethylterthienyl	48	—	—	y	wine r→mauve
Thiophene-carboxaldehyde	—	34(84)[b]	—	cr	(y[b])
2-Nitrothiophene	—	61(91)[b]	—	cr	(pl bl[b])
2-Benzoylthiophene	—	39	—	cr	vt (120°)
2-Acetylthiophene	—	25	—	cr	vt (120°)
Thiophene-carboxylic acid	—	—	65	cr-vt	gy-bl (120°)
β-(α-Thienyl)-acrylic acid	—	—	57	cr-vt	vt
4-(α-Thienyl)-butyric acid	—	—	60	cr-vt	o
Bithienyl-5-carboxylic acid	—	—	63	d-vt	y
2,2′-Bithienyl-methylamine hydrochloride	—	—	35	pl vt	vt

[a] Volatile samples run at 4 °C.
[b] When run at the same concentration on alumina and eluted with benzene-chloroform (9 : 1), these samples gave the results indicated (*e.g.* yellow).

Layer L1 = Alumina G.
L2 = Silica gel G; both activated 120 °C for 1 hour.
Solvent S1 = petroleum ether (b.p. 40–60 °C).
S2 = benzene-chloroform (9 : 1).
S3 = methanol.
Detection See Section II.I, Volume Two.
Technique Plates equilibrated in tank for 30 minutes before developing for 30 minutes at room temperature.
Literature 1. Curtis, R. F. and G. T. Phillips, *J. Chromatogr.*, **9**, 366–368 (1962).

TABLE TLC 105
TERPENE ESTERS

Layer	L1	L1		
Solvent	S1	S2		
Detection			D1	D1
Literature	1	1		
			Color	
Compound	$R_F \times 100$		5 min.	24 hrs.
Benzyl formate	63	38	wk	—
Linalyl formate	70	41	dk bl	grass gr
Geranyl formate	70	43	dk bl	grass gr
Citronellyl formate	72	44	dk gy	gy-gr
Nerolidyl formate	75	39	dk bl	grass gr
Terpinyl formate	—	39	dk bl	bl
Citronellyl acetate	55	32	dk gy	gy-gr
Rhodinyl acetate	55	32	dk gy	gy-gr
Geranyl acetate	55	31	dk bl	grass gr
Neryl acetate	55	31	dk bl	grass gr
Carvyl acetate	56	32	p	dk bl
Terpinyl acetate	56	34	dk bl	bl
Bornyl acetate	60	33	wk	—
Cedryl acetate	62	34	pk	pk
Nerolidyl acetate	62	33	dk bl	grass gr
Isobutyl acetate	50	31	—	—
3-Hexen-1-yl acetate	52	—	dim	—
Amyl acetate	54	31	—	—
Hexyl acetate	57	32	—	—
Iso-octyl acetate	57	32	—	—
Octyl acetate	57	33	—	—
Nonyl acetate	57	34	—	—
Decyl acetate	58	35	—	—
Isoamyl propionate	62	38	—	—
Geranyl propionate	67	38	dk bl	grass gr
Neryl propionate	67	38	dk bl	grass gr
Linalyl propionate	67	40	dk bl	grass gr
Carvyl propionate	73	41	p	dk bl
Phenethyl butyrate	57	27	wk	—
Isoamyl butyrate	62	39	—	—
Octyl butyrate	68	42	—	—
Hexyl butyrate	68	41	—	—
Nonyl butyrate	68	43	—	—
Geranyl butyrate	69	40	dk bl	grass gr
Linalyl butyrate	70	41	dk bl	grass gr
Terpinyl butyrate	70	42	dk bl	bl
Citronellyl butyrate	71	43	dk gy	gy-gr
Terpinen-4-yl butyrate	71	41	dk bl	bl
Ethyl caproate	61	—	—	—
Methyl caprylate	63	40	—	—
Ethyl caprylate	65	41	—	—
Propyl caprylate	68	43	—	—
Butyl caprylate	73	46	—	—
Isoamyl caprylate	80	56	—	—
Hexyl caprylate	80	52	—	—
Octyl caprylate	83	54	—	—
Decyl caprylate	85	57	—	—
Ethyl caprate	70	43	—	—
Isoamyl caprate	84	—	—	—
Octyl caprate	84	57	—	—
Dimethyl anthranilate	40	37	y	y
Methyl anthranilate	47	13	y	y
Ethyl anthranilate	50	16	y	y
Linalyl benzoate	81	47	dk bl	grass gr
Nerolidyl isobutyrate	74	45	dk bl	grass gr

Layer L1 = Silica gel G, air-dried 30 minutes, 110 °C, overnight.
Solvent S1 = benzene.
S2 = trifluorotrichloroethane-methylene chloride (6 : 4).
Detection D1 = 5% vanillin in 96% H_2SO_4.
Technique Ascending, 15 cm.
Literature 1. Attaway, J. A., Wolford, R. W., and G. J. Edwards, *Anal. Chem.*, **37**, 75 (1965).

TABLE TLC 106
TERPENES

Layer	L1	L2			
Solvent	S1	S1			
Detection	—	—	D1	D1	
Technique	T1	T1			
Literature	1	1			
			Color		
Compound	$R_F \times 100$		1 min.	30 min.	4 hr.
Azulene	48	77	r	dk r	dk r
γ-Bisabolene	71	93	p	br	br gr
δ-Cadinene	72	93	sky bl	dk bl	p
Camphene	83	95	lt tan	lt o	weak
δ-3-Carene	85	95	russet	gr	bl
β-Carotene	16	43	dk gr	dk gr	dk bl
β-Caryophyllene	74	94	r	p	p
α-Cedrene	89	96	r	p	p
***p*-Cymene**	67	89	none	none	none
Guaiazulene	46	80	pk	dk pk	r
Guaiene	76, 82	94	p-mn	dk bl	dk p
α-Humulene	65	90	p-mn	lv	br
***p*-Isopropenyltoluene**	62	84	lv	pk	lv
***d*-Limonene**	76	93	gr br	bl gr	bl
2,4(8)-Menthadiene	76	93	gr br	bl gr	bl
β-Myrcene	74	92	p	bl	y-gr
α-Ocimene	71	91	p	bl	y-gr
***trans*-β-Ocimene**	71	91	p	bl	y-gr
β-Phellandrene	79	93	p	bl	bl
α-Pinene	90	96	gr br	bl	bl
β-Pinene	88	95	gr br	bl	bl
Pinene dimer	87	95	p	p	bl
α-Pyronene	80	94	dk pk	bl	bl
β-Pyronene	80	94	dk pk	bl	bl
Sabinene	75	93	gr br	bl	bl
β-Santalene	78	94	bl p	dk bl	dk gr
β-Selinene	78	94	dk gr	dk gr	dk bl
Squalene	48	85	dk pk	tan	tan
α-Terpinene	76	93	gr br	bl gr	bl
γ-Terpinene	76	93	gr br	bl gr	bl
Terpinolene	75	92	gr br	bl gr	bl
α-Thujene	80	95	dk bl	bl	bl
Valencene	76	93	p	mn	dk gy
Verbenene	81	95	lv	bl	bl
Ylangene	88	96	dk bl	lv	dk pk
β-Zingaberene	75	94	bl gr	dk gr	gr

Layer L1 = Silica gel G.
L2 = Aluminum oxide G.
Solvent S1 = skellysolve B (hexanes).
Detection D1 = vanillin-H_2SO_4 spray reagent.
Technique 10 cm development distance.
Literature
1. Attaway, J. A., Barabas, L. J., and R. W. Wolford, *Anal. Chem.*, **37**, 1289 (1965).

TABLE TLC 107
VITAMINS, FAT-SOLUBLE

Layer	L1	L1	L1	L1	L1	L1	L1	L1	L1	L1	L1	L1	L1		
Solvent	S1	S2	S3	S4	S5	S6	S7	S8	S9	S10	S11	S12	S13		
Literature	1	1	1	1	1	1	1	1	1	1	1	1	1		
														Color	
Compound	$R_F \times 100$													Perchloric acid	H_2SO_4
α-Carotene	76	87	90	90	93	92	75	50	87	88	89	94	95	—	—
β-Carotene	—	—	89	89	—	—	64	50	—	—	—	—	90	—	—
Vitamin A	78	71	90	89	90	88	24	13	91	91	91	93	63	vt	vt
Vitamin D_2	80	91	93	98	79	98	5	0	24	17	12	58	9	o-br	o-r
Vitamin E	81	98	91	92	90	100	5	0	93	69	72	87	54	br	br
Vitamin K_1	78	89	92	89	85	90	31	21	94	90	91	94	74	y-br	y-br
Vitamin K_2	71	84	91	92	85	94	21	11	88	88	87	95	70	y-br	y-br
Vitamin K_3	93	91	92	93	80	98	29	10	74	71	72	92	49	y-br	y-br

Layer L1 = Loose-layer alumina, particle size 0.3 mm and Brockmann grade III–IV.
Solvent S1 = methyl alcohol.
S2 = anhydrous ethyl alcohol.
S3 = *n*-butyl alcohol.
S4 = benzyl alcohol.
S5 = hexane.
S6 = cyclohexane.
S7 = petroleum ether.
S8 = petrol.
S9 = benzene.
S10 = toluene.
S11 = xylene.
S12 = chloroform.
S13 = carbon tetrachloride.
Detection $HClO_4$ or H_2SO_4.
Technique 30 cm, ascending, about 45–60 minutes.
Literature 1. Davidek, J. and J. Blattná, *J. Chromatogr.*, 7, 204 (1962).

TABLE TLC 108
VITAMIN A AND RELATED COMPOUNDS

Layer	L1	L1	L1	L1	L1	L1	L1
Solvent	S1	S2	S3	S4	S5	S6	S7
Literature	1	1	1	1	1	1	1
Compound	$R_F{}^a \times 100$						
Anhydrovitamin A_1	63	90	97	—	—	—	—
β-Carotene	6	80	—	—	—	—	—
***retro*-Vitamin A_1 acetate**	0	36	90	—	—	—	—
Vitamin A_1 acetate	—	19	88	—	—	—	—
Vitamin A_1 aldehyde	—	0	66	—	—	—	—
Vitamin A_2 aldehyde	—	—	59	100	—	—	—
***retro*-Vitamin A_1 alcohol**	—	—	12	36	42	—	—
Vitamin A_1 alcohol	—	—	6	16	28	45	—
Vitamin A_1 epoxide	—	—	3	12	—	32	—
Vitamin A_2 alcohol	—	—	0	8	26	28	58
Vitamin A_1 acid	—	—	—	0	—	0	5

[a] $R_F \pm 5\%$.

TABLE TLC 108—(Continued)
VITAMIN A AND RELATED COMPOUNDS

Layer L1 = Alumina G.
Solvent S1 = cyclohexane.
S2 = 5% benzene in cyclohexane.
S3 = 0.25% methanol in cyclohexane.
S4 = 1% methanol in cyclohexane.
S5 = 3% methanol in cyclohexane.
S6 = 3% ethanol in cyclohexane.
S7 = 8% ethanol in cyclohexane.
Detection Carr-Price reagent.
Technique 3–4″ Development distance, 20–30 minutes.
Literature 1. Varma, T. N. R., Panalks, T., and T. K. Murray, *Anal. Chem.*, **36**, 1864 (1964).

TABLE TLC 109
VITAMIN B_6 COMPOUNDS

Layer	L1	L1	L1[a]	L1	L1	L2	L2	L2	L3	L3
Solvent	S1	S2	S2	S3	S3	S4	S5	S6	S4	S7
Technique	na	na	na	na	T1	T1	T1	T1	T1	T1
Literature	1	1	1	1	1	1	1	1	1	1
Compound	$R_F \times 100$									
Pyridoxol	62	47	[d]	—	—	—	—	—	—	—
Pyridoxal	68	56	36	—	—	—	—	—	—	—
Pyridoxamine	12	5	5	—	—	—	—	—	—	—
Pyridoxal ethyl acetal	54	84	83	—	—	—	—	—	—	—
4-Pyridoxic acid	91	49	—	—	—	—	—	—	—	—
4-Pyridoxic acid lactone	91	18	—	—	—	—	—	—	—	—
Pyridoxol phosphate	95	0	—	30	18	47	68	70	21	69[b]
Pyridoxal phosphate	95	0	—	54[c]	33[c]	64	80	78	37	79[b]
Pyridoxamine phosphate	96	0	—	41	28	7	36	62	1	57[b]
Pyridoxal phosphate phenylhydrazone	—	—	—	76	55	—	—	—	—	—

[a] H_3BO_3-treated plate 2 cm from starting point. [b] Spots show tailing. [c] Phenylhydrazine test is negative.
[d] Pyridoxol remains near origin in boric acid-treated region.

Layer L1[a] = Silica gel HF_{254}.
L2 = Cellulose MN-300G.
L3 = Eastman Kodak Chromagram cellulose sheets.
Solvent S1 = 0.2% NH_4OH in water (1 : 139 v/v conc. NH_4OH-H_2O).
S2 = chloroform-methanol (75 : 25 v/v).
S3 = methyl ethyl ketone-ethanol-conc. NH_4OH-H_2O (15 : 5 : 5 : 5 v/v).
S4 = *n*-butanol saturated with 1*N* HCl, upper layer.
S5 = *n*-butanol-conc. HCl-H_2O (25 : 5 : 10 v/v).
S6 = *n*-butanol-acetic acid-water (15 : 10 : 10 v/v).
S7 = methanol-butanol-benzene-water-triethylamine (20 : 10 : 10 : 10 : 5 v/v).
Detection See Section II.I, Volume Two.
Technique Chamber saturation.
Literature 1. Ahrens, H. and W. Korytnyk, *Methods of Enzymology*, Vol. XVIII, Pt. A., Colowick, S. P. and N. O. Kaplan, Editors-in-chief, Academic Press, N.Y., 1970, pp. 489–494.

TABLE TLC 110
VITAMINS, TOCOPHEROLS

Layer	L1	L1	L1	L1	L2	L2					
Solvent	S1	S2	S1	S2	S1	S2					
Detection							D1	D2	D3	D4	D1
Technique	T1	T1	na	na	na	na					
Literature	1	1	2	2	2	2	2	2	2	2	2
Compound	$R_F \times 100$						Color				Sensitivity μg
α-Tocopherol	52	26	35	56	37	62	o-r	y	y-br	y-br	0.2
α-Tocopheryl acetate	—	—	40	76	60	80	—	y	lt br	y-br	—
α-Tocopheryl succinate	—	—	0	0	0	0	—	lt y	lt br	lt br	—
β-Tocopherol	37	20	22	42	18	35	o-r	y	y-br	y-br	0.2
γ-Tocopherol	37	18	22	42	18	35	o-r	y	y-br	r-br	0.2
δ-Tocopherol	24	15	16	22	8	20	o-r	y	y-br	r-br	0.2

Layer L1 = Silica gel G.
L2 = Alumina.

Solvent S1 = benzene.
S2 = chloroform.

Detection D1 = Emmerie-Engel reagent.
D2 = H_2SO_4.
D3 = $HClO_4$.
D4 = HNO_3.

Technique T1 = ascending, saturated tank, 15 cm development.

Literature
1. Dilley, R. A. and F. L. Crane, *Anal. Biochem.*, **7**, 240 (1964).
2. Ishikawa, S. and G. Katsui, *Bitamin*, **29**, 203 (1964); *Chem. Abstr.*, **61**, 15932b (1964).

TABLE TLC 111
ALCOHOLS, DNB[b] AND ONPU[c] DERIVATIVES

Layer	L1	L2	L3	L4
Solvent	S1	S2	S3	S4
Technique	T1	T1	T1	T1
Literature	1	1	1	1
Compound	$R_Y^a \times 100$			
DNB[b]				
1-Propanol	68	—	201	—
1-Butanol	82	—	190	—
1-Pentanol	96	—	171	—
1-Hexanol	104	—	157	—
1-Heptanol	108	—	142	—
1-Octanol	115	—	128	—
1-Nonanol	119	—	117	—
1-Decanol	124	—	103	—
1-Undecanol	128	—	91	—
1-Dodecanol	135	—	78	—
2-Propanol	79	—	208	—
2-Butanol	92	—	202	—
2-Pentanol	105	—	189	—
2-Heptanol	115	—	177	—
2-Octanol	123	—	154	—
3-Pentanol	105	—	191	—
4-Heptanol	142	—	172	—
2-Methyl-1-butanol	101	—	176	—
3-Methyl-1-butanol	96	—	182	—
3-Hexen-1-ol	94	—	173	—
Phenylethyl alcohol	61	—	154	—
Cinnamyl alcohol	66	—	131	—
α-Terpineol	115	—	117	—
Geraniol	104	—	144	—
Citronellol	108	—	140	—
Linaloöl	107	—	146	—
Menthol	146	—	144	—
ONPU[c]				
1-Propanol	—	98	205	119
1-Butanol	—	108	194	111
1-Pentanol	—	121	180	104
1-Hexanol	—	121	161	100
1-Heptanol	—	123	145	91
1-Octanol	—	128	130	84
1-Nonanol	—	130	112	75
1-Decanol	—	130	100	66
1-Undecanol	—	130	87	52
1-Dodecanol	—	130	75	41
2-Propanol	—	106	230	126
2-Butanol	—	109	222	119
2-Pentanol	—	115	216	107
2-Hexanol	—	123	186	105
2-Heptanol	—	125	170	98
2-Octanol	—	131	154	86
4-Heptanol	—	127	189	93
2-Methyl-1-butanol	—	114	196	110
3-Methyl-1-butanol	—	116	182	113
3-Hexen-1-ol	—	128	176	102
Phenylethyl alcohol	—	89	170	119
α-Terpineol	—	123	135	84
Geraniol	—	120	138	73
Citronellol	—	127	134	63
Linaloöl	—	117	155	88
Menthol	—	128	135	72
Fenchyl alcohol	—	145	135	88

[a] $R_Y = R_F$ relative to butter yellow dye.
[b] 3,5-Dinitrobenzoates.
[c] *o*-Nitrophenylurethanes.

Layer L1 = Silica gel G with 2′,7′-dichlorofluorescein.
L2 = Silica gel G.
L3 = Polyamide.
L4 = Silica gel G coated with 5% vaseline.

Solvent S1 = benzene–petroleum ether (38–50 °C) (1 : 1).
S2 = heptane–acetone (4 : 1).
S3 = methanol–water (9 : 1).
S4 = methanol–water (95 : 5).

Detection UV or daylight.

Technique T1 = ascending, tank saturation, 10 cm development distance.

Literature
1. Minyard, J. P., Tumlinson, J. H., Thompson, A. C., and P. A. Hedin, *J. Chromatogr.*, **29**, 88–93 (1967).

TABLE TLC 112
ALCOHOLS, GLYCOL ESTERS

Layer	L1	L1	L1	L1	L1
Solvent	S1	S2	S3	S4	S5
Technique	T1	T1	T1	T1	T1
Literature	1	1	1	1	1
Compound[a]	$R_F \times 100$				
2-Butyne-1,4-diol	0	0	0	5	8
Diacetate	8	22	23	41	48
Dibutyrate	18	36	45	64	65
Diformate	12	27	24	48	51
Dipropionate	13	27	38	59	60
Divalerate	23	44	51	69	69
Monoacetate	0	4	7	20	27
Monobutyrate	3	4	13	32	38
Monoformate	0	4	7	24	28
Monopropionate	0	4	10	28	34
Monovalerate	3	5	14	35	40

[a] Esters of 2 butyne-1,4-diol.

Layer L1 = Silica gel.
Solvent S1 = benzene.
S2 = chloroform.
S3 = isopropyl ether–isooctane (1 : 1).
S4 = isopropyl ether.
S5 = ethyl acetate–isooctane (1 : 1).
Detection Potassium permanganate.
Technique T1 = 15 cm development distance, 27 ± 1 °C, paper-lined tank.
Literature 1. Naff, M. B., Naff, A. S., and J. A. Strite, *J. Chromatogr.*, **11**, 496 (1963).

TABLE TLC 113
ALCOHOLS, POLYALCOHOLS

Layer	L1	L2	L3	L4	L4
Solvent	S1	S2	S3	S4	S5
Literature	1	1	1	2	2
Compound	$R_F \times 100$				
2,2-Bishydroxymethyl-1-butanol	19	66	29	—	—
2,2-Bishydroxymethyl-1-propanol	12	57	19	—	—
1,3-Butanediol	50	77	84	55	3
1,4-Butanediol	35	77	45	47	2
2,3-Butanediol	—	—	—	62	4
Diethylene glycol	33	52	92	—	—
2,2-Dimethyl-1,3-propanediol	60	87	80	—	—
Dipropylene glycol-1,2	66	72	91	—	—
Ethylene glycol	13	56	52	25	0
Glycerin	2	31	16	—	—
1,6-Hexanediol	42	77	68	—	—
1,2,6-Hexanetriol	75	83	62	—	—
2-Methyl-2,4-pentanediol	80	90	93	84	21
Pentaerythritol	0	27	0	—	—
1,5-Pentanediol	—	—	—	63	4
1,2-Propanediol	36	67	72	38	2
1,3-Propanediol	26	64	55	—	—
Tetramethylcyclohexanol	4	47	9	—	—
Triethylene glycol	40	42	94	—	—

Layer L1 = Aluminum oxide G.
L2 = Silica gel G.
L3 = Polyamide-impregnated Kieselguhr (0.2 mm thickness).
L4 = Loose-layer Silica gel (0.2 mm thickness).
Solvent S1 = chloroform-toluene-formic acid (80 : 17 : 3).
S2 = *n*-BuOH saturated with 1.5*N* NH_4OH.
S3 = chloroform.
S4 = ether–ethanol (99 : 1) saturated with H_2O.
S5 = hexane–acetone (4 : 1).
Detection See Section II.I, Volume Two.
Literature 1. Knappe, E., Peteri, D., and I. Rohdewald, *Z. Anal. Chem.* **199**, 270 (1963).
2. Kucera, J., *Collect. Czech. Chem. Commun.* **28**, 134 (1963).

TABLE TLC 114
ALKALOIDS, ERGOT ALKALOIDS AND HYDROGENATED DERIVATIVES

Layer	L1	L1	L1	L2	L3	L3	L3	L3	L2	L1	L1	L1	L1	L4	L4	
Solvent	S1	S2	S3	S4	S5	S6	S7	S8	S9	S10	S11	S12	S13	S14 S15	S16	
Technique	na	na	na	na	T1	T1	T1	T1	T2	T2	T2	T3	T4	T5	na	
Literature	1	1	1	1	2	2	2	2	3	3	3	4	4	5	6	1
Compound	$R_F \times 100$															Fluorescence
Agroclavine	39	51	67	63	60	48	26	41	—	—	—	—	—	—	—	–
Chanoclavine-(I)	5	11	12	3	—	—	—	—	—	—	—	—	—	—	—	–
Chanoclavine-(II)	6	7	10	3	—	—	—	—	—	—	—	—	—	—	—	–
Costaclavine	5	66	24	56	—	—	—	—	—	—	—	—	—	—	—	–
Dihydroergocornine	—	—	—	—	54	46	31	46	—	—	—	—	—	—	38	
Dihydroergocristine	—	—	—	—	60	49	31	46	—	—	—	—	14	—	30	
Dihydroergocryptine	—	—	—	—	—	—	—	—	—	—	—	—	—	—	50	
Dihydroergotamine	—	—	—	—	52	39	11	30	—	—	—	—	11	—	9	
α-Dihydrolysergol	8	9	22	18	—	—	—	—	—	—	—	—	—	—	—	–
DL-4-Dimethylallyltryptophan	0	0	0	0	—	—	—	—	—	—	—	—	—	—	—	–
Elymoclavine	17	11	34	29	34	16	3	8	—	—	—	—	—	—	—	–
Ergine	—	—	—	—	49	45	23	41	—	—	—	—	—	—	—	
Ergocornine	75	41	87	71	66	61	43	53	9.5	58	59	28	69	73	50	+
Ergocorninine	82	56	90	71	—	—	—	—	31	83	73	45	81	93	—	+
Ergocristine	75	37	87	72	69	61	43	53	9.3	54	56	28	69	67	41	+
Ergocristinine	83	57	91	72	80	72	62	66	38	80	74	45	81	90	—	+
Ergocryptine	74	38	88	71	66	64	44	55	15	60	55	28	69	83	61	+
Ergocryptinine	81	57	90	72	—	—	—	—	46	85	75	45	81	97	—	+
Ergometrine	27	2	40	21	30	16	2	10	—	—	—	—	17	—	—	+
Ergometrinine	39	20	66	46	41	37	22	35	8	44	38	—	30	—	—	+
Ergonovine	—	—	—	—	0	17	12	—	—	—	—	—	—	—	—	
Ergosine	63	10	73	58	64	55	42	51	1.6	35	31	13	81	11	17	+
Ergosinine	76	45	89	66	—	—	—	—	11.5	75	68	34	75	65	—	+
Ergostine	72	18	80	63	—	—	—	—	—	—	—	—	—	—	—	+
Ergostinine	81	56	88	64	—	—	—	—	—	—	—	—	—	—	—	+
Ergotamine	65	9	75	57	56	44	16	31	1.0	31	31	13	51	6	11	+
Ergotaminine	78	44	90	71	77	67	51	58	6.5	68	64	34	75	50	—	+
Festuclavine	23	48	51	60	—	—	—	—	—	—	—	—	—	—	—	–
Fumigaclavine A	57	60	78	71	—	—	—	—	—	—	—	—	—	—	—	–
Fumigaclavine B	13	45	39	58	—	—	—	—	—	—	—	—	—	—	—	–
Fumigaclavine C	72	78	91	75	—	—	—	—	—	—	—	—	—	—	—	–
D-Isolysergic acid amide	51	22	68	39	—	—	—	—	—	—	—	—	—	—	—	+
Isolysergol	28	28	61	52	—	—	—	—	—	—	—	—	—	—	—	+
Isopenniclavine	41	10	56	8	—	—	—	—	—	—	—	—	—	—	—	+
Isosetoclavine	54	38	68	60	—	—	—	—	—	—	—	—	—	—	—	+
Lysergene	58	44	74	63	—	—	—	—	—	—	—	—	—	—	—	+

TABLE TLC 114—(Continued)

ALKALOIDS, ERGOT ALKALOIDS AND HYDROGENATED DERIVATIVES

Layer / Solvent	L1 S1	L1 S2	L1 S3	L2 S4	L3 S5	L3 S6	L3 S7	L3 S8	L2 S9	L1 S10	L1 S11	L1 S12	L1 S13	L4 S14 S15	L4 S16	
Technique / Literature	na 1	na 1	na 1	na 1	T1 2	T1 2	T1 2	T1 2	T2 3	T2 3	T2 3	T3 4	T4 4	T5 5	na 6	1
Compound	$R_F \times 100$															Fluorescence
D-Lysergic acid	0	0	0	0	—	—	—	—	—	—	—	—	—	—	—	+
D-Lysergic acid amide	22	3	34	25	—	—	—	—	—	—	—	—	—	—	—	+
D-Lysergic acid methyl carbinolamide	36	5	43	26	—	—	—	—	—	—	—	—	—	—	—	+
Lysergine	42	44	69	63	—	—	—	—	—	—	—	—	—	—	—	+
Lysergol	16	9	35	28	—	—	—	—	—	—	—	—	—	—	—	+
D-Lysergyl-L-valine methyl ester	70	43	80	65	—	—	—	—	—	—	—	—	—	—	—	+
6-Methyl-$\Delta^{8,9}$-ergolene-8-carboxylic acid	0	0	0	0	—	—	—	—	—	—	—	—	—	—	—	−
Nor-agroclavine	16	30	34	23	—	—	—	—	—	—	—	—	—	—	—	−
Penniclavine	21	7	35	5	33	53	4	10	—	—	—	—	—	—	—	+
Pyroclavine	20	59	63	68	—	—	—	—	—	—	—	—	—	—	—	−
Setoclavine	48	31	66	56	—	—	—	—	—	—	—	—	—	—	—	+

Layer
L1 = Silica gel G.
L2 = Aluminum oxide G.
L3 = Kieselgel GF_{254} (Merck); activated 30 minutes at 105 °C.
L4 = Cellulose impregnated with dimethylformamide.

Solvent
S1 = methanol-chloroform (2 : 8).
S2 = diethylamine-chloroform (1 : 9).
S3 = methanol-chloroform-conc. NH_3 (20 : 80 : 0.2).
S4 = ethanol-chloroform (4 : 96).
S5 = dichloromethane-methanol (92.7 : 7.3 wt.%).
S6 = chloroform-ethanol (92.0 : 8.0 wt.%).
S7 = chloroform-2-butanone (17.0 : 83.0 wt%).
S8 = acetone-cyclohexane (67.5 : 32.5 wt%).
S9 = chloroform-ethanol-water (3 : 1 : 1).
S10 = ethyl acetate-*N,N*-dimethylformamide-ethanol (13 : 1.9 : 0.1).
S11 = benzene-*N,N*-dimethylformamide (13 : 2).
S12 = chloroform-ethanol (19 : 1).
S13 = chloroform-ethanol (9 : 1).
S14 = benzene-heptane-chloroform (6 : 5 : 3).
S15 = benzene-heptane (6 : 5).
S16 = ethyl acetate-heptane-diethylamine (5 : 6 : 0.02).

Detection Technique
See Part Section II.I, Volume Two.
T1 = development dist = 13 cm, temperature 23 ± 2°C.
T2 = 17 cm development dist.
T3 = 10 cm development dist, 1st development in S12, 2nd in ethyl acetate-ethanol-dimethylformamide (85 : 10 : 5), same direction.
T4 = 10 cm development dist.
T5 = 15 cm development dist with S14, 2nd development 15 cm with S15.

Literature

1. Agurell, S., *Acta Pharm. Suec.*, **2**, 362 (1965).
2. Röder, E., Mutschler, E., and H. Rochelmeyer, *Z. Anal. Chem.*, **244**, 47 (1969).
3. McLaughlin, J. L., Goyan, J. E., and A. G. Paul, *J. Pharm. Sci.*, **53**, 308 (1964).
4. Klavehn, M. and H. Rochelmeyer, *Deut. Apotheker-Ztg.*, **101**, 477 (1961) and Klavehn, M., Rochelmeyer, H. and J. Seyfried, *Deut. Apotheker-Ztg.*, **101**, 75 (1961).
5. Teichert, K., Mutschler, E., and H. Rochelmeyer, *Deut. Apotheker-Ztg.*, **100**, 283 (1960).
6. Hohmann, T. and H. Rochelmeyer, *Arch. Pharm.*, **297/69**, 186 (1964).

TABLE TLC 115

ALKALOIDS, RAUWOLFIA

Layer	L1	L2	L2	L2
Solvent	S1	S2	S3	S4
Technique	T1	T2	T2	T2
Literature	1	2	2	2
Compound	$R_F \times 100$			
Ajmalicine	—	77	—	—
Ajmaline	28	2.4	87	51
Rescinnamine	51	—	—	—
Reserpine	59	60	—	89
Reserpinine	89	—	—	—
Sarpagine	3.0	—	—	—
Serpentine	6.0	2.4	75	34
Serpentinine	—	—	86	73
Yohimbine	33	—	—	—

Layer L1 = Formamide impregnated cellulose layers.
L2 = Loose layers of alumina.

Solvent S1 = heptane-methyl ethyl ketone (1 : 1).
S2 = chloroform-acetone (85 : 15).
S3 = absol. ethanol.
S4 = chloroform-ethanol-acetone (90 : 5 : 5).

Technique T1 = development in NH_3 atmosphere; removal of formamide before spraying.
T2 = 33 cm development distance, ascending.

Literature
1. Teichert, K., Mutschler, E., and H. Rochelmeyer, *Deut. Apotheker-Ztg.*, **100**, 477 (1960).
2. Ikram, M., Miana, G. A., and M. Islam, *J. Chromatogr.*, **11**, 260 (1963).

TABLE TLC 116

ALKALOIDS, OPIUM ALKALOIDS AND SYNTHETIC MORPHINE DERIVATIVES

Layer	L1	L1	L1	L2
Solvent	S1	S2	S3	S4
Detection	na	na	na	D1
Literature	1	1	1	1
Compound	$R_F \times 100$			
Acedicon	59	24	—	90
Codeine	33	12	41	37
Dicodid	28	10	34	63
Dilaudid	13	5	27	6
Dionin	37	14	44	57
Eucodal (Oxycodone)	70	47	79	75
Morphine	10	2	27	0
Narcotine	81	78	92	94
Papaverine	78	74	86	89
Thebaine	—	—	—	85

Layer L1 = Silica gel G.
L2 = Cellulose layers impregnated with 20% formamide in acetone.

Solvent S1 = chloroform-ethanol (8 : 2).
S2 = chloroform-ethanol (9 : 1).
S3 = dimethylformamide-diethylamine-ethanol-ethyl acetate (5 : 2 : 20 : 75).
S4 = benzene-heptane-chloroform-diethylamine (6 : 5 : 1 : 0.02).

Detection D1 = Modified Dragendorff.

Literature
1. Teichert, K., Mutschler, E., and H. Rochelmeyer, *Deut. Apotheker-Ztg.*, **100**, 477 (1960) and *Z. Anal. Chem.* **181**, 325 (1961).

TABLE TLC 117
ALKALOIDS, TROPINE

Layer	L1	L1	L1	L1	L2	L3
Solvent	S1	S2	S3	S4	S5	S6, S7
Technique	na	na	na	na	na	T1
Literature	1	1	2	3	3	3

Compound	$R_F \times 100$					
Apoatropine	52	72	—	44	44	74
Aposcopolamine	70	90	—	—	—	—
Atropine	37	58	89	37	36	15
Belladonnine	—	—	—	26	17	69
Genatropine	0	19	—	—	—	—
Genohomatropine	0	12	—	—	—	—
Genoscopolamine	0	3.0	—	—	—	—
Genotropanol	0	24	—	—	—	—
Homatropine	36	58	87	—	—	—
Methylhomatropine	—	—	84	—	—	—
Methyltropine	—	—	38	—	—	—
Scopolamine	54	74	—	73	83	53
Scopolanol	58	79	—	—	—	—
Tropanol	20	26	—	—	—	—
Tropine	—	—	43	16	3	13

Layer L1 = Silica gel G.
L2 = Alkaline (0.5*N* KOH) Silica gel G layer.
L3 = Cellulose layer impregnated with formamide.

Solvent S1 = chloroform-acetone-diethylamine (5: 4 : 1).
S2 = chloroform-methanol-diethylamine (18 : 2 : 1).
S3 = ethanol-pyridine-water (10 : 60 : 40).
S4 = dimethylformamide-diethylamine-ethanol-ethyl acetate (1 : 1 : 6 : 12).
S5 = 70% ethanol-25% NH_3 (99 : 1).
S6 = heptane-diethylamine (500 : 1).
S7 = benzene-heptane-chloroform-diethylamine (6 : 5 : 1 : 0.02).

Detection See Section II.I, Volume Two.

Technique T1 = 15 cm development distance with S6, then 10 cm in same direction with S7.

Literature

1. Kaess, A. and C. Mathis, *Ann. Pharm. Fr.*, **23**, 270 (1965).
2. Guven, K. C. and A. Hincal, *Istanbul Univ. Eczacilik Fa. Mecmuasi*, **1**, 1965; *Chem. Abstr.*, **65**, 8668d, 1966.
3. Teichert, K., Mutschler, E., and H. Rochelmeyer, *Deut. Apotheker-Ztg.*, **100**, 477 (1960).

TABLE TLC 118
AMINO ACIDS, TWO-DIMENSIONAL

Layer	L1	L1
Solvent	S1	S2
Technique	T1	T2
Literature	1	1

Compound	$R_F \times 100$ 1st dimension	2nd dimension	Color
DL-Alanine	43–53	42–52	bt p
β-Alanine	35–41	37–48	br-ish-p
γ-Aminobutyric acid	36–46	40–50	bt p
Arginine	21–30	21–31	bt p
Asparagine	35–45	38–48	bl-ish
Aspartic acid	13–23	13–23	pe bl
Cysteic acid	20–50	23–52	very dk p
Cystine	33–43	32–35	br-ish-p
L-Cystine	34–44	31–34	br-ish-p
Glutamic acid	17–27	17–27	dk p
Glycine	40–50	40–50	du p
Histidine	57–67	55–66	du br
Hydroxyproline	39–47	40–50	my
Isoleucine	81–91	80–90	bt p
L-Leucine	83–93	81–90	bt p
L-Lysine	45–55	45–55	du p
Methionine	73–83	80–90	bt p
Phenylalanine	90–96	88–96	du br
L-Proline	46–56	46–56	y with p margin
DL-Serine	50–60	50–60	bt p
L-Serine	60–70	60–70	bt p
DL-Threonine	78–88	80–90	bt p
Tryptophan	85–95	83–93	du br
Tyrosine	64–74	63–74	d br-ish p
L-Valine	64–73	60–70	bt p

Layer L1 = Cellulose MN-300.

Solvent S1 = butanol-acetone-ammonia-water (10 : 10 : 5 : 2).
S2 = isopropanol-formic acid-water (20 : 1 : 5).

Detection Fresh solution of 0.2% ninhydrin in absolute alcohol, 10 ml of acetic acid and 2 ml of 2,4,6-collidine, plates heated 40–60 °C for 10–15 minutes.

Technique T1 = 1st dimension, saturated chamber, ascending technique, development distance = 15 cm, air-dried.
T2 = 2nd dimension, ascending, saturated chamber, development distance = 15 cm. Average of 15–20 determinations.

Literature

1. Pillay, D. T. N. and Rafia Mehdi, *J. Chromatogr.*, **47**, 121 (1970).

TABLE TLC 119
AMINES, AROMATIC

Layer	L1	L1	L2	L2	L2	L2	L3	L3	L4	L4	L4	L4		
Solvent	S1	S2	S3	S4	S5	S6	S6	S7	S8	S9	S10	S4		
Detection													D1	D2
Literature	1	1	2	2	2	2	2	2	2	2	2	2	1	2
Compound	$R_F \times 100$												Color	
N-Acetyl-2,5-dichloro-1,4-phenylenediamine	—	44	—	—	—	—	—	—	—	—	—	—	—	—
N-Acetyl-1,3-phenylenediamine	36	—	—	—	—	—	—	—	—	—	—	—	y-br	—
N-Acetyl-1,4-phenylenediamine	0	11	—	—	—	—	—	—	—	—	—	—	y	—
4-Aminoacetophenone	17	—	—	—	—	—	—	—	—	—	—	—	y	—
1-Aminoanthracene	47	—	—	—	—	—	—	—	—	—	—	—	o	—
2-Aminoanthracene	43	—	—	—	—	—	—	—	—	—	—	—	o	—
2-Aminobenzoic acid	—	—	—	—	—	—	72	—	—	—	—	—	—	cr
3-Aminobenzoic acid	—	—	—	—	—	—	64	—	—	—	—	—	—	ru
4-Aminobenzoic acid	—	—	—	—	—	—	69	—	—	—	—	—	—	cr
4-Aminobenzoic acid methyl ester	15	—	—	—	—	—	—	—	—	—	—	—	y	—
4-Aminodiphenyl	42	—	21	44	54	69	—	—	—	—	—	—	y	bor
1-Amino-3-methoxy-4-acetaminobenzene	—	19	—	—	—	—	—	—	—	—	—	—	—	—
2-Amino-6-methoxybenzothiazole	2	23	—	—	—	—	—	—	—	—	—	—	y	—
2-Amino-6-methylbenzothiazole	—	25	—	—	—	—	—	—	—	—	—	—	—	—
1-Amino-2-naphthol	—	—	36	58	—	—	—	—	—	—	—	—	—	lt br[b]
2-Amino-5-nitrophenol	—	13	—	—	—	—	—	—	—	—	—	—	—	—
2-Aminophenol	—	—	3	11	17	56	59	77	—	—	—	—	—	ru
3-Aminophenol	—	—	2	10	17	53	58	76	—	—	—	—	—	ru
4-Aminophenol	—	—	—	—	3	7	25	59	—	—	—	—	—	ru
1,2-Aminophenol	5	—	—	—	—	—	—	—	—	—	—	—	r	—
1,3-Aminophenol	15	—	—	—	—	—	—	—	—	—	—	—	y	—
1,4-Aminophenol	1	7	—	—	—	—	—	—	—	—	—	—	y	—
2-Aminophenol-4-diethylaminosulphonamide	—	27	—	—	—	—	—	—	—	—	—	—	—	—
Aniline	41	—	13	28	38	64	68	—	58	44	58	28	y	r
6-Amino-2-phenylbenzothiazole	26	—	—	—	—	—	—	—	—	—	—	—	o	—
Benzidine	—	—	—	5	10	—	59	78	54	31	57	5	—	vt
3-Bromoaniline	57	—	—	—	—	—	—	—	—	—	—	—	y	—
4-Bromoaniline	54	—	31	59	61	69	—	—	36	15	50	59	y	bor
2-Chloroaniline	71	—	62	80	83	—	—	—	13	6	44	80	y	cr
3-Chloroaniline	54	—	45	67	67	—	—	—	33	13	53	67	y	cr
4-Chloroaniline	49	—	30	51	58	69	—	—	45	23	53	51	y	cr
2-Chloro-1,4-diaminobenzene	—	—	—	8	10	30	49	78	62	48	—	8	—	bl
4-Chloro-1,2-diaminobenzene	—	—	15	31	36	54	61	[a]	62	41	—	31	—	gy-vt
4-Chloro-1,3-diaminobenzene	—	—	8	17	25	61	—	84	58	42	59	17	—	r-vt
3-Chloro-4-methylaniline	53	—	—	—	—	—	—	—	—	—	—	—	y	—
2-Chloro-4-nitroaniline	37	—	—	—	—	—	—	—	—	—	—	—	y	—
2-Chloro-1,4-phenylenediamine	12	—	—	—	—	—	—	—	—	—	—	—	y-br	—

TABLE TLC 119—(Continued)
AMINES, AROMATIC

Layer	L1	L1	L2	L2	L2	L2	L3	L3	L4	L4	L4	L4		
Solvent	S1	S2	S3	S4	S5	S6	S6	S7	S8	S9	S10	S4		
Detection													D1	D2
Literature	1	1	2	2	2	2	2	2	2	2	2	2	1	2
Compound	$R_F \times 100$												Color	
1,2-Diaminobenzene	—	—	4	10	16	35	45	—	—	—	—	—	—	ft-vt
1,3-Diaminobenzene	—	—	0	2	5	18	29	71	—	—	—	—	—	r-vt
1,4-Diaminobenzene	—	—	—	—	0	5	12	37	—	—	—	—	—	bl
2,2′-Diaminodiphenyl	23	—	—	—	—	—	—	—	—	—	—	—	y	—
4,4′-Diaminodiphenyl	49	—	—	—	—	—	—	—	—	—	—	—	y	—
2,5-Dichloro-4-acetaminoaniline	5	—	—	—	—	—	—	—	—	—	—	—	y	—
2,4-Dichloroaniline	78	—	62	79	80	—	—	—	2	0	14	79	y	r-vt
2,5-Dichloroaniline	78	—	67	83	—	—	—	—	2	—	7	83	y	r-vt
3,3′-Dichlorobenzidine	—	—	56	73	—	—	—	—	3	0	22	73	—	v
2,2′-Dichloro-5,5′-dimethoxybenzidine	—	—	44	65	75	—	—	—	6	—	33	65	—	bl-vt
2,6-Dichloro-4-nitroaniline	69	—	—	—	—	—	—	—	—	—	—	—	ft-y	—
2,5-Dimethoxyaniline	25	—	20	41	50	69	71	—	—	—	—	—	y	r-vt
3,3′-Dimethoxybenzidine	—	—	1	9	18	56	64	79	44	21	38	9	—	bl-vt
2,4-Dimethoxy-5-chloroaniline	33	—	10	27	34	57	—	—	39	15	26	27	o	br-vt
2,5-Dimethoxy-4-chloroaniline	36	—	39	58	62	68	—	—	18	4	31	58	o	vt
2,5-Dimethoxy-1,4-diaminobenzene	—	—	—	—	—	9	26	58	—	—	—	—	—	bl
3,3′-Dimethoxy-4,4′-diaminodiphenyl	17	—	—	—	—	—	—	—	—	—	—	—	o	—
2,3-Dimethylaniline	44	—	—	—	—	—	—	—	—	—	—	—	y	—
2,4-Dimethylaniline	41	—	14	23	33	55	62	—	—	—	—	—	y	sc
2,5-Dimethylaniline	49	—	18	35	46	65	—	—	—	—	—	—	y	sc
2,6-Dimethylaniline	60	—	40	59	65	72	—	—	35	14	35	59	y	o-y
3,4-Dimethylaniline	37	—	—	—	—	—	—	—	—	—	—	—	y	—
3,3′-Dimethylbenzidine	—	—	—	11	20	—	—	—	—	—	—	—	—	r-vt
2,5-Dimethyl-1,4-diaminobenzene	—	—	—	2	3	6	17	46	—	—	—	—	—	bl
2,6-Dimethyl-1,4-diaminobenzene	—	—	20	40	51	69	70	83	—	—	—	—	—	bl
4,6-Dimethyl-1,3-diaminobenzene	—	—	—	2	3	11	25	62	—	—	—	—	—	r-vt
3,3′-Dimethyl-4,4′-diaminodiphenyl	13	—	—	—	—	—	—	—	—	—	—	—	o	—
2,5-Dimethyl-1,4-phenylenediamine	—	22	—	—	—	—	—	—	—	—	—	—	—	—
2,4-Dinitroaniline	17	—	—	—	—	—	—	—	—	—	—	—	y	—
2,4-Dinitro-6-chloroaniline	51	—	—	—	—	—	—	—	—	—	—	—	y	—
2,6-Dinitro-4-methylaniline	68	—	—	—	—	—	—	—	—	—	—	—	y	—
2-Ethoxyaniline	47	—	20	40	—	—	—	—	—	—	—	—	y	ru
4-Ethoxyaniline	21	—	6	9	—	—	—	—	—	—	—	—	y	ru
3-Iodoaniline	60	—	—	—	—	—	—	—	—	—	—	—	y	—
2-Methoxyaniline	42	—	17	35	50	65	—	—	—	—	—	—	y	cr
3-Methoxyaniline	28	—	—	—	—	—	—	—	—	—	—	—	y	—

TABLE TLC 119—(Continued) AMINES, AROMATIC

Layer	L1	L1	L2	L2	L2	L2	L3	L3	L4	L4	L4	L4		
Solvent	S1	S2	S3	S4	S5	S6	S6	S7	S8	S9	S10	S4		
Detection													D1	D2
Literature	1	1	2	2	2	2	2	2	2	2	2	2	1	2
Compound	$R_F \times 100$												Color	
4-Methoxyaniline	19	—	4	9	16	23	42	—	—	—	—	—	y	ru
2-Methoxy-4-chloroaniline	64	—	—	—	—	—	—	—	—	—	—	—	o	—
2-Methoxy-5-chloroaniline	61	—	49	71	76	—	—	—	13	7	39	71	y	r-vt
4-Methoxy-3-chloroaniline	—	—	9	30	34	58	65	—	54	29	50	30	—	cr
2-Methoxy-5-chloro-1,4-diaminobenzene	—	—	—	5	—	25	52	76	—	—	—	—	—	bl
2-Methoxy-4-chloro-5-methylaniline	48	—	36	58	60	72	72	84	18	3	21	58	yo	br-vt
2-Methoxy-1,4-diaminobenzene	—	—	—	—	—	3	12	39	—	—	—	—	—	bl
4-Methoxy-1,2-diaminobenzene	—	—	6	11	13	21	27	[a]	—	—	—	—	—	gy-vt
4-Methoxy-1,3-diaminobenzene	—	—	—	2	3	9	18	46	—	—	—	—	—	r-vt
2-Methoxy-5-methylaniline	38	—	13	31	46	66	—	—	—	—	—	—	y	r-vt
4-Methoxy-2-methylaniline	—	—	2	7	12	32	50	75	—	—	—	—	—	y-r
2-Methoxy-5-methyl-1,4-diaminobenzene	—	—	—	—	—	4	13	42	—	—	—	—	—	bl
2-Methoxy-4-nitroaniline	35	—	—	—	—	—	—	—	—	—	—	—	y	—
2-Methoxy-5-nitroaniline	31	—	—	—	—	—	—	—	—	—	—	—	y	—
2-Methylaniline	49	—	21	39	50	68	—	—	—	—	—	—	y	sc
3-Methylaniline	42	—	12	25	35	62	67	—	—	—	—	—	y	y-r
4-Methylaniline	35	—	6	16	23	49	63	—	—	—	—	—	y	r
2-Methyl-3-chloroaniline	62	—	47	67	69	—	—	—	20	5	41	67	y	y-r
2-Methyl-4-chloroaniline	56	—	38	59	62	69	—	—	25	9	42	59	y	ru
2-Methyl-5-chloroaniline	65	—	50	70	70	—	—	—	12	5	39	70	y	ru
2-Methyl-6-chloroaniline	75	—	—	—	—	—	—	—	—	—	—	—	y	—
2-Methyl-1,4-diaminobenzene	—	—	—	—	—	7	14	40	—	—	—	—	—	bl
4-Methyl-1,2-diaminobenzene	—	—	8	14	15	26	36	[a]	—	—	—	—	—	gy-vt
4-Methyl-1,3-diaminobenzene	—	—	—	2	7	13	25	64	—	—	—	—	—	r-vt
2-Methyl-3-nitroaniline	55	—	—	—	—	—	—	—	—	—	—	—	y	—
2-Methyl-4-nitroaniline	27	—	—	—	—	—	—	—	—	—	—	—	y	—
2-Methyl-5-nitroaniline	34	—	—	—	—	—	—	—	—	—	—	—	y	—
4-Methyl-1,3-phenylenediamine	6	—	—	—	—	—	—	—	—	—	—	—	y-br	—
1-Naphthylamine	47	—	34	54	67	74	—	—	—	—	—	—	y	—
2-Naphthylamine	43	—	21	42	54	70	—	—	—	—	—	—	y	—
2-Nitroaniline	45	—	45	67	—	—	—	—	—	—	—	—	y	br-vt
3-Nitroaniline	30	—	30	47	—	—	—	—	—	—	—	—	y	cr
4-Nitroaniline	20	—	21	40	—	—	—	—	—	—	—	—	o	r-vt
2-Nitro-4-chloroaniline	49	—	—	—	—	—	—	—	—	—	—	—	y	—
2-Nitro-4-cyanoaniline	14	—	—	—	—	—	—	—	—	—	—	—	y	—
2-Nitro-4,4′-diaminodiphenyl	6	—	—	—	—	—	—	—	—	—	—	—	o	—

TABLE TLC 119—(Continued)
AMINES, AROMATIC

Layer	L1	L1	L2	L2	L2	L2	L3	L3	L4	L4	L4	L4		
Solvent	S1	S2	S3	S4	S5	S6	S6	S7	S8	S9	S10	S4		
Detection													D1	D2
Literature	1	1	2	2	2	2	2	2	2	2	2	2	1	2
Compound	$R_F \times 100$												Color	
2-Nitro-4-methylaniline	41	—	—	—	—	—	—	—	—	—	—	—	o	—
3-Nitro-4-methylaniline	24	—	—	—	—	—	—	—	—	—	—	—	y	—
2-Nitro-1,4-phenylenediamine	12	—	—	—	—	—	—	—	—	—	—	—	o	—
3-Nitro-1,2-phenylenediamine	19	—	—	—	—	—	—	—	—	—	—	—	r-vt	—
4-Nitro-1,2-phenylenediamine	2	20	—	—	—	—	—	—	—	—	—	—	y	—
4-Nitro-1,3-phenylenediamine	5	—	—	—	—	—	—	—	—	—	—	—	y	—
1,2-Phenylenediamine	0	25	—	—	—	—	—	—	—	—	—	—	y	—
1,3-Phenylenediamine	7	—	—	—	—	—	—	—	—	—	—	—	y-br	—
1,4-Phenylenediamine	8	—	—	—	—	—	—	—	—	—	—	—	r	—
1,2,4-Triaminobenzene	—	—	—	—	—	0	4	16	—	—	—	—	—	r-vt
2,4,5-Trichloroaniline	—	—	67	82	—	—	—	—	—	—	2	82	—	r-vt
2,4,6-Trimethylaniline	—	—	23	41	52	68	—	—	—	—	—	—	—	o-y

[a] Spot of oxidized compound.
[b] Light blue fluorescence before spraying.

Layer
L1 = Aluminum oxide neutral, activity III, without binder (Reanal, Budapest).
L2 = Kieselgel G (Merck) with 0.2*M* moles of sodium acetate added.
L3 = Kieselgel G (Merck) with 1.0*M* moles of sodium acetate added.
L4 = Kieselgel G (Merck) activated at 85 °C for 30 minutes, impregnated with 20% 1-bromonaphthalene in hexane.

Solvent
S1 = benzene.
S2 = chloroform.
S3 = benzene-acetic acid (85 : 15).
S4 = benzene-butyl acetate-acetic acid (65 : 20 : 15).
S5 = benzene-butyl acetate-acetic acid (60 : 30 : 10).
S6 = butyl acetate-acetic acid (9 : 1).
S7 = methyl ethyl ketone-acetic acid (9 : 1).
S8 = 10% acetic acid
S9 = 2% isobutyric acid.
S10 = 5% dichloroacetic acid.

Detection
D1 = *p*-dimethylaminobenzaldehyde.
D2 = *p*-dimethylaminocinnamaldehyde.

Literature
1. Gemzová, I. and J. Gasparič, *Collect. Czech. Chem. Commun.*, **31**, 2527, 2529 (1966).
2. Bassl, A., Heckemann, H. J., and E. Baumann, *J. Prakt. Chem.*, **36**, 271, 278 (1967).

TABLE TLC 120
AMINES AND IMINES

Layer Solvent Detection Literature	L1 S1 1	L2 S2 2	L2 S3 2	D1 1	D2 1	D3 1	D4 1	D5 1
Compound	$R_F \times 100$			Color[a]				
1-Acetylpyrene	—	—	—	bl	bl	q	q	bl
Acridine	74	—	—	bl	bl	q	bl	ft bl
Acridone	75	—	—	bl	bl	lt bl	bl	bl
7-Aminobenz[*a*]anthracene	34	—	—	bl	bl	bl	q	q
6-Aminobenzo[*a*]pyrene	17	—	—	gr	ro	gr	q	q
6-Aminochrysene	35	—	—	bl	yo	bl	q	bl vt
2-Aminodibenzofuran	66	—	—	ft vt	bl	q	q	dk vt
2-Aminodibenzoselenophene	streak	—	—	ft bl vt	ro	q	q	q
2-Aminodibenzothiophene	38	—	—	ft bl vt	p	q	q	q
3-Aminofluoranthene	43	—	—	gr	r	gr	q	gr
8-Aminofluoranthene	19	—	—	gr	o	gr	q	gr
1-Aminofluorene	67	—	—	lt bl	b gr	q	lt y vt	lt bl
2-Aminofluorene	streak	—	—	lt bl	bl gr	q	lt bl	q
3-Aminofluorene	streak	—	—	lt bl	dk y	q	lt bl	q
4-Aminofluorene	65	—	—	ft vt	bl gr	q	q	dk r vt
1-Aminopyrene	50	—	—	bl	gr	bl	lt bl	bl
1-Anthramine	61	—	—	lt b gr	d bl	b	q	q
2-Anthramine	48	—	—	b gr	d bl	b	q	q
9-Anthramine	68	—	—	bl	d bl	lt bl	lt bl	q
Benz[*c*]acridine	26	—	—	bl	bl	q	bl gr	ft bl
Benzathrone	41	—	—	bl	bl	q	y	bl
4H-Benzo[*d,e,f*]carbazole	46	49	60	bl	bl	q	q	q
5H-Benzo[*b*]carbazole	18	39	<2	bl	y gr	lt bl	q	q
7H-Benzo[*c*]carbazole	36	37	<2	bl	bl	lt bl	q	q
11H-Benzo[*a*]carbazole	34	46	<2	bl	bl	q	q	q
Benzo[*a*]pyrene	7	—	—	bl	bl	q	q	q
13H-Benzo[*g*]pyrido[2,3-*a*]carbazole	13	—	—	lt bl	gr	q	q	q
13H-Benzo[*g*]pyrido[3,2-*a*]carbazole	26	—	—	bl	gr	q	q	q
Carbazole	68	50	<2	lt bl	bl	q	q	q
β-Carboline	54	—	—	bl	bl	q	bl	bl
7H-Dibenzo[*b,g*]carbazole	12	—	—	bl	y gr	lt bl	q	q
7H-Dibenzo[*c,g*]carbazole	12	30	<2	bl	bl	lt bl	q	q
13H-Dibenzo[*a,i*]carbazole	10	—	—	bl	bl	q	q	q
5,10-Dihydroindeno[1,2-*b*]indole	31	—	—	lt bl	bl	q	bl	q
1,4-Dihydroxyanthraquinone	—	—	—	r	q	q	y	r
7,8-Dimethyl-11H-benzo[*a*]carbazole	16	—	—	lt bl	bl	lt bl	q	lt bl

TABLE TLC 120—(Continued)
AMINES AND IMINES

Layer	L1	L2	L2					
Solvent	S1	S2	S3					
Detection				D1	D2	D3	D4	D5
Literature	1	2	2	1	1	1	1	1
Compound	$R_F \times 100$			Color[a]				
2,5-Dimethyl-13H-naphtho[2,1-*a*]carbazole	0	—	—	bl	gr	q	q	q
11,12-Dimethyl-14H-phenanthro[3,2-*a*]-carbazole	0	—	—	bl	y	lt bl	q	q
Fluoranthene	26	—	—	bl	bl	q	lt bl	bl
2-Methylcarbazole	46	—	—	bl	bl	q	q	q
9-Methylcarbazole	39	—	—	bl	bl	q	q	q
5-Methyl-5,10-dihydroindeno[1,2-*b*]indole	15	—	—	lt bl	bl	q	q	q
5H-Naphtho[2,3-*c*]carbazole	11	—	—	bl	y	lt bl	q	q
1-Naphthylamine	81	—	—	b	y gr	b	q	q
2-Naphthylamine	78	—	—	b	y gr	b	q	q
Perylene	7	—	—	bl	bl	q	q	bl
Phenanthridone	69	—	—	bl	bl	lt bl	lt bl	lt bl
Phenothiazine	51	—	—	bl	bl	q	q	q
9-Phenylcarbazole	9	—	—	bl	bl	q	q	q
***N*-Phenylphenothiazine**	3	—	—	bl	dk bl	q	q	q
9H-Pyrido[2,3-*b*]indole	69	—	—	bl	bl	q	bl	bl
1,2,4-Trimethyl-13H-dibenzo[*a*,g]carbazole	0	—	—	lt bl	bl	lt bl	q	lt bl
1,2,4-Trimethyl-13H-dibenzo[*a*,*i*]carbazole	0	—	—	lt bl	bl	q	q	q

[a] For abbreviations, see inside cover.
q = quenched.

Layer L1 = Cellulose powder.
L2 = Aluminum oxide G, activated at 110 °C for 60 minutes.

Solvent S1 = dimethylformamide-water (35 : 65).
S2 = pentane-chloroform (3 : 2).
S3 = NH_4OH.

Detection D1 = UV light.
D2 = tetraethylammonium hydroxide, UV light.
D3 = 3-methylthioaniline, UV light.
D4 = NO_2 vapors and trifluoroacetic acid.
D5 = D4 plus CS_2.

Literature

1. Sawicki, E., Johnson, H., and K. Kosinski, *Microchem. J.*, **10**, 78 (1966).
2. Bender, D. F., Sawicki, E., and R. M. Wilson, Jr., *Anal. Chem.*, **36**, 1011 (1964).

TABLE TLC 121
AMINES, CATECHOLAMINES

Layer	L1
Solvent	S1
Literature	1
Compound	**$R_F \times 100$**
Adrenaline	56
Dihydroxymandelic acid	11
3,4-Dihydroxyphenylacetic acid	27
Dihydroxyphenylalanine	29
Homovanillic acid	63
5-Hydroxyindolylacetic acid	22
3-Hydroxytyramine	51
3-Methoxyadrenaline	80
3-Methoxy-4-hydroxymandelic acid	35
3-Methoxy-4-hydroxyphenylethylamine	75
3-Methoxynoradrenaline	69
Noradrenaline	42

Layer L1 = Polyamide (Woelm), air-dried for two hours.
Solvent S1 = isobutanol-acetic acid-cyclohexane (80 : 7 : 10).
Detection Diazotized-*p*-nitroaniline, ethylenediamine and UV.
Technique Saturated chamber.
Literature
1. Segura-Cardona, R. and K. Soehring, *Med. Exp.*, **10**, 251–257 (1964); from *Chem. Abstr.* **61**, 5992d (1965).

TABLE TLC 122
AMINES, UREAS

Layer	L1	L2
Solvent	S1	S2
Literature	1	1
Compound	**$R_F \times 100$**	
Benzylurea	37	24
***N,N′*-Dimethylurea**	28	49
***N,N′*-Dimethylurea**	16	21
Ethyleneurea	31	62
Ethylurea	23	24
Methylurea	15	18
Phenylurea	42	15
***m*-Tolylurea**	59	20
***o*-Tolylurea**	65	39
***p*-Tolylurea**	58	20
Urea	10	10

Layer L1 = MN Silica gel N-HR.
L2 = MN Polyamide-acetylated.
Solvent S1 = carbon tetrachloride-dichloromethane-ethyl acetate-formic acid (70 : 50 : 50 : 10).
S2 = carbon tetrachloride-ethyl acetate-ethanol (100 : 5 : 2).
Detection Dimethylaminobenzaldehyde.
Literature
1. Knappe, E. and I. Rohdewald, *Z. Anal. Chem.*, **217**, 111 (1966); from *J. Chromatogr.*, **28**, D3 (1967).

TABLE TLC 123
AMINO ACIDS, 1-DIMETHYLAMINONAPHTHALENE-5-SULFONYLAMINO

Layer	L1	L1	L1	L1	L1	L1	L1	L1	L1	L2	L2	L2	L1	L1	L1	L1
Solvent	S1	S2	S3	S4	S5	S6	S7	S8	S9	S2	S4	S5	S3	S10	S11	S12
Technique	T1	T1	T1	T1	T1	T1	T1	T1	T1	T2	T2	T2	T3	T3	T3	T3
Literature	1	1	1	1	1	1	1	1	1	2	2	2	3	3	3	3
Compound	$R_F \times 100$												R_{DNS-NH_2}[a] $\times 100$			
DNS-Alanine	48	54	47	39	88	—	60	86	—	—	—	—	61	89	96	29
DNS-α-Alanine	—	—	—	—	—	—	—	—	—	37	60	88	—	—	—	—
DNS-Arginine	0	0	—	1	58	—	—	—	—	0	0	58	0	0	0	0
DNS-Asparagine	—	—	—	—	—	—	—	—	—	12	5	—	2	2	10	2
DNS-Aspartic acid	3	0	—	—	38	—	—	—	—	2	21	38	6	2	60	0
DNS-*S*-Carboxymethylcysteine	—	0	—	—	—	—	—	—	—	—	—	—	5	2	70	0
Di-DNS-Cystine	—	—	—	—	—	—	—	—	—	8	7	—	—	—	—	—
DNS-Glutamic acid	9	0	—	5	71	—	—	—	—	1	30	71	18	42	85	8
DNS-Glutamine	—	—	—	—	—	—	—	—	—	8	5	—	3	7	14	0
N^6-β-Glyceryllysine	—	—	—	—	—	—	—	—	—	—	—	—	2	0	0	6
DNS-Glycine	28	23	29	22	79	88	46	77	—	23	33	79	38	65	93	12
DNS-Histidine	—	—	—	—	—	—	—	—	—	47	22	95	27	59	91	9
Di-DNS-Histidine	19	52	—	22	95	—	—	—	—	—	—	—	—	—	—	—
Di-DNS-Hydroxyproline	—	—	—	—	—	—	—	—	—	15	20	—	—	—	—	—
DNS-Isoleucine	—	74	73	—	—	82	—	—	—	62	83	—	92	100	100	89
DNS-Leucine	73	72	73	67	97	92	70	92	94	60	78	97	89	100	99	63
ε-DNS-Lysine	—	—	—	—	—	—	—	—	—	—	—	—	2	8	5	4
Di-DNS-Lysine	0	2	—	0	50	—	—	—	—	40	41	50	—	—	—	—
α,ε-Di-DNS-Lysine	—	—	—	—	—	—	—	—	—	—	—	—	48	93	96	33
DNS-Methionine	59	48	—	46	94	—	—	—	—	48	64	94	65	92	99	33
DNS-NH_2	—	—	—	—	—	—	—	—	—	—	—	—	100	100	100	100
DNS-Phenylalanine	70	50	—	55	96	—	—	—	—	50	76	96	69	94	99	42
DNS-Proline	65	35	—	52	89	—	68	—	—	23	70	89	79	93	96	50
DNS-Serine	7	36	11	4	70	63	15	59	74	36	17	67	15	20	52	2
DNS-Threonine	6	55	—	7	78	—	—	67	78	50	18	78	18	25	57	2
DNS-Tryptophan	41	35	—	33	97	—	—	—	—	29	55	97	34	87	95	25
DNS-Tyrosine	—	—	—	—	—	—	—	—	—	55	67	—	55	82	92	23
Di-DNS-Tyrosine	62	55	—	60	95	—	—	—	—	—	—	—	—	—	—	—
DNS-Valine	70	65	66	54	97	87	67	—	—	58	75	97	85	100	97	75

[a] Migration relative to DNS-NH_2.

TABLE TLC 123—(Continued)
AMINO ACIDS, 1-DIMETHYLAMINONAPHTHALENE-5-SULFONYLAMINO

Layer
L1 = Silica gel G.
L2 = ITLC-SA.

Solvent
S1 = chloroform-benzyl alcohol-glacial acetic acid (100 : 30 : 5).
S2 = 2-butanone-triethylamine-dimethylsulfoxide-benzene (4 : 2 : 1 : 3), 75 °C.
S3 = benzene-pyridine-glacial acetic acid (80 : 20 : 2).
S4 = chloroform-benzyl alcohol-glacial acetic acid (30 : 10 : 1).
S5 = 2-butanone-propionic acid-distilled water (15 : 5 : 6).
S6 = 2-butanone-glacial acetic acid (13 : 5).
S7 = 2-butanone-glacial acetic acid-chloroform (13 : 5 : 15).
S8 = 2-butanone-propionic acid (2 : 1).
S9 = 2-butanone-propionic acid-water (5 : 5 : 1).
S10 = chloroform-*tert*-amyl alcohol-acetic acid (70 : 30 : 3).
S11 = chloroform-*tert*-amyl alcohol-formic acid (70 : 30 : 1).
S12 = chloroform-*tert*-amyl alcohol-acetic acid (70 : 30 : 0.5).

Detection
UV light (350–360 nm).

Technique
T1 = ascending, saturated chamber.
T2 = Gelman chamber (saturated), ascending, 10 cm migration.
T3 = ascending, 15–16 cm solvent migration.

Literature
1. David, I. B., French, T. C., and J. M. Buchanan, *Techniques of Thin-Layer Chromatography in Amino Acid and Peptide Chemistry*, Rev. Ed., Pataki, G., Ann Arbor Science Publishers, Inc., Ann Arbor, Mich., 1968, p. 148.
2. Haer, F. C., *An Introduction to Chromatography on Impregnated Glass Fiber*, Ann Arbor Science Publishers, Inc., Ann Arbor, Mich., 1969, p. 104.
3. Morse, D. and B. L. Horecker, *Anal. Biochem.*, **14**, 432 (1966).

TABLE TLC 124
AMINO ACIDS, 3-PHENYL-2-THIOHYDANTOIN (PTH)

Layer	L1	L1	L1	L1	L1	L2	L3	L3	L4	L5	L5	L5	L6	L6	L6
Solvent	S1	S2	S3	S4	S5	S6	S7	S8	S9	S10	S11	S12	S13	S14	S15
Literature	1	1	1	1	2	3	3	3	3	3	3	3	4	4	4
PTH of	$R_F \times 100$														
Alanine	18	77	44	~100	11	41	—	45	46	51	66	—	—	—	—
α-Aminobutyric acid	26	79	54	—	—	—	—	—	—	—	—	—	—	—	—
α-Aminocaprylic acid	44	84	67	—	—	—	—	—	—	—	—	—	86	67	39
Arginine	0	1	0	24	0	—	—	—	—	0	0	0	—	—	—
Asparagine	0	34	9	~100	0	3	67	2	—	0	0	32	0	0	0
Aspartic acid	0	2	16	70	0	0	17	0	—	0	0	20	—	—	—
Citrulline	0	34	8	—	—	—	—	—	—	—	—	—	—	—	—
Cystine	—	—	—	—	—	—	—	—	—	—	—	69	—	—	—
Cysteic acid	0	0	0	—	—	—	—	—	—	—	—	—	—	—	—
Diphenylthiourea	42	82	71	—	22	—	—	—	—	—	—	—	—	—	—
Glutamic acid	1	5	18	75	0	1	25	0	—	0	0	32	11	19	0
Glutamine	0	40	11	~100	0	2	65	2	—	0	0	36	39	7	0
Glycine	11	68	35	90	5	30	—	30	34	28	45	78	—	—	—
Histidine	1	40	1	~100	2	1	58	0	—	0	0	4	—	—	—
Hydroxyproline	5	64	28	—	—	—	—	—	—	—	—	—	65	28	3
α-Isobutyric acid	27	80	56	—	—	—	—	—	—	—	—	—	—	—	—
Isoleucine	39	83	62	100	37	64	—	66	71	78	80	—	85	64	31
Leucine	39	84	63	~100	37	69	—	70	72	78	80	—	85	65	35
Lysine	12	78	34	~100	3	28	—	25	23	28	62	—	—	—	—
Methionine	34	81	54	~100	13	52	—	51	57	58	76	—	—	—	—
Methionine sulfone	2	59	17	—	—	—	—	—	—	—	—	—	—	—	—
Methylglutamic acid	23	82	50	—	—	—	—	—	—	—	—	—	—	—	—
Methylserine	1	51	18	—	—	—	—	—	—	—	—	—	—	—	—
Methionine sulfoxide	0	54	15	~100	1	—	—	—	—	—	—	—	—	—	—
Monophenylthiourea	12	65	32	—	3	—	—	—	—	—	—	—	—	—	—
Norleucine	40	83	62	—	—	—	—	—	—	—	—	—	—	—	—
Norvaline	34	81	57	—	—	—	—	—	—	—	—	—	—	—	—
Ornithine	7	72	30	—	—	—	—	—	—	—	—	—	—	—	—
Phenylalanine	30	81	54	~100	18	56	—	53	60	60	74	—	—	—	—
Proline	60	89	70	~100	21	75	—	61	81	91	91	—	88	75	62
Serine	1	43	10	—	—	6	79	7	5	0	0	48	—	—	—
Threonine	1	58	17	~100	0	11	83	14	9	0	0	51	—	—	—
Tryptophan	14	71	41	~100	10	—	—	—	—	28	52	—	76	47	6
Tyrosine	3	59	22	~100	1	19	—	35	16	0	8	64	—	—	—
Valine	33	81	58	~100	23	59	—	61	61	72	76	—	—	—	—

TABLE TLC 124—(Continued)
AMINO ACIDS, 3-PHENYL-2-THIOHYDANTOIN (PTH)

Layer

L1 = Silica gel G layers, air-dried.
L2 = Silica gel G, dried 2 hours at 130–140 °C.
L3 = Silica gel G, dried 2 hours at 40 °C.
L4 = Woelm Silica gel. Small portions of 85 g Silica gel (Woelm) are placed into a mortar filled with 15 ml water; 10 g of the de-activated adsorbent are vigorously shaken in an Erlenmeyer flask with 30 ml ethyl acetate; 5 ml of this suspension are spread on a 5 × 20 cm plate. After 20–30 minutes of drying in air, the plates are used immediately.
L5 = aluminum oxide (Fluka). 10 g aluminum oxide, Fluka, 2 hours drying at 150 °C, and 80 ml isopropanol, dried over Na_2SO_4, are vigorously shaken in an Erlenmeyer flask, 5.5 ml of the suspension are spread on a 5 × 20 cm plate. After 20–30 minutes of drying in air, the plates are used immediately.
L6 = ITLC-SAF.

Solvent

S1 = chloroform (stabilized with 1.5% alcohol), 18 cm development distance.
S2 = chloroform (stabilized with 1.5% alcohol)-methanol (9 : 1), 18 cm development distance.
S3 = chloroform (stabilized with 1.5% alcohol)-formic acid (100 : 5), 18 cm development distance.
S4 = chloroform (stabilized with 1.5% alcohol)-methanol-formic acid (70 : 30 : 2), 11 cm development distance.
S5 = *n*-heptane-ethylene chloride-formic acid-propionic acid (90 : 30 : 21 : 18). 100 ml of the upper phase are used, 15 cm development distance.
S6 = chloroform (Pharmacopoea Helvetica V)-isopropanol-water (28 : 8 : 1).
S7 = ethyl acetate-pyridine-water (7 : 2 : 1).
S8 = chloroform (Pharmacopoea Helvetica V)-ethyl acetate-water (6 : 3 : 1).
S9 = chloroform (Pharmacopoea Helvetica V)-ethyl acetate (95 : 5).
S10 = chloroform (Pharmacopoea Helvetica V).
S11 = chloroform (Pharmacopoea Helvetica V)-ethanol (98 : 2).
S12 = chloroform (Pharmacopoea Helvetica V)-isopropanol-formic acid (35 : 4 : 1).
S13 = chloroform-methanol (9 : 1).
S14 = chloroform-formic acid (100 : 5).
S15 = chloroform.

Detection Chlorine-toluidine reaction, UV light.

Technique Ascending development in saturated chambers.

Literature

1. Brenner, M., Niederwieser, A., and G. Pataki, *Experientia* (Basel) **17**, 145 (1961).
2. Pataki, G., Dissertation, Universität Basel, 1962.
3. Cherbuliez, E., Baehler, Br., Marszalek, J., Sussman, N. A. R., and J. Rabinowitz, *Helv. Chim. Acta*, **46**, 2446 (1963) and Cherbuliez, E., Baehler, Br., and J. Rabinowitz, *Helv. Chim. Acta*, **47**, 1350 (1964).
4. Haer, F. C., *An Introduction to Chromatography on Impregnated Glass Fiber*, Ann Arbor Science Publishers, Inc., Ann Arbor, Mich., 1969, p. 102.

TABLE TLC 125
ANTIOXIDANTS

Layer	L1	L2	L2	L2	L2	L2	L2	L2	L2	L2	L3	L3	L4	L5	L5	L5	L5	L5	L6	L6		
Solvent	S1	S2	S3	S4	S5	S6	S7	S8	S9	S10	S9	S11	S12	S9	S13	S14	S15	S16	S17	S18		
Detection																					D1	D2
Technique	na	T1	T1	T1	T1	T1	T1	na	T2	T2	T2	T2	T2	T3	T3	T3	T3	T3	T4	T4		
Literature	1	2	2	2	2	2	2	3	4	4	4	4	5	6	6	6	6	6	7	7	7	7
Compound	$R_F \times 100$																				Color[a]	
Amyl gallate	—	—	—	—	—	—	—	—	—	—	—	—	—	—	—	—	—	—	17	49	br	y
Butyl gallate	40	—	—	—	—	—	—	—	—	—	—	—	—	—	—	—	—	—	—	—	—	—
Butylhydroxyanisole	—	—	—	—	—	—	—	67	52	66	31	31	65	42	69	83	95	100	96	46	br	y
Butylhydroxytoluene	—	—	—	—	—	—	—	—	89	70	79	75	12	—	—	—	—	—	—	—	—	—
Cetyl gallate	—	—	—	—	—	—	—	—	—	—	—	—	—	—	—	—	—	—	48	20	br	y
Dibutylhydroxytoluene	—	—	—	—	—	—	—	—	—	—	—	—	—	87	87	92	95	100	98	33	br	y
Dodecyl gallate	83	—	—	—	—	—	—	—	—	—	—	—	—	—	—	—	—	—	—	—	—	—
Ethyl gallate	19	70	26	44	67	39	13	—	—	—	—	—	—	0	—	43	68	90	—	—	—	—
Ethylprotecatechuate	—	—	—	—	—	—	—	—	—	—	—	—	—	—	—	—	—	—	31	57	br	y
Gallic acid	3	28	5	8	31	6	4	—	—	—	—	—	—	—	—	—	—	—	—	—	—	—
Isoamyl gallate	—	—	—	—	—	—	—	—	0	61	0	0	—	—	—	—	—	—	—	—	—	—
Lauryl gallate	—	89	80	85	80	80	67	—	—	—	—	—	20	0	—	68	85	97	37	30	br	y
Methyl gallate	10	60	19	33	53	29	7	—	—	—	—	—	—	0	—	34	60	60	—	—	—	—
Norhydroguaiaretic acid	—	—	—	—	—	—	—	27	0	45	—	—	—	0	12	43	77	100	—	—	—	—
Octyl gallate	67	89	62	73	80	63	45	—	—	—	—	—	50	0	—	64	81	95	—	—	—	—
Propyl gallate	29	68	44	47	55	46	28	56	0	61	0	0	82	0	—	51	70	91	9	53	br	y
Stearyl gallate	—	—	—	—	—	—	—	—	—	—	—	—	—	—	—	—	—	—	64	8	br	y

[a] Sprays are used successively. Colors given are at each stage of spraying.

Layer

L1 = 10% acetylated cellulose.
L2 = Polyamide powder, 1 mm thickness.
L3 = Silica gel.
L4 = Silica gel impregnated with paraffin oil.
L5 = Silica gel CH with 13% gypsum.
L6 = Polyamide sheets 15 × 15 cm.

Solvent

S1 = Shell Sol A-propanol-acetic acid-formic acid (15 : 2 : 1 : 2).
S2 = butanol-acetic acid-water (4 : 1 : 5).
S3 = carbon tetrachloride-ethanol (7 : 3).
S4 = carbon tetrachloride-ethanol (3 : 2).
S5 = ethanol.
S6 = carbon tetrachloride-methanol (7 : 3).
S7 = diethyl ether.
S8 = methanol-actone-H_2O (6 : 1 : 3).
S9 = $CHCl_3$.
S10 = MeOH.
S11 = C_6H_6.
S12 = MeOH-H_2O (3 : 1).
S13 = chloroform-acetic acid (9 : 1).
S14 = chloroform-acetic acid (7.5 : 2.5).
S15 = chloroform-acetic acid (6 : 4).
S16 = chloroform-acetic acid (2.5 : 7.5).
S17 = petroleum ether (30–70 °C)-benzene-acetone (8 : 2 : 5).
S18 = acetone-ethyl alcohol-water (4 : 1 : 2).

Detection

D1 = 5% v/v bromide solution. Dissolve 5 ml bormine water in 100 ml CCl_4.
D2 = 0.25% w/v sodium fluorescein solution. Dissolve 250 mg sodium fluorescein in 5 ml *N,N*-dimethylformamide, then add anhydrous ethanol to 100 ml.

Technique

T1 = ascending (20–30°C angle), 30 cm development distance.
T2 = 10 cm development distance.
T3 = 12 cm development distance, 30 minutes.
T4 = Stahl's ascending technique.

Literature

1. Salo, T. and K. Salminen, *Z. Lebensm.-Unters.-Forsch.*, **125**, 167 (1964).
2. Davidek, J., *J. Chromatogr.*, **9**, 363 (1962).
3. Davidek, J. and J. Pokorný, *Z. Lebensm-Unters.-Forsch.*, **115**, 113 (1961).
4. Ishikawa, S. and G. Katsui, *Bitamin*, **30**, 203 (1964).
5. Jonas, J., *J. Pharm. Belg.*, **17**, 103 (1962).
6. Janicek, J. and E. Davidkova, *Z. Lebensm.-Unters.-Forsch.*, **131**, 345–347 (1967); from *Z. Anal. Chem.*, **234**, 211 (1968).
7. Wang, R-T. and S. S. Chou, *J. Chromatogr.*, **43**, 522–23 (1969).

TABLE TLC 126
CARBOHYDRATES, MONO- AND DISACCHARIDES

Layer Solvent Technique Literature	L1 S1 T1 1	L1 S2 T1 1	L1 S3 T2 2	L1 S4 T2 2	L2 S5 T3 3	L3 S6 T4 4	L3 S7 T4 4	L4 S8 T5 5	L5 S9 T6 6	L5 S10 T6 6	L6 S11 na 7	L6 S12 na 7	L7 S13 T7 8	L7 S14 T7 8	L3 S6 T4 4	L3 S7 T4 4	L4 S8 T5 5
Compound	$R_F \times 100$														Colors		
D-Arabinose	61	65	28	39	—	—	—	—	—	—	28	47	—	—	bl gr	bl gr	—
L-Arabinose	61	65	28	40	57	62	42	28	35	28	—	—	49	42	bl gr	bl gr	y-gr
D-Fructose	65	52	31	42	51	52	31	25	33	26	26	46	47	42	bl r	r-br	vt
D-Galactose	62	67	23	34	—	55	32	18	24	16	16	30	42	35	bl v	gy gr	gy gr
D-Galacturonic acid	—	—	—	—	—	36	10	—	—	—	—	—	—	—	bl	bl gr	—
D-Glucose	66	67	28	42	49	63	42	17	31	24	22	41	45	39	br	v	l bl
D-Glucuronic acid	—	—	—	—	—	48	34	—	—	—	—	—	—	—	bl	bl	—
Lactose	61	61	12	23	9	56	25	4	—	—	6	22	34	25	rv	gy bl	gr
Maltose	59	63	18	30	29	—	—	6	—	—	12	35	—	—	—	—	vt
D-Mannose	66	66	32	46	55	58	32	23	36	26	28	49	47	41	l bl	bl gr	gr
Raffinose	64	43	9	20	13	—	—	—	—	—	4	12	32	26	—	—	—
L-Rhamnose	67	65	50	64	—	67	52	62	58	49	55	72	60	—	gr	bl gr	gr
D-Ribose	58	60	34	45	59	—	—	49	—	—	37	58	—	—	—	—	bl
L-Sorbose	66	55	31	45	—	51	24	26	—	—	—	—	—	—	r	r br	vt
Sucrose (Saccharose)	70	56	22	40	25	63	29	8	—	—	14	36	44	39	r	r br	v
Trehalose	62	50	20	32	—	—	—	—	—	—	—	—	—	—	—	—	—
D-Xylose	66	65	36	53	57	59	39	39	40	37	40	58	—	—	l bl	bl	gy

Layer Solvent Technique Literature	L8 S15 T7 8	L8 S14 T7 8	L9 S16 T7 9	L10 S16 T7 9	L11 S16 T7 9	L12 S16 T7 9	L13 S17 T8 10	L13 S18 T8 10	L14 S19 T9 11	L14 S20 T9 11	L15 S21 T9 11	L16 S22 T9 11	L17 S21 T9 11	L3 S6 T4 4	L3 S7 T4 4	L4 S8 T5 5
Compound	$R_F \times 100$													Colors		
D-Arabinose	—	—	—	—	—	—	61	50	—	—	—	—	—	bl gr	bl gr	—
L-Arabinose	43	42	—	—	—	—	—	—	—	—	—	—	—	bl gr	bl gr	y-gr
D-Fructose	43	42	—	—	—	—	61	47	65	40	67	44	34	bl r	r-br	vt
D-Galactose	37	35	27	22	25	21	53	40	60	27	64	46	26	bl v	gy gr	gy gr
D-Galacturonic acid	—	—	—	—	—	—	—	—	—	—	—	—	—	bl	bl gr	—
D-Glucose	39	39	22	18	18	35	47	35	66	32	68	47	28	br	v	l bl
D-Glucuronic acid	—	—	—	—	—	—	—	—	—	—	—	—	—	bl	bl	—
Lactose	26	26	16	0	7	11	39	37	18	5	52	22	12	rv	gy bl	gr

TABLE TLC 126—(Continued)
CARBOHYDRATES, MONO- AND DISACCHARIDES

Layer	L8	L8	L9	L10	L11	L12	L13	L13	L14	L14	L15	L16	L17	L3	L3	L4
Solvent	S15	S14	S16	S16	S16	S16	S17	S18	S19	S20	S21	S22	S21	S6	S7	S8
Technique	T7	T7	T7	T7	T7	T7	T8	T8	T9	T9	T9	T9	T9	T4	T4	T5
Literature	8	8	9	9	9	9	10	10	11	11	11	11	11	4	4	5
Compound	$R_F \times 100$													Colors		
Maltose	—	—	—	—	—	—	48	34	32	9	60	32	17	—	—	vt
D-Mannose	43	42	—	—	—	—	60	40	62	34	68	52	30	1 bl	bl gr	gr
Raffinose	20	23	—	—	—	—	—	—	—	—	—	—	—	—	—	—
L-Rhamnose	—	55	—	—	—	—	—	—	—	—	—	—	—	gr	bl gr	gr
D-Ribose	—	—	—	—	—	—	69	57	82	64	75	72	43	—	—	bl
L-Sorbose	—	—	—	—	—	—	—	—	—	—	—	—	—	r	r br	vt
Sucrose (Saccharose)	32	39	11	18	15	19	67	37	37	14	61	37	19	r	r br	v
Trehalose	—	—	—	—	—	—	—	—	—	—	—	—	—	—	—	—
D-Xylose	—	—	—	—	—	—	65	41	—	—	—	—	—	1 bl	bl	gy

Layer

L1 = Magnesium Silicate (Woelm), air-dried.
L2 = Silica gel G dried at 135 °C.
L3 = Silica gel G prepared with 0.1*N* boric acid.
L4 = Kieselguhr G prepared with 0.02*M* sodium acetate.
L5 = Acidic aluminum oxide.
L6 = Silicic acid prepared with 1.15*M* phosphate buffer, pH 8.
L7 = activated Silica gel G + alumina.
L8 = Silica gel G + alumina (inactive layers).
L9 = Silicic acid (Merck, 18–25 mesh) with 10% anhydrous calcium sulfate.
L10 = Alumina (Merck, 18–25 mesh) with 10% anhydrous calcium sulfate.
L11 = Silica gel (Merck, 18–25 mesh) with 13% anhydrous $CaSO_4$.
L12 = Cellulose powder (Merck, 18–25 mesh) with 10% anhydrous $CaSO_4$.
L13 = Cellulose.
L14 = ITLC-SG.
L15 = ITLC-SA.
L16 = ITLC-SA impregnated with 10% (w/v) H_3BO_3.
L17 = ITLC-SA impregnated with 0.1*M* KH_2PO_4.

Solvent

S1 = *n*-propanol-H_2O (5 : 5).
S2 = *n*-propanol-H_2O-*n*-propylamine (5 : 3 : 2).
S3 = *n*-propanol-H_2O-chloroform (6 : 2 : 1).
S4 = *n*-propanol-H_2O-methyl ethyl ketone (2 : 1 : 1).
S5 = *n*-butanol-acetic acid-ethyl ether-H_2O (9 : 6 : 3 : 1).
S6 = benzene-acetic acid-methanol (1 : 1 : 3).
S7 = methyl ethyl ketone-acetic acid-methanol (3 : 1 : 1).
S8 = ethel acetate-isopropanol-H_2O (130 : 57 : 23).
S9 = *n*-butanol-acetone-H_2O (4 : 5 : 1).
S10 = *n*-butanol-acetone-H_2O (7 : 2 : 1).
S11 = *n*-butanol-acetone-H_2O (4 : 5 : 1).
S12 = *n*-butanol-dioxane-H_2O (4 : 5 : 1).
S13 = propanol-ethyl acetate-H_2O-25% ammonia (5 : 1 : 3 : 1).
S14 = propanol-ethyl acetate-H_2O-25% ammonia (6 : 1 : 3 : 1).
S15 = *n*-butanol-acetic acid-H_2O (6 : 3 : 1).
S16 = butyl alcohol-glacial acetic acid-H_2O (4 : 1 : 5).
S17 = isopropyl alcohol-pyridine-acetic acid-H_2O (8 : 8 : 1 : 4).
S18 = 90% aq. phenol-water (8 : 1) plus 0.002% 8-hydroxyquinoline.
S19 = ethyl acetate-pyridine-H_2O (8 : 2 : 1).
S20 = chloroform-methanol-acetic acid (70 : 30 : 15).

TABLE TLC 126—(Continued)
CARBOHYDRATES, MONO- AND DISACCHARIDES

Solvent

S21 = chloroform-glacial acetic acid-H_2O (30 : 35 : 5).
S22 = ethyl acetate-methanol-glacial acetic acid-H_2O (60 : 15 : 15 : 10).

Detection Technique

See Section II.I, Volume Two.

T1 = dry two hours in air + one hour at 130 °C, development distance = 10 cm unsaturated chamber.
T2 = unsaturated chamber, 10 cm development distance.
T3 = ascending in pre-equilibrated jars.
T4 = 1 μg–200 μg samples used, development distance = 10 cm.
T5 = 20 × 20 cm plates, tank saturation, development distance = 10 cm.
T6 = development distance = 10 cm.
T7 = saturated tanks.
T8 = ascending 15 cm development distance, 23 °C.
T9 = Gelman chamber, ascending development, 10 cm development distance.

Literature

1. Grasshof, H., *J. Chromatogr.*, **14**, 513 (1964).
2. Grasshof, H., *Deut-Apotheker-Ztg.*, **103**, 1396 (1963).
3. Hay, G. W., Lewis, B. A., and F. Smith, *J. Chromatogr.*, **11**, 479 (1963).
4. Pastuska, G., *Z. Anal. Chem.*, **179**, 427 (1961).
5. Stahl, E. and Kaltenbach, *J. Chromatogr.*, **5**, 351 (1961).
6. Stroh, H. H. and W. Schueler, *Z. Chem.*, **4**, 188 (1964).
7. Ragazzi, E. and G. Vernonese, *Farmaco* (*Pavia Ed. Pract.*), **18**, 152, (1963).
8. Waldi, E. D., "*Thin Layer Chromatography*", E. Stahl, Ed. Academic Press, New York (1965), 465.
9. Shimi, I. S. and G. M. Iman, *Analyst*, **94**, 65–67 (1969).
10. Vomhof, D. W. and T. C. Tucker, *J. Chromatogr.*, **17**, 300 (1965).
11. Haer, F. C., "*Introduction to Chromatography on Impregnated Glass Fiber*", Ann Arbor Science Publishers, Inc., Ann Arbor, Michigan (1969) p. 134.

TABLE TLC 127
CARBOHYDRATES, SUGAR ALCOHOLS

Layer	L1	L2	L1	L1	L3	L3	L3	L3	L3	L4	L4	L5	L5	L6	L7	L7	L7	L7
Solvent	S1	S2	S3	S4	S5	S6	S7	S8	S9	S10	S11	S11	S12	S13	S14	S15	S16	S17
Technique	T1	T1	T1	T1	T1	T1	T1	T1	T1	T2	T2	T2	T2	T3	T3	T3	T3	T3
Literature	1	1	1	1	2	2	2	2	2	3	3	3	3	4	5	5	6	6
Compound	$R_F \times 100$																	
N-Acetyl galactosamine	—	—	34	49	42	47	79	29	96	—	—	—	—	—	—	—	—	—
N-Acetyl glucosamine	—	—	36	55	39	63	86	63	79	—	—	—	—	—	—	—	—	—
N-Acetyl-D-mannosamine	—	—	—	—	34	53	94	49	93	—	—	—	—	—	—	—	—	—
Adonitol	45	60	—	—	—	—	—	—	—	52	51	46	42	—	—	—	—	—
Arabitol	36	55	—	—	—	—	—	—	—	—	—	—	—	—	—	—	—	—
L-Arabitol	—	—	—	—	—	—	—	—	—	—	—	—	—	—	70	55	36	52
Dulcitol	41	56	—	—	—	—	—	—	—	45	40	38	36	—	67	50	28	43
Erythritol	15	22	—	—	—	—	—	—	—	61	57	52	47	—	—	—	—	—
Fructose	—	—	—	—	—	58	57	—	—	—	—	—	—	—	—	—	—	—
Galactosamine	—	—	14	6	—	9	29	—	—	—	—	—	—	—	—	—	—	—
Glucosamine	—	—	16	16	29	34	69	49	—	—	—	—	—	—	—	—	—	—
D-Glucose	—	—	—	—	—	—	71	—	—	—	—	—	—	—	—	—	—	—
Glucuronolactone	—	—	—	—	93	89	22	25	5	—	—	—	—	—	—	—	—	—
D-Glucuronic acid	—	—	—	—	8	16	21	—	—	—	—	—	—	—	—	—	—	—
Inositol	73	71	—	—	—	—	—	—	—	29	21	—	—	—	—	—	—	—
Mannitol	42	49	—	—	—	—	—	—	—	46	41	42	39	—	—	—	—	—
D-Mannitol	—	—	—	—	—	—	—	—	—	—	—	—	—	50	70	51	30	46
Ribose	—	—	—	—	—	73	80	—	—	—	—	—	—	—	—	—	—	—
Sorbitol	61	58	—	—	—	—	—	—	—	41	35	38	37	—	—	—	—	—
D-Sorbitol	—	—	—	—	—	—	—	—	—	—	—	—	—	50	65	48	27	43
Xylitol	80	75	—	—	—	—	—	—	—	—	—	—	—	—	—	—	—	—

Layer
L1 = ITLC-SA.
L2 = ITLC-SA impregnated with 10% (w/v) H_3BO_3.
L3 = ITLC-SG.
L4 = Active Alusil (a mixture of acidic Silica gel G and basic alumina oxide G), dried at 110 °C for 30 minutes.
L5 = inactive Alusil (air-dried at room temperature).
L6 = Silica gel G.
L7 = Mg Silicate (Woelm).

Solvent
S1 = *n*-butanol-glacial acetic acid-diethyl ether-H_2O (9 : 6 : 3 : 1).
S2 = ethyl acetate-methanol-glacial acetic acid-H_2O (60 : 15 : 15 : 10).
S3 = chloroform-glacial acetic acid-acetone (30 : 35 : 4).
S4 = *n*-propanol-ethyl acetate-H_2O (70 : 20 : 10).
S5 = butyl alcohol.
S6 = isopropyl alcohol.
S7 = ethyl alcohol.
S8 = phenol.
S9 = pyridine.
S10 = *n*-propanol-ethyl acetate-water-25% ammonia (50 : 10 : 30 : 10 v/v).
S11 = S10 in proportion (60 : 10 : 30 : 10 v/v).
S12 = *n*-butanol-glacial acetic acid-water (60 : 30 : 10).
S13 = *n*-butanol-water (9 : 1).
S14 = *n*-propanol-water (1 : 1).
S15 = *n*-propanol-water-*n*-propylamine (5 : 3 : 2).
S16 = *n*-propanol-water-chloroform (6 : 2 : 1).
S17 = *n*-propanol-water-ethyl methyl ketone (2 : 1 : 1).

TABLE TLC 127—(Continued)
CARBOHYDRATES, SUGAR ALCOHOLS

Detection See Section II.I, Volume Two.

Technique T1 = convert sugars to alcohols by reduction with an equal amount of sodium borohydride for 3 hours at room temperature; Gelman chambers, ascending, 10 cm development distance, saturated chamber.
T2 = 50 μg of 1% aqueous solution sugars applied, tank saturation.
T3 = 10 cm development distance, unsaturated chamber.

Literature

1. Barker, S. A., Bourne, E. J., O'Maut, D. M., and M. Stacey, *J. Chem. Soc.*, **1957**, 2448 and Haer, F. C., *An Introduction to Chromatography on Impregnated Glass Fiber*, Ann Arbor Science Publishers, Inc., Ann Arbor, Mich., 1969, pp. 138–39.
2. Chan, D., Hull, E. W., Fields, M., and E. McNall, *Clin. Chem.*, **13**, 506 (1967) and Haer, *op. cit.*
3. Waldi, D., "*Thin-Layer Chromatography*", E. Stahl, Ed., Springer-Verlag, Germany, 1965, p. 466.
4. Prey, V., Berbalk, H., and M. Kausz, *Mikrochim. Acta*, **1962**, 449.
5. Grasshof, H., *J. Chromatogr.*, **14**, 513 (1964).
6. Grasshof, H., *Deut. Apotheker-Ztg.*, **103**, 1396 (1963).

TABLE TLC 128
CARBOHYDRATES, SUGAR PHOSPHATES

Layer	L1	L1
Solvent	S1	S2
Literature	1	1
Compound	$R_F \times 100$	
Fructose-1,6-diphosphate	14	6
Fructose-1-phosphate	26	19
Fructose-6-phosphate	33	10
Glucose-1-phosphate	26	19
Glucose-6-phosphate	26	13
Glycerol-α-phosphate	40	21
Glycerol-β-phosphate	36	26
Phosphoenolpyruvic acid	33	19
3-Phosphoglyceric acid	20	18
Phosphoric acid	39	6

Layer L1 = Kieselgel H (Merck) + 0.5% sodium acetate + 0.5% tylose.

Solvent S1 = ethyl acetate-acetic acid-water-ammonia (6 : 6 : 2 : 1).
S2 = isopropanol-ammonia-water-acetic acid (5 : 3 : 1 : 1).

Detection Ammonium molybdate-perchloric acid reagent.

Literature

1. Bancher, E. and J. Washüttl, *Mikrochim. Acta.*, **1967**, p. 225; from *J. Chromatogr.*, **29**, D1 (1967).

TABLE TLC 129
CARBOXYLIC ACIDS, AMINO BENZOATES AND SALICYLATES

Layer Solvent	L1 S1
Compound	$R_F \times 100$
o-Aminobenzoic acid methyl ester	74
o-Aminobenzoic acid isobutyl ester	73*
	66
	43
o-Aminobenzoic acid ethyl ester	61
m-Aminobenzoic acid ethyl ester	51
Benzoic acid	50
o-Aminobenzoic acid	48
4-N-Dimethylaminobenzoic acid	46
p-Aminobenzoic acid isopropyl ester	44
p-Aminobenzoic acid isobutyl ester	41
p-Aminobenzoic acid ethyl ester	37
p-N-Methylaminobenzoic acid	34*
	16
p-Aminobenzoic acid methyl ester	33
2-Hydroxy-4-aminobenzoic acid	33
p-Aminobenzoic acid	19
m-Aminobenzoic acid	13
2-Butoxy-4-aminobenzoic acid diethylaminoethyl ester	0
Octyl salicylate	0
2-Ethylhexyl salicylate	0
Salicylaldehyde	88
o-Salicylsalicylic acid	48
5-Methyl-2-hydroxybenzoic acid	46
Salicylic acid (o-hydroxybenzoic acid)	44
Acetylsalicylic acid	42
m-Cresotinic acid	41
Salicylic acid methylamide	37
Salicylamide	33
m-Hydroxybenzoic acid	19
p-Hydroxybenzoic acid	19
Salicyluric acid	13

* When more than one spot was observed the asterisk indicates the major component.

Layer L1 = non-activated Merck pre-coated SGF thin layer plates.
Solvent S1 = petroleum ether (b.p. 30–70 °C)-chloroform-methanol-glacial acetic acid-benzine (70 : 10 : 5 : 5 : 10).
Detection Short-wave UV light.
Technique Ascending.
Literature Khemani, L. and I. W. French, *J. Chromatogr.*, **41**, 274–275 (1969).

TABLE TLC 130
CARBOXYLIC ACIDS

Layer Solvent	L1 S1	L2 S2
Compound	$R_F \times 100$	
	1st direction	2nd direction
Methyl esters of saturated acids		
Dodecanoic	78	80[a]
Tetradecanoic	67	80[a]
Hexadecanoic	56	80[a]
Octadecanoic	46	80[a]
Eicosanoic	36	80[a]
Docosanoic	28	80[a]
Methyl esters of unsaturated acids		
C_{16}		
cis-Hexadecen-9-oic	71	39
cis-Hexadecen-11-oic	71	46
C_{18}		
cis-Octadecen-7-oic	60	34
cis-Octadecen-9-oic	60	42
cis-Octadecen-11-oic	60	51
trans-Octadecen-9-oic	60	64
C_{20}		
cis-Eicosen-11-oic	50	52
C_{22}		
cis-Docosen-5-oic	40	39
cis-Docosen-11-oic	40	52
C_{26}		
cis-Hexacosen-9-oic	25	47
Octadeca-9,12-dienoic	71	8
Octadeca-9,12,15-trienoic	80	3
Eicosa-5,8,11,14-tetraenoic	89	0

[a] Layer not impregnated with silver nitrate.

Layer L1 = Gypsum bound silica gel layers impregnated with 10% solution of dodecane in hexane.
L2 = Silica gel impregnated with 20% silver nitrate.
Solvent S1 = acetonitrile-acetone (1 : 1), 90% saturated with dodecane.
S2 = dipropyl ether-hexane (2 : 3).
Detection See Section II.I, Volume Two.
Technique Two-dimensional development; development time 70–90 minutes with S1.
Literature Bergel'son, L. D., Dyatlovitskaya, E. V., and V. V. Voronkova, *J. Chromatogr.*, **15**, 191 (1964).

TABLE TLC 131
CARBOXYLIC ACIDS, LICHEN SUBSTANCES (ALIPHATIC LICHEN ACIDS)

Layer	L1	L1
Solvent	S1	S2
Literature	1	1
Compound	$R_F \times 100$	
Acaranoic acid	60	[a]
Acarenoic acid	48	[a]
Caperatic acid	8	28
Lichesterinic acid	62	[a]
Nephromopsinic acid	70	[a]
Norrangiformic acid	10	36
Protolichesterinic acid	66	[a]
Rangiformic acid	61	[a]
Roccellic acid	75	[a]

[a] Travels with secondary front.

Layer L1 = Eastman Chromagram sheets, type K301R2 (6.7 cm).
Solvent S1 = diethyl ether-butyric acid (30 : 1).
S2 = dichloromethane-glacial acetic acid (9 : 1).
Detection Bromocresol green.
Technique Ascending, 5 cm.
Literature 1. Santesson, J., *Acta Chem. Scand.*, **21**, 1169 (1967).

TABLE TLC 132
CARBOXYLIC ACIDS, HYDROXAMIC ACID, DERIVATIVES OF FATTY ACIDS

Layer	L1	L2
Solvent	S1	S1
Literature	1	1
Compound	$R_F \times 100$	
Hydroxamic acid of		
Acetic acid	15	9
Propionic acid	25	13
Butyric acid	34	19
***n*-Valeric acid**	43	26
Caproic acid	52	33
Heptoic acid	63	43
Caprylic acid	73	53
Nonylic acid	84	63
Capric acid	92	76
Undecylic acid	94	87
Lauric acid	96	96

Layer L1 = Diethylene glycol adipate-impregnated kieselguhr (1 : 3).
L2 = Triethylene glycol adipate-impregnated kieselguhr (1 : 3).
Solvent S1 = diisopropyl ether-petroleum ether-carbon tetrachloride-formic acid-water (50 : 20 : 20 : 8 : 1).
Technique See Section II.I, Volume Two, for preparation of derivative.
Literature 1. Knappe, E. and K. G. Yekundi, *Z. Anal. Chem.*, **203**, 87 (1964).

TABLE TLC 133
CARBOXYLIC ACIDS, PHENOLIC ACIDS

Layer	L1	L1
Solvent	S1	S2
Literature	1	1
Acid	R_{GA}[a] × 100	
Chlorogenic	0	0
Isochlorogenic	0	0
Caffeic	105	120
Hydrocaffeic	100	110
Protocatechuic	90	105
Ferulic	160	185
Isoferulic	150	175
***o*-Coumaric**	140	170
***p*-Coumaric**	140	165
Sinapic	140	170
Gentisic	100	100
***p*-Hydroxybenzoic**	115	140
Gallic	105	55
Salicylic	190	150
Veratric	170	155
Piperonylic	185	170
Mandelic	145	100
Cinnamic	170	175
Ellagic	120	40
Chinic	0	0
Shikimic	25	35

[a] R_F relative to genistic acid.

Layer L1 = Silica gel G (Merck).
Solvent S1 = toluene-ethyl formate-formic acid (5 : 4 : 1).
S2 = benzene-methanol-acetic acid (45 : 8 : 4).
Detection UV with or without $FeCl_3$ spray.
Technique Ascending.
Literature 1. Hiller, K., *Pharmazie* **20**, 354–55 (1965); from *J. Chromatogr.*, **24**, D16 (1966); *Chem. Abstr.*, **63**, 10651 g (1965).

TABLE TLC 134
CARBOXYLIC ACIDS

Layer	L1	L1	L1	L2	L2
Solvent	S1	S1	S1	S2	S3
Technique	T1	T2	T3	na	na
Literature	1	1	1	2	2
Compound	R_F × 100				
Formic	5	7	3	52	64
Acetic	10	13	6	58	66
Propionic	15	30	15	—	—
***n*-Butyric**	24	40	22	—	—
***n*-Valeric**	39	50	30	—	—
***n*-Hexanoic**	52	57	34	—	—
***n*-Heptanoic**	55	60	39	—	—
***n*-Octanoic**	58	66	43	—	—
***n*-Nonanoic**	61	69	45	—	—
Trimethylacetic	57	71	47	—	—
α-Methylbutyric (*dl*)	39	65	39	—	—
β-Methylbutyric	34	53	31	—	—
Isobutyric	27	57	32	—	—
Lactic	—	—	—	63	51
Pyruvic	—	—	—	48	42

Layer L1 = Silica gel.
L2 = Silica gel G.
Solvent S1 = methyl acetate-2.5% ammonia (95 : 5).
S2 = pyridine-petroleum ether (1 : 2).
S3 = ethanol-NH_3-H_2O (80 : 4 : 16).
Detection Methyl red indicator.
Technique T1 = double run, fresh solvent.
T2 = double run, solvent aged 24 hours.
T3 = single run, solvent aged 24 hours.
Literature 1. Lynes, A., *J. Chromatogr.* **15**, 108 (1964).
2. Prey, V., Berbalk, H., and M. Kausz, *Mikrochim. Acta*, **1962**, 449.

TABLE TLC 135
CARBOXYLIC ACIDS, PHENOLIC DICARBOXYLIC ACIDS

Layer	L1	L2	L3	L4	L1	L2	L3	L4	L1	L2	L3	L4	L1	L2
Solvent	S1	S1	S1	S1	S2	S2	S2	S2	S3	S3	S3	S3	S4	S4
Technique	na	na	na	na	na	na	na	na	na	na	na	na	na	na
Literature	1	1	1	1	1	1	1	1	1	1	1	1	1	1
Acid	$R_F \times 100$													
Vanillic	92	95	88	82	73	91	50	42	72	80	47	56	45	44
Protocatechuic	95	76	59	57	38	32[a]	11	15	31	31[a]	4	2	38	10
Guaiacyl acetic	93	89	80	78	60	84	34	43	54	71	38	47	29	22
Homoprotocatechuic	91	63	61	51	36	37[a]	10	13	22	31[a]	3	3	23	6
Guaiacyl propionic	92	87	84	77	57	84	34	42	50	68	37	43	34	29
Dihydrocaffeic	95	69	91	64	39	42[a]	19	16	27	39	11	10	27	14
Ferulic	96	87	90	82	61	81	39	49	53	73	41	49	35	34
Caffeic	95	61	85	61	42	47[a]	12	16	25	34	2	5	31	14
Isoferulic	93	85	89	82	64	82	41	53	53	65	39	45	30	25
Chlorogenic	50	10	6	22	2	2	0	0	0	0	0	0	1	0
Salicylic	—	—	—	—	—	—	—	—	—	—	—	—	—	—
Gentisic	—	—	—	—	—	—	—	—	—	—	—	—	—	—
2,4,6-Trihydroxybenzoic	—	—	—	—	—	—	—	—	—	—	—	—	—	—
p-Hydroxyphenylacetic	—	—	—	—	—	—	—	—	—	—	—	—	—	—
m-Hydroxyphenylacetic	—	—	—	—	—	—	—	—	—	—	—	—	—	—
2,5-Dihydroxyphenylacetic	—	—	—	—	—	—	—	—	—	—	—	—	—	—
α-Resorcylic	—	—	—	—	—	—	—	—	—	—	—	—	—	—
β-Resorcylic	—	—	—	—	—	—	—	—	—	—	—	—	—	—
γ-Resorcylic	—	—	—	—	—	—	—	—	—	—	—	—	—	—
Syringic	—	—	—	—	—	—	—	—	—	—	—	—	—	—
m-Hydroxybenzoic	—	—	—	—	—	—	—	—	—	—	—	—	—	—
p-Hydroxybenzoic	—	—	—	—	—	—	—	—	—	—	—	—	—	—
Gallic	—	—	—	—	—	—	—	—	—	—	—	—	—	—
o-Coumaric	—	60	21	—	—	—	—	—	—	—	—	—	—	—
p-Coumaric[b]	—	—	—	—	—	—	—	—	—	—	—	—	—	—

TABLE TLC 135—(Continued)
CARBOXYLIC ACIDS, PHENOLIC DICARBOXYLIC ACIDS

Layer	L3	L4	L1	L2	L3	L4	L5	L5	L5	L6	L6	L7	L7
Solvent	S4	S4	S5	S5	S5	S5	S6	S7	S8	S9	S10	S11	S12
Technique	na	na	na	na	na	na	T1	T1	T1	T2	T2	T3	T3
Literature	1	1	1	1	1	1	2	2	2	3	3	4	4
Acid	$R_F \times 100$												
Vanillic	39	34	82	93	38	53	46	76	29	54	61	75	48
Protocatechuic	24	10	81	71	15	4	32	55	10	32	39	35	42
Guaiacyl acetic	24	18	59	69	19	21[a]	—	—	—	—	—	—	—
Homoprotocatechuic	13	5	55	66	13	3	—	—	—	—	—	—	—
Guaiacyl propionic	32	25	56	84	29	40	—	—	—	—	—	—	—
Dihydrocaffeic	19	9	57	80	23	3	—	—	—	—	—	46	51
Ferulic	36	28	63	87	37	50	42	76	16	50	58	79	30
Caffeic	22	8	65	85	31[a]	2	—	—	—	—	—	35	25[c],41[d]
Isoferulic	31	25	60	86	33	45	—	—	—	—	—	75	26[c],44[d]
Chlorogenic	0	0	8	14	10	0	—	—	—	—	—	—	—
Salicylic	—	—	—	—	—	—	72	88	48	—	—	84	37
Gentisic	—	—	—	—	—	—	35	65	17	30	40	32	35
2,4,6-Trihydroxybenzoic	—	—	—	—	—	—	15	50	10	—	—	—	—
p-Hydroxyphenylacetic	—	—	—	—	—	—	35	72	14	—	—	—	—
m-Hydroxyphenylacetic	—	—	—	—	—	—	32	65	19	—	—	—	—
2,5-Dihydroxyphenylacetic	—	—	—	—	—	—	67	85	43	—	—	—	—
α-Resorcylic	—	—	—	—	—	—	21	61	73	—	—	—	—
β-Resorcylic	—	—	—	—	—	—	57	85	19	54	52	—	—
γ-Resorcylic	—	—	—	—	—	—	10	15	10	—	—	—	—
Syringic	—	—	—	—	—	—	33	57	16	48	60	82	59
m-Hydroxybenzoic	—	—	—	—	—	—	—	—	—	49	51	—	—
p-Hydroxybenzoic	—	—	—	—	—	—	—	—	—	—	—	60	46
Gallic	—	—	—	—	—	—	—	—	—	18	23	11	37
o-Coumaric	—	—	—	—	—	—	—	—	—	—	—	60	21
p-Coumaric[b]	—	—	—	—	—	—	—	—	—	49	52	64	26[c],40[d]

[a] Elongated spots.
[b] These give 2 spots.
[c] Main spot.
[d] Secondary spot.

TABLE TLC 135—(Continued)
CARBOXYLIC ACIDS, PHENOLIC DICARBOXYLIC ACIDS

Layer
L1 = Kieselgel G.
L2 = L1 treated with Na_2MoO_4
L3 = L1 treated with Na_2WO_4.
L4 = L1 treated with borax.
L5 = Chromostrips of silicic acid with starch binder, 0.5 mm layer.
L6 = Silica gel G, 0.5 mm layer.
L7 = Polyamide with starch binder, 0.15 mm layer.

Solvent
S1 = organic layer of *n*-butanol-acetic acid-water (4 : 1 : 5).
S2 = benzene-methanol-acetic acid (45 : 8 : 4).
S3 = benzene-dioxane-acetic acid (90 : 25 : 4).
S4 = *n*-butyl ether (saturated with water)-acetic acid (10 : 1).
S5 = ethyl acetate-isopropanol-water (65 : 24 : 11).
S6 = ether-Skellysolve B (7 : 3).
S7 = ethyl acetate-Skellysolve B (3 : 1).
S8 = acetone-Skellysolve B (1 : 3).
S9 = benzene-dioxane-acetic acid (90 : 25 : 4).

Solvent
S10 = benzene-methanol-acetic acid (90 : 16 : 8).
S11 = ethyl acetate-acetic acid (95 : 5).
S12 = acetic acid-water (30 : 70).

Technique
T1 = 10–11 cm development distance.
T2 = 10 cm development distance.
T3 = 15 cm development distance. Solvents can be used for two-dimensional development.

Literature

1. Halmekoski, J., *Suomen Kemistilehti*, **35B**, 39 (1962) from Kirchner, J. G., *Thin-Layer Chromatography, Technique of Organic Chemistry*, XII, Perry, E. S. and A. Weissberger, Eds., Interscience Publishers, Division of John Wiley & Sons, New York, 1967, p. 45.
2. Lyman, R. L., Livingston, A. L., Bickoff, E. M., and J. N. Booth, *J. Org. Chem.*, **23**, 756 (1958).
3. Pastuska, G., *Z. Anal. Chem.*, **179**, 355 (1961).
4. Urion, E., Metche, M., and J. P. Haluk, *Brauwissenschaft*, **16**, 211 (1963).

TABLE TLC 136
DRUGS, ANIMAL

Compound	Trivial Name														Detection
Layer		L1	L2	L2	L3	L4	L3	L1	L1	L1	L3	L3	L3	L3	
Solvent		S1	S2	S3	S4	S5	S5	S6	S7	S8	S9	S10	S11	S12	
Technique		T1	T2	T3	T2	T4	T4	T4	T5	T5	T1	T6	T6	T5	
Literature		1	1	1	1	1	1	1	1	1	1	1	1	1	
		$R_F \times 100$													
Phenylarsonic acids															
p-Aminobenzenearsonic acid	Arsanilic acid	69	—	—	—	—	—	—	25	35	51	—	—	—	D1 or D4
p-Ureidobenzenearsonic acid	Carbarsone®	54	—	—	—	—	—	—	—	—	—	20	—	45	D1 or D4
4-Nitrophenylarsonic acid	—	46	—	—	—	—	—	—	—	—	—	—	—	—	D1 or D4
3-Nitro-4-hydroxyphenylarsonic acid	—	36	—	—	—	—	—	—	—	—	—	—	—	—	D1 or D4
Nitrofuraldehydes															
3-(5-Nitrofurfurylideneamino)-2-oxazolidone	Furazolidone	—	84	—	—	—	—	—	—	72	33	—	28	—	D2 then D1
5-Nitro-2-furaldehyde acetylhydrazone	Nihydrazone	—	64	—	—	—	—	—	—	—	—	—	—	—	D2, D1
1-Ethyl-3-(5-nitro-2-thiazolyl)-urea	Nithiazide	—	50	—	—	—	—	—	—	—	—	—	—	—	D2, D1
5-Nitro-2-furaldehyde semicarbarzone	Nitrofurazone	—	43	—	—	—	—	—	—	—	—	—	38	—	D2, D1
Heterocyclic compounds															
Thiodiphenylamine	Phenothiazine	—	—	81	—	—	—	—	—	—	—	—	—	—	D3, D5, D1
2-Sulfanilamidoquinoxaline	Sulfaquinoxaline	—	—	68	—	—	—	—	—	—	—	—	—	—	D3, D5, D1
Hexahydropyrazine	Piperazine	—	—	58	—	—	—	—	—	—	—	—	—	—	D3, D5, D1
1-(4-Amino-2-n-propyl-5-pyrimidinylmethyl)-2-picolinium chloride hydrochloride	Amprolium	—	—	16	—	—	—	35	87	—	—	—	—	—	D3, D5, D1
Nitrophenyl compounds															
m,m′-Dinitrophenyl disulfide	Nitrophenide	—	—	—	95	—	—	—	—	—	—	80	—	—	D2
3,5-Dinitrobenzamide		—	—	—	73	—	—	—	—	—	—	—	—	—	D2
3,5-Dinitro-o-toluamide	Zoalene®	—	—	—	73	—	—	—	—	—	—	—	—	—	D2
4,4′-Dinitrocarbanilide and 2-hydroxy-4,6-dimethylpyrimidine	Nicarbazin	—	—	—	25*	—	—	—	—	—	—	—	—	—	D2
		—	—	—	50*	—	—	—	—	—	—	—	—	—	—
Aryl esters															
Diacetate of 3,4-bis-(p-hydroxyphenyl)-2,4-hexadiene	Dienestrol diacetate	—	—	—	—	60	57	—	—	—	—	—	77	73	D3
Methyl 4-acetamido-2-ethoxybenzoate	Ethopabate	—	—	—	—	25	15	80	—	—	—	—	—	—	D3
Miscellaneous															
Acetyl-(p-nitrophenyl)-sulfanilamide		—	—	—	—	—	—	—	—	—	—	70	—	—	—

* R_F 25 is of a freshly prepared nicarbazin standard; R_F 50 is of a nicarbazin standard approximately 2 weeks old.

TABLE TLC 136—(Continued)
DRUGS, ANIMAL

Layer

L1 = Polyamide G or GF.
L2 = Alumina G or GF.
L3 = Silica gel G or GF.
L4 = Silica gel GF (flexible plastic sheet).

Solvent

S1 = *n*-butanol-water-acetic acid (80 : 20 : 2.5), upper layer.
S2 = acetone-acetic acid (100 : 2).
S3 = ethanol-ammonium hydroxide (80 : 20).
S4 = diethyl ether-dimethyl sulfoxide-*n*-butanol-acetic acid (98 : 1 : 1 : 0.2).
S5 = hexane-acetone-*n*-butanol (21 : 2 : 2).
S6 = diethyl ether-ethanol-acetic acid (96 : 3 : 1).
S7 = acetone-*n*-butanol-acetic acid (78 : 20 : 2).
S8 = diethyl ether-*n*-butanol-acetic acid (73 : 25 : 2).
S9 = *n*-butanol-water-acetic acid (80 : 20 : 5).
S10 = ethyl acetate-acetic acid (75 : 25).
S11 = pyridine-benzene (50 : 50).
S12 = *n*-butanol-acetic acid (75 : 25).

Detection

D1 = 5% (w/v) solution potassium hydroxide, UV (254 mμ) when necessary.
D2 = saturated solution barium-diphenylamine-4-sulfonate, and UV (254 mμ).
D3 = UV (254 mμ).
D4 = 4-methylumbelliferone, UV (254 mμ) and ammonia vapor, if necessary.
D5 = vanillin.

Technique

For all systems: 0.5–1.5 μg applied; saturated chamber, ascending development, 10 cm.
T1 = 90 minutes development time.
T2 = 15 minutes development time.
T3 = 80 minutes development time.
T4 = 25 minutes development time.
T5 = 20 minutes development time.
T6 = 60 minutes development time.

Literature

1. Antkowiak, J. J. and A. L. Spatorico, *J. Chromatogr.*, **29**, 277–282 (1967).

TABLE TLC 137
DRUGS, BARBITALS

Layer	L1	L1	L1	L1	L2	L2	L3	L3	L3	L3	L3	L3	L3	L3	L4	L5	L1	L6	L1	L1
Solvent	S1	S2	S3	S4	S5	S6	S7	S8	S9	S10	S1	S6	S10	S11	S12	S13	S1	S14	S1	S1
Detection	—	—	—	—	—	—	—	—	—	—	—	—	—	—	—	—	—	—	D1	D2
Technique	na	T1	T1	T2	T3	T3	na	na	na	na	na	na	na	na	na	T4	T2	T5	—	—
Literature	1	2	2	3	4	4	5	5	5	5	5	5	5	5	6	7	8	9	8	3
Compound	$R_F \times 100$																R_p[c]		Color	
Acetylcarbromal	—	—	—	—	27	87	52	—	—	54	—	—	21	32	—	—	—	—	—	—
Allobarbital	43	43	48	—	66	48	35	28	37	30	34	37	—	—	34	—	130	—	bl	—
Allylbarbituric acid (Butalbital)	48	—	—	—	—	—	—	—	—	—	—	—	—	—	—	—	142	48	bl	—
5-Allyl-5-cyclopentyl-BA[b]	—	—	—	—	—	—	45	36	44	34	32	54	—	—	—	—	—	—	—	—
Allylisobutylbarbital	—	—	—	—	76	67	—	—	—	—	—	—	—	—	—	—	—	—	—	—
5-Allyl-5-isopropyl-1-methyl-BA[b]	—	—	—	—	—	—	70	78	68	74	60	68	—	—	—	—	—	—	—	—
Amobarbital (Amylobarbitone)	48	42	52	—	68	69	56	48	52	60	35	55	—	—	58	75	144	—	vt	—
Aprobarbital	43	—	—	—	65	61	50	39	48	46	33	43	—	—	—	—	130	43	vt	—
Barbital (Barbitone)	33	38	40	25	52	49	35	33	35	33	28	37	—	—	31	20	100	—	vt	gy
Bemegride	—	—	—	—	—	—	69	—	—	83	—	—	65	70	—	—	—	—	—	—
Brallobarbital	—	—	—	—	74	46	—	—	—	—	—	—	—	—	—	—	—	—	—	—
Bromisoralum	—	—	—	60	30	90	55	—	—	67	—	—	32	39	—	—	—	—	br	—
Butabarbital (Butabarbitone)	42	49	52	—	66	66	52	44	52	56	28	53	—	—	53	60	128	42	vt	—
Butallylon	—	—	—	—	—	—	52	42	48	51	36	48	—	—	—	—	—	—	—	—
Butallylonal	50	—	—	—	—	—	—	—	—	—	—	—	—	—	—	—	150	—	bl	—
Butethal	47	—	—	—	—	—	—	—	—	—	—	—	—	—	—	—	141	—	vt	—
Cabromal	—	—	—	75	58	94	68	—	—	81	—	—	55	62	—	—	—	—	—	w
Cyclobarbital (Cyclobarbitone)	—	55	38	20	58	53	41	37	40	40	35	44	—	—	30	40	+[a]	—	bl	gy
Cyclopal	46	—	—	—	—	—	—	—	—	—	—	—	—	—	—	—	138	—	bl	—
Dihydroprylon	—	—	—	40	—	—	—	—	—	—	—	—	—	—	—	—	—	—	—	—
Dihyprylon	—	—	—	—	—	—	32	—	—	45	—	—	23	34	—	—	—	—	—	gy
Ethchlorvynol	—	32	95	—	—	—	—	—	—	—	—	—	—	—	—	—	—	—	—	—
Ethinamal	—	—	—	—	—	—	63	—	—	68	—	—	40	46	—	—	—	—	—	—
Ethinamate	—	65	55	45	—	—	—	—	—	—	—	—	—	—	—	—	—	100	—	r
5-Ethyl-5-crotyl-BA[b]	—	—	—	—	—	—	52	40	46	48	33	46	—	—	—	—	—	—	—	—
5-Ethyl-5-piperidino-BA[b]	—	—	—	—	—	—	27	22	24	22	20	30	—	—	—	—	—	—	—	—
5-Furfuryl-5-isopropyl-BA[b]	—	—	—	—	—	—	43	35	44	43	33	52	—	—	—	—	—	—	—	—
Glutethimide	—	54	99	90	76	96	75	—	—	86	—	—	73	76	—	—	—	—	gy	—
Heptabarbital	43	—	—	20	32	27	42	35	39	42	28	48	—	—	—	—	132	—	vt	gy
Hexethal	52	—	—	—	—	—	—	—	—	—	—	—	—	—	—	—	156	—	vt	—
Hexobarbital	68	49	58	—	76	75	68	65	59	66	43	63	—	—	77	—	206	—	vt	—
Itobarbital (Butalbital)	—	48	50	—	—	—	—	—	—	—	—	—	—	—	—	—	—	—	—	—

TABLE TLC 137—(Continued)
DRUGS, BARBITALS

Layer	L1	L1	L1	L1	L2	L2	L3	L3	L3	L3	L3	L3	L3	L3	L4	L5	L1	L6	L1	L1
Solvent	S1	S2	S3	S4	S5	S6	S7	S8	S9	S10	S1	S6	S10	S11	S12	S13	S1	S14	S1	S1
Detection	—	—	—	—	—	—	—	—	—	—	—	—	—	—	—	—	—	—	D1	D2
Technique	na	T1	T1	T2	T3	T3	na	na	na	na	na	na	na	na	na	T4	T2	T5	—	—
Literature	1	2	2	3	4	4	5	5	5	5	5	5	5	5	6	7	8	9	8	3
Compound	$R_F \times 100$																R_p[c]		Color	
Mephobarbital	79	85	60	—	—	—	—	—	—	—	—	—	—	—	64	—	237	—	vt	—
Meprobarbital	—	—	—	—	—	—	56	46	47	52	56	57	—	—	—	—	—	—	—	—
Methaqualone	—	—	—	—	—	—	77	—	—	84	—	—	80	82	—	—	—	—	—	—
Metharibtal	78	53	66	—	—	—	—	—	—	—	—	—	—	—	—	—	235	—	vt	—
Methitural	—	—	—	—	—	—	72	64	55	62	61	62	—	—	—	—	—	—	—	—
3-Methyl-3,4-dihydroxy-4-phenylbutyne-1	—	—	—	—	—	—	55	—	—	73	—	—	40	52	—	—	—	—	—	—
Methylphenobarbital	—	—	—	—	85	64	—	—	—	—	—	—	—	—	—	—	—	—	—	—
2-Methyl-3-*o*-tolyl-4-(3H)-quinazolinone	—	—	—	80	—	—	—	—	—	—	—	—	—	—	—	—	—	—	—	w
Methyprylon	—	81	45	50	—	—	43	—	—	58	—	—	36	43	—	—	—	—	—	r
Nealbarbitone	—	—	—	—	—	—	—	—	—	—	—	—	—	—	—	70	—	—	—	—
Pentobarbital (Pentobarbitone)	47	40	49	—	69	73	59	51	56	63	33	56	—	—	66	80	130	—	bl	—
Pentothiobarbital	—	—	—	—	96	75	—	—	—	—	—	—	—	—	—	—	—	—	—	—
Phenobarbital (Phenobarbitone)	33	36	26	10	53	34	24	16	26	17	27	27	—	—	20	30	100	31	vt	gy
Phenytoin	—	—	—	—	40	68	—	—	—	—	—	—	—	—	—	—	—	—	—	
Propallylonal	—	—	—	—	—	—	40	27	38	36	36	52	—	—	—	—	—	—	—	gy
Pyrithyldione	—	—	—	—	—	—	56	—	—	65	—	—	17	25	—	—	—	—	—	—
Secobarbital (Seconal, quinalbarbitone)	55	46	54	—	76	73	60	54	59	63	41	60	—	—	63	85	167	—	bl	—
Sigmodal	59	—	—	—	—	—	—	—	—	—	—	—	—	—	—	—	177	54	vt	—
Talbutal	47	—	—	—	—	—	—	—	—	—	—	—	—	—	—	—	141	—	bl	—
Thalidomide	—	—	—	—	—	—	65	—	—	61	—	—	42	50	—	—	—	—	—	—
Thiamylal	—	80	70	—	—	—	—	—	—	—	—	—	—	—	—	—	+[a]	81	bl	—
Thiopental	—	73	70	—	—	—	—	—	—	—	—	—	—	—	—	—	+[a]	—	—	—
Vinbarbital	36	—	—	—	—	—	—	—	—	—	—	—	—	—	—	—	119	—	vt	—

[a] Runs with solvent front.
[b] BA = barbituric acid.
[c] R_p = R_F relative to phenobarbital.

TABLE TLC 137—(Continued)
DRUGS, BARBITALS

Layer

L1 = Silica gel G.
L2 = Silica gel GF-254.
L3 = Silica gel HF-254 (DC-Fertigplatte Merck).
L4 = Silica gel (with or without CA-504).
L5 = cellulose plates with fluorescent indicator Merck, or Kodak Chromagram Sheet Type 6065. The plates were dipped in 10% sodium orthophosphate ($Na_3PO_4 \cdot 12\,H_2O$) aqueous solution which was then drained off and the plates then blotted. They were oven-dried at 100 °C for half an hour and after cooling dipped in acetone-water (75 : 25) and the acetone allowed to evaporate by gently warming.
L6 = ITLC-SG activated at 110 °C for 30 minutes.

Solvent

S1 = chloroform-acetone (9 : 1).
S2 = benzene-acetic acid (9 : 1).
S3 = dioxane-benzene-ammonium hydroxide (4 : 15 : 1).
S4 = piperidine-petroleum ether, 50–70 °C (1 : 5).
S5 = chloroform-ether (75 : 25).
S6 = isopropanol-chloroform-25% ammonia (45 : 45 : 10).
S7 = benzene-dioxane-acetone-ammonia (25 : 10 : 14 : 1).
S8 = chloroform-*n*-butyl alcohol-ammonia (14 : 8 : 1).
S9 = cyclohexane-dioxane-ethyl acetate-ethyl alcohol-ammonia (3 : 3 : 8 : 4 : 2).
S10 = petroleum ether-dioxane-acetone-ammonia (3 : 3 : 3 : 1).
S11 = cyclohexane-chloroform-diethylamine (2 : 7 : 1).
S12 = ethyl acetate, hexane, ammonium hydroxide (20 : 9 : 10).

Solvent

S13 = *n*-amyl methyl ketone.
S14 = isooctane-glacial acetic acid (100.4 : 100.7 depending on relative humidity).

Detection

D1 = $KMnO_4$, $HgSO_4$, diphenylcarbazone and Na fluorescein.
D2 = $Hg(NO_3)_2$-eosin.

Technique

T1 = development distance = 10 cm.
T2 = saturated atmosphere, 10 cm development distance.
T3 = unsaturated chamber; solvents S5 and S6 can be used for 2-dimensional developments in that order.
T4 = 10 minutes development time.
T5 = saturated chamber, ascending development, 10–15 cm distance.

Literature

1. Sunshine, I., *Amer. J. Clin. Pathol.*, **40**, 578 (1963).
2. Cochin, J. and J. W. Daly, *J. Pharmacol. Exp. Ther.*, **139**, 154 (1963).
3. Eberhardt, H., Freund, K. J., and J. W. Langbein, *Arzneimittel-Forschung*, **12**, 1087 (1962).
4. deZeeuw, R. A. and M. T. Fietsma, *Pharm. Weekbl.*, **101**, 969 (1966).
5. Waldi, D., *Med. Monatsspiegel*, **No. 4**, 94 (1966).
6. Comer, J. and I. Comer, *J. Pharm. Sci.*, **56**, 423 (1967).
7. Curry, A. S. and R. H. Fox, *Analyst*, **93**, 834 (1968).
8. Sunshine, I., Rose, E., and J. LeBeau, *Clin. Chem.*, **9**, 312 (1963).
9. Haer, F. C., "*An Introduction to Chromatography on Impregnated Glass Fiber*", Ann Arbor Science Publishers, Inc., Ann Arbor Mich., 1969, p. 122.

TABLE TLC 138
DRUGS, NARCOTICS, ANALGESICS

Layer	L1	L1	L1	L1	L1	L1	L1	L2	L2	L3	L3	L3	L3
Solvent	S1	S2	S3	S4	S5	S6	S7	S4	S5	S3	S8	S9	S10
Technique	T1	T1	T1	T2	T1	T1	T1	T1	T1	T3	T3	T3	T3
Literature	1	1	1	1	1	1	1	1	1	2	2	2	2
Compound	$R_F \times 100$												
l-Acetylmethadol	64	60	99	40	38	52	62	—	—	—	—	—	—
dl-Alphaprodine	39	40	93	34	20	42	40	—	—	—	—	—	—
Alphaprodine HCl	—	—	—	—	—	—	—	—	—	83	79	87	25
Anileridine	—	—	—	—	—	—	—	—	—	95	95	93	77
Codeine	30	29	39	25	8	53	30	86	91	28	42	40	5
Dihydrocodeinone	17	25	41	19	str.	42	23	76	94	—	—	—	—
Dihydrohydroxycodeinone	46	24	87	29	16	32	34	—	—	—	—	—	—
Dihydrohydroxymorphinone	46	29	34	24	10	45	28	63	81	—	—	—	—
Dihydromorphine	15	21	10	10	str.	43	29	67	73	9	16	17	1
Dihydromorphinone	11	21	17	13	str.	41	25	65	85	—	—	—	—
Ethylmorphine	33	25	46	27	8	53	33	84	96	—	—	—	—
Ethylmorphine HCl (Dionin®)	—	—	—	—	—	—	—	—	—	40	46	47	5
Heroin	37	35	76	35	15	61	32	90	95	74	67	73	19
Hydrocodone bitartrate	—	—	—	—	—	—	—	—	—	36	35	42	5
Hydromorphone HCl	—	—	—	—	—	—	—	—	—	18	19	21	2
l-3-Hydroxy-*N*-allylmorphinan	65	70	98	41	44	64	73	—	—	—	—	—	—
2′-Hydroxy-5,9-dimethyl-2-cyclopropylmethyl-6 : 7-benzomorphan	45	71	92	15	16	55	67	—	—	—	—	—	—
2′-Hydroxy-5,9-dimethyl-2-(3,3-dimethylallyl)-6 : 7-benzomorphan	73	81	96	25	34	65	77	—	—	—	—	—	—
dl-2′-Hydroxy-5,9-dimethyl-2-phenethyl-6 : 7-benzomorphan	88	87	97	82	70	76	77	—	—	—	—	—	—
l-3-Hydroxy-*N*-methylmorphinan	11	47	80	10	7	51	60	—	—	—	—	—	—
l-3-Hydroxymorphinan	5	68	19	10	8	72	80	—	—	—	—	—	—
l-2′-Hydroxy-2,5,9-trimethyl-6 : 7-benzomorphan	12	36	56	8	5	43	51	—	—	—	—	—	—
Ketobemidone	31	39	47	24	12	42	40	—	—	—	—	—	—
Levorphanol tartrate	—	—	—	—	—	—	—	—	—	57	54	44	20
Methadone	—	—	—	—	—	—	—	—	—	79	81	82	58
dl-Methadone	34	59	99	17	17	55	62	—	—	—	—	—	—
l-3-Methoxy-*N*-methylmorphinan	13	43	91	8	7	55	59	—	—	—	—	—	—
l-3-Methoxymorphinan	7	65	38	str.	str.	66	81	—	—	—	—	—	—
Methyldihydromorphinone (Metopon)	16	24	25	15	str.	45	26	76	92	28	30	34	3
6-Monoacetylmorphine	38	40	64	29	19	37	37	—	—	50	55	56	11
Morphine	29	27	11	21	7	54	34	86	85	18	20	23	2
Nalorphine	71	55	35	67	25	59	41	88	96	—	—	—	—
Narcotine (Noscapine)	—	—	—	—	—	—	—	—	—	80	83	85	72
Norcodeine	12	50	13	9	6	63	49	59	56	—	—	—	—

TABLE TLC 138—(Continued)

DRUGS, NARCOTICS, ANALGESICS

Layer	L1	L1	L1	L1	L1	L1	L1	L2	L2	L3	L3	L3	L3
Solvent	S1	S2	S3	S4	S5	S6	S7	S4	S5	S3	S8	S9	S10
Technique	T1	T1	T1	T2	T1	T1	T1	T1	T1	T3	T3	T3	T3
Literature	1	1	1	1	1	1	1	1	1	2	2	2	2
Compound	$R_F \times 100$												
Normorphine	8	48	4	7	str.	66	62	48	25	—	—	—	—
Norpethidine	12	65	51	10	11	58	63	—	—	—	—	—	—
Oxymorphone HCl	—	—	—	—	—	—	—	—	—	56	56	58	15
Papaverine	—	—	—	—	—	—	—	—	—	73	77	77	53
Peronin	—	—	—	—	—	—	—	—	—	53	51	52	9
Pethidine (Meperidine)	42	41	97	36	20	46	44	—	—	63	70	70	27
Phenazocine	—	—	—	—	—	—	—	—	—	96	96	95	80
Piminodine	88	73	99	85	76	69	58	—	—	—	—	—	—
Piminodine ethanesulfonate	—	—	—	—	—	—	—	—	—	97	98	97	93
***d*-Propoxyphene**	73	68	97	54	53	61	—	—	—	—	—	—	—
Thebaine	—	—	—	—	—	—	—	—	—	65	73	74	23

Layer
L1 = Silica gel G (Merck).
L2 = MN-cellulose powder 300G (Machery-Nagel & Co.).
L3 = Silica gel (with or without $CaSO_4$).

Solvent
S1 = ethanol-pyridine-dioxane-water (50 : 20 : 25 : 5).
S2 = ethanol-glacial acetic acid-water (60 : 30 : 10).
S3 = ethanol-dioxane-benzene-ammonium hydroxide (5 : 40 : 50 : 5).
S4 = methanol-*n*-butanol-benzene-water (60 : 15 : 10 : 15).
S5 = *tert*-amyl alcohol-*n*-butyl ether-water (80 : 7 : 13).
S6 = *n*-butanol-glacial acetic acid-water (4 : 1 : 2).
S7 = *n*-butanol-conc. HCl, saturated with water (90 : 10).
S8 = chloroform-dioxane-ethyl acetate-ammonium hydroxide (25 : 60 : 10 : 5).
S9 = ethanol-chloroform-dioxane-petroleum ether (30–60 °C)-benzene-ammonium hydroxide-ethyl acetate (5 : 10 : 50 : 15 : 10 : 5 : 5).
S10 = ethyl acetate-benzene-ammonium hydroxide (60 : 35 : 5).

Detection
K iodoplatinate.

Technique
T1 = solvent migration 10 cm at 23 ± 2 °C.
T2 = 1st dimension with S4; 2nd dimension with S4, S5, or S6.
T3 = solvent migration 10 cm, 30 minutes.

Literature
1. Mulé, S. J., *Anal. Chem.*, **36**, 1907 (1964).
2. Steele, J. A., *J. Chromatogr.*, **19**, 300 (1965).

TABLE TLC 139
DRUGS, PHENOTHIAZINES

Layer	L1	L1	L1	L1	L1	L1	L1	L1	L1	L1	L1	L1	L2	L3	L4	L5	L6	L6	L7	L8	L9	L9	L9	L10
Solvent	S1	S2	S3	S4	S5	S6	S7	S8	S9	S10	S11	S12	S13	S13	S14	S15	S15	S16	S17	S18	S19	S20	S21	S22
Detection	—	—	—	—	—	—	—	—	—	—	—	—	—	—	—	—	—	—	—	—	—	—	—	—
Technique	na	na	na	na	na	na	T1	T1	T1	T1	T2	T1	T3	T3	na	na	na	na	na	T4	na	na	na	na
Literature	1	1	1	1	1	1	2	2	2	2	3	4	5	5	6	7	7	7	7	8	9	9	9	10
Compound							$R_F \times 100$											$R_F \times 100$						
Acetophenazine	12	36	4	52	18	38	—	—	—	—	—	—	—	—	—	—	—	—	—	—	—	—	—	—
Butylrylperazine	8	31	6	28	10	45	—	—	—	—	—	—	—	—	—	—	—	—	—	—	—	—	—	10
Chlorpromazine	37	44	28	14	23	66	94	70	30	—	5	49–50	72	62	32	31	64	85	46	41	21	21	12	38
Chlorpromazine sulfoxide	5	10	1	9	5	27	26	47	19	86	—	—	—	—	—	—	—	—	—	—	—	—	—	—
Chlorprothixene (chlorprothixenum)	34	78	—	20	36	88	80	73	46	—	69	55–57	—	—	—	—	—	—	—	64	—	—	—	—
***N*-Demethylchlorpromazine sulfoxide**	—	—	—	—	—	—	20	56	10	68	—	—	—	—	—	—	—	—	—	—	—	—	—	—
Diethazine	—	—	—	—	—	—	—	—	—	—	81	53–55	—	—	16	60	93	95	55	—	31	10	7	—
***N,N*-Didemethylchlorpromazine sulfoxide**	—	—	—	—	—	—	23	68	18	72	—	—	—	—	—	—	—	—	—	—	—	—	—	—
Ethopropazine	—	—	—	—	—	—	—	—	—	—	94	57–59	70	64	—	—	—	—	—	—	—	—	—	—
Fluphenazine	37	56	9	68	27	57	34	58	68	—	—	—	48	40	—	—	—	—	—	72	—	—	—	—
Levomepropazine	—	—	—	—	—	—	—	—	—	—	72	62–64	—	—	—	21	83	90	62	52	32	9	9	47
Librium	—	—	—	—	—	—	—	—	—	—	—	84–86	56	—	—	—	—	—	—	—	—	—	—	—
Mepazine	29	46	13	13	13	62	88	57	29	—	45	—	—	—	—	—	—	—	—	33	—	—	—	22
Mepazine sulfoxide	—	—	—	—	—	—	16	59	24	74	—	—	—	—	—	—	—	—	—	—	—	—	—	—
Methdilazine	—	—	—	—	—	—	40	61	16	—	—	—	—	—	—	—	—	—	—	—	—	—	—	—
Methoxypromazine	15	26	12	12	9	45	74	58	24	—	—	—	—	—	—	—	—	—	—	—	—	—	—	—
Omca	—	—	—	—	—	—	—	—	—	—	12	—	—	—	—	—	—	—	—	—	—	—	—	—
Perazine	—	—	—	—	—	—	—	—	—	—	17	—	—	—	—	—	—	—	—	—	—	—	—	7
Perphenazine	28	57	7	48	24	53	36	44	48	—	9	—	47	38	—	—	—	—	—	56	14	14	19	—
Phenothiazine	—	—	—	—	—	—	—	—	—	—	—	89–91	—	—	—	—	—	—	—	—	—	—	—	—
Pipamazine	38	71	—	41	37	79	12	72	56	—	—	—	12	9	—	—	—	—	—	—	—	—	—	—
Prochlorperazine	10	31	6	24	8	55	70	27	32	—	26	—	70	59	12	—	—	—	—	22	10	7	7	—
Prochlorperazine sulfoxide	—	—	—	—	—	—	13	28	13	88	—	—	—	—	—	—	—	—	—	—	—	—	—	—
Proketazine	19	45	6	53	25	44	—	—	—	—	—	—	—	—	—	—	—	—	—	—	—	—	—	—
Promazine	16	31	12	11	11	50	62	38	37	—	37	37–39	—	—	—	20	54	70	57	27	11	8	10	17
Promazine sulfoxide	—	—	—	—	—	—	17	46	15	70	—	—	—	—	—	—	—	—	—	—	—	—	—	—
Promethazine	—	—	—	—	—	—	70	59	22	—	52	55–57	—	—	35	35	69	79	60	—	23	12	18	27
Promethazine sulfoxide	—	—	—	—	—	—	22	57	30	78	—	—	—	—	—	—	—	—	—	—	—	—	—	—
Prothypendyl	16	21	11	11	8	44	—	—	—	—	—	—	—	—	—	—	—	—	—	—	—	—	—	—
Thiethylperazine	14	50	—	23	13	61	—	—	—	—	—	—	—	—	—	—	—	—	—	—	—	—	—	—
Thiopropazate	—	—	—	—	—	—	97	67	70	—	—	—	—	—	—	—	—	—	—	—	77	30	9	—
Thiopropazine	64	79	49	65	53	81	—	—	—	—	—	—	—	—	—	—	—	—	—	—	—	—	—	—
Thioridazine	24	39	14	24	15	64	97	65	20	—	45	47–49	72	63	29	—	—	—	—	—	—	—	—	—

TABLE TLC 139—(Continued)
DRUGS, PHENOTHIAZINES

Layer	L1	L1	L1	L1	L1	L1	L1	L1	L1	L1	L1	L1	L2	L3	L4	L5	L6	L6	L7	L8	L9	L9	L9	L10
Solvent	S1	S2	S3	S4	S5	S6	S7	S8	S9	S10	S11	S12	S13	S13	S14	S15	S15	S16	S17	S18	S19	S20	S21	S22
Detection	—	—	—	—	—	—	—	—	—	—	—	—	—	—	—	—	—	—	—	—	—	—	—	—
Technique	na	na	na	na	na	na	T1	T1	T1	T1	T2	T1	T3	T3	na	na	na	na	na	T4	na	na	na	na
Literature	1	1	1	1	1	1	2	2	2	2	3	4	5	5	6	7	7	7	7	8	9	9	9	10
Compound	$R_F \times 100$												$R_F \times 100$											
Thioridazine sulfoxide	3	13	1	9	3	33	—	—	—	—	—	—	—	—	—	—	—	—	—	—	15	12	8	30
Trifluoperazine	18	33	9	34	12	51	69	33	40	—	—	—	68	60	—	—	—	—	—	—	—	—	—	—
Trifluoperazine sulfoxide	3	11	2	17	3	31	—	—	—	—	—	—	—	—	—	—	—	—	—	—	14	12	5	—
Triflupromazine	36	50	48	22	22	72	95	79	40	—	62	—	—	—	—	—	—	—	—	—	—	—	—	—
Triflupromazine sulfoxide	—	—	—	—	—	—	26	48	24	93	—	—	—	—	—	—	—	—	—	48	30	14	17	54
Trimeprazine	—	—	—	—	—	—	96	64	30	—	—	—	—	—	—	—	—	—	—	—	—	—	—	—
Imipramine	—	—	—	—	—	—	—	—	—	—	—	44–46	69	64	—	—	—	—	—	—	—	—	—	—
Acepromazine	—	—	—	—	—	—	—	—	—	—	—	33–35	—	—	—	—	—	—	—	36	10	8	10	—
Profenamine hydrochloride	—	—	—	—	—	—	—	—	—	—	—	—	—	—	—	—	—	—	—	—	—	—	—	—
Chlordiazepoxide	—	—	—	—	—	—	—	—	—	—	—	—	—	—	—	—	—	—	—	—	—	—	—	—
Sordinol	—	—	—	—	—	—	—	—	—	—	—	—	—	—	30	—	—	—	—	—	—	—	—	—
Histantin	—	—	—	—	—	—	—	—	—	—	—	—	—	—	63	—	—	—	—	—	—	—	—	—
Pecazine	—	—	—	—	—	—	—	—	—	—	—	—	—	—	38	—	—	—	—	—	13	8	15	—
Aminopromazine	—	—	—	—	—	—	—	—	—	—	—	—	—	—	—	13	42	63	60	—	—	—	—	—

Layer	—	—	—	—	—	—	—	—	—	—	—	—
Solvent	—	—	—	—	—	—	—	—	—	—	—	—
Detection	D1	D2	D3	D4	D5	D6	D7	D8	D9	D10	D11	D12
Technique	—	—	—	—	—	—	—	—	—	—	—	—
Literature	1	2	3	4	4	4	10	10	10	10	10	10
Compound	Color											
Acetophenazine	o-pk	—	—	—	—	—	—	—	—	—	—	—
Butylrylperazine	o-pk	—	—	—	—	—	y	r-br	y	y	y	y-br → y
Chlorpromazine	—	r-pk	—	lt bl	o	r-br	r	dk r	r	r	y-gr	r → dk br
Chlorpromazine sulfoxide	—	r-pk	—	—	—	—	—	—	—	—	—	—
Chlorprothixene (chlorprothixenum)	—	lt o	—	bl	o	y-ish	—	—	—	—	—	—
N-Demethylchlorpromazine sulfoxide	—	r-pk	—	—	—	—	—	—	—	—	—	—
Diethazine	—	—	r	bl	o	vt	—	—	—	—	—	—
N,N-Didemethylchlorpromazine sulfoxide	—	r-pk	—	—	—	—	—	—	—	—	—	—
Ethopropazine	—	—	r	—	—	—	—	—	—	—	—	—

TABLE TLC 139—(Continued)
DRUGS, PHENOTHIAZINES

Layer	—	—	—	—	—	—	—	—	—	—	—	—
Solvent	—	—	—	—	—	—	—	—	—	—	—	—
Detection	D1	D2	D3	D4	D5	D6	D7	D8	D9	D10	D11	D12
Technique	—	—	—	—	—	—	—	—	—	—	—	—
Literature	1	2	3	4	4	4	10	10	10	10	10	10
Compound	Color											
Fluphenazine	—	o	—	—	—	vt	—	—	—	—	—	—
Levomepropazine	—	—	bl	bl	o	vt	vt	vt	vt	vt	y	vt-y
Librium	—	—	—	—	—	—	—	—	—	—	—	—
Mepazine	—	o	—	—	—	—	r-br	pk	br	r-br	y	r-br → y
Mepazine sulfoxide	—	r-pk	—	—	—	—	—	—	—	—	—	—
Methdilazine	—	r	—	—	—	—	—	—	—	—	—	—
Methoxypromazine	—	p	—	—	—	—	—	—	—	—	—	—
Omca	—	—	y	—	—	—	r-br	br	br	br	y	br-y
Perazine	—	—	yo	—	—	—	—	—	—	—	—	—
Perphenazine	—	r-pk	—	dk	r-vt	bl	—	—	—	—	—	—
Phenothiazine	—	—	—	—	—	—	—	—	—	—	—	—
Pipamazine	—	r-pk	—	—	—	—	—	—	—	—	—	—
Prochlorperazine	—	r-pk	—	—	—	—	—	—	—	—	—	—
Prochlorperazine sulfoxide	—	pk	—	—	—	—	—	—	—	—	—	—
Proketazine	o-pk	—	—	—	—	—	—	—	—	—	—	—
Promazine	—	—	r-o	bl	o	vt	r-br	pk	br	r-br	y → gr	r-br → y
Promazine sulfoxide	—	r-pk	—	—	—	—	—	—	—	—	—	—
Promethazine	—	o	—	bl	o	vt	pk	r	weak r	pk	y → gr	pk → y
Promethazine sulfoxide	—	r-pk	—	—	—	—	—	—	—	—	—	—
Prothypendyl	y	—	—	—	—	—	—	—	—	—	—	—
Thiethylperazine	bl	—	—	—	—	—	—	—	—	—	—	—
Thiopropazate	—	r-pk	—	—	—	—	—	—	—	—	—	—
Thiopropazine	—	—	—	—	—	—	—	—	—	—	—	—
Thioridazine	—	gr-bl	—	lt bl	r	r-br	—	—	—	—	—	—
Thioridazine sulfoxide	bl	—	—	—	—	—	gr	gr	gr	gr	y	gr → r-vt
Trifluoperazine	—	o	—	—	—	—	—	—	—	—	—	—
Trifluoperazine sulfoxide	o	—	—	—	—	—	—	—	—	—	—	—
Triflupromazine	—	o	—	—	—	—	br	br	weak br	weak br	y-br	bt br → y
Triflupromazine sulfoxide	—	o	—	—	—	—	—	—	—	—	—	—
Trimeprazine	—	o	—	—	—	—	—	—	—	—	—	—
Imipramine	—	—	—	lt bl	o	wh	—	—	—	—	—	—
Acepromazine	—	—	—	o	r	r-br	—	—	—	—	—	—
Profenamine hydrochloride	—	—	—	bl	o	vt	—	—	—	—	—	—
Chlordiazepoxide	—	—	—	br-ish	o	y	—	—	—	—	—	—
Sordinol	—	—	—	—	—	—	—	—	—	—	—	—
Histantin	—	—	—	—	—	—	—	—	—	—	—	—
Pecazine	—	—	—	—	—	—	—	—	—	—	—	—
Aminopromazine	—	—	—	—	—	—	—	—	—	—	—	—

TABLE TLC 139—(Continued)
DRUGS, PHENOTHIAZINES

Layer

L1 = Silica gel G.
L2 = Eastman Chromagram sheets (Type K 301 R2), silica gel without fluorescent indicator.
L3 = Eastman Chromagram sheets (Type K 301 R), silica gel with fluorescent indicator.
L4 = Kieselgel G_{254} Merck.
L5 = Woelm neutral alumina.
L6 = Woelm basic alumina.
L7 = cellulose layers.
L8 = Kieselgel G.
L9 = Silica gel (100/200 mesh) with 10% gypsum.
L10 = Silica gel.

Solvent

S1 = *tert*-butanol-1*N* ammonium hydroxide (9 : 1).
S2 = *n*-propanol-1*N* ammonium hydroxide (22 : 3).
S3 = diethyl ether saturated with water, overrun.
S4 = methanol-water (7 : 3).
S5 = *n*-propanol-water (17 : 3).
S6 = *n*-butanol saturated with 1*N* ammonium hydroxide.
S7 = benzene-dioxane-ammonium hydroxide (12 : 7 : 1).
S8 = ethanol-acetic acid-water (5 : 3 : 2).
S9 = methanol-butanol (3 : 2).
S10 = benzene-dioxane-ammonium hydroxide (1 : 8 : 1).
S11 = benzene-acetone-25% ammonium hydroxide (10 : 2 : 1).
S12 = methanol-acetone-triethanolamine (1 : 1 : 0.03).
S13 = chloroform-cyclohexane-diethylamine (5 : 4 : 1).
S14 = ethyl acetate-acetone (1 : 1), satd. with methylamine.
S15 = benzene-acetone (95 : 5).
S16 = benzene-acetone (90 : 10).
S17 = water-acetone (70 : 30).
S18 = butanol-pyridine-water (1 : 2 : 1).
S19 = acetone-ethyl acetate-95% ethanol (5 : 4 : 1) satd. with ammonium lactate buffer (0.5–0.8 ml/100 ml solvent). Buffer pH 7 (18 ml 85% lactic acid with 10 ml conc. ammonia and diluted to 100 ml).
S20 = S19, except pH 3.0: saturation with lactic acid.
S21 = S19, except pH 9.0: saturation with ammonia.
S22 = pyridine-petroleum ether (50–70 °C)-methanol (1 : 4.5 : 0.1).

Detection

D1 = 40% sulfuric acid or UV (263 nm).
D2 = sulfuric acid-ethanol (4 : 1); or potassium iodoplatinate reagent.
D3 = sulfuric acid-ethanol (10%).
D4 = UV.
D5 = Dragendorff reagent.
D6 = palladous chloride.
D7 = sulfuric acid-ethanol (1 : 9).
D8 = perchloric acid.
D9 = ferric chloride.
D10 = phosphomolybdic acid.
D11 = iodine-potassium iodide.
D12 = nitric acid.

Technique

T1 = 10 cm development distance.
T2 = 11.5 cm development distance.
T3 = tank type containers, 21.5 ° ± 1.5 °C, 12 cm, ascending.
T4 = treated 2 hours at 110 °C with S18.

Literature

1. Mellinger, T. J. and C. E. Keeler, *J. Pharm. Sci.*, **1169** (1962); from Kirchner, J. G., "*Thin-Layer Chromatography*", *Techniques of Org. Chem.* XII, Interscience Publishers, Division of John Wiley & Sons, New York, 1967, p. 508.
2. Cochin, J. and J. W. Daly, *J. Pharmacol. Exp. Ther.*, **139**, 160 (1963); from Kirchner, J. G., *op. cit.*, p. 508.
3. Paulus, W., Hoch, W., and R. Keymer, *Arzneimettel-Forschung*, **13**, 609 (1963); from Kirchner, J. G. *op. cit.*, p. 508.
4. Baeumler, J. and S. Rippstein, *Pharm. Acta Helv.*, **36**, 382 (1961); from Kirchner, J. G. *op. cit.*, p. 508.
5. Schweda, P., *Anal. Chem.*, **39**, 1020 (1967).
6. Maček, K. and J. Vecerková, *Pharmazie*, **20**, 611–614 (1965); from *J. Chromatogr.*, **21**, D8 (1966).
7. Eiden, F. and H. D. Stachel, *Deut. Apotheker-Ztg*, **103**, 121 (1963); from Kirchner, J. G., *op. cit.* p. 511.
8. Kinoshita, Y., Koike, H., Iwashima, K., and S. Oriyama, *Nagoya Shiritsu Daigaku Yakugakubu Kenkyu Nempo*, **12**, 33–37 (1964); *Chem. Abstr.*, **64**, 19324h.
9. Margasinski, Z., Danielak, R., Pamazanska, T., and H. Rafalowska, *Acta Pol. Pharm.*, **21**, 6, 1964; from *J. Chromatogr.*, **19**, D11 (1965).
10. Eberhardt, H., Lerbs, O. W., and Freundt, K. J., *Arzneimittel-Forschung*, **13**, 804 (1963); from Kirchner, J. G. *op. cit.* (61) p. 510.

TABLE TLC 140
DRUGS, SULFONAMIDES

Layer	L1	L1	L1	L2
Solvent	S1	S2	S3	S4/S5
Literature	1	1	1	2
Compound		$R_F \times 100$		
Acesulfamethoxypyrazine	72	75	90	—
Acesulfamethoxypyridazine	75 (50)[a]	80 (60)	92	
Acetazolamide	[b]	45	[b]	—
Altizide	40	75	90	—
Badional	—	—	—	—
Buthiazide	35	75	80	—
Carbutamide (Oranil)	—	—	—	84
Chlorexolone	75	87	90	—
Clopamide	55	42	80	—
Desaglybuzole	70[b]	95	70[b]	—
Dichlorphenamide	35[b]	70	[b]	—
Formothiazole	—	—	—	75
Glybutamide	30	82	48	—
Glybuthiazole	40	70	45	—
Glyprothiazole	40	70	35	—
Glysobuzole	55	95	70	—
Maleylsulfathiazole	0	60	0	—
Methyclothiazide	40	80	77	—
Monochlorphenamide	25	58[b]	35	—
Phthalylsulfamethizole	0	78	0	—
Phthalylsulfathiazole (Talisulfazol)	30 and 0	55	50 and 0	78
Salazosulfapyridine	0	80 (35)	0	—
Succinylsulfathiazole	0	40	0	—
Sulfacarbamide (Euvernil®)	—	—	—	0
Sulfacetamide (Albucid®)	25	60 (10)	30 and 0	18
Sulfachrysoidine	0	20	0	—
Sulfadiazine	30	60	42	67
Sulfadicramide	50	80	65	—
Sulfadimethine	—	—	—	—
Sulfadimethoxine	50	85	80	97
Sulfadimethylbenzoylamide	—	—	—	40
Sulfadimethyloxazole	—	—	—	10
Sulfadimidine	45	60	75	99
Sulfaethidole (Globucid)	—	—	—	0
Sulfafurazole (Sulfasol®)	40	70	40	38
Sulfaguanidine	10	20	20	65
Sulfamerazine	40	65	70	92
Sulfamethazine	—	—	—	—
Sulfamethizole	40	60	25	0
Sulfamethopyrazine	70	72	90	—
Sulfamethoxazole	50	75	75	—
Sulfamethoxydiazine	—	—	—	—
Sulfamethoxypyrazine	70	75	75	—
Sulfamethoxypyridazine (Depovernil)®	55	58	85	88
Sulfamethylthiazole	—	—	—	98
Sulfametoyl	40[b]	90	50	—
Sulfanilamide	30	35	50	100
Sulfaphenazole (Depocid)	—	—	—	98
Sulfaproxyline	—	—	—	31
Sulfapyridine	40	50	70	100
Sulfasomidine	—	—	—	70
Sulfasomizol (e)	40	75	45	55
Sulfathiazole	33	42	45	61
Sulfathiourea	25	77	20	—
Sulfisomidine (Sulfamethin)	30	25	40	96
Sulforthomidine	75	80	78	—
Tolbutamide (Oraket)	—	—	—	—

[a] Values in parentheses refer to impurities.
[b] R_F values are inaccurate; tailing.

Layer
- L1 = Silica gel G.
- L2 = Al_2O_3 containing 10% $CaSO_4$.

Solvent
- S1 = chloroform-methanol (10 : 1).
- S2 = chloroform-*n*-butanol-acetone-85% formic acid (4 : 1 : 1 : 1).
- S3 = chloroform-methanol-*n*-butanol-2% NH_4OH (80: 10 : 10 : 1).
- S4* = ethyl acetate-water saturated with NH_3 (80 : 1).
- S5* = S4-dioxane (7 : 1).

*(S4 and S5 are applied successively)

Detection I_2 vapors.

Literature

1. Kiger, J. L. and J. G. Kiger, *Ann. Pharm. Fr.*, **24**, 599 (1966) and *J. Chromatogr.*, **29**, D4 (1967).
2. Margasinski, Z., Danielak, R., and H. Rafalowska, *Acta Pol. Pharm.*, **22**, 423 (1965); *Chem. Abstr.*, **64**, 9509c (1966).

TABLE TLC 141
FOOD DYES, SYNTHETIC

Layer				L1	L1	L1	L1	L2	L2	L1	L1
Solvent				S1	S2	S3	S4	S5	S6	S7	S4
Technique				T1	T1	T2	T2	na	na	T2	T2
Literature				1	1	2	2	3	3	4	4
Compound				$R_F \times 100$							
DFG-8th Comm.[a]	Commercial name	Schultz No.[b]	Col. Ind. No.[c]								
Acid violet		—	42640	—	—	10	73	—	—	—	—
Black 1	Brilliant Black BN	—	28440	20[e](—)	10(—)	13	1	(0)[f]	(0)[f]	10	2
Blue 1	Indanthrene Blue RS	1228	69800	0	0	0	0	0	0	0	0
Blue 2	Indigotin I or Ia, Indigo Carmine, F.D. & Co. Blue No. 2	1309	73015	26(—)	19(7)	18	11	0	0–3	22	11
Ethyrosin		—	45430	—	—	5	86	—	—	(74)[f]	100
Guinea green		—	42085	—	—	17	73	—	—	—	—
New blue		—	42045	—	—	68	76	—	—	—	—
Orange 1	Orange GGN or GGL	—	15980	56(69)	45(—)	47	43	51	28	54	51
Orange 2	Yellow Orange S, Sunset Yellow FCF, F.D. & Co. Yellow No. 6	—	15985	55(69)	42(24)	45	41	(51)[f]	(0–28)[f]	52	48
Patent green		—	42053	—	—	86	62	—	—	—	—
Ponceau SX		—	14700	—	—	37	43	—	—	—	—
Red 1	Azorubin	208	14720	64(75)	12(—)	19	49	—	24–36	56	63
Red 2	Genuine Red E	210	16045	57(71)	20(—)	25	44	—	—	55	56
Red 3	Amaranth S, Naphtha Red S	212	16185	27(—)	31(—)	42	5	17	0	15	7
Red 4	Brilliant Ponceau 4 RC, Cochineal Red A	213	16255	33(—)	55(—)	65	20	37	12	28	23
Red 5	Ponceau 6 R, Scarlet 6 R	215	16290	16[e](—)	76(—)	—	—	—	—	—	—
Red 6	Scharlach GN	—	—	64(73)	90(94)	—	—	—	—	—	—
Scarlet GN		—	14815	—	—	81	59	55	0	—	—
Violet		—	42650	—	—	17	71	—	—	—	—
Yellow 1	Genuine Yellow (also Extra), Acid or Fast Yellow, Acidic Yellow, Jaune Solide	172	13015	53(50)[d]	58(—)[d]	58	42	51	46	51	45

TABLE TLC 141—(Continued)
FOOD DYES, SYNTHETIC

Compound				$R_F \times 100$							
Layer				L1	L1	L1	L1	L2	L2	L1	L1
Solvent				S1	S2	S3	S4	S5	S6	S7	S4
Technique				T1	T1	T2	T2	na	na	T2	T2
Literature				1	1	2	2	3	3	4	4
DFG-8th Comm.[a]	Commercial name	Schultz No.[b]	Col. Ind. No.[c]								
Yellow 2	**Tartrazine, Jaune Tartarique, Hydrazine Yellow 0, F.D. & Co. Yellow No. 5**	737	19140	25(—)	72(—)	70	9	(13–33)[f]	4	20	10
Yellow 3	**Quinoline Yellow (also Water-Soluble or Extra) Quinoline Yellow**	918	47005	39(27)	12(20)	—	—	68	56	33	18
Yellow 4	**Chrysoidine 5, Tropeoline 0, Resorcine Yellow**	186	14270	88(—)	34(—)	—	—	65	64	60	78
Yellow 5	**Yellow 27 175N**	—	—	31(—)	29(—)	—	—	—	—	—	—

[a] Communication No. 8 of the Dyestuff-Commission of DFG. 2nd Ed., published by Verlag Franz Steiner, Wiesbaden, Germany, 1957.
[b] Schultz, G., Farbe Stofftabellen (Color Index), 7th Ed., Leipzig, 1931.
[c] Color Index of the Society of Dyers and Colourists, Bradford, 1956.
[d] Minor spots ( ).
[e] Drawn out spots (tail formation).
[f] ( ) indicates streaking.

Layer
L1 = MN-Cellulose powder 300G, dried at 105 °C for 10 minutes.
L2 = Silica gel activated at 120° C for 30 minutes.

Solvent
S1 = *n*-propanol-ethyl acetate-water (60 : 10 : 30).
S2 = 2.5% aqueous sodium citrate solution-25% ammonia solution (80 : 20).
S3 = 2% sodium citrate in 5% NH_3 solution.
S4 = *tert*-butanol-propionic acid-water (50 : 12 : 38) with 0.4% KCl.
S5 = *n*-butanol-ethanol-water (20 : 20 : 5).
S6 = *n*-butanol-ethanol-water-0.3*N* HCl (20 : 20 : 5 : 1).
S7 = propanol-ethyl acetate-water (50 : 12 : 38).

Technique
T1 = ascending development time, 30 minutes.
T2 = ascending development time, 90 minutes.

Literature

1. Gaenshirt, H., Waldi, D., and E. Stahl, "*Thin-Layer Chromatography*" (Eng. ed.), E. Stahl, Ed., Academic Press, N.Y., 1965, p. 349.
2. Salo, T. and K. Salminem, *Suomen Kemistilehti*: *B*, **35**, 146 (1962).
3. Ciasea, M. A. and C. G. Casinovi, *Thin-Layer Chromatography*, G. B. Marini-Bettòlo, Eds., Elsevier, Amsterdam, 1964, p. 212.
4. Wollenweber, P., *J. Chromatogr.*, **7**, 557 (1962).

TABLE TLC 142
DYES, AZO DYESTUFFS

Layer	L1	L1	L1	L1	L1
Solvent	S1	S2	S3	S4	S5
Literature	1	1	1	1	1
Compound	$R_F \times 100$				
Ponceau 3R	86[a]	85[a]	71[a]	50[a]	0
Amaranth	69	70	52	16	2
Erythrosine	44	43	56	22	40
Ponceau SX	74	73	67	36	32
Oil Red XO	67	81	97	94	88
Ponceau 2R	76	88	74	44	37[a]
New Coccine	61	78	48	24	12
Eosine	55	58	65	31	44
Phloxine	57	66	69	34	35
Rose Bengale	37	45	53	26	25
Acid Red	64	64	78	48	42
Rodamine B	62	68	94	71	70
Coralline	80	91	81	69	56

[a] Tailing.

Layer L1 = Mixed polyamide-silica gel layers (7 : 52 w/w, prepared in 100 ml 75% formic acid).

Solvent
S1 = isopropyl alcohol-5% NH_4Cl solution (4 : 1.5).
S2 = ether-isopropyl alcohol-5% NH_4Cl solution (0.5 : 1 : 1).
S3 = $CHCl_3$-isopropyl alcohol-5% NH_4Cl solution-glacial acetic acid (0.5 : 2.5 : 1.0 : 0.5).
S4 = *n*-butanol-ethanol-5% sodium citrate solution (3 : 2 : 1.5).
S5 = $CHCl_3$-isopropyl alcohol-5% NaCl solution-glacial acetic acid (0.5 : 2.5 : 0.5 : 0.1).

Technique Ascending, 10 cm, 3–4 hours.

Literature
1. Chiang, Hung-Cheh, *J. Chromatogr.*, **40**, 189–90 (1969).

TABLE TLC 143
HYDROCARBONS, TERPENE

Layer	L1	L1	L2	L2
Solvent	S1	S2	S1	S2
Literature	1	1	1	1
Compound	$R_F \times 100$			
γ-Bisabolene	50	33	10	8
δ-Cadinene	71	35	20	12
β-Caryophyllene	65	44	30	16
α-Cedrene	73	56	44	36
Guaiene	49, 58	34, 44	18, 28	14, 21
α-Humulene	35	30	12	8
β-Santalene	64	48	20	15
β-Selinene	41	33	17	13
Valencene	51	32	21	17
Ylangene	87	64	40	32
β-Zingaberene	57	30	18	13
Azulene	19	0	5	3
Guaiazulene	10	5	0	0
Pinene dimer	54	38	22	14
Squalene	0	0	0	0
β-Carotene	0	0	0	0

Layer
L1 = Aluminum oxide G.
L2 = Silica gel G.

Solvent
S1 = perfluoroalkane (boiling range 70–80 °C).
S2 = low boiling perfluorokerosene.

Detection See Section II.I, Volume Two.

Technique Ascending 10 cm, cylindrical chamber.

Literature
1. Attaway, J. A., Barabas, L. J., and R. W. Wolford, *Anal. Chem.*, **37**, 1289 (1965).

TABLE TLC 144
GIBBERELLINS

Layer	L1	L1	L1	L2	L2	L2	L1	L1	L2	L2
Solvent	S1	S2	S3	S4	S3	S5	S6	S7	S8	S7
Technique	T1	T1	T1	T1	T2	T2	T3	na	T4	na
Literature	1	1	1	1	2	2	3	3	3	3
Compound	$R_F \times 100$									
Gibberellins										
A_1	20	49	0	0	28	49	11	0	54	26
A_2	17	40	0	0	23	37	4	0	64	30
A_3	19	54	0	0	18	40	11	0	42	18
A_4	63	95	67	67	90	100	37	82	100	100
A_5	53	87	27	45	85	90	31	35	100	88
A_6	59	87	11	33	86	84	25	21	95	76
A_7	60	90	57	45	85	91	37	70	100	100
A_8	4	30	0	0	6	10	4	0	28	6
A_9	87	95	100	100	100	100	75	100	100	100

Layer L1 = Silica gel.
L2 = Kieselguhr.

Solvent S1 = benzene-*n*-butanol-acetic acid (16 : 3 : 1).
S2 = benzene-*n*-butanol-acetic acid (14 : 5 : 1).
S3 = carbon tetrachloride-acetic acid-water (8 : 3 : 5), lower phase.
S4 = lower phase of S3 10% ethyl acetate.
S5 = lower phase of S3 20% ethyl acetate.
S6 = diisopropyl ether-acetic acid (95 : 5).
S7 = benzene-acetic acid-water (8 : 3 : 5), upper phase.
S8 = benzene-propionic acid-water (8 : 3 : 5), upper phase.

Detection UV and/or H_2SO_4.

Technique T1 = 15 cm development distance.
T2 = 15 cm development distance, 50–80 minutes.
T3 = 15 cm development distance, 25 minutes.
T4 = 15 cm development distance, 70 minutes.

Literature
1. Kagawa, T., Fukinbara, T. and Y. Sumiki, *Agr. Biol. Chem.* (Tokyo) **27**, 598 (1963).
2. Ikewawa, N., Kagawa, T., and Y. Sumiki, *Proc. Jap. Acad.*, **39**, 507 (1963); *Chem. Abstr.*, **60**, 12347c (1964).
3. MacMillan, J. and P. J. Suter, *Nature*, **197**, 790 (1963).

TABLE TLC 145
GIBBERELLINS, METHYL ESTERS OF

Layer	L1	L1	L1	L1
Solvent	S1	S2	S3	S4
Technique	T1	T1	T1	T1
Literature	1	1	2	2
Compound	$R_F \times 100$			
Methyl Esters				
A_1	33	18	31	29
A_2	44	8	23	13
A_3	26	16	35	32
A_4	100	50	73	75
A_5	100	38	60	69
A_6	100	31	66	67
A_7	100	48	71	72
A_8	10	8	17	12
A_9	100	80	98	96

Layer L1 = Silica gel.

Solvent S1 = benzene-acetic acid-water (8 : 3 : 5), upper phase.
S2 = diisopropyl ether-acetic acid (98 : 2).
S3 = ethyl ether-benzene (4 : 1).
S4 = ethyl ether-petroleum ether (4 : 1).

Detection Concentrated H_2SO_4.

Technique T1 = 15 cm development distance.

Literature
1. MacMillan, J. and P. J. Suter, *Nature*, **197**, 790 (1963).
2. Ikewawa, N., Kagawa, T., and Y. Sumiki, *Proc. Jap. Acad.*, **39**, 507 (1963).

TABLE TLC 146
HYDROCARBONS, POLYCYCLIC AROMATIC

Layer	L1	L2
Solvent	S1	S2
Literature	1	1
Compound	$R_F{}^a \times 100$	
7,12-Dimethylbenz[*a*]anthracene	74	50
4-Methylpyrene	70	70
5,6-Dimethylchrysene	69	42
3-Methylcholanthrene	69	57
Anthracene	65	57
12-Methylbenz[*a*]anthracene	63	34
7,8-Dimethylbenz[*a*]anthracene	63	42
Benzo[*c*]phenanthrene	62	54
1,2-Dihydrobenz[*e*]aceanthrylene	62	39
Cholanthrene	60	50
5-Methylchrysene	59	39
Pyrene	58	62
Fluoranthene	57	60
7-Methylbenz[*a*]anthracene	56	45
4-Methylbenzo[*a*]pyrene	51	37
Benz[*a*]anthracene	45	42
Chrysene	44	35
Triphenylene	44	49
Benzo[*a*]pyrene	40	23
Benzo[*k*]fluoranthene	40	40
Benzo[*e*]pyrene	39	56
Anthanthrene	36	32
Perylene	33	47
Benzo[*g,h,i*]perylene	33	51
Dibenz[*a,h*]anthracene	30	48
Dibenzo[*a,i*]pyrene	28	24
Dibenzo[*a,i*]phenanthrene	27	57
Dibenzo[*a,e*]pyrene	26	38
Dibenz[*b,i*]anthracene	19	54

[a] Average of five determinations.

Layer L1 = Cellulose.
L2 = Acetylated cellulose.

Solvent S1 = immobile phase, 20% DMF in ethyl ether, mobile phase, iso-octane.
S2 = ethanol-toluene-water (17 : 4 : 4).

Detection Shortwave UV light.

Technique Reverse phase chromatography.

Literature 1. White, R. H. and J. W. Howard, *J. Chromatogr.*, **29**, 108–114 (1967).

TABLE TLC 147
INORGANICS, ANIONS

Layer	L1	L1	L2	L2	L2
Solvent	S1	S2	S3	S4	S5
Literature	1	1	2	2	2
Ion	$R_F \times 100$				
SO_4^{--}	100	60[a]	—	—	—
PO_4^{3-}	0	0	—	—	—
$Fe(CN)_6^{4-}$	100	45	—	—	—
$Fe(CN)_6^{3-}$	100	90	—	—	—
MnO_4^-	0	0	—	—	—
SCN^-	100[a]	90	—	—	—
$Cr_2O_7^{--}$	80	35[a]	—	—	—
IO_3^-	85	45	6	70	58
I^-	95	90	78	67	50
Br^-	—	90	60	77	70
Cl^-	—	100	40	85	80
SO_3^{--}	100	80	—	—	—
$S_2O_3^{--}$	100	60	—	—	—
SeO_3^{--}	75	30	—	—	—
TeO_3^{--}	0	0	—	—	—
MoO_4^{--}	90	60	—	—	—
NO_2^-	90	85	—	—	—
NO_3^-	90	90	—	—	—
F^-	—	—	4	—	—
ClO_3^-	—	—	82	85	71
BO_3^-	—	—	53	78	67
ClO_4^-	—	—	97	58	50
IO_4^-	—	—	6	70	58

[a] Tailing.

Layer L1 = Starch-bound lanthanum oxide.
L2 = Corn starch.

Solvent S1 = 1*N* ammonium hydroxide.
S2 = 1*N* ammonium hydroxide-acetone (1 : 1).
S3 = acetone-3*N* ammonia (7 : 3).
S4 = 3*N* ammonia.
S5 = water.

Detection See Section II.I, Volume Two.

Literature 1. Moghissi, A., *J. Chromatogr.*, **13**, 542 (1964).
2. Petrović, S. M. and V. D. Canić, *Z. Anal. Chem.*, **228**, 340 (1967); from *J. Chromatogr.*, **33**, D17 (1968).

TABLE TLC 148
INORGANICS, METAL IONS

Layer	L1	L1	L1	L1	L1	L1	L1	L1	L1	L1	L1	L1	L1
Solvent	S1	S2	S3	S4	S5	S6	S7	S8	S9	S10	S11	S12	S13
Technique	T1	T1	T1	T2	T2	T2	T2	T1	T1	T1	T3	T3	T3
Literature	1	1	1	1	1	1	1	1	1	1	1	1	1
Ion	$R_F \times 100$												
Co(II)	84–95	80–95	79–95	73–85	66–80	63–79	56–74	82–94	81–93	62–76	89–100	89–100	88–100
Cu(II)	01–27	02–29	03–28	01–24	01–23	02–24	03–29	00–29	00–17	00–21	03– 38	02– 30	06– 41
Hg(II)	00–06	00–06	00–05	00–04	00–05	00–06	00–06	00–14	00–12	00–11	00– 23	00– 15	00– 29
Tl(I)	34–65	38–59	37–54	26–46	24–40	25–39	22–38	44–62	43–61	38–50	49– 75	63– 78	67– 79
Pb(II)	45–82	61–79	61–77	31–58	30–52	33–52	33–53	56–81	68–77	52–66	76– 94	83– 94	82– 93
Bi(III)	00–18	00–28	02–34	00–19	00–24	02–34	03–55	00–28	00–29	02–58	00– 19	00– 37	02– 55
Fe(III)	00–17	02–34[a]	19–51	05–31[b]	18–47[b]	53–76	57–74	59–95	79–94	64–80	02–100	85–100	87–100
Ag(I)	00–04	00–04	00–04	00–05	00–04	00–03	00–04	00–07	00–05	00–05	00– 05	00– 07	00– 10
Ni(II)	86–95	84–95	82–94	73–84	66–78	62–75	57–71	82–92	78–90	61–73	90–100	87– 98	85– 97
Cd(II)	87–93	86–94	84–92	79–86	73–81	71–79	67–78	86–93	84–91	71–80	90– 98	89– 96	85– 96
Al(III)	00–95	77–97	77–97	75–90	69–85	73–86	65–81	86–97	84–94	68–81	68–100	88–100	86–100
Ca(II)	84–91	82–91	79–88	68–79	61–71	57–67	48–61	83–91	82–91	66–77	89– 98	89– 98	83– 92
Zn(II)	85–94	84–96	82–94	73–84	68–79	66–78	59–74	81–92	81–91	65–77	89–100	87– 99	85– 96
Mg(II)	85–94	85–97	81–93	73–84	68–78	64–78	56–72	85–93	82–93	69–79	90–100	88– 99	85– 97
In(III)	00–93	87–96	84–95	78–91	74–88	78–91	70–86	89–97	87–97	75–87	80–100	91–100	89–100

[a] A small portion of Fe(III) was detected at R_F 88–92.
[b] A small portion of Fe(III) was detected at R_F 80–85 in EtOH–0.5*M* HNO_3 (8 : 2) and R_F 71–87 in EtOH–1*M* HNO_3 (8 : 2).

Layer
L1 = Silica gel-cellulose (5 : 2) plates impregnated with polysulfide polymer, purified.

Solvent
S1 = MeOH-0.1*M* HNO_3 (8 : 2).
S2 = MeOH-0.5*M* HNO_3 (8 : 2).
S3 = MeOH-1*M* HNO_3 (8 : 2).
S4 = EtOH-0.5*M* HNO_3 (8 : 2).
S5 = EtOH-1*M* HNO_3 (8 : 2).
S6 = EtOH-3*M* HNO_3 (8 : 2).
S7 = EtOH-5*M* HNO_3 (8 : 2).
S8 = acetone-0.5*M* HNO_3 (8 : 2).
S9 = acetone-1*M* HNO_3 (8 : 2).
S10 = acetone-3*M* HNO_3 (8 : 2).
S11 = dioxane-0.1*M* HNO_3 (7 : 3).
S12 = dioxane-0.5*M* HNO_3 (7 : 3).
S13 = dioxane-1*M* HNO_3 (7 : 3).

Detection
See Section II.I, Volume Two.

Technique
T1 = ascending, 16 cm, about 30 minutes.
T2 = ascending, 16 cm, about 55 minutes.
T3 = ascending, 16 cm, about 50 minutes.

Literature
1. Oguma, K., *J. Chromatogr.*, **42**, 96–102 (1969).

TABLE TLC 149
INORGANICS, METAL IONS

Layer	L1	L1	L1	L1	L1	L1	L1	L1	L1	L1	L1	L1	L1	L2	L2	L2
Solvent	S1	S2	S3	S4	S5	S6	S7	S8	S9	S10	S11	S12	S13	S14	S15	S16
Literature	1	1	1	1	1	1	1	1	1	1	1	1	1	1	1	1
Ion	$R_F \times 100$															
Tl(I)	13	15	(0–95)	16	31	12	16	17	23	—	—	—	—	57	46	21
Ag(I)	0	0	0	28	53	—	—	—	—	—	—	—	—	49	36	14
Co(II)	21	54	92	53	90	26	53	53	54	28	57	80	95	62	38	4
Ni(II)	20	54	92	52	90	—	—	—	—	27	56	80	95	61	38	4
Hg(II)	71	85	88	71	74	—	—	—	—	—	—	—	—	81	76	e.s.[b]
Mn(II)	33	60	95	59	95	—	—	—	—	—	—	—	—	58	38	8
Mg(II)	32	59	94	60	95	56	60	65	68	36	62	90	96	78	53	9
Cd(II)	27	61	94	39	83	12	39	41	47	36	65	9	96	61	36	3
Zn(II)	20	54	93	51	88	15	51	56	56	26	52	75	95	62	35	4
Ca(II)	6	21	n.d.[a]	22	n.d.	—	—	—	—	24	45	n.d.	n.d.	57	35	4
Sr(II)	4	5	7	5	7	—	—	—	—	6	21	41	52	55	38	6
Ba(II)	0	0	5	0	5	—	—	—	—	0	3	12	20	40	26	2
Cu(II)	2	8	25	7	20	—	—	—	—	8	26	52	68	55	28	0
Pb(II)	0	0	0	0	0	—	—	—	—	0	0	4	9	37	13	0
Bi(III)	0	0	(0–90)	0	0	—	—	—	—	0	0	0	13	84	50	2
In(III)	2	2	80	0	0	—	—	—	—	2	16	66	95	91	55	0
Th(IV)	0	0	0	0	0	—	—	—	—	0	0	0	7	—	—	—
Mo(VI)	0	0	0	0	0	—	—	—	—	0	0	0	4	—	—	—
Fe(III)	—	—	—	—	—	—	—	—	—	e.s.	e.s.	26	64	—	—	—
Al(III)	—	—	—	—	—	—	—	—	—	12	39	e.s.	e.s.	65	18	0
Ga(III)	—	—	—	—	—	—	—	—	—	3	18	47	e.s.	51	9	0
Be(II)	—	—	—	—	—	—	—	—	—	—	—	—	—	79	55	6

[a] n.d. = not determined.
[b] e.s. = elongated spot.

Layer
L1 = Alginic acid thin layers.
L2 = Carboxymethylcellulose layers.

Solvent
S1 = NaCl 0.05*M*.
S2 = NaCl 0.1*M*.
S3 = NaCl 0.3*M*.
S4 = $NaClO_4$ 0.1*M*.
S5 = $NaClO_4$ 0.3*M*.
S6 = 0.1*M* $LiNO_3$
S7 = 0.1*M* $NaNO_3$.
S8 = 0.1*M* NH_4NO_3.
S9 = 0.1*M* KNO_3.
S10 = KCl-HCl solution, pH = 2.00.
S11 = KCl-HCl solution, pH = 1.43.
S12 = KCl-HCl solution, pH = 1.10.
S13 = KCl-HCl solution, pH = .91.
S14 = 1*N* sodium acetate-1*N* acetic acid.
S15 = 0.5*N* sodium acetate-0.5*N* acetic acid.
S16 = 0.1*N* sodium acetate-0.1*N* acetic acid.

Literature

1. Cozzi, D., Desideri, P. G., and L. Lepri, *J. Chromatogr.*, **42**, 532–540 (1969).

TABLE TLC 150
INORGANIC IONS

Layer	L1	L1
Solvent	S1	S2
Literature	1	1
Ion	$R_F \times 100$	
Ag^+	50–90	30
As^{3+}	80	40
Ba^{++}	0–10	10
Be^{++}	70	60
Bi^{3+}	100	90
Cd^{++}	100	40
Ce^{4+}	10	0
Co^{++}	40	30
Cu^{++}	80	50
Hg^{++}	90	90
Mn^{++}	30	50
Ni^{++}	20	30
Pb^{++}	70	30
SeO_3^{--}	100	50
Sb^{3+}	100	90
TeO_3^-	30	20
Tl^+	0	20
UO_2^{++}	90	90
Zn^{++}	100	30

Layer L1 = Cellulose layers (MN-300).

Solvent S1 = acetone–4*N* HCl (7 : 3).
S2 = acetone–25% HNO_3 (7 : 3).

Detection See Section II.I, Volume Two.

Technique 10 cm development distance.

Literature
1. Merkus, F. W. H. M., *Pharm. Weekbl.*, **98**, 947 (1963), from Kirchner, J. G., *Thin-Layer Chromatography*, Interscience Publishers, Inc., Division of John Wiley & Sons, Inc., New York, N.Y., 1967, p. 719.

TABLE TLC 151
INORGANICS, CONDENSED PHOSPHATES

Layer	L1	L1	L2	L2	L3	L3
Solvent	S1	S2	S2	S3	S4	S5
Technique	T1	T1	T1	T1	T2	T2
Literature	1	1	1	1	2	2
Compound	$R_F \times 100$					
Monophosphate	84	84	83	55	55	34
Di-	75	64	67	72	40	15
Tri-	60	46	45	58	31	15
Tetra-	46	30	30	45	22	—
Penta-	33	18	20	30	16	—
Hexa-	23	—	12	19	11	—
Hepta-	16	—	7	11	7	—
Octa-	10	—	4	6	5	—
Trimeta-	45	19	17	34	—	6
Tetrameta-	24	10	19	18	—	47
Pentameta-	—	—	5	12	—	—
Hexameta-	—	—	—	8	—	—
Heptameta-	—	—	—	5	—	—

Layer L1 = Cellulose 140 dg (S & S).
L2 = Cellulose 142 dg (S & S).
L3 = Purified Cellulose MN-300.

Solvent S1 = 60 ml methanol, 30 ml dioxane, 30 ml (isopropanol-water) (7 : 1), 8 ml (96% acetic acid-water) (1 : 4), 40 ml of (125 g trichloroacetic acid, 35 ml 25% ammonia diluted to 1000 ml).
S2 = 70 ml dioxane; 30 ml (160 g trichloroacetic acid and 8 ml 25% ammonia diluted to 100° ml.
S3 = 75 ml methanol, 20 ml (isopropanol-water) 7 : 1), 25 ml (125 g trichloroacetic acid and 32 ml 25% ammonia diluted to 1000 ml), 6 ml (96% acetic acid-water (1 : 4).
S4 = 30 ml water, 35 ml ethanol, 15 ml isobutanol, 20 ml isopropanol-H_2O-ethanol-isobutanol-isopropanol-ammonia-trichloracetic acid (30 : 35 : 15 : 20 : 04 : 5, v/v/v/v/v/w).
S5 = ammonia-methanol-isobutanol-H_2O-formic acid (9 : 50 : 10 : 30 : 03).

Detection Molybdate, followed by sulfite.

Technique T1 = development distance, 16 cm, development time = 60 minutes.
T2 = ascending development distance, 14 cm; two dimensional chromatography may be carried out, by developing in one direction with the acidic solvent, and then in second direction with the basic solvent.

Literature
1. Roessel, T., *Z. Anal. Chem.*, **197**, 333 (1963).
2. Aurenge, J., Degeorges, M., and J. Normand, *Bull. Soc. Chim., France*, **1964**, 508 (1964).

TABLE TLC 152
LIPIDS, BRAIN

Layer	L1	L1	L1	L2	L3	L4	L5	L6	L1	L1	L7	L7	L1	L2
Solvent	S1	S2	S3	S2	S2	S1	S1	S1	S4	S5	S4	S5	S1	S1
Technique	T1	T1	T1	T1	T1	T1	T1	T1	T2	T2	T2	T2	T3	T3
Literature	1	1	1	1	1	1	1	1	2	2	2	2	2	2
Compound							$R_F \times 100$							
Cerebroside	89	96	58	44	90	61	82	77	—	—	—	—	—	—
	85	93	49	35	85	55	79	70	—	—	—	—	—	—
Phosphatidyl ethanolamine	78	90	47	59	92	45	55	65	65	60	60	60	83	75
Phosphatidyl serine	—[a]	29	11	35	57	41[a]	78[a]	65	20	47	14	15	40	35
Lysophosphatidyl ethanolamine	55	46	21	—	70	—	29	37	—	—	—	—	—	—
Lysophosphatidyl serine	—[a]	—	0	—	14	—	—	—	—	—	—	—	—	—
Phosphatidyl choline	55	63	43	45	77	35	35	45	36	36	35	30	60	53
Lysophosphatidyl choline	31	28	13	—	46	—	10	7	14	9	9	9	23	13
Sphingomyelin	47	47	24	30	58	20	19	30	—	—	—	—	—	—
Cerebroside sulfate	—	—	41	35	71	28	50	65	—	—	—	—	—	—
	—	—	36	30	67	23	35	55	—	—	—	—	—	—
Cardiolipin	—	—	—	—	—	—	—	—	91	94	91	95	92	91
Phosphatidic acid	—	—	—	—	—	—	—	—	77	71	73	71	—	—
Phosphatidyl inositol	—	—	—	—	—	—	—	—	30	64	17	60	30	24

[a] Streaking.

Layer
L1 = Silica gel G.
L2 = Silica gel G, impregnated with sodium borate.
L3 = Silica gel G, impregnated with sodium acetate.
L4 = Silica gel G, impregnated with potassium hydroxide.
L5 = Silica gel G, impregnated with oxalic acid.
L6 = Silica gel G, impregnated with ammonium sulfate.
L7 = L1 made up with borate buffer at pH 8, freshly prepared; after spreading, dry at room temperature for several hours, then activate at 100 °C for one hour.

Solvent
S1 = chloroform-methanol-water (65 : 25 : 4).
S2 = chloroform-methanol-ammonium hydroxide (62 : 25 : 4).
S3 = chloroform-methanol-ammonium hydroxide (75 : 25 : 4).
S4 = chloroform-methanol-7*N* ammonium hydroxide (60 : 35 : 5).
S5 = chloroform-methanol-7*N* ammonium hydroxide (35 : 60 : 5).

Technique
T1 = unlined but equilibrated tanks.
T2 = two-dimensional systems, S4 for 1st dimension, S5 for 2nd dimension.
T3 = one-dimensional systems.

Detection
See Section II.I, Volume Two.

Literature
1. Horrocks, L. A., *J. Amer. Oil Chem. Soc.*, **40**, 235 (1963).
2. Bunn, C. R., Keele, B. B. Jr., and G. H. Elkan, *J. Chromatogr.*, **45**, 326 (1969).

TABLE TLC 153
ANTIBIOTICS, MACROLIDE GROUP

Layer	L1	L1	L1
Solvent	S1	S2	S3
Literature	1	1	1
Compound	$R_F \times 100$		
Acumycin	66	35	82
Angolamycin	65	18	82
Tylosin	68	7	81
Carbomycin	75	40	88
Foromacidin A	32	2	59
Foromacidin B	34	5	61
Foromacidin C	37	6	64
Erythromycin	16	3	29
Narbomycin	22	12	41
Picromycin	22	7	36
Lankamycin	74	37	87

Layer L1 = Silica gel G.
Solvent S1 = methanol.
S2 = chloroform-methanol (95 : 5).
S3 = chloroform-methanol (1 : 1).
Detection Conc. sulfuric acid or bioautography.
Literature
1. Bichel, H., Gaeumann, Huetter, R., Sackman, W., Vischer, E., Voser, W., Wettstein, A., and H. Zaehner, *Helv. Chim. Acta.*, **45**, 1396 (1962); from Kirchner, J. G., "*Thin-Layer Chromatography*", Techniques of Organic Chemistry, Vol XII, Interscience Publishers, Division of John Wiley & Sons, New York, N.Y., 1967, p. 346.

TABLE TLC 154
LIPIDS, CHOLESTEROL AND METHYL ESTERS OF FATTY ACIDS

Layer	L1
Solvent	S1
Literature	1
Compound	$R_F \times 100$
Cholesteryl palmitate	75
Cholesteryl stearate	75
Cholesteryl oleate	66
Cholesteryl linoleate	57
Cholesteryl linolenate	46
Cholesteryl arachidonate	43
Methyl palmitate	80
Methyl stearate	80
Methyl oleate	70
Methyl linoleate	61
Methyl linolenate	51
Methyl arachidonate	48

Layer L1 = ITLC—SG.
Solvent S1 = isooctane.
Detection Sulfuric acid.
Technique 70% relative humidity, ascending development, saturated chamber.
Literature
1. Hamilton, J. G., Swartout, J. R., Miller, O. N., and J. E. Muldrey, *Biochem. Biophys. Res. Commun.*, **5**, 226 (1961)

TABLE TLC 155
NITROGEN HETEROCYCLICS, IMIDAZOLES

Layer	L1	L2	L1	L2	L1	L2	L1	L2	L1	L2	L1	L2
Solvent	S1	S1	S2	S2	S3	S3	S4	S4	S5	S5	S6	S6
Technique	T1	T2	T1	T2	T1	T3	T1	T3	T1	T3	T4	T5
Literature	1	1	1	1	1	1	1	1	1	1	1	1
Compound	$R_F \times 100$											
4-Amino-5-carboxamide imidazole	51	44	62	60	36	25	56	50	54	48	46	31
Anserine	—	15	24	19	9	7	37	33	9	6	32	28
Carnosine	21	15	22	17	6	4	28	23	8	5	33	21
Ergothioneine	33	25	43	39	5	4	31	28	28	25	12	8
Histamine	23	23	13[a]	9	59	48	62	53	23[a]	12	71	66
Histamine-6-*N*-acetyl	56	48	72	69	84	75	87	86	61	54	73	62
Histidine	23	18	27	23	7	6	32	28	13	6	38	33
Histidinol	24	19	25[a]	16	55	43	67	61	15[a]	9[a]	67	58
Hydroxymethyl imidazole	52	37	69	66	71	62	76	73	55	45	65	54
Imidazole	57	50	78	73	87	81	88	88	54	43	82	75
Imidazole acetic acid	46	40	42	38	13	10	40	40	53	46	39	29
Imidazole acrylic acid	53	49	61	62	15	14	42	42	67	53	46	32
Imidazole aldehyde	—	62	79	78	62	64	77	80	78	—	72	65
Imidazole carboxylic acid	36	30	42	45	11	12	34	37	35	25	39	30
Imidazole glycerol	37	28	57	51	36	29	52	46	39	24	39	26
Imidazole lactic acid	35	30	40	40	14	12	38	38	44	26	41	31
Imidazole propionic acid	45	40	42	40	17	15	45	41	58	57	46	34
Imidazole pyruvic acid	34	32	40[a]	45	7	6	38[a]	37	35[a]	27[a]	15[a]	40
3-Methyl histidine	29	21	36	33	8	6	37	32	20	10	31	21
1-Methyl histidine	21	17	29	25	9	8	36	33	12	7	27	22
4-Ureido imidazole-5-carboxylic acid	28[a]	27	50	50	10	10	35	41	27[a]	32	48	37
Acetyl histidine	45	36	45	44	17	13	40	40	35	31	43	29
Acetyl histidinol	53	44	70	60	72	67	81	79	53	48	65	58
4-Amino-imidazole-5-carboxylic acid	39	33	63	57	35	32	52	49	43	39	45	31
Imidazole-4,5-carboxylic acid	38	44	68	77	3	—	27[a]	49	45	44	21	19
1:4 Dimethyl histidine	38	28	48	62	63	52	68	60	24	15[a]	69	64
1-Histidinol phosphate ester	15	14	19	27	0	0	8	12	8	4	10	8

[a] = Streaking

Layer L1 = Cellulose MN-300.
L2 = Avicel.

Solvent S1 = *n*-butanol-acetic acid-water (60 : 15 : 25).
S2 = *n*-butanol-pyridine-water (60 : 60 : 60).
S3 = isopropanol-water-conc. ammonia (200 : 20 : 10).
S4 = *n*-propanol-0.2*N* ammonia (3 : 1).
S5 = *n*-propanol-1*N* acetic acid (3 : 1).
S6 = *n*-butanol-acetone-diethylamine-water (70 : 70 : 14 : 35).

Detection Ninhydrin, Ehrlich's reagent and diazotized sulfanilic acid.

Technique T1 = ascending development, 15 cm, 5 hours, 25 °C.
T2 = ascending development, 12 cm, 4.5 hours, 25 °C.
T3 = ascending development, 12 cm, 4.3 hours, 25 °C.
T4 = ascending development, 15 cm, 2.5 hours, 25 °C.
T5 = ascending development, 12 cm, 3 hours, 25 °C.

Literature
1. Smith, I., Rider, L. J., and R. P. Lerner, *J. Chromatogr.*, **26**, 449–455 (1967).

TABLE TLC 156
NITROGEN HETEROCYCLICS, INDOLES

Layer	L1	L1	L1	L1	L1	L1	L1					
Solvent	S1	S2	S3	S4	S5	S6	S7					
Detection								D1	D1	D2	D3	D3
Literature	1	1	2	2	2	2	2	2	1	1	1	1
Compound	$R_F \times 100$							Color	Limit of detection (μg)	Color	Color	Limit of detection (μg)
Ethyl indole-(3)-acetate	—	—	61	77	75	80	80	vt	—	—	—	—
Gramine	77	0	—	—	—	—	—	y to be	(1)	gr-bl	gr	0.5
5-Hydroxyindole-(3)-acetic acid	19	4	—	—	—	—	—	bl to vt	0.05	lt y to be	dk vt	0.05
3-Hydroxymethylindole	84	45	—	—	—	—	—	pk	0.1	y-pk	y	0.1
DL-5-Hydroxytryptophan	14	0	—	—	—	—	—	bl-gy	0.05	lt y to be	y	0.005
Indole	84	73	—	—	—	—	—	dk r to vt	0.1	pl gr	gr	0.01
Indole-3-acetaldehyde	86	46	—	—	—	—	—	r-br	1	y[b]	—	—
Indole-(3)-acetamide	—	—	16	29	59	54	57	bl to vt	—	—	—	—
Indole-3-acetic acid	31	28	0	1	29	6	11	bl-vt	0.05	y	y with gr edge	0.005
Indole-(3)-acetonitrile	85	46	49	70	75	73	78	gy	0.1	lt be	y	0.1
β-Indole-(3)-acrylic acid	33	29	—	—	—	—	—	pk	0.5	lt be	gr	0.1
Indole-3-aldehyde	81	20	29	—	—	68	—	y[a]	1	o[b]	—	—
γ-Indole-(3)-butyric acid	40	38	2	7	44	14	27	bl	0.1	y	y	0.01
Indole-3-carboxylic acid	—	—	4	10	57	17	34	r	—	—	—	—
Indole-(3)-ethanol	—	—	30	55	70	67	70	bl-gy (y rim)	—	—	—	—
β-Indole-(3)-propionic acid	38	34	1	3	37	10	19	bl	0.1	y	y	0.01
3-Indolylacetamide	83	12	—	—	—	—	—	bl	0.1	lt y to be	y	0.01
Isatin	75	27	—	—	—	—	—	stays o	1	—	—	—
DL-5-Methyltryptophan	28	0	—	—	—	—	—	bl	0.05	t	y-gr	0.005
Serotonin	65	0	—	—	—	—	—	gy	0.01	y	br	0.01
Skatole	87	78	—	—	—	—	—	bl	0.1	y	y	0.01
Tryptamine	77	0	—	—	—	—	—	bl-gr	0.1	y	y with bl edge	0.01
DL-Tryptophan	23	0	—	—	—	—	—	bl-gr	0.05	y	y with bl edge	0.005

[a] Stahl and Kaldewey (Ref. 1) obtained a pk color.
[b] 2,4-Dinitrophenylhydrazine reagent used.

Layer L1 = Silica gel G, 0.5 μg layer thickness, dried 30 minutes at 110 °C.

Solvent S1 = methyl acetate-isopropanol-25% NH_4OH (9 : 7 : 4).
S2 = chloroform-96% acetic acid (95 : 5).
S3 = chloroform-carbon tetrachloride-methanol (5 : 4 : 1).
S4 = $CHCl_3$-96% ethanol (9 : 1).
S5 = ethyl acetate-isopropanol-water (65 : 24 : 11).
S6 = chloroform-carbon tetrachloride-methanol (2 : 1 : 1).
S7 = chloroform-96% ethanol (13 : 7).

Detection D1 = Van Urk's reagent (*p*-dimethylaminobenzaldehyde reagent).
D2 = Prochazka reagent (formaldehyde reagent).
D3 = UV fluorescence with D2.

Technique Development distance = 10 cm; saturated chamber.

Literature
1. Stahl E. and H. Kaldewey, *Hoppe-Seyler's Z. Physiol. Chem.*, **323**, 182 (1961).
2. Ballin, G., *J. Chromatogr.*, **16**, 152 (1964).

TABLE TLC 157

NITROGEN COMPOUNDS, EXPLOSIVE MATERIALS, COMPONENTS OF

Layer	L1	L1	L1
Solvent	S1	S2	S3
Literature	1	1	1,2
Compound	$R_F \times 100$		
Ammonium nitrate	0	0	0
DINA[c]	16	42	56
2,4-Dinitrodiphenylamine	43	81	—
2,4′-Dinitrodiphenylamine	37	77	—
Diphenylamine	62	86	—
Dipicrylamine	0	0	6
***sym*-Hexanitrodiphenylamine**	5	5	—
Hexogen[b]	5	10	39
2-Nitrodiphenylamine	58	86	—
4-Nitrosodiphenylamine	7	28	—
***N*-Nitrosodiphenylamine**	46	81	—
Octogen[a]	4	0	23
Penthrit[e]	41	61	74
Picric acid	0	0	9
Tetryl[d]	26	46	62
Trinitrobenzene	40	62	71
Triphenylamine	75	91	—
Trotyl[f]	48	68	73

[a] Octogen = 1,3,5,7-Tetranitrotetramethylenetetramine.
[b] Hexogen = 1,3,5-Trinitrotrimethylenetriamine.
[c] DINA = Dinitroxydiethylnitramine.
[d] Tetryl = *N*-Methyl-*N*-2,4,6-tetranitroaniline.
[e] Penthrit = Pentaerythrityl tetranitrate.
[f] Trotyl = Trinitrotoluene.

Layer L1 = Silica gel G activated at 110 °C.
Solvent S1 = benzene.
S2 = chloroform.
S3 = petroleum ether-acetone (5 : 3).
Detection Diphenylamine, examine under UV.
Technique Ascending, 10 cm in about 30 minutes.
Literature
1. Hansson, J. and A. Alm, *J. Chromatogr.*, **9**, 385 (1962).
2. Hansson, J., *Explosivstoffe*, **10**, 73 (1963).

TABLE TLC 158

NITROGEN HETEROCYCLICS, PYRIDINE CARBOXYLIC ACID DERIVATIVES

Layer	L1	L1
Solvent	S1	S2
Literature	1	1
Compound	$R_F \times 100$	
Picolinic acid methyl ester	52	50
Nicotinic acid methyl ester	72	69
Isonicotinic acid ethyl ester	49	46
6-Methylpicolinic acid methyl ester	53	33
4-Methylpicolinic acid methyl ester	63	—
5-Ethylpicolinic acid methyl ester	67	48
5-Ethylpicolinic acid ethyl ester	52	15
5-Ethyl-4-chloropicolinic acid methyl ester	65	50
5-Ethylpicolinic acid amide	38	11
(−)-5-Ethylpicolinic acid α-phenylethylamide	85	65
(−)-Picolinic acid α-phenylethylamide	80	61
(−)-Isonicotinic acid α-phenylethylamide	34	22
(−)-Nicotinic acid α-phenylethylamide	53	40
(−)-2-Methylisonicotinic acid α-phenylethylamide	54	43

Layer L1 = Aluminum oxide, activity II, 50–200 mesh.
Solvent S1 = benzene-methanol (9 : 1).
S2 = benzene-methanol (25 : 1).
Detection Iodine vapor.
Literature
1. Kost, A. N., Golovleva, L. A., Terent'ev, P. B., and M. Islam, *Zh. Anal. Khim*, **21**, 860 (1966); from *J. Chromatogr.*, **26**, D4 (1967).

TABLE TLC 159
NUCLEIC ACID, FREE BASES AND NUCLEOSIDES

Layer	L1	L2	L3	L4	L4	L5	L6	L4	L6	L7	L8	L4	L4	L9	L9	L9	L9	L9	L10	L10	L10	L10
Solvent	S1	S1	S1	S2	S3	S3	S3	S4	S4	S5	S1	S6	S7	S8	S9	S10	S11	S12	S13	S14	S15	S16
Technique	T1	T1	T1	T2	T2	T2	T2	T2	T2	T3	T4	T5	T5	T6	T6	T6	T6	T6	T7	T7	T7	T7
Literature	1	1	1	2	2	2	2	2	2	3	4	5	5	6	6	6	6	6	7	7	7	7
Compound	$R_F \times 100$											R_{CMP}* × 100		$R_F \times 100$								
Adenine	29	30	38	98	20	33	44	14	13	—	8	55	98	31	88	36	40	10	75	85	54	51
Guanine	33	37	38	73	10	31	40	23	0	—	18	52	98	16	50	13	25	15	00	00	00	44
Hypoxanthine	46	55	57	—	—	—	—	—	—	—	23	—	—	—	—	—	—	—	48	52	37	74
Uracil	73	72	75	74	26	66	75	62	68	—	86	113	102	70	57	35	47	54	58	72	54	83
Cytosine	—	—	—	92	26	80	90	50	56	—	40	136	138	41	59	31	59	59	—	—	—	—
Thymine	—	—	—	83	41	74	85	52	54	—	81	—	—	80	65	57	57	44	—	—	—	—
Adenosine	56	53	75	91[a]	68	58	62	—	—	77	8	92	99	23	—	—	41	—	65	80	50	65
Guanosine	50	58	80	59[a]	13	31[b]	39[b]	—	—	—	77	84	98	22	50	9	26	27	22	17	17	74
Inosine	61	70	82	—	—	—	—	—	—	21	—	—	—	—	—	—	—	—	32	30	23	85
2-Deoxyadenosine	—	—	—	97[a]	66	70	70	—	—	—	—	—	—	27	82	37	42	9	—	—	—	—
2-Deoxyguanosine	—	—	—	73[a]	12	40	50	—	—	—	—	—	—	14	59	17	36	23	—	—	—	—
2-Deoxycytidine	—	—	—	90[a]	93	82	90	—	—	—	—	—	—	50	65	35	—	47	—	—	—	—
Cytidine	82	80	76	78[a]	96	79	94	—	—	55	55	139	138	36	65	27	—	52	30	37	23	90
Thymidine	—	—	—	81[a]	49	81	90	—	—	—	—	—	—	88	66	52	55	37	56	75	51	78
Uridine	84	81	85	63[a]	30	68	87	—	—	39	93	130	102	62	—	—	—	—	39	50	37	90
Xanthine	—	—	—	—	—	—	—	—	—	—	30	—	—	—	—	—	—	—	—	—	—	—
5-Methylcytosine	—	—	—	—	—	—	—	—	—	—	29	—	—	—	—	—	—	—	—	—	—	—
Methyldeoxycytidine	—	—	—	—	—	—	—	—	—	—	—	—	—	59	73	40	42	39	—	—	—	—
Deoxyuridine	—	—	—	—	—	—	—	—	—	—	—	—	—	77	61	33	47	45	—	—	—	—

* R_F relative to CMP (cytidine-5′-monophosphate).
[a] with 10% $CaSO_4$.
[b] Streaks.

Layer
L1 = ECTEOLA-cellulose.
L2 = Cellulose.
L3 = Silica gel G.
L4 = DEAE-cellulose.
L5 = DEAE-cellulose with 5% $CaSO_4$ binder.
L6 = DEAE-cellulose with 10% $CaSO_4$ binder.
L7 = Celite with 5% starch binder.
L8 = ITLC-SA.
L9 = Commercial grade microcrystalline cellulose (Avicel).
L10 = Polyamide.

Solvent
S1 = water (distilled).
S2 = isobutyric acid-conc. ammonium hydroxide-water (33 : 1 : 16).
S3 = 0.005*N* HCl.
S4 = saturated ammonium sulfate-1*N* sodium acetate-isopropanol (40 : 9 : 1).
S5 = isopropanol-water (9 : 1).
S6 = 75 ml of 1.0*M* NH_4HCO_3 in one chamber and 85 ml of 0.15*M* NH_4HCO_3 in the second chamber.
S7 = 75 ml of 2.0*M* ammonium formate, pH 4.2, in one chamber and 85 ml of 0.2*M* ammonium formate, pH 2.8, in the second chamber.
S8 = isopropanol-HCl-H_2O (65 : 16.7 : 18.3 v/v).
S9 = isobutyric acid-H_2O-25% NH_4OH (400 : 208 : 0.4 v/v).
S10 = isopropanol-H_2O-concentrated (28%) NH_4OH (85 : 15 : 1.3).

TABLE TLC 159—(Continued)

NUCLEIC ACID, FREE BASES AND NUCLEOSIDES

Solvent

S11 = *n*-butanol-H_2O-formic acid (77 : 13 : 10).
S12 = saturated $(NH_4)_2SO_4$-1*M* sodium acetate-isopropanol (80 : 18 : 2).
S13 = heptane-butanol-acetic acid (4 : 4 : 1), development time for 8 cm, 60 minutes.
S14 = carbon tetrachloride-acetic acid-acetone (4 : 1 : 4).
S15 = toluene-pyridine-ethylenechlorohydrin-0.8*N* NH_4OH (5 : 1 : 5 : 3 : 3 v/v).
S16 = 0.5*M* aqueous NaCl solution, 15 minutes.

Detection

UV light.

Technique

T1 = ascending development, 10 cm in 45 minutes.
T2 = 6″ development distance.
T3 = 12 cm development distance.
T4 = ascending, saturated chamber.
T5 = gradient elution, two-dimensional chromatography.
T6 = room temperature 18–22 °C, for two-dimensional, 11–13 cm, dry at 40 °C, then 8–9 cm in second direction.
T7 = Wang method.

Literature

1. Randerath, K., *Angew. Chem.*, **73**, 674 (1961) from *J. Chromatogr.*, **9**, D35 (1962).
2. Coffey, R. G. and R. W. Newburgh, *J. Chromatogr.*, **11**, 376 (1963).
3. Shasha, B. and R. L. Whistler, *J. Chromatogr.*, **14**, 532 (1964).
4. Haer, F. C., "An Introduction to Chromatography on Impregnated Glass Fiber," Ann Arbor Science Publishers, Inc., Ann Arbor, Michigan, 1969, p. 146.
5. Strickland, R. G., *Anal. Biochem.*, **10**, 116 (1965).
6. Grippo, B., Iaccarino, M., Rossi, M., and E. Scarano, *Biochim. Biophys. Acta*, **95**, 1 (1965).
7. Wang, K. T. and I. S. Y. Wang, *Biochim. Biophys. Acta*, **142**, 280 (1967).

TABLE TLC 160
NUCLEOTIDES

Layer		L1	L1	L2	L3	L4	L5	L5	L5	L5	L5	L6	L7	L7	L7	L8	L8	L8	L5	L5
Solvent		S1	S2	S3	S4	S4	S5	S2	S6	S7	S8	S9	S10	S11	S12	S13	S14	S15	S16	S17
Technique		T1	T1	T2	T3	T3	T3	T4	T4	T4	T4	T5	T5	T5	T5	T6	T6	T6	T7	T7
Literature		1	1	2	3	3	3	4	4	4	4	5,6	5,6	5,6	5,6	7	7	7	8	8
Compound		$R_F \times 100$		R_{ST}[a]								$R_F \times 100$							R_{CMP}	
Adenosine-5′-monophosphate	(5′-AMP)	57	26	—	—	—	—	45	65	—	—	38	35	52	41	—	—	—	—	—
Adenosine-monophosphate	(AMP)	—	—	42	—	—	—	—	—	—	—	—	—	—	—	22	47	54	76	94
Adenosine-3′-monophosphate	(3′-AMP)	48	—	—	4	24	63	—	—	—	—	—	—	—	28	—	—	—	—	—
Adenosine-2′-monophosphate	(2′-AMP)	—	—	—	4	32	63	—	—	—	—	—	—	—	35	—	—	—	—	—
Adenosine diphosphate	(ADP)	36	8	36	—	—	—	24	48	68	—	26	17	29	48	4	27	46	65	26
Adenosine triphosphate	(ATP)	21	—	32	—	—	—	6	11	20	56	16	8	16	56	0	16	37	63	6
Cytidine-5′-monophosphate	(5′-CMP)	74	31	—	—	—	—	46	65	—	—	34	30	48	—	—	—	—	—	—
Cytidine-3′-monophosphate	(3′-CMP)	71	—	—	5	50	51	—	—	—	—	—	—	—	—	—	—	—	—	—
Cytidine-2′-monophosphate	(2′-CMP)	—	—	—	9	61	55	—	—	—	—	—	—	—	—	—	—	—	—	—
Cytidine diphosphate	(CDP)	51	11	—	—	—	—	31	53	—	—	22	13	27	—	7	38	56	89	64
Cytidine triphosphate	(CTP)	34	—	—	—	—	—	9	13	31	64	13	7	13	—	0	21	47	86	18
Guanosine-5′-monophosphate	(5′-GMP)	55	14	—	—	—	—	—	—	—	—	—	—	—	—	—	—	—	—	—
Guanosine-3′-monophosphate	(3′-GMP)	44	—	—	—	—	—	—	—	—	—	—	—	—	49	—	—	—	—	—
Guanosine diphosphate	(GDP)	37	3	—	—	—	—	—	—	—	—	—	—	—	—	0	28	49	50	14
Guanosine triphosphate	(GTP)	17	—	—	—	—	—	—	—	—	—	—	—	—	—	0	15	35	50	2
Inosine-5′-monophosphate	(5′-IMP)	74	—	—	—	—	—	—	—	—	—	—	—	—	—	—	—	—	—	—
Thymidine-5′-monophosphate	(5′-TMP)	—	—	—	5	48	47	—	—	—	—	—	—	—	—	—	—	—	—	—
Thymidine monophosphate	(TMP)	—	—	110	—	—	—	—	—	—	—	—	—	—	—	—	—	—	—	—
Uridine-5′-monophosphate	(5′-UMP)	80	13	—	—	—	—	—	—	—	—	37	30	47	—	—	—	—	—	—
Uridine monophosphate	(UMP)	—	—	53	—	—	—	—	—	—	—	—	—	—	—	44	74	76	92	73
Uridine-3′-monophosphate	(3′-UMP)	75	—	—	2	42	36	—	—	—	—	—	—	—	—	—	—	—	—	—
Uridine-2′-monophosphate	(2′-UMP)	—	—	—	2	48	36	—	—	—	—	—	—	—	—	—	—	—	—	—
Uridine diphosphate	(UDP)	63	0	42	—	—	—	7	15	25	—	25	14	26	—	9	55	69	86	18
Uridine triphosphate	(UTP)	44	—	37	—	—	—	4	4	8	18	17	8	13	—	4	38	58	81	6
2-d-Adenosine-5′-monophosphate	(2-d-5′-AMP)	—	—	—	4	28	64	—	—	—	—	—	—	—	—	—	—	—	—	—
2-d-Guanosine-5′-monophosphate	(2-d-5′-GMP)	—	—	—	2	15	35	—	—	—	—	—	—	—	—	—	—	—	—	—
2-d-Cytidine-5′-monophosphate	(2-d-5′-CMP)	—	—	—	8	48	60	—	—	—	—	—	—	—	—	—	—	—	—	—
P′-[Adenosyl-(5′)-P² methylpyrophosphate	—	46	—	—	—	—	—	—	—	—	—	—	—	—	—	—	—	—	—	—
P′-[Adenosyl-(5′)-P² phenylpyrophosphate	—	32	—	—	—	—	—	—	—	—	—	—	—	—	—	—	—	—	—	—
2′-Guanosine monophosphate	(2′-GMP)	—	—	—	—	—	—	—	—	—	—	—	—	—	58	—	—	—	—	—
	(2′+3′-CMP)	—	—	—	—	—	—	—	—	—	—	—	—	—	73	—	—	—	—	—
	(2′+3′-UMP)	—	—	—	—	—	—	—	—	—	—	—	—	—	73	—	—	—	—	—
Uridine diphosphate glucose	(UDPG)	—	—	—	—	—	—	—	—	—	—	—	13	12	—	—	—	—	100	45
	(CMP)	—	—	—	—	—	—	—	—	—	—	—	—	—	—	34	61	64	100	100
	(GMP)	—	—	—	—	—	—	—	—	—	—	—	—	—	—	19	49	63	52	38

[a] R_{ST} values relative to inorganic phosphate.

TABLE TLC 160—(Continued)
NUCLEOTIDES

Layer

L1 = ECTEOLA-cellulose.
L2 = ECTEOLA-cellulose sprayed with 0.01*M* ammonium tetraborate (pH 9.0) and dried before development.
L3 = DEAE-cellulose with 5% $CaSO_4$ binder.
L4 = DEAE cellulose with 10% $CaSO_4$ binder.
L5 = DEAE cellulose.
L6 = MN-300 cellulose layer.
L7 = MN-300 G layer containing plaster of Paris.
L8 = ITLC-SAF impregnated with 1% polyethylenimine.

Solvent

S1 = 0.15*M* sodium chloride.
S2 = 0.01*N* HCl.
S3 = 95% ethanol-0.1*M* ammonium tetraborate pH 9.0 (3 : 2).
S4 = 0.005*N* HCl.
S5 = isobutyric acid-conc. ammonium hydroxide-water (33 : 1 : 16).
S6 = 0.02*N* HCl.
S7 = 0.03*N* HCl.
S8 = 0.04*N* HCl.
S9 = *n*-butanol-acetone-acetic acid-5% ammonium hydroxide-water (4.5 : 1.5 : 1 : 1 : 2).
S10 = *n*-butanol-acetone-acetic acid-5% ammonium hydroxide-water (3.5 : 2.5 : 1.5 : 1.5 : 1).
S11 = *tert* amyl alcohol-alcohol-formic acid-water (3 : 2 : 1).
S12 = saturated ammonium sulfate-1*M* sodium acetate-isopropanol (80 : 18 : 2).
S13 = 2*M* LiCl.
S14 = 1.0*M* LiCl.
S15 = LiCl (1.6*M*).
S16 = 75 ml of 1.0*M* NH_4HCO_3 in one chamber and 85 ml of 0.15*M* NH_4HCO_3 in the second chamber.
S17 = 75 ml of 2.0*M* ammonium formate, pH = 4.2 in one chamber and 85 ml of 0.2*N* ammonium formate, pH = 2.8 in the second chamber.

Technique

T1 = 15 minutes development time, 8–10 cm.
T2 = closed tank, 17 cm, 2 hours.
T3 = closed tanks or open trays when two-dimensional development, 6 inches.
T4 = 5–10 minutes.
T5 = 50–60 minutes.
T6 = stepwise development, S13—45 seconds, S14—6 minutes and S15 to 15 cm.
T7 = gradient elution, two-dimensional.

Detection

UV light; also see Section II.I, Volume Two.

Literature

1. Randerath, K., *Angew. Chem.*, **73**, 674 (1961).
2. Dietrich, C. P., Dietrich, S. M. C., and H. G. Pontis, *J. Chromatogr.*, **15**, 277 (1964).
3. Coffey, R. G. and R. W. Newburgh, *J. Chromatogr.*, **11**, 376 (1963).
4. Randerath, K., *Angew. Chem. Intern. Ed. Engl.*, **1**, 435 (1962).
5. Randerath, K. and H. Struck, *J. Chromatogr.*, **6**, 365 (1961).
6. Randerath, K., *Biochem. Biophys. Res. Commun.*, **6**, 452 (1961/62).
7. Haer, F. C., "*An Introduction to Chromatography on Impregnated Glass Fiber*," Ann Arbor Science Publishers, Inc., Ann Arbor, Mich., 1969, p. 169.
8. Strickland, R. G., *Anal. Biochem.*, **10**, 116 (1965).

TABLE TLC 161
ORGANOMERCURY COMPOUNDS, DITHIZONATES OF

Layer	L1	L1	L1	L1	L2	L2
Solvent	S1	S2	S3	S4	S2	S5
Literature	1	1	1	1	1	1
Dithizonates	$R_F \times 100$					
Methylmercury	64	48	57	77	89	86
Ethylmercury	64	51	62	78	91	87
Methoxyethylmercury	32	16	25	44	58	49
Ethoxyethylmercury	44	23	34	55	71	67
Phenylmercury	48	34	46	62	72	69
Tolylmercury	52	40	52	69	79	76
Mercury di-dithizonate	19	9	17	28	19	15

Layer L1 = Silica gel, 250 μ thickness.
L2 = Alumina, 250 μ thickness.

Solvent S1 = hexane-acetone (9 : 1).
S2 = hexane-acetone (19 : 1).
S3 = hexane-acetone (93 : 7).
S4 = light petroleum-acetone (9 : 1).
S5 = light petroleum-acetone (19 : 1).

Detection Dithizonates are yellow to red in color.

Literature
1. Tatton, J. O'G. and P. J. Wagstaffe, *J. Chromatogr.*, **44**, 286 (1969).

TABLE TLC 162
ORGANIC SULFUR COMPOUNDS, MERCAPTANS AND THIOPHENOLS

Layer	L1	L2
Solvent	S1	S2
Literature	1	1
Compound	$R_F \times 100$	
Amyl mercaptan	84	93
Hexyl mercaptan	84	84
Octyl mercaptan	84	69
Decyl mercaptan	84	59
Dodecyl mercaptan	84	43
Tetradecyl mercaptan	84	50
Thiophenol	71	—
Benzyl mercaptan	66	—
***p*-Thiocresol**	68	—
***o*-Thiocresol**	73	—
2-Methyl-5-isopropylthiophenol	78	—

Layer L1 = Activated aluminum oxide.
L2 = Aluminum oxide impregnated with 5% cetane.

Solvent S1 = petroleum ether (40–60 °C).
S2 = 96% acetic acid-acetonitrile (3 : 1).

Detection Bromthymol blue.

Literature
1. Prinzler, H. W., Pape, D., Tauchmann, H., Teppke, M., and C. Tzscharnke, *Ropa Uhluc.*, **8**, 14 (1966); from *J. Chromatogr.*, **32**, D11 (1968).

TABLE TLC 163
ORGANIC SULFUR COMPOUNDS, SULFIDES

Layer	L1	L1
Solvent	S1	S2
Literature	1	1
Compound	$R_F \times 100$	
Dipropyl sulfide	74	—
Dibutyl sulfide	69	—
Dioctyl sulfide	42	—
Didodecyl sulfide	21	—
Methyl octyl sulfide	67	—
Ethyl octyl sulfide	65	—
Propyl octyl sulfide	58	—
Hexyl octyl sulfide	51	—
Dioctyl sulfide	48	—
Decyl octyl sulfide	45	—
Dodecyl octyl sulfide	42	—
Methyl dodecyl sulfide	56	—
***o*-Tolyl ethyl sulfide**	—	75
***p*-Tolyl methyl sulfide**	—	79
***p*-Tolyl octyl sulfide**	—	43
2-Methyl-5-isopropylphenyl hexyl sulfide	—	42
Diphenyl sulfide	—	69
***p*-Tolyl phenyl sulfide**	—	72
2-Methyl-5-isopropylbenzyl *p*-tolyl sulfide	—	53
2-Methyl-5-isopropylbenzyl *o*-tolyl sulfide	—	51

Layer L1 = Aluminum oxide impregnated with 5% cetane.
Solvent S1 = chloroform-methanol (1 : 3).
S2 = chloroform-methanol-water (5 : 15 : 1).
Detection Potassium permanganate.
Literature 1. Prinzler, H. W., Pape, D., Tauchmann, H., Teppke, M., and C. Tzscharnke, *Ropa Uhluc*, **8**, 14 (1966); from *J. Chromatogr.*, **32**, D11 (1968).

TABLE TLC 164
ORGANIC SULFUR COMPOUNDS, SULFOXIDES

Layer	L1
Solvent	S1
Literature	1
Compound	$R_F \times 100$
Tetrahydrothiophene S-oxide	08
2-Methyl-1-thiadecalin S-oxide	21
(*cis* + *trans*)	37
1,3-Dihydroisothionaphthene S-oxide	08
Dibenzyl sulfoxide	20
Isopropyl phenyl sulfoxide	41
Allyl benzyl sulfoxide	19
***p*-Aminoazobenzene (R_F control)**	30
Dicyclohexyl sulfoxide	37
Diisoamyl sulfoxide	39
Di-*n*-hexyl sulfoxide	39
Di-*n*-decyl sulfoxide	53
Diphenyl sulfoxide	48
Di-α-naphthyl-sulfoxide	48
β-Naphthyl phenyl sulfoxide	49

Layer L1 = Aluminum oxide, activity II, loose layer.
Solvent S1 = Acetone-carbon tetrachloride (1 : 4).
Detection Iodine vapor.
Technique 1.5 hours development time.
Literature 1. Karaulova, E. N., Bobruiskaya, T. S., and G. D. Gal'pern, *Zh. Anal. Khim.*, **21**, 894 (1966); *J. Chromatogr.*, **25**, D13 (1966); *Chem. Abstr.*, **65**, 16046F (1966).

TABLE TLC 165
OXO COMPOUNDS, DERIVATIVES

Layer Solvent	L1 S1	L2 S2	L3 S3	L1 S4
Compound		R_{BY}[a] × 100		
2,4-Dinitrophenylhydrazones of:				
Hexanal	104	112	70	—
Heptanal	107	102	89	—
Nonanal	116	85	110	—
Isovaleraldehyde	95	135	64	—
2-Hexenal	93	86	54	—
2-Heptenal	100	80	65	—
2-Octenal	105	74	75	—
2-Nonenal	108	65	84	—
trans-2, *cis*-6-Nonadienal	102	81	53	—
Benzaldehyde	92	54	13	—
p-Tolualdehyde	122	33	25	—
Acetone	33	124	35	—
2-Butanone	87	133	70	—
2-Nonanone	108	104	116	—
6-Methyl-5-hepten-2-one	103	121	103	—
Acetophenone	98	141	42	—
l-Carvone	119	64	103	—
Menthone	116	84	117	—
p-Nitrophenylhydrazones of:				
Propionaldehyde	22	113	—	—
Butyraldehyde	27	110	—	—
Valeraldehyde	29	102	—	—
Hexanal	32	94	—	—
Heptanal	36	86	—	—
Nonanal	41	73	—	—
Decanal	43	63	—	—
Undecanal	46	61	—	—
Dodecanal	48	57	—	—
Isobutyraldehyde	33	115	—	—
Isovaleraldehyde	31	113	—	—
Crotonaldehyde	23	80	—	—
2-Hexenal	35	76	—	—
2-Nonenal	43	56	—	—
Benzaldehyde	31	52	—	—
p-Isopropylbenzaldehyde	41	51	—	—
Salicylaldehyde	15	35	—	—
Cinnamaldehyde	26	28	—	—
2-Butanone	22	115	—	—
2-Nonanone	35	87	—	—
6-Methyl-5-hepten-2-one	27	108	—	—
Acetophenone	31	58	—	—
l-Carvone	55	55	—	—
Menthone	50	80	—	—
2-4-Dinitrophenylsemicarbazones of:				
Valeraldehyde	—	100	—	54
Hexanal	—	88	—	59
Heptanal	—	77	—	61
Nonanal	—	68	—	65
Decanal	—	54	—	71
Undecanal	—	51	—	67
Dodecanal	—	42	—	68
Isobutyraldehyde	—	109	—	53
Isovaleraldehyde	—	111	—	56
2-Methylvaleraldehyde	—	101	—	64
Crotonaldehyde	—	82	—	51
2-Hexenal	—	69	—	62
2-Nonenal	—	54	—	68
Citral	—	61	—	80
	—	52	—	64
Citronellal	—	87	—	70
Benzaldehyde	—	27	—	52
p-Isopropylbenzaldehyde	—	21	—	57
Salicylaldehyde	—	22	—	31
Cinnamaldehyde	—	19	—	52
2-Butanone	—	84	—	54
2-Nonanone	—	54	—	67
6-Methyl-5-hepten-2-one	—	59	—	63
Acetophenone	—	27	—	58
l-Carvone	—	32	—	90
Menthone	—	59	—	80

[a] Relative to butter yellow dye.

Layer L1 = Silica gel G.
L2 = Polyamide.
L3 = Carbowax 600/Silica gel G.

Solvent S1 = benzene-petroleum ether (38–50 °C) (4 : 1).
S2 = methanol-water (95 : 5).
S3 = heptane-benzene (4 : 1).
S4 = heptane-benzene (4 : 1).

Detection Daylight and UV.

Technique Ascending, 10 cm, saturated tank.

Literature Tumlinson, J. H., Minyard, J. P., Hedin, P. A., and A. C. Thompson, *J. Chromatogr.*, **29**, 80–87 (1967).

TABLE TLC 166
OXO COMPOUNDS, DNPH-DERIVATIVES

Layer	L1	L1	L1	L1	L1	L1	L2	L2	L2	L2	L2	L2
Solvent	S1	S2	S3	S4	S5	S6	S7	S8	S9	S10	S11	S12
Technique	T1	T1	T1	T1	T1	T1	T2	T2	T2	T2	T2	T2
Literature	1	1	1	1	1	1	2	2	2	2	2	2
Compound[a]	$R_{Acetone} \times 100$							$R_{Formaldehyde} \times 100$				
2,4-DNPH of:												
Formaldehyde	70	52	52	74	80	84	100	100	100	100	100	100
Acetaldehyde	86	69	69	84	94	92	104	119	117	118	103	103
Propionaldehyde	97	95	96	96	98	100	119	170	140	154	122	127
Butyraldehyde	110	111	109	105	—	—	125	199	146	165	134	145
n-Valeraldehyde	116	119	121	108	—	—	131	214	151	182	139	158
Caproicaldehyde	122	126	126	112	—	—	—	—	—	—	—	—
Oenanthal	127	130	131	114	—	—	146	224	155	199	146	173
Caprylaldehyde	130	134	134	117	—	—	—	—	—	—	—	—
Glycolaldehyde	4	0	0	2	6	5	—	—	—	—	—	—
Glyoxal[b]	10	0	0	17	67	42	—	—	—	—	—	—
Glycollic acid	0	0	0	0	0	0	—	—	—	—	—	—
Acrylaldehyde	86	92	91	93	—	101	—	—	—	—	—	—
Glyceraldehyde	—	—	—	5	—	7	—	—	—	—	—	—
Dihydroxyacetone[b]	—	—	—	—	72	65	—	—	—	—	—	—
Acetone	100	100	100	100	100	100	100	155	119	130	115	114
Pyruvic acid	0	0	0	0	0	0	—	—	—	—	—	—
Mesoxalic acid (diethyl ester)	—	—	—	72	—	87	—	—	—	—	—	—
Isobutyraldehyde	—	—	—	—	—	—	126	207	147	169	135	152
Isovaleraldehyde	—	—	—	—	—	—	132	220	152	185	136	160
Acrolein	—	—	—	—	—	—	127	160	134	161	118	136
Crotonaldehyde	—	—	—	—	—	—	128	177	147	171	130	158
α-Methyl-β-ethylacrolein	—	—	—	—	—	—	157	243	161	207	160	186
Cyclopentanone	—	—	—	—	—	—	107	198	142	163	138	141
Cyclohexanone	—	—	—	—	—	—	126	222	147	174	146	157
Methyl ethyl ketone	—	—	—	—	—	—	131	212	134	167	134	147
Methyl propyl ketone	—	—	—	—	—	—	140	246	146	182	150	172
Methyl *tert*-butyl ketone	—	—	—	—	—	—	167	290	154	202	165	201
3-Methylcyclohexen-2-one	—	—	—	—	—	—	130	225	143	179	147	162

[a] 2,4-Dinitrophenylhydrazones.
[b] Values are for osazones.

TABLE TLC 166—(Continued)
OXO COMPOUNDS, DNPH-DERIVATIVES

Layer
L1 = Neutral alumina (Woelm).
L2 = Silica gel G.

Solvent
S1 = nitrobenzene-chloroform-hexane (1 : 2 : 8).
S2 = ether-petroleum ether (bp 40–70 °C) (2.5 : 7).
S3 = ethyl acetate-hexane (1 : 9).
S4 = nitrobenzene-carbon tetrachloride (1 : 9).
S5 = nitrobenzene-cyclohexane (1 : 2).
S6 = nitrobenzene-carbon tetrachloride (1 : 2).
S7 = benzene.
S8 = light petroleum (60–80 °C)-ether (7 : 3).
S9 = carbon tetrachloride-acetone (9 : 1).
S10 = chloroform-diethyl ether (8 : 2).
S11 = n-C_6H_{14}-methyl propionate (7 : 3).
S12 = C_6H_{14}-ethyl propionate (7 : 3).

Technique
T1 = 60 minutes development time.
T2 = 20–30 minutes development time.

Literature

1. Nano, G. M. and P. Sanein, *Experientia*, **19**, 323 (1963).
2. Zamojski, A. and F. Zamojska, *Chem. Anal.* (Warsaw) **9**, 589 (1964); from *J. Chromatogr.* **18**, D13 (1965) and *Chem. Abstr.* **62**, 8391h (1965).

TABLE TLC 167
OXO COMPOUNDS, 2,4-DINITROPHENYLHYDRAZONES OF

Layer	L1	L1
Solvent	S1	S1
Literature	1	1
Compound	$R_F \times 100$	$R_{Veratraldehyde} \times 100$
2,4-DNPH of:		
Acetosyringone	0	0
Acetovanillone	23	68
Acetylvanillin	33	97
Anisaldehyde	48	141
Benzaldehyde	60	177
Cinnamaldehyde	60	177
Coniferaldehyde	27	80
***p*-Coumaraldehyde**	33	97
2,4-Dihydroxyacetophenone	26	76
2,4-Dihydroxybenzaldehyde	23	68
2,5-Dihydroxybenzaldehyde	21	62
2,3-Dimethoxybenzaldehyde	47	138
2,4-Dimethoxybenzaldehyde	41	121
2,5-Dimethoxybenzaldehyde	45	133
3,5-Dimethoxybenzaldehyde	47	138
***p*-Ethoxybenzaldehyde**	53	156
3-Ethoxy-4-hydroxybenzaldehyde	32	94
4-Ethoxy-3-methoxybenzaldehyde	42	124
***m*-Hydroxybenzaldehyde**	32	94
***p*-Hydroxybenzaldehyde**	30	88
***p*-Hydroxybenzylacetone**	32	94
Isovanillin	23	68
***m*-Methoxybenzaldehyde**	50	147
***o*-Methoxybenzaldehyde**	49	144
***p*-Methoxybenzylacetone**	49	144
Protocatechuic aldehyde	2	6
Salicylaldehyde	48	141
Sinapaldehyde	15	44
Syringaldehyde	5	15
Vanillin	23	68
***o*-Vanillin**	32	94
Veratraldehyde	34	100

Layer L1 = Starch-bound silica gel layers.
Solvent S1 = ethyl acetate-petroleum ether (75–120 °C), (1 : 2).
Detection NH_3 vapors.
Technique 14 cm development distance.
Literature 1. Ruffini, G., *J. Chromatogr.*, **17**, 483 (1965).

TABLE TLC 168
OXO COMPOUNDS, 2-HYDRAZINEBENZOTHIAZOLE DERIVATIVES

Layer	L1		
Solvent	S1		
Detection		D1	D2
Literature	1		
Compound	$R_F \times 100$	Color	
2-Hydrazinebenzothiazole deriv. of:			
Formaldehyde	35	o	bl
Acetaldehyde	40	o	bl
Propionaldehyde	44	o	bl
Butyraldehyde	50	o	bl
Isobutyraldehyde	48	p	bl
2-Methylbutyraldehyde	52	p	bl
Hexyl aldehyde	55	o	bl
Heptyl aldehyde	56	o	bl
Octyl aldehyde	57	o	bl
Nonyl aldehyde	58	o	bl
Benzaldehyde	38	p	bl
***p*-Tolualdehyde**	36	p	bl
Cuminyl aldehyde	44	p	bl
Phenylacetaldehyde	40	o	bl
Hydrotropaldehyde	39	o	bl
Cinnamaldehyde	38	o	gr
Hydrocinnamaldehyde	36	o	bl
α-Amylcinnamaldehyde	52	o	bl
α-Hexylcinnamaldehyde	51	o	bl
Citral	49, 57	o	bl
Citronellal	54	o	bl
Hydroxycitronellal	9	o	bl
Salicylaldehyde	25	p	bl
Acrolein	35	r	bl
Isovalerylaldehyde	45	o	bl
Piperonal	23	pl gr	bl
Anisaldehyde	26	ol	bl

Layer L1 = Silica gel F_{254}, activated at 110 °C.
Solvent S1 = petroleum ether (30–40 °C)-ethyl acetate-acetic acid (88 : 10 : 2).
Detection D1 = *p*-Nitrobenzenediazonium fluoborate.
D2 = D1, then alc. KOH or NH_3.
Technique Ascending development, 15 cm distance.
Literature 1. Hunt, F. C., *J. Chromatogr.* **40**, 465–467 (1969).

TABLE TLC 169
OXO COMPOUNDS, QUINONES, NATURAL

Layer	L1	L1	L1
Solvent	S1	S2	S3
Technique	na	na	T1
Literature	1	1	2
Compound	$R_F \times 100$		
Plastoquinone A	81	74	89
Plastoquinone B	88	78	94
Plastoquinone C	0	49	58
Plastoquinone D	0	40	45
α-Tocopherylquinone	21	37	26
β-Tocopherylquinone	19	33	20
γ-Tocopherylquinone	16	25	18
δ-Tocopherylquinone	—	—	15

Layer L1 = Silica gel G.
Solvent S1 = 15% trichloroethyl acetate in benzene.
S2 = 10% ether in chloroform.
S3 = chloroform.
Detection UV absorbance (255–262 nm).
Technique T1 = 10 cm development distance, ascending, saturated tank.
Literature
1. Henninger, M. D. and F. L. Crane, *Biochemistry*, **2**, 1168 (1963); *Chem. Abstr*. **59**, 10468d (1963).
2. Dilley, R. A., *Anal. Biochem*., **7**, 240 (1964).

TABLE TLC 170
PESTICIDES, INSECTICIDES

Layer	L1	L2	L1	L2	L1	L2
Solvent	S1	S1	S2	S2	S3	S3
Literature	1	1	1	1	1	1
Compound	$R_F \times 100$					
Bromophos	55	—	83	—	47	85
Carbamult	27	83	59	76	19	36
	25	59	64	75	19	35
Carbaryl	18	44	67	76	10	24
	34	—	—	—	—	—
	32	65	71	76	20	33
Cidial	52	73	85	96	37	70
	26	54	79	80	16	44
Dicapthon	50	62	85	90	27	55
Dichlorvos	30	13	72	—	13	83
	—	57	—	—	—	—
Fenitrothion	55	73	77	78	41	70
Fenthion	31	50	74	89	18	79
	55	—	—	—	42	—
Iodofenphos	46	—	63	85	56	—
	68	—	85	—	—	—
	25	19	86	79	34	68
	48	65	—	—	—	—
	32	74	88	81	44	75
	55	—	—	—	—	—
Landrin	26	53	58	82	17	35
	25	58	55	76	17	30
Malathion	42	69	84	88	25	58
Mobam	26	34	52	81	12	21
Propoxur	26	38	52	76	14	32

Layer L1 = Silica gel G.
L2 = Aluminum oxide G.
Solvent S1 = hexane-acetone (3 : 1).
S2 = chloroform-acetone (9 : 1).
S3 = hexane-acetone (5 : 1).
Technique Saturated chamber, development distance = 12 cm; development time about 30 minutes.
Literature
1. Ramasamy, M., *Analyst*, **94**, 1078–79 (1969); from *J. Chromatogr*., **49**, D16 (1970).

TABLE TLC 171
PESTICIDES, INSECTICIDES, PHOSPHORUS

Layer	L1	L1	L1	L1	L1	L1	L1	L1	L1	L1	L1	L1	L1	L1	L1	L1	L1
Solvent	S1	S2	S3	S4	S5	S6	S7	S8	S9	S10	S11	S12	S13	S14	S15	S16	S17
Literature	1	1	1	1	1	1	1	1	1	1	1	1	1	1	1	1	1
Compound	$R_F \times 100$																
Bidrin®	0	0	0	0	0	0	0	0	41	40	19	2	0	9	1	25	—
Bromophos	34	81	83	75	68	71	65	74	78	74	78	79	69	78	88	82	46
Chlorthion	8	72	75	68	51	45	44	47	75	73	78	74	54	77	84	83	21 (27)
Cidial	3	70	69	61	39	25	29	26	77	73	76	78	63	77	86	83	29
Ciodrin®	0	14	14	10	0	0	0	0	74	69	68	41	13	62	47	76	—
Diazinon®	0	37	30	25	8	3	4	5	77	71	74	74	53	73	78	78	—
Dicapthon	9	73	77	69	53	47	47	49	73	72	77	74	55	76	84	83	21
Dichlorphos	0	41	20	16	4	2	0	2	74	66	65	44	18	61	47	72	—
Dimethoate	0	8	4	0	0	0	0	0	62	63	58	15	2	42	17	68	—
Disulfoton	10	77	75	69	55	42	46	49	78	75	77	81	70	78	88	82	51
Disulfoton S	0	8	0	0	0	0	0	0	52	56	46	8	0	31	6	60	—
Famphur	0	41	40	35	7	4	4	3	73	73	75	59	30	73	75	82	—
Fenitrothion	6	72	74	65	47	37	40	41	75	73	76	74	59	76	86	81	22
Fenthion	10	77	78	70	56	48	47	50	78	74	77	77	64	77	87	82	32
Formothion	0	27	29	26	5	2	3	3	74	72	74	54	20	72	72	80	—
Imidan®	0	52	50	43	14	8	9	7	75	72	74	65	35	72	77	80	—
Malathion	0	53	45	36	12	5	7	5	76	74	76	74	52	76	83	82	19 (29)
Meta-Systox®	6	71	70	60	40	27	31	31	78	75	77	80	66	77	86	83	39
Meta-Systox (i)	0	13	8	6	0	0	0	0	70	68	64	40	14	58	40	71	—
Meta-Systox R	0	0	0	0	0	0	0	0	17	21	6	0	0	0	0	12	—
Meta-Systox S	0	0	0	0	0	0	0	0	24	25	8	0	0	0	0	10	—
Methyl parathion	5	72	71	64	46	37	37	40	78	72	75	74	58	76	85	80	21
Mevinphos	0	9	6	4	0	0	0	0	67	65	61	32	10	53	32	69	—
Paraoxon	0	20	13	11	0	0	0	0	72	72	68	44	15	63	48	76	—
Parathion	6	73	76	66	51	40	45	45	78	75	78	79	67	78	88	81	31 (62)
Phenkapton	27	81	83	75	69	71	67	76	76	75	78	82	73	79	91	83	46 (71)
Phorate	16	76	80	75	58	52	54	58	78	74	77	81	72	78	89	80	51 (75)
Phosphamidon	0	6	0	0	0	0	0	0	63	60	51	10	0	33	7	62	—

TABLE TLC 171—(Continued)
PESTICIDES, INSECTICIDES, PHOSPHORUS

Layer	L1	L1	L1	L1	L1	L1	L1	L1	L1	L1	L1	L1	L1	L1	L1	L1	L1
Solvent	S1	S2	S3	S4	S5	S6	S7	S8	S9	S10	S11	S12	S13	S14	S15	S16	S17
Literature	1	1	1	1	1	1	1	1	1	1	1	1	1	1	1	1	1
Compound	$R_F \times 100$																
Thiocron	0	7	4	0	0	0	0	0	64	63	59	16	3	43	13	67	—
Thiometon	10	73	74	65	51	39	43	44	78	73	76	80	69	77	87	81	41 (61)
Trichlorphon	0	3	0	0	0	0	0	0	63	58	51	11	3	35	12	63	—
Vamidothion	0	3	0	0	0	0	0	0	52	52	35	3	0	16	0	47	—
Zinophos®	0	36	31	27	9	4	5	5	72	71	72	68	43	71	74	77	—

Layer L1 = Silica gel GF_{254}.

Solvent S1 = carbon tetrachloride.
S2 = chloroform + 1% ethanol.
S3 = methylene chloride.
S4 = 1,2-dichloroethane.
S5 = benzene.
S6 = chlorobenzene.
S7 = toluene.
S8 = xylene.
S9 = dioxane.
S10 = acetone.
S11 = methyl ethyl ketone.
S12 = diethyl ether.
S13 = diisopropyl ether.
S14 = ethyl acetate.
S15 = *n*-butyl acetate.
S16 = acetonitrile.
S17 = *n*-hexane-acetone (9 : 1).

Detection See Section II.I, Volume Two.

Literature
1. Guth, J. A., *Pflanzenschutz Ber.*, **35**, 138–139 (1967); from *J. Chromatogr.*, **32**, D12–D13 (1968).

TABLE TLC 172
PESTICIDES, ORGANOPHOSPHORUS

Layer	L1	L1	L1	L1	L1	L1	L1	L1	L1	L1	L1	L1	L2	L2	—	—	—	—
Solvent	S1	S2	S3	S4	S5	S6	S7	S8	S9	S10	S11	S12	S13	S14	—	—	—	—
Technique	T1	T1	T1	T1	T1	T1	T1	T1	T1	T1	T1	T1	T2	T2	—	—	—	—
Literature	1	1	1	1	1	1	1	1	1	1	1	1	2	2	2	2	2	2
															Color[a]			
Compound	$R_F \times 100$														Bromine	Fluorescein	*o*-toluidine	UV
Parathion	3	7	58	27	13	60	35	61	89	82	87	93	—	—	—	—	—	—
Methyl parathion	8	15	63	33	20	66	43	65	62	59	73	76	64	48	br	y	dk	dk gr
Sumithion® (MEP)	4	10	61	30	17	65	39	63	74	69	82	85	64	56	br	y	dk	dk gr
Baycid	1	3	49	23	10	61	33	55	91	87	91	95	—	—	—	—	—	—
ECP® (VC-13)	0	0	40	16	5	53	21	50	100	100	100	100	—	—	—	—	—	—
Cyanox (S-4084)	12	22	70	45	32	73	51	71	52	46	63	60	—	—	—	—	—	—
EPN	0	1	41	12	3	47	16	42	82	76	84	93	—	—	—	—	—	—
S-sevin	0	0	23	5	1	37	6	32	100	97	96	100	—	—	—	—	—	—
Surecide (CYP)	0	2	50	22	8	56	25	54	75	68	79	88	—	—	—	—	—	—
Methyl demeton (Meta-Systox®)	63	68	89	81	75	88	81	89	63	46	61	79	—	—	—	—	—	—
Demeton (Systox)	45	57	81	79	69	85	77	84	85	70	82	89	—	—	—	—	—	—
Disyston (Dithio-Systox)	4	8	57	36	20	64	40	65	100	100	100	100	—	—	—	—	—	—
Thiometon (Ekatin)	10	17	63	41	28	69	48	67	100	94	96	100	—	—	—	—	—	—
Cidial	3	7	57	36	18	68	41	67	94	87	91	100	—	—	—	—	—	—
Malathion	19	30	75	59	46	75	62	80	81	78	88	87	78	85	br	y	dk	dk gr
Estox	95	91	93	95	94	97	95	95	10	3	11	7	—	—	—	—	—	—
Dimethoate	60	65	82	76	69	86	78	84	9	3	9	7	—	—	—	—	—	—
Methyl paraoxon	50	61	85	73	65	87	77	86	25	15	31	26	—	—	—	—	—	—
Oxo-Sumithion	39	53	83	69	59	85	76	85	34	22	40	38	—	—	—	—	—	—
DDVP	—	—	—	—	—	—	—	—	—	—	—	—	81	92	—	—	—	br
Diazinon®	—	—	—	—	—	—	—	—	—	—	—	—	78	67	br	y	dk	dk gr
Ethyl parathion	—	—	—	—	—	—	—	—	—	—	—	—	65	46	br	y	dk	dk gr
Ethion	—	—	—	—	—	—	—	—	—	—	—	—	58	32	br	y	dk	dk gr
Lindane	—	—	—	—	—	—	—	—	—	—	—	—	45	26	—	—	—	dk br
Endrin	—	—	—	—	—	—	—	—	—	—	—	—	40	21	—	—	—	br
Aldrin	—	—	—	—	—	—	—	—	—	—	—	—	36	20	—	—	—	br

TABLE TLC 172—(Continued)
PESTICIDES, ORGANOPHOSPHORUS

Layer	L1	L1	L1	L1	L1	L1	L1	L1	L1	L1	L1	L1	L2	L2	—	—	—	—
Solvent	S1	S2	S3	S4	S5	S6	S7	S8	S9	S10	S11	S12	S13	S14	—	—	—	—
Technique	T1	T1	T1	T1	T1	T1	T1	T1	T1	T1	T1	T1	T2	T2	—	—	—	—
Literature	1	1	1	1	1	1	1	1	1	1	1	1	2	2	2	2	2	2
															Color[a]			
Compound	$R_F \times 100$														Bromine	Fluorescein	*o*-toluidine	UV
DDT	—	—	—	—	—	—	—	—	—	—	—	—	31	7	—	—	—	br
Kelthane®	—	—	—	—	—	—	—	—	—	—	—	—	21	31	—	—	—	br
Sevin®	—	—	—	—	—	—	—	—	—	—	—	—	68	55	br	y	dk	gy
Zineb	—	—	—	—	—	—	—	—	—	—	—	—	0	78	br	y	dk	dk gr

[a] — Indicates negative reaction to detection reagent; colors given are for spots at each stage when reagents are used in the indicated order.

Layer
L1 = Polyamide powder "Wakō Polyamide B-10", 0.25 mm thickness, dried at 60 °C for one hour.
L2 = Polyamide layer sheets.

Solvent
S1 = H_2O-AcOH (9 : 1).
S2 = H_2O-AcOH (8 : 2).
S3 = H_2O-AcOH (5 : 5).
S4 = H_2O-MeOH (5 : 5).
S5 = H_2O-MeOH (6 : 4).
S6 = H_2O-acetone (5 : 5).
S7 = H_2O-acetone (6 : 4).
S8 = H_2O-AcOH-MeOH (5 : 1 : 4).
S9 = *n*-hexane-AcOH (9 : 1).
S10 = *n*-hexane-AcOH (95 : 5).
S11 = petroleum ether-AcOH (95 : 5).
S12 = cyclohexane-AcOH (95 : 5).
S13 = acetone-ethanol-water (2 : 2 : 4).
S14 = ethanol-ammonia-water (5 : 2 : 4).

Detection
5% (v/v) bromine solution; 0.25% (w/v) fluorescein sodium solution; 1% (w/v) o-toluidine solution; after each spraying, dry chromatogram and observe by UV exposure.

Technique
T1 = ascending development, 24–26 °C, for 12 cm.
T2 = ascending development.

Literature

1. Nagasawa, K. and H. Yoshidome, *J. Chromatogr.*, **39**, 282–290 (1969).
2. Wang, R.-T. and S. S. Chou, *J. Chromatogr.*, **42**, 416–418 (1969).

TABLE TLC 173
PESTICIDES, UREA HERBICIDES

Layer	L1	L2	L2	L2	L2	L2	L2	L2	L2	L2	L2	L2	L2	L2	L3
Solvent	S1	S2	S3	S4	S5	S6	S7	S8	S9	S10	S11	S12	S13	S14	S15
Technique	T1	T2	T2	T2	T2	T2	T2	T2	T2	T2	T2	T2	T2	T2	T3
Literature	1	2	2	2	2	2	2	2	2	2	2	2	2	2	2
Compound	$R_F \times 100$														
Fluometuron	40	14	7	22	21	43	50	39	58	4	34	15	63	22	53
Monolinuron	—	48	24	48	58	71	63	62	70	11	44	41	81	38	58
Buturon	—	15	19	53	57	66	57	60	75	7	40	44	84	31	54
Monuron	31	12	6	15	18	42	49	33	51	3	27	11	54	22	60
Metobromuron	71	51	25	51	59	72	61	64	72	11	43	43	82	38	53
Diuron	38	14	7	16	20	42	51	39	58	4	34	17	64	22	46
Linuron	73	58	32	51	62	72	68	68	74	11	48	44	86	43	41
Metoxymarc	—	13	7	16	21	42	44	38	57	4	33	13	64	23	44
Neburon	62	38	17	50	54	68	61	59	75	10	49	38	85	32	18
Benzomarc	—	14	7	15	19	70	48	34	58	5	35	14	78	30	42
Chlorbromuron	—	56	33	52	63	72	73	68	74	10	48	44	85	46	38
3-4-DCA[a]	81	—	—	—	—	—	—	—	—	—	—	—	—	—	—
***p*-CA**[a]	77	—	—	—	—	—	—	—	—	—	—	—	—	—	—
Chloroxuron	32	—	—	—	—	—	—	—	—	—	—	—	—	—	—
3,4-DCPMOU[a]	27	—	—	—	—	—	—	—	—	—	—	—	—	—	—
Fenuron	22	—	—	—	—	—	—	—	—	—	—	—	—	—	—
3,4-DCPMU[a]	11,2	—	—	—	—	—	—	—	—	—	—	—	—	—	—
***p*-CPMU**[a]	9	—	—	—	—	—	—	—	—	—	—	—	—	—	—
***p*-CPU**[a]	1	—	—	—	—	—	—	—	—	—	—	—	—	—	—

[a] Urea Metabolites
- 3,4-DCA = 3,4-dichloroaniline
- *p*-CA = *p*-chloroaniline.
- 3,4-DCPMOU = 3-(3,4-dichlorophenyl)-1-methoxyurea
- 3,4-DCPMU = 3-(3,4-dichlorophenyl)-1-methylurea
- *p*-CPMU = 3-(*p*-chlorophenyl)-1-methylurea
- *p*-CPU = 3-(*p*-chlorophenyl)urea

Layer
- L1 = Aluminum oxide GF.
- L2 = Silica gel Merck $PF_{254-366}$.
- L3 = L2 impregnated with paraffin.

Solvent
- S1 = dioxane-butyl ether-benzene (1 : 20 : 40).
- S2 = chloroform.
- S3 = dichloromethane.
- S4 = diethyl ether.
- S5 = dichloromethane-diisopropyl ether (4 : 1).
- S6 = chloroform-ethanol (19 : 1).
- S7 = dichloromethane-ethanol (19 : 1).
- S8 = chloroform-acetone (9 : 1).
- S9 = chloroform-acetone (7 : 3).
- S10 = hexane-acetone (9 : 1).
- S11 = hexane-acetone (7 : 3).
- S12 = acetone-*n*-hexane-benzene (1 : 2 : 5).
- S13 = chloroform-nitromethane (1 : 1).
- S14 = chloroform-glacial acetic acid (60 : 1).
- S15 = ethanol-water (4 : 6).

Detection Bratton–Marshall reagent.

Technique
- **T1** = 15 cm, ascending development, saturated tank.
- **T2** = 15 cm, ascending development.
- T3 = 15 cm, reversed-phase.

Literature

1. Onley, J. H. and George, Yip. *J. Ass. Offic. Agr. Chem.*, **52**, 546 (1969).
2. Hance, R. J., *J. Chromatogr.*, **44**, 419–421 (1969).

TABLE TLC 174
PHENOLIC COMPOUNDS

Layer	L1	L1	L1	L1	L1	L1	L1	L2	L2	L2	L2	L2	L3	L3	L3	L4	L4	L5	L6	
Solvent	S1	S2	S3	S4	S5	S6	S7	S8	S9	S10	S11	S12	S3	S13	S14	S13	S15	S16	S17	
Detection	—	—	—	—	—	—	—	—	—	—	—	—	—	—	—	—	—	—	—	D1
Technique	T1	T1	T1	T1	T2	na	na	T1	T1	T1	T1	T1	na	na	na	na	na	T3	T3	
Literature	1 & 2	1 & 2	2 & 3	2 & 3	4	5	5	6[a]	6[a]	6[a]	6[a]	6[a]	7	7	7[a]	8[a]	8[a]	9[a]	9[a]	2
Compound	$R_F \times 100$																			Color
m-Aminophenol	—	—	—	—	—	—	—	—	—	—	—	—	—	—	—	28	27	—	—	—
o-Aminophenol	—	—	—	—	—	—	—	—	—	—	—	—	—	—	—	7	4	—	—	—
p-Aminophenol	—	—	—	—	—	—	—	—	—	—	—	—	—	—	—	37	33	—	—	—
Anisaldehyde	79	66	—	65	—	—	—	—	—	—	—	—	—	—	—	—	—	—	—	o
Anisic acid	84	69	—	—	—	—	—	—	—	—	—	—	—	—	—	—	—	—	—	—
Benzaldehyde	68	62	—	38	—	—	—	—	—	—	—	—	—	—	—	—	—	—	—	pl y
Benzoic acid	74	—	—	—	—	—	—	—	—	—	—	—	—	—	—	—	—	—	—	—[b]
Benzoquinone	65	69	—	—	—	—	—	—	—	—	—	—	—	—	—	—	—	—	—	—
p-Benzoquinone	—	—	—	—	—	—	—	71	80	55	65	97	—	—	—	—	—	—	—	y (own color)
m-Bromphenol	—	—	—	—	—	—	—	—	—	—	—	—	12	9	7	—	—	—	—	—
p-Bromphenol	—	—	—	—	—	—	—	—	—	—	—	—	13	12	70	—	—	—	—	—
2-*tert*-Butyl-4-hydroxyanisole	—	—	—	—	—	59	38	—	—	—	—	—	—	—	—	—	—	—	—	—
3-*tert*-Butyl-4-hydroxyanisole	—	—	—	—	—	62	45	—	—	—	—	—	—	—	—	—	—	—	—	—
Caffeic acid	24	43	—	—	—	—	—	—	—	—	—	—	—	—	—	—	—	—	—	y-ish br
Catechol	58	54	—	—	42	—	—	45	55	42	35	80	0	0	32	—	—	—	—	gy-gr
m-Chlorophenol	—	—	—	—	—	—	—	—	—	—	—	—	12	9	72	—	—	—	—	—
p-Chlorophenol	—	—	—	—	—	—	—	—	—	—	—	—	13	10	74	—	—	—	—	—
Cinnamic acid	63	65	—	—	—	—	—	—	—	—	—	—	—	—	—	—	—	—	—	—[b]
o-Coumaric acid	47	62	—	—	—	—	—	—	—	—	—	—	—	—	—	—	—	—	—	r-ish br
p-Coumaric acid	49	52	—	—	—	—	—	—	—	—	—	—	—	—	—	—	—	—	—	gy-br
m-Cresol	—	—	19	45	57	—	—	—	—	—	—	—	21	30	78	—	—	—	—	y
o-Cresol	—	—	26	52	—	—	—	—	—	—	—	—	30	37	81	—	—	—	—	y
p-Cresol	—	—	20	49	—	—	—	—	—	—	—	—	21	32	78	—	—	—	—	o-y
o,o′-Dihydroxybiphenyl	71	66	5	35	—	—	—	—	—	—	—	—	—	—	—	—	—	—	—	—
1,3-Dihydroxynaphthalene	70	80	0	28	—	—	—	—	—	—	—	—	—	—	—	—	—	—	—	—
2,3-Dimethylanisole	—	—	82	—	—	—	—	—	—	—	—	—	—	—	—	—	—	—	—	—
2,4-Dimethylanisole	—	—	83	—	—	—	—	—	—	—	—	—	—	—	—	—	—	—	—	—
2,5-Dimethylanisole	—	—	80	—	—	—	—	—	—	—	—	—	—	—	—	—	—	—	—	—
2,6-Dimethylanisole	—	—	68	—	—	—	—	—	—	—	—	—	—	—	—	—	—	—	—	—
3,4-Dimethylanisole	—	—	73	—	—	—	—	—	—	—	—	—	—	—	—	—	—	—	—	—
3,5-Dimethylanisole	—	—	75	—	—	—	—	—	—	—	—	—	—	—	—	—	—	—	—	—

TABLE TLC 174—(Continued)
PHENOLIC COMPOUNDS

Layer	L1	L1	L1	L1	L1	L1	L1	L2	L2	L2	L2	L2	L3	L3	L3	L4	L4	L5	L6	—
Solvent	S1	S2	S3	S4	S5	S6	S7	S8	S9	S10	S11	S12	S3	S13	S14	S13	S15	S16	S17	—
Detection	—	—	—	—	—	—	—	—	—	—	—	—	—	—	—	—	—	—	—	D1
Technique	T1	T1	T1	T1	T2	na	na	T1	T1	T1	T1	T1	na	na	na	na	na	T3	T3	
Literature	1 & 2	1 & 2	2 & 3	2 & 3	4	5	5	6[a]	6[a]	6[a]	6[a]	6[a]	7	7	7[a]	8[a]	8[a]	9[a]	9[a]	2
Compound	$R_F \times 100$																			Color
2,3-Dimethylphenol	—	—	27	59	—	—	—	—	—	—	—	—	—	—	—	—	—	—	—	o
2,4-Dimethylphenol	—	—	24	51	—	—	—	—	—	—	—	—	—	—	—	—	—	—	—	lt-y
2,5-Dimethylphenol	—	—	28	58	—	—	—	—	—	—	—	—	—	—	—	—	—	—	—	y
2,6-Dimethylphenol	—	—	39	67	—	—	—	—	—	—	—	—	—	—	—	—	—	—	—	lt-y
3,4-Dimethylphenol	—	—	15	44	—	—	—	—	—	—	—	—	—	—	—	—	—	—	—	br-ish y
3,5-Dimethylphenol	—	—	17	44	—	—	—	—	—	—	—	—	—	—	—	—	—	—	—	o-y
o,p-Dinitrophenol	—	—	—	—	—	—	—	—	—	—	—	—	—	—	—	0	0	—	—	—
Ethyl gallate	36	38	—	—	—	—	—	—	—	—	—	—	—	—	—	—	—	—	—	gy-br
Ethyl *m*-hydroxybenzoate	76	61	—	—	—	—	—	—	—	—	—	—	—	—	—	—	—	—	—	y
2-Ethylphenol	—	—	—	—	—	—	—	—	—	—	—	—	—	—	—	—	—	40	15	—
3-Ethylphenol	—	—	—	—	—	—	—	—	—	—	—	—	—	—	—	—	—	48	20	—
4-Ethylphenol	—	—	—	—	—	—	—	—	—	—	—	—	—	—	—	—	—	45	19	—
Ethyl protocatechuate	50	62	—	—	—	—	—	—	—	—	—	—	—	—	—	—	—	—	—	br
Ferulic acid	50	58	—	—	—	—	—	—	—	—	—	—	—	—	—	—	—	—	—	gy-ish br
Gallic acid	18	23	—	—	—	—	—	—	—	—	—	—	—	—	—	—	—	—	—	lt br
Gentisic acid	30	40	—	—	—	—	—	—	—	—	—	—	—	—	—	—	—	—	—	o
Guaicol (Guaiacol)	83	72	—	—	63	—	—	—	—	—	—	—	—	—	—	—	—	—	—	r-ish-br
α-Hydrinone	85	75	—	—	—	—	—	—	—	—	—	—	—	—	—	—	—	—	—	—
Hydroquinone	54	46	—	—	19	—	—	71	80	55	65	97	0	2	50	—	—	—	—	y
p-Hydroxybenzaldehyde	63	56	—	17	—	—	—	—	—	—	—	—	—	—	—	—	—	—	—	br
m-Hydroxybenzoic acid	49	51	—	—	—	—	—	—	—	—	—	—	—	—	—	—	—	—	—	—
p-Hydroxybenzoic acid	55	60	—	—	—	—	—	—	—	—	—	—	—	—	—	—	—	—	—	lt y
o-Hydroxybiphenyl	87	71	46	71	—	—	—	—	—	—	—	—	—	—	—	—	—	—	—	—
p-Hydroxybiphenyl	76	62	21	54	—	—	—	—	—	—	—	—	—	—	—	—	—	—	—	—
p-Iodophenol	—	—	—	—	—	—	—	—	—	—	—	—	13	11	70	—	—	—	—	—
Isoferulic (hesperitic) acid	43	66	—	—	—	—	—	—	—	—	—	—	—	—	—	—	—	—	—	carmine
m-Isopropylphenol	—	—	—	—	—	—	—	—	—	—	—	—	36	42	81	—	—	—	—	—
o-Isopropylphenol	—	—	—	—	—	—	—	—	—	—	—	—	43	47	85	—	—	—	—	—
p-Isopropylphenol	—	—	—	—	—	—	—	—	—	—	—	—	33	39	78	—	—	—	—	—
Isovanillin	56	47	—	23	—	—	—	—	—	—	—	—	—	—	—	—	—	—	—	y-ish br
Kojic acid	6	6	—	—	—	—	—	—	—	—	—	—	—	—	—	—	—	—	—	lt r
p-Methoxyphenol	—	—	—	—	—	47	24	—	—	—	—	—	—	—	—	—	—	—	—	—

TABLE TLC 174—(Continued)
PHENOLIC COMPOUNDS

Layer	L1	L1	L1	L1	L1	L1	L1	L2	L2	L2	L2	L2	L3	L3	L3	L4	L4	L5	L6	
Solvent	S1	S2	S3	S4	S5	S6	S7	S8	S9	S10	S11	S12	S3	S13	S14	S13	S15	S16	S17	
Detection	—	—	—	—	—	—	—	—	—	—	—	—	—	—	—	—	—	—	—	D1
Technique	T1	T1	T1	T1	T2	na	na	T1	T1	T1	T1	T1	na	na	na	na	na	T3	T3	
Literature	1 & 2	1 & 2	2 & 3	2 & 3	4	5	5	6[a]	6[a]	6[a]	6[a]	6[a]	7	7	7[a]	8[a]	8[a]	9[a]	9[a]	2
Compound	$R_F \times 100$																			Color
4-Methyl gentisate	71	61	—	—	—	—	—	—	—	—	—	—	—	—	—	—	—	—	—	gy
2-Methylphenol	—	—	—	—	—	—	—	—	—	—	—	—	—	—	—	—	—	62	29	—
3-Methylphenol	—	—	—	—	—	—	—	—	—	—	—	—	—	—	—	—	—	67	32	—
4-Methylphenol	—	—	—	—	—	—	—	—	—	—	—	—	—	—	—	—	—	66	32	—
α-Naphthol	81	67	28	60	63	54	31	—	—	—	—	—	14	12	71	—	—	—	—	bl-vt
β-Naphthol	79	63	18	56	58	39	17	—	—	—	—	—	12	10	68	—	—	—	—	bl-vt
o-Nitro-*p*-bromphenol	—	—	—	—	—	—	—	—	—	—	—	—	—	—	—	9	0	—	—	—
o-Nitro-*p*-hydroxyphenol	—	—	—	—	—	—	—	—	—	—	—	—	—	—	—	4	5	—	—	—
o-Nitro-*p*-methoxyphenol	—	—	—	—	—	—	—	—	—	—	—	—	—	—	—	30	16	—	—	—
o-Nitro-*p*-methylphenol	—	—	—	—	—	—	—	—	—	—	—	—	—	—	—	32	20	—	—	—
m-Nitrophenol	—	—	—	—	—	—	—	—	—	—	—	—	—	—	—	5	10	—	—	—
o-Nitrophenol	—	—	—	—	—	—	—	—	—	—	—	—	100	100	100	30	5	—	—	—
p-Nitrophenol	—	—	—	—	—	—	—	—	—	—	—	—	2	3	51	3	3	—	—	—
Phenol	76	60	17	43	70	13	2	54	72	42	12	83	16	18	75	—	—	79	50	y
o-Phenylphenol	—	—	—	—	—	21	88	—	—	—	—	—	—	—	—	—	—	—	—	—
Phlorogluernol	34	32	—	—	3	—	—	—	—	—	—	—	0	0	4	—	—	—	—	vt
Protocatechuic acid	32	39	—	—	—	—	—	—	—	—	—	—	—	—	—	—	—	—	—	br
Protocatechuic aldehyde	32	45	—	6	—	—	—	—	—	—	—	—	—	—	—	—	—	—	—	gy-br
Pyrogallol	32	45	—	—	15	—	—	—	—	—	—	—	0	0	20	—	—	—	—	dk br
Resorcinol	56	52	—	—	22	—	—	35	40	18	2	22	0	0	29	—	—	—	—	r-ish br
Resorcinol dimethyl ether	87	87	—	—	—	—	—	—	—	—	—	—	—	—	—	—	—	—	—	—
β-Resorcylic acid	54	52	—	—	—	—	—	—	—	—	—	—	—	—	—	—	—	—	—	rust r
Salicylaldehyde	82	83	—	91	—	—	—	—	—	—	—	—	—	—	—	—	—	—	—	vt[b]
Salicylic acid	64	68	—	—	—	—	—	—	—	—	—	—	—	—	—	—	—	—	—	—[c]
Syringaldehyde	60	57	—	14	—	—	—	—	—	—	—	—	—	—	—	—	—	—	—	o
Syringic acid	48	60	—	—	—	—	—	—	—	—	—	—	—	—	—	—	—	—	—	o
Thymol	—	—	—	—	—	28	6	—	—	—	—	—	64	57	87	—	—	—	—	—
Umbelliferone	55	58	—	—	—	—	—	—	—	—	—	—	—	—	—	—	—	—	—	—
Vanillic acid	54	61	—	—	—	—	—	—	—	—	—	—	—	—	—	—	—	—	—	—
Vanillin	70	64	—	27	—	—	—	—	—	—	—	—	—	—	—	—	—	—	—	r-ish br
o-Vanillin	74	59	—	53	—	—	—	—	—	—	—	—	—	—	—	—	—	—	—	y
Veratraldehyde	80	72	—	45	—	—	—	—	—	—	—	—	—	—	—	—	—	—	—	br
Veratric acid	73	70	—	—	—	—	—	—	—	—	—	—	—	—	—	—	—	—	—	r-ish vt
Veratrole	80	75	—	—	—	—	—	—	—	—	—	—	—	—	—	—	—	—	—	—

[a] Additional data presented in cited reference.
[b] Visualized with alkaline permanganate.

TABLE TLC 174—(Continued)
PHENOLIC COMPOUNDS

Layer

L1 = Silica gel G layer.
L2 = Silica gel G (Merck) impregnated with silver nitrate.
L3 = Polyamide layers.
L4 = Loose layers of aluminum oxide (Brockmann activity = III).
L5 = Cellulose powder Camlab MN300 impregnated with 0.75% (with respect to cellulose powder) ethyl oleate in diethyl ether.
L6 = Cellulose powder Camlab MN300 impregnated with 5% (with respect to cellulose powder) ethyl oleate in diethyl ether.

Solvent

S1 = benzene-dioxane-acetic acid (90 : 25 : 4).
S2 = benzene-methanol-acetic acid (45 : 8 : 4).
S3 = benzene.
S4 = benzene-methanol (95 : 5).
S5 = benzene-acetic acid (5 : 1).
S6 = cyclohexane-isopropyl ether-chloroform (7 : 1 : 7).
S7 = cyclohexane-chloroform (1 : 1).
S8 = benzene-methanol-acetic acid (8 : 1 : 1).
S9 = benzene-dioxane-acetic acid (8 : 1 : 1).
S10 = benzene-acetic acid (8 : 2).
S11 = benzene saturated with concentrated ammonium hydroxide solution.
S12 = chloroform-acetic acid (9 : 1).
S13 = chloroform.
S14 = ethyl acetate.
S15 = diethyl ether.
S16 = ethanol-water (1 : 3).
S17 = ethanol-water (3.75 : 6.25).

Detection Technique

D1 = Diazotized benzidine. See also Section II.I, Volume Two.
T1 = ascending development distance = 10 cm.
T2 = development distance = 12 cm.
T3 = saturated chamber, development time about $2\frac{1}{4}$ hours.

Literature

1. Pastuska, G., *Z. Anal. Chem.*, **179**, 355 (1961).
2. Pastuska, G. and H-J. Petrowitz, *Chemiker Ztg.*, **86**, 311 (1962).
3. Petrowitz, H-J., *Erdoel Kohle*, **14**, 923 (1961).
4. Seeboth, H., *Chem. Tech.*, (Berlin) **15**, 34 (1963).
5. Kondo, T. and I. Kawashiro, *Shokuhin Eisigabu Zasshi*, **6**, 436 (1965).
6. Aly, O. M., *Z. Anal. Chem.*, **234**, 253 (1968).
7. Wang, K. T., *J. Clin. Chem. Soc.*, (Taiwan) **8**, 241 (1961).
8. Heřmánek, S., Schwarz, V., and Z. Cekan, *Pharmazie*, **16**, 566 (1961).
9. Bark, L. S. and R. J. T. Graham, *Talanta*, **13**, 1282–83 (1966).

TABLE TLC 175
PHENOLS, RUBBER COMPOUNDING INGREDIENTS

Layer	L1
Solvent	S1
Literature	1

Compound	Trade name	$R_F \times 100$	Color
Hydroquinone monobenzyl ether	Age Rite Alba	7	bl
2,6-Di-*tert*-butylphenol	Ethyl 701	72	vt
2,6-Di-*tert*-butyl-4-methyl-phenol	Ionol	74	creamy y
2,6-Di-*tert*-butyl-α-methoxy-4-methylphenol	Ethyl 762	30	vt-y
2-α-Methylcyclohexyl-4,6-dimethylphenol	Nonox WSL	64	creamy y
Butylated hydroxyanisole	Tenox BHA®	22	bl
Butylated hydroxytoluene	Tenox BHT®	74	creamy y
4,4′-Bis(2,6-di-*tert*-butylphenol)	Ethyl 712	74	y
2,2′-Methylenebis(4-methyl-6-*tert*-butylphenol)	Antioxidant 2246	62	y
2,2′-Methylenebis(4-ethyl-6-*tert*-butylphenol)	Antioxidant 425	59	gy-bl
2,2′-Methylenebis(6-α-methyl-cyclohexyl-4-methylphenol)	Nonox WSP	62	y
4,4′-Methylenebis(6-*tert*-butyl-2-methylphenol)	Ethyl 720	41	r-vt
4,4′-Methylenebis(2,6-di-*tert*-butylphenol)	Ethyl 702	76	r-br
4,4′-Butylidenebis(6-*tert*-butyl-3-methylphenol)	Santowhite Powder	19	bl
2,2′-Thiobis(4-methyl-6-*tert*-butylphenol)	CAO-6	72	y-br
4,4′-Thiobis(6-*tert*-butyl-2-methylphenol)	Ethyl 736	45	r-br
4,4′-Thiobis(6-*tert*-butyl-3-methylphenol)	Santowhite Crystals	28	bl-vt
4,4′-Thiobis(3,6-di-*sec*-amylphenol)	Santowhite L	64	bl and gy
***n*-Butyl-*p*-aminophenol**	Tenamene 1	3	bl
Mixture of *tert*-octyl and *tert*-butyl cresols	Wingstay T	73	vt
Fortified phenol	Zalba Special	69	r-br
Alkylated phenol	Naugawhite	69, 60	y
Alkylated phenol	Nevastain B	64, 29	bl
Alkylated phenol	Stabilite White	76, 59, 50	bl-gy and br
Alkylated phenol	Age Rite Superlite	69, 54, 47, 43, 39, 30, 26, 18, 9	vt, bl and gy
Substituted styrenated phenol	Phenox	58, 26, 10	bl
Substituted styrenated phenol	Wingstay S	58, 26, 10	bl
Substituted styrenated phenol	Age Rite Spar	58, 26, 10	bl
Substituted styrenated phenol	Wingstay V	67, 61, 56, 50, 44, 28, 17, 9	bl and cream
Tri(nonylated phenyl) phosphite	Polygard	63, 52, 48, 39, 22, 12	bl, gy, salmon, w
Phosphited polyalkylpolyphenol	Age Rite Geltrol	76, 69, 63, 48, 41, 19, 12, 3	bl, gy, salmon, vt

Layer L1 = Silica gel G, layer thickness between 250 to 300 μ.
Solvent S1 = benzene.
Detection 2,6-Dichloridequinonechlorimide.
Technique Saturated tank. 15 cm development.
Literature 1. Kreiner, J. G. and W. C. Warner, *J. Chromatogr.*, **44**, 315–330 (1969).

TABLE TLC 176
PHENOLS, ANTIOXIDANTS

Layer	L1	L1
Solvent	S1	S2
Literature	1	1
Compound	$R_F \times 100$	
p-tert-Butylphenol	67	59
p-tert-Butylpyrocatechol	25	53
p-Cyclohexylphenol	70	33
3,4-Di-*tert*-butylpyrocatechol	73	63
Gallic acid	1	32
Hydroquinone	72	5
Hydroxyhydroquinone	18	95
p-Methylphenol	57	10
3-Methylpyrocatechol	24	70
4-Methylpyrocatechol	20	69
α-Naphthol	38	10
β-Naphthol	42	37
p-tert-Octylhomopyrocatechol	33	36
p-tert-Octylhydroxybenzoquinone	46	23
2-*tert*-Octylphenol	54	25
4-*tert*-Octylphenol	73	35
p-tert-Octylpyrocatechol	56	56
p-tert-Octylpyrogallol	43	29
Phenol	67	37
o-Phenylphenol	76	37
p-Phenylphenol	46	36
Phloroglucinol	54	8
Pyrocatechol	29	73
Pyrogallol	67	4
Resorcinol	50	62

Layer L1 = Loose-layer polyamide.
Solvent S1 = MeOH-CCl_4 (1 : 9.
S2 = MeOH-H_2O (6 : 4).
Literature 1. Davidek, J. and Pokorný, *Rev. Univ. Ind.* Sandtander (Colombia), **4**, 111 (1962).

TABLE TLC 177
STEROID GLYCOSIDES, DIGITALIS GLYCOSIDES AND AGLYCONES

Layer	L1	L1
Solvent	S1	S2
Technique	T1	T1
Literature	1	1
Compound	$R_D{}^a \times 100$	
Acetyldigitoxin-1 (α?)	132	195
Acetyldigitoxin-2 (β?)	142	203
Acetyldigoxin-α	100	138
Acetyldigoxin-β	113	157
Acetylgitoxin-α	104	139
Acetylgitoxin-β	117	147
Digitalinum verum	6	2
Digitonin	0	0
Digitoxigenin	141	217
Digitoxigenin-mono-digitoxose	127	141
Digitoxin	100	100
Digoxigenin	99	122
Digoxin	68	62
Gitaloxigenin	143	205
Gitaloxin	101	92
Gitoxigenin	110	128
Gitoxigenin-mono-digitoxose	102	86
Gitoxin	75	59
Glucogitaloxin	13	4
Lanadoxin	130	136
Lanatoside A	26	18
Lanatoside B	16	12
Lanatoside C	15	8
Odoroside H	76	149
Purpureaglycoside A	14	5
Purpureaglycoside B	9	2
Stropeside	50	91
Verodoxin	77	141

[a] R_F relative to digitoxin.

Layer L1 = Silica gel layers dried at 110 °C for 30 minutes.
Solvent S1 = ethyl acetate-methanol-water (16 : 1 : 1).
S2 = chloroform-pyridine (6 : 1).
Detection *p*-anisaldehyde-perchloric acid spray.
Technique T1 = Two-dimensional separation. S1 in first dimension, S2 in second dimension; heat to 90–100 °C for five minutes between developments.
Literature 1. Sjoehelm, I., *Svensk Farm. Tidskr.*, **66**, 321 (1962); *Chem. Abstr.* 57, 13888i.

TABLE TLC 178
STEROIDS

Layer	L1	L1	L1	L1	L1	L1	L1	L1							
Solvent	S1	S2	S3	S4	S5	S6	S7	S8							
Detection	—	—	—	—	—	—	—	—	D1	D1	D2	D2	D3	D3	D4
Literature	1	1	1	1	1	1	1	1							
									Color						
Compound[a]	$R_F \times 100$								UV	Daylight	UV	Daylight	UV	Daylight	Daylight
Testosterone propionate	31	65	72	75	83	86	<80	<80	lt-gr-y	gr-gy	bt-o	d-gy-br	bt-p-r	bt-bl	d-gr
Androstane-3,17-dione	30	62	72	72	80	82	<80	<80	pl-vt	y-br	bt-bl-gr	gy-gr	bt-vt	bt-bl-gr	o-y
Progesterone	28	59	71	68	77	80	<80	<80	pl-bl-vt	ft-y	bt-y	bt-y	bt-gr-bl	pl-y-br	o-br
Etiocholane-3,17-dione	17	59	69	70	79	75	<80	<80	br-y	y-br	d-y	y-gr	pl-vt	—	d-br
1-Dehydro-testosterone propionate	18	58	68	70	75	73	<80	<80	r-br	bt-r	bt-br-vt	o-r	bt-y-br	pk	r-vt
Estrone	13	56	58	63	72	65	<80	<80	bt-o	r	bt-y-o	br-r	ft-pl-vt	pk	vt
6-Chlor-6-dehydro-17α-hydroxyprogesterone acetate	20	55	70	75	75	75	<80	<80	d-gr-bl	d-gr	d-gr-bl	d-bl-br	bt-y-gr	gr-bl	p-vt
Cholesterol	25	52	69	67	—	73	<80	<80	bt-br-r	gy	bt-o-r	pk-r	bt-vt-bl	vt-bl	br-r
4-Androstene-3,17-dione	18	50	69	65	70	70	<80	<80	d-gy-gr	gr-bl	bt-r	d-br-r	pl-vt	—	br-r
Dihydro-testosterone	19	48	68	67	67	68	<80	<80	d-gr-bl	d-gr-bl	rt-r	d-gy-gr	ft-pl-gr	—	be
Etiocholane-3β-ol-17-one	16	48	55	53	66	68	<80	<80	pl-y-gr	ft-y-br	d-gy-gr	gr-br	pl-vt	bt-bl	d-gr-bl
Pregnenolone	17	41	52	58	66	66	<80	<80	bt-vt	y-br	bt-vt	vt	d-gy-bl	gy-bl	vt
4-Pregnene-3α-ol-20-one	18	40	49	60	65	67	<80	<80	vt	bt-vt-gr	pl-vt	br-gy	bt-br-r	gy-bl	d-vt
Pregnane-3,20-dione-21-ol	18	37	53	59	62	66	<80	<80	be	ft-y	y-gr	y-gr	pl-vt	—	y-br
5-Androstene-3α-ol-17-one (Dehydroandrosterone)	16	37	48	54	55	63	<80	<80	bt-vt	br-vt	p-y-o	vt-r	pk	p	r-vt
Dehydro-epiandrosterone	13	36	48	53	55	56	<80	<80	bt-vt	bt-vt	p-y-or	vt-r	pk-r	p	r-vt
Androstane-3α-ol-17-one androsterone	14	34	46	52	53	65	<80	<80	y-br	bt-be	d-gr-gy	r-br	bt-r-vt	gr-gy	vt
17α-Hydroxyprogesterone	11	31	47	60	54	64	<80	<80	d-br-y	be	bt-vt-bl	bl	bt-vt-r	y-gr	br-r
17α-Methyltestosterone	12	30	46	50	54	69	<80	<80	pl-y-br	y-br	bt-y-br	br	bt-p-r	p	br-r
Estromone (diethyl stilbestrol, no steroid)	4	28	45	45	54	31	<80	<80	d-bl	d-br-vt	br-vt	r-vt	p	d-vt-br	bl
Estradiol	5	32	35	47	45	56	<80	<80	bt-y	y-o	bt-gr	bt-r	bt-y-br	bt-bl-r	gr
11-Desoxy-corticosterone	15	28	45	52	55	70	<80	<80	d-gr-gy	d-gy-br	bt-br-r	d-gy-bl	d-br	br-r	r-vt
Etiocholane-3α-ol-17-one	9	27	38	45	50	60	<80	<80	ft-gr-gy	ft-bt-gy	d-br	pl-bt-r	pl-vt	bt-bl	gr
6-Dehydro-17α-hydroxyprogesterone	12	27	42	59	59	66	<80	<80	bt-y-gr	y-br	bt-gr-br	gy-gr	bt-r-br	gy	br-r
4-Androstene-3,11,17-trione	19	26	60	56	58	78	<80	<80	ft-gr-gy	ft-bt-gy	pl-gy	bt-y-gy	ft-pl-vt	—	be
Testosterone	15	26	45	48	50	66	<80	<80	d-gr-bl	d-gr-bl	bt-y-o	gy-br	bt-y-br	bl	br-r
Allopregnane-3α,20α-diol	10	22	39	44	46	62	71	<80	pl-pk	ft-pl-r	ft-pl-y	ft-br-r	pl-vt	bt-gy-bl	bt-gr-bl
Pregnane-3α,20α-diol	5	12	24	34	35	61	65	75	pl-gy-gr	ft-bt-gy	bt-y-gr	bt-br-gy	pl-y-gr	bt-gy	p
Cortisone acetate	—	10	30	44	58	60	55	68	vt	y-br	pl-vt	bt-br	pl-gr	—	br-r
Prednisone acetate	—	8	29	42	54	56	55	70	d-br	br	d-br-r	br-vt	pl-vt	—	br-r
16-Methylene-1-dehydro-11β,17α-dihydroxy-progesterone	—	5	18	34	45	42	43	35	r-br	pk-r	r-br	pk-r	r-vt	vt-bl	r-vt

TABLE TLC 178—(Continued)
STEROIDS

Layer	L1	L1	L1	L1	L1	L1	L1	L1							
Solvent	S1	S2	S3	S4	S5	S6	S7	S8							
Detection	—	—	—	—	—	—	—	—	D1	D1	D2	D2	D3	D3	D4
Literature	1	1	1	1	1	1	1	1							
									Color						
Compound	$R_F \times 100$								UV	Daylight	UV	Daylight	UV	Daylight	Daylight
Hydrocortisone acetate	—	4	20	36	45	55	44	38	y-br	gr-gy	bt-gr-y	y-gy	pl-y	y-br	br-r
Prednisolone acetate	—	—	15	30	38	49	36	29	gr-gy	br	d-br-r	br-vt	br	vt-br	gr-gy
Cortisone	—	—	10	20	23	31	20	15	pl-vt	bt-y	ft-y	bt-gr	ft-gr-bl	—	bt-br-r
Prednisone	—	—	8	17	18	18	16	12	d-br	d-be	d-br	vt-br	bt-br	bt-be	bt-br
Dexamethasone	—	—	6	11	8	16	10	10	d-br	br	d-br	d-bl-gr	bt-br	bt-be	bt-br
Estriol	—	—	4	9	15	17	13	19	bt-o	r-br	bt-vt-r	br-r	ft-pl-vt	bt-gr	d-vt
Hydrocortisone	—	—	3	6	8	23	8	11	y-br	y-br	d-br	bt-br	bt-br	—	bt-br-r
Prednisolone	—	—	—	4	5	14	8	8	y-br	br	d-br	br	bt-y-br	bt-br	d-gr
16-Methylene prednisolone	—	—	—	3	12	17	8	11	br-r	r-br	d-br	r-br	br	br-vt	r-vt
Aldosterone	—	—	—	6	25	12	20	22	bt-gr	y-br	d-br	br	pl-gr	—	gr

[a] Esters are printed in italics.

Layer L1 = Silica gel G.

Solvent
S1 = chloroform.
S2 = chloroform-ethyl acetate (80 : 20).
S3 = chloroform-acetone (90 : 10).
S4 = chloroform-acetone (80 : 20).
S5 = cyclohexane-chloroform-glacial acetic acid (70 : 20 : 10).
S6 = methylene chloride-acetone (80 : 20).
S7 = chloroform-glacial acetic acid (90 : 10).
S8 = methylene chloride-glacial acetic acid (90 : 10).

Detection
D1 = phosphoric acid reagent.
D2 = acetic anhydride-H_2SO_4 reagent.
D3 = antimony trichloride reagent.
D4 = vanillin-H_2SO_4 reagent.

Literature
1. Waldi, D., "Steroids", *Thin-Layer Chromatography*, Stahl, E., Academic Press, New York, 1965, pp 262–263.

TABLE TLC 179
STEROIDS

Layer	L1	L1	L1	L1	L1	L1	L1	L1	L1	L1	
Solvent	S1	S2	S3	S4	S5	S6	S7	S8	S9	S10	
Detection	—	—	—	—	—	—	—	—	—	—	D1
Literature	1	1	1	1	1	1	1	1	1	1	
Compound	$R_F \times 100$										Color
5α-Androstan-11-one	—	69	—	87	—	72	72	—	66	64	none
5β-Androstan-11-one	—	69	—	—	—	73	71	—	66	64	none
5α-Androstan-17-one	75	73	78	—	75	78	76	80	66	55	bl
5α-Androstane-3,17-dione	63	46	74	77	66	63	69	72	53	17	ol
5β-Androstane-3,17-dione	64	43	74	78	67	60	67	69	49	14	br
5α-Androst-1-ene-3,17-dione	61	38	76	75	66	58	67	67	45	13	vt bl
Androst-4-ene-3,17-dione	—	—	—	—	—	—	—	—	—	8	salmon
5α-Androst-2-ene-7,17-dione	67	52	77	81	70	61	70	74	55	27	ol
5β-Androstane-3,11,17-trione	52	25	69	70	66	50	58	56	36	4	r-br
5α-Androstan-17-ol	64	54	64	78	62	57	57	73	46	31	bl
17β-Hydroxy-5α-androstan-3-one	57	35	61	74	57	42	49	62	31	—	br-ol to bk-ol
17β-Hydroxy-5β-androstan-3-one	55	30	61	73	59	42	46	63	28	—	bt vt
17β-Hydroxy-androst-4-en-3-one	48	23	58	66	56	49	43	52	24	4	r-o to vt p
3α-Hydroxy-5α-androstan-17-one	55	33	63	74	58	43	50	62	29	—	gr
3β-Hydroxy-5α-androstan-17-one	52	30	58	72	55	39	44	61	27	—	gr
3α-Hydroxy-5β-androstan-17-one	49	24	58	71	56	38	43	57	23	—	gr
3β-Hydroxy-5β-androstan-17-one	57	36	62	73	59	44	50	66	33	—	gr
11β-Hydroxy-5α-androstane-3,17-dione	57	27	63	75	64	44	46	59	27	—	br
17β-Hydroxy-5α-androstane-3,17-dione	40	12	56	57	49	29	36	42	18	—	br
3α-Hydroxy-5β-androstane-11,17-dione	38	9	52	61	55	33	31	43	18	—	br-ol
3α-Hydroxy-5α-androstane-7,17-dione	40	12	53	54	47	27	32	40	16	—	gr-ol
3β-Hydroxy-5α-androstane-7,17-dione	33	6	50	49	45	23	27	34	13	—	gr-ol
5α-Androstane-3α,17β-diol	50	25	50	70	48	32	32	56	18	—	bl
5α-Androstane-3β,17β-diol	48	25	47	68	49	25	29	54	16	—	bl
Androst-4-ene-3β,17β-diol	50	25	47	70	50	30	32	58	17	—	dk bl
5β-Androstane-3α,17β-diol	44	17	40	65	48	23	23	50	13	—	bl
3β,11β-Dihydroxy-5α-androstan-17-one	44	14	43	68	51	28	22	51	13	—	gr
3α,11β-Dihydroxy-5β-androstan-17-one	43	12	44	66	52	29	23	48	13	—	dk-gy-bl

Layer L1 = Silica gel G.

Solvent
S1 = ethyl acetate-cyclohexane-ethanol (9 : 9 : 2).
S2 = ethyl acetate-cyclohexane (1 : 1).
S3 = chloroform-ethanol (9 : 1).
S4 = ethyl acetate-*n*-hexane-ethanol-acetic acid (72 : 13.5 : 4.5 : 10).
S5 = benzene-ethanol (4 : 1).
S6 = benzene-ethanol (9 : 1).
S7 = chloroform-ethanol (19 : 1).
S8 = ethyl acetate-*n*-hexane-acetic acid (15 : 4 : 1).
S9 = benzene-ethanol (19 : 1).
S10 = *n*-hexane-ethyl acetate (3 : 1).

Detection D1 = Anisaldehyde-sulfuric acid.

Technique 15 cm development distance, ascending, saturated tank.

Literature
1. Lisboa, B. P., *J. Chromatogr.*, **13**, 391 (1964).

TABLE TLC 180
STEROIDS, CHOLANIC ACIDS

Layer / Solvent / Literature	L1 S1 1	L1 S2 1	L1 S3 1	L1 S4 1	L1 S5 1	L1 S6 1	L1 S7 1	L1 S8 1	L1 S9 1	L1 S10 1	L1 S11 1	L1 S12 1	L1 S13 1	L1 S14 1	L1 S15 1	L1 S16 1	L1 S17 1
Reference compound[a]	R_D	R_L	R_D	R_C	R_C	R_C	R_C	R_C × 100	R_C	R_C	R_D	R_L	R_D	R_D	R_L	R_L	R_L
Reference mobility[b]	5.0	9.7	5.4	4.7	8.5	4.0	5.2	5.1	9.0	4.7	9.8	12.0	9.1	5.7	9.1	8.9	7.5
Cholanic acid[c]																	
3α,7α,12α,23ξ	—	13	4	17	12	7	13	28	6	8	—	—	—	—	—	—	—
3α,7α,23ξ	—	—	15	62	43	25	85	43	26	28	—	—	—	—	—	—	—
3α,7α,16α	—	—	67	208	141	250	214	254	162	179	88	—	72	—	—	—	—
3α,7α,12α-C$_{27}$	—	—	13	152	128	161	196	188	138	189	85	—	—	—	—	—	—
3α,7α,12α	34	34	17	100	100	100	100	100	100	100	76	50	22	16	—	—	—
3β,7α,12α	—	—	—	127	114	144	100	100	87	100	—	—	—	—	—	—	—
3α,7β,12α	—	—	30	146	117	147	159	175	122	140	—	—	—	—	—	—	—
3α,6α,7α	—	—	30	135	105	130	147	149	112	136	—	—	—	—	—	—	—
3α,6β,7α	—	—	—	115	98	117	117	114	82	83	—	—	—	—	—	—	—
3α,6β,7β	—	—	—	125	89	111	112	86	92	87	—	—	—	—	—	—	—
7α,12α,3-keto	—	—	131	282	155	294	228	302	166	204	87	81	63	126	—	—	—
3α,12α,7-keto	—	—	45	195	115	192	183	232	137	163	66	62	44	60	—	—	—
3α,7α,12-keto	—	66	45	195	115	194	185	232	137	163	66	62	48	—	—	—	—
3α,7,12-diketo	116	85	107	256	130	258	194	265	152	139	54	53	52	—	—	—	—
3α,7α	100	—	100	—	—	—	—	—	—	—	91	82	88	100	—	—	—
3β,7α	—	—	126	—	—	—	—	—	—	—	85	73	95	133	—	—	—
3α,7β	—	—	114	—	—	—	—	—	—	—	77	77	78	100	—	—	—
3α,12α	100	86	100	262	128	259	348	368	167	233	100	83	100	100	21	—	—
3β,12α	—	—	136	—	—	—	—	—	—	—	96	81	106	140	—	—	—
3α,12β	—	—	158	—	—	—	—	—	—	—	100	87	109	153	34	—	—
3α,6α	—	46	63	186	112	180	202	224	124	170	67	71	50	56	10	—	—
7α,12α	—	—	251	—	—	—	—	—	—	—	—	98	162	223	80	84	—
3,7,12-triketo	158	100	213	—	—	—	—	—	—	—	55	46	91	178	46	41	41
3α,7-keto	138	90	163	—	—	—	—	—	—	—	88	70	95	144	45	42	—
3α,12-keto	138	—	178	—	—	—	—	—	—	—	103	80	117	—	55	54	—
7α,3-keto	—	99	213	—	—	—	—	—	—	—	—	93	135	211	74	—	—
12α.3-keto	—	—	185	—	—	—	—	—	—	—	122	87	145	—	54	—	—
3,7-diketo	156	108	265	—	—	—	—	—	—	—	110	85	—	—	93	91	93
3,12-diketo	156	108	265	—	—	—	—	—	—	—	117	91	—	—	93	91	95
3α	156	100	254	—	—	—	—	—	—	—	128	100	160	227	100	100	100
3β	—	—	—	—	—	—	—	—	—	—	—	104	—	—	113	108	115
7α	—	112	320	—	—	—	—	—	—	—	—	108	—	—	135	124	140

TABLE TLC 180—(Continued)
STEROIDS, CHOLANIC ACIDS

Layer	L1	L1	L1	L1	L1	L1	L1	L1	L1	L1	L1	L1	L1	L1	L1	L1	L1
Solvent	S1	S2	S3	S4	S5	S6	S7	S8	S9	S10	S11	S12	S13	S14	S15	S16	S13
Literature	1	1	1	1	1	1	1	1	1	1	1	1	1	1	1	1	1
Reference compound[a]	R_D	R_L	**222** R_D	R_C	R_C	R_C	R_C	R_C × 100	R_C	R_C	R_D	R_L	R_D	R_D	R_L	R_L	R_L
Reference mobility[b]	5.0	9.7	5.4	4.7	8.5	4.0	5.2	5.1	9.0	4.7	9.8	12.0	9.1	5.7	9.1	8.9	7.5
7β	—	—	—	—	—	—	—	—	—	—	—	101	—	—	119	110	121
12α	—	—	—	—	—	—	—	—	—	—	—	112	171	—	137	—	140
12β	—	—	—	—	—	—	—	—	—	—	—	112	—	—	128	—	132
3-keto	162	113	310	—	—	—	—	—	—	—	—	112	—	—	150	121	135
7-keto	—	—	—	—	—	—	—	—	—	—	—	116	—	—	162	127	150
12-keto	—	—	—	—	—	—	—	—	—	—	—	116	—	—	162	127	150
Unsubst.	—	—	—	—	—	—	—	—	—	—	—	—	—	—	173	158	189

[a] R_C = Cholic acid, R_D = Deoxycholic acid, R_L = Lithocholic acid.
[b] Distance in cm traveled by the reference compound.
[c] Hydroxyl groups are indicated by Greek letters. A coprostanic acid is indicated by -C_{27}.

Layer L1 = Silica gel G.

Solvent
S1 = diethyl oxalate-dioxane (40 : 10).
S2 = diethyl oxalate-isopropyl alcohol (48 : 8).
S3 = benzene-dioxane-acetic acid (75 : 20 : 2.0).
S4 = benzene-dioxane-acetic acid (20 : 10 : 2.0).
S5 = benzene-dioxane-acetic acid (15 : 5 : 2.0).
S6 = benzene-dioxane-acetic acid (55 : 40 : 2.0).
S7 = cyclohexane-ethyl acetate-acetic acid (10 : 15 : 4.0).
S8 = cyclohexane-ethyl acetate-acetic acid (7 : 23 : 3.0).
S9 = benzene-isopropyl alcohol-acetic acid (30 : 10 : 1.0).
S10 = cyclohexane-isopropyl alcohol-acetic acid (30 : 10 : 1.0).
S11 = trimethylpentane-isopropyl alcohol-acetic acid (30 : 10 : 1.0).
S12 = trimethylpentane-isopropyl alcohol-acetic acid (60 : 20 : 0.5).
S13 = trimethylpentane-ethyl acetate-acetic acid (10 : 10 : 2.0).
S14 = trimethylpentane-ethyl acetate-acetic acid (5 : 25 : 0.2).
S15 = trimethylpentane-ethyl acetate-acetic acid (50 : 50 : 0.7).
S16 = trimethylpentane-ethyl acetate-acetic acid (10 : 10 : 0.25).
S17 = trimethylpentane-ethyl acetate-acetic acid (10 : 10 : 0.1).

Detection H_2SO_4.

Technique 17–18 cm development distance in unsaturated system.

Literature
1. Eneroth, P., *J. Lipid Res.*, **4**, 11 (1963).

TABLE TLC 181
STEROIDS, ESTROGENS

Layer / Solvent		L1 S1	L1 S2	L1 S3	L1 S4	L1 S5	L1 S6	L1 S7	L1 S8	L1 S9	L1 S10	L1 S11	L1 S12	L1 S13	L1 S14
Trivial name	Systematic name	$R_F \times 100$													
16-Desoxy-estrone	Estra-1,3,5(10)-trien-3-ol	72	63	67	84	61	56	63	76	44	—	—	—	—	—
Estratetraenol	Estra-1,3,5(10),16-tetraen-3-ol	65	42	63	78	56	51	44	67	18	—	—	—	—	—
—	Estra-1,3,5(10)-trien-3-ol,16β,17β-epoxide	69	56	66	80	59	54	47	70	27	—	—	—	—	—
—	Estra-1,3,5(10)-trien-3-ol-16α,17α-epoxide	67	54	66	79	60	51	47	69	26	—	—	—	—	—
Estrone	3-Hydroxy-estra-1,3,5(10)-trien-17-one	69	53	65	80	59	51	46	69	20	50	44	60	47	56
6-Dehydro-estrone	3-Hydroxy-estra-1,3,5(10),6-tetraen-17-one	62	—	—	75	—	—	—	—	—	—	—	—	—	—
Equilin	3-Hydroxy-estra-1,3,5(10),7-tetraen-17-one	68	—	66	78	—	—	—	—	—	—	—	60	—	55
9-Dehydro-estrone	3-Hydroxy-estra-1,3,5(10),9(11)-tetraen-17-one	65	50	64	79	56	52	45	68	17	—	—	—	—	—
Equilenin	3-Hydroxy-estra-1,3,5(10),6,8-pentaen-17-one	63	47	63	77	58	50	42	65	16	—	—	—	—	—
6-Oxo-estrone	3-Hydroxy-estra-1,3,5(10)-triene-6,17-dione	60	33	—	—	—	—	—	—	—	30	22	44	—	38
7-Oxo-estrone	3-Hydroxy-estra-1,3,5(10)-triene-7,17-dione	61	36	55	72	55	44	56	—	—	—	—	—	—	—
11-Oxo-estrone	3-Hydroxy-estra-1,3,5(10)-triene-11,17-dione	59	34	61	73	—	—	38	58	5	—	—	—	—	—
16-Oxo-estrone	3-Hydroxy-estra-1,3,5(10)-triene-16,17-dione	66	43	64	—	—	—	45	—	—	36	39	58	40	44
4-Hydroxy-estrone	3,4-Dihydroxy-estra-1,3,5(10)-trien-17-one	54	33	39	76	—	—	34	61	—	—	—	—	—	—
6α-Hydroxy-estrone	3,6α-Dihydroxy-estra-1,3,5(10)-trien-17-one	47	13	35	65	—	—	—	—	—	—	—	—	—	—
6β-Hydroxy-estrone	3,6β-Dihydroxy-estra-1,3,5(10)-trien-17-one	46	12	34	62	—	—	—	—	—	—	—	—	—	—
7α-Hydroxy-estrone	3,7α-Dihydroxy-estra-1,3,5(10)-trien-17-one	42	12	40	60	47	30	17	41	—	—	—	—	—	—
7β-Hydroxy-estrone	3,7β-Dihydroxy-estra-1,3,5(10)-trien-17-one	43	—	—	59	—	—	—	—	—	—	—	—	—	—
11α-Hydroxy-estrone	3,11α-Dihydroxy-estra-1,3,5(10)-trien-17-one	51	—	—	—	—	—	—	—	—	—	—	—	—	—
11β-Hydroxy-estrone	3,11β-Dihydroxy-estra-1,3,5(10)-trien-17-one	56	27	50	72	51	42	26	54	4	—	—	—	—	—
15α-Hydroxy-estrone	3,15α-Dihydroxy-estra-1,3,5(10)-trien-17-one	48	18	46	66	45	30	21	46	—	—	—	—	—	—
16α-Hydroxy-estrone	3,16α-Dihydroxy-estra-1,3,5(10)-trien-17-one	57	32	58	68	—	—	33	—	6	21	21	—	21	37

TABLE TLC 181—(Continued)
STEROIDS, ESTROGENS

Layer / Solvent		L1 S1	L1 S2	L1 S3	L1 S4	L1 S5	L1 S6	L1 S7	L1 S8	L1 S9	L1 S10	L1 S11	L1 S12	L1 S13	L1 S14
Trivial name	Systematic name	$R_F \times 10$													
16β-Hydroxy-estrone	3,16β-Dihydroxy-estra-1,3,5(10)-trien-17-one	55	28	—	—	—	—	—	—	—	—	—	—	—	—
3-Methoxy-estrone	2,3-Dihydroxy-estra-1,3,5(10)-trien-17-one-2-methyl ether	68	50	73	—	—	—	66	19	18	—	—	—	—	—
3-Methoxy-4-hydroxy-estrone	3,4-Dihydroxy-estra-1,3,5(10)-trien-17-one-3-methyl ether	66	53	76	79	—	—	—	—	—	—	—	—	—	—
Estradiol	Estra-1,3,5(10)-triene-3,17β-diol	61	40	52	74	53	42	30	61	10	30	24	44	25	38
Estradiol-17α	Estra-1,3,5(10)-triene-3,17α-diol	61	43	55	75	52	45	34	62	11	35	27	51	30	43
6-Dehydro-estradiol	Estra-1,3,5(10),6-tetraene-3,17β-diol	57	36	52	73	49	41	29	59	8	—	—	—	—	—
Dehydro-equilin	Estra-1,3,5(10), 7-tetraene-3,17β-diol	57	36	53	74	50	41	30	59	7	—	—	—	—	—
6-Oxo-estradiol	3,17β-Dihydroxy-estra-1,3,5(10)-trien-6-one	52	23	—	—	—	—	—	—	—	—	—	—	—	—
7-Oxo-estradiol	3,17β-Dihydroxy-estra-1,3,5(10)-trien-7-one	44	—	41	—	—	40	22	50	—	—	—	—	—	—
16-Oxo-estradiol	3,17β-Dihydroxy-estra-1,3,5(10)-trien-16-one	59	32	55	69	—	—	—	—	—	23	19	40	—	35
2-Hydroxy-estradiol	Estra-1,3,5(10)-triene-2,3,17β-triol	47	21	24	70	—	—	—	47	2	—	—	—	—	—
6α-Hydroxy-estradiol	Estra-1,3,5(10)-triene-3,6α,17β-triol	45	12	23	61	—	—	12	44	—	—	—	—	—	—
6β-Hydroxy-estradiol	Estra-1,3,5(10)-triene-3,6β,17β-triol	44	12	23	61	—	—	10	43	—	—	—	—	—	—
7α-Hydroxy-estradiol	Estra-1,3,5(10)-triene-3,7α,17β-triol	40	9	22	59	37	24	8	40	—	—	—	—	—	—
11β-Hydroxy-estradiol	Estra-1,3,5(10)-triene-3,11β,17β-triol	34	8	29	54	39	20	9	34	0	—	—	—	—	—
Estriol	Estra-1,3,5(10)-triene-3,16α,17β-triol	29	6	21	48	38	17	7	29	—	1	2	7	—	4
16-*epi*-Estriol	Estra-1,3,5(10)-triene-3,16β,17β-triol	43	18	26	54	—	—	17	—	—	—	—	—	—	—
17-*epi*-Estriol	Estra-1,3,5(10)-triene-3,16α,17α-triol	45	20	30	58	—	—	—	—	—	8	5	16	—	15
16,17-*epi*-Estriol	Estra-1,3,5(10)-triene-3,16β,17α-triol	34	8	18	50	—	—	—	—	—	—	—	—	—	—
15α-Hydroxy-estradiol	Estra-1,3,5(10)-triene-3,15α,17β-triol	26	—	16	38	—	—	—	—	—	0	2	5	—	2
2-Methoxy-estradiol	Estra-1,3,5(10)-triene-2,3,17β-triol-2-methyl ether	58	36	58	72	53	45	41	57	8	32	21	37	—	34
6-Oxo-estriol	3,16α,17β-Trihydroxy-estra-1,3,5-(10)-trien-6-one	21	—	—	34	—	—	—	18	—	—	—	—	—	—
6,7-Dihydroxy-estradiol	Estra-1,3,5(10)-triene-3,6,7,17β-tetrol	23	—	—	—	—	—	—	—	—	—	—	—	—	—
2-Hydroxy-estriol	Estra-1,3,5(10)-triene-3,6,16α,17β-tetrol	12	1	3	24	—	—	1	10	—	—	—	—	—	—
	Estra-1,3,5(10)-triene-2,3,16α,17β-tetrol	16	5	—	—	—	—	—	—	—	—	—	—	—	—
Doisynolic acid	3-Hydroxy-16,17-seco-estra-1,3,5(10)-triene-16-methyl-17-oine acid	31	—	30	78	—	—	—	—	—	—	—	—	—	—
2-Methoxy-estriol	Estra-1,3,5(10)-triene-2,3,16α,17β-tetrol-2-methyl ether	29	6	—	43	—	—	13	26	—	—	—	—	—	—

TABLE TLC 181—(Continued)
STEROIDS, ESTROGENS

Layer / Solvent		L1 S1	L S	L1 S3	L S4	L S5	L S6	L S7	L S8	L L S9	L S10	L S11	L1 S12	L1 S13	S14
Trival name	Systematic name	$R_F \times 100$													
3-Methoxy-6α-hydroxy-estradiol	**Estra-1,3,5(10)-triene-3,6α,17β-triol-3-methyl ether**	—	—	—	—	—	—	31	—	—	—	—	—	—	—
3-Methoxy-6β hydroxy-estradiol	**Estra-1,3,5(10)-triene-3,6β,17β-triol-3-methyl ether**	—	—	—	—	—	—	28	—	—	—	—	—	—	—
3-Methoxy-estriol	**Estra-1,3,5(10)-triene-3,16α,17β-triol-3-methyl ether**	—	—	—	—	—	—	—	—	—	3	2	8	—	—
3-Methoxy-6β-hydroxy-estradiol	**Estra-1,3,5(10)-triene-3,16β,17β-triol-3-methyl ether**	—	—	—	—	—	—	—	—	—	12	5	18	—	—
3-Methoxy-17-*epi*-estriol	**Estra-1,3,5(10)-triene-3,16α,17α-triol-3-methyl ether**	—	—	—	—	—	—	—	—	—	12	7	20	—	—

Layer L1 = Silica gel G.

Solvent
S1 = ethyl acetate-cyclohexane-ethanol (45 : 45 : 10).
S2 = ethyl acetate-cyclohexane (1 : 1).
S3 = chloroform-ethanol (9 : 1).
S4 = ethyl acetate-*n*-hexane-acetic acid-ethanol (72 : 13.5 : 10 : 4.5).
S5 = benzene-ethanol (4 : 1).
S6 = benzene-ethanol (9 : 1).
S7 = chloroform-ethanol (95.5).
S8 = ethyl acetate-*n*-hexane-acetic acid (15 : 4 : 1).
S9 = *n*-hexane-ethyl acetate (3 : 1).
S10 = chloroform-ether (3 : 1).
S11 = benzene-ethyl acetate (3 : 1).
S12 = benzene-ethyl acetate (1 : 1).
S13 = chloroform-ethyl acetate (3 : 1).
S14 = benzene-ether (1 : 1).

Detection See Section II.I, Volume Two.

Technique Tank saturation.

Literature Lisboa, B. P., *Clin. Chim. Acta*, **13**, 179 (1966), from Kirchner, J. G., *Thin-Layer Chromatography*, Techniques of Organic Chemistry, Vol. XII. Interscience Publishers, Inc., Division of John Wiley & Sons, Inc., New York, N.Y., 1967, pp. 581–583.

TABLE TLC 182
STEROIDS, β-SITOSTERYL ESTERS

Layer	L1	L1	L1	L1	L1	L1
Solvent	S1	S2	S3	S4	S5	S6
Literature	1	1	1	1	1	1
β-Sitosteryl ester of:			$R_F \times 100$			
Acetate	40	22	19	18	48	83
Propionate	53	33	26	25	53	84
Butyrate	55	36	29	26	56	85
Caproate	70	41	33	31	59	86
Caprylate	75	46	39	37	60	88
Pelargonate	74	48	40	36	59	87
Caprate	76	50	42	38	62	88
10-Undecenoate	70	46	35	34	61	87
Laurate	80	53	43	40	63	90
Myristate	82	54	44	41	64	90
Palmitate	85	54	47	43	65	91
Stearate	88	55	50	44	66	92
Oleate	86	53	45	42	65	91
Linoleate	89	56	52	45	67	92
Arachidonate	92	58	54	49	69	93

Layer L1 = Silica gel G.

Solvent S1 = cyclohexane-benzene (1 : 1).
S2 = cyclohexane-benzene (2 : 1).
S3 = cyclohexane-benzene (4 : 1).
S4 = carbon tetrachloride-chloroform (19 : 1).
S5 = *n*-heptane-ethyl acetate (19 : 1).
S6 = chloroform-acetone (19 : 1).

Detection $SbCl_3$ followed by phosphomolybdic acid.

Technique Ascending development; 13–15 cm; 30–90 minutes at 27 °–30 °C.

Literature
1. Bergmann, E. A., Ikau, R., and S. Harel, *J. Chromatogr.*, **15**, 204 (1964).

TABLE TLC 183
TERPENE ALCOHOLS

Layer	L1	L1	L1	L1	L1	L1	L1	L1	L1	L1	L1	L2	L2	L3	L3	L3	L3	L2	—
Solvent	S1	S2	S3	S4	S5	S6	S7	S8	S9	S10	S11	S12	S13	S14	S15	S16	S17	S1	—
Technique	T1	T1	T1	T1	T1	T1	T1	T1	T1	T1	T1	T2	T3	na	na	na	na	—	—
Literature	1	1	1	1	1	1	1	1	1	1	1	2	2	3	3	3	3	4	—
Compound	$R_F \times 100$																		Color
Borneol	—	—	—	—	—	—	—	—	—	—	—	20	—	10	17	32	48	27–30	vt
Carveol	34	44	49	38	40	60	51	45	29	0	12	—	18	—	—	—	—	—	—
Citronellol	27	39	41	33	36	56	41	41	36	0	19	27	15	—	—	—	—	—	—
Geraniol	20	40	43	34	29	55	31	30	25	0	14	13	15	—	—	—	—	—	—
Linalool	36	47	45	31	53	67	45	48	48	0	15	30	20	—	—	—	—	36–41	gr br
Menthol	42[a]	—	—	—	—	—	—	—	—	—	37[a]	25	—	12	21	28	42	—	—
Nerol	26	38	42	30	28	54	38	36	27	0	14	—	15	—	—	—	—	—	—
Nopol	27	48	52	51	59	65	60	43	41	0	12	—	—	—	—	—	—	—	—
α-Terpineol	29	34	36	22	32	56	29	35	25	0	8	—	15	—	16	27	43	24–27	gr

[a] = Ref. 2.

Layer
L1 = Silicic acid + 5% starch binder.
L2 = Silica gel G.
L3 = Al_2O_3 (Grade III) loose layer.

Solvent
S1 = 15% ethyl acetate in hexane.
S2 = 10% ethyl acetate in chloroform (alcohol free).
S3 = 15% ethyl acetate in benzene.
S4 = 5% ethyl acetate in chloroform (alcohol free).
S5 = 50% 1-nitropropane in hexane.
S6 = 30% ethyl acetate in hexane.
S7 = 15% ethyl carbonate in chloroform (alcohol free).
S8 = 30% isopropyl formate in hexane.
S9 = 50% isopropyl ether in hexane.
S10 = hexane.
S11 = chloroform (alcohol free).
S12 = 5% ethyl acetate in benzene.
S13 = methylene chloride.
S14 = 50% benzene in petroleum ether.
S15 = benzene.
S16 = 2% ethanol in benzene.
S17 = 5% ethanol in benzene.

Detection Fluorescein-Br_2 or conc. H_2SO_4.

Technique
T1 = 10 cm development distance.
T2 = 14 cm development distance.
T3 = 15 cm development distance.

Literature
1. Miller, J. M. and J. G. Kirchner, *Anal. Chem.*, **25**, 1107 (1953).
2. Ito, M., Wakamatsu, S., and H. Kawahara, *J. Chem. Soc. Japan*, Pure Chem. Section, **75**, 413 (1954).
3. Hermánek, S., Schwarz, V., and Z. Čekan, *Pharmazie*, **16**, 566 (1961).
4. Geyer, S. and R. Mayer, *Z. Chem.*, **5**, 308 (1965).

TABLE TLC 184
TERPENES, TRITERPENE ACIDS

Layer	L1	L1
Solvent	S1	S2
Literature	1	1
Acid[a]	$R_F \times 100$	
Acantholic	29	—
Betulinic[b]	68	—
Bredemolic	59[c]	62
Cochalic	18[c]	15
Emmolic	59	—
Guiiavolic	35	—
Isomasticadienonic	47	—
Machaerinic	23[c]	27
Masticadienonic	47	—
Medicagenic	29[c]	—
Morolic	59	—
Oleanic[b]	68	—
Oleanonic	68	—
Quinovic	55	—
Siaresinolic	66	—
Ursolic[b]	68	—

[a] Found in Bredemeyera floribunda, Alphitonia excelsa and Crataegus oxycantha.
[b] Can be separated on anion-exchange paper with cyclohexane-toluene (4 : 1) saturated with formic acid.
[c] Tailing.

Layer L1 = Silica gel G layers.
Solvent S1 = diisopropyl ether-acetone (5 : 2).
S2 = S1 + 0.5% pyridine.
Detection Chlorosulfonic acid.
Technique Ascending development.
Literature 1. Tschesche, R., Lampert, F., and G. Spatzke, *J. Chromatogr.*, **5**, 217 (1961).

TABLE TLC 185
TOXINS, MYCOTOXINS

Layer	L1			
Solvent	S1			
Detection	—	D1	D2	D3
Literature	1	1	1	1
Mycotoxin	$R_F{}^a \times 100$	Color		
Aspertoxin	12	lt y	gr-y	—
Ochratoxin B	20	bl	—	r-br
Secalonic acid D	23	dk	lt br	lt br
8α-(3-Methylbutyryloxy)-4β,15-diacetoxyscirp--9-en-3α-ol	28	—	lead gy	—
Aflatoxin G$_1$	30	gr	gr-gy	—
Aflatoxin B$_1$	40	bl	gr-gy	—
6β-Hydroxyrosenonolactone	44	—	o-r	—
Ochratoxin A	48	gr	—	r-br
Cyclopiazonic acid	65	dk	r-br	r-br
Zearalenone	72	ft bl	lt y	r-br
Sterigmatocystin	85	o	gr-gy	gr

[a] Average of 10 independent determinations.

Layer L1 = 0.25 mm layer of Silica gel G impregnated with oxalic acid, activated at 100 °C for 40 minutes, and kept at room temperature.
Solvent S1 = chloroform-methyl isobutyl ketone (4 : 1).
Detection D1 = fluorescence, UV (366 nm).
D2 = conc. H_2SO_4 and 110 °C for 10 minutes.
D3 = 1% ethanolic ferric chloride.
Technique Development distance about 14 cm in saturated tank.
Literature 1. Steyn, P. S., *J. Chromatogr.*, **45**, 473–475 (1969).

TABLE TLC 186
VITAMINS, FAT-SOLUBLE

Layer Solvent Detection Literature	L1 S1 — 1	L1 S2 — 1	L1 S3 — 1	L1 S4 — 1	L1 S5 — 1	L1 S6 — 1	L1 S7 — 1	L1 S8 — 1	L1 S9 — 1	L1 S10 — 1	L1 S11 — 1	L1 S12 — 1	L1 S13 — 1	L1 S14 — 1	D1	D2	D3
Compound	$R_F \times 100$														Color[a]		
Vitamin A	92	92	91	85	90	91	90	94	90	88	92	92	93	91	bk-gy (0.5)	bl-ish (1.0)	bl-ish y (1.2)
Vitamin D_2	42	42	49	56	38	59	32	66	34	23	32	55	58	32	gy (0.8)	bl-gy (0.8)	gy-ish gr (0.7)
Vitamin D_3	33	35	40	48	35	50	32	66	34	23	32	50	52	32	gy (0.7)	bl-ish gy (0.7)	gy-ish gr (0.8)
Vitamin E (α-Tocopherol)	70	73	73	74	73	78	67	84	72	54	70	77	78	71	gn-ish y (1.0)	y (1.5)	y-ish br (1.0)
Vitamin K_1	84	85	86	80	84	86	83	88	84	77	80	84	86	80	br (1.0)	y-ish br (1.0)	lt y (1.2)
Vitamin K_3	50	58	58	63	53	68	55	80	53	38	54	65	69	58	lt br (7.0)	r-ish br (8.0)	lt br (9.0)

[a] Sensitivity in μg in ( ).

Layer L1 = Silica gel D-O. Camag, without binder.

Solvent
S1 = benzene-petroleum ether (4 : 1).
S2 = cyclohexane-methyl ethyl ketone (21 : 4).
S3 = toluene-methyl ethyl ketone (99 : 1).
S4 = cyclohexane-cyclopentanone (24 : 1).
S5 = cyclohexane-ether (4 : 1)
S6 = cyclohexane-methyl isopropyl ketone (9 : 1).
S7 = toluene-chloroform (9 : 1).
S8 = cyclohexane-acetic acid (22 : 3).
S9 = benzene-cyclohexane (9 : 1).
S10 = toluene-cyclohexane (9 : 1).
S11 = cyclohexane-methanol (43 : 7).
S12 = cyclohexane-methyl ethyl ketone (21 : 4) on plates to which sulfuric acid has been added to the slurry; hence no spraying is necessary.
S13 = cyclohexane-ether (4 : 1) on plates with sulfuric acid as in S12.
S14 = cyclohexane-ether (4 : 1) on plates to which perchloric acid has been added in the slurry; hence no spraying is necessary.

Detection
D1 = 70% perchloric acid.
D2 = Concentrated sulfuric acid.
D3 = Antimony pentachloride, saturated solution in carbon tetrachloride.

Technique Circular.

Literature
1. Hashmi, M. H., Chughtai, F. R., Adil, A. S., and T. Qureshi, *Mikrochim. Acta*, **1967**, 1114; from *J. Chromatogr.*, **34**, D51 (1968).

TABLE TLC 187
VITAMINS, FAT SOLUBLE

Layer	L1	L1	L1	L1	L1	L1	L1	L1	L1	L1	L1	L1			
Solvent	S1	S2	S3	S4	S5	S6	S7	S8	S9	S10	S11	S12			
Detection													D1	D2	D3
Literature	1	1	1	1	1	1	1	1	1	1	1	1	1	1	1
Compound	$R_F \times 100$												Color[a]		
Vitamin A	97	95	95	97	94	90	95	96	97	91	97	84	gy (0.1)	bl (0.1)	bl (0.1)
Vitamin D_2	62	59	58	57	32	38	45	63	75	38	47	88	br (0.5)	lt gy (0.3)	lt br (0.4)
Vitamin D_3	53	53	52	47	30	31	45	63	59	38	47	82	br (0.6)	gy (0.3)	lt br (0.4)
Vitamin E (α-Tocopherol)	89	81	88	87	76	76	86	81	90	83	81	69	y-ish br (1.2)	y-gr (1.8)	y-gr (1.9)
Vitamin K_1	93	89	93	93	85	86	91	89	97	88	91	90	lt br (0.1)	dk br (0.1)	br (0.1)
Vitamin K_3	82	70	70	69	57	48	77	74	81	76	67	100	r-ish br (0.7)	r-ish br (0.6)	pk (0.7)

[a] Sensitivity in μg in ( ).

Layer L1 = Aluminum oxide G (E. Merck).

Solvent
S1 = benzene-petroleum ether (4 : 1).
S2 = cyclohexane-methyl ethyl ketone (21 : 4).
S3 = toluene-methyl ethyl ketone (99 : 1).
S4 = cyclohexane-cyclopentanone (24 : 1).
S5 = cyclohexane-ether (4 : 1).
S6 = cyclohexane-methyl isopropyl ketone (9 : 1).
S7 = toluene-chloroform (9 : 1).
S8 = cyclohexane-acetic acid (22 : 3).
S9 = benzene-cyclohexane (9 : 1).
S10 = toluene-cyclohexane (9 : 1).
S11 = cyclohexane-methanol (43 : 7).
S12 = methanol-glacial acetic acid (49 : 1).

Detection
D1 = Antimony pentachloride, saturated solution in carbon tetrachloride.
D2 = 70% perchloric acid.
D3 = Concentrated sulfuric acid.

Technique Circular.

Literature
1. Hashmi, M. H., Chughtai, F. R., Adil, A. S., and T. Qureshi, *Mikrochim. Acta*, **1967**, 1114; from *J. Chromatogr.*, **34**, D50 (1968).

TABLE TLC 188
VITAMINS, WATER-SOLUBLE

Layer	L1	L2	L3	L4	L2	L5	L5	L6	L1	L7	L6	L1	L7					
Solvent	S1	S1	S2	S3	S1	S4	S5	S6	S6	S6	S7	S7	S7					
Detection														D1	D2	D3	D4	D5
Technique	na	na	T1	T2	T3	T4	T4	T5	T5	T5	T5	T5	T5		—	—	—	—
Literature	1	1	2	3	4	5	5	6	6	6	6	6	6	4	4	4	4	4
Compound	$R_F \times 100$													Color				
Thiamine hydrochloride	0	54	0	0	22	0	0	40	15	86	32	10	93	bl	—	—	—	—
Thiamine mononitrate	0	54	—	—	—	—	—	—	—	—	—	—	—	—	—	—	—	—
Thiamine dicetylsulfate	0	54	—	—	—	—	—	—	—	—	—	—	—	—	—	—	—	—
Riboflavin	29	24	35	42	53	52	82	18	22	29	20	29	24	—	—	—	gr-y	—
Riboflavin phosphate	0	0	—	—	—	—	—	—	—	—	—	—	—	—	—	—	—	—
Pyridoxine hydrochloride	12	26	15	35	30	—	—	60	45	74	58	59	77	—	—	gy	—	—
Ascorbic acid	25	0	30	—	0	—	—	98	88	90	90	85	98	pale y	—	—	—	—
Niacinamide	44	62	65	70	—	37	31	—	—	—	—	—	—	—	—	—	—	—
Pantothenic acid	40	0	—	—	—	30	12	—	—	—	—	—	—	—	—	—	—	—
Biotin	55	54	80	—	53	0	0	—	—	—	—	—	—	w	—	—	—	—
Folic acid	7	0	—	—	0	0	0	—	—	—	—	—	—	—	—	br	—	—
Carnitine	3	20	—	—	—	—	—	—	—	—	—	—	—	—	—	—	—	—
Choline	2	45	—	—	—	—	—	—	—	—	—	—	—	—	—	—	—	—
Vitamin B_{12}	0	23	—	100	—	—	—	—	—	—	—	—	—	—	—	—	—	—
p-Aminobenzoic acid	60	54	—	—	58	—	—	6	64	11	65	80	78	—	p-r	—	—	—
Inositol	2	0	—	—	—	—	—	—	—	—	—	—	—	—	—	—	—	—
Orotic acid	0	0	—	—	—	—	—	0	86	0	0	90	0	—	—	—	—	—
Lipoic acid	65	0	—	—	—	—	—	—	—	—	—	—	—	—	—	—	—	—
Calcium panthothenate	40	0	57	—	—	—	—	—	—	—	—	—	—	—	—	—	—	—
Nicotinic acid (Niacin)	—	—	—	100	46	83	25	74	47	77	73	62	88	w	—	—	—	—
Choline chloride	—	—	—	—	28	94	43	—	—	—	—	—	—	vt	—	—	—	—
D-Pantothenic acid	—	—	—	—	0	—	—	—	—	—	—	—	—	vt	—	—	—	—
Cyanocobalamine	—	—	—	—	28	—	—	27	10	4	40	30	6	—	—	br	—	—
K_3 (sodium salt)	—	—	—	—	16	—	—	—	—	—	—	—	—	—	—	—	—	p-r
Pyridoxine	—	—	—	—	—	44	3	—	—	—	—	—	—	—	—	—	—	—
L-Ascorbic acid	—	—	—	—	—	92	22	—	—	—	—	—	—	—	—	—	—	—
Rutin	—	—	—	—	—	—	—	1	43[a]	30[a]	10	54[a]	5	—	—	—	—	—
Riboflavin-5′-phosphate sodium	—	—	—	—	—	—	—	43	22	79	47	29	67	—	—	—	—	—
Nicotinamide	—	—	—	—	—	—	—	53	43	60	52	55	70	—	—	—	—	—

[a] Tailing.

TABLE TLC 188—(Continued)
VITAMINS, WATER-SOLUBLE

Layer
L1 = Silica gel.
L2 = Alumina.
L3 = Silica gel G with fluorescence indicator.
L4 = Ion exchange resin; WOFATIT CP 300.
L5 = ITLC-SAF.
L6 = Polyamide-silica gel.
L7 = Polyamide.

Solvent
S1 = glacial acetic acid-acetone-methanol-benzene (1 : 1 : 4 : 14).
S2 = glacial acetic acid-acetone-methanol-benzene (5 : 5 : 20 : 70).
S3 = 10% ethanol.
S4 = distilled water.
S5 = acetone-methanol-water-acetic acid (70 : 20 : 5 : 5).
S6 = 10% aq. NaCl.
S7 = 10% sodium acetate in water.

Detection
D1 = Potassium iodoplatinate.
D2 = ninhydrin.
D3 = *o*-toluidine.
D4 = UV light.
D5 = $SbCl_5$.

Technique
T1 = 19 cm ascending in dark.
T2 = ascending.
T3 = circular.
T4 = 13 cm ascending.
T5 = 10 cm ascending in dark; saturated chamber.

Literature

1. Katsui, G., Ishikawa, S., Shimizu, M., and Y. Nishimoto, *Bitamin* **28**, 41 (1963); *Chem. Abstr.*, **60**, 9577 (1964) and Ishikawa, S. and G. Katsui, *Bitamin* **29**, 203 (1964); *Chem. Abstr.*, **61**, 15932b (1964).
2. Gaenshirt, H. and A. Malzacher, *Naturwiss.* **47**, 279 (1960).
3. Huettenrauch, R., Klotz, L., and W. Mueller, *Z. Chem.* **3**, 193 (1963); *Chem. Abstr.*, **59**, 14278 (1963).
4. Hashmi, M. H., Chughtai, F. R., Adil, A. S., and T. Qureshi, *Mikrochim. Acta.*, **1967**, 1117; *J. Chromatogr.*, **34**, D52 (1968).
5. Haer, F. C., *An Introduction to Chromatography on Impregnated Glass Fiber*, Ann Arbor Sci. Publishers Inc., Ann Arbor, Mich., 1968, p. 162.
6. Chiang, H.-C., Lin, Yu, and Y.-C. Wu, *J. Chromatogr.*, **45**, 161–162 (1969).

TABLE TLC 189
VITAMINS

Layer	L1	L1	L2	L2	L3	L3				
Solvent	S1	S2	S1	S2	S2	S3				
Detection							D1	D2	D3	D1
Literature	1	1	1	1	2	2				
Compound	$R_F \times 100$						Color			Sensitivity μg
Vitamin A alcohol	8	22	8	28	80	69	bl	bl	r-vt	0.1
Vitamin A acetate	41	69	62	78	—	—	bl	bl	vt	0.1
Vitamin A palmitate	75	94	74	82	87	41	bl	bl	vt	0.1
Vitamin A benzoate	63	88	66	78	—	—	bl	bl-vt	vt	—
Vitamin A trimethoxybenzoate	15	35	38	76	—	—	bl	bl-vt	vt	—
Vitamin A senecionate	57	82	65	78	—	—	bl	bl-vt	vt	—
Vitamin A pivalate	61	84	66	78	—	—	bl	vt	vt	—
Vitamin A aldehyde	18	38	35	68	—	—	dk bl	vt	vt	0.05
Vitamin A acid	0	0	0	0	58	56	r-vt	r→r-vt	r-vt	0.2
Vitamin A acid methyl ester	50	72	66	81	88	60	vt	r→r-vt	r-vt	0.2
β-Carotene	100	100	78	86	—	—	gr-bl	bl	bl	0.05

Layer
L1 = Silica gel G.
L2 = Aluminium oxide G.
L3 = Polyamide.

Solvent
S1 = benzene.
S2 = chloroform.
S3 = methanol.

Detection
D1 = $SbCl_3$.
D2 = H_2SO_4.
D3 = $HClO_4$.

Literature

1. Katsui, G., Ishikawa, S., Shimizu, M., and Y. Nishimoto, *Bitamin*, **28**, 41 (1963); *Chem. Abstr.*, **60**, 9577 (1964).
2. Ishikawa, S. and G. Katsui, *Bitamin*, **29**, 203 (1964); *Chem. Abstr.*, **61**, 15932b (1964).

Volume One

SECTION III.I

COMPOUND INDEX

COMPOUND INDEX

Abate®	GC	65
	TLC	84
Abietate	LC	68
Ac, amines	GC	98
Acacetin (*Flavone aglycones*)	PC	26
Acacetin (methoxyflavone)	TLC	92
Acantholic	TLC	184
Acaranoic acid	TLC	131
Acarenoic acid	TLC	131
Accelerator 552	TLC	66
Accothion®	GC	65
Accothion *O*-analog	GC	65
Acedicon	TLC	116
Acenaphthene	GC	38, 40
	TLC	51, 53
Acepromazine	TLC	139
Acesulfamethoxypyrazine	TLC	140
Acesulfamethoxypyridazine	TLC	140
Acetaldehyde	GC	7, 52, 55, 57, 125
	LC	8, 29
	PC	20
	TLC	77
Acetaldehyde (aldehydes)	GC	126
Acetaldehyde diethyl acetal, organics	GC	52
Acetaldehyde dimethyl acetal, organics	GC	52
Acetaldehyde, 2,4-DNPH of	TLC	166
Acetaldehyde, 2-hydrazinebenzothiazole derivative of	TLC	168
Acetamide	LC	73
4-Acetamido-3-hydroxybiphenyl	TLC	6
4-Acetamido-4′-hydroxybiphenyl	TLC	6
2-Acetamido-1-hydroxyfluorene	TLC	6
2-Acetamido-7-hydroxyfluorene	TLC	6
2-Acetamido-1-hydroxynaphthalene	TLC	6
p-Acetamidophenol	GC	105
1-Acetamino-4-chloro-anthraquinone	TLC	69
m-Acetaminophenol	PC	120
o-Acetaminophenol	PC	120
p-Acetaminophenol	PC	120
Acetanilide	TLC	7, 40
2-(*N*-Acetanilido)-3-Cl-1,4-naphthoquinones	PC	111
Acetate	PC	14
Acetate, *β*-sitosteryl ester of	TLC	182
Acetate, Vitamin A_1	PC	49
Acetate, Vitamin A_2	PC	49
Acetates	GC	84
Acetazolamide	TLC	140
Acetic acid	GC	49, 127
	LC	4, 34, 36, 57, 58, 61, 65
	PC	77, 79
	TLC	29, 132, 134
Acetic acid, amyl ester of	GC	24
Acetic acid, benzyl ester of	GC	24
Acetic acid, butyl ester of	GC	24
Acetic acid, cinnamyl ester of	GC	24
Acetic acid, decyl ester of	GC	24
Acetic acid, dodecyl ester of	GC	24
Acetic acid, ethyl ester of	GC	24
Acetic acid ester of butanol-1	GC	51

Acetic acid ester of butanol-2	GC	51
Acetic acid ester of ethanol	GC	51
Acetic acid ester of methanol	GC	51
Acetic acid ester of 3-methylbutanol-1	GC	51
Acetic acid ester of 2-methylpropanol-1	GC	51
Acetic acid ester of 2-methylpropanol-2	GC	51
Acetic acid ester of pentanol-1	GC	51
Acetic acid ester of propanol-1	GC	51
Acetic acid ester of propanol-2	GC	51
Acetic acid esters	GC	51
Acetic acid, hexyl ester of	GC	24
Acetic acid, isoamyl ester of	GC	24
Acetic acid, isobutyl ester of	GC	24
Acetic acid, isohexyl ester of	GC	24
Acetic acid, isopropyl ester of	GC	24
Acetic acid, nonyl ester of	GC	24
Acetic acid, propyl ester of	GC	24
α-G-Acetic(glycocyamine)guanidine	PC	9
Acetoacetic acid	PC	81
	TLC	32
4-Acetoaminobenzalazine	PC	23
4-Acetocatechol	TLC	87
4-Acetoguaiacone	TLC	87
Acetoin	LC	8
Acetone	GC	7, 52, 117, 121, 125, 127
	PC	20, 81
Acetone, 2,4-DNPH of	TLC	72, 165, 166
Acetone (hydrocarbons)	GC	118
Acetone (ketones)	GC	126
Acetone (2-propanone)	GC	47, 49, 51, 53, 55, 57, 58
	LC	8, 18, 36, 70
Acetone-SB	GC	98
Acetone TF-SB	GC	98
Acetone, (-SB), 1,1,1-trifluoroacetone, TF	GC	98
Acetonitrile	GC	51, 55, 123, 127
	LC	36, 60
Acetophenazine	TLC	139
Acetophenetidin (Phenacetin)	GC	21, 105
Acetophenone	GC	51
	LC	16
	PC	22
Acetophenone azine	PC	23
Acetophenone, 2,4-DNPH of	TLC	72, 73, 74, 165
Acetophenone, 2,4-DNPSC of	TLC	165
Acetophenone, *p*-NPH of	TLC	165
Acetophenone, 4-NPH of	TLC	74
Acetosalicylic acid	TLC	40
Acetosyringone, 2,4-DNPH of	TLC	167
Acetovanillone	PC	20
Acetovanillone, 2,4-DNPH of	TLC	167
3*β*-Acetoxy-5*α*-androstane	TLC	98, 100
16*β*-Acetoxy-5*α*-androstane	TLC	98
17*β*-Acetoxy-5*α*-androstane	TLC	98
3*α*-Acetoxy-5*α*-androstan-17-one	TLC	98
3*β*-Acetoxy-5*α*-androstan-17-one	TLC	98
3*β*-Acetoxyandrost-5-en-17-one	TLC	98

17β-Acetoxy-4-androsten-3-one	PC	124
3β-Acetoxy-5-androsten-17-one	PC	124
3β-Acetoxy-6-aza-B-homo-5α-cholestan-7-one	TLC	96
21-Acetoxy-4β-bromo-17α-hydroxy-5β-pregnane-3,11,20-trione	PC	124
3β-Acetoxy-5α-cholestane	TLC	98
3β-Acetoxycholest-5-ene	TLC	98
3β-Acetoxy-5α-conanine	TLC	98
21-Acetoxy-11β,17α-dihydroxy-1,4-pregnadiene-3,20-dione	PC	124
21-Acetoxy-3α,17α-dihydroxy-5β-pregnane-11,20-dione	PC	124
21-Acetoxy-3β,17α-dihydroxy-5α-pregnane-11,20-dione	PC	124
12α-Acetoxy-3α,7α-dihydroxy-5β-pregnan-20-one	TLC	101
21-Acetoxy-11β,17α-dihydroxy-4-pregnene-3,20-dione	PC	124
5-Acetoxy-*N,N*-dimethyltryptamine	GC	9
3β-Acetoxy-14α,15α-epoxy-5β-cardenolide	TLC	102
3β-Acetoxy-14β,15β-epoxy-5β-cardenolide	TLC	102
21-Acetoxy-9α-fluoro-11β,17α-dihydroxy-4-pregnene-3,20-dione	PC	124
21-Acetoxy-11β-hydroxy-4,17(20)-pregnadien-3-one	PC	124
21-Acetoxy-17α-hydroxy-1,4-pregnadiene-3,11,20-trione	PC	124
21-Acetoxy-17α-hydroxy-5β-pregnane-3,11,20-trione	PC	124
21-Acetoxy-11β-hydroxy-4-pregnene-3,20-dione	PC	124
21-Acetoxy-15α-hydroxy-4-pregnene-3,20-dione	PC	124
21-Acetoxy-17α-hydroxy-4-pregnene-3,20-dione	PC	124
21-Acetoxy-17α-hydroxy-4-pregnene-3,11,20-trione	PC	124
2-Acetoxykessan-8-ol	GC	141
8-Acetoxykessan-2-ol	GC	141
cis-7-Acetoxymethyl-5,6-dihydro-5,6-dihydroxy-12-methylbenz[*a*]-anthracene	TLC	50
3β-Acetoxy-B-norcholest-5-ene	TLC	98
3β-Acetoxy-15-oxo-5β,14β-cardenolide	TLC	102
3β-Acetoxy-15-oxo-5β,14β,17βH-cardenolide	TLC	102
3β-Acetoxy-5,16-pregnadien-20-one	TLC	101
21-Acetoxy-4,16-pregnadiene-3,20-dione	PC	124
3β-Acetoxy-5α-pregnan-20-one	TLC	98
11α-Acetoxy-4-pregnene-3,20-dione	PC	124
21-Acetoxy-4-pregnene-3,20-dione	PC	124
21-Acetoxy-4-pregnene-3,11,20-trione	PC	124
3β-Acetoxypregn-5-en-20-one	TLC	98
21-Acetoxy-11β,17α,20α-trihydroxy-4-pregnen-3-one	PC	124
Acetyl, Ac	GC	98
Acetylacetone	GC	7
N-Acetyl-5-acetoxytryptamine (Diacetyl-serotonin)	GC	9
N-Acetyladenosine 2′,3′,5′-triacetate	GC	134
N-Acetyl-*dl*-alanine	PC	63
5-Acetylamino-6-amino-3-methyluracil	PC	36
2-Acetylamino-1,4-naphthoquinone	PC	111
N-Acetyl-*p*-aminophenol	TLC	40
3-Acetylamino-2,4,6-triiodobenzoic acid	TLC	64
9-Acetylanthracene	TLC	75
6-Acetyl-6-aza-B-homo-5α-cholestan-3β-ol acetate	TLC	96
Acetylcaranine	GC	96
Acetylcarbromal	GC	17
	TLC	137
Acetylcholine	PC	8, 60
Acetyl-demethylchlorpromazine (acetyl-Nor, CP)	GC	103
Acetyl-demethylchlorpromazine sulfoxide (acetyl-Nor, CPSO)	GC	103
N-Acetyl-2′-deoxyadenosine 3′,5′-diacetate	GC	134
N-Acetyl-2,5-dichloro-1,4-phenylenediamine	TLC	119
Acetyl-didemethylchlorpromazine (acetyl-Nor_2 CP)	GC	103
Acetyl-didemethylchlorpromazine sulfoxide (acetyl-Nor_2 CPSO)	GC	103
16-Acetyl-diginatigenin (3β,12β,14β-Trihydroxy-16β-acetoxy-5β-cardenolide)	TLC	102
Acetyldigitoxin-α	PC	127
Acetyldigitoxin-β	PC	127
Acetyldigitoxin-1(α?)	TLC	177
Acetyldigitoxin-2(β-?)	TLC	177
Acetyldigoxin-α	PC	127
	TLC	177
Acetyldigoxin-β	PC	127
	TLC	177
N-Acetyl-3,3′-diiodothyronine	TLC	19
N-Acetyl-3,5-diiodothyronine	TLC	19
N-Acetyl-3,5-diiodotyrosine	TLC	19
1-Acetyl-2,2-dimethylhydrazine	PC	61
Acetylene	GC	29, 58, 125, 127
Acetylene (alkynes)	GC	126
Acetylene (hydrocarbons)	GC	118
Acetyl-fluphenazine	GC	103
N-Acetyl galactosamine	TLC	127
Acetyl-gitaloxin	TLC	102
Acetylgitoxin-α	PC	127
	TLC	177
Acetylgitoxin-β	PC	127
	TLC	177
N-Acetylglucosamine	TLC	127
N-Acetyl-L-glutamic acid	PC	63
N-Acetylhistamine	PC	60
Acetyl histidine	TLC	155
N-Acetyl-L-histidine	PC	63
Acetyl histidinol	TLC	155
Acetyl-7-hydroxychlorpromazine (acetyl-7-OH-CP)	GC	103
N-Acetyl-5-hydroxytryptamine	LC	77
	PC	101
N-Acetylindoxyl	PC	102, 103
N-Acetyl isatin	PC	103
3-*O*-Acetyl-1,2:4,5-di-*O*-isopropylidene-D-fructose	GC	13
N^a-Acetylkynurenine	PC	63, 103
N-Acetyl-D-mannosamine	TLC	127
l-Acetylmethadol	GC	18
	TLC	138
N-Acetyl-*p*-methoxyphenylalanine	GC	5
N-Acetyl-5-methoxytryptamine	GC	9
	LC	77
N-Acetyl-5-methoxytryptamine = melantonin	PC	101

Acetyl-β-methylcholine	PC	8
N-Acetyl-O-methyl-3,3′-diiodothyronine	TLC	19
N-Acetyl-O-methyl-3,5-diiodothyronine	TLC	19
N-Acetyl-N-methylhistamine	GC	9
N-Acetyl-O-methylthyroxine	TLC	19
N-Acetyl-O-methyl-3,3′,5′-triiodothyronine	TLC	19
N-Acetyl-O-methyl-3,5,3′-triiodothyronine	TLC	19
N-Acetyl-N-methyltryptamine	GC	9
N-Acetyl-3-monoiodohistidine	TLC	19
N-Acetyl-3-monoiodotyrosine	TLC	19
Acetylnaphthalene	GC	40
N-Acetyl-S-(2-nitrophenyl)-L-cysteine	PC	29
N-Acetyl-S-(4-nitrophenyl)-L-cysteine	PC	29
Acetyl-(p-nitrophenyl)-sulfanilamide	TLC	136
9-Acetylphenanthrene	TLC	75
N-Acetyl-p-phenetidine	TLC	40
2-Acetylphenol	TLC	89
4-Acetylphenol	TLC	89
N-Acetylphenylalanine	GC	5
N-Acetyl-1,3-phenylenediamine	TLC	119
N-Acetyl-1,4-phenylenediamine	TLC	119
Acetyl phosphate	PC	77
Acetylpromazine	GC	22
Acetylpromazine maleate	PC	85
1-Acetylpyrene	TLC	120
3-Acetylpyrene	TLC	75
2-Acetylpyridine	TLC	60
3-Acetylpyridine	PC	2
Acetylsalicylcholine	PC	8
Acetylsalicylic acid	GC	105, 106
	TLC	129
O-Acetylserine	PC	63
Acetylspermidine A	LC	66
Acetylspermidine B	LC	66
Acetylspermine	LC	66
N[1]-Acetyl-2-sulfanilamido-3-methoxypyrazine	TLC	44
N[1]-Acetyl-3-sulfanilamido-6-methoxypyridazine	TLC	44
O-Acetyl-tetraiodothyrolactic acid	TLC	19
O-Acetyl-tetraiodothyropyruvic acid	TLC	19
Acetylthiamine	PC	8
Acetylthiocholine iodide	PC	8
2-Acetylthiophene	TLC	104
N-Acetylthyroxine	TLC	19
Acetyl tributyl citrate (citric acid ester)	TLC	27
Acetyl triethyl citrate (citric acid ester)	TLC	27
Acetyl tri-2-ethyl hexyl citrate (citric acid ester)	TLC	27
O-Acetyl-3,5,3′-triiodothyrolactic acid	TLC	19
N-Acetyl-3,5,3′-triiodothyronamine	TLC	19
N-Acetyl-3,3′,5′-triiodothyronine	TLC	19
N-Acetyl-3,5,3′-triiodothyronine	TLC	19
O-Acetyl-3,5,3′-triiodothyropyruvic acid	TLC	19
N-Acetyltryptamine	GC	9
	PC	101
Acetyltryptophan	PC	103
N-Acetyltryptophan	GC	5
	PC	102, 103
N-Acetyltryptophanamide	PC	63, 101
N-Acetyltyramine	PC	60
N-Acetyltyrosine	GC	5
N-Acetyl-l-tyrosine	PC	63
5-Acetyluracil	PC	36
Acetylurea	PC	60
Acetylvanillin	TLC	167
Acidomycin	TLC	20
Acid Red	TLC	142
Acid Violet	TLC	141
Acid, Vitamin A_1	PC	49
Acids, aromatic	PC	82
Aconitic acid	PC	13
Aconitine	GC	8
	PC	3, 4, 54
	TLC	3
cis-Aconitrate acid	TLC	36
Acoschimperoside P	TLC	102
Acoschimperoside P peracetate	TLC	102
Acovenoside A	TLC	102
Acovenosigenin (1β,3β,14β-Trihydroxy-5β-cardenolide)	TLC	102
9-Acridanone	TLC	75
Acridine	GC	45
	PC	84
	TLC	58, 120
Acridone	TLC	120
Acrolein	GC	7, 55
	LC	29
	PC	20
Acrolein, 2,4-DNPH of	TLC	72, 166
Acrolein, 2-hydrazinebenzothiazole derivative of	TLC	168
Acrylaldehyde, 2,4-DNPH of	TLC	166
Acrylamide	LC	73
Acrylate, methyl, organics	GC	52
Acrylic acid	LC	58
	TLC	29
Acrylonitrile	GC	55,123, 127
Actidil® (Triprolidine)	TLC	39
Actinomycin C	TLC	20
Actinomycin J	TLC	20
Acumycin	PC	66
	TLC	153
Adenine (Ad)	GC	77
	LC	3, 13, 15, 17, 22, 31, 46, 50
	PC	11, 32, 36, 37, 108
	TLC	159
Adenine-N′-oxide	PC	36
Adenine-1-N-oxide	LC	46
Adenosine (Ao)	GC	77
	LC	3, 13, 15, 17, 22, 31, 46, 50
	PC	32, 36, 37, 108
	TLC	62, 159
Adenosine 3′:5′-cyclic phosphate	PC	108
Adenosine diphosphate (ADP)	TLC	160
Adenosine-5′-diphosphate	PC	36, 108
Adenosine-monophosphate (AMP)	LC	47
	TLC	160
Adenosine 2′-monophosphate (2′-AMP)	GC	77
	PC	108
	TLC	160

Adenosine 3′-monophosphate (3′-AMP)	GC	77
	PC	108
	TLC	160
Adenosine 5′-monophosphate (5′-AMP)	GC	77
	LC	15
	PC	108
	TLC	160
2-d-Adenosine-5′-monophosphate (2-d-5′-AMP)	TLC	160
Adenosine-3′-phosphate (AMP-3′)	PC	32
Adenosine-5′-phosphate (AMP-5′)	PC	32, 36
	TLC	62
Adenosine 2′,3′,5′-triacetate	GC	134
Adenosine triphosphate (ATP)	PC	32
	TLC	160
Adenosine-5′-triphosphate	PC	36, 108
S-Adenosylhomocysteine	LC	41
S-Adenosylmethionine	PC	36
P^1-[Adenosyl-(5′)] P^2 methylpyrophosphate	TLC	160
P^1-[Adenosyl-(5′)] P^2 phenylpyrophosphate	TLC	160
Adenylic acid	LC	13
5′-Adenylic acid	LC	50
Adenylyl-(3′-5′)-uridine	TLC	62
Adipates, esters	GC	25
Adiphenine	GC	17, 22
	TLC	39
Adiphenine hydrochloride	GC	17
Adipic acid	LC	65
	PC	13, 77
	TLC	31
Adipic acid, dimethyl ester of	GC	24
N,N′-Adipic-di-(3-amino-2,4,6-triiodobenzoic acid)	TLC	64
Adiponitrile	LC	60
Adonitol	TLC	127
Adonitoxin	TLC	102
Adonitoxigenin (3β,14β,16β-Trihydroxy-19-oxo-5β-cardenolide)	TLC	102
ADP (Adenosine diphosphate)	LC	3
	PC	31
	TLC	63
ADPG	TLC	63
Adrenaline	PC	5, 56, 58, 64
	TLC	9, 121
m-Adrenaline	PC	58
Adrenochrome	TLC	8
Adrenone	TLC	9
Adynerigenin (3β-Hydroxy-8β,14β-epoxy-5β-cardenolide)	TLC	102
Adynerin	TLC	102
Adynerin peracetate	TLC	102
Aesculetin	PC	113
Aesculin	PC	113
Aetioporphyrin	PC	33
Aflatoxin B_1	TLC	185
Aflatoxin G_1	TLC	185
Ag (Silver)	LC	85, 88
	PC	15
	TLC	52
Ag^+, Silver	PC	92, 97
Ag(I), Silver	PC	95, 96
	TLC	148,149, 150
Age-Rite Alba	TLC	175
Age Rite Geltrol	TLC	175
Age Rite Spar	TLC	175
Age Rite Superlite	TLC	175
Aglycone	PC	24
Aglycone 3,5-diglucoside	PC	24
Aglycone 3,7-diglucoside	PC	24
Aglycone 3-galactoside	PC	24
Aglycone 3-gentiobioside	PC	24
Aglycone 3-gentiotrioside	PC	24
Aglycone 3-glucoside	PC	24
Aglycone 5-glucoside	PC	24
Aglycone 7-glucoside	PC	24
Aglycone 5-glucoside-3-rutinoside	PC	24
Aglycone 5-glucoside-3-sambubioside	PC	24
Aglycone 5-glucoside-3-sophoroside	PC	24
Aglycone 7-glucoside-3-sophoroside	PC	24
Aglycone 3-(2^G-glucosylrutinoside)	PC	24
Aglycone 3-lathyroside	PC	24
Aglycone 3-rhamnoside	PC	24
Aglycone 3-rhamnoside-5-glucoside	PC	24
Aglycone 3-rutinoside	PC	24
Aglycone 3-sambubioside	PC	24
Aglycone 3-sophoroside	PC	24
Aglycones, aurone	PC	26
Aglycones, chalcone	PC	26
Aglycones, flavone	PC	26
Aglycones, flavonol	PC	26
Aglycones, flavonone	PC	26
Aglycones, isoflavone	PC	26
Agmatine	LC	66
	PC	9, 56, 64
Agmatine·H_2SO_4 (guanidobutylamine)	PC	10
Agnosterol	TLC	99
Agroclavine	TLC	114
AIR	GC	26,113, 127
Ajmalicine	TLC	115
Ajmaline	PC	2, 54
	TLC	2, 3, 115
Al^{3+}, Aluminum	PC	92, 97
Al(III), Aluminum	GC	43
	PC	95, 96
	TLC	148, 149
Alanine	GC	12, 101
	LC	2, 14, 33, 41, 56
	PC	11, 62, 63, 64, 65
	TLC	13, 15, 29
Alanine, DNAP of	TLC	14
Alanine, PTH of	TLC	124
α-Alanine	GC	101
	TLC	11, 16
β-Alanine	GC	12, 101
	LC	41
	PC	63, 64
	TLC	11, 13, 15, 16, 118
β-Alanine, DNAP of	TLC	14
DL-Alanine	TLC	12, 118
Cbo-DL-Alanine	TLC	12
Cbo-DL-Alanine-Glycine	TLC	12
DL-Alanine-Glycine	TLC	12
Cbo-DL-Alanine-Glycine ethyl ester	TLC	12
Cbo-DL-Alanine-DL-Phenylalanine	TLC	12
β-1-(Naphth-2-ol) alanine	PC	64
Alaninol	PC	57
Alanyl-leucine (*cis* + *trans*)	GC	16

Alanyl-phenylalanine (*cis*)	GC	16
Alanyl-phenylalanine (*trans*)	GC	16
Alanyl-proline (*cis* + *trans*)	GC	16
Alanyl-valine (*cis*)	GC	16
Alanyl-valine (*trans*)	GC	16
Albucid® (sulfacetamide)	TLC	140
Alcohol	PC	52
Alcohol, Vitamin A_1	PC	49
Alcohol, Vitamin A_2	PC	49
Alcohols	GC	87, 126
Alcohols, ethanol	GC	126
Alcohols, methanol	GC	126
Alcohols, propanol	GC	126
Alcohols, sugar	PC	53
Aldehyde, Vitamin A_1	PC	49
Aldehyde, Vitamin A_2	PC	49
Aldehydes	GC	126
Aldehydes, propionaldehyde	GC	126
Aldehydes, valeraldehyde	GC	126
Aldehydo and keto acids	LC	40
Aldicarb	GC	65
Aldicarb sulfone	GC	65
Aldicarb sulfoxide	GC	65
Aldobionic acids	LC	40
Aldonic acids	LC	40
Aldosterone	TLC	101, 178
Aldrin	GC	60, 61, 129
	LC	44
	PC	116
	TLC	80, 81, 82, 172
Aldrin (reference)	GC	68, 128
Alectoronic acid	PC	30
Aliphatic alcohols	PC	53
Aliphatic alcohols, glycols	PC	53
Aliphatic alcohols, polyhydroxy compounds	PC	53
Aliphatic alcohols, primary	PC	53
Aliphatic alcohols, secondary	PC	53
Aliphatic alcohols, tertiary	PC	53
Alizarin	PC	109
Alkaline hydrolysis myoinositol pentaphosphate	LC	47
Alkaloids, Genista	PC	3
Alkaloids, miscellaneous	PC	3
Alkaloids, Papaveraceae	PC	3
Alkaloids, Rauwolfia, free form	GC	17
Alkaloids, Solonaceous	PC	3
Alkylated phenol	TLC	175
Alkyl guanidines	PC	9
Alkyl-guanidino guanidines	PC	9
1-Alkyl-3-nitroguanidines	PC	9
Alkynes	GC	126
Alkynes, propyne	GC	126
Allantoin	LC	46
	PC	37, 64
Allene (hydrocarbons)	GC	118
Allethrin	TLC	85
α-DL-*trans*-Allethrin	PC	119
Allethrolone	PC	119
Allitol	GC	14
Allobarbital	TLC	137
Allobarbitone	PC	86
Allobarbitone (Allobarbital, diallyl-barbituric acid)	GC	19
Allocholesterol	TLC	99
Allocryptopine	PC	3
Allocymarin	PC	127
Alloemicymarin	PC	127
Alloglaucotoxigenin (3β,14β,15β-Trihydroxy-19-oxo-5α-cardenolide)	TLC	102
Alloglaucotoxigenin peracetate	TLC	102
Allohydroxyproline	GC	101
Alloisoleucine	GC	12, 101
	LC	41
	TLC	11
Dg-β,D-allomethyloside	TLC	102
Dg-β,D-allomethyloside peracetate	TLC	102
D-Allonic	LC	23
Allonolactone	PC	76
Alloperiplocymarin	PC	127
Alloperiplogenin acetate	PC	127
Allopregnane-3α,20α-diol	TLC	101, 178
Allopregnanolone	GC	136
Allose	LC	48
Allostrophanthidin	PC	127
Allothreonine	GC	12, 101
	TLC	11
Alloxanthine	GC	77
Allyl acetate	GC	55, 91
Allyl alcohol	GC	47
	LC	54
	PC	52
	TLC	1
Allyl alcohol (2-propen-l-ol)	GC	53, 55
	LC	37
Allyl alkyl guanidine	PC	9
Allyl-1-alkyl-3-nitroguanidine	PC	9
Allylamine	PC	56
	TLC	10
Allylamine, trifluoroacetyl derivative	GC	10
Allylbarbituric acid	GC	106
Allylbarbituric acid (Butalbital)	TLC	137
Allyl benzyl sulfoxide	TLC	164
Allyl butyrate	GC	91
Allyl chloride	GC	47
Allyl cyanide	GC	123
5-Allyl-5-cyclopentyl-Ba	TLC	137
1-Allyl-2,5-dimethylpyrrole	GC	133
Allyl esters	GC	24
Allyl formate	GC	55, 91
Allyl guanidine	PC	9
Allyl hexanoate	GC	91
Allylisobutylbarbital	TLC	137
Allyl isobutyrate	GC	91
Allyl isocyanate	GC	44
Allylisopropylacetylurea	GC	19
	TLC	40
5-Allyl-5-isopropyl-1-methyl-Ba	TLC	137
Allyl isothiocyanate	GC	85
Allyl mercaptan	GC	85
Allyl 3-methylbutyrate	GC	91
Allyl 4-methylpentanoate	GC	91
N-Allyl-4-morpholine thiocarboxamide	PC	47
1-Allyl-3-nitroguanidine	PC	9
N-Allylnormorphine	GC	8
Allyl octanoate	GC	91
Allyl pentanoate	GC	91
4-Allylphenol	GC	71
Allylphenylbarbituric acid (Alphenal)	GC	105
Allyl propionate	GC	91
1-Allylpyrrole	GC	133
2-Allylpyrrole	GC	133
N-Allylthiourea	PC	47
Aloe-emodin	PC	109
Alphanaphthol	PC	31

Alphaprodine	GC	105
	PC	4
dl-Alphaprodine	GC	17
	TLC	138
Alphaprodine HCl	TLC	138
Alphenal	GC	106
Althiomycin	TLC	20
Altizide	TLC	140
Altritol	GC	14
D-Altronic	LC	23
Altronolactone	PC	76
Altrose	LC	48
Aluminum, Al^{3+}	PC	92, 97
Aluminum, Al(III)	GC	43
	PC	95, 96
	TLC	148, 149
Amaranth	TLC	49, 142
Amaryllidaceae	GC	96
Amax	TLC	66
Ambrodryl® (Bromdiphenhydramine)	TLC	39
Amethocaine	PC	4
Ametryne®	GC	66
	TLC	81, 86
Amides, Ureides, Hydrazines, free form, salt form	GC	17
Amidithion	GC	65
Amidonomycin	TLC	20
Amidopyrine	TLC	40
Amine, free	GC	98
Amines, nitrogen compounds	GC	45
1-Amino-3-acetaminobenzene	PC	6
1-Amino-4-acetaminobenzene	PC	6
m-Aminoacetophenone	PC	78
o-Aminoacetophenone	PC	78
p-Aminoacetophenone	LC	35
4-Aminoacetophenone	PC	23
	TLC	119
4-Aminoacetophenone azine hydrochloride	PC	23
α-Aminoadipic acid	GC	101
	LC	41
	PC	64
	TLC	13
1-Aminoanthracene	TLC	119
1-Aminoanthracene pentafluoropropionamide	GC	99
2-Aminoanthracene	TLC	119
2-Aminoanthracene pentafluoropropionamide	GC	99
9-Aminoanthracene pentafluoropropionamide	GC	99
1-Aminoanthraquinone	PC	105
	TLC	69
2-Aminoanthraquinone	PC	105
	TLC	69
p-Aminoazobenzene (R_F control)	TLC	164
m-Aminobenzaldehyde	PC	78
7-Aminobenz[*a*]anthracene	TLC	120
p-Aminobenzenearsonic acid	TLC	136
2-Aminobenzoic acid	TLC	119
3-Aminobenzoic acid	TLC	119
4-Aminobenzoic acid	LC	58
	TLC	119
m-Aminobenzoic acid	LC	35
	TLC	37, 129
m-Aminobenzoic acid ethyl ester	TLC	129
o-Aminobenzoic acid	LC	35
	TLC	37, 129
o-Aminobenzoic acid ethyl ester	TLC	129
o-Aminobenzoic acid isobutyl ester	TLC	129
o-Aminobenzoic acid methyl ester	TLC	129
p-Aminobenzoic acid	LC	30, 35
	PC	64, 82
	TLC	37, 129, 188
p-Aminobenzoic acid ethyl ester	TLC	129
p-Aminobenzoic acid isobutyl ester	TLC	129
p-Aminobenzoic acid isopropyl ester	TLC	129
p-Aminobenzoic acid methyl ester	TLC	129
4-Aminobenzoic acid methyl ester	TLC	119
2-Aminobenzophenone	PC	59
6-Aminobenzo[*a*]pyrene	TLC	120
1-Amino-4-benzoylamino-anthraquinone	TLC	69
4-Aminobiphenyl	TLC	6
2-Aminobiphenyl pentafluoropropionamide	GC	99
3-Aminobiphenyl pentafluoropropionamide	GC	99
4-Aminobiphenyl pentafluoropropionamide	GC	99
1-Amino-2-bromo-anthraquinone	TLC	69
1-Amino-3-bromo-anthraquinone	TLC	69
1-Amino-4-bromo-anthraquinone	TLC	69
2-Amino-3-bromo-anthraquinone	TLC	69
2-Aminobutyric acid	TLC	29
3-Aminobutyric acid	TLC	29
4-Aminobutyric acid	TLC	29
α-Aminobutyric acid	GC	101
	TLC	16
α-Aminobutyric acid, PTH of	TLC	124
α-Amino-*n*-butyric acid	GC	12
	LC	41
	PC	64
	TLC	11, 13, 15
dl-α-Amino-*n*-butyric acid	PC	63
β-Aminobutyric acid	GC	101
	TLC	11, 16
β-Amino-*n*-butyric acid	PC	64
γ-Aminobutyric acid	GC	101
	LC	41
	TLC	11, 16, 118
DL-γ-Aminobutyric acid	PC	63
γ-Amino-*n*-butyric acid	PC	64
	TLC	13, 15
6-Aminocaproic acid	TLC	29
α-Amino-*n*-caproic acid	PC	64
ε-Aminocaproic acid	PC	63
	TLC	13
ε-Amino-*n*-caproic acid	PC	64
	TLC	11
α-Aminocaprylic acid	TLC	16
α-Aminocaprylic acid, PTH of	TLC	124
α-Amino-*n*-caprylic acid	PC	64
	TLC	15
ω-Aminocaprylic acid	PC	64
4-Amino-5-carboxamide imidazole	PC	17
	TLC	155
5-Amino-6-carboxyuracil	PC	36
1-Amino-4-chloro-anthraquinone	TLC	69
1-Amino-5-chloro-anthraquinone	TLC	69
1-Amino-8-chloro-anthraquinone	TLC	69
2-Amino-3-chloro-anthraquinone	TLC	69
1-Amino-2-chloro-4-bromo-anthraquinone	TLC	69
2-Amino-3-chloro-1,4-naphthoquinone	PC	111
2-Amino-3-chlorophenol	PC	29
2-Amino-4-chlorophenol	PC	6, 29
2-Amino-5-chlorophenol	PC	29
2-Amino-6-chlorophenol	PC	29
3-Amino-2-chlorophenol	PC	29
3-Amino-4-chlorophenol	PC	29

3-Amino-5-chlorophenol	PC	29
4-Amino-2-chlorophenol	PC	29
4-Amino-3-chlorophenol	PC	29
5-Amino-2-chlorophenol	PC	29
4-Amino-7-chloroquinoline	PC	104
2-Amino-5-chlorotoluene	PC	59
3α-Amino-5α-cholestane hydrochloride	TLC	96
6-Aminochrysene	TLC	120
1:1 Aminocyclopropane carboxylic acid	PC	64
2-Aminodibenzofuran	TLC	120
2-Aminodibenzofuran pentafluoro-propionamide	GC	99
2-Aminodibenzoselenophene	TLC	120
2-Aminodibenzothiophene	TLC	120
1-Amino-2,4-dibromo-anthraquinone	TLC	69
2-Amino-1,3-dibromo-anthraquinone	TLC	69
6-Amino-2,8-dihydroxypurine	PC	36
6-Amino-1,3-dimethyl-5-formylaminouracil	PC	36
2-Amino-4,6-dimethylpyrimidine	PC	36
6-Amino-2,4-dimethylpyrimidine	PC	36
6-Amino-1,3-dimethyluracil	PC	36
2-Amino-4,6-dinitrophenol	PC	6
OH(3-Amino-4,6-dinitrophenol), DNAP of	TLC	14
2-Amino-4,6-dioxy-5-methylpyrimidine	PC	36
2-Amino-4,6-dioxypyrimidine	PC	36
4-Aminodiphenyl	TLC	119
S-Aminoethylcysteine	LC	59
4-Amino-5-ethyl-2-methylpyridine *N*-oxide	TLC	61
2-Aminoethylphosphonic acid	LC	41
3-Aminofluoranthene	TLC	120
8-Aminofluoranthene	TLC	120
1-Aminofluorene	TLC	120
2-Aminofluorene	TLC	6, 120
2-Aminofluorene pentafluoropropionamide	GC	99
3-Aminofluorene	TLC	120
4-Aminofluorene	TLC	120
9-Aminofluorene pentafluoropropionamide	GC	99
6-Amino-5-formylamino-4-hydroxy-pyrimidine	PC	36
6-Amino-5-formylamino-1-methyl-4-oxypyrimidine	PC	36
6-Amino-5-formylamino-3-methyl-4-oxypyrimidine	PC	36
6-Amino-5-formylamino-l-methyluracil	PC	36
6-Amino-5-formylamino-3-methyluracil	PC	36
6-Amino-5-formylaminouracil	PC	36
Aminoguanidine	PC	9
Aminoguanidine H_2SO_4	PC	10
2-Aminoheptane	PC	5
	TLC	39
p-Aminohippuric acid	LC	35
	PC	64
	TLC	11, 13
o-Aminohippuric acid methyl ester	PC	80
Aminohydroxyacetic acid	PC	63
4-Amino-3-hydroxybiphenyl	TLC	6
4-Amino-4′-hydroxybiphenyl	TLC	6
2-Amino-2′-hydroxydiethylamine	PC	5
2-Amino-1-hydroxyfluorene	TLC	6
2-Amino-7-hydroxyfluorene	TLC	6
2-Amino-4-hydroxy-6-methylpyrimidine	PC	36
2-Amino-8-hydroxypurine	PC	36
6-Amino-2-hydroxypurine	PC	36
6-Amino-8-hydroxypurine	PC	36
2-Amino-4-hydroxypyrimidine	PC	36
4-Aminoimidazole-5-carboxamide	LC	46
4-Amino-5-imidazolecarboxamide	PC	16, 36

4-Amino-5-imidazolecarboxamide ribonucleoside	PC	36
4-Amino-5-imidazolecarboxamidine	PC	16
4-Aminoimidazole-5-carboxylic acid	TLC	155
4-Amino-5-imidazole carboxylic acid	TLC	13
1′-Aminoindolo[*b*-3,2]-5α-cholest-2-one	TLC	96
2-Aminoisobutyric acid	TLC	29
α-Aminoisobutyric acid	PC	64
	TLC	11, 13, 15, 16
β-Aminoisobutyric acid	GC	101
	LC	41
	PC	64
	TLC	11, 13, 15, 16
δ-Aminolevulinic acid	PC	64
	TLC	13
Δ-Aminolevulinic acid	LC	41
1-Amino-3-methoxy-4-acetaminobenzene	TLC	119
5-Amino-2-methoxyaniline	GC	102
2-Amino-6-methoxybenzothiazole	TLC	119
8-Amino-6-methoxyquinoline	PC	104
2-Amino-4-methylbenzoic acid	PC	78
2-Amino-5-methylbenzoic acid	PC	78
3-Amino-4-methylbenzoic acid	PC	78
4-Amino-3-methylbenzoic acid	PC	78
2-Amino-6-methylbenzothiazole	TLC	119
6-Amino-3-methyl-5(*N*-formyl-methylamino)uracil	PC	36
2-Amino-5-methyl-4-oxypyrimidine	PC	36
4-Amino-3-methylphenol	PC	120
2,2′-Aminomethyl-1,3-propanediol	PC	60
2-Amino-6-methylpurine	PC	36
2-Amino-4-methylpyrimidine	PC	36
5-Amino-6-methyluracil	PC	36
6-Amino-1-methyluracil	PC	36
1-Aminonaphthalene	PC	59
2-Aminonaphthalene	PC	59
1-Amino-2-naphthol	PC	105
	TLC	119
1-Amino-5-naphthol	PC	105
2-Amino-3-naphthol	PC	105
2-Amino-1,4-naphthoquinone	PC	111
o-Aminonaphtholsulfonic acid isomer 1,2,0	PC	46
o-Aminonaphtholsulfonic acid isomer 1,2,4	PC	46
o-Aminonaphtholsulfonic acid isomer 1,2,5	PC	46
o-Aminonaphtholsulfonic acid isomer 1,2,6	PC	46
o-Aminonaphtholsulfonic acid isomer 1,2,7	PC	46
o-Aminonaphtholsulfonic acid isomer 1,2,8	PC	46
o-Aminonaphtholsulfonic acid isomer 2,1,0	PC	46
o-Aminonaphtholsulfonic acid isomer 2,1,4	PC	46
o-Aminonaphtholsulfonic acid isomer 2,1,5	PC	46
2-Amino-4-nitrophenol	PC	120
2-Amino-5-nitrophenol	PC	6, 120
	TLC	119
2-Amino-6-nitrophenol	PC	121
4-Amino-2-nitrophenol	PC	120
2-Amino-octanoic acid	TLC	13

α-Amino-*n*-octanoic acid	TLC	15
Amino parathion	GC	67
3-Aminopentane	GC	97
1-Aminophenanthrene pentafluoropropionamide	GC	99
2-Aminophenanthrene pentafluoropropionamide	GC	99
3-Aminophenanthrene pentafluoropropionamide	GC	99
4-Aminophenanthrene pentafluoropropionamide	GC	99
9-Aminophenanthrene pentafluoropropionamide	GC	99
m-Aminophenol	LC	52
	PC	29, 82
	TLC	88, 174
o-Aminophenol	PC	29
	TLC	88, 174
o-Aminophenol, *N*-4-acetyl-2-nitrophenol ether of	TLC	91
p-Aminophenol	LC	52
	PC	29, 82
	TLC	88, 174
1,2-Aminophenol	TLC	119
1,3-Aminophenol	TLC	119
1,4-Aminophenol	TLC	119
2-Aminophenol	PC	6
	TLC	119
3-Aminophenol	PC	6
	TLC	119
4-Aminophenol	PC	6
	TLC	119
N-ANP-*m*-Aminophenol, 4-acetyl-2-nitrophenol ether of	TLC	91
N-ANP-*p*-Aminophenol, 4-acetyl-2-nitrophenol ether of	TLC	91
2-Aminophenol-4-diethylaminosulphonamide	TLC	119
2-Aminophenol-5-sulfonamide	PC	6
m-Aminophenyl glucoside	PC	82
6-Amino-2-phenylbenzothiazole	TLC	119
Aminophenylpyridone	PC	84
3-Aminophenyl sulfoxide	GC	139
4-Aminophenyl sulfoxide	GC	139
m-Aminophenyl sulfuric acid	PC	82
p-Aminophenyl sulfuric acid	PC	82
2-Amino-4-picoline	PC	84
2-Amino-6-picoline	PC	84
α-Amino-pimelic acid	PC	64
Aminopromazine	TLC	139
3-Aminopropanol	PC	7
Aminopropiron	TLC	40
1-(4-Amino-2-*n*-propyl-5-pyrimidinylmethyl)-2-picolinium chloride hydrochloride	TLC	136
2-Aminopurine	LC	46
	PC	36
1-Aminopyrene	TLC	120
2-Aminopyridine	PC	84
	TLC	7, 60
3-Aminopyridine	PC	84
	TLC	60
4-Aminopyridine	PC	84
	TLC	60
2-Aminopyrimidine	PC	36
Aminopyrine	GC	21
Aminopyrine (pyramidon)	PC	84
8-Aminoquinoline	PC	104
4-Aminosalicylic acid	LC	35
5-Aminosalicylic acid	LC	35
2-Aminostilbene	PC	6
2-Aminostilbene (*trans*) pentafluoropropionamide	GC	99
3-Aminostilbene (*trans*) pentafluoropropionamide	GC	99
4-Aminostilbene (*trans*) pentafluoropropionamide	GC	99
2-Aminothiazole	TLC	7
2′-Aminothiazolo[*d*-3,2]-5α-cholest-2-ene	TLC	96
2′-Aminothiazolo[*d*-3,2]-17α-methyl-5α-androst-2-en-17β-ol	TLC	96
o-Aminothiophenol	PC	121
p-Aminothiophenol	PC	121
6-Amino-2-thiouracil	PC	36
N-(3-Amino-2,4,6-triiodobenzoyl)-*N*-phenyl-β-aminopropionic acid	TLC	64
5-Aminouracil	LC	17
	PC	36, 37
6-Aminouracil	PC	36, 37
5-Aminouridine	PC	36
α-Amino-*n*-valeric acid	PC	64
α-Amino-*n*-valeric acid (Norvaline)	TLC	15
δ-Amino-*n*-valeric acid	PC	64
Amitriptyline	TLC	39
Amitrole	TLC	81
Ammarvomycin	TLC	20
Amminosidin	TLC	20
Ammonia	LC	4, 33, 41
	PC	55
Ammonium *n*-butyl *n*-propyl phosphate	PC	34
Ammonium chloride (NH_4Cl)	LC	2
Ammonium dibenzyl hydrogen phosphate	PC	34
Ammonium diethyl phosphate	PC	34
Ammonium ethyl *n*-propyl phosphate	PC	34
Ammonium hydroxide	LC	34
Ammonium nitrate	TLC	157
Amobarbital	GC	105, 106
Amobarbital (Amylobarbitone)	GC	19
Amobarbital (Amytal)	GC	21
Amobarbital (Amytal, Amylobarbitone)	TLC	137
Amobarbital metabolite	GC	106
Amolanone	TLC	39
α-Amorphene	GC	140
γ-Amorphene	GC	140
AMP (Adenosine monophosphate)	LC	3, 17
	PC	37
AMP-3′ (Adenosine-3′-phosphate)	PC	32
AMP-5′ (Adenosine-5′-phosphate)	PC	32
d-AMP	LC	3, 18
	TLC	63
2′-AMP	LC	17
2′,3′-AMP	LC	46
5′-AMP	LC	22
	TLC	63
Amphetamine	GC	9, 21, 22,23, 105
	LC	55
	TLC	10, 39, 42
Amphomycin	TLC	20
Amphotericin	TLC	20
Amphotericin A	TLC	20
Amphotericin B	TLC	21
Ampicillin Na	TLC	21
Amprolium	TLC	136

Amydricaine	PC	4
Amyl acetate	GC	57
	TLC	105
iso-Amyl acetate	GC	57
Amyl alcohol	GC	53
n-Amyl alcohol	LC	5, 11
sec-Amyl alcohol	TLC	1
tert-Amyl alcohol	GC	55,125, 126
	LC	11
n-Amyl alkyl guanidine	PC	9
Amylamine	TLC	10
iso-Amylamine	PC	55
n-Amylamine	LC	26
	PC	55, 56
n-Amylamine, trifluoroacetyl derivative	GC	10
N-iso-Amylaniline	GC	102
Amyl *n*-butyrate	GC	57
iso-Amyl *n*-butyrate	GC	57
tert-Amyl chloride	GC	53
α-Amylcinnamaldehyde, 2-hydrazine-benzothiazole derivative of	TLC	168
S-Amyl cysteine	PC	64
1-*N*-Amyl-2,5-dimethylpyrrole	GC	133
iso-Amylene (galegine) guanidine	PC	9
Amyl formate	GC	57
iso-Amyl formate	GC	57
Amyl gallate	TLC	125
iso-Amyl guanidine	PC	9
n-Amyl guanidine	PC	9
n-Amyl-*n*-heptyl sebacate	GC	25
tert-Amyl hydroperoxide	PC	27
Amyl isobutyrate	GC	57
iso-Amyl isobutyrate	GC	57
Amyl mercaptan	TLC	162
Amylobarbitone	PC	86
Amylocaine	PC	4
n-Amyl-*n*-octyl succinate	GC	25
4-*tert*-Amylphenol	GC	72
Amyl *n*-propionate	GC	57
iso-Amyl propionate	GC	57
n-Amyl sulfoxide	GC	139
α-Amyrin acetate	LC	45
Anabasine	GC	95
	PC	2
Anagyrine	PC	3
Anatabine	GC	95
Androsta-1,4-diene-3,17-dione	PC	40, 42
Androsta-4,9(11)-diene-3,17-dione	PC	40
Androsta-4,11-diene-3,17-dione	PC	40
Androsta-1,4-dien-17β-ol-3-one	PC	42
Androsta-1,4-diene-3,6,17-trione	PC	40
1,4-Androstadiene-3,17-dione	GC	136
	PC	124
	TLC	100
4,9(11)-Androstadiene-3,17-dione	PC	124
Androstane derivatives	PC	42
Androstanes (steroids)	GC	80
5α-Androstane	GC	80, 81, 137
	LC	6, 45
5β-Androstane	GC	80, 81, 137
Androstane-3α,17β-diol	GC	79
Androstane-3β,16β-diol	GC	79
Androstane-3β,17β-diol	GC	79
5α-Androstane-3α,17β-diol	GC	136, 137
	PC	42
	TLC	179

5α-Androstane-3β,17α-diol	TLC	98
5α-Androstane-3β,17β-diol	GC	80, 137
	TLC	98, 179
5β-Androstane-3α,17β-diol	GC	80, 137
	PC	42
	TLC	179
5β-Androstane-3β,17β-diol	GC	137
	TLC	98
5α-Androstane-3α,11β-diol-17-one	GC	81
5α-Androstane-3β,11β-diol-17-one	GC	136
5β-Androstane-3α,11β-diol-17-one	GC	136
	PC	42
5β-Androstane-3α,12α-diol-17-one	TLC	100
Androstane-3,16-dione	GC	79
Androstane-3,17-dione	GC	79
	PC	124
	TLC	178
5α-Androstane-3,17-dione	GC	80, 81, 137
	PC	40, 42, 124
	TLC	100,179
5β-Androstane-3,17-dione	GC	80, 81, 137
	PC	40, 42
	TLC	179
Androstane-5-en-3β-ol,17-one	GC	137
5α-Androstane-17α-methyl-17β-ol-3-one	TLC	100
5α-Androstane-3α-ol	GC	81
5α-Androstane-3β-ol	GC	81, 137
5α-Androstane-17β-ol	GC	81, 137
5β-Androstane-3α-ol	GC	80
5β-Androstane-3β-ol	GC	137
5β-Androstane-17β-ol	GC	137
5α-Androstane-3α-ol,11,17-dione	GC	81
5α-Androstane-3β-ol,3,16-dione	GC	81
Androstane-3α-ol-17-one (Androsterone)	TLC	178
5β-Androstane-3α-ol,11β-ol,17-one	GC	81
5α-Androstane-3α-ol,17-one	GC	137
5α-Androstane-3α-ol,17-one (Androsterone)	GC	81
5α-Androstane-3β-ol,17-one	GC	137
5β-Androstane-3α-ol,17-one	GC	81, 137
5β-Androstane-3β-ol,17-one	GC	81, 137
5β-Androstane-17β-ol,3-one	GC	81
5α-Androstane-3-one	GC	137
5α-Androstane-17-one	GC	137
5β-Androstane-3-one	GC	137
5β-Androstane-17-one	GC	137
5α-Androstane-3-one, 17α-ol	GC	137
5α-Androstane-3-one, 17β-ol	GC	137
5β-Androstane-3-one, 17α-ol	GC	137
5β-Androstane-3-one, 17β-ol	GC	137
5α-Androstane-3,7,17-trione (Group III)	PC	40
5α-Androstane-3,11,17-trione	GC	80
5β-Androstane-3,11,17-trione	GC	80
	PC	40, 41
	TLC	179
5α-Androstane-3,11,17-trione (Group III)	PC	40
5β-Androstane-3,11,17-trione (Group III)	PC	40, 41
5β-Androstane-3,12,17-trione	TLC	100
5α-Androstan-3α-ol	TLC	98
5α-Androstan-3β-ol	GC	80
	TLC	98

5α-Androstan-17-ol	TLC	179
5α-Androstan-17α-ol	GC	80
5α-Androstan-17β-ol	GC	80
	LC	6
	TLC	98
5α-Androstan-3α-ol-11,17-dione	GC	136
	TLC	100
5β-Androstan-3α-ol-11,17-dione	GC	136
5β-Androstan-3α-ol-12,17-dione	TLC	100
5β-Androstan-12α-ol-3,17-dione	TLC	100
Androstanolone	GC	136
	TLC	95
Androstan-3α-ol-17-one	GC	79
Androstan-3β-ol-17-one	GC	79
5α-Androstan-3α-ol-16-one	TLC	98
5α-Androstan-3β-ol-16-one	TLC	98
5α-Androstan-3α-ol-17-one	GC	136
	PC	42
	TLC	98, 100
5α-Androstan-3β-ol-17-one	GC	136
	PC	42
	TLC	98, 100
5α-Androstan-17β-ol-3-one	GC	136
	TLC	100
5β-Androstan-3α-ol-17-one	GC	136
	PC	42
	TLC	98
5β-Androstan-3β-ol-17-one	PC	42
	TLC	98
5β-Androstan-17β-ol-3-one	PC	42
	TLC	98
Androstan-17-one	GC	79
5α-Androstan-3-one	GC	80
5α-Androstan-11-one	TLC	179
5α-Androstan-17-one	GC	80
	LC	6
	PC	42
	TLC	179
5β-Androstan-11-one	TLC	179
Androstenediol	TLC	95
Androstenediol dipropionate	TLC	95
Androst-4-ene-3β,17β-diol	TLC	179
Androst-5-ene-3β,17α-diol	PC	42
	TLC	98
Androst-5-ene-3β,17β-diol	GC	80
	LC	62
	PC	42
	TLC	98
5-Androstene-3β,17α-diol	GC	81
5-Androstene-3β,17β-diol	GC	79, 81, 136
	PC	124
5-Androstene-5β,17α-diol	GC	136
Androst-4-ene-11β,17β-diol-3-one	PC	42
5-Androstene-3β,16α-diol,17-one	GC	81
5-Androstene-3β,17β-diol,16-one	GC	81
Androstenedione	GC	83
Androst-4-ene-3,16-dione	TLC	98
Androst-4-ene-3,17-dione	GC	80
	PC	40, 42
	TLC	98, 179
Δ^4-Androstene-3,17-dione	TLC	100
4-Androstene-3,16-dione	GC	79
4-Androstene-3,17-dione	GC	79, 81, 136
	PC	41, 124
	TLC	178
5α-Androst-1-ene-3,17-dione	PC	40
	TLC	179
5α-Androst-2-ene-7,17-dione	TLC	179
5α-1-Androstene-3,17-dione	TLC	100
Δ^4-Androsten-17α-ethynyl-17β-ol-3-one	TLC	100
Δ^4-Androstene-17α-methyl-17β-ol-3-one	TLC	100
4-Androstene-11α-ol-3,17-dione	GC	79
4-Androstene-11β-ol-3,17-dione	GC	79
4-Androstene-14α-ol,3,17-dione	GC	81
5-Androstene-3β-ol,16-one	GC	81
5-Androstene-3α-ol-17-one (Dehydroandrosterone)	TLC	178
5-Androstene-3β-ol,17-one (DHA)	GC	81
4-Androstene-3-one,17α-ol	GC	81
4-Androstene-3-one,17β-ol Testosterone)	GC	81
5-Androstene-3β,16α,17α-triol	GC	136
Androst-4-ene-3,6,17-trione	GC	80
	PC	40
Androst-4-ene-3,11,17-trione	GC	80
	PC	40, 42
Δ^4-Androstene-3,11,17-trione	GC	83
	TLC	100
4-Androstene-3,6,17-trione	PC	124
4-Androstene-3,11,17-trione	GC	79, 81
	PC	124
	TLC	178
5α-Androst-l-ene-3,11,17-trione	PC	40
Androst-5-en-3β-ol	GC	80
	PC	42
	TLC	98
Androst-5-en-17β-ol	GC	80
Androst-4-en-11β-ol-3,17-dione	PC	42
Androst-4-en-17α-ol-3-one	PC	42
Androst-4-en-17β-ol-3-one	PC	42
	TLC	98
Δ^4-Androsten-17β-ol-3-one	TLC	100
4-Androsten-17α-ol-3-one	GC	136
4-Androsten-17β-ol-3-one	GC	79, 136
Androst-5-en-3β-ol-17-one	PC	42
	TLC	98
5-Androsten-3β-ol-17-one	GC	79, 136
5β-Androst-9(11)-en-3α-ol-17-one	PC	42
3α-OH-4-Androstenone	TLC	97
3β-OH-4-Androstenone	TLC	97
Androst-4-en-3-one	GC	80
5α-Androst-2-en-17-one	GC	80
	PC	40
Androsterone	GC	83, 136
	TLC	97
IIβ-OH-Androsterone	TLC	97
5β-Androstrostane-3α-ol,11,17-dione	GC	81
Anethol	TLC	90
trans-Anethole	GC	87
Angolamycin	PC	66
	TLC	153
Anhydrodemethylchlortetracycline	PC	68
"α"-Anhydrodigitoxigenin	PC	127
"β"-Anhydrodigitoxigenin	PC	127
"β"-Anhydrodigoxigenin	PC	127
"β"-Anhydrodigoxigenone	PC	127
1,6-Anhydro-β-D-glucofuranose	LC	48
2,5-Anhydro-D-gluconic (chitaric)	LC	23
1,6-Anhydro-β-D-glucopyranose	LC	48
2,5-Anhydro-D-mannonic (chitonic)	LC	23
"β"-Anhydrosarmentogenin	PC	127
2,5-Anhydro-D-talonic	LC	23
Anhydrotetracycline	PC	68

1,6-Anhydro-2,3,4-tri-*O*-methyl-β-D-glucopyranose	TLC	26
"β"-Anhydrouzarigenin	PC	127
Anhydrovitamin A_1	PC	49
	TLC	108
Anhydrovitamin A_2	PC	48
Anileridine (Leritine)	GC	21, 105
	TLC	138
Aniline	GC	45
	LC	26, 36, 53
	PC	6, 59
	TLC	7,119
Aniline, trifluoroacetyl derivative	GC	10
1-Anilino-anthraquinone	TLC	69
Anisaldazine	PC	23
Anisaldehyde	PC	22
	TLC	77, 174
Anisaldehyde, 2,4-DNPH of	TLC	72, 73, 74, 167
Anisaldehyde, 2-hydrazinebenzothiazole derivative of	TLC	168
Anisaldehyde, 4-NPH of	TLC	74
p-Anisaldehyde	PC	20
Anisic acid	TLC	174
m-Anisidine	PC	121
o-Anisidine	PC	121
p-Anisidine	PC	120, 121
Anisole	GC	51, 73, 75
	LC	7, 21
Anserine	LC	41
	PC	17
	TLC	155
Antazoline	GC	20
	PC	85
Antazoline (Antistine®)	GC	21
	TLC	39, 45
Anthanthrene	GC	40
	PC	89
	TLC	51, 146
Anthracene	GC	40, 45
	PC	89
	TLC	7, 51, 53, 146
9,10-Anthradiol	PC	121
Anthragallol	PC	109
1-Anthramine	TLC	120
2-Anthramine	TLC	120
9-Anthramine	TLC	120
Anthranilic acid	LC	30, 53, 72
	PC	64, 82, 102, 103
Anthranilic acid, methyl ester of	GC	24
Anthranilic acid, phenylethyl ester of	GC	24
Anthranylglycine	PC	80
Anthraquinone	PC	110
1-Anthrol	TLC	92
Anthrone	TLC	75
Antiarigenin (3β,5β,12β,14β-tetrahydroxy-19-oxo-cardenolide)	TLC	102
β-Antiarin	TLC	102
β-Antiarin peracetate	TLC	102
Antimony, Sb	PC	15
Antimony, Sb^{3+}	PC	97
	TLC	150
Antimony, Sb(III)	PC	95, 96

Antimony pentachloride	GC	50
Antimony trichloride	GC	50
Antimycin A	TLC	20
Antioxidant 425	TLC	175
Antioxidant 2246	TLC	175
Antipyrine	GC	21
Antistine (Antazoline)	TLC	39
Antu®	GC	62
Ao (Adenosine)	PC	32
Apesan	TLC	42
Apholate	GC	65
Apigenin	LC	28
	PC	26
	TLC	92
Apigeninidin	PC	25
Apiin	PC	26
Apoatropine	GC	109
	PC	3
	TLC	2, 3, 117
Apocannoside	TLC	102
Apocannoside peracetate	TLC	102
β-Apo-8′-carotenal	TLC	94
β-Apo-8′-carotenic acid	TLC	93
Apomorphine	GC	8, 109
	PC	4
Apomorphine HCl	GC	109
Aporheine	PC	3
Aposcopolamine	TLC	117
Aprobarbital	GC	19,105, 106
	TLC	137
Aprobarbitone	PC	86
Aquatriammine cobalt tribromide $[Co(NH_3)_3(H_2O)]Br_3$	PC	94
Ar, Argon	GC	112
L-Arabinitol	GC	14, 15
Arabinose	LC	25, 29, 48
	PC	70
	TLC	23, 24
D-Arabinose	TLC	126
L-Arabinose	GC	15
	PC	75
	TLC	126
D-Arabinonic	LC	40
Arabinuronic	LC	23
Arabitol	LC	69
	PC	1
	TLC	127
L-Arabitol	GC	14
	TLC	127
Arabitol pentaacetate	PC	72
L-Arabono-γ-lactone	PC	72
Arachidate	LC	68
Arachidic acid	PC	77
	TLC	33
Arachidonate, β-sitosteryl ester of	TLC	182
Arachidyl alcohol	LC	45
Aramite®	GC	64, 129
	TLC	80
Arbutin	LC	1
	PC	121
Arcaine · H_2SO_4 (L-4-diguanidobutane)	PC	10
Arecoline	TLC	3
Arginine	GC	101
	LC	4, 33, 41, 66

Arginine (*continued*)	PC	9, 10, 11, 62, 64, 65
	TLC	11, 13, 15, 118
Arginine, DNAP of	TLC	14
Arginine, PTH of	TLC	124
Arginine hydrochloride	LC	56
	TLC	16
Argininol	PC	57
Argininosuccinic acid	LC	59
	PC	9, 64
	TLC	13
Argininosuccinic anhydride	PC	9, 64
Argon, Ar	GC	112
Aribine	GC	22
Armepavine	PC	3
Aromadendrene	GC	140
Aromatic acids	PC	82
Aromatics	GC	87, 126
Aromatics, *m*-xylene	GC	126
Aromatics, *o*-xylene	GC	126
Aromatics, *p*-xylene	GC	126
Aromatics, toluene	GC	126
Aromatics, ethylbenzene	GC	126
Arsanilic acid	TLC	136
m-Arsanilic acid	LC	35
o-Arsanilic acid	LC	35
p-Arsanilic acid	LC	35
Arsenate	PC	14, 91, 93
Arsenate, $AsO_4{}^{3-}$	PC	97
Arsenic, As^{3+}	PC	97
Arsenic, As(III)	PC	15, 95, 96
	TLC	150
Arsenic trichloride	GC	50
Arsenite	PC	14, 91, 93
As(III), Arsenic	PC	15, 95, 96
As^{3+}, Arsenic	PC	97
	TLC	150
Ascorbic acid	LC	20, 29
	PC	13
	TLC	188
L-Ascorbic acid	TLC	188
$AsO_4{}^{3-}$, Arsenate	PC	97
Asparagine	LC	41
	PC	62, 64
	TLC	11, 13, 15, 16, 118
Asparagine, PTH of	TLC	124
iso-Asparagine	PC	64
Aspartate acid	TLC	36
l-Aspartate acid	TLC	36
Aspartic acid	GC	12, 101
	LC	2, 4, 33, 41, 56
	PC	11, 62, 64, 65
	TLC	11, 13, 15, 16, 118
Aspartic acid, DNAP of	TLC	14
Aspartic acid, PTH of	TLC	124
Aspartidiol	PC	57
Aspertoxin	TLC	185
Aspidospermine	TLC	3
Aspirin	GC	104
Aspon	LC	43
Astacin	TLC	93
Astilbin	LC	28
Astragalin	PC	26
	TLC	92
Atarax®	TLC	42
Atarax®(Hydroxyzine)	TLC	39, 45
Atebrin = quinacrine	PC	84
ATP (Adenosine triphosphate)	LC	3, 13, 47
	TLC	63
d-ATP	TLC	63
Atranorin	PC	30
	TLC	34
Atratone	GC	66
	TLC	86
Atrazine	GC	62, 66, 68, 128
	TLC	81, 86
Atropine	GC	21, 96, 109, 111
	PC	2, 4
	TLC	2, 3, 39, 45, 117
Atropine (d-hyoscyamine)	GC	8
Atropine sulfate	GC	109
Au(III), Gold	PC	95, 96
Auranetin	PC	26
Aureothricin	TLC	20
Aureusidin	PC	26
Aureusin	PC	26
Aurone aglycones	PC	26
Aurone, glucosides	PC	26
Avicularin	PC	26
8-Azaadenine	LC	46
4-Aza-5-cholesten-3-one	TLC	96
Azacyclonol	GC	17
	TLC	39, 42
Azacyclonol hydrochloride	GC	17
8-Aza-2,6-diaminopurine	LC	46
8-Azaguanine	LC	46
3-Aza-A-homo-5α-cholestan-4-one	TLC	96
3-Aza-A-homo-5β-cholestan-4-one	TLC	96
7a-Aza-β-homo-5α-cholestan-3α-ol-7-one acetate	TLC	96
8-Azahypoxanthine	LC	46
Azaleatin	LC	28
	PC	26
4-Aza-5-pregnene-3,20-dione	TLC	96
6-Azathymine	PC	36
6-Azauridine	PC	36
8-Azaxanthine	LC	46
Azelaic acid	PC	13, 77
	TLC	31
Azelaic acid, dimethyl ester of	GC	24
Azinphosethyl	GC	65
	LC	43
	TLC	81, 84
Azinphosmethyl	TLC	81
Azinphosmethyl (Guthion®) (Gusathion®)	GC	65, 67
	LC	43
	TLC	84
Azobenzene	TLC	7, 48
Azodrin®	GC	65, 67
Azoxybenzene	TLC	7
Azulene	GC	40, 51
	TLC	106, 143

Adsorption of the Elements (LC Section – Figures)

Ac, Actinium	Pg	278, 279, 280, 281
Ag, Silver	Pg	278, 279, 280, 281
Al, Aluminum	Pg	278, 279, 280, 281
Am, Americium	Pg	278, 279, 280, 281
As, Arsenic	Pg	278, 279, 280, 281
At, Astatine	Pg	278, 279, 280, 281
Au, Gold	Pg	278, 279, 280, 281

Ba^{2+}, Barium	PC	92, 97
Ba(II), Barium	GC	43
	PC	95, 96
	TLC	149, 150
Bacitracin	LC	4
	TLC	20, 21
Badional	TLC	140
Baeomycesic acid	PC	30
	TLC	34
Baicalein	PC	26
Baikaiin	PC	64
Banol	GC	130
Barbatic acid	PC	30
	TLC	35
Barbatolic acid	TLC	34
Barbital (Barbitone)	GC	19, 21, 105, 106
	TLC	40, 137
Barbitone	PC	86
Barbitone (Barbital)	TLC	40, 137
Barbituric acid	PC	36
Barium, Ba^{2+}	PC	92, 97
Barium, Ba(II)	GC	43
	PC	95, 96
	TLC	149, 150
Batallylonal	GC	106
Bay 30911	GC	65
Bay 37289	GC	65
Bay 37342	GC	65
Bay 77488	GC	65
Baycid	TLC	172
Bayer 25141 (Dasanit)	TLC	80, 84
Bayer 29493 (Fenthion)	TLC	84
Bayer 37344 (Mesurol)	GC	130
Baygon	GC	130
Be(II), Beryllium	GC	43
	PC	96
Be^{++}, Beryllium	TLC	150
Be^{2+}, Beryllium	PC	92, 97
Behenate	LC	67
Behenic acid	PC	77
	TLC	33
Behenolic acid	TLC	33
Behenyl alcohol	LC	45
Belladine	GC	96
Belladonnine	PC	3
	TLC	117
Bemegride (Megimide)	TLC	137
Benactyzine	PC	85
	TLC	39
Benadryl® (Diphenhydramine)	TLC	39
Bendrofluazide	TLC	41
Bendroflumethiazide	TLC	41
Benfluralin	TLC	81
Benoxinate HCl	GC	110
Benz[*c*]acridine	TLC	120
1,2-Benzacridine	GC	45
3,4-Benzacridine	GC	45
	TLC	58
Benzalacetophenone	PC	22
Benzalazine	PC	23
Benzaldehyde	GC	7, 51, 53
	PC	20, 22
	TLC	71, 77, 174
Benzaldehyde, 2,4-DNPH of	TLC	72, 73, 74, 165
Benzaldehyde, 2,4-DNPSC of	TLC	165
Benzaldehyde, 2-hydrazinebenzothiazole derivative of	TLC	168
Benzaldehyde, *p*-NPH of	TLC	165
Benzaldehyde, 4-NPH of	TLC	74
Benzamid-residue	TLC	84
Benz(*a*)anthracene	TLC	51, 146
1,2-Benzanthracene	PC	90
Benzanthrone	TLC	75, 120
Benzene	GC	7, 31, 47, 49, 51, 53, 55, 58, 93, 120, 121, 125, 127
	LC	7, 27, 36, 42, 51
Vg-Benzene	GC	135
Benzene (aromatics)	GC	126
Benzene hexachloride	LC	75
Benzene (hydrocarbons)	GC	118
Benzene, organics	GC	52
Benzene-*p*-sulfonic acid-azo-*m*-cresol (BAC-S)	TLC	47
Benzene-*p*-sulfonic acid-azo-resorcinol (BAR-S)	TLC	47
Benzenesulfonic acid *N*-butylamide	TLC	27
11H-Benz(*b*)fluorene	TLC	51
Benz(*k*)fluoranthene	TLC	51
Benzhexol	PC	85
Benzhydrol	PC	53
Benzidine	PC	6, 59
	TLC	119
Benzidine pentafluoropropionamide	GC	99
Benzimidazole	LC	46
4,5-Benzindane	TLC	51
1,2-Benzoanthracene	GC	40
Benzocaine (Ethoform)	TLC	39, 43
4H-Benzo[*d,e,f*] carbazole	TLC	120
5H-Benzo[*b*] carbazole	TLC	120
7H-Benzo[*c*] carbazole	TLC	120
11H-Benzo[*a*] carbazole	TLC	120
7:8-Benzoflavone	TLC	75
Benzo(*g,h,i*)fluoranthene	GC	40
Benzo[*k*] fluoranthene	TLC	146
2,3-Benzofluoranthene	GC	40
3,4-Benzofluoranthene	GC	40
11,12-Benzofluoranthene	GC	40
1,2-Benzofluorene	GC	40
	TLC	53
2,3-Benzofluorene	GC	40
	PC	89
	TLC	53
3,4-Benzofluorene	GC	40
11H-Benzo[*b*] fluorene-11-one	TLC	75
Benzofuran	GC	119
2-Benzohydryl-1,3-indanedione	TLC	79
Benzoic acid	GC	4, 5
	LC	30, 51, 53, 74
	PC	82

Benzoic acid (*continued*) TLC 37, 129, 174
Benzoic acid, amyl ester of GC 24
Benzoic acid, butyl ester of GC 24
Benzoic acid, hexyl ester of GC 24
Benzoic acid, isoamyl ester of GC 24
Benzoic acid, methyl ester of GC 24
Benzoin PC 53
Benzomarc TLC 173
Benzo[*g,h,i*] perylene GC 40
TLC 146
Benzo[*c*] phenanthrene PC 90
TLC 146
3,4-Benzophenanthrene GC 40
Benzophenone PC 22
TLC 7
Benzophenone, 2,4-DNPH of TLC 73, 74
Benzophenone, 4-NPH of TLC 74
Benzophenone azine PC 23
Benzo[*a*] pyrene TLC 120, 146
Benzo[*e*] pyrene TLC 146
3,4-Benzo[*a*] pyrene TLC 51
3,4-Benzopyrene PC 89
6H-Benzo[*c,d*] pyrene-6-one TLC 75
13H-Benzo[*g*] pyrido[3,2-*a*]-carbazole TLC 120
13H-Benzo[*g*] pyrido[2,3-*a*]-carbazole TLC 120
3,4-Benzoquinoline TLC 58
5,6-Benzoquinoline TLC 58
7,8-Benzoquinoline TLC 58
Benzoquinone TLC 174
p-Benzoquinone PC 111
TLC 88, 174
Benzoquinones PC 111
g-Benzoquinoxalino[*b*-2,3]-5α-cholestane TLC 96
Benzotetraphene GC 40
1-(2′-Benzothiazolylazo)-2-naphthol (BTAN) TLC 47
4-(2′-Benzothiazolylazo)-resorcinol (BTAR) TLC 47
2-Benzothiazyl-*N,N*-diethylthio-carbamyl sulfide TLC 66
Benzothiazyl disulfide TLC 66
Benzo(*b*)thiophene GC 78
Benzoylcholine PC 8
6-Benzoylchrysene TLC 75
1-Benzoyl-2,2-dimethylhydrazine PC 61
Benzoyl peroxide PC 27
TLC 65
2-Benzoylpyrene TLC 75
2-Benzoylpyridine TLC 60
N^1-Benzoylsulfanilamide TLC 44
2-Benzoylthiophene TLC 104
Benzphetamine GC 23
TLC 39
Benz[*a*] pyrene TLC 51
Benz(*e*)pyrene TLC 51
1,2-Benzpyrene GC 40
3,4-Benzpyrene GC 40
Benzthiazide TLC 41
Benzyl LC 53
Benzyl alcohol GC 51
LC 11, 54
PC 52
TLC 1, 7
Benzyl alkyl guanidine PC 9
Benzylamine GC 45
LC 26
PC 5, 56, 60
TLC 5, 7, 10
4-Benzyl-4-aza-5-cholesten-3-one TLC 96
Benzylbromide GC 51
Benzylbutyl phthalate (phthalic acid ester) TLC 27
S-Benzyl cysteine PC 64
2-*O*-Benzyl-3,4:5,6-di-*O*-isopropylidene-D-mannose dimethylacetal GC 13
3-*O*-Benzyl-1,2:4,5-di-*O*-isopropylidene-D-fructose GC 13
Benzyl esters GC 24
Benzyl formate TLC 105
Benzyl guanidine PC 9
3,3′-Benzylidene-bis-(4-hydroxy-coumarin) TLC 79
Benzyl isocyanate GC 44
Benzyl mercaptan TLC 162
4-Benzyl-17α-methyl-4-aza-5-androsten-17β-ol TLC 96
Benzyl methyl ketone TLC 73
Benzyl octyl adipate (adipic acid ester) TLC 27
4-Benzyloxy-5-ethyl-2-methylpyridine *N*-oxide TLC 61
5-Benzyloxyindole-3-acetic acid PC 103
S-Benzyl penicillamine PC 64
Benzylpenicillin Na TLC 21
4-Benzylphenol GC 71
Benzylphosphate PC 112
2-Benzylpyridine GC 45
4-Benzylpyridine GC 45
Benzyl-α-pyridyl-phosphate PC 112
N-Benzyl-3,5-seco-4-norcholestan-5β-ol-3-amide TLC 96
N-Benzylthiourea PC 47
Benzylurea TLC 122
Benzylvanillin TLC 68
BEP ester 2,4-D (tech) GC 129
BEP ester 2,4,5-T GC 129
BEP ester 2,4,5 (tech) GC 129
Berberine GC 8
PC 2, 3
Bergapten PC 113
Beryllium, Be(II) GC 43
PC 96
TLC 149
Beryllium, Be^{2+} PC 92, 97
TLC 150
Betaine GC 8
PC 2, 8
Betanaphthol PC 31
Betazole GC 17
TLC 39
Betazole dihydrochloride GC 17
Bethanidine LC 55
Betulinic TLC 184
B_4H_{10} GC 27
B_5H_9 GC 27
B_5H_{11} GC 27
BHA (butylated hydroxyanisole) PC 28
BHC GC 129
PC 116
TLC 82
α-BHC GC 60, 68,

α-BHC (*continued*)		128
	TLC	81, 82
β-BHC	GC	60, 68, 128
	TLC	82
γ-BHC	TLC	81, 82
δ-BHC	GC	68, 128
	TLC	81
BHC (alpha)	GC	129
BHC (beta)	GC	129
BHC (delta)	GC	129
BHC (T)	TLC	82
BHC (tech)	GC	129
BHT (butylated hydroxytoluene)	PC	69
Bi(III), Bismuth	PC	95, 96
	TLC	148, 149, 150
Bi^{3+}, Bismuth	PC	92, 97
	TLC	148, 149, 150
Biacetyl	GC	55
	PC	20
Bibenzyl	TLC	7
Bicyclics	GC	87
Bidrin®	GC	65, 67
	LC	43
	TLC	84, 171
Bilirubin	PC	84
Biliverdin	PC	84
Binapacryl (Morocide®)	TLC	80
Biochanin A	PC	26
Biotin	PC	51
	TLC	188
(+)-Biotin	PC	54
Biphenyl	GC	38
	LC	21
	TLC	7, 51
Bipindogenin (3β,5β,11α,14β-tetrahydroxy-cardenolide)	TLC	102
β-Bisabolene	GC	140
	TLC	54
γ-Bisabolene	GC	140
	TLC	106, 143
2,2-Bis(*tert*-butylperoxy)-butane	TLC	65
N,N-Bis-(β-choroethyl)-3β-amino-5α-cholestane hydrochloride	TLC	96
4,4′-Bis(2,6-di-*tert*-butylphenol)	TLC	175
Dg-bisdigitoxoside	TLC	102
Dg-bisdigitoxoside peracetate	TLC	102
4,4′-Bis(dimethylamino)benzophenone	TLC	7
Bis(2-ethylbutyl)phthalate	GC	132
Bis(2-ethylhexyl)adipate	GC	132
Bis(2-ethylhexyl)isophthalate	GC	132
Bis(2-ethylhexyl)phthalate	GC	132
2,2-Bishydroxymethyl-1-butanol	TLC	113
2,2-Bishydroxymethyl-1-propanol	TLC	113
1,1′-Bis-(4-hydroxyphenyl)-cyclohexane	TLC	22
2,6-Bis-(2′-hydroxyphenyliminomethyl)-pyridine (BPOP)	TLC	47
1,1-Bis-(4-hydroxyphenyl)-2,2,2-trichloroethane	TLC	81
Bismate	TLC	66
Bis(2-methoxyethyl)phthalate	GC	132
Bismuth, Bi(III)	PC	95, 96
	TLC	148, 149, 150
Bismuth, Bi^{3+}	PC	92, 97
	TLC	148, 149, 150
Bismuth dimethyldithiocarbamate	TLC	66
Bisphenol "A" (*p,p′*)	LC	61
o,p′-Bisphenol "A"	LC	61
p,p′-Bisphenol diglycidyl ether	LC	60
2,6-Bis-(salicylideneamino)-pyridine (diSAMPy)	TLC	47
Bis(4-thiuridine)disulfide	LC	22
Bis(3,5,5-trimethylhexyl)phthalate	GC	132
Bis[tris(ethylenediamine)chromium] trisulfate $[Cr(en)_3]_2\ (So_4)_3$	PC	94
Bithienyl	TLC	104
Bithienyl-5-carboxylic acid	TLC	104
2,2′-Bithienyl-methylamine hydrochloride	TLC	104
Biuret	LC	63, 73
Biuret = carbamylurea	PC	60
Bixin	TLC	94
Black 1	TLC	141
Black PN	TLC	49
Blasticidin S	TLC	20
Blastmycin	TLC	20
Blue 1	TLC	141
Blue 2	TLC	141
Blue Dextran 2000	LC	18
Blue VRS	TLC	49
Boldine	PC	2, 54
	TLC	3
Bomyl®	TLC	84
Boninic acid	TLC	35
Borate	PC	14
Borneol	GC	87
	TLC	183
Bornyl acetate	TLC	105
Bornylene	GC	86, 87
Bornyl guanidine	PC	9
Boron trichloride	GC	26, 27
Botogenin	TLC	98
Botran	GC	62
α-Bourbonene	GC	140
β-Bourbonene	GC	140
Bovine serum albumin	LC	64
Bovogenin A (3β,14β -Dihydroxy-19-oxo-5α-bufadienolide)	TLC	102
Bovoside A	TLC	102
Bovoside A peracetate	TLC	102
Br^-, Bromine	TLC	147
Brallobarbital	TLC	137
Branched paraffins	GC	126
Brassicasterol	GC	84
	TLC	99
Brazilin	PC	54
Bredemolic	TLC	184
Brestan® (Triphenylzinc acetate)	PC	117
BrO_3^-	TLC	147
Bromacil	GC	61
	TLC	81
5(2-Bromallyl)-5-isopropyl barbituric acid	GC	19
Bromate	PC	14, 91, 93
Bromcresol green	PC	88
Bromcresol purple	PC	88
Bromide	PC	14, 91, 93
Bromide, vinyl	GC	26
Bromine$^-$, Br^-	TLC	147
Bromine pentafluoride	GC	26, 114
Bromisovalum (Bromvaletone)	TLC	137
Bromoacetic acid	PC	79
	TLC	29

p-Bromoaniline	PC	59
2-Bromoaniline	GC	102
3-Bromoaniline	GC	102
	TLC	119
4-Bromoaniline	GC	102
	PC	6
	TLC	119
Bromobenzene	GC	51, 53
	LC	75
m-Bromobenzoic acid	TLC	37
o-Bromobenzoic acid	TLC	37
p-Bromobenzoic acid	TLC	37
1-Bromobutane	GC	56
Bromobutyric	PC	79
1-Bromo-2-chloroanthraquinone	PC	110
2-Bromo-4-chloro-6-propionylphenol	TLC	89
Bromocycloheptane	GC	51
Bromocyclohexane	GC	51
	LC	36
Bromocyclooctane	GC	51
Bromocyclopentane	GC	51
5-Bromodeoxyuridine	PC	36
5-Bromo-2,4-dihydroxybenzoic acid	PC	120
1-Bromo-2,2-dimethylpropane	GC	34
2-Bromo-3,4-dimethyl-6-propionylphenol	TLC	89
Bromodiphenhydramine (Ambodryl®)	GC	21
	TLC	39
Bromoethane	GC	56
2-Bromo-4-fluoro-6-propionylphenol	TLC	89
1-Bromoheptane	GC	56
1-Bromohexane	GC	56
2-Bromohydroquinone	PC	120
5-Bromoindole	PC	101
d-Bromolysergic acid diethylamide	PC	101
4-Bromomandelic acid	PC	83
Bromomethane	GC	26
3-Bromo-4-methylaniline	GC	102
1-Bromo-2-methylbutane	GC	34
1-Bromo-3-methylbutane	GC	34
2-Bromo-2-methylbutane	GC	34
2-Bromo-3-methylbutane	GC	34
2-Bromo-4-methyl-6-propionylphenol	TLC	89
1-Bromopentane	GC	34
1-Bromopentane, organics	GC	56
2-Bromopentane	GC	34
3-Bromopentane	GC	34
Bromophenhydramine HCl	GC	20
m-Bromophenol	LC	52
o-Bromophenol	PC	31
p-Bromophenol	LC	52
	PC	31
2-Bromophenol	TLC	88
3-Bromophenol	TLC	88
4-Bromophenol	GC	71
	TLC	88
N-*p*-Bromophenylthiourea	PC	47
Bromophos	LC	43
	TLC	170, 171
Bromophos-ethyl	TLC	81
1-Bromopropane	GC	118
1-Bromopropane, organics	GC	56
2-Bromopropionic	PC	79
3-Bromopropionic	PC	79
2-Bromo-4-propionylphenol	TLC	89
2-Bromo-6-propionylphenol	TLC	89
3-Bromo-4-propionylphenol	TLC	89
3-Bromo-6-propionylphenol	TLC	89
4-Bromo-6-propionylphenol	TLC	89
6-Bromopurine	LC	46
2-Bromopyridine	TLC	60
3-Bromopyridine	TLC	60
3-Bromoquinoline	GC	78
5-Bromosalicylic acid	PC	120
m-Bromotoluene	GC	51
o-Bromotoluene	GC	51
p-Bromotoluene	GC	51
5-Bromouridine	PC	36
Bromovalerylurea	TLC	40
Bromoxynil octanoate	TLC	81
Bromoxynol	GC	63
Brompheniramine (Dimetan®)	TLC	39
Brompheniramine maleate	GC	20, 104
m-Bromphenol	TLC	174
p-Bromphenol	TLC	174
Bromphenol blue	PC	88
Bromphenol red	PC	88
2-Brom-1-phenylethane	GC	51
Bromthymol blue	PC	88
Brown FK	TLC	49
Brucine	GC	22, 96, 107
	PC	2, 3, 4
	TLC	2, 3, 45
Brucine sulfate	GC	107
Bu, amines	GC	98
1-Bromoanthraquinone	PC	110
2-Bromoanthraquinone	PC	110
Bu, butyryl	GC	98
Bucarpolate	TLC	85
Buclizine	GC	17
	TLC	39
Buclizine dihydrochloride	GC	17
Bufalin (3β,14β-Dihydroxy-5β-bufadienolide) peracetate	TLC	102
Bufotalin (3β,14β-Dihydroxy-16β-acetoxy-5β-bufadienolide)	TLC	102
Bufotalin peracetate	TLC	102
Bufotenin	PC	103
Bufotenine	TLC	59
Bufothionine	PC	102
Bulan®	GC	129
Bulbocapnine	TLC	3
e-Bulgarene	GC	140
Butabarbital (Butabarbitone)	GC	105, 106
	TLC	137
Butabarbitone (Butabarbital)	GC	19
Butacaine	TLC	39
Butacaine sulfate	GC	110
Butadiene	GC	127
Butadiene-1,3	GC	29, 125
Butadiene-1,3(olefins)	GC	126
1,3-Butadiene (hydrocarbons)	GC	118
Butadiene sulfone	GC	139
Butalbital	GC	106
Butallylon	TLC	137
Butallylonal	GC	106
	TLC	137
Butamben (Butylaminobenzoate)	TLC	39
Butanal	GC	55
Butanal, 2,4-DNPH of	TLC	76
n-Butanal, 2,4-DNPH of	TLC	72
Butane	GC	58, 125
Butane (*n*-paraffins)	GC	126
n-Butane	GC	29, 42, 121
	LC	42

n-Butane (hydrocarbons)	GC	118
1-Butane (hydrocarbons)	GC	118
1,3-Butanediol	GC	76
	LC	37
	TLC	113
1,4-Butanediol	GC	76
	LC	37
	TLC	113
2,3-Butanediol	TLC	113
Butanediols, Propanediols and Related Compounds, Free form	GC	17
2,3-Butanedione	GC	57
Butanefrin HCl	PC	5
Butane-1-sulfonic acid	TLC	29
n-Butanethiol	GC	56
1,2,4-Butanetriol	PC	1
Butanoic acid	LC	68
Butanol	GC	49, 53, 58, 125
Butanol (alcohols)	GC	126
Butanol, 2-propanol	GC	126
iso-Butanol	GC	57, 94
	LC	37
n-Butanol	GC	47, 57, 92, 94
	LC	5, 37, 54
	TLC	1
n-Butanol, organics	GC	52
sec-Butanol	LC	5, 37
sec-Butanol, organics	GC	52
tert-Butanol	GC	94
	LC	5, 37
tert-Butanol, organics	GC	52
1-Butanol	PC	52
1-Butanol (DNB)	TLC	111
1-Butanol (ONPU)	TLC	111
Butanol-1, butyric acid ester of	GC	51
Butanol-1, formic acid ester of	GC	51
Butanol-1, isobutyric acid ester of	GC	51
Butanol-1, propionic acid ester of	GC	51
2-Butanol	GC	57, 92, 94, 125
	PC	52
2-Butanol (alcohols)	GC	126
2-Butanol (DNB)	TLC	111
2-Butanol (ONPU)	TLC	111
Butanol-2, butyric acid ester of	GC	51
Butanol-2, formic acid ester of	GC	51
Butanol-2, isobutyric acid ester of	GC	51
Butanol-2, propionic acid ester of	GC	51
Butan-2-one, 2,4-DNPH of	TLC	76
Butanone (2)	GC	51
2-Butanone	GC	49, 58, 125
	LC	70
2-Butanone, 2,4-DNPH of	TLC	165
2-Butanone, 2,4-DNPSC of	TLC	165
2-Butanone, *p*-NPH of	TLC	165
2-Butanone (ketones)	GC	126
Butein	PC	26
1-Butene	GC	122
Butene-1	GC	29, 58, 125
Butene-1 (olefins)	GC	126
cis-2-Butene	GC	122
cis-2-Butene (hydrocarbons)	GC	118
cis-Butene-2	GC	29, 58, 125
cis-Butene-2 (olefins)	GC	126
trans-2-Butene	GC	122
trans-2-Butene (hydrocarbons)	GC	118
trans-Butene-2	GC	29, 58, 125
trans-Butene-2 (olefins)	GC	126
2-Butene-1-ol	LC	54
N-3-Butenylthiourea	PC	47
Butethal	GC	19, 106
	TLC	137
Butethamine	TLC	39
Butethamine HCl	GC	110
Buthiazide	TLC	140
Butin	PC	26
Butobarbitone	PC	86
2-Butoxy-4-aminobenzoic acid diethylaminoethyl ester	TLC	129
Butoxyethanol ester	PC	115
Butoxy ethanol ester 2,4-D	GC	129
Butoxy ethanol ester 2,4,5-T	GC	129
2,4-D Butoxyethyl	TLC	81
2,4,5-T Butoxyethyl	TLC	81
5-Butoxymethyluracil	PC	36
Butter Yellow	TLC	46, 85
Butter yellow (Dimethylazobenzene)	TLC	48
Buturon	TLC	173
n-Butyl	PC	9
Butyl 2,4-D	GC	60
2,4-D *n*-Butyl	TLC	81
2,4-D *sec*-Butyl	TLC	81
2,4,5-T *n*-Butyl	TLC	81
N-*tert*-Butylacrylamide	LC	73
Butyl acetate	GC	2, 49
iso-Butyl acetate	GC	57
n-Butyl acetate	GC	53, 55, 57, 125
n-Butyl acetate (esters)	GC	126
n-Butyl acetate, organics	GC	52
sec-Butyl acetate	GC	121
tert-Butyl acetate	GC	57
Butyl acetylricinoleate	TLC	27
Butyl acrylate	GC	91
n-Butyl alcohol	GC	55
sec-Butyl alcohol	GC	53, 55
	TLC	1
tert-Butyl alcohol	GC	55,125, 126, 127
tert-Butyl alcohol (2-methyl-2-propanol)	GC	53
	LC	54
sec-Butyl allyl sulfide	GC	85
Butylamine	TLC	10
n-Butylamine	GC	11, 97
	LC	26
	PC	7, 55, 56
	TLC	5
iso-Butylamine	PC	55
sec-Butylamine	GC	97
	LC	26
	PC	55
	TLC	10
tert-Butylamine	GC	97
	LC	26
	PC	55
	TLC	10
n-Butylamine, trifluoroacetyl derivative	GC	10
sec-Butylamine, trifluoroacetyl derivative	GC	10

tert-Butylamine, trifluoroacetyl derivative	GC	10
N-Butyl-*p*-aminophenol	TLC	175
N-*n*-Butylaniline	GC	102
Butylated hydroxyanisole (BHA)	PC	28
	TLC	175
Butylated hydroxytoluene (BHT)	PC	69
	TLC	175
n-Butylbenzene	GC	53,119, 120
	LC	42
sec-Butylbenzene	GC	53, 119
	LC	42
	TLC	51
tert-Butylbenzene	GC	53
	LC	42
	TLC	51
n-Butylbenzene sulfonamide	TLC	27
p-*tert*-Butylbenzoic acid	TLC	37
N-*tert*-Butyl-2-benzothiazole sulfenamide	TLC	66
n-Butyl bromide	GC	47
sec-Butyl bromide	GC	47
Butyl butyrate	GC	2
iso-Butyl *n*-butyrate	GC	57
iso-Butyl *iso*-butyrate	GC	57
n-Butyl *iso*-butyrate	GC	57
n-Butyl *n*-butyrate	GC	57
Butyl caprylate	TLC	105
tert-Butyl chloride	GC	53
Butyl crotonate	GC	91
n-Butyl cyanide	GC	56
n-Butylcyclohexane	GC	35
S-Butyl cysteine	PC	64
S-*n*-Butyl-L-cysteine	LC	59
1-*n*-Butyl-2,5-dimethylpyrrole	GC	133
1,3-Butyleneglycol	PC	1
1,4-Butyleneglycol	PC	1
2,3-Butyleneglycol	PC	1
1,2-Butylene oxide, organics	GC	52
cis-2,3-Butylene oxide, organics	GC	52
trans-2,3-Butylene oxide, organics	GC	52
Butyl esters	GC	24
n-Butyl ester 2,4-D	GC	129
n-Butyl ester 2,4,5-T	GC	129
p-Butyl ester	PC	115
Butyl ethers	GC	116
tert-Butyl ethyl hydrogen phosphate	PC	34
Butyl formate	GC	2
n-Butyl formate	GC	47, 55, 57
Butyl gallate	TLC	125
Butyl glycidyl ether	LC	60
iso-Butyl guanidine	PC	9
Butyl hexanoate	GC	2
n-,*iso*-,*sec*-Butyl hydroperoxide	PC	27
sec-Butylhydroperoxide	TLC	65
tert-Butylhydroperoxide	PC	27
	TLC	65
Butylhydroxyanisole	TLC	125
2-*tert*-Butyl-4-hydroxyanisole	TLC	174
3-*tert*-Butyl-4-hydroxyanisole	TLC	174
n-Butyl *p*-hydroxybenzoate	LC	74
tert-Butyl 2-hydroxyethylperoxide	PC	27
Butylhydroxytoluene	TLC	125
4,4′-Butylidene-bis(6-*tert*-butyl-3-methylphenol)	TLC	175
3,3′-Butylidene-bis-(4-hydroxycoumarin)	TLC	79
Butyl isobutyrate	GC	2
n-Butyl isocyanate	GC	44
sec-Butyl isocyanate	GC	44
tert-Butyl isocyanate	GC	44
n-Butylmercaptan	GC	85
4-Butylmercapto-5-ethyl-2-methylpyridine *N*-oxide	TLC	61
Butyl methacrylate	GC	91
Butyl 3-methylbutyrate	GC	2
Butyl 4-methylpentanoate	GC	2
1-*n*-Butyl-3-nitroguanidine	PC	9
1-*tert*-Butyl-3-nitroguanidine	PC	9
Butyl pentanoate	GC	2
tert-Butylperacetate	TLC	65
tert-Butylperbenzoate	PC	27
	TLC	65
tert-Butylperoctoate	TLC	65
tert-Butyl peroxyacetate	TLC	65
tert-Butyl peroxybenzoate	TLC	65
4-*sec*-Butylphenol	GC	71
o-*tert*-Butylphenol	PC	120
p-*tert*-Butylphenol	GC	73
	LC	61
	TLC	176
4-*tert*-Butylphenol	GC	71, 72
tert-Butyl phosphate	PC	34
Butyl phthalyl butyl glycolate	TLC	27
Butyl propionate	GC	2
p-*tert*-Butylpyrocatechol	TLC	176
1-prim-Butylpyrrole	GC	133
2-prim-Butylpyrrole	GC	133
1-*sec*-Butylpyrrole	GC	133
2-*sec*-Butylpyrrole	GC	133
2-*tert*-Butylpyrrole	GC	133
Butyl stearate	TLC	27
n-Butyl sulfoxide	GC	139
sec-Butyl sulfoxide	GC	139
tert-Butyl sulfoxide	GC	139
Butylthiostannous acid	TLC	67
N-DL-*sec*-Butylthiourea	PC	47
N-*n*-Butylthiourea	PC	47
N-*tert*-Butylthiourea	PC	47
Butyltin trichloride	TLC	67
tert-Butylurea	LC	63
iso-Butyl *iso*-valerate	GC	57
Butyl Zimate	TLC	66
Butyne-1	GC	29, 125
Butyne-1 (alkynes)	GC	126
Butyne-2	GC	29
2-Butyne-1,4-diol	TLC	112
Butyraldehyde	GC	49, 57, 58, 125
	LC	8
Butyraldehyde (aldehydes)	GC	126
Butyraldehyde dimethyl acetal, organics	GC	52
Butyraldehyde, 2,4-DNPH of	TLC	166
Butyraldehyde, 2-hydrazinebenzothiazole derivative of	TLC	168
Butyraldehyde, *p*-NPH of	TLC	165
n-Butyraldehyde	GC	7, 53
	LC	70
	PC	20
n-Butyraldehyde, organics	GC	52
Butyrate, ethyl, organics	GC	52
Butyrate, methyl, organics	GC	52
Butyrate, *β*-sitosteryl ester of	TLC	182

Butyrate, vinyl, organics GC 52
Butyric acid LC 7
PC 77
TLC 29, 132
Butyric acid, benzyl ester of GC 24
Butyric acid esters GC 51
Butyric acid, ethyl ester of GC 24
Butyric acid, hexyl ester of GC 24
Butyric acid, methyl ester of GC 24
n-Butyric acid LC 57, 65
PC 11, 79
TLC 134
α-G-*n*-Butyric carboxyalkyl guanidine PC 9
γ-G-*n*-Butyric carboxyalkyl guanidine PC 9
Butyronitrile GC 51, 55
Butyrophenone GC 51
PC 22
Butyrophenone, 2,4-DNPH of TLC 74
Butyrophenone, 4-NPH of TLC 74
Butyrophenone azine PC 23
Butyryl, Bu GC 98
Butyrylcholine PC 8
Butyrylperazine TLC 139
2-Butyrylphenol TLC 89
4-Butyrylphenol TLC 89
Butyryl phosphate PC 77

Adsorption of the Elements (LC Section – Figures)

B, Boron Pg 278, 279, 280, 281
Ba, Barium Pg 278, 279, 280, 281
Be, Beryllium Pg 278, 279, 280, 281
Bi, Bismuth Pg 278, 279, 280, 281
Br, Bromine Pg 278, 279, 280, 281

Ca, Calcium LC 81, 82, 83, 85, 86, 87, 88, 89
PC 15
TLC 52
Ca(II), Calcium GC 43
PC 95, 96
TLC 148, 149
Ca^{2+}, Calcium PC 92, 97
p-CA TLC 173
Cabromal GC 19
TLC 137
Cadaverine LC 66
PC 64
TLC 5, 13
γ-Cadinene GC 140
δ-Cadinene GC 140
TLC 106, 143
Cadmate TLC 66
Cadmium, Cd LC 81, 82, 83, 85, 86, 87, 88, 89
PC 15
TLC 52
Cadmium, Cd(II) GC 43
PC 95, 96
TLC 148,149, 150
Cadmium, Cd^{2+} PC 92, 97
Cadmium diethyldithiocarbamate TLC 66

Caffeic acid PC 13, 82, 113
TLC 133,135, 174
Caffeine GC 8, 19, 21, 96,104 105,107, 111
LC 31, 46
PC 2, 3, 37
TLC 39, 40, 45
Caffeine citrated GC 107
Caffeine HB-R GC 107
Caffeine HCl GC 107
Caffeoylglycine PC 80
Caffeyl-glucose PC 70
Calamenene GC 140
Calarene GC 140
Calcium, Ca LC 81, 82, 83, 85, 86, 87, 88, 89
PC 15
TLC 52
Calcium, Ca(II) GC 43
PC 95, 96
TLC 148, 149
Calcium, Ca^{2+} PC 92, 97
Calcium panthothenate TLC 188
Calycin TLC 28
Calycotomine PC 3
Camatan (*N*-Trichloromethylthio-tetrahydrophthalimide) PC 117
Campestanol TLC 99
Campesterol GC 84
LC 45
TLC 99
Camphene GC 86, 87
TLC 51, 54, 106
Camphor TLC 77
Camphor, 2,4-DNPH of TLC 76
Canaline PC 64
Canaridigitoxoside TLC 102
Canaridigitoxoside peracetate TLC 102
Canarigenin (3β,14β-Dihydroxy-cardadien-4,20(22)-olide) TLC 102
Canarigenin peracetate TLC 102
L-Canavanine LC 59
Canavanine·H_2SO_4 PC 10
Canavanine sulfate TLC 11
Canavanosuccinic acid PC 64
Cannabiscitrin PC 26
Cannogenin (3,14β-Dihydroxy-19-oxo-5β-cardenolide) TLC 102
Cannogenin peracetate TLC 102
Cannogenol peracetate TLC 102
Cannogenol (3β, 14β, 19-Trihydroxy-5β-cardenolide) TLC 102
Canthaxanthin TLC 93
Canthaxin TLC 94
CAO-6 TLC 175
Capaxanthin TLC 94
Capensinidin PC 25
Caperatic acid TLC 131
Caprate LC 67
Caprate, β-sitosteryl ester of TLC 182

Capric acid	PC	77
	TLC	132
n-Capric acid	LC	7
Caproaldehyde	GC	57
n-Caproaldehyde	PC	20
Caproate, β-sitosteryl ester of	TLC	182
Caproic acid	LC	61
	TLC	29, 132
n-Caproic acid	LC	7
	PC	77, 79
Caproicaldehyde, 2,4-DNPH of	TLC	166
α-G-*n*-Caproic guanidine	PC	9
α-G-*n*-Caproic carboxyalkyl guanidine	PC	9
ε-G-*n*-Caproic carboxyalkyl guanidine	PC	9
ε-G-*n*-Caproic guanidine	PC	9
α-G-*iso*-Caproic guanidine	PC	9
Capronitrile	GC	51
2-Caproylphenol	TLC	89
4-Caproylphenol	TLC	89
Caproyl phosphate	PC	77
Caprylaldehyde	TLC	77
Caprylaldehyde, 2,4-DNPH of	TLC	166
Caprylate, β-sitosteryl ester of	TLC	182
Caprylic acid	PC	77
	TLC	29, 132
n-Caprylic acid	LC	7
2-Caprylphenol	TLC	89
4-Caprylphenol	TLC	89
2-Caprylylphenol	TLC	89
4-Caprylylphenol	TLC	89
Capsanthin	TLC	93
Capsorubin	TLC	93
Captan	GC	60, 62, 64, 65, 68, 129
	TLC	80, 81, 82
Captodiamine	GC	17
	PC	85
Captodiamine (Covatix)	TLC	45
Captodiamine hydrochloride	GC	17
Carbadrine	TLC	9
Carbaminoylcholine	PC	8
Carbamult	TLC	170
S-Carbamyl-L-cysteine	LC	59
Carbamyl-*N*′,2′-hydroxyethylpyridinium bromide	PC	18
3-Carbamyl-*N*′-ribosidylpyridinium bromide	PC	18
3-Carbamyltetraacetyl-*N*′-D-glucosidylpyridinium bromide	PC	18
Carbarsone®	TLC	136
Carbaryl	GC	130
	TLC	81, 83, 170
Carbazole	GC	45
	TLC	7, 58, 120
Carbetapentane	TLC	39
N^4-Carbethoxysulfoethyl-thiodiazole	TLC	44
Carbinoxamine	GC	20
Carbinoxamine (Clistin®)	TLC	39
Carbinoxamine maleate	GC	20
N-Carbobenzoxy-3,3′-diiodothyronine	TLC	19
N-Carbobenzoxy-3,5-diiodothyronine	TLC	19
N-Carbobenzoxy-3,5-diiodotyrosine	TLC	19
N-Carbobenzoxy-3-monoiodohistidine	TLC	19
N-Carbobenzoxy-3-monoiodotyrosine	TLC	19
N-Carbobenzoxythyroxine	TLC	19
N-Carbobenzoxy-3,3′,5′-triiodothyronine	TLC	19
N-Carbobenzoxy-3,5,3′-triiodothyronine	TLC	19
β-Carboline	TLC	120
Carbomycin	PC	66
	TLC	20, 153
Carbonate	PC	93
Carbon dioxide	GC	112, 127
Carbon dioxide (hydrocarbons)	GC	118
Carbon dioxide (inorganic)	GC	126
Carbon disulfide	LC	60
Carbon disulfide, organics	GC	52
Carbon monoxide	GC	112
Carbon monoxide (hydrocarbons)	GC	118
Carbon tetrachloride	GC	26, 32, 47, 50, 53, 93
	LC	36, 42, 75
Carbon tetrachloride, organics	GC	52
Carbon tetrafluoride	GC	26
Carbophenothion	GC	64, 65, 67, 128, 129
	LC	43
Carbophenothion O-analog	GC	65
Carbophenothion oxygen analog	GC	67
Carbophenothion oxygen analog sulfone	GC	67
Carbophenothion, sulfon	GC	129
Carbophenothion oxygen analog sulfoxide	GC	67
Carbophenothion sulfone	GC	67
Carbophenothion sulfoxide	GC	64, 67, 129
Carbophenothion, thiol	GC	64
Carbophenothion, thiolsulfone	GC	63
Carbophenothion, thiolsulfoxide	GC	64, 129
Carbophenothion (Trithion®)	TLC	80, 82, 84
Carboxyalkyl guanidines (guanidino-acids)	PC	9
4-Carboxybenzalazine	PC	23
2-(6-Carboxyhexylamino)-ethylguanidine	LC	55
2-Carboxyphenol	LC	51
3-Carboxyphenol	LC	51
4-Carboxyphenol	LC	51
6-Carboxypurine	LC	46
	PC	36
5-Carboxyvanillin	PC	20
Carbutamide (Oranil)	TLC	140
Cardiolipin	TLC	152
δ-3-Carene	TLC	106
3-Carene	GC	87
Δ^3-Carene	GC	86
Δ^4-Carene	GC	86
Carisoprodol	GC	17, 105, 108
	TLC	38
Carminic acid	PC	105
Carmoisene	TLC	49
Carnitine	TLC	188
Carnosine	LC	41
	PC	17
	TLC	155
α-Carotene	TLC	94, 107
β-Carotene	LC	45
	PC	49
	TLC	93, 94, 106, 107, 108, 143, 189
β-Carotene (C)	PC	48

ε-Carotene	TLC	94
γ-Carotene	TLC	94
δ-Carotene	TLC	94
Carotenes	PC	122
Carpogenin (3α,14β-Dihydroxy-19-oxo-5β-cardenolide)	TLC	102
Carpogenin peracetate	TLC	102
Carvacrol	TLC	90
Carveol	TLC	183
Carvone	TLC	77
Carvone, 2,4-DNPH of	TLC	76
d-Carvone, 2,4-DNPH of	TLC	72
l-Carvone, 2,4-DNPH of	TLC	72, 165
l-Carvone, 2,4-DNPSC of	TLC	165
l-Carvone, *p*-NPH of	TLC	165
Carvotane acetate, 2,4-DNPH of	TLC	76
Carvyl acetate	TLC	105
Carvyl propionate	TLC	105
Caryophyllene	GC	140
α-Caryophyllene	TLC	54
β-Caryophyllene	TLC	54,106, 143
γ-Caryophyllene	TLC	54
Catechin	LC	1, 28
	PC	30
d-Catechin	LC	28
D-Catechin	TLC	92
L-*epi*-Catechin	PC	30
L-*epi*-Catechin gallate	PC	30
Catechol	LC	1, 11
	PC	82
	TLC	87, 88, 174
Catenulin	TLC	20
Cd, Cadmium	LC	81, 82, 83, 85, 86, 87, 88, 89
	PC	15
	TLC	52
Cd(II), Cadmium	GC	43
	PC	95, 96
	TLC	148,149, 150
Cd^{2+},Cadmium	PC	92, 97
CDAA	GC	128, 129
CDEC	GC	68, 128
CDP	TLC	63
CDPG	TLC	63
Ce (Cerium)	LC	79, 80, 81, 82, 85, 86, 88, 90, 91
Ce(III), Cerium	GC	43
	PC	15, 96
Ce^{3+}, Cerium	PC	92, 97
Ce^{4+},Cerium	TLC	150
Cedrene	TLC	51, 54
α-Cedrene	GC	140
	TLC	106, 143
β-Cedrene	GC	140
Cedryl acetate	TLC	105
Cellobionic	LC	40
Cellobiose	LC	25, 48
	PC	70
	TLC	24
Cellohexaose	LC	48
Cellopentaose	LC	48
Cellotetraose	LC	48
Cellotriose	LC	48
Celocurine	PC	60
Cephaeline	TLC	3
Cephalin	LC	68
Cephaloridine	TLC	21
Cerberigenin (3β,14β-Dihydroxy-11β,12β-epoxy-5β-cardenolide)	TLC	102
Cerberigenin peracetate	TLC	102
Cerbertin	TLC	102
Cerebroside	TLC	152
Cerebroside sulfate	TLC	152
Cerenox (Quinoneoximebenzoyl hydrazone)	PC	117
Cerenox-Special (Phenyl mercuric chloride)	PC	117
Ceresan (Methoxyethyl mercuric chloride)	PC	117
Ceres Red G	TLC	48
Ceres Yellow	TLC	48
Cerium (Ce)	LC	79, 80, 81, 82, 85, 86, 88, 90, 91
Cerium, Ce(III)	GC	43
	PC	15, 96
Cerium, Ce^{3+}	PC	92, 97
Cerium, Ce^{4+}	TLC	150
Cernuoside	PC	26
Ceryl alcohol	LC	45
Cesium, Cs	LC	83, 85, 86, 88
	PC	15
	TLC	52
Cesium, Cs^+	PC	97
Cetyl alcohol	LC	45
Cetyl gallate	TLC	125
Cevadine	PC	3
n-$C_{28}H_{58}$	GC	142
Chalcone	LC	1
Chalcone, aglycones	PC	26
Chalcone glucosides	PC	26
Chanoclavine-(I)	TLC	114
Chanoclavine-(II)	TLC	114
Chavicol methyl ether	TLC	90
Chelerythrine	PC	3
Chelidonine	PC	2, 3
Chelilutine	PC	3
Chelirubine	PC	3
Chenodeoxycholic acid	PC	39
Chenodeoxycholic acid (3α,7α-OH)	LC	45
Chinic acid	TLC	133
Chipman RP-11783	GC	65
Chloranil	GC	129
Chlorate	PC	14, 91, 93
Chlorazin	TLC	86
Chlorazine	GC	66
Chlorbenside	GC	129
	TLC	82
Chlorbenzide	GC	60, 65, 128
Chlorbenzilate	TLC	82
Chlorbromuron	TLC	173
Chlorcyclizine	GC	17, 20, 22
	PC	85
Chlorcyclizine HCl	GC	20, 54, 104
Chlorcyclizine (Perazil)	GC	17, 21
Chlorcyclizine (Perazil) (Di-Paralene®)	TLC	39

Chlordane	GC	60, 128, 129
	PC	116
	TLC	80, 81, 82
Chlordane (alpha)	GC	129
Chlordane (beta)	GC	129
Chlordane (gamma)	GC	129
Chlordane (T)	TLC	82
Chlordane (tech)	GC	129
6-Chlor-6-dehydro-17α-hydroxy-progesterone acetate	TLC	178
Chlordiazepoxide	TLC	139
Chlorexolone	TLC	140
Chlorfenvinphos	LC	43
Chlorgenic	PC	13
Chloride	PC	14, 91, 93
Chloride, Cobalt $CoCl_2$	PC	94
Chloride, cobalt, dichlorobis (ethylenediamine) $[Co(en)_2Cl_2]Cl$	PC	94
Chloride, cobalt, hexammine $[Co(NH_3)_6]Cl_3$	PC	94
Chloride, methylene	GC	32
Chloride, nickel, $NiCl_2$	PC	94
Chloride, vinyl	GC	26, 32
Chloride, vinylidene	GC	32
Chlorin a_2 (methyl ester)	PC	33
Chlorine, Cl	GC	26, 114
Chlorine, Cl^-	TLC	147
Chlorine, Cl_2	GC	113
Chlorine dioxide	GC	114
Chlorine monofluoride	GC	26, 114
Chlorine monoxide	GC	114
Chlorine oxidifluoride	GC	114
Chlorine pentafluoride	GC	114
Chlorine trifluoride	GC	26, 114
Chlormethazanone (Chlormezanone)	GC	17, 108
Chloroacetic acid	LC	58
	PC	79
	TLC	29
1-Chloro-2-aminoethane	PC	7
3β-Chloro-androst-5-ene-17-one (Group III)	PC	40
m-Chloroaniline	PC	29
o-Chloroaniline	PC	29
p-Chloroaniline	PC	29
2-Chloraniline	GC	102
	PC	6
	TLC	119
3-Chloroaniline	GC	102
	PC	6
	TLC	119
4-Chloroaniline	GC	102
	PC	6
	TLC	119
1-(*o*-Chloroanilino)-anthraquinone	TLC	69
1-Chloroanthraquinone	PC	110
2-Chloroanthraquinone	PC	110
Chloroatranorin	TLC	34
Chlorobenside	GC	68
Chlorobenzene	GC	31, 47, 51, 53
	LC	36, 75
Chlorobenzilate®	GC	60,128, 129
	PC	116
m-Chlorobenzoic acid	TLC	37
o-Chlorobenzoic acid	TLC	37
p-Chlorobenzoic acid	TLC	37
3-(*m*-Chlorobenzoyl)-*N*-methylcarbazole	TLC	7
4-Chlorobenzoylperoxide	TLC	65
1-Chloro-2,2-bis(4-chlorophenyl)-ethylene	TLC	81
1-Chloro-2-bromoanthraquinone	PC	110
1-Chloro-5-bromoanthraquinone	PC	110
2-Chloro-3-bromoanthraquinone	PC	110
2-Chloro-4-bromo-6-propionylphenol	TLC	89
2-Chloro-6-bromo-4-propionylphenol	TLC	89
2-Chloro-1,3-butadiene	GC	32
1-Chlorobutane	GC	34, 56
1-Chlorobutane (*n*-butyl chloride)	GC	53
2-Chlorobutane (*sec*-butyl chloride)	GC	53
cis-1-Chloro-1-butene	GC	32
trans-1-Chloro-1-butene	GC	32
trans-1-Chloro-2-butene	GC	32
2-Chloro-1-butene	GC	32
cis-2-Chloro-2-butene	GC	32
trans-2-Chloro-2-butene	GC	32
DL-3-Chloro-1-butene	GC	32
2-Chlorobutyric acid	TLC	29
3-Chlorocatechol	PC	29
4-Chlorocatechol	PC	29
4-Chlorocinnamic acid	PC	83
4-Chloro-*m*-cresol	PC	31
4-Chloro-*o*-cresol	PC	31
6-Chloro-*m*-cresol	PC	31
6-Chloro-*o*-cresol	PC	31
Chlorocycloheptane	GC	51
Chlorocyclohexane	GC	51
Chlorocyclooctane	GC	51
Chlorocyclopentane	GC	51
Chlorocyclizine dihydrochloride	GC	17
1-Chlorodecane	GC	34
2-Chloro-3-*n*-decylamino-1,4-naphthoquinone	PC	111
3β-Chloro-dehydroepiandrosterone	TLC	97
2-Chloro-1,4-diaminobenzene	TLC	119
4-Chloro-1,2-diaminobenzene	TLC	119
4-Chloro-1,3-diaminobenzene	TLC	119
Chlorodiborane	GC	27
3-Chloro-2,4-dibromo-6-propionylphenol	TLC	89
3-Chloro-2,6-dibromo-4-propionylphenol	TLC	89
2-Chloro-3-dimethylamino-1,4-naphthoquinone	PC	111
4-Chloro-2,3-dimethylphenol	PC	31
4-Chloro-2,5-dimethylphenol	PC	31
4-Chloro-3,5-dimethylphenol	PC	31
6-Chloro-3,4-dimethylphenol	PC	31
2-Chloro-3,4-dimethylphenyl *N*-methyl carbamate	GC	130
1-Chloro-2,2-dimethylpropane	GC	34
Chloroethane	GC	34
2-Chloro-3-ethylamino-1,4-naphthoquinone	PC	111
4-Chloro-5-ethyl-3-methylphenol	PC	31
4-Chloro-5-ethyl-2-methylpyridine	TLC	61
4-Chloro-5-ethyl-2-methylpyridine *N*-oxide	TLC	61
1-Chloro-2-fluoranthraquinone	PC	110
1-Chloro-5-fluoranthraquinone	PC	110
2-Chloro-3-fluoranthraquinone	PC	110
Chlorofluoromethane	GC	26
Chloroform	GC	26, 32, 47, 53, 93
	LC	36, 42, 75
Chloroform, organics	GC	52
Chlorogenic acid	LC	1, 46
	PC	113
	TLC	133, 135

Chlorogenin	TLC	98, 103
1-Chloroheptane, organics	GC	56
1-Chlorohexane	GC	34
1-Chlorohexane, organics	GC	56
2-Chloro-6-hydroxypyridine	PC	120
5-Chloroindole-2-carboxylic acid	PC	101
1-Chloro-2-iodoanthraquinone	PC	110
1-Chloro-5-iodoanthraquinone	PC	110
2-Chloro-3-iodoanthraquinone	PC	110
Chloromethane	GC	26
Chloromethapyrilene chlorothen	GC	20
3-Chloro-4-methoxyaniline	PC	6
3-Chloro-6-methoxyaniline	PC	6
6-Chloro-9-methoxymethylpurine	LC	46
2-Chloro-*n*-methylaniline	GC	102
2-Chloro-4-methylaniline	TLC	7
2-Chloro-6-methylaniline	TLC	7
3-Chloro-2-methylaniline	PC	6
3-Chloro-4-methylaniline	TLC	119
3-Chloro-6-methylaniline	PC	6
4-Chloro-2-methylaniline	PC	6
	TLC	7
1-Chloro-2-methylbutane	GC	34
1-Chloro-3-methylbutane	GC	34
2-Chloro-2-methylbutane	GC	34
2-Chloro-3-methylbutane	GC	34
2-Chloro-2-methylpropane	GC	34
4-Chloro-3-methylpyridine	TLC	61
4-Chloro-3-methylpyridine *N*-oxide	TLC	61
2-Chloro-4-methylquinoline	GC	78
2-Chloro-4-nitroaniline	TLC	119
Chloro-nitrophenol	TLC	84
2-Chloro-3-nitrophenol	PC	29
2-Chloro-4-nitrophenol	PC	29
2-Chloro-5-nitrophenol	PC	29
2-Chloro-6-nitrophenol	PC	29
3-Chloro-2-nitrophenol	PC	29
3-Chloro-4-nitrophenol	PC	29
3-Chloro-5-nitrophenol	PC	29
4-Chloro-2-nitrophenol	PC	29
4-Chloro-3-nitrophenol	PC	29
5-Chloro-2-nitrophenol	PC	29
1-Chlorononane	GC	34
1-Chlorooctane	GC	34
1-Chlorooctane, organics	GC	56
1-Chloro-1,1,2,2,2-pentafluoroethane	GC	26
1-Chloropentane	GC	34
1-Chloropentane, organics	GC	56
2-Chloropentane	GC	34
3-Chloropentane	GC	34
m-Chloro(peroxybenzoic acid)	PC	107
o-Chloro(peroxybenzoic acid)	PC	107
p-Chloro(peroxybenzoic acid)	PC	107
m-Chlorophenol	GC	74
	TLC	174
o-Chlorophenol	GC	74
	LC	1, 11
	PC	31
p-Chlorophenol	GC	74
	PC	31
	TLC	174
2-Chlorophenol	GC	71
3-Chlorophenol	GC	71
4-Chlorophenol	GC	71
o-Chorophenol, 4-acetyl-2-nitrophenol ether of	TLC	91
Chlorophenol red	PC	88
2-Chlorophenothiazine (2-Cp)	GC	103
2-Chlorophenothiazine sulfoxide (2-CpSO)	GC	103
2-Chloro-1,4-phenylenediamine	TLC	119
L-2-Chlorophenylmercapturic acid	PC	29
L-3-Chlorophenylmercapturic acid	PC	29
L-4-Chlorophenylmercapturic acid	PC	29
4-Chorophenyl phosphate	PC	112
4-Chlorophenyl sulfoxide	GC	139
Chlorophyll *a*	LC	45
	PC	122
Chlorophyll *b*	LC	45
	PC	122
Chlorophyll *c*	PC	122
Chlorophylls (Chl) *a* + *b*	PC	48
2-Chloroprocaine	TLC	39
Chloroprocaine HCl	GC	110
1-Chloropropane	GC	34
1-Chloropropane, organics	GC	56
2-Chloropropane	GC	34
2-Chloropropionic acid	PC	79
	TLC	29
3-Chloropropionic acid	PC	79
2-Chloro-4-propionylphenol	TLC	89
2-Chloro-6-propionylphenol	TLC	89
3-Chloro-4-propionylphenol	TLC	89
3-Chloro-6-propionylphenol	TLC	89
4-Chloro-6-propionylphenol	TLC	89
Chloropropylate	GC	129
6-Chloropurine	LC	46
2-Chloropyridine	GC	131
	TLC	60
3-Chloropyridine	TLC	60
Chloroquin	GC	21
Chloroquinol	PC	29
4-Chlororesorcinol	PC	29
Chlorotestosterone	TLC	95
Chlorotetracycline HCl	TLC	21
Chlorothen (Chloromethapyrilene)	GC	20
Chlorothen citrate	GC	20, 54
Chlorothen (Tagathen®)	TLC	39
Chlorothiazide	TLC	41
Chlorothion	GC	129
2-Chlorothiophenol	PC	29
3-Chlorothiophenol	PC	29
4-Chlorothiophenol	PC	29
m-Chlorotoluene	GC	51
o-Chlorotoluene	GC	51
p-Chlorotoluene	GC	51
2-Chlorovinyldiethyl phosphate	GC	65
Chloroxuron	TLC	173
Chlorphenesin carbamate	TLC	38
Chlorpheniramine	GC	20, 22
Chlorpheniramine (Chlor-Trimeton®)	GC	21
	TLC	39
Chlorpheniramine maleate	GC	20, 53, 104
Chlorphentermine	GC	22
Chlorpicrin	GC	129
Chlorpromazine	GC	17, 20, 22,105, 108, 109
	PC	84, 85
	TLC	139
Chlorpromazine (CP)	GC	103
Chlorpromazine hydrochloride	GC	17
Chlorpromazine (Largactil®)	TLC	45
Chlorpromazine-*N*-oxide (CPNO)	GC	103
Chlorpromazine-*N*-oxide sulfoxide (CPNSO)	GC	103
Chlorpromazine sulfoxide (CPSO)	GC	103
	TLC	139

Chlorpromazine (Thorazine®)	GC	21
Chlorprotixene (chlorprothixenum)	TLC	139
Chlortestosterone acetate	TLC	95
Chlortetracycline	PC	68
iso-Chlortetracycline	PC	68
Chlorthion®	GC	64, 65
	LC	43
	TLC	80, 84, 171
Chlorzoxazone	GC	17
Chocolate Brown FB	TLC	49
Chocolate Brown HF	TLC	49
5α-Cholane	GC	80
	LC	45
5β-Cholane	GC	80
5α-Cholane-3α,24-diol	LC	45
5α-Cholane-3α,7α,12α,24-tetrol	LC	45
5α-Cholane-3α,12α,24-triol	LC	45
Cholanes (steroids)	GC	80
Cholanic acid	TLC	180
5α-Cholan-12α-ol	GC	80
5α-Cholan-24-ol	LC	45
5β-Cholan-12α-ol	GC	80
5α-Cholan-12-one	GC	80
5β-Cholan-12-one	GC	80
Cholanthrene	TLC	146
5,7-Cholestadien-3β-ol	GC	136
3,5-Cholestadien-7-one	TLC	96
Cholesta-3,5-dien-7-one	LC	62
Cholestane	LC	45
Cholestane, 5-en-3β-acetate	LC	6
Cholestane, 5-en-3β-OCH_3	LC	6
5α-Cholestane	GC	80, 81
	LC	6
5α-Cholestane-3,6-dione	GC	80
5α-Cholestane-3β-ol	GC	81
	LC	6
5α-Cholestane-3-one	GC	81
	LC	6
5α-Cholestane-3β,5,6β-triol	LC	62
5β-Cholestane	GC	80, 81
5β-Cholestane-3α,12α-diol	LC	6
5β-Cholestane-3,12-dione	LC	6
5β-Cholestane-3α-ol	GC	81
5β-Cholestane-3β-ol	GC	81
	LC	6
5β-Cholestane-3-one	GC	81
	LC	6
5β-Cholestane-3α,7α,12α-triol	LC	6
5β-Cholestane-3,7,12-trione	LC	6
Cholestane-3β-ol	LC	24
Cholestanes (steroids)	GC	80
Cholestanol	GC	79, 136
	TLC	99
β-Cholestanol	PC	126
5α-Cholestan-3α-ol	GC	80
	TLC	98
5α-Cholestan-3β-ol	GC	80, 84, 136
	TLC	98
5β-Cholestan-3α-ol	GC	80
	TLC	98
5β-Cholestan-3β-ol	GC	80
	TLC	98
Cholestan-3-one	GC	79
	LC	45
5α-Cholestan-3-one	GC	80
	TLC	98
5α-Cholestan-6-one	GC	80
5β-Cholestan-3-one	GC	80
	TLC	98
Cholestanyl acetate	GC	79
Cholestanyl methyl ether	GC	79
Cholestanyl trifluoracetate	GC	79
Cholestanyl trimethylsilyl ether	GC	79
Cholest-3,5-diene-3β-ol	LC	24
Cholest-5,7-diene-3β-ol	LC	24
Cholest-5,24-diene-3β-ol	LC	24
Cholest-7,9-diene-3β-ol	LC	24
Cholest-7,14-diene-3β-ol	LC	24
Cholest-8,24-diene-3β-ol	LC	24
Cholest-4-ene-3,6-dione	GC	80
	LC	45
Cholest-5-ene	GC	80
5α-Cholest-2-ene	GC	80
5-Cholestene (Cholestene)	GC	81
Cholest-5-ene-3β,7α-diol	LC	62
Cholest-5-ene-3β,7β-diol	LC	62
Cholest-5-ene-3β,20α-diol	LC	62
Cholest-5-ene-3β,25-diol	LC	62
5-Cholestene-3β-ol (Cholesterol)	GC	81
Cholest-5-ene-3β-ol	LC	24
Cholest-7-ene-3β-ol	LC	24
Cholest-8(14)-ene-3β-ol	LC	24
Cholest-5-ene-3β,20α,22R-triol	LC	62
5-Cholesten-3β-ol	GC	136
7-Cholesten-3β-ol	GC	79
Δ^7-Cholesten-3β-ol	GC	84
Cholest-5-en-3β-ol	GC	80
	TLC	98
Cholest-5-en-3β-ol (cholesterol)	LC	62
Cholestenone	GC	83
	PC	39
4-Cholesten-3-one	GC	79
Cholest-4-en-3-one	GC	80
	TLC	98
Cholest-5-en-3-one	GC	80
Cholest-5-en-3β-yl (cholesteryl) acetate	LC	62
Cholesterol	GC	79, 83, 84, 136
	LC	27, 45
	PC	53, 126
	TLC	56, 57, 99, 178
Cholesterol acetate	PC	126
	TLC	57
Cholesterol oleate	TLC	57
Cholesterol palmitate	TLC	57
Cholesteryl acetate	LC	45
	TLC	56
Cholesteryl arachidonate	TLC	154
Cholesteryl benzoate	LC	45
Cholesteryl butyrate	LC	45
Cholesteryl chloride	LC	45
Cholesteryl laureate	LC	27
Cholesteryl linoleate	TLC	154
Cholesteryl linolenate	TLC	154
Cholesteryl oleate	TLC	154
Cholesteryl palmitate	LC	27, 45
	TLC	56, 154
Cholesteryl stearate	LC	27
	TLC	154
Cholesteryl TMSi	LC	45
Cholic acid	PC	39
Cholic acid (3α,7α,12α-OH)	LC	45

Choline	PC	2,8, 47, 99
	TLC	188
Choline chloride	TLC	188
Chondrosamine (galactosamine)	PC	64
Chromate	PC	14, 91, 93
Chromium, Cr	LC	83, 84, 85, 86, 87, 89
	PC	15
Chromium, Cr(III)	GC	43
	PC	95, 96
Chromium, Cr^{3+}	PC	97
Chromium oxychloride	GC	50
Chromium sulfate, potassium	PC	94
Chromium, tetra, hexachloride, hexahydroxyhexakis (ethylenediamine) $(Cr_4(en)_6(OH)_6Cl_6$	PC	94
Chromomycin A_3	TLC	20
cis-Chrysanthemic acid	PC	119
trans-Chrysanthemic acid	PC	119
Chrysanthemum dicarboxylic acid	PC	119
Chrysene	GC	40
	LC	27
	PC	89
	TLC	7, 51, 53, 146
Chrysin	PC	26
	TLC	92
Chrysoeriol	PC	26
Chrysophanol	PC	109
Chymotrypsinogen	LC	19
Ciba 10573 (*O*-(1,3-dithiolan-2-yl)-phenyl-*N*-methylcarbamate	GC	61
Ciba C-2307	GC	65
Ciba C-7019	GC	65
Ciba C-8874	GC	65
Ciba C-9491	GC	65
Ciba C-9491 O-analog	GC	65
Ciba C-14421	GC	65
Cidial	TLC	170,171, 172
Cinchocaine	PC	4
Cinchocaine HCl (Dibucaine HCl)	TLC	43
Cinchomeronic acid	PC	18
Cinchomeronic acid hydrazide	PC	18
Cinchonamine	PC	2
Cinchonidine	GC	21, 107
	PC	2
Cinchonidine (*l*-cinchonine)	GC	8
Cinchonine	GC	21, 96, 107, 111
	PC	2, 3
	TLC	2, 3, 39
Cinchonine (*d*-cinchonine)	GC	8
Cinchonine sulfate	GC	107
1,4-Cineole	GC	87
1,8-Cineole	GC	87
Cinnamaldehyde	PC	20, 22
	TLC	77
Cinnamaldehyde, 2,4-DNPH of	TLC	73, 74
Cinnamaldehyde, 2,4-DNPSC of	TLC	165
Cinnamaldehyde, 2-hydrazinebenzothiazole derivative of	TLC	168
Cinnamaldehyde, *p*-NPH of	TLC	165
Cinnamaldehyde, 4-NPH of	TLC	74
Cinnamein sulfanilamide	TLC	44
Cinnamic acid	GC	4, 5
	LC	53
	PC	82
	TLC	133, 174
Cinnamic acid, benzyl ester of	GC	24
Cinnamic acid, cinnamyl ester of	GC	24
Cinnamic acid, ethyl ester of	GC	24
Cinnamic acid, isobutyl ester of	GC	24
Cinnamic acid, methyl ester of	GC	24
Cinnamic acid, phenylethyl ester of	GC	24
Cinnamic alcohol	PC	52
Cinnamoylformic acid	PC	81
Cinnamyl alcohol	PC	53
Cinnamyl alcohol (DNB)	TLC	111
Cinnamylephedrine	PC	60
Cinnamyl esters	GC	24
Ciodrin®	GC	61, 65, 67
	TLC	84, 171
CIPC	GC	129
	TLC	82
Citraconic anhydride, *p*-anisidide of	TLC	30
Citral	TLC	77
Citral, 2,4-DNPH of	TLC	72, 76
Citral, 2,4-DNPSC of	TLC	165
Citral, 2-hydrazinebenzothiazole derivative of	TLC	168
Citramalate acid	TLC	36
Citrate acid	TLC	36
Citric acid	LC	34, 58, 65
	PC	13, 77
	TLC	31
Citronellal, 2,4-DNPH of	TLC	76
Citronellal, 2,4-DNPSC of	TLC	165
Citronellal, 2-hydrazinebenzothiazole derivative of	TLC	168
Citronellol	TLC	77, 183
Citronellol (DNB)	TLC	111
Citronellol (ONPU)	TLC	111
Citronellyl acetate	TLC	105
Citronellyl butyrate	TLC	105
Citronellyl formate	TLC	105
Citrostadienol	GC	84
Citrulline	LC	41
	PC	64, 102
	TLC	11, 13, 15, 16
Citrulline, PTH of	TLC	124
Cl (Chlorine)	GC	26, 114
Cl^-, Chlorine	TLC	147
Cl_2, Chlorine	GC	113
Clemizole	TLC	39
Clemizole (Allercur®)	TLC	45
Clemizole HCl	GC	20
Cliradon	PC	2
Clistin® (Carbinoxamine)	TLC	39
ClO_3^-	TLC	147
ClO_4^-	TLC	147
Clopamide	TLC	41, 140
Cloxacillin Na	TLC	21
Cm, Curium	LC	90
Cm, Curium(II)	PC	15, 95, 96
CMP	LC	3
	PC	37
	TLC	160
d-CMP	LC	18
	TLC	63

2′ + 3′-CMP	TLC	160
5′-CMP	LC	22, 46
	TLC	63
Co(Cobalt)	LC	81, 82, 83, 85, 86, 87, 88, 89
	PC	15
Co_2, Cobalt	GC	113
Co(II), Cobalt	GC	43
	PC	95, 96
	TLC	148,149, 150
Co^{2+}, Cobalt	PC	92, 97
Co(III), Cobalt	GC	43
Cobalt, Co	LC	81, 82, 83, 85, 86, 87, 88, 89
	PC	15
Cobalt, Co(II)	GC	43
	PC	95, 96
	TLC	148,149, 150
Cobalt, Co(III)	GC	43
Cobalt, Co^{2+}	PC	92, 97
Cobalt chloride $CoCl_2$	PC	94
Cobalt chloride, dichlorobis (ethylene diamine) $[Co(en)_2Cl_2]Cl$	PC	94
Cobalt chloride, hexammine $[Co(NH_3)_6]Cl_3$	PC	94
Cobalt hexammine $Co(NH_3)6^{3+}$	PC	92
Cobalt protoporphyrin	PC	33
Cobalt, tetra, hexachloride, hexahydroxyhexakis (ethylenediamine) $[Co_4(en)_6(OH)_6Cl_6$	PC	94
Cobalt tribromide, aquatriammine $[Co(NH_3)_3(H_2O)]Br_3$	PC	94
Cobalt, trichloride tris (ethylenediamine) $[Co(en)_3Cl_3]$	PC	94
Cobalt tris (ethylenediamine) $Co(en)3^{3+}$	PC	92
Cobefrin·HCl	PC	5
Cocaine	GC	8, 96, 107, 111
	PC	3, 4, 54
	TLC	2, 3, 39, 45
Cocaine HCl	GC	107
	TLC	43
Cochalic	TLC	184
Codeine	GC	8, 18, 21, 96, 107, 111
	PC	1, 2, 54
	TLC	2, 3, 4, 39, 40, 45, 116, 138
Codeine hydrochloride	GC	107
	LC	20
Codeine phosphate	GC	107
Codethyline	TLC	4
$Co(en)3^{3+}$, Cobalt tris (ethylenediamine)	PC	92
Coenzyme-A	LC	18
Coenzyme Q_6	PC	50
Coenzyme Q_7	PC	50
Coenzyme Q_8	PC	50
Coenzyme Q_9	PC	50
Coenzyme Q_{10}	PC	50
Coenzyme Q_{10} (Q_{10})	PC	48
Coenzyme Q_6 hydroquinone	PC	50
Coenzyme Q_7 hydroquinone	PC	50
Coenzyme Q_8 hydroquinone	PC	50
Coenzyme Q_9 hydroquinone	PC	50
Coenzyme Q_{10} hydroquinone	PC	50
Colarnine	PC	2
Colchicine	GC	8, 111
	PC	3
	TLC	2, 3
α-Collatolic acid	PC	30
γ-Collidine	LC	34
Compound 4072	GC	65
Conac T	TLC	66
Conanin-3-one	TLC	98
Conessine	PC	2
	TLC	98
Confluentinic acid	TLC	35
Congo red	PC	88
$Co(NH_3)_6{}^{3+}$, Cobalt hexammine	PC	92
Coniferaldehyde, 2,4-DNPH of	TLC	167
Conjugated phenols	PC	82
Convallamarogenin	TLC	103
Convallatoxin	TLC	102
Convallatoxin peracetate	TLC	102
Convallatoxol	TLC	102
Convallatoxol peracetate	TLC	102
Copaene	TLC	54
α-Copaene	GC	140
β-Copaene	GC	140
Copper (Cu)	LC	81, 82, 83, 85, 86, 87, 88
Copper, Cu(II)	GC	43
	TLC	148, 149
Copper, Cu^{2+}	PC	92
Copper, Cu^{++}	TLC	52, 150
Copper dimethyldithiocarbamate	TLC	22, 66
Copper oxychloride	PC	117
Coproporphyrin	PC	33
Coproporphyrin I	PC	123
Coproporphyrin III	PC	123
Coprostanol	GC	79
	TLC	99
Coprostane	LC	45
Coptisine	PC	3
Co-Ral®	GC	64, 129
Co-Ral® (Coumphos)	PC	28
	TLC	80, 84
Co-Ral oxygen analog	GC	61
Coralline	TLC	142
Corbasil	PC	5
Corchoroside	TLC	102
Corchoroside peracetate	TLC	102
Coreopsin	PC	26
Coroglaucigenin (3β,14β-19-Trihydroxy-5α-cardenolide)	TLC	102
Coroglaucigenin peracetate	TLC	102
Coronaridine	GC	96
Coronene	GC	40
	LC	27
Corotoxigenin (3β,14β-Dihydroxy-19-oxo-5α-cardenolide)	TLC	102
Corotoxigenin peracetate	TLC	102
Corticosterone	TLC	101
Cortisone	PC	124
	TLC	101, 178
Cortisone acetate	TLC	178

Corydaline	GC	96
Corynantheine	PC	2
Corynanthine	PC	54
Corysamine	PC	3
Cosmetin	PC	26,
Costaclavine	TLC	114
Cotarnine	GC	8, 107
	TLC	3
Cotinine	GC	95
Coumaphos (Co-Ral®)	GC	65, 67
	LC	43
	TLC	81, 84
Coumaphos *O*-analog	GC	65
	TLC	84
p-Coumaraldehyde, 2,4-DNPH of	TLC	167
o-Coumaric acid	GC	90
	PC	82, 113
	TLC	133,135, 174
p-Coumaric acid	GC	90
	PC	113
	TLC	133,135, 174
Coumarin	PC	113
	TLC	68, 75, 79
7-Coumarinylglucoside	PC	82
7-Coumarinyl sulfuric acid	PC	82
Covatin	TLC	39
p-CPMU	TLC	173
p-CPU	TLC	173
Cr, Chromium	LC	83, 84, 85, 86, 87, 89
	PC	15
Cr(III), Chromium	GC	43
	PC	95, 96
Cr^{3+}, Chromium	PC	97
Creatine	PC	9, 10, 60
	TLC	11
Creatine phosphate	PC	9, 10
Creatine phosphate (Cr-P)	PC	32
Creatine phosphate kinase	LC	19
Creatinine	LC	46, 50
	PC	9, 10
	TLC	11
Cresol	TLC	90
m-Cresol	GC	71, 73
	LC	52
	PC	31
	TLC	88, 174
o-Cresol	GC	71, 73
	LC	1, 11
	PC	31, 82
	TLC	88, 174
p-Cresol	GC	71, 73
	LC	30, 52, 61
	PC	82
	TLC	88, 174
m-Cresol, 4-acetyl-2-nitrophenol ether of	TLC	91
o-Cresol, 4-acetyl-2-nitrophenol ether of	TLC	91
p-Cresol, 4-acetyl-2-nitrophenol ether of	TLC	91
p-Cresolmonomethylol	PC	82
Cresol red	PC	88

m-Cresotinic acid	TLC	129
Crinine	GC	96
$Cr_2O_7^{--}$	TLC	147
Crocetin	TLC	94
Crotonaldehyde	GC	7, 53, 55
	LC	29
	PC	20
Crotonaldehyde, 2,4-DNPH of	TLC	72, 166
Crotonaldehyde, 2,4-DNPSC of	TLC	165
Crotonaldehyde, *p*-NPH of	TLC	165
Crotonaldehyde, organics	GC	52
Crotonic acid	LC	58
	TLC	29
Crotonic acid, methyl ester of	GC	24
Crotononitrile	GC	55
cis-Crotononitrile	GC	123
trans-Crotononitrile	GC	123
Crotoxyphos	TLC	81
N-trans-Crotylthiourea	PC	47
Crufomate	LC	43
Cryptocapsin	TLC	94
Cryptochlorophaeic acid	TLC	35
Cryptofauronol	GC	141
Cryptograndoside A	TLC	102
Cryptograndoside A peracetate	TLC	102
Cryptopine	GC	96, 107
	PC	2, 3
	TLC	4
Cryptopine oralate	GC	107
Cryptoxanthin	TLC	93, 94
R-2 Crystals	TLC	66
Cs, Cesium	LC	83, 85, 86, 88
	PC	15
	TLC	52
Cs^{+}, Cesium	PC	97
CTP	TLC	63
d-CTP	TLC	63
Cu (Copper)	LC	81, 82, 83, 85, 86, 87, 88
Cu(II), Copper	GC	43
	TLC	148, 149
Cu^{2+}, Copper	PC	92
Cu^{++}, Copper	TLC	52, 150
Cubebene	GC	140
2-*p*-Cumaroyl-glucose	PC	71
3-*p*-Cumaroyl-glucose	PC	71
6-*p*-Cumaroyl-glucose	PC	71
1-*p*-Cumaroyl-β-D-glucose	PC	71
4-*p*-Cumaroyl-rhamnose	PC	71
Cumate	TLC	66
Cumene	GC	87
	TLC	51
Cumene hydroperoxide	PC	27
	TLC	65
Cumene (isopropylbenzene)	GC	53
Cuminaldehyde	TLC	77
Cuminaldehyde, 2,4-DNPH of	TLC	73
Cuminyl aldehyde, 2-hydrazinebenzothiazole derivative of	TLC	168
Cumyl-2-hydroxy-2-phenylethylperoxide	PC	27
Cuparene	TLC	54
Cupreine	TLC	3
α-Curcumene	GC	140
β-Curcumene	GC	140
γ-Curcumene	GC	140
Curium (Cm)	LC	90

Curium, Cm(II)	PC	15, 95, 96
Cyanamide	PC	9
Cyanidin	PC	25
	TLC	78
Cyanidin chloride	PC	54
Cyanidin-3,5-diglucoside	TLC	78
Cyanidin glycosides, 3-diglucoside	PC	24
Cyanidin glycosides, 3,5-diglucoside	PC	24
Cyanidin glycosides, 3-monogalactoside	PC	24
Cyanidin glycosides, 3-monoglucoside	PC	24
Cyanidin glycosides, 3-rhamnoglucoside	PC	24
Cyanidin glycosides, 3-rhamnoglucosido-5-glucoside	PC	24
Cyanidin glycosides, 3-xyloglucoside	PC	24
Cyanidin-3-monoglucoside	TLC	78
Cyanidin-3-triglucoside	TLC	78
Cyanoacetic acid	PC	64
Cyanocobalamine	TLC	188
Cyanoethyl pT	PC	19
m-Cyanophenol	PC	31
p-Cyanophenol	PC	31, 121
6-Cyanopurine	LC	46
Cyanox (S-4084)	TLC	172
Cyanuric acid	PC	60
Cyclazine	GC	20
Cyclic dimer	LC	49
Cyclic hexamer	LC	49
Cyclic 66 linear trimer	LC	49
Cyclic monomer	LC	49
Cyclic pentamer	LC	49
Cyclic tetramer	LC	49
Cyclic thiuram	TLC	66
Cyclic trimer	LC	49
Cyclizine	LC	55
	TLC	39
Cyclizine HCl	GC	20, 54
Cycloartenol	GC	84
	LC	45
Cyclobarbital	GC	106
Cyclobarbital (cyclobarbitone)	GC	19
	TLC	137
Cyclobarbitone	PC	86
Cyclobutanone	GC	55
Cyclocopacamphene	GC	140
Cyclodecane	GC	51
Cyclododecane	GC	51
Cyclododecanol	GC	76
Cyclofenchene	GC	87
Cyclohept, cycloheptanone	GC	98
Cyclohept-SB	GC	98
Cyclohept-SB, cycloheptanone	GC	98
Cycloheptane	GC	42, 51
Cycloheptane (hydrocarbons)	GC	118
Cycloheptanol	GC	51
Cycloheptanone	GC	51
Cycloheptanone, cyclohept	GC	98
Cycloheptanone, 2,4-DNPH of	TLC	76
Cycloheptene	GC	51
Cyclohex-SB	GC	98
Cyclohex-SB, cyclohexanone	GC	98
Cyclohexane	GC	29, 35, 42, 47, 51, 53, 55, 93, 125, 127
	LC	21, 27, 36, 42
Cyclohexane (hydrocarbons)	GC	118
Cyclohexane hydroperoxide	PC	27
Cyclohexane, methylcyclohexane	GC	126
Cyclohexane, organics	GC	52
Cyclohexanes	GC	126
Cyclohexanediol-1,2 *trans*	GC	76
Cyclohexanediol-1,3 *cis*	GC	76
Cyclohexanediol-1,3 *trans*	GC	76
Cyclohexanediol-1,4 *trans*	GC	76
Cyclohexanol	GC	51, 76
	LC	11, 36
	PC	52
	TLC	1
Cyclohexanone	GC	7, 51
	LC	8, 46
	PC	20
Cyclohexanone, cyclohex-SB	GC	98
Cyclohexanone, 2,4-DNPH of	TLC	72, 76, 166
Cyclohexanone peroxide Component A	TLC	65
Cyclohexanone peroxide Component B	TLC	65
Cyclohexanone peroxide Component C	TLC	65
Cyclohexene	GC	47, 51, 53, 55, 122
	LC	42
Cyclohexene (hydrocarbons)	GC	118
Cyclohexene hydroperoxide	PC	27
2-Cyclohexene-1-one	GC	53
Cyclohexylamine	TLC	10
Cyclohexylammonium benzyl *n*-butyl phosphate	PC	34
Cyclohexylammonium isopropyl *n*-propyl phosphate	PC	34
Cyclohexylammonium isopropyl *n*-propyl pyrophosphate	PC	34
N-Cyclohexylaniline	TLC	7
N-Cyclohexyl-2-benzothiazole sulfenamide	TLC	22, 66
Cyclohexyl isocyanate	GC	44
p-Cyclohexylphenol	TLC	176
4-Cyclohexylphenol	GC	71
α-Cyclohexyl-α-phenyl-l-piperdine-propanol (Artane®)	GC	21
N-Cyclohexylpiperidine	TLC	7
Cyclomethycaine	TLC	39
Cyclononane	GC	51
Cyclooctane	GC	51
Cyclooctane (hydrocarbons)	GC	118
Cyclooctanol	GC	51
Cyclooctanone	GC	51
Cyclooctanone, 2,4-DNPH of	TLC	76
Cyclooctene	GC	51
Cyclopal	GC	19, 106
	TLC	137
Cyclopent-SB	GC	98
Cyclopent-SB, cyclopentanone	GC	98
Cyclopentadiene	GC	29
Cyclopentadiene (hydrocarbons)	GC	118
Cyclopentamine	GC	23
	TLC	39
Cyclopentane	GC	29, 36, 42, 51, 53, 55, 93, 125, 127
	LC	42
Cyclopentane (hydrocarbons)	GC	118
Cyclopentane, methylcyclopentane	GC	126
Cyclopentanes	GC	126

Cyclopentanol GC 51, 76
PC 52
Cyclopentanone GC 51, 55
LC 8
PC 20
Cyclopentanone, cyclopent-SB GC 98
Cyclopentanone, 2,4-DNPH of TLC 76, 166
Cyclopentene GC 29, 47, 51, 122
LC 42
Cyclopentene (hydrocarbons) GC 118
Cyclopenthiazide TLC 41
Cyclopentylamine TLC 10
17β-Cyclopentylpropionoxy-4-androsten-3-one PC 124
Cyclopiazonic acid TLC 185
Cyclopropane (hydrocarbons) GC 118
Cyclosativene GC 140
d-Cycloserine (*d*-4-amino-3-isoxalidone) PC 54
Cyclothiazide TLC 41
Cycloundecane GC 51
Cymarin PC 127
TLC 102
Cymarin acetate PC 127
Cymarin acid TLC 102
Cymarin acid methylester TLC 102
Cymarin acid methylester peracetate TLC 102
Cymarin acid peracetate TLC 102
Cymarin peracetate TLC 102
Cymarol PC 127
TLC 102
Cymarol peracetate TLC 102
m-Cymene GC 87
o-Cymene GC 86, 87
p-Cymene GC 51, 86, 87
TLC 7, 51, 54, 106
Cyolane TLC 81
Cyperene GC 140
Cystathionine LC 41
PC 11, 64
DL+ *allo*-Cystathionine LC 32
Cysteamine PC 64
Cysteamine-glutathione disulfide LC 32
Cysteic acid LC 33, 41
PC 11, 62, 63, 64, 65, 105, 106
TLC 11, 13, 15, 16, 118
Cysteic acid, PTH of TLC 124
Cysteine GC 101
PC 11, 64
TLC 13
L-Cysteine LC 32
L-Cysteine-cysteamine disulfide LC 32
L-Cysteine-glutathione disulfide LC 32
Cysteine-homocysteine mixed disulfide LC 41
L-Cysteine-L-homocysteine disulfide LC 32
L-Cysteine-L-homocysteine trisulfide LC 32
Cysteine methyl ester PC 11
L-Cysteine-D-penicillamine disulfide LC 32
L-Cysteine-D-penicillamine trisulfide LC 32
Cysteine sulfinic acid LC 41
PC 63, 64
L-Cysteine trisulfide LC 32
Cystine GC 101
LC 33, 41
PC 3, 11, 62, 64, 65
TLC 11, 13, 15, 118
L-Cystine LC 32
TLC 118
L-Cystine disulfoxide LC 59
meso-Cystine LC 59
Cystine-HCl TLC 16
Cythioate GC 67
Cytidine GC 77
LC 3, 13, 15, 22, 31, 46, 50
PC 36, 37, 108
TLC 62, 159
MO-cytidine GC 77
Cytidine diphosphate (CDP) TLC 160
Cytidine-5′-diphosphate PC 108
TLC 62
Cytidine-2′-monophosphate (2′-CMP) TLC 160
Cytidine-2′:3′-monophosphate PC 108
Cytidine-2′(3′)-monophosphate GC 77
Cytidine-3′-monophosphate (3′-CMP) TLC 160
Cytidine-2′ and 3′-monophosphates PC 108
Cytidine-5′-monophosphate (5′-CMP) PC 108
TLC 160
2-d-Cytidine-5′-monophosphate (2-d-5′-CMP) TLC 160
Cytidine-3′(2′)-phosphate LC 15
Cytidine-5′-phosphate PC 36
TLC 62
Cytidine triphosphate (CTP) PC 32
TLC 160
Cytidine-5′-triphosphate PC 108
Cytidylic acid LC 13
Cytisine GC 96
PC 3
Cytochrome-C LC 64
Cytochrome C (horse heart) LC 19
Cytosine GC 77
LC 3, 13, 15, 17, 22, 31, 46, 50
PC 36, 37, 108
TLC 159
Cytosine-5-carboxylic acid PC 36

Adsorption of the Elements (LC Section – Figures)

C, Carbon Pg 278, 279, 280, 281
Ca, Calcium Pg 278, 279, 280, 281
Cd, Cadmium Pg 278, 279, 280, 281
Cl, Chlorine Pg 278, 279, 280, 281
Cm, Curium Pg 278, 279, 280
Co, Cobalt Pg 278, 279, 280, 281
Cr, Chromium Pg 278, 279, 280, 281
Cs, Cesium Pg 278, 279, 280, 281
Cu, Copper Pg 278, 279, 280, 281

D_2, Deuterium	GC	115
2,4-D	PC	116, 118
2,4-D(butoxyethanol ester)	GC	128
2,4-D(butyl ester)	GC	128
2,4-D(IPE)	GC	68
2,4-D(isooctyl ester)	GC	128
2,4-D(isopropyl ester)	GC	128
2,4-D(ME)	GC	68
2,4-D(P.G.B. ether ester)	GC	128
Daconil 2787	GC	128
Dacthal®	GC	68, 129
	TLC	82
Dacthal mono acid	GC	61
Dactylin	PC	26
Daidzein	PC	26
Dalapon	PC	118
DAO (2,3-Diacetyloleandomycin)	PC	66
1,2-DAO (1,2-Diacetyloleandomycin)	PC	66
1,3-DAO (Diacetyloleandomycin)	PC	66
Daphnetin	PC	113
Darvon (Propoxyphene)	TLC	138
Dasanit®	GC	65, 67
Dasanit® (Bayer 25141)	TLC	80, 84
Dasanit sulfone	GC	67
Dasanit O-analog	GC	65
Dasanit oxygen analog	GC	67
Dasanit oxygen analog sulfone	GC	67
Dazomet	GC	65
	TLC	81
2,4-DB	PC	118
2,4-D(BE)I	GC	68
2,4-D(BE)II	GC	68
3,4-DCA	TLC	173
3,4-DCPMOU	TLC	173
3,4-DCPMU	TLC	173
DDA(ME)	GC	68
DDA (methyl ester)	GC	128
DDD	LC	44
DDD (TDE)	TLC	80, 82
o,p'-DDD	GC	60, 62, 68,128, 129
p,p'-DDD (*p,p'*-TDE)	GC	60, 61, 62, 68,128, 129
	TLC	82
p,p'-DDD olefin	GC	129
DDE	GC	129
	PC	116
	TLC	81
o,p'-DDE	GC	68,128, 129
p,p'-DDE	GC	60, 61, 62, 68,128, 129
	TLC	80, 82
DDT	TLC	81, 172
DDT (Dichlorodiphenyltrichloroethane)	GC	21
DDT (tech)	GC	129
o,p'-DDT	GC	60, 62, 68,128, 129
	TLC	80, 82
p,p'-DDT	GC	61
	LC	44
	TLC	80, 82
p,p'-DDT, pesticides	GC	60, 129
DDVP	GC	129
	TLC	84, 172
Deacetyldigilanid A	PC	127
Deacetyldigilanid B	PC	127
Deacetyldigilanid C	PC	127
Decacyclene	LC	27
Decahydro-9-naphthyl hydroperoxide	TLC	65
Decalin	LC	27, 45
Decalin (*cis*)	GC	45
Decalin (*trans*)	GC	45
cis-Decalin	GC	51
trans-Decalin	GC	51
Decamethylenediamine, trifluoroacetyl derivative	GC	10
Decanal, 2,4-DNPH of	TLC	72, 76
Decanal, 2,4-DNPSC of	TLC	165
Decanal, *p*-NPH of	TLC	165
Decane	GC	49
n-Decane	LC	21, 38
	GC	53, 55, 120, 121
n-Decane, organics	GC	52
Decanoic acid	PC	77
Decanoic acid, ethyl ester of	GC	24
Decanoic acid, methyl ester of	GC	24
n-Decanoic acid	LC	57
n-Decanol	GC	92
	LC	54
n-Decanol (*n*-Decyl alcohol)	TLC	1
1-Decanol	GC	92
1-Decanol (DNB)	TLC	111
1-Decanol (ONPU)	TLC	111
Decanone-2	GC	51
Decapryn® (Doxylamine)	TLC	39
Decarboxythamnolic acid	TLC	34
2-Decen-1-al, 2,4-DNPH of	TLC	72
Decene-1	GC	53
	LC	21
Dec-1-ene	GC	56
1-Decene	GC	120
cis-2-Decene	GC	120
trans-2-Decene	GC	120
cis-3-Decene	GC	120
trans-3-Decene	GC	120
cis-4-Decene	GC	120
trans-4-Decene	GC	120
cis-5-Decene	GC	120
trans-5-Decene	GC	120
Decyl acetate	TLC	105
Decyl aldehyde	PC	20
n-Decyl alcohol	LC	11
Decylamine	TLC	10
n-Decylamine	PC	5, 55
Decyl caprylate	TLC	105
Decyl esters	GC	24
n-Decyl isocyanate	GC	44
Decyl mercaptan	TLC	162
Decyl octyl sulfide	TLC	163
n-Decyl-3-thienyl sulfide	TLC	104
DEF®	GC	65, 67
Dehydroabietate	LC	67
Dehydroacetic acid	LC	74
Dehydroadonirubin	TLC	93
Dehydroandrosterone (iso)	TLC	95
Dehydrochlortetracycline	PC	68
7-Dehydrocholesterol	GC	136
	PC	126
	TLC	56, 99
Dehydrocholic acid	PC	38
Dehydrocorydaline	PC	3
Dehydroepiandrosterone	GC	83, 136
	TLC	97, 178

Dehydroepiandrosterone or (3β-Hydroxy-5-androsten-17-one)	PC	41
Dehydro-equilin	TLC	181
6-Dehydro-estradiol	TLC	181
6-Dehydro-estrone	TLC	181
9-Dehydro-estrone	TLC	181
6-Dehydro-17α-hydroxyprogesterone	TLC	178
Dehydroisoandrosterone	PC	124
6-Dehydro-progesterone	GC	81
1-Dehydro-testololactone (Group II)	PC	40
1-Dehydro-testosterone propionate	TLC	178
Dehydrovanillin	PC	20
Delnav (Diozathion)	PC	28
	TLC	80, 84
Delphenine	PC	2
Delphinidin	PC	25
	TLC	78
Delphinidin-3,5-diglucoside	TLC	78
Delphinidin glycosides	PC	24
Delphinidin glycosides, 3,5-diglucoside	PC	24
Delphinidin glycosides, 3,7-diglucoside	PC	24
Delphinidin glycosides, 3-gentiobioside	PC	24
Delphinidin glycosides, 3-gentiotrioside	PC	24
Delphinidin glycosides, 5-glucoside-3-rutinoside	PC	24
Delphinidin glycosides, 5-glucoside-3-sambubioside	PC	24
Delphinidin glycosides, 5-glucoside-3-sophoroside	PC	24
Delphinidin glycosides, 7-glucoside-3-sophoroside	PC	24
Delphinidin glycosides, 3-monoglucoside	PC	24
Delphinidin glycosides, 3-rhamnoglucoside	PC	24
Delphinidin-3-monoglucoside	TLC	78
Delta Keto "153"	GC	129
	TLC	82
Delta Trichloro "237"	GC	129
Demecolcine	PC	3
Demethylchlorpromazine (Nor,CP)	GC	103
Demethylchlorpromazine sulfoxide (Nor,CPSO)	GC	103
N-Demethylchlorpromazine sulfoxide	TLC	139
Demethylchlortetracycline	PC	68
Demethylchlortetracycline HCl	TLC	21
Demethyltetracycline	PC	68
Demeton-S	LC	43
Demeton	GC	60, 64, 65
Demeton-methyl	TLC	81
Demeton-*O*-methyl	LC	43
	TLC	84
Demeton-S-methyl	LC	43
	TLC	84
Demeton-S-methylsulfone	LC	43
Demeton-S-methylsulfoxide	LC	43
Demeton (Systox)	TLC	80, 84, 172
Demeton thiol	GC	67
Demeton, thiol isomer	TLC	80
Demeton, thiol isomer sulfone	TLC	80
Demeton, thiol isomer sulfoxide	TLC	80
Demeton thiol sulfone	TLC	84
Demeton thiono	GC	67
Demeton thiono isomer	TLC	80
Demeton thiono oxygen analog	GC	67
Demeton thiono sulfone	GC	67
Deoxyadenosine	GC	77
	LC	3, 31
	PC	19, 36, 37, 108
2-Deoxyadenosine	TLC	159
2′-Deoxyadenosine	LC	46
Deoxyadenosine-5′-diphosphate	PC	108
Deoxyadenosine-5′-monophosphate	GC	77
	PC	108
Deoxyadenosine-5′-phosphoric acid trichloroethyl ester	TLC	62
Deoxyadenosine-5′-triphosphate	PC	108
2′-Deoxyadenosine 3′,5′-diacetate	GC	134
Deoxyadenylic acid	PC	19
Deoxy-AMP (Adenosine monophosphate)	PC	37
2-Deoxy-D-arabino-hexonic	LC	23
Deoxycholic acid	PC	39
Deoxycholic acid (3α,12α-OH)	LC	45
Deoxy-CMP	PC	37
Deoxycytidine	GC	77
	LC	31
	PC	19, 36, 37, 108
2-Deoxycytidine	TLC	159
Deoxycytidine-5′-monophosphate	PC	108
Deoxycytidine-5′-phosphate	PC	36
Deoxycytidylic acid	PC	19, 37
Deoxycytosine	LC	3
Deoxyephedrine	GC	9
2-Deoxy-DL-erythro-pentonic	LC	23
3-Deoxy-D-erythro-pentonic	LC	23
5-Deoxy-5-S-ethyl-1,2-*O*-isopropylidene-L-arabinose	GC	13
5-Deoxy-5-S-ethyl-1,2-*O*-isopropylidene-D-xylose	GC	13
2-Deoxygalactinol	GC	14
6-Deoxy-D-galactonic	LC	23
2-Deoxy-D-galactose	LC	48
2-Deoxyglucitol	GC	14
6-Deoxy-gluco-hepturonic	LC	23
Deoxyglucose	PC	70
2-Deoxyglucose	GC	15
2-Deoxy-D-glucose	LC	48
6-Deoxy-D-glucose	LC	48
Deoxy-GMP	PC	37
Deoxyguanosine	GC	77
	LC	3, 31
	PC	19, 36, 37, 108
2-Deoxyguanosine	TLC	159
2′-Deoxyguanosine	LC	46
Deoxyguanosine-5′-diphosphate	PC	108
Deoxyguanosine-5′-monophosphate	GC	77
	PC	108
Deoxyguanosine-5′-triphosphate	PC	108
Deoxyguanylic acid	PC	19
3-Deoxy-2-C-hydroxymethyl-tetronic	LC	23
Deoxyinosine	GC	77
	LC	3, 31
	PC	36, 37, 108
2′-Deoxyinosine	LC	46
5-Deoxy-1,2-*O*-isopropylidene-L-arabinose	GC	13
2-Deoxy-D-lyxo-hexonic	LC	23
3-Deoxy-D-lyxo-hexonic	LC	23
6-Deoxy-D-mannonic	LC	40
Deoxynucleosides	PC	108
4-Deoxypyridoxol	GC	88
2-Deoxyribitol	GC	14
Deoxyribonucleosides	GC	77
Deoxyribose	GC	15
	PC	70

2-Deoxy-D-ribose	LC	48
6-Deoxy-L-talitol	GC	14
6-Deoxy-L-talose	PC	75
Deoxythymidine	TLC	62
Deoxythymidine-5′-phosphate	TLC	62
Deoxythymidine-5′-phosphoric acid benzhydryl ester	TLC	62
Deoxythymidine-5′-phosphoric acid trichloroethyl ester	TLC	62
Deoxythymidylyl-(3′-5′)-deoxythymidine	TLC	62
Deoxyuridine	LC	3, 31
	PC	36, 37, 108
	TLC	159
2′-Deoxyuridine 3′,5′-diacetate	GC	134
Deoxyuridine-5′-phosphate	PC	36
3-Deoxy-D-xylo-hexonic	LC	23
Depocid (sulfaphenazole)	TLC	140
Depovernil® (sulfamethoxypyridazine)	TLC	140
Derivatives, androstane	PC	42
Derivatives, pregnane	PC	42
Desacetyl-bufotalin (3β,14β,16β-Trihydroxy-5β-bufadienolide)	TLC	102
Desacetyl-cerbertin	TLC	102
Desacetyl-cerbertin peracetate	TLC	102
Desacetyl-protanghinin	TLC	102
Desacetyl-rhodexin B	TLC	102
Desacetyl-tanghinin	TLC	102
Desacetyl-tanghinin peracetate	TLC	102
Desaglybuzole	TLC	140
Deserpidine	GC	17
Deserpine	PC	2
Desethylphosphamidon	TLC	84
Desgluco-cheirotoxin	TLC	102
Desgluco-cheirotoxin peracetate	TLC	102
Desgluco-ericordin	TLC	102
Desgluco-ericordin peracetate	TLC	102
Desglucohellebrin	TLC	102
Desglucohellebrin peracetate	TLC	102
Desgluco-uzarin	TLC	102
Desgluco-uzarin peracetate	TLC	102
Desimpramine	TLC	39
Desmosterol	GC	79
	TLC	99
Dg-2-desoxy-β,D-alloside	TLC	102
Dg-2-desoxy-β,D-alloside peracetate	TLC	102
11-Desoxy-corticosterone	TLC	178
Desoxycortone acetate	TLC	95
14-Desoxy-14αH-digitoxigenin (3β-Hydroxy-5β,14α-cardenolide)	TLC	102
14-Desoxy-17βH-digitoxigenin (3β-Hydroxy-5β,14β,17βH-cardenolide)	TLC	102
Desoxyephedrine (Methamphetamine)	GC	21
dl-Desoxyephedrine	PC	60
16-Desoxy-estrone	TLC	181
Dg-2-desoxy-β,D-glucoside	TLC	102
Dg-2-desoxy-β,D-glucoside peracetate	TLC	102
dl-Desthiobiotin (5-methyl-2-imidazolidone-4-caproic acid)	PC	54
Deuterium, D_2	GC	115
Deuteroporphyrin	PC	33
Deuteroporphyrin IX	PC	123
Dexamethasone	TLC	101, 178
Dexbrompheniramine maleate (brompheniramine maleate)	GC	20, 54
Dexedrine	PC	5
Dextroamphetamine	GC	17
Dextroamphetamine sulfate	GC	17

Dextrocaine (Psicaine)	TLC	43
Dextromethorphan	PC	4
	TLC	4
Dextromethorphan HBr	GC	20, 54, 104
Dextromoramide	PC	4
	TLC	4
Diacetate	TLC	112
Diacetate of 3,4-bis-(*p*-hydroxyphenyl)-2,4-hexadiene	TLC	136
Diacetone alcohol	TLC	1
Diacetone alcohol (4-methyl-4-hydroxy-2-pentanone)	GC	53
	LC	8
Diacetyl	LC	8
Diacetyl, 2,4-DNPH of	TLC	72
Diacetyl (bis)	GC	7
3,5-Di-*O*-acetyl-2,4-*O*-benzylidene-D-xylose dimethylacetal	GC	13
N,N′-Diacetyl-3,5-diamino-2,4,6-triiodobenzoic acid	TLC	64
O,N-Diacetyl-3,3′-diiodothyronine	TLC	19
O,N-Diacetyl-3,5-diiodothyronine	TLC	19
1,2-Diacetyl-1,2-dimethylhydrazine	PC	61
N,N-Diacetylhistamine	GC	9
1,2-Diacetylhydrazine	PC	61
3,6-Di-*O*-acetyl-5-*O*-methyl-1,2-*O*-isopropylidene-D-glucose	GC	13
Diacetylmorphine (Heroin)	GC	8, 21, 107
	TLC	39, 138
Diacetylmorphine HCl (Heroin HCl)	GC	107
1,2-Diacetyl oleandomycin	PC	66
1,3-Diacetyl oleandomycin	PC	66
2,3-Diacetyl oleandomycin	PC	66
Diacetyl, organics	GC	52
Diacetylperoxide	PC	27
O,O-Diacetyl-tetraiodothyrolactic acid	TLC	19
O,N-Diacetylthyroxamine	TLC	19
O,N-Diacetylthyroxine	TLC	19
O,O-Diacetyl-3,5,3′-triiodothyrolactic acid	TLC	19
O,N-Diacetyl-3,3′,5′-triiodothyronine	TLC	19
O,N-Diacetyl-3,5,3′-triiodothyronine	TLC	19
Diadinoxanthin	PC	122
Cbo-DL-Dialanine	TLC	12
Diallylamine	TLC	10
Diallylamine, trifluoroacetyl derivative	GC	10
4-Diallylamino-3,5-methylphenyl *N*-methylcarbamate	GC	130
Diallylbarbituric acid (Allobarbital)	GC	105, 106
Di-allyl disulfide	GC	85
Di-allyl sulfide	GC	85
5,5-Diallythiobarbituric acid	PC	46
Di-allyl trisulfide	GC	85
1,2-Diamino-anthraquinone	TLC	69
1,3-Diamino-anthraquinone	TLC	69
1,4-Diamino-anthraquinone	TLC	69
1,5-Diamino-anthraquinone	TLC	69
1,6-Diamino-anthraquinone	TLC	69
1,7-Diamino-anthraquinone	TLC	69
1,8-Diamino-anthraquinone	TLC	69
2,3-Diamino-anthraquinone	TLC	69
2,6-Diamino-anthraquinone	TLC	69
2,7-Diamino-anthraquinone	TLC	69
1,2-Diaminobenzene	TLC	119

1,3-Diaminobenzene	TLC	119
1,4-Diaminobenzene	TLC	119
3,4-Diaminobenzoic acid	PC	78
1,4-Diamino-2-bromo-anthraquinone	TLC	69
1,4-Diaminobutane	GC	11
1,4-Diaminobutane (putrescine)	PC	56
α,γ-Diaminobutyric acid	TLC	11
α,γ-Diamino-*n*-butyric acid	PC	64
2,4-Diaminobutyric acid	TLC	13
L-2,4-Diaminobutyric acid	LC	59
β,ε-Diamino-*n*-caproic acid	PC	64
2,5-Diamino-4,6-deoxypyrimidine	PC	36
4,5-Diamino-2,6-dihydroxypyrimidine	PC	16
2,2′-Diaminodiphenyl	TLC	119
4,4′-Diaminodiphenyl	TLC	119
Diaminoethane	GC	11
1,2-Diaminoethane	PC	56
2,7-Diaminofluorene	PC	59
2,6-Diamino-5-formylamino-4-hydroxy-pyrimidine	PC	36
2,6-Diamino-5-formylamino-4-methoxy-pyrimidine	PC	36
4,6-Diamino-5-formylamino-2-methyl-pyrimidine	PC	36
α,γ-Diaminoglutaric acid	PC	64
1,7-Diaminoheptane	GC	11
1,6-Diaminohexane	GC	11
	PC	56
3,4-Diamino-6-hydroxypyrimidine	PC	16
4,5-Diamino-2-hydroxypyrimidine	PC	16
2,6-Diamino-7-methylpurine	PC	36
4,5-Diamino-6-methylpyrimidine	PC	36
1,5-Diaminonaphthalene	PC	59
2,7-Diaminonaphthalene	PC	59
1,2-Diamino-4-nitrobenzene	PC	6
1,3-Diamino-4-nitrobenzene	PC	6
1,8-Diaminooctane	GC	11
2,4-Diamino-6-oxypyrimidine	PC	36
4,6-Diamino-2-oxypyrimidine	PC	36
1,5-Diaminopentane	GC	11
1,5-Diaminopentane (cadaverine)	PC	56
α,α-Diaminopimelic acid	TLC	11
α,β-Diaminopimelic acid	PC	64
α,ε-Diaminopimelic acid	LC	59
2,6-Diaminopimelic acid	TLC	13
1,2-Diaminopropane	GC	11
1,3-Diaminopropane	GC	11
	LC	66
	PC	56
α,β-Diaminopropionic acid	PC	64
2,6-Diaminopurine	LC	46
2,6-Diaminopurine hemisulfate	PC	36
2,3-Diaminopyridine	PC	60
4,6-Diaminopyrimidine	PC	36
4,6-Diamino-2-thiopyrimidine	PC	36
Diammonium P^1,P^2-di-*n*-butyl pyrophosphate	PC	34
Diammonium P^1,P^2-diethyl pyrophosphate	PC	34
Diamorphine	PC	4
Di-*n*-amyl adipate	GC	25
Diamylamine	TLC	10
Di-*n*-amylamine	GC	45
Di-*n*-amylamine, trifluoroacetyl derivative	GC	10
Di-*n*-amyl disulfide	GC	85
Diamyl ether	GC	51
Di-*n*-amyl ether	LC	7
Di-*n*-amyl sebacate	GC	25
Di-*n*-amyl suberate	GC	25
Di-*n*-amyl succinate	GC	25
Dianhydro-gitoxigenin (3β-Hydroxy-5β-cardatrien-14,16,20(22)-olide)	TLC	102
o-Dianisidine	PC	120
N,N′-Diaryl-*p*-phenylenediamine	TLC	22
Diatoxanthin	PC	122
Diazepam	GC	22
	TLC	39
Diazine residue	TLC	84
Diazinon®	GC	60, 64, 65, 68, 128, 129
	LC	43
	PC	28
	TLC	80, 81, 84, 171, 172
Diazinon oxygen analog	GC	67
Diazinoxon	TLC	84
Diazoxon	GC	65
Dibenz[*a,h*]anthracene	TLC	146
Dibenz[*b,i*]anthracene	TLC	146
1,2,3,4-Dibenzanthracene	GC	40
1,2:5,6-Dibenzanthracene	PC	89
1,2,5,6-Dibenzanthracene	GC	40
1,2,7,8-Dibenzanthracene	GC	40
7H-Dibenzo[*b,g*]carbazole	TLC	120
7H-Dibenzo[*c,g*]carbazole	TLC	120
13H-Dibenzo[*a,i*]carbazole	TLC	120
Dibenzo[*a,i*]phenanthrene	TLC	146
Dibenzo[*a,e*]pyrene	TLC	146
Dibenzo[*a,i*]pyrene	TLC	146
1,2:3,4-Dibenzopyrene	PC	89
1,2,3,4-Dibenzopyrene	GC	40
1,2:4,5-Dibenzopyrene	PC	89
1,2,4,5-Dibenzopyrene	GC	40
1,2:6,7-Dibenzopyrene	PC	89
3,4,8,9-Dibenzopyrene	GC	40
3,4:9,10-Dibenzopyrene	PC	89
3,4,9,10-Dibenzopyrene	GC	40
Dibenzothiazolyl disulfide	TLC	22
Dibenzothiophene sulfone	GC	139
7H-Dibenzo[*c,h*]xanthen-7-one	TLC	75
Dibenzylamine	TLC	7, 10
Dibenzyl sulfoxide	TLC	164
Diborane	GC	27
Dibrom (Naled)	GC	61
	LC	43
	PC	28
	TLC	80, 84
Dibromoacetic	PC	79
p-Dibromobenzene	LC	75
Dibromofluorescein	PC	88
2,3-Dibromopropionic	PC	79
2,4-Dibromo-6-propionylphenol	TLC	89
2,6-Dibromo-4-propionylphenol	TLC	89
DIBS	TLC	66
Dibucaine (Nupercaine)	GC	8, 21
Dibutylamine	TLC	7, 10
Di-*n*-butylamine	GC	97
	PC	55
Di-*n*-butylamine, trifluoroacetyl derivative	GC	10
Di-*sec*-butylamine	GC	97
Di-*sec*-butylamine, trifluoroacetyl derivative	GC	10

Di-*n*-butyl disulfide	GC	85
Dibutyl ether	GC	51
Di-*n*-butyl ether	GC	125, 126
	LC	7
Di-*n*-butyl ether, organics	GC	52
N,N'-Dibutylguanidine	PC	10
Dibutylhydroxytoluene	TLC	125
Dibutyl isophthalate	GC	132
Dibutyl ketone	GC	117
2,6-Di-*tert*-butyl-α-methoxy-4-methylphenol	TLC	175
2,6-Di-*tert*-butyl-4-methylphenol	GC	72
	TLC	175
Di-*tert*-butylperoxide	TLC	65
2,4-Di-*tert*-butylphenol	GC	73
2,6-Di-*tert*-butylphenol	GC	69, 73
	PC	120
	TLC	175
3,5-Di-*tert*-butylphenol	GC	71
Dibutyl phthalate	GC	105, 132
Dibutyl phthalate (phthalic acid ester)	TLC	27
Di-*n*-butyl phthalate	GC	25
3,4-Di-*tert*-butylpyrocatechol	TLC	176
Dibutyl sebacate	GC	132
	TLC	27
Di-*n*-butyl sebacate	GC	25
Di-*n*-butyl succinate	GC	25
Dibutyl sulfide	GC	135
	TLC	163
N',N''-Dibutylthiourea	PC	47
Dibutyltin dichloride	TLC	67
Dibutyrate	TLC	112
1,2-Dibutyrylhydrazine	PC	61
Dicapthon (Dicapton)	GC	65
	TLC	80, 84, 170, 171
O,N-Dicarbobenzoxy-3,3'-diiodothyronine	TLC	19
O,N-Dicarbobenzoxy-3,5-diiodothyronine	TLC	19
O,N-Dicarbobenzoxy-3,5-diiodotyrosine	TLC	19
O,N-Dicarbobenzoxy-3,3,',5'-triiodothyronine	TLC	19
O,N-Dicarbobenzoxy-3,5,3'-triiodothyronine	TLC	19
O,N-Dicarbobenzoxythyroxine	TLC	19
Dichlofenthion	LC	43
	TLC	81
Dichlone	GC	62, 68, 129
	TLC	82
Dichloral urea	GC	129
2,5-Dichloro-4-acetaminoaniline	TLC	119
Dichloroacetic acid	PC	79
	TLC	29
2,3-Dichloroaniline	GC	102
2,4-Dichloroaniline	GC	102
	PC	6, 59
	TLC	7, 119
2,5-Dichloroaniline	PC	6
	TLC	119
2,6-Dichloroaniline	TLC	7
3,4-Dichloroaniline	GC	102
	TLC	81
m-Dichlorobenzene	GC	51
o-Dichlorobenzene	GC	51
	LC	78
p-Dichlorobenzene	GC	51, 129
	LC	75
1,2-Dichlorobenzene	GC	31
1,3-Dichlorobenzene	GC	31
1,4-Dichlorobenzene	GC	31
3,3'-Dichlorobenzidine	PC	6
	TLC	119
2,4-Dichlorobenzoic acid	TLC	37
2,6-Dichlorobenzoic acid	PC	78
3,4-Dichlorobenzoic acid	TLC	37
3,5-Dichlorobenzoic acid	TLC	37
4,4'-Dichlorobenzophenone	TLC	81
2,5-Dichloro-*p*-benzoquinone	PC	111
2,4-Dichlorobenzoyl peroxide	TLC	65
5,5'-Dichlorobithienyl	TLC	104
Dichloroborine	GC	27
1,4-Dichlorobutane	LC	75
3,4-Dichlorocatechol	PC	29
3,5-Dichlorocatechol	PC	29
4,5-Dichlorocatechol	PC	29
4,6-Dichloro-*m*-cresol	PC	31
4,6-Dichloro-*o*-cresol	PC	31
2,2'-Dichloro-5,5'-dimethoxybenzidine	TLC	119
5,5'-Dichloro-3,3'-dimethoxybenzidine	PC	6
2,4-Dichloro-3,5-dimethylphenol	PC	31
2,6-Dichloro-3,4-dimethylphenol	PC	31
o,p'-Dichlorodiphenylmonochloroethane	GC	129
p,p'-Dichlorodiphenylmonochloroethane	GC	129
Dichloroethane	LC	42
1,1-Dichloroethane	GC	32, 49, 58
1,1-Dichloroethane, organics	GC	52
1,2-Dichloroethane	GC	49, 58, 93
1,2-Dichloroethane, organics	GC	52
cis-1,2-Dichloroethene	GC	93
trans-1,2-Dichloroethene	GC	93
cis-Dichloroethylene	GC	32, 47
trans-Dichloroethylene	GC	32, 47
1,1-Dichloroethylene, organics	GC	52
cis-1,2-Dichloroethylene	GC	49, 52, 58
trans-1,2-Dichloroethylene	GC	49, 58
trans-1,2-Dichloroethylene, organics	GC	52
Dichlorobis(ethylenediamine)cobalt chloride [Co(en)$_2$Cl$_2$]Cl	PC	94
2,4-Dichloro-5-ethyl-3-methylphenol	PC	31
3,5-Dichloro-4-hydroxybenzoic acid	PC	120
O-(2,5-Dichloro-4-iodophenyl)-*O,O*-dimethyl phosphorothioate (Iodofenphos)	TLC	170
Dichloromethane	GC	26
Dichloromethane (CH_2Cl_2) (methylene chloride)	GC	47, 49, 58
2,3-Dichloro-1,4-naphthoquinone	PC	111
2,6-Dichloro-4-nitroaniline	TLC	119
2,3-Dichlorophenol	GC	74
	PC	29
2,4-Dichlorophenol	GC	71, 74
	LC	51
	PC	29
	TLC	81
2,5-Dichlorophenol	GC	74
	PC	29

2,6-Dichlorophenol	GC	74
	PC	29
3,4-Dichlorophenol	GC	74
	PC	29
3,5-Dichlorophenol	PC	29
2,4-Dichlorophenoxyethanol	TLC	81
2,3-Dichlorophenylmercapturic acid	PC	29
3,4-Dichlorophenylmercapturic acid	PC	29
L-2,4-Dichlorophenylmercapturic acid	PC	29
L-2,6-Dichlorophenylmercapturic acid	PC	29
L-3,5-Dichlorophenylmercapturic acid	PC	29
Di(*p*-chlorophenyl)pyrophosphate	PC	112
2,4-Dichloro-6-propionylphenol	TLC	89
2,5-Dichloropyridine	GC	131
2,6-Dichloropyridine	GC	131
3,5-Dichloropyridine	GC	131
2,3-Dichloroquinol	PC	29
2,5-Dichloroquinol	PC	29
2,6-Dichloroquinol	PC	29
1,2-Dichloro-1,1,2,2-tetrafluoroethane	GC	26
2,4-Dichlorothiophenol	PC	29
2,6-Dichlorothiophenol	PC	29
3,5-Dichlorothiophenol	PC	29
2,2-Dichlorovinyl dimethyl phosphate (Dichlorvos)	TLC	170
Dichlorphenamide	TLC	140
Dichlorphos	TLC	171
Dichlorvos	GC	65, 67
	LC	43
	TLC	84, 170
Dichromate	PC	93
Dicloran	TLC	81
Dicloxacillin Na	TLC	21
Dicodid	PC	2
	TLC	116
Dicofol	GC	60, 128
	TLC	81
Dicofol (Kelthane®)	TLC	80, 82
Di-condensed phosphate	TLC	151
Dicryl	GC	62
Dicumyl peroxide	LC	21
Dicyanodiamide	LC	63
Dicyclohexylamine	TLC	7
Dicyclohexyl phthalate (phthalic acid ester)	TLC	27
Dicyclohexyl sulfoxide	TLC	164
Dicyclopentadiene	TLC	51
Didecyl phthalate (phthalic acid ester)	TLC	27
Di-*n*-decyl sulfoxide	TLC	164
Didemethylchlorpromazine (Nor$_2$CP)	GC	103
Didemethylchlorpromazine sulfoxide (Nor$_2$CPSO)	GC	103
N,N-Didemethylchlorpromazine sulfoxide	TLC	139
2,6-Dideoxy-D-ribo-hexonic	LC	23
N,N'-Di-DNP-L-cysteine	TLC	17
Di-*S,N*-DNP-L-cysteine	TLC	17
Di-DNP-cystine	TLC	18
Di-DNP-histidine	TLC	18
Di-DNP-L-histidine	TLC	17
Di-DNP-lysine	TLC	18
Di-*N,N'*-DNP-L-lysine	TLC	17
Di-DNP-tyrosine	TLC	18
Di-*O,N*-DNP-L-tyrosine	TLC	17
Di-DNS-cystine	TLC	123
Di-DNS-histidine	TLC	123
Di-DNS-hydroxyproline	TLC	123
Di-DNS-lysine	TLC	123
α,ε-Di-DNS-lysine	TLC	123
Di-DNS-tyrosine	TLC	123
(Di) dodecyl ether ($(n\text{-}C_{12}H_{25})_2O$)	LC	21
Didodecyl sulfide	TLC	163
Dieldrin	GC	60, 61, 68, 128, 129
	PC	116
	TLC	80, 81, 82
Dienestrol	PC	124
Dienestrol diacetate	TLC	136
Diethanolamine	LC	26
	PC	7
Diethazine	TLC	139
Diethoxyalanine	PC	64
Diethyl adipate	GC	1, 25
Diethylamine	GC	97
	LC	26
	PC	55
	TLC	5, 7, 10
Diethylamine, trifluoroacetyl derivative	GC	10
2-Diethylaminoethanethiole	PC	60
N,N-Diethylaniline	GC	102
	TLC	7
2,6-Diethylaniline	TLC	7
Diethyl azelate	GC	1
Diethylbarbituric acid	LC	34
m-Diethylbenzene	LC	42
o-Diethylbenzene	LC	42
p-Diethylbenzene	LC	42
1,2-Diethylbenzene	GC	119, 120
Diethyl carbonate	LC	60
O,O-Diethyl *O*-(2-chloro-α-cyanobenzylideneamino)phosphorothioate)	TLC	170
O,O-Diethyl *O*-(α-cyanobenzylideneamino)phosphorothioate	TLC	170
1,3-Diethylcyclohexane	GC	35
1,4-Diethylcyclohexane	GC	35
N,N'-Diethyldiaminoethane	GC	11
N,N-Diethyl-2,4-dimethylaniline	TLC	7
Diethyl disulfide	GC	85
N,N-Diethylethylenediamine	PC	7
Diethylene glycol	GC	116
	PC	1, 53
	TLC	113
Diethylene glycol, butyl ether of	GC	116
Diethylene glycol dibenzoate	GC	132
Diethylene glycol (min)	GC	116
Diethylene glycol, ethyl ether of	GC	116
Diethylene glycol, methyl ether of	GC	116
Diethylene glycol, propyl ether of	GC	116
Diethylene glycol monomethyl ether	LC	54
Diethylenetriamine	LC	26
	PC	7
	TLC	7
Diethyl esters	GC	24
Diethyl ether	GC	27, 47, 51, 125, 126, 127, 131
	LC	21
Di-2-ethylhexyl adipate	TLC	27
Di-(2-ethylhexyl)-phthalate (phthalic acid ester)	TLC	27

Di-2-ethylhexyl sebacate	TLC	27
Di-2-ethylhexyl thiobutyrate	TLC	27
Di-2-ethylhexyltin dichloride	TLC	67
Diethyl fumarate	GC	1
Diethyl glutarate	GC	1
N,N′-Diethylguanidine	PC	10
1,2-Diethylhydrazine·2 HCl	PC	61
N,N-Diethylhydroxylamine	TLC	7
Diethyl ketone	GC	7, 55, 117, 121
Diethyl ketone (see 3-Pentanone)	LC	8
Diethyl ketone, organics	GC	52
Diethylnicotinamide	PC	18
Diethyl 4-nitrophenyl phosphate (Paraoxon)	TLC	84
Diethyl malate	GC	1
Diethyl maleate	GC	1
Diethyl malonate	GC	1
N,N-Diethyl-3-methylaniline	GC	102
N,N-Diethyl-4-methylaniline	GC	102
Diethyl orthophosphate	PC	35
3,3-Diethylpentane	GC	121
Diethyl phthalate	GC	25, 132
Diethyl phthalate (phthalic acid ester)	TLC	27
2,6-Diethylphenol	GC	72
3,5-Diethylphenol	GC	72
Diethyl pimelate	GC	1
Diethylpropion	TLC	39
Diethyl pyrophosphate	PC	35
Diethyl sebacate	GC	25
α,β-Diethylstilbestrol	PC	124
Diethyl suberate	GC	1, 25
Diethyl succinate	GC	1, 25
N′,N′-Diethylsulfanilamide	PC	84
Diethyl sulfide	GC	85, 135
5,5-Diethylthiobarbituric acid	PC	47
N′,N″-Diethylthiourea	PC	47
Diethyltin chloride	TLC	67
Diethyl triphosphate	PC	35
N,N-Diethyltryptamine	PC	101, 103
N,N-Diethyltryptamine (as HCl)	PC	102
Diffractaic acid	PC	30
	TLC	35
Difluoroacetic acid	TLC	29
Difluoroacetylene	GC	26
2,4-Difluoroaniline	GC	102
2,5-Difluoroaniline	GC	102
2,4-Difluorobutylphenol	GC	73
Difluorodibromomethane	GC	26
Difluorodichloromethane	GC	26
1,1-Difluoroethene	GC	26
Difluoromethane	GC	26
Difolatan®	GC	65
Diformate	TLC	112
1,2-Diformyl-1,2-dimethylhydrazine	PC	61
1,2-Diformylhydrazine	PC	61
Digallium hexachloride	GC	50
Digilanid A	PC	127
Digilanid B	PC	127
Digilanid C	PC	127
Diginatigenin peracetate	TLC	102
Diginatigenin (3β,12β,14β,16β-Tetrahydroxy-5β-cardenolide)	TLC	102
Diginatin	TLC	102
Diginatin peracetate	TLC	102
Diginin	PC	127
Digiproside	TLC	102
Digiproside peracetate	TLC	102
Digitalinum verum	PC	127
	TLC	177

Digitonin	TLC	177
Digitoxigenin	PC	127
	TLC	177
Digitoxigenin acetate	PC	127
Digitoxigenin(Dg) (3β,14β-Dihydroxy-5β-cardenolide)	TLC	102
Digitoxigenin(Dg) peracetate	TLC	102
Digitoxigenin glycoside	PC	127
Digitoxigenin-mono-digitoxose	TLC	177
Digitoxigenon (14β-Hydroxy-3-oxo-5β-cardenolide)	TLC	102
Digitoxin	PC	127
	TLC	102, 177
Digitoxin percetate	TLC	102
Digitoxose	LC	48
Dg-α,D-digitoxoside	TLC	102
Dg-α,D-digitoxoside percetate	TLC	102
Cbo-Diglycine	TLC	12
Cbo-Diglycine ethyl ester	TLC	12
Digoxigenin	PC	127
	TLC	177
Digoxigenin-bisdigitoxoside	TLC	102
Digoxigenin-bisdigitoxoside peracetate	TLC	102
Digoxigenin diacetate	PC	127
Digoxigenin-monodigitoxoside	TLC	102
Digoxigenin-monodigitoxoside peracetate	TLC	102
Digoxigenin peracetate	TLC	102
Digoxigenin (Dxg) (3β,12β,14β-Trihydroxy-5β-cardenolide)	TLC	102
Digoxigenone	PC	127
Digoxin	PC	127
	TLC	102, 177
Digoxin peracetate	TLC	102
Digoxoside	TLC	102
Digoxoside peracetate	TLC	102
α:ε-Diguanidino-caproic acid	PC	9
α,ω-Diguanidino-deca-alkane	PC	9
α,ω-Diguanidino-dialkane	PC	9
α,ω-Diguanidino-hexa-alkane	PC	9
α,ω-Diguanidino-penta-alkane	PC	9
α,ω-Diguanidino-tetra-alkane (arcaine)	PC	9
α,ω-Diguanidino-tri-alkane	PC	9
α:δ-Diguanidino-valeric acid	PC	9
Di-*n*-heptyl adipate	GC	25
Di-*n*-heptylperoxide	PC	27
Di-*n*-heptyl phthalate	GC	25
Di-*n*-heptyl suberate	GC	25
Di-*n*-heptyl succinate	GC	25
Di-*n*-hexyl adipate	GC	25
Di-*n*-hexyl disulfide	GC	85
Dihexyl phthalate (phthalic acid ester)	TLC	27
Di-*n*-hexyl phthalate	GC	25
Di-*n*-hexyl sebacate	GC	25
Di-*n*-hexyl suberate	GC	25
Di-*n*-hexyl succinate	GC	25
Di-*n*-hexyl sulfoxide	TLC	164
Dihexyltin dichloride	TLC	67
Dihydroagnosterol	TLC	99
Dihydroandrosterone	GC	136
1,2-Dihydrobenz[*e*]aceanthrylene	TLC	146
Dihydrocaffeic acid	TLC	135
Dihydrocanthaxanthin	TLC	93
Dihydrocanthaxanthin, red	TLC	93
Dihydrocarvone, 2,4-DNPH of	TLC	76
Dihydrocholesterol	TLC	99
Dihydrocodeine	TLC	3, 4
Dihydrocodeinone (hydrocodone)	GC	8, 18, 107, 111
	PC	4

Dihydrocodeinone (hydrocodone) (*continued*)	TLC	3, 39, 138
Dihydrocodeinone bitartrate (hydrocodone bitartrate)	GC	107
Dihydrocoumarin	TLC	68
10,11-Dihydro-5H-dibenzo[*a,d*]-cycloheptene-5-one	TLC	75
cis-5,6-Dihydro-5,6-dihydroxy-7,12-dimethylbenz[*a*]-anthracene	TLC	50
trans-5,6-Dihydro-5,6-dihydroxy-7,12-dimethylbenz[*a*]-anthracene	TLC	50
8,9-Dihydro-8,9-dihydroxy-7,12-dimethylbenz[*a*]anthracene	TLC	50
cis-5,6-Dihydro-5,6-dihydroxy-7-hydroxymethyl-12-methylbenz[*a*]-anthracene	TLC	50
8,9-Dihydro-8,9-dihydroxy-7-hydroxy-methyl-12-methylbenz[*a*] anthracene	TLC	50
Dihydroergocornine	TLC	2, 114
Dihydroergocristine	TLC	2, 3, 114
Dihydroergocryptine	TLC	2, 114
Dihydroergotamine	TLC	2, 3, 114
Dihydroeugenol	TLC	90
Dihydrohydroxycodeinone (Oxycodone)	GC	8, 18, 111
	TLC	138
Dihydrohydroxymorphinone (Oxymorphone)	GC	18
	TLC	138
5,10-Dihydroindeno[1,2-*b*]indole	TLC	120
11,12-Dihydroindeno[2,1-*a*]fluorene-11,12-dione	TLC	75
1,3-Dihydroisothionaphthene S-oxide	TLC	164
Dihydrolanosterol	LC	45
	TLC	99
α-Dihydrolysergal	TLC	114
Dihydromorphine	GC	18, 107
	TLC	138
Dihydromorphinone (Hydromorphone)	GC	8, 18, 107
	PC	4
	TLC	3, 39, 138
Dihydromorphinone HCl (hydromorphone HCl)	GC	107
5,12-Dihydronaphthacen-5-one	TLC	75
1,2-Dihydronaphthalene	TLC	53
Dihydronicotyrine	PC	2
1,1′-Dihydroperoxy-dicyclohexyl peroxide	TLC	65
1,1′-Dihydroperoxy-2,2′-dimethyldicyclohexyl peroxide	TLC	65
1,1′-Dihydroperoxy-3,3′-dimethyldicyclohexyl peroxide	TLC	65
1,1′-Dihydroperoxy-4,4′-dimethyldicyclohexyl peroxide	TLC	65
Dihydrophenanthrene	LC	27
9,10-Dihydrophenanthrene	TLC	53
Dihydroprylon	TLC	137
Dihydroquercetin	LC	1
Dihydrorhodoxanthin	TLC	93
Dihydrostreptomycin	PC	67
	TLC	20
cis-Dihydro-α-terpineol	GC	87
trans-Dihydro-α-terpineol	GC	87
Dihydro-testosterone	TLC	178
Dihydrotetracene	LC	27
Dihydrouracanic acid	PC	83
Dihydroxyacetone	GC	15
	LC	29
	TLC	23
Dihydroxyacetone, 2,4-DNPH of	TLC	166
Dihydroxyacetone phosphate (DHAP)	PC	32
2,4-Dihydroxyacetophenone, 2,4-DNPH of	TLC	167
2,4-Dihydroxyacetophenone azine	PC	21
2,5-Dihydroxyacetophenone azine	PC	23
2,6-Dihydroxyacetophenone azine	PC	21
3α,11β-Dihydroxy-5α-androstan-17-one	PC	40, 41
3α,11β-Dihydroxy-5β-androstan-17-one	TLC	179
3α,11β-Dihydroxy-5β-androstan-17-one (Group III)	PC	40
3β,11β-Dihydroxy-5β-androstan-17-one	PC	41
3β,11β-Dihydroxy-5α-androstan-17-one	PC	40, 41
	TLC	179
3β,11β-Dihydroxy-androst-5-en-17-one (Group III)	PC	40
6β,17β-Dihydroxy-androst-4-en-3-one (Group II)	PC	40
11β,17β-Dihydroxy-androst-4-en-3-one (Group II)	PC	40
6β,17β-Dihydroxy-4-androsten-3-one	PC	124
11α,17β-Dihydroxy-4-androsten-3-one	PC	124
11β,17β-Dihydroxy-4-androsten-3-one	PC	124
14α,17β-Dihydroxy-4-androsten-3-one	PC	124
1,8-Dihydroxyanthracene (chrysazol)	PC	105
1,4-Dihydroxyanthraquinone	TLC	92, 120
1,5-Dihydroxyanthraquinone (anthrarufin)	PC	105
1,8-Dihydroxyanthraquinone (chrysazin)	PC	105
2,4-Dihydroxybenzaldehyde	TLC	71
2,4-Dihydroxybenzaldehyde, 2,4-DNPH of	TLC	167
2,5-Dihydroxybenzaldehyde	TLC	71
2,5-Dihydroxybenzaldehyde, 2,4-DNPH of	TLC	167
3,4-Dihydroxybenzaldehyde	TLC	71
1,2-Dihydroxybenzene	TLC	37
1,3-Dihydroxybenzene	TLC	37
1,4-Dihydroxybenzene	TLC	37
2,3-Dihydroxybenzoic acid	GC	5
	LC	1
	TLC	37
2,4-Dihydroxybenzoic acid	GC	5
	LC	1
	TLC	37
2,5-Dihydroxybenzoic acid	GC	5
	LC	1
	TLC	37
2,6-Dihydroxybenzoic acid	GC	5
	LC	1
	TLC	37
3,4-Dihydroxybenzoic acid	GC	5
	LC	1
	TLC	37
3,4-Dihydroxybenzoic acid ethyl ester	PC	121
3,5-Dihydroxybenzoic acid	GC	5
	LC	1
	TLC	37
2,4-Dihydroxybenzophenone azine	PC	23
2,5-Dihydroxy-*p*-benzoquinone	TLC	88
o,o′-Dihydroxybiphenyl	TLC	174
2,2′-Dihydroxybiphenyl	TLC	88
2,3-Dihydroxybiphenyl	TLC	88
2,4-Dihydroxybiphenyl	TLC	88
2,4′-Dihydroxybiphenyl	TLC	88
2,5-Dihydroxybiphenyl	TLC	88
2,6-Dihydroxybiphenyl	TLC	88
3,3′-Dihydroxybiphenyl	TLC	88
3,4-Dihydroxybiphenyl	TLC	88

4,4′-Dihydroxybiphenyl	TLC	88
2,4-Dihydroxybutyric	LC	40
3,4-Dihydroxybutyric	LC	40
D-*threo*-2,3-Dihydroxybutyric	LC	40
3,4-Dihydroxy-β-carotene	TLC	93
3,4-Dihydroxycinnamic acid	GC	4
3α,6α-Dihydroxy cholanic acid	TLC	180
3α,7α-Dihydroxy cholanic acid	TLC	180
3α,12α-Dihydroxy cholanic acid	TLC	180
3α,7β-Dihydroxy cholanic acid	TLC	180
3α,12β-Dihydroxy cholanic acid	TLC	180
7α,12α-Dihydroxy cholanic acid	TLC	180
3β,7α-Dihydroxy cholanic acid	TLC	180
3β,12α-Dihydroxy cholanic acid	TLC	180
Dihydroxy compounds	GC	136
β,β-Di-(4-hydroxy-3-coumarin)-ethyl chloride	TLC	79
1,1′-Dihydroxydicyclohexyl peroxide	TLC	65
1β,7β-Dihydroxy-digitoxigenin peracetate	TLC	102
1β,7β-Dihydroxy-digitoxigenin (1β,3β,7β,14β-Tetrahydroxy-5β-cardenolide)	TLC	102
1β,7β-Dihydroxy-digitoxigenon peracetate	TLC	102
1β,7β-Dihydroxy-digitoxigenon (1β,7β,14β-Trihydroxy-3-oxo-5β-cardenolide)	TLC	102
2,2′-Dihydroxy-5,5′-dimethylbiphenyl	TLC	88
4,4′-Dihydroxy-3,3′-dimethylbiphenyl	TLC	88
D-2,4-Dihydroxy-3,3-dimethylbutyric	LC	23
1,1′-Dihydroxy-2,2′-dimethyl-dicyclohexyl peroxide	TLC	65
1,1′-Dihydroxy 3,3′-dimethyldicyclohexyl peroxide	TLC	65
1,1′-Dihydroxy-4,4′-dimethyldicyclohexyl peroxide	TLC	65
2,2′-Dihydroxydiphenyl	PC	31
2,2′-Dihydroxydiphenylmethane	TLC	88
4,4′-Dihydroxydiphenylmethane	TLC	88
6,7-Dihydroxy-estradiol	TLC	181
2,3-Dihydroxyestra-1,3,5(10)-triene-17-one	GC	82
3,6β-Dihydroxyestra-1,3,5(10)-triene-17-one	GC	82
3,7α-Dihydroxyestra-1,3,5(10)-triene-17-one	GC	82
3,11β-Dihydroxyestra-1,3,5(10)-triene-17-one	GC	82
3-16α-Dihydroxyestra-1,3,5(10)-triene-17-one	GC	82
3,4-Dihydroxy-estra-1,3,5(10)-trien-17-one	TLC	181
3,6α-Dihydroxy-estra-1,3,5(10)-trien-17-one	TLC	181
3,6β-Dihydroxy-estra-1,3,5(10)-trien-17-one	TLC	181
3,7α-Dihydroxy-estra-1,3,5(10)-trien-17-one	TLC	181
3,7β-Dihydroxy-estra-1,3,5(10)-trien-17-one	TLC	181
3,11α-Dihydroxy-estra-1,3,5(10)-trien-17-one	TLC	181
3,11β-Dihydroxy-estra-1,3,5(10)-trien-17-one	TLC	181
3,15α-Dihydroxy-estra-1,3,5(10)-trien-17-one	TLC	181
3,16α-Dihydroxy-estra-1,3,5(10)-trien-17-one	TLC	181
3,16β-Dihydroxy-estra-1,3,5(10)-trien-17-one	TLC	181
3,17β-Dihydroxy-estra-1,3,5(10)-trien-6-one	TLC	181
3,17β-Dihydroxy-estra-1,3,5(10)-trien-7-one	TLC	181
3,17β-Dihydroxy-estra-1,3,5(10)-trien-16-one	TLC	181
2,3-Dihydroxy-estra-1,3,5(10)-trien-17-one-2-methyl ether	TLC	181
3,4-Dihydroxy-estra-1,3,5(10)-trien-17-one-3-methyl ether	TLC	181
10ε,17β-Dihydroxy-4-estren-3-one	PC	124
11α,17β-Dihydroxy-4-estren-3-one	PC	124
Di-2-hydroxyethylperoxide	PC	27
2,3-Dihydroxyhippuric acid	PC	80
2,5-Dihydroxyhippuric acid	PC	80
Dihydroxyindole	PC	103
2,6-Dihydroxyisonicotinic acid	PC	120
1,3-Dihydroxyisoquinoline	PC	104
1,5-Dihydroxyisoquinoline	PC	105
Dihydroxymandelic acid	TLC	8, 121
3,4-Dihydroxymandelic acid	GC	5
dl-3,4-Dihydroxymandelic acid	PC	120
3,4-Dihydroxy-5-methoxybenzaldehyde	PC	120
	TLC	71
1,2-Dihydroxy-3-methoxybenzene	TLC	37
1,3-Dihydroxy-2-methoxybenzene	TLC	37
3,4-Dihydroxy-5-methoxybenzoic acid	TLC	37
3,5-Dihydroxy-4-methoxybenzoic acid	TLC	37
Dihydroxymethylanthraquinone	PC	109
7,12-Dihydroxymethylbenz(*a*)anthracene	TLC	50
3,5-Dihydroxy-4-methylbenzoic acid	PC	121
Dihydroxymethylperoxide	PC	27
6,8-Dihydroxy-2-methylpurine	PC	36
DL-3,5-Dihydroxy-3-methylvaleric	LC	23
Dihydroxymonoketone compounds	GC	136
6,11-Dihydroxynaphthacenequinone	PC	121
	TLC	92
1,3-Dihydroxynaphthalene	TLC	174
2,8-Dihydroxy-3-naphthoic acid	PC	105
3,5-Dihydroxy-2-naphthoic acid	PC	105
3,4-Dihydroxynorephedrine	GC	9
3,4-Dihydroxyphenethylamine	LC	77
3,4-Dihydroxyphenethylglycol	LC	72
2,3-Dihydroxyphenol	LC	51
3,5-Dihydroxyphenol	LC	51
Dihydroxyphenylacetic acid	TLC	8
2,3-Dihydroxyphenylacetic acid	PC	121
2,5-Dihydroxyphenylacetic acid	GC	5
	TLC	135
2,5-Dihydroxyphenylacetic, lactone	GC	5
3,4-Dihydroxyphenylacetic acid	GC	5
	LC	72
	TLC	121
Dihydroxyphenylalanine	PC	5
	TLC	8, 11, 16, 121
3,4-Dihydroxyphenylalanine	LC	41, 72
p-(2,4-Dihydroxyphenyl)azophenol	PC	82
p-(2,4-Dihydroxyphenyl)azophenyl sulfuric acid	PC	82
3,4-Dihydroxyphenylcinnamic acid	GC	5
3,4-Dihydroxyphenylglycol	PC	121
3-(3,4-Dihydroxyphenyl)propionic acid	GC	4
3,4-Dihydroxyphenylpropionic acid = hydrocaffeic acid	PC	120
3,4-Dihydroxyphenylpyruvic acid	GC	4
2,5-Dihydroxyphenylpyruvic, lactone	GC	5

3,6-Dihydroxyphthalimide	PC	120
11β,21-Dihydroxy-4,17(20)-pregnadien-3-one	PC	124
3α,21-Dihydroxypregnane-11,20-dione	TLC	101
17α,21-Dihydroxypregnane-3,20-dione	TLC	101
6β,11α-Dihydroxy-5β-pregnane-3,20-dione	PC	124
17α,21-Dihydroxy-4-pregnane-3,20-dione	PC	124
3α,17-Dihydroxy-5β-pregnane-11,20-dione	PC	44
3α,17α-Dihydroxy-5β-pregnane-11,20-dione	PC	124
3α,21-Dihydroxy-5β-pregnane-11,20-dione	PC	45
3β,21-Dihydroxy-5α-pregnane-11,20-dione	PC	45
17α,21-Dihydroxypregnane-3,11,20-trione	TLC	101
17,21-Dihydroxy-5α-pregnane-3,11,20-trione	PC	45
17,21-Dihydroxy-5β-pregnane-3,11,20-trione	PC	45
17α,21-Dihydroxy-5β-pregnane-3,11,20-trione	PC	124
17α,21-Dihydroxy-5β-pregnane-3,11,20-trione-3,20-ethylene glycol diketal	PC	124
3β,11α-Dihydroxy-5α-pregnan-20-one	PC	124
3α,11α-Dihydroxy-5β-pregnan-20-one	PC	124
3α,17-Dihydroxy-5β-pregnan-20-one	PC	44
11β,21-Dihydroxy-pregn-4-en-18-al-3,20-dione (aldosterone)	PC	125
11β,21-Dihydroxy-17-iso-pregn-4-en-18-al-3,20-dione (17-iso-aldosterone)	PC	125
3β,21-Dihydroxy-pregn-5-en-20-one (BT+)	PC	44
6β,11α-Dihydroxy-4-pregnene-3,20-dione	PC	124
6β,17-Dihydroxy-pregn-4-ene-3,20-dione	PC	43
6β,17α-Dihydroxy-4-pregnene-3,20-dione	PC	124
6β,21-Dihydroxy-pregn-4-ene-3,20-dione	PC	125
6β,21-Dihydroxy-4-pregnene-3,20-dione	PC	124
11α,17α-Dihydroxy-4-pregnene-3,20-dione	PC	124
11β,17α-Dihydroxy-4-pregnene-3,20-dione	PC	124
11α,21-Dihydroxy-pregn-4-ene-3,20-dione	PC	125
11α,21-Dihydroxy-4-pregnene-3,20-dione	PC	124
11β,21-Dihydroxy-pregn-4-ene-3,20-dione	PC	125
11β,21-Dihydroxy-4-pregnene-3,20-dione	PC	124
14α,21-Dihydroxy-4-pregnene-3,20-dione	PC	124
15α,21-Dihydroxy-4-pregnene-3,20-dione	PC	124
16α,21-Dihydroxy-pregn-4-ene-3,20-dione	PC	125
17,21-Dihydroxy-pregn-4-ene-3,20-dione	PC	125
19,21-Dihydroxy-pregn-4-ene-3,20-dione	PC	125
11β,21-Dihydroxy-pregn-4-ene-3,20-dione-18-acid-lactone(18→11)	PC	125
6β,21-Dihydroxy-pregn-4-ene-3,11,20-trione	PC	125
17,21-Dihydroxy-pregn-4-ene-3,11,20-trione (cortisone)	PC	125
17α,21-Dihydroxy-4-pregnene-3,11,20-trione	PC	124
18,21-Dihydroxy-pregn-4-ene-3,11,20-trione	PC	125
17α,21-Dihydroxy-5-pregnene-3,11,20-trione-3,20-ethylene glycol diketal	PC	124
2,4-Dihydroxypropiophenone	PC	120
3,4-Dihydroxypropiophenone	PC	121
2,6-Dihydroxypurine (Xanthine)	PC	16
2,8-Dihydroxypurine	PC	16
6,8-Dihydroxypurine	LC	46
	PC	36
2,3-Dihydroxypyridine	PC	121
2,5-Dihydroxypyridine	PC	121
2,6-Dihydroxypyridine	PC	120
3,4-Dihydroxypyridine	PC	121
2,4-Dihydroxypyrimidine-6-methylsulfone	PC	36
2,3-Dihydroxyquinoline	PC	106
2,4-Dihydroxyquinoline	PC	106
2,6-Dihydroxyquinoline	PC	106
4,6-Dihydroxyquinoline	PC	106
4,4′-Dihydroxystilbene	TLC	88
2,3-Dihydroxytoluene	TLC	88
2,5-Dihydroxytoluene	TLC	88
2,6-Dihydroxytoluene	TLC	88
3,4-Dihydroxytoluene	TLC	88
3,5-Dihydroxytoluene	TLC	88
2,5-Dihydroxyvaleric	LC	23
Dihyprylon	TLC	137
3,5-Diiodobenzaldehyde	LC	76
1,2-Diiodoethane	GC	37
Diiodomethane	GC	37
1,3-Diiodopropane	GC	37
3,5-Diiodo-4-pyridone-*N*-acetic acid	TLC	64
3,5-Diiodo-4-pyridone-*N*-acetic acid *n*-propyl ester	TLC	64
3,5-Diiodothyroacetic acid	TLC	19
3,5-Diiodothyroacetic acid methyl ester	TLC	19
3,3′-Diiodothyronine	TLC	19
3,3′-Diiodothyronine methyl ester	TLC	19
3,5-Diiodothyronine	LC	76
	PC	63
	TLC	19
3,5-Diiodothyronine methyl ester	TLC	19
3,5-Diiodo-L-thyronine	TLC	64
3′,5′-Diiodothyronine	LC	76
Diiodotyrosine	TLC	11
3,5-Diiodotyrosine	LC	76
	TLC	15, 19
3,5-Diiodotyrosine methyl ester	TLC	19
3,5-Diiodo-L-tyrosyl-glycine	PC	114
Diisoamylamine, trifluoroacetyl derivative	GC	10
Diisoamyl ether	LC	7
Diisoamyl sulfoxide	TLC	164
Diisobutyl adipate (Adipic acid ester)	TLC	27
Diisobutylamine	GC	97
	TLC	10
Diisobutylamine, trifluoroacetyl derivative	GC	10
Diisobutyl azelate	GC	132
Diisobutylene	GC	53
Diisobutyl ketone	GC	53
Diisobutyl phthalate	GC	132
Diisobutyl phthalate (phthalic acid ester)	TLC	27
Diisodecyl phthalate (phthalic acid ester)	TLC	27
Diisononyl phthalate	TLC	27

Diisoctyl adipate	GC	132
Diisooctyl phthalate	TLC	27
Diisopropanolamine	PC	7
Diisopropylamine	GC	97
Diisopropylamine, trifluoroacetyl derivative	GC	10
Diisopropylbenzene	LC	27
1,2-Diisopropylbenzene	GC	119
1,4-Diisopropylbenzene	GC	119
N,N-Diisopropyl-2-benzothiazole sulfenamide	TLC	66
Diisopropyl ether	GC	51,125, 126
	LC	7
Diisopropyl ether, organics	GC	52
2,3:4,5-Di-*O*-isopropylidene-1,6-di-*O*-methane-sulfonylgalactitol	GC	13
1,2,4,5-Diisopropylidenefructopyranose	TLC	26
2,3,4,5-Diisopropylidenefructopyranose	TLC	26
1,2:4,5-Di-*O*-isopropylidene-D-fructose	GC	13
2,3:4,5-Di-*O*-isopropylidene-D-fructose	GC	13
1,2:4,5-Di-*O*-isopropylidenegalactitol	GC	13
2,3:4,5-Di-*O*-isopropylidenegalactitol	GC	13
1,2,5,6-Diisopropylideneglucofuranose	TLC	26
1,2:5,6-Di-*O*-isopropylidene-D-glucose	GC	13
3,4:5,6-Di-*O*-isopropylidene-D-glucose dimethylacetal	GC	13
1,2,4,5-Diisopropylidene-3-*O*-methylfructopyranose	TLC	26
2,3,4,5-Diisopropylidene-1-*O*-methylfructopyranose	TLC	26
1,2,5,6-Diisopropylidene-3-*O*-methylglucofuranose	TLC	26
1,2:4,5-Di-*O*-isopropylidene-D-mannitol	GC	13
1,2,5,6-Di-*O*-isopropylidene-D-mannitol	GC	13
2,3:5,6-Di-*O*-isopropylidene-D-mannono-1→4-lactone	GC	13
2,3:5,6-Di-*O*-isopropylidene-D-mannose	GC	13
1,2:4,5-Di-*O*-isopropylidene-3-*O*-methanesulfonyl-D-fructose	GC	13
1,2:3,4-Di-*O*-isopropylidene-6-*O*-methanesulfonyl-D-galactose	GC	13
2,3:4,5-Di-*O*-isopropylidene-1-*O*-toulene-*p*-sulfonyl-D-fructose	GC	13
1,2:5,6-Di-*O*-isopropylidene-3-*O*-toluene-*p*-sulfonyl-D-glucose	GC	13
2,3:5,6-Di-*O*-isopropylidene-4-*O*-toluene-*p*-sulfonyl-D-glucose dimethylacetal	GC	13
1,2:4,5-Di-*O*-isopropylideneribitol	GC	13
2,6-Diisopropylphenol	GC	72, 73
	LC	61, 77
3,4-Diisopropylphenyl *N*-methylcarbamate	GC	130
3,5-Diisopropylphenyl *N*-methylcarbamate	GC	130
Diisopropyl-sulfide	GC	125
3,7-Diketo cholanic acid	TLC	180
3,12-Diketo cholanic acid	TLC	180
3α,7,12-Diketo hydroxy cholanic acid	TLC	180
1-(1,3-Diketo-2-indonyl)-1-phenyl-2-nitropropane	TLC	79
Diketone compounds	GC	136
Dilan®	GC	128, 129
Dilan I	GC	68
Dilan II	GC	68
Dilaudid	PC	2
	TLC	116
Dilauryl thiodipropionate	TLC	22
Dilituric acid (5-nitrobarbituric acid)	PC	60
Dimefox	LC	43
Dimetane® (Brompheniramine)	TLC	39
Dimethoate	GC	60, 61, 64, 65, 67, 68,128, 129
	LC	43
	TLC	81,171, 172
Dimethoate oxygen analog	GC	61, 67
	TLC	80, 84
Dimethoate (Rogor®)	TLC	80, 84, 171
Dimethocaine (Larocaine®)	TLC	43
Dimethoxon	TLC	84
2′,4′-Dimethoxyacetophenone	GC	48
2,5-Dimethoxyaniline	GC	102
	PC	121
	TLC	119
2,3-Dimethoxybenzalazine	PC	23
2,3-Dimethoxybenzaldehyde	TLC	71
2,3-Dimethoxybenzaldehyde, 2,4-DNPH of	TLC	167
2,4-Dimethoxybenzaldehyde	TLC	68, 71
2,4-Dimethoxybenzaldehyde, 2,4-DNPH of	TLC	167
2,5-Dimethoxybenzaldehyde	TLC	71
2,5-Dimethoxybenzaldehyde, 2,4-DNPH of	TLC	167
3,4-Dimethoxybenzaldehyde	TLC	71
3,5-Dimethoxybenzaldehyde	TLC	71
3,5-Dimethoxybenzaldehyde, 2,4-DNPH of	TLC	167
3,3′-Dimethoxybenzidine	PC	6
	TLC	119
2,4-Dimethoxybenzoic acid	PC	120
2,5-Dimethoxybenzoic acid	PC	120
3,4-Dimethoxybenzoic acid	GC	5
2,6-Dimethoxy-1,4-benzoquinone	PC	120
2,4-Dimethoxy-5-chloroaniline	TLC	119
2,5-Dimethoxy-4-chloroaniline	TLC	119
2,4-Dimethoxycinnamic acid	GC	4
3,4-Dimethoxycinnamic acid	GC	4
3,5-Dimethoxycinnamic acid	GC	4
2,5-Dimethoxy-1,4-diaminobenzene	TLC	119
3,3′-Dimethoxy-4,4′-diaminodiphenyl	TLC	119
Dimethoxy ethyl phthalate	TLC	27
3,5-Dimethoxy-4-hydroxybenzoic acid	TLC	37
3,5-Dimethoxy-4-hydroxycinnamic acid	GC	4
3,5-Dimethoxy-4-hydroxy-cinnamic acid = sinapic acid	PC	120
Dimethoxymethane (methylal)	GC	53
2,4-Dimethoxy-5-methylpyrimidine	PC	36
2,3-Dimethoxy-1,4-naphthoquinone	PC	111
2,3-Dimethoxyphenol, 4-acetyl-2-nitrophenol ether of	TLC	91
2,6-Dimethoxyphenol	TLC	37
2,6-Dimethoxyphenol, 4-acetyl-2-nitrophenol ether of	TLC	91
2,5-Dimethoxyphenylacetic acid	GC	4
3,4-Dimethoxyphenylacetic acid	GC	4, 5
β-(3,4-Dimethoxyphenyl)-ethylamine	PC	60
3,4-Dimethoxyphenylpyruvic acid	GC	4
5,6-Dimethoxytryptamine	PC	103
5,6-Dimethoxytryptamine (as HCl)	PC	102
4,8-Dimethoxyxanthurenic acid	PC	104
N,N-Dimethylacetamide	LC	73

2,4-Dimethylacetophenone	PC	22
2,4-Dimethylacetophenone, 2,4-DNPH of	TLC	74
2,4-Dimethylacetophenone, 4-NPH of	TLC	74
2,4-Dimethylacetophenone azine	PC	23
2,5-Dimethylacetophenone	PC	22
2,5-Dimethylacetophenone, 2,4-DNPH of	TLC	74
2,5-Dimethylacetophenone, 4-NPH of	TLC	74
2,5-Dimethylacetophenone azine	PC	23
3,4-Dimethylacetophenone	PC	22
3,4-Dimethylacetophenone, 2,4-DNPH of	TLC	74
3,4-Dimethylacetophenone, 4-NPH of	TLC	74
3,4-Dimethylacetophenone azine	PC	23
N,N-Dimethylacroylsulfanilamide	TLC	44
Dimethylacrylsulfanilamide	TLC	44
6-Dimethyladenine	GC	77
N^6-Dimethyladenosine	LC	22
Dimethyl adipate	GC	1, 25
	LC	70
DL-4-Dimethylallyltryptophan	TLC	114
Dimethylamine	GC	97
	PC	55, 56
	TLC	5, 10
Dimethylamine, trifluoroacetyl derivative	GC	10
4-Dimethylaminobenzalazine	PC	23
4-*N*-Dimethylaminobenzoic acid	TLC	129
3α-Dimethylamino-5α-cholestane	TLC	98
3β-Dimethylamino-5α-cholestane	TLC	98
3α-Dimethylamino-5β-cholestane	TLC	98
3β-Dimethylamino-5β-cholestane	TLC	98
3α-Dimethylamino-cholest-5-ene	TLC	98
3β-Dimethylamino-cholest-5-ene	TLC	98
4-(Dimethylaminoethyl)-4-aza-5-pregnene-3,20-dione methiodide	TLC	96
4-(Dimethylaminoethyl)-17α-methyl-4-aza-5-androsten-17β-ol-3-one	TLC	96
2-Dimethylamino-6-hydroxypurine	PC	36
β-(3-Dimethylamino-methyleneamino-2,4,6-triiodophenyl)-propionic acid	TLC	64
3-Dimethylamino-4-methylphenyl *N*-methylcarbamate	GC	130
2-Dimethylaminomethylpyrrole	PC	60
2-Dimethylamino-1,2-naphthoquinone	PC	111
6-Dimethylaminopurine	LC	46
	PC	36
6-Dimethylaminopurine ribonucleoside	PC	36
4-Dimethylamino-3,5-xylenol	GC	130
	TLC	83
N,N-Dimethylaniline	GC	102
	PC	6, 59
	TLC	7
N,N-Dimethylaniline oxide	TLC	7
2,3-Dimethylaniline	GC	45, 46, 100, 102
	PC	6
	TLC	119
2,3-Dimethylaniline, trifluoroacetyl derivative	GC	10
2,4-Dimethylaniline	GC	46,100, 102
	PC	6
	TLC	7, 119
2,4-Dimethylaniline, trifluoroacetyl derivative	GC	10
2,5-Dimethylaniline	GC	46,100, 102
	PC	6
	TLC	119
2,5-Dimethylaniline, trifluoroacetyl derivative	GC	10
2,6-Dimethylaniline	GC	46,100, 102
	PC	6
	TLC	119
2,6-Dimethylaniline, trifluoroacetyl derivative	GC	10
3,4-Dimethylaniline	GC	46,100, 102
	TLC	119
3,4-Dimethylaniline, trifluoroacetyl derivative	GC	10
3,5-Dimethylaniline	GC	46, 100
2,3-Dimethylanisole	GC	75
	TLC	174
2,4-Dimethylanisole	GC	75
	TLC	174
2,5-Dimethylanisole	GC	75
	TLC	174
2,6-Dimethylanisole	GC	75
	TLC	174
3,4-Dimethylanisole	GC	75
	TLC	174
3,5-Dimethylanisole	GC	75
	TLC	174
9,10-Dimethylanthracene	TLC	53
Dimethyl anthranilate	TLC	105
3,5-Dimethylanthranilic acid	PC	78
2,3-Dimethylarabinose	PC	73
2,4-Dimethylarabinose	PC	73
2,5-Dimethylarabinose	PC	73
DL-β,β-Dimethylaspartic acid	PC	11
4,17α-Dimethyl-4-aza-5-androsten-17β-ol-3-one	TLC	96
Dimethyl azelate	GC	1
7,8-Dimethylbenz[*a*]anthracene	TLC	146
7,12-Dimethylbenz(*a*)anthracene	TLC	50, 146
7,12-Dimethyl-1,2-benzanthracene	GC	40
1,2-Dimethylbenzene	GC	119
1,3-Dimethylbenzene	GC	119
1,4-Dimethylbenzene (*p*-xylene)	GC	119
3,3′-Dimethylbenzidine	TLC	119
7,8-Dimethyl-11H-benzo[*a*]carbazole	TLC	120
N^1-3,4-Dimethylbenzoyl sulfanilamide	TLC	44
3,5-Dimethylbiphenyl	TLC	51
4,4′-Dimethylbiphenyl	TLC	51
5,5′-Dimethylbithienyl	TLC	104
1,1-Dimethylborane	GC	27
N,N-Dimethyl-4-bromoaniline	GC	102
2,3-Dimethyl-4-bromo-6-propionylphenol	TLC	89
2,3-Dimethyl-6-bromo-4-propionylphenol	TLC	89
2,5-Dimethyl-6-bromo-4-propionylphenol	TLC	89
3,5-Dimethyl-4-bromo-6-propionylphenol	TLC	89
2,3-Dimethylbutadiene	GC	118
2,2-Dimethylbutane	GC	30, 42, 118, 121
2,2-Dimethylbutane (neohexane)	GC	47, 53
	LC	42
2,3-Dimethylbutane	GC	30, 42, 47, 51, 53, 118, 121
	LC	42
2,3-Dimethylbutane, paraffins	GC	33
3,3-Dimethyl-2-butanone	GC	118
3,3-Dimethylbutanone-2	GC	51
2,3-Dimethyl-1-butene	GC	118, 122
2,3-Dimethylbutene-1	LC	42
2,3-Dimethyl-2-butene	GC	118, 122
3,3-Dimethyl-1-butene	GC	118, 122
N,N-Dimethyl-*n*-butylamine	GC	97

N,N-Dimethyl-*sec*-butylamine	GC	97
N,N-Dimethyl-*tert*-butylamine	GC	97
2,2-Dimethylbutyric acid	TLC	29
3,3-Dimethylbutyric acid	TLC	29
Dimethyl carbonate	GC	47
N,N-Dimethyl-2-chloroaniline	GC	102
4,4-Dimethylcholestane-3β-ol	LC	24
4,4-Dimethylcholest-5-ene-3β-ol	LC	24
4,4-Dimethylcholest-7-ene-3β-ol	LC	24
4,4-Dimethylcholest-8(14)-ene-3β-ol	LC	24
5,6-Dimethylchrysene	TLC	146
1,1-Dimethylcyclohexane	GC	35, 42, 121
cis-1,2-Dimethylcyclohexane	GC	35
1-*cis*-2-Dimethylcyclohexane	GC	118
1-*trans*-2-Dimethylcyclohexane	GC	118
trans-1,2-Dimethylcyclohexane	GC	35, 121
1-*cis*-3-Dimethylcyclohexane	GC	118
1-*trans*-3-Dimethylcyclohexane	GC	118
cis-1,3-Dimethylcyclohexane	GC	35
trans-1,3-Dimethylcyclohexane	GC	35
1-*cis*-4-Dimethylcyclohexane	GC	118
1-*trans*-4-Dimethylcyclohexane	GC	118
cis-1,4-Dimethylcyclohexane	GC	35
trans-1,4-Dimethylcyclohexane	GC	35
1-*cis*-2-Dimethylcyclohexane	GC	42
1-*trans*-2-Dimethylcyclohexane	GC	42
1-*cis*-3-Dimethylcyclohexane	GC	42
1-*trans*-3-Dimethylcyclohexane	GC	42
1-*cis*-4-Dimethylcyclohexane	GC	42
1-*trans*-4-Dimethylcyclohexane	GC	42
2,2-Dimethylcyclohexanone	PC	20
N,N-Dimethylcyclohexylamine salt of dibutyldithiocarbamic acid	TLC	66
Dimethylcyclohexyl phthalate	TLC	27
1,1-Dimethylcyclopentane	GC	36, 42
1,2-*cis*-Dimethylcyclopentane	GC	36
1,2-*trans*-Dimethylcyclopentane	GC	36
1,3-*cis*-Dimethylcyclopentane	GC	36
1,3-*trans*-Dimethylcyclopentane	GC	36
1-*cis*-2-Dimethylcyclopentane	GC	42
1-*trans*-2-Dimethylcyclopentane	GC	42
1-*cis*-3-Dimethylcyclopentane	GC	42
1-*trans*-3-Dimethylcyclopentane	GC	42
trans-1,2-Dimethylcyclopentane	GC	121
cis-1,3-Dimethylcyclopentane	GC	121
trans-1,3-Dimethylcyclopentane	GC	121
1,1-Dimethylcyclopropane	GC	118
Dimethylcysteine	TLC	11
2,3-Di-*O*-methyl-6-deoxyallose (mycinose)	LC	48
1,3-Dimethyl-2,4-diamino-anthraquinone	TLC	69
2,5-Dimethyl-1,4-diaminobenzene	TLC	119
2,6-Dimethyl-1,4-diaminobenzene	TLC	119
4,6-Dimethyl-1,3-diaminobenzene	TLC	119
3,3′-Dimethyl-4,4′-diaminodiphenyl	TLC	119
N,N′-Dimethyldiaminoethane	GC	11
N,N-Dimethyl-1,3-diaminopropane	GC	11
N,N′-Dimethyl-1,3-diaminopropane	GC	11
3,5-Dimethyl-2,4-dibromo-6-propionyl-phenol	TLC	89
2,2-Dimethyl-4-[2,5-dimethylpyrrolyl-(1)]-propane (1-neopentyl-2,5-dimethyl-pyrrole)	GC	133
4,4′-Dimethyldiphenyl	TLC	53
Dimethyl disulfide	GC	85
Dimethyl dodecanedioate	GC	1
1,3:2,4-Di-*O*-methylene-D-glucitol	GC	13
1,3:2,4-Di-*O*-methyleneribitol	GC	13
Dimethyl esters	GC	24
Dimethyl ether	GC	52
N,N-Dimethylethylamine	GC	97
1,2-Dimethyl-3-ethylbenzene	GC	119
1,2-Dimethyl-4-ethylbenzene	GC	119
1,3-Dimethyl-4-ethylbenzene	GC	120
1,3-Dimethyl-5-ethylbenzene	GC	119
1,4-Dimethyl-2-ethylbenzene	GC	119
3,3-Dimethyl-2-ethyl-butene	GC	118
2,3-Dimethyl-4-ethylthiophene	TLC	104
N,N-Dimethylformamide	GC	53
Dimethyl fumarate	GC	1
2,5-Dimethylfuran	GC	55
2,3-Dimethylgalactose	PC	73
2,4-Dimethylgalactose	PC	73
2,5-Dimethylgalactose	PC	73
3,4-Dimethylgalactose	PC	73
2,4-Di-*O*-methyl-D-glucose	TLC	26
2,6-Di-*O*-methyl-D-glucose	TLC	26
4,6-Di-*O*-methyl-D-glucose	TLC	26
Dimethyl glutarate	GC	1
N,N-Dimethylguanidine	PC	10
N,N′-Dimethylguanidine	PC	10
N^2-Dimethylguanine	LC	46
1,7-Dimethylguanine	PC	36
2-Dimethylguanine	GC	77
N^2-Dimethylguanosine	LC	22
1,7-Dimethylguanosine	LC	22
2,2-Dimethylheptane	GC	30, 42
2,3-Dimethylheptane	GC	30
2,4-Dimethylheptane	GC	30
2,5-Dimethylheptane	GC	30
2,6-Dimethylheptane	GC	30
2,6-Dimethylheptanone-4	GC	51
N,N-Dimethylhexadecylamine oxide	TLC	7
2,5-Dimethyl-1,5-hexadiene	GC	118
2,2-Dimethylhexane	GC	30, 42
2,2-Dimethylhexane, paraffins	GC	33
2,3-Dimethylhexane	GC	30, 42
2,3-Dimethylhexane, paraffins	GC	33
2,4-Dimethylhexane	GC	30, 42, 118, 121
	LC	42
2,4-Dimethylhexane, paraffins	GC	33
2,5-Dimethylhexane	GC	30, 42, 118
2,5-Dimethylhexane, paraffins	GC	33
2,5-Dimethylhexane-2,5-dihydroperoxide	PC	27
3,3-Dimethylhexane	GC	30, 42, 121
3,3-Dimethylhexane, paraffins	GC	33
3,4-Dimethylhexane	GC	30, 42, 121
3,4-Dimethylhexane, paraffins	GC	33
2,3-Dimethyl-2-hexene	GC	39
2,4-Dimethyl-1-hexene	GC	39
2,5-Dimethyl-1-hexene	GC	39, 118
2,5-Dimethyl-2-hexene	GC	39
3,4-Dimethyl-1-hexene	GC	39
3,4-Dimethyl-*x*-hexene	GC	39
4,5-Dimethyl-1-hexene	GC	39
5,5-Dimethyl-*cis*-2-hexene	GC	39
2,3-Dimethyl-*cis-trans*-3-hexene	GC	39
2,4-Dimethyl-*cis-trans*-3-hexene	GC	39
3,4-Dimethyl-*cis-trans*-2-hexene	GC	39
3,5-Dimethyl-*cis-trans*-2-hexene	GC	39
2,2-Dimethyl-*trans*-3-hexene	GC	39
5,5-Dimethyl-*trans*-2-hexene	GC	39
1:4 Dimethyl histidine	TLC	155

1,1-Dimethylhydrazine·HCl	PC	61
1,2-Dimethylhydrazine-2HCl	PC	61
N,N-Dimethyl-5-hydroxytryptamine	LC	77
N,N-Dimethyl-5-hydroxytryptamine (as acetate)	PC	102
1,7-Dimethylhypoxanthine	PC	36
1,2-Dimethylindole	GC	45
2,3-Dimethylindole	GC	45
2,5-Dimethylindole	GC	45
2,7-Dimethylindole	GC	45
N,N-Dimethylisopropylamine	GC	97
1,3-Dimethylisoquinoline	TLC	58
Dimethyl isosebacate	GC	1
Dimethyl itaconate	GC	1
Dimethyl maleate	GC	1
Dimethyl malonate	GC	1
Dimethylmannose	PC	73
N,N-Dimethyl-2-methylaniline	GC	102
N,N-Dimethyl-3-methylaniline	GC	102
N,N-Dimethyl-4-methylaniline	GC	102
3,5-Dimethyl-4-methylthiophenol	GC	71
2-(2,6-Dimethyl-4-morpholinothio)-benzothiazole	TLC	66
Dimethylnaphthalene	LC	27
1,2-Dimethylnaphthalene	GC	38
	TLC	51
1,3-Dimethylnaphthalene	GC	38
	TLC	51, 53
1,4-Dimethylnaphthalene	GC	38
	LC	7
	TLC	51, 53
1,5-Dimethylnaphthalene	GC	38
	TLC	51, 53
1,6-Dimethylnaphthalene	GC	38
	TLC	51, 53
1,7-Dimethylnaphthalene	GC	38
	TLC	51, 53
1,8-Dimethylnaphthalene	GC	38
2,3-Dimethylnaphthalene	GC	38
	TLC	51, 53
2,6-Dimethylnaphthalene	GC	38
	TLC	51, 53
2,7-Dimethylnaphthalene	GC	38
	TLC	51
2,5-Dimethyl-13H-naphtho[2,1-*a*]-carbazole	TLC	120
N,N-Dimethyl-1-naphthylamine	TLC	7
Dimethyloctylamine	TLC	7
2,4-Dimethylol-*o*-cresol	PC	82
2,6-Dimethylol-*p*-cresol	PC	82
2,4-Dimethylolphenol	PC	82
2,6-Dimethylolphenol	PC	82
Dimethyl orthophthalate	GC	1
Dimethyl oxalate	GC	1
2,2-Dimethylpentane	GC	30, 42, 118
	LC	42
2,2-Dimethylpentane, paraffins	GC	33
2,3-Dimethylpentane	GC	30, 42, 53, 118, 121
	LC	42
2,3-Dimethylpentane, paraffins	GC	33
2,4-Dimethylpentane	GC	30, 42, 47, 53, 118, 121
	LC	42
2,4-Dimethylpentane, paraffins	GC	33
3,3-Dimethylpentane	GC	30, 42, 118, 121

2,2-Dimethyl-3-pentanone	GC	118
2,4-Dimethylpentanone-2	GC	51
2,3-Dimethyl-1-pentene	GC	39, 118, 122
2,4-Dimethyl-1-pentene	GC	39, 118, 122
3,3-Dimethylpentene	GC	118
3,3-Dimethyl-1-pentene	GC	122
3,4-Dimethyl-1-pentene	GC	122
4,4-Dimethyl-1-pentene	GC	122
2,3-Dimethyl-2-pentene	GC	39, 118, 122
2,4-Dimethyl-2-pentene	GC	39, 122
cis-3,4-Dimethyl-2-pentene	GC	122
trans-3,4-Dimethyl-2-pentene	GC	122
cis-4,4-Dimethyl-2-pentene	GC	122
trans-4,4-Dimethyl-2-pentene	GC	122
3,4-Dimethyl-*cis*-2-pentene	GC	118
3,4-Dimethyl-*trans*-2-pentene	GC	39, 118
4,4-Dimethyl-*cis*-2-pentene	GC	118
4,4-Dimethyl-*trans*-2-pentene	GC	39, 118
11,12-Dimethyl-14H-phenanthro[3,2-*a*]-carbazole	TLC	120
2,3-Dimethylphenetole	GC	75
2,4-Dimethylphenetole	GC	75
2,5-Dimethylphenetole	GC	75
2,6-Dimethylphenetole	GC	75
3,4-Dimethylphenetole	GC	75
3,5-Dimethylphenetole	GC	75
2,3-Dimethylphenol	GC	69, 72, 75
	PC	31
	TLC	174
2,3-Dimethylphenol, 4-acetyl-2-nitrophenol ether of	TLC	91
2,4-Dimethylphenol	GC	69, 72, 75
	LC	51
	TLC	174
2,4-Dimethylphenol, 4-acetyl-2-nitrophenol ether of	TLC	91
2,5-Dimethylphenol	GC	69, 72, 75
	PC	31
	TLC	174
2,5-Dimethylphenol, 4-acetyl-2-nitrophenol ether of	TLC	91
2,6-Dimethylphenol	GC	69, 72, 75
	LC	51
	TLC	174
3,4-Dimethylphenol	GC	69, 72, 75
	PC	31
	TLC	174
3,4-Dimethylphenol, 4-acetyl-2-nitrophenol ether of	TLC	91
3,5-Dimethylphenol	GC	69, 72, 75, 130
	PC	31
	TLC	174
N,N-Dimethyl-*p*-phenyl-azoaniline	TLC	81
N,N-Dimethyl-*p*-phenylenediamine	TLC	7
2,5-Dimethyl-1,4-phenylenediamine	TLC	119
3,5-Dimethylphenyl-*N*-methylcarbamate	GC	130
Dimethyl phthalate	GC	25, 132
	TLC	27
Dimethyl pimelate	GC	1
2,6-Dimethylpiperidine	TLC	7

Compound	Method	Page
2,2-Dimethylpropane	GC	42
2,2-Dimethyl-1,3-propanediol	LC	37
	TLC	113
2,2-Dimethylpropionic acid	TLC	29
2,3-Dimethyl-4-propionylphenol	TLC	89
2,3-Dimethyl-6-propionylphenol	TLC	89
2,4-Dimethyl-6-propionylphenol	TLC	89
2,5-Dimethyl-4-propionylphenol	TLC	89
2,6-Dimethyl-4-propionylphenol	TLC	89
3,4-Dimethyl-6-propionylphenol	TLC	89
3,5-Dimethyl-6-propionylphenol	TLC	89
N,N-Dimethyl-*n*-propylamine	GC	97
2,3-Dimethylpyridine	GC	131
2,4-Dimethylpyridine	GC	131
	TLC	60
2,5-Dimethylpyridine	GC	131
2,6-Dimethylpyridine	GC	131
	TLC	60
3,5-Dimethylpyridine	GC	131
4,6-Dimethylpyrimidine	PC	36
2,4-Dimethylpyrrole	GC	133
2,5-Dimethylpyrrole	GC	133
3-[2,5-Dimethylpyrrolyl-(1)]-pentane	GC	133
Dimethylpyruvic acid	PC	81
2,3-Dimethylquinoline	TLC	58
2,4-Dimethylquinoline	GC	45, 78
	TLC	58
2,6-Dimethylquinoline	GC	45, 78
	TLC	58
2,8-Dimethylquinoline	GC	78
	TLC	58
4,6-Dimethylquinoline	GC	78
4,8-Dimethylquinoline	GC	78
3,4-Dimethylrhamnose	PC	73
Dimethyl sebacate	GC	1, 25
	LC	70
Dimethyl silane	GC	27
Dimethyl suberate	GC	1
Dimethyl succinate	GC	1, 25
2,4-Dimethyl-6-sulfanilamido-1,3-diazine	TLC	44
Dimethyl sulfide	GC	85, 135
2,4-Dimethylsulfolane	GC	139
2,4-Dimethyl-3-sulfolene sulfone	GC	139
Dimethyl terephthalate	LC	70
5,5″-Dimethylterthienyl	TLC	104
2,5-Dimethylthiophene	TLC	104
N′-Dimethylthiourea	PC	47
N′,*N*″-Dimethylthiourea	PC	47
Dimethyltin chloride	TLC	67
N,N-Dimethyl-*o*-toluidine	TLC	7
N,N-Dimethyl-*p*-toluidine	TLC	7
Di-methyl trisulfide	GC	85
α,α-Dimethyltryptamine	PC	103
α,α-Dimethyltryptamine (as acetate)	PC	102
β,β-Dimethyltryptamine (as oxalate)	PC	102
N,N-Dimethyltryptamine	GC	9
	LC	77
	PC	101, 103
N,N-Dimethyltryptamine (as acetate)	PC	102
1,3-Dimethyluracil	PC	36, 37
5,6-Dimethyluracil	PC	36
N,N-Dimethylurea	TLC	122
N,N′-Dimethylurea	TLC	122
1,3-Dimethylurea	PC	60
1,3-Dimethyluric acid	PC	36
1,7-Dimethyluric acid	PC	36
3,7-Dimethyluric acid	PC	36
2,2-Dimethylvaleric acid	TLC	29
Dimethylxanthine	LC	13

Compound	Method	Page
1,3-Dimethylxanthine	PC	36
1,7-Dimethylxanthine	LC	46
	PC	36
3,7-Dimethylxanthine	PC	36
3,8-Dimethylxanthine	PC	36
2,3-Dimethylxylose	PC	73
2,3-Di-*O*-methyl-D-xylose	TLC	26
Dimetilan	GC	130
	TLC	83
Dimite®	GC	129
DINA	TLC	157
2,2′-Dinaphthyl	TLC	53
N,N′-Di-β-naphthyl-*p*-phenylenediamine	TLC	22
Di-α-naphthyl sulfoxide	TLC	164
2,4-Dinitroaniline	TLC	18, 119
2,3-Dinitroanisole	GC	70
2,4-Dinitroanisole	GC	70
2,5-Dinitroanisole	GC	70
2,6-Dinitroanisole	GC	70
3,4-Dinitroanisole	GC	70
3,5-Dinitroanisole	GC	70
3,5-Dinitrobenzamide	TLC	136
2,5-Dinitrobenzoic acid	PC	78
Dinitrobutylphenol (DNOSBP)	GC	61, 62
4,4′-Dinitrocarbanilide and 2-hydroxy-4,6-dimethylpyrimidine	TLC	136
1,4-Dinitro-6-chloroaniline	TLC	119
Dinitrocresol (DNOC)	GC	61, 63
Dinitrocyclohexylphenol (DNOCHP)	GC	63
2,4-Dinitrodiphenylamine	TLC	157
2,4′-Dinitrodiphenylamine	TLC	157
2,6-Dinitro-4-methylaniline	TLC	119
3,5-Dinitro-4-methylaniline	PC	59
2,4-Dinitro-1-naphthol	PC	105
Dinitrophenol	PC	65
o,p-Dinitrophenol	TLC	174
2,3-Dinitrophenol	GC	70
2,4-Dinitrophenol	GC	70
	LC	51
	TLC	18
2,5-Dinitrophenol	GC	70
2,6-Dinitrophenol	GC	70
3,4-Dinitrophenol	GC	70
3,5-Dinitrophenol	GC	70
m,m′-Dinitrophenyl disulfide	TLC	136
2,4-Dinitrophenylhydrazine	PC	20
	TLC	32
2,4-Dinitrophenylsulfenyl chloride	TLC	59
2,4-Dinitrosoresorcinol	PC	120
2,5-Dinitro-*o*-toluamide	TLC	136
3,5-Dinitrotyrosine	PC	63
Dinocap	TLC	81
Dinonyl adipate (Adipic acid ester)	TLC	27
Dinonyl phthalate	TLC	27
Di-*n*-nonyl succinate	GC	25
Dinoxanthin	PC	122
Diocaine	TLC	43
(Di)octadecyl ether $(n\text{-}C_{18}H_{37})_2O$	LC	21
Dioctyl adipate (Adipic acid ester)	TLC	27
Di-*n*-octyl adipate	GC	25
Dioctyl azelate	TLC	27
Dioctylbenzene	LC	27
Dioctyl ether	LC	21
Dioctyl phthalate	TLC	27
Dioctyl sebacate	TLC	27
Di-*n*-octyl succinate	GC	25
Dioctyl sulfide	TLC	163

Dioctyltin dichloride	TLC	67
3,5-Diodothyroacetic acid	LC	76
1,3-Diolein	TLC	56
Dionin® (Ethylmorphine HCl)	TLC	45,116, 138
Dionyl	PC	4
Diosgenin	TLC	98, 103
Diosgenin 3-acetate	TLC	103
Diosmetin	PC	26
Dioxan	GC	47
Dioxan, organics	GC	52
1,4-Dioxane	GC	55
Dioxathion	GC	64, 65, 67, 129
	TLC	81
Dioxathion (Delnav)	TLC	80, 84
3,20-Dioxo-11β-hydroxy-eti-4-enic acid-lactone (18 → 11)	PC	43
Dioxymonovinylhydroxy-ethyldeuteroporphyrin	PC	33
Dioxyprotoporphyrin	PC	33
1,3-Dipalmitin	TLC	56
α,β-Dipalmitolein	PC	99
Dipalmitoyl-phosphatidylcholine	TLC	57
Di-Paralene® (Chlorcyclizine)	TLC	39
Diparcol®	TLC	42
Dipentamethylenethiuram tetrasulfide	TLC	66
Dipentene	GC	87
Dipentyl ketone	GC	117
Diphenhydramine	PC	85
Diphenhydramine (Benadryl®)	GC	20, 21
	TLC	39, 45
Diphenhydramine HCl	GC	104
Diphenolic acid	PC	121
Diphenoquinone	TLC	88
Diphenoxyethane	LC	21
Diphenyl	GC	40
	LC	27
	TLC	53
Diphenylamine	TLC	7, 157
1,4-Diphenylbenzene	TLC	53
5,5′-Diphenylbithienyl	TLC	104
Diphenylcresyl phosphate (Phosphoric acid ester)	TLC	27
Diphenylene oxide	TLC	88
Diphenyl ether	LC	7
β,β-Diphenylethylamine	PC	56
Diphenylfulvene	TLC	7
Diphenylguanidine	TLC	22
Diphenylhydantoin	GC	105
Diphenylhydantoin (Dilantin®)	GC	21
Diphenylhydramine HCl	GC	20
Diphenylmethane	LC	21
Diphenylmethane derivatives	GC	17
Diphenyloctyl phosphate (Phosphoric acid ester)	TLC	27
N,N′-Diphenyl-*p*-phenylenediamine	TLC	22
Diphenylpyraline	GC	17
	TLC	39
Diphenylpyraline hydrochloride	GC	17, 20, 54, 104
Diphenylpyraline (Mepiben)	TLC	45
Diphenyl pyrophosphate	PC	112
Diphenyl sulfide	TLC	163
Diphenyl sulfoxide	TLC	164
Diphenylthiourea, PTH of	TLC	124
Diphenyltin dichloride	TLC	67
2,3-Diphosphoglycerate	PC	12

2,3-Diphosphoglyceric acid (2,3-PPGA)	PC	32
Diphosphopyridine nucleotide	LC	13
Diphosphopyridine nucleotide (DPN)	PC	32
Dipicrylamine	TLC	157
Dipipanone	PC	4
Dipropargylamine	TLC	10
Dipropionate	TLC	112
1,2-Dipropionylhydrazine	PC	61
Di-*n*-propyl adipate	GC	25
Dipropylamine	TLC	10
Di-*n*-propylamine	GC	97
	PC	55
Di-*n*-propylamine, trifluoroacetyl derivative	GC	10
Di-*n*-propylbarbituric acid	GC	105
Di-*n*-propyl disulfide	GC	85
Dipropylene glycol	LC	54
Dipropylene glycol-1,2	TLC	113
Dipropylenetriamine	PC	7
Dipropyl ether	GC	51
Di-*n*-propyl ether	LC	7
Di-*n*-propyl ether, organics	GC	52
Di-*n*-propyl ketone	PC	20
2,6-Di-*n*-propylphenol	GC	72
Di-*n*-propyl phthalate	GC	25
P^1,P^2-Di-*n*-propyl pyrophosphate	PC	34
Di-*n*-propyl sebacate	GC	25
Di-*n*-propyl suberate	GC	25
Di-*n*-propyl succinate	GC	25
Dipropyl sulfide	GC	135
	TLC	163
Di-*n*-propyl sulfide	GC	85
Di-*n*-propyl trisulfide	GC	85
Dipterex	PC	28
Dipterex (Trichlorfon) (Dylox®)	TLC	80, 84
2,3′-Dipyridyl	GC	95
	PC	2
Di-α-pyridyl phosphate	PC	112
Distearin (2-OH)	LC	45
Distearin (3-OH)	LC	45
Disulfides (saturated)	GC	85
Disulfides (unsaturated)	GC	85
Disulfoton	GC	65, 67
	LC	43
Disulfoton (Di-Syston)	TLC	171
Disulfoton O-analog	GC	65
Disulfoton O-analog sulfone	GC	65
Disulfoton O-analog sulfoxide	GC	65
Disulfoton oxygen analog sulfone or demeton thiol sulfone	GC	67
Disulfoton S	TLC	171
Disulfoton sulfone	GC	65, 67
Disulfoton sulfoxide	GC	65
Di-Syston®	GC	60
Di-Syston® (Disulfoton)	PC	28
	TLC	80, 84
Disyston (Dithio-Systox)	TLC	172
Di-Syston, sulfone	TLC	80
Di-Syston, sulfoxide	TLC	80
Dithianon	TLC	81
Di(α-G-β-thiol propionic acid)	PC	9
Dithiouracil	PC	36, 37
Dition®	GC	65
Di-*o*-tolylguanidine	TLC	22
Ditolyl pyrophosphate	PC	112
Diuron	GC	129
	TLC	81, 82, 173

Divalerate	TLC	112
Divaric acid	PC	30
Divaricatic acid	PC	30
	TLC	35
Divaricatinic acid	PC	30
Djenkolic acid	TLC	13
dl-Djenkolic acid	PC	63
L-Djenkolic acid	LC	59
$DMTrTpa^{Bz}$	PC	19
$DMTrTpA^{Bz}$-OAc	PC	19
$DMTrTpA^{Bz}pG^{Ac}$-OAc	PC	19
DNOPP, Pesticides	GC	63
DNP-Alanine	TLC	18
DNP-β-Alanine	TLC	18
DNP-L-Alanine	TLC	17
DNP-α-Aminobutyric acid	TLC	18
DNP-α-Aminocaprylic acid	TLC	18
DNP-L-Arginine	TLC	17
DNP-L-Asparagine	TLC	17
DNP-Aspartate	TLC	18
DNP-L-Aspartic acid	TLC	17
DNP-Glutamate	TLC	18
DNP-DL-Glutamic acid	TLC	17
DNP-L-Glutamine	TLC	17
DNP-Glycine	TLC	17, 18
DNP-Isoleucine	TLC	18
DNP-L-Isoleucine	TLC	17
DNP-Leucine	TLC	18
DNP-L-Leucine	TLC	17
ε-DNP-L-Lysine	TLC	17
DNP-Methionine	TLC	18
DNP-DL-Methionine	TLC	17
DNP-Methionine sulfone	TLC	18
DNP-DL-Methionine sulfoxide	TLC	17
DNP-Norleucine	TLC	18
DNP-Norvaline	TLC	18
DNP-Ornithine	TLC	18
DNP-Phenylalanine	TLC	18
DNP-L-Phenylalanine	TLC	17
DNP-Proline	TLC	18
DNP-L-Proline	TLC	17
DNP-Sarcosine	TLC	18
DNP-Serine	TLC	18
DNP-L-Serine	TLC	17
DNP-Threonine	TLC	18
DNP-L-Threonine	TLC	17
DNP-Tryptophan	TLC	18
DNP-L-Tryptophan	TLC	17
DNP-Valine	TLC	18
DNP-L-Valine	TLC	17
DNS	LC	15
DNS-Alanine	TLC	123
DNS-α-Alanine	TLC	123
DNS-Arginine	TLC	123
DNS-Asparagine	TLC	123
DNS-Aspartic acid	TLC	123
DNS-*S*-Carboxymethylcysteine	TLC	123
DNS-Glutamic acid	TLC	123
DNS-Glutamine	TLC	123
DNS-Glycine	TLC	123
DNS-Histidine	TLC	123
DNS-Isoleucine	TLC	123
DNS-Leucine	TLC	123
ε-DNS-Lysine	TLC	123
DNS-Methionine	TLC	123
DNS-NH_2	TLC	123
DNS-Phenylalanine	TLC	123
DNS-Proline	TLC	123
DNS-Serine	TLC	123
DNS-Threonine	TLC	123
DNS-Tryptophan	TLC	123
DNS-Tyrosine	TLC	123
DNS-Valine	TLC	123
Docosanoic acid, methyl ester of	TLC	130
cis-Docosen-5-oic acid, methyl ester of	TLC	130
cis-Docosen-11-oic acid, methyl ester of	TLC	130
Dodecamethylenediamine, trifluoroacetyl derivative	GC	10
Dodecanal, dimethyl acetal of	GC	6
Dodecanal, 2,4-DNPSC of	TLC	165
Dodecanal, *p*-NPH of	TLC	165
Dodecane	GC	49
n-Dodecane	GC	53, 55
	LC	21, 27, 38
Dodecanoic acid	LC	68
Dodecanoic acid, methyl ester of	GC	24
	TLC	130
n-Dodecanoic acid	LC	57
n-Dodecanol	GC	92
1-Dodecanol	GC	92
1-Dodecanol (DNB)	TLC	111
1-Dodecanol (ONPU)	TLC	111
Dodecenal, 2,4-DNPH of	TLC	72, 76
2-Dodecen-1-al, 2,4-DNPH of	TLC	72
Dodecene-1	GC	53
Dodecyl alcohol	TLC	1
n-Dodecyl alcohol	LC	11
n-Dodecylamine	PC	55
Dodecylbenzene	TLC	53
Dodecyl esters	GC	24
Dodecyl gallate	TLC	125
Dodecyl mercaptan	TLC	162
Dodecyl octyl sulfide	TLC	163
Doisynolic acid	TLC	181
Dolantin	PC	2
Dolophin (methadone)	TLC	39, 138
DOPA	PC	58
DOPA 2:3	PC	64
DOPA 2:4	PC	64
DOPA 2:5	PC	64
DOPA 2:6	PC	64
3,4-DOPA	TLC	13
DOPA 3:4	PC	64
DOPA 3:5	PC	64
Dopamine	GC	9
	PC	58
	TLC	8
Dopamine 2:5	PC	64
Dopamine 3:4	PC	64
Doriden (Glutethimide)	TLC	137
Dow ET-14 (Ronnel) (Fenchlorphos)	TLC	82, 84
Doxylamine	GC	20
Doxylamine (Decapryn®)	TLC	39
Doxylamine succinate	GC	20
DPN	TLC	63
DPN (Diphosphopyridine nucleotide)	PC	18
DPNH	TLC	63
DPPD (*N,N'*-diphenyl-*p*-phenylenediamine)	PC	69
Dromoran	PC	2, 84
Drugs, Free form	GC	17
DT	GC	115
DTBH (2,5-di-*tert*-butylhydroquinone)	PC	68
Dulcitol	PC	1, 53, 70
	TLC	127

Dursban® GC 65, 67
TLC 81, 84
Dursban O-analog GC 65
Dursban oxygen analog GC 67
Dy (Dysprosium) LC 39, 79, 89, 91
Dyclonine TLC 39
Dyclonine HCl GC 110
Dyfonate® GC 65, 67
Dyfonate O-analog GC 65
Dyfonate oxygen analog GC 67
Dylox®(Dipterex) (Trichlorfon) TLC 80, 84
Dyrene GC 68, 128, 129
TLC 82
Dysprosium (Dy) LC 39, 79, 89, 91

ECP® (VC-13) TLC 172
Ectylurea GC 17, 108
Eicosane GC 45
n-Eicosane LC 38
Eicosanoic acid, methyl ester of TLC 130
Eicosa-5,8,11,14-tetraenoic acid, methyl ester of TLC 130
cis-Eicosen-11-oic acid, methyl ester of TLC 130
Elaidic acid PC 77
β-Elemene GC 140
TLC 54
β-Eleostearic acid PC 77
Ellagic acid TLC 92, 133
Elymoclavine TLC 114
Emetine GC 8, 107
PC 3, 4
TLC 3
Emicymarin PC 127
Emicymarin acetate PC 127
Emmolic acid TLC 184
Emodin PC 109
Emylcamate TLC 38
Enanthic acid PC 77
TLC 29
Endocrocin PC 109
Endomethylenetetrahydrophthalic anhydride, *p*-anisidide of TLC 30
Endosulfan (Thiodan®) GC 60, 64, 129
TLC 80, 82
Endosulfan I GC 65, 129
TLC 82
Endosulfan II GC 65, 129
TLC 82
Endosulfan alcohol GC 129
Endosulfan, high m.p. TLC 80
Endosulfan, low m.p. TLC 80
Endosulfan sulfate GC 65, 129
Endrin GC 60, 68, 128, 129
LC 44
PC 116
TLC 80, 81, 82, 172
Endrin alcohol GC 129
TLC 82
Endrin aldehyde GC 129
TLC 82
Enolase LC 19
Enteromycin TLC 20

Eosine TLC 142
Epanorin TLC 28
Ephedrine GC 9, 21, 22, 23, 105
LC 55
PC 2, 3, 56, 64
TLC 2, 3, 10, 39, 45
Epiandrosterone GC 83, 136
TLC 97
Epianhydrodemethylchlortetracycline PC 68
Epianhydrotetracycline PC 68
Epichlortetracycline PC 68
Epicholestanol GC 79
Epicholesterol TLC 99
11-Epicortisol TLC 101
11-Epicortisol-21-acetate TLC 101
11-Epicortisol-11,21-diacetate TLC 101
Epidemethylchlortetracycline PC 68
Epiestriol (Estriol) TLC 95
16-*epi*-Estriol TLC 181
16,17-*epi*-Estriol TLC 181
17-*epi*-Estriol TLC 181
Epietiocholanolone TLC 97
Epilupinine PC 3
Epimethyltetracycline PC 68
Epinephrine GC 9, 23
TLC 8, 13, 39
L-Epinephrine LC 77
Epinine PC 5
Epioxytetracycline PC 68
Epipregnanolone GC 136
Epitestosterone GC 83, 136
TLC 97
Epitestosterone acetate GC 83
Epitetracycline PC 68
Epithiazide TLC 41
EPN GC 64, 65, 67
PC 28
TLC 80, 84, 172
4*α*,5*α*-Epoxypregnane-3,20-dione TLC 96
4*β*,5*β*-Epoxypregnane-3,20-dione TLC 96
Equilenin GC 79
PC 124
TLC 181
Equilin PC 124
TLC 181
Equinenone TLC 93
Eradex TLC 80
Er (Erbium) LC 39, 79, 85, 86, 90, 91
Ergine TLC 114
Ergocornine TLC 114
Ergocorninine TLC 114
Ergocristine TLC 3, 114
Ergocristinine TLC 3, 114
Ergocryptine TLC 114
Ergocryptinine TLC 114
Ergometrine (Ergonovine) TLC 2, 3, 114
Ergometrinine TLC 3, 114
Ergonovine TLC 114
Ergosine TLC 114
Ergosinine TLC 114

Ergosterol	GC	84, 136
	LC	24
	PC	126
	TLC	57, 99
Ergostine	TLC	114
Ergostinine	TLC	114
Ergotamine	PC	101
	TLC	2, 3, 114
Ergotaminine	TLC	3, 114
Ergothioneine	PC	17
	TLC	155
Eriochrome Black T	TLC	47
Eriodictyol	LC	28
	PC	26
Erucic acid	PC	77
	TLC	33
Erycanoside	TLC	102
Erycanoside peracetate	TLC	102
Erysimoside	TLC	102
Erysimoside peracetate	TLC	102
Erythritol	GC	14, 15
	LC	69
	PC	1
	TLC	127
Erythro-dihydro-sphingosine	TLC	57
DL-Erythro-2,3-dihydroxybutyric	LC	23
D-Erythro-L-galactooctitol	GC	14
Erythromycin	PC	66
	TLC	20, 153
D-Erythronic	LC	40
Erythrose	LC	48
	TLC	23
D-Erythrose	GC	15
Erythrose-4-phosphate	PC	32
Erythrosine	TLC	142
Erythrosine BS	TLC	49
Erythrosphingosine	TLC	57
Escholtzxanthin	TLC	93
Esculin	TLC	92
Eserine	PC	2, 3
Esters, ethyl acetate	GC	126
Esters, methyl acetate	GC	126
Esters, *n*-propyl acetate	GC	126
Estetrol	GC	82
Estox	TLC	172
Estradiol	GC	79
	PC	124
	LC	6
	TLC	178, 181
Estradiol-17α	PC	124
	TLC	181
Estradiol-17β	GC	83
	PC	124
α-Estradiol	TLC	95
β-Estradiol	TLC	95
17β-Estradiol	GC	136
Estragole	GC	87
Estra-1,3,5(10),6-tetraene-3,17β-diol	TLC	181
Estra-1,3,5(10),7-tetraene-3,17β-diol	TLC	181
Estra-1,3,5(10)11(12)-tetraene-3,17α-diol	GC	82
Estratetraenol	TLC	181
Estra-1,3,5(10),16-tetraen-3-ol	TLC	181
1,3,5(10)-Estratriene-3,16α-diol	GC	81
1,3,5(10)-Estratriene-3,16β-diol	GC	81
Estra-1,3,5(10)-triene-3,17α-diol	GC	82
	TLC	181
1,3,5(10)-Estratriene-3,17α-diol	GC	81
Estra-1,3,5(10)-triene-3,17β-diol (Estradiol)	GC	82
	TLC	181
1,3,5(10)-Estratriene-3,17β-diol	GC	136
1,3,5(10)-Estratriene-3,17β-diol (Estradiol-17β)	GC	81
Estra-1,3,5(10)-triene-3,17β-diol-6-one	GC	82
Estra-1,3,5(10)-triene-3,17β-diol-11-one	GC	82
Estra-1,3,5(10)-triene-3,17β-diol-16-one	GC	82
1,3,5(10)Estratriene-3,17β-diol, 16-one	GC	81, 136
1,3,5(10)-Estratriene-3-methoxy-16,17-diketo	GC	81
1,3,5(10)-Estratriene-3-methoxy-16α,17β-diol	GC	81
1,3,5(10)-Estratriene-3-methoxy-17β-ol	GC	81
1,3,5(10)-Estratriene-2-methoxy-3-ol,17-one	GC	81
1,3,5(10)-Estratriene-3-methoxy-17-one	GC	81
1,3,5(10)-Estratriene-2-methoxy-3,16α,17β-triol	GC	81
1,3,5(10)-Estratriene-3-ol,16,17-dione	GC	81
1,3,5(10)-Estratriene-3-ol,16-one	GC	81
1,3,5(10)-Estratriene-3-ol,17-one (Estrone)	GC	81
1,3,5(10)-Estratrienes (Steroids)	GC	81
Estra-1,3,5(10)-triene-2,3,16α,17β-tetrol	GC	82
	TLC	181
1,3,5(10)-Estratriene-2,3,16α,17β-tetrol	GC	81
Estra-1,3,5(10)-triene-2,3,16α,17β-tetrol-2-methyl ether	TLC	181
Estra-1,3,5(10)-triene-3,6,7,17β-tetrol	TLC	181
Estra-1,3,5(10)-triene-3,6α,16α,17β-tetrol	GC	82
	TLC	181
Estra-1,3,5(10)-triene-2,3,17β-triol	GC	82
	TLC	181
Estra-1,3,5(10)-triene-2,3,17β-triol-2-methyl ether	TLC	181
Estra-1,3,5(10)-triene-3,6α,17β-triol	GC	82
	TLC	181
Estra-1,3,5(10)-triene-3,6α,17β-triol-3-methyl ether	TLC	181
Estra-1,3,5(10)-triene-3-6β,17β-triol	GC	82
	TLC	181
Estra-1,3,5(10)-triene-3,6β,17β-triol-3-methyl ether	TLC	181
Estra-1,3,5(10)-triene-3,7α,17β-triol	GC	82
	TLC	181
Estra-1,3,5(10)-triene-3,11β,17β-triol	GC	82
	TLC	181

Compound	Method	Pages
Estra-1,3,5(10)-triene-3,15α,17β-triol	GC	82
	TLC	181
Estra-1,3,5(10)-triene-3,16α,17α-triol	TLC	181
1,3,5(10)-Estratriene-3,16α,17α-triol	GC	81
Estra-1,3,5(10)-triene-3,16α,17α-triol-3-methyl ether	TLC	181
Estra-1,3,5(10)-triene-3,16α-17β-triol (Estriol)	GC	82
	TLC	181
Estra-1,3,5(10)-triene-3,16α,17β-triol-3-methyl ether	TLC	181
1,3,5(10)-Estratriene-3,16α,17β-triol (Estriol)	GC	81, 136
Estra-1,3,5(10)-triene-3,16β,17α-triol	GC	82
	TLC	181
1,3,5(10)-Estratriene-3,16β,17α-triol	GC	81
Estra-1,3,5(10)-triene-3-16β-17β-triol	GC	82
	TLC	181
1,3,5(10)-Estratriene-3,16β,17β-triol	GC	81
Estra-1,3,5(10)-triene-3,16β,17β-triol-3-methyl ether	TLC	181
Estra-1,3,5(10)-trien-3-ol	TLC	181
Estra-1,3,5(10)-trien-3-ol-16α,17α-epoxide	TLC	181
Estra-1,3,5(10)-trien-3-ol,16β,17β-epoxide	TLC	181
1,3,5(10)-Estratrien-3-ol-17-one	GC	136
4-Estrene-3,17-dione	PC	124
Estriol	GC	136
	PC	124
	TLC	178, 181
Estromone(diethyl stilbestrol, no steroid)	TLC	178
Estrone	GC	79, 83, 136
	LC	6, 45
	PC	124
	TLC	95,178, 181
Estrone acetate	TLC	95
Estrone benzoate	TLC	95
Esucos®	TLC	42
Etamycin	TLC	20
Ethanal, 2,4-DNPH of	TLC	72, 76
Ethane	GC	27, 42, 58,112,118, 125,126, 127
1,2-Ethanediol	LC	37
Ethanethiol	GC	56
Ethanol	GC	7, 47, 49, 51, 53, 55, 57, 58, 92, 93, 94, 125
	LC	5, 37, 54
	PC	52
	TLC	1
Ethanol, alcohols	GC	126
Ethanol, butyric acid ester of	GC	51
Ethanol, formic acid of	GC	51
Ethanol, isobutyric acid ester of	GC	51
Ethanol, organics	GC	52
Ethanol, propionic acid ester of	GC	51
Ethanolamine	LC	41
	PC	56, 57, 60, 64
	TLC	5, 13, 15
Ethchlorvynol	TLC	137
Ethchlorvynol (Placidyl®)	GC	21
Ether, di-*n*-butyl	GC	126
Ether, diethyl	GC	126
Ether, diisopropyl	GC	126
Ethide	GC	129
Ethinamal	TLC	137
Ethinamate	GC	19, 105
	TLC	38, 137
Ethinylestradiol	TLC	95
17-Ethinylestradiol	PC	124
17α-Ethinyl-17β-hydroxy-4-androsten-3-one	PC	124
17α-Ethinylpregn-4-en-17β-ol-3-one	TLC	98
Ethiocholanolone	GC	136
Ethion	GC	60, 64, 65, 67, 68, 128, 129
	LC	43
	PC	28
	TLC	80, 81, 84, 172
Ethionine	GC	101
	PC	64
	TLC	13
DL-Ethionine	PC	63
L-Ethionine	LC	59
Ethisterone	TLC	95
Ethoate-methyl	LC	43
Ethoform (Benzocaine)	TLC	43
Ethoheptazine	TLC	39
Ethopabate	TLC	136
Ethopropazine	GC	17, 20, 108, 109
	TLC	139
Ethopropazine hydrochloride	GC	17
Ethotoin	GC	105
Ethoxyacetic acid	TLC	29
2-Ethoxyaniline	PC	6
	TLC	119
4-Ethoxyaniline	GC	102
	TLC	119
p-Ethoxybenzaldehyde, 2,4-DNPH of	TLC	167
Ethoxybenzamide	TLC	40
S-(α-Ethoxycarbonylbenzyl)*O,O*-dimethyl phosphoro-dithioate (Cidial)	TLC	170
2-Ethoxy-6,9-diaminoacridine=rivanol	PC	84
6-Ethoxy-1,2-dihydro-2,2,4-trimethylquinoline	TLC	7
2′-*O*-Ethoxyethyleneuridylyl-(3′-5′)-2′,3′-*O*-ethoxymethyleneuridine	TLC	62
Ethoxyethylmercury dithizonate	TLC	161
3-Ethoxy-4-hydroxybenzaldehyde, 2,4-DNPH of	TLC	167
4-Ethoxy-2-hydroxybenzoic acid	PC	120
4-Ethoxy-3-methoxybenzaldehyde, 2,4-DNPH of	TLC	167

5-Ethoxymethyl-6-methyluracil	PC	36
5-Ethoxymethyluracil	PC	36
5-Ethoxymethyluridine	PC	36
6-Ethoxy-2,2,4-trimethyl-1,2-dihydroquinoline	TLC	22
2,4-D ethyl	TLC	81
2,4,5-T ethyl	TLC	81
Ethyl 701	TLC	175
Ethyl 702	TLC	175
Ethyl 712	TLC	175
Ethyl 720	TLC	175
Ethyl 736	TLC	175
Ethyl 762	TLC	175
Ethylac	TLC	66
Ethyl acetate	GC	2, 47, 49, 53, 55, 57, 58, 121, 125
	LC	70
Ethyl acetate, esters	GC	126
Ethyl acetate, organics	GC	52
Ethyl acrylate	GC	91
Ethyl alcohol	GC	121, 127
Ethyl alkyl guanidine	PC	9
Ethyl-1-alkyl-3-nitroguanidine	PC	9
Ethyl allyl disulfide	GC	85
Ethylamine	GC	97
	LC	26
	PC	7, 55, 56
	TLC	5, 10
Ethylamine, trifluoroacetyl derivative	GC	10
2-Ethyl-1-amino-4-benzoylamino-anthraquinone	TLC	69
N-Ethylaniline	GC	102
	TLC	7
o-Ethylaniline, trifluoroacetyl derivative	GC	9
p-Ethylaniline	GC	10
2-Ethylaniline	GC	46
4-Ethylaniline	GC	45, 46, 100
2-Ethylanisole	GC	75
3-Ethylanisole	GC	75
4-Ethylanisole	GC	75
Ethyl anthranilate	TLC	105
Ethyl β-apo-8′-carotenoate	TLC	94
Ethylbenzene	GC	49, 51, 53, 55, 118, 119, 120, 125
	LC	21, 27, 36, 42
Ethylbenzene, aromatics	GC	126
Ethyl bromide	GC	47
	LC	75
2-Ethyl-1,3-butadiene	GC	118
2-Ethylbutanol-1	GC	51
2-Ethyl-1-butene	GC	118, 122
Ethyl *n*-butyl ether	LC	7
Ethyl butyl ketone	GC	117
Ethyl butyl sulfide	GC	135
Ethyl *sec*-butyl sulfide	GC	135
2-Ethylbutyraldehyde	GC	49, 58
Ethyl butyrate	GC	2, 53
Ethyl *n*-butyrate	GC	57
Ethyl butyrate, organics	GC	52
2-Ethylbutyric acid	TLC	29
Ethyl caprate	TLC	105
Ethyl caproate	TLC	105
Ethyl *n*-caproate	GC	57
Ethyl caprylate	TLC	105
4-Ethylcatechol	TLC	87
Ethyl chloride	GC	127
Ethyl chloride, organics	GC	52
Ethyl chloroformate	GC	47
5-Ethyl-4-chloropicolinic acid methyl ester	TLC	158
24β-Ethyl-5-cholestaen-3β-ol	GC	136
3-Ethylcoumarin	TLC	68
Ethyl crotonate	GC	91
5-Ethyl-5-crotyl-BA	TLC	137
Ethyl cyanide	GC	56
Ethylcyclohexane	GC	35, 42, 118
	LC	42
1-Ethylcyclohexane	GC	41
3-Ethylcyclohexene	GC	41
4-Ethylcyclohexene	GC	41
Ethylcyclopentane	GC	36, 42, 118
1-Ethylcyclopentene	GC	41
3-Ethylcyclopentene	GC	41
4-Ethylcyclopentene	GC	41
S-Ethylcysteine	LC	59
	PC	11, 64
1-Ethyl-1-*cis*-2-dimethylcyclopropane	GC	118
1-Ethyl-1-*trans*-2-dimethylcyclopropane	GC	118
3-Ethyl-2,2-dimethylpentane	GC	42
1-Ethyl-2,5-dimethylpyrrole	GC	133
Ethyl 2,4-dinitrophenylsulfenate	TLC	59
Ethylene (ethene)	GC	26, 29, 58, 112, 118, 122, 125, 126, 127
Ethylenediamine	LC	26
	PC	5, 7
Ethylene dibromide	GC	129
Ethylene dichloride	GC	32, 93, 129
	LC	75
Ethylene glycol	GC	116, 127
	LC	36, 68
	PC	1, 53
	TLC	113
Ethylene glycol (min)	GC	116
Ethylene glycol, butyl ether of	GC	116
Ethylene glycol, ethyl ether of	GC	116
Ethylene glycol, methyl ether of	GC	116
Ethylene glycol, monoisoamyl ether of	GC	116
Ethylene glycol, monoisobutyl ether of	GC	116
Ethylene glycol, monoisopropyl ether of	GC	116
Ethylene glycol, propyl ether of	GC	116
Ethylene oxide	GC	127
Ethylene oxide, organics	GC	52
N-Ethylenethiourea	PC	47
Ethyleneurea	TLC	122
Ethyl esters	GC	24
Ethylestrenol	TLC	95
Ethyl ether	GC	116, 118, 121

Ethyl formate	GC	2, 47, 53, 55, 57, 121
	LC	36
Ethyl formate, organics	GC	52
Ethyl gallate	TLC	125, 174
4-Ethylguaiacol	TLC	87
Ethyl guanidine	PC	9
Ethyl *n*-heptanoate	GC	57
Ethyl heptyl ketone	GC	117
3-Ethylhexane	GC	30, 42, 118, 121
Ethyl hexanoate	GC	2
2-Ethylhexanol-1	GC	51
3-Ethyl-3-hexanol	GC	118
3-Ethyl-2-hexene	GC	118
3-Ethyl-3-hexene	GC	118
2-Ethylhexyl adipate (Adipic acid ester)	TLC	27
2-Ethylhexyl alcohol	TLC	1
2-Ethylhexylamine	TLC	7
2-Ethylhexyl diphenyl phosphate (Phosphoric acid ester)	TLC	27
Ethyl hexyl ester 2,4-D	GC	129
2-Ethylhexyl *p*-hydroxybenzoate	TLC	27
Ethyl hexyl ketone	GC	117
2-Ethylhexyl salicylate	TLC	129
Ethylhexyl sebacate	TLC	27
Ethyl *p*-hydrobenzoate	LC	74
Ethyl hydroperoxide	PC	27
Ethyl *m*-hydroxybenzoate	TLC	174
Ethyl-2-hydroxyethyl sulfoxide	GC	139
Ethylidene-bis(*tert*-butyl) peroxide	PC	27
3,3′-Ethylidene-bis-(4-hydroxycoumarin)	TLC	79
Ethylidenecyclohexane	GC	41
Ethylidenecyclopentane	GC	41
4,6-*O*-Ethylidene-1,2-*O*-isopropylidene-*O*- galactose	GC	13
4,6-*O*-Ethylidene-2,3-*O*-isopropylidene-*D*-mannose	GC	13
3,4-*O*-Ethylidene-1,2-*O*-isopropylidene-6-*O*-toluene-*p*-sulfonyl-D-galactose	GC	13
4,6-*O*-Ethylidene-1,2-*O*-isopropylidene-3-*O*-toluene-*p*-sulfonyl-D-galactose	GC	13
Ethylindole-3-acetate	TLC	156
Ethyl 3-indoleacetate	TLC	59
Ethylindolylacetate	PC	102
Ethyl iodide	GC	47
Ethyl isobutrazine	GC	20, 109
Ethyl isobutyl sulfide	GC	135
Ethyl isobutyrate	GC	2, 57
Ethyl isocyanate	GC	44
N-Ethylisopropylamine	GC	97
Ethyl isopropyl sulfide	GC	135
Ethyl Ledate	TLC	66
Ethyl mercaptan	GC	85
Ethyl mercury dithizonate	TLC	161
Ethyl methacrylate	GC	91
N-Ethyl-*N*-methylaniline	GC	102
N-Ethyl-2-methylaniline	GC	102
N-Ethyl-3-methylaniline	GC	102
N-Ethyl-4-methylaniline	GC	102
2-Ethyl-3-methylaniline	GC	100
2-Ethyl-4-methylaniline	GC	100
2-Ethyl-5-methylaniline	GC	100
3-Ethyl-4-methylaniline	GC	100
3-Ethyl-4-methyl-4-aza-2,5-cholestadiene	TLC	96
2-Ethyl-3-methyl-1-butene	GC	122
Ethyl 3-methylbutyrate	GC	2
1-Ethyl-1-methylcyclopentane	GC	42
1-Ethyl-*cis*-2-methylcyclopentane	GC	42
1-Ethyl-*trans*-2-methylcyclopentane	GC	42
1-Ethyl-*cis*-3-methylcyclopentane	GC	42
1-Ethyl-*trans*-3-methylcyclopentane	GC	42
3-Ethyl-2-methylpentane	GC	42
3-Ethyl-3-methylpentane	GC	42
Ethyl 4-methylpentanoate	GC	2
2-Ethyl-4-methylphenol	GC	69
2-Ethyl-5-methylphenol	GC	69, 72
3-Ethyl-2-methylphenol	GC	72
3-Ethyl-5-methylphenol	GC	69, 71, 72
4-Ethyl-2-methylphenol	GC	69, 72
4-Ethyl-3-methylphenol	GC	69, 72
5-Ethyl-2-methylphenol	GC	69, 72
5-Ethyl-3-methylphenol	PC	31
5-Ethyl-3-methyl-5-phenylhydantoin (Mephenytoin)	GC	105
3-Ethyl-4-methylpyridine	GC	131
Ethylmorphine	GC	8, 18
	PC	4
	TLC	4, 138
Ethylmorphine (Dionin®)	GC	21, 107
Ethylmorphine HCl (Dionin®)	GC	107
	TLC	138
N-Ethylmorpholine	TLC	7
N-Ethyl-4-morpholine thiocarboxamide	PC	47
1-Ethylnaphthalene	GC	38
	TLC	53
2-Ethylnaphthalene	GC	38
	TLC	51
1-Ethyl-3-nitroguanidine	PC	9
1-Ethyl-3-(5-nitro-2-thiazolyl)-urea	TLC	136
Ethyl nonyl ketone	GC	117
Ethyl octyl ketone	GC	117
Ethyl octyl sulfide	TLC	163
Ethyl ortophosphate	PC	35
Ethyl parathion	GC	68, 128
	TLC	172
3-Ethylpentane	GC	42,118, 121
	LC	42
3-Ethylpentane, paraffins	GC	33
Ethyl pentanoate	GC	2
2-Ethyl-1-pentene	GC	122
3-Ethyl-1-pentene	GC	122
3-Ethyl-2-pentene	GC	122
2-Ethylphenetole	GC	75
3-Ethylphenetole	GC	75
4-Ethylphenetole	GC	75
m-Ethylphenol	GC	73
o-Ethylphenol	GC	73
p-Ethylphenol	LC	52
	PC	121
2-Ethylphenol	GC	69, 72, 75
	TLC	174

3-Ethylphenol	GC	69, 72, 75
	TLC	174
4-Ethylphenol	GC	69, 71, 72, 75
	TLC	87, 174
N',*N''*-Ethylphenyl thiourea	PC	47
Ethylphthalyl ethyl glycolate	TLC	27
5-Ethylpicolinic acid amide	TLC	158
5-Ethylpicolinic acid ethyl ester	TLC	158
5-Ethylpicolinic acid methyl ester	TLC	158
(-)-5-Ethylpicolinic acid α-phenylethylamide	TLC	158
5-Ethyl-5-piperidino-BA	TLC	137
Ethyl propionate	GC	2, 47, 55, 57
2-Ethylpropionic acid	TLC	29
N-Ethyl-*n*-propylamine	GC	97
Ethyl *n*-propyl hydrogen phosphate	PC	34
Ethylprotecatechuate	TLC	125
Ethyl protocatechuate	TLC	174
2-Ethylpyridine	GC	131
	TLC	60
4-Ethylpyridine	GC	131
Ethyl pyrophosphate	PC	35
1-Ethylpyrrole	GC	133
2-Ethylpyrrole	GC	133
N-Ethyl-1-pyrrolidine thiocarboxamide	PC	47
Ethylquinine carbonate	TLC	40
Ethyl styryl ketone	TLC	73
Ethyl sulfoxide	GC	139
5-Ethylthiobarbituric acid	PC	47
2-Ethylthiophene	TLC	104
N-Ethylthiourea	PC	47
m-Ethyltoluene	GC	120
o–Ethyltoluene	GC	120
p-Ethyltoluene	GC	120
α-Ethyl-β-(2,4,6-triiodo-3-hydroxyphenyl)-propionic acid	TLC	64
α-Ethyltryptamine	PC	103
α-Ethyltryptamine (as acetate)	PC	102
N-Ethyltryptamine	PC	102, 103
N-Ethyltryptamine (as HCl)	PC	102
Ethyl Tuads	TLC	66
Ethylurea	LC	63
	TLC	122
Ethyl *n*-valerate	GC	57
Ethyl vanillin, 2,4-DNPH of	TLC	73
Ethyl vanillin (4-hydroxy-3-ethoxy-benzaldehyde)	TLC	68
Ethyl Zimate	TLC	66
Ethynodiol diacetate	TLC	95
Ethynylcyclohexyl carbamate (Valmid®)	GC	21
Ethynylestradiol acetate	TLC	95
Ethynylestradiol benzoate	TLC	95
Ethyrosin	TLC	141
Etiocholane-3,17-dione	TLC	178
Etiocholane-3α-ol-17-one	TLC	178
Etiocholane-3β-ol-17-one	TLC	178
Etiocholanolone	GC	83
	TLC	97
11β-OH-Etiocholanolone	TLC	97
Etioporphyrin	PC	33
Eucaine	GC	107
Eucaine HCl	GC	107
β-Eucaine	TLC	43
Eucodal (Oxycodone)	TLC	116
Eugenol	GC	71
	TLC	90
Eugenol benzyl ether	TLC	90
Eugenol methyl ether	TLC	90
Euglenanone	TLC	93
Eukodal®	PC	2
Eu (Europium)	LC	39, 79, 90, 91
Eu(III), Europium	PC	96
Euvernil® (Sulfacarbamide)	TLC	140
Evatromonoside	TLC	102
Evatromonoside peracetate	TLC	102
Evernic acid	PC	30
	TLC	35
Everninic acid	PC	30
EXD	GC	65
F (Fluorine)	GC	26
	LC	85, 86, 87, 88
	TLC	147
FAC	TLC	84
Famphur	GC	65
	TLC	171
β-Farnesene	GC	140
cis, trans-Farnesol	LC	45
trans, trans-Farnesol	LC	45
Fast Acid Red E	TLC	49
Fauronyl acetate	GC	141
Fe (Iron)	LC	81, 82, 83, 84, 85, 86, 87, 88, 89
Fe(II), (III), Iron	GC	43
	PC	15, 95, 96
	TLC	148, 149
Felinine	PC	64
α-Fenchene	GC	86, 87
Fenchlorphos	LC	43
Fenchone	GC	87
	TLC	77
Fenchone, 2,4-DNPH of	TLC	76
Fenchyl alcohol (ONPU)	TLC	111
α-Fenchyl alcohol	GC	87
β-Fenchyl alcohol	GC	87
Fenitrothion	LC	43
	TLC	170, 171
Fenoprop butyl	TLC	81
Fenoprop methyl	TLC	81
Fenson	GC	65
Fenthion	GC	65, 67, 128
	LC	43
Fenthion (Bayer 29493)	TLC	84, 170, 171
Fenthion O-analog	GC	65
Fenthion O-analog sulfone	GC	65
Fenthion O-analog sulfoxide	GC	65
Fenthion sulfone	GC	65
Fenthion sulfoxide	GC	65
Fenuron	TLC	81, 173
Ferbam (Ferric dimethyldithiocarbamate)	PC	117
Ferric-2-chloro-3-hydroxy-1,4-naphthoquinone	PC	111
Ferric dimethyldithiocarbamate (Ferbam)	PC	117

Ferricyanide	PC	14, 91, 93
Ferrocyanide	PC	14, 91, 93
Ferulic acid	GC	90
	PC	58, 71, 113
	TLC	133,135, 174
Ferulic acid, 4-acetyl-2-nitrophenol ether of	TLC	91
1-Feruloyl-β-D-glucose	PC	71
2-Feruloylglucose	PC	71
3-Feruloylglucose	PC	71
4-Feruloylglucose	PC	71
6-Feruloylglucose	PC	71
Feruoylglycine	PC	80
Festuclavine	TLC	114
Fisetin	LC	1, 28
	PC	26
	TLC	92
Flavanone	LC	28
	TLC	92
Flavanone aglycones	PC	26
Flavone	LC	28
Flavone, aglycones	PC	26
Flavone, glycosides	PC	26
Flavonol aglycones	PC	26
Flavonol glycosides	PC	26
Flumethiazide	TLC	41
Fluometuron	TLC	81, 173
Fluoranthene	GC	40
	PC	89
	TLC	7, 51, 53, 120, 146
Fluoranthrene	LC	27
Fluorene	GC	40
	TLC	7, 51, 53
Fluoren-9-one	TLC	75
Fluorescein	PC	88
Fluoride	PC	14
Fluorine	GC	26
	LC	85, 86, 87, 88
	TLC	147
Fluorine monoxide	GC	114
Fluoroacetate acid	TLC	36
2-Fluoroaniline	GC	102
3-Fluoroaniline	GC	102
4-Fluoroaniline	GC	102
1-Fluoroanthraquinone	PC	110
2-Fluoroanthraquinone	PC	110
Fluorobenzene	GC	51, 53
2-Fluorobutylphenol	GC	73
1-Fluoro-2-chloroanthraquinone	PC	110
5-Fluorodeoxyuridine	PC	36
Fluorodichloromethane	GC	26
Fluoroethylene (vinyl fluoride)	GC	26
Fluoromethane	GC	26
3-Fluoro-4-methylaniline	GC	102
4-Fluoro-2-methylaniline	GC	102
o-Fluoroperoxybenzoic acid	PC	107
p-Fluoroperoxybenzoic acid	PC	107
m-Fluorophenol	LC	52
p-Fluorophenol	LC	52
4-Fluorophenol	GC	71
4-Fluoro-6-propionylphenol	TLC	89
2-Fluoropyridine	TLC	60
Fluorotrichloromethane	GC	26
5-Fluorotryptamine	PC	101
5-Fluorouracil	PC	36
Fluoxymestrone	TLC	95
Fluphenazine	GC	17, 20, 108, 109
	TLC	139
Fluphenazine dihydrochloride	GC	17
FMN	LC	18
Folic acid	LC	30
	PC	51, 84
	TLC	188
Folpet	GC	65, 68, 129
	TLC	81
Formaldehyde	GC	49
	LC	8
	PC	20
Formaldehyde diethyl acetal, organics	GC	52
Formaldehyde dimethyl acetal, organics	GC	52
Formaldehyde, 2,4-DNPH of	TLC	166
Formaldehyde, 2-hydrazinebenzothiazole derivative of	TLC	168
Formamide	LC	71
N-Formamidino-hydroxyproline	PC	9
N-Formamidino-proline	PC	9
Formate	PC	14
Formic acid	GC	127
	LC	34, 65, 71
	PC	77, 79
	TLC	29, 134
Formic acid, benzyl ester of	GC	24
Formic acid, cinnamyl ester of	GC	24
Formic acid, ethyl ester of	GC	24
Formic acid, hexyl ester of	GC	24
Formiminoglycine	TLC	13
Formononetin	PC	26
Formothiazole	TLC	140
Formothion	TLC	171
5-Formylamino-3-methyl-6-methylaminouracil	PC	36
5-Formylcytosine	PC	36
5-Formyldeoxyuridine	PC	36
1-Formyl-2,2-dimethylhydrazine	PC	61
2-Formylphenol	TLC	89
4-Formylphenol	TLC	89
16-Formyl-strophadogenin peracetate	TLC	102
16-Formyl-strophadogenin-(3β,5β,14β-Trihydroxy-16β-formyloxy-19-oxo-cardenolide	TLC	102
N-Formyl-*dl*-tryptophan	PC	63, 101
N-Formyl-*l*-tyrosine	PC	63
5-Formyluracil	PC	36
5-Formyluridine	PC	36
5-Formylvanillin	PC	20
Foromacidin A	PC	66
	TLC	153
Foromacidin B	PC	66
	TLC	153
Foromacidin C	PC	66
	TLC	153
Fortified phenol	TLC	175

Fradiomycin (sulfate) TLC 20
Fraxetin PC 113
Fraxin PC 113
Free Amine GC 98
Frenguel (Azacyclonol HCl) GC 108
Freon 22 ($CHClF_2$) GC 26
Freon 23 GC 26
Fructose LC 25, 29, 47, 48, 71
PC 32, 70
TLC 24, 26, 127
Fructose-1,6-diphosphate PC 32
TLC 128
Fructose-1-phosphate PC 32
TLC 128
Fructose-6-phosphate PC 32
TLC 128
D-Fructose GC 15
TLC 126
D-Fructose-1,6-diphosphate PC 12
D-Fructose-6-phosphate PC 12
Fructuronic acid PC 76
Frugoside TLC 102
Frugoside peracetate TLC 102
Fucitol LC 69
D-Fucitol GC 14
Fucose LC 25, 48
PC 70
TLC 23
L-Fucose GC 15
PC 75
Fucosterol GC 84
29-*iso*Fucosterol GC 84
29-*iso*Fucosterol-
(Δ^5 *Avena sterol*) GC 84
Fucosyllactose, reduced,
trisilyl derivative GC 124
Fucoxanthin PC 122
TLC 93
Fumarate acid TLC 36
Fumaric acid LC 58, 65
PC 113
Fumarprotocetraric acid PC 30
TLC 34
Fumigaclavine A TLC 114
Fumigaclavine B TLC 114
Fumigaclavine C TLC 114
Furan GC 55
LC 60
Furazolidone TLC 136
Furfural (Furfuraldehyde) PC 20
TLC 77
Furfural, 2,4-DNPH of TLC 72, 73, 74
Furfural, 4-NPH of TLC 74
6-Furfurylaminopurine LC 46
PC 36
5-Furfuryl-5-isopropyl-BA TLC 137
Furoic acid, methyl ester of GC 24
2-Furoic acid LC 40

Adsorption of the Elements
(LC Section – Figures)

F, Fluorine Pg 278, 279, 280, 281
Fe, Iron Pg 278, 279, 280, 281
Fr, Francium Pg 278, 279, 280, 281

Ga(III), Gallium GC 43
LC 82, 83, 84, 85, 86, 87
PC 15, 96
TLC 149
Gadolinium (Gd) LC 39, 79, 83, 85, 86, 90, 91
Galactitol GC 14
LC 69
D-Galactitol GC 76
D-Galactone γ-lactone PC 72
D-Galactonic LC 40
Galactonolactone PC 76
Galactosamine LC 41
PC 64
TLC 127
Galactose LC 25, 29, 48, 71
PC 70, 73
TLC 23, 24
D-Galactose GC 15
TLC 126
Galacturonic A PC 70, 76
Galacturonic acid PC 72
D-Galacturonic acid LC 40
TLC 126
Galanthamine GC 96
Galanthine GC 96
Galgal NH_2 GC 124
Gallic acid PC 30, 82
TLC 125,133, 135,174, 176
Gallic acid methyl ester PC 121
Gallium, Ga(III) GC 43
LC 82, 83, 84, 85, 86, 87
PC 15, 96
TLC 149
Gallium, Ga^{3+} PC 97
Gallocatechin-α-gallate PC 30
DL-Gallocatechin PC 30
L-*epi*-Gallocatechin PC 30
L-*epi*-Gallocatechin gallate PC 30
Galuteolin (*Flavone Glycosides*) PC 26
Gamabufotalin peracetate TLC 102
Gamabufotalin (3β,11α,14β-
Trihydroxy-5β-bufadienolide) TLC 102
Gardona® GC 65
Gd (Gadolinium) LC 39, 79, 83, 85, 86, 90, 91
GDP (Guanosine diphosphate) TLC 63
GDPM TLC 63
Ge(IV), Germanium LC 83
PC 96
Geigy G-28029 GC 65
Gelsemine GC 8, 107
PC 2
Gelsemine sulfate GC 107
Genatropine TLC 117
Genista alkaloids PC 3
Genistein PC 26
Genite® GC 65
TLC 80
Genohomatropine TLC 117
Genoscopolamine TLC 117

Genotropanol	TLC	117
Gentamicin	TLC	20
Gentianose	LC	48
Gentiobiose	GC	124
Gentisic acid	PC	82
	TLC	133,135, 174
Gentrogenin	TLC	103
Gentrogenin 3-acetate	TLC	103
Geraniol	TLC	183
Geraniol (DNB)	TLC	111
Geraniol (ONPU)	TLC	111
Geranyl acetate	TLC	105
Geranyl butyrate	TLC	105
Geranyl formate	TLC	105
Geranyl propionate	TLC	105
Germanium, Ge(IV)	LC	83
	PC	96
Germanium tetrachloride	GC	50
Gibberellin A_1	LC	45
	TLC	144
Gibberellin A_2	TLC	144
Gibberellin A_3	LC	45
	TLC	144
Gibberellin A_4	LC	45
	TLC	144
Gibberellin A_5	LC	45
	TLC	144
Gibberellin A_6	TLC	144
Gibberellin A_7	LC	45
	TLC	144
Gibberellin A_8	TLC	144
Gibberellin A_9	LC	45
	TLC	144
Gibberellin A_1, methyl ester of	TLC	145
Gibberellin A_2, methyl ester of	TLC	145
Gibberellin A_3, methyl ester of	TLC	145
Gibberellin A_4, methyl ester of	TLC	145
Gibberellin A_5, methyl ester of	TLC	145
Gibberellin A_6, methyl ester of	TLC	145
Gibberellin A_7, methyl ester of	TLC	145
Gibberellin A_8, methyl ester of	TLC	145
Gibberellin A_9, methyl ester of	TLC	145
Girard Reagent	PC	20
Gitaloxin	PC	127
	TLC	102, 177
Gitaloxin peracetate	TLC	102
Gitaloxigenin	TLC	177
Gitaloxigenin (3β,14β-Dihydroxy-16β-formyloxy-5β-cardenolide)	TLC	102
Gitaloxigenin peracetate	TLC	102
Gitogenin	TLC	98, 103
Gitorin	PC	127
Gitoside	TLC	102
Gitoside peracetate	TLC	102
Gitoxigenin	PC	127
	TLC	177
Gitoxigenin diacetate	PC	127
Gitoxigenin glycoside	PC	127
Gitoxigenin-mono-digitoxose	TLC	177
Gitoxigenin peracetate	TLC	102
Gitoxigenin (3β,14β,16β-Trihydroxy-5β-cardenolide)	TLC	102
16-*epi*-Gitoxigenin (3β,14β, 16α-Trihydroxy-5β-cardenolide)	TLC	102
Gitoxin	PC	127
	TLC	102, 177
16-*epi*-Gitoxin	TLC	102
Gitoxin peracetate	TLC	102
Glaucamine	PC	3
Glebomycin	TLC	20
Globucid (sulfaethidol)	TLC	140
γ-Globulin (human)	LC	19, 64
Glucitol	LC	69
D-Glucitol	GC	14, 15
Gluco-Dg-bisdigitoxoside	TLC	102
Gluco-Dg-bisdigitoxoside peracetate	TLC	102
Glucocanaridigitoxoside	TLC	102
Glucocanaridigitoxoside peracetate	TLC	102
Glucodigifucoside	TLC	102
Glucodigifucoside peracetate	TLC	102
Glucoevatromonoside	TLC	102
Glucoevatromonoside peracetate	TLC	102
Glucogitaloxin	PC	127
	TLC	102, 177
Glucogitaloxin peracetate	TLC	102
Glucogitoroside	TLC	102
Glucogitoroside peracetate	TLC	102
D-Glucoheptono-γ-lactone	PC	72
α-D-Glucoisosaccharinic	LC	40
β-D-Glucoisosaccharinic	LC	40
Glucolanadoxin	TLC	102
Glucolanadoxin peracetate	TLC	102
α-D-Glucometasaccharinic	LC	40
β-D-Glucometasaccharinic	LC	40
Dg-β,D-glucomethyloside	TLC	102
Dg-β,D-glucomethyloside peracetate	TLC	102
D-Gluconic	LC	23, 40
Gluconolactone	PC	76
D-Glucono-γ-lactone	PC	72
D-Glucono-δ-lactone	PC	72
Glucopyranoside, methyl ethers	GC	28
1-β-D-Glucopyranosyl-4-methoxy-2(1*H*)-pyrimidone 2′,3′,4′,6′-tetraacetate	GC	134
1-β-D-Glucopyranosyl-3-methylthymine 2′,3′,4′,6′-tetraacetate	GC	134
1-β-D-Glucopyranosyl-3-methyluracil 2′,3′,4′,6′-tetraacetate	GC	134
1-β-D-Glucopyranosylthymine 2′,3′,4′,6′-tetraacetate	GC	134
1-β-D-Glucopyranosyluracil 2′,3′,4′,6′-tetraacetate	GC	134
α-D-Glucosaccharinic	LC	40
Glucosamine	LC	4, 41
	PC	56, 64
	TLC	13, 20, 127
Glucosaminic acid	PC	64
Glucose	LC	4, 25, 29, 48

Glucose (*continued*) PC 32, 71, 74
TLC 23, 24, 26
D-Glucose LC 64, 71
PC 72
TLC 126, 127
L-Glucose LC 71
α-D-Glucose GC 15
β-D-Glucose GC 15
Glucose-1-phosphate PC 32
TLC 128
Glucose-6-phosphate PC 32
TLC 128
D-Glucose-1-phosphate PC 12
D-Glucose-6-phosphate PC 12
Dg-β,D-glucoside TLC 102
Dg-β,D-glucoside peracetate TLC 102
Glucosides, aurone PC 26
Glucosides, chalcone PC 26
Glucosides, isoflavone PC 26
Glucosyl-(1→2)-6-*p*-cumaroyl-β-D-glucose from *Raphanus* PC 71
Glucosyl-(1→2)-6-feruloyl-β-D-glucose from *Raphanus* PC 71
Gluco-verodoxin TLC 102
Gluco-verodoxin peracetate TLC 102
Glucuronic A PC 70
Glucuronic acid LC 71
PC 72, 76
D-Glucuronic acid LC 40
TLC 126, 127
Glucuronolactone PC 76
TLC 127
Glutamate acid TLC 36
l-Glutamate acid TLC 36
Glutamic acid GC 12, 101
LC 2, 4, 12, 33, 41, 56
PC 11, 62, 64, 65
TLC 11, 13, 15, 16, 118
Glutamic acid, DNAP of TLC 14
Glutamic acid, PTH of TLC 124
Glutamidiol PC 57
Glutamine LC 41
PC 11, 62, 64
TLC 11, 13, 15, 16
iso-Glutamine PC 64
Glutamine, PTH of TLC 124
Glutaric acid LC 65
PC 13, 77
TLC 31
α-G-Glutaric acid PC 9
Glutaric acid, dimethyl ester of GC 24
Glutathione PC 11
Glutathione (oxidized) LC 32, 41
Glutathione (reduced) LC 41
Glutethimide GC 17, 19, 105
TLC 137
Glutethimide (Doriden®) GC 21
Gly-Asp LC 4
Glybutamide TLC 140
Glybuthiazole TLC 140
Glyceraldehyde TLC 23
Glyceraldehyde (dimer) LC 29
Glyceraldehyde, 2,4-DNPH of TLC 166
Glyceraldehyde (monomer) LC 29
Glyceraldehyde 3-phosphate (GAP) PC 32
D-Glyceraldehyde-3-phosphate PC 12
DL-Glyceraldehyde GC 15
Glycerate (Ca) PC 13
Glyceric acid LC 40
PC 13
TLC 29
Glycerin TLC 113
meso-Glycero-alloheptitol GC 14
D-Glycero-D-altroheptitol GC 14
D-Glycero-D-galactoheptitol GC 14
D-Glycero-D-glucoheptitol GC 14
L-Glycero-D-glucoheptitol GC 14
meso-Glycero-guloheptitol GC 14
meso-Glycero-idoheptitol GC 14
D-Glycero-D-mannoheptitol GC 14
Glycerol LC 5, 69
PC 1, 14, 53, 76
TLC 1
Glycerol-α-phosphate TLC 128
Glycerol-β-phosphate TLC 128
Glycerol-1-phosphate PC 12, 32
Glycerol-2-phosphate PC 32
Glycerol phosphoryl myoinositol LC 47
Glycerol triacetate TLC 27
Glycerol tripropionate GC 132
β-Glycerophosphate PC 99
Glyceryl guaiacolate GC 104
N^6-β-Glyceryllysine TLC 123
Glycerylphosphorylcholine PC 99
Glycine GC 12, 101
LC 2, 4, 12, 14, 33, 34, 41, 58
PC 11, 62, 63, 64, 65
TLC 11, 12, 13, 15, 16, 29, 118
Glycine, DNAP of TLC 14
Glycine, PTH of TLC 124
Cbo-Glycine TLC 12
Cbo-Glycine-DL-Alanine TLC 12
Cbo-Glycine-DL-Alanine-Glycine TLC 12
Cbo-Glycine-L-Glutamic acid TLC 12
Glycine-Glycine TLC 12
Cbo-Glycine-L-Isoleucine TLC 12
Glycine-L-Leucine TLC 12
Cbo-Glycine-L-Leucine TLC 12
Cbo-Glycine-Glycine-L-Leucine TLC 12
Cbo-Glycine-DL-Phenylalanine TLC 12
Cbo-Glycine-Glycine-DL-Phenylalanine ethyl ester TLC 12
Cbo-Glycine-DL-Phenylalanine-Glycine ethyl ester TLC 12
Glycoaldehyde (dimer) LC 29
Glycoaldehyde (monomer) LC 29
Glycochenodeoxycholic PC 39
Glycocholic PC 39
Glycocyamine TLC 11
Glycodeoxycholic PC 39
Glycol GC 76

Glycolaldehyde GC 15
PC 20
Glycolaldehyde, 2,4-DNPH of TLC 166
Glycol-bis-cyanoethylether LC 68
Glycolate acid TLC 36
Glycolate (K) PC 13
Glycolic acid LC 40
PC 13
TLC 29
Glycollic acid, 2,4-DNPH of TLC 166
Glycosides, Cyanidin PC 24
Glycosides, Delphinidin PC 24
Glycosides, Flavone PC 26
Glycosides, Flavonol PC 26
Glycosides, Malvidin PC 24
Glycosides, Pelargonidin PC 24
Glycosides, Peonidin PC 24
Glycosides, Petunidin PC 24
Glycyl-alanine GC 16
Glycyl-L-3,5-diiodotyrosine PC 114
Glycyl-3,5-diiodo-L-tyrosyl-glycine PC 114
Glycylglycine GC 16
PC 63
Glycyl-phenylalanine GC 16
Glycyl proline GC 16
Glycyl-L-thyroxine PC 114
Glycyl-L-thyroxyl-glycine PC 114
Glycyl-L-tyrosine PC 114
Glycyl-L-tyrosyl-glycine PC 114
Glycyl-valine PC 114
Glyoxal (bis) PC 20
Glyoxal, 2,4-DNPH of TLC 166
Glyoxalate acid TLC 36
Glyoxylic acid LC 40
PC 81
TLC 29, 32
Glyprothiazole TLC 140
Glysobuzole TLC 140
GMP LC 3
PC 37
TLC 160
dGMP LC 18
TLC 63
5′-GMP LC 22, 46
TLC 63
Gnoscopine GC 96
Gold, Au LC 80, 83, 86
PC 15
TLC 55
Gold, Au(III) PC 95, 96
Gossypetin PC 26
Gossypin PC 26
Gossypitrin PC 26
Gramicidin TLC 21
Gramine PC 102, 103
TLC 156
Gramine-*N*-oxide PC 101
Grayanic acid TLC 35
Green S TLC 49
Griseofulvin TLC 21
GTP TLC 63
d-GTP TLC 63
Guaiacol LC 1
TLC 87, 90
Guaiacol, 4-acetyl-2-nitrophenol ether of TLC 91
Guaiacyl acetic acid TLC 135
Guaiacyl propionic acid TLC 135
Guaiazulene TLC 106, 143
Guaicol (Guaiacol) TLC 174
Guaiene TLC 106, 143
Guanethidine LC 55
TLC 39
Guanethidine N-oxide LC 55
Guanidines, alkyl PC 9
Guanidines, carboxyalkyl (guanidino-acids) PC 9
Guanidines, hydroxy-*n*-alkyl PC 9
Guanidines, miscellaneous PC 9
Guanidino-acids PC 9
Guanidino-alkyl guanidines (α,ω-diguanidino-alkanes) PC 9
1-Guanidino-2(4-imidazolyl)-propionic acid PC 17
Guanine GC 77
LC 3, 15, 17, 22, 31, 46, 50
PC 36, 37, 108
TLC 159
Guanosine GC 77
LC 3, 13, 15, 22, 31, 46, 50
PC 36, 37, 108
TLC 62, 159
Guanosine diphosphate (GDP) PC 32
TLC 160
Guanosine 5′-diphosphate PC 108
Guanosine 2′ and 3′-monophosphates GC 77
PC 108
Guanosine-2′-monophosphate (2′-GMP) TLC 160
Guanosine-3′-monophosphate (3′-GMP) TLC 160
Guanosine-5′-monophosphate (5′-GMP) GC 77
LC 15
PC 108
TLC 160
2-d-Guanosine-5′-monophosphate (2-d-5′-GMP) TLC 160
Guanosine-3′-phosphate (GMP-3′) PC 32
Guanosine-5′-phosphate (GMP-5′) PC 32, 36
TLC 62
Guanosine triphosphate (GTP) PC 32
TLC 160
Guanosine 5′-triphosphate PC 108
Guanylic acid LC 13
Guiiavolic TLC 184
Guinea green TLC 141
D-Gulonic acid LC 40
Gulonolactone PC 76
Gulose LC 48
Guluronic acid PC 72
L-Guluronic acid LC 40
α-Gurjunene GC 140
TLC 54
Gusathion (Guthion® and Azinphosmethyl) TLC 80, 84
Guthion® GC 60, 61, 64, 68, 128, 129

Guthion® (Azinphosmethyl) (Gusathion®)	PC	28
	TLC	80, 84
Guthion oxygen analog	GC	61
Gyrophoric acid	TLC	35

Adsorption of the Elements (LC Section – Figures)

Ga, Gallium	Pg	278, 279, 280, 281
Ge, Germanium	Pg	278, 279, 280, 281

Haematoporphyrin IX	PC	123
Haematoporphyrin IX dimethyl ester	PC	123
Haemin	PC	33
Hafnium (Hf)	LC	80, 84, 85, 86, 88, 89
Haloperidol	TLC	42
Haloxon	LC	43
	TLC	81
Harmalol	PC	101
Harmane	PC	101
Harmine	PC	101
Harmol	PC	101
HCl-Glycine ethyl ester	TLC	12
HCl-DL-Phenylalanine ethyl ester	TLC	12
Hecogenin	TLC	98, 103
Hecogenin acetate	TLC	98
Hecogenin 3-acetate	TLC	103
Helium	GC	112
Hellebrigenin peracetate	TLC	102
Hellebrigenin (3β,5β,14β-Trihydroxy-19-oxo-bufadienolide)	TLC	102
Hellebrigenol peracetate	TLC	102
Hellebrigenol (3β,5β,14β,19-Tetrahydroxy-bufadienolide)	TLC	102
Heloniogenin	TLC	103
Helveticoside	TLC	102
Helveticoside peracetate	TLC	102
Helveticosol	TLC	102
Helveticosol peracetate	TLC	102
Hematoporphyrin	PC	33
Heminporphyrin	PC	33
Hempa	GC	65
Heptabarbital (Medomin)	GC	19, 106
	TLC	137
Heptabarbitone	PC	86
Heptachlor	GC	60, 68, 128, 129
	PC	116
	TLC	80, 82
Heptachlor epoxide	GC	60, 68, 128, 129
	PC	116
	TLC	80, 82
Hepta-condensed phosphate	TLC	151
Heptadecanal, dimethyl acetal of	GC	6
Heptadecanoic acid	PC	77
Heptafluorobutyryl, HFBu	GC	98
n-Heptaldehyde	LC	70
Heptameta-condensed phosphate	TLC	151
Heptamethylenediamine, trifluoroacetyl derivative	GC	10
Heptanal	GC	57
Heptanal, 2,4-DNPH of	TLC	72, 76, 165
Heptanal, 2,4-DNPSC of	TLC	165
Heptanal, *p*-NPH of	TLC	165
n-Heptan-3,3'-dihydroperoxide	PC	27
Heptane	GC	49, 51, 58, 125, 126
n-Heptane	GC	29, 42, 47, 53, 55, 118, 120, 121
	LC	21, 27, 38, 42
n-Heptane, organics	GC	52
n-Heptane, paraffins	GC	33
n-Heptanethiol	GC	56
Heptanoic acid	LC	61
	PC	11
Heptanoic acid, allyl ester of	GC	24
Heptanoic acid, ethyl ester of	GC	24
Heptanoic acid, methyl ester of	GC	24
n-Heptanoic acid	LC	57
	PC	79
	TLC	134
Heptanol	GC	49, 53
n-Heptanol	GC	92
	LC	37, 54
1-Heptanol	PC	52
	TLC	1
1-Heptanol (DNB)	TLC	111
1-Heptanol (ONPU)	TLC	111
2-Heptanol (DNB)	TLC	111
2-Heptanol (ONPU)	TLC	111
3-Heptanol	LC	54
4-Heptanol (DNB)	TLC	111
4-Heptanol (ONPU)	TLC	111
Heptan-2-one, 2,4-DNPH of	TLC	76
2-Heptanone	GC	51, 53, 125
2-Heptanone, ketones	GC	126
3-Heptanone	GC	51, 53
4-Heptanone	GC	51, 53
2,2',3,3',4',6,6'-Hepta-*O*-acetyl-α-maltose	TLC	25
2,2',3,3',4',6,6'-Hepta-*O*-acetyl-β-maltose	TLC	25
2-Heptenal, 2,4-DNPH of	TLC	165
Heptene	TLC	51
Heptene-1	GC	51, 53, 56
	LC	21, 42
1-Heptene	GC	118, 120, 122
cis-2-Heptene	GC	118, 120, 122
trans-2-Heptene	GC	118, 120, 122
cis-3-Heptene	GC	118, 120, 122
trans-3-Heptene	GC	118, 120, 122
Heptoic acid	TLC	132
D-*glycero*-D-*gulo*-Heptonic	LC	40
D-*glycero*-L-*manno*-Heptonic	LC	40
n-Heptyl alcohol	LC	11
Heptyl aldehyde, 2-hydrazinebenzothiazole derivative of	TLC	168
n-Heptylamine	PC	5, 56
n-Heptylamine, trifluoroacetyl derivative	GC	10
1-*n*-Heptyl-2,5-dimethylpyrrole	GC	133
Heptyl esters	GC	24

n-Heptyl hydroperoxide, 3-methylhexyl-3-hydroperoxide	PC	27
n-Heptyl-2-hydroxyethylperoxide	PC	27
n-Heptyl-*n*-nonyl succinate	GC	25
Hercules 14503	GC	67
Herniarin	PC	113
Heroin (Diacetylmorphine)	GC	18, 111
	PC	3, 54
	TLC	39, 138
Hesperetin	LC	1, 28
Hesperidin	LC	1, 28
Hesperitin	PC	26
Hetacillin K	TLC	21
Heterocyclic acids	LC	40
2′,3,3′,4′,6,6′-Hexaacetyl-*β*-maltose	TLC	25
Hexa-*O*-acetylxylobiose	TLC	25
Hexacarboxylic porphyrin	PC	33
Hexachlorobenzene	GC	31, 129
	TLC	81, 82
Hexachlorocyclopentadiene	GC	129
Hexachloroendomethylenetetrahydrophthalic anhydride, *p*-anisidide of	TLC	30
Hexa-condensed phosphate	TLC	151
cis-Hexacosen-9-oic acid, methyl ester of	TLC	130
Hexadecanal (palmitaldehyde), dimethyl acetal of	GC	6
n-Hexadecane	GC	38
	LC	27
Hexadecanoic acid	GC	24
	LC	68
Hexadecanoic acid, methyl ester of	TLC	130
Hexadecanol	TLC	56
n-Hexadecanol	GC	92
Hexadecenal (palmitoleic aldehyde), dimethyl acetal of	GC	6
2-Hexadecenoic acid (16)	PC	77
9-Hexadecenoic acid (16)	PC	77
cis-Hexadecen-9-oic acid, methyl ester of	TLC	130
cis-Hexadecen-11-oic acid, methyl ester of	TLC	130
n-Hexadecylamine	PC	55
Hexadecylbenzene	TLC	53
Hexadecyl palmitate	TLC	56
1,5-Hexadiene	LC	42
Hexafluorocyclopropane	GC	26
sym.-Hexafluorodiethyl ether, organics	GC	52
Hexafluoroethane	GC	26
Hexafluoropropene	GC	26
Hexahydrophthalic anhydride, *p*-anisidide of	TLC	30
Hexahydropyrazine	TLC	136
Hexahydroxyhexakis (ethylenediamine) tetra cobalt hexachloride $[Co_4(en)_6(OH)_6]Cl_6$	PC	94
Hexameta-condensed phosphate	TLC	151
Hexamethylbenzene	TLC	7
Hexamethylenediamine	PC	5, 7
Hexamethylenediamine, trifluoroacetyl derivative	GC	10
Hexamethylenetetramine	TLC	22
Hexa-*O*-methylmyricitin	LC	28
Hexammine cobalt chloride $[Co(NH_3)_6]Cl_3$	PC	94
Hexanal, 2,4-DNPH of	TLC	72, 76, 165
Hexanal, 2,4-DNPSC of	TLC	165
Hexanal, *p*-NPH of	TLC	165
Hexane	GC	49, 58, 125, 126, 127
n-Hexane	GC	29, 42, 47, 53, 55, 115, 120, 121, 122
	LC	21, 38, 42
n-Hexane, paraffins	GC	33
1,6-Hexanediol	TLC	113
2,5-Hexanedione (acetonylacetone)	GC	53
	LC	8
n-Hexanethiol	GC	56
1,2,6-Hexanetriol	TLC	113
sym-Hexanitrodiphenylamine	TLC	157
Hexanoic acid	LC	68
Hexanoic acid, allyl ester of	GC	24
Hexanoic acid, ethyl ester of	GC	24
Hexanoic acid, hexyl ester of	GC	24
Hexanoic acid, methyl ester of	GC	24
n-Hexanoic acid	LC	57
	TLC	134
Hexanol	GC	49, 53
n-Hexanol	GC	57, 92
	LC	37
	TLC	1
1-Hexanol	GC	92
	PC	52
1-Hexanol (DNB)	TLC	111
1-Hexanol (ONPU)	TLC	111
2-Hexanol (ONPU)	TLC	111
3-Hexanol	GC	57
2-Hexanone	GC	7, 51
Hexan-2-one, 2,4-DNPH of	TLC	76
3-Hexanone	GC	51
Hexanonitrile	LC	60
Hexanophenone	PC	21
	TLC	70
n-Hexatriacontane	LC	27, 38, 45
2-Hexenal	GC	7, 57
2-Hexenal, 2,4-DNPH of	TLC	165
2-Hexenal, 2,4-DNPSC of	TLC	165
2-Hexenal, *p*-NPH of	TLC	165
trans-Hexen-2-al-1, 2,4-DNPH of	TLC	72
1-Hexene	GC	47, 118, 120, 122
Hexene-1	GC	51, 53, 125, 126
	LC	42
Hex-1-ene	GC	56
cis-2-Hexene	GC	118, 120, 122, 125, 126
trans-2-Hexene	GC	118, 120, 122, 125, 126
cis-3-Hexene	GC	118, 120, 122
trans-3-Hexene	GC	118, 120, 122
Hexene-2-al-1, 2,4-DNPH of	TLC	72
1-Hexen-3-ol	GC	57
cis-3-Hexen-1-ol	GC	57
trans-3-Hexen-1-ol	GC	57
3-Hexen-1-ol	TLC	1
3-Hexen-1-ol (DNB)	TLC	111
3-Hexen-1-ol (ONPU)	TLC	111
3-Hexen-1-yl acetate	TLC	105

Hexethal GC 19, 106; TLC 137
Hexobarbital (Hexobarbitone) GC 19,105, 106; PC 86; TLC 137
Hexobarbitone (Hexobarbitol) GC 19,105, 106; PC 86; TLC 137
Hexoestrol PC 124
Hexogen TLC 157
Hexyl acetate GC 2; TLC 105
n-Hexyl acetate GC 57
Hexyl acrylate GC 91
n-Hexyl alcohol LC 11
Hexyl aldehyde, 2-hydrazinebenzothiazole derivative of TLC 168
n-Hexyl allyl disulfide GC 85
Hexylamine TLC 10
n-Hexylamine GC 11, 97; PC 5, 55
n-Hexylamine, trifluoroacetyl derivative GC 10
n-Hexyl *n*-amyl disulfide GC 85
Hexylbenzene TLC 53
n-Hexylbenzene LC 42
Hexyl butyrate GC 2; TLC 105
Hexyl caprylate TLC 105
α-Hexylcinnamaldehyde, 2-hydrazinebenzothiazole derivative of TLC 168
1-*n*-Hexyl-2,5-dimethylpyrrole GC 133
Hexyl esters GC 24
Hexyl formate GC 2
n-Hexyl formate GC 57
n-Hexyl-*n*-heptyl suberate GC 25
Hexyl hexanoate GC 2
Hexyl isobutyrate GC 2
n-Hexyl isocyanate GC 44
Hexyl mercaptan TLC 162
n-Hexyl mercaptan GC 85
Hexyl methacrylate GC 91
Hexyl 3-methylbutyrate GC 2
Hexyl 4-methylpentanoate GC 2
Hexyl octyl sulfide TLC 163
Hexyl pentanoate GC 2
Hexyl propionate GC 2
n-Hexyl sulfoxide GC 139
Hexyne TLC 51
Hf (Hafnium) LC 80, 84, 85, 86, 88, 89
HFBu, amines GC 98
HFBu, heptafluorobutyryl GC 98
Hg (Mercury) LC 80, 83, 85, 86, 87, 88
Hg(II), Mercury PC 15, 95, 96; TLC 52,148, 149, 150
Hg^{+}, Mercury PC 97; TLC 52
Hg^{2+}, Mercury PC 92
Hiascic acid TLC 35
High temperature reaction product of diphenylamine and acetone TLC 22
α-Himachalene GC 140; TLC 54
β-Himachalene GC 140; TLC 54
ar-Himachalene TLC 54
Himbacine GC 96
Hippuric acid GC 5; LC 30, 46; PC 82, 102
Hirsuitidin PC 25
Histamine PC 17, 56, 64; TLC 5, 13, 155
Histamine β-*N*-acetyl PC 17; TLC 155
Histamine β-*N*-diethyl PC 17
Histamine β-*N*-dimethyl PC 17
Histamine β-*N*-ethyl PC 17
Histamine 1-methyl PC 17
Histantin TLC 139
Histidine PC 11, 17, 62, 64; TLC 11, 13, 15,118, 155
Histidine HCl (Hydrochloride) LC 12, 56; TLC 16
Histidine, N-α-DNAP- TLC 14
Histidine (*bis*-DNP) PC 65
Histidine (*im*-DNP) PC 65
Histidine, PTH of TLC 124
Histidinol PC 17, 57, 64; TLC 155
1-Histidinol phosphate ester TLC 155
Ho (Holmium) LC 39, 79, 90, 91
Holocaine (Phenacaine) TLC 39, 43, 45
Homatropine GC 8, 107; PC 4; TLC 3, 117
Homatropine HBr GC 107
Homoarginine PC 9, 64; TLC 13
Homocarnosine LC 41
Homocitrulline LC 41
Homocysteic acid LC 41; PC 11, 64
DL-Homocysteic acid LC 59
Homocysteine PC 64
dl-Homocysteine PC 63
L-Homocysteine-D-penicillamine disulfide LC 32
l-Homocysteine thiolactone PC 63
L-Homocysteine trisulfide LC 32
Homocystine LC 41; PC 11, 64
dl-Homocystine PC 63
L-Homocystine LC 32, 59
DL + *meso*-Homocystine LC 59
Homocystine thiolacetone PC 11
Homoeriodictyol PC 26
Homogentisic acid LC 30; PC 82; TLC 8
Homomycin TLC 20
Homoprotocatechuic acid TLC 135
Homosekikaic acid TLC 35

Homoserine	LC	41
	PC	11, 64
	TLC	13
4-Homosulfanilamide salt of 1-sulfanilyl-2-thiourea	TLC	44
Homotaurine	LC	41
Homovanillic acid	LC	30
	PC	58
	TLC	121
Homovanillylamine	GC	9
Homoveratrylamine	GC	9
Honghelin	TLC	102
Honghelin peracetate	TLC	102
Hordenine	GC	9, 22
	PC	2, 3
	TLC	3
Hostocaine	TLC	43
Humulene	GC	140
α-Humulene	TLC	106, 143
Hydantoin-5-acetic acid	PC	63
Hydrallazine	PC	85
Hydrastine	GC	8,107, 111
Hydrastine HCl	GC	107
(-)-β-Hydrastine	PC	3
Hydrastinine	GC	107
	PC	2, 3
	TLC	3
Hydrastinine HCl	GC	107
Hydratropic aldehyde, 2,4-DNPH of	TLC	73
Hydrazine	PC	18
Hydrazines, Amides, Ureides, free form, salt form	GC	17
Hydrazine·2 HCl	PC	61
p-Hydrazinobenzoic acid	PC	78
α-Hydrinone	TLC	174
Hydrocaffeic acid	TLC	123
Hydrocarbons	GC	121
Hydrocarbons, nitrogen compounds	GC	45
Hydrochloric acid	LC	34, 58
Hydrochlorothiazide	TLC	41
Hydrocinnamaldehyde	GC	51
Hydrocinnamaldehyde, 2-hydrazine-benzothiazole derivative of	TLC	168
p-Hydrocinnamic acid, 4-acetyl-2-nitrophenol ether of	TLC	91
Hydrocodone bitartrate	TLC	138
Hydrocodone (Dihydrocodeinone)	TLC	2, 4, 39, 138
Hydrocortisone	TLC	101, 178
Hydrocortisone acetate	TLC	101, 178
Hydroflumethiazide	TLC	41
Hydrogen	GC	27,112, 115
Hydrogenated ε-tocopherol	PC	50
Hydrogen chloride	GC	26, 27, 113
Hydrogen fluoride	GC	26, 114
Hydrogen peroxide	PC	27
	TLC	65
Hydrogen sulfide	GC	85
Hydromorphone (Dihydromorphinone)	TLC	2, 4
Hydromorphone (Dilaudid®)	TLC	45
Hydromorphone HCl	TLC	4, 138
Hydroquinone	LC	1
	PC	82
	TLC	88, 90, 174, 176

Hydroquinone, di(4-acetyl-2-nitrophenol) ether of	TLC	91
Hydroquinone dimethyl ether	TLC	90
Hydroquinone monobenzyl ether	TLC	175
Hydrotropaldehyde, 2-hydrazinebenzo-thiazole derivative of	TLC	168
4-(*N*-Hydroxyacetamido)-biphenyl	TLC	6
2-(*N*-Hydroxyacetamido)-fluorene	TLC	6
2-(*N*-Hydroxyacetamido)-naphthalene	TLC	6
N-Hydroxyacetanilide	TLC	6
2-Hydroxyacetanilide	TLC	6
2-Hydroxyacetophenone	PC	22
2-Hydroxyacetophenone azine	PC	23
2-Hydroxyacetophenone, 4-NPH of	TLC	74
3-Hydroxyacetophenone	PC	22
3-Hydroxyacetophenone, 4-NPH of	TLC	74
4-Hydroxyacetophenone	PC	22
4-Hydroxyacetophenone azine	PC	23
4-Hydroxyacetophenone, 4-NPH of	TLC	74
Hydroxy-*n*-alkyl guanidines	PC	9
l-3-Hydroxy-*N*-allylmorphinan (Levallorphan)	GC	18
	TLC	138
2-Hydroxy-4-aminobenzoic acid	TLC	129
4-(Hydroxyamino)-biphenyl	TLC	6
2-(Hydroxyamino)-fluorene	TLC	6
Hydroxy-α-aminopimelic acid	PC	64
Hydroxyamphetamine	GC	23
	TLC	39
5-Hydroxyamyl guanidine (*hydroxy-n-alkyl guanidine*)	PC	9
5-Hydroxyamyl (*hydroxy-n-alkyl guanidine*)	PC	9
3-Hydroxyamylobarbitone	GC	19
17β-Hydroxy-1,4-androstadien-3-one	PC	124
17β-Hydroxy-androsta-1,4-dien-3-one (Group II)	PC	40
3α-Hydroxy-5α-androstane-7, 17-dione	TLC	179
3α-Hydroxy-5α-androstane-11, 17-dione (Group III)	PC	40
3α-Hydroxy-5β-androstane-11, 17-dione	GC	80
	PC	40, 41
	TLC	179
3β-Hydroxy-5α-androstane-6, 17-dione (Group III)	PC	40
3β-Hydroxy-5α-androstane-7, 17-dione	TLC	179
3β-Hydroxy-5α-androstane-11, 17-dione (Group III)	PC	40
3β-Hydroxy-5β-androstane-11, 17-dione (Group III)	PC	40
11β-Hydroxy-5α-androstane-3,17-dione	PC	40, 41
	TLC	179
17β-Hydroxy-5α-androstane-3,17-dione	TLC	179
3α-Hydroxy-5α-androstan-17-one	GC	80
	PC	40, 41
	TLC	179
3α-Hydroxy-5β-androstan-17-one	GC	80
	PC	40, 41
	TLC	179
3β-Hydroxy-5α-androstan-17-one	GC	80
	PC	41
	TLC	179

3β-Hydroxy-5α-androstan-17-one (Group III)	PC	40
3β-Hydroxy-5β-androstan-17-one	PC	40, 41
	TLC	179
6-Hydroxy-1-androstan-17-one (Group III)	PC	40
17β-Hydroxy-5α-androstan-3-one	GC	80
	TLC	179
17β-Hydroxy-5β-androstan-3-one	GC	80
	TLC	179
3β-Hydroxy-androst-5-ene-11,17-dione (Group III)	PC	40
6α-Hydroxy-androst-4-ene-3,17-dione (Group I)	PC	40
6β-Hydroxy-4-androstene-3,17-dione	PC	124
6β-Hydroxy-androst-4-ene-3,17-dione (Group I)	PC	40
11α-Hydroxy-4-androstene-3,17-dione	PC	124
11α-Hydroxy-androst-4-ene-3,17-dione (Group I)	PC	40
11β-Hydroxy-4-androstene-3,17-dione	PC	124
11β-Hydroxy-androst-4-ene-3,11-dione (Group II)	PC	40
11β-Hydroxyandrost-4-ene-3,17-dione	GC	80
11β-Hydroxy-androst-4-ene-3,17-dione (Group I)	PC	40
12α-Hydroxy-androst-4-ene-3,17-dione (Group I)	PC	40
14α-Hydroxy-4-androstene-3,17-dione	PC	124
14α-Hydroxy-androst-4-ene-3,17-dione (Group I)	PC	40
15α-Hydroxy-4-androstene-3,17-dione	PC	124
15α-Hydroxy-androst-4-ene-3,17-dione (Group I)	PC	40
17α-Hydroxy-androst-4-ene-3,11-dione (Group II)	PC	40
17β-Hydroxy-4-androstene-3,11-dione	PC	124
19-Hydroxy-androst-4-ene-3,17-dione (Group I)	PC	40
3α-Hydroxy-5α-androst-9-ene-17-one (Group III)	PC	40
3β-Hydroxy-androst-5-ene-17-one (Group III)	PC	40
3α-Hydroxy-5β-androst-9-en-17-one	PC	40, 41
3β-Hydroxyandrost-5-en-17-one	GC	80
	LC	62
3β-Hydroxy-5-androsten-17-one	PC	41, 124
17α-Hydroxyandrost-4-en-3-one	GC	80
17α-Hydroxy-androst-4-en-3-one (Group II)	PC	40
17β-Hydroxyandrost-4-en-3-one	GC	80
	TLC	179
17β-Hydroxy-androst-4-en-3-one (Group II)	PC	40
17β-Hydroxy-4-androsten-3-one	PC	124
p-Hydroxyanisole	TLC	90
9-Hydroxyanthracene	PC	121
3-Hydroxyanthranilic acid	LC	30
	PC	64, 82, 102, 103
5-Hydroxyanthranilic acid	PC	102,103, 120
3-Hydroxyanthranylglycine	PC	80
1-Hydroxyanthraquinone	TLC	92
2-Hydroxybenzalazine	PC	23
m-Hydroxybenzaldehyde	LC	52
	PC	22
	TLC	68, 71
m-Hydroxybenzaldehyde, 2,4-DNPH of	TLC	167
o-Hydroxybenzaldehyde	LC	1
	TLC	71
p-Hydroxybenzaldehyde	LC	52
	PC	20, 22
	TLC	68, 71, 174
p-Hydroxybenzaldehyde, 4-acetyl-2-nitrophenol ether of	TLC	91
p-Hydroxybenzaldehyde, 2,4-DNPH of	TLC	73, 167
3-Hydroxybenzaldehyde, 4-NPH of	TLC	74
4-Hydroxybenzaldehyde, 4-NPH of	TLC	74
m-Hydroxybenzoic acid	GC	4, 5, 90
	LC	1, 30
	PC	31, 82
	TLC	37,129, 135, 174
o-Hydroxybenzoic acid	GC	5
	LC	1, 30
	TLC	37
o-Hydroxybenzoic (salicylic acid)	PC	82
p-Hydroxybenzoic acid	GC	4, 5, 90
	LC	1, 30, 52, 74
	PC	31, 82
	TLC	37,129, 133,135, 174
p-Hydroxybenzoic acid, 4-acetyl-2-nitrophenol ether of	TLC	91
p-Hydroxybenzoic acid, methyl ester of	GC	24
3-Hydroxybenzoic acid methyl ester	PC	120
m-Hydroxybenzophenone	PC	22
m-Hydroxybenzophenone, 4-NPH of	TLC	74
o-Hydroxybenzophenone	PC	22
o-Hydroxybenzophenone, 4-NPH of	TLC	74
p-Hydroxybenzophenone	PC	22
p-Hydroxybenzophenone, 4-NPH of	TLC	74
4-Hydroxybenzophenone azine	PC	23
p-Hydroxybenzyl acetone, 2,4-DNPH of	TLC	167
m-Hydroxybenzyl alcohol	PC	82
o-Hydroxybenzyl alcohol	PC	82
p-Hydroxybenzyl alcohol	PC	82
4-Hydroxybenzyl alcohol	TLC	88
4-Hydroxybenzylideneacetophenone	TLC	92
o-Hydroxybiphenyl	TLC	174
p-Hydroxybiphenyl	TLC	174
12β-Hydroxy-bufalin peracetate	TLC	102
12β-Hydroxy-bufalin (3β,12β,14β-Trihydroxy- 5β-bufadienolide)	TLC	102
2-Hydroxybutyl guanidine	PC	9
4-Hydroxybutyl guanidine	PC	9

2-Hydroxybutyl (hydroxy-*n*-alkyl guanidine)	PC	9
4-Hydroxybutyl (hydroxy-*n*-alkyl guanidine)	PC	9
β-Hydroxybutyrate acid	TLC	36
2-Hydroxybutyric acid	LC	23
	TLC	29
3-Hydroxybutyric acid	LC	23
	TLC	29
4-Hydroxybutyric acid	LC	23
	TLC	29
α-G-β-Hydroxybutyric acid	PC	9
β-Hydroxybutyric acid, methyl ester of	GC	24
2-Hydroxycaproic acid	TLC	29
2-Hydroxycarbazole	TLC	92
3β-Hydroxy-5α,14α,17βH-cardenolide	TLC	102
3β-Hydroxy-5α,14α,17βH-cardenolide peracetate	TLC	102
1-Hydroxychlordene	GC	68, 129
3α-Hydroxy cholanic acid	TLC	180
3β-Hydroxy cholanic acid	TLC	180
7α-Hydroxy cholanic acid	TLC	180
7β-Hydroxy cholanic acid	TLC	180
12α-Hydroxy cholanic acid	TLC	180
12β-Hydroxy cholanic acid	TLC	180
3β-Hydroxycholest-5-ene-17α-hydroperoxide	LC	62
3β-Hydroxycholest-5-ene-20α-hydroperoxide	LC	62
3β-Hydroxycholest-5-ene-25-hydroperoxide	LC	62
3β-Hydroxy-5α-cholest-6-ene-5-hydroperoxide	LC	62
3β-Hydroxychol-5-en-24-ic-acid	LC	62
3β-Hydroxycholest-5-en-7-one	LC	62
6-Hydroxychrysene	TLC	92
m-Hydroxycinnamic acid	GC	5
	PC	82
m-Hydroxycinnamic acid, methyl ester of	GC	24
o-Hydroxycinnamic acid	GC	5
	PC	82
o-Hydroxycinnamic acid, methyl ester of	GC	24
p-Hydroxycinnamic acid	GC	5
	LC	30
	PC	82
p-Hydroxycinnamic acid, methyl ester of	GC	24
Hydroxycitronellal, 2-hydrazine-benzothiazole derivative of	TLC	168
10-Hydroxycodeine	TLC	4
4-Hydroxycoumarin	TLC	79, 92
4-Hydroxycoumarin-3-carboxylic acid ethyl ester	TLC	79
7-Hydroxycoumarin	PC	84
	TLC	79, 92
7-Hydroxycoumarin (umbelliferone)	PC	82
1-Hydroxy-2,4-diamino-anthraquinone	TLC	69
2-Hydroxydibenzofuran	TLC	92
5-Hydroxydeoxyuridine	PC	36
7β-Hydroxy-digitoxigenin peracetate	TLC	102
7β-Hydroxy-digitoxigenin (3β,7β,14β-Trihydroxy-5β-cardenolide)	TLC	102
15α-Hydroxy-digitoxigenin peracetate	TLC	102
15α-Hydroxy-digitoxigenin (3β,14β,15α-Trihydroxy-5β-cardenolide)	TLC	102
15β-Hydroxy-digitoxigenin peracetate	TLC	102
15β-Hydroxy-digitoxigenin (3β,14β,15β-Trihydroxy-5β-cardenolide)	TLC	102
7β-Hydroxy-digitoxin	TLC	102
7β-Hydroxy-digitoxin peracetate	TLC	102
6′-Hydroxy-2′,3′-dimethoxyacetophenone	GC	48
4-Hydroxy-3,5-dimethoxybenzaldehyde	TLC	71
4-Hydroxy-3,5-dimethoxybenzoic	GC	5
3-Hydroxy-7,12-dimethylbenz[*a*]-anthracene	TLC	50
4-Hydroxy-7,12-dimethylbenz[*a*]-anthracene	TLC	50
2′-Hydroxy-5,9-dimethyl-2-cyclopropylmethyl-6,7-benzomorphan	GC	18
	TLC	138
2′-Hydroxy-5,9-dimethyl-2-(3,3-dimethylallyl)-6,7-benzomorphan	GC	18
	TLC	138
2-Hydroxy-4′,5-dimethyldiphenyl ether	TLC	88
4-Hydroxy-2′,3-dimethyldiphenyl ether	TLC	88
dl-2′-Hydroxy-5,9-dimethyl-2-phenethyl-6,7-benzomorphan	GC	18
	TLC	138
5-Hydroxy-*N,N*-dimethyltryptamine (bufotenin)	GC	9
o-Hydroxydiphenyl	PC	31
	TLC	92
p-Hydroxydiphenyl	LC	1
	TLC	92
2-Hydroxydiphenyl ether	TLC	88
3-Hydroxydiphenyl ether	TLC	88
4-Hydroxydiphenyl ether	TLC	88
2-Hydroxy-estradiol	TLC	181
6α-Hydroxy-estradiol	TLC	181
6β-Hydroxy-estradiol	TLC	181
7α-Hydroxy-estradiol	TLC	181
11β-Hydroxy-estradiol	TLC	181
15α-Hydroxy-estradiol	TLC	181
3-Hydroxy-estra-1,3,5(10),6,8-pentaen-17-one	TLC	181
3-Hydroxy-estra-1,3,5(10),6-tetraen-17-one	TLC	181
3-Hydroxy-estra-1,3,5(10),7-tetraen-17-one	TLC	181
3-Hydroxy-estra-1,3,5(10),9(11)-tetraen-17-one	TLC	181
3-Hydroxy-estra-1,3,5(10)-triene-6,17-dione	TLC	181
3-Hydroxy-estra-1,3,5(10)-triene-7,17-dione	TLC	181
3-Hydroxy-estra-1,3,5(10)-triene-11,17-dione	TLC	181
3-Hydroxy-estra-1,3,5(10)-triene-16,17-dione	GC	82
	TLC	181

3-Hydroxyestra-1,3,5(10)-triene-17-one (Estrone)	GC	82
3-Hydroxy-estra-1,3,5(10)-trien-17-one	TLC	181
11α-Hydroxy-4-estrene-3,17-dione	PC	124
17β-Hydroxy-4-estren-3-one	PC	124
2-Hydroxy-estriol	TLC	181
4-Hydroxy-estrone	TLC	181
6α-Hydroxy-estrone	TLC	181
6β-Hydroxy-estrone	TLC	181
7α-Hydroxy-estrone	TLC	181
7β-Hydroxy-estrone	TLC	181
11α-Hydroxy-estrone	TLC	181
11β-Hydroxy-estrone	TLC	181
15α-Hydroxy-estrone	TLC	181
16α-Hydroxy-estrone	TLC	181
16β-Hydroxy-estrone	TLC	181
11-Hydroxyethiocholanolone	GC	136
2-Hydroxy-3-ethoxybenzaldehyde	TLC	71
3-Hydroxy-4-ethoxybenzaldehyde	TLC	71
4-Hydroxy-3-ethoxybenzaldehyde	TLC	71
4-Hydroxy-3-ethoxybenzoic	GC	5
2-Hydroxyethyl	GC	139
2-Hydroxyethylamine	PC	5
4-(β-Hydroxyethyl)-4-aza-5-cholesten-3-one	TLC	96
4-(β-Hydroxyethyl)-4-aza-5-pregnen-20β-ol-3-one	TLC	96
N-(2-Hydroxyethyl)-ethylenediamine	PC	7
2-Hydroxyethylhydroperoxide	PC	27
2-Hydroxyethyl (hydroxy-*n*-alkyl guanidine)	PC	9
o-Hydroxyethylguaiacol	PC	53
2-Hydroxyethyl guanidine	PC	9
N-(β-Hydroxyethyl)-3,5-seco-4-norcholestan-5β-ol-3-amide	TLC	96
3-Hydroxyflavone	LC	1
	TLC	92
2-Hydroxyfluorene	TLC	92
3-Hydroxyfluorene	TLC	92
Hydroxyglutamic acid	PC	64
	TLC	11
Hydroxyglutamic acid allo	PC	64
α-Hydroxyglutarate acid	TLC	36
α-Hydroxyglutaryl lactone acid	TLC	36
8-Hydroxyguanine	LC	46
2-Hydroxyhexanoic acid	TLC	29
m-Hydroxyhippuric acid	GC	5
	PC	82
o-Hydroxyhippuric acid	GC	5
	LC	30
	PC	82
p-Hydroxyhippuric acid	PC	82
1-Hydroxy-1′-hydroperoxy-dicyclohexyl peroxide	TLC	65
1-Hydroxy-1′-hydroperoxy-2,2′-dimethyldicyclohexyl peroxide	TLC	65
1-Hydroxy-1′-hydroperoxy-3,3′-dimethyldicyclohexyl peroxide	TLC	65
1-Hydroxy-1′-hydroperoxy-4,4′-dimethyldicyclohexyl peroxide	TLC	65
Hydroxyhydroquinone	TLC	176
17β-Hydroxy-4-(β-hydroxyethyl)-17α-methyl-4-aza-5-androstene	TLC	96
3-Hydroxy-7-hydroxymethyl-12-methyl-benz[*a*]anthracene	TLC	50
4-Hydroxy-7-hydroxymethyl-12-methyl-benz[*a*]anthracene	TLC	50

3-Hydroxy-2-hydroxymethylpyridine	PC	121
4-Hydroxyindole	PC	101
5-Hydroxyindole	PC	101, 120
5-Hydroxyindoleacetic acid	LC	72
	PC	82, 103
5-Hydroxy-3-indoleacetic acid	TLC	59
5-Hydroxyindole-(3)-acetic acid	TLC	156
7-Hydroxyindoleacetic acid	PC	103
5-Hydroxyindole-3-acetylglycine	PC	80
5-Hydroxy-3-indolylacetamide	PC	101, 120
3-(5-Hydroxyindolyl)acetic acid	GC	5
5-Hydroxyindolylacetic acid	PC	102
	TLC	121
5-Hydroxy-3-indolylacetic acid methyl ester	PC	101
5-Hydroxyindolylacetic acid 5-sulfate ester	PC	102
2-Hydroxyisobutyric acid	LC	23
2-Hydroxyisobutyric acid, methyl ester of	GC	24
5-Hydroxyisophthalic acid	PC	120
5-Hydroxyisoquinoline	PC	105
	TLC	92
2-Hydroxyisovaleric	LC	23
3-Hydroxykynurenine	PC	103
D-3-Hydroxykynurenine	PC	103
L-3-Hydroxykynurenine	PC	103
5-Hydroxykynurenine	PC	63
D-5-Hydroxykynurenine	PC	103
L-5-Hydroxykynurenine	PC	103
Hydroxyleucine	PC	64
Hydroxylupanine	PC	3
13-Hydroxylupanine	GC	96
Hydroxylysine	LC	41
	PC	64
	TLC	11, 13
allo-Hydroxylysine	LC	33
n-Hydroxylysine	LC	33
m-Hydroxymandelic acid	PC	82
o-Hydroxymandelic acid	PC	82
p-Hydroxymandelic acid	GC	5
	LC	30
	PC	82
dl-*p*-Hydroxymandelic acid	PC	120
2-Hydroxy-4-methoxyacetophenone, 4-acetyl-2-nitrophenol ether of	TLC	91
2′-Hydroxy-4′-methoxy-acetophenone	GC	48
4-Hydroxy-2-methoxyacetophenone, 4-acetyl-2-nitrophenol ether of	TLC	91
4-Hydroxy-3-methoxyacetophenone,4-acetyl-2-nitrophenol ether of	TLC	91
2-Hydroxy-3-methoxybenzalazine	PC	23
2-Hydroxy-3-methoxybenzaldehyde	PC	22
	TLC	71
2-Hydroxy-3-methoxybenzaldehyde, 4-NPH of	TLC	74
3-Hydroxy-4-methoxybenzaldehyde	TLC	71
4-Hydroxy-3-methoxybenzaldehyde	TLC	71
3-Hydroxy-4-methoxybenzoic acid	GC	5
	TLC	37
4-Hydroxy-3-methoxybenzoic acid	GC	5
	TLC	37
2-Hydroxy-4-methoxybenzophenone azine	PC	23
2-Hydroxy-4-methoxybenzophenone hydrazone	PC	23
3-Hydroxy-4-methoxybenzyl alcohol (isovanillyl alcohol)	PC	121

3-Hydroxy-4-methoxycinnamic acid	GC	5
	PC	58,120, 121
4-Hydroxy-3-methoxycinnamic acid	GC	4, 5
3-Hydroxy-4-methoxyhippuric	GC	5
4-Hydroxy-3-methoxyhippuric	GC	5
3-Hydroxy-4-methoxymandelic acid	PC	58, 120
4-Hydroxy-3-methoxymandelic acid	GC	5
	PC	120
3-Hydroxy-4-methoxyphenylacetic	GC	5
β-(3-Hydroxy-4-methoxyphenyl)-hydracrylic	GC	5
β-(3-Hydroxy-4-methoxyphenyl)-propionic	GC	5
4-Hydroxy-3-methoxyphenylacetic	GC	5
β-(4-Hydroxy-3-methoxyphenyl)-lactic	GC	5
β-(4-Hydroxy-3-methoxyphenyl)-propionic	GC	5
4-Hydroxy-3-methoxyphenylpyruvic	GC	5
4-Hydroxy-6-methoxyquinoline	PC	104
4-Hydroxy-6-methoxy-3-quinoline-carboxylic acid	PC	104
4-Hydroxy-7-methoxy-3-quinoline-carboxylic acid	PC	104
4-Hydroxy-8-methoxyquinoline-carboxylic acid	PC	104
6-Hydroxy-2-methylaminopurine	PC	36
2-Hydroxy-5-methylaniline	PC	121
3-Hydroxy-4-methylaniline	PC	121
2-Hydroxy-5-methylbenzophenone	PC	22
2-Hydroxy-5-methylbenzophenone, 4-NPH of	TLC	74
2-Hydroxy-5-methylbenzophenone azine	PC	23
2-Hydroxy-2-methylbutyric	LC	23
7-Hydroxy-4-methylcoumarin	PC	82
	TLC	92
5-Hydroxymethylcytosine	PC	36
5-Hydroxymethyldeoxycytidine	PC	36
5-Hydroxymethyldeoxyuridine	PC	36
Hydroxymethylfurfural	LC	29
5-Hydroxymethylfurfural, 2,4-DNPH of	TLC	73
8-Hydroxy-7-methylguanine	LC	46
	PC	36
Hydroxymethylhydroperoxide	PC	27
17α-Hydroxymethyl-17aα-hydroxy-D-homo-4-androstene-3,17-dione (D-homo compound S)	PC	124
Hydroxymethyl imidazole	PC	17
	TLC	155
3-Hydroxymethylindole	TLC	156
5-Hydroxy-2-methylindole	PC	101
2-Hydroxymethylisobutyric	LC	23
4-Hydroxy-2-methylmercaptopyrimidine	PC	36
7-Hydroxymethyl-12-methylbenz(*a*)-anthracene	TLC	50
12-Hydroxymethyl-7-methylbenz[*a*]-anthracene	TLC	50
5-Hydroxymethyl-6-methyluracil	PC	36
l-3-Hydroxy-*N*-methylmorphinan (Levorphanol)	GC	18
	TLC	138
5 Hydroxymethylorotic acid	PC	36
5-Hydroxymethylorotic acid lactone	PC	36
3-Hydroxy-*N*-methylpiperidine	PC	60
Hydroxymethylproline	PC	64
2-Hydroxy-6-methylpurine	PC	36
2-Hydroxy-4-methylpyrimidine	PC	36
4-Hydroxy-6-methylpyrimidine	PC	36
3-Hydroxy-2-methyl-1,4-pyron(maltol)	LC	23
4-Hydroxy-*N*-methyl-2-quinoline	PC	104
5-Hydroxymethyluracil	PC	36
5-Hydroxymethyluridine	PC	36
	TLC	62
2-Hydroxy-3-methylvaleric	LC	23
l-3-Hydroxymorphinan	GC	18
	TLC	138
2-Hydroxymyristic acid	PC	77
2-Hydroxy-3-naphthoic acid hydrazide	PC	105
2-Hydroxy-6-naphthoic acid	PC	105
2-Hydroxy-1,4-naphthoquinone	PC	111
5-Hydroxy-1,4-naphthoquinone	PC	105
1-Hydroxy-(2′-naphthylidene-amino)-pyridine (NAMP)	TLC	47
3-Hydroxy-4-nitrobenzoic acid	PC	120
Hydroxy-nor-leucine allo	PC	64
Hydroxy-nor-leucine threo	PC	64
10β-Hydroxy-19-nor-periplogenin peracetate	TLC	102
10β-Hydroxy-19-nor-periplogenin (3β,5β,10β,14β-Tetrahydroxy-19-nor-cardenolide)	TLC	102
Hydroxy-nor-valine allo	PC	64
Hydroxy-nor-valine threo	PC	64
3β-Hydroxy-15-oxo-5β,14α-cardenolide	TLC	102
3β-Hydroxy-15-oxo-5β,14α-cardenolide peracetate	TLC	102
2-Hydroxypalmitic acid	PC	77
Hydroxyphenamate	TLC	38
Hydroxyphenamate (Listica®)	GC	21
p-Hydroxyphenobarbital	GC	105
2-Hydroxyphenol	LC	51
3-Hydroxyphenol	LC	51
4-Hydroxyphenol	LC	51
m-Hydroxyphenylacetic acid	GC	4, 5
	LC	30
	PC	82, 120
	TLC	135
o-Hydroxyphenylacetic acid	GC	4, 5
	LC	30
	PC	58, 82
o-Hydroxyphenylacetic acid, methyl ester of	GC	24
p-Hydroxyphenylacetic acid	GC	4, 5
	LC	30
	TLC	135
p-Hydroxyphenylacetic acid, methyl ester of	GC	24
4-Hydroxyphenylacetic acid	LC	72
m-Hydroxyphenylethanolamine	PC	5
β-Hydroxyphenylethylamine	GC	9
β-(*m*-Hydroxyphenyl)-hydracrylic	GC	5
2-(2′-Hydroxyphenyliminomethyl)-pyridine (POAP)	TLC	47
m-Hydroxyphenyllactic acid	PC	82
o-Hydroxyphenyllactic acid	PC	82
p-Hydroxyphenyllactic acid	LC	30
	PC	82
β-(*p*-Hydroxyphenyl)-lactic acid	GC	4, 5
α-G-β-Hydroxyphenylpropionic acid	PC	9
m-Hydroxyphenylpropionic acid	PC	82, 120
o-Hydroxyphenylpropionic acid	PC	82
p-Hydroxyphenylpropionic acid	PC	82

β-(*m*-Hydroxyphenyl)-propionic acid	GC	5
β-(*p*-Hydroxyphenyl)-propionic acid	GC	5
3-(*p*-Hydroxyphenyl)-propionic acid	GC	4
m-Hydroxyphenylpyruvic acid	GC	4
	PC	82
o-Hydroxyphenylpyruvic acid	PC	81, 82
o-Hydroxyphenylpyruvic, lactone	GC	5
	PC	82
p-Hydroxyphenylpyruvic acid	GC	4, 5
	PC	81
m-Hydroxyphenylsulfamic acid	PC	82
4-Hydroxyphenyl sulfoxide	GC	139
5-Hydroxypipecolinic acid	PC	64
3β-Hydroxy-5,16-pregnadien-20-one	PC	124
11α-Hydroxy-4,16-pregnadiene-3,20-dione	PC	124
3α-Hydroxy-5α-pregnane-11,20-dione	PC	44
3α-Hydroxy-5β-pregnane-11,20-dione	PC	44, 124
3β-Hydroxy-5α-pregnane-11,20-dione	GC	80
	PC	44, 124
3β-Hydroxy-5β-pregnane-11,20-dione	PC	44, 124
11α-Hydroxy-5α-pregnane-3,20-dione	PC	124
11α-Hydroxy-5β-pregnane-3,20-dione	PC	124
11β-Hydroxy-5β-pregnane-3,20 dione	PC	44, 124
17α-Hydroxy-5β-pregnane-3,11,20-trione	PC	44, 124
3α-Hydroxy-5α-pregnan-20-one	PC	124
3α-Hydroxy-5β-pregnan-20-one	GC	80
	PC	44, 124
3β-Hydroxy-5α-pregnan-20-one	GC	80
	PC	44, 124
3β-Hydroxy-5β-pregnan-20-one	GC	80
	PC	44
20β-Hydroxy-5α-pregnan-3-one	GC	80
6β-Hydroxy-4-pregnene-3,20-dione	PC	124
6β-Hydroxy-pregn-4-ene-3,20-dione	PC	43
11α-Hydroxy-4-pregnene-3,20-dione	PC	124
11α-Hydroxypregn-4-ene-3,20-dione	GC	80
11β-Hydroxy-4-pregnene-3,20-dione	PC	124
11β-Hydroxypregn-4-ene-3,20-dione	GC	80
	PC	43
14α-Hydroxy-4-pregnene-3,20-dione	PC	124
14α-Hydroxypregn-4-ene-3,20-dione	PC	43
15α-Hydroxy-4-pregnene-3,20-dione	PC	124
15β-Hydroxy-4-pregnene-3,20-dione	PC	124
16α-Hydroxy-4-pregnene-3,20-dione	PC	124
16α-Hydroxypregn-4-ene-3,20-dione	PC	43
17-Hydroxypregn-4-ene-3,20-dione	PC	43
17α-Hydroxy-4-pregnene-3,20-dione	PC	124
18-Hydroxypregn-4-ene-3,20-dione	PC	43
21-Hydroxy-4-pregnene-3,20-dione	PC	124
21-Hydroxypregn-4-ene-3,20-dione	PC	125
21-Hydroxy-pregn-4-en-18-al-11,20-dione	PC	125
6β-Hydroxy-4-pregnene-3,11,20-trione	PC	124
6β-Hydroxy-pregn-4-ene-3,11,20-trione	PC	43
17-Hydroxy-pregn-4-ene-3,11,20-trione	PC	43
17α-Hydroxy-4-pregnene-3,11,20-trione	PC	124
21-Hydroxy-4-pregnene-3,11,20-trione	PC	124
3β-Hydroxy-5-pregnen-20-one	PC	124
3β-Hydroxypregn-5-en-20-one	GC	80
	LC	62
	PC	44
11α-Hydroxy-5α-pregn-2-en-20-one	GC	80
20α-Hydroxypregn-4-en-3-one	GC	83
	PC	43
20β-Hydroxypregn-4-en-3-one	GC	83
	PC	43
17-Hydroxyprogesterone	GC	136
17α-Hydroxyprogesterone	TLC	101, 178
17α-Hydroxyprogesterone caproate	TLC	101
Hydroxyproline	GC	101
	LC	29, 33, 41, 56
	PC	11, 64
	TLC	11, 13, 16, 118
Hydroxyproline, PTH of	TLC	124
Hydroxyproline allo	PC	64
allo-Hydroxy-D-proline	LC	59
α-G-β-Hydroxypropionic acid	PC	9
2-Hydroxypropionic (lactic)	LC	40
3-Hydroxypropionic acid	LC	23
	TLC	29
3-Hydroxypropyl(hydroxy-*n*-alkyl guanidine)	PC	9
2-Hydroxypurine	LC	46
	PC	16
6-Hydroxypurine (Hypoxanthine)	PC	16
8-Hydroxypurine	PC	16, 36
1-Hydroxypyrene	TLC	92
1-Hydroxypyridine-2-thione	PC	60
2-Hydroxypyridine	TLC	60
3-Hydroxypyridine	TLC	60
4-Hydroxypyridine	TLC	60
4-Hydroxypyrimidine	PC	36
Hydroxypyruvic acid	TLC	32
4-Hydroxyquinazoline	PC	104
Hydroxyquinol	PC	82

2-Hydroxyquinoline-4-carboxylic acid	PC	105
3-Hydroxyquinoline	PC	82, 106
4-Hydroxyquinoline	TLC	92
5-Hydroxyquinoline	TLC	92
6-Hydroxyquinoline	PC	82, 106
7-Hydroxyquinoline	PC	82
8-Hydroxyquinoline	PC	82
	TLC	92
6-Hydroxyquinolyl-5-sulfuric acid	PC	106
6β-Hydroxyrosenonolactone	TLC	185
3-Hydroxy-16,17-seco-estra-1,3,5(10)-triene-16-methyl-17-oine acid	TLC	181
4-Hydroxyskatole	PC	101
6-Hydroxyskatole	PC	101
Hydroxysparteine	PC	3
13-Hydroxysparteine	GC	96
2-Hydroxystearic acid	PC	77
C_{19} 17β-Hydroxysteroids	PC	124
C_{21}-Hydroxysteroids	PC	124
Hydroxystreptomycin	PC	67
	TLC	20
6-Hydroxytetrahydroharman	PC	101
5-Hydroxy-1-tetralone	PC	105
2′-Hydroxy-2,3′,4′,6′-tetramethoxyacetophenone	GC	48
6′-Hydroxy-2,2′,3′,4′-tetramethoxyacetophenone	GC	48
2′-Hydroxy-2,4′,6′-trimethoxyacetophenone	GC	48
2′-Hydroxy-3′,4′,6′-trimethoxyacetophenone	GC	48
6′-Hydroxy-2′,3′,4′-trimethoxyacetophenone	GC	48
6′-Hydroxy-2′,3′,4′-trimethoxyacetophenone (min.)	GC	48
l-2′-Hydroxy-2,5,9-trimethyl-6,7-benzomorphan	GC	18
	TLC	138
4-Hydroxytryptamine(as creatinine sulfate)	PC	102
5-Hydroxytryptamine	PC	64, 103
5-Hydroxytryptamine(as creatinine sulfate)	PC	102
6-Hydroxytryptamine	PC	103
6-Hydroxytryptamine(as creatinine sulfate)	PC	102
DL-5-Hydroxytryptophan	TLC	156
5-Hydroxytryptophan	LC	41
	PC	102, 103
	TLC	8
Hydroxytyramine	PC	5
3-Hydroxytyramine	PC	64
	TLC	9, 121
Hydroxyurea	PC	60
2-Hydroxyvaleric acid	LC	23
	TLC	29
4-Hydroxyvaleric acid	LC	23
Hydroxyvaline	PC	64
β-Hydroxyvaline	TLC	11
4-Hydroxyanthrone	TLC	92
Hydroxyzine	GC	17
	PC	85
Hydroxyzine (Atarax®)	TLC	39, 45
Hydroxyzine dihydrochloride	GC	17
Hygroton	TLC	41
Hyodeoxycholic acid	PC	38
Hyodeoxycholic acid(3α,6α-OH)	LC	45
Hyoscine	PC	4
Hyoscyamine	PC	3, 4
l-Hyoscyamine	GC	8
Hyperin	PC	26
Hypoglycin A	PC	64
Hypotaurine	LC	41
Hypothamnolic acid	TLC	35
Hypoxanthine	GC	77
	LC	3, 13, 15, 22, 31, 46, 50
	PC	32, 36, 37, 108
	TLC	159

Adsorption of the Elements (LC Section – Figures)

Hf, Hafnium	Pg	278, 279, 280, 281
Hg, Mercury	Pg	278, 279, 280, 281

I^-, Iodine	TLC	147
Ibogaine	GC	96
	PC	54
Ibogamine	GC	96
D-Iditol	GC	14
D-Idonic	LC	23
Idonolactone	PC	76
IDP (Inosine diphosphate)	LC	18
	PC	32
	TLC	63
L-Iduronic	LC	40
Imidan®	GC	61, 64, 65, 67, 129
	TLC	81, 84, 171
Imidan oxygen analog	GC	67
Imidazole	LC	46
	PC	17
	TLC	10, 155
Imidazole acetic acid	PC	17
	TLC	155
Imidazole-4-acetic acid	PC	83
Imidazole acetol	PC	17
Imidazole acrylic acid	PC	17
	TLC	155
Imidazole aldehyde	PC	17
	TLC	155
Imidazole carboxylic acid	PC	17
	TLC	155
Imidazole-4,5-carboxylic acid	TLC	155
4,5-Imidazoledicarboxylic acid	PC	83
Imidazole glycerol	PC	17
	TLC	155
Imidazole lactic acid	PC	17
	TLC	155
Imidazole propionic acid	PC	17
	TLC	155
Imidazole pyruvic acid	PC	17
	TLC	155
α-G-β-Imidazolyl propionic acid	PC	9
Imidoxon	GC	61, 65
Iminobispropylamine	LC	66
Iminodibenzyl	TLC	7
Imipramine	GC	17
	TLC	39, 139
Imipramine hydrochloride	GC	17

Imipramine (Tofranil®) TLC 45
IMP (Inosine monophosphate) LC 3, 18
PC 32, 37
5′-IMP TLC 63
Imperatorin PC 113
Imuran (Azathioprine) LC 46
In (Indium) LC 81, 82, 83, 84, 85, 86, 87, 89
In(II), (III), Indium GC 43
PC 15, 96
TLC 148, 149
In^{3+}, Indium PC 97
Indan GC 119
LC 27
4-Indanol GC 72
5-Indanol GC 72
1-Indanone TLC 75
Indene GC 45, 119
TLC 51
Indican PC 101, 103
Indican glucoside PC 103
Indigo carmine PC 88
TLC 49
Indigo disulfonate PC 103
Indigo tetrasulfate PC 103
Indigotin PC 103
Indium, In(II), (III) GC 43
PC 15, 96
TLC 148, 149
Indium, In^{3+} PC 97
Indole GC 45
LC 77
PC 102, 103
TLC 7, 58, 156
Indole-3-acetaldehyde TLC 156
3-Indoleacetaldehyde PC 101
Indole-3-acetamide PC 103
TLC 156
3-Indoleacetamide TLC 59
Indole-3-acetic acid LC 72
PC 103
TLC 156
3-Indoleacetic acid TLC 59
3-Indoleacetone TLC 59
Indole-3-acetonitrile PC 103
TLC 156
3-Indoleacetonitrile TLC 59
3-Indoleacetylaspartic acid TLC 59
Indole-3-acetylglycine PC 80
3-Indoleacetyl-ϵ-L-lysine TLC 59
Indole-3-acrylglycine PC 80
β-Indole-(3)-acrylic acid TLC 156
Indole-3-acrylic acid PC 103
Indole-3-aldehyde PC 102, 103
TLC 156
Indole alkaloids GC 96
Indole-3-butyric acid PC 103
γ-Indole-(3)-butyric acid TLC 156
3-Indole-γ-butyric acid TLC 59
Indole-3-butyronitrile PC 103
Indole-3-carbinol PC 101
Indole-2-carboxylic acid PC 101, 102
Indole-3-carboxylic acid PC 102, 103
TLC 156
Indole-5-carboxylic acid PC 101
Indole-(3)-ethanol TLC 156
3-Indole-L-lactic acid TLC 59
Indole-3-lactic acid PC 103
Indole-3-propionic acid PC 103
β-Indole-(3)-propionic acid TLC 156
3-Indole-β-propionic acid TLC 59
Indole-3-pyruvic acid PC 103
Indoles, nitrogen compounds GC 45
Indoline TLC 7
Indolo[b-3,4]-5β-cholest-3-ene TLC 96
Indolon-2 PC 101
β-Indolpyruvic acid PC 81
Indolylacetamide PC 102
3-Indolylacetamide TLC 156
3-Indolylacetic acid GC 5
PC 102
3-Indolylacetic acid hydrazide PC 101
3-Indolylacetic acid methyl ester PC 83, 101
Indolylacetonitrile PC 103
3-Indolylacetonitrile PC 102
Indolylacetyl-asparagine PC 102
Indolylacetyl-aspartic acid PC 102
Indolylacetyl-glutamic acid PC 102
Indolylacetyl-glutamine PC 102
Indolylacetyl-glycine PC 102
3-Indolylacrylic acid PC 101, 102
γ-(3-Indolyl)butyric acid PC 102
3-Indolylcarboxylic acid GC 5
C-(3-Indolyl)-glycine PC 102
3-Indolyl-glycollic acid PC 102
3-Indolylglyoxylamide PC 101
3-Indolylglyoxyldimethylamide PC 101
3-Indolyl-glyoxylic acid PC 83,101, 102
β-(3-Indolyl)lactic acid GC 5
3-Indolyl-lactic acid PC 102
Indolyl-propane 1,2-diol PC 102
3-(3-Indolyl)propionic acid GC 5
β-(3-Indolyl)propionic acid PC 81
α-G-β-Indolylpropionic acid PC 9
3-Indolylpyruvic acid GC 5
PC 83, 101
Indomethacin TLC 40
Indophenol TLC 7, 85
Indoxyl O-acetate PC 102, 103
Indoxyl O-phosphate PC 102
Indoxyl sulfate PC 102,103
Inorganic compounds, water GC 126
Inorganic phosphate PC 112
Inorganic pyrophosphate PC 112
Inorganic trimetaphosphate PC 112
Inosine (I) GC 77
LC 3, 13, 15, 22, 31, 46, 50
PC 32, 36, 37, 108
TLC 62, 159
Inosine diphosphate (IDP) PC 32
Inosine 5′-diphosphate PC 108
Inosine monophosphate (IMP) PC 32
Inosine 5′-monophosphate GC 77
PC 108
TLC 160
Inosine-5′-phosphoric acid morpholidate TLC 62
Inosine triphosphate (ITP) PC 32
Inosine 5′-triphosphate PC 108
Inosinic acid LC 50

Compound	Method	Pages
5′-Inosinic acid	LC	50
Inositol	GC	15
	PC	1, 47, 70
	TLC	127, 188
Insidon	TLC	42
IO_3^-	TLC	147
IO_4^-	TLC	147
Iodate	PC	14, 91, 93
Iodide	PC	14, 91, 93
	TLC	19
Iodine$^-$	TLC	147
Iodoacetic acid	PC	79
	TLC	29
2-Iodoaniline	GC	102
3-Iodoaniline	GC	102
	TLC	119
4-Iodoaniline	GC	102
1-Iodoanthraquinone	PC	110
2-Iodoanthraquinone	PC	110
Iodobenzene	GC	51, 53
	LC	36, 75
1-Iodobutane	GC	37, 56
Iodobutyric acid	PC	79
1-Iodo-2-chloroanthraquinone	PC	110
1-Iododecane	GC	37
1-Iodo-3,3-dimethylbutane	GC	37
2-Iodo-2,3-dimethylbutane	GC	37
2-Iodo-3,3-dimethylbutane	GC	37
2-Iodo-2,4-dimethylpentane	GC	37
1-Iodo-2,2-dimethylpropane	GC	37
Iodoethane	GC	37, 56
Iodofenphos	TLC	170
Iodogorgoic acid (3,5-diiodo-L-tyrosine)	TLC	64
1-Iodoheptane	GC	37, 56
1-Iodohexane	GC	37, 56
5-Iodohexene-1	GC	37
Iodomethane	GC	37, 56
1-Iodo-2-methylbutane	GC	37
1-Iodo-3-methylbutane	GC	37
2-Iodo-2-methylbutane	GC	37
2-Iodo-3-methylbutane	GC	37
1-Iodo-2-methylpropane	GC	37
2-Iodo-2-methylpropane	GC	37
1-Iodononane	GC	37
1-Iodooctane	GC	37
1-Iodopentane	GC	37, 56
2-Iodopentane	GC	37
3-Iodopentane	GC	37
o-Iodophenol	PC	31
p-Iodophenol	LC	52
	PC	31
	TLC	174
4-Iodophenol	GC	71
N-*p*-Iodophenylthiourea	PC	46
1-Iodopropane	GC	37, 56
2-Iodopropane	GC	37
3-Iodopropane	GC	37
Iodopropionic acid	PC	79
6-Iodopurine	LC	46
3-Iodopyridine	TLC	60
3-Iodothyronine	LC	76
1-Iodo-2,2,4-trimethylpentane	GC	37
3-Iodotyrosine	LC	76
	PC	63
	TLC	15
3-Iodo-L-tyrosine	TLC	64
Ionol	LC	54
	TLC	175
α-Ionone, 2,4-DNPH of	TLC	72, 76
β-Ionone, 2,4-DNPH of	TLC	72, 76
Ioxynil	GC	63
Ipazine	GC	66
	TLC	86
Iproniazid	GC	17
	TLC	39, 42
Iproniazid phosphate	GC	17
Ir (Iridium)	LC	83, 84
Ir(III), Iridium	GC	43
Iridin	PC	26
Iridium	TLC	55
Iridium (Ir)	LC	83, 84
Iridium, Ir(III)	GC	43
Irigenin	PC	26
Irigenol	PC	26
Iron, Fe	LC	81, 82, 83, 84, 85, 86, 87, 88, 89
Iron, Fe(II), (III)	GC	43
	PC	15, 95, 96
	TLC	148, 149
Iron, Fe^{3+}	PC	92
Isatin	PC	102, 103
	TLC	156
Isoamyl acetate	GC	53, 55
Isoamyl alcohol (3-methyl-1-butanol)	GC	53
Isoamylamine	PC	56
	TLC	5, 10
Isoamylamine, trifluoroacetyl derivative	GC	10
Isoamyl butyrate	TLC	105
Isoamyl caprate	TLC	105
Isoamyl caprylate	TLC	105
Isoamyl esters	GC	24
Isoamyl gallate	TLC	125
Isoamylhydrocupreine	TLC	39
Isoamyl propionate	TLC	105
Isoamyl sulfoxide	GC	139
Isoanhydrovitamin A_1	PC	49
Isobarbituric acid	PC	36
Isobenzan	TLC	81
Isoborneol	GC	87
Isobucaine HCl	GC	110
Isobutanal	GC	55
	TLC	1
Isobutanal, 2,4-DNPH of	TLC	72
Isobutane	GC	58, 118
Isobutanol	GC	49, 58
	LC	5
Isobutanol, organics	GC	52
Isobutene	GC	58, 125, 126
2,4-D Isobutyl	TLC	81
2,4,5–T Isobutyl	TLC	81
Isobutyl acetate	GC	49, 55
	TLC	105
Isobutyl alcohol	GC	55
Isobutyl alcohol (2-methyl-1-propanol)	GC	53
Isobutylamine	GC	97
	TLC	5, 10

Compound	Method	Page
Isobutylamine, trifluoroacetyl derivative	GC	10
Isobutylbenzene	GC	53, 119
Isobutyl *sec*-butyl sulfide	GC	135
Isobutylcyclohexane	GC	35
Isobutylcyclopentane	GC	121
Isobutylene	GC	29, 118
Isobutylene oxide, organics	GC	52
Isobutyl ester 2,4-D	GC	129
Isobutyl esters	GC	24
Isobutyl formate	GC	121
Isobutyl isocyanate	GC	44
Isobutyl methyl ketone	LC	70
Isobutyl palmitate	GC	132
Isobutyl sulfone	GC	139
Isobutyl sulfoxide	GC	85
N-Isobutylthiourea	PC	46
Isobutyraldehyde	GC	7, 49, 57, 58
	PC	20
Isobutyraldehyde, 2,4-DNPH of	TLC	166
Isobutyraldehyde, 2,4-DNPSC of	TLC	165
Isobutyraldehyde, 2-hydrazinebenzo-thiazole derivative of	TLC	168
Isobutyraldehyde, *p*-NPH of	TLC	165
Isobutyric acid	LC	65
	TLC	134
Isobutyric acid, cinnamyl ester of	GC	24
Isobutyric acid esters	GC	51
Isobutyric acid, ethyl ester of	GC	24
Isobutyric acid, hexyl ester of	GC	24
Isobutyric acid, isobutyl ester of	GC	24
Isobutyric acid, methyl ester of	GC	24
α-Isobutyric acid, PTH of	TLC	124
Isobutyronitrile	GC	55
Isocamphane	GC	87
Isocarboxazid	TLC	39
Isocarboxazide	GC	17
Isochavibetol	TLC	90
Isochlorogenic acid	TLC	133
Isochroman hydroperoxide	PC	27
Isochroman peroxide	PC	27
dl-Isocitrate acid	TLC	36
Isocitric acid	PC	13
Isocytosine	PC	37
Isodrin	GC	129
Isoeugenol	TLC	90
Isoeugenol benzyl ether	TLC	90
Isoeugenol methyl ether	TLC	90
Isoferulic acid	TLC	133, 135
Isoferulic(hesperitic) acid	TLC	174
Isoflavone, aglycones	PC	26
Isoflavone glucosides	PC	26
29-*iso*Fucosterol	GC	84
Isoguanine	LC	46
Isohexanoic acid, methyl ester of	GC	24
Isohexyl esters	GC	24
Isojasmon, 2,4-DNPH of	TLC	73
Isolan®	GC	130
	TLC	83
Isoleucine	GC	12, 101
	LC	4, 12, 33, 41
	PC	11, 62
	TLC	11, 13, 15, 16, 29, 118
Isoleucine, DNAP of	TLC	14
Isoleucine, PTH of	TLC	124
L-Isoleucine	LC	32
Isoleucinol	PC	57
Isoleucinyl-proline (*cis*)	GC	16
Isoleucyl-proline (*trans*)	GC	16
Isolongifolene	TLC	54
D-Isolysergic acid amide	TLC	114
Isolysergol	TLC	114
Isomaltoheptaose	PC	74
Isomaltohexaose	PC	74
Isomaltopentaose	PC	74
Isomaltose	LC	48
	PC	70, 74
Isomaltotetraose	PC	74
Isomaltotriose	PC	70, 74
Isomasticadienonic	TLC	184
Isomenthone, 2,4-DNPH of	TLC	76
Isomescaline	PC	60
Isomyristicin	TLC	90
Isoniazid = isonicotinoylhydrazine	PC	84
Isonicotinic acid	PC	2, 18
Isonicotinic acid ethyl ester	TLC	158
Isonicotinic acid hydrazide	PC	18
(-)-Isonicotinic acid α-phenylethylamide	TLC	158
Isooctane	GC	126
Isooctane (see 2,2,4-Trimethylpentane)	LC	21
Iso-octyl acetate	TLC	105
Iso-octyl ester	PC	115
Iso-octyl ester 2,4-D	GC	129
Iso-octyl ester 2,4,5-T	GC	129
Isopenniclavine	TLC	114
Isopentanal, 2,4-DNPH of	TLC	72
Isopentane	GC	29, 47, 58, 118, 121
Isopentanol	GC	49
	TLC	1
Isopentylamine	GC	97
Isopentylcyclohexane	GC	35
Isopentylcyclopentane	GC	36
N-Isopentylthiourea	PC	46
Isopetasin	LC	45
Isopetasol	LC	45
Isopetasone	LC	45
Isopetasyl acetate	LC	45
Isopimarate	LC	67
Isopregnenone	TLC	95
Isoprenaline	TLC	9
Isoprene	GC	29, 51, 55, 118
Isopropanol	GC	57
	LC	5, 71
Isopropanol, organics	GC	52
Isopropanolamine	PC	7
Isopropenyl acetate	GC	91
Isopropenylacetylene	GC	29
Isopropenyl butyrate	GC	91
Isopropenyl hexanoate	GC	91
Isopropenyl isobutyrate	GC	91
Isopropenyl 3-methylbutyrate	GC	91
Isopropenyl 4-methylpentanoate	GC	91
Isopropenyl pentanoate	GC	91
Isopropenyl propionate	GC	91
p-Isopropenyltoluene	TLC	106
2-Isopropoxyphenol	GC	71
2-Isopropoxyylphenyl methyl-carbamate (Propoxur)	TLC	170

N^1-Isopropoxysulfanilamide	TLC	44
Isopropyl 2,4-D	GC	60
	TLC	81
Isopropyl 2,4,5-T	GC	60
	TLC	81
Isopropyl acetate	GC	2, 49, 53, 55, 58, 121
Isopropyl acetate, organics	GC	52
Isopropyl acrylate	GC	91
Isopropyl alcohol	GC	55, 127
	TLC	1
Isopropyl alcohol (2-propanol)	GC	53
	LC	54
Isopropylamine	GC	97
	LC	26
	PC	56
	TLC	10
Isopropylamine, trifluoroacetyl derivative	GC	10
Isopropylantipyrine	TLC	40
p-Isopropylbenzaldehyde, 2,4-DNPSC of	TLC	165
p-Isopropylbenzaldehyde, *p*-NPH of	TLC	165
Isopropylbenzene	GC	51, 55, 119,120, 125
Isopropylbenzene (see Cumene)	LC	42
Isopropylbenzene, aromatics	GC	126
Isopropyl butyrate	GC	2
Isopropyl crotonate	GC	91
Isopropylcyclohexane	GC	121
1-Isopropylcyclohexene	GC	41
3-Isopropylcyclohexene	GC	41
4-Isopropylcyclohexene	GC	41
Isopropylcyclopentane	GC	42, 121
Isopropylcyclopropane	GC	118
Isopropyl ester	PC	115
Isopropyl ester 2,4-D	GC	129
Isopropyl ester 2,4,5-T	GC	129
Isopropyl esters	GC	24
Isopropyl ether	GC	121
Isopropyl formate	GC	2, 55, 57, 121
Isopropyl hexanoate	GC	2
2′,3′-*O*-Isopropylidineadenosine	GC	134
	TLC	62
Isopropylidenecyclohexane	GC	41
2′,3′-*O*-Isopropylidenecytidine	TLC	62
1,2-*O*-Isopropylidene-D-fructose	GC	13
3,4-*O*-Isopropylidene-*β*-D-galactosan	GC	13
1,2-*O*-Isopropylidene-*O*-glucofuranose	GC	13
1,2-*O*-Isopropylidene-*α*-D-glucofuranose 5,6-carbonate	GC	13
2′,3′-*O*-Isopropylideneguanosine	TLC	62
2′,3′-*O*-Isopropylideneinosine	TLC	62
2,3-*O*-Isopropylidene-L-rhamnose	GC	13
2,3-*O*-Isopropylidene-D-ribono-1 → 4-lactone	GC	13
2′,3′-*O*-Isopropylidene-4-thiouridine	TLC	62
2′,3′-*O*-Isopropylideneuridine	GC	134
	TLC	62
1,2-*O*-Isopropylidene-D-xylulose	GC	13
Isopropyl isobutyl sulfide	GC	135
Isopropyl isobutyrate	GC	2
Isopropyl isocyanate	GC	44
Isopropyl 3-methylbutyrate	GC	2
p-Isopropyl-*α*-methylhydrocinnamaldehyde, 2,4-DNPH of	TLC	73
Isopropyl methyl ketone	GC	55
1-Isopropyl-3-methyl-4-oximino-2-pyrazolin-5-one	TLC	83
Isopropyl 4-methylpentanoate	GC	2
1-Isopropyl-3-methylpyrazolon	GC	130
	TLC	83
N-Isopropylnoradrenaline	PC	60
Isopropyl pentanoate	GC	2
2-Isopropyl-1-pentene	GC	39
m-Isopropylphenol	TLC	174
o-Isopropylphenol	TLC	174
p-Isopropylphenol	TLC	174
2-Isopropylphenol	GC	69, 72
3-Isopropylphenol	GC	72
4-Isopropylphenol	GC	69, 71, 72
3-Isopropylphenyl methylcarbamate	TLC	170
Isopropylphenyl sulfoxide	TLC	164
Isopropyl propionate	GC	2
4-Isopropylpyridine	GC	131
4-Isopropylpyrocatechol	PC	120
1-Isopropylpyrrole	GC	133
2-Isopropylpyrrole	GC	133
Isopropyl sulfoxide	GC	139
N-Isopropylthiourea	PC	46
Isopulegone, 2,4-DNPH of	TLC	76
Isopyrethrin I	TLC	85
Isopyrethrin I peroxide	TLC	85
Isopyrethrin II	TLC	85
Isopyrethrin II peroxide	TLC	85
Isoquercitrin	PC	26
Isoquinoline	GC	45, 78
	TLC	7, 58
Isoquinolines, nitrogen compounds	GC	45
Isorhamnetin	LC	28
	PC	26
Isorhodeasapogenin	TLC	103
Isosafrol	TLC	90
Isosativene	GC	140
Isosetoclavine	TLC	114
α-Isosparteine	GC	96
l-*α*-Isosparteine	PC	3
Isostrophanthidin	PC	127
Isothymol	TLC	90
Isovaleraldehyde	GC	7, 55, 121
Isovaleraldehyde, 2,4-DNPH of	TLC	165, 166
Isovaleraldehyde, 2,4-DNPSC of	TLC	165
Isovaleraldehyde, *p*-NPH of	TLC	165
Isovalerylaldehyde	PC	20
Isovalerylaldehyde, 2-hydrazinebenzothiazole derivative of	TLC	168
Isovaleric acid, cinnamyl ester of	GC	24
Isovaleric acid, ethyl ester of	GC	24
Isovaleric acid, methyl ester of	GC	24
Isovanillin	TLC	77, 174
Isovanillin, 2,4-DNPH of	TLC	167
Isozeaxanthin	TLC	93, 94
Itaconic acid	PC	13

Itobarbital (butalbital)	GC	105
	TLC	137
ITP (Inosine triphosphate)	PC	32
	TLC	63

Adsorption of the Elements (LC Section – Figures)

I, Iodine	Pg	278, 279, 280, 281
In, Indium	Pg	278, 279, 280, 281
Ir, Iridium	Pg	278, 279, 280, 281

Jatrorrhizine	PC	3
Juloidine	TLC	7
K (Potassium)	LC	83, 85, 86,
	TLC	52
K(I), Potassium	GC	43
	PC	15, 95, 96
K^+, Potassium	PC	97
K_3 (sodium salt)	TLC	188
Kampferol	PC	26
Kanamycin	PC	67
	TLC	20
Kanamycin A	PC	67
Kanamycin A (sulfate)	TLC	20
Kanamycin B	PC	67
Kanamycin B (sulfate)	TLC	20
Kanamycin C (sulfate)	TLC	20
Kanokonol	GC	141
Kanokonyl acetate	GC	141
Karathane®	TLC	80
Kelthane®	GC	129
	TLC	80, 82, 172
Kelthane® (Dicofol)	PC	116
Kelthane (tech)	GC	129
p,p'-Kelthane	GC	129
Kemithal	GC	106
Kempferol	LC	28
Kepone®	GC	68,128, 129
	TLC	80
Kessane	GC	141
Kessanol	GC	141
Kessanyl acetate	GC	141
Kessoglycol	GC	141
Kessoglycol diacetate	GC	141
Kessyl acetate	GC	141
α-Kessyl alcohol	GC	141
α-Keto-ε-aminocaproic acid	PC	81
11-Ketoandrosterone	GC	136
Ketobemidone	GC	18
	TLC	4, 138
α-Ketobutyric acid	PC	81
	TLC	32
α-Ketocaproic acid	TLC	32
3-Keto cholanic acid	TLC	180
7-Keto cholanic acid	TLC	180
12-Keto cholanic acid	TLC	180
12α,3-Keto cholanic acid	TLC	180
3α,7α,12-Keto dihydroxy cholanic acid	TLC	180
3α,12α,7-Keto dihydroxy cholanic acid	TLC	180
7α,12α,3-Keto dihydroxy cholanic acid	TLC	180
11-Ketoethiocholanolone	GC	136
2-Ketogluconic acid	PC	72
2-Keto-D-gluconic acid	LC	40
5-Ketogluconic acid	PC	72
5-Keto-D-gluconic acid	LC	40
α-Ketoglutarate acid	TLC	36
α-Ketoglutaric acid	PC	13
	TLC	32
3α,7-Keto hydroxy cholanic acid	TLC	180
3α,12-Keto hydroxy cholanic acid	TLC	180
7α,3-Keto hydroxy cholanic acid	TLC	180
α-Ketoisocaproic acid	PC	81
	TLC	32
α-Ketoisovaleric acid	PC	81
	TLC	32
α-Keto-γ-methiolbutyric acid	PC	81
α-Keto-β-methylvaleric acid	TLC	32
D-α-Keto-β-methylvaleric acid	PC	81
Ketones and ethers	GC	87
Ketones, 2-heptanone	GC	126
Ketones, 2-pentanone	GC	126
2-Keto-octanoic acid methyl ester of	GC	24
Ketoprogesterone P^4-17α-ol-3,11,20-one	LC	10
Ketoprogesterone P^4-17α, 21-ol-3,11,20-one	LC	10
Ketoprogesterone P^4-17α, 21-ol,3,20-one	LC	10
Ketoprogesterone P^4-20α-ol-3-one	LC	10
Ketoprogesterone P^4-11β-ol-3,20-one	LC	10
Ketoprogesterone P^4-11β, 17α-ol-3,20-one	LC	10
Ketoprogesterone P^4-11β, 17α,21-ol-3,20-one	LC	10
Ketoprogesterone P^4-11β, 21-ol-3,20-one	LC	10
Ketoprogesterone P^4-20β-ol-3-one	LC	10
Ketoprogesterone P^4-21-ol-3,11,20-one	LC	10
Ketoprogesterone P^4-3,20-one	LC	10
Ketoprogesterone 6α-Me-P^4-17α,21-ol-3,20-one	LC	10
Ketoprogesterone 6α-Me-P^4-11β-ol-3,20-one	LC	10
Ketoprogesterone 6α-Me-P^4-3,11,20-one	LC	10
1-Ketose	LC	48
α-Keto-δ-G-valeric acid	PC	9
Kinetin	GC	77
Kitigenin	TLC	103
Kogagenin	TLC	103
Kojic acid	LC	29
	TLC	174
Korlan (Trolene)	PC	28
Kryptogenin	TLC	98, 103
Krypton	GC	112
Kryptopine	TLC	4
Kynuramine	PC	103
Kynurenic acid	LC	72
	PC	64, 82, 102, 103
Kynurenine	LC	41

Kynurenine (*continued*)	PC	64,102, 103
	TLC	11, 13
D-Kynurenine	PC	103
L-Kynurenine	PC	103

Adsorption of the Elements (LC Section – Figures)

K, Potassium	Pg	278, 279, 280, 281

La (Lanthanum)	LC	39, 79, 80, 83, 85, 86, 89, 90, 91
La(III), Lanthanum	GC	43
	PC	15, 96
La^{3+}, Lanthanum	PC	92
$LacNH_2$, reduced, trisilyl derivative	GC	124
Lactate acid	TLC	36
l-Lactate acid	TLC	36
Lactic acid	LC	58, 65
	PC	13
	TLC	29, 134
Lactic acid, ethyl ester of	GC	24
Lactobionic acid	LC	40
Lactose	LC	25, 29
	PC	70
	TLC	23, 24, 126
Lactose, reduced, trimethylsilyl derivative	GC	124
Lactulose	PC	70
Lactylcholine	PC	8
Laervorphanol	TLC	4
Lanadoxin	TLC	102, 177
Lanadoxin peracetate	TLC	102
Lanatoside A	TLC	102, 177
Lanatoside A peracetate	TLC	102
Lanatoside B	TLC	102, 177
Lanatoside B peracetate	TLC	102
Lanatoside C	TLC	102, 177
Lanatoside C peracetate	TLC	102
Landrin	TLC	170
Lankamycin	TLC	153
Lanosterol	LC	45
	TLC	99
Lanthanum (La)	LC	39, 79, 80, 83, 85, 86, 89, 90, 91
Lanthanum, La(III)	GC	43
	PC	15, 96
Lanthanum, La^{3+}	PC	92
Lanthionine	PC	64
	TLC	11
DL+*meso*-Lanthionine	LC	32
Largactil®	TLC	42
Larocaine® (Dimethocaine)	TLC	43
Laudanidine	TLC	4
Laudanine	TLC	4
Laudanosine	GC	96, 107
	PC	54
	TLC	4
Laurate·	LC	67
Laurate, β-sitosteryl ester of	TLC	182
Lauric acid	LC	45
	PC	77
	TLC	33, 132
Lauric acid, ethyl ester of	GC	24
Lauric aldehyde	TLC	77
Lauroyl peroxide	TLC	65
Lauryl gallate	TLC	125
$Lead^{++}$	TLC	52,148, 149, 150
Lead, Pb(II)	GC	43
	PC	15, 95, 96
Lead, Pb^{2+}	PC	92
Lead diethyldithiocarbamate	TLC	66
Lebaycid	TLC	84
Lecanoric acid	PC	30
Lecithin	LC	45, 68
Lecithin (dimyristyl-)	TLC	56
Lecithin (dipalmitoleyl-)	TLC	56
Lecithin (distearyl-)	TLC	56
Lecithins (from egg)	TLC	56
Lecithin (saturated)	PC	99
Lecithin unsaturated (dipalmitoleyl)	PC	99
Leptosidin	PC	26
Leptosin	PC	26
Leritine (Anileridine)	TLC	138
Lethane A-70®	GC	65
Lethane 384®	GC	65
Leucine	GC	12, 101
	LC	4, 12, 14, 33, 41, 56
	PC	11, 62, 64, 65
	TLC	11, 13, 14, 15, 16, 29
Leucine, PTH of	TLC	124
L-Leucine	TLC	118
iso-Leucine	PC	64
Leucinol	PC	57
Leucomycin	TLC	20
Leucyl-phenylalanine (*cis*+*trans*)	GC	16
Leu-Gly-Gly	LC	4
Levallorphan	TLC	39
Levarterenol (*l*-norepinephrine)	GC	23
Levomepromazine (Nozinan)	TLC	45
Levomepromazine-5-sulfoxide	PC	84
Levomepropazine	TLC	139
Levopimarate	LC	67
Levorphanol	PC	4
Levorphanol tartrate	TLC	138
Levulinic acid	LC	29, 40
	PC	13
	TLC	29, 32
Levulose	TLC	23
Li (Lithium)	LC	83, 85, 86
	TLC	52
Li(I), Lithium	PC	96
Li^+, Lithium	PC	97
Librium®	TLC	42, 139
Librium (chlordiazepoxide)	GC	21, 108
Librium (Methaminodiazepoxide)	TLC	45

Lichesterinic acid TLC 131
Lidocaine HCl GC 110
Lidocaine (Xylocaine®) GC 21
TLC 39, 43, 45
Lignocaine PC 4
Lignocerate LC 67
Lignoceric acid LC 45
PC 77
Lignoceryl alcohol LC 45
Limettin PC 113
Limonene GC 86
TLC 51, 54
d-Limonene TLC 106
Linaloöl TLC 183
Linaloöl (DNB) TLC 111
Linaloöl (ONPU) TLC 111
Linalyl benzoate TLC 105
Linalyl butyrate TLC 105
Linalyl formate TLC 105
Linalyl propionate TLC 105
Linarin PC 26
Lincomycin HCl TLC 21
Lindane GC 60, 61, 68, 128, 129
LC 44
PC 116
TLC 80, 82, 172
Lindane (gamma BHC) GC 129
Linear dimer LC 49
Linear heptamer LC 49
Linear hexamer LC 49
Linear monomer LC 49
Linear octamer LC 49
Linear pentamer LC 49
Linear tetramer LC 49
Linear trimer LC 49
Linoleate LC 67
Linoleate, β-sitosteryl ester of TLC 182
Linoleic acid PC 77
TLC 33
Linolenate LC 67
Linolenic acid PC 77
TLC 33
Linuron TLC 81, 173
Lipoic acid TLC 188
Lithium LC 83, 85, 86
TLC 52
Lithium, Li(I) PC 96
Lithium, Li^+ PC 97
Lithocholic acid (3α-OH) LC 45
Lobaric acid PC 30
Lobeline PC 3
TLC 3
Lokundjoside TLC 102
Lokundjoside peracetate TLC 102
Longicyclene GC 140
TLC 54
Longifolene GC 140
TLC 54
α-Longipinene GC 140
Lu (Lutetium) LC 39, 79, 89, 90, 91
Lumipyrethrin TLC 85
Lumisterol GC 136
PC 126
Lupanine GC 96
PC 3
Lupin alkaloids GC 96
Lupinine GC 96
PC 3
Lutein PC 122
TLC 93, 94
Luteinepoxide TLC 94
Luteolin LC 28
PC 26
Luteolinidin PC 25
Lutetium (Lu) LC 39, 79, 89, 90, 91
Luvigenin TLC 103
Lycopene TLC 93, 94
Lycopersene TLC 94
Lycophyll TLC 93
Lycorenine GC 96
Lycoxanthin TLC 93
Lynestrenol TLC 95
D-Lysergic acid PC 101
TLC 114
D-Lysergic acid amide TLC 114
D-Lysergic acid methyl carbinolamide TLC 114
Lysergine TLC 114
Lysergol TLC 114
D-Lysergyl-L-valine methyl ester TLC 114
Lysine GC 101
LC 4, 33, 41
PC 11, 62, 64
TLC 11, 13, 15
Lysine (*bis*-DNP) PC 65
Lysine (ε-DNP) PC 65
Lysine, PTH of TLC 124
Lysine·HCl LC 12, 56
TLC 16
L-Lysine TLC 118
Lysinol PC 57
Lysolecithin (2-palmitoleyl-) TLC 56
Lysolecithin (saturated) PC 99
Lysolecithin (2-stearyl) TLC 56
Lysolecithin (unsaturated) PC 99
Lysolecithins TLC 56
Lysophosphatidyl choline TLC 152
Lysophosphatidyl ethanolamine TLC 152
Lysophosphatidyl serine TLC 152
D-Lyxonic LC 40
Lyxose LC 25, 29, 48
TLC 23
Lyxuronic LC 23

Adsorption of the Elements (LC Section – Figures)

Li, Lithium Pg 278, 279, 280, 281

α-Maaliene GC 140
Maaliol GC 141
Macarpine PC 3
Machaerinic TLC 184

Compound	Method	Pages
Magnesium (Mg)	LC	81, 82, 83, 85, 86, 87, 89
	TLC	52
Magnesium, Mg(II)	GC	43
	PC	15, 95, 96
	TLC	148, 149
Magnesium, Mg^{2+}	PC	92
Malaoxon	GC	65
	TLC	84
Malate acid	TLC	36
l-Malate acid	TLC	36
Malathion	GC	60, 61, 64, 65, 67, 68, 77,128, 129
	LC	43
	PC	28
	TLC	80, 81, 84,170,171, 172
Malathion oxygen analog	GC	61, 67
Malathion-O-analogue	LC	43
Maleic acid	LC	58, 65
	PC	13
Maleic anhydride, *p*-anisidide of	TLC	30
Maleylsulfanilamide	TLC	44
Maleyl sulfathiazole	TLC	140
Malic acid	LC	58, 65
	PC	13
	TLC	31
Mallogenin (3β,11β,14β-Trihydroxy-5α-cardenolide)	TLC	102
Malloside	TLC	102
Malloside peracetate	TLC	102
Malonate acid	TLC	36
Malonic acid	LC	65
	PC	13, 77
	TLC	31
Malonic acid, diethyl ester of	GC	24
Malonic acid, dimethyl ester of	GC	24
Maltobionic	LC	40
Maltoheptaose	PC	74
Maltohexaose	PC	74
Maltopentaose	PC	74
Maltose	LC	25
	PC	70, 74
	TLC	23, 24, 126
Maltose, trimethylsilyl derivative	GC	124.
Maltotetraose	PC	74
Maltotriose	LC	48
	PC	70, 74
Malvidin	PC	25
	TLC	78
Malvidin-3,5-diglucoside	TLC	78
Malvidin, glycosides	PC	24
Malvidin glycosides, 3,5-diglucoside	PC	24
Malvidin glycosides, 3-monoglucoside	PC	24
Malvidin glycosides, 3-rhamnoglucosido-5-glucoside	PC	24
Malvidin-3-monoglucoside	TLC	78
Malvin	TLC	92
Mandelic acid	GC	5
	PC	82
	TLC	133
Maneb	PC	117
Manganese (Mn)	LC	81, 82, 83, 85, 86, 87, 88, 89
Manganese, Mn(II)	GC	43
	PC	15, 95, 96
	TLC	149, 150
Manganese, Mn^{2+}	PC	92
Manganese, Mn(III)	GC	43
Mannitol	LC	69
	PC	1, 53, 70
	TLC	127
D-Mannitol	GC	14
	TLC	127
Mannitol hexaacetate	PC	72
Mannobiose	LC	48
Mannoheptulose	LC	25
	TLC	23
D-Mannonic	LC	40
Mannonolactone	PC	76
D-Mannono-γ-lactone	PC	72
Mannose	LC	25, 29, 48, 71
	PC	76
	TLC	23, 24
D-Mannose	GC	15
	TLC	126
Mannosidodihydrostreptomycin	PC	67
Mannosidostreptomycin	PC	67
Mannuronic acid	PC	72
D-Mannuronic acid	LC	40
Mansonin	TLC	102
Mansonin peracetate	TLC	102
MAO (3-Monoacetyloleandomycin)	PC	66
1-MAO (1-Monoacetyloleandomycin)	PC	66
2-MAO (2-Monoacetyloleandomycin)	PC	66
Mapazine	TLC	42
Marein	PC	26
Maritimein	PC	26
Maritimetin	PC	26
Marplan®	TLC	42
Martius Yellow	TLC	48
Masticadienonic	TLC	184
Matacil	GC	130
Matrina	GC	96
MBT	TLC	66
MBTS	TLC	66
MCPA	PC	118
MCPB	PC	118
Mebutamate	GC	105
	TLC	38
Mecambrine	PC	3
Mecarbam	LC	43
Meclizine	GC	17, 20
Meclizine (Bonamine)	GC	21
	TLC	39
Meclizine dihydrochloride	GC	17
Meclofenoxate	GC	22
Meclozine	PC	85
Medicagenic	TLC	184
Megestrol acetate	TLC	95
Megimide (Bemigride)	GC	19
Melatonin	PC	103
	TLC	59
Melezitose	LC	25
Melibionic	LC	40

Compound	Method	References
Melibiose	PC	70
	TLC	23
Melibiose, reduced, trimethylsilyl derivative	GC	124
Melibitol (min), trimethylsilyl derivative	GC	124
Melleril	TLC	42
Menabegenin (3β,14β-Dihydroxy-5β,17βH-cardenolide)	TLC	102
Menabegenin peracetate	TLC	102
Menadione	PC	51
Menaxon	GC	65
Menazon	PC	28
	TLC	80, 84
2,4(8)-Menthadiene	TLC	106
2,4(8)-*p*-Menthadiene	GC	87
3,8-*p*-Menthadiene	GC	87
cis-p-Menthane	GC	87
trans-p-Menthane	GC	87
1-*p*-Menthene	GC	41, 87
trans-2-*p*-Menthene	GC	41
3-*p*-Menthene	GC	41, 87
8-*p*-Menthene	GC	41
1(7)-*p*-Menthene	GC	41
4(8)-*p*-Menthene	GC	41, 87
Menthol	TLC	183
Menthol (DNB)	TLC	111
Menthol (ONPU)	TLC	111
Menthone, 2,4-DNPH of	TLC	76, 165
Menthone, 2,4-DNPSC of	TLC	165
Menthone, *p*-NPH of	TLC	165
Mepacrine	PC	4
Mepazine	GC	17, 20, 108, 109
	TLC	139
Mepazine hydrochloride	GC	17
Mepazine sulfoxide	TLC	139
Meperidine	GC	8
	TLC	39
Meperidine (Demerol®)	GC	21
Mephenesin	GC	17
	TLC	42
Mephenesin carbamate	TLC	38
Mephentermine	GC	22, 23
	TLC	39
Mephobarbital	GC	105, 106
Mephobarbital (Methylphenobarbitone)	GC	19
	TLC	137
Mepivacaine	TLC	39
Mepivacaine HCl	GC	110
Meprobamate	GC	17, 105, 108
	TLC	38, 42
Meprobamate (Miltown®)	GC	21
Meprobarbital	TLC	137
Meprylcaine HCl	GC	110
Mepyramine maleate	PC	85
2-Mercaptobenzimidazole	TLC	22
2-Mercaptobenzothiazole	TLC	22, 66
Mercaptodimethur	TLC	81
2-Mercaptoimidazoline	TLC	22
8-Mercapto-6-methylpurine	PC	36
6-Mercaptopurine	LC	46
	PC	37
2-Mercaptopyridine	PC	120
4-Mercaptopyridine	PC	120
β-Mercaptopyruvic acid	PC	81
2-Mercaptothiazoline	TLC	66
Mercury (Hg)	LC	80, 83, 85, 86, 87, 88
Mercury, Hg(I)	TLC	52
Mercury, Hg(II)	PC	15, 95, 96
	TLC	52,148, 149, 150
Mercury, Hg^+	PC	97
Mercury, Hg^{2+}	PC	92
Mercury di-dithizonate	TLC	161
Merphos	GC	65, 67
Mescaline	GC	8, 9
	TLC	10
Mesitylene	TLC	7, 51
Mesitylene (1,3,5-trimethylbenzene)	GC	53
	LC	7
Mesityl oxide (4-methyl-3-penten-2-one)	GC	53
Mesocaine (Trimecaine)	TLC	43
Mesochlorin	PC	33
Meso Cystine	PC	64
Meso-erythritol	GC	76
Mesoporphyrin	PC	33
Mesoporphyrin IX	PC	123
Mesoxalic acid (diethyl ester), 2,4-DNPH of	TLC	166
Mestranol	TLC	95
Mesurol® (Bayer 37344) (*N*-methyl-3,5-dimethyl-4-thiomethylphenylcarbamate)	GC	130
Mesurol®, phenol	GC	65
Mesurol sulfone	GC	65
Mesurol sulfoxide	GC	65
Metabutethamine HCl	GC	110
Metabutoxycaine	TLC	39
Metabutoxycaine HCl	GC	110
Metacresol purple	PC	88
Metagenin	TLC	103
Metanephrine	GC	9
	PC	64
	TLC	8
DL-Metanephrine	LC	77
dl-Metanephrine	PC	60
Metanicotine	GC	95
	PC	2
Metaoxedrin	TLC	9
Metaphosphate	PC	14
Metaraminol	GC	22, 23
	TLC	39
Meta-Systox® (Methyl demeton)	TLC	84, 171
Meta-Systox® R	GC	129
	TLC	171
Meta-Systox S	TLC	171
Meta-Systox (I)	TLC	171
Meteloidine	PC	3
Meteogenin	TLC	103
Metepa	GC	65
Methacrolein	GC	55
Methacrylonitrile	GC	55, 123
Methadone	GC	105
	PC	4
Methadone (Dolophin®)	GC	21
	LC	55
	TLC	4, 39, 138
Methadone (Heptadone)	TLC	45
dl-Methadone	GC	8, 18
	TLC	138

Methallatal	GC	106
N-α-Methallylthiourea	PC	47
Methamphetamine	GC	23, 105
	TLC	39
Methamphetamine (Pervitin®)	TLC	45
Methanal, 2,4-DNPH of	TLC	72, 76
Methandienone	TLC	95
Methandriol	TLC	95
Methane	GC	42, 58, 112,118,125, 126, 127
Methanesulfonamide	GC	98
Methanol	GC	47, 49, 52, 53, 55, 58, 92, 94, 125
	LC	5, 37, 54, 71
	PC	52
	TLC	1
Methanol, alcohols	GC	126
Methanol, butyric acid ester of	GC	51
Methanol, formic acid ester of	GC	51
Methanol, isobutyric acid ester of	GC	51
Methanol, propionic acid ester of	GC	51
Methaphenilene	TLC	39
Methapyrilene	GC	20, 105
Methapyrilene (Histadyl®)	GC	21
Methapyrilene (Thenylpyramine)	TLC	39
Methapyrilene HCl	GC	20, 54, 104
Methaqualone	GC	19
	TLC	137
Methaqualone (Revonal)	TLC	45
Metharbital	GC	19, 106
	TLC	137
Methdilazine	GC	20, 109
	TLC	139
Methdilazine HCl	GC	20
Methenamine	TLC	39
Methicillin Na	TLC	21
α-G-γ-Methiol-butyric acid	PC	9
Methionine	GC	12, 101
	LC	4, 12, 33, 41, 56
	PC	11, 62, 64, 65
	TLC	11, 13, 15, 16, 118
Methionine, DNAP of	TLC	14
Methionine, PTH of	TLC	124
Methionine sulfone	GC	101
	PC	11, 64
	TLC	11, 13, 15
Methionine sulfone, PTH of	TLC	124
Methionine sulfoxide	GC	101
	LC	41
	PC	11, 64
	TLC	11, 13, 15
Methionine sulfoxide, PTH of	TLC	124
DL-Methionine	PC	63
dl-Methionine methylsulfonium chloride	PC	63
dl-Methionine sulfone	PC	63
dl-Methionine sulfoxide	PC	63
DL-Methionine sulfoximine	PC	63
Methiotepa	GC	65

Methitural	TLC	137
Methobarbitone	PC	86
Methocarbamol	GC	17
	TLC	38
Methomyl	TLC	81
Methopromazine	GC	109
	PC	84
Methopromazine (Methoxypromazine)	GC	20
Methoxamine	TLC	39
Methoxinine	PC	64
3-Methoxyadrenaline	TLC	121
2-Methoxy-1-amino-anthraquinone	TLC	69
2-Methoxy-3-amino-anthraquinone	TLC	69
2-Methoxyaniline	GC	102
	PC	6
	TLC	119
3-Methoxyaniline	GC	102
	TLC	119
4-Methoxyaniline	GC	102
	TLC	119
p-Methoxyazobenzene	TLC	48
2-Methoxybenzalazine	PC	21
m-Methoxybenzaldehyde	TLC	71
m-Methoxybenzaldehyde, 2,4-DNPH of	TLC	167
o-Methoxybenzaldehyde	PC	20, 121
	TLC	68, 71
o-Methoxybenzaldehyde, 2,4-DNPH of	TLC	167
p-Methoxybenzaldehyde	TLC	68, 71
2-Methoxybenzaldehyde, 4-NPH of	TLC	74
m-Methoxybenzoic acid	TLC	37
o-Methoxybenzoic acid	TLC	37
p-Methoxybenzoic acid	GC	5
	TLC	37
p-Methoxybenzylacetone, 2,4-DNPH of	TLC	167
p-Methoxybenzyl alcohol	PC	53
p-Methoxybenzylamine	PC	60
Methoxychlor	GC	60, 68, 128, 129
	PC	116
	TLC	80, 82
p,p'-Methoxychlor	GC	129
p,p'-Methoxychlor, dichloroethane analog	GC	129
p,p'-Methoxychlor, monochloroethylene analog	GC	129
p,p'-Methoxychlor olefin	GC	129
Methoxychlor (tech)	GC	129
2-Methoxy-4-chloroaniline	TLC	119
2-Methoxy-5-chloroaniline	PC	6
	TLC	119
4-Methoxy-3-chloroanaline	TLC	119
2-Methoxy-5-chloro-1,4-diaminobenzene	TLC	119
2-Methoxy-4-chloro-5-methylaniline	TLC	119
m-Methoxycinnamic acid	GC	4
	PC	120
o-Methoxycinnamic acid	GC	4
	PC	120
p-Methoxycinnamic acid	GC	4, 5
2-Methoxy-1,4-diaminobenzene	TLC	119
4-Methoxy-1,2-diaminobenzene	TLC	119
4-Methoxy-1,3-diaminobenzene	TLC	119
5-Methoxy-*N*-dimethyltryptamine	PC	103
5-Methoxy-*N,N*-dimethyltryptamine	GC	9
3-Methoxyestra-1,3,5(10)11(12)-tetraene-3,17α-diol	GC	82
2-Methoxyestra-1,3,5(10)-triene-3,17β-diol	GC	82
3-Methoxyestra-1,3,5(10)-triene-3,17β-diol	GC	82

2-Methoxyestradiol	TLC	181
2-Methoxyestriol	TLC	181
3-Methoxyestriol	TLC	181
3-Methoxy-17-*epi*-estriol	TLC	181
3-Methoxyestrone	TLC	181
Methoxyethylmercury dithizonate	TLC	161
2-Methoxyethyl palmitate	GC	132
5-Methoxygramine	PC	101
p-Methoxyhippuric acid	GC	5
2-Methoxy-3-hydroxy-1,4-naphthoquinone	PC	111
3-Methoxy-4-hydroxybenzoic acid (vanillic acid)	PC	82
3-Methoxy-4-hydroxycinnamic acid	LC	30
	PC	82
3-Methoxy-6α-hydroxyestradiol	TLC	181
3-Methoxy-6β-hydroxyestradiol	TLC	181
3-Methoxy-4-hydroxyestrone	TLC	181
3-Methoxy-4-hydroxyhippuric acid	PC	82
3-Methoxy-4-hydroxymandelic acid	LC	30, 72
	PC	82
	TLC	121
3-Methoxy-4-hydroxyphenethylamine	LC	77
3-Methoxy-4-hydroxyphenethylglycol	LC	72
3-Methoxy-4-hydroxyphenylacetic acid	LC	72
	PC	82
3-Methoxy-4-hydroxyphenylacetic acid (homovanillic acid)	PC	120
3-Methoxy-4-hydroxyphenylethylamine	TLC	121
β-(3-Methoxy-4-hydroxyphenyl)-ethylamine	PC	60
3-Methoxy-4-hydroxyphenylglycol	TLC	8
3-Methoxy-4-hydroxyphenyl-propionic acid	PC	82
6-Methoxyindole	PC	101
5-Methoxyindole-2-carboxylic acid ethyl ester	PC	101
5-Methoxy-3-indolylacetamide	PC	101, 120
5-Methoxyindolyl acetic acid	PC	102
5-Methoxy-3-indolylacetic acid	PC	101, 120
m-Methoxymandelic acid	PC	120
o-Methoxymandelic acid	PC	120
p-Methoxymandelic acid	GC	5
	PC	120
2-Methoxy-5-methylaniline	PC	6
	TLC	119
4-Methoxy-2-methylaniline	PC	121
	TLC	119
2-Methoxy-5-methyl-1,4-diaminobenzene	TLC	119
3-Methoxy-4,5-methylenedioxy-amphetamine	PC	60
l-3-Methoxy-*N*-methylmorphinan	GC	18
	TLC	138
5-Methoxymethyluracil	PC	36
5-Methoxymethyluridine	PC	36
l-3-Methoxymorphinan	GC	18
	TLC	138
2-Methoxy-1,4-naphthoquinones	PC	111
2-Methoxy-3-OH-1,4-naphthoquinones	PC	111
2-Methoxy-4-nitroaniline	TLC	119
2-Methoxy-5-nitroaniline	TLC	119
2-Methoxy-4-nitro-5-methylaniline	PC	121
3-Methoxynoradrenaline	TLC	121
o-Methoxyphenaceturic acid	PC	80
p-Methoxyphenaceturic acid	GC	5
Methoxyphenamine	TLC	39
p-Methoxyphenol	LC	52
	TLC	174
2-Methoxyphenol	GC	71
	TLC	37
3-Methoxyphenol	GC	69
3-Methoxyphenol, 4-acetyl-2-nitrophenol ether of	TLC	91
m-Methoxyphenylacetic acid	GC	4
o-Methoxyphenylacetic acid	PC	120
p-Methoxyphenylacetic acid	GC	4, 5
p-Methoxyphenylacetylglutamic acid	GC	5
β-(*p*-Methoxyphenyl)-ethylamine	PC	60
4-Methoxy-β-phenylethylamine	GC	9
β-(*p*-Methoxyphenyl)lactic acid	GC	4, 5
β-(*p*-Methoxyphenol)propionic acid	GC	5
3-(*o*-Methoxyphenyl)propionic acid	GC	4
3-(*p*-Methoxyphenyl)propionic acid	GC	4
o-Methoxyphenylpyruvic acid	GC	4
p-Methoxyphenylpyruvic acid	GC	4, 5
Methoxypromazine	GC	17, 108
	TLC	139
Methoxypromazine maleate	GC	17
6-Methoxypurine	LC	46
	PC	36
6-Methoxypurine ribonucleoside	PC	36
6-Methoxypurine riboside	TLC	62
3-Methoxypyrocatechol	PC	121
6-Methoxy-4-quinoline carboxylic acid (quininic acid)	PC	104
6-Methoxy-9-β-D-ribofuranosyl-purine 2′,3′,5′-triacetate	GC	134
6-Methoxytetrahydroharman	PC	101
4′-Methoxy-3′,5,7,-trihydroxyflavanone	TLC	92
5-Methoxytryptamine	GC	9
	LC	77
	PC	101, 103
5-Methoxytryptamine (as HCl)	PC	102
6-Methoxytryptamine	PC	101, 103
6-Methoxytryptamine (as HCl)	PC	102
5-Methoxytryptophan	PC	102
dl-5-Methoxytryptophan	PC	63, 101
3-Methoxy-*dl*-tyrosine	PC	63
4-Methoxyxanthurenic acid	PC	104
8-Methoxyxanthurenic acid	PC	104
Methyclothiazide	TLC	41, 140
2,4-D Methyl	TLC	81
2,4,5-T Methyl	TLC	81
N-Methylacetamide	LC	73
Methyl 4-acetamido-2-ethoxybenzoate	TLC	136
Methyl acetate	GC	2, 49, 52, 55, 57, 58,121, 125
Methyl acetate, esters	GC	126
Methyl acetate, organics	GC	52, 55
4′-Methylacetophenone	PC	23
	TLC	70
Methyl acetoricinoleate	TLC	27
2-Methyl-21-acetoxy-11β,17α-dihydroxy-4-pregnene-3,20-dione	PC	124
2-Methyl-21-acetoxy-9α-fluoro-11β,17α-dihydroxy-4-pregnene-3,20-dione	PC	124
Methylacetylene	GC	118
2-Methyl-3-acetyl-1,4-naphthoquinone	PC	111
10-Methylacridinium chloride	TLC	7

Methyl acrylate	GC	52, 55, 91
Methyl acrylate, organics	GC	52
1-Methyladenine	LC	46
	PC	36
2-Methyladenine	PC	36
2-Methyladenine sulfate	PC	36
3-Methyladenine	PC	36
6-Methyladenine	GC	77
7-Methyladenine	PC	36
9-Methyladenine	PC	36
N[6]-Methyladenosine	LC	22
1-Methyladenosine	LC	22
	PC	36
Methyl adrenaline	PC	5
Methyl alcohol	GC	127
Methyl alkyl guanidine	PC	9
Methyl-1-alkyl-3-nitroguanidine	PC	9
Methyl allyl disulfide	GC	85
Methyl allyl sulfide	GC	85
5-Methyl-5-allylthiobarbituric acid	PC	47
Methyl allyl trisulfide	GC	85
Methylamine	GC	97
	LC	26
	PC	7, 55, 56
	TLC	5, 10
1-Methylamino-anthraquinone	TLC	69
2-Methyl-1-amino-anthraquinone	TLC	69
2-Methyl-3-amino-anthraquinone	TLC	69
2-Methyl-4-amino-anthraquinone	TLC	69
2-Methyl-5(8)-amino-anthraquinone	TLC	69
2-Methyl-6(7)-amino-anthraquinone	TLC	69
4-Methyl-1-amino-anthraquinone	TLC	69
4-Methyl-2-amino-anthraquinone	TLC	69
p-*N*-Methylaminobenzoic acid	TLC	129
2-Methyl-1-amino-4-benzoylamino-anthraquinone	TLC	69
2-Methyl-1-amino-3-bromo-anthraquinone	TLC	69
2-Methyl-1-amino-4-bromo-anthraquinone	TLC	69
2-Methyl-1-amino-3-chloro-anthraquinone	TLC	69
3β-Methylamino-5α-conanine	TLC	98
3β-Methylamino-con-5-enine	TLC	98
6-Methylamino-5-(*N*-formylmethyl-amino)uracil	PC	36
p-Methylaminophenol	PC	120
6-Methylaminopurine	LC	31, 46
	PC	36
6-Methylaminopurine deoxyribonucleoside	PC	36
6-Methylaminopurine ribonucleoside	PC	36
4-Methylamino-1-(3-pyridyl)-1-butanol	PC	2
5-Methylaminouracil	PC	36
Methylamphetamine	GC	22
N-Methyl-*n*-amylamine, trifluoroacetyl derivative	GC	10
Methyl *n*-amyl ketone	LC	16
N-Methylanabasine	PC	2
17α-Methyl-5α-androstane-3β,17β-diol	PC	42
17α-Methyl-androst-5-ene-3β,17β-diol	PC	42
17α-Methyl-5-androstene-3β,17β-diol	GC	79
17α-Methylandrost-4-en-17β-ol-3-one	PC	42
	TLC	98
Methylandrostenolone acetate	TLC	95
N-Methylaniline	GC	102
	PC	6, 59
	TLC	7
2-Methylaniline	GC	46,100, 102
	PC	6
	TLC	119
3-Methylaniline	GC	46,100, 102
	TLC	119
4-Methylaniline	GC	46,100, 102
	TLC	119
2-Methylanisole	GC	75
3-Methylanisole	GC	75
4-Methylanisole	GC	75
Methyl anthracene	LC	27
2-Methylanthracene	GC	40
	TLC	50, 53
9-Methylanthracene	GC	40
	TLC	53
Methyl anthranilate	TLC	105
N-Methylanthranilic acid	PC	78
Methyl β-apo-8′-carotenoate	TLC	94
α-D-Methylarabinopyranoside	GC	76
β-D-Methylarabinopyranoside	GC	76
Methyl β-L-arabinopyranoside	PC	75
2-Methylarabinose	PC	73
2-C-Methyl-L-arabinose	PC	75
Methyl α-L-arabinoside	PC	72
Methyl β-L-arabinoside	PC	72
Methyl arachidate	LC	45
Methyl arachidonate	TLC	154
DL-*threo*-β-Methylaspartic acid	PC	11
Methylated aldonic acids	LC	40
17α-Methyl-4-aza-5α-androstan-17β-ol-3-one	TLC	96
17α-Methyl-4-aza-5-androsten-17β-ol-3-one	TLC	96
4-Methyl-4-aza-5α-cholestan-3-one	TLC	96
4-Methyl-4-aza-5-cholestene	TLC	96
4-Methyl-4-aza-5-cholesten-3-one	TLC	96
6-Methyl-6-aza-B-homo-5α-cholestan-3β-ol	TLC	96
6-Methyl-6-aza-B-homo-5α-cholestan-3β-ol methiodide	TLC	96
4-Methyl-4-aza-5-pregnene-3,20-dione	TLC	96
4-Methyl-4-aza-5-pregnen-20β-ol-3-one	TLC	96
5-Methylbarbituric acid	PC	36
Methyl behenate	LC	45
1′-Methyl-1,2-benzanthracene	PC	90
2′-Methyl-1,2-benzanthracene	PC	90
3-Methyl-1,2 benzanthracene	PC	90
3′-Methyl-1,2-benzanthracene	PC	90
4-Methyl-1,2-benzanthracene	PC	90
4′-Methyl-1,2-benzanthracene	PC	90
5-Methyl-1,2-benzanthracene	PC	90
6-Methyl-1,2-benzanthracene	PC	90
7-Methylbenz[*a*]anthracene	TLC	146

7-Methyl-1,2-benzanthracene	GC	40
	PC	90
9-Methyl-1,2-benzanthracene	PC	90
10-Methyl-1,2-benzanthracene	PC	90
12-Methylbenz[*a*]anthracene	TLC	146
12-Methylbenz[*a*]anthracene-7-carboxaldehyde	TLC	50
12-Methylbenz(*a*)anthracene-7-carboxylic acid	TLC	50
Methylbenzene(toluene)	GC	119
m-Methylbenzoic acid	TLC	37
o-Methylbenzoic acid	TLC	37
p-Methylbenzoic acid	TLC	37
1-Methylbenzo(*c*)phenanthrene	PC	90
2-Methylbenzo(*c*)phenanthrene	PC	90
3-Methylbenzo(*c*)phenanthrene	PC	90
4-Methylbenzo(*c*)phenanthrene	PC	90
5-Methylbenzo(*c*)phenanthrene	PC	90
6-Methylbenzo(*c*)phenanthrene	PC	90
4-Methylbenzo[*a*]pyrene	TLC	146
Methyl 4,6-*O*-benzylidene-2,3-di-*O*-methyl-α-D-glucoside	GC	13
2-Methyl-4-bromo-6-propionylphenol	TLC	89
2-Methyl-6-bromo-4-propionylphenol	TLC	89
2-Methyl-1,3-butadiene	GC	53
2-Methylbutadiene-1,3	GC	125
2-Methylbutane	GC	42, 125
2-Methylbutane (isopentane)	GC	53, 126
	LC	42
2-Methyl-1-butane	GC	47
2-Methyl-2-butane	GC	47
2-Methyl-1-butanal	GC	7
2-Methyl-1-butanol	GC	57, 92, 94
	PC	52
2-Methyl-1-butanol (DNB)	TLC	111
2-Methyl-1-butanol (ONPU)	TLC	111
2-Methyl-2-butanol	GC	57, 92
3-Methyl-1-butanol	GC	57, 92, 94
	PC	52
3-Methyl-1-butanol (DNB)	TLC	111
3-Methyl-1-butanol (ONPU)	TLC	111
3-Methylbutanol-1, butyric acid ester of	GC	51
3-Methylbutanol-1, formic acid ester of	GC	51
3-Methylbutanol-1, isobutyric acid ester of	GC	51
3-Methylbutanol-1, propionic acid ester of	GC	51
3-Methyl-2-butanol	GC	55, 92, 94
3-Methyl-2-butanone	GC	7
3-Methylbutanone-2	GC	51
2-Methyl-1-butene	GC	118, 122
2-Methylbutene-1	GC	29, 51
	LC	42
2-Methyl-2-butene	GC	39,118, 122
2-Methylbutene-2	GC	29, 51
3-Methyl-1-butene	GC	118, 122
3-Methylbutene-1	GC	29, 51
3-Methyl-3-butenyl methyl ether	GC	53
3-Methylbutyl acetate	GC	2
N-Methyl-*n*-butylamine	GC	97
N-Methyl-*n*-butylamine, trifluoroacetyl derivative	GC	10
N-Methyl-*sec*-butylamine	GC	97
1-Methyl-*n*-butylamine	GC	97
5-(Methylbutyl)barbituric acid	GC	105
3-Methylbutyl butyrate	GC	2
3-Methylbutyl crotonate	GC	91
3-Methylbutyl formate	GC	2
3-Methylbutyl hexanoate	GC	2
3-Methylbutyl isobutyrate	GC	2
Methyl butyl ketone	GC	57, 117
Methyl *n*-butyl ketone	LC	16
	PC	20
Methyl *tert*-butyl ketone, 2,4-DNPH of	TLC	166
3-Methylbutyl-methacrylate	GC	91
3-Methylbutyl 3-methylbutyrate	GC	2
3-Methylbutyl 4-methylpentanoate	GC	2
3-Methylbutyl pentanoate	GC	2
3-Methylbutyl propionate	GC	2
Methyl *sec*-butylsulfide	GC	135
N′,*N″*-Methyl butyl thiourea	PC	47
2-Methylbutyraldehyde, 2-hydrazine-benzothiazole derivative of	TLC	168
Methyl butyrate	GC	2, 52, 53, 55
Methyl butyrate, organics	GC	52
α-Methylbutyric(*dl*) acid	TLC	134
β-Methylbutyric acid	TLC	134
2-Methylbutyric acid	TLC	29
2-Methylbutyric acid, methyl ester of	GC	24
8α-(3-Methylbutyryloxy)-4β,15-diacetoxyscrip-9-en-3α-ol	TLC	185
Methyl *n*-caproate	GC	57
Methyl caprylate	TLC	105
Methyl carbamate	GC	98
N-Methylcarbazole	TLC	7
2-Methylcarbazole	TLC	58, 120
9-Methylcarbazole	TLC	120
4-Methylcatechol	TLC	87
Methyl chenodeoxycholate	LC	45
Methyl chloride	GC	52, 127
N-Methyl-4-chloroaniline	GC	102
2-Methyl-3-chloroaniline	TLC	119
2-Methyl-4-chloroaniline	TLC	119
2-Methyl-5-chloroaniline	PC	6
	TLC	119
2-Methyl-6-chloroaniline	TLC	119
3-Methyl-3-chlorobutyl methyl ether	GC	53
4-Methyl-3-chlorobutyl methyl ether	GC	53
Methylchloroform	LC	75
Methylcholanoate	GC	138
3-Methylcholanthrene	GC	40
	TLC	146
20-Methylcholanthrene	GC	40
	PC	89
4α-Methylcholestane-3β-ol	LC	24
4β-Methylcholestane-3β-ol	LC	24
14α-Methylcholestane-3β-ol	LC	24
24β-Methyl-5,7,22-cholestatrien-3β-ol	GC	136
24β-Methyl-5,7,22(9β,10α)-cholestatrien-3β-ol	GC	136
4α-Methylcholest-7,9-diene-3β-ol	LC	24
14α-Methylcholest-7,9-diene-3β-ol	LC	24
4α-Methylcholest-5-ene-3β-ol	LC	24
4β-Methylcholest-5-ene-3β-ol	LC	24

4α-Methylcholest-7-ene-3β-ol LC 24
14α-Methylcholest-7-ene-3β-ol LC 24
14α-Methylcholest-8-ene-3β-ol LC 24
3-Methylchromone TLC 92
5-Methylchrysene TLC 146
Methyl citrate GC 89
N-Methyl corbasil PC 5
3-Methylcoumarin TLC 68
4-Methylcoumarin-7-yl sulfuric acid PC 82
6-Methylcoumarin TLC 68
Methyl cyanide GC 56
Methylcycloheptane GC 51, 118
1-Methylcycloheptanol-1 GC 51
1-Methylcycloheptene-1 GC 51
Methylcyclohexane GC 35, 42, 51, 52, 53, 55, 93,118, 121, 125
LC 42
Methylcyclohexane, cyclohexane GC 126
1-Methylcyclohexanol-1 GC 51
2-Methylcyclohexanone GC 51
3-Methylcyclohexanone GC 51
4-Methylcyclohexanone GC 51
1-Methylcyclohexene-1 GC 51
3-Methylcyclohexen-2-one, 2,4-DNPH of TLC 166
2-α-Methylcyclohexyl-4,6-dimethylphenol TLC 175
Methylcyclooctane GC 51
1-Methylcyclooctanol-1 GC 51
1-Methylcyclooctene-1 GC 51
1-Methylcyclopentadiene GC 118
2-Methylcyclopentadiene GC 118
5-Methylcyclopentadiene GC 118
Methylcyclopentane GC 36, 42, 51, 53, 55, 93,118,121, 125
LC 42
Methylcyclopentane, cyclopentanes GC 126
1-Methylcyclopentanol-1 GC 51
1-Methylcyclopentene GC 41,118, 122
1-Methylcyclopentene-1 GC 51
3-Methylcyclopentene GC 41,118, 122
4-Methylcyclopentene GC 41, 122
α-Methylcysteine PC 64
N-Methylcysteine PC 11
S-Methylcysteine GC 101
LC 59
PC 11, 64
S-Methylcysteine S-oxide PC 11
S-Methylcysteine sulfoxide PC 11
S-Methyl-*l*-cysteine PC 63
3-Methylcytidine LC 22
5-Methylcytidine PC 36
Methylcytisine GC 96
PC 3
Methylcytisine *N*-oxide GC 96
1-Methylcytosine PC 36
5-Methylcytosine GC 77
LC 15, 46
PC 36
TLC 159
4-Methyldaphnetine TLC 92
Methyl *n*-decanoate GC 3
Methyl demeton (Meta-Systox®) TLC 171, 172
Methyl deoxycholate LC 45
Methyldeoxycytidine TLC 159
5-Methyldeoxycytidine PC 36
5-Methyldeoxycytidine-5′-phosphate PC 36
2-Methyl-1,3-diamino-anthraquinone TLC 69
2-Methyl-1,4-diamino-anthraquinone TLC 69
2-Methyl-1,5-diamino-anthraquinone TLC 69
2-Methyl-1,6-diamino-anthraquinone TLC 69
2-Methyl-1,7-diamino-anthraquinone TLC 69
2-Methyl-1,8-diamino-anthraquinone TLC 69
2-Methyl-3,6(7)-diamino-anthraquinone TLC 69
4-Methyl-1,3-diamino-anthraquinone TLC 69
4-Methyl-1,5-diamino-anthraquinone TLC 69
4-Methyl-1,6-diamino-anthraquinone TLC 69
4-Methyl-1,7-diamino-anthraquinone TLC 69
4-Methyl-1,8-diamino-anthraquinone TLC 69
2-Methyl-1,4-diaminobenzene TLC 119
4-Methyl-1,2-diaminobenzene TLC 119
4-Methyl-1,3-diaminobenzene TLC 119
N-Methyldiaminoethane GC 11
N-Methyl-1,3-diaminopropane GC 11
Methyldiborane GC 27
3-Methyl-2,4-dibromo-6-propionylphenol TLC 89
3-Methyl-2,6-dibromo-4-propionylphenol TLC 89
N-Methyldiethanolamine PC 7
N-Methyldiethylamine GC 97
1-Methyl-3,4-diethylbenzene GC 119
1-Methyl-3,5-diethylbenzene GC 119
5-Methyl-5,10-dihydroindeno-[1,2-*b*]indole TLC 120
Methyldihydromorphinone (Metopon) GC 18
TLC 138
17α-Methyl-11α,17β-dihydroxy-4-androsten-3-one PC 124
3-Methyl-3,4-dihydroxy-4-phenylbutyne-1 TLC 137
2-Methyl-2,3-dihydroxypropionic LC 40
O-Methyl-3,5-diiodothyroacetic acid TLC 19
O-Methyl-3,5-diiodothyronine TLC 19
Methyl 2,3:4,6-di-*O*-isopropylidene-α-D-mannopyranoside GC 13
Methyl-3,12-diketocholanate PC 39
Methyl-7,12-diketocholanate PC 39
Methyl-3,11-diketo-3,17(20)-(*cis*)-pregnadien-21-oate PC 124
Methyl 2,3-di-*O*-methyl-4,6-*O*-ethylidene-β-D-mannopyranoside GC 13
Methyl 2,3-di-*O*-methyl-4,6-*O*-isopropylidene-α-D-glucopyranoside GC 13
2-Methyl-3-[2,5-dimethylpyrrolyl-(1)]-butane GC 133
2-Methyl-*N*,*N*-dimethyltryptamine PC 101
3-Methyldiphenyl TLC 53
4-Methyldiphenyl TLC 53
N-Methyldiphenylamine TLC 7
N-Methyldiphenylamine oxide TLC 7
Methyl Δ^{13}-docosaenoate GC 3
Methyl docosahexaenoate GC 3
Methyl-*n*-docosanoate GC 3
Methyl-*n*-dodecanoate GC 3
10-Methyldodecanol, dimethyl acetal of GC 6
11-Methyldodecanol, dimethyl acetal of GC 6
Methyl dodecyl sulfide TLC 163
Methylendomethylenetetrahydrophthalic anhydride, *p*-anisidide of TLC 30

α-Methyl-DOPA	PC	58
Methyl *n*-eicosanoate	GC	3
Methyl eicosapentaenoate	GC	3
Methyl $\Delta^{5,8,11,14}$-eicostatetraenoate	GC	3
4,4′-Methylenebis-(6-*tert*-butyl-2-methylphenol)	TLC	175
4,4′-Methylenebis-(2,6-di-*tert*-butylphenol)	TLC	175
2,2′-Methylenebis-(4-ethyl-6-*tert*-butylphenol)	TLC	175
3,3′-Methylenebis-(4-hydroxycoumarin)	TLC	79
3,3′-Methylenebis-(4-hydroxy-7-methoxycoumarin)	TLC	79
2,2′-Methylenebis-(4-methyl-6-*tert*-butylphenol)	TLC	175
2,2′-Methylenebis(6-α-methylcyclohexyl-4-methylphenol)	TLC	175
3,3′-Methylenebis-(4-propionoxycoumarin)	TLC	79
Methylene chloride	GC	52, 53, 93, 127
Methylene chloride, CH_2Cl_2 (see dichloromethane)	LC	42
Methylenecyclopentane	GC	41
16-Methylene-1-dehydro-11β,17α-dihydroxyprogesterone	TLC	101, 178
3,4-Methylene-dihydroxybenzaldehyde	TLC	71
3,4-Methylenedioxyphenol	TLC	90
1,1-Methylene-dipiperidine-carbon disulfide reaction product	TLC	66
γ-Methylene glutamic acid	PC	64
γ-Methylene glutamine	PC	64
4,5-Methylenephenanthrene	GC	40
16-Methyleneprednisolone	TLC	101, 178
6-Methyl-$\Delta^{8,9}$-ergolene-8-carboxylic acid	TLC	114
Methyl ester 2,4-D	GC	129
Methyl ester DDA	GC	129
Methyl esters	GC	24
3-Methyl estradiol 17β	GC	83
Methylestrenolone	TLC	95
Methyl ethers	GC	116
α-Methyl-β-ethylacrolein, 2,4-DNPH of	TLC	166
N-Methyl-ethylamine, trifluoroacetyl derivative	GC	10
2-Methyl-3-ethyl-5-aminoindole	PC	101
2-Methyl-3-ethylaniline	GC	46, 100
2-Methyl-4-ethylaniline	GC	46, 100
2-Methyl-5-ethylaniline	GC	46, 100
2-Methyl-6-ethylaniline	GC	46, 100
3-Methyl-2-ethylaniline	GC	46
3-Methyl-4-ethylaniline	GC	46, 100
4-Methyl-2-ethylaniline	GC	46
4-Methyl-3-ethylaniline	GC	46
5-Methyl-2-ethylaniline	GC	46
1-Methyl-2-ethylbenzene	GC	119
1-Methyl-3-ethylbenzene	GC	119
1-Methyl-4-ethylbenzene	GC	119
3-Methyl-2-ethyl-1-butene	GC	118
2-Methyl-2-ethylbutyric acid	TLC	29
1-Methyl-1-ethylcyclopentane	GC	36, 121
cis-1-Methyl-2-ethylcyclopentane	GC	36
trans-1-Methyl-2-ethylcyclopentane	GC	36
1-Methyl-*cis*-2-ethylcyclopentane	GC	118
1-Methyl-*trans*-2-ethylcyclopentane	GC	118
cis-1-Methyl-3-ethylcyclopentane	GC	36
trans-1-Methyl-3-ethylcyclopentane	GC	36
1-Methyl-2-ethylcyclopentene	GC	118
2-Methyl-3-ethyl-5-dimethylamino-indole	PC	101
2-Methyl-3-ethylhexane	GC	30
Methyl 4,6-*O*-ethylidene-2,3-di-*O*-methyl-β-D-galactopyranoside	GC	13
Methyl ethyl ketone	GC	7, 52, 55, 93,117, 121
	PC	20
Methyl ethyl ketone (2-butanone)	GC	47, 53, 55, 57
	LC	8
Methyl ethyl ketone, 2,4-DNPH of	TLC	72, 166
Methyl ethyl ketone peroxide, Component A	TLC	65
Methyl ethyl ketone peroxide, Component B	TLC	65
3-Methyl-3-ethylpentane	GC	30, 121
3-Methyl-5-ethylphenol	GC	73
2-Methyl-5-ethylpyridine	GC	78, 131
	TLC	61
2-Methyl-5-ethylpyridine *N*-oxide	TLC	61
4-Methyl-3-ethylpyridine	GC	78
Methyl ethyl sulfide	GC	135
N′,*N″*-Methyl ethyl thiourea	PC	47
1-Methylfluorene	GC	40
2-Methylfluorene	GC	40
2-Methyl-9α-fluoro-11β,17α,21-trihydroxy-4-pregnene-3,20-dione	PC	124
Methyl formate	GC	2, 47, 52, 53, 55, 57
1-*O*-Methylfructose	TLC	26
3-*O*-Methylfructose	TLC	26
6-*O*-Methylfructose	TLC	26
Methyl α-L-fucopyranoside	PC	75
Methyl α-L-fucoside	PC	72
Methyl fumarate	GC	89
2-Methylfuran	GC	55
5-Methylfurfural, 2,4-DNPH of	TLC	73
5-Methyl-2-furfural	PC	20, 22
5-Methyl-2-furfural, 4-NPH of	TLC	74
2-Methylgalactose	PC	73
Methyl α-D-galactoside	PC	72
Methyl β-D-galactoside	PC	72
Methyl gallate	TLC	125
4-Methyl gentisate	TLC	174
N-Methylglucamine	PC	64
Methyl β-D-glucofuranoside	PC	72
α-D-Methylglucopyranoside	GC	76
β-D-Methylglucopyranoside	GC	76
2-*O*-Methylglucose	TLC	26
3-*O*-Methylglucose	PC	70
	TLC	26
6-*O*-Methylglucose	TLC	26
α-Methylglucosides	TLC	26
Methyl α-D-glucoside	PC	72
α-Methyl-D-glucoside	TLC	23
Methyl β-D-glucoside	PC	72
Methylglutamic acid, PTH of	TLC	124
γ-Methylglutamic acid	PC	64
γ-Methylglutamine	PC	64
S-Methylglutathione	PC	11
Methylglyoxal	LC	29
	PC	20
4-Methylguaiacol	TLC	87

Methylguanidine	PC	9, 10
N^2-Methylguanine	LC	46
1-Methylguanine	LC	17, 31, 46
	PC	36
1-Methylguanine sulfate	PC	36
3-Methylguanine	PC	36
7-Methylguanine	LC	22, 31, 46
	PC	36
N^2-Methylguanosine	LC	22
1-Methylguanosine	PC	36
7-Methylguanosine	LC	22, 46
	PC	36
Methyl *n*-heptadecanoate	GC	3
2-Methylheptane	GC	30, 33, 42, 118
3-Methylheptane	GC	30, 33, 42, 47, 121
4-Methylheptane	GC	30, 33, 42
Methyl *n*-heptanoate	GC	3
2-Methyl-2-heptene	GC	39
2-Methyl-*x*-heptene	GC	39
3-Methyl-*x*-heptene	GC	39
3-Methyl-*cis-trans*-2-heptene	GC	39
4-Methyl-*x*-heptene	GC	39
Methyl heptenol	TLC	1
Methylheptenone	TLC	77
Methylheptenone, 2,4-DNPH of	TLC	76
6-Methyl-5-hepten-2-one, 2,4-DNPH of	TLC	165
6-Methyl-5-hepten-2-one, 2,4-DNPSC of	TLC	165
6-Methyl-5-hepten-2-one, *p*-NPH of	TLC	165
1-Methylheptylamine	TLC	7
Methyl heptyl ketone	GC	117
	PC	20
Methyl *n*-heptyl ketone	LC	16
14-Methylhexadecanal, dimethyl acetal of	GC	6
15-Methylhexadecanal, dimethyl acetal of	GC	6
Methyl hexadecanoate (palmitate), dimethyl acetal of	GC	6
Methyl *n*-hexadecanoate	GC	3
Methyl *cis*-Δ^9-hexadecenoate	GC	3
Methyl *trans*-Δ^9-hexadecenoate	GC	3
2-Methylhexane	GC	30, 42, 121
	LC	42
3-Methylhexane	GC	30, 33, 42, 47, 118
Methylhexaneamine	TLC	39
Methyl hexanoate	GC	2
5-Methylhexanone-2	GC	51
2-Methyl-1-hexene	GC	122
2-Methyl-2-hexene	GC	39, 122
2-Methyl-*trans*-3-hexene	GC	39
trans-2-Methyl-3-hexene	GC	122
3-Methyl-1-hexene	GC	122
cis-3-Methyl-2-hexene	GC	122
trans-3-Methyl-2-hexene	GC	122
3-Methyl-*cis*-3-hexene	GC	39
3-Methyl-*trans*-3-hexene	GC	39
cis-3-Methyl-3-hexene	GC	122
trans-3-Methyl-3-hexene	GC	122
4-Methyl-1-hexene	GC	39, 122
4-Methyl-*trans*-2-hexene	GC	39
cis-4-Methyl-2-hexene	GC	122
trans-4-Methyl-2-hexene	GC	122
5-Methyl-1-hexene	GC	122
trans-5-Methyl-2-hexene	GC	122
Methyl *n*-hexyl ketone	LC	16
1-Methylhistamine	PC	60
Methyl histidine	PC	17
N-Methylhistidine	PC	64
1-Methylhistidine	LC	41
	TLC	13, 15, 155
3-Methylhistidine	LC	41
	TLC	13, 15, 155
Methylhomatropine	TLC	117
Methylhydrazine sulfate	PC	61
Methyl hydroperoxide	PC	27
Methylhydroquinone, di(4-acetyl-2-nitrophenol) ether of	TLC	91
17α-Methyl-17β-hydroxy-androst-4-ene-3,11-dione (Group II)	PC	40
17α-Methyl-17β-hydroxy-4-androstene-3,11-dione	PC	124
17α-Methyl-17β-hydroxy-androst-4-en-3-one (Group II)	PC	40
17α-Methyl-17β-hydroxy-4-androsten-3-one	PC	124
5-Methyl-3 hydroxybarbituric acid	PC	36
Methyl-*p*-hydroxybenzoic acid	GC	19
5-Methyl-2-hydroxybenzoic acid	TLC	129
4-Methyl-6-hydroxycoumarin	PC	121
Methyl-3-hydroxy-7,12-diketocholanate	PC	39
Methyl-12-hydroxy-3,7-diketocholanate	PC	39
3-Methyl-5-hydroxypyrazol-5-dimethylcarbamate	GC	130
	TLC	83
N-Methyl-5-hydroxytryptamine (as oxalate)	PC	102
N-ω-Methyl-5-hydroxytryptamine	PC	101
Methyl hyodeoxycholate	LC	45
1-Methylhypoxanthine	LC	46
	PC	36
2-Methylhypoxanthine	PC	36
3-Methylhypoxanthine	PC	36
7-Methylhypoxanthine	LC	46
	PC	36
8-Methylhypoxanthine	PC	36
9-Methylhypoxanthine	LC	46
	PC	36
1-Methylimidazole	PC	17
1-Methylimidazole-4-acetic acid	PC	83
1-Methylindan	GC	119
4-Methylindan	GC	119
5-Methylindan	GC	119
1-Methylindene	GC	119
2-Methylindene	GC	119
3-Methylindene	GC	119
1-Methylindole	GC	45
2-Methylindole	GC	45
	PC	103
	TLC	58
3-Methylindole	GC	45
	PC	103
	TLC	58
5-Methylindole	GC	45
	TLC	58
7-Methylindole	TLC	58
7-Methyl-3-indolealdehyde	PC	101

1-Methylindolo[*b*-3,2]-5α-cholest-2-ene TLC 96
1-Methylindolo[*b*-3,4]-5β-cholest-3-ene TLC 96
Methyl indolyl-acetate PC 102
7-Methyl-3-indolylacetonitrile PC 102
1-Methylinosine LC 22
PC 36
TLC 62
1-Methylinosine 2′,3′,5′-triacetate GC 134
7-Methylinosine LC 22, 46
PC 36
Methyl iodide LC 75
α-Methylionone, 2,4-DNPH of TLC 76
β-Methylionone, 2,4-DNPH of TLC 76
5-Methylisatin PC 101
N-Methyl-isoamylamine, trifluoroacetyl derivative GC 10
N-Methyl-isobutylamine, trifluoroacetyl derivative GC 10
Methyl isobutyl ketone LC 16
Methyl isobutyl ketone peroxide, Component A TLC 65
Methyl isobutyl ketone peroxide, Component B TLC 65
Methyl isobutyl sulfide GC 135
Methyl isobutyrate GC 2, 52, 57
Methyl isocyanate GC 44
(-)-2-Methylisonicotinic acid α-phenylethylamide TLC 158
cis-1-Methyl-4-isopentylcyclohexane GC 35
trans-1-Methyl-4-isopentylcyclohexane GC 35
N-Methylisopropylamine GC 97
N-Methylisopropylamine, trifluoroacetyl derivative GC 10
1-Methyl-2-isopropylbenzene GC 119
1-Methyl-3-isopropylbenzene GC 119
1-Methyl-4 isopropylbenzene GC 119
LC 42
2-Methyl-5-isopropylbenzyl *o*-tolyl sulfide TLC 163
2-Methyl-5-isopropylbenzyl *p*-tolyl sulfide TLC 163
3-Methyl-2-isopropyl-1-butene GC 39
Methyl 4,6-*O*-isopropylidene-2,3-di-*O*-benzoyl-α-D-glucoside GC 13
Methyl 4,6-*O*-isopropylidene-2,3-di-*O*-methanesulfonyl-α-D-glucoside GC 13
3-*O*-Methyl-1,2-*O*-isopropylidene-D-fructose GC 13
Methyl 3,4-*O*-isopropylidene-2-*O*-methanesulfonyl-β-D-arabinoside GC 13
Methyl 3,4-*O*-isopropylidene-2-*O*-toluene-*p*-sulfonyl-β-D-arabinoside GC 13
Methyl isopropyl ketone GC 52
LC 8
PC 20
2-Methyl-5-isopropylphenol GC 72
4-Methyl-2-isopropylphenol GC 72
5-Methyl-2-isopropylphenol GC 72
2-Methyl-5-isopropylphenylhexyl sulfide TLC 163
3-Methyl-5-isopropylphenylmethyl carbamate (Carbamult) TLC 170
2-Methyl-5-isopropylthiophenol TLC 162
1-Methylisoquinoline GC 78
TLC 58
3-Methylisoquinoline GC 45, 78
TLC 58
α-Methylisoserine PC 64
Methyl *iso*-valerate GC 57
Methyl-12-keto-3-acetoxycholanate PC 39
Methyl-12-ketocholanate PC 39
Methyl-7-keto-3,12-dihydroxycholanate PC 39
Methyl α-ketoglutarate GC 89
Methyl-12-keto-3-hydroxycholanate PC 39
Methyl lactate GC 89
Methyl β-lactoside PC 72
β-Methyl lanthionine PC 64
Methyl laurate GC 89
LC 45
Methyl lignocerate LC 45
Methyl linoleate TLC 154
Methyl linolenate TLC 154
Methyl malate GC 89
Methyl β-maltoside PC 72
Methyl α-D-mannofuranoside PC 72
α-D-Methylmannopyranoside GC 76
4-Methylmannose PC 73
α-Methyl-D-mannoside TLC 23
Methyl α-D-mannoside PC 72
Methyl mercaptan GC 85
2-Methylmercapto-1,4-naphthoquinone PC 111
6-Methylmercaptopurine LC 46
PC 36
Methylmercury dithizonate TLC 161
N-Methylmescaline PC 60
Methyl methacrylate GC 55, 91
LC 21
S-Methyl methionine PC 64
2-Methyl-5-methoxy-3-indolylacetamide PC 101
1-Methyl-6-methylamino-5-(*N*-formylmethylamino)-uracil PC 36
3-Methyl-6-methylamino-5-(*N*-formylmethylamino)-uracil PC 36
3-Methyl-6-methylaminouracil PC 36
N-Methyl-2-methylaniline GC 102
N-Methyl-3-methylaniline GC 102
Methyl-2-C-methyl-β-L-arabinopyranoside PC 75
Methyl-12-methylbenz(*a*)anthracene-7-carboxylate TLC 50
Methyl 3-methylbutyrate GC 2
Methyl 8-methyldecanoate GC 3
Methyl 10-methyldodecanoate GC 3
N-1-Methyl-2-methyl-3-ethyl-5-dimethylamino-indole PC 101
Methyl 6-methylheptanoate GC 3
Methyl 14-methylhexadecanoate GC 3
Methyl 4-methylhexanoate GC 3
Methyl 8-methylnonanoate GC 3
Methyl 6-methyloctanoate GC 3
Methyl 14-methylpentadecanoate GC 3
Methyl 4-methylpentanoate GC 2
Methyl 12-methyltetradecanoate GC 3
O-(3-Methyl-4-methylthiophenyl)-*O,O*-dimethyl phosphorothioate (Fenthion) TLC 170
Methyl 10-methylundecanoate GC 3
N-Methyl-4-morpholine thiocarboxamide PC 47
N-Methylmyosmine PC 2

Methyl myristate LC 45
β-Methylnaphthalene LC 7
1-Methylnaphthalene GC 38
TLC 51, 53
2-Methylnaphthalene GC 38
TLC 51, 53
2-Methyl-1,4-naphthoquinones PC 111
2-Methyl-1-naphthylamine pentafluoropropionamide GC 99
3-Methyl-2-naphthylamine pentafluoropropionamide GC 99
Methyl α-naphthyl ketone, 2,4-DNPH of TLC 73
Methyl β-naphthyl ketone, 2,4-DNPH of TLC 73
N-Methylnicotinamide GC 95
N'-Methylnicotinamide chloride PC 18
N'-Methylnicotinuric acid betaine PC 18
2-Methyl-3-nitroaniline TLC 119
2-Methyl-4-nitroaniline TLC 119
2-Methyl-5-nitroaniline TLC 119
1-Methyl-3-nitroguanidine PC 9
Methyl-*p*-nitrophenyl GC 139
N-Methyl-*N*-nitrosoaniline TLC 7
O-(3-Methyl-4-nitrophenyl)-*O,O*-dimethyl phosphorothioate (Fenitrothion) TLC 170
2-Methyl-5-nitrotryptamine PC 101
4-Methylnonane GC 30
Methyl *n*-nonanoate GC 3
Methyl nonyl ketone GC 117
Methyl *n*-nonyl ketone LC 16
Methyl *cis*-$\Delta^{9,12}$-octadecadienoate GC 3
Methyl *n*-octadecanoate GC 3
Methyl $\Delta^{9,12,15}$-octadecatrienoate GC 3
Methyl *cis*-Δ^{4}-octadecenoate GC 3
Methyl *trans*-Δ^{4}-octadecenoate GC 3
Methyl *cis*-Δ^{6}-octadecenoate GC 3
Methyl *cis*-Δ^{9}-octadecenoate GC 3
Methyl *trans*-Δ^{9}-octadecenoate GC 3
2-Methyloctane GC 30
3-Methyloctane GC 30
4-Methyloctane GC 30
Methyl *n*-octanoate GC 3
Methyl octyl ketone GC 117
Methyl *n*-octyl ketone LC 16
Methyl octyl sulfide TLC 163
Methyl oleate TLC 154
Methyl Orange PC 88
5-Methylorotic acid PC 36
Methyl orsellate PC 30
17α-Methyl-4-oxa-5-androsten-17β-ol-3-one acetate TLC 96
Methyl palmitate TLC 154
Methyl paraoxon GC 67
TLC 172
Methyl parathion GC 60, 64, 65, 67, 68, 128, 129
PC 28
TLC 80, 84, 171, 172
13-Methyl pentadecanal, dimethyl acetal of GC 6
14-Methyl pentadecanal, dimethyl acetal of GC 6
Methyl *n*-pentadecanoate GC 3
2-Methylpentadiene GC 118
3-Methyl-1,3-pentadiene GC 118
2-Methylpentane GC 30, 33, 42, 47, 51, 53, 58, 118, 121, 125, 126
LC 42
3-Methylpentane GC 30, 33, 42, 47, 53, 118, 121
LC 38, 42
2-Methyl-2,4-pentanediol TLC 113
Methyl pentanoate GC 2
Methyl *n*-pentanoate GC 3
4-Methyl-2-pentanone GC 53
4-Methylpentanone-2 GC 51
2-Methyl-1-pentene GC 118, 122
2-Methylpentene-1 GC 29
LC 21, 38, 42
2-Methyl-2-pentene GC 118, 122
3-Methyl-1-pentene GC 118, 122
3-Methylpentene-1 LC 42
cis-3-Methyl-2-pentene GC 122
trans-3-Methyl-2-pentene GC 122
3-Methyl-*cis*-2-pentene GC 39, 118
3-Methyl-*trans*-2-pentene GC 118
4-Methyl-1-pentene GC 118, 122
4-Methylpentene-1 LC 21, 38
cis-4-Methyl-2-pentene GC 122
trans-4-Methyl-2-pentene GC 122
4-Methyl-*cis*-2-pentene GC 118
4-Methyl-*trans*-2-pentene GC 118
cis-4-Methylpentene-2 GC 51
trans-4-Methylpentene-2 GC 51
4-Methyl-3-penten-2-one GC 7
Methyl pentyl ketone GC 117
Methylpentynol carbamate TLC 38
m-Methylperoxybenzoic acid PC 107
p-Methylperoxybenzoic acid PC 107
1-Methylphenanthrene GC 40
TLC 51
Methylphenethylhydrazine TLC 39
α-Methylphenethylhydrazine GC 17
α-Methylphenethylhydrazine hydrochloride GC 17
2-Methylphenetole GC 75
3-Methylphenetole GC 75
4-Methylphenetole GC 75
Methylphenidate GC 17
TLC 39
Methylphenidate hydrochloride GC 17
Methylphenobarbital TLC 137
Methylphenobarbitone PC 86
p-Methylphenol TLC 176
2-Methylphenol GC 69, 72, 75
LC 51
TLC 174
3-Methylphenol GC 69, 75, 130
LC 51
TLC 174
4-Methylphenol GC 69, 75
TLC 87, 174
4-Methyl-3-phenyl-4-aza-2,5-cholestadiene TLC 96
4-Methyl-1,3-phenylenediamine TLC 119
1-Methyl-1-phenylethyl hydroperoxide TLC 65
1-Methyl-1-phenylethyl xanthydryl peroxide TLC 65

3-Methyl-3-phenylglutaric acid	PC	83
1-Methyl-3-phenylisoquinoline	TLC	58
3-Methylphenyl-*N*-methylcarbamate	GC	130
Methylphenyl phosphate	PC	112
1-Methyl-1-phenylpropyl hydroperoxide	TLC	65
2-Methyl-3-phenylquinoline	TLC	58
N',*N''*-Methylphenyl thiourea	PC	47
4-Methylpicolinic acid methyl ester	TLC	158
6-Methylpicolinic acid methyl ester	TLC	158
N-Methyl-2-piperidinecarboxylic acid	PC	63
N-Methyl-1-piperidine thiocarboxamide	PC	47
6α-Methylprednisolone	TLC	101
16α-Methyl-pregn-4-ene-3,20-dione	PC	42
16α-Methyl-5α-pregn-2-ene-11,20-dione	GC	80
16α-Methyl-5α-pregn-2-en-20-one	GC	80
2-Methylpropane	GC	42,122, 125, 126
2-Methyl-1-propanol	GC	92
	PC	52
2-Methylpropanol-1, butyric acid ester of	GC	51
2-Methylpropanol-1, formic acid ester of	GC	51
2-Methylpropanol-1, isobutyric acid ester of	GC	51
2-Methylpropanol-1, propionic acid ester of	GC	51
2-Methyl-2-propanol	GC	92, 118
2-Methylpropanol-2, butyric acid ester of	GC	51
2-Methylpropanol-2, formic acid ester of	GC	51
2-Methylpropanol-2, isobutyric acid ester of	GC	51
2-Methylpropanol-2, propionic acid ester of	GC	51
Methyl propionate	GC	2, 52, 55, 57
2-Methylpropionic acid	TLC	29
2-Methyl-4-propionylphenol	TLC	89
2-Methyl-6-propionylphenol	TLC	89
3-Methyl-4-propionylphenol	TLC	89
3-Methyl-6-propionylphenol	TLC	89
4-Methyl-6-propionylphenol	TLC	89
2-Methylpropyl acetate	GC	2
2-Methylpropyl acrylate	GC	91
N-Methyl-*n*-propylamine, trifluoroacetyl derivative	GC	10
1-Methyl-2-*n*-propylbenzene	GC	119
1-Methyl-3-*n*-propylbenzene	GC	119
1-Methyl-4-*n*-propylbenzene	GC	119
2-Methylpropyl butyrate	GC	2
3-Methylpropyl crotonate	GC	91
Methyl *n*-propyl disulfide	GC	85
2-Methylpropyl formate	GC	2
2-Methylpropyl hexanoate	GC	2
2-Methylpropyl isobutyrate	GC	2
Methyl propyl ketone	GC	57, 117
	LC	8
Methyl propyl ketone, 2,4-DNPH of	TLC	72, 166
Methyl *n*-propyl ketone	GC	47, 52
	PC	20
2-Methylpropyl methacrylate	GC	91
2-Methylpropyl 3-methylbutyrate	GC	2
2-Methylpropyl 4-methylpentanoate	GC	2
2-Methylpropyl pentanoate	GC	2

2-Methyl-6-*n*-propylphenol	GC	72
2-Methylpropyl propionate	GC	2
Methyl *n*-propyl trisulfide	GC	85
6-Methylpurine	LC	46
	PC	36
1-Methylpyrene	GC	40
3-Methylpyrene	GC	40
	TLC	53
4-Methylpyrene	TLC	146
Methyl pyrethrate	PC	119
2-Methylpyridine	GC	131
	TLC	60, 61
3-Methylpyridine	GC	131
	TLC	60
4-Methylpyridine	GC	131
	TLC	60
6-Methylpyridine-2-aldehyde	PC	2
2-Methylpyridine-5-carboxylic acid	PC	2
6-Methylpyridine-2-carboxylic acid	PC	2
2-Methylpyridine *N*-oxide	TLC	61
3-Methylpyridine *N*-oxide	TLC	61
N-Methyl-2-pyridone-5-carboxamide	PC	60
N-Methyl-2-pyridone-3-carboxylic acid	PC	18
N'-Methyl-2-pyridone-5-carboxylic acid	PC	18
N-Methyl-2-pyridone-5-carboxylic amide	PC	18
2-Methyl-6-(3-pyridyl)-tetrahydro-1,2-oxazine·HCl	PC	2
3-Methylpyrocatechol	TLC	176
4-Methylpyrocatechol	TLC	176
1-Methylpyrrole	GC	133
2-Methylpyrrole	GC	133
N-Methyl-1-pyrrolidinethiocarboxamide	PC	47
Methyl pyruvate	GC	89
3-*O*-Methylquercitin	LC	28
2-Methylquinizarin	PC	109
2-Methylquinoline	GC	78
	TLC	58
3-Methylquinoline	TLC	58
4-Methylquinoline	GC	45, 78
	TLC	58
5-Methylquinoline	GC	78
6-Methylquinoline	GC	45, 78
	TLC	58
7-Methylquinoline	GC	78
	TLC	58
8-Methylquinoline	GC	45, 78
	TLC	58
Methyl red	PC	88
Methyl reserpate	PC	2
2-Methylresorcinol, di(4-acetyl-2-nitrophenol) ether of	TLC	91
Methyl α-L-rhamnopyranoside	PC	75
α-L-Methylrhamnopyranoside	GC	76
3-Methylrhamnose	PC	73
Methyl α-L-rhamnoside	PC	72
Methyl-β-D-ribopyranoside	PC	75
β-D-Methylribopyranoside	GC	76
Methyl-2-C-methyl-β-L-ribopyranoside	PC	75
2-C-Methyl-L-ribose	PC	75
Methyl salicylate	GC	105
17α-Methyl-3,5-seco-4-nor-androstan-17β-ol-5-on-3-oic acid	TLC	96
Methyl Selenac	TLC	66

Methylserine, PTH of	TLC	124
α-Methylserine	PC	64
N-Methylserotonin	PC	103
Methylsilane	GC	27
Methyl stearate	LC	45
	TLC	57, 154
α-Methylstyrene	GC	119
β-Methylstyrene (*trans*)	GC	119
Methyl succinate	GC	89
3-Methyl-6-succino-1,4-naphthoquinones	PC	111
5-Methyl-3-sulfanilamidoisoxazole	TLC	44
3-Methylsulfolane	GC	139
3-Methylsulfolene	GC	139
Methylsulfonal	GC	139
Methyl sulfoxide	GC	139
N-Methyltaurine	PC	63
Methyltestosterone	TLC	95
17-Methyltestosterone	PC	124
17α-Methyltestosterone	TLC	178
Methyltestosterone acetate	TLC	95
12-Methyltetradecanal, dimethyl acetal of	GC	6
13-Methyltetradecanal, dimethyl acetal of	GC	6
Methyl *n*-tetradecanoate	GC	3
Methyl *cis*-Δ^9-tetradecenoate	GC	3
O-Methyl-tetraiodothyroacetic acid	TLC	19
Methyl 2,3,4,6-tetra-O-methyl-α-D-galactopyranoside	GC	28
Methyl 2,3,4,6-tetra-O-methyl-α-D-glucopyranoside	GC	28
Methyl 2,3,4,6-tetra-O-methyl-α-D-mannopyranoside	GC	28
2-Methyl-1-thiadecalin S-oxide (*cis* + *trans*)	TLC	164
Methyl-3-thienyl sulfide	TLC	104
Methyl thiolcarbamate	GC	98
2-Methylthiophene	TLC	104
3-Methylthiophene	TLC	104
p-(Methylthio)-phenol	PC	121
4-Methylthiosemicarbazide	PC	47
6-Methylthiouracil	PC	37
6-Methyl-2-thiouracil	PC	36
N-Methylthiourea	PC	47
3-Methyl-4-thiouridine	TLC	62
3-Methylthymidine	GC	134
	TLC	62
3-Methylthymidine 3′,5′-diacetate	GC	134
1-Methylthymine	PC	36
O-Methylthyroxine	TLC	19
5-Methyltocol	GC	142
N-Methyl-*o*-toluidine	TLC	7
2-Methyl-3-*o*-tolyl-4(3H)-quinazolinone	TLC	137
2-Methyl-1,4,5(8)-triaminoanthraquinone	TLC	69
4-Methyl-1,5,8-triaminoanthraquinone	TLC	69
2-Methyl-4,5,6-triaminopyrimidine sulfate	PC	36
11-Methyltridecanal, dimethyl acetal of	GC	6
12-Methyltridecanal, dimethyl acetal of	GC	6
Methyl *n*-tridecanoate	GC	3
Methyl 2,3,4-tri-*O*-ethyl-α-D-glucopyranoside	TLC	26
Methyl 2,3,4-tri-*O*-ethyl-β-D-glucopyranoside	TLC	26
2-Methyl-11β,17α,21-trihydroxy-4-pregnene-3,20-dione	PC	124
O-Methyl-3,5,3′-triiodothyroacetic acid	TLC	19
O-Methyl-3,5,3′-triiodothyronine	TLC	19
Methyl-3,7,12-triketocholanate	PC	39
Methyl 2,3,4-tri-O-methyl-β-L-arabopyranoside	GC	28
Methyl 2,3,4-tri-*O*-methyl-α-D-glucopyranoside	TLC	26
Methyl 2,3,4-tri-*O*-methyl-β-D-glucopyranoside	TLC	26
Methyl 2,3,4-tri-*O*-methyl-β-D-xylopyranoside	GC	28
Methyl Trithion®	GC	60, 65, 67
	TLC	84
Methyltropine	TLC	117
N-Methyltryptamine	GC	9
	PC	101, 103
	TLC	59
N-Methyltryptamine (as HCl)	PC	102
α-Methyltryptamine	PC	101, 103
α-Methyltryptamine (as acetate)	PC	102
1-Methyltryptamine	PC	103
1-Methyltryptamine (as HCl)	PC	102
2-Methyltryptamine	PC	103
2-Methyltryptamine (as creatinine sulfate)	PC	102
5-Methyltryptamine	PC	101
7-Methyltryptamine	PC	101
N-Methyltryptophan	PC	102, 103
DL-5-Methyltryptophan	TLC	156
6-Methyltryptophan	PC	63, 101
Methyl Tuads	TLC	66
dl-α-Methyl-*m*-tyrosine	PC	63
Methylumbelliferone	TLC	84
Methyl *n*-undecanoate	GC	3
1-Methyluracil	PC	36
3-Methyluracil	PC	36
6-Methyluracil	LC	17
	PC	36, 37
Methylurea	LC	63
	TLC	122
1-Methyluric acid	PC	36
3-Methyluric acid	PC	36
7-Methyluric acid	PC	36
3-Methyluridine	GC	134
	LC	22
	PC	36
3-Methyluridine 2′,3′,5′-triacetate	GC	134
5-Methyluridine	GC	77
	PC	36
5-Methyluridine (thymidine)	TLC	62
5-Methyluridine-5′-phosphate (thymidine-5′-phosphate)	TLC	62
5-Methyluridylyl-(3′,5′)-uridine	TLC	62
Methyl ursodeoxycholate	LC	45
2-Methylvaleraldehyde, 2,4-DNPSC of	TLC	165
4-Methylvaleric acid	TLC	29
α-G-β-Methylvaleric carboxyalkylguanidine	PC	9
α-G-β-Methylvaleric guanidine	PC	9
Methylvanillin (3,4-dimethoxybenzaldehyde)	TLC	68

Methyl violet	PC	88
1-Methylxanthine	LC	46
	PC	36
3-Methylxanthine	PC	36
7-Methylxanthine	GC	77
	LC	46
	PC	36
α-D-Methylxylopyranoside	GC	76
2-Methylxylose	PC	73
Methyl α-D-xyloside	PC	72
Methyl Zimate	TLC	66
Methymycin	PC	66
Methyprylon	GC	105
Methyprylon (Noludar®)	GC	21
	TLC	137
Methyprylone	GC	19
Metobromuron	TLC	173
Metopon (Methyldihydromorphinone)	TLC	138
Metoxymarc	TLC	173
Metyrapone	GC	22
Mevinphos	GC	65, 67
	LC	43
Mevinphos (Phosdrin®)	TLC	84, 171
Mexogenin	TLC	98
Mg (Magnesium)	LC	81, 82, 83, 85, 86, 87, 89
	TLC	52
Mg(II), Magnesium	GC	43
	PC	15, 95, 96
	TLC	148, 149
Mg^{2+}, Magnesium	PC	92
Microphyllic acid	PC	30
Mimosine	PC	63
Mirex	GC	60, 68, 128, 129
	TLC	80
Mitomycin C	TLC	20
Mitox	PC	116
Mixture of *tert*-octyl and *tert*-butyl cresols	TLC	175
MMTrThymidine	PC	19
MMTrTpC^{An}-OAc	PC	19
MMTrTpG^{Ac}-OAc	PC	19
MMTrTpT-OAc	PC	19
Mn (Manganese)	LC	81, 82, 83, 85, 86, 87, 88, 89
Mn(II), Manganese	GC	43
	PC	15, 95, 96
	TLC	149, 150
Mn^{2+}, Manganese	PC	92
Mn(III), Manganese	GC	43
MnO_4^-	TLC	147
Mo (Molybdenum)	LC	80, 81, 82, 83, 84, 85, 86, 87
Mo(VI), Molybdenum	PC	96
	TLC	149
MoO_2, Molybdenum	GC	43
MoO_4^{2-}, Molybdate	PC	97
MoO_4^{--}	TLC	147
MoO_4^{-2}, Molybdenum	PC	95
Mobam	TLC	170
Mobam (4-benzothienyl *N*-methylcarbamate)	GC	61
Mobam®, phenol	GC	65
MO-Cytidine	GC	77
MO-Deoxycytidine	GC	77
Moditen (or Sevinol)	TLC	42
Molinate	GC	65
Molybdate	PC	14, 91, 93
Molybdate, MoO_4^{2-}	PC	97
Molybdenum (Mo)	LC	80, 81, 82, 83, 84, 85, 86, 87
Molybdenum, Mo(VI)	PC	96
	TLC	149
Molybdenum, MoO_2	GC	43
Molybdenum, MoO_4^{-2}	PC	95
Molybdenum dioxydichloride	GC	50
6-Monacetylmorphine	GC	18
Monethyl phosphate	PC	112
Monoacetate	TLC	112
3,O-Monoacetylmorphine	GC	107
6-Monoacetylmorphine	TLC	138
6,O-Monoacetylmorphine	GC	107
1-Monoacetyloleandomycin	PC	66
2-Monoacetyloleandomycin	PC	66
3-Monoacetyloleandomycin	PC	66
Monoanhydro-gitoxigenin (3β,16β-Dihydroxy-5β-cardadien-14,20(22)-olide)	TLC	102
Monobutyrate	TLC	112
Mono(*p*-chlorophenyl)phosphate	PC	112
Monochlorphenamide	TLC	140
Mono-condensed phosphate	TLC	151
Monocyclics	GC	87
O-Mono-DNAP-Tyrosine	TLC	14
Monoethanolamine	LC	26
	PC	7
Monoformate	TLC	112
Monohydroxy compounds	GC	136
Monohydroxydiketone compounds	GC	136
Monohydroxymonoketone compounds	GC	136
3-Monoiodohistidine	TLC	19
3-Monoiodohistidine methyl ester	TLC	19
Monoiodomethanesulfonic acid	TLC	64
3-Monoiodotyrosine	TLC	19
3-Monoiodotyrosine methyl ester	TLC	19
Monoisopropanolamine	PC	7
Monolinuron	TLC	173
1-Monoölein	TLC	56
α-Monopalmitin	PC	99
1-Monopalmitin	TLC	56
Monophenyl phosphate	PC	112
Monophenyl pyrophosphate	PC	112
Monophenylthiourea, PTH of	TLC	124
Monophosphate $(PO_4)^{-3}$	PC	98
Monopropionate	TLC	112
Monostearin (1,2-diOH)	LC	45
Monotolyl phosphate	PC	112
Monovalerate	TLC	112
Monovinylmonohydroxyethyl-deuteroporphyrin IX	PC	33, 123
Monuron	GC	61, 129
	TLC	81, 82, 173
Morestan®	GC	65
Morin	LC	1, 28
	PC	26
	TLC	92
Morocide® (Binapacryl)	TLC	80
Morolic	TLC	184

Morphine GC 8, 18, 21, 96,105, 107, 111
LC 55
PC 2, 3, 4
TLC 2, 3, 4, 39, 45, 116, 138
Morphine hydrochloride LC 20
Morphine sulfate GC 107
Morpholine LC 26
TLC 10
Morphothion LC 43
2-MT TLC 66
Muramine PC 3
ϵ-Muurolene GC 140
γ-Muurolene GC 140
Myoglobin (horse) LC 19, 64
Myoinositol LC 47
Myoinositol diphosphate LC 47
Myoinositol hexaphosphate LC 47
Myoinositol monophosphate LC 47
Myoinositol pentaphosphate (chicken blood) LC 47
Myoinositol tetraphosphate LC 47
Myoinositol triphosphate LC 47
Myoinositol tripyrophosphate LC 47
Myosmine GC 95
PC 2
Myrcene GC 86, 87
TLC 51
β-Myrcene TLC 106
Myricetin PC 26
TLC 92
Myricitin LC 28
Myricitrin PC 26
Myristate LC 67
Myristate, β-sitosteryl ester of TLC 182
Myristic acid LC 57
PC 77
TLC 33, 57
Myristic acid, ethyl ester of GC 24
Myristic acid, methyl ester of GC 24
Myristicin TLC 90
Myristyl acetate LC 45
Myristyl alcohol LC 45

Adsorption of the Elements (LC Section – Figures)

Mg, Magnesium Pg 278, 279, 280, 281
Mn, Manganese Pg 278, 279, 280, 281
Mo, Molybdenum Pg 278, 279, 280, 281

Na(I) (Sodium) GC 43
LC 83, 85, 86, 88
PC 95, 96
TLC 52
Nabam PC 117
Nacpaine HCl GC 110
NADase I LC 64
NADase II LC 64
Nafcillin Na TLC 21
Naled (Dibrom) GC 60, 65, 67
TLC 80, 84
Nalorphine GC 18
PC 4
TLC 138
Nalorpine GC 107
Naphazoline TLC 39
Naphthacene GC 40
9-Naphthacenone TLC 92
2-Naphthaldehyde PC 20
2-Naphthaldehyde, 4-NPH of TLC 74
Naphthalene GC 38, 40, 45, 51, 119
LC 7, 21, 27, 42, 45
TLC 7, 51, 53
Naphthalene derivative 1-sulfonic acid PC 46
Naphthalene derivative 2-sulfonic acid PC 46
Naphthalene derivative 1,5-disulfonic acid PC 46
Naphthalene derivative 1,6-disulfonic acid PC 46
Naphthalene derivative 1,7-disulfonic acid PC 46
Naphthalene derivative 2,6-disulfonic acid PC 46
Naphthalene derivative 2,7-disulfonic acid PC 46
Naphthalene derivative 1,3,5,7-tetrasulfonic acid PC 46
Naphthalene derivative 1,3,5-trisulfonic acid PC 46
Naphthalene derivative 1,3,6-trisulfonic acid PC 46
Naphthalene derivative 1,3,7-trisulfonic acid PC 46
5H-Naphtho[2,3-*c*]carbazole TLC 120
1-Naphthoic acid TLC 37
2-Naphthoic acid TLC 37
α-Naphthol PC 31
TLC 81, 83, 90,174, 176
β-Naphthol PC 31
TLC 90,174, 176
α-Naphthol, 4-acetyl-2-nitrophenol ether of TLC 91
β-Naphthol, 4-acetyl-2-nitrophenol ether of TLC 91
1-Naphthol GC 71, 130
TLC 92
2-Naphthol GC 71
TLC 92
β-Naphthol ethyl ether TLC 90
β-1-(Naphth-2-ol)alanine PC 64
β-Naphthol isobutyl ether TLC 90
β-Naphthol methyl ether TLC 90
Naphthol Yellow TLC 49
1,2-Naphthoquinone derivative 1-diazide PC 46
1,2-Naphthoquinone derivative 2-diazide PC 46
1,2-Naphthoquinone derivative 4-sulfonic acid 1-diazide PC 46
1,2-Naphthoquinone derivative 4-sulfonic acid 2-diazide PC 46
1,2-Naphthoquinone derivative 5-sulfonic acid 1-diazide PC 46

1,2-Naphthoquinone derivative 5-sulfonic acid 2-diazide	PC	46
1,2-Naphthoquinone derivative 6-sulfonic acid 1-diazide	PC	46
1,2 Naphthoquinone derivative 7-sulfonic acid 1-diazide	PC	46
1,2-Naphthoquinone derivative 8-sulfonic acid 1-diazide	PC	46
1,2-Naphthoquinone-3-sulfonic acid	PC	111
1,4-Naphthoquinones	PC	111
1-Naphthylamine	GC	45
	PC	6
	TLC	7,119, 120
1-Naphthylamine pentafluoropropionamide	GC	99
2-Naphthylamine	GC	45
	PC	6
	TLC	119, 120
2-Naphthylamine pentafluoropropionamide	GC	99
1-Naphthyl methyl carbamate (Carbaryl)	TLC	170
β-Naphthyl phenyl sulfoxide	TLC	164
β-Naphthylphosphate	PC	112
β-Naphthyl-α-pyridyl-phosphate	PC	112
Narbomycin	PC	66
	TLC	153
Narceine	TLC	3, 4
Narconumal	GC	106
Narcotine	TLC	45
Narcotine (Noscapine)	GC	111
	PC	2, 3
	TLC	2, 3, 4,116, 138
Narcotoline	TLC	4
Naringenin	LC	1, 28
	PC	26
Naringin	LC	28
	PC	26
Naugawhite	TLC	175
Nb (Niobium)	LC	83, 84, 85, 86
Nd(III) (Neodymium)	GC	9, 43
	LC	39, 79, 80, 90, 91
NDGA (Nordihydroguaiaretic acid)	PC	69
Nealbarbitone	GC	19
	PC	86
	TLC	137
Neamine (Neomycin A)	PC	67
Neburon	GC	129
	TLC	173
Nemacide® (VC-13)	GC	65
	PC	28
Neoabietate	LC	67
Neo-Antergan® (Pyrilamine)	TLC	45
Neodiadinoxanthin	PC	122
Neo-digoxin	TLC	102
Neo-digoxin peracetate	TLC	102
Neodymium (Nd)(III)	GC	9, 43
	LC	39, 79, 80, 90, 91
Neofucoxanthin A and B	PC	122
Neo-gluco-digifucoside	TLC	102
Neo-gluco-digifucoside peracetate	TLC	102
Neo-hesperidin	PC	26
Neohetramine® (Thonzylamine)	TLC	39
Neohexane	GC	29
Neometeogenin	TLC	103
Neomycin	TLC	20
Neomycin A (Neamine)	PC	67
Neomycin B	PC	67
	TLC	20
Neomycin C	PC	67
Neomycin complex	PC	67
Neomycin (sulfate)	TLC	21
Neon	GC	112
Neo-odorobioside G peracetate	TLC	102
Neopentane	GC	118, 126
Neopentanol	GC	94
Neopentyl alcohol (2,2-dimethyl-1-propanol)	GC	53
Neopentylamine	GC	97
Neoperidinin	PC	122
Neophyllin M	TLC	40
Neopine	GC	96, 107
	TLC	4
Neopine HBR	GC	107
Neostigmine	PC	4
Neoxanthin	PC	122
	TLC	93, 94
Nephromopsinic acid	TLC	131
Neptunium (Np)	LC	80
Neriifolin peracetate	TLC	102
Nerol	TLC	183
Nerolidol	LC	45
Nerolidyl acetate	TLC	105
Nerolidyl formate	TLC	105
Nerolidyl isobutyrate	TLC	105
Nervifolin	PC	127
Neryl acetate	TLC	105
Neryl propionate	TLC	105
Netropsin	TLC	20
Neutral red	PC	88
Nevastain B	TLC	175
New blue	TLC	141
New Coccine	TLC	142
NH_2 (4,6-Dinitro-*m*-phenylenediamine), DNAP of	TLC	14
Ni(II), Nickel	GC	9, 43
	LC	81, 82, 83, 85, 86, 87, 88, 89
	PC	15, 95, 96
	TLC	148,149, 150
Ni^{++}, Nickel	PC	92
	TLC	150
Niacin	PC	51
Niacinamide	TLC	188
Nialamide	GC	17, 105
	TLC	39
Niamide	TLC	42
Nicarbazin	TLC	136
Nickel, Ni(II)	GC	9, 43
	LC	81, 82, 83, 85, 86, 87, 88, 89
	PC	15, 95, 96
	TLC	148,149, 150
Nickel, Ni^{++}	PC	92
	TLC	150

Nickel chloride $NiCl_2$	PC	94
Nickel dibutyldithiocarbamate	TLC	22
Nickel isopropyl xanthate	TLC	22
Nicotiflorin	PC	26
Nicotinamide	LC	20
	PC	2, 18, 51
	TLC	188
Nicotine	GC	8, 21, 95, 105
	PC	2, 4, 54
	TLC	2, 39, 45
Nicotinic acid	PC	2, 18
Nicotinic acid (Niacin)	TLC	188
Nicotinic acid hydrazide	PC	18
Nicotinic acid methyl ester	TLC	158
(-)-Nicotinic acid α-phenylethylamide	TLC	158
Nicotinuric acid	PC	18
Nicotyrine	GC	95
	PC	2
Nigrescigenin peracetate	TLC	102
Nigrescigenin (=*Sarmentosigenin A*) (3β,5β,11α,14β-Tetrahydroxy-19-oxo-cardenolide)	TLC	102
Nihydrazone	TLC	136
Nikethamide	TLC	39
Niobium (Nb)	LC	83, 84, 85, 86
Nithiazide	TLC	136
Nitrate	PC	14, 93
Nitrite	PC	14, 91, 93
m-Nitroaniline	LC	35
	PC	59
o-Nitroaniline	LC	35
	PC	59
p-Nitroaniline	LC	35
	PC	59
2-Nitroaniline	GC	102
	PC	6
	TLC	119
3-Nitroaniline	PC	6
4-Nitroaniline	PC	6
	TLC	119
1-(*p*-Nitroanilino)-anthraquinone	TLC	69
2-Nitro-*p*-anisidine	PC	120
4-Nitro-*o*-anisidine	PC	120
5-Nitro-*o*-anisidine	PC	120
2-Nitroanisole	GC	70
3-Nitroanisole	GC	70
4-Nitroanisole	GC	70
1-Nitroarginine-3-nitroguanidine	PC	9
o-Nitrobenzaldehyde	PC	78
Nitrobenzene	GC	51
	LC	60
m-Nitrobenzoic acid	TLC	37
o-Nitrobenzoic acid	TLC	37
p-Nitrobenzoic acid	PC	100
	TLC	37
m-Nitrobenzyl alcohol	PC	78
o-Nitro-*p*-bromphenol	TLC	174
1-Nitrobutane	GC	51
2-Nitro-4-chloroaniline	PC	6
	TLC	119
4-Nitro-2-chloroaniline	PC	6
2-Nitro-*p*-cresol	PC	121
4-Nitro-*m*-cresol	PC	121
2-Nitro-4-cyanoaniline	TLC	119
Nitrocyclohexane	GC	51
2-Nitro-4,4′-diaminodiphenyl	TLC	119
1-Nitro-2,3-dimethylbenzene	GC	46
1 Nitro-2,5-dimethylbenzene	GC	46
1-Nitro-2,6-dimethylbenzene	GC	46
2-Nitrodiphenylamine	TLC	157
Nitroethane	GC	51, 127
	LC	60
p-Nitroethylbenzene	PC	100
1-Nitro-2-ethylbenzene	GC	46
4-Nitro-5-ethyl-2-methylpyridine	TLC	61
4-Nitro-5-ethyl-2-methylpyridine *N*-oxide	TLC	61
Nitrofen	TLC	81
5-Nitro-2-furaldehyde acetylhydrazone	TLC	136
5-Nitro-2-furaldehyde semicarbazone	TLC	136
Nitrofurazone	TLC	136
3-(5-Nitrofurfurylideneamino)-2-oxazolidinone	TLC	136
Nitrogen	GC	26, 112
Nitrogen trifluoride	GC	26
Nitroguanidine	LC	63
1-H-3-Nitroguanidine	PC	9
2-Nitro-3-hydroxyacetophenone	PC	120
o-Nitro-*p*-hydroxyphenol	TLC	174
3-Nitro-4-hydroxyphenylarsonic acid	TLC	136
5-Nitroindole	PC	101
Nitromethane	GC	47, 51
2-Nitro-4-methoxyaniline	PC	6
4-Nitro-2-methoxyaniline	PC	6
o-Nitro-*p*-methoxyphenol	TLC	174
2-Nitro-4-methylaniline	PC	6, 59
	TLC	119
2-Nitro-6-methylaniline	PC	59
3-Nitro-4-methylaniline	TLC	119
3-Nitro-5-methylaniline	PC	59
3-Nitro-6-methylaniline	PC	6, 59
4-Nitro-2-methylaniline	PC	6
4-Nitro-6-methylaniline	PC	59
1-Nitro-2-methyl-3-ethylbenzene	GC	46
1-Nitro-2-methyl-4-ethylbenzene	GC	46
1-Nitro-2-methyl-5-ethylbenzene	GC	46
1-Nitro-2-methyl-6-ethylbenzene	GC	46
1-Nitro-3-methyl-2-ethylbenzene	GC	46
1-Nitro-3-methyl-4-ethylbenzene	GC	46
1-Nitro-4-methyl-2-ethylbenzene	GC	46
1-Nitro-4-methyl-3-ethylbenzene	GC	46
1-Nitro-5-methyl-2-ethylbenzene	GC	46
o-Nitro-*p*-methylphenol	TLC	174
4-Nitro-2-methylpyridine	TLC	61
4-Nitro-2-methylpyridine *N*-oxide	TLC	61
4-Nitro-3-methylpyridine *N*-oxide	TLC	61
1-Nitropentane	GC	51
o-Nitroperoxybenzoic acid	PC	107
Nitrophenide	TLC	136
Nitrophenol	TLC	84
m-Nitrophenol	PC	29, 31
	TLC	174
m-Nitrophenol, 4-acetyl-2-nitrophenol ether of	TLC	91
o-Nitrophenol	LC	1, 11
	PC	29, 31
	TLC	174
o-Nitrophenol, 4-acetyl-2-nitrophenol ether of	TLC	91

p-Nitrophenol	GC	19
	PC	29, 31
	TLC	174
p-Nitrophenol, 4-acetyl-2-nitrophenol ether of	TLC	91
2-Nitrophenol	GC	70, 71
	LC	51
3-Nitrophenol	GC	70, 71
	LC	51
4-Nitrophenol	GC	19, 70, 71
	LC	51
	TLC	81
p-Nitrophenolsulfate	PC	121
p-Nitrophenylacetic acid	PC	100
4-Nitrophenylarsonic acid	TLC	136
2-Nitro-1,4-phenylenediamine	TLC	119
3-Nitro-1,2-phenylenediamine	TLC	119
4-Nitro-*o*-phenylenediamine	PC	59
4-Nitro-1,2-phenylenediamine	TLC	119
4-Nitro-1,3-phenylenediamine	TLC	119
1-(*p*-Nitrophenyl)ethanol	PC	100
2-(*p*-Nitrophenyl)ethanol	PC	100
3-Nitrophthalic acid	PC	78
1-Nitropropane	GC	51
4-Nitropyridine *N*-oxide	TLC	61
3-Nitropyrocatechol	PC	120
4-Nitropyrocatechol	PC	121
4-Nitropyrocatecholsulfate	PC	121
4-Nitrosalicylic acid	PC	121
5-Nitrosalicylic acid	PC	120
Nitrosobenzene	TLC	7
N-Nitrosodiphenylamine	TLC	157
4-Nitrosodiphenylamine	TLC	157
1-Nitrosoindolo[*b*-3,2]-5α-cholest-2-ene	TLC	96
1′-Nitrosoindolo[*b*-3,4]-5β-cholest-3-ene	TLC	96
2-Nitroso-1-naphthol	TLC	83
4-Nitrosoresorcinol	PC	120
2-Nitrothiophene	TLC	104
m-Nitrotoluene	GC	51
o-Nitrotoluene	GC	51
p-Nitrotoluene	GC	51
	PC	100
2-Nitrotoluene	GC	46
4-Nitrotoluene	GC	46
	TLC	7
1-Nitro-2,3,4-trimethylbenzene	GC	46
1-Nitro-2,3,5-trimethylbenzene	GC	46
1-Nitro-2,3,6-trimethylbenzene	GC	46
1-Nitro-2,4,5-trimethylbenzene	GC	46
1-Nitro-2,4,6-trimethylbenzene	GC	46
1-Nitro-3,4,5-trimethylbenzene	GC	46
3-Nitrotyrosine	PC	63
5-Nitrouracil	LC	17
	PC	37
5-Nitrovanillin	PC	121
Nogiragenin	TLC	103
Nonadecanoic acid	PC	77
trans-2,*cis*-6-Nonadienal, 2,4-DNPH of	TLC	165
Nonamethylenediamine, trifluoroacetyl derivative	GC	10
Nonanal, 2,4-DNPH of	TLC	72, 76, 165
Nonanal, 2,4-DNPSC of	TLC	165
Nonanal, *p*-NPH of	TLC	165
Nonane	GC	49, 51
n-Nonane	GC	53, 55
	LC	21, 38, 42
Nonanoate	LC	67
Nonanoic acid	PC	11, 77
n-Nonanoic acid	LC	57
	PC	79
	TLC	134
Nonanoic acid, methyl ester of	GC	24
Nonanol	GC	49
n-Nonanol	GC	92
	TLC	1
1-Nonanol (DNB)	TLC	111
1-Nonanol (ONPU)	TLC	111
Nonan-2-one, 2,4-DNPH of	TLC	76
2-Nonanone, 2,4-DNPH of	TLC	165
2-Nonanone, 2,4-DNPSC of	TLC	165
2-Nonanone, *p*-NPH of	TLC	165
Nonanone-2	GC	51
Nonanone-5	GC	51
2-Nonenal, 2,4-DNPH of	TLC	165
2-Nonenal, 2,4-DNPSC of	TLC	165
2-Nonenal, *p*-NPH of	TLC	165
n-Nonene	GC	120
Nonene-1	GC	53
1-Nonene	GC	120
cis-2-Nonene	GC	120
trans-2-Nonene	GC	120
cis-3-Nonene	GC	120
trans-3-Nonene	GC	120
4-Nonene	TLC	51
cis-4-Nonene	GC	120
trans-4-Nonene	GC	120
Nonox WSL	TLC	175
Nonox WSP	TLC	175
Nonyl acetate	TLC	105
n-Nonyl alcohol	LC	11
Nonyl aldehyde	PC	20
Nonyl aldehyde, 2-hydrazine-benzothiazole derivative of	TLC	168
n-Nonylbenzene	LC	42
Nonyl butyrate	TLC	105
Nonyl esters	GC	24
Nonylic acid	TLC	132
2-(5-Nonyl)pyridine	GC	45
4-(5-Nonyl)pyridine	GC	45
Nopol	TLC	183
Noradrenaline	PC	5, 58
	TLC	9, 121
Noradrenaline·HCl	PC	5
Nor-*m*-adrenaline	PC	58
Noradrenalone	PC	60
Nor-agroclavine	TLC	114
19-Nor-4-androstene-3,17-dione	GC	79
B-Norandost-5-en-3β-ol-17-one	TLC	98
19-Nor-androst-4-en-17β-ol-3-one	PC	42
B-Nor-3,5-cholestadiene	TLC	96
21-Norcholesterol	TLC	99
Norcholic	PC	39
Norcodeine	GC	18
	TLC	138
Norephedrine	GC	9
	PC	5
Norepinephrine	GC	9
	TLC	8, 39
DL-Norepinephrine	LC	77
Norethisterone	TLC	95
Norethynodrel	TLC	95
Norharman	PC	101

Norhydroguaiaretic acid	TLC	125
Norhyoscyamine	PC	3
Norleucine	GC	12, 101
	LC	4, 33, 41
	PC	11
	TLC	11, 13, 16
Norleucine, DNAP of	TLC	14
Norleucine, PTH of	TLC	124
dl-Norleucine	PC	63
Normetanephrine	GC	9
	PC	64
	TLC	8
dl-Normetanephrine	PC	60
DL-Normetanephrine	LC	77
Normorphine	GC	18
	TLC	138
Nornicotine	GC	95
	PC	2
Nornicotyrine	GC	95
	PC	2
Norpethidine	GC	18
	TLC	138
Norrangiformic acid	TLC	131
Norstictic acid	PC	29
	TLC	34
Nortangeretin	PC	26
Nortestosterone	TLC	95
Nortestosterone acetate	TLC	95
Nortestosterone benzoate	TLC	95
Nortestosterone decanoate	TLC	95
Nortestosterone phenylpropionate	TLC	95
Nortriptyline	GC	22
Norvaline	GC	12, 101
	PC	11
	TLC	11, 16, 29
Norvaline, DNAP of	TLC	14
Norvaline PTH of	TLC	124
dl-Norvaline	PC	63
Norwogonin	PC	26
Noscapine	GC	107
	TLC	40
Noscapine (Narcotine)	TLC	138
Novaminosulfone	TLC	40
Novesine (Dorsacaine®)	TLC	45
Novobiocin	TLC	20, 21
Nozinan	TLC	42
Np (Neptunium)	LC	80
Nucleotides	GC	77
Nupercaine® hydrochloride (Dibucaine)	TLC	45
Nylidrin	TLC	39
Nylon 12 linear dimer	LC	49
Nylon 12 linear heptamer	LC	49
Nylon 12 linear hexamer	LC	49
Nylon 12 linear monomer	LC	49
Nylon 12 linear pentamer	LC	49
Nylon 12 linear tetramer	LC	49
Nylon 12 linear trimer	LC	49
Nylon 66 cyclic dimer	LC	49
Nylon 66 cyclic monomer	LC	49
Nylon 66 cyclic trimer	LC	49
Nylon 66 linear dimer	LC	49
Nylon 66 linear monomer	LC	49
Nylon 66 linear tetramer	LC	49
Nystatin	TLC	20
Nystose	LC	48

Adsorption of the Elements
(LC Section – Figures)

N, Nitrogen	Pg	278, 279, 280, 281
Na, Sodium	Pg	278, 279, 280, 281
Nb, Niobium	Pg	278, 279, 280, 281
Ni, Nickel	Pg	278, 279, 280, 281
Np, Neptunium	Pg	278, 279, 280, 281

O (oxygen)	GC	26, 112
N Oblivon	TLC	42
Obtusatic acid	PC	30
Ochratoxin A	TLC	185
Ochratoxin B	TLC	185
Ocimene	GC	86
α-Ocimene	TLC	106
trans-β-Ocimene	TLC	106
Octa-*O*-acetyl-β-cellobiose	TLC	25
Octa-*O*-acetyl-gentiobiose	TLC	25
Octa-*O*-acetyl-β-laminaribose	TLC	25
Octa-*O*-acetyl-α-maltose	TLC	25
Octa-*O*-acetyl-β-maltose	TLC	25
Octa-*O*-acetyl-sucrose	TLC	25
Octa-condensed phosphate	TLC	151
n-Octacosane	LC	38
Octacosyl acetate	LC	45
Octadeca-9,12-dienoic acid, methyl ester of	TLC	130
cis,cis-Octadeca-9,12-dienol (linoleylaldehyde), dimethyl acetal of	GC	6
cis-Octadeca-9-dienol (oleylaldehyde), dimethyl acetal of	GC	6
Octadeca-9,12,15-enal (linolenylaldehyde), dimethyl acetal of	GC	6
Octadecanal (stearaldehyde), dimethyl acetal of	GC	6
n-Octadecane	LC	38
Octadecanoic acid	LC	68
Octadecanoic acid, methyl ester of	GC	24
	TLC	130
n-Octadecanol	GC	92
Octadeca-9,12-15-trienoic acid, methyl ester of	TCL	130
Octadecenal, dimethyl acetal of	GC	6
Octadecene	TLC	51
2-Octadecenoic acid	PC	77
cis-Octadecen-7-oic acid, methyl ester of	TLC	130
cis-Octadecen-9-oic acid, methyl ester of	TLC	130
trans-Octadecen-9-oic acid, methyl ester of	TLC	130
cis-Octadecen-11-oic acid, methyl ester of	TLC	130
n-Octadecylamine	PC	55
2-Octadecylnaphthalene	TLC	53
Octafluorobutene-1	GC	26
cis-Octafluorobutene-2	GC	26
trans-Octafluorobutene-2	GC	26
Octafluorocyclobutane	GC	26
Octafluoroisobutene	GC	26
Octafluoropropane	GC	26
Octahydrophenanthrene	LC	27
Octamethylenediamine, trifluoroacetyl derivative	GC	10
Octamylamine (Octinum D®)	TLC	45
Octanal	GC	57
Octanal, 2,4-DNPH of	TLC	72, 76

Octane	GC	49, 58, 125, 126
n-Octane	GC	42, 47, 53, 55, 118, 120
	LC	38, 42
n-Octane, organics	GC	52
n-Octane, paraffins	GC	33
Octanediol-1,8	GC	76
Octanoic acid	LC	68
	PC	11, 77
Octanoic acid, allyl ester of	GC	24
Octanoic acid, ethyl ester of	GC	24
Octanoic acid, methyl ester of	GC	24
n-Octanoic acid	LC	57
	PC	79
	TLC	134
Octanol	GC	49
n-Octanol	GC	92
	TLC	1
1-Octanol	PC	53
1-Octanol (DNB)	TLC	111
1-Octanol (ONPU)	TLC	111
2-Octanol	GC	92
	TLC	1
2-Octanol (DNB)	TLC	111
2-Octanol (ONPU)	TLC	111
Octan-2-one, 2,4-DNPH of	TLC	76
Octanone-2	GC	51
Octanone-3	GC	51
Octaphosphate $(P_8O_{25})^{-10}$	PC	98
Octaverine	GC	22
2-Octenal, 2,4-DNPH of	TLC	165
1-Octene	GC	118, 120
Octene-1	GC	51, 53
	LC	42
Oct-1-ene, organics	GC	56
cis-2-Octene	GC	120
trans-2-Octene	GC	120
cis-3-Octene	GC	120
trans-3-Octene	GC	120
cis-4-Octene	GC	120
trans-4-Octene	GC	120
Octogen	TLC	157
Octopamine	GC	9
DL-Octopamine	LC	77
Octopine	PC	9, 10
Octyl acetate	GC	2
	TLC	105
n-Octyl alcohol	LC	11
Octyl aldehyde, 2-hydrazinebenzothiazole derivative of	TLC	168
Octylamine	TLC	10
n-Octylamine	PC	5, 55
	TLC	7
n-Octylamine, trifluoroacetyl derivative	GC	10
tert-Octylamine	TLC	7
Octylbenzene	TLC	53
n-Octylbenzene	LC	42
Octyl butyrate	GC	2
	TLC	105
Octyl caprate	TLC	105
Octyl caprylate	TLC	105
Octyl esters	GC	24
Octyl formate	GC	2
Octyl gallate	TLC	125
Octyl hexanoate	GC	2
p-tert-Octylhomopyrocatechol	TLC	176
p-tert-Octylhydroxybenzoquinone	TLC	176
Octyl isobutyrate	GC	2
Octyl mercaptan	TLC	162
n-Octyl mercaptan	GC	85
Octyl methacrylate	GC	91
Octyl 3-methylbutyrate	GC	2
Octyl 4-methylpentanoate	GC	2
Octyl pentanoate	GC	2
2-*tert*-Octylphenol	TLC	176
4-*tert*-Octylphenol	GC	71
	TLC	176
Octyl propionate	GC	2
p-tert-Octylpyrocatechol	TLC	176
p-tert-Octylpyrogallol	TLC	176
Octyl salicylate	TLC	129
Odorobioside G	TLC	102
Odorobioside G peracetate	TLC	102
Odoroside A	TLC	102
Odoroside A peracetate	TLC	102
Odoroside B	TLC	102
Odoroside B peracetate	TLC	102
Odoroside H	TLC	102, 177
Odoroside H peracetate	TLC	102
Oenanthal, 2,4-DNPH of	TLC	166
Oestrone	PC	42
	TLC	98
Oil Orange SS	TLC	48
Oil Red OS	TLC	46, 48
Oil Red XO	TLC	142
Okanin	PC	26
Oleandomycin	PC	66
Oleandrigenin	PC	127
Oleandrigenin(Og) (3β,14β-Dihydroxy-16β-acetoxy-5β-cardenolide)	TLC	102
Oleandrigenin-tridigitoxoside	TLC	102
Oleandrin	PC	127
	TLC	102
Oleandrin peracetate	TLC	102
Oleanic	TLC	184
Oleanonic	TLC	184
Oleate	LC	67
Oleate, β-sitosteryl ester of	TLC	182
Olefins	GC	126
Oleic acid	PC	77, 99
	TLC	33, 57
Olendomycin	TLC	20
Oligosaccharides	PC	73
Olivetol-carboxylic acid	PC	30
Olivetoric acid	PC	30
Omca	TLC	139
Orabet (Tolbutamide)	TLC	140
Oracet Orange (Ciba)	TLC	7
Oracet Red (Ciba)	TLC	7
Orange 1	TLC	141
Orange 2	TLC	141
Orange G	TLC	49
Orange RN	TLC	49
Orange SS	TLC	46
Oranil (Carbutamide)	TLC	140
Orasol Scarlet 2B (Ciba)	TLC	7
Orcin	TLC	90
Orcinol	LC	1
	PC	82
Orcinol, di(4-acetyl-2-nitrophenol) ether of	TLC	91
β-Orcinolcarboxylic acid	PC	30
β-Orcinolcarboxylic acid dimethyl ether	PC	30
Orcinoldicarboxylic acid monomethyl ether	PC	30
Ornithine	GC	101

Ornithine (*continued*) LC 4, 41; PC 64; TLC 11, 13, 15, 16
Ornithine, PTH of TLC 124
Orobol PC 26
Orotic acid GC 77; LC 3, 46; PC 36, 37; TLC 188
Orotidine LC 3; PC 36
Orotidine monophosphate PC 36
Orphenadrine GC 22; TLC 39
Orsellic acid PC 30
Orthocaine® (Orthoform) TLC 43
Orthoform (orthocaine®) TLC 43
Orthophosphate LC 47; PC 12, 14, 32, 91
Orthophosphoric acid (H_2PO_4) PC 35
Os (osmium) TLC 55
Oscine PC 3
Osmium (Os) TLC 55
Osthol PC 113
Ostruthin PC 113
Ovalbumin LC 19, 64
Ovex GC 62, 64, 65, 129; TLC 82
4-Oxa-5α-cholestane TLC 96
4-Oxa-5α-cholestan-3-one TLC 96
Oxacillin Na TLC 21
Oxalacetate acid TLC 36
Oxalacetic acid PC 81
Oxalate PC 14, 93
Oxalic acid LC 58, 65; PC 13, 77; TLC 31
Oxalic acid, diethyl ester of GC 24
Oxalic acid, dimethyl ester of GC 24
Oxaloacetic, 2,4-DNPH of TLC 32
Oxamphetamine TLC 9
4-Oxa-5α-pregnane-3,20-dione TLC 96
Oxedrine TLC 9
7-Oxo-digitoxigenon (14β-Hydroxy-3,7-dioxo-5β-cardenolide) TLC 102
15-Oxo-digitoxigenon (14β-hydroxy-3,15-dioxo-5β-cardenolide) TLC 102
3-Oxo-14α,15α-epoxy-5β-cardenolide TLC 102
3-Oxo-11β,18-epoxy-eti-4-enic acid-lactone (20 → 18) (aldosterone lactone) PC 43
6-Oxo-estradiol TLC 181
7-Oxo-estradiol TLC 181
16-Oxo-estradiol TLC 181
6-Oxo-estriol TLC 181
6-Oxo-estrone TLC 181
7-Oxo-estrone TLC 181
11-Oxo-estrone TLC 181
16-Oxo-estrone TLC 181
3-Oxo-18-hydroxy-eti-4-enic acid-lactone (20 → 18) PC 43
17-Oxosparteine PC 3
Oxo-Sumithion TLC 172
Oxycodone (Dihydrohydroxycodeinone) TLC 2, 138
Oxycodone HCl TLC 4
Oxydemetonmethyl LC 43; TLC 84
Oxydemetonmethyl sulfone GC 65
N-Oxydiethylene-2-benzothiazole sulfenamide TLC 22, 66
Oxygenated compounds GC 121
Oxygen (O) GC 26, 112
Oxymorphone (Dihydrohydroxymorphinone) TLC 138
Oxymorphone HCl TLC 138
Oxynicotine PC 2
Oxyphenbutazone TLC 40
Oxytetracycline PC 68
epi-Oxytetracycline PC 68
Oxytetracycline HCl TLC 21

Adsorption of the Elements (LC Section – Figures)

O, Oxygen Pg 278, 279, 280, 281
Os, Osmium Pg 278, 279, 280, 281

Pa (Protactinium) LC 80
4-PABSA TLC 10
Paeonidin TLC 78
Paeonidin-3,5-diglucoside TLC 78
Palatinose PC 70
Palladium (Pd) LC 80, 83, 85, 86, 87; TLC 55
Palladium, Pd(II) GC 43; PC 15, 95, 96
Palmatine PC 3
Palmitate LC 68
Palmitate, β-sitosteryl ester of TLC 182
Palmitate, Vitamin A_1 PC 49
Palmitic acid LC 57, 61; PC 77; TLC 33
Palmitic acid, ethyl ester of GC 24
Palustrate LC 67
Pannarin TLC 34
Panogenin (3β,11β,14β,19-Tetrahydroxy-5α-cardenolide) TLC 102
Panoside TLC 102
Panoside peracetate TLC 102
Pantothenic acid PC 51; TLC 188
D-Pantothenic acid TLC 188
Papaveraceae GC 96
Papaveraceae alkaloids PC 3
Papaverine GC 8, 22, 96, 107, 111; PC 2, 3, 4, 54; TLC 2, 3, 4, 39, 40, 45, 116, 138
Papaverine HCl GC 107
Paraacetamole TLC 40
Parabromdylamone maleate (Brompheniramine maleate) GC 20
Paraffins GC 33
n-Paraffins GC 126
Paraldehyde GC 7
Paraoxon GC 65, 129; TLC 84, 171

Parathion GC 60, 61, 64, 65, 67, 129
LC 43
PC 28
TLC 80, 81, 84, 171, 172
Parathion-methyl LC 43
Parathion oxygen analog GC 61, 67
LC 43
Parathion-t_r(min) GC 65
Paredrin PC 5
Parethoxycaine HCl GC 110
Paromomycin PC 67
TLC 20
Paromomycin (sulfate) TLC 20
Parsidol TLC 42
Patent Green TLC 141
Patuletin PC 26
Pb (lead) LC 80, 81, 82, 85, 86, 87, 88, 89
Pb(II), Lead GC 43
PC 15, 95, 96
TLC 148, 149
Pb^{2+}. Lead PC 92
Pb^{++} TLC 52, 150
PCNB GC 62, 129
PC 116
TLC 82
Pd (Palladium) LC 80, 83, 85, 86, 87
TLC 55
Pd(II), Palladium GC 43
PC 15, 95, 96
Pebulate GC 65
Pecazine PC 85
TLC 139
Pelargonate, β-sitosteryl ester of TLC 182
Pelargonic acid PC 77
TLC 33
Pelargonidin PC 25
TLC 78
Pelargonidin, glycosides PC 24
Penicillamine PC 64
TLC 13
Penicillamine disulfide PC 64
D-Penicillamine LC 32, 59
D-Penicillamine disulfide LC 32
D-Penicillamine tetrasulfide LC 32
D-Penicillamine trisulfide LC 32
dl-Pencillamine(β-mercaptovaline) PC 63
Penniclavine TLC 114
Pennogenin TLC 103
Penta-*O*-acetyl-α-D-galactofuranose TLC 25
Penta-*O*-acetyl-β-D-galactofuranose TLC 25
Penta-*O*-acetyl-α-D-galactopyranose TLC 25
Penta-*O*-acetyl-β-D-galactopyranose TLC 25
Penta-*O*-acetyl-α-D-glucopyranose TLC 25
Penta-*O*-acetyl-β-D-glucopyranose TLC 25
Penta-*O*-acetyl-β-D-mannopyranose TLC 25
Pentachlorobenzene GC 31
Pentachoroethane LC 75
Pentachlorophenol GC 129
TLC 81
Pentachlorophenyl acetate TLC 81
Penta-condensed phosphate TLC 151
Pentadecanal, dimethyl acetal of GC 6
Pentadecanoic acid PC 77
2,4-Pentadienal GC 7
1,3-Pentadiene GC 53, 55
cis-1,3-Pentadiene GC 29
trans-1,3-Pentadiene GC 29
1-*cis*-3-Pentadiene GC 118
1-*trans*-3-Pentadiene GC 118
1,4-Pentadiene GC 118
Pentadiene-1,4 GC 29
Pentaerythritol GC 14, 76
PC 53
TLC 113
Pentafluoropropionyl, PFPr GC 98
Pentameta-condensed phosphate TLC 151
2′,2′,3′,4,6′-Pentamethoxyacetophenone GC 48
Pentamethylbenzene GC 119
TLC 53
Pentamethylenediamine PC 7
Pentamethylenediamine, trifluoroacetyl derivative GC 10
Pentamethylphenol GC 72
Penta-O-methylquercitin LC 28
Pentamycin TLC 20
Pentanal, 2,4-DNPH of TLC 76
Pentanal-1, isobutyric acid ester of GC 51
n-Pentanal, 2,4-DNPH of TLC 72
Pentane GC 49, 51, 58, 125, 127
n-Pentane GC 7, 29, 34, 42, 47, 53, 118, 120, 121
LC 21, 38, 42
n-Pentane, paraffins GC 33
1,5-Pentanediol TLC 113
2,3-Pentanedione LC 8
2,4-Pentanedione LC 8
TLC 7
n-Pentanethiol GC 56
Pentanol GC 49
n-Pentanol GC 47, 57, 92, 94
LC 37
TLC 1
tert-Pentanol GC 94
1-Pentanol PC 52
1-Pentanol (DNB) TLC 111
1-Pentanol (ONPU) TLC 111
Pentanol-1, butyric acid ester of GC 51
Pentanol-1, formic acid ester of GC 51
Pentanol-1, propionic acid ester of GC 51
2-Pentanol GC 55, 57, 92, 94
PC 52
2-Pentanol (DNB) TLC 111
2-Pentanol (ONPU) TLC 111
3-Pentanol GC 55, 57, 92, 94
3-Pentanol (DNB) TLC 111
Pentanone-2 GC 51
2-Pentanone GC 7, 53, 125
2-Pentanone, ketones GC 126
Pentan-2-one, 2,4-DNPH of TLC 76
Pentanone-3 GC 51
3-Pentanone GC 7, 118

3-Pentanone (diethyl ketone)	GC	53
3-Pentanone (diethyl ketone), organics	GC	53
Pentaquine	GC	22
n-Pentatriacontane	LC	38
1-Pentene	GC	47,118, 122
Pent-1-ene	GC	56
Pentene-1	GC	29, 51, 53, 58,125, 126
	LC	42
Pentene-2 (*trans*)	GC	53
cis-Pentene-2	GC	29, 58, 125, 126
trans-Pentene-2	GC	29, 58, 125, 126
cis-2-Pentene	GC	118, 122
trans-2-Pentene	GC	118, 122
Pentex	TLC	66
Penthrit	TLC	157
Pentobarbital	GC	21,105, 106
Pentobarbital (Pentobarbitone)	GC	19
	TLC	137
Pentobarbitone	PC	86
Pentothiobarbital	TLC	137
Pentyl acetate	GC	2
Pentyl acrylate	GC	91
n-Pentylamine	GC	11, 97
tert-Pentylamine	GC	97
n-Pentylbenzene	LC	42
Pentyl butyrate	GC	2
α-Pentylcinnamaldehyde, 2,4-DNPH of	TLC	73
Pentyl crotonate	GC	91
n-Pentyl cyanide	GC	56
n-Pentylcyclohexane	GC	35
n-Pentylcyclopentane	GC	36
Pentylenetetrazol	TLC	39
Pentyl formate	GC	2
iso-Pentyl formate	GC	47
Pentyl hexanoate	GC	2
Pentyl isobutyrate	GC	2
Pentyl methacrylate	GC	91
Pentyl 3-methylbutyrate	GC	2
Pentyl 4-methylpentanoate	GC	2
Pentyl pentanoate	GC	2
4-*tert*-Pentylphenol	GC	71
Pentyl propionate	GC	2
2-(*n*-Pentyl)pyridine	GC	45
4-(*n*-Pentyl)pyridine	GC	45
Pentyne-1	GC	29
Pentyne-2	GC	29
Peonidin	PC	25
Peonidin, glycosides	PC	24
Peonidin glycosides, 3,5-diglucoside	PC	24
Peonidin glycosides, 3-monoglucoside	PC	24
Peonidin glycosides, 3-rhamnoglucosido-5-glucoside	PC	24
PEP (Phospho-enol pyruvate)	PC	32
Pepsin	LC	19
Peptide Ala-Gly	LC	14
Peptide α-Asp-Ala	LC	2
Peptide α-Asp-Asn	LC	2
Peptide β-Asp-Asn	LC	2
Peptide α-Asp-Gln	LC	2
Peptide β-Asp-Gln	LC	2
Peptide α-Asp-Gly	LC	2
Peptide β-Asp-Gly	LC	2
Peptide α-Asp-Gly-Ala	LC	2
Peptide β-Asp-Gly-Ala	LC	2
Peptide α-Asp-Gly-Asn	LC	2
Peptide β-Asp-Gly-Asn	LC	2
Peptide α-Asp-Gly-Gln	LC	2
Peptide β-Asp-Gly-Gln	LC	2
Peptide α-Asp-Gly-Gly	LC	2
Peptide β-Asp-Gly-Gly	LC	2
Peptide α-Asp-Gly-Pro	LC	2
Peptide β-Asp-Gly-Pro	LC	2
Peptide α-Asp-Gly-Val	LC	2
Peptide β-Asp-Gly-Val	LC	2
Peptide α-Asp-Ile	LC	2
Peptide β-Asp-Ile	LC	2
Peptide α-Asp-Leu	LC	2
Peptide β-Asp-Leu	LC	2
Peptide α-Asp-Ser	LC	2
Peptide β-Asp-Ser	LC	2
Peptide β-Asp-Thr	LC	2
Peptide α-Asp-Val	LC	2
Peptide β-Asp-Val	LC	2
Peptide Gly-Ala	LC	14
Peptide Gly-Gly-Phe	LC	14
Peptide Gly-Leu	LC	14
Peptide Gly-Phe	LC	14
Peptide Gly-Pro	LC	14
Peptide Leu-Gly	LC	14
Peptide Phe-Gly	LC	14
Peptide Phe-Gly-Gly	LC	14
Peptide Pro-Gly	LC	14
Peptide Z·Gly-Gly·OMe	LC	9
Peptide Z·Gly-Gly-Gly·OMe	LC	9
Peptide Z·Gly-Gly-Gly-Gly·OMe	LC	9
Peptide Z·Val-Lys(BOC)·OH	LC	9
Peptide Z·Lys(BOC)-Val-Tyr-Pro·O*t*Bu	LC	9
Peptide Z·Val-Lys(BOC)-Val-Tyr-Pro·(O*t*Bu)	LC	9
Peptide H·Lys(BOC)-Val-Tyr-Pro·O*t*Bu	LC	9
Peptide H·Val-Lys(BOC)-Val-Tyr-Pro·O*t*Bu	LC	9
Peptide H·Tyr(*t*Bu)-Ser(*t*Bu)·O*t*Bu	LC	9
Peptide Z·Val-Tyr(*t*Bu)-Ser-(*t*Bu)·O*t*Bu	LC	9
Peptide Z·Leu-Val-Tyr(*t*Bu)-Ser(*t*Bu)·O*t*Bu	LC	9
Peptide Z·Leu-Leu-Val-Tyr(*t*Bu)-Ser(*t*Bu)·O*t*Bu	LC	9
Peptide Z·His-Leu-Leu-Val-Tyr-(*t*Bu)-Ser(*t*Bu)·O*t*Bu	LC	9
Peptide H·His-Leu-Leu-Val-Tyr(*t*Bu)-Ser(*t*Bu)·O*t*Bu	LC	9
Peptide Z·Val-Tyr-Val-His-Pro-Phe·OMe	LC	9
Peptide Z·Arg(NO_2)-Val-Tyr-Val-His-Pro-Phe·OMe	LC	9
Peptide H·Arg(NO_2)-Val-Tyr-Val-His-Pro-Phe·OMe	LC	9
Peptide Z·Asp(OBzl)-Arg(NO_2)-Val-Tyr-Val-His-Pro-Phe·OMe	LC	9
Peptide Z·Asp(O*t*Bu)-Glu(NH_2)-Leu-Ala-Glu(O*t*Bu)-Ala-Phe-Pro-Leu-Glu(O*t*Bu)-Phe·O*t*Bu	LC	9
Peptide Z·Glu(O*t*Bu)-Asp(O*t*Bu)-Glu(NH_2)-Leu-Ala-Glu(O*t*Bu)-Ala-Phe-Pro-Leu-Glu(O*t*Bu)-Phe·O*t*Bu	LC	9

Peracetic acid	PC	27
Perazine	TLC	139
Per-*n*-butyric acid, per-isobutyric acid	PC	27
Perchlorate	PC	14
Perchloroethane	LC	75
Perchloroethylene, organics	GC	52
Percrotonic acid	PC	27
Per-*n*-hexoic acid	PC	27
Peridinin	PC	122
Perilla-aldehyde, 2,4-DNPH of	TLC	76
Periodate	PC	14, 91, 93
Periplocymarin	PC	127
Periplogenin	PC	127
Periplogenin acetate	PC	127
Periplogenin peracetate	TLC	102
Periplogenin (3β,5β,14β-Trihydroxy-cardenolide)	TLC	102
Periplogenon(5β,14β-Dihydroxy-3-oxo-cardenolide)	TLC	102
Periplorhamnoside	TLC	102
Periplorhamnoside peracetate	TLC	102
Perlatolic acid	PC	30
Peronin	TLC	138
Peroxidase-I	LC	19
Perphenazine	GC	17, 20, 108, 109
	PC	84, 85
	TLC	139
Perphenazine (Decentan)	TLC	45
Perpropionic acid	PC	27
Perrhenate	PC	14, 93
Perrillaldehyde	TLC	77
Persulphate	PC	14
Perthane®	GC	60, 62, 68, 128, 129
	PC	116
	TLC	82
Perthane olefin	GC	129
	TLC	82
Perthane, trichloroethane analog	GC	129
Per-*n*-valeric acid	PC	27
Pervitin	PC	2
Perylene	GC	40
	TLC	51, 53, 120, 146
Petasin	LC	45
S-Petasin	LC	45
Pethidine	PC	4
Pethidine (Dolantin)	TLC	45
Pethidine (Meperidine)	GC	18, 105
	LC	55
	TLC	2, 138
Petroselinic acid	TLC	33
Petunidin	PC	25
	TLC	78
Petunidin glycosides	PC	24
Petunidin glycosides, 3,5-diglucoside	PC	24
Petunidin glycosides, 3-monoglucoside	PC	24
Petunidin glycosides, 3-rhamnoglucoside	PC	24
Petunidin glycosides, 3-rhamnoglucosido-5-glucoside	PC	24
Petunidin-3-monoglucoside	TLC	78
Pevitin	TLC	42
PFPr amines	GC	98
PFPr, pentafluoropropionyl	GC	98
PG (Propyl gallate)	PC	69
6-PG (6-Phosphogluconic acid)	PC	32
2-PGA (2-Phosphoglycerate)	PC	32
3-PGA (3-Phosphoglycerate)	PC	32
α-Phellandral, 2,4-DNPH of	TLC	76
α-Phellandrene	GC	86, 87
β-Phellandrene	GC	86, 87
	TLC	106
Phenacaine (Holocaine)	TLC	39, 43, 45
Phenacetin	GC	19, 104, 106
	TLC	40
Phenaceturic	GC	5
Phenaglycodol	GC	17, 108
Phenalene-1-one	TLC	75
Phenanthrene	GC	40, 45
	LC	27
	TLC	7, 51, 53
Phenanthridine	GC	45
Phenanthridone	TLC	120
6(5H)-Phenanthridinone	TLC	75
Phenazocine	TLC	138
Phenazone	GC	19
Phenazone (Antipyrine)	TLC	40
Phendimetrazine	TLC	39
Phenelzine	GC	17
	TLC	39
Phenelzine dihydrogen sulfate	GC	17
Phenergan®	TLC	42
Phenethicillin K	TLC	21
Phenethylamine	GC	21
α-Phenethylamine	TLC	10
β-Phenethylamine	TLC	10
(±)-1-Phenethylamine, trifluoroacetyl derivative	GC	10
2-Phenethylamine, trifluoroacetyl derivative	GC	10
Phenethyl butyrate	TLC	105
p-Phenetidine	GC	105
Phenetole	GC	51, 75
	LC	21
Phenindamine (Theophorin®)	TLC	39
Phenindamine tartrate	GC	20, 54, 104
	PC	85
Pheniramine	GC	20, 22
Pheniramine (Avil®)	TLC	45
Pheniramine (Trimeton®)	GC	21
	TLC	39
Phenirimine maleate	GC	20, 54, 104
Phenkapton	GC	67
	LC	43
	TLC	84, 171
Phenmetrazine	GC	22
Phenmetrazine (Preludin®)	TLC	39, 45
Phenobarbital	GC	21, 105, 106
Phenobarbital (Phenobarbitone)	GC	19
	TLC	137
Phenobarbitone	PC	86
Phenobarbitone (Phenobarbital)	TLC	137
Phenol	GC	69, 71, 73, 74, 75, 130
	LC	1, 11, 30, 51, 52, 53, 61
	PC	31, 82

Phenol (*continued*)	TLC	37, 87, 88, 90,174, 176
Phenol, 4-acetyl-2-nitrophenol ether of	TLC	91
Phenolphthalein	PC	88
Phenol red	PC	88
Phenols, conjugated	PC	82
Phenothiazine	GC	65, 103
	PC	84
	TLC	81,120, 136, 139
Phenothiazine derivatives	GC	17
Phenothiazine-5-oxide	PC	84
Phenox	TLC	175
2-Phenoxyethanol	PC	53
Phenoxy methylpenicillin K	TLC	21
Phentermine	GC	22
Phenylacetaldehyde	GC	51
Phenylacetaldehyde, 2,4-DNPH of	TLC	73
Phenylacetaldehyde, 2-hydrazinebenzo-thiazole derivative of	TLC	168
Phenylacetic acid	GC	4, 5
	PC	82
Phenylacetic acid, methyl ester of	GC	24
Phenylacetonitrile	GC	51
4′-Phenylacetophenone	PC	23
	TLC	70
Phenylacetylene	LC	60
Phenylacetylglutamic	GC	5
Phenylacetylglutamine	GC	5
Phenylalanine	GC	9, 12, 101
	LC	4, 14, 33, 41, 56
	PC	11, 62, 64, 65
	TLC	13, 15, 16, 118
Phenylalanine, DNAP of	TLC	14
Phenylalanine, PTH of	TLC	124
α-Phenylalanine	TLC	11
DL-Phenylalanine	TLC	12
Cbo-DL-Phenylalanine	TLC	12
Cbo-DL-Phenylalanine-Glycine ethyl ester	TLC	12
l-Phenylalanine methyl ester	PC	63
Phenylalaninol	PC	57
l-Phenylalaninol	PC	63
6-Phenylaminopurine	LC	46
4-Phenyl-4-aza-5-cholesten-3-one	TLC	96
p-Phenylazoaniline	TLC	81
Phenylazo-*m*-cresol (BAmC)	TLC	47
4-Phenylazodimethylaniline	TLC	7
1-Phenylazo-2-naphthol (Sudan Yellow)	TLC	7
4-(Phenylazo)-resorcinol (BAR)	TLC	47
Phenylbenzyl phosphate	PC	112
α-Phenyl-α-bithienyl	TLC	104
1-Phenyl-1,3-butanedione	PC	23
	TLC	70
1-Phenylbutanone-2	GC	51
1-Phenylbutanone-3	GC	51
Phenylbutazone	TLC	40
trans-4-Phenyl-3-buten-2-one	PC	23
	TLC	70
N-Phenylcarbazole	TLC	7
9-Phenylcarbazole	TLC	120
Phenylcyanide	GC	51
Phenylcyclohexane	TLC	53
m-Phenylenediamine	PC	59

o-Phenylenediamine	GC	102
	PC	59
p-Phenylenediamine	PC	59
1,2-Phenylenediamine	TLC	119
1,3-Phenylenediamine	TLC	119
1,4-Phenylenediamine	TLC	119
Phenylephrine	GC	9, 23
	TLC	39
L-Phenylephrine	PC	58
1-Phenylethanol-1	GC	51
2-Phenylethanol	GC	51
	PC	52
Phenylethyl alcohol (DNB)	TLC	111
Phenylethyl alcohol (ONPU)	TLC	111
Phenylethylamine	TLC	5
β-Phenylethylamine	GC	9
	PC	9, 64
Phenylethyl esters	GC	24
1-Phenylethyl hydroperoxide	TLC	65
N-β-Phenylethylthiourea	PC	47
2-Phenylglutaramide	GC	105
Phenyl glycidyl ether	LC	60
Phenylglycine	PC	64
α-Phenylglycine	TLC	11
β-Phenyl-β-hydroxyethylamine	PC	56
Phenyl isocyanate	GC	44
β-Phenylisopropylmethylamine	TLC	10
N-Phenyl-*N*′-isopropyl-*p*-phenylenediamine	TLC	22
β-Phenyllactic acid	GC	4, 5
Phenyl mercury dithizonate	TLC	161
Phenyl-*N*-methyl carbamate	GC	130
Phenyl-3-methyl-4-oximino-2-pyrazolin-5-one	TLC	83
1-Phenyl-3-methylpyrazolon	GC	130
	TLC	83
1-Phenylnaphthalene	TLC	53
2-Phenylnaphthalene	TLC	51, 53
N-Phenyl-α-naphthylamine	TLC	22
N-Phenyl-β-naphthylamine	TLC	22
o-Phenylphenol	TLC	174, 176
p-Phenylphenol	TLC	176
2-Phenylphenol	GC	72
N-Phenylphenothiazine	TLC	120
2-Phenylpropanol-2	GC	51
3-Phenyl-1-propanol	PC	52, 53
3-Phenylpropanol-1	GC	51
Phenylpropanolamine	GC	23
	TLC	39
Phenylpropanolamine HCl	GC	20, 54, 104
1-Phenylpropanone-2	GC	51
Phenyl-2-propanone	PC	23
	TLC	70
Phenylpropionaldehyde, 2,4-DNPH of	TLC	73
Phenylpropionic acid	PC	82
β-Phenylpropionic acid	GC	5
α-G-β-Phenylpropionic acid	PC	9
3-Phenylpropionic acid	GC	4
β-Phenyl-β-propylthioethylamine	PC	60
Phenyl-α-pyridyl phosphate	PC	112
Phenylpyruvic acid	GC	4, 5
	PC	81
	TLC	8
Phenylpyruvic acid, 2,4-DNPH of	TLC	32
Phenylpyruvic (Isomer 1), 2,4-DNPH of	TLC	32
Phenylpyruvic (Isomer 2), 2,4-DNPH of	TLC	32
Phenyl salicylate	GC	105
Phenylserine allo	PC	64

Phenylserine 4-Cl allo	PC	64
Phenylserine 4-Cl threo	PC	64
Phenylserine 4-NO_2 allo	PC	64
Phenylserine 4-NO_2 threo	PC	64
Phenylserine threo	PC	64
Phenyl sulfoxide	GC	139
Phenyl-2-thienyl ketone	PC	21
	TLC	70
1-Phenylthiosemicarbazide	PC	47
Phenyltoloxamine	GC	17
	TLC	39
Phenyltoloxamine hydrogen citrate	GC	17, 20, 54
Phenylurea	TLC	122
Phenyramidol	TLC	39
Phenytoin	TLC	137
Pheophytin (Ph)	PC	48
Phloretic	LC	30
Phloroglucinol	LC	1
	PC	82, 113
	TLC	90, 176
Phlorogluernol	TLC	174
Phloxine	TLC	142
Pholcodeine	PC	4
Phorate	GC	65, 67, 128
	LC	43
Phorate (Thimet®)	TLC	80, 84, 171
Phorate O-analog	GC	65
	LC	43
Phorate O-analog sulfone	GC	65
Phorate O-analog sulfoxide	GC	65
Phorate oxygen analog	GC	67
	TLC	84
Phorate oxygen analog sulfone	GC	67
	TLC	84
Phorate sulfone	GC	65, 67
	LC	43
Phorate sulfoxide	GC	65, 67
	LC	43
Phosalone	GC	65, 67
	LC	43
Phosdrin®	GC	60, 61, 68, 128, 129
	PC	28
	TLC	80, 84
Phosfon®	GC	65
Phosgene	GC	26
Phosphamidon	GC	61, 65, 67
	LC	43
	TLC	80, 84, 171
Phosphate	PC	93
Phosphates, esters	GC	25
Phosphatidic acid	LC	45
	PC	99
	TLC	152
Phosphatidyl choline	TLC	152
Phosphatidylethanolamine	TLC	57, 152
Phosphatidylethanolamines	TLC	56
Phosphatidyl inositol	TLC	152
Phosphatidyl serine	TLC	152
Phosphite	PC	14
Phosphited polyalkylpolyphenol	TLC	175
Phospho-enol-pyruvate (PEP)	PC	12, 32
Phosphoenolpyruvic acid	TLC	128
Phosphoethanolamine	LC	41
	PC	64
	TLC	13
O-Phosphoethanolamine	PC	60
6-Phosphogluconic acid (6-PG)	PC	32
2-Phosphoglycerate	PC	12
2-Phosphoglycerate (2-PGA)	PC	32
3-Phosphoglycerate	PC	12
3-Phosphoglycerate (3-PGA)	PC	32
3-Phosphoglyceric acid	TLC	128
Phosphoribulose pyrophosphates	PC	32
Phosphoric acid	TLC	128
Phosphorus oxychloride	GC	50
Phosphorus trichloride	GC	50
Phosphorylcholine	PC	99
5′-*O*-Phosphoryldeoxythymidylyl-(3′-5′)-deoxythymidylyl-(3′-5′)-deoxythymidine	TLC	62
dl-O-Phosphoserine	PC	63
Phosphothreonine	LC	41
dl-O-Phosphothreonine	PC	63
Phostex	TLC	80
Phosvitin	LC	47
Phthalates, esters	GC	25
Phthalic acid	LC	53
Phthalic anhydride, *p*-anisidide of	TLC	30
Phthalylsulfacetamide	TLC	44
Phthalylsulfamethizole	TLC	140
Phthalylsulfathiazole (Talisulfazol)	TLC	44, 140
Physaliene	TLC	93
Physcion	PC	109
Physodalic acid	PC	30
	TLC	34
Physostigmine	GC	107
	PC	4
	TLC	2, 3
Physostigmine salicylate	GC	107
Phytoene	TLC	94
Phytofluene	TLC	94
Phytol	LC	45
Phytosterol from rapeseed or olive oil	PC	126
Picene	GC	40
	LC	27
β-Picoline	PC	2
Picolinic acid	PC	18
α-Picolinic acid	PC	2
Picolinic acid hydrazide	PC	18
Picolinic acid methyl ester	TLC	158
(-)-Picolinic acid α-phenylethylamide	TLC	158
Picrate	PC	93
Picric acid	LC	53
	TLC	157
Picromycin	PC	66
	TLC	20, 153
Pilocarpine	GC	8, 107
	PC	2, 3, 4
	TLC	2, 3
Pilocarpine HCl	GC	107
Pimarate	LC	68
Pimaricin	TLC	20
Pimelic acid	PC	13, 77
	TLC	31
Pimelic acid, dimethyl ester of	GC	24
Piminodine	TLC	138
Piminodine ethanesulfonate	TLC	138
Pimonodine	GC	18

cis-Pinane GC 87
Pinastric acid TLC 28
α-Pinene GC 86, 87
TLC 51, 54, 106
β-Pinene GC 86, 87
TLC 51, 54, 106
trans-Pinene GC 87
Pinene dimer TLC 106, 143
Pinocembrin PC 26
Pipamazine GC 20, 109
TLC 139
Pipecolic acid GC 101
TLC 13
Piperazine TLC 7, 136
Piperidine-2-carboxylic acid PC 64
Piperidine-3-carboxylic acid PC 64
Piperidine-4-carboxylic acid PC 64
4-Piperidinecarboxylic acid (isonipecotic acid) PC 63
Piperidine-2,6-dicarboxylic acid PC 64
Piperidine pentamethylenedithiocarbamate TLC 22
Piperidine, trifluoroacetyl derivative GC 10
Piperidinium pentamethylenedithiocarbamate TLC 66
Piperidolate TLC 39
Piperilate GC 17
Piperilate hydrochloride GC 17
Piperine GC 8, 96, 111
PC 2
Piperitone, 2,4-DNPH of TLC 76
Piperocaine TLC 39
Piperocaine HCl GC 110
Piperonal PC 22
Piperonal azine PC 23
Piperonal, 2,4-DNPH of TLC 73
Piperonal (Heliotropin) TLC 68, 77
Piperonal, 2-hydrazinebenzothiazole derivative of TLC 168
Piperonal, 4-NPH of TLC 74
β-Piperonylacrylic acid PC 121
Piperonyl butoxide TLC 85
Piperonylic acid TLC 133
Pipradol GC 17
Pipradol hydrochloride GC 17
Pipradol (Pipradrol) TLC 39
Pipradrol TLC 42
Pirazinon GC 65
Pivalaldehyde GC 55
Pivalylbenzhydrazine TLC 39
Placidyl (Ethchlorvynol) TLC 137
Planteose LC 48
Plantvax® GC 65
Plastoquinone A TLC 169
Plastoquinone A (PQ_9) PC 48
Plastoquinone B TLC 169
Plastoquinone C TLC 169
Plastoquinone D TLC 169
Platinum (Pt) LC 83, 86
TLC 55
Platinum, Pt(II) GC 43
Platinum, Pt(IV) PC 15, 95, 96
Plegicil® TLC 42
Plutonium (Pu) LC 80, 88
$Po_4{}^{3-}$ TLC 147
Podospicatin PC 26
Polamidon PC 2
$Polonium_4{}^{3-}$ TLC 147
Polyester adipate (Adipic acid ester) TLC 27
Polyethylenethiuram disulfide (Polyram) PC 117
Polygard TLC 175
Polymer of 2,2,4-trimethyl-1,2-dihydroquinoline TLC 22
Polymyxin B sulfate TLC 21
Polyram (Polyethylenethiuram disulfide) PC 117
Polystyrene LC 78
Polythiazide TLC 41
Ponceau 2R TLC 142
Ponceau 3R TLC 49, 142
Ponceau 4R TLC 49
Ponceau MX TLC 49
Ponceau SX TLC 49, 141, 142
Porfiomycin TLC 20
Porphobilinogen PC 102
(All) Porphyrin esters PC 33
Porphyroxine TLC 4
Potasan® GC 65
TLC 84
Potassium (K) LC 83, 85, 86
TLC 52
Potassium, K(I) GC 43
PC 15, 95, 96
Potassium, K^+ PC 97
Potassium chromium sulfate PC 94
Potassium hydroxide LC 34
Potassium iodate TLC 64
Potassium iodide TLC 64
Powelline GC 96
Pr (Praseodymium) GC 98
LC 39, 79, 80, 90, 91
Pr(III), Praseodymium GC 43
Praseodymium (Pr) GC 98
LC 39, 79, 80, 90, 91
Praseodymium, Pr(III) GC 43
Pramoxine TLC 39
Pratensein PC 26
Prazine® TLC 42
Prednisilone TLC 101
Prednisolone TLC 178
Prednisolone acetate TLC 178
Prednisone TLC 101, 178
Prednisone acetate TLC 101, 178
Prefar GC 61
Pregna-1,4-diene-3,20-dione PC 43
1,4-Pregnadiene-3,20-dione PC 124
4,6-Pregnadiene-3,20-dione PC 124
Pregna-4,9(11)-diene-3,20-dione PC 43
4,9(11)-Pregnadiene-3,20-dione PC 124
Pregna-4,16-diene-3,20-dione PC 43
4,16-Pregnadiene-3,20-dione PC 124
Pregna-5,16-dien-3β-ol-20-one PC 42
TLC 98, 101
5α-Pregnane GC 80, 81
LC 6, 45
5β-Pregnane GC 80, 81
Pregnane derivatives PC 42

Pregnanediol	GC	136
	TLC	101
Pregnane-3α,20α-diol	TLC	101, 178
5α-Pregnane-3α,20α-diol	GC	81
	PC	42
5α-Pregnane-3β,20α-diol	GC	79, 80
	TLC	101
5α-Pregnane-3β,20β-diol	GC	79, 80, 81, 136
	TLC	101
5β-Pregnane-3α,20α-diol	GC	80, 81, 136
	PC	42
5β-Pregnane-3α,20β-diol	GC	80, 81
5α-Pregnane-3β,16α-diol-20-one	TLC	101
Pregnane-3,20-dione	GC	79
4-Pregnane-3,20-dione (Progesterone)	GC	81
5α-Pregnane-3,20-dione	GC	79, 80
	PC	42, 44, 124
	TLC	101
5β-Pregnane-3,20-dione	GC	80, 81
	PC	42, 44, 124
Pregnane-3,20-dione-21-ol	TLC	101, 178
5α-Pregnan-2-en-20-one	GC	80
5α-Pregnane-3β-ol	GC	81
5β-Pregnane-3α-ol	GC	81
5α-Pregnane-3β-ol,20-one	GC	81
5β-Pregnane-3α-ol,20-one	GC	81
5β-Pregnane-3β-ol,20-one	GC	81
5β-Pregnane-3-one	GC	81
Pregnanes (Steroids)	GC	80, 81
Pregnanetriol	GC	136
5α-Pregnane-3α,17α,20α-triol	PC	42
5β-Pregnane-3α,17α,20α-triol	GC	81, 136
5β-Pregnane-3β,16β,20α-triol	PC	42
5α-Pregnane-3,11,20-trione	GC	79, 80
	PC	44, 124
5β-Pregnane-3,11,20-trione	GC	80
	PC	44, 124
5α-Pregnan-3α-ol	GC	80
5α-Pregnan-3β-ol	GC	80
	PC	42
	TLC	98
5β-Pregnan-3α-ol	GC	80
5β-Pregnan-3β-ol	GC	80
	PC	42
5α-Pregnan-20α-ol	GC	80
5α-Pregnan-20β-ol	GC	80
Pregnan-21-ol-3,20-dione	GC	79
5α-Pregnan-3β-ol-11,20-dione	GC	79
5α-Pregnan-3α-ol-20-one	PC	42
5α-Pregnan-3β-ol-20-one	GC	79, 136
5α-Pregnan-20β-ol-3-one	GC	79
5β-Pregnan-3α-ol-20-one	GC	136
	PC	42
5α-Pregnan-3-one	GC	80
	TLC	98
5α-Pregnan-11-one	GC	80
5α-Pregnan-20-one	GC	80
5β-Pregnan-20-one	GC	80
Pregn-5-ene-3β,20α-diol	LC	62
	PC	42
5-Pregnene-3β,20α-diol	GC	136
Pregn-5-ene-3β,20β-diol	LC	62
	PC	42
Pregn-4-ene-17α,21-diol-3,20-dione	TLC	101
Pregn-4-ene-17α,21-diol-3,20-dione-17,21-diacetate	TLC	101
Pregn-4-ene-17α,21-diol-3,20-dione-21-acetate	TLC	101
Pregn-5-ene-3β,17α-diol-20-one	TLC	101
Pregn-4-ene-17α,21-diol-3,11,20-trione-21-acetate	TLC	101
Pregn-4-ene-3,20-dione	GC	80
	PC	42
	TLC	98
4-Pregnene-3,20-dione	GC	79, 136
	PC	124
5α-Pregn-2-ene-11,20-dione	GC	80
4-Pregnene-3,20-dione,17α-ol	GC	81
Pregn-4-ene-21-ol-3,20-dione	TLC	101
4-Pregnene-3α-ol-20-one	TLC	178
5-Pregnene-3β-ol,20-one	GC	81
Pregn-4-ene-21-ol-3,11,20-trione	TLC	101
Pregn-4-ene-3,6,11,20-tetrone	PC	43
Pregn-5-ene-3β,17α,21-triol-20-one	TLC	101
Pregn-4-ene-3,6,20-trione	GC	80
	PC	43
Pregn-4-ene-3,11,20-trione	PC	42
4-Pregnene-3,11,20-trione	PC	124
Pregn-4-ene-3,18,20-trione	PC	43
Pregn-5-en-3β-ol	LC	62
4-Pregnen-17α-ol-3,20-dione	GC	136
Pregn-4-en-21-ol-3,20-dione acetate	TLC	101
Pregnenolone	GC	83
	TLC	101, 178
Pregn-4-en-3α-ol-20-one	TLC	101
4-Pregnen-20α-ol-3-one	GC	136
Pregn-4-en-20β-ol-3-one	PC	42
4-Pregnen-20β-ol-3-one	GC	136
Pregn-5-en-3β-ol-20-one	PC	42
	TLC	98
5-Pregnen-3β-ol-20-one	GC	136
5α-Pregn-16-en-3β-ol-20-one	TLC	101
Pregn-4-en-3-one	GC	80
Preludin®	TLC	42
Prethcamide (Micoren®)	TLC	45
Probarbital (Ipral®)	GC	19, 21, 106
Procaine	GC	8, 21
	TLC	39
Procaine HCl	GC	110
	TLC	43
Procaine (Novocaine)	TLC	45
Prochlorperazine	GC	17, 20, 22,108, 109
	PC	85
	TLC	139
Prochlorperazine dihydrochloride	GC	17
Prochlorperazine dimaleate	GC	17
Prochlorperazine ethane disulfonate	GC	17
Prochlorperazine sulfoxide	TLC	139
Procyclidine	TLC	39
Profenamine hydrochloride	TLC	139
Progesterone	GC	83, 136
	LC	6, 10, 45
	PC	124
	TLC	95,101, 178
Progesterone 11β-ol	LC	10
Progesterone 11β,17α-ol	LC	10
Progesterone 11β,21-ol	LC	10
Progesterone 11β,17α,21-ol	LC	10
Progesterone 17α-ol	LC	10

Progesterone 17α,21-ol LC 10
Progesterone 21-ol LC 10
Proketazine TLC 139
Prolan GC 129
Proline GC 12, 101
LC 2, 4, 14, 33, 41, 56
PC 11, 62, 64, 65
TLC 11, 13, 15, 16
Proline, DNAP of TLC 14
Proline, PTH of TLC 124
L-Proline TLC 118
Prolinol PC 57
Prolixin GC 22
Promazine GC 17, 20, 103,105,108, 109
PC 85
TLC 139
Promazine hydrochloride GC 17
Promazine (Sparine®) GC 21
Promazine sulfoxide TLC 139
Promethazine GC 17, 20, 108, 109
TLC 139
Promethazine-8-chlorotheophyllinate PC 85
Promethazine hydrochloride GC 17
Promethazine (Phenergan®) TLC 45
Promethazine sulfoxide TLC 139
Prometon(e) GC 66
TLC 86
Prometryn(e)® GC 65, 66
TLC 86
Prominal (Mebaral®) GC 21
Propadiene GC 29,125, 126
Propallylonal GC 106
TLC 137
Propanal, 2,4-DNPH of TLC 72, 76
Propanal-2, isobutyric acid ester of GC 51
Propane GC 42, 58, 118,125,126, 127
Propanediols, Butanediols and Related compounds, free form GC 17
1,2-Propanediol LC 37
TLC 113
1,3-Propanediol TLC 113
n-Propanethiol GC 56
Propanol GC 49, 53, 55, 58, 125
Propanol, alcoholic GC 126
iso-Propanol LC 37
n-Propanol GC 57, 92, 94
LC 5, 37, 54, 71
n-Propanol, organics GC 52
1-Propanol GC 55
PC 52
TLC 1
1-Propanol (DNB) TLC 111
1-Propanol (ONPU) TLC 111
Propanol-1, butyric acid ester of GC 51
Propanol-1, formic acid ester of GC 51
Propanol-1, isobutyric acid ester of GC 51
Propanol-1, propionic acid ester of GC 51
2-Propanol GC 92, 94, 125
PC 52
2-Propanol, butanol GC 126
2-Propanol (DNB) TLC 111
2-Propanol (ONPU) TLC 111
Propanol-2, butyric acid ester of GC 51
Propanol-2, formic acid ester of GC 51
Propanol-2, propionic acid ester of GC 51
Propan-2-one, 2,4-DNPH of TLC 76
Propargyl chloride LC 60
1-Propargylpyrrole GC 133
Propazine GC 66, 129
TLC 81, 86
Propenylguaethol TLC 90
Propenylguaethol (vanitrope, 1-ethoxy-2-hydroxy-4-propenylbenzene) TLC 68
4-Propiocatechol TLC 87
4-Propioguaiacone TLC 87
Propionaldehyde GC 7, 53, 57, 125
LC 8
TLC 77
Propionaldehyde, 2,4-DNPH of TLC 166
Propionaldehyde, 2-hydrazinebenzothiazole derivative of TLC 168
Propionaldehyde, p-NPH of TLC 165
Propionaldehyde, aldehydes GC 126
Propionaldehyde dimethyl acetal, organics GC 52
Propionaldehyde, organics GC 52
n-Propionaldehyde PC 20
Propionate, β-sitosteryl ester of TLC 182
Propionic acid GC 127
LC 57, 65
PC 77, 79
TLC 29,132, 134
Propionic acid, benzyl ester of GC 24
Propionic acid, cinnamyl ester of GC 24
Propionic acid esters GC 51
Propionic acid, ethyl ester of GC 24
Propionic acid, methyl ester of GC 24
α-G-Propionic carboxyalkylguanidine (guanidino acid) PC 9
β-G-Propionic carboxyalkylguanidine (guanidino acid) PC 9
α-G-Propionic guanidine PC 9
β-G-Propionic guanidine PC 9
Propionitrile GC 51, 55, 123
LC 60
17β-Propionoxy-4-androsten-3-one PC 124
Propionyl, Pr GC 98
Propionylcholine PC 8
2-Propionylphenol TLC 89
4-Propionylphenol TLC 89
Propionyl phosphate PC 77
Propiophenone GC 51
PC 22
Propiophenone, 4-NPH of TLC 74
Propiophenone azine PC 23
Propiopromazine GC 20, 109
Propoxur TLC 170
Propoxycaine HCl GC 110
Propoxyphene (Darvon®) GC 21
TLC 39, 138
d-Propoxyphene GC 18, 105

2,4,5-T-n-propyl TLC 81
Propyl acetate GC 2, 49, 58
n-Propyl acetate GC 53, 55, 57,121, 125
n-Propyl acetate, esters GC 126
n-Propyl acetate, organics GC 52
Propyl acrylate GC 91
n-Propyl alcohol GC 127
n-Propyl alkyl guanidine PC 9
n-Propyl-1-alkyl-3-nitroguanidine PC 9
n-Propyl allyl sulfide GC 85
n-Propyl allyl trisulfide GC 85
Propylamine TLC 10
n-Propylamine GC 11, 97
LC 26
PC 7, 55, 56
TLC 5
n-Propylamine, trifluoroacetyl derivative GC 10
N-*n*-Propylaniline GC 102
Propylbenzene GC 51, 53
n-Propylbenzene GC 55,119, 120
LC 42
1-*m*-2-*n*-Propylbenzene GC 120
1-*m*-3-*n*-Propylbenzene GC 120
n-Propyl bromide GC 47
Propyl butyl ketone GC 117
Propyl butyl sulfide GC 135
Propyl butyrate GC 2
n-Propyl *n*-butyrate GC 57
Propyl caprylate TLC 105
4-Propylcatechol TLC 87
n-Propyl chloride GC 47
Propyl crotonate GC 91
n-Propyl cyanide GC 56
n-Propylcyclopentane GC 42
S-Propyl cysteine PC 64
n-Propyl dihydrogen phosphate PC 34
1-*n*-Propyl-2,5-dimethylpyrrole GC 133
Propylene GC 29, 58, 118,122,125, 126, 127
Propylene carbonate LC 60
Propylene dichloride GC 32
Propylene glycol GC 127
LC 5
1,2-Propylene glycol PC 1
1,3-Propylene glycol PC 1
Propylene glycol butyl ether PC 115
Propylene glycol butyl ether ester 2,4-D GC 129
Propylene oxide LC 60
Propylene oxide, organics GC 52
Propyl esters GC 24
Propyl ethers GC 116
Propyl formate GC 2. 57
n-Propyl formate GC 47, 55, 121
Propyl gallate TLC 125
4-Propylguaiacol TLC 87
iso-Propyl guanidine PC 9
n-Propyl guanidine PC 9
Propyl hexanoate GC 2
Propylhexedrine GC 23
TLC 39
Propylhexedrine (Eventine®) TLC 45
n-Propyl-*n*-hexyl adipate GC 25
n-Propyl-*n*-hexyl phthalate GC 25
n-Propyl-*n*-hexyl sebacate GC 25
n-Propyl-*n*-hexyl suberate GC 25
n-Propyl-*n*-hexyl succinate GC 25
n-Propyl *p*-hydroxybenzoate LC 74
3,3′-Propylidene-bis-(4-hydroxycoumarin) TLC 79
n-Propyl iodide GC 47
Propyl isobutyrate GC 2
n-Propyl isocyanate GC 44
Propyl-isopropyl-sulfide GC 135
Propyl methacrylate GC 91
Propyl 3-methylbutyrate GC 2
n-Propyl methyl ketone GC 55
Propyl 4-methylpentanoate GC 2
1-*iso*-Propyl-3-nitroguanidine PC 9
1-*n*-Propyl-3-nitroguanidine PC 9
Propyl octyl sulfide TLC 163
Propyl pentanoate GC 2
2-*n*-Propylphenol GC 72
4-Propylphenol TLC 87
4-*n*-Propylphenol GC 72
Propylphenyl phosphate PC 112
Propyl propionate GC 2
n-Propyl *n*-propionate GC 57
2-*n*-Propylpyridine TLC 60
4-*n*-Propylpyridine GC 131
1-*n*-Propylpyrrole GC 133
2-*n*-Propylpyrrole GC 133
4-*n*-Propylresorcinol PC 121
n-Propyl sulfide LC 60
n-Propyl sulfoxide GC 139
Propyl thiopyrophosphate GC 65
Propyl-2-thiouracil PC 36
N-*n*-Propylthiourea PC 47
1-Propyl-2,4,6-trimethylbenzene TLC 53
Propyne GC 29, 125
Propyne, alkynes GC 126
Protactinium (Pa) LC 80
Protanghinin TLC 102
Protanghinin peracetate TLC 102
Prothipendyl (Dominal) TLC 45
Prothypendyl TLC 139
Protocatechuic acid GC 90
PC 82
TLC 133,135, 174
Protocatechuic aldehyde TLC 87, 174
Protocatechuic aldehyde, 2,4-DNPH of TLC 167
Protocetraric acid PC 30
TLC 34
Protolichesterinic acid TLC 131
Protopine GC 96
PC 3
TLC 4
Protoporphyrin PC 33
Protoporphyrin I PC 123
Protoprophyrin III PC 123
Protoporphyrin IX PC 123
Pr, propionyl GC 98
Prunetin PC 26
Pseudocumene (1,2,4-trimethylbenzene) GC 53
Pseudoephedrine GC 23
Pseudoionone TLC 77
Pseudoionone, 2,4-DNPH of TLC 76
Pseudouridine LC 22, 31, 46
α-Pseudouridine GC 77

β-Pseudouridine	GC	77
Psicaine (Dextrocaine)	TLC	43
Psicaine new	TLC	3
Psilocin	PC	103
Psoromic acid	PC	30
	TLC	34
Psylocybin	PC	101
Pt (Platinum)	LC	83, 86
	TLC	55
Pt(II), Platinum	GC	43
Pt(IV), Platinum	PC	15, 95, 96
Pt-OAc	PC	19
Pu (Plutonium)	LC	80, 88
Pulchellidin	PC	25
Pulegone	TLC	77
Pulegone, 2,4-DNPH of	TLC	76
Pulvic acid	TLC	28
Pulvic dilactone	TLC	28
Purine	LC	15, 46
	PC	16, 36, 37
Purine and pyrimidine bases	PC	108
Purines and pyrimîdines	GC	77
Purpurea glycoside A	PC	127
	TLC	102, 177
Purpurea glycoside B	PC	127
	TLC	102, 177
Purpurin	PC	109
Putrescine	LC	66
	TLC	5, 13
Putrescine (1,4-Diaminobutane)	PC	64
Pyocyanine chloride	PC	54
Pyrathiazine	GC	17, 20, 108, 109
Pyrathiazine hydrochloride	GC	17
Pyrazole	LC	46
	TLC	10
Pyrazolo[3,4-*d*] pyrimidine, 4-amino-	LC	46
Pyrazolo[3,4-*d*] pyrimidine, 4,6-dihydroxy-	LC	46
Pyrazolo[3,4-*d*] pyrimidine, 4-hydroxy-	LC	46
Pyrene	GC	40
	LC	27
	PC	89
	TLC	7, 51, 53, 146
Pyrethrin I	PC	119
	TLC	85
Pyrethrin I peroxide	TLC	85
Pyrethrin II	PC	119
	TLC	85
Pyrethrin II peroxide	TLC	85
Pyribenzamine (Tripelennamine)	TLC	39
Pyribenzamine citrate (tripelennamine citrate)	GC	20, 54
Pyridine	GC	131
	LC	26, 34, 46
	TLC	7, 60, 61
Pyridine-2-aldehyde	TLC	60
Pyridine-3-aldehyde	TLC	60
Pyridine-4-aldehyde	TLC	60
α-Pyridinealdehyde	PC	2
β-Pyridinealdehyde	PC	2
γ-Pyridinealdehyde	PC	2
Pyridine-2-carbinol	TLC	60
Pyridine-3-carbinol	TLC	60
Pyridine-4-carbinol	TLC	60
Pyridine-2-carboxylic acid	TLC	60
Pyridine-3-carboxylic acid	TLC	60
Pyridine-4-carboxylic acid	TLC	60
2,6-Pyridinedialdehyde	PC	2
2,3-Pyridinedicarboxylic acid	PC	2
2,4-Pyridinedicarboxylic acid	PC	2, 18
2,4-Pyridinedicarboxylic acid hydrazide	PC	18
2,5-Pyridinedicarboxylic acid	PC	2
Pyridine-2,6-dicarboxylic acid	TLC	60
2,6-Pyridinedicarboxylic acid	PC	2
3,4-Pyridinedicarboxylic acid	PC	2
3-Pyridinemethanol	TLC	39
Pyridine oxide	TLC	7
Pyridine *N*-oxide	TLC	61
Pyridines, nitrogen compounds	GC	45
9H-Pyrido[2,3-*b*] indole	TLC	120
Pyridomycin	TLC	20
Pyridoxal	GC	88
	TLC	109
Pyridoxal ethyl acetal	GC	88
	TLC	109
Pyridoxal-P	LC	18
Pyridoxal phosphate	TLC	109
Pyridoxal phosphate phenylhydrazone	TLC	109
Pyridoxamine	GC	88
	TLC	109
Pyridoxamine-P	LC	18
Pyridoxamine phosphate	TLC	109
4-Pyridoxic acid	TLC	109
4-Pyridoxic acid lactone	GC	88
	TLC	109
5-Pyridoxic acid lactone	GC	88
Pyridoxine	GC	88
	PC	18, 51
	TLC	188
Pyridoxine hydrochloride	LC	20
	TLC	188
Pyridoxol	GC	88
	TLC	109
Pyridoxol phosphate	TLC	109
3-Pyridylacetic acid	PC	83
1-(2′-Pyridylazo)-2-naphthol (PAN)	TLC	47
4-(2′-Pyridylazo)-resorcinol (PAR)	TLC	47
3-Pyridyl ethyl ketone	GC	95
3-Pyridyl methyl ketone	GC	95
3-Pyridyl *n*-propyl ketone	GC	95
Pyrilamine (Neo-Antergan®)	GC	20
	TLC	39
Pyrilamine maleate	GC	20, 54, 104
Pyrimidine	LC	46
Pyrimidine and purine bases	PC	108
Pyrimidines and purines	GC	77
Pyrimithate	LC	43
Pyrithyldione	TLC	137
Pyrocatechol	TLC	90, 176
Pyrocatechol, di(4-acetyl-2-nitrophenol) ether of	TLC	91
Pyroclavine	TLC	114
Pyrogallol	LC	1
	PC	82
	TLC	90,174, 176
Pyrogallol carboxylic acid	PC	82
Pyrolan	GC	130
	TLC	83
Pyromellitic anhydride, *p*-anisidide of	TLC	30

α-Pyronene	GC	87
	TLC	106
β-Pyronene	GC	87
	TLC	106
Pyronil® (Pyrrobutamine)	TLC	39
Pyrophosphate	LC	47
	PC	12, 32
Pyrophosphoric acid ($H_4P_2O_7$)	PC	35
α-Pyrovetivene	GC	140
Pyrrobutamine	GC	17
Pyrrobutamine diphosphate	GC	17
Pyrrobutamine (Pyronil®)	TLC	39
Pyrrocaine HCl	GC	110
Pyrrochlorin	PC	33
Pyrrole	GC	133
	LC	46
Pyrrole-2-carboxylic acid	PC	83, 102
Pyrrole, trifluoroacetyl derivative	GC	10
Pyrrolidine, trifluoroacetyl derivative	GC	10
Pyrrolidinomethyltetracycline	TLC	21
Pyrroline, trifluoroacetyl derivative	GC	10
Pyrroporphyrin	PC	33
Pyruvate acid	TLC	36
Pyruvate (K)	PC	13
Pyruvic acid	PC	77, 81
	TLC	29, 134
Pyruvic acid, 2,4-DNPH of	TLC	32, 166
Pyruvic (Isomer 1), 2,4-DNPH of	TLC	32
Pyruvic (Isomer 2), 2,4-DNPH of	TLC	32

Adsorption of the Elements (LC Section – Figures)

P, Phosphorus	Pg	278, 279, 280, 281
Pa, Protactinium	Pg	278, 279, 280, 281
Pb, Lead	Pg	278, 279, 280, 281
Pd, Palladium	Pg	278, 279, 280, 281
Po, Polonium	Pg	278, 279, 280, 281
Pt, Platinum	Pg	278, 279, 280, 281
Pu, Plutonium	Pg	278, 279, 280, 281

α-Quaterthienyl	TLC	104
Quercetagetin	PC	26
Quercetin	LC	1
	PC	26
Quercetrin	PC	26
Quercimeritrin	PC	26
Quercitin	LC	28
Quercitin pent-acetate	LC	28
Quercitrin	LC	1, 28
Quinalbarbitone (Secobarbital) (Seconal)	PC	86
	TLC	137
Quinaldylglycine	PC	80
Quinethazone	TLC	41
Quinic acid	LC	23
Quinidine	GC	21, 107
	PC	2, 3, 4, 54
	TLC	2, 3
Quinidine sulfate	GC	107
Quinine	GC	8, 21, 22, 96, 107, 111
	PC	2, 3, 4, 54
	TLC	2, 3, 39, 40, 45
Quinine ethyl carbonate	GC	107
Quinine HCl	GC	107
Quinine sulfate	LC	20
Quinizarin	PC	109
Quinol	LC	52
Quinoline	GC	28, 45, 78
	TLC	7, 58
Quinolines, nitrogen compounds	GC	45
Quinoline Yellow SS	TLC	48
Quinolinic acid	PC	18, 82
Quinolinic acid hydrazide	PC	18
2-Quinolinol	TLC	92
2-Quinolonyl-6-glucuronide	PC	106
4-Quinolonyl-6-glucuronide	PC	106
8-Quinolyl glucoside	PC	82
Quinolyl-6-glucuronide	PC	106
3-Quinolyl sulfuric acid	PC	82
6-Quinolyl sulfuric acid	PC	82
7-Quinolyl sulfuric acid	PC	82
8-Quinolyl sulfuric acid	PC	82
Quinovic	TLC	184
Quinoxalino[*b*-2,3]-5α-cholestane	TLC	96
Quintozene	TLC	81
3-Quinuclidinol	PC	54
Raffinose	LC	25, 71
	PC	70
	TLC	23, 126
Rangiformic acid	TLC	131
Rauwolfia alkaloids, free form	GC	17
Rauwolscine	PC	2
	TLC	3
Rb (Rubidium)	LC	83, 85, 86
	PC	15
	TLC	52
Rb^+, Rubidium	PC	97
Re (Rhenium)	LC	80
Reaction product of phenyl-β-naphthylamine and acetone	TLC	22
Reactivan	TLC	45
Recinoleic acid	PC	77
Red 1	TLC	141
Red 2	TLC	141
Red 2G	TLC	49
Red 3	TLC	141
Red 4	TLC	141
Red 5	TLC	141
Red 6	TLC	141
Red 6B	TLC	49
Red FB	TLC	49
Red 10B	TLC	49
Redax	GC	62
Remalinolic acid	TLC	35
ReO_4^-, Rhenium tetraoxide	PC	92
Rescinnamine	GC	17
	PC	2
	TLC	115
Reserpine	GC	17, 108
	PC	2
	TLC	2, 3, 115
Reserpinine	TLC	115
Resibufogenin (3β-Hydroxy-14β,15β-epoxy-5β-bufadienolide)	TLC	102
Resibufogenin peracetate	TLC	102
Resokempferol	LC	28

Resorcinol	LC	1, 52
	PC	82
	TLC	88, 90, 174, 176
Resorcinol, di(4-acetyl-2-nitrophenol) ether of	TLC	91
Resorcinol dimethyl ether	TLC	174
α-Resorcylic acid	LC	30
	PC	82
	TLC	135
β-Resorcylic acid	PC	82
	TLC	135, 174
γ-Resorcylic acid	PC	82
	TLC	135
Retamine	PC	3
Reticuline	TLC	4
Reynoutrin	PC	26
Rh, Rhodium	LC	80, 83, 84, 85, 86
	PC	15
	TLC	55
Rh(III), Rhodium	GC	43
Rhamnetin	LC	28
	PC	26
Rhamnitol	LC	69
L-Rhamnitol	GC	14
L-Rhamnono-γ-lactone	PC	72
Rhamnose	GC	15
	LC	25, 29, 48, 71
	PC	70, 73
	TLC	23, 24
L-Rhamnose	PC	75
	TLC	126
Dg, α,D-rhamnoside	TLC	102
Dg, α,D-rhamnoside peracetate	TLC	102
Dg, β,D-rhamnoside	TLC	102
Dg, β,D-rhamnoside peracetate	TLC	102
Rhein	PC	109
Rhenium (Re)	LC	80
Rhenium tetraoxide, ReO_4	PC	92
Rhizocarpic acid	TLC	28
Rhizonic acid	PC	30
Rhodamine B	TLC	7
Rhodeasapogenin	TLC	103
Rhodexin A	TLC	102
Rhodexin A peracetate	TLC	102
Rhodexin B	TLC	102
Rhodexin B peracetate	TLC	102
Rhodexin C	TLC	102
Rhodexin C peracetate	TLC	102
Rhodinyl acetate	TLC	105
Rhodium, Rh	LC	80, 83, 84, 85, 86
	PC	15
	TLC	55
Rhodium, Rh(III)	GC	43
Rhodochlorin	PC	33
Rhodocladonic acid	PC	109
Rhodoporphin	PC	33
Rhodoxanthin	TLC	93, 94
Rhoeadine	PC	3
Rhoeagenine	PC	3
Rhoifolin	PC	26
Ribitol	GC	14, 15
	LC	69
Ribitol (Adonitol)	PC	1
D-Ribitol	GC	76
Riboflavin	LC	17, 18, 20
	PC	51
	TLC	188
Riboflavin phosphate	LC	17
	TLC	188
Riboflavin-5′-phosphate sodium	TLC	188
3-Ribohexulose	LC	48
D-Ribonic	LC	40
Ribonuclease	LC	4
Ribonucleic acid	LC	13
Ribonucleosides	GC	77
Ribose	LC	25, 29, 48
	PC	32
	TLC	23, 24, 127
Ribose-5-phosphate (R-5-P)	PC	32
D-Ribose	GC	15
	PC	75
	TLC	126
D-Ribose-1-phosphate	PC	12
D-Ribose-5-phosphate	PC	12
5-Ribosyluracil	PC	36
Ribothymidine	LC	22
Ribulose	PC	70
D-Ribulose	GC	15
Ribulose-1,5-diphosphate (Ru-1,5-PP)	PC	32
Ribulose-5-phosphate (Ru-5-P)	PC	32
Riburonic	LC	23
Rilatin	TLC	42
Rimifon®	TLC	42
Ristocetin	TLC	21
Ritalin (Rilatin)	TLC	45
RNS	LC	15
Robinetin	LC	1, 28
	PC	26
Robinin	LC	28
	PC	26
	TLC	92
Robinucleosides	PC	108
Roccellic acid	TLC	131
Rodamine B	TLC	142
Roemeridine	PC	3
Rogor® (Dimethoate)	PC	28
	TLC	84
Rogor-oxon	PC	28
Ronnel (Fenchlorphos) (Dow ET-14)	GC	60, 64, 65, 67, 68, 129
	PC	28
	TLC	82, 84
Ronnel oxygen analog	GC	67
Rose Bengale	TLC	142
Rosinidin	PC	25
Rothane	PC	116
Ru (Ruthenium)	LC	80
	TLC	55
Ru (III), Ruthenium	GC	43
Ru-1,5-PP (Ribulose-1,5-diphosphate)	PC	32
Rubiadin	PC	109
Rubidium (Rb)	LC	83, 85, 86
	PC	15
	TLC	52
Rubidium, Rb^+	PC	97
Rubixanthin	TLC	93
Rubrene	LC	27

Ruelene® GC 65, 67
TLC 84
Rufigallic acid PC 109
Ruthenium, (Ru) LC 80
TLC 55
Ruthenium, Ru(III) GC 43
Rutin LC 1, 28
PC 26
TLC 92, 188
Rutonal GC 19
RZ-50A TLC 66

Adsorption of the Elements
(LC Section – Figures)

Ra, Radium Pg 278, 279, 280, 281
Rb, Rubidium Pg 278, 279, 280, 281
Re, Rhenium Pg 278, 279, 280, 281
Rh, Rhodium Pg 278, 279, 280, 281
Ru, Ruthenium Pg 278, 279, 280, 281

S 421 TLC 85
Sabinene GC 86
TLC 106
Saccharinic acids LC 40
Safrol TLC 90
Sakuranetin PC 26
Salazinic acid PC 30
TLC 34
Salazosulfapyridine TLC 140
Salicylacetic acid LC 30
Salicyl alcohol LC 53
Salicylaldazine PC 23
Salicylaldehyde PC 20, 22
TLC 40, 129, 174
Salicylaldehyde, 2,4-DNPH of TLC 73, 167
Salicylaldehyde, 2,4-DNPSC of TLC 165
Salicylaldehyde, 2-hydrazinebenzothiazole derivative of TLC 168
Salicylaldehyde, *p*-NPH of TLC 165
Salicylaldehyde, 4-NPH of TLC 74
Salicylaldoxime PC 120
Salicylamide GC 19, 104, 105
PC 120
TLC 38, 40, 129
Salicylcholine PC 8
Salicylic acid GC 105
LC 53, 74
TLC 40, 133, 135, 174
Salicylic acid, amyl ester of GC 24
Salicylic acid, ethyl ester of GC 24
Salicylic acid, hexyl ester of GC 24
Salicylic acid hydrazide PC 121
Salicylic acid (*o*-hydroxybenzoic acid) TLC 129
Salicylic acid methylamide TLC 129
Salicylic acid, methyl ester of GC 24
2-(Salicylideneamino)-pyridine (SAMPy) TLC 47
o-Salicylsalicylic acid TLC 129
Salicyluric acid TLC 129
Saligenin LC 1
Salt form, drugs GC 17
Samarium (Sm) LC 39, 79, 85, 86, 90, 91
Sanguilutine PC 3
Sanguinarine PC 3
Sanguirubine PC 3
Santalene (major) GC 140
Santalene (minor) GC 140
β-Santalene TLC 106, 143
Santene GC 86
Santocure TLC 66
Santocure NS TLC 66
Santocure 26 TLC 66
Santoquin (6-ethoxy-2,2,4-trimethyl-1,2-dihydroquinoline) PC 69
Santowhite Crystals TLC 175
Santowhite L TLC 175
Santowhite Powder TLC 175
Saponaretin PC 26
Saponarin PC 26
Saponin PC 127
Sarcosine GC 101
LC 41
PC 64
TLC 11, 13, 15, 16
Sarmentocymarin PC 127
Sarmentogenin PC 127
Sarmentogenin acetate PC 127
Sarmentogenin diacetate PC 127
Sarmentogenin peracetate TLC 102
Sarmentogenin (3*β*,11*α*,14*β*-Trihydroxy-5*β*-cardenolide) TLC 102
Sarmentoside A PC 127
TLC 102
Sarmentoside A peracetate TLC 102
Sarnovide TLC 102
Sarnovide mono(?) acetate PC 127
Sarnovide peracetate TLC 102
Sarpagine TLC 3, 115
Sarsasapogenin TLC 103
Sarsasapogenin acetate TLC 98
Sarsasapogenin 3-acetate TLC 103
Sarverogenin PC 127
Sarverosid PC 127
Sativene GC 140
Sb (Antimony) LC 86, 88
PC 15
Sb(III), Antimony PC 95, 96
Sb^{3+}, Antimony PC 97
TLC 150
Sc, Scandium LC 79, 83, 84, 85, 86, 87, 88, 90, 91
PC 15
Sc(III), Scandium GC 43
Sc^{3+}, Scandium PC 92
Scandium, Sc LC 79, 83, 84, 85, 86, 87, 88, 90, 91
PC 15
Scandium, Sc(III) GC 43
Scandium, Sc^{3+} PC 92
Scarlet GN TLC 141
Schradan® GC 65, 67
LC 43
TLC 80
Scillaren A TLC 102
Scillaren A peracetate TLC 102
Scillarenin A (3*β*,14*β*-Dihydroxy-bufatrien-4,20,22-olide) TLC 102

Scillarenin A peracetate	TLC	102
Scillarenin-β,D-glucoside	TLC	102
Scillarenin-β,D-glucoside peracetate	TLC	102
Scilliglaucosidin (3β,14β-Dihydroxy-19-oxo-bufatrien-4,20,22-olide)	TLC	102
Scillirosid	TLC	102
Scillirosid peracetate	TLC	102
Scillirosidin (3β,8β,14β-Trihydroxy-6β-acetoxy-bufatrien-4,20,22-olide)	TLC	102
SCN	TLC	147
Scopolamine	GC	8, 21, 107
	PC	3
	TLC	2, 3, 39, 117
Scopolamine HBr	GC	107
Scopolanol	TLC	117
Scopoletin	PC	113
Scopolin	PC	113
Scopoline	TLC	3
SD-7438	GC	67
Se (Selenium)	LC	83, 84, 85, 86
	TLC	55
Se(III), Selenium	PC	96
Sebacates, esters	GC	25
Sebacic acid	PC	13, 77
	TLC	31
Sebacic acid, dimethyl ester of	GC	24
Sebacic acid polyester	TLC	27
Secalonic acid D	TLC	185
Secbutobarbitone	PC	86
Secobarbital (Seconal®) (Quinalbarbitone)	GC	19, 21, 105, 106
	TLC	137
2,3-Seco-5α-cholestane-2,3-diol	TLC	96
Seconal® (Secobarbital, Quinalbarbitone)	GC	21
	TLC	137
Secondary front	TLC	35
3,5-Seco-4-norandrostan-17β-ol-5-on-3-oic acid	TLC	96
3,5-Seco-4-norcholestane-3,5β-diol	TLC	96
3,5-Seco-4-norpregnan-20β-ol-5-on-3-oic acid	TLC	96
D-Sedoheptulosan	GC	15
Sedoheptulose	TLC	23
Sedoheptulose-7-phosphate (S-7-P)	PC	32
Sekikaic acid	TLC	35
Selenate	PC	14, 91
Selenite	PC	14, 91, 93
Selenium (Se)	LC	83, 84, 85, 86
	TLC	55
Selenium, Se(III)	PC	96
Selenium diethyldithiocarbamate	TLC	22
Selenium dimethyldithiocarbamate	TLC	66
Selenomethionine	LC	41
Selina-3,7(11)-diene	GC	140
Selina-4(14),7-diene	GC	140
Selina-4(14),7(11)-diene	GC	140
β-Selinene	GC	140
	TLC	54, 106, 143
δ-Selinene	GC	140
SeO_3^{--}	TLC	147, 150
Serine	GC	12, 101
	LC	2, 4, 33, 41, 56
	PC	11, 64, 65
	TLC	11, 13, 15, 16
Serine, DNAP of	TLC	14
Serine, PTH of	TLC	124
DL-Serine	PC	63
	TLC	118
L-Serine	TLC	118
Serinol	PC	57
Serotonin	PC	103
	TLC	8, 59, 156
Serotonin (5-hydroxytryptamine)	LC	77
Serpentine	GC	96
	PC	2
	TLC	3, 115
Serpentinine	TLC	3, 115
Serum albumin (bovine)	LC	19
Serum albumin dimer (bovine)	LC	19
Setoclavine	TLC	114
Sevin®	TLC	80, 172
S-sevin	TLC	172
Shell SD-8280	GC	65
Shell SD-8436	GC	65
Shell SD-8448	GC	65
Shikimic acid	LC	23
	TLC	133
Siaresinolic	TLC	184
Sigmodal	TLC	137
Silane	GC	27
Silicon compounds, methyl phenyl polysiloxanes	PC	38
Silicon tetrachloride	GC	26, 50
Silicon tetrafluoride	GC	26
Silver, Ag	PC	15
	TLC	52
Silver, Ag(I)	PC	95, 96
	TLC	148, 149, 150
Silver, Ag^+	PC	92, 97
Simazine	GC	66, 68, 128, 129
	TLC	81, 86
Simetone	GC	66
	TLC	86
Simetryne®	GC	66
	TLC	86
Sinapaldehyde, 2,4-DNPH of	TLC	167
Sinapic acid	TLC	133
Sinaxar®	TLC	42
Siphonaxanthin	PC	122
Siphonein	PC	122
β-Sitostanol	TLC	99
Sitosterol	LC	24
β-Sitosterol	GC	84
	LC	45
	PC	126
	TLC	57, 99
β-Sitosterol acetate	TLC	57
Skatole	PC	102, 103
	TLC	156
Skimmin	PC	113
Sm (Samarium)	LC	39, 79,

Sm (Samarium) (*continued*) 85, 86, 90, 91
Smilagenin TLC 98
Smilagenin 3-acetate TLC 103
Sn (Tin) LC 80, 83, 86, 87
Sn(II), (IV), Tin PC 15, 95, 96
Sn^{2+}, Tin PC 97
SO_2 GC 113
SO_3^{--} TLC 147
SO_4^{--} TLC 147
$S_2O_3^{--}$ TLC 147
Sodium (Na) LC 83, 85, 86, 88
TLC 52
Sodium, Na(I) GC 43
PC 95, 96
Sodium acetate LC 34
Sodium bicarbonate LC 34
Sodium carbonate LC 34
Sodium citrate LC 34
Sodium dibutyldithiocarbamate TLC 22
Sodium dimethyldithiocarbamate TLC 22
Sodium hydroxide LC 34
Sodium noramidopyrine methanesulfonate TLC 40
Sodium phosphate LC 34
Sodium sulfadiazine PC 87
Sodium sulfamerazine PC 87
Sodium sulfamethiazine PC 87
Sodium sulfapyridine PC 87
Sodium sulfathiazole PC 87
Sodium tetraborate LC 34
Solanidan-3β-ol GC 79
Solanidine GC 96
Solasodine GC 96
Solonaceous alkaloids PC 3
Solvent Yellow TLC 48
Solvent Yellow 7 TLC 48
Somalin TLC 102
Somalin peracetate TLC 102
Sorbic acid LC 74
Sorbitol GC 15
PC 1, 53, 70
TLC 127
D-Sorbitol TLC 127
Sorbose LC 25, 29, 48
PC 70
TLC 23, 24
D-Sorbose GC 15
L-Sorbose TLC 126
Sordinol TLC 45, 139
Sparteine GC 8, 96, 107
PC 2, 3, 4
TLC 2, 3
Sparteine sulfate GC 107
Spermidine LC 66
Spermine LC 66
PC 56
Sphaerophorin PC 30
TLC 35
Sphaerophorol-carboxylic acid PC 30
Sphingomyelin TLC 57, 152
Sphingomyelins TLC 56
α-Spinasterol GC 84
Spiraeoside PC 26
Spiramycin TLC 20
Spironolactone TLC 41
Spongothymidine PC 36
Squalene LC 27, 45, 68
TLC 56, 106, 143
Squamatic acid PC 30
TLC 35
Sr (Strontium) LC 79, 81, 83, 85, 86, 87, 88, 89, 90
TLC 52
Sr(II), Strontium GC 43
PC 15, 95, 96
TLC 149
Stabilite white TLC 175
Stachyose LC 48, 71
Stanazolol TLC 95
Stannic chloride GC 50
Stauffer N-2788 GC 65
Stearate LC 68
Stearate, β-sitosteryl ester of TLC 182
Stearic acid LC 57
PC 77
TLC 33, 57
Stearic acid, ethyl ester of GC 24
Stearic acid, linoleic acid TLC 56
Stearolic acid PC 77
TLC 33
Stearyl alcohol LC 45
Stearyl gallate TLC 125
Stemetil® TLC 42
Sterigmatocystin TLC 185
Steroidal alkaloids GC 96
C_{18} steroids PC 124
C_{21} steroids PC 124
Steroids, C_{19} PC 124
Steroids, C_{18} PC 124
Steroids, C_{21} PC 124
Stictic acid PC 30
TLC 34
Stigmasterol LC 24, 45
GC 84
PC 126
TLC 99
Stigmasteryl acetate LC 45
Stilbene TLC 51
trans-Stilbene GC 40
Stillopsidin PC 26
Stillopsin PC 26
Strdn-β,D-glucoside TLC 102
Strdn peracetate TLC 102
Strdn-α,D-rhamnoside TLC 102
Strdn-β,D-rhamnoside TLC 102
Strdn-β,D-rhamnoside peracetate TLC 102
Strdn-3-tetrahydropyranylether TLC 102
Strdn-β,D-xyloside TLC 102
Strdn-β,D-xyloside peracetate TLC 102
Streptomycin PC 67
TLC 20
Streptothricin PC 67
TLC 20
Streptothricin I PC 67
Streptothricin II PC 67
Streptothricin III PC 67
Streptothricin (sulfate) TLC 20

Strobane®	GC	60, 91, 128, 129
	PC	116
	TLC	82
Strontium (Sr)	LC	79, 81, 83, 85, 86, 87, 88, 89, 90
	TLC	52
Strontium, Sr(II)	GC	43
	PC	15, 95, 96
	TLC	149
Strontium, Sr^{2+}	PC	92
Stropeside	TLC	177
Strophadogenin peracetate	TLC	102
Strophadogenin (3β,5β,14β,16β-Tetrahydroxy-19-oxo-cardenolide)	TLC	102
Strophanthidin	PC	127
Strophanthidin acetate	PC	127
Strophanthidin acid (3β,5β,14β-Trihydroxy-cardenolide-19-acid)	TLC	102
Strophanthidin acid methylester (3β,5β,14β-Trihydroxy cardenolide-19-acid methylester)	TLC	102
Strophanthidin peracetate	TLC	102
Strophanthidin (Strdn) (3β,5β,14β-Trihydroxy-19-oxo-cardenolide)	TLC	102
Strophanthidol	PC	127
Strophanthidol acetate	PC	127
Strophanthidol peracetate	TLC	102
Strophanthidol (3β,5β,14β,19-Tetrahydroxy-cardenolide)	TLC	102
Strophanthojavoside	TLC	102
Strophanthojavoside peracetate	TLC	102
Strophothevoside	TLC	102
Strophothevoside peracetate	TLC	102
Strospeside	PC	127
	TLC	102
Strospeside peracetate	TLC	102
Strychnine	GC	8, 21, 22, 96, 107, 111
	LC	55
	PC	2, 3, 4
	TLC	2, 3, 39, 45
Strychnine nitrate	GC	107
Stylophylline	PC	3
Stylopine	PC	3
Styramate	TLC	38
Styrenated phenol	TLC	22
Styrene	GC	51, 118
	LC	21, 36
	TLC	51
Suberates, esters	GC	25
Suberic acid	PC	13
	TLC	31
Suberic acid, dimethyl ester of	GC	24
Substituted styrenated phenol	TLC	175
Succinate acid	TLC	36
Succinates, esters	GC	25
Succinic acid	LC	58, 65
	PC	13, 77
	TLC	31
Succinic acid, diethyl ester of	GC	24
Succinic acid, dimethyl ester of	GC	24
α-G-Succinic acid	PC	9
α-G-Succinic anhydride	PC	9
Succinic anhydride, *p*-anisidide of	TLC	30
6-Succinoaminopurine	LC	46
	PC	36
Succinyldicholine iodide	PC	8
Succinyl peroxide	PC	27
Succinylsulfanilamide (sodium salt)	TLC	44
Succinylsulfathiazole	TLC	140
Sucrose	LC	25, 71
	PC	70
	TLC	22, 126
Sudan I	TLC	46, 48
Sudan Black B	TLC	46
Sudan Blue G	TLC	46
Sudan G (Ceres Orange GN)	TLC	48
Sudan GN	TLC	46
Sudan II	TLC	46, 48
Sudan III	TLC	46, 48
Sudan IV	TLC	46, 48
Sudan Orange	TLC	49
Sudan Orange RR	TLC	46
Sudan R	TLC	46
Sudan Red	TLC	7
Sudan Red G	TLC	46, 85
Sudan Red G acid	TLC	36
Sudan Violet BR	TLC	46
Sudan Yellow 3G	TLC	46
Sugar alcohols	PC	53
Sulfacarbamide (Euvernil®)	TLC	140
Sulfacetamide (Albucid®)	PC	87
	TLC	44, 140
Sulfachrysoidine	TLC	140
Sulfadiazine	PC	87
	TLC	44, 140
Sulfadicramide	TLC	140
Sulfadimethine	TLC	140
Sulfadimethoxine	TLC	44, 140
Sulfa-4,6-dimethoxypyrimidine	TLC	44
Sulfadimethylbenzoylamide	TLC	140
Sulfadimethyloxazole	TLC	140
Sulfadimidine	TLC	140
Sulfaethidole (Globucid)	TLC	140
Sulfaethylthiadiazole	TLC	44
Sulfafurazole (Sulfasol®)	TLC	140
Sulfaguanidine	PC	87
	TLC	44, 140
Sulfamerazine	PC	87
	TLC	44, 140
Sulfamethazine	PC	87
	TLC	44, 140
Sulfamethin (Sulfisomidine)	TLC	44, 140
Sulfamethizole	TLC	140
Sulfamethopyrazine	TLC	140
Sulfamethoxazole	TLC	140
Sulfamethoxydiazine	TLC	140
Sulfamethoxypyrazine	TLC	140
Sulfamethoxypyridazine (Depovernil®)	TLC	140
Sulfa-5-methylthiadiazole	TLC	44
Sulfamethylthiazole	TLC	140
Sulfametoyl	TLC	140
4′-Sulfamyl-2,4-diaminoazobenzene·HCl	TLC	44
Sulfanilamide	LC	35
	PC	84, 87
	TLC	7, 44, 140
Sulfanilamido-4,5-dimethyloxazole	TLC	44
2-Sulfanilamido-5-methoxydiazine	TLC	44
2-Sulfanilamido-3-methoxypyrazine	TLC	44

3-Sulfanilamido-6-methoxypyridazine	TLC	44
2-Sulfanilamido-5-methylpyrimidine	TLC	44
Sulfanilamido-2-phenylpyrazole	TLC	44
2-Sulfanilamidoquinoxaline	TLC	136
Sulfanilic acid	LC	53
	PC	82
Sulfaphenazole (Depocid)	TLC	140
Sulfaproxyline	TLC	44, 140
Sulfapyridine	PC	87
	TLC	44, 140
Sulfaquinoxaline	TLC	136
Sulfasol® (Sulfafurazole)	TLC	140
Sulfasomidine	TLC	140
Sulfasomizol(e)	TLC	140
Sulfate	PC	14, 93
Sulfathiazol(e)	PC	87
	TLC	44, 140
Sulfathiazole-formaldehyde	TLC	44
Sulfathiourea	TLC	44, 140
Sulfide	PC	93
Sulfisomidine (Sulfamethin)	TLC	44, 140
Sulfisoxazole	TLC	44
Sulfolane	GC	139
Sulfolene	GC	139
Sulfonamidodimethoxytriazine	TLC	44
β-Sulfopyruvic acid	PC	81
Sulforthomidine	TLC	140
Sulfosalicylic acid	GC	105
Sulfotepp	GC	67
	LC	43
	TLC	80
Sulfoxide	GC	65
Sulfur chloride (mono)	GC	50
Sulfur dichloride	GC	50
Sulfur dioxide	GC	85
Sulfurein	PC	26
Sulfuretin	PC	26
Sulfur hexafluoride	GC	26
Sulfuryl chloride	GC	50
Sulphenone	GC	64, 65, 129
	TLC	80, 82
Sulphite	PC	14
Sulpyrin	TLC	40
Sumithion®	GC	67
	TLC	172
Sunset Yellow	TLC	49
Suprasterol II	PC	126
Surecide (CYP)	TLC	172
Surmontil®	TLC	42
Sursum	TLC	42
Synephrine	GC	9, 23
Synopen (Halopyramine)	TLC	45
Syringaldehyde	PC	20
	TLC	174
Syringaldehyde, 2,4-DNPH of	TLC	73, 167
Syringic acid	GC	90
	LC	1, 30
	PC	13, 82
	TLC	135, 174
Syringic acid, 4-acetyl-2-nitrophenol ether of	TLC	91
Syringin	PC	121
Systox (Demeton)	PC	28
	TLC	80, 84
Systox-S-methylsulfone	PC	28
Systox-S-methylsulfoxide	PC	28

Adsorption of the Elements (LC Section – Figures)

S, Sulfur	Pg	278, 279, 280, 281
Sb, Antimony	Pg	278, 279, 280, 281
Sc, Scandium	Pg	278, 279, 280, 281
Se, Selenium	Pg	278, 279, 280, 281
Si, Silicon	Pg	278, 279, 280, 281
Sn, Tin	Pg	278, 279, 280, 281
Sr, Strontium	Pg	278, 279, 280, 281

T_2	GC	115
2,4,5-T (butoxyethanol ester)	GC	128
2,4,5-T (IPE)	GC	68
2,4,5-T (isooctyl ester)	GC	128
2,4,5-T (isopropyl ester)	GC	128
2,4,5-T (ME)	GC	68
Ta (Tantalum)	LC	84
Tachysterol	PC	126
Tagatose	LC	48
Tagaturonic acid	PC	76
Talbutal	PC	86
	TLC	137
Talisulfazol (Phthalylsulfathiazole)	TLC	44, 140
D-Talitol	GC	14
D-Talonic	LC	40
Talonolactone	PC	76
Talose	LC	48
Tamarixetin	LC	28
Tanghinigenin (3β,19β-Dihydroxy-7β,8β-epoxy-5β-cardenolide)	TLC	102
Tannic acid	PC	82
Tantalum (Ta)	LC	84
Taractan®	TLC	42
Taractan® (Truxal®)	TLC	45
Tareoxanthin	TLC	93
Tariric acid	TLC	33
Tartaric acid	LC	58, 65
	PC	13, 77
	TLC	31
Tartrazine	TLC	49
Tasettine	GC	96
Taurine	LC	12, 41
	PC	11, 63, 64
	TLC	11, 13, 15, 16
Taurochenodeoxycholic	PC	38
Taurocholic	PC	38
Taurocyanine	PC	9, 10
Taurodeoxycholic	PC	38
Taxifolin	LC	28
Tb (Terbium)	LC	39, 79, 90, 91
Tb(III), Terbium	GC	43
Tc (Technetium)	LC	80
TCNB	GC	129
	TLC	82
TDE	TLC	81
TDE (DDD)	TLC	80, 82
o,p'-TDE	GC	129
	TLC	82
o,p'-TDE olefin	GC	129
	TLC	82
p,p'-TDE	GC	129
p,p'-TDE olefin	GC	129
	TLC	82

TDP	LC	3
Te (Tellurium)	LC	85, 86
	TLC	55
Technetium (Tc)	LC	80
Techtochrysin	TLC	92
Tectiridin	PC	26
Tectirigenin	PC	26
Tectochrysin	PC	26
Tedion® (Tetradifon)	GC	62, 68
	TLC	80, 82
Tellurac	TLC	66
Tellurite	PC	14, 91, 93
Tellurium, Te	LC	85, 86
	TLC	55
Tellurium diethyldithiocarbamate	TLC	22, 66
Telocinobufagin peracetate	TLC	102
Telocinobufagin (3β,5β,14β-Trihydroxy-bufadienolide)	TLC	102
Telodrin®	GC	129
	TLC	82
Telomycin	TLC	20
Temik	GC	61
Tenamene 1	TLC	175
Tenox BHA®	TLC	175
Tenox BHT®	TLC	175
TeO_3^-	TLC	150
TeO_3^{--}	TLC	147
Tepa	GC	65
Tepanil (Regenon®)	TLC	45
TEPP	GC	65, 129
	LC	43
	PC	28
Terbacil	GC	61
Terbium (Tb)	LC	39, 79, 90, 91
Terbium, Tb(III)	GC	43
Terephthaladazine	PC	21
p-Terphenyl	LC	27
	TLC	7
α-Terpinene	GC	86, 87
	TLC	106
γ-Terpinene	GC	86, 87
	TLC	106
Terpinen-4-yl butyrate	TLC	105
α-Terpineol	GC	87
	TLC	183
α-Terpineol (DNB)	TLC	111
α-Terpineol (ONPU)	TLC	111
β-Terpineol	GC	87
Terpinolene	GC	86, 87
	TLC	51, 54, 106
Terpinyl acetate	TLC	105
Terpinyl butyrate	TLC	105
Terpinyl formate	TLC	105
α-Terthienyl	TLC	104
Tertiomycin A	TLC	20
Tertiomycin B	TLC	20
Testololactone	PC	40, 124
Testosterone	GC	83
	TLC	95, 97, 178
Testosterone acetate	GC	83
	TLC	95
Testosterone isobutyrate	TLC	95
Testosterone propionate	TLC	95, 178
cis-Testosterone	PC	124
	TLC	95
cis-Testosterone acetate	TLC	95
trans-Testosterone	PC	124
Tetra-*O*-acetyl-α-D-xylopyranose	TLC	25
Tetra-*O*-acetyl-β-D-xylopyranose	TLC	25
1,4,5,8-Tetraamino-anthraquinone	TLC	69
Tetraborate	PC	93
Tetrabromofluorescein	PC	88
4,6,4′,6′-Tetra-*tert*-butyl-2,2′-dihydroxybiphenyl	LC	61
Tetrabutylthiuram disulfide	TLC	22
Tetrabutylthiuram monosulfide	TLC	66
Tetrabutyltin	TLC	67
Tetracaine HCl	GC	110
	TLC	43
Tetracaine (Pontocaine®)	GC	21
	TLC	39, 45
1,2,3,4-Tetrachlorobenzene	GC	31
1,2,3,5-Tetrachlorobenzene	GC	31
1,2,4,5-Tetrachlorobenzene	GC	31
2,3,5,6-Tetrachloro-*p*-benzoquinone	PC	111
3,4,5,6-Tetrachloro-*o*-benzoquinone	PC	111
Tetrachloroethane	LC	75
Tetrachloromethane (CCl_4)	GC	49, 58
Tetrachorophthalic anhydride, *p*-anisidide of	TLC	30
Tetrachlorotetrabromofluorescein	PC	88
Tetrachlorothiophene	GC	65
Tetra-condensed phosphate	TLC	151
Tetracycline	PC	68
Tetracycline HCl	TLC	21
iso-Tetracycline	PC	68
Tetradecanal (myristaldehyde)	GC	6
Tetradecane	GC	49
n-Tetradecane	GC	38
	LC	38, 45
Tetradecanoic acid, methyl ester of	TLC	130
n-Tetradecanol	GC	92
1-Tetradecanol	GC	92
2-Tetradecenoic acid	PC	77
9-Tetradecenoic acid	PC	77
n-Tetradecylamine	PC	55
Tetradecyl mercaptan	TLC	162
Tetradifon (Tedion®)	GC	64, 65, 128, 129
	TLC	80, 81
Tetraethylene glycol	LC	54
Tetraethylene glycol, butyl ether of	GC	116
Tetraethylene glycol, ethyl ether of	GC	116
Tetraethylene glycol, methyl ether of	GC	116
Tetraethylene glycol, propyl ether of	GC	116
Tetraethylenepentamine	LC	26
Tetra-2-ethylhexyltin	TLC	67
Tetraethyl pyrophosphate	PC	35
Tetraethylthiuram disulfide	TLC	22, 66
Tetraethyltin	TLC	67
Tetraethyl triphosphate	PC	35
Tetrafluoroethene	GC	26
Tetraglycine	LC	4
Cbo-Tetraglycine	TLC	12
Tetrahexyltin	TLC	67
Tetrahydroabietate (8β,9α,13α-H)	LC	67
Tetrahydroberberine	PC	3
Tetrahydrofuran	GC	51, 55
	LC	60

Tetrahydrofuran hydroperoxide PC 27
Tetrahydrofuran, organics GC 52
2-Tetrahydrofuroic LC 23
Tetrahydroharman PC 101
1,2,3,4-Tetrahydroisoquinoline GC 45
Tetrahydronaphthalene TLC 51, 53
1,2,3,4-Tetrahydronaphthalene GC 119
1,2,3,4-Tetrahydro-1-naphthyl hydroperoxide TLC 65
Tetrahydrophthalic anhydride, *p*-anisidide of TLC 30
Tetrahydropyran GC 51, 55
Dg-3-Tetrahydropyranylether TLC 102
1,2,5,6-Tetrahydropyridine, trifluoroacetyl derivative GC 10
1,2,3,4-Tetrahydroquinoline GC 45
TLC 7
Tetrahydrothiophene S-oxide TLC 164
Tetrahydroxyanthraquinone PC 109
1,2,5,8-Tetrahydroxyanthraquinone (quinalizarin) PC 105
Tetrahydroxy-*p*-benzoquinone PC 120
TLC 88
2,3,2′,3′-Tetrahydroxybiphenyl TLC 88
2,4,2′,4′-Tetrahydroxybiphenyl TLC 88
2,5,2′,5′-Tetrahydroxybiphenyl TLC 88
2,6,2′,6′-Tetrahydroxybiphenyl TLC 88
3,4,3′,4′-Tetrahydroxybiphenyl TLC 88
3,4,3′,4′-Tetrahydroxy-β-carotene TLC 93
3α,7α,12α,23ξ-Tetrahydroxy cholanic acid TLC 180
3α,11β,17,21-Tetrahydroxy-5α-pregnan-20-one PC 45
3β,11β,17,21-Tetrahydroxy-5α-pregnan-20-one PC 45
3α,11β,17,21-Tetrahydroxy-5β-pregnan-20-one PC 45
3α,11β,17α,21-Tetrahydroxy-pregnan-20-one TLC 101
2α,11β,17,21-Tetrahydroxy-pregn-4-ene-3,20-dione PC 125
6β,11β,17,21-Tetrahydroxy-pregn-4-ene-3,20-dione PC 125
11β,17α,20α,21-Tetrahydroxy-4-pregnen-3-one PC 124
11β,17α,20β,21-Tetrahydroxy-4-pregnen-3-one PC 124
Tetraiodoethylene GC 129
Tetraiodofluorescein PC 88
3,5,3′,5′-Tetraiodothyroacetic acid TLC 19
Tetraiodothyroacetic acid methyl ester TLC 19
3,5,3′,5′-Tetraiodothyrolactic acid TLC 19
Tetraiodothyrolactic acid methyl ester TLC 19
3,5,3′,5′-Tetraiodothyronine (thyroxine) LC 76
3,5,3′,5′-Tetraiodothyropyruvic acid TLC 19
Tetraiodothyropyruvic acid methyl ester TLC 19
Tetralin LC 27, 42, 45
Tetralin hydroperoxide PC 27
Tetrameta-condensed phosphate TLC 151
Tetrametaphosphate PC 98
2′,3′,4′,6′-Tetramethoxy-acetophenone GC 48
N,N-2,4-Tetramethylaniline TLC 7
N,N-2,6-Tetramethylaniline TLC 7
2,3,5,6-Tetramethylaniline GC 46, 100
2,4,5,6-Tetramethylaniline GC 46, 100
1,2,3,4-Tetramethylbenzene GC 119
1,2,3,5-Tetramethylbenzene GC 119
1,2,4,5-Tetramethylbenzene GC 119
TLC 51, 53
N,N,N′,N′-Tetramethylbenzidine TLC 7
2,2,3,3-Tetramethylbutane GC 30, 42
2,2,4,4-Tetramethylcaproic acid TLC 29
1,2,4,5-Tetramethylcyclohexane GC 35
Tetramethylcyclohexanol TLC 113
1,1,3,3-Tetramethylcyclopentane GC 42
1,1,*cis*-2-*cis*-4-Tetramethylcyclopentane GC 42
1,1,*cis*-2-*trans*-4-Tetramethylcyclopentane GC 42
1,1,*cis*-3-*cis*-4-Tetramethylcyclopentane GC 42
1,1,*cis*-3-*trans*-4-Tetramethylcyclopentane GC 42
N,N,N′,N′-Tetramethyl-1,3-diaminobutane GC 11
N,N,N′,N′-Tetramethyl-1,4-diaminobutane GC 11
N,N,N′,N′-Tetramethyl-diaminoethane GC 11
N,N,N′,N′-Tetramethyl-1,3-diaminopropane GC 11
4,4′,6,6′-Tetramethyl-2,2′-dihydroxybiphenyl LC 61
Tetramethylenediamine LC 26
PC 7
Tetramethylenediamine, trifluoroacetyl derivative GC 10
Tetramethylgalactose PC 73
2,3,4,6-Tetra-*O*-methyl-D-galactose TLC 26
2,3,4,6-Tetra-*O*-methyl-D-gluconic LC 40
2,3,4,6-Tetra-*O*-methyl-D-glucose TLC 26
Tetramethylmannose PC 73
2,2,3,4-Tetramethylpentane GC 42, 118
2,2,4,4-Tetramethylpentane GC 30, 42
2,2,3,4-Tetramethylpentene GC 118
2,3,4,5-Tetramethylphenol GC 72
2,3,5,6-Tetramethylphenol GC 72
2,2,6,6-Tetramethylpiperidine TLC 7
2,3,5,6-Tetramethylpyridine GC 78
2,4,6,8-Tetramethylquinoline GC 45
Tetramethylsilane GC 27
Tetramethylthiophene TLC 104
Tetramethylthiourea PC 47
Tetramethylthiuram disulfate (Thiram) PC 117
Tetramethylthiuram disulfide TLC 22, 66
Tetramethylthiuram monosulfide TLC 22, 66
1,3,7,9-Tetramethyluric acid PC 36
Tetraphenylethylene GC 105
Tetraphenyltin TLC 67
Tetraphosphate PC 98
Tetrone A TLC 66
Tetryl TLC 157
TFAc GC 98
TFAc, trifluoroacetyl GC 98
Th, Thorium LC 80, 81, 82, 83, 84, 85, 86, 87, 88, 89
PC 15
Th(IV), Thorium GC 43
TLC 149

Th^{4+}, Thorium PC 97
Thalidomide TLC 137
Thallium (Tl) LC 80, 83, 85, 86, 88
Thallium, Tl(I) GC 43
TLC 52,148, 149, 150
Thallium, Tl(III) PC 15
Thallium, Tl^{+} PC 92
Thallium, Tl^{3+} PC 97
Thamnolic acid PC 30
TLC 34
Thanite® GC 65
Thebacone TLC 4
Thebaine GC 8, 96, 107, 111
PC 2, 3, 4
TLC 2, 3, 4,116, 138
Thenalidine (Sandosten®) TLC 45
Thenyldiamine GC 20
TLC 39
Thenylpyramine (Methapyrilene) TLC 39
Theobromine GC 8,105, 111
LC 31, 46
PC 2, 3, 37
TLC 45
Theophorin® (Phenindamine) TLC 39
Theophylline GC 77, 105
LC 31, 46
PC 2, 3, 37
TLC 45
Theophylline-7-acetic acid PC 36
Thevebioside TLC 102
Thevebioside peracetate TLC 102
Thiabendazole TLC 81
Thiabutazide TLC 41
Thialbarbitone PC 86
Thiamine LC 31
PC 51
Thiamine dicetylsulfate TLC 188
Thiamine hydrochloride LC 20
TLC 188
Thiamine mononitrate TLC 188
Thiamylal GC 19, 106
TLC 137
Thiazolo[*d*-3,2]-5α-cholest-2-ene TLC 96
2-(2′-Thiazolylazo)-*p*-cresol (TAC) TLC 47
1-(2′-Thiazolylazo)-2-naphthol (TAN) TLC 47
(2′-Thiazolylazo)-1-naphthol-3,6-disulfonic acid (TANα36S) TLC 47
1-(2′-Thiazolylazo)-2-naphthol-3,6-disulfonic acid (TAN-β-36S) TLC 47
1-(2′-Thiazolylazo)-2-naphthol-6,8-disulfonic acid (TANβ68S) TLC 47
(2′-Thiazolylazo)-1-naphthol-5-sulfonic acid (TANα5S) TLC 47
1-(2′-Thiazolylazo)-2-naphthol-6-sulfonic acid (TANβ6S) TLC 47
1-(2′-Thiazolylazo)-2-naphthol-7-sulfonic acid (TANβ7S) TLC 47
1-(2′-Thiazolylazo)-2-naphthol-8-sulfonic acid(TANβ8S) TLC 47
2-(2′-Thiazolylazo)-1-naphthol-4-sulfonic acid (TANα4S) TLC 47
4-(2′-Thiazolylazo)-phenol (TAP) TLC 47
2-(2′-Thiazolylazo)-phenol-4-sulfonic acid (TAP-S) TLC 47
2-(2′-Thiazolylazo)-phloroglucinol (TAPh) TLC 47
2-(2′-Thiazolylazo)-pyrogallol (TAPyg) TLC 47
4-(2′-Thiazolylazo)-resorcinol (TAR) TLC 47
β-(α-Thienyl)-acrylic acid TLC 104
4-(α-Thienyl)-butyric acid TLC 104
Thiethylperazine TLC 139
Thimet® GC 61, 64, 68, 129
Thimet® (Phorate) PC 28
TLC 80, 84, 171
Thimet oxygen analog GC 61
Thimet oxygen analog sulfone GC 61
Thimet oxygen analog sulfoxide GC 61
Thimet sulfone GC 61
TLC 84
Thimet sulfoxide GC 61
TLC 84
Thioacetamide PC 47
Thiobarbituric acid PC 47
2-Thiobarbituric acid PC 36
Thiobenzamide PC 47
Thiobenzoic acid PC 78
4,4′-Thiobis(6-*tert*-butyl-2-methylphenol) TLC 175
4,4′-Thiobis-(6-*tert*-butyl-3-methylphenol) TLC 22, 175
4,4′-Thiobis(3,6-di-*sec*-amylphenol) TLC 175
3,3′-Thiobis-(4-hydroxycoumarin) TLC 79
2,2′-Thiobis(4-methyl-6-*tert*-butylphenol) TLC 175
o-Thiocresol TLC 162
p-Thiocresol TLC 162
Thiocron TLC 171
Thiocyanate PC 14, 93
2-Thiocytosine PC 36
Thiodan® GC 128
PC 116
TLC 80, 82
Thiodan I GC 68
Thiodan II GC 68
Thiodibutyric acid di-(2-ethylhexyl ester) TLC 27
Thiodiphenylamine TLC 136
Thioglycolic acid TLC 29
6-Thioguanine LC 46
Thiol demeton-disulfoton oxygen analog GC 67
2-Thiol histidine PC 17
L-2-Thiolhistidine LC 59
PC 63, 64
2-Thiol imidazole PC 17
2-Thiol imidazole propionic acid PC 17
2-Thiol-4-methyl imidazole PC 17
Thiolutin TLC 20
5′-Thiomethyladenosine PC 36
5′-Thiomethylinosine PC 36
4-Thiomethylpyrimidin-2-one-1-(β-D-ribofuranoside) TLC 62
Thiometon LC 43
TLC 84,171, 172

Thionazin	LC	43
	TLC	84, 171
Thionazin-O-analogue	LC	43
Thiono demeton	GC	64, 67
Thionyl chloride	GC	50
2-Thio-orotic acid	PC	36
Thiopental	GC	19, 105, 106
	TLC	137
Thiopentone	PC	86
Thiophene	GC	55
Thiophene-carboxaldehyde	TLC	104
Thiophene-carboxylic acid	TLC	104
Thiophenol	TLC	162
Thioproline (thiazolidine-4-carboxylic acid)	PC	63
Thiopropazate	GC	17, 20, 108, 109
	PC	85
	TLC	139
Thiopropazate dihydrochloride	GC	17
Thiopropazine	TLC	139
Thioproperazine	TLC	42
Thioridazine	GC	17, 20, 108, 109
Thioridazine hydrochloride	GC	17
Thioridazine (Melleril)	TLC	45, 139
Thioridazine sulfoxide	TLC	139
Thiosemicarbazide	PC	47
Thiosulfate	PC	14, 91, 93
Thiotepa	GC	65
2-Thiothymine	PC	36
Thiouracil	PC	37
4-Thiouracil	LC	22
Thiourea	LC	63
	PC	60, 64
4-Thiouridine	LC	22
	TLC	62
Thioxanthen-9-one	TLC	75
Thiram	GC	60, 61, 65
	TLC	81
Thiram (Tetramethylthiram disulfate)	PC	117
Thonzylamine	GC	17, 20
Thonzylamine (Anahist®)	GC	17, 20, 21
Thonzylamine HCl	GC	17, 20, 54
Thonzylamine (Neohetramine®) (Anahist)	TLC	39
Thorium, Th	LC	80, 81, 82, 83, 84, 85, 86, 87, 88, 89
	PC	15
Thorium, Th(IV)	GC	43
	TLC	149
Thorium, Th^{4+}	PC	97
Threitol	LC	69
D-Threitol	GC	14
DL-Threitol	GC	76
D-Threo-L-galactooctitol	GC	14
D-Threonic	LC	40
Threonine	GC	12, 101
	LC	2, 4, 33, 41
Threonine (*continued*)	PC	11, 62, 64, 65
	TLC	11, 13, 15, 16
Threonine, DNAP of	TLC	14
Threonine, PTH of	TLC	124
Threonine allo	PC	64
DL-Threonine	PC	63
	TLC	118
Threoninol	PC	57
Threose	LC	48
α-Thujene	GC	86
	TLC	106
α-Thujone, 2,4-DNPH of	TLC	76
β-Thujone, 2,4-DNPH of	TLC	76
Thujopsene	GC	140
	TLC	54
Thullium (Tm)	LC	39, 79, 90, 91
Thymidine	GC	77
	LC	3, 15, 31, 46
	PC	19, 36, 37, 108
	TLC	159
Thymidine 3′,5′-diacetate	GC	134
Thymidine 5′-diphosphate	PC	108
Thymidine monophosphate (TMP)	TLC	160
Thymidine 5′-monophosphate	LC	15
	PC	108
	TLC	160
Thymidine-5′-phosphate	PC	36
Thymidine 5′-triphosphate	PC	108
DMTrThymidine	PC	19
TrThymidine	PC	19
Thymidylic acid	PC	19
Thymine	LC	3, 15, 17, 22, 46
	PC	36, 37, 108
	TLC	159
Thymine riboside	LC	3
Thymol	GC	71
	TLC	90, 174
Thymol, 4-acetyl-2-nitrophenol ether of	TLC	91
Thymol blue	PC	88
Thymolphthalein	PC	88
Thymylphosphate	PC	112
Thymyl-α-pyridyl-phosphate	PC	112
Thyronine	LC	76
	TLC	11
Thyroxine	TLC	11, 19
Thyroxine methyl ester	TLC	19
DL-Thyroxine	TLC	64
L-Thyroxyl-glycine	PC	114
Ti, Titanium	LC	83, 84, 85, 86, 87, 88, 89
	PC	15
Ti(I), Titanium	GC	43
Ti^{4+},Titanium	PC	97
$TiCl_2$	GC	43
TiO, Titanium	GC	43
Tigloidine	PC	3
Tigogenin	GC	79
	TLC	98, 103
Tigogenin acetate	TLC	98

Tigogenin 3-acetate	TLC	103
Tin, Sn(II)	PC	15
Tin, Sn(II),(IV)	PC	15, 95, 96
Tin, Sn^{2+}	PC	97
Titanium, Ti	LC	83, 84, 85, 86, 87, 88, 89
	PC	15
Titanium, Ti(I)	GC	43
Titanium, Ti^{4+}	PC	97
Titanium, TiO	GC	43
Titanium tetrachloride	GC	50
Tl (Thallium)	LC	80, 83, 85, 86, 88
Tl(I), Thallium	GC	43
	TLC	52,148, 149, 150
Tl(III), Thallium	PC	15
Tl^{+}, Thallium	PC	92
Tl^{3+}, Thallium	PC	97
Tm (Thulium)	LC	39, 79, 90, 91
TMP	LC	3
	PC	37
	TLC	63
T-OAc	PC	19
Tocol	GC	142
	PC	50
n-Tocopherol	GC	142
α-Tocopherol	GC	142
	PC	48, 50, 69, 126
	TLC	56, 110
β-Tocopherol	GC	142
	PC	50
	TLC	110
γ-Tocopherol	GC	142
	PC	50
	TLC	110
δ-Tocopherol	GC	142
	PC	50
	TLC	110
ϵ-Tocopherol	PC	50
ζ_2-Tocopherol	GC	142
α-Tocopheryl acetate	TLC	56, 110
α-Tocopherylquinone	PC	48, 50
	TLC	169
β-Tocopherylquinone	TLC	169
γ-Tocopherylquinone	TLC	169
δ-Tocopherylquinone	TLC	169
α-Tocopheryl succinate	TLC	110
Tofranil®	TLC	42
Tokorogenin	TLC	103
Tolazoline (Priscol®)	TLC	45
Tolbutamide (Oraket)	TLC	140
o-Tolidine	PC	59, 68
m-Tolualdehyde	GC	51
p-Tolualdehyde	GC	51
p-Tolualdehyde, 2,4-DNPH of	TLC	165
p-Tolualdehyde, 2-hydrazinebenzothiazole derivative of	TLC	168
Toluene	GC	47, 49, 51, 53, 55, 58, 118,120, 121, 125
	LC	7, 21, 36, 42
Toluene, aromatics	GC	126
m-Toluidine	PC	59
m-Toluidine, trifluoroacetyl derivative	GC	10
o-Toluidine	PC	59
	TLC	7
o-Toluidine, trifluoroacetyl derivative	GC	10
p-Toluidine	GC	45
	PC	59
	TLC	7
p-Toluidine, trifluoroacetyl derivative	GC	10
p-Toluquinone	PC	111
p-Toluylaldehyde, 2,4-DNPH of	TLC	73
o-Tolylbiguanidine	TLC	22
o-Tolyl ethyl sulfide	TLC	163
m-Tolyl isocyanate	GC	44
o-Tolyl isocyanate	GC	44
p-Tolyl isocyanate	GC	44
Tolylmercury dithizonate	TLC	161
p-Tolylmethyl sulfide	TLC	163
1-Tolyl-3-nitroguanidine	PC	9
p-Tolyl octyl sulfide	TLC	163
p-Tolyl phenyl sulfide	TLC	163
N-m-Tolythiourea	PC	47
N-o-Tolylthiourea	PC	47
N-p-Tolylthiourea	PC	47
m-Tolylurea	TLC	122
o-Tolylurea	TLC	122
p-Tolylurea	TLC	122
Tomatidine	GC	96
	TLC	103
Torularhodin	TLC	93
1-(*p*-Tosylamino)-3-bromo-anthraquinone	TLC	69
Toxaphene	GC	60, 128, 129
	PC	116
	TLC	80, 82
TPN	TLC	63
TPN (Triphosphopyridine nucleotide)	PC	18
TPNH	TLC	63
Transferrin (human)	LC	19
Tranylcypromine	TLC	39
Trehalose	LC	48
	TLC	23, 126
Trehalose, trisilyl derivative	GC	124
Triacetin	TLC	27
1,3,5-Tri-*O*-acetyl-2,4-*O*-benzylidenexylitol	GC	13
Tri(4-acetyl-2-nitrophenol) ether of phloroglucinol	TLC	91
Tri(4-acetyl-2-nitrophenol) ether of 2,4,6-trihydroxyacetophenone	TLC	91
Triacetyloleandomycin	PC	66
Triamcinolone	TLC	101
1,2,3-Triamino-anthraquinone	TLC	69
1,2,4-Triamino-anthraquinone	TLC	69
1,2,4-Triaminobenzene	TLC	119
2,4,5-Triaminopyrimidine	PC	16
4,5,6-Triaminopyrimidine	PC	16
4,5,6-Triaminopyrimidine sulfate	PC	36
Triamterene	TLC	41
Triarachidin	LC	78
Tribenzylamine	TLC	7

Tribromoacetic acid	PC	79
	TLC	29
2,4,6-Tribromophenol	PC	121
2,3,4-Tribromo-6-propionylphenol	TLC	89
2,3,6-Tribromo-4-propionylphenol	TLC	89
Tributylamine	TLC	7
Tri-*n*-butylamine	GC	97
Tributyl citrate	TLC	27
2,4,6-Tri-*tert*-butylphenol	GC	73
Tributyl phosphate	TLC	27
Tri-*n*-butyl phosphate	GC	25
Tributyltin chloride	TLC	67
Tricaprin	LC	78
Tricarballylic acid	PC	13, 118
Tricetinidin	PC	25
Trichlorethyl phosphate (Phosphoric acid esters)	TLC	27
Trichlorfon	GC	67
Trichlorfon (Dipterex) (Dylox®)	TLC	80, 81, 84
Trichlormethiazide	TLC	41
Trichloroacetic acid	LC	58
	PC	79
	TLC	29
2,4,5-Trichloroaniline	PC	6
	TLC	119
2,4,6-Trichloroaniline	TLC	7
1,2,3-Trichlorobenzene	GC	31
1,2,5-Trichlorobenzene	GC	31
1,3,5-Trichlorobenzene	GC	31
2,4,6-Trichloro-*m*-cresol	PC	31
1,1,1-Trichloroethane	GC	32, 93
1,1,1-Trichloroethane, organics	GC	52
1,1,2-Trichloroethane, organics	GC	52
1,2,2-Trichloroethene	GC	93
Trichloroethyl pT	PC	19
Trichloroethyl pT-OAc	PC	19
Trichloroethylene	GC	32, 49, 58
Trichloroethylene, organics	GC	52
5′-*O*-Trichloroethylphosphoryldeoxy-adenylyl-(3′-5′)-deoxyadenosine	TLC	62
5′-*O*-Trichloroethylphosphoryldeoxy-thymidylyl-(3′-5′)-deoxyguanylyl-(3′-5′)-deoxyguanosine	TLC	62
5′-*O*-Trichloroethylphosphoryl-deoxythymidylyl-(3′-5′)-deoxythymidylyl-(3′-5′)-deoxyadenosine	TLC	62
5′-*O*-Trichloroethylphosphoryl-deoxythymidylyl-(3′-5′)-deoxythymidylyl-(3′-5′)-deoxyguanosine	TLC	62
5′-*O*-Trichloroethylphosphoryl-deoxythymidylyl-(3′-5′)-deoxythymidylyl-(3′-5′)-deoxythymidine	TLC	62
5′-*O*-Trichloroethylphosphoryl-deoxythymidylyl-(3′-5′)-thymidine	TLC	62
Trichlorolactic acid	TLC	29
Trichloromethane (see chloroform)	GC	49, 58
N-Trichloromethylthio-tetrahydrophthalimide (Camatan)	PC	117
Trichlorophenol	TLC	81
2,4,5-Trichlorophenol	GC	71, 74
2,4,6-Trichlorophenol	GC	74
	LC	51
2,4,6-Trichlorophenol, 4-acetyl-2-nitrophenol ether of	TLC	91
1,1,2-Trichloro-1,2,2-trifluoroethane	GC	26
Trichlorphon	LC	43
Trichlorphon (Trichlorfon) (dylox) (dipterex)	TLC	171
Trichomycin	TLC	20
Tricin	PC	26
Tricresyl phosphate (phosphoric acid esters)	TLC	27
Tricyclene	GC	86, 87
Tridecanal, dimethyl acetal of	GC	6
Tridecane	GC	49
n-Tridecane	LC	38
Tridecanoic acid	PC	77
2-Tridecanone	PC	21
	TLC	70
Tridecylbenzene	LC	27
Trietazine	GC	66
	TLC	86
Triethanolamine	LC	26
	PC	7
Triethylamine	GC	97
	LC	26, 34
1,3,5-Triethylbenzene	GC	119
	LC	42
	TLC	53
Triethyl citrate	TLC	27
Triethylene diamine	GC	11
Triethylene glycol	GC	116
	LC	68
	PC	1, 53
	TLC	113
Triethylene glycol, butyl ether of	GC	116
Triethylene glycol, ethyl ether of	GC	116
Triethylene glycol, methyl ether of	GC	116
Triethylene glycol, propyl ether of	GC	116
Triethylenetetramine	LC	26
	PC	7
Tri-2-ethylhexyltin chloride	TLC	67
Triethyl phosphate	GC	25, 132
Triethyl pyrophosphate	PC	35
Triethyltin chloride	TLC	67
Triethyl triphosphate	PC	35
Triflorperazine	PC	85
Triflorpromazine	PC	85
Trifluomeprazine	GC	20, 109
Trifluoperazine	GC	17, 20, 108, 109
	TLC	139
Trifluoperazine dihydrochloride	GC	17
Trifluoperazine sulfoxide	TLC	139
Trifluoroacetic acid	TLC	29
Trifluoroacetyl, TFAc	GC	98
Trifluorobromomethane	GC	26
Trifluorochloroethylene	GC	26
Trifluorochloromethane	GC	26
Trifluorochloromethane, gases	GC	26
Trifluoromethane	GC	26
Trifluperazine	TLC	42
Triflupromazine	GC	17, 20, 108, 109
	TLC	139
Triflupromazine hydrochloride	GC	17
Triflupromazine sulfoxide	TLC	139
Trifluralin	TLC	81
Triformylnorcholic	PC	38
Triglycerides (unsaturated)	TLC	56
Cbo-Triglycine	TLC	12

Cbo-Triglycine ethyl ester TLC 12
Trigonelline PC 2, 18
Tri-*n*-hexyl phosphate GC 25
Trihexyltin chloride TLC 67
3β,11,21-Trihydroxyallopregnan-20-one TLC 101
1,2,3-Trihydroxybenzene TLC 37, 88
1,2,4-Trihydroxybenzene PC 120
TLC 88
1,3,5-Trihydroxybenzene TLC 37, 88
2,3,4-Trihydroxybenzoic acid LC 1
TLC 37
2,4,6-Trihydroxybenzoic acid PC 120
TLC 37, 135
3,4,5-Trihydroxybenzoic acid LC 1
TLC 37
3,4,4′-Trihydroxy-β-carotene TLC 93
3α,6α,7α-Trihydroxy cholanic acid TLC 180
3α,7α,12α-Trihydroxy cholanic acid TLC 180
3α,7α,16α-Trihydroxy cholanic acid TLC 180
3α,7α,23ξ-Trihydroxy cholanic acid TLC 180
3α,6β,7α-Trihydroxy cholanic acid TLC 180
3α,6β,7β-Trihydroxy cholanic acid TLC 180
3α,7β,12α-Trihydroxy cholanic acid TLC 180
3β,7α,12α-Trihydroxy cholanic acid TLC 180
Trihydroxy compounds GC 136
3α,7α,12α-Trihydroxy coprostanic acid TLC 180
3,16α,17β-Trihydroxy-estra-1,3,5-(10)-trien-6-one TLC 181
4′,5,7-Trihydroxyflavanone TLC 92
11β,17α,21-Trihydroxy-1,4-pregnadiene-3,20-dione PC 124
3α,11β,21-Trihydroxy-5β-pregnan-18-al-20-one PC 45
3α,17α,21-Trihydroxypregnane-11,20-dione TLC 101
3α,17,21-Trihydroxy-5β-pregnane-11,20-dione PC 45
3β,17,21-Trihydroxy-5α-pregnane-11,20-dione PC 45
11β,17α-21-Trihydroxypregnane-3,20-dione TLC 101
11β,17,21-Trihydroxy-5α-pregnane-3,20-dione PC 45
11β,17,21-Trihydroxy-5β-pregnane-3,20-dione PC 45
3α,17α,21-Trihydroxypregnan-20-one TLC 101
3α,11β,21-Trihydroxy-5β-pregnan-20-one PC 45
3β,11β,21-Trihydroxy-5α-pregnan-20-one PC 45
3α,17,21-Trihydroxy-5α-pregnan-20-one PC 45
3α,17,21-Trihydroxy-5β-pregnan-20-one PC 45
3β,17,21-Trihydroxy-5α-pregnan-20-one PC 45
6β,11β,21-Trihydroxypregn-4-en-3,20-dione PC 125
11β,17,21-Trihydroxypregn-4-en-18-al-3,20-dione (17α-hydroxyaldosterone) PC 125
20,20,21-Trihydroxypregn-4-en-3-one-18-acid-lactone (18→20) PC 125
6β,17,21-Trihydroxypregn-4-ene-3,20-dione PC 125
6β,17α,21-Trihydroxy-4-pregnene-3,20-dione PC 124
11α,17α,21-Trihydroxy-4-pregnene-3,20-dione PC 124
11α,17,21-Trihydroxypregn-4-ene-3,20-dione PC 125
11β,17α,21-Trihydroxy-4-pregnene-3,20-dione PC 124
11β,18,21-Trihydroxypregn-4-ene-3,20-dione PC 125
11β,19,21-Trihydroxypregn-4-ene-3,20-dione PC 125
17,19,21-Trihydroxypregn-4-ene-3,20-dione PC 125
11β,17α,21-Trihydroxy-5-pregnene-3,20-dione-3,20-ethylene glycol diketal PC 124
2,6,8-Trihydroxypurine (Uric Acid) PC 16
D-*threo*-2,4,5-Trihydroxyvaleric LC 40
α-(2,4,6-Triiodophenoxy)-butyric acid TLC 64
3,5,3′-Triiodothyroacetic acid LC 76
TLC 19
3,5,3′-Triiodothyroacetic acid methyl ester TLC 19
3,5,3′-Triiodothyrolactic acid TLC 19
3,5,3′-Triiodothyrolactic acid methyl ester TLC 19
3,5,3′-Triiodothyronamine TLC 19
3,3′,5-Triiodo-L-thyronine TLC 64
3,3′,5′-Triiodothyronine LC 76
TLC 19
3,3′,5′-Triiodothyronine methyl ester TLC 19
3,5,3′-Triiodothyronine LC 76
TLC 19
3,5,3′-Triiodothyronine methyl ester TLC 19
3,5,3′-Triiodothyropyruvic acid TLC 19
3,5,3′-Triiodothyropyruvic acid methyl ester TLC 19
Triiodotyrosine TLC 8
Triisopropanolamine PC 7
1,2:3,4:5,6-Tri-*O*-isopropylidene-D-mannitol GC 13
3,7,12-Triketo cholanic acid TLC 180
Trilafon® TLC 42
Trilaurin LC 78
Trimecaine (Mesocaine) TLC 43
Trimellitic anhydride, *p*-anisidide of TLC 30
Trimeprazine GC 17, 20, 108, 109
TLC 139
Trimeprazine tartrate GC 17
Trimeta-condensed phosphate TLC 151
Trimetaphosphate PC 98
3,4,5-Trimethoxyaniline PC 120
3,4,5-Trimethoxybenzalazine PC 23
3,4,5-Trimethoxybenzaldehyde TLC 71
3,4,5-Trimethoxybenzyl alcohol PC 121
3,4,5-Trimethoxycinnamic acid GC 4
3,4,5-Trimethoxyphenylacetic acid GC 4
Trimethylacetic acid TLC 134
Trimethylamine GC 97
Trimethylamine oxide TLC 7

6-Trimethylammoniumpurinide	LC	46
2,3,4-Trimethylaniline	GC	46, 100
2,3,5-Trimethylaniline	GC	46, 100
2,3,6-Trimethylaniline	GC	46, 100
2,4,5-Trimethylaniline	GC	46, 100
2,4,6-Trimethylaniline	GC	46, 100
	TLC	119
3,4,5-Trimethylaniline	GC	46, 100
Trimethylanilinium benzenesulfonate	TLC	7
2,3,4-Trimethylarabinose	PC	73
2,3,5-Trimethylarabinose	PC	73
1,2,3-Trimethylbenzene	GC	49,119, 120
1,2,4-Trimethylbenzene	GC	119, 120
	TLC	53
1,2,4-Trimethylbenzene (see Pseudocumene)	GC	49
	LC	42
1,3,5-Trimethylbenzene	GC	119, 120
1,3,5-Trimethylbenzene (see Mesitylene)	LC	42
Trimethylboron	GC	27
2,2,3-Trimethylbutane	GC	30, 33, 42, 118
2,3,3-Trimethyl-l-butene	GC	118, 122
4,4,14α-Trimethylcholestane-3β-ol	LC	24
4,4,14α-Trimethylcholest-7,9-diene-3β-ol	LC	24
4,4,14α-Trimethylcholest-7,24-diene-3β-ol	LC	24
4,4,14α-Trimethylcholest-8,24-diene-3β-ol	LC	24
4,4,14α-Trimethylcholest-7-ene-3β-ol	LC	24
4,4,14α-Trimethylcholest-8-ene-3β-ol	LC	24
4,4,14α-Trimethylcholest-7,9,24-triene-3β-ol	LC	24
1,*cis*-3,*cis*-5-Trimethylcyclohexane	GC	42
1,1,3-Trimethylcyclohexane	GC	121
1,2,3-Trimethylcyclohexane (*cis*)	GC	35
1,2,3-Trimethylcyclohexane (*trans*)	GC	35
1,2,4-Trimethylcyclohexane (*cis*)	GC	35
1,2,4-Trimethylcyclohexane (*trans*)	GC	35
1,3,5-Trimethylcyclohexane (*cis*)	GC	35
1,3,5-Trimethylcyclohexane (*trans*)	GC	35
1,1,2-Trimethylcyclopentane	GC	36, 42
1,1,3-Trimethylcyclopentane	GC	36, 42
1,*cis*-2,*cis*-3-Trimethylcyclopentane	GC	42
1,*cis*-2,*trans*-3-Trimethylcyclopentane	GC	42
1,*trans*-2,*cis*-3-Trimethylcyclopentane	GC	42
1,*cis*-2,*trans*-4-Trimethylcyclopentane	GC	42
1,*trans*-2,*cis*-4-Trimethylcyclopentane	GC	42
1,2-*cis*-(1)3-*trans*-Trimethylcyclopentane	GC	36
1,2-*trans*-(1)3-*cis*-Trimethylcyclopentane	GC	36
1,2-*cis*-(1)4-*cis*-Trimethylcyclopentane	GC	36
1,2-*cis*-(1)4-*trans*-Trimethylcyclopentane	GC	36
1,2-*trans*-(1)4-*cis*-Trimethylcyclopentane	GC	36
1,1,2-Trimethylcyclopropane	GC	118
N,N,N'-Trimethyldiaminoethane	GC	11
1,2,4-Trimethyl-13H-dibenzo[*a,g*]-carbazole	TLC	120
1,2,4-Trimethyl-13H-dibenzo[*a,i*]-carbazole	TLC	120
Trimethylenediamine	PC	7
Trimethylenediamine, trifluoroacetyl derivative	GC	10
sym-Trimethylene trisulfone	GC	139
2,3,5-Tri-*O*-methyl-D-galactonic	LC	40
2,3,4-Trimethylgalactose	PC	73
2,3,4-Tri-*O*-methyl-D-galactose	TLC	26
2,3,5-Trimethylgalactose	PC	73
2,3,6-Trimethylgalactose	PC	73
2,3,6-Tri-*O*-methyl-D-galactose	TLC	26
2,4,6-Trimethylgalactose	PC	73
2,4,6-Tri-*O*-methyl-D-galactose	TLC	26
3,5,6-Tri-*O*-methyl-D-gluconic	LC	40
2,3,4-Trimethylglucose	PC	73
2,3,4-Tri-*O*-methyl-D-glucose	TLC	26
2,3,6-Tri-*O*-methyl-D-glucose	TLC	26
2,4,6-Tri-*O*-methyl-D-glucose	TLC	26
3,4,6-Tri-*O*-methyl-D-glucose	TLC	26
1,1,3-Trimethylhexane	GC	30
2,2,3-Trimethylhexane	GC	42
2,2,4-Trimethylhexane	GC	30, 33, 42
2,2,5-Trimethylhexane	GC	30, 33, 42, 121
2,3,5-Trimethylhexane	GC	30, 33, 42
2,4,4-Trimethylhexane	GC	42
2,5,5-Trimethyl-*x*-hexene	GC	39
Trimethylhydrazine·HCl	PC	61
3,4,6-Trimethylmannose	PC	73
Trimethyl naphthalene	LC	27
2,3,5-Trimethylnaphthalene	TLC	51
2,4,6-Trimethylnaphthalene	TLC	53
2,4,6-Trimethylolphenol	PC	82
2,2,3-Trimethylpentane	GC	30, 33, 42, 118
2,2,4-Trimethylpentane	GC	30, 33, 42, 51, 118, 121
2,2,4-Trimethylpentane (isooctane)	GC	51, 53
	LC	38, 42
2,2,4-Trimethylpentane (isooctane), organics	GC	53
2,3,3-Trimethylpentane	GC	30, 33, 42, 118
2,3,4-Trimethylpentane	GC	30, 33, 42, 118
2,3,4-Trimethylpentane, organics	GC	53
2,2,4-Trimethyl-1,3-pentanediol	LC	37
2,2,3-Trimethyl-3-pentanol	GC	118
2,2,4-Trimethylpentanol	LC	54
2,2,3-Trimethyl-3-pentanone	GC	118
2,3,3-Trimethyl-1-pentene	GC	39, 118
2,3,4-Trimethyl-1-pentene	GC	39, 118
2,3,4-Trimethyl-2-pentene	GC	39, 118
2,4,4-Trimethyl-1-pentene	GC	39, 118
2,4,4-Trimethylpentene-1	LC	42
2,4,4-Trimethyl-2-pentene	GC	39, 118
3,4,4-Trimethyl-1-pentene	GC	39, 118
3,4,4-Trimethyl-2-pentene	GC	118
3,4,4-Trimethyl-*cis-trans*-2-pentene	GC	39
2,3,4-Trimethylphenol	GC	72
2,3,5-Trimethylphenol	GC	69, 72, 73

2,3,6-Trimethylphenol GC 72
2,4,5-Trimethylphenol GC 69, 72
2,4,6-Trimethylphenol GC 69, 72
3,4,5-Trimethylphenol GC 69, 72, 73
2,3,5-Trimethylphenyl methylcarbamate TLC 170
3,4,5-Trimethylphenyl methylcarbamate (Landrin) TLC 170
2,3,4-Trimethylpyridine GC 78
2,3,5-Trimethylpyridine GC 78
2,4,6-Trimethylpyridine GC 78, 131
TLC 7, 60
1,2,5-Trimethylpyrrole GC 133
2,4,6-Trimethylquinoline GC 78
1,1,4-Trimethylsemicarbazide PC 47
Trimethyl silane GC 27
Trimethylsilyl ethers GC 84
2,3,5-Trimethylthiophene TLC 104
Trimethylthiourea PC 47
Trimethylxanthine LC 13
1,3,7-Trimethylxanthine PC 36
1,3,8-Trimethylxanthine PC 36
2,3,4-Trimethylxylose PC 73
2,3,4-Tri-*O*-methyl-D-xylose TLC 26
Trimyristin LC 78
Trinitrobenzene TLC 157
2,4,6-Trinitrophenol PC 31
Tri(nonylated phenyl) phosphite TLC 175
Tri(nonylphenyl) phosphite TLC 22
Trioctyl phosphate (phosphoric acid ester) TLC 27
Triolein LC 45
TLC 57
Tripalmatin LC 45, 68, 78
TLC 56, 57
Tripelennamine GC 20
Tripelennamine (Pyribenzamine®) GC 20, 21
TLC 39, 45
Tripelennamine base GC 20
Tripelennamine citrate GC 20
Triperidol TLC 42
Triphenylamine TLC 7, 157
Triphenylcarbinol LC 54
Triphenylene GC 40
TLC 51, 146
Triphenylmethanol PC 53
Triphenyl phosphate GC 132
Triphenyl phosphate (phosphoric acid ester) TLC 27
Triphenyltin acetate TLC 67
Triphenylzinc acetate (Brestan®) PC 117
Triphosphate PC 98
Triphosphopyridine nucleotide (TPN) PC 32
Triprolidine (Actidil®) TLC 39
Triprolidine HCl GC 20
Tri-*n*-propylamine GC 97
Tripropylene glycol LC 54
Tripropyltin acetate TLC 67
Tris(ethylenediamine)cobalt trichloride $[Co(en)_3Cl_3]$ PC 94
Tris(ethylenediamine)nickel dichloride $[Ni(en_3)]Cl_2$ PC 94
Tris(*m*-tolyl) phosphate GC 132
Tristearin LC 78
Trisulfides GC 85
Trithion® GC 60, 68
Trithion® (Carbophenothion) PC 28
TLC 80, 82, 84
Trolene (Korlan) PC 28
Tropacocaine TLC 3
Tropanol TLC 17
Tropine PC 3
TLC 117
Trotyl TLC 157
TrTpT PC 19
Truxene LC 27
Trypsin LC 19
Trypsin inhibitor (from soya bean) LC 19
Tryptamine GC 9
LC 77
PC 56, 102, 103
TLC 5, 59, 156
Tryptizol TLC 45
Tryptophan GC 9, 101
LC 4, 41, 56
PC 11, 62, 64, 65, 102, 103
TLC 8, 11, 13, 15, 16, 118
Tryptophan, DNAP of TLC 14
Tryptophan, PTH of TLC 124
DL-Tryptophan TLC 156
Tryptophan-4-carboxylic acid PC 102
L-Tryptophan LC 46
L-Tryptophan methyl ester PC 101
Tryptophanamide PC 102
Tryptophanol PC 102
dl-Tryptophanol PC 63, 101
Tryptophol PC 101
TLC 59
TTP LC 3
TLC 63
Tuaminoheptane GC 23
d-Tubocurarine GC 8
Tungstate PC 14
Tungsten (W) LC 83, 84, 86
Tungsten, W(VI) PC 96
Tungsten, WO_4^{-2} PC 95
Turanose LC 29
TLC 23
Tutocaine (Butamine) TLC 43, 45
Tybamate GC 19, 105
TLC 38
Tylosin PC 66
TLC 20, 153
Tyramine GC 9
LC 77
PC 56, 58, 64
TLC 5, 8
Tyrisine PC 62
Tyrocidine HCl TLC 21
Tyrosine GC 9, 101
LC 4, 12, 33, 41, 56, 72, 76
PC 11, 62, 64, 65
TLC 8, 11, 13, 15, 16, 118

Tyrosine, DNAP of TLC 14
Tyrosine, PTH of TLC 124
Tyrosine, O-mono-DNAP- TLC 14
l-Tyrosine methyl ester PC 63
Tyrosinol PC 57
l-Tyrosinol PC 63
L-Tyrosyl-glycine PC 114
Tyrothricin TLC 21

Adsorption of the Elements
(LC Section – Figures)

Ta, Tantalum Pg 278, 279, 280, 281
Tc, Technetium Pg 278, 279, 280, 281
Te, Tellurium Pg 278, 279, 280, 281
Th, Thorium Pg 278, 279, 280, 281
Ti, Titanium Pg 278, 279, 280, 281
Tl, Thallium Pg 278, 279, 280, 281

U (Uranium) LC 80, 83, 84, 85, 86, 88
U(IV), U(VI), Uranium PC 15, 96
U(IV), UO_2, Uranium GC 43
UO_2 (II), Uranium PC 95
UDP TLC 63
UDPAG TLC 63
UDPG (Uridine diphosphate glucose) PC 36; TLC 63
UDPGA TLC 63
Umbelliferone (7-Hydroxycoumarin) PC 113; TLC 174
UMP LC 3, 17; PC 37
dUMP LC 18
2′+3′-UMP TLC 160
5′-UMP LC 22, 46; TLC 63
Unads GC 62; TLC 66
Unamycin A TLC 20
Undecanal, 2,4-DNPH of TLC 76
Undecanal, 2,4-DNPSC of TLC 165
Undecanal, *p*-NPH of TLC 165
Undecane GC 49, 51
n-Undecane GC 55, 120
n-Undecane (Hendecane) GC 53; LC 38
Undecanoic acid PC 77; TLC 33
Undecanoic acid, methyl ester of GC 24
n-Undecanoic acid LC 57
1-Undecanol (DNB) TLC 111
1-Undecanol (ONPU) TLC 111
Undecan-2-one, 2,4-DNPH of TLC 76
Undecanone-2 GC 51
Undecanone-6 GC 51
Undecene-1 GC 53
10-Undecenoate, *β*-sitosteryl ester of TLC 182
Undecenoic acid TLC 33
10-Undecenoic acid PC 77
n-Undecyl alcohol LC 11
Undecylic acid PC 77; TLC 132
Undecynoic acid TLC 33
Union Carbide UC-8305 GC 65
UO_2 (II) LC 81, 82, 87, 89
UO_2^{++} TLC 150
UO_2^{2+}, Uranium oxide PC 97
Uracil GC 77; LC 3, 13, 15, 22, 31, 46, 50; PC 36, 37, 108; TLC 159
Uracil-6-acetic acid PC 36
Uracil-5-carboxylic acid PC 36
Uramil (5-aminobarbituric acid) PC 60
Uranium, U LC 80, 83, 84, 85, 86, 88
Uranium, U(IV), U(VI) PC 15, 96
Uranium, U(IV), UO_2 GC 43
Uranium, UO_2 (II) PC 95
Uranium hexafluoride GC 26, 114
Uranium oxide, UO_2^{2+} PC 97
p-Urazine (diurea) PC 60
Urbazit (Methyl arsinebisdimethyldithiocarbamate) PC 117
Urea LC 4, 41, 46, 63, 73; PC 64, 102; TLC 15, 38, 122
Ureides, amides, hydrazines, free form, salt form GC 17
p-Ureidobenzenearsonic acid TLC 136
4-Ureidoimidazole-5-carboxylic acid PC 17; TLC 155
Urethane TLC 38
Uric acid GC 77; LC 3, 31, 46, 50; PC 36, 37
Uridine GC 77; LC 3, 13, 15, 22, 31, 46, 50; PC 36, 37, 108; TLC 62, 159
Uridine diphosphate (UDP) TLC 160
Uridine-5′-diphosphate PC 36, 108
Uridine diphosphate glucose (UDPG) PC 32, 36; TLC 160
Uridine monophosphate (UMP) TLC 160
Uridine-2′-monophosphate (2′-UMP) TLC 160
Uridine 2′:3′-monophosphate PC 108
Uridine-2′(3′)monophosphate GC 77
Uridine 2′ and 3′-monophosphates PC 108
Uridine-3′-monophosphate (3′-UMP) TLC 160
Uridine-5′-monophosphate GC 77; LC 15; PC 108; TLC 160
Uridine-5′-phosphate PC 36; TLC 62
Uridine 2′,3′,5′-triacetate GC 134
Uridine triphosphate (UTP) TLC 160
Uridine-5′-triphosphate PC 108
Uridylic acid LC 13
Uridylyl-(3′-5′)-6-methoxypurine riboside TLC 62
Uridylyl-(3′-5′)-1-methylinosine TLC 62
Uridylyl-(3′-5′)-uridine TLC 62

Uronic acids	LC	40
Uroporphyrin	PC	33
Uroporphyrin I	PC	123
Uroporphyrin III	PC	123
Ursodeoxycholic acid (3α,7β-OH)	LC	45
Ursolic acid	TLC	184
Usnic acid	PC	30
	TLC	28
UTP	TLC	63
Uzarigenin	PC	127
Uzarigenin (Uzg) (3β,14β-Dihydroxy-5α-cardenolide)	TLC	102
Uzarigenin peracetate	TLC	102
Uzg-canaroside	TLC	102
Uzg-canaroside peracetate	TLC	102
Uzg-β,D-digitoxoside	TLC	102
Uzg-β,D-digitoxoside peracetate	TLC	102
Adsorption of the Elements (LC Section – Figures)		
U, Uranium	Pg	278, 279, 280, 281
V (Vanadium)	LC	81, 82, 83, 84, 85, 86, 87
V(IV), (V), Vanadium	PC	15, 96
Valencene	GC	140
	TLC	106, 143
Valeraldehyde	GC	55, 125
Valeraldehyde (pentanol)	GC	53
Valeraldehyde, 2,4-DNPSC of	TLC	165
Valeraldehyde, *p*-NPH of	TLC	165
n-Valeraldehyde	GC	7
	PC	20
n-Valeraldehyde, 2,4-DNPH of	TLC	166
Valeranone	GC	141
Valeric acid	GC	24
	PC	77
	TLC	29
Valeric acid, methyl ester of	GC	24
iso-Valeric acid	PC	77
n-Valeric acid	LC	57, 65
	PC	79
	TLC	132, 134
α-G-*n*-Valeric carboxyalkyl guanidine	PC	9
δ-G-*n*-Valeric carboxyalkyl guanidine	PC	9
α-G-*iso*-Valeric guanidine	PC	9
α-G-*n*-Valeric guanidine	PC	9
δ-G-*n*-Valeric guanidine	PC	9
Valeroidine	PC	3
Valeronitrile	GC	51
Valerylcholine	PC	8
2-Valerylphenol	TLC	89
4-Valerylphenol	TLC	89
Valeryl phosphate	PC	77
Valine	GC	12, 101
	LC	2, 4, 33, 41
	PC	11, 62, 64, 65
	TLC	11, 13, 15, 16, 29
Valine, DNAP of	TLC	14
Valine, PTH of	TLC	124
L-Valine	TLC	118
Valinol	PC	57
Valium	TLC	45
Vallaroside	TLC	102
Vallaroside peracetate	TLC	102
Val-Try-Arg	LC	4
Valyl-proline (*cis*)	GC	16
Valyl-proline (*trans*)	GC	16
Valyl-valine (*cis*)	GC	16
Valyl-valine (*trans*)	GC	16
Vamidothion	LC	43
	PC	28
	TLC	171
Vanadate	PC	14, 91
Vanadium, V	LC	81, 82, 83, 84, 85, 86, 87
Vanadium, VO	GC	43
Vanadium, VO(II)	GC	43
	PC	95
Vanadium, V(IV), (V)	PC	15, 96
Vanadium oxychloride	GC	50
Vanadium tetrachloride	GC	50
Vancide PB	GC	62
Vancomycin	PC	67
Vancomycin HCl	TLC	21
Vanillic acid	GC	90
	LC	1, 30, 72
Vanillic acid (3-Methoxy-4-hydroxy-benzoic)	PC	58
	TLC	174
Vanillic acid, 4-acetyl-2-nitrophenol ether of	TLC	91
Vanillic acid diethylamide (Ethamivan)	GC	19
Vanillin	PC	20, 22
	TLC	77, 87, 184
Vanillin, 4-acetyl-2-nitrophenol ether of	TLC	91
Vanillin, 2,4-DNPH of	TLC	73, 167
Vanillin, 4-NPH of	TLC	74
Vanillin (4-hydroxy-3-methoxybenzaldehyde)	TLC	68
Vanillin azine	PC	23
o-Vanillin	TLC	174
o-Vanillin, 2,4-DNPH of	TLC	167
o-Vanillin (2-hydroxy-3-methoxybenzaldehyde)	TLC	68
Vanillovanillone	PC	20
Vanilloylglycine	PC	80
Vanillylamine	GC	9
Vanilmandelic acid	PC	58
	TLC	8
Variolaric acid	TLC	35
VC-13®	TLC	80
Vegadex®	GC	129
	TLC	82
Velsicol UCS-506	GC	65
Velsicol UCS-506 O-analog	GC	65
Veneniferin	TLC	102
Veratraldazine	PC	23
Veratraldehyde	PC	22
	TLC	174
Veratraldehyde, 2,4-DNPH of	TLC	73, 167
Veratraldehyde, 4-NPH of	TLC	74
Veratric acid	TLC	133, 174
Veratridine	PC	3
Veratrine	PC	2
Veratrole	TLC	174
Verbenene	TLC	106
Verodoxin	PC	127
	TLC	102, 177

Verodoxin peracetate	TLC	102
β-Vetivenene	GC	140
Vinbarbital	GC	19, 106
	TLC	137
Vincamine	PC	3
Vinyl acetate	GC	53, 91
Vinyl acetate, organics	GC	52
Vinyl bromide	GC	26
Vinyl butyrate	GC	91
Vinyl butyrate, organics	GC	52
Vinyl chloride	GC	26, 32, 127
Vinyl chloride, organics	GC	52
Vinylcyclohexane	GC	41
Vinyl hexanoate	GC	91
Vinylidene chloride	GC	32
Vinyl isobutyrate	GC	91
Vinyl 3-methylbutyrate	GC	91
Vinyl methyl ketone	GC	55
Vinyl 4-methylpentanoate	GC	91
Vinyl pentanoate	GC	91
Vinyl propionate	GC	91
N-Vinyl-2-pyrrolidone	LC	73
Violaxanthin	PC	122
	TLC	93, 94
Violet	TLC	141
Violet BNP	TLC	49
Viomycin	PC	67
	TLC	20
Viomycin (sulfate)	TLC	20
Vitamin A	PC	48, 51, 126
	TLC	107,186, 187
Vitamin A acetate	PC	126
	TLC	189
Vitamin A acid	TLC	189
Vitamin A acid methyl ester	TLC	189
Vitamin A alcohol	TLC	56, 189
Vitamin A aldehyde	TLC	189
Vitamin A benzoate	TLC	189
Vitamin A palmitate	TLC	56, 189
Vitamin A pivalate	TLC	189
Vitamin A senecionate	TLC	189
Vitamin A trimethoxybenzoate	TLC	189
Vitamin A_1 acetate	PC	49, 126
	TLC	108
retro-Vitamin A_1 acetate	TLC	108
Vitamin A_1 acid	PC	49
	TLC	108
Vitamin A_1 alcohol	PC	49
	TLC	108
retro-Vitamin A_1 alcohol	TLC	108
Vitamin A_1 aldehyde	PC	49
	TLC	108
Vitamin A_1 epoxide	TLC	108
Vitamin A_1 palmitate	PC	49
Vitamin A_2 acetate	PC	49
Vitamin A_2 alcohol	PC	49
	TLC	108
Vitamin A_2 aldehyde	PC	49
	TLC	108
Vitamin B_{12}	LC	17, 18
	PC	51
	TLC	188
Vitamin C	PC	51
Vitamin D_2	PC	48, 126
	TLC	107,186, 187
Vitamin D_3	PC	126
	TLC	56,186, 187
Vitamin E (α-Tocopherol)	TLC	107,186, 187
Vitamin K_1	PC	48, 50
	TLC	107,186, 187
Vitamin K_2	PC	50
	TLC	107
Vitamin K_3	TLC	107,186, 187
Vitamins, n-$C_{28}H_{58}$	GC	142
Vitamins, 5-Methyltocol	GC	142
Vitamins, Tocol	GC	142
Vitamins, α-Tocopherol	GC	142
Vitamins, β-Tocopherol	GC	142
Vitamins, γ-Tocopherol	GC	142
Vitamins, δ-Tocopherol	GC	142
Vitamins, ζ_2-Tocopherol	GC	142
Vitamins, *n*-Tocopherol	GC	142
Vitavax®	GC	65
Vitexin	PC	26
Voacangine	GC	96
Vulpinic acid	PC	54
	TLC	27

Adsorption of the Elements (LC Section – Figures)

V, Vanadium	Pg	278, 279, 280, 281
W (Tungsten)	LC	83, 84, 86
W(VI), Tungsten	PC	96
WO_4^{-2}, Tungsten	PC	95
Wallichoside	TLC	102
Wallichoside peracetate	TLC	102
Warfarin	GC	21
	TLC	81
Water	GC	53,118, 127
Water, inorganic compounds	GC	126
Wingstay S	TLC	175
Wingstay T	TLC	175
Wingstay V	TLC	175

Adsorption of the Elements (LC Section – Figures)

W, Tungsten	Pg	278, 279, 280, 281
Xanthen-9-one	TLC	75
Xanthine	GC	77
	LC	3, 13, 15, 17, 31, 46
	PC	36, 37, 108
	TLC	159
Xanthophylls (X)	PC	48
Xanthosine	LC	3, 15, 31, 46
	PC	36, 37, 108
Xanthosine-5′-monophosphate	GC	77
Xanthotoxin	PC	113

Xanthurenic acid LC 72
PC 82, 102, 103
Xanthurenic acid-8-methyl ether PC 103
Xenon GC 112
XMP LC 3
Xylene TLC 51
m-Xylene GC 49, 51, 53, 55, 118, 120, 125
LC 7, 36, 42
m-Xylene, aromatics GC 126
o-Xylene GC 49, 51, 53, 55, 118, 120, 125
LC 36, 42
o-Xylene, aromatics GC 126
p-Xylene GC 49, 51, 53, 118, 120, 121, 125
LC 21, 36, 42
p-Xylene, aromatics GC 126
2,3-Xylenol GC 73
2,4-Xylenol GC 73
2,5-Xylenol GC 71, 73
2,6-Xylenol GC 73
3,4-Xylenol GC 71, 73
3,5-Xylenol GC 71, 73
Xylitol GC 14, 15
LC 69
PC 1
TLC 127
D-Xylitol GC 76
Xylobiose LC 48
Xylocaine® (Lidocaine) TLC 43
D-Xylonic LC 40
D-Xylono-γ-lactone PC 72
p-Xyloquinone PC 111
Xylose LC 25, 29, 48, 71
PC 70
TLC 23, 24
D-Xylose GC 15
TLC 126
Xylulose LC 29
PC 70
Xylulose-5-phosphate PC 32
Xyluronic LC 23
Xysmalogenin (3β,14β-Dihydroxy-cardadien-5,20(22)-olide) TLC 102
Xysmalogenin peracetate TLC 102

Y, Yttrium LC 79, 83, 85, 86, 89, 90, 91
PC 15
Y(III), Yttrium GC 43
PC 96
Y^{3+}, Yttrium PC 92
Yamogenin TLC 98
Yb, Ytterbium LC 39, 79, 83, 84, 85, 86, 88, 90, 91

Yellow 1 TLC 141
Yellow 2 TLC 141
Yellow 2G TLC 49
Yellow 3 TLC 141
Yellow 4 TLC 141
Yellow 5 TLC 141
Yellow AB TLC 46, 48
Yellow OB TLC 46, 48
Yellow RFS TLC 49
Yellow RY TLC 49
Yellow XP TLC 48
Ylangene TLC 106, 143
α-Ylangene GC 140
β-Ylangene GC 140
Yohimbine GC 8
PC 2, 4
TLC 3, 39, 115
Yonogenin TLC 103
Ytterbium, Yb LC 39, 79, 83, 84, 85, 86, 88, 90, 91
Yttrium, Y LC 79, 83, 85, 86, 89, 90, 91
PC 15
Yttrium, Y(III) GC 43
PC 96
Yttrium, Y^{3+} PC 92
Yuccagenin TLC 98

Adsorption of the Elements
(LC Section – Figures)

Y, Yttrium Pg 278, 279, 280, 281

Zalba Special TLC 175
Zearalenone TLC 185
Zeaxanthin TLC 93, 94
Zectran® GC 130
TLC 80, 83
Zetax TLC 66
Zinc, Zn LC 81, 82, 83, 85, 86, 87, 88, 89
PC 15
TLC 52
Zinc, Zn(II) GC 43
PC 95
TLC 148, 149, 150
Zinc, Zn^{2+} PC 92
Zinc benzothiazyl sulfide TLC 66
Zinc dibutyldithiocarbamate TLC 66
Zinc di-*n*-butyldithiocarbamate TLC 22
Zinc diethyldithiocarbamate TLC 22, 66
Zinc dimethyldithiocarbamate PC 117
TLC 66
Zinc ethylphenyldithiocarbamate TLC 22
Zinc salt of 2-mercaptobenzimidazole TLC 22
Zinc salt of 2-mercaptobenzothiazole TLC 22
Zincb TLC 172
Zineb (Zinc ethylenebisdithiocarbamate) PC 117
α-Zingaberene GC 140
β-Zingaberene TLC 106, 143
Zinophos® (Thionazin) GC 65, 67
TLC 84, 171
Ziram (Zinc dimethyldithiocarbamate) PC 117

Zirconium, Zr	LC	80, 82, 83, 84, 85, 86, 87, 89
	PC	15
Zirconium, Zr(IV)	GC	43
	PC	95, 96
Zirconium, Zr^{4+}	PC	92
Zizaene	GC	140
Zn, Zinc	LC	81, 82, 83, 85, 86, 87, 88, 89
	PC	15
	TLC	52
Zn(II), Zinc	GC	43
	PC	95
	TLC	148,149, 150
Zn^{2+}, Zinc	PC	92
Zoalene®	TLC	136
Zoxazolamine	TLC	39
Zr, Zirconium	LC	80, 82, 83, 84, 85, 86, 87, 89
	PC	15
	GC	43
Zr(IV), Zirconium	PC	95, 96
Zr^{4+}, Zirconium	PC	92
Zygomycin A	TLC	20
Zygomycin (sulfate)	TLC	20
Zymosterol	PC	126
	TLC	99
Zytron	GC	65, 67
	TLC	84

Adsorption of the Elements (LC Section – Figures)

Zn, Zinc	Pg	278, 279, 280, 281
Zr, Zirconium	Pg	278, 279, 280, 281